Stem Cell-Free Therapy

This book provides a comprehensive review of the advancements in cell-free therapeutic applications, focusing on the biological and biomedical potential of mesenchymal stem cell (MSC)-derived secretome and extracellular vesicles (EVs) in disease treatment and tissue repair. It examines the mechanistic functions of MSC-derived biomolecules and the regulatory landscape shaping their development. The discussion then progresses to innovative strategies for enhancing the MSC secretome through priming techniques, optimizing its biological activity for therapeutic applications. The book examines the application of exosome-based and EV-driven therapies across a broad range of biomedical fields, including neuroscience, cardiology, ophthalmology, gastroenterology, and oncology. Further, it highlights the role of MSC-derived secretome in treating neurocognitive disorders, autoimmune diseases, diabetes complications, and inflammatory conditions, with a special focus on innovations in drug delivery and tissue repair. Additionally, it examines the emerging use of EVs as theranostic tools in cancer therapy and their application in novel treatment strategies for COVID-19. The book also addresses key challenges in transitioning these cell-free therapies from bench to bedside, covering critical aspects such as formulation, large-scale production, regulatory approvals, and clinical implementation.

Key Features:

- Explores the biomedical applications of MSC-derived secretome and EVs across different diseases.
- Discusses the biological and therapeutic potential of exosome-based treatments for neurodegenerative, cardiovascular, ophthalmological, and metabolic diseases.
- Examines innovations in extracellular vesicle-based cell-free therapies for targeted tissue repair and disease intervention.
- Reviews the challenges of translating EV-based therapeutics into clinical practice, including regulatory and manufacturing considerations.
- Highlights cutting-edge applications of EVs in oncology, infectious diseases, and theranostics.

This book serves as an important resource for researchers and students of biomedical engineering, biotechnology, and stem cell biology.

Stem Cell-Free Therapy

Engineering Exosomes and Secretomes for Next-Generation Therapeutics

Edited by
Khawaja H. Haider

CRC Press
Taylor & Francis Group
Boca Raton London New York

CRC Press is an imprint of the
Taylor & Francis Group, an **informa** business

Designed cover image: Khawaja H. Haider

First edition published 2027
by CRC Press
2385 NW Executive Center Drive, Suite 320, Boca Raton FL 33431

and by CRC Press
4 Park Square, Milton Park, Abingdon, Oxon, OX14 4RN

CRC Press is an imprint of Taylor & Francis Group, LLC

ISBN: 978-1-041-10100-0 (hbk)
ISBN: 978-1-041-10103-1 (pbk)
ISBN: 978-1-003-65338-7 (ebk)

DOI: 10.1201/9781003653387

Typeset in Times
by Deanta Global Publishing Services, Chennai, India

Contents

About the Editor

Khawaja H. Haider is a Professor of Cellular and Molecular Pharmacology (Stem Cell and Gene Therapy) and Chairman of the Basic Sciences Department (Medical Program) at Sulaiman AlRajhi University, Saudi Arabia. Before his current assignment, he served in several prestigious institutions in various parts of the world. He has also served as Principal Investigator (PI) and Co-PI on multiple National Institute of Health (NIH) funded stem cell research projects. He has been on the editorial boards and invited reviewer panels for several respected international journals. His research focuses on using, either alone or in combination, DNA, miRNAs, and stem cells as "drugs," a topic that has gained popularity in regenerative medicine. He has extensively published on cellular and sub-cellular preconditioning of stem cells and using their conditioned media for cell-free therapy. He has published nearly 350 book chapters, abstracts, and research papers in various books and leading research journals, including *Circulation*, *Circulation Research*, *Cardiovascular Research*, *Journal of Cellular and Molecular Medicine*, the *Journal of Biological Chemistry*, *Cell Cycle*, *Basic Research in Cardiology*, and *Antioxidant Redox Signaling*. He has also given numerous presentations and edited ten books covering various facets of stem cells and their applications from drug to drug development.

1 Living Biodrugs MSCs-Derived Exosomes and Secretome as Advanced Therapeutic Products for the Treatment of COVID-19

Syed Mohammad Habib[1], Tasneem Komaqli[1], Mohammad Fateh Martini[1], Reem Ali AL-Hussain[1], Yacoub Nail Hamdan Abu-Hamdan[1], Omar Essa[1], and Khawaja H. Haider[1]

[1]College of Medicine, Sulaiman Al-Rajhi University, Al-Bukayriyah, Qassim, Saudi Arabia

1.1 LIST OF ABBREVIATIONS

ACE2	Angiotensin-converting enzyme 2
ALI	Acute lung injury
Ang1	Angiopoietin-1
ARDS	Acute respiratory distress syndrome
ASCs	Adipose-derived stem cells
Akt	Protein kinase B
ASC-CS	Adipose-derived stem cell–conditioned secretome
Alix	Apoptosis-linked gene 2–interacting protein X
AA	Arachidonic acid
AMPK	AMP-activated protein kinase
α-SMA	Alpha–smooth muscle actin
AlgXO	Alginate–exosome hybrid nanoparticle
BFGF	Basic fibroblast growth factor
BLM	Bleomycin
BM-MSC	Bone marrow-derived mesenchymal stem cells
BM-MSC EVs	Bone marrow mesenchymal stem cell–extracellular vesicles
C3	Complement 3
CD	Cluster of differentiation
CXCL	C-X-C motif chemokine ligand
circRNA	circular RNA
CM	Conditioned medium
COVID-19	Coronavirus disease 2019

DOI: 10.1201/9781003653387-1

CS	Cytokine storm
Cas	CRISPR-associated protein
cGAS	Cyclic GMP–AMP synthase
ctDNA	Circulating tumor DNA
CL-6B	Cross-linked 6B sepharose
cyclic RGD (cRGD)	Cyclic arginine–glycine–aspartic acid peptide
CRP	C-reactive protein
CRISPR	Clustered regularly interspaced short palindromic repeats
DLD	Deterministic lateral displacement
dsDNA	Double-stranded DNA
DUC	Differential ultracentrifugation
DGU	Density gradient ultracentrifugation
DPSCs	Dental pulp stem cells
DMSO	Dimethyl sulfoxide
ECs	Endothelial cells
EMSCs	Ectodermal mesenchymal stem cells
EPM	Exosome–polyethyleneimine matrix
ESCRT	Endosomal sorting complex required for transport
EVs	Extracellular vesicles
Exos	exosomes
EV-DNA	Extracellular vesicle–derived DNA
EGF	Epidermal growth factor
ELISA	Enzyme-linked immunosorbent assay
Exo-hASC	Exosomes derived from human adipose-derived stem cells
ExoFlo	Bone marrow-derived mesenchymal stem cell exosome preparation.
FBS	Fetal bovine serum
FN	Fibronectin
FDA	Food and drug administration
FTC	Federal trade commission
gDNA	Genomic DNA
GVHD	Graft-versus-host disease
HGF	Hepatocyte growth factor
HRS	Hepatocyte growth factor–regulated tyrosine kinase substrate
HIF-1α	Hypoxia-inducible factor 1 alpha
HSP	Heat shock protein
hESC-MSC-IMRC-CM	Human embryonic stem cell-derived msc-like immune and matrix regulatory cell–conditioned medium
hUC-MSCs	Human umbilical cord mesenchymal stem cells
hUC-MSCs-CM	Human umbilical cord mesenchymal stem cell–conditioned medium
HE	Hybrid exosome
HG	Hydrogel
hCMEC	Human cerebral microvascular endothelial cell.
haMSC-Exos	Human adipose-derived mesenchymal stem cell exosomes
IAV	Influenza A virus
IFN-γ	Interferon-gamma
IL-1β	Interleukin-1 beta
IL-6	Interleukin-6
IL-10	Interleukin-10
ILVs	Intraluminal vesicles
iMSCs	Induced pluripotent stem cell-derived mesenchymal stem cells
iMSCs-Exo	Induced pluripotent stem cell-derived msc exosomes

iPSCs	Induced pluripotent stem cells
iECs	Induced pluripotent stem cell-derived endothelial cells
IV	Intravenous
ICU	Intensive care unit
IgG	Immunoglobulin G
ISCT	International Society for Cell & Gene Therapy (formerly International Society for Cellular Therapy)
ISSCR	International Society for Stem Cell Research
ISEV	International Society for Extracellular Vesicles
KGF	Keratinocyte growth factor
lncRNA	Long non-coding RNA
LPS	Lipopolysaccharide
LFB	Luxol fast blue
LSC-Exo	Lung spheroid cell-derived exosomes
MenSCs	Menstrual blood stromal cells
MMSC	Muscle mesenchymal stem cell
miRNAs	MicroRNAs
mRNA	Messenger RNA
MSCs	Mesenchymal stem cells
MSC-Exos	Mesenchymal stem cell-derived exosomes
MSC-SECR/MSC-Sec	Mesenchymal stem cell secretome
MVBs	Multivesicular bodies
MWCO	Molecular weight cutoff
MyD88	Myeloid differentiation primary response 88
mtDNA	Mitochondrial DNA
mtRNA	Mitochondrial RNA
MSC-SCE	Mesenchymal stem cell-derived secretome
mTOR	Mechanistic target of rapamycin
MHC	Major Histocompatibility complex
MV	Microvesicle
MSC-NTF	Mesenchymal stem cell-derived neurotrophic factors
MIP	Macrophage inflammatory protein
TLR4	Toll-like receptor 4
MSC-CM	Mesenchymal stem cell-conditioned medium
MSC-Exos	Mesenchymal stem cell-derived exosomes
ncRNA	Non-coding RNA
Nox4	NADPH oxidase 4
NF-κB	Nuclear factor kappa B
nSMase2	Neutral sphingomyelinase 2
NP	nanoparticles
ncRNAs	Non-coding RNAs
NK	Natural killer (cell)
NO	Nitric oxide
Nrf2	Nuclear factor erythroid 2–related factor 2
NLRP3	NOD-like receptor pyrin domain–containing 3
PBS	Phosphate-buffered saline
PEG	Polyethylene glycol
PGE2	Prostaglandin E2
PF	Pulmonary fibrosis
PI3K	Phosphoinositide 3-kinase
PS	Phosphatidylserine

PUFAs	Polyunsaturated fatty acids.
PBMC	Peripheral blood mononuclear cell(s)
PGE2	Prostaglandin E2
QC	Quality control
RBD	Receptor binding domain
RBD-Exo	Receptor binding domain-coated exosomes
RNA	Ribonucleic acid
ROS	Reactive oxygen species
rRNA	Ribosomal RNA
RZC	Rate-zonal centrifugation
RT-qPCR	Reverse transcription quantitative polymerase chain reaction
RT-PCR	Reverse transcription polymerase chain reaction
SARS-CoV-2	Severe acute respiratory syndrome coronavirus 2
SCE/SECR	Secretome
SQ	Subcutaneous
sEVs	Small extracellular vesicles
siRNA	Small interfering RNA
shRNA	Short hairpin RNA
SNAP	Synaptosomal-associated protein
ssDNA	Single-stranded DNA
snoRNAs	Small nucleolar RNAs
SOFA	Sequential organ failure assessment
snRNAs	Small nuclear RNAs
SEC	Size-exclusion chromatography.
STING	Stimulator of interferon genes
SNARE	Soluble N-ethylmaleimide-sensitive factor attachment protein receptor
SAEs	Serious adverse events
TGF-β	Transforming growth factor beta
TLR	Toll-like receptor
TNF-α	Tumor necrosis factor alpha
TM	Transmembrane (protein)
TP	Triptolide
TLR4	Toll-like receptor 4
t-SNARE	target SNAREs
tRNA	Transfer RNA
TSP1	Thrombospondin-1
TSG	Tumor susceptibility gene
TRAIL	Tumor necrosis factor–related apoptosis-inducing ligand
UC-MSCs	Umbilical cord mesenchymal stem cells
UC-MSC-CM	Umbilical cord mesenchymal stem cell–conditioned medium
VAMP	Vesicle-associated membrane protein
VEGF	Vascular endothelial growth factor
VEGFA	Vascular endothelial growth factor A
v-SNARE	vesicle-associated SNAREs
vWF	von Willebrand factor
XOs	Exosomes
XLs	Targeting exosomes

1.2 INTRODUCTION

1.2.1 Background

Severe Acute Respiratory Syndrome Coronavirus 2 (SARS-CoV-2) is the causative agent of Coronavirus Disease 2019 (COVID-19). It is a positive-sense, single-stranded RNA virus first identified in Wuhan, China, in December 2019 (Chung et al., 2024). It belongs to the beta-coronavirus genus and has genetic similarities with COVID-19 and MERS-CoV, which have earlier caused epidemics (Soy et al., 2021). On March 11, 2020, the World Health Organization declared COVID-19 (Ochani et al., 2021). COVID-19 is transmitted mainly through respiratory droplets and aerosols, and sometimes by fomites (Rothan et al., 2020). The global burden includes over 100 million confirmed cases and millions of deaths, besides overwhelming economic loss.

1.2.2 Pathogenesis of COVID-19

The COVID-19 pathogenesis begins with viral entry into host cells via angiotensin-converting enzyme 2 (ACE2) receptors, which are widely distributed across various tissues, including the lungs, heart, kidneys, intestines, liver, and the central nervous system (Bhalerao et al., 2021). Once inside the host cell, the virus replicates using host cell machinery and initiates a series of complex innate and adaptive immune responses. ARDS mediated by COVID-19 is characterized by alveolar epithelial cell injury, immune cell infiltration, including macrophages and neutrophils, and injury to the alveolar-capillary membrane, leading to pulmonary edema, impaired gas exchange, and hypoxemia (Zheng et al., 2022; Bhalerao et al., 2021). The histologic findings are characterized by diffuse alveolar damage and fibrin deposition, which are responsible for the compromised lung compliance leading to respiratory failure in severe cases.

The cytokine storm is a significant determinant of disease severity, characterized by a hyperinflammatory syndrome marked by the increased release of pro-inflammatory cytokines, including IL-6, IL-1β, TNF-α, and interferon-γ. The downstream effects include increased vascular permeability, disseminated intravascular coagulation, and diffuse tissue injury. The cytokine storm also facilitates lung injury, playing a crucial role in the pathogenesis of ARDS. The storm is due to the overactivation of macrophages, T cells, and other immune effectors, triggering a systemic inflammatory response syndrome (Soy et al., 2021) (Figure 1.1). The downstream effects include increased vascular permeability, disseminated intravascular coagulation, and widespread tissue damage. The cytokine storm also causes lung damage and is a primary mediator of ARDS pathogenesis.

Apart from pulmonary involvement, COVID-19 is linked with multi-organ failure, particularly in critically ill and severe patients. The systemic involvement is multifactorial and involves direct viral cytopathic effects, endothelial damage, thrombo-inflammation, and immune-mediated tissue injury (Machado-Curbelo et al., 2021). Organs often involved include the heart, kidneys, liver, and central nervous system. The heart can be affected by myocardial injury, arrhythmias, and thrombotic complications, which may result from direct infection of cardiomyocytes and exposure to pro-inflammatory cytokines. In the liver, transaminases may be elevated, and hepatic dysfunction may occur due to hypoxia, systemic inflammation, or direct hepatocyte injury. The kidneys are susceptible to acute damage resulting from hemodynamic instability, direct viral infection, and cytokine-mediated inflammation. The central nervous system could be affected by stroke, encephalopathy, and chronic cognitive impairment. These also might lead not only to in-hospital mortality but also to the rising burden of post-acute sequelae of COVID-19 infection (PASC), which is also known as "long COVID." The survivors can have chronic fatigue, dyspnea, neurocognitive impairment, and cardiovascular impairment weeks after viral clearance (Bhalerao et al., 2021). Understanding the intricate interplay between viral pathogenesis, host immune response, and systemic complications is crucial for guiding successful therapeutic interventions and enhancing long-term survival (Figure 1.1).

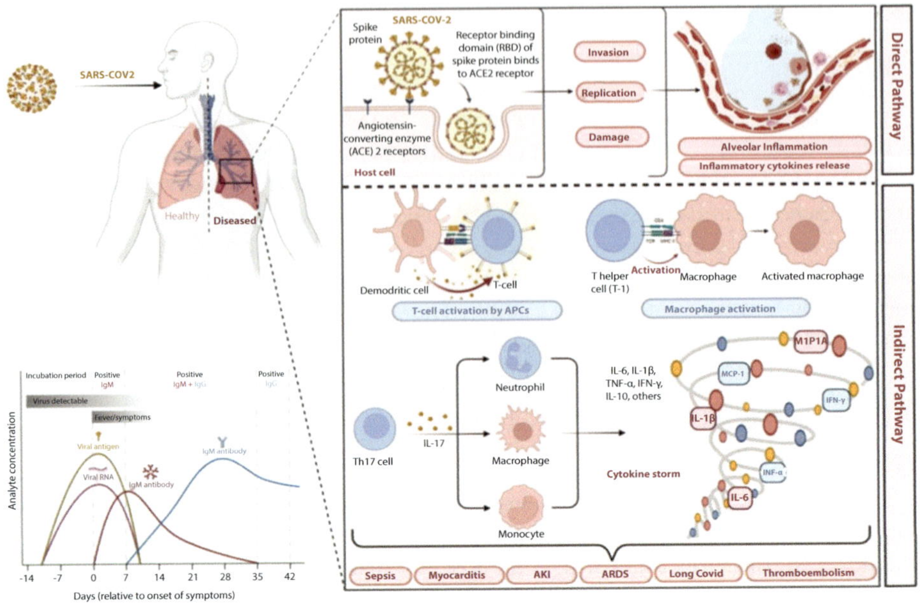

FIGURE 1.1 Direct and indirect immunopathological pathways of COVID-19 and associated serological timeline. Habib et al. (2025) BioRender.

1.2.3 Present Approach to the Management of COVID-19

The current management approaches for COVID-19 depend on disease severity. Treatments include antivirals (such as protease inhibitors), immunosuppressive agents (like corticosteroids, interleukin-6 inhibitors, and Janus kinase inhibitors), anticoagulants, vaccination, and supportive care. However, these drugs may exhibit variable responses and carry potential side effects, especially in individuals with advanced disease (COVID-19 Treatment Guidelines Panel, 2024; Soy et al., 2021).

Nowadays, vaccination is the primary method for preventing COVID-19, but due to the emergence of new viral variants and breakthrough cases, there is a need for adjuvant therapeutic strategies. Furthermore, many patients experience chronic symptoms (commonly referred to as long COVID), which highlights a continued need for therapeutic innovation (Chung et al., 2024).

1.3 EXPLORING THE REALM OF ACELLULAR REGENERATIVE THERAPIES

Regenerative medicine, particularly stem cells, has emerged as a promising approach to addressing COVID-19-induced immune responses and tissue damage (Bhalerao et al., 2021; Habib et al., 2025). Preclinical trials show considerable potential for MSCs, which exhibit immunomodulatory, anti-inflammatory, and reparative properties. However, it raises concerns about immunogenicity and tumorigenicity. Because of that, attention has shifted to cell-free therapy such as secretomes and exosomes, which offer similar therapeutic effects and benefits, but without the disadvantages associated with cell-based therapy (Habib et al., 2025).

The soluble fraction of the cell secretome is rich in bioactive molecules, including cytokines, growth factors, and nucleic acids. In contrast, the insoluble fraction of extracellular vesicles carries a specific cargo of bioactive molecules that may modulate inflammation, promote angiogenesis, and facilitate tissue repair (Habib et al., 2025). Among the insoluble fraction, exosomes are considered nanosized vesicles that mediate intercellular communication through paracrine signaling and represent a targeted, safer therapeutic platform (Mushtaq et al., 2024). These new approaches have paved the way for cell-free therapy for the treatment of COVID-19 and its associated complications

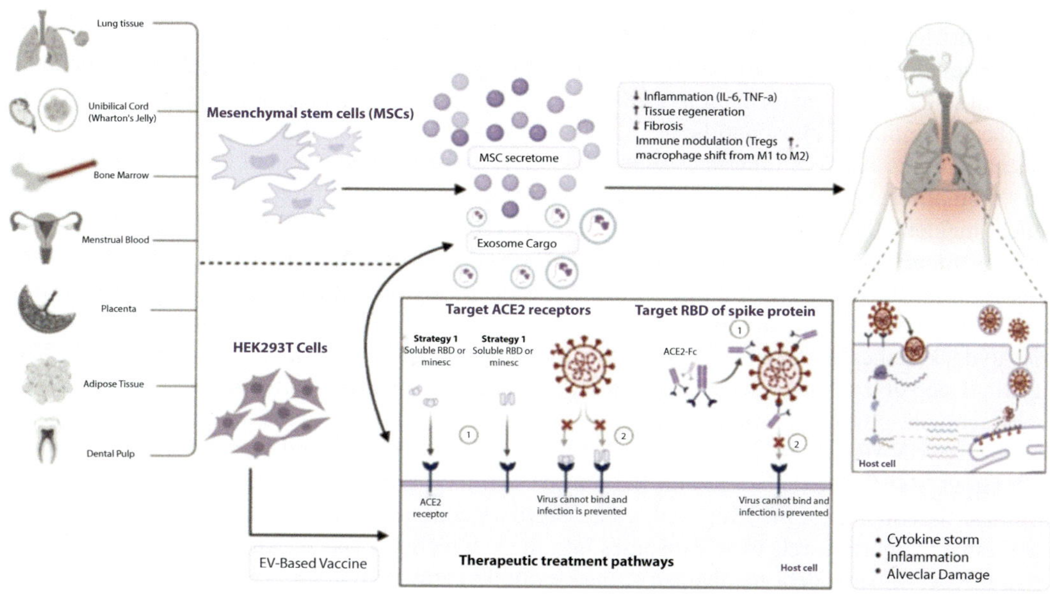

FIGURE 1.2 Therapeutic strategies using mesenchymal stem cell (MSC)-derived cell-free therapy molecules and vaccine-based strategies against COVID-19. Habib et al. (2025). BioRender.

(Figure 1.2). Additional research and development are needed to assess their therapeutic potential fully.

This chapter delves into the potential of MSC-derived exosomes (MSC-Exos) and secretome (MSC-Sec) for COVID-19 treatment. Being cell-derived products, these biologically active substances offer novel, safer treatment approaches for COVID-19-infected patients. The chapter also describes techniques for secretome and exosome biogenesis and purification, discusses preclinical and clinical data, and highlights translational impediments. Lastly, we aim to highlight the pros and cons and promise scalable, cell-free therapies for COVID-19 (Figure 1.2).

1.4 INTRODUCTION TO EXOSOMES

1.4.1 Biogenesis of Exosomes

Exosomes are a type of EV that range in size from 30 to 150 nm and play a critical role in intercellular communication (Yuan et al., 2023). Their biogenesis begins when the cell's proteins, lipids, and other small molecules are taken into the cell via endocytosis, forming early endosomes. Early endosomes mature into late endosomes, which subsequently form multivesicular bodies (MVBs). These MVBs contain internal intraluminal vesicles (ILVs). Over time, they can either fuse with lysosomes and be degraded or fuse with the plasma membrane, releasing their ILVs as exosomes into the extracellular space (Théry et al., 2002; Beshbishy et al., 2020). The two primary pathways regulating exosome formation are ESCRT-dependent and ESCRT-independent pathways (Zhang et al., 2019). The ESCRT-dependent pathway involves four multiprotein complexes, namely ESCRT-0, -I, -II, and -III, which contribute to cargo selection, facilitate vesicular budding into the endosomal lumen, and ultimately lead to the formation of ILVs within MVBs (Hade et al., 2021). The first step begins when ESCRT-0, which is composed of the hepatocyte growth factor (HGF)-regulated tyrosine kinase substrate (HRS) and a signal-transducing adaptor molecule, recognizes and binds mono-ubiquitinated cargo proteins. Thereafter, they get clustered into microdomains on the endosomal membrane (Han et al., 2022). Subsequently, ESCRT-I and ESCRT-II complexes cause membrane deformation, inward budding of the endosomal membrane, and cargo selection (Lee et al., 2024). Finally, ESCRT-III mediates membrane constriction and scission, completing ILV formation (Liu et al., 2024).

Another closely linked pathway to ESCRT is the Syndecan-Syntenin-Alix pathway, which plays a crucial role in exosome formation (Lee et al., 2024). Regulated by ARF6, PLD2, etc., which regulate the generation of phosphatidic acid, aiding syntenin in binding the membrane and promoting vesicle budding. Alix also interacts with the lipid lysobisphosphatidic acid to facilitate late endosome formation and with tetraspanins such as CD9 and CD63, enabling their sorting into exosomes, which, incidentally, serve as exosomal biomarkers (Han et al., 2022). Exosomes produced by this pathway are resistant to lysosomal degradation (Dyball et al., 2022). In the next phase, they must be transported to the plasma membrane for release, which is driven by members of the Ras superfamily of small GTPases, specifically Rab-27a, Rab-27b, Rab-11, and Rab-35. These proteins help the vesicles to dock, tether, and traffic to the plasma membrane (Han et al., 2022; Hessvik NP et al., 2018). The final step of exosome release is primarily mediated by soluble N-ethylmaleimide-sensitive factor attachment protein receptor (SNARE) proteins. This process begins when vesicle-associated SNAREs (v-SNAREs), such as VAMP2, VAMP3, and VAMP7, located on MVB membranes, pair with target SNAREs (t-SNAREs) on the plasma membrane, including the syntaxin and SNAP25 families. Together, these proteins form a stable ternary complex that drives the fusion of MVBs with the plasma membrane, releasing exosomes into the extracellular space (Xu et al., 2023a).

In ESCRT-independent mechanisms, the ceramide pathway plays a critical role. It involves the lipid ceramide, derived from sphingomyelin via a process dependent on the enzyme neutral sphingomyelinase 2 (nSMase2). Ceramide promotes the negative membrane curvature required for vesicle budding and formation (Liu et al., 2024; Trajkovic et al., 2008). It has been hypothesized that MVBs produced by ceramide-dependent and ESCRT-dependent mechanisms may have different fates; for example, ESCRT-produced exosomes tend to be degraded in lysosomes (Dyball et al., 2022).

1.4.2 Characteristics of iPSC-Derived Exosomes

Induced pluripotent stem cells (iPSCs) are essentially somatic cells reprogrammed into a surrogate embryonic stem cell by the introduction of the classical quartet of transcription factors Oct4, Sox2, Klf4, and c-Myc (Ibrahim et al., 2016). They can be further differentiated into any cell type of the three germ layers, i.e., MSCs (iMSCs) or endothelial cells (iECs), both of which are capable of immunomodulation and regeneration (Wang et al., 2021; Hu et al., 2015). For the isolation of iPSC-derived exosomes, iPSCs are cultured in serum-free conditions for about 48 hours, and the CM is then collected and sequentially processed by centrifugation and filtration to remove cellular debris. Finally, the resulting precipitate is washed with PBS and stored for further analysis and characterization (Ye et al., 2019).

In an experimental mouse model of ischemic limbs, iMSC-derived exosomes (iMSCs-Exo) and iPSC-endothelial cell-derived exosomes (iECs-Exo) have both shown vigorous proangiogenic activity. Mechanistic studies showed that iMSCs-Exo promoted angiogenesis by upregulating proangiogenic genes and proteins in endothelial cells, including VEGFA, TGFB1, HIF-1α, bFGF, VEGF, and angiogenin. It is also believed that angiogenesis can be mediated by paracrine signaling, which involves the delivery of miRNAs such as miR-126, miR-132, miR-21, and miR-210 (Hu et al., 2015). On the other hand, iECs-Exo stimulated angiogenesis by delivering miR-199b-5p, which inhibited the Jagged1/Notch1 signaling pathway, leading to increased VEGFR2 expression and enhanced sprouting angiogenesis in endothelial cells (Ye et al., 2019). iPSC-Exos have also shown significant therapeutic promise in treating myocardial infarction. Wang et al. reported that mouse iPSC-Exos were effectively taken up by cardiomyocytes and provided notable cytoprotection against oxidative stress, decreasing caspase 3/7 activation in vitro and in mouse models of ischemic myocardium by reducing active caspase-3 protein levels. Additionally, exosomes also transferred cardioprotective miRNAs, specifically miR-21 and miR-210, which were regulated by Nanog and HIF-1α, respectively, to protect cardiomyocytes against apoptosis (Wang et al., 2015).

1.5 EXOSOME ENGINEERING FORMATION OF INNOVATIVE NANO DRUG DELIVERY SYSTEMS

Many nanotechnological and bioengineering techniques have been developed to engineer exosome-mimetic vesicles that enhance their therapeutic efficacy. Exosome engineering is primarily intended to control their size, composition, loading efficiency, modify the cargo profile, and alter surface characteristics for targeted delivery, while retaining the beneficial biological characteristics of naïve exosomes. The most commonly employed methodological approaches include top-down, bottom-up, and biohybrid approaches (Maeki et al., 2025).

1.5.1 Cell Deformation (Top-Down Approaches)

The top-down approach involves mechanical breakdown of intact cells into nanovesicles; therefore, the membrane composition of the resulting exosomes is similar to that of natural exosomes. The process of cell deformation involves forcing the cells through membrane pores or microfluidic devices (Li et al., 2021). Jo et al. (2014) reported a device that employs centrifugal force and a micro-sized polycarbonate filter to deform cells. During extrusion, the polycarbonate filter creates surface tension, disrupting cells as they pass through the pores and forming nanovesicles. This strategy produces nanovesicles containing various cellular components, such as RNAs and proteins, and yields roughly 250 times more than natural exosome secretion, doubling the drug-loading capacity (Maeki et al., 2025). On the other hand, in microfluidic devices, cells are moved through hydrophilic microchannels. There, shear stress and pressure gradients cause the cells to elongate, then break, and reassemble as nanovesicles rich in membrane proteins and cytoplasmic contents (Zhu et al., 2018).

1.5.2 The Biohybrid Approach

The biohybrid approach primarily combines natural exosomes with synthetic lipid-based nanocarriers to create hybrid vesicles that leverage exosomes' innate targeting properties and synthetic systems' structural flexibility and drug-delivery potential. This is achieved using freeze–thaw fusion, incubation, polyethylene glycol (PEG)-mediated fusion, and membrane extrusion. In freeze–thaw fusion, vesicles undergo several rounds of freezing and thawing, which cause exosomes and liposomes to merge by forming temporary ice crystals that break lipid bilayers (Singh et al., 2021). Sato et al. (2016) demonstrated liposome fusion with HER2-expressing exosomes derived from genetically engineered CMS7 cells, enabling the incorporation of tumor-targeting membrane proteins, such as HER2, into the hybrid vesicles and maintaining their functional activity, albeit with potential stability and membrane integrity issues. On the other hand, the incubation technique is gentler, promoting fusion between exosomes and liposomes via electrostatic and hydrophobic interactions, thereby alleviating unnecessary shear stress and minimizing structural damage; however, it has low fusion efficiency (Lin et al., 2018; Liu A et al., 2022a). Lin et al. incubated exosomes with liposomes to generate hybrid nanoparticles that efficiently encapsulated large plasmids, including the CRISPR–Cas9 expression vector, for the genetic manipulation of MSCs (Lin et al., 2018). The fusion method generally uses PEG as a fusogen to achieve liposome fusion with exosomes, with the advantage that the PEG-modified fusion enhances circulation time but with lower cellular uptake (Liu A et al., 2022a). Piffoux et al. generated mTHPC-containing hybrid EVs through PEG-mediated fusion of EVs with Foslip liposomes to investigate the ability of hybrid EVs to deliver drugs (Piffoux et al., 2018). Hybrid EVs loaded with the anti-tumor photosensitizer mTHPC demonstrated better delivery and cytotoxicity in cancer cells. Lastly, the membrane extrusion method efficiently creates uniformly sized hybrid vesicles with high fusion efficiency by passing exosomes and liposomes through a size-regulated membrane pore. However, it may damage the EV's membrane (Maeki et al., 2025).

1.5.3 The Bottom-Up Approach

Bottom-up approach creates nanomaterials from small molecules (Xu et al., 2023b). A typical example of a bottom-up strategy is a liposome, a lipid-based NP often considered to resemble exosomes closely. Specific lipids are used to form lipid bilayers, which are then modified with chemical groups or engineered with surface proteins (Li et al., 2021). Maeki et al. prepared artificial exosomes using a microfluidic device by rapidly mixing an aqueous solution containing proteins and RNAs with a lipid solution comprising several lipid types mimicking cell-derived exosomes. The resulting particles ranged in diameter from 130 to 140 nm and displayed exosomal surface markers, including CD9, CD63, CD81, and integrin α6β4. The approach is significant because it enables more precise control over particle size and surface features than top-down methods.

1.6 THERAPEUTIC POTENTIAL OF ENGINEERED EXOSOMES AND EXOSOME-MIMETICS

Rayamajhi et al. developed a hybrid exosome (HE) by fusing small EVs with synthetic liposomes loaded with water-soluble doxorubicin, which demonstrated increased toxicity against cancer cells due to their efficient uptake and pH-dependent drug release in the acidic tumor microenvironment (Rayamajhi et al., 2019). Interestingly, HEs interacted more selectively with cancer cells than with normal cells, thereby reducing off-target effects and enhancing their safety profile. On the same note, Li et al. engineered an HE preparation by fusing CD47-expressing tumor-derived exosomes with cRGD-modified liposomes to co-deliver miR-497 and triptolide (TP) to treat cisplatin-resistant ovarian cancer. Treatment of ovarian cancer cells abrogated PI3K/Akt/mTOR signaling, leading to apoptosis and reduced tumor cell growth (Li et al., 2022). Treatment also helped to overcome chemoresistance, besides reducing systemic toxicity. Ning et al. (2021) have used a fusion approach to develop a diagnostic test for COVID-19. They fused liposomes containing detection reagents with EVs isolated directly from patient plasma, thereby sensitively amplifying and detecting a SARS-CoV-2 gene target, thus overcoming the common limitations of plasma-based RT-qPCR, such as low viral RNA stability and dilution (Ning et al., 2021).

1.6.1 Essential Constituents of Exosome Cargo

1.6.1.1 Proteins

Exosomes carry a diverse range of proteins involved in cellular communication, disease progression, and therapeutic applications. Among the vast array of proteins in their cargo, tetraspanins—CD9, CD63, CD81, CD82, and CD37—are the key contributors to many exosomal functions, including recipient cell penetration, invasion, fusion, and cargo selection during payload packaging (Zhang et al., 2019). They are important biomarkers for the isolation and purification of exosomes (Rech et al., 2022). Heat shock proteins, including HSP70, HSP90, Hsc70, Hsp60, Hsp20, and Hsp27, constitute another important group that contributes to exosome release and cell signaling (Mashouri et al., 2019; Tenchov et al., 2022). Alix and TSG 101 are essential for multivesicular body formation and exosome biogenesis (Lee et al., 2024). Adhesion molecules, such as integrins and P-selectin, facilitate exosome uptake by recipient cells, while glycoproteins contribute to exosome targeting and cellular interactions. Further, Rab GTPases, flotillin, and dynamin mediate membrane transport, fusion, exosomal secretion, and cellular uptake (Rech et al., 2022). It was also demonstrated that growth factors and cytokines, such as TGF-β, TNF-α, and TRAIL, mediate the signaling process and exosome uptake. Additionally, some exosome proteins can regulate the tumor immune response, thereby preventing the immune system's response to tumors and promoting tumor evasion (Chen et al., 2024). Exosomes also have immunomodulation functions, as they carry major histocompatibility complex (MHC) of Class I and II molecules, enabling antigen presentation (Gurung et al., 2021).

1.6.1.2 Lipids

Lipids are essential components of exosomes, providing structural integrity and enhancing their function. A well-known example is ceramides, composed of sphingosine and fatty acids. Due to their special shape, they can facilitate negative membrane curvature, thus enabling inward budding of ceramide-containing membrane, which facilitates EV formation (Trajkovic et al., 2008). Additionally, ceramide can generate exosomes independently of the ESCRT-based pathway (Fyfe et al., 2023). Cholesterol is another key lipid in EVs, providing structural rigidity to the phospholipid bilayer of the exosome membrane. It can also facilitate the loading of vesicles from late endosomes into multivesicular bodies (Ghadami et al., 2023). It has also been established that cholesterol can participate in exosome secretion (Rocha et al., 2009).

Phosphatidylserine can introduce exosomes to macrophages, while phosphatidic acid is another critical lipid that is derived from lysophosphatidic acid and diacylglycerol or phosphatidylcholine through the enzyme phospholipase D (Kooijman et al., 2005). Like ceramide, it initiates spontaneous negative membrane curvature, which allows budding and exosome formation. Additionally, it has been reported to interact with syntenin, which recruits syndecan, CD63, and ALIX. Interestingly, EVs can also serve as reservoirs for bioactive lipids, including esterified fatty acids, such as arachidonic acid (AA), and polyunsaturated fatty acids (PUFAs). These substrates can be enzymatically converted into eicosanoids, which regulate inflammation, immunity, and hemostasis. Phuyal et al. (2015) have reported that increased levels of hexadecylglycerol (HG) in cells were associated with elevated exosome release compared with control cells. This can be explained by the fact that HG increases MVB fusion with the plasma membrane, thereby enabling exosome release.

1.6.1.3 Deoxyribonucleic Acids

The presence of DNA in exosomes plays a crucial role in cellular communication and has diagnostic significance. DNA can be identified in multiple forms within EVs, including single-stranded DNA (ssDNA), double-stranded DNA (dsDNA), genomic DNA (gDNA), and mitochondrial DNA (mtDNA) (Ghanam et al., 2022). It is pertinent to mention that the DNA packaged into EVs plays a vital role in cellular hemostasis (Elzanowska et al., 2021). Therefore, if it remained there without being loaded into exosomes, it could activate the cytoplasmic inflammatory sensing machinery, cGAS/STING, due to nuclear DNA accumulation in the cytoplasm (Ghanam et al., 2022). Moreover, the DNA cargo of apoptotic bodies can also participate in horizontal gene transfer, thereby influencing recipient cells and modulating signaling pathways within them (Malkin et al., 2020). From a clinical perspective, EV-associated DNA may be an excellent biomarker in liquid biopsies. In comparison with circulating tumor DNA (ctDNA), which is rapidly degraded in plasma, EV-DNA is encapsulated within a lipid bilayer that protects it from nucleases, offering exceptional stability (Herrero et al., 2023).

1.6.1.4 Ribonucleic Acids

RNA cargo in exosomes may exist in diverse biotypes that reflect the source cell's content (O'Brien et al., 2020). EVs predominantly contain small non-coding RNAs (ncRNAs), alongside fragments of messenger RNAs (mRNAs), ribosomal RNAs (rRNAs), long non-coding RNAs (lncRNAs), transfer RNA (tRNA), mitochondrial RNAs (mtRNAs), circular RNAs (circRNAs), small nucleolar RNA (snoRNAs), and small nuclear RNAs (snRNAs) (Lee et al., 2024). Among these, microRNAs have been extensively studied for their critical role in various cellular functions, i.e., cell development, proliferation, and differentiation; apoptosis; and immune regulation (Kim et al., 2017). Also, miRNA can function in gene repression (Narang et al., 2022). Transfer RNA (tRNA) fragments are the second-most-abundant RNA biotype after miRNAs and play essential roles in regulating gene expression and epigenetic processes, particularly in immunomodulation and cancer progression (Narang et al., 2022). Furthermore, long non-coding RNAs (lncRNAs) have diverse biological functions; for example, serving as scaffolds for the formation and localization of protein complexes that regulate gene expression and transcription (Di Liegro et al., 2017). Further, circRNAs can act as miRNA sponges, in addition to their function in regulating splicing and transcription. It has also been reported that mRNAs in EVs can be translated into functional proteins in recipient cells.

1.7 THE ROLE OF EXOSOMES AND SECRETOME IN PARACRINE SIGNALING

Exosomes, after their release, are internalized by target cells via endocytosis, either clathrin-dependent or clathrin-independent (Isaac et al., 2021). Alternatively, phagocytosis is another mechanism of exosome internalization by macrophages and dendritic cells, which depends on the actin cytoskeleton, phosphatidylinositol 3-kinase (PI3K), and dynamin 2 (McKelvey et al., 2015). The exosome-bound FasL, TRAIL, and TNF target the interacting cell to initiate pro-apoptotic signaling. MSC-derived exosomes (MSC-Exos) from MSCs can modulate immunity by polarizing macrophages, shifting pro-inflammatory M1 macrophages toward an anti-inflammatory M2 phenotype. This transition can be facilitated by specific exosomal miRNAs that regulate Toll-like receptor and NF-κB signaling pathways (Yi et al., 2025). Furthermore, they decrease pro-inflammatory factors such as TNF-α, IL-6, IFN-γ, and IL-17, and increase anti-inflammatory factors, i.e., IL-4, IL-10, and TGF-β (Jayaramayya et al., 2020). In addition, they inhibit T cell proliferation and cytotoxic functions and promote the generation of Tregs, thereby maintaining immune tolerance (Yi et al., 2025). In the study by Mohammadi et al. (2021), they evaluated the ability of umbilical cord MSC-derived exosomes (XOs) in hybrid alginate microcapsules (AlgXO) to mitigate immune rejection of implanted rat islets in diabetic mice. Interestingly, AlgXO-treated mice maintained euglycemia for over 170 days, demonstrating long-term graft survival. Also, exosomes suppressed CD3/CD28-activated human peripheral blood mononuclear cells (PBMCs) and reduced cytokine secretion, including IL-2, IL-6, IL-12p70, IL-22, and TNFα (Mohammadi et al., 2021). It has been reported that Exo-hASCs enhanced the inhibition of CD4+ and CD8+ T cell proliferation, impaired their differentiation into effector-memory (CD4+) and terminally differentiated (CD8+) cells, and decreased IFN-γ production in both T cell subsets, thereby suppressing inflammatory disorders (Blázquez et al., 2014).

On the other hand, MSC-SCE contains many components, such as cytokines, chemokines, growth factors, and extracellular vesicles, that promote tissue repair, modulate the immune response, and support tissue regeneration (Prado-Yupanqui et al., 2025). One critical immunomodulatory factor is PGE2, which suppresses NK cell and T cell proliferation while promoting macrophage polarization toward the M2 phenotype. Additional components, such as galectin-1 and -9, inhibit cytokine release and T- and B-cell activity (Muñoz-Pérez et al., 2021). MSC-SCE also expresses thrombospondin 1 (TSP1), which suppresses TGF-β/Smad signaling, thereby reducing NK cell proliferation and cytotoxicity. It was also suggested that the MSC-SCE could act in multifactorial mechanisms to mitigate COVID-19 infection by restoring lung homeostasis by secreting angiopoietin 1 (Ang1) and keratinocyte growth factor (KGF), which play an essential role in alveolar capillary barrier repair (Mahajan et al., 2021). In a study by Kudinov et al., it was demonstrated that freeze-dried secretomes from 2D- and 3D-cultured placental multipotent MSCs (MMSCs) and lung fibroblasts (LFBs) significantly improved the outcomes in a lethal acute lung injury (ALI) mouse model when administered via inhalation. Multiple mechanisms mediated the therapeutic benefits; for example, elevating human KGF protein levels promoted the repair of alveolar epithelial cells (Kudinov et al., 2022).

1.8 ISOLATION TECHNIQUES OF EXTRACELLULAR VESICLES

EVs can be found in various body fluids, such as blood, urine, and saliva. They can be classified into three main groups based on size and origin. The smallest are exosomes, measuring 30–150 nm, generated by inward budding of endosomal membranes and released upon fusion of multivesicular bodies with the plasma membrane. Isolating exosomes with nanosized particles from different body fluids remains challenging due to their small size and the presence of contaminating proteins. Hence, various isolation protocols have been developed to address these problems using different techniques, which are discussed in the following section along with their underlying principles.

1.8.1 Ultracentrifugation-Based Methods

Ultracentrifugation is the most commonly used technique for isolating exosomes, employing high centrifugal force to separate particles based on size, density, and shape (Li et al., 2017). It includes two main types: differential ultracentrifugation and density gradient ultracentrifugation. Differential ultracentrifugation (DUC) separates particles based on size and density. It involves multiple sequential steps, ranging from a low-speed spin at 300 g for 10 minutes, followed by a high-speed spin at 2,000 g for 10 minutes to remove large cells and debris. Thereafter, an intermediate step is applied at 10,000 g for 30 min to remove MVs and particles larger than 100 nm. Finally, it ends with high-speed ultracentrifugation at 100,000 g for 70 min to sediment small EVs smaller than 100 nm, in which the exosome size lies (Livshits et al., 2015). DUC offers many benefits, including its low cost and ability to process large sample volumes with a high exosome yield. However, the technique has limitations, including purity, labor intensity, and a time-consuming nature. Also, due to the high centrifugal force, exosomes can be damaged, thereby affecting their structural integrity and functional properties.

Another commonly used technique is density-gradient ultracentrifugation (DGU), which separates particles based on both density and size. Two primary modalities are used: isopycnic (equilibrium) and moving-zone (rate-zonal) centrifugation (Liu A et al., 2022a). Isopycnic ultracentrifugation separates particles based on their densities. Firstly, a density gradient medium is prepared in layers of increasing density from top to bottom, encompassing exosome densities of 1.10 to 1.21 g/mL. During centrifugation, particles, including exosomes, move until they reach a static position in a medium of density similar to that of each component (Yang et al., 2020). On the other hand, rate-zonal density gradient ultracentrifugation (RZC) separates particles based on both their density and size. In this method, the gradient density medium has a lower density than that of any of the solutes. Therefore, after centrifugation, all solutes will be separated based on their different sizes rather than solely on density, allowing the separation of vesicles of similar density but different sizes. DGU offers several advantages, notably its capacity to isolate exosomes with high purity, efficiency, and integrity. The limitations include time intensity and the complexity of the protocol. Additionally, it cannot process large sample volumes, resulting in low throughput (Jia et al., 2022).

1.8.2 Size-Based Techniques

Size-exclusion chromatography (SEC) is a widely used technique for isolating EVs, particularly exosomes, based on their size or hydrodynamic radius. This technique consists of two main phases—the mobile phase and the stationary phase. It works when the sample, i.e., the mobile phase, is passed through the porous gel-filtration polymer, which serves as the stationary phase. As the sample flows through the column, larger particles, such as exosomes, are eluted first because they can traverse only a few pores, whereas smaller solutes diffuse into the pores and elute more slowly (Liu WZ et al., 2022b). The efficiency of SEC is highly dependent on column characteristics, including the resin type, pore size, and column height (Lane et al., 2019; Guo et al., 2021). Resins with smaller pore sizes, such as CL-6B and Sepharose 4B, yielded higher EV yields (Ter-Ovanesyan et al., 2021). Additionally, increasing the column height from 10 to 20 mL enhanced the separation of EVs from free proteins and albumin, thereby increasing EV purity.

SEC has many advantages over other techniques. It can maintain the biological activity and integrity of exosomes, preserving their biophysical and functional characteristics, unlike differential ultracentrifugation, which can damage them. In addition, the SEC showed low levels of contamination of the exosomes with other particles, such as proteins, indicating high purity (Gámez-Valero et al., 2016). Additionally, it is considered a fast procedure, consuming only minutes compared to other procedures that can last for hours. However, it has significant limitations, including its inability to process large amounts of samples, which leads to low yields and limits its application in clinical settings.

Ultrafiltration uses membrane filters with defined molecular weight cutoffs (MWCOs) to separate particles, including exosomes, based on size (Liu A et al., 2022a). Firstly, the large particles are separated using membrane filters with pore diameters of 0.22 and 0.45 µm. Subsequently, the fluid is filtered again through a filter with an MWCO of 5 to 100 kDa or a pore diameter of 0.22 to 0.1 µm, eliminating minor contaminations from the exosomes (Konoshenko et al., 2018). This method offers many advantages for exosome isolation, as it is fast and requires no specialized equipment, making it suitable for clinical use. However, it has potential disadvantages, including the potential to damage exosomes through shear stress and to cause exosomal loss due to clogging and membrane trapping of vesicles.

1.8.3 Microfluidic Technologies for Exosome Isolation

Microfluidic technologies are widely employed and integrate multiple steps, including exosome isolation, detection, and analysis, using a single lab-on-a-chip device (Jia et al., 2022). These devices can separate exosomes based on either their physical or immunoaffinity properties. They have many examples that can be listed, such as filtration, Deterministic Lateral Displacement (DLD), Acoustofluidics, and Viscoelastic flow (Ozcelik et al., 2023). Filtration isolates exosomes based on their size, using nanomembranes that allow the passage of vesicles smaller than the pore size. This method, although suitable for isolating large samples with high yields, can cause device clogging, leading to device malfunction. DLD is another method that utilizes gradient pillar arrays to separate particles as they flow through them, allowing large particles to be pumped laterally as they pass. In contrast, smaller particles will follow a straight path without lateral displacement (Contreras-Naranjo et al., 2017). This method, however, poses a high risk of clogging. Further, acoustofluidic devices use ultrasound waves to generate size-dependent acoustic forces that trap particles within microfluidic channels. It is a contact-free approach that separates exosomes with minimal to no damage. Another microfluidic approach that yields high-purity exosomes is the viscoelastic flow-based device, which separates exosomes based on elastic lift forces that increase with particle volume. On the other hand, immunoaffinity-based microfluidics targets exosomal markers with specific antibodies immobilized on hard surfaces, enabling the isolation of high-purity, specific exosomal subpopulations (Contreras-Naranjo et al., 2017). In conclusion, despite the numerous advantages of microfluidic technology, the lack of standardization can introduce a significant limitation, hindering further improvement, as many microfluidic tools remain at the proof-of-concept stage (Ozcelik et al., 2023).

Immunoaffinity-based exosome isolation works by allowing specific binding between antibodies coated on magnetic beads to target exosome surface markers such as CD9, CD63, and CD81. This method enables precise exosome isolation, including specific exosomal subtypes, yielding high purity. However, some exosomes may lack these markers, leading to lower exosome yields. Furthermore, this method is time-consuming due to its long incubation period. Additionally, it is considered an expensive approach due to the high cost of the antibodies (Chen et al., 2022). Moreover, the need for exosome elution can cause damage to the exosomes.

Lastly, precipitation-based exosome isolation alters exosome solubility. It works by adding water-excluding polymers, such as the most commonly used, PEG, which draws water away from the particles, causing the less soluble components, including exosomes, to aggregate during low-speed centrifugation (Popović et al., 2017). This method offers several advantages, including simplicity, low time consumption, and low cost, making it highly practical for both clinical and research applications (Sani et al., 2024). However, a significant limitation is the high risk of contamination with proteins and polymeric substances (Konoshenko et al., 2018).

1.9 ISOLATION AND OPTIMIZATION OF MSC-SECR

MSC-SECR contains all bioactive molecules secreted by MSCs, including soluble factors—growth factors, cytokines, chemokines, and hormones—as well as insoluble factors, such as EVs. These paracrine factors facilitate tissue repair, immunomodulation, and angiogenesis (Singh et al., 2025). MSC-SECR are being utilized in cell-free therapy as an alternative to MSC transplantation due to their reduced immunogenicity, simplicity of use, and lower cost (Smolinská et al., 2023). The tissue source of MSCs is a critical determinant of the composition of their derivative secretome. For example, umbilical cord-derived MSCs (UC-MSCs) exhibit significant immunosuppressive effects by releasing indoleamine 2,3-dioxygenase (Trigo et al., 2025). However, adipose-derived stem cells (ASCs) can also induce immunomodulation, primarily by inhibiting lymphocyte proliferation, thereby shifting macrophages toward the M2 anti-inflammatory phenotype. Moreover, it downregulates the pro-inflammatory cytokines such as TNF-α and IL-1β. Their secretome is rich in bioactive factors, including VEGF, HGF, TGF-β1, and interleukins such as IL-1Ra, IL-6, and IL-10 (Al-Ghadban et al., 2020). Gupta et al. (2024) used human cerebral microvascular endothelial cells (hCMEC) for exposure to ionizing radiation. Prior treatment of the cells with iMSC-CM improved cell viability, enhanced angiogenesis, and suppressed pro-inflammatory cytokine release. These data showed that CM contained multiple factors with both proangiogenic and immunosuppressive effects, in addition to pro-survival factors.

1.9.1 Factors That Influence the Production of Secretomes

The process of secretome production starts by culturing MSCs in a medium supplemented with fetal bovine serum (FBS) until they reach 60–80% confluency. Then, cells are thoroughly washed with phosphate-buffered saline (PBS) and incubated again in serum-free medium for 12 to 48 hours before CM is harvested (Andreev et al., 2025).

1.9.1.1 Two-Dimensional and Three-Dimensional Culture Systems

The traditional two-dimensional (2D) monolayer culture has been widely used for secretome production due to its simplicity and ease of use. However, 2D culturing does not fully mimic their natural habitat (Kapałczyńska et al., 2018). Alternatively, the three-dimensional (3D) culture systems are superior for secretome production. These systems include spheroids, extracellular matrices, and synthetic scaffolds. They mimic the cells' physiological environment, thus supporting the natural paracrine activity. It has been demonstrated that spheroid cultures can restrict oxygen entry to the inner cells, resulting in a hypoxic environment that enhances the reduction of inflammation and improves tissue regeneration properties (Singh et al., 2025). In addition, it was shown to promote immunomodulation and reduce apoptosis compared to 2D culture.

1.9.1.2 Role of Oxygen Conditions

Oxygen levels also play a critical role in shaping secretome composition. Standard practice in cell culture employs normoxia (21% O2). However, these levels exceed the physiological O2 levels experienced by most stem cells in the body, with bone marrow oxygen concentrations ranging from 1–7% to about 12% in blood (Trigo et al., 2025). Therefore, hypoxic culture conditions (0.5–5% O_2) have been used to enhance MSC proliferation and migration. It was demonstrated that hypoxia (1% O_2) increased MSC secretion of proangiogenic factors, such as VEGF and BFGF. Additionally, a 2% concentration was found to reduce tumorigenic risk (Hu et al., 2018) while 5% O_2 has been shown to enhance autophagy by modulating the AMPK/mTOR signaling pathway (Yang et al., 2022). In the study by Chacko et al. (2010), MSCs were cultured under hypoxia (0.5% O2) for 24h, which stabilized HIF-1α expression and led to enhanced release of angiogenic factors, such as VEGF.

1.9.1.3 Impact of Pro-Inflammatory Cytokine Preconditioning

Exposing MSCs to pro-inflammatory cytokines, especially the most commonly used ones—tumor necrosis factor alpha (TNF-α) and interferon-gamma (IFN-γ)—can significantly alter their immunomodulatory properties. A study by Liang YC et al. (2019) clearly showed that MSCs stimulated with TNF-α upregulate miR-146a, which has strong anti-inflammatory properties. IFN-γ, on the other hand, downregulates the expression of miR149, which is involved in inflammation, while increasing miR210 levels (Song et al., 2025). In addition, IFN-γ increases the expression of IDO and other immunomodulatory molecules, such as TGF-β, NO, and IL-10, thereby suppressing natural killer cell cytotoxicity and reducing lymphocyte proliferation (Uberti et al., 2022).

1.9.2 Concentration and Lyophilization of MSC-SECR

After collection of the CM, it should be centrifuged at 3,500 × g for 10 minutes to remove apoptotic bodies and cell debris from the supernatant. Then, the collected supernatant is subjected to MSC-SECR purification using methods such as ultrafiltration or tangential flow filtration, thereby improving the isolation of secretomes (Chouaib et al., 2023). Once harvested, lyophilization enables easy storage and long-term stability of MSCs by maintaining them in a dry state (Paresishvili et al., 2024). Nevertheless, this strategy has some disadvantages, including disruption of cellular membrane integrity during freeze-drying and cellular stress during rehydration, which can reduce viability.

1.9.3 Importance of Quality Control

Quality control (QC) is a critical step that ensures the safety, purity, and efficacy of exosomes, thereby minimizing risks to patients and guaranteeing safe clinical use. There are multiple QC tests considered essential, which can be categorized into EV characterization tests and safety and purity testing (Zheng et al., 2024). Characterization of EVs involves methods that detect and quantify them. These can be either molecular-composition-based or physicochemical-property-based methods. The composition analysis involves testing EVs to detect the presence of positive exosomal markers to eliminate other impurities. On the other hand, the primary goal of physicochemical analysis is to characterize EVs based on their size and quantity. It is also vital to ensure the sterility of the EVs by eliminating viruses and bacteria. Therefore, it is mandatory to test for bacterial endotoxins, mycoplasmas, and viruses (Lau et al., 2022).

1.10 REPARATIVE MECHANISMS OF CELL-FREE THERAPY MOLECULES AND VIRAL-HOST INFECTIVE MECHANISMS

1.10.1 A Comprehensive Review of the Immunomodulation

The exact pathogenesis of COVID-19 is not yet fully understood. Within the alveoli, COVID-19 primarily infects type II alveolar cells because they express high levels of ACE2, a cellular receptor that facilitates viral entry. The infection itself leads to upregulation of ACE2 expression in the heart and small intestine. Thus, it explains the complex clinical spectrum of the disease, particularly regarding lung injury, cardiac dysfunction, and gastrointestinal manifestations. An increase in pro-inflammatory cytokines induces ACE2 expression, which might help stop a cytokine storm by suppressing inflammatory signaling pathways (Costela-Ruiz et al., 2020). However, studies show that the symptoms and complications of COVID-19 are mainly linked to the virus's attack on specific cells in the respiratory tract, such as epithelial cells, vascular ECs, and macrophages, all of which express ACE2, which acts as a receptor. This immunological reaction is associated with EC dysfunction, leading to the loss of barrier integrity and increased levels of inflammatory cytokines that attract immune cells, such as monocytes and T lymphocytes. This results in blood vessel inflammation and plays a significant role in airway damage. In severe cases, T cells undergo exhaustion, leading to lymphopenia and functional impairment; macrophages become hyperactivated and release large amounts of pro-inflammatory cytokines, including IL-6, IL-1β, and TNF-α. It is also well known that this viral infection can trigger ARDS, which may result in fatal outcomes in severe cases, due to excessive activation and release of immune molecules, including interleukins (IL-1, IL-6, IL-10, IL-17), interferon-gamma (IFN-γ), and tumor necrosis factor alpha (TNF-α). Such uncontrolled immune activation not only impairs adequate viral clearance but also leads to widespread tissue damage and multi-organ failure (Costela-Ruiz et al., 2020) (Figure 1.1).

EVs can either promote or suppress inflammatory responses and play a significant role in modulating ARDS by expressing Runx1 isoforms (Runx1p66/p52), which promote EC proliferation and tissue repair. Stem cell-derived EVs, particularly those from MSCs, exert potent immunomodulatory effects. Due to their diverse roles, EVs represent a double-edged sword in the context of viral infections, such as COVID-19 (Xia et al., 2021). Pathologically, the hyperinflammatory state leads to epithelial cell shedding, hyaline membrane formation, and interstitial and vascular edema, impairing gas exchange, causing hypoxia and respiratory distress, and increasing susceptibility to secondary infections (Costela-Ruiz et al., 2020).

Throughout the COVID-19 pandemic, several therapeutic approaches were explored to reduce mortality. One promising intervention was ExoFlo, an exosome-rich fluid derived from bone marrow MSCs. Clinical trials demonstrated its potential in managing COVID-19 complications, including lung injury and ARDS. Exosomes are crucial mediators of intercellular communication by transferring bioactive molecules to the target cells, thereby exerting anti-inflammatory and immunomodulatory effects (Li G et al., 2024). Animal studies have shown that MSC-derived exosomes contribute to tissue repair and inflammation attenuation by decreasing levels of IL-1β, TNF-α, IL-6, MIP-1, MIP-2, and CXCL2, and by enhancing IL-10 production. Additionally, exosomes alleviate oxidative stress, reduce pulmonary fibrosis, and support lung tissue remodeling. Notably, they also appear to restore mitochondrial function in damaged lung tissue. Moreover, exosome therapy has been shown to promote endothelial regeneration and restore vascular function—both critical components of COVID-19 recovery. While current evidence supports the therapeutic potential of exosomes, further preclinical and clinical studies are necessary to confirm their safety, efficacy, and mechanisms of action (Li G et al., 2024).

1.10.2 Role in Tissue Regeneration and Reparative Properties

The MSC-SECR may represent a new therapy for COVID-19 by its potent anti-inflammatory and anti-fibrotic mechanisms. In addition to other key features such as immune modulation, tissue repair, and regeneration, as well as proangiogenic activity, fibroids and epithelial-mesenchymal transition represent vital repair mechanisms for damaged tissue (Gazzaniga et al., 2024). Evidence from clinical trials indicates that MSCs have shown efficacy in reducing COVID-19 complications and suppressing fibrotic development (Li Z et al., 2020). Data from experimental models suggest that MSC-Exos can prevent and reverse pulmonary fibrosis by modulating monocyte phenotypic states, suppressing monocyte infiltration, and repressing profibrotic gene expression (e.g., IL-10, *Col1a1*) (Dinh et al., 2020). Secretome therapy has been associated with multiple advantages, including its availability as an off-the-shelf product, ease of handling and outpatient use, and improved safety due to a lower risk of tumorigenicity compared with living cells (Bari et al., 2021). It is anticipated that pluripotent stem cell-derived secretions could significantly ease the economic burden for patients.

Although MSCs and MSC-Exos are capable of immune modulation, tissue repair, and regeneration, they can also be actively targeted to the site of injury and prevent fibrotic progression in COVID-19. MSCs also promote the repair of damaged alveolar epithelial structure by secreting TGF-α, TGF-β, HGF, EGF, and angiopoietin (Gazzaniga et al., 2024). When MSCs were cultured with alveolar epithelial cells in vitro, they could differentiate into type II alveolar epithelial cells and facilitate the repair of damaged alveolar structures. In addition, MSCs secrete proangiogenic factors that activate both the extrinsic and intrinsic coagulation pathways, thereby promoting hemostasis. In animal experiments, macrophages can secrete crucial bioactive molecules that reduce inflammation and enhance the release of cellular vesicles from MSCs, resulting in a 43% reduction in extravascular lung water, which, in turn, improves lung function by lowering lung permeability. During ARDS, MSCs accumulate in the lungs, secrete anti-inflammatory and antimicrobial effects, and exhibit anti-apoptotic effects, significantly enhancing respiratory function after a single injection without side effects (Li Z et al., 2020). Published data in regenerative medicine demonstrate that MSC-derived exosomes represent a viable, cell-free candidate therapy for tissue regeneration and anti-fibrotic effects.

1.10.3 Potential for Exosomal RNA Delivery in Antiviral Strategies

The unique ability of exosomes to transport functional molecules to diseased cells further supports their use as therapeutic agents, enabling them to cross multiple biological barriers to reach target cells (Zhang et al., 2019). In addition, they serve as mediators in paracrine and endocrine communication; all these qualities make exosomes a promising and feasible therapeutic option (Sengupta et al., 2020). Indeed, the structural composition of exosomes, notably lipids and proteins, influences their pharmacokinetics and may enhance bioavailability while mitigating side effects. Exosomes obtained from tissues or bodily fluids exhibit favorable biodistribution and biocompatibility. Thus, exosomes are crucial mediators in both physiological and pathological mechanisms (Kalluri et al., 2020). The transfer of functional mitochondrial or mitochondrial DNA from MSCs to recipient cells can restore cellular aerobic respiration in defective mitochondria or modulate T cell function. The study demonstrated that EVs from MSCs, particularly during lung injury, transport metabolites, DNA, miRNA, mRNA, and proteins to cells, thereby promoting lung healing and functional restoration. miRNAs have been established as key players in physiological processes, including genetic regulation. Furthermore, exosomes have been shown to capture membrane-active virulence factors, such as pore-forming toxins, thereby preventing viral uptake by host cells (Zhang et al., 2019). By encapsulating mRNA within their bilipid layers, exosomes protect it from degradation by extracellular enzymes, thus ensuring its stability during transfer to target cells. Exosomes enter target cells primarily via endocytosis, membrane fusion, or receptor-mediated uptake. Additionally, their immunomodulatory functions are facilitated by miRNA-mediated regulation of gene expression and by the activation of several signaling pathways via surface ligands on extracellular vesicles (Kalluri et al., 2020). Ang-1mRNA (an angiogenic trophic factor) found in extracellular vesicles secreted by MSCs is a factor essential for leukocyte-endothelial interactions and vascular barrier maintenance. Ang-1 mRNA is considered critical for endothelial cell stability. Exosome quantity and molecular composition may change in response to environmental or cellular stress, altering cell behavior and activating the immune system (El Safadi et al., 2024).

1.11 PRECLINICAL STUDIES USING CELL-FREE-BASED APPROACHES FOR COVID-19

1.11.1 Exosome-Based Cell-Free Therapy for COVID-19 and ARDS in Animal Models

Exosome-based therapy contains a variety of immunomodulatory and protective components, but only a subset demonstrated significant therapeutic and immune-protective effects. Examples of those components include SARS-CoV-2 spike and nucleocapsid protein, neurotrophic factors (AREG, LIF, HGF, TSG-6), siRNA molecules, ACE2 receptors, and receptor-binding domain (RBD) antigens. Those factors demonstrated substantial improvements in immune responses, antiviral protection, and reduced inflammation, as shown in the studies in Table 1.1. For example, Cacciottolo et al. administered exosomes displaying the SARS-CoV-2 spike protein via intramuscular injection

TABLE 1.1
Clinical Trials Involving COVID-19 ARDS Patients

Title and ID	Condition	Intervention	Country	Sample Size	Phase/Status	ROA	Dose
A Pilot Clinical Study on Inhalation of MSCs Exosomes for Treating Severe Novel Coronavirus Pneumonia (NCT04276987)	Coronavirus	Biological MSCs-derived exosomes	China	Experimental group n =7 No placebo.	1 (Completed)	Aerosol inhalation	Five times aerosol inhalation of MSCs-derived exosomes (2.0 × 10E8 nano vesicles/3 ml at Day 1, 2, 3, 4, 5).
Evaluation of Safety and Efficiency of the Method of Exosome Inhalation in SARS-CoV-2 Associated Pneumonia (NCT0449124)	COVID-19, SARS-CoV-2 Pneumonia	Drug EXO 1 inhalation, Drug EXO 2 inhalation, Drug Placebo inhalation	Russia	N = 30; Experimental group 1 = 10 Experimental group 2 = 10 Placebo group = 10	1 and 2 (Completed)	Inhalation	Twice a day for 10 days, inhalation of 3 ml of a special solution containing 0.5 2×10^{10} 10^{10} nanoparticles (exosomes) of the Exo 1 (first type) and Exo 2 (second type), and Placebo (3 ml free of exosomes).
Title A Clinical Study of MSC Exosomes Nebulizer for the Treatment of ARDS (NCT04602104)	Acute Respiratory Distress Syndrome (ARDS)	Low dose hMSC-Exos; Medium dose hMSC-Exos; High dose hMSC-Exos	China	Not provided	1 and 2 (Completed)	Aerosol inhalation	Basic treatment and 7 times aerosol inhalation of hMSC-Exos (Low dose 2.0×10^8, Medium dose 8.0×10^8, High dose 16.0×10^8 particles at Day 1, 2, 3, 4, 5, 6, 7). Basic treatment and 7 times aerosol inhalation of hMSC-Exos (Dosage 1 a quarter of MTD/day, Dosage 2 MTD/day on days 1, 2, 3, 4, 5, 6, 7). No hMSC-derived exosomes: Basic treatment and 7 times aerosol inhalation of normal saline (at Days 1, 2, 3, 4, 5, 6, 7).

(Continued)

TABLE 1.1 CONTINUED
Clinical Trials Involving COVID-19 ARDS Patients

Title and ID	Condition	Intervention	Country	Sample Size	Phase/Status	ROA	Dose
Extracellular Vesicle Infusion Treatment for COVID-19 Associated ARDS (NCT04493242)	COVID-19, ARDS	Biological ExoFlo, or normal saline	USA	N = 102	2 (Completed)	IV	Dose 1 Normal saline 90 mL and ExoFlo 10 mL; Dose 2 Normal saline 85 mL and ExoFlo 15 mL; Placebo Normal saline 100 mL.
Intravenous Infusion of CAP-1002 in Patients With COVID-19 (NCT04623671)	COVID-19	Biological CAP-1002 or Placebo	USA	Not provided	2 (Completed)	Peripheral Infusion	CAP-1002 100-mL (total volume) infusion of 150M Cardiosphere-Derived Cells (CDCs) in 5% Human Serum Albumin (HSA).
Zofin (Organicell Flow) for Patients With COVID-19 (NCT 0484445)	Coronavirus Infection, COVID-19, SARS	Biological Zofin or Placebo	USA	N = 20 Experimental group = 10 Placebo group = 10	1 and 2 (Completed)	IV	The Zofin dose will be diluted in 100 mL of sterile saline. The placebo dose will be diluted in 100 mL of sterile saline. Both will be administered IV with 1 mL in addition to standard care on days 0, 4, and 8.
Treatment of Severe COVID-19 Patients Using Secretome of Hypoxia-MSCs in Indonesia (NCT 04753476)	Covid-19 Cytokine storm	Biological MSC-SECR Or Standard COVID-19 treatment.	Indonesia	N = 48 Experimental group = 24 Placebo group = 24	2 (Unknown)	I.M	Injection of Hypoxic MSCs-SECR: Day 1 1 cc every 12 hours; Day 2 1 cc every 12 hours Day 3 1 cc every 12 hours; Additionally, Patients will receive standard COVID-19 treatment according to the National protocol. CControl group patients will receive the Standard treatment for COVID-19, as per the National protocol.
Zofin to Treat COVID-19 Long Haulers (NCT05228899)	COVID-19	Drug Zofin or Placebo	USA	N = 30 Experimental group = 15; Placebo group = 15	1 and 2 (Completed)	IV	Group 1 1 mL of Zofin diluted with 100ml of sterile saline on day 0, day 4, and day 8, containing 2-5 × 10^{11} particles/mL IV Group 2 1mL of placebo diluted with 100ml of sterile saline on day 0, day 4, and day 8, by IV.

Title and ID	Condition	Intervention	Country	Sample Size	Phase/Status	ROA	Dose
MSCs Secretome In Severe Cases of COVID-19 (NCT05122234)	COVID-19	Biological MSC-SECR Or Placebo Or Standard COVID-19 treatment	Indonesia	N = 40 Experimental group = 20 Placebo group = 20	3 (Completed)	IV	MSC-SECR was administered once at a dose of 15 mL in 100 mL of normal saline. The administration is done IV for 60 minutes. Drug Standard treatment of COVID-19 Standard therapy of COVID-19 based on national protocol. Placebo Normal saline Drug Standard treatment of COVID-19 Standard therapy of COVID-19 based on national protocol.
Menstrual Blood Stem Cells in Severe Covid-19 (NCT05019287)	Covid19 Cytokine Storm	Allogeneic human MSC-SECR or IV saline	Iran	N = 39	1 and 2 (Completed)	IV	Not provided
MSC Therapy in Patients With COVID-19 Pneumonia (NCT04713878)	Coronavirus Disease 2019 (COVID-19) Pneumonia	MSCs	Turkey	N = 21	NA (Completed)	IV	Dosage of MSCs: 1 million cells/kg IV on day 0; 1 million cells/kg IV on day 2; 1 million cells/kg IV on day 4 No control group.
A Study of ADR-001 in Patients With Severe Pneumonia Caused by SARS-CoV-2 Infection (COVID-19) (NCT04888949)	SARS-CoV-2 Infection (COVID-19)	MSCs or Placebo	Japan	N = 15	2 (Completed)	IV	1×10^8 MSCs are administered once a week, for a total of 4 intravenous administrations. Placebo Commercially available saline.

(Continued)

TABLE 1.1 CONTINUED
Clinical Trials Involving COVID-19 ARDS Patients

Title and ID	Condition	Intervention	Country	Sample Size	Phase/Status	ROA	Dose
A Proof of Concept Study for the DNA Repair Driven by MSCs in Critical COVID-19 Patients (NCT04898088)	COVID-19 Pneumonia	MSCs	Turkey	N = 30	NA (Completed)	IV	Not provided
Treatment of Severe COVID-19 Pneumonia With Allogeneic MSCs (NCT04361942)	COVID-19 Pneumonia	MSCs or Placebo	Spain	N = 24	2 (Completed)	IV	IV injection of 1 million MSCs (MSV cells)/Kg in 100 ml saline solution. Other Names MSV (GMP-compliant MSCs manufactured by IBGM in Valladolid) placebo IV injection of 100 ml physiological saline solution containing no cells (Saline).
Study Evaluating the Safety and Efficacy of Autologous Non-Hematopoietic Peripheral Blood Stem Cells in COVID-19 (NCT04473170)	Coronavirus Disease (2019) (COVID-19)	Autologous Non-Hematopoietic Peripheral Blood Stem Cells (NHPBSC) or COVID-19 standard care	UAE	N = 146	1 and 2 (Completed)	Jet Nebulization	Not provided

Abbreviations AE (Adverse reaction), ARDS (Acute Respiratory Distress Syndrome), BGA (Blood Gas Analisis), CDCs (Cardiosphere-Derived Cells), COVID-19 (Coronavirus Disease, 2019), CRP (C-reactive protein), ECG (Electrocardiogram), HSA (Human Serum Albumin), hMSC-Exos (human Mesenchymal Stem Cell Exosomes), ICU (Intensive Care Unit), IL-6 (Interleukin 6), IMV (Invasive Mechanical Ventilation), LDH (Lactic Acid Dehydrogenase), LIF (Leukemia Inhibitory Factor), MSCs (Mesenchymal Stem Cells), MTD (Maximum Tolerated Dose), NCT (National Clinical Trial), NHPBSC (Non-Hematopoietic Peripheral Blood Stem Cells), ROA (Route of Administration), RT-PCR (Reverse Transcription Polymerase Chain Reaction), SAE (Severe Adverse Reaction), SARS (Severe Acute Respiratory Syndrome), SARS-CoV-2 (Severe Acute Respiratory Syndrome Coronavirus 2), SOFA (Sequential Organ Failure Assessment), SpO2 (Peripheral Oxygen Saturation), TTCI (Time to Clinical Improvement), TTCR (Time to Clinical Recovery), VEGF (Vascular Endothelial Growth Factor).

Source: Clinicaltrials.gov.

into mice. They demonstrated potent neutralizing antibody responses against both the Delta and Omicron BA.1 and BA.5 variants, alongside robust activation of CD4+ and CD8+ T cells, achieving significant virus neutralization with minimal protein dosage (Cacciottolo et al., 2023a). Another study by Cacciottolo et al. used a multivalent exosome vaccine containing the spike protein and additional viral antigens. They observed broad-spectrum immune responses, including potent humoral and cellular immunity, in immunized mice and rabbits. Wallen et al. developed a novel exosome–polyethyleneimine matrix (EPM), derived from bovine colostrum, capable of delivering plasmid DNA or siRNA to simulate viral protein expression without handling live virus. EPM-mediated delivery of spike (S1) or nucleocapsid (N) induced sustained antibody production. The EPM system demonstrated efficient nucleic acid encapsulation and low cytotoxicity, thereby enhancing the vaccine safety profile compared with polyethyleneimine alone. It also eliminated the need for high-containment laboratory facilities (Wallen et al., 2022).

A study by Wang et al. demonstrated that human lung spheroid cells–derived exosomes expressing ACE2 (LSC-Exo) can be delivered as an inhaled therapy against SARS-CoV-2 infection. The inhaled LSC-Exo effectively bound and neutralized SARS-CoV-2 in the lungs of mice, blocking viral entry into host cells. In a hamster model, LSC-Exo treatment significantly reduced viral load and protected against disease symptoms upon viral challenge. In addition, LSC-Exo showed neutralizing activity against both wild-type SARS-CoV-2 and pseudovirus variants. Finally, the safety assessments revealed no significant toxicity in treated animals (Wang et al., 2024). Wang et al. preferred a novel inhalable vaccine, RBD-Exo, composed of human lung spheroid cell exosomes coated with the SARS-CoV-2 receptor-binding domain (RBD). Using a murine experimental model, they reported that inhalation of RBD-Exo triggered a strong immune response, including RBD-Exo-specific systemic IgG and mucosal IgA, and a robust lung-resident population of $CD4^+$ and $CD8^+$ T cells with a Th1 cytokine profile (Wang et al., 2022). Moreover, two doses of the inhaled vaccine reduced lung inflammation and lowered viral loads after live SARS-CoV-2 exposure in a hamster model and significantly reduced severe pneumonia symptoms (Wang et al., 2022). Guo et al. developed an exosome-based nanovesicle vaccine system that displayed viral antigens by engineering exosomes, which were strongly immunogenic, eliciting neutralizing antibodies and active T cells, thereby demonstrating their strong immunogenic potential for safe yet effective vaccination (Guo et al., 2024). Moving away from cell-derived exosomes, Zhang et al. used bovine milk-derived exosomes encapsulating SARS-CoV-2 mRNA encoding the receptor-binding domain (RBD) to develop an mRNA-based vaccine. The vaccine was administered intraduodenally in a murine model. The use of non-cell-derived exosomes improved the vaccine preparation's safety profile and provided a cost-effective approach (Zhang et al., 2023). Kaspi et al. compared naïve MSC-Exos with engineered exosomes from MSCs stimulated to release elevated levels of neurotrophic and immunomodulatory factors (Exo MSC-NTF). Mice treated with intratracheal Exo MSC-NTF showed greater improvement than those treated with naïve preparations. It reduced alveolar wall thickening, fibrin deposition, and neutrophil infiltration, while also enhancing blood oxygen saturation. In addition, Exo MSC-NTF reversed the inflammatory imbalance by remarkably lowering the levels of IFN-γ, IL-6, and TNF-α (Kaspi et al., 2021). These data strongly support the potential of exosomes as an effective, safe, and low-cost therapy for SARS-CoV-2.

1.11.2 Secretome-Based Therapy for COVID-19, ALI, and PF in Animal Models

Acute lung injury (ALI) and pulmonary fibrosis (PF) are among the most severe complications of COVID-19, contributing significantly to morbidity and mortality. MSC-SECR, which contains EVs, cytokines, and growth factors, is being investigated as a promising therapeutic approach due to its potent immunomodulatory and tissue-repair properties. However, direct preclinical studies evaluating secretome-based therapies for COVID-19 in small animal models remain scarce. To address this gap, we have compiled evidence from multiple studies investigating the application of secretome-based therapy in small animal models of ALI, PF, and virus-induced lung damage, which share pathophysiological features with severe COVID-19 (Table 1.2). These studies demonstrated that MSC-CM or MSC-SECR can mitigate lung injury by reducing inflammation through multiple mechanisms and restoring vascular integrity. While only one study, by Pezzotta et al. (2025), directly examined the effect of IV injection of CM derived from human umbilical cord (UC)-MSCs in ACE2-humanized mice with acute lung injury induced by IV injection of SARS-CoV-2 antigens. The findings showed that CM significantly reduced lung injury and fibrosis by exerting an anti-fibrotic effect, decreasing the expression of fibronectin (FN) and α-smooth muscle actin (α-SMA), which are typically associated with pulmonary fibrosis. In addition, CM reduced vascular dysfunction by restoring thrombomodulin (TM) expression, which has anti-thrombogenic properties, and by reducing vWF expression in the lung vasculature. Furthermore, CM inhibited complement activation and endothelial C3 deposition. It was also demonstrated that treatment with CM significantly reduced total NF-κB expression in lung tissues, thereby reducing neutrophil and macrophage infiltration and the inflammatory response (Pezzotta et al., 2025) (Table 1.2). In another related study by Bogatcheva et al. (2021), they evaluated the application of Adipose Stromal Cell (ASC) concentrated secretome (CS) administered subcutaneously (SQ) in the treatment of influenza A virus (IAV)-induced lung injury in murine models. ASC-CS mitigated lung injury by reducing Angpt2 release, which is associated with vascular injury and leak (Bogatcheva et al., 2021) (Table 1.2). Furthermore, in a study by Tan J et al. (2023), they explored the effects of aerosolized ectodermal MSCs (EMSCs)-derived secretomes (EMSC-Sec) in treating LPS-induced acute lung injury in mice. EMSC-Sec was shown to significantly reduce the degree of pulmonary edema, as well as reverse the alveolar destruction. Additionally, EMSC-Sec inhibited the NF-κB (p50/p65)/NLRP3 pathway, which plays a crucial role in inflammation, and increased the expression of anti-inflammatory cytokines, such as IL-10, indicating strong immunomodulatory properties (Tan et al., 2023) (Table 1.2). Moreover, in a study by Kudinov et al. (2022), the authors analyzed the effects of inhaled placental multipotent MSCs (MMSCs) on mice with lethal endotoxin-induced acute lung injury. It was observed that MSC-sec treatment restored lung architecture, with a significant reduction in peribronchial edema, and most alveoli had cleared lumens and showed no thickened alveolar walls. Moreover, secretome-treated groups showed less fibrin deposition than the control group, leading to enhanced alveolar fluid clearance and decreased neutrophil activation (Kudinov et al., 2022) (Table 1.2). Additionally, Su et al. (2019) investigated the effect of tail vein-delivered MSC-CM in mice with endotoxin-induced ALI. CM treatment reduced alveolar neutrophil infiltration only after 24 h. Moreover, CM induced neutrophil apoptosis. The treated group showed a significant reduction in pro-inflammatory markers like IL-6, MIP-2, and NF-κB p65 (Su et al., 2019) (Table 1.2). Tang et al. (2021) used hUC-MSCs-CM (hUC-MSCs-CM) to attenuate LPS-induced acute lung injury in mice by tail vein injection. It was shown that hUC-MSCs-CM significantly reduced pulmonary edema. Furthermore, it improved the lung tissue inflammatory response by reducing the number of inflammatory cells and neutrophil and macrophage infiltration. Additionally, it led to decreased cytokine production, including IL-1β, IL-6, and TNF-α (Tang et al., 2021). Hu et al. assessed hESC-derived MSC-like immune and matrix regulatory cell-derived CM (hESC-MSC-IMRC-CM) to alleviate pulmonary fibrosis in a bleomycin (BLM)-injured mice model. Interestingly, hESC-MSC-IMRC-CM significantly inhibited BLM-induced pulmonary fibrosis in mice by reducing ROS and pro-inflammatory cytokine production, thereby regulating the Nox4/Nrf2 and Tlr4/MyD88 signaling pathways (Hu et al., 2023).

1.12 CLINICAL TRIALS INSIGHTS

1.12.1 Overview of Studies Using Exosomes and Secretomes in COVID-19 Patients

According to the National Library of Medicine's trial registry and database (Clinicaltrials.gov), about 474 trials are registered under exosomal therapies and 87 under secretome therapy for diverse pathologies. A few examples of the clinical trials on exosomes include Metastatic Melanoma,

TABLE 1.2

Preclinical Studies of Secretomes Cell-Free Therapy for COVID-19 and ARDS in Small Animal Models

Author/year	Model	Condition/Injury	Secretomes Source	Administration	Condition Medium	Findings
Pezzotta et al. (2025) (1)	Mice	Acute lung injury induced by SARS-CoV-2 Spike protein	Human umbilical cord-derived MSC secretome	IV	Not individually specified, contains MSC paracrine factors including cytokines and extracellular vesicles	• Attenuated lung injury by reducing edema and fibrosis. • Secretomes reduced lung injury, swelling, and scarring, and protected the blood vessel lining. • Restored thrombomodulin, lowered von Willebrand factor and complement buildup, and reduced inflammation signals and immune cell entry.
Su et al. (2019) (2)	Mice	LPS-induced acute lung injury	Human BM-MSC-SECR	IV	N/A	• MSC secretomes reduced histological lung injury and neutrophil infiltration. • Enhanced neutrophil apoptosis and lowered IL-6 and MIP-2. • Decreased NF-κB p65 activation in lung tissue neutrophils.
Ding et al. (2020) (3)	Mice	LPS-induced acute lung injury	BM–MSC-SECR	Intratracheal instillation	MSC paracrine factors (exosomes, cytokines, miRNAs), enriched in miR-214	• Enhanced lung fluid clearance. • Restored alveolar fluid clearance to near-normal levels. • miR-214 transfer upregulated alveolar ENaC expression/activity, enhancing lung fluid clearance.
Peng et al. (2017) (4)	Male Sprague–Dawley rats	Acute lung injury induced by 30 min lung ischemia + 90 min reperfusion.	NCSCs from adult rat hair follicle	IV	No specific factors quantified	• Reduced lung edema and permeability. • Decreased BAL fluid TNF-α and IL-6 levels. • Lower histopathology injury score and neutrophil infiltration. • Inhibited NF-κB activation. • Pulmonary arterial pressure levels were decreased.
Bogatcheva and Coleman (2021) (5)	Mice	Influenza A virus infection	Human ASC.	Subcutaneous injection	Contains a broad spectrum of proteins/cytokines (not Specified)	• Secretome improved weight loss, O_2 saturation, and lung histology. • Lowered BAL cytokines IL-6, MCP-1, IFNγ, and TNFα. • Pre-treating ASC secretomes with TNF-α and IFN-γ improved the treatment effectiveness. • Lowered lung markers (Angpt2 and PDL1) without affecting virus clearance.

(Continued)

TABLE 1.2 CONTINUED
Preclinical Studies of Secretomes Cell-Free Therapy for COVID-19 and ARDS in Small Animal Models

Author/year	Model	Condition/Injury	Secretomes Source	Administration	Condition Medium	Findings
Kudinov et al. (2022) (6)	Mice	Lethal acute lung injury (ARDS) induced by inhalation of 10 mg/kg LPS	Human placenta–derived MSCs; also compared to human LFB secretome	Inhalation (nebulized aerosol)	EVs, growth factors, cytokines, chemokines, angiogenesis proteins, immune regulation, and growth factor signaling	• Inhalation therapy with secretomes resulted in complete survival in all treatment groups. • Histology showed restoration of normal alveoli and marked reduction of inflammatory cell infiltration. • Fibrin deposition was significantly decreased. • Significantly reduced lung injury scores.
González et al. (2023) (7)	Adult male Sprague–Dawley rats	E. coli–induced pneumonia (ALI/ARDS model via intratracheal instillation of bacteria)	Human BM–derived MSCs	Intratracheal delivery by nebulizer (inhalation) vs. direct Intratracheal instillation.	N/A	• Secretomes administered through intratracheal instillation or nebulization significantly improved lung function and reduced bacterial load in lungs. • The treatment reduced the inflammatory infiltration cytokine levels in BAL fluid, including IL-1β and IL-6. • Treated animals showed reduced lung inflammation and improved respiratory parameters compared with untreated controls.
Tan et al. (2023) (8)	Mice	LPS-induced acute lung injury	Ectodermal MSC (nasal mucosa–derived; neural crest origin)	Aerosol inhalation	Secretome containing IL-10, TGF-β, SHH; exosome markers CD9, CD63	• Treated animals have shown a reduction in lung edema and tissue damage. • The inhalation decreased pro-inflammatory cytokines (TNF-α, MPO, IL-17). • The Secretome treatment increased the anti-inflammatory IL-10 and suppressed the activation of NF-κB p50/p65 complex and the NLRP3 inflammasome pathway.

Abbreviations: Angiopoietin-2 (Angpt2), Adipose Stromal Cell (ASC), Bronchoalveolar Lavage (BAL), Epithelial Sodium Channels (ENaC), Extracellular Vesicle (EVs), Interferon-Gamma (IFN-γ), Interleukin (IL), Intravenous (IV), Lung Fibroblast (LFB), Lipopolysaccharide (LPS), Macrophage Inflammatory Protein-2 (MIP-2), Myeloperoxidase (MPO), Mesenchymal Stem Cells (MSC), Neural Crest–derived Stem Cells (NCSCs), Nuclear Factor Kappa B (NF-κB), NOD-like Receptor Protein 3 (NLRP3), Programmed Cell Death Ligand 1 (PDL1), and Tumor Necrosis Factor Alpha (TNF-α).

Periodontitis, Hair Loss, Neurodegenerative Diseases, Umbilical Cord (UC), Diabetes Mellitus Type 1, Long COVID-19 Syndrome, Alopecia Androgenica, and Knee Osteoarthritis (Clinical trials identifier NCT02310451, NCT04270006, NCT06932393, NCT06853522, NCT02138331, NCT05808400, NCT06930326, NCT05060107). Similarly, to name a few clinical trials that utilized secretome therapy for a variety of different conditions, we have Epilepsy, Drug-Resistant Epilepsy, Lupus Erythematosus, Knee Osteoarthritis, Rheumatoid Arthritis, Skin Rejuvenation, and graft vs host disease (GVHD) (Clinical trials identifier NCT06638970, NCT05921058, NCT04314661, NCT05925647, NCT06217627, NCT06469411). A summary of all the identified clinical trials on COVID-19 has been shown in Table 1.3.

1.12.2 Safety and Efficacy of Clinical Trials Using Cell-Free Therapy for COVID-19

Several clinical trials have been conducted assessing the safety and efficacy of the cell-free approach. For example, one Zofin (formerly Organically Flow) is an investigational, acellular biologic derived from human amniotic fluid and containing exosome fractions. Zofin was administered to three patients with severe COVID-19-related multi-organ complications, concurrent with standard inpatient care. All patients were followed for 28 days post-intervention, during which SOFA scores were evaluated, chest radiography was performed, and inflammatory biomarker tests were performed. The primary safety outcome was that no adverse events related to Zofin were identified. Following treatment, the three patients showed clinical improvement in the ICU, concurrent improvement in respiratory status, and full recovery from acute delirium. Preliminary data indicated that inflammatory marker levels decrease following treatment, suggesting acceptable safety and clinical benefits. This is the first clinical evidence that amniotic fluid-derived nanoparticle vesicles are a potential treatment for COVID-19 respiratory failure (Mitrani et al., 2021a). The term "long haulers" refers to patients who experience prolonged symptoms after contracting COVID-19. In a separate study, Zofin was administered to a single patient with COVID-19 who was suffering from protracted respiratory distress. On the 28th day following treatment, a chest x-ray revealed fully expanded lungs and no acute cardiopulmonary pathology. By day 60, no chest x-ray was deemed necessary, and the patient had resumed full activity and returned to work. Laboratory results were predominantly normal and remained so throughout the follow-up period, with no recorded side effects or safety concerns (Mitrani et al., 2021b). Patients with severe, multi-organ COVID-19 exhibited higher levels of both pro-inflammatory and anti-inflammatory cytokines than those with milder illness.

An excessive cytokine and chemokine release has been implicated in progression to multi-organ failure and worse clinical outcomes, including more extended hospitalization and increased mortality. Given the secretome's well-recognized advantages over MSCs, a study evaluated its effectiveness and safety for treating acute COVID-19 cases. In this randomized, double-blind, placebo-controlled, multicenter study, 40 participants were enrolled at three major referral hospitals for COVID-19. The primary endpoints include changes in inflammatory markers and the pro-inflammatory-to-anti-inflammatory ratio. In contrast, the secondary endpoints were clinical outcomes, laboratory results, imaging, RT-PCR conversion, and assessment of MSC/secretome safety. Compared with the intervention arm, the control group showed greater inflammatory upregulation at day 14 and elevated IL-6 and ferritin at day 7. The intervention arm did not demonstrate an increase in pro-anti-inflammatory ratio at days 7 or 14. No adverse events were reported, supporting the preliminary safety and efficacy of MSC-SECR therapy on patients with severe COVID-19 (NCT05122234). MSC-SECR can promote rapid lung recovery. Three patients with critical, rare conditions were admitted to ICUs in Indonesia. All treated patients demonstrated reduced CRP levels, which paralleled improvements in radiologic findings, arterial blood gases, and hematologic assessments (Putra et al., 2021). A double-masked, placebo-controlled trial assessed the safety and efficacy of secretome from allogeneic menstrual blood stromal cells (MenSCs) in patients with severe COVID-19. Five daily IV doses of MenSCs or a placebo were monitored for 28 days. Follow-up for safety, labs, symptoms, oxygenation, lymphocytes, hospital stay, and chest imaging. Oxygenation improved in 64% of patients

TABLE 1.3
Preclinical Studies of Exosomes Cell-Free Therapy for COVID-19 and ARDS in Small Animal Models

Author/ Year	Model	Exosomes Source	EVs Size	EVs Markers	Administration	EVs Cargo	Dose	Findings
Cacciottolo et al. (2023a).	Mouse	Engineered exosomes from human cells.	NA	CD81	IM	SARS-CoV-2 spike protein	Two injections on day 1 and day 21, each containing 3–10 ng.	• High level of antibody neutralization. • Antigen-specific T cell activation. • Decreased weight loss. • Reduced viral loads. • Improved lung histopathology.
Kaspi et al. (2021).	Female BALB/c mice (LPS-induced ARDS)	Bone marrow-derived MSCs: Exo-MSC and Exo-MSC-NTF.	Exo-MSC ~146nm, Exo-MSC-NTF ~114nm	CD9, CD63, CD81, plus MSC markers CD44 and CD29.	Intratracheal (endotracheal tube)	LIF, AREG, HGF, and TSG-6	2×10^{10} vesicles/ml per dose, 50 µl; Exo MSC/Exo MSC-NTF, administered at 3 h, 24 h, and 48 h post-LPS instillation.	• Significant Histological reduction in epithelial fibrosis and luminal obliteration. • Less infiltration of CD3 + T cells and macrophages. • Decrease expression of TNF-α, IL-1β, IFN-γ. • Increased expression of IL-10.
Cacciottolo et al. (2023b).	Mice and Rabbits	Engineered exosomes from human cells.	~144.8nm and ~140.1nm	CD81, CD63, CD9.	IM	SARS-CoV-2 antigens Spike and Nucleocapsid.	Low dose 1×10^{12} exosomes/mL with 3 ng Spike + 3 ng Nucleocapsid. High dose 1×10^{12} exosomes/mL with 10 ng Spike + 10 ng Nucleocapsid.	• Increased levels of Anti-Spike and Anti-Nucleocapsid IgG. • Plasma antibodies neutralization against Delta and Omicron variants of the virus. • CD4+ and CD8+ T-cell activation in response to antigen stimulations. • No adverse effects (weight loss or tissue abnormalities).

Author/ Year	Model	Exosomes Source	EVs Size	EVs Markers	Administration	EVs Cargo	Dose	Findings
Wallen et al. (2022)	Mouse	Bovine colostrum-derived exosomes.	N/A	N/A	IV	SARS-CoV-2 proteins (Spike, Nucleocapsid, replicase)	Three doses of the EPM formulation contained 75 µg of exosome protein and 40 µg of PEI over three days.	• Robust humoral immune response induced by strong Spike-specific IgG responses. • Elevated T cell activation (CD8+, CD4+), increased secretion of cytokines (IFN-γ, TNF-α, and IL-2). • Exosomes induced antigen-specific CTL activity. • Vaccinated groups showed a reduction in viral loads and lung inflammation.
Zhenzhen Wang et al. (2024)	Syrian golden hamsters	Human lung spheroid cells (LSC-Exo).	~178 ± 3.5nm	CD63, CD9, CD81.	Inhalation (nebulization/ aerosol delivery)	Surface-enriched human ACE2	Multiple doses on days 1, 2, 3 post-infections. (did not mention does)	• LSC-Exo bound SARS-CoV-2 spike protein, effectively blocking viral binding. • Inhaled LSC-Exo maintained body weight and showed significantly lower viral loads. • LSC-Exo had good/improved the efficacy of blocking viral entry. • The lung gene expression profiles in the LSC-Exo group resembled normalization of cytokine, stress response, and oxidative phosphorylation.
Guo et al. (2024)	Hamsters infected with SARS-CoV-2	Exosomes from engineered cells	N/A	N/A	IM	Spike protein on the exosome surface	Low dose exosomes 10^9 + 0.5 ng spike. High dose exosomes 10^{10} + 5 ng spike.	• Exosomes conferred high neutralization of Delta pseudo viruses. • Inhaled LSC-Exo significantly reduced viral loads and prevented severe lung pathology. • Exosomes reduced pro-inflammatory signaling, restored oxidative stress pathways, and activated antioxidants. • Exosomes showed decreased inflammation by a reduction of white blood cells, neutrophils, and lymphocytes counts.

(Continued)

TABLE 1.3 CONTINUED
Preclinical Studies of Exosomes Cell-Free Therapy for COVID-19 and ARDS in Small Animal Models

Author/ Year	Model	Exosomes Source	EVs Size	EVs Markers	Administration	EVs Cargo	Dose	Findings
Zhenzhen Wang (2022)	Syrian golden hamsters	Human lung spheroid cells (LSC-Exo).	~180 ± 5nm	CD9, CD63, CD81, TSG101, Alix + RBD	Inhalation (aerosol), using a lyophilized vaccine delivered to the lungs.	Surface-bound RBD protein; no nucleic acids.	Two inhaled doses—amounts not specified	• Inhaled RBD-Exo induced a strong RBD-specific IgG response systemically and mucosal IgA response in the lungs. • Mice developed potent $CD4^+$ and $CD8^+$ T cell responses in the lung with a biased cytokine profile (e.g., IFN-γ, IL-2), favoring antiviral immunity. • Vaccinated mice efficiently cleared a SARS-CoV-2 pseudovirus from lungs. • Inhaled doses of RBD-Exo significantly attenuated severe pneumonia, reduced inflammatory infiltrates, and lung damage.
Zhang et al. (2023)	BALB/c mice	Bovine milk-derived exosomes	30–150 nm	CD9, CD63, CD81, and TSG101	Oral, intraduodenal administration	RBD mRNA.	Three doses, no specific amount.	• RBD mRNA–loaded exosomes induced significant neutralizing antibodies. • No adverse effects or weight loss reported. • The use of milk-derived edible exosomes offers a promising, scalable, and cost-effective strategy for vaccine delivery.

within five days and increased survival to 57% versus 28% in controls. CRP and ferritin decreased, with no therapy-related adverse events, indicating potential application in severe inflammatory lung diseases (NCT05019287). In patients with severe COVID-19, secretome derived from menstrual blood stromal cells (MenSCs) showed promising results. Receiving 14 doses of (BM-MSC EVs) over two months resulted in almost complete lung recovery, improved quality of life, and discharge after three months.

In another reported clinical study (NCT05354141), repeated administration of BM-MSC EVs was well-tolerated and without adverse events. On the same note, another study evaluated the safety and efficacy of an exosome-based therapeutic product (ExoFlo) in patients with severe COVID-19 and moderate-to-severe ARDS. The primary research question was whether two doses of ExoFlo could safely reduce mortality compared to a placebo. Patients received a placebo, 10 mL ExoFlo, or 15 mL ExoFlo. Sixty-day mortality decreased with 15 mL ExoFlo and increased ventilation-free days in adults aged 18–65 years. No treatment-related adverse events (NCT04493242). In the MEXCOVID trial, a phase 2a, single-arm, open-label study, seven COVID-19 patients received daily doses of human adipose-derived MSCs-Exos (haMSC-Exos) for five consecutive days. Primary outcomes included inhalation-associated events and serious adverse events (SAEs), while secondary observations encompassed demographics, clinical data, lymphocyte counts, D-dimer levels, IL-6 levels, and chest imaging results. The therapy was safe and well-tolerated, associated with early CT improvement and a mild increase in lymphocytes, with no adverse events (NCT04276987).

1.13 CELL-FREE THERAPY APPROACH AND CHALLENGES

1.13.1 Immunogenicity and Safety

Since exosome and secretome therapies are being considered as therapeutic modalities to treat COVID-19 and its associated pathologies, their immunocompatibility and safety are being extensively studied. A few characteristics of this spike protein include increased anti-exosome IgG antibodies in plasma samples from COVID-19 patients, which facilitate the delivery of therapeutic XLs (targeting exosomes) into the mononuclear phagocytic system to accelerate clearance through Fc receptor-mediated phagocytosis, and potentially involve immune complex formation that could lead to more inflammation across different organ systems. This antibody-mediated neutralization mechanism presents a significant barrier to the development of efficacious dosing regimens for repeated exosome administration. Several advanced methods are being developed to enhance immunocompatibility and mitigate this mismatch (Fathi-Kazerooni et al., 2022). Remodeling the surface glycan architecture by enzymatic removal of terminal sialic acid residues has emerged as an additional promising approach, because this modification potently prevents complement activation (by limiting C3 deposition and ultimately membrane attack complex [MAC] formation). In addition, combination strategies, including the use of artificial self-markers such as CD47, exosomes from low-immunogenicity sources (e.g., immature DPSCs), or temporary immunosuppressive protocols, could enhance therapeutic persistence and reduce toxicity (Seoane-Vazquez et al., 2019). Another critical factor influencing immunogenicity is the inherent variability of the source cell, which affects the arrangement of donor cells and the production of exosomes. Reflection in surface proteins, lipid raft organization, and cargo (miRNA, cytokines) all play significant roles in how the recipient immune system perceives exosomes and responds to them. For example, inflammatory-primed haMSC-Exos may contain higher levels of immunomodulatory factors (e.g., TGF-β, PGE2) than those from naive cells, which affects both treatment efficacy and potential side effects. Implementing standards for donor selection, isolation protocols, and exosome characterization will ensure batch-to-batch consistency and, hence, predictable immune system compatibility (Fathi-Kazerooni et al., 2022).

For the treatment of COVID-19, the cell-free biologics have a strong immunomodulatory effect by inhibiting key pro-inflammatory factors (IL-6, TNF-α, and IL-1β, etc.) using microRNA regulation or binding with receptors to block them from participating in cytokine storms, which lead directly to acute lung injury (Aldoghachi et al., 2023). Therefore, when treating critically ill patients with COVID-19, strict dosage adjustment combined with real-time monitoring of coagulation indices is essential to maximize therapeutic benefit while minimizing the risk of thrombosis.

1.13.2 Ethical Issues, Regulatory Frameworks, and International Institutional Efforts to Manage and Control Therapeutic Claims

Many companies have claimed to offer cell-based treatments for COVID-19. Unfortunately, such claims have led to violations of scientific and ethical standards, with potentially serious consequences. While rapid therapeutic development is a critical goal, it must not come at the expense of scientific rigor and moral principles. According to recent studies, stem cell-derived exosomes are among the most widely marketed of these unproven products (Fujita et al., 2024). Without evidence to support their safety and efficacy, such interventions pose both financial and physical risks. Moreover, some patients who receive these unapproved therapies may believe they are fully protected, leading them to decline vaccination and ignore public safety and health measures. To counter this, healthcare professionals and companies should avoid overstating the benefits of stem cell-based COVID-19 treatments until sound research outcomes support such claims. The International Society for Cell & Gene Therapy (ISCT) and the International Society for Stem Cell Research (ISSCR) have also raised concerns about companies advertising and distributing unproven cell-based treatments that lack scientific evidence and safety validation. Regulatory agencies have taken some steps to address this issue. For example, the US Food and Drug Administration (FDA) and the Federal Trade Commission (FTC) have jointly issued 22 warning letters to companies selling unverified and unapproved cellular therapies for COVID-19. Suggested regulatory measures include imposing financial penalties on offending companies and medical professionals, requiring the return of profits to patients, and revoking medical licenses (Fujita et al., 2024). Protecting public health requires stricter rules, more vigorous enforcement, and decisive action against the spread of misinformation and the exploitation of patients.

1.13.3 Optimal Storage Conditions and Shelf-Life Challenges

EV-based drug delivery offers several advantages over artificial systems, such as liposomes, including reduced immunogenicity and toxicity, superior vascular and tissue stability, and enhanced immunological safety (Lener et al., 2015). Several studies have identified EVs as promising multi-target therapeutic agents. However, before large-scale human application, researchers must establish reliable production methods, determine optimal dosing, and select effective delivery routes (Khalaj et al., 2020). Recently, the International Society for Extracellular Vesicles (ISEV) and the International Society for Cell & Gene Therapy emphasized the need to identify the most appropriate EV source and to ensure optimal storage conditions, reproducible isolation procedures, and quality control. Multiple isolation techniques exist, as discussed earlier, with differential ultracentrifugation being the most common. Regardless of whether current or novel methods are used, EV isolation protocols must be standardized to ensure reproducibility, maximize purity, minimize contamination, and preserve vesicle properties. Importantly, the process that achieves the highest purity does not necessarily yield vesicles with the most significant therapeutic efficacy (Khalaj et al., 2020).

All steps, reagents, and laboratory materials—whether plastic or glass—must be certified for therapeutic human use. Although exosomes are routinely stored at −80°C, some studies have shown that storage at room temperature combined with freeze–thaw cycles can still preserve specific biological signals. However, storing at −80°C is impractical for large-scale transport or long-term distribution. Morphological changes have been observed in exosomes stored at −80°C for only four days compared with freshly isolated samples, and biological activity declines significantly after 28 days (Charoenviriyakul et al., 2018). Cryopreservation is widely used to maintain EV integrity, preserving concentration, morphology, size, and biological activity. Cryoprotectants such as dimethyl sulfoxide (DMSO) or glycerol are commonly used to minimize freezing damage. Short-term storage at 4°C for up to 14 days can maintain EV morphology, RNA integrity, and biological activity. Despite their advantages over whole-cell therapies, purified EVs may have reduced therapeutic potency compared to their parent cells due to the loss of certain paracrine factors and their relatively short half-life. This limitation may necessitate repeated dosing or combination with cell therapy to sustain therapeutic effects.

1.13.4 Delivery Methods and Their Efficacy

EVs have demonstrated excellent biocompatibility, low immunogenicity, efficient uptake by diverse tissues, and the ability to penetrate biological membranes more effectively than liposomes or polymer-based systems, making them promising vehicles for drug delivery (Deng et al., 2024). In the case of COVID-19, the complexity of the disease necessitates a careful evaluation of the optimal route of administration and dosage. EVs can be delivered either intravenously or via airway administration, each with distinct advantages and limitations. The administration route significantly influences EV biodistribution, pharmacokinetics, and therapeutic efficacy (Tolomeo et al., 2023).

Intratracheal delivery of EVs directly to the lungs avoids systemic dilution. However, aerosol delivery often delivers high concentrations to the proximal airways, with insufficient penetration into distal regions, which are usually more severely affected in COVID-19. Intravenous (IV) administration allows repeated dosing and has been successfully applied in regenerative medicine. In severe COVID-19, where multi-organ failure and sepsis are common, IV EV-based therapy may support the recovery of multiple organs in addition to lung repair. As COVID-19 patients are also at risk of cardiac dysfunction, microvascular injury, and plaque destabilization, EVs may be effective in supporting the recovery of multiple organs beyond lung repair. However, IV administration in COVID-19 carries the risk of worsening the disease's pro-thrombotic state, including disseminated intravascular coagulation without bleeding, which is associated with poor prognosis (Khalaj et al., 2020). In an acute lung injury (ALI) mouse model, human umbilical cord MSC-derived EVs administered intravenously, intranasally, or via aerosol inhalation all reduced pulmonary inflammation at 4 and 24 hours, except in mice receiving low-dose aerosol therapy. ELISA showed decreased levels of pro-inflammatory cytokines in serum, with IV delivery yielding the most effective results (Chen et al., 2025). Another study has also suggested a preference for IV administration to rapidly deliver the therapeutic agent into systemic circulation in COVID-19 patients. However, for patients with early-stage disease, oral treatment is preferred due to its lower cost, convenience, and better compliance.

1.14 INSIGHTS INTO THE FUTURE OF CELL-FREE THERAPY

1.14.1 Comparative Analysis of Exosome and Secretome Sources

Secretomes and exosomes have been extensively studied due to their reparability and immunomodulatory roles. While exosomes are nanovesicles (30–150 nm) secreted by nearly all cell types and carry a specific payload profile (Li G et al., 2024), the secretome comprises an array of bioactive molecules released by cells as discussed earlier. MSC-derived Exos have been the most extensively studied for therapeutic use. They can modulate the immune response by downregulating pro-inflammatory cytokines and enhancing regulatory T cell function (Li G et al., 2024). For example, miR-146a in MSC-derived Exos has been linked to inhibition of pro-inflammatory signaling pathways, including Tumor necrosis factor-α (TNF-α) and interleukin-6 (IL-6). Also, ACE2-expressing exosomes isolated from genetically engineered cells have been shown to neutralize SARS-CoV-2 virions by decoy-mediated reduction in viral entry into host cells. On the other hand, the MSC-Secr offers a more comprehensive profile of action. In preclinical models of COVID-19, MSC-Secr treatment reduced pro-inflammatory cytokines, including IL-6, TNF-α, and IL-1β, and induced epithelial repair and angiogenesis (Pezzotta et al., 2025). However, heterogeneity in cell culture conditions, donors, and passage numbers hinders the standardization of secretome composition. In terms of the delivery systems, exosomes offer a more stable, uniform means of delivery. Their lipid bilayer protects their contents from enzymatic degradation and allows them to reach the target tissues (Li G et al., 2024). They are thus well-suited for drug delivery systems and long-term storage. On the contrary, the secretome lacks membrane protection, leading to faster degradation and reduced control over tissue targeting for site-directed delivery (Pezzotta et al., 2025). In COVID-19 applications, both methods have been tested with promising results. Exosomes from various sources—including MSCs and lung spheroid cells—demonstrate antiviral and anti-inflammatory effects, thereby improving oxygenation and reducing lung edema in animal models of SARS-CoV-2-induced ARDS (Ghiasi et al., 2024). Intratracheal or intravenous delivery of secretomes stimulated alveolar regeneration, reduced neutrophilic invasion, and cytokine storm suppression (Yao et al., 2025). Both can mimic the paracrine effects of stem cells; they are cost-intensive to manufacture, and effective large-scale isolation methods are under development.

1.14.2 Future Perspectives of Cell-Free Therapies

Exosome engineering is advancing rapidly and opening new directions in precision therapy. One of the most desirable approaches is the genetic or pharmacological preconditioning of donor cells with specific RNAs, proteins, or ligands that target inflamed tissue or viral entry mechanisms. For instance, ACE2-overexpressing cells release decoy exosomes that trap SARS-CoV-2 and prevent infection of the host cells (Li L et al., 2024). Other research areas focus on new delivery platforms, including inhalable aerosols, injectable hydrogels, and surface-modified vesicles with tissue-selective ligands, which enable lung targeting and increase bioavailability (Ghiasi et al., 2024). This is especially applicable in COVID-19, where targeting the lung is crucial. Additionally, loading siRNAs, anti-inflammatory drugs, or CRISPR/Cas components into exosomes is also being explored for targeted gene control in viral or inflammatory disease. These bioengineered vesicles enhance therapeutic specificity and minimize side effects by limiting systemic exposure, making them promising tools in next-generation medicine (Schürz et al., 2025) (Figure 1.2). The impact of exosome or secretome therapy on long-term immune memory, lung fibrosis, and recurrent infections remains unknown. Post-COVID syndromes, such as persistent fatigue, neuroinflammation, or cardiovascular dysfunction, may be assisted by continuous immunomodulation and tissue regeneration, but robust clinical evidence is lacking. There is also little consensus on the optimal dosing, route of administration, and timing of therapy initiation for these products. Another challenge to clinical translation, however, is the absence of standardized protocols for secretome and exosome isolation, quantification, and quality control. Laboratory-to-laboratory and batch-to-batch variability is a persistent limitation to reproducibility and regulatory licensure (Schürz et al., 2025). To bridge these gaps through large-scale, multicenter clinical trials and shared manufacturing standards is imperative for bringing these therapies from the bench to the bedside.

REFERENCES

Aldoghachi AF, Loh JK, Wang ML, Yang YP, Chien CS, Teh HX, Omar AH, Cheong SK, Yeap SK, Ho WY, Ong AH. Current developments and therapeutic potentials of exosomes from induced pluripotent stem cell-derived mesenchymal stem cells. *J Chin Med Assoc.* 2023 Apr 1;86(4):356–365. doi: 10.1097/JCMA.0000000000000899.

Al-Ghadban S, Bunnell BA. Adipose tissue-derived stem cells immunomodulatory effects and therapeutic potential. *Physiology* (Bethesda). 2020 Mar 1;35(2):125–133. doi: 10.1152/physiol. 00021.2019.

Andreev E, Kravchenko E, Zavyalova E, Eremin P, Rzyanina A, Markov P, Nechaev A. Comparative analysis of enriched mesenchymal stem cells conditioned medium fractions obtained by ultrafiltration. *Nat Sci Rev.* 2025;2:100401. doi: 10.54546/NaturalSciRev.100401.

Bari E, Ferrarotti I, Saracino L, Perteghella S, Torre ML, Richeldi L, Corsico AG. Mesenchymal stromal cell secretome for post-COVID-19 pulmonary fibrosis a new therapy to treat the long-term lung sequelae? *Cells.* 2021 May 14;10(5):1203. doi: 10.3390/cells10051203.

Beshbishy AM, Alghamdi S, Onyiche TE, Zahoor M, Rivero-Perez N, Zaragoza-Bastida A, Ghorab MA, Meshaal AK, El-Esawi MA, Hetta HF, et al. Biogenesis, biologic function, and clinical potential of exosomes in different diseases. *Appl Sci.* 2020;10(13):4428. doi: 10.3390/app10134428.

Bhalerao A, Raut S, Noorani B, Mancuso S, Cucullo L. Molecular mechanisms of multi-organ failure in COVID-19 and potential of stem cell therapy. *Cells.* 2021 Oct 25;10(11):2878. doi: 10.3390/cells10112878.

Blázquez R, Sánchez-Margallo FM, de la Rosa O, Dalemans W, Álvarez V, Tarazona R, Casado JG. Immunomodulatory potential of human adipose mesenchymal stem cells derived exosomes on in vitro stimulated T cells. *Front Immunol.* 2014 Nov 4;5:556. doi: 10.3389/fimmu.2014.00556.

Bogatcheva NV., Coleman ME. Concentrated secretome of adipose stromal cells limits influenza a virus-induced lung injury in mice. *Cells.* 2021 Mar 24;10(4):720.

Cacciottolo M, Li Y, Nice JB, LeClaire MJ, Twaddle R, Mora CL, Adachi SY, Young M, Angeles J, Elliott K, Sun M. Nanograms of SARS-CoV-2 spike protein delivered by exosomes induce potent neutralization of both delta and omicron variants. *PLoS One.* 2023a Aug 22;18(8):e0290046. doi: 10.1371/journal.pone.0290046.

Cacciottolo M, Nice JB, Li Y, LeClaire MJ, Twaddle R, Mora CL, et al. Exosome-based multivalent vaccine achieving potent immunization, broadened reactivity, and strong T-Cell responses with nanograms of proteins. *Microbiol Spectr* [Internet]. 2023b Jun 15 [cited 2025 Jul 22];11(3). https://doi.org/10.1128/spectrum.00503-23.

Chacko SM, Ahmed S, Selvendiran K, Kuppusamy ML, Khan M, Kuppusamy P. Hypoxic preconditioning induces the expression of prosurvival and proangiogenic markers in mesenchymal stem cells. *Am J Physiol Cell Physiol.* 2010 Dec;299(6):C1562–1570. doi: 10.1152/ajpcell. 00221.2010.

Charoenviriyakul C, Takahashi Y, Nishikawa M, Takakura Y. Preservation of exosomes at room temperature using lyophilization. *Int J Pharm.* 2018 Dec 20;553(1–2):1–7. doi 10.1016/j.ijpharm.2018.10.032. Erratum in Int J Pharm. 2019 Mar 25;559:427–428. doi: 10.1016/j.ijpharm.2019.02.037.

Chen J, Li P, Zhang T, Xu Z, Huang X, Wang R, Du L. Review on strategies and technologies for exosome isolation and purification. *Front Bioeng Biotechnol.* 2022 Jan 5;9:811971. doi: 10.3389/fbioe.2021.811971.

Chen J, Liu S, Zou J, Wang Y, Ge H, Hui Y, Huang S, Li W, Na W, Huang X, Bai L, Huang Y, Qu D. Comparison of efficacy of exosomes derived from human umbilical cord blood mesenchymal stem cells in treating mouse acute lung injury via different routes. *Front Pediatr.* 2025 May 29;13:1560915. doi: 10.3389/fped.2025.1560915.

Chen YF, Luh F, Ho YS, Yen Y. Exosomes a review of biologic function, diagnostic and targeted therapy applications, and clinical trials. *J Biomed Sci.* 2024 Jul 11;31(1):67. doi: 10.1186/s12929-024-01055-0.

Chouaib B, Haack-Sørensen M, Chaubron F, Cuisinier F, Collart-Dutilleul PY. Towards the standardization of mesenchymal stem cell secretome-derived product manufacturing for tissue regeneration. *Int J Mol Sci.* 2023 Aug 9;24(16):12594. doi: 10.3390/ijms241612594.

Chung YS, Lam CY, Tan PH, Tsang HF, Wong SC. Comprehensive review of COVID-19 epidemiology, pathogenesis, advancement in diagnostic and detection techniques, and post-pandemic treatment strategies. *Int J Mol Sci.* 2024 Jul 26;25(15):8155. doi: 10.3390/ijms25158155.

Contreras-Naranjo JC, Wu HJ, Ugaz VM. Microfluidics for exosome isolation and analysis Enabling liquid biopsy for personalized medicine. *Lab Chip.* 2017 Oct 25;17(21):3558–3577.

Coronavirus Disease 2019 (COVID-19) Treatment Guidelines [Internet]. Bethesda (MD): National Institutes of Health; 2021 Apr 21–2024 Mar 1.

Costela-Ruiz VJ, Illescas-Montes R, Puerta-Puerta JM, Ruiz C, Melguizo-Rodríguez L. SARS-CoV-2 infection The role of cytokines in COVID-19 disease. *Cytokine Growth Factor Rev.* 2020 Aug;54:62–75. doi: 10.1016/j.cytogfr.2020.06.001.

Deng J, Wang Y, Li Y, Zhang Y, Li Y, et al. Novel administration strategies for tissue-specific delivery of extracellular vesicles a comprehensive review. *Adv Drug Deliv Rev.* 2024;190:114539. doi: 10.1016/j.addr.2023.114539.

Di Liegro CM, Schiera G, Di Liegro I. Extracellular vesicle-associated RNA as a carrier of epigenetic information. *Genes* (Basel). 2017 Sep 22;8(10):240. doi: 10.3390/genes8100240.

Ding Y, Cui Y, Hou Y, Nie H. Bone marrow mesenchymal stem cell-conditioned medium facilitates fluid resolution via miR-214-activating epithelial sodium channels. *MedComm* (Beijing) [Internet]. 2020 Dec 1 [cited 2025 Aug 12];1(3):376.

Dinh PC, Paudel D, Brochu H, Popowski KD, Gracieux MC, Cores J, Huang K, et al. Inhalation of lung spheroid cell secretome and exosomes promotes lung repair in pulmonary fibrosis. *Nat Commun*. 2020 Feb 28;11(1):1064.

Dyball LE, Smales CM. Exosomes Biogenesis, targeting, characterization, and their potential as "plug & play" vaccine platforms. *Biotechnol J*. 2022;17(5):e2100646.

El Safadi D, Mokhtari A, Krejbich M, Lagrave A, Hirigoyen U, Lebeau G, Viranaicken W, Krejbich-Trotot P. Exosome-mediated antigen delivery unveiling novel strategies in viral infection control and vaccine design. *Vaccines* (Basel). 2024 Mar 7;12(3):280.

Elzanowska J, Semira C, Costa-Silva B. DNA in extracellular vesicles biological and clinical aspects. *Mol Oncol*. 2021 Jun;15(6):1701–1714. doi: 10.1002/1878-0261.12777.

Fathi-Kazerooni M, Fattah-Ghazi S, Darzi M, Makarem J, Nasiri R, Salahshour F, Dehghan-Manshadi SA, et al. Safety and efficacy study of allogeneic human menstrual blood stromal cells secretome to treat severe COVID-19 patients clinical trial phase I & II. *Stem Cell Res Ther*. 2022 Mar 7;13(1):96. doi: 10.1186/s13287-022-02771-w.

Fujita M, Hatta T, Ikka T, Onishi T. There is an urgent need for clear and concise regulations on exosome-based interventions. *Stem Cell Reports*. 2024 Nov 12;19(11):1517–1519. doi: 10.1016/j.stemcr.2024.09.008.

Fyfe J, Casari I, Manfredi M, Falasca M. Role of lipid signalling in extracellular vesicles-mediated cell-to-cell communication. *Cytokine Growth Factor Rev*. 2023 Oct;73:20–26. doi: 10.1016/j.cytogfr.2023.08.006.

Gámez-Valero A, Monguió-Tortajada M, Carreras-Planella L, Franquesa M, Beyer K, Borràs FE. Size-exclusion chromatography-based isolation minimally alters Extracellular Vesicle characteristics compared to precipitating agents. *Sci Rep*. 2016 Sep 19;6:33641.

Gazzaniga G, Voltini M, Carletti A, Lenta E, Meloni F, Briganti DF, Avanzini MA, Comoli P, Belliato M. Potential application of mesenchymal stromal cells as a new therapeutic approach in acute respiratory distress syndrome and pulmonary fibrosis. *Respir Res*. 2024 Apr 18;25(1):170. doi: 10.1186/s12931-024-02795-1.

Ghadami S, Dellinger K. The lipid composition of extracellular vesicles applications in diagnostics and therapeutic delivery. *Front Mol Biosci*. 2023 Jul 13;10:1198044. doi: 10.3389/fmolb.2023.1198044.

Ghanam J, Chetty VK, Barthel L, Reinhardt D, Hoyer PF, Thakur BK. DNA in extracellular vesicles from evolution to its current application in health and disease. *Cell Biosci*. 2022 Mar 28;12(1):37. doi: 10.1186/s13578-022-00771-0.

Ghiasi M, Kheirandish Zarandi P, Dayani A, Salimi A, Shokri E. Potential therapeutic effects and nano-based delivery systems of mesenchymal stem cells and their isolated exosomes to alleviate acute respiratory distress syndrome caused by COVID-19. *Regen Ther*. 2024 Apr 16;27:319–328.

González HE, McCarthy SD, Masterson C, Laffey JG, MacLoughlin R, O'Toole D. Nebulized mesenchymal stem cell-derived conditioned medium attenuates Escherichia coli-induced pneumonia in acoli-induced. *Front Med* (Lausanne) [Internet]. 2023 [cited 2025 Aug 12];10:1162615.

Gulick RM, Pau AK, Daar E, Evans L, Gandhi RT, Tebas P, Ridzon R, Masur H, Lane HC; NIH COVID-19 Treatment Guidelines Panel; Adimora AA, Baker J, Kreuziger LB, Bedimo R, Belperio P, Bhalla A, Burgess T, Campbell D, Cantrill S, Chew K, Chiotos K, Coopersmith C, Davey R, Dzierba A, Eisnor D, Eschenauer G, Francis J, Gallagher J, Glidden D, Goldenberg N, Grund B, Han A, Hardy E, Harrison C, Henderson L, Higgs E, Hinkson C, Hughes B, Johnson S, Keller M, Kim A, Knight R, Kuriakose S, Lennox J, Lerner A, Levy M, Li J, MacBrayne C, Martin G, Nadig N, Nason M, Patel P, Pavia A, Proschan M, Schulert G, Seam N, Sheikh V, Simpson S, Singh K, Swindells S, Tien P, Uyeki T, Waghmare A, Wolfe C, Yazdany J, Aberg J. National Institutes of Health COVID-19 treatment guidelines panel: Perspectives and lessons learned. *Ann Intern Med*. 2024 Nov;177(11):1547–1557. doi: 10.7326/ANNALS-24-00464.

Guo C, Sachithanandham J, Zhong W, Craney M, Villano J, Pekosz A, Gould SJ. Antigen-display exosomes provide adjuvant-free protection against SARS-CoV-2 disease at nanogram levels of spike protein. *bioRxiv* [Preprint]. 2024 Jan 5. doi: 10.1101/2024.01.04.574272.

Guo J, Wu C, Lin X, Zhou J, Zhang J, Zheng W, Wang T, Cui Y. Establishment of a simplified dichotomic size-exclusion chromatography for isolating extracellular vesicles toward clinical applications. *J Extracell Vesicles*. 2021 Sep;10(11):e12145. doi: 10.1002/jev2.12145.

Gupta K, Perkerson RB, Parsons TM, Angom R, Amerna D, Burgess JD, Ren Y, McLean PJ, Mukhopadhyay D, Vibhute P, Wszolek ZK, Zubair AC, Quiñones-Hinojosa A, Kanekiyo T. Secretome from iPSC-derived MSCs exerts proangiogenic and immunosuppressive effects to alleviate radiation-induced vascular endothelial cell damage. *Stem Cell Res Ther.* 2024;15(1):230. doi: 10.1186/s13287-024-03735-6.

Gurung S, Perocheau D, Touramanidou L, Baruteau J. The exosome journey from biogenesis to uptake and intracellular signalling. *Cell Commun Signal.* 2021 Apr 23;19(1):47. doi: 10.1186/s12964-021-00730-1.

Habib SM, Martini MF, Al Ahmed O, et al., Mesenchymal stem cell secretomes for cell-free therapy of the broken heart. In: KH Haider, editor. *Handbook of Regenerative Medicine: Cell-Free Therapy Approach.* Boca Raton (FL): CRC Press, Taylor & Francis; 2025. p. 274–294. (Chapter 18). doi: 10.1201/9781003611509-18.

Hade MD, Suire CN, Suo Z. Mesenchymal stem cell-derived exosomes Applications in regenerative medicine. *Cells.* 2021;10(8):1959. doi: 10.3390/cells10081959.

Han QF, Li WJ, Hu KS, Gao J, Zhai WL, Yang JH, Zhang SJ. Exosome biogenesis machinery, regulation, and therapeutic implications in cancer. *Mol Cancer.* 2022 Nov 1;21(1):207. doi: 10.1186/s12943-022-01671-0.

Herrero C, Ferreirós A, Pérez-Fentes D, León-Mateos L, López-López R, Abal M, Alonso-Alconada L. Extracellular vesicles' genetic cargo as noninvasive biomarkers in cancer: a pilot study using ExoGAG technology. *Biomedicines.* 2023;11(2):404. doi: 10.3390/biomedicines11020404.

Hessvik NP, Llorente A. Current knowledge on exosome biogenesis and release. *Cell Mol Life Sci.* 2018 Jan;75(2):193–208. doi: 10.1007/s00018-017-2595-9.

Hu C, Li L. Preconditioning influences mesenchymal stem cell properties in vitro and in vivo. *J Cell Mol Med.* 2018 Mar;22(3):1428–1442. doi: 10.1111/jcmm.13492.

Hu GW, Li Q, Niu X, Hu B, Liu J, Zhou SM, Guo SC, Lang HL, Zhang CQ, Wang Y, Deng ZF. Exosomes secreted by human-induced pluripotent stem cell-derived mesenchymal stem cells attenuate limb ischemia by promoting angiogenesis in mice. *Stem Cell Res Ther.* 2015;6:10. doi: 10.1186/scrt546.

Hu W, Yang J, Xue J, Ma J, Wu S, Wang J, Xu R, Wei J, Wang Y, Wang S, Liu X. Secretome of hESC-derived MSC-like immune and matrix regulatory cells mitigate pulmonary fibrosis through antioxidant and anti-inflammatory effects. *Biomedicines.* 2023 Feb 5;11(2):463. doi: 10.3390/biomedicines11020463.

Ibrahim AY, Mehdi Q, Abbas AO, Alashkar A, Haider KH. Induced pluripotent stem cells: Next generation cells for tissue regeneration. *J Biomed Sci Eng.* 2016;9(4):226–244.

Isaac R, Reis FCG, Ying W, Olefsky JM. Exosomes as mediators of intercellular crosstalk in metabolism. *Cell Metab.* 2021 Sep;33(9):1744–1762. doi: 10.1016/j.cmet.2021.08.006.

Jayaramayya K, Mahalaxmi I, Subramaniam MD, Raj N, Dayem AA, Lim KM, Kim SJ, An JY, Lee Y, Choi Y, Raj A, Cho SG, Vellingiri B. Immunomodulatory effect of mesenchymal stem cells and mesenchymal stem-cell-derived exosomes for COVID-19 treatment. *BMB Rep.* 2020 Aug;53(8):400–412. doi: 10.5483/BMBRep. 2020.53.8.121.

Jia Y, Yu L, Ma T, Xu W, Qian H, Sun Y, et al. Isolation and separation of small extracellular vesicles current techniques, open questions, and clinical applications. *Theranostics.* 2022;12(15):6548–6575. doi: 10.7150/thno.74305.

Jo W, Jeong D, Kim J, Cho S, Jang SC, Han C, Kang JY, Gho YS, Park J. Microfluidic fabrication of cell-derived nanovesicles as endogenous RNA carriers. *Lab Chip.* 2014 Apr 21;14(7):1261–1269. doi: 10.1039/c3lc51235e.

Kalluri R, LeBleu VS. The biology, function, and biomedical applications of exosomes. *Science.* 2020 Feb 7;367(6478):eaau6977. doi: 10.1126/science.aau6977.

Kapałczyńska M, Kolenda T, Przybyła W, Zajączkowska M, Teresiak A, Filas V, Ibbs M, Bliźniak R, Łuczewski Ł, Lamperska K. 2D and 3D cell cultures - a comparison of different types of cancer cell cultures. *Arch Med Sci.* 2018 Jun;14(4):910–919. doi: 10.5114/aoms.2016.63743.

Kaspi H, Semo J, Abramov N, Dekel C, Lindborg S, Kern R, Lebovits C, Aricha R. MSC-NTF (NurOwn®) exosomes a novel therapeutic modality in the mouse LPS-induced ARDS model. *Stem Cell Res Ther.* 2021 Jan 19;12(1):72. doi: 10.1186/s13287-021-02143-w.

Khalaj K, Figueira RL, Antounians L, Lauriti G, Zani A. Systematic review of extracellular vesicle-based treatments for lung injury are EVs a potential therapy for COVID-19? *J Extracell Vesicles.* 2020 Aug 6;9(1):1795365. doi: 10.1080/20013078.2020.1795365.

Kim KM, Abdelmohsen K, Mustapic M, Kapogiannis D, Gorospe M. RNA in extracellular vesicles. *Wiley Interdiscip Rev RNA.* 2017 Jul;8(4). doi: 10.1002/wrna.1413.

Konoshenko MY, Lekchnov EA, Vlassov AV, Laktionov PP. Isolation of extracellular vesicles general methodologies and latest trends. *Biomed Res Int.* 2018 Jan 30;2018:8545347. doi: 10.1155/2018/8545347.

Kooijman EE, Chupin V, Fuller NL, Kozlov MM, de Kruijff B, Burger KN, Rand PR. Spontaneous curvature of phosphatidic acid and lysophosphatidic acid. *Biochemistry*. 2005 Feb 15;44(6):2097–2102. doi: 10.1021/bi0478502.

Kudinov VA, Artyushev RI, Zurina IM, Zorina ES, Lapshin RD, Snopova LB, Mukhina IV, Saburina IN. Inhaled placental mesenchymal stromal cell secretome from two- and three-dimensional cell cultures promotes survival and regeneration in acute lung injury model in mice. *Int J Mol Sci*. 2022 Mar 22;23(7):3417. doi: 10.3390/ijms23073417.

Lane RE, Korbie D, Trau M, Hill MM. Optimizing size-exclusion chromatography for extracellular vesicle enrichment and proteomic analysis of clinically relevant samples. *Proteomics*. 2019 Apr;19(8):e1800156.

Lau HC, Han DW, Park J, Lehner E, Kals C, Arzt C, Bayer E, Auer D, Schally T, Grasmann E, Fang H, Lee JY, Lee HS, Han J, Gimona M, Rohde E, Bae S, Oh SW. GMP-compliant manufacturing of biologically active cell-derived vesicles produced by extrusion technology. *J Extracell Biol*. 2022 Dec 25;1(12):e70. doi: 10.1002/jex2.70.

Lee YJ, Shin KJ, Chae YC. Regulation of cargo selection in exosome biogenesis and its biomedical applications in cancer. *Exp Mol Med*. 2024 Apr;56(4):877–889. doi: 10.1038/s12276-024-01209-y.

Lener T, Gimona M, Aigner L, Börger V, Buzas E, Camussi G, Chaput N, et al. Applying extracellular vesicles-based therapeutics in clinical trials - an ISEV position paper. J Extracell Vesicles. 2015 Dec 31;4:30087. doi: 10.3402/jev.v4.30087.

Li G, Zhang S, Zou Y, Ai H, Zheng X, Qian K, Lei C, Fu W. The therapeutic potential of exosomes in immunotherapy. *Front Immunol*. 2024 Jul 8;15:1424081. doi: 10.3389/fimmu.2024.1424081.

Li L, Yang Z, Li J. Exosomes and SARS-CoV-2 infection. *Front Immunol*. 2024 Nov 26;15:1467109. doi: 10.3389/fimmu.2024.1467109.

Li P, Kaslan M, Lee SH, Yao J, Gao Z. Progress in exosome isolation techniques. *Theranostics*. 2017 Jan 26;7(3):789–804. doi: 10.7150/thno.18133.

Li YJ, Wu JY, Liu J, Xu W, Qiu X, Huang S, Hu XB, Xiang DX. Artificial exosomes for translational nanomedicine. *J Nanobiotechnology*. 2021 Aug 12;19(1):242. doi: 10.1186/s12951-021-00986-2.

Li Z, Niu S, Guo B, Gao T, Wang L, Wang Y, Wang L, Tan Y, Wu J, Hao J. Stem cell therapy for COVID-19, ARDS, and pulmonary fibrosis. *Cell Prolif*. 2020 Dec;53(12):e12939. doi: 10.1111/cpr.12939.

Liang YC, Wu YP, Li XD, Chen SH, Ye XJ, Xue XY, Xu N. TNF-α-induced exosomal miR-146a mediates mesenchymal stem cell-dependent suppression of urethral stricture. *J Cell Physiol*. 2019 Dec;234(12):23243–23255. doi: 10.1002/jcp.28891.

Lin Y, Wu J, Gu W, Huang Y, Tong Z, Huang L, Tan J. Exosome-liposome hybrid nanoparticles deliver CRISPR/Cas9 system in MSCs. *Adv Sci* (Weinh). 2018 May 16;5(4):1700611. doi: 10.1002/advs.201700611.

Liu A, Yang G, Liu Y, Liu T. Research progress in membrane fusion-based hybrid exosomes for drug delivery systems. *Front Bioeng Biotechnol*. 2022a;10:939441. doi: 10.3389/fbioe.2022.939441.

Liu M, Wen Z, Zhang T, Zhang L, Liu X, Wang M. The role of exosomal molecular cargo in exosome biogenesis and disease diagnosis. *Front Immunol*. 2024 Jun 25;15:1417758. doi: 10.3389/fimmu.2024.1417758.

Liu WZ, Ma ZJ, Kang XW. Current status and outlook of advances in exosome isolation. *Anal Bioanal Chem*. 2022b;414(9):2607–2627

Livshits MA, Khomyakova E, Evtushenko EG, Lazarev VN, Kulemin NA, Semina SE, et al. Isolation of exosomes by differential centrifugation theoretical analysis of a commonly used protocol. *Sci Rep*. 2015 Nov 20;5:17319.

Machado C, González-Quevedo A. Hypoxemia and cytokine storm in COVID-19: clinical implications. *MEDICC Rev*. 2021 Jul–Oct;23(3–4):54–59. doi: 10.37757/MR2021.V23.N3.10.

Maeki M, Tokeshi M. Engineered and artificial exosomes for non-viral drug delivery nanocarriers. In: Baba Y, Hanayama R, Akita H, Yasui T, editors. *Extracellular Fine Particles*. Singapore: Springer Nature Singapore Pte Ltd; 2025. p. 275–290.

Mahajan A, Bhattacharyya S. A brief review on potential application of mesenchymal stem cells and secretome in combating mortality and morbidity in COVID-19 patients. *Biomed J*. 2021 Mar;44(1):63–73. doi: 10.1016/j.bj.2020.09.003.

Malkin EZ, Bratman SV. Bioactive DNA from extracellular vesicles and particles. *Cell Death Dis*. 2020 Jul 27;11(7):584. doi: 10.1038/s41419-020-02803-4.

Mashouri L, Yousefi H, Aref AR, Ahadi AM, Molaei F, Alahari SK. Exosomes composition, biogenesis, and mechanisms in cancer metastasis and drug resistance. *Mol Cancer*. 2019 Apr 2;18(1):75. doi: 10.1186/s12943-019-0991-5.

McKelvey KJ, Powell KL, Ashton AW, Morris JM, McCracken SA. Exosomes mechanisms of uptake. *J Circ Biomark*. 2015 Jul 17;4:7. doi: 10.5772/61186.

Mitrani MI, Bellio MA, Meglin A, Khan A, Xu X, Haskell G, Arango A, Shapiro GC. Treatment of a COVID-19 long-hauler with an amniotic fluid-derived extracellular vesicle biologic. *Respir Med Case Rep.* 2021a;34:101502. doi: 10.1016/j.rmcr.2021.101502.

Mitrani MI, Bellio MA, Sagel A, Saylor M, Kapp W, VanOsdol K, Haskell G, et al. Report administration of amniotic fluid-derived nanoparticles in three severely Ill COVID-19 patients. *Front Med* (Lausanne). 2021b Mar 17;8:583842. doi: 10.3389/fmed.2021.583842.

Mohammadi MR, Rodriguez SM, Luong JC, Li S, Cao R, Alshetaiwi H, Lau H, et al. Exosome-loaded immunomodulatory biomaterials attenuate the local immune response in immunocompetent diabetic mice after islet xenotransplantation. *Commun Biol.* 2021 Jun 3;4(1):685. doi: 10.1038/s42003-021-02229-4.

Muñoz-Pérez E, González-Pujana A, Igartua M, Santos-Vizcaíno E, Hernández RM. Mesenchymal stromal cell secretome for the treatment of immune-mediated inflammatory diseases latest trends in isolation, content optimization, and delivery avenues. *Pharmaceutics.* 2021 Oct 27;13(11):1802. doi: 10.3390/pharmaceutics13111802.

Mushtaq M, Zineldeen DH, Mateen MA, Haider KH. Mesenchymal stem cells' "garbage bags" at work: Treating radial nerve injury with mesenchymal stem cell–derived exosomes. *World J Stem Cells.* 2024;16(5):467–478. doi: 10.4252/wjsc.v16.i5.467.

Narang P, Shah M, Beljanski V. Exosomal RNAs in diagnosis and therapies. *Noncoding RNA Res.* 2022 Jan 14;7(1):7–15. doi: 10.1016/j.ncrna.2022.01.001.

Ning B, Huang Z, Youngquist BM, Scott JW, Niu A, Bojanowski CM, Zwezdaryk KJ, et al. Liposome-mediated detection of SARS-CoV-2 RNA-positive extracellular vesicles in plasma. *Nat Nanotechnol.* 2021 Sep;16(9):1039–1044. doi: 10.1038/s41565-021-00939-8.

O'Brien K, Breyne K, Ughetto S, Laurent LC, Breakefield XO. RNA delivery by extracellular vesicles in mammalian cells and its applications. *Nat Rev Mol Cell Biol.* 2020 Oct;21(10):585–606. doi: 10.1038/s41580-020-0251-y.

Ochani R, Asad A, Yasmin F, Shaikh S, Khalid H, Batra S, Sohail MR, et al. COVID-19 pandemic from origins to outcomes. A comprehensive review of viral pathogenesis, clinical manifestations, diagnostic evaluation, and management. *Infez Med.* 2021 Mar 1;29(1):20–36.

Ozcelik A, Cevik O. Microfluidic methods used in exosome isolation. *BIOCELL.* 2023 Apr 7;47(4):959–964. doi: 10.32604/biocell. 2023.028371.

Paresishvili T, Kakabadze Z. Freeze-dried mesenchymal stem cells from bench to bedside. *Adv Biol.* 2024 Feb;8(2):2300155. doi: 10.1002/adbi.202300155.

Peng CK, Wu SY, Tang SE, Li MH, Lin SS, Chu SJ, et al. Protective effects of neural crest-derived stem cell-conditioned media against ischemia-reperfusion-induced lung injury in rats. *Inflammation* [Internet]. 2017 Oct 1 [cited 2025 Aug 12];40(5):1532. Available from https://pmc.ncbi.nlm.nih.gov/articles/PMC7102066/

Pezzotta A, Bovio A, Imberti B, Locatelli M, Corna D, Cerullo D, Gastoldi S, Benigni A, Remuzzi G, Morigi M, Perico L. Mesenchymal stromal cell secretome reduces lung injury and thrombo-inflammation induced by SARS-CoV-2 spike protein. *Stem Cell Res Ther.* 2025 Jul 1;16(1):324. doi: 10.1186/s13287-025-04472-6.

Phuyal S, Skotland T, Hessvik NP, Simolin H, Øverbye A, Brech A, Parton RG, Ekroos K, Sandvig K, Llorente A. The ether lipid precursor hexadecylglycerol stimulates the release and changes the composition of exosomes derived from PC-3 cells. *J Biol Chem.* 2015 Feb 13;290(7):4225–4237. doi: 10.1074/jbc.M114.593962.

Piffoux M, Silva AKA, Wilhelm C, Gazeau F, Tareste D. Modification of extracellular vesicles by fusion with liposomes for the design of personalized biogenic drug delivery systems. *ACS Nano.* 2018 Jul 24;12(7):6830–6842. doi: 10.1021/acsnano.8b02053.

Popović M, de Marco A. Canonical and selective approaches in exosome purification and their implications for diagnostic accuracy. *Transl Cancer Res.* 2017 Aug;6(Suppl 2) S209–S225. doi: 10.21037/tcr.2017.08.44.

Prado-Yupanqui JW, Ramírez-Orrego L, Cortez D, Vera-Ponce VJ, Chenet SM, Tejedo JR, Tapia-Limonchi R. The hidden power of the secretome therapeutic potential on wound healing and cell-free regenerative medicine- a systematic review. *Int J Mol Sci.* 2025 Feb 23;26(5):1926. doi: 10.3390/ijms26051926.

Putra A, Widyatmoko A, Ibrahim S, Amansyah F, Berlian MA, Retnaningsih R, Pasongka Z, et al. Case series of the first three severe COVID-19 patients treated with the secretome of hypoxia-mesenchymal stem cells in Indonesia. *F1000Res.* 2021 Mar 22;10:228. doi: 10.12688/f1000research.51191.3.

Rayamajhi S, Nguyen TDT, Marasini R, Aryal S. Macrophage-derived exosome-mimetic hybrid vesicles for tumor-targeted drug delivery. *Acta Biomater.* 2019 Sep 1;88:504–520. doi: 10.1016/j.actbio.2019.05.054.

Rech J, Getinger-Panek A, Gałka S, Bednarek I. Origin and composition of exosomes as crucial factors in designing drug delivery systems. *Appl Sci*. 2022 Nov 30;12(23):12259. doi: 10.3390/app122312259.

Rocha N, Kuijl C, van der Kant R, Janssen L, Houben D, Janssen H, Zwart W, Neefjes J. Cholesterol sensor ORP1L contacts the ER protein VAP to control Rab7-RILP-p150 glued and late endosome positioning. *J Cell Biol*. 2009 Jun 29;185(7):1209–1225. doi: 10.1083/jcb.200811005.

Rothan HA, Byrareddy SN. The epidemiology and pathogenesis of the coronavirus disease (COVID-19) outbreak. *J Autoimmun*. 2020 May;109:102433. doi: 10.1016/j.jaut.2020.102433.

Sani F, Shafiei F, Dehghani F, Mohammadi Y, Khorraminejad-Shirazi M, Anvari-Yazdi AF, Moayedfard Z, Azarpira N, Sani M. Unveiling exosomes Cutting-edge isolation techniques and their therapeutic potential. *J Cell Mol Med*. 2024 Oct;28(20):e70139. doi: 10.1111/jcmm.70139.

Sato YT, Umezaki K, Sawada S, Mukai SA, Sasaki Y, Harada N, Shiku H, Akiyoshi K. Engineering hybrid exosomes by membrane fusion with liposomes. *Sci Rep*. 2016;6:21933. doi: 10.1038/srep21933.

Schürz M, Pagani I, Klinglmayr E, Melo Benirschke H, Mayora Neto M, Galietta LJV, Venturini A, et al. Quantitative characterisation of extracellular vesicles designed to decoy or compete with SARS-CoV-2 reveals a differential mode of action across variants of concern and highlights the diversity of Omicron. *Cell Commun Signal*. 2025 Jul 2;23(1):323. doi: 10.1186/s12964-025-02223-x.

Sengupta V, Sengupta S, Lazo A, Woods P, Nolan A, Bremer N. Exosomes derived from bone marrow mesenchymal stem cells as treatment for severe COVID-19. *Stem Cells Dev*. 2020 Jun 15;29(12):747–754. doi: 10.1089/scd.2020.0080.

Seoane-Vazquez E, Shukla V, Rodriguez-Monguio R. Innovation and competition in advanced therapy medicinal products. *EMBO Mol Med*. 2019 Mar;11(3):e9992. doi: 10.15252/emmm.201809992.

Singh A, Raghav A, Shiekh PA, Kumar A. Transplantation of engineered exosomes derived from bone marrow mesenchymal stromal cells ameliorates diabetic peripheral neuropathy under electrical stimulation. *Bioact Mater*. 2021 Jul;6(7):2231–2249. doi: 10.1016/j.bioactmat.2021.01.008.

Singh N, Choonara YE, Kumar P. Method standardization of secretome production, collection, and characterization New insights and challenges. *Regen Ther*. 2025;29:466–473. doi: 10.1016/j.reth.2025.04.005.

Smolinská V, Boháč M, Danišovič Ľ. Current status of the applications of conditioned media derived from mesenchymal stem cells for regenerative medicine. *Physiol Res*. 2023 Oct 27;72(S3) S233–S245. doi: 10.33549/physiolres. 935186.

Song Y, Liang F, Tian W, Rayhill E, Ye L, Tian X. Optimizing therapeutic outcomes preconditioning strategies for MSC-derived extracellular vesicles. *Front Pharmacol*. 2025 Feb 10;16:1509418. doi: 10.3389/fphar. 2025.1509418.

Soy M, Keser G, Atagündüz P. Pathogenesis and treatment of cytokine storm in COVID-19. *Turk J Biol*. 2021 Aug 30;45(4):372–389. doi: 10.3906/biy-2105-37.

Su VY, Lin CS, Hung SC, Yang KY. Mesenchymal stem cell-conditioned medium induces neutrophil apoptosis associated with inhibition of the NF-κB pathway in endotoxin-induced acute lung injury. *Int J Mol Sci*. 2019 May 5;20(9):2208. doi: 10.3390/ijms20092208.

Tan J, Zhuo Z, Wang X, Zhang Y, Qian Y, Lin F. Secretome of EMSCs neutralizes LPS-induced acute lung injury via aerosol administration. *Int J Mol Med*. 2023 Nov;52(5):104. doi: 10.3892/ijmm.2023.5307.

Tang Y, Ding F, Wu C, Liu B. hucMSC-conditioned medium ameliorates lipopolysaccharide-induced acute lung injury by suppressing oxidative stress and inflammation via the Nrf2/NF-κB signaling pathway. *Anal Cell Pathol* (Amst). 2021;2021:6653681. doi: 10.1155/2021/6653681.

Tenchov R, Sasso JM, Wang X, Liaw WS, Chen CA, Zhou QA. Exosomes—Nature's lipid nanoparticles, a rising star in drug delivery and diagnostics. *ACS Nano*. 2022 Nov 22;16(11):17802–17846. doi: 10.1021/acsnano.2c08774.

Ter-Ovanesyan D, Norman M, Lazarovits R, Trieu W, Lee JH, Church GM, et al. Framework for rapid comparison of extracellular vesicle isolation methods. *Elife*. 2021 Nov 16;10:e70725.

Théry C, Zitvogel L, Amigorena S. Exosomes composition, biogenesis and function. *Nat Rev Immunol*. 2002 Aug;2(8):569–579. doi: 10.1038/nri855.

Tolomeo AM, Zuccolotto G, Malvicini R, De Lazzari G, Penna A, Franco C, Caicci F, Magarotto F, Quarta S, Pozzobon M, Rosato A, Muraca M, Collino F. Biodistribution of intratracheal, intranasal, and intravenous injections of human mesenchymal stromal cell-derived extracellular vesicles in a mouse model for drug delivery studies. *Pharmaceutics*. 2023 Feb 7;15(2):548. doi: 10.3390/pharmaceutics15020548.

Trajkovic K, Hsu C, Chiantia S, Rajendran L, Wenzel D, Wieland F, Schwille P, Brügger B, Simons M. Ceramide triggers budding of exosome vesicles into multivesicular endosomes. *Science*. 2008 Feb 29;319(5867):1244–1247. doi: 10.1126/science.1153124.

Trigo CM, Rodrigues JS, Camões SP, Solá S, Miranda JP. Mesenchymal stem cell secretome for regenerative medicine where do we stand? *J Adv Res.* 2025;70:103–124. doi: 10.1016/j.jare.2024.05.004.

Uberti B, Plaza A, Henríquez C. Pre-conditioning strategies for mesenchymal stromal/stem cells in inflammatory conditions of livestock species. *Front Vet Sci.* 2022 Mar 16;9:806069. doi: 10.3389/fvets.2022.806069.

Wallen M, Aqil F, Kandimalla R, Jeyabalan J, Auwardt S, Tyagi N, Schultz DJ, Spencer W, Gupta RC. A model system for antiviral siRNA therapeutics using exosome-based delivery. *Mol Ther Nucleic Acids.* 2022 Sep 13;29:691–704. doi: 10.1016/j.omtn.2022.08.011. Erratum in Mol Ther Nucleic Acids. 2022 Dec 13;30:28–29. doi: 10.1016/j.omtn.2022.09.002.

Wang AYL. Human-induced pluripotent stem cell-derived exosomes as a new therapeutic strategy for various diseases. *Int J Mol Sci.* 2021;22(4):1769. doi: 10.3390/ijms22041769.

Wang Y, Zhang L, Li Y, Chen L, Wang X, Guo W, Zhang X, Qin G, He S, Zimmerman A, et al. Exosomes/microvesicles from induced pluripotent stem cells deliver cardioprotective miRNAs and prevent cardiomyocyte apoptosis in the ischemic myocardium. *Int J Cardiol.* 2015;192:61–69. doi: 10.1016/j.ijcard.2015.05.020.

Wang Z, Hu S, Popowski KD, Liu S, Zhu D, Mei X, Li J, Hu Y, Dinh PC, Wang X, Cheng K. Inhalation of ACE2-expressing lung exosomes provides prophylactic protection against SARS-CoV-2. *Nat Commun.* 2024 Mar 12;15(1):2236. doi: 10.1038/s41467-024-45628-x.

Wang Z, Popowski KD, Zhu D, de Juan Abad BL, Wang X, Liu M, Lutz H, De Naeyer N, DeMarco CT, Denny TN, Dinh PC, Li Z, Cheng K. Exosomes decorated with a recombinant SARS-CoV-2 receptor-binding domain as an inhalable COVID-19 vaccine. *Nat Biomed Eng.* 2022 Jul;6(7):791–805. doi: 10.1038/s41551-022-00902-5.

Xia X, Yuan P, Liu Y, Wang Y, Cao W, Zheng JC. Emerging roles of extracellular vesicles in COVID-19 a double-edged sword? *Immunology.* 2021 Aug;163(4):416–430. doi: 10.1111/imm.13329.

Xu M, Ji J, Jin D, Wu Y, Wu T, Lin R, Zhu S, Jiang F, Ji Y, Bao B, Li M, Xu W, Xiao M. The biogenesis and secretion of exosomes and multivesicular bodies (MVBs) Intercellular shuttles and implications in human diseases. *Genes Dis.* 2023a;10(5):1894–1907. doi: 10.1016/j.gendis.2022.03.021.

Xu X, Xu L, Wen C, Xia J, Zhang Y, Liang Y. Programming assembly of biomimetic exosomes An emerging theranostic nanomedicine platform. *Mater Today Bio.* 2023b Aug 11;22:100760. doi: 10.1016/j.mtbio.2023.100760.

Yang D, Zhang W, Zhang H, Zhang F, Chen L, Ma L, et al. Progress, opportunities, and perspectives on exosome isolation—efforts toward efficient exosome-based theranostics. *Theranostics.* 2020 Jan 1;10(8):3684–3707.

Yang Y, Lee EH, Yang Z. Hypoxia-conditioned mesenchymal stem cells in tissue regeneration application. *Tissue Eng Part B Rev.* 2022 Oct;28(5):966–977. doi: 10.1089/ten.teb.2021.0145.

Yao W, Li B, Jiang Y, Yuan Q, Wu W, Hou R, Qi Q, Dong H, Zhang Y, Zhang Y. Salvage therapy utilizing human umbilical cord-derived mesenchymal stem cells is being explored for the treatment of critically ill patients with COVID-19. *Front Immunol.* 2025 Jul 4;16:1594373. doi: 10.3389/fimmu.2025.1594373.

Ye M, Ni Q, Qi H, Qian X, Chen J, Guo X, Li M, Zhao Y, Xue G, Deng H, et al. Exosomes derived from human induced pluripotent stem cells-endothelial cells promote postnatal angiogenesis in mice bearing ischemic limbs. *Int J Biol Sci.* 2019;15(2):158–168. doi: 10.7150/ijbs.28392.

Yi YF, Fan ZQ, Liu C, Ding YT, Chen Y, Wen J, Jian XH, Li YF. Immunomodulatory effects and clinical application of exosomes derived from mesenchymal stem cells. *World J Stem Cells.* 2025 Mar 26;17(3):103560. doi: 10.4252/wjsc.v17.i3.103560.

Yuan YG, Wang JL, Zhang YX, Li L, Reza AMMT, Gurunathan S. Biogenesis, composition, and potential therapeutic applications of mesenchymal stem cells derived exosomes in various diseases. *Int J Nanomedicine.* 2023;18:3801–3852. doi: 10.2147/IJN.S407029.

Zhang Q, Wang M, Han C, Wen Z, Meng X, Qi D, Wang N, Du H, Wang J, Lu L, Ge X. Intraduodenal delivery of exosome-loaded SARS-CoV-2 RBD mRNA induces a neutralizing antibody response in mice. *Vaccines* (Basel). 2023 Mar 16;11(3):673. doi: 10.3390/vaccines11030673.

Zhang Y, Liu Y, Liu H, Tang WH. Exosomes biogenesis, biologic function and clinical potential. *Cell Biosci.* 2019 Feb 15;9:19. doi: 10.1186/s13578-019-0282-2.

Zheng J, Miao J, Guo R, Guo J, Fan Z, Kong X, Gao R, Yang L. Mechanism of COVID-19 causing ARDS exploring the possibility of preventing and treating SARS-CoV-2. *Front Cell Infect Microbiol.* 2022 Jun 14;12:931061. doi: 10.3389/fcimb.2022.931061.

Zheng X, Ai H, Qian K, Li G, Zhang S, Zou Y, Lei C, Fu W, Hu S. Small extracellular vesicles purification and scale-up. *Front Immunol*. 2024 Feb 26;15:1344681. doi: 10.3389/fimmu.2024.1344681.

Zhu Q, Heon M, Zhao Z, He M. Microfluidic engineering of exosomes editing cellular messages for precision therapeutics. *Lab Chip*. 2018 Jun 21;18(11):1690–1703. doi: 10.1039/C8LC00246K.

2 Paracrining the Heart with Stem Cell Secretome

Mechanism of Action, Ethical Considerations, and Regulatory Framework for Development as Advanced Treatment Medicinal Products

Sarah Dayyati[1], Rafah Al-Sumaani[1], Sarah Alluhaydan[1], Yara Baradeh[1], Rania Aziz[1], Gul Abdur Rahman[2], and Khawaja H. Haider[1]

[1]Department of Basic Sciences, Sulaiman AlRajhi Medical School, Kingdom of Saudi Arabia

[2]Withybush General Hospital, Fishguard Road, Haverfordwest, UK

2.1 LIST OF ABBREVIATIONS

AD-MSCs	Adipose tissue-derived MSCs
Akt-MSCs	Akt-overexpressing MSCs
Ang-1	Angiopoietin-1
BM	Bone marrow
CPCs	Cardiac progenitor cells
CSCs	Cardiac stem cells
ECs	Endothelial cells
EPCs	Endothelial progenitor cells
ESCs	Embryonic stem cells
G-CSF	Granulocyte colony-stimulating factor
GM-CSF	Granulocyte-macrophage colony-stimulating factor
hCSCs	Human cancer stem cells
HGF	Hepatocyte growth factor
IGF-1	Insulin-like growth factor
IL	Interleukin
iPSCs	Induced pluripotent stem cells
LVEF	Left ventricular ejection fraction
MCP-1	Monocyte chemoattractant protein-1

DOI: 10.1201/9781003653387-2

MMP	Matrix metalloproteinase
MSCs	Mesenchymal stem cells
MMSCs	Mouse MSCs
NGF	Nerve growth factor
NYHA	New York Heart Association
PBMCs	Peripheral blood mononuclear cells
PDGF	Platelet-derived growth factor
PlGF	Placenta growth factor
Sca-1	Stem cell antigen 1
SkMs	Skeletal myoblasts
TIMP	Tissue inhibitor of metalloproteinase
TEM	Transmission electron microscopy
UC-MSCs	Umbilical cord-derived MSCs
VEGF	Vascular endothelial growth Factor

2.2 INTRODUCTION

Cardiovascular diseases alone are the most prevalent cause of death and morbidity worldwide (Martin et al., 2025). Despite significant reductions in mortality rates over the last two decades due to surgical and pharmacological advancements, as well as improved patient management protocols, the urgent need for a new therapeutic approach for patients at risk of ventricular remodeling and dysfunction remains evident (Tamargo et al., 2023; Leela et al., 2025). The need for developing novel treatment approaches is accentuated by the limited endogenous regenerative capacity of the cardiomyocytes to replenish the damaged myocardium. This underscores the importance and relevance of modern-day research in exploring the potential of cell-based and cell-free therapy treatment approaches for cardiac repair (Liang et al., 2025). Both these novel therapeutic approaches offer a biological complement to pharmacological and surgical therapeutics for patients with cardiac disease, offering hope for the future of cardiac repair.

Although the safety profile of cell-based therapy is favorable and its efficacy has been well-documented, though inconsistent and, overall, modest, it has advanced to late clinical phase assessment and even to the launching of some MSC-based products after approval from drug regulating authorities of different countries, including the FDA of the USA (Bhatti et al., 2025). Multiple cell types have been used in preclinical animal models and in patients to repair the injured heart, including: SkMs (Menasché et al., 2001; Sim et al., 2005; Ye et al., 2007), endothelial progenitor cells (EPCs) (Kawamoto et al., 2001; Menasche et al., 2008), mesenchymal stem cells (MSCs) (Nguyen et al., 2010; Ye et al., 2013; Ahmed et al., 2024; Safwan et al., 2024), embryonic stem cells (ESCs)(Dai et al., 2007; Laflamme et al., 2007; Rufaihah et al. 2010), induced pluripotent stem cells (iPSCs) (Ahmed et al., 2011; Buccini et al., 2012; Fakoya et al., 2022) and cardiac stem cells (CSCs) (Beltrami et al., 2003; Takamiya et al., 2011; Mehanna et al., 2022). The therapeutic benefits of cell-based therapy have been multifactorial, primarily attributed to paracrine mechanisms, with donor cell engraftment and cardiomyogenic differentiation as secondary mechanisms that sustain myocardial pump function and support the regenerative process of the injured myocardium (Haider & Aziz, 2017). This chapter summarizes recent advances in the secretome-based approach to myocardial repair, using both the soluble (secretomes) and insoluble (exosomes) fractions of paracrine secretions from stem cells. It delves into providing mechanistic insights into how the secretome-based cell-free therapy approach confers cytoprotection and facilitates repair of the infarcted myocardium, thereby preserving global cardiac function. The chapter first discusses soluble factors (the secretome) and then their regulatory requirements. It offers a future perspective for further research in this area, a topic sure to stimulate our readers' interest.

2.3 OVERVIEW OF CELL TRANSPLANTATION FOR CARDIAC REPAIR

There has been a long-standing dogma that cardiomyocytes can proliferate during embryogenesis and the early postnatal period (Günthel et al., 2018); however, they permanently withdraw from the cell cycle a few months into postnatal life and stay as terminally differentiated tissue (Dimasi et al., 2023). Any further cardiac muscle development occurs only through myocyte hypertrophy and growth of the non-myogenic resident cells within the cardiac muscle (Hudlicka and Brown, 1996). However, there is mounting evidence challenging this long-standing dogma by showing data that the cardiomyocytes can re-enter the cell cycle under conditions of myocardial ischemia and injury (Beltrami et al., 2001; Kajstura et al., 1998). Also, resident CSCs have been reported to participate in the repair process following myocardial damage, replacing dead or injured cardiomyocytes with morpho-functionally competent cardiomyocytes, albeit at a very low rate of proliferation and differentiation (Mollova et al., 2013). Due to a very low rate of proliferation, cardiomyocyte turnover is insufficient to repair damaged cardiomyocytes and prevent extensive scar formation, eventually leading to loss of contractile function. Given the human heart's limited regenerative capacity, stem cell-based therapy offers a more realistic approach to regenerate lost cardiomyocytes through transdifferentiation of transplanted cells (Rajab et al., 2019). This approach aims to repopulate post-infarction scar tissue with myogenic cells that may be functionally integrated into the host tissue (Ali et al., 2025). Different cell types from diverse tissue sources, ranging from extraembryonic to embryonic and adult tissues, have been used, some of which offer the opportunity to use autologous tissue-source-derived cells to overcome immunological rejection of the graft without immunosuppression. Some of these cell types have advanced to the clinical phase after in vitro and in vivo characterization and assessment in experimental animal models, as discussed in the following section.

2.4 MOST EXTENSIVELY STUDIED CELL TYPES AND THEIR SECRETOME-BASED STUDIES

2.4.1 Skeletal Myoblasts and Paracrine Secretome Activity

Soonpaa et al. were the first to demonstrate the feasibility of the cellular cardiomyoplasty concept, showing that mouse fetal cardiomyocytes had intercalated disks connected with host myocytes (Soonpaa et al., 1994). Following this report, neonatal (Scorsin et al., 1996) and fetal rat cardiomyocytes (Li et al., 1996, 1997) were reported, demonstrating that fetal rat cardiomyocyte transplantation formed cardiac tissue within the myocardial scar, limited scar expansion, and improved heart function. After pioneering SkM-based cellular cardiomyoplasty in 1995 by Chiu et al. (Chiu et al., 1995) and a subsequent study in a rabbit model of cryoinjured heart by Doris Taylor's group (Taylor et al., 1998), Phillip Menasche and his colleagues from Hôpital Bichat Claude Bernard, Paris, France, reported the first-in-man study data using autologous SkMs in a 57-year-old man (Menasché et al., 2001). Seventeen and a half months later, when the patient died due to a cell therapy-unrelated cause, post-mortem histological studies on his heart revealed that the cardiac milieu influenced the differentiation of SkMs into cardiac-like muscle cells. Encouraged by the results, the same group of researchers conducted another study involving nine additional patients (Menasché et al., 2003). The authors reported that the mean New York Heart Association (NYHA) functional class improved from 2.7 ± 0.2 preoperatively to 1.6 ± 0.1 postoperatively, and the left ventricular ejection fraction (LVEF) improved from $24 \pm 1\%$ to $32 \pm 1\%$ ($p < 0.02$). A blinded echocardiographic analysis showed that 63% of the cell-implanted scars (14 of 22) showed improved systolic thickening. These data set the ball rolling for the clinical assessment of cell-based therapy using other cell types.

SkMs are one of the most robust and abundant cell types available for tissue repair that can be easily expanded in vitro, have been used as a sole therapy or synergized with an angiogenic factor for cardiac repair, with encouraging results (Chazaud et al., 2003; Haider et al., 2004; Brasselet et al., 2005; Ye et al., 2007, 2008). Due to their myogenic lineage commitment, they are among the most widely used cells for cardiac repair. SkMs from various species, such as mice, rats, rabbits, sheep, dogs, and pigs, have been investigated for their potential to repair damaged myocardium and in human clinical trials (Menasché et al., 2001, 2008; Sim et al., 2005; Ye et al., 2006, 2007). It has been shown that SkM transplantation can improve both systolic and diastolic function of the injured heart (Taylor et al., 1998; Atkins et al., 1999; McCue et al., 2008; Ye et al., 2007) and limit left ventricular (LV) remodeling by decreasing LV diastolic dimension (Thompson et al., 2003) and increasing myocardial wall thickness (Tambara et al., 2003). Niagara et al. demonstrated that preconditioning rat SkMs with 200 μM diazoxide for 30 min could protect SkMs from injury induced by 100 μM H2O2, as shown by reduced lactate dehydrogenase release, preserved mitochondrial membrane potential, and decreased apoptosis (Niagara et al., 2007). The pretreatment enhanced cell engraftment after intramyocardial transplantation in MI rats. Elevated phospho-Akt (pAkt), bFGF, HGF, and cyclooxygenase-2 contribute to these benefits.

In addition to their direct effect on forming skeletal muscle fibers in the recipient heart, SkMs also exert paracrine effects on the recipient cardiac tissue. These paracrine actions are mediated by several cytokines and growth factors that regulate angiogenesis, cell proliferation, chemotaxis, inflammation, matrix remodeling, migration, and apoptosis (Peterson & Pizza, 2009). Perez-Ilzarbe et al. analyzed the cell culture supernatant of SkMs. They identified the release of proangiogenic growth factors, including VEGF, placenta growth factor (PLGF), angiogenin, angiopoietin, HGF, and platelet-derived growth factor (PDGF), matrix metalloproteinase 2, 9, and 10 (MMP2, MMP9, and MMP10), and their inhibitors, i.e., tissue inhibitor of metalloproteinase (TIMP) (Pérez-Ilzarbe et al., 2008). Priming the cells with mechanical stretch results in the release of HGF and accelerated entry into the cell cycle when the medium pH is between 7.1 and 7.4 (Tatsumi et al., 2002). Similarly, mechanical strain regimens of negative pressure increased the production of IL-8, granulocyte-macrophage colony-stimulating factor (GM-CSF), granulocyte colony-stimulating factor (G-CSF), monocyte chemoattractant protein-1 (MCP-1), and IL-6 in conditioned media, which, in turn, induced neutrophil chemotaxis (Peterson & Pizza, 2009). Henningsen et al. documented the identification and quantitative analysis of 635 secreted proteins, including 35 growth factors, 40 cytokines, and 36 metallopeptidases, using C_2C_{12} SkMs during differentiation (Henningsen et al., 2010). Comparative analyses of selected secreted proteins revealed limited correlation between mRNA and protein levels, suggesting pronounced post-transcriptional regulation. Norheim et al. identified 236 proteins in medium conditioned by cultured human myotubes from strength-training individuals (Norheim et al., 2011). However, 17 of 235 proteins identified by proteomic analysis are potentially secreted via the classical secretory pathway into the media from cultured human skeletal muscle cells. The two studies highlighted the importance of the SkMs as a prominent secretory tissue.

2.4.2 Resident Cardiac Stem/Progenitor Cells and Paracrine Activity

The second type of cells reaching the clinical phase of assessment are cardiac progenitor cells (CPCs), which are self-renewing, clonogenic, and multipotent. It is proposed that c-Kit$^+$/Lin$^-$ CSCs comprise a heterogeneous population of lineage-committed progenitor cells (Sandstedt et al., 2014). Beltrami et al. (2003) were the first to document the presence of c-Kit$^+$/Lin$^-$ cells, which were relatively rare (one out of 10,000 myocytes) (Beltrami et al., 2003). They were typically found in small clusters and expressed the transcription factors Gata4, Nkx2.5, and Mef2 when isolated and cultured. c-Kit$^+$/Lin$^-$ cells could differentiate into cardiomyocytes, smooth muscle cells, and endothelial cells in vitro. Since the publication of this data, c-Kit$^+$/Lin$^-$ CPCs have been reported in different mammalian species, including rats (Miyamoto et al., 2010; Smith et al., 2014), dogs (Linke et al., 2005), pigs (Ellison et al., 2011), and humans (Bearzi et al., 2007; Urbanek et al., 2005). Bearzi et al. developed methods for isolating and expanding c-Kit$^+$ human cancer stem cells (hCSCs) from small myocardial specimens (Bearzi et al., 2007). They found that hCSCs differentiated predominantly into cardiomyocytes, and to a lesser extent, into smooth muscle cells and endothelial cells.

In addition to directly dividing into cardiac cells, CSCs secrete many growth factors (VEGF, IGF, HGF, and IGF-1, etc.) (Chimenti et al., 2010). The conditioned media exerted anti-apoptotic effects on neonatal rat ventricular myocytes and proangiogenic effects on human umbilical vein endothelial cells. Maxeiner et al. reported that the CPC supernatant contained a diverse array of cytokines. It was found that, for rats, 8 (juvenile) or 7 (adult) cytokines out of 19 were significantly released into the cell culture medium (Maxeiner et al., 2010), and, for humans, 29 (juvenile) or 18 (adult) cytokines out of 507 were significantly released into the cell culture medium (Maxeiner et al., 2010). The supernatant enhanced the contractility of isolated adult cardiomyocytes in a concentration-dependent manner after 24 hours of incubation, regardless of the age of the CPC donors. Furthermore, the supernatant could normalize angiotensin-II-induced contractile dysfunction.

Another cell type reported as CPCs expresses stem cell antigen-1 (Sca-1 or Ly6A), a member of the Ly6 family of glycosylphosphatidylinositol (GPI)-anchored cell surface proteins. Sca-1 has long been associated with murine stem and progenitor cells (Holmes & Stanford, 2007). Recently, Sca-1 cells have been found in multiple organs (Wang et al., 2006). Gojo et al. demonstrated that $CD34^{low}$ $c\text{-}Kit^{+}$ $CD140a^{+}Sca\text{-}1^{high}$ cells could contribute to vasculature formation when injected into skeletal muscle and the lung (Gojo et al., 2003). Liang et al. demonstrated that Sca-1+CD31+ CPCs can migrate into damaged myocardium from non-ischemic areas of the heart and differentiate into cardiomyocyte- and endothelial-like cells following acute ischemic injury via the SDF-1/CXCR4 system in a murine model of myocardial ischemia (Liang et al., 2010). Wang et al. documented that murine heart-derived $Sca\text{-}1^{+}CD31^{-}$ cells improved cardiac function in a mouse model of MI. $Sca\text{-}1^{+}CD31^{-}$ cells could be induced to express both ECs and cardiomyocyte markers in vitro (Wang et al., 2006). Transplantation of $Sca\text{-}1^{+}CD31^{-}$ cells into the acutely infarcted mouse heart improved pump function and neovascularization, limited structural remodeling, and increased cardiomyocyte regeneration. Remarkably improved myocardial bioenergetic characteristics accompanied the attenuation of LV remodeling. It was suggested that the beneficial effects of Sca-1+CD31-transplanted cells appeared to primarily depend on the paracrine effects of these cells on new vessel formation and native cardiomyocyte function. The same group further demonstrated that $Sca\text{-}1^{+}CD31^{-}$ had high gene expression levels of a variety of growth factors, including FGF11, IGF1, Kit ligand, NGF, PLGF, TGFβ1 and β2, and VEGF-A, -B, and -C (Wang et al., 2014). IGF and HGF treatment significantly altered the mRNA expression of PDGF, FGF1, FGF3, FGF6, VEGF-A, and VEGF-B. Simultaneous injection of IGF+HGF and Sca-1+CD31—into the mouse MI model increased the cell engraftment rate, promoted the transplanted cell viability, enhanced angiogenesis, and minimally stimulated endogenous cardiomyocyte regeneration in vivo.

2.4.3 Mesenchymal Stem Cells and Paracrine Activity

Bone marrow-derived MSCs are multipotent, lowly immunogenic, easy to expand in vitro, robust, and genetically modifiable without losing their stemness for cellular cardiomyoplasty (Álvarez-Viejo & Haider, 2022). They were first reported in 1999 by Tomita et al., who showed that the transplanted cells differentiated into cardiac-like myocytes in vitro and in vivo in ventricular scar tissue and improved global myocardial function (Tomita et al., 1999). Incidentally, MSCs derived from different mesenchymal tissues are among the most well-characterized and studied cell types for myocardial repair and regeneration (Álvarez-Viejo & Haider, 2022; Kamal et al., 2022). They have progressed to advanced phases of clinical assessment and are currently approved and available as living biodrugs for patient use. Several clinical studies have shown their safety and efficacy in the treatment of patients with heart failure and myocardial infarction (Ahmed et al., 2024; Safwan et al., 2024; Ahmed et al., 2025).

One characteristic feature of MSCs is their paracrine activity, which has been recognized as a primary contributor to their therapeutic effects (Haider & Aziz, 2017). The impact of paracrine secretions is multifactorial, encompassing at least seven mechanisms reported in the published literature, comprising: anti-apoptotic effects on myocardial cells, including the cardiomyocytes, proangiogenic effects, anti-inflammatory and immunomodulatory effects, attenuation of oxidative stress, antifibrotic effects, reduction in extracellular matrix remodeling, and myocardial repair and regeneration by involving resident cardiac stem/progenitor cells (Habib et al., 2026). Figure 2.1 depicts and summarizes the mechanisms underlying secretome-based therapy.

Chen et al. determined the growth factor profile of human MSCs (hMSCs) conditioned with rat cerebral tissue extract following traumatic brain injury (Chen et al., 2002). The authors observed a time-dependent increase in the expression of VEGF, HGF, BDNF, and nerve growth factor (NGF) in the conditioned medium. Kinnaird et al. reported that VEGF, bFGF, PlGF, and monocyte chemoattractant protein-1 (MCP-1) were released by murine MSCs (mMSCs) (Kinnaird et al., 2004). Cell injection into a mouse ischemic limb disease model improved limb function, reduced the incidence of amputation, and attenuated muscle atrophy and fibrosis, accompanied by increased collateral development. However, the injected mMSCs dispersed between muscle fibers but were not incorporated into mature collaterals, suggesting that mMSCs can contribute to collateral remodeling through a paracrine mechanism. On the same note, hypoxia and Akt-gene induction not only enhanced the secretion of stem cell factor (SCF) to promote cardiomyocyte proliferation but also protected against ischemia-induced cardiomyocyte death through protein kinase C epsilon (Huang et al., 2014).

Various strategies have been developed to prime MSCs to enhance their survival, paracrine activity, and reparability (Haider, 2024a, Haider, 2024b). For example, intramyocardial transplantation of Akt-overexpressing MSCs (Akt-MSCs) into the ischemic rat myocardium inhibited cardiac remodeling, reduced intramyocardial inflammation, collagen deposition, and cardiomyocyte hypertrophy, regenerated 80–90% of lost myocardial volume, and completely normalized systolic and diastolic cardiac function (Mangi et al., 2003). Rat Akt-MSCs restored a fourfold greater myocardial volume than control MSCs. The same group demonstrated that conditioned medium from

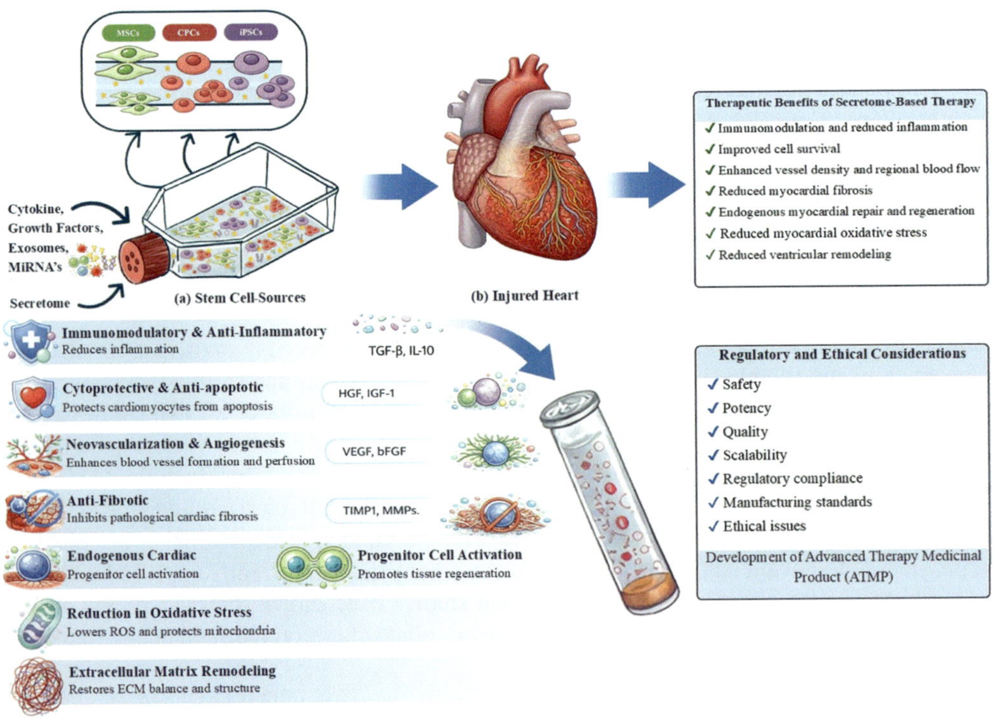

FIGURE 2.1 A schematic illustration showing the multifactorial mechanism of stem cell-derived secretome for HF treatment.

hypoxic Akt-MSCs markedly inhibited hypoxia-induced apoptosis and triggered vigorous spontaneous contraction of adult rat cardiomyocytes *in vitro* (Gnecchi et al., 2006). The authors reported the release of VEGF, bFGF, HGF, IGF-I, and thymosin β4 (TB4), which are potential mediators of the effects of the Akt-MSC conditioned medium and were significantly upregulated in Akt-MSCs. When injected into infarcted hearts, the conditioned medium from Akt-MSCs significantly delimited infarct size and improved ventricular function. Akt-MSCs also released Frizzled-related protein 2 (Sfrp2) that could dramatically attenuate Wnt3a-induced caspase activity in a concentration-dependent manner during hypoxia-reoxygenation (Zhang et al., 2009). Wang et al. reported that HSP20, via enhancing Akt activation, improved MSC engraftment rate at day four after transplantation into the infarcted heart (Wang et al., 2009). Increased secretion of growth factors, including VEGF, FGF-2, and IGF-1, was observed in the genetically modified MSCs. They can secrete a variety of paracrine factors, including angiogenic growth factors (VEGF, bFGF, HGF, IGF, PDGF, epiregulin, endothelin) (Wang et al., 2009; Windmolders et al., 2014; Nguyen et al., 2010) and anti-apoptotic factors (Galectin-3, Smad-5, sRFP: Secreted frizzled-related protein 1 (sRFP-1), and sRFP-4) (Nguyen et al., 2010), which contribute to the beneficial effects on cardiac repair by promoting angiogenesis, anti-apoptosis, and anti-fibrosis, limiting left ventricular (LV) remodeling, and improving cardiac function.

2.4.4 Other Cell Types Used for Myocardial Repair and Regeneration

Endothelial progenitor cells (EPCs), also known as angioblasts, were first reported by Kocher et al., demonstrating that BM-derived angioblasts can revascularize infarcted myocardium (Kocher et al., 2001). Gojo et al. showed that bone marrow-derived CD34low c-Kit+ CD140a+ Sca-1high cells could differentiate into cardiomyocytes, ECs, and SMCs upon direct injection into the adult heart (Gojo et al., 2003). Matsuura et al. reported that Sca-1$^+$ cells in adult mouse hearts exhibited some features of stem cells and could differentiate into cardiomyocyte-like cells (Matsuura et al., 2004). Some other cell types, such as pluripotent stem cells, derived cells (Xiong et al., 2012; Ye et al., 2014; Yeghiazarians et al., 2012; Zhang et al., 2015), umbilical cord and cord blood-derived stem cells (Acosta et al., 2013), very small embryonic/epiblast-like stem cells (VSELs) (Dawn et al., 2008), adipose-derived stromal cells (Hamdi et al., 2011), etc., have been studied to assess their paracrine behavior and reparability.

Pluripotent stem cells, including ESCs and induced pluripotent stem cells (iPSCs), have the potential to repair cardiac structure and significantly improve cardiac function and bioenergetics after injury. Zhang et al. found that conditioned medium from hiPSC-MSCs (CdM-hiPSC-MSCs) could limit reactive oxidative stress (ROS) production and cell apoptosis *in vitro* (Zhang et al., 2015). Transplantation of CdM-BM-MSCs can remarkably attenuate left ventricular (LV) dysfunction and dilatation after anthracycline-induced cardiomyopathy *in vivo*. Analysis of the conditioned medium revealed that macrophage migration inhibitory factor (MIF) and growth differentiation factor-15 (GDF-15) may be partially responsible for the cardioprotective effects of CdM-iPSC-MSCs (Zhang et al., 2015). Xiong et al. analyzed the paracrine factor profile of human ESC-derived ECs (hESC-ECs) and SMCs (hESC-SMCs) under normoxia and hypoxia (Xiong et al., 2012). It was found that both hESC-ECs expressed high protein levels of angiogenin, epithelial-derived neutrophil-activating peptide 78, IL-6, IL-8, TIMP-1, and TIMP-2, and low levels of epithelial growth factor (EGF), IGF, PDGF-BB, thrombopoietin, VEGF-A, and -D. In contrast, hESC-SMCs expressed only high levels of TIMP-1 and low levels of angiogenin, IGF-1, IL-8, MCP-1, and TIMP-2. Culture under hypoxic conditions accentuated the expression of angiogenin and TIMP-1 in the conditioned media from hESC-ECs (CdM-hESC-ECs) and hESC-SMCs (CdM-hESC-ECs).

Ye et al. have reported the paracrine factor profile of hiPSC-derived cardiac cells, including endothelial cells (hiPSC-ECs), smooth muscle cells (hiPSC-SMCs), and cardiomyocytes (hiPSC-CMs) under normoxic conditions (Ye et al., 2014). hiPSC-ECs not only expressed high protein levels of angiogenin, IL-6, IL-8, TIMP-1, and TIMP-2, but also expressed high levels of angiopoietin-2, growth-regulated protein, and MCP-1, and low levels of angiopoietin-1, PDGF-BB, RANTES, MCP-3, and uPAR. hiPSC-SMCs expressed high levels of angiogenin, angiopoietin-1, MCP-1, TIMP-1, and TIMP-2, and low levels of IL-6, growth-regulated protein, IL-8, RANTES, and uPAR. Similarly, hiPSC-CMs expressed high levels of angiogenin, IL-6, growth-regulated protein, MCP-1, MCP-3, TIMP-1, and TIMP-2, and low levels of angiopoietin-1, angiopoietin-2, uPAR, and VEGF. They inhibited apoptosis (angiogenin, angiopoietin, IL-6, MMP-1, PDGF-BB, TIMP-1, uPAR, and VEGF), induced cell migration or homing (angiogenin, angiopoietin, IL-8, MCP-1, MCP-3, MMP-9, uPAR, and VEGF), and promoted cell division (angiogenin, angiopoietin, PDGF-BB, and VEGF). The conditioned medium of hiPSC-ECs (CdM-hiPSC-ECs) and hiPSC-SMCs (CdM-hiPSC-SMCs) protected hiPSC-CM from hypoxic injury (in a 1% O_2 incubator for 48 hrs), suggesting a paracrine mechanism through which hiPSC-ECs and hiPSC-SMCs could promote CM survival. A summary of the studies reported that a diverse array of paracrine factors released by different cell types has been reported in Tables 2.1 and 2.2.

2.5 CHALLENGES, LIMITATIONS, AND FUTURE DIRECTIONS

Secretome-mediated processes in myocardial repair can be divided into discrete, yet overlapping, categories. A minireview by the Haider group elegantly reviews the therapeutic potential of the stem cell secretome, which exerts pleiotropic effects on cardiac repair and regeneration (Habib et al., 2026). Providing mechanistic insight, the authors describe that secretome-based treatment involves seven primary modules: angiogenesis and tissue perfusion; apoptosis inhibition; reduction of oxidative stress; modulation of inflammatory responses; facilitation of cardiomyogenesis; and reduction of myocardial fibrosis. Although many cell-based products are now available on the market (Bhatti & Haider, 2025), no secretome-based therapy has been approved by the FDA for routine clinical use. Currently, early-phase clinical trials are underway to assess the safety and feasibility of some investigational new drugs. For example, Zofin™ is an investigational acellular biologic derived from full-term amniotic fluid, enriched in cytokines, chemokines, and EVs. It is in IND-approved Phase I/II trials for the treatment of moderate-to-severe COVID-19. Produced from rigorously screened donors by scheduled cesarean delivery and processed to eliminate cells while preserving bioactive factors, Zofin™ is designed to be antiviral and immunomodulatory. In early clinical use, no disastrous or undesirable effects were noted. Preclinical studies have shown that Zofin™ and its EV fraction significantly suppress SARS-CoV-2 replication, elevate interferon-β expression, and inhibit proinflammatory cytokines, including TNF. In an experimental murine infection model, treatment with Zofin™ enhanced survival, limited inflammatory cell infiltration, and decreased viral load in the lungs. These data showed that Zofin™ successfully reduced inflammation and improved the antiviral effects (Bellio et al., 2021).

Another secretome-based product in the clinical assessment phase is APOSEC™, derived from stressed peripheral blood mononuclear cells (PBMCs) (Simader et al., 2017). The product is rich in bioactive proteins, lipids, EVs, and antimicrobial peptides. Phase I study (MARSYAS I; ClinicalTrials.gov Identifier: NCT02284360), autologous APOSEC was topically administered in a randomized, double-masked, placebo-controlled manner to minor, artificially inflicted dermal wounds in 10 healthy male volunteers; both low-dose and high-dose treatments were well tolerated, with no therapy-related serious undesirable events and no effect on wound closure. These data form the basis of the ongoing Phase I/II trial (MARSYAS II; ClinicalTrials.gov Identifier: NCT04277598), a randomized, double-blinded, placebo-controlled, parallel-group, dose-ranging, multicenter study involving approximately 132 patients with diabetic foot ulcers. APOSEC in a hydrogel mixture (APO-2) is applied topically three times a week for 4 weeks at three escalating doses versus placebo (Hafner et al., 2020). An 8-week follow-up revealed a significant reduction in wound size at four weeks as the primary endpoint. Promising preclinical studies suggest therapeutic value in cardiovascular disease (e.g., acute myocardial ischemia), which further warrants development outside the realm of wound healing uses.

TABLE 2.1

Secretome Profiling and Therapeutic Benefits of the Paracrine Factors Released from Adult and Pluripotent Stem Cells

Cell Type	Paracrine Factors Secreted	Treatment Effects	Reference
Adult Stem Cells Derived Secretome			
Cardiac progenitor cell	VEGF, IGF, HGF, and IGF-1	Promote endogenous regeneration, myocardial protection.	Chimenti et al., 2010
	Rat: 8 (one juvenile and seven adult) cytokines. Human: 29 (18 juvenile, 11 adult).	Improve cardiomyocyte contractility and function.	Maxeiner et al., 2010
	IL-6, IL-8, VEGF, and bFGF	Mechanical stress increases the release of these factors.	Kurazumi et al., 2011
Cardiac Explants	IL-6, IL-8, GRO, MCP-1, bFGF, EGF, HGF, IGFBP-1, and TGF-β	Promote myocardial differentiation.	Schittini et al., 2010
Human Aortic $CD133^+$ Progenitor Cells	VEGF, IL-8	Activate the Wnt signaling pathway.	Barcelos et al., 2009
Hematopoietic progenitor cell	Galectin-3, gelsolin, IL-6, IL-13, MCP-1, MCP-3, MIP-1α, MIP-1β	Promote tissue regeneration.	Luecke et al., 2010
Bone marrow-derived progenitor cell.	HGF-1	Reduces cardiac fibrosis and improves heart function.	Kishore et al., 2013
Endothelial progenitor cells	TGFβ1, IGF-1	TGFβ1 promotes cardiac.	Doyle et al., 2008
Skeletal myoblast	VEGF, PLGF, angiogenin, Ang HGF, PDGF-BB, MMP2, MMP9, MMP10, and TIMPs.	Promote angiogenesis and anti-apoptosis.	Pérez-Ilzarbe et al., 2008
	HGF-1	Stretch triggers an intracellular Signaling pathway leading to NOS activity and HGF release.	Tatsumi et al., 2002
	bFGF and HGF	Preconditioned rat SkMs promoted cell-enhanced release of paracrine factors.	Niagara et al., 2007
	IL-8, GM-CSF, G-CSF, MCP-1	Mechanical strain increases chemokine release.	Peterson and Pizza, 2009
	635 secreted proteins, including 35 growth factors, 40 cytokines, and 36 metallopeptidases	These proteins can act as potent signaling mediators for other cells.	Henningsen et al., 2010
Sca-1	FGF11, IGF1, Kit ligand, Nerve growth factor, PDGF, TGFβ1 and β2, and VEGF-A,-B, and -C.	Simultaneous IGF+HGF and $Sca\text{-}1^+CD31^-$ transplantation achieved better cardiac function improvement.	Wang et al., 2014
Pluripotent Stem Cell-Derived Secretome			
hiPSC-ECs	Ang-1, VEGF-A, VEGF-C, and PDGF-AA	Increase capillary density and improve blood perfusion.	Rufaihah et al., 2011

(Continued)

TABLE 2.1 CONTINUED
Secretome Profiling and Therapeutic Benefits of the Paracrine Factors Released from Adult and Pluripotent Stem Cells

Cell Type	Paracrine Factors Secreted	Treatment Effects	Reference
Irradiated ESCs	Unknown	Improve cardiac function and decrease infarct size.	Burt et al., 2012.
hESC-ECs	Ang, ENA78, IL-6, IL-8, TIMP-1&2, EGF, IGF-1, PDGF-BB, thrombopoietin, VEGF-A, and VEGF-D, TIMP-1, angiogenin, IGF-1, IL-8, MCP-1, and TIMP-2	Anti-apoptosis and cytoprotection.	Xiong et al., 2012
	Angiogenin, IL-6, IL-8, TIMP-1, TIMP-2, Ang-2, MCP-1, and PDGF-BB, growth-regulated protein, MCP-3, Ang-1, RANTES, and uPAR.	Anti-apoptosis and cytoprotection, enhance angiogenesis and blood flow, promote cell migration and division.	Ye et al., 2014
hiPSC-MSCs	MIF and GDF-15.	Attenuated LV dysfunction and dilatation.	Zhang et al., 2015
MSC-derived exosomes	miR-210, VEGF	Enhanced angiogenesis and reduced apoptosis.	Li et al., 2017
hiPSC-CM secretome	VEGF-A, HGF, IGF-1	Improved myocardial repair and reduced fibrosis	Shen et al., 2019
Cardiac MSC secretome	Angiopoietins, cytokines, EV-miRNAs	Improves LV function and vascular regeneration.	Matsumoto et al., 2020
MSC-derived exosomes	Exosomal miR-21, miR-126, miR-146a; VEGF, HGF	Pro-angiogenic, anti-apoptotic, improves cardiac repair.	Zheng et al., 2021
Wharton's Jelly MSC secretome	VEGF, HGF, angiopoietins, cytokines, EVs	Enhanced angiogenesis and immunomodulation.	Drobiova et al., 2023
MSC-Exos (clinical translation)	VEGF, HGF, EV-miRNAs	Documented safety and benefit signals in early clinical studies.	Lotfy et al., 2013
Hypoxia-preconditioned MSC secretome	↑VEGF, angiogenin, pro-angiogenic miRNAs	Improved wound healing and cardiac function.	Yang et al., 2023; Zhou et al., 2024
Hypoxia-primed MSCs	↑VEGF | altered cytokine profile	Enhanced neuro-cardiac regeneration in preclinical models.	De Palma, 2024

Abbreviations: Ang: angiopoietin-1; BDNF: brain-derived neurotrophic factor; ENA78: epithelial neutrophil-activating protein 78; EGF: epithelial growth factor; bFGF: basic fibroblast growth factor; G-CSF: granulocyte-colony stimulating factor; GDF-15: growth differentiation factor-15; GM-CSF: granulocyte macrophage colony-stimulating factor; GRO: growth-regulated oncogene; HASF: hypoxia and Akt induced Stem cell Factor; HGF: hepatocyte growth factor; IGF-I: Insulin-like growth factor 1; IGFBP-1: insulin-like growth factor binding protein-1; IL: interleukin 6; MCP: monocyte chemoattractant protein 1; MIF: migration inhibitory factor; MIP: macrophage inflammatory protein; MMP: matrix metallopeptidase; NGF: nerve growth factor; PDGF: platelet-derived growth factor; PlGF: placental growth factor; RANTES: regulated on activation, normal T cell expressed and secreted; sRFP: secreted frizzled-related protein; TB4: thymosin β4; TGF-β: Transforming growth factor β; TIMP: tissue inhibitor of metalloproteinase; uPAR: urokinase receptor; VEGF: vascular endothelial growth factor.

TABLE 2.2
Mesenchymal Stem Cell-Derived Secretome Profiling and Their Therapeutic Effects

Cell Type	Paracrine Factors Secreted	Effects	Reference
Naïve MSCs			
	BDNF, NGF, bFGF, VEGF, and HGF	Neuroprotection and angiogenesis.	Chen et al., 2002
	VEGF, bFGF, PlGF, and MCP-1	Collateral remodeling.	Kinnaird et al., 2004
	Unknown	Inhibit apoptosis and improve cardiac function.	Timmers et al., 2007
	VEGF, endothelin, epiregulin, Galectin-3, Smad-5, and sRFP-1, and sRFP-4	Promote angiogenesis, anti-apoptosis.	Nguyen et al., 2010
	Paracrine factors	Paracrine factors prolong the repolarization of neonatal cells.	Askar et al., 2013
	PDGF	Exerts a pro-migratory effect.	Windmolders et al., 2014
	HASF is a stem cell paracrine factor	Anti-apoptotic and reduces fibrosis.	Huang et al., 2014
	VEGF, IL-6, HGF	Improved neurovascular repair.	Park et al., 2016
	EV-miRNAs, cytokines	Anti-inflammatory + anti-fibrotic effects.	Rahman et al., 2018
Genetically Modified Cells			
AKT modified MSC	VEGF, bFGF, HGF, IGF-I, and TB4	Anti-apoptotic, limits infarction, and improves heart function.	Gnecchi et al., 2006
	Frizzled-related protein 2	Anti-apoptosis.	Zhang et al., 2009
HSP-20 modified MSC	VEGF, bFGF, and IGF-1	Improve cell engraftment rate and cardiac function.	Wang et al., 2009
Umbilical cord-derived MSCs	VEGF, angiopoietins, HGF, bFGF, TGF-β, PDGF	Promote angiogenesis and mitosis, and anti-apoptosis.	Nascimento et al., 2014
AKT-enhanced MSCs (2017)	VEGF, HGF, IGF-1	Enhanced cardiac regeneration.	Lopez et al., 2017
HIF-1α modified MSCs (2020)	↑VEGF, ↑Ang-1	Improved perfusion and angiogenesis.	Cheng et al., 2020
miR-21 overexpressing MSCs (2021)	miR-21, VEGF	Reduced cell death and improved repair.	Zhou et al., 2021
MSC-derived exosomes	Exosomal miR-21, miR-126, miR-146a; VEGF, HGF	Pro-angiogenic, anti-apoptotic, improves cardiac repair.	Zheng et al., 2021
Wharton's Jelly MSC secretome	VEGF, HGF, angiopoietins, cytokines, EVs	Enhanced angiogenesis and immunomodulation.	Drobiova et al., 2023

(Continued)

TABLE 2.2 CONTINUED

Mesenchymal Stem Cell-Derived Secretome Profiling and Their Therapeutic Effects

Cell Type	Paracrine Factors Secreted	Effects	Reference
MSC-Exos (clinical translation)	VEGF, HGF, EV-miRNAs	Documented safety and benefit signals in early clinical studies.	Lotfy et al., 2013
Hypoxia-preconditioned MSC secretome	↑VEGF, angiogenin, pro-angiogenic miRNAs	Improved wound healing and cardiac function.	Yang et al., 2023; Zhou et al., 2024
Hypoxia-primed MSCs	↑VEGF ǀ altered cytokine profile	Enhanced neuro-cardiac regeneration in preclinical models.	De Palma ST, 2024

Abbreviations: Ang: angiopoietin-1; BDNF: brain-derived neurotrophic factor; ENA78: epithelial neutrophil-activating protein 78; EGF: epithelial growth factor; bFGF: basic fibroblast growth factor; G-CSF: granulocyte-colony stimulating factor; GDF-15: growth differentiation factor-15; GM-CSF: granulocyte macrophage colony-stimulating factor; GRO: growth-regulated oncogene; HASF: hypoxia and Akt induced Stem cell Factor; HGF: hepatocyte growth factor; IGF-I : insulin-like growth factor 1; IGFBP-1: insulin-like growth factor binding protein-1; IL: interleukin 6; MCP: monocyte chemoattractant protein 1; MIF: migration inhibitory factor; MIP: macrophage inflammatory protein; MMP: matrix metallopeptidase; NGF: nerve TGF-β: transforming growth factor β; TIMP: tissue inhibitor of metalloproteinase; uPAR: urokinase receptor; VEGF: vascular endothelial growth growth factor; PDGF: platelet-derived growth factor; PlGF: placental growth factor; RANTES: regulated on activation, normal T cell expressed and secreted; sRFP: secreted frizzled-related protein; TB4: thymosin β4.

2.6 QUALITY CONTROL MEASURES, INCLUDING ASSAYS AND TECHNIQUES TO ENSURE CONSISTENCY AND EFFICACY

Like for any drug development and manufacturing process, product quality assurance is also an integral part of biopharmaceuticals, as well as stem cells and their secretome-based product development. There is mounting evidence that the therapeutic benefits of stem cells are primarily due to their paracrine activity rather than to their differentiation, long-term engraftment in the heart, or development into cardiac tissue (Aslam & Haider, 2019). The released bioactive factors modulate cell survival, inflammation, and angiogenesis in the injured heart by recruiting cells from the intrinsic/resident stem/progenitor pool and supporting the repair and regenerative processes. For example, Kompa et al. have reported that the secretome from cardiac stem cells has shown a strong pro-survival, pro-angiogenic, and pro-migratory effect on cardiomyocytes and endothelial cells, reinforcing the potential of cell-free regenerative therapy (Kompa et al., 2021). A recent systematic review by Mabotuwana et al., including 86 published studies, has emphasized that MSCs primarily exert their cardioprotective effects through the secretion of paracrine mediators, including growth factors, cytokines, and EVs, which act on surrounding cells to promote cardiac regeneration (Mabotuwana et al., 2022). The authors identified 234 individual protective factors secreted by MSCs as part of their paracrine activity, with HGF, VEGF, and FGF2 among the most frequently reported bioactive molecules. On the same note, Pires et al. have compared the secretome from MSCs derived from three different human tissue sources and reported that BM-MSCs, AD-MSCs, and UC-MSCs differed in their neurotrophic, neurogenic, axon guidance, axon growth, and neurodifferentiative protein contents, as well as proteins with neuroprotective actions against oxidative stress, apoptosis, and excitotoxicity (Pires et al., 2016). Du et al. reported that the three cell types differed in the secretion of angiogenic growth factors from their secretome (Du et al., 2016). Such variations are not only observed in cells from diverse tissue sources but also in MSCs isolated from the same tissue and cultured under divergent conditions.

Given the complex composition of the secretome, which is incidentally dependent on numerous factors, the production of secretome-based ATMP and its use in clinical settings require rigorous, robust quality control at every stage, encompassing cell selection through collection, purification, and patient availability. These quality control measures are essential for the safety, effectiveness, and reproducibility of secretome-based therapeutics, ensuring batch-to-batch consistency, homogeneous, standardized bioactive content, and standardized functional potency. All of these are critical to therapeutic reliability, regulatory approval, and clinical translation. Without strict quality assurance systems, the secretome's inherent compositional variability, driven by factors such as donor differences, cell passage number, or culture conditions, may lead to diminished efficacy, erratic, non-homogeneous, and unpredictable clinical outcomes, or even serious therapeutic adverse events (Chouaib et al., 2023). Therefore, well-defined optimized protocols for secretome-based products must encompass several key domains: source cell validation, secretome composition analysis, potency assays, safety and purity testing, and stability evaluation. These measures are not only necessary for meeting regulatory standards but are also indispensable for building trust in the clinical use of cell-free, paracrine-based therapeutics (Bari et al., 2019).

A recent review has highlighted significant challenges in developing secretome-based GMP-grade pharmaceuticals with the desired potency and efficacy. One of these critical challenges is the development of an appropriate potency assay that reflects the product's mechanism of action and the establishment of workflows that regulatory authorities can adopt. Confirming the identity of stem and progenitor cells, such as MSCs, involves evaluating surface marker expression, as well as assessing viability, senescence, and passage number to prevent drift in secretome output. Additionally, maintaining consistent culture conditions—including cell density, medium composition, conditioning period, and oxygen tension—is essential, as fluctuations in these parameters can lead to substantial alterations in the secreted profile (Turlo et al., 2023). Without optimizing these parameters, avoiding batch-to-batch variability will be difficult, and the resulting variability in the end product will undermine efficacy or safety. A detailed analytical characterization of the purified secretome is required, from total protein content determination (e.g., BCA or Bradford assay) to quantification of growth factors/cytokines, their concentrations, and their contributions to cardiovascular effects. For the insoluble paracrine constituents of extracellular EVs, additional tests such as particle size and concentration must be measured (i.e., by nanoparticle tracking analysis, NTA), morphology assessed by transmission electron microscopy (TEM), and EV-specific markers (CD63, CD81, TSG101) confirmed by Western blot or other immunoassay up to their cargo profiling (Sagaradze et al., 2023).

Another critical aspect of secretome characterization is determining secretome potency. Because secretome preparations for cardiovascular applications are generally intended to support cardiomyocyte survival, promote angiogenesis, attract resident cardiac stem/progenitor cells to migrate and home into injured myocardium to participate in the repair process, and alleviate fibrosis, the use of potency assays will provide mechanistic insight into secretome-based treatment. The most commonly used potency assays include tube formation, cardiomyocyte survival, and cell migration assays, among others, for assessing the angiogenic and pro-survival effects of the secretome. Similarly, a fibroblast activation assay is performed to determine the anti-remodeling or anti-fibrotic potential of the product. Importantly, these assays are both qualitative and quantitative for each batch of the secretome product. A reference for each potency assay must be established to benchmark and define the minimum potency assessment (Sagaradze et al., 2023). Another critical aspect of evaluation for each batch is rigorous sterility testing, e.g., culture-based or rapid microbial methods, mycoplasma detection, and endotoxin measurement (typically by the Limulus Amebocyte Lysate assay, or LAL). Additionally, the secretome product must be cell- and cell debris-free to ensure that contaminating material is absent and to reduce the immunogenicity risk of the product (Maumus et al., 2020). One of the essential advantages of secretome-based ATMPs is their logistical advantage over cell-based living biodrugs and their potential for ready-to-use, off-the-shelf availability. However, the optimized storage conditions and safety and potency profile are retained after thawing for clinical use. It is therefore imperative that secretome-based ATMPs undergo stability studies to establish their expiry dates and define storage conditions that ensure the products retain their safety and potency during storage.

It is pertinent to note that this series of proposed assays must be documented, with standard operating protocols established to govern manufacturing from start to finish, including the source and quality of the cells, down to isolation and purification of the secretome, its standardization, storage, distribution, and post-thaw analysis report. In conclusion, the development of secretome-based ATMPs must follow a multi-tiered QC framework to meet cGMP requirements, similar to pharmaceutical development, albeit with additional tests and assays discussed above.

2.7 CONCLUSION AND FUTURE PERSPECTIVE

Each cell type secretes a wide range of cytokines and growth factors, constituting a cell-specific paracrine secretome under the given conditions. The composition of the secretome may be engineered by modifying culture conditions or by priming or preconditioning the cells (Haider, 2025). The composition of paracrine secretomes includes both soluble and insoluble fractions, both of which are therapeutically active and beneficial. These fractions, used either individually or in combination, exert their therapeutic effect via at least seven well-studied mechanisms: inflammation; inhibition of cell necrosis and apoptosis; promotion of cell viability, migration, homing, and proliferation; and induction of angiogenesis and arteriogenesis. These may directly or indirectly enhance the engraftment rate of transplanted donor cells and promote endogenous cardiac repair, thereby improving the efficacy of cellular myoplasty. Therefore, using the conditioned medium for pretreatment or co-transplantation of donor cells would be a realistic and straightforward approach for cardiac repair and regeneration. With current advances in targeting strategies, non-invasive delivery systems, artificial intelligence-guided optimization, and advanced biomaterial platforms, stem cell paracrine secretion-based therapy, in general and for HF in particular, offers a more accurate, safe, scalable, and efficacious treatment option, either alone or as an adjunct to routine treatment interventions. Combining these advancements—using engineered secretomes and exosomes containing cardiac-specific bioactive molecules—positions cell-free therapeutics at the forefront of next-generation regenerative medicine for cardiovascular disease.

REFERENCES

Acosta SA, Franzese N, Staples M, Weinbren NL, Babilonia M, Patel J, Merchant N, Simancas AJ, Slakter A, Caputo M, et al. Human umbilical cord blood for transplantation therapy in myocardial infarction. *J Stem Cell Res. Ther.* 2013; (Suppl 4), S4-005.

Ahmed OTF, Ahmed ZT, Dairi AW, Zain Al-Abeden MS, Alkahlot MH, Alkahlot RH, Al Jowf GI, et al. The inconclusive superiority debate of allogeneic versus autologous MSCs in treating patients with HFrEF: a systematic review and meta-analysis of RCTs. *Stem Cell Res Ther.* 2025; 16(1), 175. doi: 10.1186/s13287-025-04209-5.

Ahmed RPH, Haider KH, Buccini S, Shujia J, Ashraf M. Reprogramming of skeletal myoblasts for induction of pluripotency for tumor-free cardiomyogenesis in the infarcted heart. *Circulation Res.* 2011; 109, 60–70.

Ahmed ZT, Zain Al-Abeden MS, Al Abdin MG, Muqresh MA, Al Jowf GI, Eijssen LMT, Haider KH. The dose-response relationship of MSCs as living Bio-drugs in HFrEF patients is a systematic review and meta-analysis of RCTs. *Stem Cell Res Ther.* 2024; 15, 165.

Ali SA, Mahmood Z, Mubarak Z, Asad M, Sarfraz Chaudhri MT, Bilal L, Ashraf T, et al. Assessing the potential benefits of stem cell therapy in cardiac regeneration for patients with ischemic heart disease. *Cureus.* 2025; 17(1), e76770. doi: 10.7759/cureus.76770.

Álvarez-Viejo M, Haider KH. Mesenchymal stem cells. In: KH Haider, editor. *Handbook of Stem Cell Therapy.* Singapore: Springer; 2022. doi: 10.1007/978-981-16-6016-0_6-1

Askar SF, Ramkisoensing AA, Atsma DE, Schalij MJ, de Vries AA, Pijnappels DA. Engraftment patterns of human adult mesenchymal stem cells expose electrotonic and paracrine proarrhythmic mechanisms in myocardial cell cultures. *Circulation Arrhythmia Electrophysiol.* 2013; 6, 380–391.

Aslam M, Haider KH. Adrenomedullins and their emerging role in regenerative medicine. In: KH Haider, editor. *Stem Cells: From Myth to Reality and Evolving.* Berlin (Germany): De Gruyter; n.d. (Published, 2019).

Atkins BZ, Hueman MT, Meuchel J, Hutcheson KA, Glower DD, Taylor DA. Cellular cardiomyoplasty improves diastolic function in the injured heart. *J Surgical Res.* 1999; 85, 234–242.

Barcelos LS, Duplaa C, Krankel N, Graiani G, Invernici G, Katare R, Siragusa M, et al. Human CD133+ progenitor cells promote the healing of diabetic ischemic ulcers through paracrine stimulation of angiogenesis and activation of the Wnt signaling pathway. *Circ Res.* 2009; 104, 1095–1102.

Bari E, Ferrarotti I, Torre ML, Corsico AG, Perteghella S. Mesenchymal stem/stromal cell secretome for lung regeneration: The long way through "pharmaceuticalization" for the best formulation. *J Control Release.* 2019 Sep 10; 309, 11–24. doi: 10.1016/j.jconrel.2019.07.022.

Bearzi C, Rota M, Hosoda T, Tillmanns J, Nascimbene A, De Angelis A, Yasuzawa-Amano S, et al. Human cardiac stem cells. *Proc Natl Acad Sci U S A.* 2007; 104, 14068–14073.

Bellio MA, Golub HL, Shapiro GC, Santos I, Bennett C, Leider H. Proof-of-concept trial of an amniotic fluid-derived extracellular vesicle biologic for treating high-risk patients with mild-to-moderate acute COVID-19 infection. *Front Immunol.* 2021; 12, 977809. doi: 10.3389/fimmu.2022.977809

Beltrami AP, Barlucchi L, Torella D, Baker M, Limana F, Chimenti S, Kasahara H, et al. Adult cardiac stem cells are multipotent and support myocardial regeneration. *Cell.* 2003; 114, 763–776.

Beltrami AP, Urbanek K, Kajstura J, Yan SM, Finato N, Bussani R, Nadal-Ginard B, et al. Evidence that human cardiac myocytes divide after myocardial infarction. *N Engl J Med.* 2001; 344, 1750–1757.

Bhatti A, Haider KH. Mesenchymal stem cells as living bio-drugs: adding a novel dimension to cardiology and cardiovascular sciences. *J Cardiol Cardiovasc Sci.* 2025; 6(1), 30–35.

Brasselet C, Morichetti MC, Messas E, Carrion C, Bissery A, Bruneval P, Vilquin JT, et al. (2005). Skeletal myoblast transplantation through a catheter-based coronary sinus approach: an effective means of improving function of infarcted myocardium. *European Heart J.* 2005; 26, 1551–1556.

Buccini S, Haider KH, Ahmed RPH, Jiang S, Ashraf M. Cardiac progenitors derived from reprogrammed mesenchymal stem cells contribute to angiomyogenic repair of the infarcted heart. *Basic Res Cardiol.* 2012; 107(6), 301–314.

Burt RK, Chen YH, Verda L, Lucena C, Navale S, Johnson J, Han X, et al. Mitotically inactivated embryonic stem cells can be used as an in vivo feeder layer to nurse damaged myocardium after acute myocardial infarction: a preclinical study. *Circ Res.* 2012; 111, 1286–1296.

Chazaud B, Hittinger L, Sonnet C, Champagne S, Le Corvoisier P, Benhaiem-Sigaux N, Unterseeh T, et al. Endoventricular porcine autologous myoblast transplantation can be successfully achieved with minor mechanical cell damage. *Cardiovasc Res.* 2003; 58, 444–450.

Chen X, Katakowski M, Li Y, Lu D, Wang L, Zhang L, Chen J, et al. Human bone marrow stromal cell cultures conditioned by traumatic brain tissue extracts: growth factor production. *J Neurosci Res.* 2002; 69, 687–691.

Cheng Y, Wang D, Wang B, Li H, Xiong J, Xu S, Chen Q. HIF-1α-modified mesenchymal stem cells enhance angiogenesis and cardiac repair. *J Mol Cell Cardiol.* 2020; 138, 1–12.

Chimenti I, Smith RR, Li TS, Gerstenblith G, Messina E, Giacomello A, Marban E. Relative roles of direct regeneration versus paracrine effects of human cardiosphere-derived cells transplanted into infarcted mice. *Circulation Res.* 2010; 106, 971–980.

Chiu RC, Zibaitis A, Kao RL. Cellular cardiomyoplasty: myocardial regeneration with satellite cell implantation. *Ann Thorac Surg.* 1995; 60, 12–18.

Chouaib B, Cuisinier F, Collart-Dutilleul P-Y, et al. Towards the standardization of mesenchymal stem cell secretome-derived product manufacturing for tissue regeneration. *Int J Mol Sci.* 2023; 24(16), 12594.

Dai W, Field LJ, Rubart M, Reuter S, Hale SL, Zweigerdt R, Graichen RE, et al. (2007). Survival and maturation of human embryonic stem cell-derived cardiomyocytes in rat hearts. *J Mol Cell Cardiol.* 2007; 43, 504–516.

Dawn B, Tiwari S, Kucia MJ, Zuba-Surma EK, Guo Y, Sanganalmath SK, Abdel-Latif A, et al. (2008). Transplantation of bone marrow-derived very small embryonic-like stem cells attenuates left ventricular dysfunction and remodeling after myocardial infarction. *Stem Cells.* 2008; 26, 1646–1655.

De Palma ST, Ricciardi M, Perrotta I, Forte G. Hypoxia-primed mesenchymal stem cells enhance neurocardiac regeneration. *Cells.* 2024; 13(2), 198.

Dimasi CG, Darby JRT, Morrison JL. A change of heart: understanding the mechanisms regulating cardiac proliferation and metabolism before and after birth. *J Physiol.* 2023 Apr; 601(8), 1319–1341. doi: 10.1113/JP284137.

Doyle B, Sorajja P, Hynes B, Kumar AH, Araoz PA, Stalboerger PG, Miller D, et al. Progenitor cell therapy in a porcine acute myocardial infarction model induces cardiac hypertrophy, mediated by paracrine secretion of cardiotrophic factors, including TGFbeta1. *Stem Cells Dev.* 2008; 17, 941–951.

Drobiova H, Antalík M, Kucerova L, Matuskova M, Rosocha J. Wharton's Jelly-derived mesenchymal stromal cell secretome enhances angiogenesis and immunomodulation. *Cells.* 2023;12(3), 421.

Du WJ, Chi Y, Yang ZX, Li ZJ, Cui JJ, Song BQ, Li X, et al. Heterogeneity of proangiogenic features in mesenchymal stem cells derived from bone marrow, adipose tissue, umbilical cord, and placenta. *Stem Cell Res Ther.* 2016; 7, 163. doi: 10.1186/s13287-016-0418-9.

Ellison GM, Torella D, Dellegrottaglie S, Pérez-Martínez C, Pérez de Prado A, Vicinanza C, Purushothaman S, et al. Endogenous cardiac stem cell activation by intracoronary injection of insulin-like growth factor-1 and hepatocyte growth factor fosters survival and regeneration of the infarcted pig heart. *J Am Coll Cardiol.* 2011; 58, 977–986.

Fakoya AOJ, Omole AE, Satyadev N, Haider KH. Induced pluripotent stem cells. In: KH Haider, editor. *Handbook of Stem Cell Therapy.* Singapore: Springer; 2022. doi: 10.1007/978-981-19-2655-6_40

Gnecchi, M., He, H., Noiseux, N., Liang, O.D., Zhang, L., Morello, F., Mu, H., et al. Evidence supporting the paracrine hypothesis for Akt-modified mesenchymal stem cell-mediated cardiac protection and functional improvement. *Faseb J.* 2006; 20, 661–669.

Gojo S, Gojo N, Takeda Y, Mori T, Abe H, Kyo S, Hata J, et al. In vivo cardiovasculogenesis by direct injection of isolated adult mesenchymal stem cells. *Exp Cell Res.* 2003; 288, 51–59.

Günthel M, Barnett P, Christoffels VM. Development, proliferation, and growth of the mammalian heart. *Mol Ther.* 2018; 26(7), 1599–1609. doi: 10.1016/j.ymthe.2018.05.022.

Habib SM, Martini MF, Hamdan YNA, Shrebaty OMM, Haider KH. Mesenchymal stem cell secretome and exosomes as potential advanced therapy medicinal products for treating a failing heart. *World J Stem Cells.* 2026; 18(3), 113871. doi: 10.4252/wjsc.v18.i3.113871

Hafner F, Riedl R, Beer R, Brodmann M, Gary T, Hackl G, Hafner S, et al. MARSYAS II – Topical application of secretome of stressed peripheral blood mononuclear cells in diabetic foot ulcers: protocol for a randomized, double-masked, placebo-controlled, dose-finding trial. *Trials.* 2020; 21, 548. doi: 10.1186/s13063-020-04948-1

Haider H, Ye L, Jiang S, Ge R, Law PK, Chua T, Wong P, et al. Angiomyogenesis for cardiac repair utilizes human myoblasts as carriers of vascular endothelial growth factor. *J Mol Med* (Berl). 2004; 82, 539–549.

Haider KH, Aziz S. Paracrine hypothesis and cardiac repair. *Int J Stem Cell Res Transplant.* 2017; 5(1), 265–267.

Haider KH. Melatonin-based priming of stem cells to alleviate oxidative stress. *World J Stem Cells.* 2024b; 16(11), 985–989. doi: 10.4252/wjsc.v16.i11.985.

Haider KH. Priming mesenchymal stem cells to develop "super stem cells." *World J Stem Cells.* 2024a; 16(6), 623–640. doi: 10.4252/wjsc.v16.i6.623.

Hamdi H, Planat-Benard V, Bel A, Puymirat E, Geha R, Pidial L, Nematalla H, et al. Epicardial adipose stem cell sheets result in more remarkable post-infarction survival than intramyocardial injections. *Cardiovasc Res.* 2011: 91, 483–491.

Henningsen J, Rigbolt KT, Blagoev B, Pedersen BK, Kratchmarova I. Dynamics of the skeletal muscle secretome during myoblast differentiation. *Mol Cell Proteomics.* 2010; 9, 2482–2496.

Holmes C, Stanford WL. Concise review: stem cell antigen-1: expression, function, and enigma. *Stem Cells.* 2007; 25, 1339–1347. doi: 10.1201/9781003611509.

Huang J, Guo J, Beigi F, Hodgkinson CP, Facundo HT, Zhang Z, Espinoza-Derout J, et al. HASF is a stem cell paracrine factor that activates PKC epsilon-mediated cytoprotection. *J Mol Cell Cardiol.* 2014; 66, 157–164.

Hudlicka O, Brown MD. Postnatal growth of the heart and its blood vessels. *J Vasc Res.* 1996; 33, 266–287.

Kajstura J, Leri A, Finato N, Di Loreto C, Beltrami CA, Anversa P. Myocyte proliferation in end-stage human cardiac failure. *Proc Natl Acad Sci U S A.* 1998; 95, 8801–8805.

Kamal M, Kassem D, Haider KH. Sources and therapeutic strategies of mesenchymal stem cells in regenerative medicine. In: KH Haider, editor. *Handbook of Stem Cell Therapy.* Singapore: Springer; 2022. doi: 10.1007/978-981-19-2655-6_2

Kawamoto A, Gwon HC, Iwaguro H, Yamaguchi JI, Uchida S, Masuda H, Silver M, et al. Therapeutic potential of ex vivo expanded endothelial progenitor cells for myocardial ischemia. *Circulation.* 2001; 103, 634–637.

Kinnaird T, Stabile E, Burnett MS, Shou M, Lee CW, Barr S, Fuchs S, et al. Local delivery of marrow-derived stromal cells augments collateral perfusion through paracrine mechanisms. *Circulation.* 2004; 109, 1543–1549.

Kishore R, Verma SK, Mackie AR, Vaughan EE, Abramova TV, Aiko I, Krishnamurthy P. Bone marrow progenitor cell therapy-mediated paracrine regulation of cardiac miRNA-155 modulates fibrotic response in diabetic hearts. *PLoS One.* 2013; 8, e60161.

Kocher AA, Schuster MD, Szabolcs MJ, Takuma S, Burkhoff D, Wang J, Homma S, et al. Neovascularization of ischemic myocardium by human bone-marrow-derived angioblasts prevents cardiomyocyte apoptosis, reduces remodeling, and improves cardiac function. *Nat Med.* 2001; 7, 430–436.

Kompa AR, Greening DW, Kong AM, McMillan PJ, Fang H, Saxena R, Wong RCB, et al. Sustained subcutaneous delivery of the secretome of human cardiac stem cells promotes cardiac repair following myocardial infarction. *Cardiovasc Res.* 2021; 117(3), 918–929. doi: 10.1093/cvr/cvaa088.

Kurazumi H, Kubo M, Ohshima M, Yamamoto Y, Takemoto Y, Suzuki R, Ikenaga S, et al. The effects of mechanical stress on the growth, differentiation, and paracrine factor production of cardiac stem cells. *PLoS One* 2011; 6, e28890.

Laflamme MA, Chen KY, Naumova AV, Muskheli V, Fugate JA, Dupras SK, Reinecke H, et al. Cardiomyocytes derived from human embryonic stem cells, when treated with pro-survival factors, enhance the function of infarcted rat hearts. *Nat Biotecnol.* 2007; 25: 1015–1024.

Leela K, Sushmitha A, Sendilkumar B, Gowtham D, Shree Harini V, Parkash J. Recent advances in cardiac surgery: minimally invasive techniques and robotic assistance. *IJPSR* 2025; 12, 1847–1856. doi: 10.13040/IJPSR.0975-8232.16(7).1847-56.

Li RK, Jia ZQ, Weisel RD, Mickle DA, Zhang J, Mohabeer MK, Rao V, et al. Cardiomyocyte transplantation improves heart function. *Ann Thoracic Surg.* 1996; 62, 654–660; discussion 660–651.

Li RK, Mickle DA, Weisel RD, Mohabeer MK, Zhang J, Rao V, Li G, et al. Natural history of fetal rat cardiomyocytes transplanted into adult rat myocardial scar tissue. *Circulation.* 1997; 96, II179–186; discussion 186–177.

Li Y, Zhang J, Shi J, Liu K, Wang X, Jia Y, He T, Shen K. Exosomes derived from mesenchymal stem cells promote angiogenesis and reduce apoptosis after myocardial infarction. *Stem Cell Res Ther.* 2017; 8, 210.

Liang SX, Tan TY, Gaudry L, Chong B. Differentiation and migration of Sca1+/CD31- cardiac side population cells in a murine myocardial ischemic model. *International J Cardiol.* 2010; 138, 40–49.

Liang Y, Xia J, Xu L, Zhang Y, Xu X, Haider KH. Cell-free and cell-based therapy in regenerative medicine. In: KH Haider, editor. *Handbook of Regenerative Medicine: Cell-Free Therapy Approach.* 1st ed. Boca Raton (FL): CRC Press; 2025, Chapter 24.

Linke A, Müller P, Nurzynska D, Casarsa C, Torella D, Nascimbene A, Castaldo C, et al. (2005). Stem cells in the dog heart are self-renewing, clonogenic, and multipotent, and they regenerate infarcted myocardium, thereby improving cardiac function. *Proc Natl Acad Sci U S A.* 2005;102: 8966–8971.

Lopez Y, Lutjemeier B, Seshareddy K, Trevino E, Prockop DJ. Akt-enhanced mesenchymal stem cells promote cardiac regeneration. *Stem Cells.* 2017; 35(4), 1006–1017.

Lotfy A, Salama M, El-Shaer A, Rashed L. Clinical translation of mesenchymal stem cell-derived exosomes: safety and therapeutic potential. *Stem Cell Res Ther.* 2013; 14, 312.

Luecke N, Templin C, Muetzelburg MV, Neumann D, Just I, Pich A. Secreted proteome of the murine multipotent hematopoietic progenitor cell line DKmix. *Rapid Commun Mass Spectrom.* 2010; 24, 561–570.

Mabotuwana NS, Rech L, Lim J, Hardy SA, Murtha LA, Rainer PP, Boyle AJ. Paracrine factors released by stem cells of mesenchymal origin and their effects in cardiovascular disease: a systematic review of preclinical studies. *Stem Cell Rev Rep.* 2022; 18(8), 2606–2628. doi: 10.1007/s12015-022-10429-6.

Mangi AA, Noiseux N, Kong D, He H, Rezvani M, Ingwall JS, Dzau VJ. Mesenchymal stem cells modified with Akt prevent remodeling and restore cardiac function in infarcted hearts. *Nat Med.* 2003; 9, 1195–1201.

Martin SS, Aday W, Allen NB, Almarzooq ZI, Anderson CAM, Arora P, Avery CL, et al. (on behalf of the American Heart Association Council on Epidemiology and Prevention Statistics Committee and Stroke Statistics Committee). 2025 disease and stroke statistics: a report of US and global data from the American Heart Association. *Circulation.* 2025; 151(8). doi: 10.1161/CIR.0000000000001303

Matsuura K, Nagai T, Nishigaki N, Oyama T, Nishi J, Wada H, Sano M, et al. Sca-1+ cells differentiate into beating cardiomyocytes. *J Biol Chem*. 2004; 279, 11384–11391.

Maumus M, Rozier P, Boulestreau J, Jorgensen C, Noël D. Mesenchymal stem cell-derived extracellular vesicles: opportunities and challenges for clinical translation. *Front Bioeng Biotechnol*. 2020; 8, 997. doi: 10.3389/fbioe. 2020.00997.

Maxeiner H, Krehbiehl N, Müller A, Woitasky N, Akinturk H, Müller M, Weigand MA, et al. New insights into the paracrine mechanisms of human cardiac progenitor cells. *Eur J Heart Fail*. 2010; 12, 730–737.

McCue JD, Swingen C, Feldberg T, Caron G, Kolb A, Denucci C, Prabhu S, et al. The real estate of myoblast cardiac transplantation: adverse remodeling is associated with location. *J Heart Lung Transplant*. 2008; 27, 116–123.

Mehanna RA, Essawy MM, Barkat MA, Awaad AK, Thabet EH, Hamed HA, Elkafrawy H, et al. Cardiac stem cells: current knowledge and prospects. *World J Stem Cells*. 2022; 14(1), 1–40. doi: 10.4252/wjsc.v14.i1.1.

Menasche P, Alfieri O, Janssens S, McKenna W, Reichenspurner H, Trinquart L, Vilquin JT, et al. The Myoblast Autologous Grafting in Ischemic Cardiomyopathy (MAGIC) trial: first randomized placebo-controlled study of myoblast transplantation. *Circulation*. 2008; 117, 1189–1200.

Menasché P, Hagège AA, Scorsin M, Pouzet B, Desnos M, Duboc D, Schwartz K, et al. Myoblast transplantation for heart failure. *Lancet*. 2001; 357(9252), 279–80. doi: 10.1016/S0140-6736(00)03617-5.

Menasché P, Hagège AA, Vilquin JT, Desnos M, Abergel E, Pouzet B, Bel A, et al. Autologous skeletal myoblast transplantation for severe postinfarction left ventricular dysfunction. *J Am Coll Cardiol*. 2003; 41(7), 1078–1083. doi: 10.1016/s0735-1097(03)00092-5.

Miyamoto S, Kawaguchi N, Ellison GM, Matsuoka R, Shin'oka T, Kurosawa H. Characterization of long-term cultured c-kit+ cardiac stem cells derived from adult rat hearts. *Stem Cells Dev*. 2010; 19, 105–116.

Mollova M, Bersell K, Walsh S, Savla J, Das LT, Park SY, Silberstein LE, et al. Cardiomyocyte proliferation contributes to heart growth in young humans. *Proc Natl Acad Sci U S A*. 2013 Jan 22; 110(4), 1446–51. doi: 10.1073/pnas.1214608110.

Nascimento DS, Mosqueira D, Sousa LM, Teixeira M, Filipe M, Resende TP, Araújo AF, et al. Human umbilical cord tissue-derived mesenchymal stromal cells attenuate remodeling following myocardial infarction through pro-angiogenic, anti-apoptotic, and endogenous cell-activation mechanisms. *Stem Cell Res Ther*. 2014; 5, 5.

Nguyen BK, Maltais S, Perrault LP, Tanguay JF, Tardif JC, Stevens LM, Borie M, et al. Improved function and myocardial repair of the infarcted heart by intracoronary injection of mesenchymal stem cell-derived growth factors. *J Cardiovasc Transl Res*. 2010; 3, 547–558.

Niagara MI, Haider H, Jiang S, Ashraf M. Pharmacologically preconditioned skeletal myoblasts are resistant to oxidative stress and promote angiomyogenesis via the release of paracrine factors in the infarcted heart. *Circ Res*. 2007;100: 545–555.

Norheim F, Raastad T, Thiede B, Rustan AC, Drevon CA, Haugen F. Proteomic identification of secreted proteins from human skeletal muscle cells and expression in response to strength training. *Am J Physiol Endocrinol Metab*. 2011; 301, E1013–E1021.

Park H, Kim J, Na K, Woo DG. Mesenchymal stem cell secretome enhances neurovascular regeneration. *J Tissue Eng Regen Med*. 2016; 10(10), 870–882.

Pérez-Ilzarbe M, Agbulut O, Pelacho B, Ciorba C, San José-Eneriz E, Desnos M, Hagege AA, et al. Characterization of the paracrine effects of human skeletal myoblasts transplanted in infarcted myocardium. *Eur J Heart Fail*. 2008; 10, 1065–1072.

Peterson JM, Pizza FX. Cytokines derived from cultured skeletal muscle cells after mechanical strain promote neutrophil chemotaxis in vitro. *J Appl Physiol*. 2009; 106, 130–137.

Pires AO, Mendes-Pinheiro B, Teixeira FG, Anjo SI, Ribeiro-Samy S, Gomes ED, Serra SC, et al. Unveiling the differences of secretome of human bone marrow mesenchymal stem cells, adipose tissue-derived stem cells, and human umbilical cord perivascular cells: A proteomic analysis. *Stem Cells Dev*. 2016; 25, 1073–1083. doi: 10.1089/scd.2016.0048.

Rahman MS, Akhtar N, Jamil HM, Banik RS, Asaduzzaman SM. TGF-β/BMP signaling and MSC-derived extracellular vesicles in tissue repair. *Stem Cells International*. 2018, 1–17.

Rajab AM, Rajab TM, AlJundi S, Haider KH. Bone stem cell therapy in the clinical perspective: a focus on nonrandomized and randomized trials. In: KH Haider, editor. *Stem Cells: From Myth to Reality and Evolving*. Berlin (Germany): De Gruyter; 2019.

Rufaihah AJ, Haider KH, Heng BC, Ye L, Tan RS, Toh WS, Tian XF, et al. Therapeutic angiogenesis by transplantation of human embryonic stem cell-derived CD133+ endothelial progenitor cells for cardiac repair. *Regenerative Med.* 2010; 5, 231–244.

Rufaihah AJ, Huang NF, Jame S, Lee JC, Nguyen HN, Byers B, De A, Okogbaa J, et al. Endothelial cells derived from human induced pluripotent stem cells (iPSCs) increase capillary density and improve perfusion in a mouse model of peripheral arterial disease. *Arterioscler Thromb Vasc Biol*, 2011; 31, e72–79.

Safwan M, Bourgleh MS, ALDoush M, Haider KH. Tissue-source effect on the safety and efficacy of MSCs as living biodrugs to treat heart failure: a systematic review and meta-analysis of RCTs. *World J Cardiology.* 2024; 16(8): 469–483 .

Sagaradze G, Monakova A, Efimenko A. Potency assays for mesenchymal stromal cell secretome-based products for tissue regeneration. *Int J Mol Sci.* 2023; 24(11), 9379. doi: 10.3390/ijms24119379.

Sandstedt J, Jonsson M, Dellgren G, Lindahl A, Jeppsson A, Asp J. Human C-kit+CD45- cardiac stem cells are heterogeneous and display both cardiac and endothelial commitment, as determined by single-cell qPCR analysis. *Biochem Biophys Res Commun.* 2014; 443, 234–238.

Schittini AV, Celedón PF, Stimamiglio MA, Krieger M, Hansen P, da Costa FD, Goldenberg S, et al. Human cardiac explant-conditioned medium: soluble factors and cardiomyogenic effect on mesenchymal stem cells. *Exp Biol Med.* (Maywood). 2010; 235, 1015–1024.

Scorsin M, Marotte F, Sabri A, Le Dref O, Demirag M, Samuel JL, Rappaport L, et al. Can grafted cardiomyocytes colonize peri-infarct myocardial areas? *Circulation.* 1996; 94, II337–340.

Shen X, Pan B, Zhou H, Liu L, Lv T, Zhu J, Huang X, Tian J. Paracrine effects of hiPSC-derived cardiomyocytes improve myocardial repair. *Theranostics* 2019; 9(13), 3729–3747.

Sim EKW, Haider KH, Aziz S, Ooi OC, Law PK. Myoblast transplantation on beating heart. *Surg International* 2005; 90(3), 148–150.

Simader E, Traxler D, Kasiri MM, Hofbauer H, Wolzt M, Glogner C, Storka A, et al. Safety and tolerability of topically administered autologous, apoptotic PBMC secretome (APOSEC) in dermal wounds: a randomized Phase 1 trial (MARSYAS I). *Sci Rep.* 2017; 7(1), 6216. doi: 10.1038/s41598-017-06223-x.

Smith AJ, Lewis FC, Aquila I, Waring CD, Nocera A, Agosti V, Nadal-Ginard B, et al. Isolation and characterization of resident endogenous c-Kit+ cardiac stem cells from the adult mouse and rat heart. *Nature Protocols.* 2014; 9, 1662–1681.

Soonpaa MH, Koh GY, Klug MG, Field LJ. Formation of nascent intercalated disks between grafted fetal cardiomyocytes and host myocardium. *Science.* 1994; 264, 98–101.

Takamiya M, Haider KH, Ashraf M. Identification and characterization of a novel multipotent sub-population of sca-1+ cardiac progenitor cells for myocardial regeneration. *PLoS One.* 2011; 6(9), e25265.

Tamargo J, Agewall S, Borghi C, Ceconi C, Cerbai E, Dan GA, Ferdinandy P, et al. New pharmacological agents and novel cardiovascular pharmacotherapy strategies in 2022. *Eur Heart J Cardiovasc Pharmacother.* 2023; 9(4), 353–370. doi: 10.1093/ehjcvp/pvad034.

Tambara K, Sakakibara Y, Sakaguchi G, Lu F, Premaratne GU, Lin X, Nishimura K, et al. Transplanted skeletal myoblasts can fully replace the infarcted myocardium when they survive in the host in large numbers. *Circulation.* 2003; 108 Suppl 1, II259–263.

Tatsumi R, Hattori A, Ikeuchi Y, Anderson JE, Allen RE. Release of hepatocyte growth factor from mechanically stretched skeletal muscle satellite cells and role of pH and nitric oxide. *Mol Biol Cell.* 2002; 13, 2909–2918.

Taylor DA, Atkins BZ, Hungspreugs P, Jones TR, Reedy MC, Hutcheson KA, Glower DD, et al. Regenerating functional myocardium: improved performance after skeletal myoblast transplantation. *Nat Med.* 1998; 4(8), 929–933. doi: 10.1038/nm0898-929.

Thompson RB, Emani SM, Davis BH, van den Bos EJ, Morimoto Y, Craig D, Glower D, et al. Comparison of intracardiac cell transplantation: autologous skeletal myoblasts versus bone marrow cells. *Circulation.* 2003; 108 Suppl 1, II264–271.

Timmers L, Lim SK, Arslan F, Armstrong JS, Hoefer IE, Doevendans PA, Piek JJ, et al. Reduction of myocardial infarct size by human mesenchymal stem cell conditioned medium. *Stem Cell Res.* 2007; 1, 129–137.

Tomita S, Li RK, Weisel RD, Mickle DA, Kim EJ, Sakai T, Jia Z. Autologous bone marrow cell transplantation enhances heart function in damaged hearts. *Circulation.* 1999; 100, II247–256.

Turlo AJ, Hammand De, Ramsbottom Ka, Soul J, Gillen A, McDonald J, Pfeffer MJ. Mesenchymal stromal cell secretome is affected by donor age, sex, and tissue source. Cells. 2023; 41(11): 1047–1059.

Urbanek K, Rota M, Cascapera S, Bearzi C, Nascimbene A, De Angelis A, Hosoda T, et al. Cardiac stem cells possess growth factor-receptor systems that, after activation, regenerate the infarcted myocardium, improving ventricular function and long-term survival. *Circulation Res.* 2005; 97, 663–673.

Wang X, Hu Q, Nakamura Y, Lee J, Zhang G, From AH, Zhang J. The role of the sca-1+/CD31- cardiac progenitor cell population in postinfarction left ventricular remodeling. *Stem Cells.* 2006; 24, 1779–1788.

Wang X, Li Q, Hu Q, Suntharalingam P, From AH, Zhang J. Intra-myocardial injection of both growth factors and heart-derived Sca-1+/CD31- cells attenuates post-MI LV remodeling more than does cell transplantation alone: neither intervention enhances functionally significant cardiomyocyte regeneration. *PLoS One.* 2014; 9, e95247.

Wang X, Zhao T, Huang W, Wang T, Qian J, Xu M, Kranias EG, et al. Hsp20-engineered mesenchymal stem cells are resistant to oxidative stress via enhanced activation of Akt and increased secretion of growth factors. *Stem Cells.* 2009; 27, 3021–3031.

Windmolders S, De Boeck A, Koninckx R, Daniels A, De Wever O, Bracke M, Hendrikx M, et al. Mesenchymal stem cell-secreted platelet-derived growth factor exerts a pro-migratory effect on resident Cardiac Atrial appendage stem cells. *J Mol Cell Cardiol.* 2014; 66, 177–188.

Xiong Q, Ye L, Zhang P, Lepley M, Swingen C, Zhang L, Kaufman DS, et al. Bioenergetic and functional consequences of cellular therapy: activation of endogenous cardiovascular progenitor cells. *Circulation Res.* 2012; 111, 455–468.

Yang Y, Li Y, Chen X, Cheng L, Wang J. Hypoxia-preconditioned mesenchymal stem cell secretome improves cardiac repair via pro-angiogenic mechanisms. *Int J Mol Sci.* 2023; 24, 11345.

Ye L, Chang YH, Xiong Q, Zhang P, Zhang L, Somasundaram P, Lepley M, et al. Cardiac repair in a porcine model of acute myocardial infarction with human induced pluripotent stem cell-derived cardiovascular cells. *Cell Stem Cell.* 2014; 15, 750–761.

Ye L, Haider H, Sim EK. Adult stem cells for cardiac repair: a choice between skeletal myoblasts and bone marrow stem cells. *Exp Biol Med.* (Maywood). 2006; 231, 8–19.

Ye L, Haider H, Tan R, Su L, Law PK, Zhang W, Sim EK. Angiomyogenesis uses liposome-based transfection of vascular endothelial growth factor-165 into skeletal myoblasts for cardiac repair. *Biomaterials.* 2008; 29, 2125–2137.

Ye L, Haider H, Tan R, Toh W, Law PK, Tan W, Su L, et al. Transplantation of nanoparticle-transfected skeletal myoblasts overexpressing vascular endothelial growth factor-165 for cardiac repair. *Circulation.* 2007; 116, I113–120.

Ye L, Zhang P, Duval S, Su L, Xiong Q, Zhang J. Thymosin beta4 increases the potency of transplanted mesenchymal stem cells for myocardial repair. *Circulation.* 2013; 128, S32–41.

Yeghiazarians Y, Gaur M, Zhang Y, Sievers RE, Ritner C, Prasad M, Boyle A, Bernstein HS. Myocardial improvement with human embryonic stem cell-derived cardiomyocytes enriched by p38MAPK inhibition. *Cytotherapy.* 2012; 14, 223–231.

Zhang Y, Liang X, Liao S, Wang W, Wang J, Li X, Ding Y, Liang Y, Gao F, Yang M, et al. Potent paracrine effects of human induced pluripotent stem cell-derived mesenchymal stem cells attenuate doxorubicin-induced cardiomyopathy. *Sci Rep.* 2015; 5, 11235.

Zhang Z, Deb A, Zhang Z, Pachori A, He W, Guo J, Pratt R, Dzau VJ. Secreted frizzled-related protein 2 protects cells from apoptosis by blocking the effect of canonical Wnt3a. *J Mol Cell Cardiol.* 2009; 46, 370–377.

Zheng D, Huo M, Li B, Wang W, Piao H, Wang Y, Zhu Z, et al. MSC-derived exosomes promote cardiac repair by enhancing angiogenesis and inhibiting apoptosis. *Stem Cell Res Ther.* 2021; 12, 151.

Zhou Q, Zhang Y, Li Z, Wang L, Sun H. Hypoxia-induced extracellular vesicles from MSCs enhance angiogenesis and tissue regeneration. *Front Cell and Dev Biol.* 2024; 12, 1298743.

Zhou Y, Yamamoto Y, Xiao Z, Ochiya T, Takahashi K. miR-21-overexpressing mesenchymal stem cells reduce apoptosis and improve tissue repair. *Stem Cell Res Ther.* 2021.

3 Cell-Free Therapy

Next Step in Regenerative Medicine

Hui Ma[1, 2], *Wing Sum Siu*[1, 2], *and Ping Chung Leung*[1, 2]

[1]Institute of Chinese Medicine, The Chinese University of Hong Kong, Shatin, New Territories, Hong Kong SAR, China

[2]Guangdong-Hong Kong-Macau Joint Laboratory for Pharmacodynamic Constituents of TCM and New Drugs Research, The Chinese University of Hong Kong, Shatin, New Territories, Hong Kong SAR, China

3.1 LIST OF ABBREVIATIONS

AD	Alzheimer's disease
ahaMSCs-Exos	Allogenic human adipose MSC-derived exosomes
ASCs	Adipose-derived stem cells
BBB	Blood-brain barrier
BDNF	Brain-derived neurotrophic factor
cGMP	Current Good Manufacturing Practice
COMT	Catechol-O-methyltransferase
CM	Conditioned medium
CXCR4	C–X–C chemokine receptor type 4
DAergic	Dopaminergic
ECM	Extracellular matrix
EVs	Extracellular vesicles
FBS	Fetal bovine serum
FGF	Fibroblast growth factor
GC-MS	Gas chromatography-mass spectrometry
GDNF	Glial cell-derived neurotrophic factor
GM-CSF	Granulocyte-macrophage colony-stimulating factor
GVHD	Graft-versus-host disease
HGF	Hepatocyte growth factor
HIFs	Hypoxia-inducible factors
HSCs	Hematopoietic stem cells
HSCT	HSC transplantation
Hypo-EVs	Extracellular vesicles from hypoxia-conditioned cells
ICH	International Council for Harmonization of Technical Requirements for Pharmaceuticals for Human Use
IFN-γ	Interferon-gamma
IGF-1	Insulin-like growth factor 1
IL-1β	Interleukin-1 beta

DOI: 10.1201/9781003653387-3

IL-6	Interleukin-6
IL-10	Interleukin-10
LC-MS	Liquid chromatography-mass spectrometry
lncRNAs	Long non-coding RNAs
LVEF	Left ventricular ejection fraction
MAO-B	Monoamine oxidase B
MCP-1	Monocyte chemoattractant protein-1
MHC	Major histocompatibility complex
miRNAs	microRNAs
MI	Myocardial infarction
MMPs	Matrix metalloproteinases
MSCs	Mesenchymal stem cells
MS	Mass spectrometry
NGF	Nerve growth factor
NGS	Next-generation sequencing
NK cells	Natural killer cells
Nor-EVs	Extracellular vesicles from normoxia-conditioned hMSCs
NSAIDs	Nonsteroidal anti-inflammatory drugs
OA	Osteoarthritis
OARSI	Osteoarthritis Research Society International
PD	Parkinson's disease
PDGF	Platelet-derived growth factor
PGE2	Prostaglandin E2
qPCR	Quantitative polymerase chain reaction
RA	Rheumatoid arthritis
ROS	Reactive oxygen species
SCI	Spinal cord injury
SCF	Stem cell factor
TBI	Traumatic brain injury
TGF-β	Transforming growth factor-beta
TNF-α	Tumor necrosis factor-alpha
TSG-6	Tumor necrosis factor-stimulated gene six
UC-MSC	Umbilical cord mesenchymal stem cell
VEGF	Vascular endothelial growth factor
2D	Two-dimensional
2D-GE	Two-dimensional gel electrophoresis
3D	Three-dimensional

3.2 INTRODUCTION

3.2.1 Traditional Stem Cell Therapies and Their Limitations

Traditional stem cell therapies primarily focus on adult stem cells, especially hematopoietic stem cells (HSCs) and mesenchymal stem cells (MSCs). These therapies have been successfully applied in various medical fields, particularly for treating blood disorders, certain cancers, cardiovascular and bone and joint diseases, and many more. HSC transplantation (HSCT) is frequently used for conditions such as leukemia, lymphoma, and aplastic anemia, in which infused HSCs regenerate the blood and immune systems.

In contrast, MSCs are multipotent stem cells capable of trilineage differentiation into osteoblasts (bone cells), chondrocytes (cartilage cells), and adipocytes (fat cells) (Álvarez-Viejo and Haider, 2022). MSCs are primarily derived from bone marrow but can also be obtained from diverse mesenchymal tissues, including adipose tissue, umbilical cord tissue, dental pulp, etc. (Kamal et al., 2022). Their ability to migrate to injury sites and secrete bioactive molecules that modulate inflammation and promote healing makes them crucial for tissue repair and regeneration (Habib et al., 2025a). Additionally, their immunomodulatory properties reduce the risk of graft-versus-host disease (GVHD), potentially enabling the use of allogeneic (donor-derived) MSCs (Mousaei Ghasroldasht et al., 2022).

MSCs have been investigated in various therapeutic contexts. In orthopedics, they are used to treat conditions such as osteoarthritis and fractures, promoting cartilage regeneration and bone healing. In neurology, MSC therapy shows promise for spinal cord injuries and neurodegenerative diseases, as these cells may help reduce inflammation and promote neuronal survival (Yusuf and Haider, 2025). Current research is also exploring the use of MSCs in cardiovascular diseases, where they may help repair damaged heart tissue after a myocardial infarction (Rajab et al., 2019; Mei et al., 2024).

Despite their potential, traditional stem cell therapies face several limitations. A significant challenge is the limited availability of suitable donors for HSCT, particularly for individuals with rare tissue types. GVHD remains a vital complication, wherein transplanted immune cells attack the host's tissues, resulting in serious health issues. Furthermore, these therapies may not be effective for all diseases, especially neurodegenerative disorders.

Regarding MSC-based therapy, while these cells are versatile and promising, their efficacy can vary depending on the donor source, donor age, and cell culture conditions (Safwan et al., 2024a). Although they exhibit immunomodulatory features, MSCs may not entirely evade the host immune system, potentially leading to rejection. Additionally, there is a risk of tumor formation if MSCs are not appropriately differentiated or if exposed to conditions that promote uncontrolled growth. Standardizing the isolation, expansion, and application protocols for MSCs remains a challenge, affecting the reproducibility and safety of these therapies.

Therefore, traditional stem cell therapies, including HSCs and MSCs, hold significant promise for treating various conditions (Safwan et al., 2024b). MSCs, in particular, represent a versatile tool in regenerative medicine with applications across multiple diseases and have progressed to the market as living biodrugs after approval by the FDA of various countries (Bhatti and Haider, 2025). However, addressing the limitations of both traditional stem cell therapies and MSC therapies is essential to optimize safety and efficacy, ensuring these treatments can be effectively utilized in clinical settings for the benefit of patients (Mousaei Ghasroldasht et al., 2022; Mei et al., 2024).

3.3 SECRETOME AS AN INNOVATIVE ALTERNATIVE

The secretome, defined as the collection of biologically active molecules secreted by cells, was first described by Gnecchi et al. (2005). Their work demonstrated that the therapeutic potential of MSCs is primarily mediated by secreted factors rather than cell replacement, thus giving rise to the paracrine hypothesis (Haider and Aziz, 2017). Today, the secretome has emerged as a promising alternative in regenerative medicine and therapeutic development (Habib et al., 2025b). Unlike conventional stem cell therapies, which rely on direct cell transplantation, secretome-based approaches leverage paracrine signaling mechanisms to facilitate tissue repair and regeneration (Haider and Aziz, 2017). This innovative strategy exploits natural intercellular communication pathways, highlighting the therapeutic potential of extracellular bioactive molecules.

The secretome comprises a diverse array of proteins, cytokines, growth factors, and exosomes released by cells in response to physiological or pathological stimuli. These components play pivotal roles in modulating inflammation, enhancing cell survival, and promoting tissue regeneration. Notably, the MSC secretome has attracted significant interest due to its abundance of anti-inflammatory cytokines and growth factors, which can aid tissue recovery while mitigating risks associated with cell transplantation, such as immune rejection and tumorigenesis. A significant advantage of secretome-based therapies lies in their potential to reduce complications commonly associated with traditional cell-based treatments. By harnessing the therapeutic effects of secreted factors, researchers aim to develop safer and more effective interventions for a wide range of conditions, including cardiovascular diseases, neurodegenerative disorders, and autoimmune diseases (da Silva et al., 2023). As scientific understanding of the secretome and its mechanisms of action continues to evolve, this innovative approach holds transformative potential for regenerative medicine, offering novel strategies to enhance healing and improve clinical outcomes.

This chapter aims to provide a comprehensive overview of the components and various engineering approaches of the secretome. It will also elucidate the complex biological mechanisms by which secretomes exert therapeutic effects across a range of fields, including cardiovascular therapies, neurological disorders, wound healing, and regenerative orthopedics. Additionally, the chapter seeks to review the current and emerging clinical applications of the secretome. Finally, the chapter aims to identify potential future directions in secretome research, discussing the challenges, opportunities, and areas that warrant further investigation. It encourages interdisciplinary collaboration and innovation, ultimately driving the development of safer, more effective therapies that can significantly improve patient care and enhance quality of life.

3.3.1 Composition and Functional Mechanisms of Secretome

The secretome is the comprehensive collection of bioactive molecules that are essential for intercellular communication and play significant roles in various biological processes, including tissue repair, immune responses, and cell differentiation. The composition of the secretome can vary widely depending on factors such as cell type, environmental conditions, and the physiological or pathological state of the cells. This variability renders the secretome a rich area for research, particularly in the context of regenerative medicine and therapeutic applications.

3.3.2 Key Components of Secretome

3.3.2.1 Growth Factors

One of the primary components of the secretome is the growth factors, proteins that stimulate cellular growth, proliferation, and differentiation. For instance, vascular endothelial growth factor (VEGF) is a key regulator of angiogenesis. VEGF promotes endothelial cell proliferation and migration, playing a crucial role in restoring regional blood flow to injured or ischemic tissues. This is particularly vital in conditions such as myocardial infarction or stroke, where enhanced blood flow is essential for tissue survival and repair. Another crucial constituent of the secretome is fibroblast growth factor (FGF), a family of proteins involved in numerous biological processes. FGFs are critical for wound healing, as they stimulate fibroblast activity, promote the synthesis of extracellular matrix components, and facilitate angiogenesis by stabilizing blood vessel conduits and reducing their leakiness. Transforming growth factor-beta (TGF-β) is a multifunctional cytokine that regulates various cellular processes, including growth, differentiation, and immune responses. TGF-β is particularly significant in tissue repair, as it promotes the differentiation of fibroblasts into myofibroblasts, which are essential for wound contraction and extracellular matrix deposition. However, dysregulation of TGF-β can lead to fibrosis, highlighting its dual role in both healing and pathological conditions (Ma et al., 2023a).

3.3.2.2 Cytokines and Chemokines

Another significant category of secretome components comprises cytokines and chemokines. These small signaling proteins are essential for regulating immune responses and inflammation. Interleukin-6 (IL-6) is a well-known pro-inflammatory cytokine that plays a significant role in the acute phase response. Produced by various cell types, including macrophages and T cells, IL-6 promotes B cell differentiation and the production of acute-phase proteins. Elevated levels of IL-6 are often associated with chronic inflammatory conditions, making it a potential therapeutic target. In contrast, Interleukin-10 (IL-10) functions as an anti-inflammatory cytokine that helps regulate immune responses. It inhibits the production of pro-inflammatory cytokines, thereby maintaining tissue homeostasis and preventing excessive inflammation, which can lead to tissue damage. Interferon-gamma (IFN-γ) is another crucial cytokine, primarily produced by activated T cells and natural killer (NK) cells. IFN-γ plays a vital role in immune responses, particularly in activating macrophages and enhancing antigen presentation, which is essential for effective adaptive immunity (Ma et al., 2023b).

3.3.2.3 Non-coding RNAs

Among non-coding RNAs, microRNAs (miRNAs) are small, 18–22-nucleotide non-coding RNAs that play a key role in post-transcriptional regulation of gene expression (Haider and Najimi, 2022). By binding to complementary sequences in target mRNAs, miRNAs can induce mRNA degradation or inhibit its translation. This regulatory mechanism is essential in various biological processes, including development, differentiation, and cellular stress responses. Dysregulation of miRNAs is frequently implicated in diseases such as cancer and cardiovascular conditions, underscoring their potential as therapeutic targets. Long non-coding RNAs (lncRNAs) represent another class of non-coding RNAs that have emerged as essential regulators of gene expression. These longer RNA molecules do not encode proteins but can influence gene expression at multiple levels, including chromatin remodeling and transcriptional regulation. The expression of lncRNAs can vary in response to different stimuli, making them significant in understanding disease mechanisms and potential therapeutic interventions (Ma et al., 2023b).

3.3.2.4 Metabolites and Enzymes

In addition to proteins and RNAs, metabolites are essential components of the secretome. These small molecules, produced during cellular metabolism, act as signaling molecules that regulate cellular functions and intercellular communication. For instance, metabolites such as lactate and adenosine have been demonstrated to modulate immune responses. Lactate, once regarded merely as a metabolic byproduct, can promote angiogenesis and inhibit the production of pro-inflammatory cytokines, thereby providing protective effects in tissues during inflammation. Similarly, adenosine exerts immunosuppressive impacts that can help protect tissues during inflammatory responses, making it an attractive target for therapeutic strategies.

Enzymes also constitute a critical component of the secretome. Secreted enzymes, particularly matrix metalloproteinases (MMPs), play vital roles in remodeling the extracellular matrix (ECM). MMPs are involved in the degradation and reorganization of ECM components, which is essential for processes such as tissue repair and wound healing. Breaking down ECM barriers facilitates cell migration and proliferation. However, dysregulation of MMP activity can lead to pathological conditions, including fibrosis and cancer metastasis. Other secreted enzymes, such as proteases and lipases, further contribute to tissue remodeling and metabolic regulation, emphasizing the multifunctional nature of the secretome.

The diverse components of the secretome—spanning growth factors, cytokines, non-coding RNAs, metabolites, and enzymes—highlight their integral role in mediating intercellular communication and orchestrating various biological processes. Understanding these components and their specific functions is crucial for harnessing the therapeutic potential of the secretome in regenerative medicine and other clinical applications. As research in this field continues to progress, the insights gained will pave the way for innovative therapies that leverage the body's natural healing mechanisms, potentially transforming patient care and enhancing health outcomes across a wide range of medical conditions (Ma et al., 2023b).

3.3.3 Variability in Secretome Components

The composition and functionality of the secretome can vary significantly based on the source of the cells used and the conditions under which they are cultured. This variability is crucial for understanding how the secretome can be effectively utilized in therapeutic applications, particularly in regenerative medicine and tissue engineering.

3.3.3.1 Different Cell Source

Different cell types produce distinct secretomes that reflect their specific functions and roles in tissue dynamics (Safwan et al., 2024a). For example, MSCs' secretome is mainly rich in growth factors, cytokines, and extracellular matrix components that facilitate tissue repair and modulate immune responses. At the same time, endothelial cells primarily secrete factors involved in angiogenesis, and the immune cells, such as macrophages, have a secretome rich in cytokines that regulate inflammation. The differentiation status of a cell also influences its secretome. Undifferentiated stem cells may produce a different set of factors compared to their differentiated counterparts. For instance, adipose-derived stem cells (ASCs) may exhibit distinct secretome profiles when cultured under conditions that promote adipocyte differentiation, compared with when they are maintained in an undifferentiated state. This variability can significantly affect the therapeutic potential of these cells in clinical settings. Additionally, the tissue of origin can influence the composition of cells' secretomes. For example, cardiac stem cells may secrete factors that promote cardiac repair, while neural stem cells may release neuroprotective substances. This tissue specificity is crucial for tailoring secretome-based therapies to address particular injuries or diseases (da Silva et al., 2023).

3.3.3.2 Different Culture Conditions

The composition of the culture medium can significantly affect the secretome profile. Nutrient availability, growth factors, and the presence of signaling molecules can all influence what cells secrete. For instance, incorporating specific growth factors into the culture medium can enhance the production of particular cytokines or proteins, thereby modifying the secretome. Oxygen levels in the culture environment are a critical determinant of secretome variability. Hypoxic conditions, which simulate the microenvironment of injured tissues, can stimulate cells to secrete factors that promote angiogenesis and tissue repair. For example, hypoxia-inducible factors (HIFs) can increase the secretion of VEGF and other growth factors, resulting in a secretome that is more conducive to wound healing.

Mechanical factors, such as shear stress or substrate stiffness, can further influence the cell secretome (Wang et al., 2024b). Cells exposed to dynamic mechanical environments may produce different levels of extracellular matrix components and signaling molecules compared to those cultured under static conditions. This variability is particularly relevant in engineering tissues that must withstand mechanical forces, such as cartilage or vascular tissues. Additionally, the density at which cells are cultured can impact secretome production (Kim et al., 2022). Higher cell densities may increase cell-cell interactions, thereby modulating the secretion of various factors. Furthermore, crowded conditions can create localized environments that affect cell-to-cell communication, thereby altering the composition of the secretome (da Silva et al., 2023).

3.3.3.3 Engineering Approaches

The engineering of secretomes is an innovative field focused on optimizing the production, composition, and delivery of secreted factors to enhance their therapeutic efficacy in regenerative medicine. This approach involves selecting specific cell types known for their strong secretory capabilities, such as MSCs, endothelial cells, and immune cells. By choosing appropriate cellular sources, researchers can significantly enhance both the quality and quantity of the secretome produced. Additionally, modifying culture conditions—such as adjusting oxygen levels, nutrient availability, and physical cues—can profoundly affect the composition of the secretome. For example, inducing hypoxic conditions can boost the secretion of angiogenic factors, which are essential for promoting blood vessel formation and tissue repair (da Silva et al., 2023).

Bioengineering techniques further enhance the potential applications of secretomes (Gerini et al., 2025). The use of bioreactors and three-dimensional (3D) culture systems improves cell viability and secretome yield by more closely mimicking physiological conditions than traditional two-dimensional (2D) cultures. This 3D environment supports cell-cell interactions and enhances the biological activity of the secreted factors (Gerini et al., 2025). Genetic engineering methods, such as CRISPR and viral vectors, can also be employed to modify cells, enabling them to produce specific therapeutic proteins or growth factors in greater quantities (Hazrati et al., 2022). Moreover, advanced techniques for isolating and characterizing exosomes—small extracellular vesicles that carry proteins, RNAs, and lipids—are crucial for refining secretome applications. Techniques such as ultracentrifugation, size-exclusion chromatography, and affinity-based methods are increasingly used to isolate and analyze these vesicles (Carter et al., 2019).

3.4 COMPONENT ANALYSIS

Component analysis of the secretome involves identifying and characterizing the various bioactive molecules secreted by cells into the extracellular environment. This analysis is essential for understanding the functional roles of the secretome in biological processes. Different techniques are employed in the study of secretome components, each tailored to identify specific types of molecules.

One of the primary methods used is proteomics, which includes several advanced techniques. Mass spectrometry (MS) is a powerful tool for identifying and quantifying proteins in complex mixtures, enabling detailed analysis of the secretome, including the detection of post-translational modifications that can influence protein function (Batschkus et al., 2017). Another significant method in proteomics is two-dimensional gel electrophoresis (2D-GE), which separates proteins based on their isoelectric point and molecular weight, enabling researchers to visualize protein expression profiles (Villatoro et al., 2021; Murphy et al., 2023). In addition to proteomics, cytokine arrays are used to measure multiple cytokines simultaneously in a secretome. These arrays contain capture antibodies specific to target cytokines, enabling high-throughput analysis of the secretome's cytokine content (Ma et al., 2023b).

For the analysis of non-coding RNAs, such as miRNAs and lncRNAs, techniques such as quantitative polymerase chain reaction (qPCR) and next-generation sequencing (NGS) are used. qPCR quantifies specific RNA molecules, while NGS provides comprehensive profiles of the secretome's RNA landscape, offering insights into regulatory mechanisms (Villatoro et al., 2021). Furthermore, the field of metabolomics employs techniques such as gas chromatography-mass spectrometry (GC-MS) and liquid chromatography-mass spectrometry (LC-MS) to analyze metabolites in the secretome. These methods help elucidate the metabolic signatures associated with different cell types or conditions, revealing critical biochemical pathways (Sayed et al., 2022). Lastly, lipidomics focuses on the thorough analysis of lipid species within secretomes. Techniques such as LC-MS are commonly used in this field to profile lipid classes and explore their signaling roles, thereby further enhancing our understanding of the functional diversity in secretomes.

3.5 FUNCTIONAL MECHANISMS

The therapeutic effects of the secretome are mediated through a variety of complex mechanisms. The secretome contains a diverse array of growth factors and cytokines that activate signaling cascades in target cells, promoting essential processes such as cell proliferation, differentiation, and survival. These signaling pathways are critical for orchestrating the healing process. Moreover, the anti-inflammatory components of the secretome can inhibit pro-inflammatory signaling pathways, thereby reducing tissue damage and fostering a healing environment (Ma et al., 2021). By enhancing angiogenesis and stimulating the recruitment of resident stem cells to injury sites, the secretome plays a vital role in facilitating tissue repair. Additionally, it influences the composition and organization of the ECM, which is essential for maintaining tissue structure and function, thus supporting overall tissue integrity.

3.6 SECRETOME IN CARDIOVASCULAR DISEASES

Cardiovascular diseases are a significant cause of mortality and morbidity worldwide. Given that injured cardiac tissue cannot repair itself, there is an urgent need to develop alternative therapies. Despite notable advances in the medical treatment of cardiovascular diseases—such as drug pre-treatment, interventional therapies, and surgical options—inducing cardiomyocyte differentiation and proliferation remains a challenge. The primary difficulty lies in restoring the function of infarcted and nonfunctional cardiomyocytes (Mittal et al., 2019). Although some studies have reported positive outcomes in patients with heart disease treated with stem cells, the limitations of cell therapy hinder its broader application (Ni et al., 2019).

The secretome derived from stem cells circumvents some of these limitations and shows therapeutic potential against cardiovascular diseases through anti-apoptotic, anti-inflammatory, anti-cardiac remodeling, and pro-angiogenic effects (He et al., 2020; Haider and Najimi et al., 2022). Apoptosis, a programmed form of cell death, plays a critical role in regulating the growth of early morphogenetic heart structures and the late developmental differentiation of cardiac tissue (Sheng et al., 2023). It can be categorized into intrinsic and extrinsic apoptosis based on different mechanisms. Inherent apoptosis is regulated by the Bcl-2 protein family, which includes Bax, Bak, and Bcl-x. A recent study demonstrated that long non-coding RNA UCA1, secreted from hypoxia-conditioned human MSCs (hMSCs), protects against myocardial injury by increasing the anti-apoptotic protein BCL2 (Sun et al., 2020). Extrinsic apoptosis is initiated primarily by receptors on the plasma membrane, such as the Fas and TNF receptors. Another study reported that exosomal miR-25-3p from MSCs mitigates myocardial infarction by targeting pro-apoptotic proteins (FASL and PTEN) and EZH2 (Peng et al., 2020).

Chronic inflammation and abnormal immune responses are integral to the development of cardiovascular diseases. Various anti-inflammatory agents, including anti-TNF-α therapy (infliximab), IL-1 and IL-6 inhibitors, and T- and B-cell inhibitors (such as Rituximab), have shown promising results. However, the use of these inhibitors can increase other health risks, such as elevated cholesterol and triglyceride levels, as well as alterations in hematopoiesis (Mittal et al., 2018). Therefore, alternative methods targeting inflammation are needed. Macrophages play a crucial role throughout the inflammatory process, continuously switching their molecular phenotypes from classically activated pro-inflammatory M1-type (M1) to alternatively activated anti-inflammatory M2-type macrophages (M2). Recent studies have shown that the secretome, enriched in cytokines and chemokines, can regulate M1/M2 macrophage polarization (Su et al., 2023).

Cardiac remodeling alters the morphology and organizational structure of the ventricle. This compensatory response to myocardial hypertrophy and fibrosis can progress to infarct-related heart failure and even sudden cardiac arrest if not managed appropriately (Di Palo et al., 2020). It has been reported that MSC-derived secretome can improve cardiac function and remodeling in a rat model of acute myocardial infarction (Alrefai et al., 2019). Angiogenesis, the formation of new blood microvessels, is essential for tissue generation and repair and is stimulated by pro-angiogenic factors (Haider and Aslam, 2018). More than a dozen pro-angiogenic factors have been identified, many of which are small peptides derived from ischemic tissues (Zeng et al., 2025). VEGF is one of the most well-studied factors. In previous research, we analyzed the components of the MSC-derived secretome. We found it to be rich in VEGF (Ma et al., 2021), which could protect myocytes against vascular remodeling during ischemic processes (Figure 3.1).

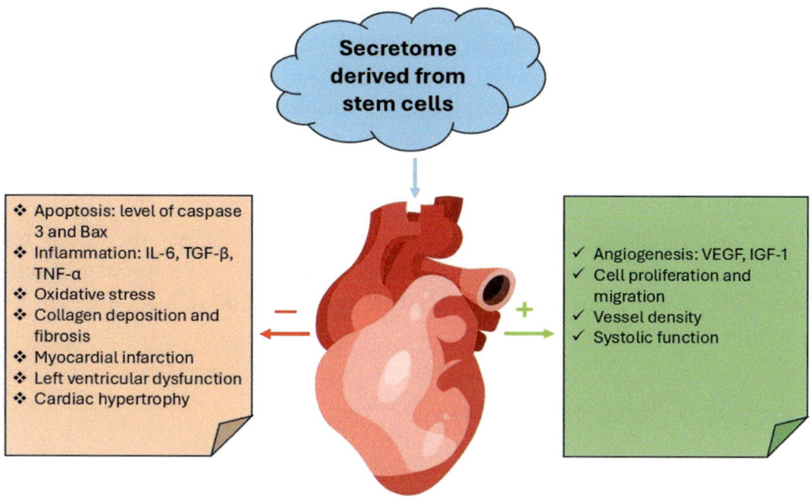

FIGURE 3.1 Secretome in cardiovascular diseases.

3.6.1 Secretome in Atherosclerosis

Atherosclerosis is an inflammatory disorder characterized by the thickening of arteries due to the formation of plaques composed of fatty acids, cholesterol, calcium, fibrin, cellular debris, and waste products in the sub-endothelium. Anti-inflammatory therapy has emerged as a promising treatment option (Soehnlein et al., 2021). Several components of the secretome can modulate macrophage function by regulating M1/M2 polarization. Currently, most research on atherosclerosis has focused on stem cell-derived exosomes. One study demonstrated that exosomes derived from MSCs (MSCs) attenuated the progression of atherosclerosis in ApoE-/- mice by promoting miR-let7-mediated infiltration and polarization of M2 macrophages, leading to increased M2 polarization and decreased expression of M1 markers in macrophages (Li et al., 2019).

In addition to inflammation, endothelial cell functionality plays a critical role in atherosclerosis. Damage or dysfunction of endothelial cells is a key event in the initiation of atherosclerotic plaque formation. The ability of arterial endothelial cells to adapt to disturbed blood flow is crucial in determining susceptibility at arterial branching sites, where atherosclerosis tends to develop preferentially. Another study showed that MSC-derived exosomes containing microRNA-342-5p protect endothelial cells against atherosclerosis in a cellular model (Xing et al., 2020). Therefore, stem cell-derived exosomes represent a potential therapeutic option for treating atherosclerosis (Al-Omar et al., 2022).

3.6.2 Seretome in Myocardial Infarction

Myocardial infarction (MI) occurs due to the occlusion of a coronary artery, which deprives cardiomyocytes of oxygen and nutrients. As a result, cell death (including apoptosis, pyroptosis, and necrosis) begins within 20 minutes, with complete necrosis occurring 2 to 4 hours after prolonged coronary artery occlusion. The death of cardiomyocytes, along with inflammatory stimuli, triggers a pro-fibrotic response in resident fibroblasts, leading to irreversible cardiac remodeling, increased arrhythmogenicity, and ultimately heart failure (Han et al., 2022). Heart failure following MI is one of the leading causes of death and disability worldwide. Therefore, therapeutic strategies aimed at repairing the heart to prevent adverse left ventricular remodeling and reduce the risk of heart failure after MI are a critical unmet clinical need.

In recent years, numerous studies have demonstrated the therapeutic effects of the secretome in treating MI. These secretomes can reduce cell apoptosis and regulate the inflammatory response after MI by modulating macrophage polarization, T cell activation, NK cell activity, and dendritic cell function, while also decreasing pyroptosis in non-immune cells (Shao et al., 2022). One study showed that sustained subcutaneous delivery of the secretome from human cardiac stem cells promotes cardiac repair following myocardial infarction. In this study, rats receiving sustained secretome infusion exhibited preserved left ventricular ejection fraction (LVEF), increased cardiomyocyte number, and enhanced vessel density (Kompa et al., 2021). Another study reported a secretome derived from preconditioned stem cells, revealing that hypoxic preconditioning increased HIF-1α levels, which could impair dendritic cell differentiation and reduce NK cell-mediated cytotoxicity (Ejtehadifar et al., 2015).

3.6.3 Secretome in Cardiac Hypertrophy

Cardiac hypertrophy is a condition marked by the enlargement and thickening of the heart muscle, particularly in the ventricular walls. This adaptation can arise from various physiological and pathological stimuli, such as increased workload, hypertension, and specific cardiac diseases. While cardiac hypertrophy initially serves as a compensatory mechanism to enhance cardiac output, prolonged hypertrophy can lead to adverse effects, including heart failure and a heightened risk of cardiovascular events (Nakamura et al., 2018).

Extracellular vesicles (EVs) secreted by stem cells could reduce the expression of molecules associated with cardiac hypertrophy in a model of human-induced pluripotent stem cell-derived cardiomyocytes (Ahmed et al., 2025). Following treatment with EVs, hypertrophic cardiomyocytes exhibited decreased protein levels of hypertrophy-specific markers (ANF, MIF, cTnI, COL1A1) and reduced IL-6 gene expression, a molecule involved in inflammatory processes linked to cardiac hypertrophy. Additionally, there was a reduction in the expression of transcription factors SMAD2, SMAD3, cJUN, and cFOS, which are implicated in the AngII- and TGF-β1-induced cardiomyocyte hypertrophic response (Constantin et al., 2022).

Hypoxic preconditioning also enhances the therapeutic effects of stem cells and their derived secretome (Kim et al., 2022; Lu et al., 2023), as oxidative stress plays a role in the pathophysiology of congestive heart failure and cardiac hypertrophy. In a study comparing extracellular vesicles from normoxia-conditioned hMSCs (Nor-EVs) with those from hypoxia-conditioned cells (Hypo-EVs), it was found that Hypo-EVs had a more pronounced inhibitory effect on cardiac hypertrophy by mitigating mitochondrial dysfunction and preventing ATRAP degradation (Lu et al., 2023).

Beyond these conditions, the secretome derived from stem cells also shows therapeutic potential in other cardiac diseases, such as dilated cardiomyopathies and aneurysms. A recent study indicated that intrapericardial injection of hydrogels containing ASCs and their secretome improved ventricular hemodynamics and reduced cardiac remodeling in dilated cardiomyopathies (Liguori et al., 2025). In the context of aneurysms, exosomal miR-17-5p from adipose-derived MSCs has been shown to inhibit the progression and inflammation of abdominal aortic aneurysms via the TXNIP-NLRP3 signaling pathway (Hu et al., 2022).

3.7 SECRETOME IN NEUROLOGICAL DISORDERS

The secretome of stem cells is an emerging focus of research in neurological diseases. It encompasses a collection of bioactive molecules, including proteins, cytokines, growth factors, and extracellular vesicles, that stem cells secrete into their surroundings. This secretome plays a pivotal role in mediating the therapeutic effects of stem cells, especially in the treatment of neurological disorders.

Key components of the stem cell secretome include various cytokines and growth factors. Neurotrophic factors such as brain-derived neurotrophic factor (BDNF), nerve growth factor (NGF), and glial cell-derived neurotrophic factor (GDNF) support neuronal survival, differentiation, and growth, thereby promoting neuroprotection and regeneration in injured tissues. Furthermore, anti-inflammatory cytokines such as IL-10 and TGF-β can modulate inflammatory responses, thereby reducing neuroinflammation associated with multiple neurological diseases (da Silva et al., 2023). EVs are also significant components of the stem cell secretome. These vesicles transport proteins, lipids, and nucleic acids that facilitate intercellular communication, influencing the behavior of recipient cells and promoting repair mechanisms in the nervous system (Yusuf and Haider, 2025). Exosomes, a subtype of EVs, carry microRNAs and proteins that can modulate gene expression, further enhancing neuroprotection and regeneration (Tran et al., 2015).

The secretome of stem cells exerts its effects through several mechanisms. A primary function is neuroprotection, in which factors released by stem cells shield neurons from apoptosis and oxidative stress, particularly in conditions such as stroke and neurodegenerative diseases (Baez-Jurado et al., 2018; Wala Eman paper ref). By enhancing cellular survival, the secretome helps maintain neuronal function. Additionally, it can modulate the immune response, decreasing the secretion of pro-inflammatory cytokines and fostering an anti-inflammatory environment in the brain (Szekiova et al., 2018). The secretome also promotes the repair of damaged neural tissue by encouraging the proliferation and differentiation of neural progenitor cells and enhancing angiogenesis, which is essential for restoring blood supply to injured areas (Teixeira et al., 2015). Moreover, components of the secretome can influence neighboring cells, thereby enhancing the overall regenerative capacity of the nervous system through intercellular signaling (Chang et al., 2021). The secretome has demonstrated promising outcomes in various neurological diseases (Figure 3.2).

3.7.1 Secretome in Neurodegenerative Diseases

Neurodegenerative diseases include Alzheimer's disease (AD) and Parkinson's disease (PD). AD is the most prevalent type of neurodegenerative disorder, accounting for 60% to 80% of all cases. It is characterized by cognitive decline and the presence of β-amyloid plaques in the brain. Although numerous therapeutic strategies have been investigated over the decades, no curative treatment currently exists. Existing drug treatments for AD, such as anticholinesterase inhibitors and anti-glutaminergics, are symptomatic and aim to slow the progression of cognitive and behavioral symptoms

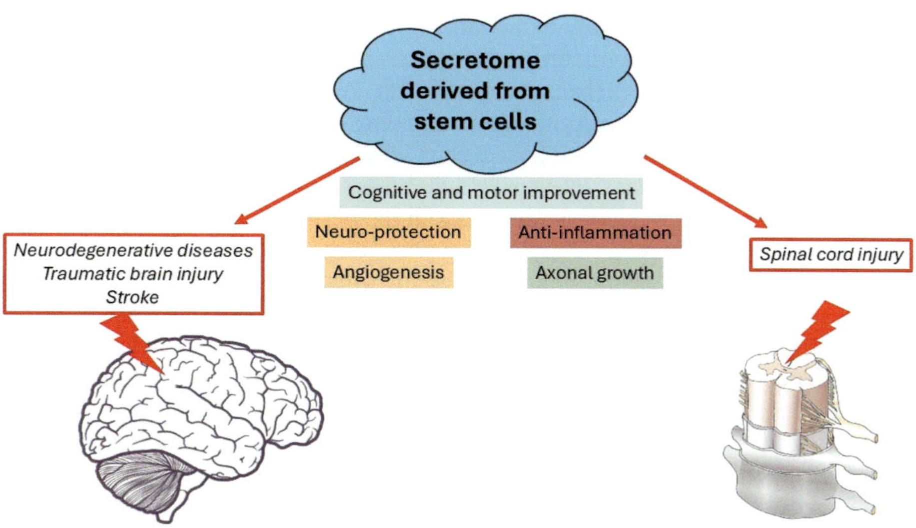

FIGURE 3.2 Secretome in neurological diseases.

associated with dementia (Passeri et al., 2022). These medications may be more effective in the early asymptomatic stages, but do not reverse neurodegeneration.

The therapeutic effects of the secretome in AD have been highlighted in various preclinical studies. One study found that the secretome derived from bovine umbilical MSCs exhibited neuroprotective effects by upregulating BDNF and NGF while downregulating tumor necrosis factor-alpha (TNF-α) and interleukin-1 beta (IL-1β) (Wihadmadyatami et al., 2025). Another study demonstrated the therapeutic potential of MSC secretome delivered intranasally in an AD mouse model, where it reduced plaque density, decreased microglial activation, increased neuronal density, and promoted cognitive and motor functional recovery (Santamaria et al., 2021).

PD is the second most common neurodegenerative disease globally, with its prevalence expected to double over the next 30 years. It results from the loss of dopaminergic (DAergic) neurons in the substantia nigra, leading to a range of motor and non-motor disorders. The clinical hallmark of Parkinson's disease includes a motor syndrome characterized by bradykinesia, resting tremor, rigidity, and changes in posture and gait. Non-motor symptoms can consist of hyposmia, constipation, urinary dysfunction, orthostatic hypotension, memory loss, depression, pain, and sleep disturbances (Tolosa et al., 2021). Drug therapy remains the most effective and widely used treatment for PD, including levodopa, DA agonists, amantadine, monoamine oxidase B (MAO-B) inhibitors, and catechol-O-methyltransferase (COMT) inhibitors (Church et al., 2021). However, these therapies can have side effects and do not halt the progression of PD.

The secretome derived from stem cells has the potential to counteract neuroinflammation, enhance antioxidant capacity, and increase the expression of neurotrophic factors. Several in vivo and in vitro studies have demonstrated the positive effects of the secretome in combating PD. A recent study showed that the secretome from neural-induced human adipose tissue-derived stem cells improved motor balance and coordination in a rotenone-induced PD rat model by reducing tyrosine hydroxylase-positive dopaminergic neurons, inhibiting neuroinflammation and apoptosis, and restoring autophagic flux properties (Ramalingam et al., 2025). Another study found that the neural stem cell secretome could protect neurons from mitochondrial damage in PD. In this study, the injection of secretome into the striatum and substantia nigra improved both motor and non-motor deficits in PD rats and reduced the loss of dopaminergic neurons. In cellular studies, it reduced oxidative stress and exacerbated mitochondrial dysfunction via the Sirtl pathway (Ni et al., 2022) (Figure 3.2).

3.7.2 Secretome in Stroke

Stroke is a leading cause of disability and death worldwide. It is most commonly caused by an abrupt blockage of an artery (ischemic stroke). However, it can also occur due to bleeding into brain tissue when a blood vessel ruptures (hemorrhagic stroke). The predominant mechanism of ischemic stroke is embolism, with the vast majority of emboli being blood clots formed in the heart (cardioembolism) as a result of cardiac disease (Feske et al., 2021). This embolism initiates the formation of the infarct core, which consists of irreversibly injured necrotic tissue within the affected vascular region, followed by late-phase injury development in the peri-infarct area—a potentially salvageable region surrounding the core.

The pathophysiology of ischemic brain damage involves the activation of a series of harmful signaling cascades, including hypoxia and glucose deprivation, excessive release of the excitatory neurotransmitter glutamate, and oxidative and nitrosative stress. These factors lead to an inflammatory response characterized by the activation of brain-intrinsic microglia, increased permeability of the blood-brain barrier (BBB), and infiltration of peripheral immune cells (Paul et al., 2021). There is an urgent need for neuroprotection strategies to safeguard brain cells from ischemic damage.

Numerous studies have highlighted the bioactive role of stem cell secretome in promoting angiogenesis, neurogenesis, and neuroprotection against ischemic stroke. A recent study demonstrated that secretome derived from human amniotic membrane could prevent brain damage due to cerebral ischemia/reperfusion during subacute, acute, and chronic phases in a rat stroke model. In this study, intraventricular injection of the secretome reduced cerebral edema and apoptosis, enhanced BBB integrity, and improved sensorimotor function (Razavi-Toosi et al., 2023). Another study also reported positive outcomes of secretome in treating brain ischemic injury using an experimental stroke model. Infusion of the secretome was found to upregulate neurotrophic factors, stimulate neurogenesis and angiogenesis, and concurrently inhibit inflammation, apoptosis, and neuronal loss in ischemic brains (Asgari Taei et al., 2022) (Figure 3.2).

3.7.3 Secretome in Traumatic Brain Injury

Traumatic brain injury (TBI) results in brain damage through both primary and secondary injury mechanisms. Primary injury leads to localized brain damage, while secondary injury is characterized by inflammatory activity that disrupts the BBB, allows infiltration of peripheral blood cells, causes brain edema, and releases various immune mediators, including chemotactic factors and interleukins. TBI disrupts molecular signaling, alters cellular structure, and impairs cellular function. In addition to tissue damage such as axonal injury, contusions, and hemorrhage, TBI generally disrupts brain physiology, impacting cognition, decision-making, memory, attention, and speech capabilities (Thapa et al., 2021).

The secretome possesses immunoregulatory properties and is considered a practical approach to mitigate both primary damage and secondary neuroinflammation resulting from TBI. The long-term safety and efficacy of secretome delivery have been demonstrated in a focal cranial blast TBI model, in which primary vision deficits caused by the explosion improved after 5 months of secretome treatment (Rasiah et al., 2022). Another study confirmed that intravenous infusion of hypoxia-preconditioned secretome reduces neuroinflammation, cerebral edema, and neurological deficits following TBI by lowering IL-6 and TNF-α levels while increasing TGF-β and tumor necrosis factor-stimulated gene six protein (TSG-6) (Xu et al., 2020) (Figure 3.2).

3.7.4 Secretome in Spinal Cord Injury

Spinal cord injury (SCI) is a severe neurological and pathological condition that leads to significant motor, sensory, and autonomic dysfunctions. The initial phase following the injury is termed primary injury, characterized by bone fragments and tearing of spinal ligaments. This phase triggers secondary injury, including ischemia, oxidative stress, inflammatory responses, and apoptotic pathways, ultimately resulting in locomotor dysfunction. Current treatments for SCI aim to reduce neuroinflammation, promote axonal growth, enhance myelination, and decrease cavity size (Anjum et al., 2020).

The secretome contains a variety of anti-inflammatory molecules and growth factors, indicating its therapeutic potential in treating SCI. One study demonstrated that intrathecal delivery of secretome derived from Wharton's Jelly-derived MSCs could improve functional recovery, enhance expression of axonal growth-related genes, promote axonal sprouting, and reduce the number of reactive astrocytes at the injury site (Chudickova et al., 2019). A recent study also highlighted the therapeutic potential of secretome from human dental pulp stem cells in a rat SCI model. Intraperitoneal injection of this secretome was shown to promote recovery of sensory and motor functions, decrease the expression of microglial pyroptosis markers such as NLRP3, GSDMD, caspase-1, and IL-1β, promote axonal and myelin regeneration, and inhibit the formation of glial scars. In vitro studies have confirmed that the NLRP3/caspase-1/IL-1β pathway is associated with the anti-pyroptosis effects of the secretome in microglia (Liu et al., 2024) (Figure 3.2).

3.8 SECRETOME IN WOUND HEALING

A skin wound occurs when the integrity of the epidermal layer is disrupted. Wound healing involves a series of well-coordinated cellular and molecular events, including hemostasis, inflammation, proliferation, and remodeling. Hemostasis is initiated immediately after the wound occurs to prevent blood loss. This is followed by an inflammatory phase that lasts 2 to 5 days, typically peaking around 72 hours. During this phase, neutrophils and macrophages are activated and secrete numerous growth factors, cytokines, and chemokines, including TGF-α, TGF-β, bFGF, platelet-derived growth factor (PDGF), VEGF, and reactive oxygen species (ROS). These factors help resolve inflammation and recruit endothelial cells and fibroblasts, thereby initiating the proliferative phase.

In the proliferative phase, neovascularization, fibroblast migration, and re-epithelialization occur in a well-regulated manner. Ultimately, granulation tissue is replaced by scar tissue during the remodeling phase. Standard wound care practices include debridement, cleaning, infection management, and dressing. Traditional approaches, such as plant extracts and herbal medicines, have been used to promote wound healing. In modern therapeutic practices, the use of secretome is emerging as a promising method to enhance wound healing. It addresses limitations associated with traditional methods, such as contamination and tissue irritation, while promoting tissue regeneration (Ma et al., 2023a).

Trophic factors in the secretome contribute to various phases of wound healing. A key step in the inflammatory phase is the shift in macrophage phenotype from M1 pro-inflammatory to M2 anti-inflammatory. Our previous study demonstrated that MSC-derived secretome suppressed the release of IL-6, TNF-α, and monocyte chemoattractant protein-1 (MCP-1) from M1 macrophages (Ma et al., 2021). Another study showed that conditioned medium (CM) from bone marrow-derived stem cells reduced the infiltration of inflammatory cells (neutrophils and macrophages) in vivo at various time points.

During the proliferative phase, fibroblasts and epithelial cells play critical roles. Application of MSC-derived secretome enhanced fibroblast proliferation and migration and increased ECM deposition, facilitating wound healing (Ma et al., 2021). TGF-β is also vital in this phase, as it promotes local ECM secretion, regulates melanin production, and activates keratinocyte differentiation and maturation. It has been reported that TGF-β, in conjunction with MMP-9, promotes wound healing in diabetic ulcers. Additionally, angiogenesis is a crucial factor in wound healing. The secretome contains a variety of angiogenic factors, including bFGF, TGF-β, PDGF, VEGF, hepatocyte growth factor (HGF), and MCP-1, which promote neovascularization both in vitro and in vivo.

In the remodeling phase, characterized by collagen synthesis and maturation, cytokines and growth factors such as HGF and prostaglandin E2 (PGE2) exhibit anti-fibrotic properties by inhibiting myofibroblast differentiation and preventing epithelial-mesenchymal transition. Furthermore, the secretome stimulates the proliferation, migration, and differentiation of local progenitor cells into new keratinocytes, which is essential for tissue remodeling. It can also modulate the behavior of other cells, including endothelial cells and fibroblasts, by promoting cell survival and enhancing ECM production (Ma et al., 2023a) (Figure 3.3).

In addition to cutaneous wounds, the secretome is also utilized for corneal wound healing. It can be delivered via eye drops, a commonly used therapeutic method that requires frequent topical administration due to rapid turnover through the lacrimal system. One study found that continuous or multiple secretome treatments were more effective at suppressing inflammation than a single treatment during corneal wound healing. Another study demonstrated that secretome delivered via a 3D electrospun fiber scaffold enabled continuous corneal application. The results indicated that this 3D delivery system could promote faster epithelialization with less scarring and suppress myofibroblastic transformation (Carter et al., 2019) (Figure 3.3).

3.9 SECRETOME IN REGENERATIVE ORTHOPAEDICS

The secretome, which comprises bioactive molecules secreted by cells, plays a crucial role in regenerative orthopedics. This field focuses on the repair and regeneration of musculoskeletal tissues, including bones, cartilage, tendons, and ligaments. It has emerged as a promising therapeutic

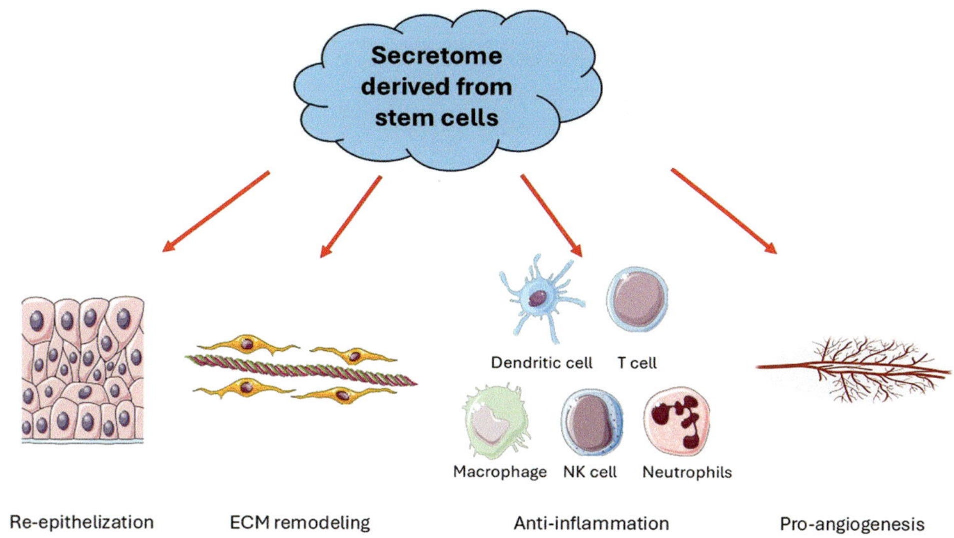

FIGURE 3.3 Secretome in wound healing.

strategy for enhancing tissue repair and promoting healing in various orthopaedic conditions. For bone regeneration, the secretome is employed to stimulate bone healing; when combined with biodegradable scaffolds, it supports new bone formation. In cartilage repair, the secretome is used to address cartilage damage, particularly in conditions such as osteoarthritis. Additionally, it can enhance the healing of tendons and ligaments following injuries (Chen et al., 2021).

The secretome exerts its effects through several mechanisms. First, growth factors stimulate the proliferation of progenitor and stem cells, increasing the pool of cells available for tissue repair. Second, cytokines modulate the inflammatory response by recruiting essential immune cells and promoting their resolution, which is vital for healing. Third, factors such as VEGF promote the formation of new blood vessels, enhancing blood supply to damaged tissues and facilitating nutrient delivery. Finally, the secretome influences ECM remodeling, which is critical for tissue regeneration and involves the degradation of damaged matrix components and the deposition of new ones (Benavides-Castellanos et al., 2020).

3.9.1 Secretome in Bone Defect

Bone defects, or bone discontinuities, caused by trauma, infection, tumors, and other diseases do not regenerate spontaneously, even with prolonged healing. Currently, treatment options are primarily limited to reconstructive trauma surgery involving bone grafts and biomaterials. Standard clinical practices include the use of autologous bone grafts, vascularized bone grafts, anastomosed vascular bone grafts, and autologous bone marrow components. However, larger bone defects often require the use of biomaterials for bone tissue engineering to facilitate repair and promote bone regeneration. Despite advancements in treatment, managing bone defects remains challenging, particularly in cases of non-unions and chronic defects (Wang et al., 2024a).

The secretome, which contains a variety of factors that exert multiple biological effects, has emerged as a novel strategy for treating bone defects. A previous study indicated that the secretome from human bone marrow-derived MSCs is rich in growth factors, including insulin-like growth factor 1 (IGF-1), VEGF, and TGF-β1, which accelerate bone regeneration (Takeuchi et al., 2019). In a rat calvaria bone defect model, the secretome from bone marrow MSCs has been shown to promote bone formation by regulating the polarization of M1/M2 macrophages

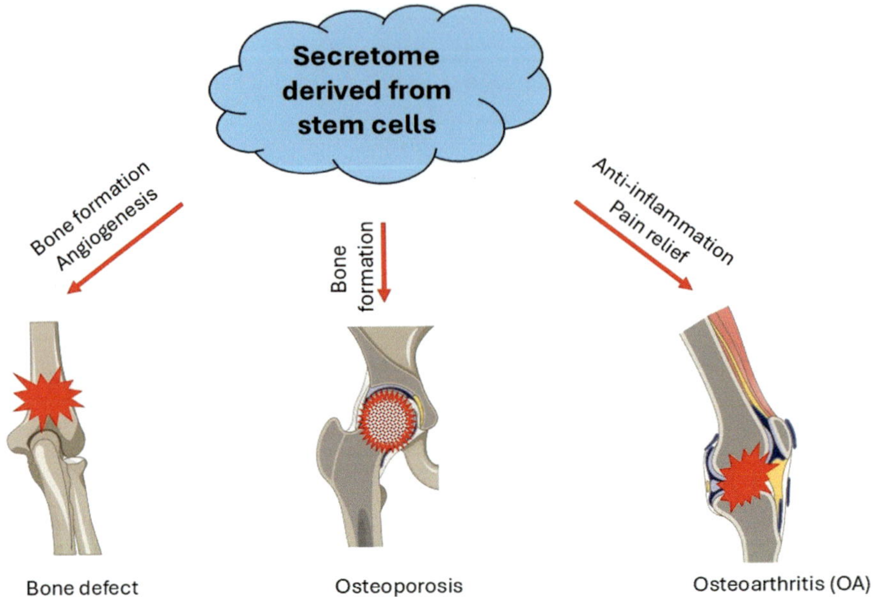

FIGURE 3.4 Secretome in regenerative orthopedics.

(Katagiri et al., 2022). In the same model, exosomes derived from the bone marrow MSC secretome also demonstrated positive outcomes, promoting bone formation, increasing the number of osteoblast-like cells and vascular endothelial cells, and enhancing angiogenesis (Takeuchi et al., 2019) (Figure 3.4).

3.9.2 Secretome in Osteoporosis

Osteoporosis is a disease characterized by low bone mass and disruption of bone architecture, leading to decreased bone strength and an increased risk of fractures. Often referred to as a silent disease, osteoporosis typically presents no symptoms until a fracture occurs. Several risk factors can increase the likelihood of developing osteoporosis. Gender is a significant factor, with women, particularly post-menopausal women, being at a higher risk. Age also plays a crucial role, as the risk of osteoporosis rises significantly with advancing age. Individuals with smaller body frames may face a greater risk due to lower peak bone mass. Lifestyle factors such as smoking, excessive alcohol consumption, and a calcium-deficient diet can further contribute to bone density loss. Treatment options for osteoporosis include medications, nutritional support, exercise, and lifestyle changes (Johnston et al., 2020).

Current therapies primarily focus on inhibiting bone resorption, which can be effective to some extent; however, the serious adverse effects associated with long-term use should be a concern. Additionally, these therapies do not promote bone formation or address the underlying causes of osteoporosis. The secretome has been shown to effectively stimulate bone regeneration, as it contains a diverse range of proteins, angiogenic factors, growth factors, hormones, cytokines, and extracellular matrix proteins. In a rat model of age-related osteoporosis, a secretome derived from human umbilical cord-derived MSCs, when loaded into silk fibroin-based hydrogels, increased bone density (Liang et al., 2019). In another osteoporosis model induced by ovariectomy, secretome from MSCs encapsulated in nanoparticles labeled with C–X–C chemokine receptor type 4 (CXCR4) was shown to accumulate in bone, inhibit osteoclast differentiation, promote osteogenic proliferation, and reduce bone mass loss (Zhang et al., 2022) (Figure 3.4).

3.9.3 Secretome in Osteoarthritis

Osteoarthritis (OA) is the most prevalent form of arthritis and a leading cause of disability globally. According to the Osteoarthritis Research Society International (OARSI), OA is a disorder of movable joints characterized by cellular stress and degradation of the extracellular matrix. It is a whole-joint disease that affects all joint components, commonly affecting the knees, hips, hands, and spine. Symptoms can vary in severity and may include joint pain, stiffness, swelling, reduced range of motion, and a grating sensation during movement.

Several factors influence the development of OA. Age is a significant contributor, as the risk of OA increases with advancing years; cartilage naturally deteriorates over time. Joint injuries, such as fractures or ligament tears, can also elevate the likelihood of developing OA in the affected joint. Obesity is another significant risk factor, as excess weight places additional stress on weight-bearing joints, particularly the knees, leading to cartilage breakdown. Genetic predisposition may also play a role, with a family history of OA heightening susceptibility. Moreover, repetitive stress from occupations or activities involving excessive joint movement can accelerate cartilage degeneration. Other medical conditions, such as rheumatoid arthritis, metabolic disorders, and diabetes, can further increase the risk of OA. While there is no cure for osteoarthritis, various treatment options can help manage symptoms and improve quality of life. Pain relievers, including over-the-counter options like acetaminophen and nonsteroidal anti-inflammatory drugs (NSAIDs), can alleviate pain and reduce inflammation. Topical treatments, such as NSAID-containing creams or gels, may provide localized relief. Physical therapy can create exercise programs to strengthen the muscles around the joint, improve flexibility, and reduce pain. Lifestyle changes, including weight loss and regular low-impact activities such as swimming or cycling, can help maintain joint function. Assistive devices, such as braces, shoe inserts, or walking aids, can relieve pressure on affected joints. In advanced cases where conservative treatments fail, surgical options may be considered, including arthroscopy, osteotomy, or joint replacement (Tang et al., 2025).

During the onset and progression of OA, a cascade of events occurs within joint structures, leading to systemic inflammation, fibrosis, ECM degradation, neovascularization, and neural sensitization. The proteins and bioactive molecules found in the stem cell secretome offer diverse therapeutic benefits. In a mouse model of OA, a single injection of secretome derived from human adipose tissue-derived MSCs produced rapid and lasting antihyperalgesic and antiallodynic effects, with intravenous infusion proving to be the most effective delivery method. This approach significantly reduced pain by decreasing the inflammatory response (Amodeo et al., 2021). Another in vitro study identified the mechanisms by which the secretome treats OA. It was found to reduce inflammation by inhibiting the NF-κB pathway, promote cartilage repair, and maintain chondrocyte phenotype by suppressing hypertrophy markers such as RUNX2 and COL10A1, while increasing COL2A1 and ACAN production and upregulating SOX9 (González-Cubero et al., 2024) (Figure 3.4).

In addition to the diseases above, stem cell secretome is also applied in other joint conditions, such as rheumatoid arthritis (RA) and rotator cuff injuries. RA is a chronic autoimmune disease characterized by progressive joint destruction, the production of autoantibodies, a predominance of pro-inflammatory cells, and dysregulation of peripheral immune tolerance. As there is currently no cure, novel therapeutic approaches are essential. The secretome exhibits immunomodulatory and anti-inflammatory effects, which are crucial for alleviating clinical symptoms in RA patients. One study found that the secretome from 3D-cultured MSCs derived from umbilical cord tissue expressed higher levels of anti-inflammatory cytokines, such as IL-10 and LIF, as well as trophic factors that support tissue regeneration, including PDGF-BB, FGF-2, I-309, stem cell factor (SCF), and granulocyte-macrophage colony-stimulating factor (GM-CSF). In rat models, this secretome reduced the manifestations of arthritis (Miranda et al., 2019).

Rotator cuff repair is a standard surgical procedure in sports medicine, where torn tendons are reattached to the bony surface. While most patients experience positive outcomes, some may suffer re-tears due to the inability to reconstruct the normal tendon-bone enthesis. The stem cell secretome can be used to enhance tendon-bone healing and promote enthesis regeneration. One study showed that secretome from human bone marrow-derived stem cells could facilitate tendon-bone healing in a rat rotator cuff model by modulating macrophage polarization, inhibiting the M1 phenotype, and promoting the M2 phenotype. This modulation may be related to the Smad2/3 signaling pathway (Chen et al., 2021).

3.10 CLINICAL APPLICATION OF SECRETOME

The secretome has garnered significant interest in regenerative medicine and tissue repair due to its essential roles in cellular communication and tissue regeneration. Despite certain clinical limitations, numerous clinical trials have been registered and conducted worldwide at www.clinicaltrials.gov.

In the realm of neurological diseases, two completed clinical trials have been published, focusing on Alzheimer's disease and ischemic stroke. One phase I/II trial evaluated the clinical safety and efficacy of allogenic human adipose mesenchymal stromal cell-derived exosomes (ahaMSCs-Exos) in patients with mild to moderate Alzheimer's disease (NCT04388982). Nine patients participated in the study, which demonstrated that intranasal administration of ahaMSCs-Exos was safe and well-tolerated. A medium dose of 4 × 10^8 particles appeared to delay cognitive decline and reduce hippocampal volume (Xie et al., 2023). Another study involved five patients assessing the safety of intraparenchymal injections of allogenic placenta mesenchymal stem cell-derived exosomes in those undergoing decompressive craniectomy after a malignant middle cerebral artery infarct (NCT03384433). This study reported no adverse effects in the five ischemic stroke patients, and two showed improvements in neurological status (Dehghani et al., 2022).

In the context of orthopedic diseases, a completed clinical trial with published results is available (NCT05579665). This randomized, open-label study involved 30 patients with knee osteoarthritis, divided into two groups that received either umbilical cord-MSC (UC-MSC) secretome or hyaluronic acid. The findings indicated that intra-articular injections of secretome reduced pain, decreased the MMP3 ratio, and increased TGF-β1 levels without side effects (Partan et al., 2023).

Additionally, a completed phase I clinical trial utilized conditioned medium from Wharton's Jelly-derived MSCs for chronic ulcers (NCT04134676), enrolling 38 patients. Preliminary results suggested that the conditioned medium could reduce edema and erythema, promote granulation tissue formation, and decrease ulcer size. Another study on hypertrophic scars involved fourteen patients treated with human stem cell-conditioned medium, which resulted in reduced skin erythema and water loss without other side effects (NCT05004779) (Joo et al., 2023).

Beyond these conditions, several clinical trials have been registered using MSC-based cell-free products for other diseases, including alopecia (NCT05658094, NCT05296863), otitis media (NCT05402267), type 1 diabetes (NCT02138331), nasopharyngeal cancer (NCT05261750), and dry eye diseases (NCT04213248, NCT05738629). Since the outbreak of SARS-CoV-2 in 2019, a considerable number of clinical trials have been registered using MSC-derived secretome to address respiratory symptoms. One study involving nebulized exosomes from allogenic adipose tissue mesenchymal stromal cells in seven patients with severe COVID-19 confirmed the safety of the secretome (NCT04276987) (Zhu et al., 2022). More extensive studies are needed further to validate its safety and efficacy (Table 3.1).

3.11 LIMITATIONS AND FUTURE PERSPECTIVES

While no adverse effects of secretome have been reported to date, several significant challenges remain regarding its application in clinical practice. Understanding the identification and bioactivities of secretome components is still incomplete. The composition of the secretome is influenced by various factors, including the characteristics of the donor cells or tissue, the conditions employed during cell culture (such as culture duration, cell density, number of passages, culture media, and

TABLE 3.1
Clinical Trials Registered Using MSC-Based Cell-Free Products for Different Diseases

Disease	Cell-Free Products	Administration Routine	ClinicalTrials.gov Identifier
Neurological Diseases			
Alzheimer's disease	AD-MSC-exo	nasal drip	NCT04388982
Ischemic stroke	MSC-exo	i.v.	NCT03384433
Cerebral palsy	UC-MSC-CM	intrathecal injection	NCT04314687
Stroke	UC-MSC-CM	intranasal	NCT05008588
ALS and Other Motor Disorders	AlloEx exosomes	intranasal	NCT07105371
ALS	SHED-CM	i.v.	NCT06889857
ALS	SHED-CM	i.v.	NCT06608719
Orthopedic Diseases			
Osteoarthritis	MSC-EVs	intra-articular injection	NCT05060107
	AD-MSC-CM	N.A.	NCT04223622
	UC-MSC-CM	intra-articular injection	NCT04314661
	UC-MSC-CM	intra-articular injection	NCT05579665
	UC-MSC-CM	intra-articular injection	NCT06688318
Degenerative Meniscal Injury	Synovial Fluid-MSC-exo	intra-articular injection	NCT05261360
Bone graft	AD-MSC-CM	bone graft	NCT04998058
Ligament injury	AD-MSC-CM	N.A	NCT04889963
Skin Diseases			
Wound	AD-MSC-exo	topical	NCT05475418
	Plasma-exo	topical	NCT02565264
	WJ-MSC-CM	topical	NCT04134676
	MSC-CM	topical	NCT04235296
Psoriasis	Commercial MSC-exo	topical	NCT05523011
Dystrophic Epidermolysis Bullosa	Commercial MSC-EVs	topical	NCT04173650
Skin aging	AD-MSC-CM	topical	NCT05508191
Hypertrophic scar	MSC-CM	topical	NCT05004779
Keloid	UC-MSC-CM	intra-lesion injection	NCT04326959
Melasma	AD-MSC-CM	intradermal injection	NCT06516419
Pilonidal Sinus	MSC-exo	intra-lesion injection	NCT06391307
Respiratory Diseases			
Covid-19	MSC-exo	aerosol inhalation	NCT04276987
	MSC-exo	i.v.	NCT05216562
	BM-EVs	i.v.	NCT04493242
	MSC-exo	aerosol inhalation	NCT04491240
	Commercial MSC-exo	i.v.	NCT04798716
	Commercial WJ-MSC-exo, UC-MSC-exo	i.v.	NCT05387278
	MSC-CM	i.v.	NCT05122234
	MSC-CM	i.m.	NCT04753476
Pulmonary infection	AD-MSC-exo	aerosol inhalation	NCT04544215
ARDS	hMSC-exo	aerosol inhalation	NCT04602104
Eye Diseases			
Macular holes	MSC-exo	intravitreal injection	NCT03437759

(Continued)

TABLE 3.1 CONTINUED
Clinical Trials Registered Using MSC-Based Cell-Free Products for Different Diseases

Disease	Cell-Free Products	Administration Routine	ClinicalTrials.gov Identifier
Dry eye	UC-MSC-exo	eye drop	NCT04213248
	PSC-MSC-exo	eye drop	NCT05738629
Retinitis pigmentosa	WJ-MSC-exo	subtenon injection	NCT05413148
	UC-MSC-CM	Peribulbar injection	NCT05909488
Ocular surface disease	MSC-CM	topical	NCT05204329
Other Diseases			
Alopecia	Amniotic-MSC-exo	local injection	NCT05658094
	AD-MSC-CM	local injection	NCT05296863
Perianal Fistula	Placental-MSC-exo	local injection	NCT05499156
T1DM	UCB-MSC-EVs	i.v.	NCT02138331
PCOS	UC-MSC-CM	i.v.	NCT05279768
Nasopharyngeal cancer	Dendritic-CM	i.m.	NCT05261750

Abbreviations: AD-MSC: adipose-derived MSCs; Exo: exosomes; WJ-MSC: Wharton's jelly-derived MSC; MSC-EVs: MSC-derived extracellular vesicles; BM-EVs: bone merrow derived extracellular vesicles; PSC-MSC: pluripotent stem cell-derived MSCs; UCB-MSC: umbilical cord blood-derived MSCs; SHED-CM: Stem Cell from Human Exfoliated Deciduous teeth Conditioned Media; ARDS: acute respiratory distress syndrome; T1DM: type 1 diabetes mellitus; PCOS: polycystic ovary syndrome; ALS: amyotrophic lateral sclerosis; i.v.: intravenous; i.m.: intramuscular.

Source: Clinicaltrials.gov.

supplements), and the protocols for preparing and concentrating the secretome. These compositional variations ultimately impact the therapeutic effects observed.

Although "omics" technologies provide avenues for comprehensive and quantitative analysis of biomolecules within the secretome, adherence to current Good Manufacturing Practice (cGMP) protocols is essential. Such protocols ensure the proper preparation of secretome, its stability during storage, and the establishment of quality control parameters necessary to guarantee safety and efficacy.

The secretome may contain biomolecular materials, such as EVs, that contribute to tissue regeneration. However, this raises concerns about immunogenicity, particularly the potential transfer of major histocompatibility complex (MHC) molecules. Standardization of preparation protocols, characterization methods, and dosage analysis is crucial. The absence of standardized protocols poses a significant barrier to regulatory approval from entities such as the International Council for Harmonization of Technical Requirements for Pharmaceuticals for Human Use (ICH). Insights can be gained from the characterization and identification processes seen in traditional Chinese medicine. Establishing a standardized list of quantified secretome biomolecules, along with identifying markers of therapeutic potential, could help mitigate some of these limitations.

In addition to the need for standardized production protocols, the economic aspects associated with producing the large volumes required for therapeutic applications present a challenge. The secretome contains substantial amounts of bioactive molecules, which necessitate careful consideration of its stability and half-life during manufacturing, storage, and delivery. Comprehensive studies are required to evaluate various application protocols, including preparation methods, concentrations, and delivery routes. Furthermore, many studies have utilized fetal bovine serum (FBS) for cell cultivation, which may not be suitable for clinical applications due to risks of disease transmission and potential xenogeneic immune reactions. Consequently, the use of human serum or serum replacement is necessary, although this significantly increases the costs of cell-free products.

Despite the above limitations, the secretome holds tremendous potential in regenerative medicine. Ongoing research aims to clarify the specific mechanisms through which secretome components exert their effects. Gaining insight into the signaling pathways and interactions underlying the actions of secretome components will facilitate the development of targeted therapies, thereby maximizing therapeutic outcomes. This knowledge could pave the way for novel treatments addressing a range of conditions, including degenerative diseases and injuries.

The secretome also has the potential to enhance personalized medicine by tailoring treatments to individual patients. For instance, preconditioning stem cells with hypoxic conditions may enhance their efficacy in treating ischemic diseases and improve healing responses. Additionally, integrating secretome therapies with existing treatments—such as stem cell therapy, gene therapy, or conventional medications—could yield synergistic effects, amplifying therapeutic benefits. Research into secretome-enriched therapies may lead to more effective strategies for tissue repair and regeneration, particularly in complex conditions like cardiovascular diseases and neurodegenerative disorders.

Moreover, the secretome can serve as a valuable source of biomarkers for monitoring disease progression and treatment response. Identifying specific components within the secretome, such as exosomes or proteins found in patients' blood or body fluids, associated with particular conditions could enhance diagnostic capabilities and inform treatment decisions, ultimately improving patient outcomes.

The future of secretome in regenerative medicine appears promising, with numerous avenues for exploration and advancement. As research advances, the potential for secretome-based therapies to redefine treatment strategies, enhance healing, and improve patient outcomes continues to expand. By addressing current challenges—such as standardization, economic feasibility, and safety concerns—and leveraging scientific discoveries, the secretome may play a pivotal role in the evolution of regenerative medicine. Continued collaboration among researchers, clinicians, and regulatory bodies will be essential in realizing the full potential of secretome therapies, ultimately leading to innovative treatments that can significantly impact patient care.

As we move forward, the integration of secretome into clinical practice could revolutionize the management of various health conditions, offering new hope for patients and advancing the field of regenerative medicine.

REFERENCES

Ahmed LA, Al-Massri KF, Haider KH. Therapeutic efficacy of mesenchymal stem cells and their derived exosomes in different neurodegenerative disorders: current status and future perspectives. Handbook of regenerative medicine. *Cell-free Therapy Approach*. 2025;55–67.

Al-Omar MT, Alnajjar MT, Ahmed ZT, Salaas FMI, Alrefaei TSM, Haider KH. Endothelial progenitor cell-derived small extracellular vesicles for myocardial angiogenesis and revascularization. *Journal of Clinical and Translational Research*. 2022;8(6):1–12.

Alrefai MT, Tarola CL, Raagas R, Ridwan K, Shalal M, Lomis N, Paul A, et al. Functional assessment of pluripotent and mesenchymal stem cell-derived secretome in heart disease. *Annals of Stem Cell Research*. 2019;2(1):29–36.

Álvarez-Viejo M, Haider KH. Mesenchymal stem cells. In: KH Haider, editor. *Handbook of Stem Cell Therapy*. Singapore: Springer; 2022. https://doi.org/10.1007/978-981-16-6016-0_6-1

Amodeo G, Niada S, Moschetti G, Franchi S, Savadori P, Brini AT, Sacerdote P. Secretome of human adipose-derived mesenchymal stem cell relieves pain and neuroinflammation independently of the route of administration in experimental osteoarthritis. *Brain, Behavior, and Immunity*. 2021;94:29–40.

Anjum A, Yazid MD, Fauzi Daud M, Idris J, Ng AMH, Selvi Naicker A, Ismail OHR, et al. Spinal cord injury: pathophysiology, multimolecular interactions, and underlying recovery mechanisms. *International Journal of Molecular Sciences*. 2020;21(20):7533.

Asgari Taei A, Dargahi L, Khodabakhsh P, Kadivar M, Farahmandfar M. Hippocampal neuroprotection mediated by secretome of human mesenchymal stem cells against experimental stroke. *CNS Neuroscience & Therapeutics*. 2022;28(9):1425–1438.

Baez-Jurado E, Vega GG, Aliev G, Tarasov VV, Esquinas P, Echeverria V, Barreto GE. Blockade of neuroglobin reduces protection of conditioned medium from human mesenchymal stem cells in human astrocyte model (T98G) under a scratch assay. *Molecular Neurobiology*. 2018:55(3):2285–2300.

Batschkus S, Atanassov I, Lenz C, Meyer-Marcotty P, Cingöz G, Kirschneck C, Urlaub H, et al. Mapping the secretome of human chondrogenic progenitor cells with mass spectrometry. *Annals of Anatomy = Anatomischer Anzeiger: Official Organ of the Anatomische Gesellschaft*. 2017;212:4–10.

Benavides-Castellanos MP, Garzón-Orjuela N, Linero I. Effectiveness of mesenchymal stem cell-conditioned medium in bone regeneration in animal and human models: a systematic review and meta-analysis. *Cell Regeneration* (London, England). 2020;9(1):5.

Bhatti A, Haider KH. Mesenchymal stem cells as living bio-drugs: adding a novel dimension to cardiology and cardiovascular sciences. *Journal of Cardiology and Cardiovascular Sciences*. 2025;6(1):30–35.

Carter K, Lee HJ, Na KS, Fernandes-Cunha GM, Blanco IJ, Djalilian A, Myung D. Characterizing the impact of 2D and 3D culture conditions on the therapeutic effects of human mesenchymal stem cell secretome on corneal wound healing in vitro and ex vivo. *Acta Biomaterialia*. 2019;99:247–257.

Chang C, Yan J, Yao Z, Zhang C, Li X, Mao HQ. Effects of mesenchymal stem cell-derived paracrine signals and their delivery strategies. *Advanced Healthcare Materials*. 2021;10(7):e2001689.

Chen W, Sun Y, Gu X, Cai J, Liu X, Zhang X, Chen J, et al. Conditioned medium of human bone marrow-derived stem cells promotes tendon-bone healing of the rotator cuff in a rat model. *Biomaterials*. 2021; 271:120714.

Chudickova M, Vackova I, Machova Urdzikova L, Jancova P, Kekulova K, Rehorova M, Turnovcova K, et al. The effect of wharton jelly-derived mesenchymal stromal cells and their conditioned media in the treatment of a rat spinal cord injury. *International Journal of Molecular Sciences*. 2019;20(18):4516.

Church FC. Treatment options for motor and non-motor symptoms of Parkinson's disease. *Biomolecules*. 2021;11(4):612.

Constantin A, Comarița IK, Alexandru N, Filippi A, Bojin F, Gherghiceanu M, Vîlcu A, et al. Stem cell-derived extracellular vesicles reduce the expression of molecules involved in cardiac hypertrophy in a model of human-induced pluripotent stem cell-derived cardiomyocytes. *Frontiers in Pharmacology*. 2022;13:1003684.

Da Silva AV, Serrenho I, Araújo B, Carvalho AM, Baltazar G. Secretome as a tool to treat neurological conditions: are we ready? *International Journal of Molecular Sciences*. 2023;24(22):16544.

Dehghani L, Khojasteh A, Soleimani M, Oraee-Yazdani S, Keshel SH, Saadatnia M, Saboori M, et al. Safety of intraparenchymal injection of allogenic placenta mesenchymal stem cells derived exosome in patients undergoing decompressive craniectomy following malignant middle cerebral artery infarct, a pilot randomized clinical trial. *International Journal of Preventive Medicine*. 2022;13:7.

Di Palo KE, Barone NJ. Hypertension and heart failure: prevention, targets, and treatment. *Heart Failure Clinics*. 2020;16(1):99–106.

Ejtehadifar M, Shamsasenjan K, Movassaghpour A, Akbarzadehlaleh P, Dehdilani N, Abbasi P, Molaeipour Z, Saleh M. The effect of hypoxia on mesenchymal stem cell biology. *Advanced Pharmaceutical Bulletin*. 2015;5(2):141–149.

Feske SK. Ischemic stroke. *The American Journal of Medicine*. 2021;134(12):1457–1464.

Gerini G, Mari E, Pontocorvi P, Camero S, Romano E, Ranieri D, Megiorni F, et al. 3D culturing as a promising strategy to enhance the angiogenic potential of adipose stem cell-derived secretome: insights into the role of miR-145-5p/ANGPT2 axis. *Stem Cell Research & Therapy*. 2025;16(1):153.

Gnecchi M, He H, Liang OD, Melo LG, Morello F, Mu H, Noiseux N, et al. Paracrine action accounts for marked protection of the ischemic heart by Akt-modified mesenchymal stem cells. *Nature Medicine*. 2005;11(4):367–368.

González-Cubero E, González-Fernández ML, Esteban-Blanco M, Pérez-Castrillo S, Pérez-Fernández E, Navasa N, Aransay AM, et al. The therapeutic potential of adipose-derived mesenchymal stem cell secretome in osteoarthritis: a comprehensive study. *International Journal of Molecular Sciences*. 2024;25(20):11287.

Habib SM, Martini MF, Al Ahmed O, Qassab AR, Haider KH. Mesenchymal stem cell secretomes for cell-free therapy of the broken heart. In: KH Haider, editor. *Handbook of Regenerative Medicine: Cell-free Therapy*. Boca Raton (FL): CRC Press; 2025a, pp. 274–293, Chapter 18.

Habib SM, Martini MF, Al Ahmed O, Qassab AR, Haider KH. The potential of mesenchymal stem cell–derived exosomes in cell-free therapy for myocardial repair. In: KH Haider, editor. *Handbook of Regenerative Medicine: Cell-free Therapy*. Boca Raton (FL): CRC Press; 2025b, pp. 372–390, Chapter-24.

Haider KH, Aslam M. Cell-free therapy with stem cell secretions: protection, repair, and regeneration of the injured myocardium. In: KH Haider, S Aziz, editors. *Stem Cells: From Hype to Real Hope*. Berlin: De Gruyter; 2018.

Haider KH, Aziz S. Paracrine hypothesis and cardiac repair. *International Journal of Stem Cell Research & Transplantation*. 2017; 5(1): 265–267.

Haider KH, Najimi M. Exosome-based cell-free therapy in regenerative medicine for myocardial repair. In: Haider KH, editor. *Handbook of Stem Cell Therapy*. Singapore: Springer; 2022. https://doi.org/10.1007/978-981-16-6016-0_9-1.

Han C, Yang J, Sun J, Qin G. Extracellular vesicles in cardiovascular disease: biological functions and therapeutic implications. *Pharmacology & Therapeutics*. 2022;233:108025.

Hazrati A, Malekpour K, Soudi S, Hashemi SM. CRISPR/Cas9-engineered mesenchymal stromal/stem cells and their extracellular vesicles: a new approach to overcoming cell therapy limitations. *Biomedicine & Pharmacotherapy*. 2022;156:113943.

He N, Zhang Y, Zhang S, Wang D, Ye H. Exosomes: cell-free therapy for cardiovascular diseases. *Journal of Cardiovascular Translational Research*. 2020;13(5):713–21.

Hu J, Jiang Y, Wu X, Wu Z, Qin J, Zhao Z, Li B, et al. Exosomal miR-17-5p from adipose-derived mesenchymal stem cells inhibits abdominal aortic aneurysm by suppressing TXNIP-NLRP3 inflammasome. *Stem Cell Research & Therapy*. 2022;13(1):349.

Johnston CB, Dagar M. Osteoporosis in older adults. *The Medical Clinics of North America*. 2020;104(5):873–884.

Joo SY, Cho YS, Lee SY, Seo CH. Regenerative effect of combined laser and human stem cell-conditioned medium therapy on hypertrophic burn scar. *Burns: Journal of the International Society for Burn Injuries*. 2023;49(4):870–876.

Kamal M, Kassem D, Haider KH. Sources and therapeutic strategies of mesenchymal stem cells in regenerative medicine. In: KH Haider, editor. *Handbook of Stem Cell Therapy*. Singapore: Springer; 2022. https://doi.org/10.1007/978-981-19-2655-6_2

Katagiri W, Takeuchi R, Saito N, Suda D, Kobayashi T. Migration and phenotype switching of macrophages at the early phase of bone formation by secretomes from bone marrow-derived mesenchymal stem cells using a rat calvaria bone defect model. *Journal of Dental Sciences*. 2022;17(1):421–429.

Kim JH, Green DS, Ju YM, Harrison M, Vaughan JW, Atala A, Lee SJ, et al. Identification and characterization of stem cell secretome-based recombinant proteins for wound healing applications. *Frontiers in Bioengineering and Biotechnology*. 2022;10:954682.

Kompa AR, Greening DW, Kong AM, McMillan PJ, Fang H, Saxena R, Wong RCB, et al. Sustained subcutaneous delivery of the secretome of human cardiac stem cells promotes cardiac repair following myocardial infarction. *Cardiovascular Research*. 2021;117(3):918–929.

Liang M, Liu W, Peng Z, Lv S, Guan Y, An G, Zhang Y, et al. The therapeutic effect of secretome from human umbilical cord-derived mesenchymal stem cells in age-related osteoporosis. *Artificial Cells, Nanomedicine, and Biotechnology*. 2019;47(1):1357–1366.

Li J, Xue H, Li T, Chu X, Xin D, Xiong Y, Qiu W, et al. Exosomes derived from mesenchymal stem cells attenuate atherosclerosis progression in ApoE-/- mice by promoting miR-let7-mediated infiltration and polarization of M2 macrophages. *Biochemical and Biophysical Research Communications*. 2019;510(4):565–572.

Liguori TTA, Liguori GR, Sinkunas V, Correia CJ, dos Santos Coutinho E Silva R, Zanoni FL, Aiello VD, et al. Intrapericardial injection of hydrogels with ASC and their secretome to treat dilated cardiomyopathies. *Scientific Reports*. 2025;15(1): 3529.

Liu T, Ma Z, Liu L, Pei Y, Wu Q, Xu S, Liu Y, et al. Conditioned medium from human dental pulp stem cells treats spinal cord injury by inhibiting microglial pyroptosis. *Neural Regeneration Research*. 2024;19(5):1105–1111.

Lu Y, Zhang J, Han B, Yu Y, Zhao W, Wu T, Mao Y, et al. Extracellular vesicles derived from DJ-1 in hypoxia-conditioned hMSCs alleviate cardiac hypertrophy by suppressing mitochondrial dysfunction and preventing ATRAP degradation. *Pharmacological Research*. 2023;187:106607.

Ma H, Lam PK, Siu WS, Tong CSW, Lo KKY, Koon CM, Wu XX, et al. Adipose tissue-derived mesenchymal stem cells (ADMSCs) and ADMSC-derived secretome expedited wound healing in a rodent model - a preliminary study. *Clinical, Cosmetic, and Investigational Dermatology*. 2021;14:753–764.

Ma H, Siu WS, Koon CM, Wu XX, Li X, Cheng W, Shum WT, et al. The application of adipose tissue-derived mesenchymal stem cells (ADMSCs) and a Twin-Herb formula to the rodent wound healing model: use alone or together? *International Journal of Molecular Sciences*. 2023a;24(2):1372.

Ma H, Siu WS, Leung PC. The potential of MSC-based cell-free therapy in wound healing: a thorough literature review. *International Journal of Molecular Sciences*. 2023b;24(11):9356.

Mei R, Wan Z, Yang C, Shen X, Wang R, Zhang H, Yang R, et al. Advances and clinical challenges of mesenchymal stem cell therapy. *Frontiers in Immunology*. 2024;15:1421854.

Miranda JP, Camões SP, Gaspar MM, Rodrigues JS, Carvalheiro M, Bárcia RN, Cruz P, et al. The secretome derived from 3D-cultured umbilical cord tissue MSCs counteracts manifestations typifying rheumatoid arthritis. *Frontiers in Immunology*. 2019;10:18.

Mittal R, Jhaveri VM, Kay SS, Greer A, Sutherland KJ, McMurry HS, Lin N, et al. Recent advances in understanding the pathogenesis of cardiovascular diseases and development of treatment modalities. *Cardiovascular & Hematological Disorders Drug Targets*. 2019;19(1):19–32.

Mittal R, Jhaveri VM, McMurry HS, Kay SS, Sutherland KJ, Nicole L, Mittal J, et al. Recent treatment modalities for cardiovascular diseases with a focus on stem cells, aptamers, exosomes, and nanomedicine. *Artificial Cells, Nanomedicine, and Biotechnology*. 2018;46(Suppl 1):831–840.

Mousaei Ghasroldasht M, Seok J, Park HS, Liakath Ali FB, Al-Hendy A. Stem cell therapy: from idea to clinical practice. *International Journal of Molecular Sciences*. 2022;23(5):2850.

Murphy S, Ohlendieck K. Protein Digestion for 2D-DIGE Analysis. *Methods in Molecular Biology* (Clifton, N.J.). 2023;2596:339–349.

Nakamura M, Sadoshima J. Mechanisms of physiological and pathological cardiac hypertrophy. *Nature Reviews Cardiology*. 2018;15(7):387–407.

Ni J, Sun Y, Liu Z. The potential of stem cells and stem cell-derived exosomes in treating cardiovascular diseases. *Journal of Cardiovascular Translational Research*. 2019;12(1): 51–61.

Ni W, Zhou J, Ling Y, Lu X, Niu D, Zeng Y, Qiu Y, et al. The neural stem cell secretome protects mitochondrial function against damage in a Parkinson's disease model. *Brain Research*. 2022;1790:147978.

Partan RU, Putra KM, Kusuma NF, Darma S, Reagan M, Muthia P, Radiandina AS, et al. Umbilical cord mesenchymal stem cell secretome improves clinical outcomes and changes biomarkers in knee osteoarthritis. *Journal of Clinical Medicine*. 2023;12(22):7138.

Passeri E, Elkhoury K, Morsink M, Broersen K, Linder M, Tamayol A, Malaplate C, et al. Alzheimer's disease: treatment strategies and their limitations. *International Journal of Molecular Sciences*. 2022; 23(22):13954.

Paul S, Candelario-Jalil E. Emerging neuroprotective strategies for the treatment of ischemic stroke: an overview of clinical and preclinical studies. *Experimental Neurology*. 2021;335:113518.

Peng Y, Zhao JL, Peng ZY, Xu WF, Yu GL. Exosomal miR-25-3p from mesenchymal stem cells alleviates myocardial infarction by targeting pro-apoptotic proteins and EZH2. *Cell Death & Disease*. 2020;11(5):317.

Rajab AM, Rajab TM, AlJundi S, Haider KH. The clinical perspective of bone stem cell therapy focuses on nonrandomized and randomized trials. In: KH Haider, editor. *Stem Cells: From Myth to Reality and Evolving*. Berlin (Germany): De Gruyter; n.d. (Published, 2019).

Ramalingam M, Jang S, Hwang J, Cho HH, Kim BC, Jeong HS. Neural-induced human adipose tissue-derived stem cell secretome exerts neuroprotection against rotenone-induced Parkinson's disease in rats. *Stem Cell Research & Therapy*. 2025;16(1):193.

Rasiah PK, Jha KA, Gentry J, Del Mar NA, Townsend T, Torgbe KE, Reiner A, et al. A long-term safety and efficacy report on intravitreal delivery of adipose stem cells and secretome on visual deficits after traumatic brain injury. *Translational Vision Science & Technology*. 2022;11(10):1.

Razavi-Toosi SMT, Asadi Y, Aboutaleb N, Faezi M. Conditioned medium derived from the human amniotic membrane prevents brain damage against cerebral ischemia/reperfusion in subacute, acute, and chronic phases in a rat model of stroke. *Basic and Clinical Neuroscience*. 2023;14(6):843–856.

Safwan M, Bourgleh MS, ALDoush M, Haider KH. Tissue-source effect on the safety and efficacy of MSCs as living biodrugs to treat heart failure: a systematic review and meta-analysis of RCTs. *World Journal of Cardiology*. 2024b;16(8):469–483. https://doi.org/10.4330/wjc.v16.i8.469

Safwan M, Bourgleh MS, Alshakaki H, Molhem A, Haider KH. Morbid cell status and donor age significantly alter mesenchymal stem cell functionality and reparability. In: KH Haider, editor. *Handbook of Stem Cell Applications*. Singapore: Springer Nature Singapore Pte Ltd; 2024a, pp. 1359–1387. https://doi.org/10.1007/978-981-99-7119-0_62

Santamaria G, Brandi E, Vitola P, Grandi F, Ferrara G, Pischiutta F, Vegliante G, et al. Intranasal delivery of mesenchymal stem cell secretome restores brain function in Alzheimer's mice. *Cell Death and Differentiation*. 2021;28(1):203–218.

Sayed AME, Omar FA, Emam MMA, Farag MA. UPLC-MS/MS and GC-MS based metabolites profiling of Moringa oleifera seed with its anti-Helicobacter pylori and anti-inflammatory activities. *Natural Product Research*. 2022;36(24):6433–6438.

Shao L, Shen Y, Ren C, Kobayashi S, Asahara T, Yang J. Inflammation in myocardial infarction: roles of mesenchymal stem cells and their secretome. *Cell Death Discovery*. 2022;8(1):452.

Sheng SY, Li JM, Hu XY, Wang Y. Regulated cell death pathways in cardiomyopathy. Acta pharmacologica *Sinica*. 2023;44(8):1521–1535.

Soehnlein O, Libby P. Targeting inflammation in atherosclerosis - from experimental insights to the clinic. *Nature Reviews Drug Discovery*. 2021;20(8): 589–610.

Su Y, Ai S, Shen Y, Cheng W, Xu C, Sui L, Zhao Y. Regulatory effects of three-dimensional cultured lipopolysaccharide-pretreated periodontal ligament stem cell-derived secretome on macrophages. *International Journal of Molecular Sciences*. 2023;24(8):6981.

Sun L, Zhu W, Zhao P, Wang Q, Fan B, Zhu Y, Lu Y, et al. Long noncoding RNA UCA1 from hypoxia-conditioned hMSC-derived exosomes: a novel molecular target for cardioprotection through miR-873-5p/XIAP axis. *Cell Death & Disease*. 2020;11(8): 696.

Szekiova E, Slovinska L, Blasko J, Plsikova J, Cizkova D. The neuroprotective effect of rat adipose tissue-derived mesenchymal stem cell-conditioned medium on cortical neurons using an in vitro model of SCI inflammation. *Neurological Research*. 2018;40(4):258–267.

Takeuchi R, Katagiri W, Endo S, Kobayashi, T. Exosomes from conditioned media of bone marrow-derived mesenchymal stem cells promote bone regeneration by enhancing angiogenesis. *PLoS One*. 2019;14(11):e0225472.

Tang S, Zhang C, Oo WM, Fu K, Risberg MA, Bierma-Zeinstra SM, Neogi T, et al. Osteoarthritis. *Nature Reviews Disease Primers*. 2025;11(1):10.

Teixeira FG, Carvalho MM, Neves-Carvalho A, Panchalingam KM, Behie LA, Pinto L, Sousa N, et al. The secretome of mesenchymal progenitors from the umbilical cord modulates neural and glial proliferation and differentiation. *Stem Cell Reviews and Reports*. 2015;11(2):288–297.

Thapa K, Khan H, Singh TG, Kaur A. Traumatic brain injury: mechanistic insight on pathophysiology and potential therapeutic targets. *Journal of Molecular Neuroscience*. 2021;71(9):1725–1742.

Tran C, Damaser MS. Stem cells as drug delivery methods: application of stem cell secretome for regeneration. *Advanced Drug Delivery Reviews*. 2015;82–83:1–11.

Tolosa E, Garrido A, Scholz SW, Poewe W. Challenges in the diagnosis of Parkinson's disease. *The Lancet Neurology*. 2021;20(5):385–397.

Villatoro AJ, Martín-Astorga MDC, Alcoholado C, Sánchez-Martín MDM, Becerra J. Proteomic analysis of the secretome and exosomes of feline adipose-derived mesenchymal stem cells. *Animals: An Open Access Journal from MDPI*. 2021;11(2):295.

Wang J, Liu M, Yang C, Pan Y, Ji S, Han N, Sun G. Biomaterials for bone defect repair: types, mechanisms, and effects. *The International Journal of Artificial Organs*. 2024a;47(2):75–84.

Wang W, Zheng X, Wang H, Zuo B, Chen S, Li J. Mechanical unloading promotes osteoclastic differentiation and bone resorption by modulating the MSC secretome to favor inflammation. *Cell Transplantation*. 2024b;33:9636897241236584.

Wihadmadyatami H, Zulfikar MA, Herawati H, Karnati S, Saragih GR, Aliffia D, Pratama DAOA, et al. Neuroprotection effect of bovine umbilical mesenchymal stem cell-conditioned medium on the rat model of Alzheimer's disease mediated by upregulation of BDNF and NGF and downregulation of TNF-α and IL-1β. *Open Veterinary Journal*. 2025;15(1):151–161.

Xie X, Song Q, Dai C, Cui S, Tang R, Li S, Chang J, et al. Clinical safety and efficacy of allogenic human adipose mesenchymal stromal cells-derived exosomes in patients with mild to moderate Alzheimer's disease: a phase I/II clinical trial. *General Psychiatry*. 2023;36(5):e101143.

Xing X, Li Z, Yang X, Li M, Liu C, Pang Y, Zhang L, et al. Adipose-derived mesenchymal stem cells-derived exosome-mediated microRNA-342-5p protects endothelial cells against atherosclerosis. *Aging*. 2020;12(4):3880–3898.

Xu C, Diao YF, Wang J, Liang J, Xu HH, Zhao ML, Zheng B, et al. Intravenously infusing the secretome of adipose-derived mesenchymal stem cells ameliorates neuroinflammation and neurological functioning after traumatic brain injury. *Stem Cells and Development*. 2020;29(4):222–234.

Yusuf R, Haider KH. Theragnostic applications of exosomes in Alzheimer's disease. In: KH Haider, editor. *Handbook of Regenerative Medicine: Cell-Free Therapy*. Boca Raton (FL): CRC Press, Taylor & Francis Group; 2025, pp. 390–409, Chapter-255.

Zeng Y, Fu BM. Angiogenesis and Microvascular Permeability. *Cold Spring Harbor Perspectives in Medicine*. 2025;15(1):a041163.

Zhang C, Zhang W, Zhu D, Li Z, Wang Z, Li J, Mei X, et al. Nanoparticles functionalized with stem cell secretome and CXCR4-overexpressing endothelial membrane for targeted osteoporosis therapy. *Journal of Nanobiotechnology*. 2022; 20(1):35.

Zhu YG, Shi MM, Monsel A, Dai CX, Dong X, Shen H, Li SK, et al. Nebulized exosomes derived from allogenic adipose tissue mesenchymal stromal cells in patients with severe COVID-19: a pilot study. *Stem Cell Research & Therapy*. 2022; 13(1):220.

4 Harnessing the Secretome to Repair Electrical Conductivity in the Heart

Juan Antonio Rojas-Murrillo[1], Michelle Giovanna Santoyo-Suarez[1], David Andrés de la Garza-Kalife[1], Geraldina Mayela del Ángel-Martinez[2], Elsa Nancy Garza-Trevino[1], Gerardo R. Padilla-Rivas[1], and José Francisco Islas[1]

[1]Department of Biochemistry and Molecular Biology, Faculty of Medicine, Autonomous University of Nuevo León. Ave. Francisco I. Madero and Dr Eduardo Aguirre Pequeño, Mitras Centro, No Number, Monterrey, N.L. Mexico

[2]Laboratory of Molecular Biology, Health Research Department, Medicine School, Autonomous University of Coahuila, Saltillo, Mexico

4.1 LIST OF ABBREVIATIONS

3D	Three-dimensional
Akt	Protein kinase B
ADAM	A desintegrin and metalloprotease
AD-MSCs	Adipose tissue-derived MSC
AP-1	Activator Protein 1
BMP-2	Bone morphogenetic protein-2
BMP-4	Bone morphogenetic protein-4
c-MET	Hepatocyte growth factor receptor
CPCs	Cardiac progenitor cells
CSCs	Human cardiac stem cells
CVDs	Cardiovascular diseases
Cx	Connexin
CXCR4	C-X-C chemokine receptor type 4
DNA	Deoxyribonucleic acid
ECs	Endothelial cells
ECM	Extracellular matrix
EPs	Epicardial cells
ESCRT	Endosomal sorting complex required for transport
EVs	Extracellular vesicles
ERK	Extracellular signal-regulated kinase

DOI: 10.1201/9781003653387-4

FGF-2	Fibroblast growth factor 2
HMGB1	High mobility group box 1
hBM-MSC	Human bone marrow-derived MSC
HGF	Hepatocyte growth factor
HSP	Heat-shock protein
hUCB-EPCs	Human umbilical cord blood-derived endothelial progenitor cells
hUC-MSCs	Human umbilical cord-derived MSC
IGF-1	Insulin-like growth factors
iPSCs	Induced pluripotent stem cells
iPSC-CMs	Induced pluripotent stem cell-derived cardiomyocytes
LVEF	Left ventricle ejection fraction
MAPK	MAP Kinases
miR	microRNA
MMPS	Matrix metalloproteinases
MSCs	Mesenchymal stromal cells or Mesenchymal stem cells
nm	nanometers
μm	micrometers
mRNA	messenger RNA
NGF	Neural growth factor
NF-κβ	Nuclear factor kappa-light-chain-enhancer of activated B cells
PDCD4	Programmed cell death 4
PKA	Protein kinase A
PTEN	Phosphatase and tensin homolog
PI3K	Phosphoinositide 3 kinase
PDGF	Platelet-derived growth factor
RNA	Ribonucleic acid
SCF	Stem cell factor
SDF-1	Stromal cell-derived factor-1α
SFRP1	Secreted frizzled-related protein 1
STAT3	Signal transducer and activator of transcription 3
TGF-β	Transforming growth factor beta
TIMP	Tissue inhibitor of metalloproteases
UC-MSC	Umbilical cord-derived MSC
UCB	Umbilical cord blood
VEGF	Vascular endothelial growth factor

4.2 INTRODUCTION

The human heart is a complex organ that primarily functions to distribute oxygenated blood and nutrients throughout the entire body. As a dual pump, the heart supports both the pulmonary and systemic circulations. Its function begins when deoxygenated blood enters the right atrium through three main vessels: the superior vena cava, the inferior vena cava, and the coronary sinus. From there, the blood flows through the tricuspid valve into the right ventricle and is subsequently pumped into the pulmonary artery, traveling toward the lungs. Once oxygenated, the blood returns to the left atrium via the pulmonary veins, crosses the mitral valve into the left ventricle, and is then propelled into the aorta and systemic circulation (Rehman & Rehman, 2023) Cardiac rhythm, over 2.5 billion coordinated beats during an average lifetime, is orchestrated by the cardiac conduction system, comprising the sinoatrial node, atrio-ventricular node, His bundle, bundle branches, and Purkinje fibers (Christoffels et al., 2010; Goodyer et al., 2019; Park & Fishman, 2011).

Cardiovascular diseases (CVDs) remain the foremost cause of death worldwide, accounting for roughly 17.9 million fatalities each year. Complementary to these fatalities, about 64 million people live with heart failure, ranking CVDs among the top 10 leading causes of death in the United States alone (Şahin & İlgün, 2022; Shahim et al., 2023). Following a myocardial infarction or other forms of cardiac insult, the adult heart mounts a reparative response dominated by fibrotic scar formation. Due to the nature of mature cardiomyocytes, they seldom re-enter the cell cycle (Senyo et al., 2014), thereby dying and promoting cardiac fibrosis. As scarring continues, both the contractile performance and electrical conductivity, two key determinants of synchronous pumping, degrade (Monteiro et al., 2017; Tajabadi et al., 2022). The recognition that limited endogenous regeneration underlies post-infarction dysfunction has driven efforts to replace, reprogram, or protect cardiomyocytes using growth factors, transcription factors, and cell-based therapies (Testa et al., 2021).

Recently, attention has shifted to the secretome, the spectrum of extracellular vesicles (EVs), soluble proteins, nucleic acids, and lipids that cells release to mediate paracrine signaling. The secretome derives from mesenchymal stromal cells (MSCs) or other progenitors, conferring anti-fibrotic, pro-survival, and angiogenic effects while circumventing immune rejection, tumorigenic risk, and poor engraftment that limit live-cell transplantation (Busatto et al., 2020; Daneshmandi et al., 2020; Rody et al., 2024; Teixeira & Salgado, 2020). Early evidence suggests that secretome preparations modulate gap-junction integrity and action-potential propagation, directly addressing the electrophysiological deficits that follow injury (Tajabadi et al., 2022). To enhance local retention and on-target activity, bioresorbable conductive materials, including hydrogels, nanoparticles, micelles, and 3D-printed scaffolds, are currently being engineered for controlled, stimulus-responsive delivery of secretome cargo (González-King et al., 2024; Rodness et al., 2016). By integrating advances in paracrine biology with tissue-engineering platforms, secretome-based strategies hold promise for restoring both the structural and electrical integrity of the damaged myocardium. The mechanistic basis and translational potential of this emerging paradigm are examined in the following sections.

4.3 UNDERSTANDING THE HEART'S SECRETOME

The cardiac secretome can be viewed as a repertoire of bioactive molecules released by resident cardiac cells (cardiomyocytes, endothelial cells, fibroblasts, and various progenitor populations) to govern tissue homeostasis, coordinate repair, and support impulse propagation (Arrell et al., 2021; Garmany et al., 2024). The secretome comprises soluble proteins, such as cytokines, chemokines, and growth factors, as well as diverse lipid species and EVs that ferry regulatory RNAs, including microRNAs and long non-coding RNAs (Balbi et al., 2019; Garmany et al., 2024). The exact composition depends on cell identity, developmental stage, metabolic status, and the local biomechanical and inflammatory milieu (Schulte, 2015; Uhlén et al., 2019).

One prominent example of such changes occurs after a myocardial infarction or sustained ischemia, when a massive loss of cardiomyocytes triggers an inflammatory cascade that initiates a fibrotic response. The result is ECM remodeling and a fibrotic scar that assists in preserving the chamber's integrity, but with reduced contractility and disrupted electrical conduction (Tran et al., 2023). Adult cardiomyocytes exhibit minimal proliferative capacity; hence, endogenous regeneration is insufficient to replace lost tissue (Senyo et al., 2014).

As a result, various strategies have been explored to repopulate the injured myocardium, including the use of embryonic stem cells, induced pluripotent stem cells (iPSCs), or adult stem cells; yet, these tactics have been hindered by manufacturing complexity, immune compatibility, incomplete maturation, poor electromechanical integration, and potential genomic instability (Gnecchi et al., 2012; Testa et al., 2021). Also, it is essential to note that numerous benefits observed in cell-transplantation studies appear to arise from paracrine signaling rather than long-term engraftment, a finding that has redirected attention toward cell-free secretome formulations (Gnecchi et al., 2016; Haider & Aziz, 2017). For analytical and therapeutic purposes, the heart's secretome is often divided into two fractions. The soluble fraction consists of freely diffusing proteins, peptides, and small EVs that act through receptor binding or direct membrane permeation. Meanwhile, the insoluble fraction includes larger microvesicles, exosomes, and apoptotic bodies that shield complex cargo and deliver it to target cells by fusion or endocytosis (Haider & Najimi, 2022) (Figure 4.1).

These two arms operate synergistically to modulate fibrosis, preserve cardiomyocyte viability, regenerate gap junctions, and stimulate neovascularization, topics that are examined in the following sections (Habib et al., 2025b). A comparative summary of the major cardiac and progenitor cell types, their secreted factors, exosomal contents, and biological functions is provided in Table 4.1.

4.3.1 Soluble Secretome

The soluble secretome comprises an ensemble of freely diffusing proteins, peptides, and other molecules (Xia et al., 2021). These components are typically embedded in nanometric-sized vesicles released by cardiac cells. Among these components are cytokines, chemokines, growth factors, hormones, proteases, and coagulation factors, together with exosomes and microvesicles that ferry regulatory RNAs such as microRNAs and long non-coding RNAs (Blackwood et al., 2020). Through receptor binding or membrane permeation, these molecules sustain autocrine, paracrine, and endocrine signals that preserve cardiac structure and the electrical syncytium. Developmental morphogens that guide cardiogenesis reappear after injury, as early Wnt activation followed by its down-regulation, plus coordinated inputs from BMP-2, BMP-4, Activin A, and FGF-2, re-engage pathways that favor cardiomyocyte survival and limited proliferation, while Nodal, Sonic Hedgehog, Notch, and Retinoic Acid renew chamber identity and proliferative balance (Lyra-Leite et al., 2022; Mummery et al., 2012; Parikh et al., 2015).

Several soluble factors exert direct cytoprotective and anti-fibrotic effects. For example, Neuregulin-1 activates ErbB receptors and promotes modest cardiomyocyte cell-cycle re-entry, thereby improving conduction velocity in the infarct border (Yester & Kühn, 2017). Insulin-like growth factor-1 combined with hepatocyte growth factor, when delivered from affinity-binding

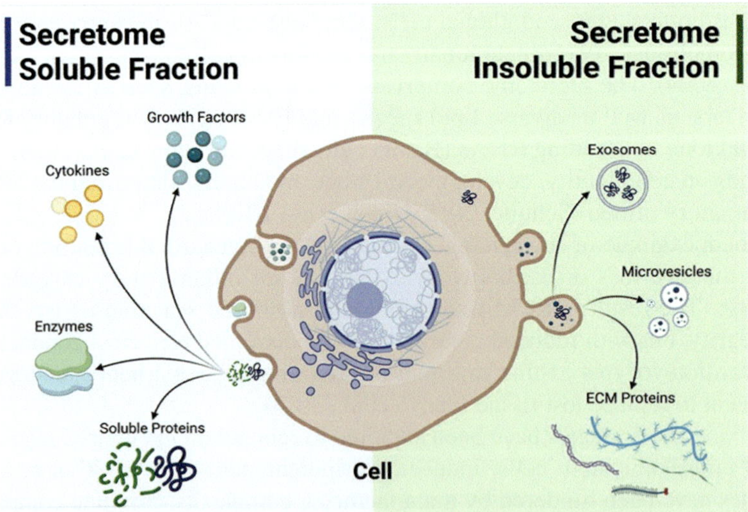

FIGURE 4.1 Schematic overview of the cellular secretome, partitioned into its soluble and insoluble fractions. In the soluble fraction (blue shading), the cell secretes diffusible bioactive molecules—including cytokines, growth factors, enzymes, and other soluble proteins—via classical exocytic pathways. In the insoluble fraction (green shading), the cell releases particulate components: exosomes generated inside multivesicular bodies, microvesicles shed directly from the plasma membrane, and structural ECM proteins that integrate into the extracellular matrix. This dual representation highlights how cells simultaneously modulate their microenvironment through both freely diffusible factors and vesicle- or matrix-associated elements.

TABLE 4.1

Secretome Composition, Exosomal Cargo, and Functional Effects of Distinct Cardiac and Progenitor Cell Types

Cell Type	Secretome	Exosomes	Functions	Ref.
BM-MSC	Release paracrine factors such as VEGF, HGF, IGF-1, FGF-2, IL-10, PDGF, SDF-1, Ang-1, TGF, INF and IL-6	Enriched in mitochondria (EV>200nm) Notch2, ADAM10, and CA2D1. miR125b, miR125-5p and miR-144	Inhibit cardiomyocyte apoptosis, promote angiogenesis, and suppress fibrosis. Immunoregulation is achieved by inhibiting T cells, NK cells, and dendritic cells, and by regulating macrophage polarization.	Fujii et al., 2018; Heo et al., 2016; Lin et al., 2013; O'Brien et al., 2021; Wang et al., 2023; Wen et al., 2020; Wu et al., 2017; Xiao et al., 2018; Xin et al., 2020; Zhu et al., 2018
AD-derived MSC)	VEGF, HGF, IGF (1 and 2), NGF, SDF-1α, MCP-1, TIMP-1, TIMP-4, FGF-2, CXCL-12, Angptl4, and SOD3	miR-671, miR-221, miR-222, miR-17-5p	Inhibit cardiomyocyte apoptosis and promote angiogenesis. Suppress the function of T cells and dendritic cells.	Hu et al., 2022; Lee et al., 2021; Park et al., 2018; Wang YC et al., 2018, Weng X, et al., 2021; Yan et al., 2017, 2020; Zhao et al., 2018
UC-derived MSCs	IL-6, TGF, PDGF, FGF2, SCF, GM-CSF, MCP-1 and IL-21	PAI (33246507), cirBBS2	Increase angiogenesis, limit inflammation, and preserve CX43 gap junctions. Inhibit T cell proliferation.	Hong et al., 2023; Lim et al., 2018; Miranda et al., 2019; Sengun et al., 2023; Wang et al., 2014, Wang Z et al., 2020; Wu et al., 2022
iPSC-CMs	VEGF and IP-10	miR-30, miR-100-5p, miR-21-5p and miR-100-5p	Promote re-musculization and attenuate remodeling and cardiac function.	Ai et al., 2023; Gao et al., 2020; Kawamura et al., 2012; Li et al., 2013; Ong et al., 2015; Roca-Alonso et al., 2015
IPSC-derived ECs	EGF, HGF, and VEGF	miR-1-3p, miR-100-5p miR-199-5p	Promote revascularization. Induce an immune rejection response mediated by T cells and NK cells.	Cheng et al., 2023; Clayton et al., 2018; Gao et al., 2020; Li et al., 2023; Park et al., 2018; Yan et al., 2024
IPSC-derived EPs	Intelectin 1 and retinoic acid	N/A	Enhance macrophage polarization and survival, as well as angiogenesis of cardiomyocytes.	(Luo et al., 2023)
Cardiac fibroblasts	FGF21, GDFs, BNP, ANP, HMGB1, C-C motif chemokine, SFRP1, TNF-α and VEGF.	PCSK6 miR-21, miR-24, miR-144, miR-210, miR-34a	PCSK6 was identified as a profibrotic protein in the transforming growth factor-β pathway and cellular communication, enhancing collagen production in fibroblasts.	Gao et al., 2021; Liu et al., 2021; Park et al., 2020

Abbreviations: AD-MSCs: Adipose tissue-derived MSC; HMGB1: High Mobility Group Box 1 (HMGB1); iPSC-CMs: Induced pluripotent stem cell-derived cardiomyocytes; EPs: Epicardial cells; ECs: Endothelial cells; SFRP1: Secreted frizzled-related protein 1; UC-derived MSCs: Umbilical cord-derived MSCs.

hydrogels, activates PI3K–Akt and c-Met pathways, reduces apoptosis, and maintains Cx43 expression (Ruvinov et al. 2011; Konoplyannikov et al., 2013). Stromal cell-derived factor-1α (SDF-1α) recruits progenitors via CXCR4, STAT3, and Akt (Haider et al., 2008). Controlled release from biomimetic hydrogels decreases cell death, augments capillary density, and preserves conduction properties (Xu et al., 2025). Interleukin-10 couples anti-inflammatory and anti-apoptotic actions, suppressing matrix metalloproteinase-9 and limiting scar thinning to maintain electrical continuity (Krishnamurthy et al., 2009).

MicroRNAs in the soluble fraction fine-tune these responses by regulating gene expression in cells (Haider et al., 2015). Exosomal miR-21 derived from oxidative-stressed cardiac progenitor cells (CPCs) down-regulates PDCD4, protecting cardiomyocytes, and supports progenitor viability through the PTEN–PI3K–Akt axis (J. Xiao et al., 2016). miR-30d-5p targets integrin α5 and MAP4K4, attenuating TGF-β/Smad signaling thereby reducing apoptosis and fibrosis (Long & Cheng, 2024). miR-126 lowers reactive oxygen species, stabilizes mitochondrial potential, and promotes microvascular growth, indirectly safeguarding gap-junction integrity (Luo et al., 2018).

A significant obstacle to achieving the maximum secretome effect is the rapid washout of soluble components. Hence, bioresorbable hydrogels, conductive patches, and nanoparticle carriers delay local retention, shield labile proteins from proteases, and permit stimulus-responsive release, as seen with temperature-sensitive hydrogels that undergo a sol-to-gel transition at body temperature and form in situ depots (Zhou Q, et al., 2021). By minimizing apoptosis, restraining fibroblast activation, enhancing angiogenesis, and preserving connexin expression, the soluble secretome maintains contiguous low-resistance pathways for action-potential propagation, thus complementing the reparative actions of the insoluble secretome that are discussed in the next section.

4.3.2 Insoluble Secretome

The insoluble secretome comprises membrane-enclosed particles and protein assemblies that cardiac and non-cardiac cells export into the extracellular space to convey complex molecular information (Ahmed et al., 2023). These EVs appear principally as apoptotic bodies (1 to 5 μm), microvesicles (50 to 1000 nm), and exosomes (30 to 150 nm), each distinguished by its biogenesis route and size distribution (Chen et al., 2021; Feng et al., 2024). By encapsulating proteins, lipids, nucleic acids, and occasionally entire organelles, vesicles protect cargo from degradation and deliver it intact to recipient cells, influencing signal-transduction pathways that regulate cell fate, proliferation, and survival (Leitolis et al., 2019).

Apoptotic bodies form when dying cells fragment their cytoplasm and nucleus into large blebs, which package histones, DNA fragments, and organelles within a phospholipid membrane. In the heart, cardiomyocyte-derived apoptotic bodies recruit CPCs, increase precursor proliferation, promote differentiation toward a cardiac lineage, and raise contraction rates in newly formed myocytes, thus converting a marker of cell death into a cue for tissue renewal (Chistiakov et al., 2016; Leitolis et al., 2019).

Microvesicles bud directly from the plasma membrane in response to calcium influx or mechanical stress. Under pro-thrombotic conditions, platelets shed microvesicles rich in clotting factors and endothelial-stabilizing proteins, then shift their cargo toward anticoagulant mediators once a thrombus has formed, restraining excessive clot growth (Chistiakov et al., 2016). In the myocardium, fibroblast-derived microvesicles transport metalloproteinases and cytokines that remodel the ECM and influence scar stiffness, thereby affecting impulse propagation.

Exosomes originate from the inward budding of the plasma membrane, which produces endosomes that mature into multivesicular bodies. The fusion of these bodies with the cell membrane releases 30 to 150 nm vesicles (Chen et al., 2021; Zhang et al., 2020). Their surfaces display tetraspanins such as CD9, CD63, and CD81, as well as Rab GTPases, heat-shock proteins, and ESCRT-associated proteins like Tsg101 and Alix, which are markers that not only reveal their origin but also mediate uptake through membrane fusion, receptor-guided endocytosis, macropinocytosis, or phagocytosis (Liu Y et al., 2023).

In the healthy heart, endothelial exosomes help regulate vascular tone and barrier integrity, whereas cardiomyocyte exosomes transfer metabolic enzymes and gap junction proteins that synchronize contraction. Under hypoxia, inflammation, or pressure overload, vesicle release increases, and their cargo shifts toward miR-21, miR-133, HSP20, and VEGF, molecules that can either protect myocardium by promoting angiogenesis and survival or aggravate disease by fostering fibrosis and arrhythmia (Bollini et al., 2018). Therapeutically, the insoluble secretome provides a cell-free avenue for cardiac repair. Exosomes enriched with cardioprotective factors have improved ventricular function, reduced scar size, and attenuated inflammation in multiple infarct models, all without the immunogenicity or tumorigenic risk associated with transplanted cells (Davidson et al., 2021; Lin et al., 2023). By selecting vesicles with tailored cargo and engineering their surfaces for precise homing, researchers aim to develop off-the-shelf therapeutics capable of sustained modulation of injured myocardium and restoration of electrical continuity.

4.4 MECHANISMS FOR REPAIRING ELECTRICAL CONDUCTIVITY

Secretome-based therapy aims to restore cardiac electrical conduction through four interconnected biological processes: reducing fibrosis to preserve low-resistance pathways, promoting cardiomyocyte survival to maintain an excitable cell mass, regenerating gap junctions to synchronize cellular depolarization, and stimulating neovascularization to provide the metabolic support required for sustained electrical activity.

4.4.1 Reduction of Fibrosis

Myocardial fibrosis develops when resident fibroblasts differentiate into α-smooth muscle actin-expressing myofibroblasts that secrete large amounts of collagen I and III together with fibronectin and periostin, leading to a stiffer ECM, altered ventricular geometry, and discontinuities in cell-to-cell electrical coupling (Pakshir et al., 2020; Tai et al., 2021). Excess matrix lengthens action-potential pathways, promotes Cx43 lateralization, and introduces regions of slow conduction where re-entry circuits can form, so reversing or attenuating fibrosis is a prerequisite for expected impulse propagation. In large-animal optical-mapping studies, even a modest reduction in collagen volume fraction is sufficient to raise conduction velocity and suppress ventricular tachycardia, underscoring the therapeutic leverage that antifibrotic interventions can provide (Carta-Bergaz et al., 2022)

Secretome preparations contain several growth factors with potent antifibrotic profiles. Hepatocyte growth factor suppresses canonical TGF-β signaling through cMet-dependent Akt activation, lowers Smad2/3 phosphorylation, and diminishes transcription of COL1A1 and COL3A1 in cultured cardiac fibroblasts (Okayama et al., 2012). When bound to pharmacologically active microcarriers and delivered with EVs rich in Sonic Hedgehog, HGF improved capillary density and reduced scar thickness by 30% in rat infarction, although intracoronary administration 48 hours after ligation failed to benefit swine, highlighting the importance of dose, route, and timing (Blanco-Blázquez et al., 2023; Riaud et al., 2021). Bone morphogenetic protein-7 similarly antagonizes TGF-β1. Repeated intraperitoneal injections in mice subjected to left anterior descending artery ligation halve the interstitial collagen density and downregulate transcripts for Smad2, Smad3, fibronectin, and TGF-β1. In diabetic cardiomyopathy, intravenous BMP-7 lessens both interstitial and perivascular fibrosis and improves LVEF (Jin et al., 2018). Relaxin enhances the activity of matrix metalloproteinases (MMPs), reduces myofibroblast persistence, and promotes angiogenesis. Continuous infusion in diabetic rodents reduced myocardial hydroxyproline content by 30% and improved diastolic compliance, whereas venous delivery of relaxin-expressing C2C12 myoblasts in swine remodeled mature scar, augmented vascular density, and recovered systolic function (Bani et al., 2009; Du et al., 2009).

Antifibrotic activity is reinforced by microRNAs packaged within cardiac or MSC-derived EVs. Exosomes enriched in miR-30d-5p silence integrin α5 and MAP4K4, curbing TGF-β/Smad feedback and sharply lowering fibroblast activation, while co-delivery of miR-21 and miR-146a further decreases caspase-3 activity, reduces infarct size, and raises ejection fraction beyond what either microRNA achieves alone (Long & Cheng, 2024). Vesicular miR-21 also dampens PTEN activity, indirectly amplifying PI3K-Akt survival signals in border-zone cardiomyocytes and limiting secondary fibrosis triggered by dying cells (Xiao et al., 2016). These RNA-mediated effects converge with protein-based cues to normalize the fibroblast phenotype, rebalance collagen turnover, and restore Cx43 localization at intercalated discs.

Because soluble factors and vesicles are rapidly diluted by coronary flow and ventricular contraction, delivery technologies are crucial (discussed in section 4). Temperature-sensitive or enzyme-responsive hydrogels retain secretome components for weeks, releasing them in phase with the resolution of inflammation and tissue remodeling. Conductive polymer patches, such as polyaniline-coated polycaprolactone meshes, not only prolong the payload's residence time but also provide electrical cues that favor the organized alignment of collagen and enhance gap-junction formation. In rodent infarction models, HGF- and miRNA-loaded hydrogels have been shown to decrease the collagen volume fraction by up to 40% and increase longitudinal conduction velocity by 20% within 4 weeks (Sonnenberg et al., 2015). Microspheres composed of poly(lactic-co-glycolic acid) or collagen–chitosan hybrids enable pulsatile, step-wise release that matches the sequential needs of the healing myocardium. Early-phase large-animal trials are testing catheter-based epicardial spraying to reduce the need for multiple injections. Continued optimization of carrier porosity, degradation kinetics, and surface charge aims to maximize antifibrotic efficacy while minimizing off-target biodistribution.

4.4.2 Promotion of Cardiomyocyte Survival

Ischemia deprives cardiomyocytes of oxygen and metabolic substrates, leading to ATP depletion, calcium overload, mitochondrial dysfunction, and the production of excess reactive oxygen species, which trigger necrosis or apoptosis. The resulting loss of viable myocyte fragments the electrical syncytium, Cx43 becomes redistributed or degraded, and conduction slows, creating regions that favor re-entry arrhythmias (Peters et al., 1997; Saffitz, 2006; Saffitz et al., 2000). Secretome-based interventions counter this cascade by delivering a concerted mix of growth factors, cytokines, chaperone proteins, and regulatory microRNAs that activate canonical survival pathways and stabilize cell-to-cell coupling.

Among soluble mediators, IGF-1, combined with HGF, engages the PI3K-Akt and c-Met axes, curbs caspase activation, and preserves sarcolemmal integrity. Sequential release from affinity-binding alginate or ureido-pyrimidinone hydrogels has expanded Ki-67 and GATA-4 positive cardiomyocyte clusters and improved left-ventricular ejection fraction in rodent and porcine infarction models (Koudstaal et al., 2014; Sonnenberg et al., 2015). VEGF not only promotes angiogenesis but also limits apoptosis through Akt and Bcl-2 signaling, and epicardial delivery on poly-L-lactic-acid scaffolds has reduced TUNEL staining in peri-infarct myocardium, thereby maintaining conduction velocity (Chung et al., 2015). FGF-1, especially when combined with the Wnt activator small-molecule CHIR99021, boosts cardiomyocyte cell-cycle entry sixfold and enhances wall thickness and contractility, effects reproduced in human iPSC-derived cardiomyocyte patches (Fan et al., 2020; Formiga et al., 2014). SDF-1α signals through CXCR4, STAT3, and Akt to recruit endogenous progenitors; collagen-binding or nanoparticle carriers' lower apoptosis and help preserve gap-junction integrity (Prokoph et al., 2012; Elmadbouh et al., 2007; Sun et al., 2016). Interleukin-10 provides a dual anti-inflammatory and antiapoptotic signal, and sustained release from supramolecular hydrogels has polarized macrophages toward an M2 phenotype while reducing scar formation (Krishnamurthy et al., 2009; Wang et al., 2023).

Secretome proteins with direct chaperone activity reinforce these cytokine effects. For example, HSP27 fused to a TAT peptide restores cardiomyoblast proliferation under hypoxic conditions. Additionally, antibody-targeted HSP27 nanoparticles have been shown to diminish fibrosis and dysfunction after ischemia-reperfusion (Kim et al., 2020; Lee et al., 2009). Cardiac-specific expression of HSP20 attenuates apoptosis in diabetic hearts, and exosomes rich in HSP20 carry phosphorylated Akt, survivin, and superoxide dismutase 1 as additional survival cargo (Wang et al., 2016).

Exosomes enriched in miR-21 downregulate PDCD4 and amplify PI3K-Akt signaling, shielding both cardiomyocytes and resident progenitors from oxidative stress (Xiao et al., 2016). Vesicles carrying miR-30d-5p target integrin α5 and MAP4K4, dampening TGF-β/Smad-driven apoptosis and fibrosis (Long & Cheng, 2024) miR-126 decreases reactive oxygen species, stabilizes mitochondrial potential, and promotes microvascular growth, while combined delivery of miR-21 and miR-146a further reduces cleaved caspase-3 and strengthens ventricular function (Luo et al. 2018; Huang et al., 2016). Because these survival cues are swiftly diluted by coronary flow and continuous myocardial motion, specialized delivery platforms are essential. Injectable hydrogels that remain liquid at room temperature and solidify inside the heart in response to body temperature or local enzymatic activity can retain proteins and vesicles for weeks; conductive nanofiber-collagen patches release bioactive factors for over thirty-five days; and biodegradable polymeric nanoparticles provide staggered dosing that matches the evolving metabolic demands of the healing myocardium (Luo et al., 2018; Kerignard et al., 2022). Collectively, growth factors, chaperones, and microRNAs converge on Akt, ERK, STAT3, and mitochondrial targets to limit cell death, maintain connexin organization, and preserve the excitable substrate, thereby preparing the ground for the structural and vascular repair mechanisms discussed in the subsequent sections.

4.4.3 Regeneration of Gap Junction

Gap-junction channels are composed of connexins and provide low-resistance pathways that synchronize depolarization across cardiac muscle. In working atrial and ventricular myocytes, Cx43 is predominant, while Cx40 is enriched in Purkinje fibers, and Cx45 dominates in the atrioventricular node and His bundle (Severs et al., 2008). To fully integrate the unit, six connexin subunits assemble into a connexon; the docking of two connexons from adjacent cells completes a pore that allows the transference of ions and signaling metabolites smaller than one kilodalton (Ma et al., 2017). After myocardial infarction, Cx43 protein falls by roughly half, plaques shrink, and the remaining channels migrate from intercalated discs to lateral membranes, a combination that slows longitudinal conduction and facilitates re-entry arrhythmias (Camman et al., 2022; Duffy, 2012). Pannexins, a related channel family that carries ATP and calcium, are also disrupted, further destabilizing impulse propagation (Sosinsky et al., 2011). Restoring gap-junction integrity, therefore, ranks with fibrosis reduction and cell survival as a central goal of secretome therapy.

Secretome-delivered growth factors engage signaling cascades that upregulate connexin transcription and stabilize plaque architecture. Insulin-like growth factor 1 and hepatocyte growth factor activate PI3K, driving Akt phosphorylation that enhances β-catenin binding to the Cx43 promoter and maintains expression; excessive Akt produces pathological hypertrophy, whereas Akt deficiency activates JNK and interleukin-22, accelerating connexin degradation, so precise dosing is critical (Matsui et al., 2002; Liang et al., 2019). Downstream kinases ERK1/2, p38, and GSK3β modulate plaque stability, and secretome fractions rich in MAPK-activating peptides reinforce this balance (Choukroun et al., 1999; Haq et al., 2000). β-Adrenergic agonists present in cardiomyocyte vesicles stimulate PKA and ERK, phosphorylating the cAMP response element-binding protein, which recruits NF-κB, AP-1, and SP-1 to the connexin promoter and enlarges junctional plaques (Geimonen et al., 1996; Gupta et al., 2019). Electrical remodeling is also tempered by heat-shock proteins, such as HSP20 and HSP27, which chaperone newly synthesized Cx43 and protect it from calpain- and caspase-mediated cleavage under oxidative stress (Kim et al., 2020; Wang et al., 2016).

Notch ligands packaged in endothelial or epicardial vesicles bind Notch receptors on the border-zone cardiomyocytes, liberating an intracellular domain that induces Cntn2, Nkx2-5, and Tbx5, all of which drive conduction lineage maturation and boost Cx40 and Cx45 without suppressing Cx43 (Gude et al., 2008; Rentschler et al., 2012; Torregrosa-Carrión et al., 2019). Calcium influx generated by vesicular ATP and purinergic signaling activates the calcineurin-NFAT axis; dephosphorylated NFATc1–c4 translocate to the nucleus and raise transcription of Cx43, Cx40, Cx45, Plakophilin-2, Desmoglein-2, ZO-1, and N-cadherin, thereby coupling mechanical and electrical junctions (Crabtree & Olson, 2002; Karkache et al., 2021). Secretome components also attenuate connexin lateralization: tissue inhibitor of metalloproteinase-3 and miR-92a, contained in MSC-derived exosomes, prevent excessive ECM degradation at intercalated discs, keeping plaques anchored to the cytoskeleton (Totland et al., 2020).

Vesicles enriched in miR-21 and miR-146a suppress PTEN, amplifying Akt signaling and improving Cx43 density in peri-infarct cardiomyocytes (Huang et al., 2016). miR-30d-5p targets integrin α5 and MAP4K4, limiting TGF-β/Smad activity that otherwise down-regulates connexin expression during fibroblast-to-myofibroblast transition (Long & Cheng, 2024). miR-1-loaded cardiac stem-cell exosomes, when delivered early after infarction, normalize Cx43 phosphorylation status and shorten QRS duration in rat models, demonstrating that miRNA cargo can affect conduction at the whole-heart level.

Sustained delivery enhances these molecular effects. Conductive graphene-oxide silk patches seeded with UC-MSC-derived secretome concentrate Cx43 at intercalated discs raise conduction velocity by 20% within four weeks in rat infarction. Injectable hydrogels that co-release miR-30d-5p vesicles and Akt stimulators prevent Cx43 lateralization and normalize anisotropic conduction in mice. Poly(lactic-co-glycolic acid) microspheres, designed for the two-stage liberation of Notch peptides and β-adrenergic agonists, restore Purkinje fiber Cx40 and shorten QRS duration in porcine models. By reinstating connexin abundance, correcting plaque localization, and balancing isoform ratios, secretome-based strategies rebuild the cellular syncytium that underlies coordinated impulse propagation, setting the stage for the metabolic and vascular support provided by neovascularization (Long & Cheng, 2024).

4.4.4 Neovascularization

Cardiac neovascularization is essential for maintaining oxygen delivery, nutrient exchange, and waste clearance in the infarcted myocardium, supporting the metabolic demands of uninterrupted impulse propagation (Ferrara et al., 2003; Zhao et al., 2022). Ischemia stabilizes hypoxia-inducible factor 1α in ECs, upregulates VEGF, loosens pericyte coverage, and initiates basement-membrane degradation; tip cells extend filopodia, stalk cells proliferate to lengthen the sprout, and fusion of opposing tips forms perfused lumens that mature through signaling by PDGF, transforming growth factor β1, and angiopoietins (Akbarian et al., 2022; Deveza et al., 2012; Ferrara et al., 2003). Secretome preparations amplify each phase of this sequence: exosomes derived from EPC or MSCs deliver messenger RNA for VEGF, FGF-2, and angiopoietin 1, increasing capillary density by approximately 40% in rodent infarction models and doubling microvascular flow, while vesicles enriched in HSP 20 activate PI3K–Akt in ECs and enhance tube formation under hypoxia (Arslan et al., 2013; Omar et al., 2022; Liang et al., 2024; Moslem et al., 2025; Ottaviani et al., 2016).

Adequate vessel maturation depends on a carefully balanced interplay between pro- and anti-angiogenic signals. Secretome fractions that combine VEGF with PDGF in a defined ratio recruit pericytes and smooth-muscle cells, generating microvessels with thicker abluminal coverage and fewer microaneurysms than those produced with VEGF alone (Haider & Aslam, 2018). In contrast, excessive PDGF skews the response toward fibrosis (Liu ZL et al., 2023; Richardson et al., 2001). MicroRNAs within the vesicular cargo refine this balance: exosomal miR-126 targets SPRED1 and the PI3K regulatory subunit 2 to heighten endothelial sensitivity to VEGF, miR-92a inhibition increases integrin α5 and stabilizes nascent tubes, miR-222 accelerates tip-cell selection, and miR-150 from monocytes promotes sprout elongation; by contrast, miR-21 can be either deleterious or beneficial depending on its pairing with miR-146a and the cellular origin of the vesicle (Ding et al., 2017; Kao & Papoutsakis, 2019; Li et al., 2021; Ottaviani et al., 2016; Radmanesh et al., 2021; Van Rooij & Olson, 2012).

Low-intensity electrical stimulation polarizes ECs, enhances MAPK–ERK signaling, and elevates endogenous VEGF and interleukin 8 (Bai et al., 2011; Geng et al., 2019). Conductive biomaterials harness this phenomenon: reduced-graphene-oxide silk fibroin patches loaded with UC-MSC-derived secretome increase capillary density, restore conduction velocity, and shorten QRS duration in rats, whereas injectable polyaniline-coated hyaluronic acid hydrogels release VEGF and PDGF over two weeks through embedded conductive networks and improve left-ventricular ejection fraction by 15% compared with nonconductive controls (He et al., 2023; Li et al., 2017; Zhao et al., 2022)

Thermoresponsive hydrogels that gel in situ trap vesicles and proteins within the infarct border zone and extend VEGF half-life from minutes to days. Porous three-dimensional printed scaffolds with graded pore sizes permit endothelial ingrowth while releasing exosomal cargo in a spatially controlled manner, resulting in homogeneous perfusion and a reduced arrhythmogenic substrate (Gil-Cabrerizo et al., 2023; He et al., 2023). Microsphere systems engineered for pulse-release kinetics mimic the natural sequence of VEGF, followed by PDGF and angiopoietin 1, and generate larger, more branched arteries in porcine infarction studies (Liang et al., 2024; Richardson et al., 2001).

4.5 INNOVATIVE DELIVERY METHODS FOR SECRETOME TO REPAIR HEART/CARDIAC INJURIES

Secretome-based therapies have emerged as an auspicious cell-free approach for cardiac repair, addressing many limitations associated with direct cell transplantation. Myocardial regeneration and repair are stimulated by these therapies, which use a complex mixture of bioactive molecules, such as cytokines, growth factors, proteins, lipids, and EVs (Villa et al., 2021). Despite their significant potential, several critical challenges must be overcome to optimize their clinical efficacy.

One major challenge is the inherent instability and rapid clearance of secretome components in the dynamic, hostile microenvironment of the injured heart. Cardiac injury creates conditions characterized by intense inflammation, oxidative stress, and elevated proteolytic enzymatic activity, all of which speed up the degradation of therapeutic molecules within the secretome (Tang et al., 2021; Villa et al., 2021). Consequently, these factors drastically reduce the bioactivity and functional lifespan of secretome components, limiting their regenerative potential and complicating accurate dosing. The heterogeneous nature of the secretome composition presents significant variability in biological activity, which further complicates reproducibility and clinical standardization (Villa et al., 2021). Differences in cell sources, culture conditions, isolation protocols, and storage methods contribute significantly to this variability. Addressing these inconsistencies is crucial for ensuring predictable therapeutic outcomes, reliable dosing, and effective clinical translation.

A substantial advantage of secretome-based therapies over traditional cell-based methods is their lower risk of immunogenicity and tumorigenicity, attributed to their acellular nature and lack of replicative potential (Sandonà et al., 2021; Liang et al., 2025). Secretome treatments circumvent significant limitations in cell survival, engraftment, and potential malignant transformation associated with cell transplantation (Siraj et al., 2023). Furthermore, secretome formulations are easier to store and standardize, enabling off-the-shelf availability and facilitating clinical use. However, their clinical success depends heavily on the chosen method of administration (Jihwaprani et al., 2024), as the therapeutic efficacy of secretome formulations relies on their retention, stability, and sustained release within myocardial tissue (Tang et al., 2021). When directly injected into the myocardium, secretome components are prone to rapid dispersion due to dynamic interstitial fluid flow and the constant mechanical contraction of the heart, resulting in insufficient local retention and suboptimal therapeutic efficacy (Balbi & Vassalli, 2020; Kompa et al., 2025; Sípos & Műzes, 2022). Therefore, enhancing secretome retention has become a critical design consideration for the development of effective delivery platforms. Without strategies to maintain a localized reservoir of bioactive molecules, the full therapeutic potential of secretome-based interventions may remain unrealized (Tang et al., 2021; Villa et al., 2021).

Given these challenges, the development of advanced delivery methods that protect, stabilize, and extend the bioactivity of the secretome is essential. Innovative delivery systems must be engineered to mitigate rapid clearance, control release kinetics, and shield therapeutic molecules from degradation in the cardiac microenvironment. Thus, refining secretome delivery methods represents a pivotal step toward the successful translation and clinical application of secretome-based therapies for cardiac injury repair.

4.5.1 Direct Injection

Direct injection remains the most straightforward and appealing method for delivering therapeutic agents into cardiac tissue, primarily because of its minimally invasive nature and ease of implementation (Prendiville et al., 2014; Kalou et al., 2021). Using fine needles under imaging guidance, secretome formulations can be precisely administered into myocardial regions quickly and efficiently, making this technique especially valuable in clinical scenarios requiring immediate intervention (Lederman et al., 2002; Prendiville et al., 2014). A significant advantage of this method is the immediate availability of therapeutic molecules at the injury site, enabling rapid initiation of the repair process.

Despite its procedural simplicity and cost-effectiveness, direct injection suffers from critical limitations. Primarily, the absence of a protective carrier leaves the secretome vulnerable to rapid degradation and dissemination from the targeted area, significantly reducing its therapeutic effectiveness over time (Arifka et al., 2022; Zhang et al., 2015). Due to dynamic interstitial fluid flow and continuous myocardial contractions, injected secretome components tend to disperse rapidly, resulting in insufficient retention and suboptimal therapeutic concentrations at the injury site. Direct injection often leads to an uneven distribution of secretome within myocardial tissue (Waters et al., 2018). The localized concentration of therapeutic agents may vary, with some regions potentially receiving subtherapeutic levels, while others become overloaded, which may lead to adverse effects (Hashimoto et al., 2018). Mechanical forces generated by cardiac contractions further exacerbate this uneven distribution, reducing overall efficacy. Consequently, the lack of effective localization and sustained delivery represents a significant challenge for direct injection methods.

Another critical concern associated with direct injection is the potential for physical trauma and damage to myocardial tissue, particularly with repeated administration. Needle insertions into the myocardium can lead to local hemorrhage, tissue necrosis, inflammation, and exacerbation of the original cardiac injury, ultimately undermining therapeutic outcomes (Lv et al., 2019; Tang et al., 2021; Waters et al., 2018). Given these risks, clinicians must carefully weigh the benefits of procedural simplicity and immediacy against the potential for tissue injury.

While direct injection offers technical ease and immediate delivery, its limited retention and uneven distribution significantly reduce its long-term therapeutic efficiency. Rapid dispersion and inadequate retention diminish overall regenerative responses, creating unpredictable biological effects that can complicate clinical outcomes (Lv et al., 2019). These limitations underscore the need for alternative methodologies and innovative modifications to enhance safety, efficacy, and patient adherence to therapy. Advanced delivery strategies capable of providing sustained, localized, and uniform secretome concentrations are thus imperative for optimizing clinical outcomes in cardiac repair.

4.5.2 Hydrogels

Hydrogel-based delivery systems have emerged as promising platforms to overcome many of the limitations associated with direct injection methods for secretome administration in cardiac repair (Sedighim et al., 2022; Zhou J et al., 2021). Hydrogels are three-dimensional, highly hydrated polymeric networks that can encapsulate therapeutic secretome components and release them in a controlled, sustained manner. This controlled delivery significantly enhances local therapeutic concentrations and reduces rapid dispersion compared to traditional direct injection approaches (Umar, 2022). One of the principal benefits of hydrogel delivery systems is their exceptional biocompatibility, making them particularly suitable for cardiac tissue repair applications. Designed to mimic the native ECM closely, hydrogels promote cell adhesion, proliferation, and successful integration into host tissues (Peng et al., 2023; Zhou Q et al., 2021). The high water content and porous architecture of hydrogels facilitate the controlled diffusion of encapsulated biomolecules, thereby maintaining sustained bioavailability of secretome components at the injury site.

Advanced hydrogel formulations can also exhibit stimulus-responsive properties, adjusting their physical or chemical states in response to specific local stimuli, such as changes in pH, temperature, or enzymatic activity. For example, smart hydrogels engineered to undergo sol-to-gel transitions upon injection form stable depots at the administration site, ensuring robust initial retention of the secretome and enabling dynamic modulation of its release aligned with the evolving conditions of cardiac tissue repair (Eil Bakhtiari et al., 2021; Sönmezer & Latifoglu, 2023). Hydrogels demonstrate a remarkable capacity to protect encapsulated secretome components from the harsh enzymatic and oxidative conditions characteristic of injured myocardial tissue, thereby substantially prolonging their bioactivity and effectiveness. By encapsulating secretome within polymeric matrices, hydrogels effectively shield therapeutic molecules, enhancing their stability and duration of action in the infarcted myocardium (Umar, 2022). Preclinical studies using hydrogel systems have consistently demonstrated significant improvements in cardiac function parameters, including increased LVEF, highlighting their efficacy in promoting cardiac regeneration (Arifka et al., 2022; Lv et al., 2019; Paul et al., 2017). Although hydrogels have promising attributes, they also present several challenges that need careful consideration. Precise control of hydrogel degradation rates is crucial; misalignment with the cardiac repair timeline can lead to either premature secretome release or prolonged hydrogel persistence, which can impede natural tissue regeneration (Hasan et al., 2015; Wang L et al., 2018). Additionally, optimizing the hydrogel crosslinking density is critical; excessive crosslinking may limit secretome diffusion, whereas insufficient crosslinking may lead to rapid, uncontrolled release, mimicking the limitations of direct injection (Lee et al., 2017; Li et al., 2023). Balancing the stiffness of hydrogels to match the mechanical environment of the contracting heart is another essential consideration, as incompatible mechanical properties can adversely affect the kinetics of secretome release and cardiac function (Jeffords et al., 2015).

Hydrogels specifically engineered for cardiac applications frequently emulate the myocardial ECM closely, enhancing tissue integration and regeneration (Gholamali et al., 2024; Ho et al., 2022). Structural and biochemical mimicry of the native extracellular matrix (ECM) not only provides reservoirs for secretome components but also actively modulates the local cardiac microenvironment to enhance regenerative responses. Incorporating bioactive motifs or cellular adhesion ligands further enhances the functionality of hydrogels, thereby maximizing therapeutic efficacy in cardiac repair contexts (Koh et al., 2022; Waters et al., 2018). Recent studies underscore the promise of alginate-based hydrogels in cardiac applications, demonstrating notable improvements in cardiac outcomes. Alginate hydrogels, often supplemented with other biocompatible materials such as melanin, have consistently shown improved cardiac function metrics, including enhanced LVEF. The sustained local presence of secretome, enabled by alginate hydrogels, contributes significantly to the modulation of inflammation and subsequent regeneration processes (Zhu et al., 2021). Its reduced immunogenicity, a result of alginate's inherent natural properties, makes it especially appealing for cardiac repair therapies. Thus, alginate-based hydrogels represent a highly promising strategy for overcoming the rapid clearance and degradation issues historically associated with direct injection approaches, thereby significantly advancing secretome-based cardiac therapy (Kwon et al., 2023; Lv et al., 2019).

4.5.3 Scaffold-Based Delivery

Scaffold-based delivery methods represent an innovative advancement in secretome therapy, leveraging three-dimensional porous structures to provide mechanical support and controlled therapeutic release simultaneously (Mehrotra et al., 2023; Yıldırım et al., 2024). Scaffolds fabricated using advanced technologies such as 3D printing closely replicate the native cardiac tissue architecture, facilitating enhanced integration and uniform distribution of therapeutic agents at the myocardial injury site (Chingale et al., 2021; Tortajada et al., 2021).

The defined spatial architecture of scaffold-based systems offers distinct advantages, such as precise control over pore sizes and interconnectivity, essential for effective nutrient diffusion and sustained secretome release (Kaiser & Coulombe, 2015). By entrapping secretome within a structured physical framework, scaffolds ensure prolonged availability of therapeutic molecules, promoting a homogeneous and sustained reparative response across the infarcted myocardium (Tortajada et al., 2021; Waters et al., 2018). Additionally, scaffolds degrade harmoniously with natural tissue regeneration, providing temporary mechanical support and minimizing adverse remodeling and excessive scar formation.

A key strength of scaffold-based delivery systems is their capability to integrate seamlessly with engineered cardiac tissues, thereby enhancing cellular viability and overall functional recovery (Sonnenberg et al., 2015). Alignment of scaffold mechanical properties with those of native myocardium is crucial for accommodating the heart's dynamic mechanical environment while ensuring prolonged regenerative capacity (Perea-Gil et al., 2018). However, the complexities of scaffold fabrication and surgical implantation pose significant challenges, especially regarding biodegradability, mechanical compatibility, and immune responses. Despite these advantages, scaffold-based methods have inherent drawbacks, including the invasiveness of surgical implantation and potential complications, such as infection, inflammation, and mechanical injury. Moreover, achieving precise synchronization between the timelines of scaffold degradation and myocardial regeneration remains a significant challenge in bioengineering. Consequently, careful scaffold design, rigorous preclinical testing, and optimization are crucial to fully realize their therapeutic potential (Chingale et al., 2021; Mehrotra et al., 2023; Waters et al., 2018).

4.6 CURRENT CHALLENGES AND FUTURE DIRECTIONS

The use of secretome-based therapies for cardiac repair presents a promising opportunity to harness the paracrine effects of a complex mixture of bioactive molecules, thereby avoiding the risks associated with direct cell transplantation. However, several critical issues hinder the practical clinical application of secretomes for the repair of heart injury (Habib et al., 2025b). Three significant challenges include standardization, targeted delivery, and the longevity of therapeutic effects.

One primary issue is the lack of standardization in secretome production. The secretome comprises a myriad of components—including proteins, cytokines, growth factors, lipids, and EVs—whose composition largely depends on the source cells, culture conditions, preconditioning regimens, and isolation (Arrell et al., 2021; Harman et al., 2021). This inherent heterogeneity results in batch-to-batch variability, making it challenging to define a reproducible therapeutic product. Consequently, the absence of well-established quality control assays and robust regulatory guidelines creates uncertainty about dosing, safety, and therapeutic efficacy (Arrell et al., 2021; Garmany et al., 2024; Kukumberg et al., 2021). Addressing these concerns through the development of standardized protocols and scalable manufacturing procedures is identified as a necessary future direction.

Another significant challenge lies in achieving effective targeted delivery. Once administered, secretome components may undergo rapid systemic clearance and fail to localize efficiently within the injured myocardium (Arifka et al., 2022; Nawaz et al., 2023). This challenge is compounded by the dynamic and often hostile post-infarct environment. To overcome these obstacles, advanced drug delivery systems—such as encapsulation within polymeric hydrogels or incorporation into acellular scaffolds—are being explored to protect bioactive molecules, improve retention at the target site, and facilitate controlled release (Aquinas Liguori et al., 2021; Arifka et al., 2022; Venditti et al., 2020). Bioengineering approaches that leverage lipid nanoparticle systems have been proposed to enhance the targeted delivery and bioavailability of therapeutic factors (Ong et al., 2023). Future research will likely focus on optimizing these delivery vehicles to ensure that sufficient and sustained concentrations of active molecules reach the damaged cardiac tissue.

The final challenge is the limited longevity of secretome effects after administration. Many biochemical signals delivered through secretome-based therapies have a transient half-life in vivo, which may not translate to long-term, durable functional improvements in the heart (Aquinas Liguori et al., 2021; Arifka et al., 2022; Kukumberg et al., 2021; Saghebasl et al., 2022). This issue causes repeated administration or the use of sophisticated sustained-release strategies. In this context, controlled-release bioformulations that can prolong the myocardium's exposure to therapeutic factors have been investigated, alongside preconditioning techniques aimed at enhancing the stability and bioactivity of secretome components (Mancuso et al., 2020; Nawaz et al., 2023). Modulation of the secretome's composition through cellular preconditioning or bioengineering to enhance cardioprotective signaling offers a promising strategy to address these temporal limitations (Arrell et al., 2021; Garmany et al., 2024).

4.7 CLINICAL TRANSLATION

Over the past several years, the concept of using the secretome—the collection of molecules that cells release naturally—has attracted significant interest for aiding heart recovery after injury. This cell-free approach holds real promise for transitioning from the lab to more practical, clinical settings. Table 4.2 summarizes representative preclinical studies exploring the therapeutic application of the secretome in cardiac and related tissue injury models, including details on cell sources, delivery methods, and observed outcomes. This table provides a snapshot of the growing body of work in this area, highlighting different strategies researchers have used to apply the secretome in models of heart damage.

One of the most common approaches involves using MSC secretome, particularly in the form of exosomes, which are small particles that convey beneficial signals between cells. Studies such as those by Shao et al. (2017) and Qu et al. (2020) demonstrate that MSC-derived exosomes can reduce scarring in the heart and facilitate the regrowth of blood vessels after injury. These results suggest that MSC secretome can help create a healing environment by calming inflammation and encouraging tissue repair. Secretome from CPCs has also shown great potential. For example, Gray et al. (2015) found that exosomes from CPCs preconditioned under low-oxygen conditions facilitate neoangiogenesis after a heart attack. Similarly, VEGF-A mRNA-loaded delivery systems have been utilized to stimulate blood vessel growth, even in hearts without apparent damage, demonstrating that secretome therapy facilitates the prevention and recovery from heart disease (Nawaz et al., 2023).

Most of the studies in Table 4.2 use direct heart injection as the delivery method. It is simple, targeted, and effective. But some teams are thinking beyond the needle (Tang et al., 2017). For instance, researchers have developed synthetic microparticles that mimic cells and slowly release the secretome where it is needed (Kawamura et al., 2015). The search for a novel route is becoming not just a fascinating research topic but a real possibility for treating heart disease in the future.

TABLE 4.2
Experimental Applications of Secretome Therapy in Cardiac and Non-cardiac Injury Models

Approach	Injury Model	Secretome Source	Secretome Fraction	Delivery Method	Relevant Outcomes	Ref.
MSC-derived exosomes reduce post-infarction cardiac fibrosis.	Rat MI Cellular model: H9c2 cardiomyoblasts under H_2O_2-induced oxidative stress.	MSCs; H9C2 cardiomyoblasts	Soluble fraction	Direct injection	MSC-Exo promoted cardiac repair by inhibiting fibrosis and inflammation.	Shao et al., 2017
Exosomal miRNAs from hypoxia-treated CPCs promoted endothelial tube formation, thereby enhancing neovascularization.	Mouse I/R injury model.	CPCs	Both soluble and insoluble fractions	Exosomes injection after I/R injury.	The secretome improved cardiac function and reduced fibrosis after injury.	Gray et al., 2015
VEGF-A mRNA delivery via lipid nanoparticles and EVs to stimulate angiogenesis in cardiac tissue.	Mouse model, direct myocardial injection of VEGF-A mRNA into a healthy heart.	CPCs	Both soluble and insoluble fractions	Direct injection. Other (Lipid nanoparticles)	CPC-secretome enhanced angiogenesis and minimized inflammation in cardiac repair.	Nawaz et al., 2023
MSC-derived exosomes in murine femur fracture healing, which involved accelerated callus formation.	Mouse (CD9−/− strain) and femur fracture.	hBM-MSCs	Both soluble and insoluble fractions	Direct injection	Exosomes significantly enhanced tissue repair processes.	Furuta et al., 2016
Neovascularization by showing that EPC-derived exosomes accelerate angiogenesis via Erk1/2 signaling.	Rat model with streptozotocin-induced diabetic cutaneous wound.	hUCB-EPCs	Soluble fraction	Direct injection.	Enhanced angiogenesis and wound healing via paracrine effects.	Zhang J et al., 2016
MSC-derived exosomes accelerated reendothelialization by enhancing ECs' proliferation and migration.	Rat and vein graft intimal hyperplasia.	hUC-MSCs	Both soluble and insoluble fractions	Direct injection	hUC-MSC exosomes promoted EC proliferation and migration, facilitating cardiac tissue repair.	Qu et al., 2020
Secretome-based therapies using synthetic cell-mimetic microparticles effectively enhanced cardiac repair by preserving viable myocardium and improving cardiac function.	Mouse MI model.	hCSCs	Both soluble and insoluble fractions	Direct injection	Synthetic microparticles enhance myocardial recovery and improve heart function by modulating paracrine signaling and membrane interactions.	Tang et al., 2017

Approach	Injury Model	Secretome Source	Secretome Fraction	Delivery Method	Relevant Outcomes	Ref.
MSC exosomes mediate paracrine trophic effects, enabling osteochondral regeneration, and show the potential of secretome therapy.	Mouse myocardial I/R. Porcine MI model. Cellular model: H9c2 cardiomyoblasts and hypoxia/ reoxygenation.	Human embryonic MSCs	Both soluble and insoluble fractions	Direct injection	The secretome promoted cardiac tissue repair through paracrine factors.	Zhang S et al., 2016
UCB-derived paracrine factors, including exosomes, mediate tissue repair through endogenous anti-inflammatory and regenerative responses, highlighting the potential use of the secretome for cardiac repair.	Neonatal rat and hypoxic-ischaemic brain injury.	hUC-MSCs	Both soluble and insoluble fractions	Direct injection	UCB-derived therapies enhanced cardiac repair through paracrine mechanisms, improving heart function after injury.	Nakanishi et al., 2017
Promotion of cardiomyocyte survival by enhancing transplanted MSC viability through rapamycin-induced autophagy.	Rat I/R injury model. Cellular model: rat BM-MSCs under hypoxia with serum deprivation.	Rat BM- MSCs	Soluble and insoluble fractions	Direct injection	Enhanced secretion of reparative factors boosts myocardial regeneration.	Li et al., 2020
MSCs injection to achieve a reduction of fibrosis.	New Zealand rabbit and bladder tuberculosis.	Did not specify the cell types used to derive secretomes	Did not adequately address the specific secretome fractions	Direct injection	Secretome enhances cardiac healing through paracrine signaling.	Yudintceva et al., 2018
Cardiomyocyte survival via enhanced MSC-to-cardiomyocyte differentiation with 5-AZA and low-dose FGF/IGF.	Animal model: rat with no induced cardiac injury.	hBM-MSCs and AD-MSCs	Both soluble and insoluble fractions	Delivery method not specified	Secretome enhances cardiomyocyte differentiation and supports tissue regeneration.	(Farag et al., 2024)
MSC sheet transplantation promoted neovascularization by enhancing vascular density in the infarct border zone.	Porcine ischemic cardiomyopathy using ameroid constrictor-mediated LAD coronary artery occlusion.	hBM-MSCs	Both soluble and insoluble fractions	Scaffold-based delivery	The secretome from BM-MSCs enhances cardiac function and reduces LV remodeling via paracrine signaling.	Kawamura et al., 2015

(Continued)

TABLE 4.2 CONTINUED
Experimental Applications of Secretome Therapy in Cardiac and Non-cardiac Injury Models

Approach	Injury Model	Secretome Source	Secretome Fraction	Delivery Method	Relevant Outcomes	Ref.
Promotion of cardiomyocyte survival.	Human iPSC-derived CPCs (no induced injury condition).	Human iPSCs-CPCs, Naïve mouse heart CPCs. Human iPSC-cardiomyocytes	Both soluble and insoluble fractions	The delivery method is not explicitly specified	FGF16 enhances CPC proliferation, supporting effective cardiac repair and regeneration.	Jennbacken et al., 2019

Abbreviations: CPCs: cardiac progenitor cells; hBM-MSC: Human bone marrow-derived MSC; hUCB-EPCs: human umbilical cord blood-derived endothelial progenitor cells; hUC-MSCs: Human umbilical cord-derived MSC; CSCs: human cardiac stem cells; UCB: umbilical cord blood; AD-MSCs: Adipose tissue-derived MSC.

4.8 CONCLUSION

Secretome-based therapy represents a transformative approach to cardiac repair, offering a potent alternative to traditional cell-based interventions (Habib et al., 2025b). By harnessing the paracrine effects of bioactive molecules, including cytokines, growth factors, microRNAs, and EVs, secretome therapies promote cardiomyocyte survival, reduce fibrosis, restore gap junction function, and stimulate neovascularization, all of which are essential for restoring electrical conductivity in the injured myocardium.

This cell-free strategy circumvents many of the limitations of live-cell transplantation, such as immune rejection, tumorigenicity, and poor engraftment. Moreover, it leverages the regenerative and immunomodulatory potential of the secretome while enabling greater control over dosage and delivery. Innovative biomaterials, including hydrogels, nanoparticles, and scaffold-based systems, have significantly enhanced the stability, retention, and targeted release of secretome components. These delivery platforms not only protect bioactive cargo from the harsh cardiac microenvironment but also allow for sustained therapeutic action aligned with the tissue repair timeline.

Despite the promising outcomes observed in preclinical models, several challenges remain before secretome-based therapies can be widely translated to clinical practice (Habib et al., 2025b). Standardization of secretome production, improved targeting, and prolonged therapeutic effects are critical hurdles that must be addressed to achieve optimal outcomes. The development of scalable, reproducible, and clinically compliant manufacturing and delivery systems will be essential to unlock the therapeutic potential of the secretome fully. As research continues to elucidate the mechanisms of action and optimize delivery strategies, secretome-based therapies stand poised to become a cornerstone in the next generation of regenerative medicine for heart disease, offering new hope for restoring cardiac function and electrical integrity following myocardial injury.

REFERENCES

Ahmed ZT, Alkahlot MH, Haider KH. MSC-derived exosomes: advances in cell-free therapy. In: KH Haider, editor. *Handbook of stem cell applications*. Singapore: Springer; 2023. https://doi.org/10.1007/978-981-99-0846-2_60-1

Ai X, Yan B, Witman N, Gong Y, Yang L, Tan Y, et al. Transient secretion of VEGF protein from transplanted hiPSC-CMs enhances engraftment and improves rat heart function post-MI. *Molecular Therapy*. 2023;31(1):211–29.

Akbarian M, Bertassoni LE, Tayebi L. Biological aspects in controlling angiogenesis: current progress. *Cell Mol Life Sci*. 2022;79(7):1–35.

Al-Omar MT, Alnajjar MT, Ahmed ZT, Salaas FMI, Alrefaei TSM, Haider KH. Endothelial progenitor cell-derived small extracellular vesicles for myocardial angiogenesis and revascularization. *J Clin Transl Res*. 2022;8(6):1–12.

Aquinas Liguori TT, Liguori GR, van Dongen JA, Pinho Moreira LF, Harmsen MC. Bioactive decellularized cardiac extracellular matrix-based hydrogel as a sustained-release platform for human adipose tissue-derived stromal cell-secreted factors. *Biomed Mater*. 2021;16(2):025022.

Arifka M, Wilar G, Elamin KM, Wathoni N. Polymeric hydrogels as mesenchymal stem cell secretome delivery system in biomedical applications. *Polymers (Basel)*. 2022;14(6):1218.

Arrell DK, Crespo-Diaz RJ, Yamada S, Jeon R, Garmany A, Park S, et al. Secretome signature of cardiopoietic cells echoed in rescued infarcted heart proteome. *Stem Cells Transl Med*. 2021;10(9):1320–8.

Arslan F, Lai RC, Smeets MB, Akeroyd L, Choo A, Aguor ENE, et al. Mesenchymal stem cell-derived exosomes increase ATP levels, decrease oxidative stress, and activate the PI3K/Akt pathway to enhance myocardial viability and prevent adverse remodeling after myocardial ischemia/reperfusion injury. *Stem Cell Res*. 2013;10(3):301–12.

Bai H, Forrester J V., Zhao M. DC electric stimulation upregulates angiogenic factors in endothelial cells through activation of VEGF receptors. *Cytokine*. 2011;55(1):110–15.

Balbi C, Lodder K, Costa A, Moimas S, Moccia F, van Herwaarden T, et al. Reactivating endogenous mechanisms of cardiac regeneration via paracrine boosting using the human amniotic fluid stem cell secretome. *Int J Cardiol*. 2019 Jul;287:87–95.

Balbi C, Vassalli G. Exosomes: beyond stem cells for cardiac protection and repair. *Int J Cell Cloning.* 2020;1387–99.

Bani D, Nistri S, Formigli L, Meacci E, Francini F, Zecchi-Orlandini S. Prominent role of relaxin in improving postinfarction heart remodeling. *Ann N Y Acad Sci.* 2009;1160(1):269–77.

Blackwood EA, Thuerauf DJ, Stastna M, Stephens H, Sand Z, Pentoney A, et al. Proteomic analysis of the cardiac myocyte secretome reveals extracellular protective functions for the ER stress response. *J Mol Cell Cardiol.* 2020;143:132–44.

Blanco-Blázquez V, Báez-Díaz C, Sánchez-Margallo FM, González-Bueno I, Martín H, Blázquez R, et al. Intracoronary administration of microencapsulated HGF in a reperfused myocardial infarction swine model. *J Cardiovasc Dev Dis.* 2023;10(2):86.

Bollini S, Smits AM, Balbi C, Lazzarini E, Ameri P. Triggering endogenous cardiac repair and regeneration via extracellular vesicle-mediated communication. *Front Physiol.* 2018 Oct;9:1–14.

Busatto S, Zendrini A, Radeghieri A, Paolini L, Romano M, Presta M, et al. The nanostructured secretome. *Biomater Sci.* 2020;8(1):39–63.

Camman M, Joanne P, Agbulut O, Hélary C. 3D models of dilated cardiomyopathy: shaping the chemical, physical, and topographical properties of biomaterials to mimic the cardiac extracellular matrix. *Bioact Mater.* 2022;7:275–91.

Carta-Bergaz A, Ríos-Muñoz GR, Crisóstomo V, Sánchez-Margallo FM, Ledesma-Carbayo MJ, Bermejo-Thomas J, et al. Intrapericardial cardiosphere-derived cells hinder epicardial dense scar expansion and promote electrical homogeneity in a porcine post-infarction model. *Front Physiol.* 2022;13:1041348.

Chen H, Wang L, Zeng X, Schwarz H, Nanda HS, Peng X, et al. Exosomes, a new star for targeted delivery. *Front Cell Dev Biol.* 2021 Oct 8;9;751079.

Cheng YC, Hsieh ML, Lin CJ, Chang CMC, Huang CY, Puntney R, et al. Combined treatment of human induced pluripotent stem cell-derived cardiomyocytes and endothelial cells regenerates the infarcted heart in mice and non-human primates. *Circulation.* 2023;148(18):1395–409.

Chingale M, Zhu D, Cheng K, Huang K. Bioengineering technologies for cardiac regenerative medicine. *Front Bioeng Biotechnol.* 2021;9:1–11.

Chistiakov DA, Orekhov AN, Bobryshevy YV. Cardiac extracellular vesicles in normal and infarcted hearts. *Int J Mol Sci.* 2016;17(1):1–18.

Choukroun G, Hajjar R, Fry S, del Monte F, Haq S, Guerrero JL, et al. Regulation of cardiac hypertrophy in vivo by the stress-activated protein kinases/c-Jun NH2-terminal kinases. *J Clin Invest.* 1999;104(4):391–8.

Christoffels VM, Smits GJ, Kispert A, Moorman AFM. Development of the pacemaker tissues of the heart. *Circ Res.* 2010;106(2):240–54.

Chung HJ, Kim JT, Kim HJ, Kyung HW, Katila P, Lee JH, et al. Epicardial delivery of VEGF and cardiac stem cells, guided by a 3-dimensional PLLA matrix, enhances cardiac regeneration and angiogenesis in acute myocardial infarction. *J Control Release.* 2015;205:218–30.

Clayton ZE, Tan RP, Miravet MM, Lennartsson K, Cooke JP, Bursill CA, et al. Induced pluripotent stem cell-derived endothelial cells promote angiogenesis and accelerate wound closure in a murine excisional wound healing model. *Biosci Rep.* 2018;38(4):1–11.

Crabtree GR, Olson EN. NFAT signaling. *Cell.* 2002;109(2):S67–79.

Daneshmandi L, Shah S, Jafari T, Bhattacharjee M, Momah D, Saveh-Shemshaki N, et al. Emergence of the stem cell secretome in regenerative engineering. *Trends Biotechnol.* 2020;38(12):1373–84.

Davidson SM, Padró T, Bollini S, Vilahur G, Duncker DJ, Evans PC, et al. Progress in cardiac research: from rebooting cardiac regeneration to a complete cell atlas of the heart. *Cardiovasc Res.* 2021;117(10):2161–74.

Deveza L, Choi J, Yang F. Therapeutic angiogenesis for treating cardiovascular diseases. *Theranostics.* 2012;2(8):801–14.

Ding S, Huang H, Xu Y, Zhu H, Zhong C. MiR-222 in Cardiovascular diseases: physiology and pathology. *Biomed Res Int.* 2017;2017(1):4962426. https://doi.org/10.1155/2017/4962426

Du X, Xu Q, Lekgabe E, Gao X, Kiriazis H, Moore X, et al. Reversal of cardiac fibrosis and related dysfunction by relaxin. *Ann N Y Acad Sci.* 2009;1160(1):278–84.

Duffy HS. The molecular mechanisms of gap junction remodeling. *Heart Rhythm.* 2012;9(8):1331–4.

Eil Bakhtiari SS, Bakhsheshi-Rad HR, Karbasi S, Razzaghi M, Tavakoli M, Ismail AF, et al. 3-Dimensional printing of hydrogel-based nanocomposites: a comprehensive review on the technology description, properties, and applications. *Adv Eng Mater.* 2021;1–21.

Elmadbouh I, Khawaja Husnain Haider, Shujia J, Niagara MI, Gang Lu, Ashraf M. Ex vivo delivered stromal cell-derived factor-1α promotes stem cell homing and induces angiomyogenesis in the infarcted myocardium. *J Mol Cell Cardiol.* 2007;42:792–803.

Fan C, Oduk Y, Zhao M, Lou X, Tang Y, Pretorius D, et al. Myocardial protection by nanomaterials formulated with CHIR99021 and FGF1. *JCI Insight.* 2020;5(12):1–19.

Farag A, Koung Ngeun S, Kaneda M, Aboubakr M, Tanaka R. Optimizing cardiomyocyte differentiation: comparative analysis of bone marrow and adipose-derived mesenchymal stem cells in rats using 5-azacytidine and low-dose FGF and IGF treatment. *Biomedicines.* 2024;12(8):1–23.

Feng Y, Wang Y, Li L, Yang Y, Tan X, Chen T. Exosomes induce crosstalk between multiple types of cells and cardiac fibroblasts: therapeutic potential for remodeling after myocardial infarction. *Int J Nanomedicine.* 2024;19:10605–21.

Ferrara N, Gerber HP, LeCouter J. The biology of VEGF and its receptors. *Nature Med.* 2003 9:6. 2003;9(6):669–76.

Formiga FR, Pelacho B, Garbayo E, Imbuluzqueta I, Díaz-Herráez P, Abizanda G, et al. Controlled delivery of fibroblast growth factor-1 and neuregulin-1 from biodegradable microparticles promotes cardiac repair in a rat myocardial infarction model by activating endogenous regeneration. *J Control Release.* 2014;173:132–9.

Fujii S, Miura Y, Fujishiro A, Shindo T, Shimazu Y, Hirai H, et al. Graft-versus-host disease amelioration by human bone marrow mesenchymal stromal/stem cell-derived extracellular vesicles is associated with peripheral preservation of naive T cell populations. *Stem Cells.* 2018;36(3):434–45.

Furuta T, Miyaki S, Ishitobi H, Ogura T, Kato Y, Kamei N, et al. Mesenchymal stem cell-derived exosomes promote fracture healing in a mouse model. *Stem Cells Transl Med.* 2016;5(12):1620–30.

Gao L, Wang L, Wei Y, Krishnamurthy P, Walcott GP, Menasché P, et al. Exosomes secreted by hiPSC-derived cardiac cells improve recovery from myocardial infarction in swine. *Sci Transl Med.* 2020;12(561):1–13.

Gao X, Gao M, Gorecka J, Langford J, Liu J, Luo J, et al. Human-induced pluripotent stem-cell-derived smooth muscle cells increase angiogenesis to treat Hindlimb Ischemia. *Cells* 2021;10(4):792.

Garmany A, Arrell DK, Yamada S, Jeon R, Behfar A, Park S, et al. Decoded cardiopoietic cell secretome linkage to heart repair biosignature. *Stem Cells Transl Med.* 2024;13(11):1144–59.

Geimonen E, Jiang W, Ali M, Fishman GI, Garfield RE, Andersen J. Activation of protein kinase C in human uterine smooth muscle induces connexin-43 gene transcription through an AP-1 site in the promoter sequence. *J Biol Chem.* 1996;271(39):23667–74.

Geng K, Wang J, Liu P, Tian X, Liu H, Wang X, et al. Electrical stimulation facilitates the angiogenesis of human umbilical vein endothelial cells through the MAPK/ERK signaling pathway by stimulating the secretion of FGF2. *Am J Physiol Cell Physiol*. 2019;317(2):C277–86.

Gholamali I, Vu TT, Jo S, Park S, Lim KT. Exploring the progress of hyaluronic acid hydrogels: synthesis, characteristics, and wide-ranging applications. *Materials.* 2024;17(10):2439.

Gil-Cabrerizo P, Scacchetti I, Garbayo E, Blanco-Prieto MJ. Cardiac tissue engineering for myocardial infarction treatment. *Eur J Pharm Sci.* 2023;185:106439.

Gnecchi M, Danieli P, Cervio E. Mesenchymal stem cell therapy for heart disease. *Vascul Pharmacol.* 2012;57(1):48–55.

Gnecchi M, Danieli P, Malpasso G, Ciuffreda MC. Paracrine mechanisms of mesenchymal stem cells in tissue repair. *Methods Mol Biol.* 2016;1416:123–46.

González-King H, Rodrigues PG, Albery T, Tangruksa B, Gurrapu R, Silva AM, et al. Head-to-head comparison of relevant cell sources of small extracellular vesicles for cardiac repair: superiority of embryonic stem cells. *J Extracell Vesicles.* 2024;13(5):1–16.

Goodyer WR, Beyersdorf BM, Paik DT, Tian L, Li G, Buikema JW, et al. Transcriptomic profiling of the developing cardiac conduction system at single-cell resolution. *Circ Res.* 2019;125(4):379–97.

Gray WD, French KM, Ghosh-Choudhary S, Maxwell JT, Brown ME, Platt MO, et al. Identification of therapeutic covariant microRNA clusters in hypoxia-treated cardiac progenitor cell exosomes using systems biology. *Circ Res.* 2015 Jan 16;116(2):255–63.

Gude NA, Emmanuel G, Wu W, Cottage CT, Fischer K, Quijada P, et al. Activation of notch-mediated protective signaling in the myocardium. *Circ Res.* 2008 May 9;102(9):1025–35.

Guo X, Bai Y, Zhang L, Zhang B, Zagidullin N, Carvalho K, et al. Cardiomyocyte differentiation of mesenchymal stem cells from bone marrow: New regulators and their implications. *Stem Cell Res Ther.* 2018;9(1):1–12.

Gupta A, Leser JM, Gould NR, Buo AM, Moorer MC, Stains JP. Connexin43 regulates osteoprotegerin expression via ERK1/2 -dependent recruitment of Sp1. *Biochem Biophys Res Commun.* 2019;509(3):728–33.

Habib SM, Haider KH, Martini MF. The potential of mesenchymal stem cell-derived exosomes in cell-free therapy for myocardial repair. In: KH Haider, editor. *Handbook of regenerative medicine: cell-free therapy approach.* 1st ed. CRC Press; 2025a, Chapter 24. https://doi.org/10.1201/9781003611509.

Habib SM, Martini MF, Al-Ahmed O, Qassab AR, Haider KH. Mesenchymal stem cell secretomes for cell-free therapy of the broken heart. In: KH Haider, editor. *Handbook of regenerative medicine: cell-free therapy*. Boca Raton (FL): CRC Press; 2025b. p. 274–93, Chapter 18.

Haider KH, Aslam M (2018) Cell-free therapy with stem cell secretions: protection, repair and regeneration of the injured myocardium. In: KH Haider, S Aziz, editors. *Stem cells: From hype to real hope*. Berlin: De Gruyter; 2018.

Haider KH, Aziz S. Paracrine hypothesis and cardiac repair. *Int J Stem Cell Res Transplant*. 2017;5(1):265–67.

Haider KH, Jiang S, Niagara MI, Ashraf M. IGF-1-overexpressing mesenchymal stem cells accelerate bone marrow stem cell mobilization via paracrine activation of SDF-1α/CXCR4 signaling to promote myocardial repair. *Circulation Res*. 2008;103:1300–8.70.

Haider KH, Khan M, Sen CK. Chapter 15 - MicroRNAs with mega functions in cardiac remodeling and repair: the micromanagement of matters of the heart. In: Sen CK, editor. *MicroRNA in regenerative medicine*. 2nd ed. Academic Press; 2015. p. 397–428.

Haider KH, Najimi M. Exosome-based cell-free therapy in regenerative medicine for myocardial repair. In: Haider KH, editor. *Handbook of stem cell therapy*. Singapore: Springer; 2022. https://doi.org/10.1007/978-981-16-6016-0_9-1.

Haq S, Choukroun G, Kang Z Bin, Ranu H, Matsui T, Rosenzweig A, et al. Glycogen synthase kinase-3β is a negative regulator of cardiomyocyte hypertrophy. *J Cell Biol*. 2000;151(1):117–30.

Harman RM, Marx C, Van Walle GR. Translational animal models provide insight into mesenchymal stromal cell (MSC) secretome therapy. *Front Cell Dev Biol*. 2021:1–23.

Hasan A, Khattab A, Islam MA, Hweij KA, Zeitouny J, Waters R, et al. Injectable hydrogels for cardiac tissue repair after myocardial infarction. *Adv Sci*. 2015 Nov 1;2(11):201500122 [cited 2025 Jun 10]. https://doi.org/10.1002/advs.201500122

Hashimoto H, Olson EN, Bassel-Duby R. Therapeutic approaches for cardiac regeneration and repair. *Nat Rev Cardiol*. 2018;15(10):585–600. [cited 2025 Jun 10]. Available from: https://www.nature.com/articles/s41569-018-0036-6

He S yi, Long L yu, Wang Z cun, Liu W qi, Zhang W, Hu C, et al. Combining the HUMSC secretome with a conductive hydrogel enhances angiogenesis and electrical transmission at myocardial infarct sites, supporting cardiac repair. *Chem Eng J*. 2023;474:145877

Heo JS, Choi Y, Kim HS, Kim HO. Comparison of molecular profiles of human mesenchymal stem cells derived from bone marrow, umbilical cord blood, placenta, and adipose tissue. *Int J Mol Med*. 2016;37(1):115–25.

Ho T, Chang CC, Chan HP, Chung T, Shu C, Chuang K, et al. Hydrogels: properties andapplications in biomedicine. *Molecules*. 2022:1–29.

Hong T, Zhao T, He W, Xia J, Huang Q, Yang J, et al. Exosomal circBBS2 inhibits ferroptosis by targeting miR-494 to activate SLC7A11 signaling in ischemic stroke. *The FASEB Journal*. 2023;37(9):e23152.

Hu J, Jiang Y, Wu X, Wu Z, Qin J, Zhao Z, et al. Exosomal miR-17-5p from adipose-derived mesenchymal stem cells inhibits abdominal aortic aneurysm by suppressing TXNIP-NLRP3 inflammasome. *Stem Cell Res Ther*. 2022;13(1):1–19.

Huang W, Tian SS, Hang PZ, Sun C, Guo J, Du ZM. Combination of microRNA-21 and microRNA-146a attenuates cardiac dysfunction and apoptosis during acute myocardial infarction in mice. *Mol Ther Nucleic Acids*. 2016;5:e296.

Jeffords ME, Wu J, Shah M, Hong Y, Zhang G. Tailoring material properties of cardiac matrix hydrogels to induce endothelial differentiation of human mesenchymal stem cells. *ACS Appl Mater Interfaces*. 2015.7(20):11053–61.

Jihwaprani MC, Sula I, Charbat MA, Haider KH. Establishing delivery route-dependent safety and efficacy of living biodrug mesenchymal stem cells in heart failure patients. *World J Cardiol* 2024;16(6):339–354. https://doi.org/10.4330/wjc.v16.i6.339.

Jin Y, Cheng X, Lu J, Li X. Exogenous BMP-7 facilitates the recovery of cardiac function after acute myocardial infarction through counteracting TGF-β1 signaling pathway. *Tohoku J Exp Med*. 2018;244(1):1–6.

Kaiser NJ, K. Coulombe KL. Physiologically inspired cardiac scaffolds for tailored *in vivo* function and heart regeneration. *Biomed Mater*. 2015.

Kalou YMN, Hashemi ASA, Joudeh RM, Aramini B, Haider KH. Mesenchymal stem cell-based heart cell therapy: the effect of route of cell delivery in the clinical perspective. In: KH Haider, editor. *Stem cells: From potential to promise*. Singapore: Springer; 2021. p. 151–91. https://doi.org/10.1007/978-981-16-0301-36.

Kao CY, Papoutsakis ET. Extracellular vesicles: exosomes, microparticles, their parts, and their targets to enable their biomanufacturing and clinical applications. *Curr Opin Biotechnol.* 2019;60:89–98.

Karkache IY, Damodaran JR, Molstad DHH, Bradley EW. Serine/threonine phosphatases in osteoclastogenesis and bone resorption. *Gene.* 2021;771:145362.

Kawamura M, Miyagawa S, Fukushima S, Saito A, Toda K, Daimon T, et al. Xenotransplantation of bone marrow-derived human mesenchymal stem cell sheets attenuates left ventricular remodeling in a porcine ischemic cardiomyopathy model. *Tissue Eng Part A.* 2015;21(15–16):2272–80.

Kawamura M, Miyagawa S, Miki K, Saito A, Fukushima S, Higuchi T, et al. Feasibility, safety, and therapeutic efficacy of human induced pluripotent stem cell-derived cardiomyocyte sheets in a porcine ischemic cardiomyopathy model. *Circulation.* 2012;126(11 SUPPL.1);S29–S37.

Kerignard E, Bethry A, Falcoz C, Nottelet B, Pinese C. Design of hybrid polymer nanofiber/collagen patches releasing IGF and HGF to promote cardiac regeneration. *Pharmaceutics.* 2022;14(9):1854.

Kim N, Ullah I, Chung K, Lee D, Cha MJ, Ban H, et al. Targeted delivery of recombinant heat shock protein 27 to cardiomyocytes promotes recovery from myocardial infarction. *Mol Pharm.* 2020;17(6):2034–43.

Koh K, Wang JK, Yuan Chen JX, Hiew SH, Cheng HS, Gabryelczyk B, et al. Squid suckerin-spider silk fusion protein hydrogel for delivery of mesenchymal stem cell secretome to chronic wounds. *Adv Healthc Mater.* 2022.12(1):e2201900.

Kompa AR, Greening DW, Lees JG, Kong AM, Cross J, Nowland A, et al. Device encapsulated MSCs for adaptive secretome therapy to effectively target ischaemic heart injury. 2025 17(1):22.

Konoplyannikov M, Haider KH, Ahmed RPH, Jiang S, Vien KL, Ashraf M. Activation of diverse signaling pathways by delivery of multiple cytokines for myocardial repair. *Stem Cell Dev.* 2013;22(2):204–15.

Koudstaal S, Bastings MMC, Feyen DAM, Waring CD, van Slochteren FJ, Dankers PYW, et al. Sustained delivery of insulin-like growth factor-1/hepatocyte growth factor stimulates endogenous cardiac repair in the chronic infarcted pig heart. *J Cardiovasc Transl Res.* 2014;7(2):232–41.

Krishnamurthy P, Rajasingh J, Lambers E, Qin G, Losordo DW, Kishore R. IL-10 Inhibits inflammation and attenuates left ventricular remodeling after myocardial infarction via activation of STAT3 and suppression of HuR. *Circ Res.* 2009;104(2):e9–18.

Kukumberg M, Phermthai T, Wichitwiengrat S, Wang X, Arjunan S, Chong SY, et al. Hypoxia-induced amniotic fluid stem cell secretome augments cardiomyocyte proliferation and enhances cardioprotective effects under hypoxic-ischemic conditions. *Sci Rep.* 2021;11(1):163.

Kwon JW, Savitri C, An B, Yang SW, Park K. Mesenchymal stem cell-derived secretomes-enriched alginate/extracellular matrix hydrogel patch accelerates skin wound healing. *Biomater Res.* 2023;27:107.

Lederman RJ, Guttman MA, Peters DC, Thompson RB, Sorger JM, Dick AJ, et al. Catheter-based endomyocardial injection with real-time magnetic resonance imaging. *Circulation.* 2002;105(11):1282–4. https://doi.org/10.1161/circ.105.11.1282?download=true

Lee J, Tan CY, Lee SK, Kim YH, Lee KY. Controlled delivery of heat shock protein using an injectable microsphere/hydrogel combination system for the treatment of myocardial infarction. *J Control Release.* 2009;137(3):196–202.

Lee S, Serpooshan V, Tong X, Venkatraman S, Lee M, Lee J, et al. Contractile force generation by 3D hiPSC-derived cardiac tissues is enhanced by rapid establishment of cellular interconnection in matrix with muscle-mimicking stiffness. *Biomaterials.* 2017;131:111–20.

Lee TL, Lai TC, Lin SR, Lin SW, Chen YC, Pu CM, et al. Conditioned medium from adipose-derived stem cells attenuates ischemia/reperfusion-induced cardiac injury through the microRNA-221/222/PUMA/ETS-1 pathway. *Theranostics.* 2021;11(7):3131–49.

Leitolis A, Robert AW, Pereira IT, Correa A, Stimamiglio MA. Cardiomyogenesis modeling using pluripotent stem cells: the role of microenvironmental signaling. *Front Cell Dev Biol.* 2019;7:164.

Li H, Wang L, Ma T, Liu Z, Gao L. Exosomes secreted by endothelial cells derived from human induced pluripotent stem cells improve recovery from myocardial infarction in mice. *Stem Cell Res Ther.* 2023;14(1):1–20.

Li LF, Liu YY, Yang CT, Chien Y, Twu NF, Wang ML, et al. Improvement of ventilator-induced lung injury by IPS cell-derived conditioned medium via inhibition of PI3K/Akt pathway and IP-10-dependent paracrine regulation. *Biomaterials.* 2013;34(1):78–91.

Li P, Hu J, Wang J, Zhang J, Wang L, Zhang C. The role of hydrogel in cardiac repair and regeneration for myocardial infarction: recent advances and future perspectives. *Bioengineering.* 2023 10(2):165.

Li Y, Chen X, Jin R, Chen L, Dang M, Cao H, et al. Injectable hydrogel with MSNs/microRNA-21-5p delivery enables both immunomodification and enhanced angiogenesis for myocardial infarction therapy in pigs. *Sci Adv.* 2021;7(9):6740–64.

Li Y, Li X, Zhao R, Wang C, Qiu F, Sun B, et al. Enhanced adhesion and proliferation of human umbilical vein endothelial cells on conductive PANI-PCL fiber scaffold by electrical stimulation. *Mater Sci Eng C*. 2017;72:106–12.

Li Z hua, Wang Y li, Wang H jie, Wu J hong, Tan Y zhen. Rapamycin-preactivated autophagy enhances survival and differentiation of mesenchymal stem cells after transplantation into infarcted myocardium. *Stem Cell Rev Rep*. 2020;16(2):344–56.

Liang J, Chen P, Li C, Li D, Wang J, Xue R, et al. IL-22 Down-regulates Cx43 expression and decreases gap junctional intercellular communication by activating the JNK pathway in psoriasis. *J Investigative Dermatol*. 2019;139(2):400–11.

Liang T, Liu J, Liu F, Su X, Li X, Zeng J, et al. Application of pro-angiogenic biomaterials in myocardial infarction. *ACS Omega*. 2024;9(36):37505–29. https://doi.org/10.1021/acsomega.4c04682.

Liang Y, Xia J, Xu L, Zhang Y, Xu X, Haider KH. Cell-free and cell-based therapy in regenerative medicine. In: KH Haider, editor. *Handbook of regenerative medicine: cell-free therapy approach*. 1st ed. Boca Raton (FL): CRC Press; 2025, Chapter 24. https://doi.org/10.1201/9781003611509.

Lim M, Wang W, Liang L, Han ZB, Li Z, Geng J, et al. Intravenous injection of allogeneic umbilical cord-derived multipotent mesenchymal stromal cells reduces the infarct area and ameliorates cardiac function in a porcine model of acute myocardial infarction. *Stem Cell Res Ther*. 2018;9(1):1–17.

Lin R, Ma H, Ding Z, Shi W, Qian W, Song J, et al. Bone marrow-derived mesenchymal stem cells favor the immunosuppressive T cells skewing in a Helicobacter pylori model of gastric cancer. *Stem Cells Dev*. 2013;22(21):2836–48.

Lin YN, Miguel-dos-Santos R, Cingolani E. Biological modification of arrhythmogenic substrates by cell-free therapeutics. *Heart Lung Circ*. 2023;32(7):844–51.

Liu D, Gu G, Gan L, Yan W, Zhang Z, Yao P, et al. Identification of a CTRP9 C-terminal polypeptide capable of enhancing bone-derived mesenchymal stem cell cardioprotection through promoting angiogenic exosome production. *Redox Biol*. 2021;41:101929.

Liu Y, Lyu X, Tan S, Zhang X. Research progress of exosomal non-coding RNAs in cardiac remodeling. *Int J Med Sci*. 2023;20(11):1469–78.

Long M, Cheng M. Small extracellular vesicles associated with miRNA in myocardial fibrosis. *Biochem Biophys Res Commun*. 2024;727:150336.

Luo Q, Guo D, Liu G, Chen G, Hang M, Jin M. Exosomes from MiR-126-overexpressing adscs are therapeutic in relieving acute myocardial ischaemic injury. *Cell Physiol Biochem*. 2018;44(6):2105–16.

Luo XL, Jiang Y, Li Q, Yu XJ, Ma T, Cao H, et al. hESC-derived epicardial cells promote repair of infarcted hearts in mouse and swine. *Adv Sci*. 2023;10(27):2300470.

Lv K, Li Q, Zhang L, Wang Y, Zhong Z, Zhao J, et al. Incorporation of small extracellular vesicles in sodium alginate hydrogel as a novel therapeutic strategy for myocardial infarction. *Theranostics*. 2019;9(24):7403–16.

Lyra-Leite DM, Gutiérrez-Gutiérrez Ó, Wang M, Zhou Y, Cyganek L, Burridge PW. A review of protocols for human iPSC culture, cardiac differentiation, subtype specification, maturation, and direct reprogramming. *STAR Protocol*. 2022;3(3);101560.

Ma Y, Iyer RP, Jung M, Czubryt MP, Lindsey ML. Cardiac fibroblast activation post-myocardial infarction: current knowledge gaps. *Trends Pharmacol Sci*. 2017;38(5):448–58.

Mancuso T, Barone A, Salatino A, Molinaro C, Marino F, Scalise M, et al. Unravelling the biology of adult cardiac stem cell-derived exosomes to foster endogenous cardiac regeneration and repair. *Int J Mol Sci*. 2020;21(10). https://doi.org/10.3390/ijms21103725

Matsui T, Li L, Wu JC, Cook SA, Nagoshi T, Picard MH, et al. Phenotypic spectrum caused by transgenic overexpression of activated akt in the heart. *J Biol Chem*. 2002;277(25):22896–901.

Mehrotra S, Dey S, Sachdeva K, Mohanty S, Mandal BB. Recent advances in tailoring stimuli-responsive hybrid scaffolds for cardiac tissue engineering and allied applications. *J Mater Chem-B*. 2023:1–35.

Miranda JP, Camões SP, Gaspar MM, Rodrigues JS, Carvalheiro M, Bárcia RN, et al. The secretome derived from 3D-cultured MSCs derived from umbilical cord tissue counteracts manifestations typical of rheumatoid arthritis. *Front Immunol*. 2019 Feb;10:425923.

Monteiro LM, Vasques-Nóvoa F, Ferreira L, Pinto-do-Ó P, Nascimento DS. Restoring heart function and electrical integrity: closing the circuit. *NPJ Regen Med*. 2017;2(1):9.

Moslem F, Shafaghi M, Karamikhaghan P, Hosseini M, Madani H, Amini ZH, et al. The functional recovery of ischemic human pluripotent stem cell-derived cardiomyocytes by the secretome of heat shock-preconditioned umbilical cord mesenchymal stem cells. *Biomedicine Advances*. 2025;2(1):25–33.

Mummery CL, Zhang J, Ng ES, Elliott DA, Elefanty AG, Kamp TJ. Differentiation of human embryonic stem cells and induced pluripotent stem cells to cardiomyocytes: A methods overview. *Circ Res.* 2012;111(3):344–58.

Nakanishi K, Sato Y, Mizutani Y, Ito M, Hirakawa A, Higashi Y. Rat umbilical cord blood cells attenuate hypoxic-ischemic brain injury in neonatal rats. *Sci Rep.* 2017;7:44111.

Nawaz M, Heydarkhan-Hagvall S, Tangruksa B, González-King Garibotti H, Jing Y, Maugeri M, et al. Lipid nanoparticles deliver the therapeutic VEGFA mRNA in vitro and in Vivo and transform extracellular vesicles for their functional extensions. *Adv Sci.* 2023;10(12):e2206187.

O'Brien CG, Ozen MO, Ikeda G, Vaskova E, Jung JH, Bayardo N, et al. Mitochondria-rich extracellular vesicles rescue patient-specific cardiomyocytes from doxorubicin injury: insights into the SENECA trial. *JACC CardioOncol.* 2021;3(3):428–40

Okayama K, Azuma J, Dosaka N, Iekushi K, Sanada F, Kusunoki H, et al. Hepatocyte growth factor reduces cardiac fibrosis by inhibiting endothelial-mesenchymal transition. *Hypertension.* 2012;59(5):958–65.

Ong LP, Bargehr J, Knight-Schrijver V, Lee J, Colzani M, Bayraktar S, et al. Epicardially secreted fibronectin drives cardiomyocyte maturation in 3D-engineered heart tissues. *Stem Cell Rep.* 2023;18(4):936–51.

Ong SG, Huber BC, Lee WH, Kodo K, Ebert AD, Ma Y, et al. Microfluidic single-cell analysis of transplanted human induced pluripotent stem cell-derived cardiomyocytes after acute myocardial infarction. *Circulation.* 2015;132(8):762–71.

Ottaviani L, De Windt LJ, Da Costa Martins PA. Exosomes: scytales in the damaged heart. *Ann Transl Med.* 2016;4(11):222.

Pakshir P, Noskovicova N, Lodyga M, Son DO, Schuster R, Goodwin A, et al. The myofibroblast at a glance. *J Cell Sci.* 2020;133(13):1–10.

Parikh A, Wu J, Blanton RM, Tzanakakis ES. Signaling pathways and gene regulatory networks in cardiomyocyte differentiation. *Tissue Eng Part B Rev.* 2015;21(4):377–92.

Park BW, Jung SH, Das S, Lee SM, Park JH, Kim H, et al. In vivo priming of human mesenchymal stem cells with hepatocyte growth factor–engineered mesenchymal stem cells promotes therapeutic potential for cardiac repair. *Sci Adv.* 2020;6(13):1–12.

Park DS, Fishman GI. The cardiac conduction system. *Circulation.* 2011;123(8):904–15.

Park SR, Kim JW, Jun HS, Roh JY, Lee HY, Hong IS. Stem cell secretome and its effect on cellular mechanisms relevant to wound healing. *Molecular Therapy.* 2018;26(2):606–17.

Paul A, Waters R, Pacelli S, Chakravarti AR. Abstract 195: Engineering shear-thinning injectable biomimetic hydrogel to repair damaged cardiac tissue. *Circ Res.* 2017(Supp_1).

Peng T, Shi Q, Chen M, Yu W, Yang T. Antibacterial-based hydrogel coatings and their application in the biomedical field—a review. *J Funct Biomater.* 2023;14:243.

Perea-Gil I, Gálvez-Montón C, Prat-Vidal C, Jorba I, Segú-Vergés C, Roura S, et al. Head-to-head comparison of two engineered cardiac grafts for myocardial repair: from scaffold characterization to pre-clinical testing. *Sci Rep.* 2018;8:6708.

Peters NS, Coromilas J, Severs NJ, Wit AL. Disturbed Connexin43 gap junction distribution correlates with the location of reentrant circuits in the epicardial border zone of healing canine infarcts that cause ventricular tachycardia. *Circulation.* 1997;95(4):988–96.

Prendiville TW, Ma Q, Lin Z, Zhou P, He A, Pu WT. Ultrasound-guided transthoracic intramyocardial injection in mice. *J Vis Exp.* 2014;(90):51566.

Prokoph S, Chavakis E, Levental KR, Zieris A, Freudenberg U, Dimmeler S, et al. Sustained delivery of SDF-1α from heparin-based hydrogels to attract circulating pro-angiogenic cells. *Biomaterials.* 2012;33(19):4792–800.

Qu Q, Pang Y, Zhang C, Liu L, Bi Y. Exosomes derived from human umbilical cord mesenchymal stem cells inhibit vein graft intimal hyperplasia and accelerate reendothelialization by enhancing endothelial function. *Stem Cell Res Ther.* 2020;11(1):133.

Radmanesh F, Sadeghi Abandansari H, Ghanian MH, Pahlavan S, Varzideh F, Yakhkeshi S, et al. Hydrogel-mediated delivery of microRNA-92a inhibitor polyplex nanoparticles induces localized angiogenesis. *Angiogenesis.* 2021;24(3):657–76.

Rehman I, Rehman A. *Anatomy, thorax, heart.* StatPearls; 2023.

Rentschler S, Yen AH, Lu J, Petrenko NB, Lu MM, Manderfield LJ, et al. Myocardial notch signaling reprograms cardiomyocytes to a conduction-like phenotype. *Circulation.* 2012;126(9):1058–66.

Riaud M, Hilairet G, Sindji L, Perdomo L, Montero-Menei CN, Martinez MC. Pharmacology active microcarriers delivering HGF associated with extracellular vesicles for myocardial repair. *Eur J Pharm Biopharm.* 2021;169:268–79.

Richardson TP, Peters MC, Ennett AB, Mooney DJ. Polymeric system for dual growth factor delivery. *Nat Biotechnol.* 2001;19(11):1029–34.

Roca-Alonso L, Castellano L, Mills A, Dabrowska AF, Sikkel MB, Pellegrino L, et al. Myocardial MiR-30 downregulation triggered by doxorubicin drives alterations in β-adrenergic signaling and enhances apoptosis. *Cell Death & Disease* 2015;6(5):e1754.

Rodness J, Mihic A, Miyagi Y, Wu J, Weisel RD, Li RK. VEGF-loaded microsphere patch for local protein delivery to the ischemic heart. *Acta Biomater.* 2016;45:169–81.

Rody E, Zwaig J, Derish I, Khan K, Kachurina N, Gendron N, et al. Evaluating the reparative potential of secretome from patient-derived induced pluripotent stem cells during ischemia–reperfusion injury in human cardiomyocytes. *Int J Mol Sci.* 2024;25(19):10279.

Ruvinov E, Leor J, Cohen S. The promotion of myocardial repair by the sequential delivery of IGF-1 and HGF from an injectable alginate biomaterial in a model of acute myocardial infarction. *Biomaterials.* 2011;32(2):565–78.

Saffitz JE. Connexins, conduction, and atrial fibrillation. *N Engl J Med.* 2006;354(25):2712–4.

Saffitz JE, Laing JG, Yamada KA. Connexin expression and turnover. *Circ Res.* 2000;86(7):723–8.

Saghebasl S, Akbarzadeh A, Gorabi AM, Nikzamir N, Seyed-Sadjadi M, Mostafavi E. Biodegradable functional macromolecules as promising scaffolds for cardiac tissue engineering. *Polym Adv Technol.* 2022;33(7):2044–68.

Şahin B, İlgün G. Risk factors of deaths related to cardiovascular diseases in World Health Organization (WHO) member countries. *Health Soc Care Community.* 2022;30(1):73–80.

Sandonà M, Pietro L Di, Esposito F, Ventura A, Silini AR, Parolini O, et al. Mesenchymal stromal cells and their secretome: new therapeutic perspectives for skeletal muscle regeneration. *Front Bioeng Biotechnol.* 2021;9:652970.

Schulte C. Diagnostic and prognostic value of circulating microRNAs in heart failure with preserved and reduced ejection fraction. *World J Cardiol.* 2015;7(12):843.

Sedighim S, Chen Y, Xu Z, Mohindra R, Liu H, Agrawal DK, et al. Carboxymethyl cellulose–alginate interpenetrating hydroxy ethyl methacrylate crosslinked polyvinyl alcohol reinforced hybrid hydrogel templates with improved biological performance for cardiac tissue engineering. *Biotechnol Bioeng.* 2022;120(3):819–35.

Sengun E, Wolfs TGAM, van Bruggen VLE, van Cranenbroek B, Simonetti ER, Ophelders D, et al. Umbilical cord-mesenchymal stem cells induce a memory phenotype in CD4+ T cells. *Front Immunol.* 2023;14:1128359.

Senyo SE, Lee RT, Kühn B. Cardiac regeneration based on mechanisms of cardiomyocyte proliferation and differentiation. *Stem Cell Res.* 2014;13(3):532–41.

Severs NJ, Bruce AF, Dupont E, Rothery S. Remodelling of gap junctions and connexin expression in diseased myocardium. *Cardiovasc Res.* 2008;80(1):9–19.

Shahim B, Kapelios CJ, Savarese G, Lund LH. Global public health burden of heart failure: an updated review. *Card Fail Rev.* 2023;9:e11.

Shao L, Zhang Y, Lan B, Wang J, Zhang Z, Zhang L, et al. MiRNA-sequence indicates that mesenchymal stem cells and exosomes have a similar mechanism to enhance cardiac repair. *Biomed Res Int.* 2017;2017:4150705.

Sípos F, Műzes G. Controversies in therapeutic application of mesenchymal stem cell-derived secretome. *Biocell.* 2022;46(4):903–6.

Siraj YA, Galderisi U, Alessio N. Senescence induces fundamental changes in the secretome of mesenchymal stromal cells (MSCs): implications for the therapeutic use of MSCs and their derivatives. *Front Bioeng Biotechnol.* 2023;11:1148761.

Sönmezer D, Latifoğlu F. Production of hydrogel with alginate and pericardial fluid for use in tissue engineering applications. *Çukurova Üniversitesi Mühendislik Fakültesi Dergisi.* 2023;38(4):1077–82.

Sonnenberg S, Rane AA, Liu CJ, Rao N, Agmon G, Suarez SL, et al. Delivery of an engineered HGF fragment in an extracellular matrix-derived hydrogel prevents negative LV remodeling post-myocardial infarction. *Biomaterials.* 2015;45:56–63.

Sosinsky GE, Boassa D, Dermietzel R, Duffy HS, Laird DW, MacVicar B, et al. Pannexin channels are not gap junction hemichannels. *Channels* (Austin). 2011;5(3):193–7.

Sun J, Zhao Y, Li Q, Chen B, Hou X, Xiao Z, et al. Controlled release of collagen-binding SDF-1α improves cardiac function after myocardial infarction by recruiting endogenous stem cells. *Sci Rep.* 2016;6(1):26683.

Tai Y, Woods EL, Dally J, Kong D, Steadman R, Moseley R, et al. Myofibroblasts: function, formation, and scope of molecular therapies for skin fibrosis. *Biomolecules.* 2021;11(8):1095.

Tajabadi M, Goran Orimi H, Ramzgouyan MR, Nemati A, Deravi N, Beheshtizadeh N, et al. Regenerative strategies for the consequences of myocardial infarction: chronological indication and upcoming visions. *Biomed Pharmacother.* 2022;146:112584.

Tang J, Cui X, Zhang Z, Xu Y, Guo J, Soliman BG, et al. Injection-free delivery of MSC-derived extracellular vesicles for myocardial infarction therapeutics. *Adv Healthc Mater.* 2021;11:2100312.

Tang J, Shen D, Caranasos TG, Wang Z, Vandergriff AC, Allen TA, et al. Therapeutic microparticles functionalized with biomimetic cardiac stem cell membranes and secretome. *Nat Commun.* 2017;8:13724.

Teixeira F, Salgado A. Mesenchymal stem cells secretome: current trends and future challenges. *Neural Regen Res.* 2020;15(1):75.

Testa G, Di Benedetto G, Passaro F. Advanced technologies to target cardiac cell fate plasticity for heart regeneration. *Int J Mol Sci.* 2021;22(17):9517.

Torregrosa-Carrión R, Luna-Zurita L, García-Marqués F, D'Amato G, Piñeiro-Sabarís R, Bonzón-Kulichenko E, et al. NOTCH Activation promotes valve formation by regulating the endocardial secretome. *Mol Cell Proteomics.* 2019;18(9):1782–95.

Tortajada M, Prat-Vidal C, Font-Morón M, Clos-Sansalvador M, Calle A, Gastelurrutia P, et al. Local administration of porcine immunomodulatory, chemotactic, and angiogenic extracellular vesicles using engineered cardiac scaffolds for myocardial infarction. *Bioact Mater.* 2021.

Totland MZ, Rasmussen NL, Knudsen LM, Leithe E. Regulation of gap junction intercellular communication by connexin ubiquitination: physiological and pathophysiological implications. *Cell Mol Life Sci.* 2020;77(4):573–91.

Tran T, Cruz C, Chan A, Awad S, Rajasingh J, Deth R, et al. Mesenchymal stem cell-derived long noncoding RNAs in cardiac injury and repair. *Cells.* 2023;12(18):2268.

Uhlén M, Karlsson MJ, Hober A, Svensson AS, Scheffel J, Kotol D, et al. The human secretome. *Sci Signal.* 2019;12(609):eaaz0274.

Umar AbdK. Stem cell's secretome delivery systems. *Adv Pharm Bull.* 2022;13(2):244–58.

Van Rooij E, Olson EN. MicroRNA therapeutics for cardiovascular disease: opportunities and challenges. *Nat Rev Drug Discov.* 2012;11(11):860–72.

Venditti I, Iucci G, Fratoddi I, Cipolletti M, Montalesi E, Marino M, et al. Direct conjugation of resveratrol on hydrophilic gold nanoparticles: structural and cytotoxic studies for biomedical applications. *Nanomaterials.* 2020;10(10):1898.

Villa F, Bruno S, Costa A, Li M, Russo M, Cimino JJ, et al. The human fetal and adult stem cell secretome can exert cardioprotective paracrine effects against cardiotoxicity and oxidative stress from cancer treatment. *Cancers* (Basel). 2021;13(15):3729.

Wang D, Feng X, Lu L, Konkel JE, Zhang H, Chen Z, et al. A CD8 T cell/indoleamine 2,3-dioxygenase axis is required for mesenchymal stem cell suppression of human systemic lupus erythematosus. *Arthritis Rheumatol.* 2014;66(8):2234–45.

Wang D, Hu Y, Zhang L, Cai H, Wang Y, Zhang Y. Dual delivery of an NF-κB inhibitor and IL-10 through supramolecular hydrogels polarizes macrophages and promotes cardiac repair after myocardial infarction. *Acta Biomater.* 2023;164:111–23.

Wang D, Zhu H, Yang Q, Sun Y. Effects of relaxin on cardiac fibrosis, apoptosis, and tachyarrhythmia in rats with myocardial infarction. *Biomed Pharmacother.* 2016;84:348–55.

Wang L, Chung J, Li EC, Uman S, Atluri P, Burdick JA. Injectable and protease-degradable hydrogel for siRNA sequestration and triggered delivery to the heart. *J Control Release.* 2018;10:152–61.

Wang X, Zhu Y, Wu C, Liu W, He Y, Yang Q. Adipose-derived mesenchymal stem cells-derived exosomes carry MicroRNA-671 to alleviate myocardial infarction through inactivating the TGFBR2/Smad2 Axis. *Inflammation.* 2021;44(5):1815–30.

Wang YC, Chen RF, Brandacher G, Lee WPA, Kuo YR. The suppression effect of dendritic cells maturation by adipose-derived stem cells through the TGF-β1-related pathway. *Exp Cell Res.* 2018;370(2):708–17.

Wang Z, He Z yi, Liang S, Yang Q, Cheng P, Chen A min. Comprehensive proteomic analysis of exosomes derived from human bone marrow, adipose tissue, and umbilical cord mesenchymal stem cells. *Stem Cell Res Ther.* 2020;11(1):1–11.

Waters R, Alam P, Pacelli S, Chakravarti AR, Ahmed RPH, Paul A. Stem cell-inspired secretome-rich injectable hydrogel to repair injured cardiac tissue. *Acta Biomater.* 2018;69:95–106.

Wen Z, Mai Z, Zhu X, Wu T, Chen Y, Geng D, et al. Mesenchymal stem cell-derived exosomes ameliorate cardiomyocyte apoptosis in hypoxic conditions through microRNA144 by targeting the PTEN/AKT pathway. *Stem Cell Res Ther.* 2020;11(1):1–17.

Wu XQ, Yan T zhong, Wang Z wei, Wu X, Cao G hui, Zhang C. BM-MSCs-derived microvesicles promote allogeneic kidney graft survival through enhancing micro-146a expression of dendritic cells. *Immunol Lett.* 2017;191:55 62.

Wu Y, Zhang H, Wang S, Li L, Wang R, Jiang S. Human umbilical cord-derived stem cell sheets improve left ventricular function in rat models of ischemic heart failure. *Eur J Pharmacol.* 2022;925:174994.

Xia J, Arai S, Haider KH. Research and applications of stem cell secretome. In: KH Haider, editor. *Stem cells: From potential to promise.* Singapore: Springer; 2021. https://doi.org/10.1007/978-981-16-0301-3_7.

Xiao C, Wang K, Xu Y, Hu H, Zhang N, Wang Y, et al. Transplanted mesenchymal stem cells reduce autophagic flux in infarcted hearts via the exosomal transfer of miR-125b. *Circ Res.* 2018;123(5):564–78.

Xiao J, Pan Y, Li XH, Yang XY, Feng YL, Tan HH, et al. Cardiac progenitor cell-derived exosomes prevent cardiomyocyte apoptosis through exosomal miR-21 by targeting PDCD4. *Cell Death Dis.* 2016;7(6):e2277.

Xin Y, Gao J, Hu R, Li H, Li Q, Han F, et al. Changes of immune parameters of T lymphocytes and macrophages in EAE mice after BM-MSCs transplantation. *Immunol Lett.* 2020;225:66–73.

Xu J, Brown J, Shaik R, Soto-Garcia L, Liao J, Nguyen K, et al. Injectable myocardium-derived hydrogels with SDF-1α release for cardiac repair. *Biomater Adv.* 2025;170:214203.

Yan W, Guo Y, Tao L, Lau WB, Gan L, Yan Z, et al. C1q/tumor necrosis factor-related protein-9 regulates the fate of implanted mesenchymal stem cells and mobilizes their protective effects against ischemic heart injury via multiple novel signaling pathways. *Circulation.* 2017;136(22):2162–77.

Yan W, Lin C, Guo Y, Chen Y, Du Y, Lau WB, et al. N-Cadherin overexpression mobilizes the protective effects of mesenchymal stromal cells against ischemic heart injury through a β-catenin-dependent manner. *Circ Res.* 2020;126(7):857–74.

Yan W, Xia Y, Zhao H, Xu X, Ma X, Tao L. Stem cell-based therapy in cardiac repair after myocardial infarction: promise, challenges, and future directions. *J Mol Cell Cardiol.* 2024;188:1–14.

Yester JW, Kühn B. Mechanisms of cardiomyocyte proliferation and differentiation in development and regeneration. *Curr Cardiol Rep.* 2017;19(2):1–9.

Yıldırım Y, Degener L, Reuter L, Petersen J, Gabel L, Sommer A, et al. Evaluation of cell survival in different 3D-printed geometric shapes of human iPSC-derived engineered heart tissue. *Artif Organs.* 2024;48(11):1251-1263.

Yudintceva NM, Bogolyubova IO, Muraviov AN, Sheykhov MG, Vinogradova TI, Sokolovich EG, et al. Application of the allogenic mesenchymal stem cells in the therapy of bladder tuberculosis. *J Tissue Eng Regen Med.* 2018;12(3):e1580–93.

Zhang J, Chen C, Hu B, Niu X, Liu X, Zhang G, et al. Exosomes derived from human endothelial progenitor cells accelerate cutaneous wound healing by promoting angiogenesis through the Erk1/2 signaling pathway. *Int J Biol Sci.* 2016;12(12):1472–87.

Zhang S, Chu WC, Lai RC, Lim SK, Hui JHP, Toh WS. Exosomes derived from human embryonic mesenchymal stem cells promote osteochondral regeneration. *Osteoarthr Cartil.* 2016;24(12):2135–40.

Zhang Y, Bi J, Huang J, Tang Y, Du S, Li P. Exosome: a review of its classification, isolation techniques, storage, diagnostic and targeted therapy applications. *Int J Nanomedicine.* 2020;15:6917–34.

Zhang Y, Liang X, Liao S, Wang W, Wang J, Li X, et al. Potent paracrine effects of human induced pluripotent stem cell-derived mesenchymal stem cells attenuate doxorubicin-induced cardiomyopathy. *Sci Rep* 2015;5(1):1–17.

Zhao G, Feng Y, Xue L, Cui M, Zhang Q, Xu F, et al. Anisotropic conductive reduced graphene oxide/silk matrices promote post-infarction myocardial function by restoring electrical integrity. *Acta Biomater.* 2022;139:190–203.

Zhao H, Shang Q, Pan Z, Bai Y, Li Z, Zhang H, et al. Exosomes from adipose-derived stem cells attenuate adipose inflammation and obesity through polarizing M2 macrophages and beiging in white adipose tissue. *Diabetes.* 2018;67(2):235–47.

Zhou J, Liu W, Zhao X, Xian Y, Wu W, Zhang X, et al. Natural melanin/alginate hydrogels achieve cardiac repair through ROS scavenging and macrophage polarization. *Adv Sci.* 2021;8(20):e2100505.

Zhu D, Li Z, Huang K, Caranasos TG, Rossi JS, Cheng K. Minimally invasive delivery of therapeutic agents by hydrogel injection into the pericardial cavity for cardiac repair. *Nat Commun.* 2021;12(1). doi.org/10.1038/s41467-021-21682-7

Zhu LP, Tian T, Wang JY, He JN, Chen T, Pan M, et al. Hypoxia-elicited mesenchymal stem cell-derived exosomes facilitate cardiac repair through miR-125 b-mediated prevention of cell death in myocardial infarction. *Theranostics.* 2018;8(22):6163–77.

5 The MSC-Derived Secretome-Based Therapy for Neuroinflammation in Traumatic Brain Injury

Yurshev YuA[1, 2], *Kostryukov PA*[3], *Basalova NA*[1], *Popov VS*[1], *Novikova MR*[3], *Aniol VA*[3], *Shilova AA*[1], *Komoltsev IG*[3], *Gulyaev MV*[1], *Pavlova OS*[1], *Manolova AO*[3], *Efimenko AYu*[1], *Gulyaeva NV*[3], *Akopyan ZhA*[1], *Malyshev AYu*[3], *Tkachuk VA*[1], *and Karagyaur MN*[1]

[1]Medical Research and Education Institute, Federal State Budget Educational Institution of Higher Education, M.V.Lomonosov Moscow State University, Moscow, Russia,

[2]Neurosurgery department, Sklifosovsky Research Institute for Emergency Medicine, Moscow, Russia

[3]Federal State Budget Institution, Institute of Higher Nervous Activity and Neurophysiology, Russian Academy of Science,Moscow, Russia

5.1 LIST OF ABBREVIATIONS

BBB	Blood-brain barrier
BIAA	Brain Injury Association of America
DAI	Diffuse axonal injury
DAPI	4′,6diamidino2phenylindole
DMEM	Dulbecco's Modified Eagle's Medium
FBS	Fetal bovine serum
ICP	Intracranial pressure
IL1β	Interleukin 1 beta
IL6	Interleukin 6
iMSC	Immortalized mesenchymal stem/stromal cells
ISCT	International society for cell therapy
LHI	Lateral hydrodynamic impact
MRI	Magnetic resonance imaging
MSCs	Mesenchymal stem/stromal cells
SASP	Senescence-associated secretory phenotype
SBI	Secondary brain injury
TBI	Traumatic brain injury
TNFα	Tumor necrosis factor

DOI: 10.1201/9781003653387-5

5.2 INTRODUCTION

Traumatic brain injury (TBI) is one of the leading causes of health loss and disability worldwide. According to various estimates, the annual number of people affected by TBI ranges from 27 to 69 million (Dewan et al., 2018; Guan et al., 2019). The overall mortality rate for severe TBI reaches 65–70%, while 50% of TBI victims lose their ability to work to some degree (Zhu et al., 2012). Many patients with TBI live the rest of their lives suffering from severe neurological deficits, reducing their quality of life and inducing significant socio-economic burden. The total global losses, including the loss of working capacity and the cost of providing medical care to victims, are estimated at approximately $100 billion per year (Zhu et al., 2012). TBI, as defined by the Brain Injury Association of America (BIAA), is a disruption of brain function or other signs of brain pathology resulting from external forces (Menon et al., 2008). The definition notes that clinical symptoms of TBI may not appear immediately or may be absent altogether, and that "other evidence of brain pathology" may include imaging or laboratory test results.

5.3 PATHOPHYSIOLOGY OF TBI, CURRENT AND PROMISING THERAPEUTIC APPROACHES

The pathophysiology of brain damage caused by external factors involves multiple mechanisms mediating primary and secondary brain injury. This concept underlies major modern clinical approaches to treating patients with TBI. The primary brain damage occurs in the acute period of trauma. Most often, it includes a combination of focal contusions and intracerebral hemorrhages, as well as diffuse axonal injury (DAI) along with focal and diffuse cerebral edema. TBI-induced DAI occurs due to the displacement of brain tissue, appearing pathologically and on neuroimaging as multiple small lesions in the white matter tracts. DAI typically affects the transition zone between gray and white matter in the telencephalon, with more severe cases often involving the corpus callosum and brainstem structures (Mishra et al., 2022). Patients with severe DAI are usually in a deep coma without increased intracranial pressure (ICP) and often have a poor prognosis.

Focal brain contusions are the most common injuries in TBI. Contusions are usually observed in the basal frontal and temporal regions, which are particularly sensitive to direct impact in acceleration/deceleration injuries. Damage to the integrity of the blood vessels in the brain can lead to intracerebral hemorrhage, which, together with injured brain tissue, forms the primary lesion site. Hemorrhages are dangerous not only because they disrupt the cytoarchitecture of the brain and cause cell death, but also because they provoke the so-called mass effect, i.e., the appearance of a foreign mass in the closed space of the skull, which leads to an increase in intracranial pressure. The spillage of blood into the tissue triggers the blood coagulation system. Its components, primarily thrombin, affect the endothelium, increase the permeability of the blood-brain barrier (BBB), promote the diffusion of serum proteins into the brain parenchyma, and induce cerebral tissue edema, which, in turn, impairs brain blood supply and cerebrospinal fluid outflow (Butcher et al., 2004). Seventy-four days following TBI, the edema surrounding the hematoma continues to expand due to the progression of delayed TBI-induced effects, such as neuroinflammation and secondary brain injury (SBI) (Balami et al., 2012; Koliaraki et al., 2020).

SBI is a cascade of molecular damage mechanisms initiated at the time of the initial injury and continuing for hours or days after trauma. SBI is thought to be the common pathological sequela of any focal brain damage (TBI, stroke, intracerebral hemorrhage, etc.). The primary mechanisms that trigger SBI include excitotoxicity (resulting from excessive release of excitatory neurotransmitters), mitochondrial dysfunction due to disturbances in brain tissue trophism (secondary to vasospasm, focal microvascular occlusion, and vascular damage), damage to cell membranes by free radicals, and neuroinflammation. Maintaining and propagating brain tissue damage, these mechanisms contribute to the spread of the lesion. Neuronal cell death by necroptosis, ferroptosis, or apoptosis amplifies the release of even more inflammatory factors in the injury and penumbra foci areas (Lu et al., 2021; Orr et al., 2024), which, together with the enhanced release of matrix metalloproteinases, maintains high BBB permeability, stimulates the expression of endothelial cell adhesion molecules, and attracts a wide range of immune cells to the lesion site.

Leukocytes that have infiltrated the lesioned brain tissue promote the resorption of hemorrhage and necrotic masses (Chang et al., 2018). Still, overproduction of pro-inflammatory cytokines (IL-1β, IL-6, TNF-α, etc.) also damages neural cells in the penumbra and stimulates prolonged, self-sustaining activation of microglial cells (Taran et al., 2022; Wafaisade et al., 2010). In response to damage and inflammation, the expression of anti-inflammatory molecules is activated, and the subsequent cascade of processes is primarily determined by the ratio of pro-inflammatory to anti-inflammatory signals.

Highly undesirable, though common consequences of TBI are deferred cognitive impairment (dementia) and/or affective disorders (depression), pathologies caused (at least partially) by impaired hippocampal function. The hippocampus is a complex and heterogeneous structure in the brain's limbic system that integrates mental and emotional responses. Its dorsal part is primarily involved in learning and memory, while the ventral part is involved in stress responses and emotions (Rao et al., 1997). Damage to the hippocampus may be secondary and distant in nature and is presumably caused by the prolonged action of excess glucocorticoids released after focal brain trauma (Komoltsev and Gulyaeva, 2022). Glucocorticoids interact with corticosteroid receptors, which are abundant in the hippocampus, and activate signaling pathways that stimulate neuroinflammation, thereby impairing neurogenesis in the subgranular niche and causing hippocampal neurodegeneration (Shin et al., 2006; Podgorny and Gulyaeva, 2021; Komoltsev et al., 2021; Gulyaeva, 2023). These changes induced by focal brain damage may affect not only the ipsilateral hippocampus (on the side of the damage) but also the contralateral hippocampus and other brain structures involved in neurodegenerative and psychiatric diseases (Gulyaeva et al., 2021). Hippocampal vulnerability to distant, glucocorticoid-mediated damage associated with neuroinflammation remains a serious and unresolved problem in modern neurology, psychiatry, and rehabilitation medicine.

5.4 CONTEMPORARY THERAPEUTIC OPTIONS FOR TBI

Existing therapeutic strategies for treating TBI primarily aim to maintain optimal oxygen delivery to the brain, thereby preventing secondary neuronal death. To this end, parameters grouped under the acronym "THE MANTLE" are monitored in these patients: Temperature, Hemoglobin, Electrolytes and acid-base status, Metabolism, Arterial blood pressure, Nutrition and glucose, target oxygen, lung-protective ventilation, Edema, and ICP control (Godoy et al., 2023). Body temperature is maintained between 36°C and 37°C, since hyperthermia exacerbates neuroinflammation, while hypothermia has several side effects. Hemoglobin levels in TBI patients are maintained at > 80 g/L, and oxygen saturation is assessed using jugular and cerebral oximetry. Maintaining normal electrolyte levels and acid-base balance ensures the conditions for proper hemoglobin oxidation and reduction, and stability of ICP. Systolic blood pressure is maintained above 90–110 mmHg to maintain adequate cerebral perfusion. Glucose is maintained within the range of 4–10 mmol/L to avoid both deficiency and excess, which may cause inflammation and edema. Sedation and control of epileptic activity reduce the release of excitatory neurotransmitters, slowing brain metabolism and thereby reducing its need for oxygen and nutrients, which, in turn, decreases the likelihood of excitotoxicity. Intracranial pressure and brain edema are monitored and managed using sensors and comprehensive methods, including medication and, if necessary, surgical measures. All parameters require individual adjustments depending on the patient's condition.

In cases of moderate TBI, systemic antifibrinolytic therapy, such as tranexamic acid, may be used to reduce the probability of intracerebral hemorrhage. Attempts have been made to stimulate neuroprotection of the damaged brain tissue using progesterone, cytidine diphosphate-choline, and erythropoietin, but these have not improved outcomes after severe TBI. Natural antioxidants such as curcumin, resveratrol, and catechins show potential but have not been clinically proven. Several other molecules exhibit neuroprotective effects. These include molecular hydrogen, minocycline, hydrogen sulfide, thiazolidinediones, glibenclamide, amantadine, and extracts of animal blood or brain tissues. Their claimed effects include decreased production of reactive oxygen species (ROS), reduced cell apoptosis, decreased inflammation, improved mitochondrial function, inhibition of caspase-3, decreased microglial activation, and changes in the synthesis of several proteins, among others. However, as noted above, none of them are currently widely used in clinical practice (Zehtabchi et al., 2008).

Attempts have been made to suppress inflammatory processes in the brain with glucocorticoids; however, they have been found to have a limited therapeutic effect and significant delayed negative consequences, including pro-inflammatory and pro-neurodegenerative impact, which have led to their abandonment. Severe brain traumas accompanied by skull fractures and vascular damage require surgical treatment to remove bone fragments, evacuate massive hematomas, stop cerebrospinal fluid leakage, and close the wound defect. However, a review of surgical methods for treating TBI is beyond the scope of this study.

The lack of effective treatment methods stimulates the search for new drugs with pleiotropic action. The above-mentioned mechanisms of secondary damage to brain tissue after TBI can potentially be mitigated by stimulating direct neuroprotection of brain cells, increasing their resistance to damaging and toxic factors, stabilizing the barrier function of the BBB, and suppressing neuroinflammation, with the best therapeutic effect being achieved through a complex interplay of these mechanisms. To date, there are virtually no registered drugs with such complex activity capable of preventing the development of undesirable acute and postponed consequences of TBI. This prompts the need to search for them among drug molecules and compositions with potential neuroprotective and anti-inflammatory activity.

5.5 CELL-BASED AND CELL-FREE THERAPY APPROACHES FOR TBI

We have previously shown that the secretome of multipotent human mesenchymal stromal cells (MSCs) can stimulate the survival of experimental animals, restore their neurological parameters, and reduce the volume of brain tissue damage in a model of intracerebral hematoma (hemorrhagic stroke) in rats (Dzhauari et al., 2023; Karagyaur et al., 2021). MSCs are considered key regulators of organ and tissue renewal and regeneration processes (Alvarez-Viejo and Haider, 2022). One of the primary mechanisms of their activity is believed to be the production of a diverse range of growth factors, cytokines, matrix proteins, microRNAs, and other molecules, which exhibit neuroprotective, proangiogenic, and anti-inflammatory properties (Haider and Aziz, 2017) (Figure 5.1).

Since the pathogenic mechanisms of TBI and hemorrhagic stroke are broadly similar, we hypothesized that the MSC secretome may have neuroprotective activity and reduce adverse consequences in TBI. Previous studies have shown that MSCs or their secretome both stimulate neural cell survival, stabilize the BBB, and suppress neuroinflammation (Hasan et al., 2017; Hosseini et al., 2018).

In open-access sources, we have also found information on several clinical trials investigating the therapeutic efficacy and safety of mesenchymal stem/stromal cells (MSCs) following traumatic brain injury (TBI). Two of the identified studies have the status "Completed":

1. NCT04063215 "A Clinical Trial to Determine the Safety and Efficacy of Hope Biosciences Autologous Mesenchymal Stem Cell Therapy for the Treatment of Traumatic Brain Injury and Hypoxic-Ischemic Encephalopathy" (USA, Phase I/II): "three infusions of autologous adipose tissue-derived MSCs (2 × 10^8 total cells per dose) in adult patients with sub-acute or chronic neurological injury" (ClinicalTrials.Gov, A). Despite the study's completion in 2024, no open-access information is available on the efficacy and safety of autologous MSCs in treating post-TBI conditions. Based on the study design, it can be assumed that the expected efficacy may be lower than could potentially be achieved. This is because TBI is an acute pathology, with the most severe brain damage occurring within hours and days after injury. In contrast, the procedure for isolating, expanding, and preparing autologous MSCs for transplantation takes weeks. With the current study

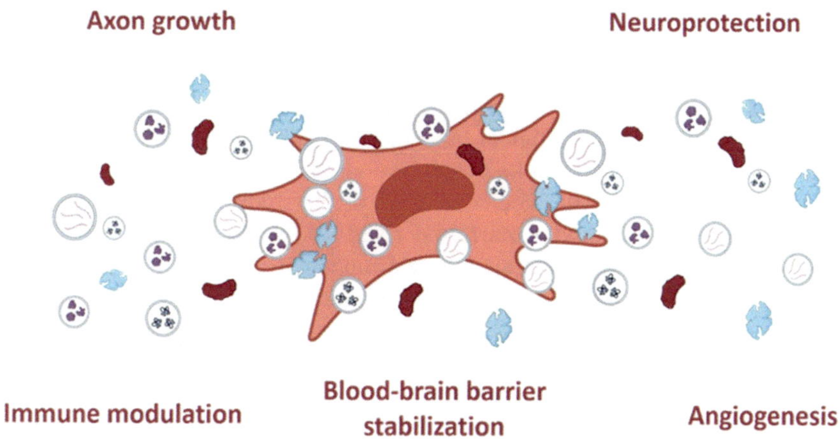

FIGURE 5.1 Potential spectrum of activity of the secretome of multipotent mesenchymal stromal cells in the central nervous system.

design, MSCs would influence the pathophysiology of TBI at the stage of spontaneous resolution of neuroinflammation and gliosis, with minimal efficacy. For this reason, a more promising approach, in our view, is the use not of cells themselves, but of their secretome, which can be obtained "off-the-shelf" and administered in the first hours after TBI to stimulate neuroprotection and suppress neuroinflammation.

To be fair, it should be noted that these same autologous MSCs have demonstrated promising efficacy and safety in another clinical trial for treating a chronic brain disease—multiple sclerosis (USA, NCT05116540, Phase 2 trial). Thus, patients treated with autologous MSCs showed significant improvements in physical and mental health quality-of-life scores compared to placebo at 52 weeks, meeting the primary efficacy endpoint. The therapy was well tolerated with no severe adverse events, and further research is suggested to evaluate its long-term efficacy in multiple sclerosis and potentially other autoimmune diseases (NeurologyLive.Com).

2. NCT02742857 "Non-randomized, Open-labeled, Interventional, Single Group, Proof of Concept Study With Multi-modality Approach in Cases of Brain Death Due to Traumatic Brain Injury Having Diffuse Axonal Injury" (India, Phase I): "multi-modality approach (using intrathecal bioactive peptides, stem cells, laser and transcranial IV laser and median nerve stimulation as adjuvants) in cases of brain death due to traumatic brain injury having diffuse axonal injury to document possibility of reversal of brain death." The study is also marked as "Completed" at 2020; however, the authors do not publish detailed study design or results (ClinicalTrials.Gov, B).

Additionally, the website https://clinicaltrials.gov contains information on five more Phase I/II clinical trials that are at the stages "Active, not recruiting" or "Recruiting," aiming at the use of autologous and allogeneic MSCs for TBI treatment (NCT02525432, NCT06146062, NCT05018832, NCT06163833, NCT05951777); results for these studies are yet to be obtained.

At the same time, there is a substantial amount of information on the results of clinical trials using MSCs or their vesicles for treating other acute brain injuries (strokes, intracerebral hemorrhages) in open-access sources:

1. NCT01287936 (USA, Phase I/IIa). Transplantation of genetically modified bone marrow-derived MSCs (with Notch1-encoding plasmid) intracerebrally, once, in patients with chronic ischemic stroke (n = 18), 6–60 months post-stroke, at doses of $2.5–10 \times 10^6$ MSCs. No serious adverse effects or deaths were observed within 12 months post-treatment. Some patients experienced mild to moderate symptoms related to the surgical procedure:

headache (77.8%), nausea (33.3%), vomiting (22.2%), depression (22.2%), muscle spasticity (22.2%), fatigue (16.7%), elevated blood glucose (16.7%), elevated C-reactive protein (16.7%), and subdural hematoma (5.6%). No association was found between MSC dose and symptom frequency or severity. Administration of genetically modified MSCs led to significant improvement in neurological outcomes in patients with chronic ischemic stroke, as assessed by the European Stroke Scale, National Institutes of Health Stroke Scale, and Fugl-Meyer Scale ($p < 0.001$). The conclusion of the study is "Intracerebral administration of genetically modified bone marrow-derived MSCs in patients with chronic ischemic stroke is generally safe, well-tolerated, and stimulates significant improvement in neurological function at 12 months" (ClinicalTrials.Gov, C; Steinberg et al., 2016).

2. No number (China, no phase, clinical case). Intra-arterial injection (into the middle cerebral artery) of allogeneic umbilical cord blood-derived MSCs after ischemic ($n = 3$) and hemorrhagic ($n = 1$) strokes. Four participants: single injection of 20×10^6 MSCs over 20 minutes, 11–50 days post-stroke. Safety and efficacy were assessed over 6 months. No serious consequences (stroke or death) were observed within 6 months post-treatment, and there was no fever or rash. Muscle strength improved in two of the three patients with ischemic stroke. In two patients with ischemic stroke, modified Rankin Scale scores improved at 90 and 180 days post-MSC therapy (Jiang et al., 2013).
3. No number (South Korea, no phase). Fifty-two participants with severe ischemic stroke in the middle cerebral artery basin, followed for 5 years—2 groups: autologous bone marrow-derived MSCs ($n = 16$) and control ($n = 36$). MSCs were administered intravenously twice, at 5 and 7 weeks post-stroke, at a dose of 50×10^6 MSCs. Over 5 years, four (25%) patients in the MSC group died compared to 21 (58.3%) in the control group, with cumulative survival at week 260 of 0.72 in the MSC group and 0.34 in the control group (log-rank; $p = 0.058$). No significant adverse effects were observed post-MSC administration. Incidence of comorbidities, including seizures and recurrent vascular events, did not differ between groups. Patients in the MSC group showed reduced severity of neurological deficits (Lee et al., 2010).
4. No number (China, no phase, clinical case). Co-transplantation of allogeneic neural stem progenitor cells (NSPCs) and mesenchymal stromal cells (MSCs) in patients ($n = 6$) with ischemic stroke in the middle or anterior cerebral artery basin, 1 week to 2 years post-stroke. Two groups: Group 1—four intravenous injections of allogeneic MSCs at 5×10^5/kg body weight ($n = 2$); Group 2—one intravenous injection of allogeneic MSCs at 5×10^5/kg body weight followed by three injections of 5×10^6/patient of allogeneic MSCs and 6×10^6/patient of allogeneic NSPCs into the basal cistern ($n = 4$). Follow-up: 2 years. Adverse effects in the post-transplantation period included low-grade fever, typically lasting 2–4 days after each MSC injection, and mild dizziness in one patient. All adverse effects occurred within the first 2–24 hours post-transplantation and resolved without specific treatment, with no signs of neurological deterioration or neurological infection. No tumor formation was detected at 2-year follow-up. Disability severity decreased, neurological functions improved, and patients' daily living abilities enhanced (Qiao et al., 2014).
5. NCT04657458 (USA, Phase II). Extracellular vesicles from allogeneic bone marrow-derived MSCs in severe COVID-19: 24 participants, single intravenous administration of exosomes derived from 10^6–10^7 MSCs/kg body weight. Treatment safety was assessed daily for two weeks post-administration. No adverse effects were observed within 72 hours after infusion. Participant survival was 83% with 70% of patients recovered, 13% remained in critical but stable condition, and 16% died from causes unrelated to treatment. Post-administration, the oxygenation index increased by an average of 192%. Additionally, C-reactive protein decreased by 77%, ferritin by 43%, and D-dimer by 42% (Sengupta et al., 2020).

Given the high similarity in pathogenetic mechanisms between TBI and other acute brain disorders (strokes, intracerebral hemorrhages), it can be assumed that MSCs themselves and their secretion products, which exhibit therapeutic activity in strokes and brain hemorrhages, may help stimulate neuroprotection and suppress the expansion of the TBI lesion.

The use of MSC secretomes offers several advantages over MSCs themselves, including storage, delivery, safety, and reduced variability in therapeutic response in brain injury (Cui et al., 2022; MRC CRASH Trial Collaborators et al., 2008; Xia et al., 2021). A meta-analysis of 31 preclinical studies published before June 2021 revealed that treating TBI with MSC secretome enhances structural and neurological recovery, reduces inflammation, and increases the production of anti-inflammatory cytokines, thereby promoting tissue repair and functional recovery (Mugenyi et al., 2023). Other studies, both in vitro and in vivo, demonstrate that the MSC secretome is a promising agent for treating traumatic brain injury. However, further research is needed to understand the mechanisms of action of the secretome and to develop standardized treatment protocols, especially for elderly patients (Moore et al., 2021; Pischiutta et al., 2022).

A significant disadvantage of using primary MSC cultures as secretome producers is their rapid aging and the accumulation of components of the senescence-associated secretory phenotype (SASP) in the secretome (Samsonraj et al., 2023). SASP components are pro-inflammatory molecules that prevent tissue regeneration, cause and maintain tissue inflammation and senescence (Coppé et al., 2010). An alternative to using primary MSCs is the use of MSC cultures with prolonged proliferative potential. Previously, we generated a line of immortalized human mesenchymal stromal cells by ectopic expression of the telomerase catalytic subunit gene. This extended the culture's proliferative potential to ~30–40 passages while maintaining the qualitative and quantitative composition and biological activity of the secretome they produce (Primak et al., 2024a, 2024b). Moreover, these cell cultures revealed a delayed onset of SASP component production (up to 20 passages) compared to that observed in primary MSC cultures (typically, 8–10 passages). Altogether, these immortalized MSC (iMSC) cultures appear to be a promising source of clinically relevant amounts of a highly active and pro-regenerative secretome. The primary objective of this study was to investigate the ability of iMSC-derived secretome to stimulate neuroprotection and suppress neuroinflammation in a rat model of traumatic brain injury.

5.6 IMMORTALIZED MSCS SECRETOME FOR TBI TREATMENT (LATERAL HYDRODYNAMIC IMPACT MODEL)

In one of our recent experiments, we used 36 male Wistar rats aged 6–6.5 months, weighing 400–500 g. The animals were purchased from the Stolbovaya subsidiary of the Federal State Budgetary Institution "Scientific Center for Biomedical Technologies of FMBA." The animals were kept in groups of three per cage in the vivarium of the Faculty of Basic Medicine under standard conditions, with free access to food and water. Immediately before the experiment, the animals underwent a 2-week adaptation to the experimental conditions. The animals were handled in the vivarium for 5 minutes per individual to accustom them to human contact. Handling was performed as follows: the animals were picked up by the chest with their tails secured, then placed on the researcher's forearms and shoulders. The procedure was performed several times on different days during the adaptation period. The cages were numbered to identify the animals. Each cage held a maximum of three animals, so each animal was marked with one or two stripes or left unmarked. All rats were also marked with an individual number. This marking was made on the tail with a permanent marker. When working with animals, standard precautions and the vivarium's internal rules were strictly adhered to, including the use of individual gloves and laboratory clothing, and adherence to the principles of asepsis and antisepsis during surgical interventions.

The animals were kept and used in experiments in full compliance with Directive 2010/63/EU of the European Parliament and of the Council of the European Union of September 22, 2010, on the protection of animals used for scientific purposes and the recommendations of the Bioethics Committee of Lomonosov Moscow State University (protocols No. 3.4 (March 21, 2021) and No. 3.5 (March 17, 2022)). This study was approved by the Bioethics Committee of Moscow State University (application No. 173-zh approved, Bioethics Commission meeting number: 160-d-z, meeting date: March 21, 2024).

Primary adipose-derived MSCs isolated from human subcutaneous fat were obtained from the biobank of the Center of Regenerative Medicine of the Medical Research and Education Institute, Lomonosov Moscow State University (https://human.depo.msu.ru). MSCs were characterized as cells adherent to plastic, expressing CD73, CD90, and CD105, lacking CD14, CD19, CD34, CD45, and HLA-DR, and able to differentiate in vitro into adipogenic, osteogenic, and chondrogenic lineages, which altogether meet the criteria established by the International Society for Cell Therapy (ISCT) (Vishnubhatla et al., 2014). Primary MSCs were immortalized via overexpression of the human telomerase catalytic subunit gene (Primak et al., 2024a, 2024b). The procedures for assembling vector and lentiviral particles, as well as cell modification, selection, and iMSC characterization, were described previously (Primak et al., 2024a, 2024b). The obtained iMSC cultures were cultured in DMEM/F12 medium (Gibco, #11320033) supplemented with 10% fetal bovine serum (FBS) (ThermoFisher Scientific, #26140079) and 1% antibiotic/antimycotic mixture (Gibco, #15240062). The culture medium was changed every 3–4 days.

iMSC-derived secretome was obtained as described earlier (Sagaradze et al., 2019). Briefly, subconfluent cultures of iMSC (13–15 passages) were thoroughly washed with Hank's solution (PanEco, #P020P) and then cultured for 7 days in DMEM low glucose medium (Gibco, #31885049) supplemented with GlutaMAX™, pyruvate, and 100 U/mL penicillin-streptomycin (Gibco, #15240062). Then, the conditioned medium of iMSC cultures (the so-called secretome) was aspirated from the cells and centrifuged for 10 minutes at 300 g to remove cell debris. Afterward, the medium was concentrated 10-fold using a centrifugal ultrafiltration membrane with a molecular weight cutoff of 10 kDa. All procedures performed with patient tissue samples complied with the World Medical Association's Declaration of Helsinki, "Ethical Principles for Medical Research Involving Human Subjects," and were approved by the Ethics Committee of Lomonosov Moscow State University (IRB00010587), protocol No. 4 (2018).

There are many models of TBI. Among them are falling cargo, pneumatic impact, penetrating impact, blast wave, and hydrodynamic impact. Lateral hydrodynamic impact (LHI) is the most commonly used model (Xi et al., 2003). It has several advantages, including good reproducibility, controllable impact, high validity, and correspondence with the clinical course of TBI. This model reproduces virtually all aspects of brain tissue damage: concussion, contusion, axonal damage, and hemorrhage. The model's disadvantages include the need for craniotomy, the absence of impact damage to the skull, and a high mortality rate (Xi et al., 2003; Komoltsev et al., 2022). However, this model has been demonstrated to have high validity, reliability, and clinical significance, confirming its suitability for TBI research and the potential application of the data obtained in clinical practice (Alder et al., 2011). Moreover, this model reproduces key features of TBI, including epileptiform activity and hippocampal damage, making it a valuable tool for studying its mechanisms and consequences, as well as for developing new therapeutic strategies. Thus, data obtained using the LHI model can be successfully translated into clinical practice (Komoltsev et al., 2022).

For LHI modeling, each rat was weighed and anesthetized with 2–3% isoflurane (Laboratorios Carizo, Spain) delivered by inhalation. Then, they were placed in the stereotaxic apparatus with Blepharogel (Geltek, Russia) applied to their eyes. The scalp was shaved and treated with 70% ethyl alcohol and 5% iodine alcohol solution (Obnovlenie, Russia). Under aseptic conditions, the scalp and aponeurosis of the skull were dissected. Soft tissues were laterally moved, and hemostasis was achieved with 3% hydrogen peroxide (Ecotex, Russia) and a hemostatic collagen sponge (Belkozin, Russia). A wound retractor was installed. The skull was dried with 96% ethyl alcohol. A 3 mm diameter trephination hole was made 3 mm caudal to the bregma and 3 mm lateral to the midline on the right side using a hand trephine. A visual assessment of the integrity of the dura mater was performed. A Luer-Lock tip was glued to the craniotomy site with cyanoacrylate adhesive gel (Loctite 454, Henkel, Ireland) and then fixed with dental cement, composed of Acrodent powder (Stoma, Ukraine) and Villacryl SP liquid (EVERALL7, Poland). The rats were then connected to a device for applying LHI, the Fluid Percussion Device with a PC-Based Pressure Measurement Unit, Model FP302 (AmScien Instruments, USA), using an 80-centimeter polyvinyl chloride tube with a 3 mm diameter. The target impact force was set at 3.00 atmospheres and monitored after each impact. After TBI or sham surgery, the bone plate was returned, fixed with BF-6 glue (Tula Pharmaceutical Factory, Russia). A mixture of ceftriaxone powder (Sintez, Russia) and 0.05% chlorhexidine bigluconate solution (Samaramedprom, Russia) was used to sterilize the operating field. Then, the surgical wound was sutured with 3/0 polypropylene on a cutting atraumatic needle (PTO Medtekhnika, Russia). The experimental setup is illustrated in Figure 5.2.

In the early postoperative period, animals were placed on a heated surface at 38°C to improve survival. The exclusion criteria used at this stage were: unsuccessful surgery, damage to the dura mater, presence of glue on the dura mater, and death within 24 hours after the TBI.

The animals were randomly divided into four groups: TBI + treatment, TBI (control for treatment), sham-operated, and intact. The first two groups underwent surgery and injury. The third group underwent the same procedure, but without the direct TBI. The intact animals were not injured. The first group received the iMSC-derived secretome, the second group received empty cell culture medium (DMEM/F12 without phenol red, Gibco, #21041025), and the third group received a saline solution (Solofarm, Russia). The preparations for the experiment were stored on ice before administration. The microcentrifuge tubes containing the iMSC-derived secretome or control cell culture medium were labeled, so the researcher who conducted the injections was blinded to which preparation he administered.

All animals were randomized before administration. The drugs were administered into the lateral tail vein using U-100 insulin syringes with a 27G needle 1 hour after the stroke. The rats were immobilized using an immobilization cylinder. To improve vein visibility, the rats' tails were immersed in warm water immediately before administration. The volume of all infusions was 100 μl. After the central part of the experiment, the animals were transported to the vivarium, where they were maintained under the strict temperature regime.

The animals were observed for 14 days after TBI modeling, with daily recording of rat deaths and neurological status assessments on days 3 and 10 post-TBI modeling. The McGraw stroke scale,

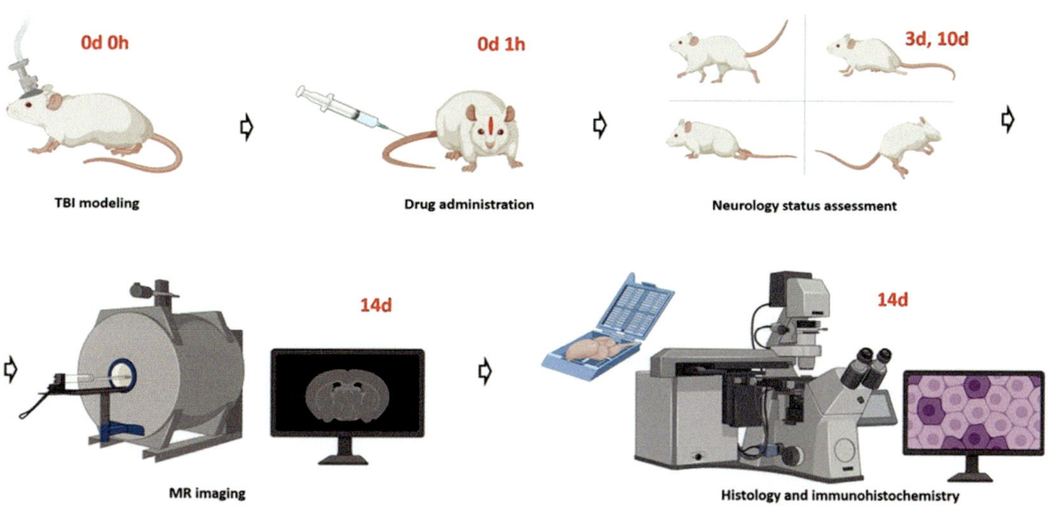

FIGURE 5.2 The pipeline of the experiment.

modified for rodents by I.V. Ganushkina (McGraw et al., 1976; Gannushkina et al., 2000), was used to assess neurological status. In short, "visually healthy animals" had no signs of neurological deficit. Animals with signs of lethargy, limb weakness, tremor, ptosis, and/or hemiparesis were considered "mildly affected." Animals with signs of paresis and/or limb paralysis, impaired coordination, or in a comatose state were considered "severely affected." For statistical purposes, the animals were divided into four groups: those with no abnormalities, those with mild abnormalities, those with severe abnormalities, and those that had died. The researchers conducting the neurological testing were blinded to each rat's treatment method.

On day 14, a subset of animals underwent MRI examination. To obtain in vivo brain images, a 7 Tesla magnetic resonance tomograph (BioSpec 70/30 USR; Bruker BioSpin, Germany) equipped with ParaVision® 5.1 software and a gradient system with a maximum amplitude of 105 mT/m was used. Animals were anesthetized with a gas mixture of oxygen (flow rate: 1 L/min) and isoflurane (4% for induction in a chamber and 1.5–2% for maintenance via a nose mask) throughout the MRI procedure.

Axial T2-weighted images were obtained using a two-dimensional Turbo spin echo (TSE) pulse sequence with the following scanning parameters: field of view (FOV) = 2.6×2.6 cm^2, matrix = 200×200, in-plane resolution = 0.13×0.13 mm^2, slice thickness = 0.8 mm, number of slices = 20, repetition time (TR) = 5 s, first echo time (TE_{first}) = 15.3 ms, adequate echo time (TE_{eff}) = 61.2 ms, echo train length (ETL) = 8, bandwidth (BW) = 25 kHz. The total scanning time for each rat was approximately 6 minutes and 15 seconds.

Each image containing a visible lesion was included in the analysis. The following criteria were used for evaluation: cerebral edema, tissue defects, hippocampal sclerosis, and intracranial hemorrhage. For each of the listed criteria, 0, 1, or 2 points were assigned depending on the severity of the criterion in the image. The scores were then summed and analyzed statistically, both as a total and for each criterion.

Fourteen days after the TBI, all animals were euthanized by slowly increasing the carbon dioxide concentration in the inhaled air. They were then decapitated, and their brains were removed using surgical instruments and fixed in buffered formalin (HistoSave, Russia) at a rate of 20 ml per sample. The formalin was changed three times every three days. To prepare histological samples, brain specimens were cut at the site of impact, placed in histological cassettes, and embedded in paraffin. Next, the blocks were cut using a microtome, the sections were deparaffinized, stained with hematoxylin and eosin, and fixed on glass. This staining allows visualization and measurement of necrotic focus size, leukocyte infiltration, and gliosis.

The same procedure was used to prepare immunohistochemical samples, but after deparaffinization, the antigens were unmasked. The sections were stained with antibodies to CD68, a marker of active phagocytosis and activated microglia (Abcam, #ab125212, USA), followed by staining with secondary fluorescently labeled antibodies (Invitrogen, #A11034, USA). The nuclei were stained with DAPI solution (Sigma, #MBD0015-10ML, USA).

The obtained samples were studied, and the pictures were taken using an inverted wide-field microscope Nikon Eclipse Ti2 equipped with a CFI Plan Fluor DLL 10X/0.3 (Nikon, Tokyo, Japan) and a Kinetix digital monochrome camera (Teledyne Photometrics, USA) and a Ri2 color camera (Nikon, Tokyo, Japan). The images obtained were analyzed using NIS-Elements AR 5.40.02 and ImageJ v. 1.54c (https://imagej.net/ij/index.html).

The histological samples were analyzed as follows. The samples were evaluated for necrosis, gliosis, and leukocyte infiltration. Each sample was assessed for each criterion on a scale of 0 to 2, where 0 indicated no manifestations and 2 indicated severe manifestations. Next, the sum of the signs and the signs themselves were separately statistically evaluated.

Immunohistochemical samples were analyzed by searching for microglial activation foci in the impact area and manually counting cells within them. Statistical processing was performed using the methods described above. Statistical analysis was performed using STATISTICA 10 (StatSoft, Germany). For clarity, scatter plots were constructed. Next, for the use of descriptive statistics in the case of normal distribution, the notation "M ± S" was used, where M is the mean value and S is the standard deviation. Otherwise, the data were described using the median (Me) and the 25th and 75th percentiles (Q1 and Q3, respectively) in the format Me [Q1, Q3]. Quantitative data were assessed for normality using the Shapiro-Wilk test. The hypothesis of normality was accepted at the $p > 0.05$ significance level. Equality of variances was assessed using the Levene test. At $p < 0.05$, the equality of variances between the groups was concluded. To compare two independent groups under the assumption of normality, we used Student's t-test or t-tests with separate variance estimates if no conclusion was made about the equality of variances between the groups. If the data were considered nonparametric or if this issue was debatable due to the sample size, the nonparametric U-test or Mann-Whitney test was used. Univariate survival analysis was performed using the Kaplan-Meier method. Due to the experimental design, the data were censored. The level of statistical significance was set at a probability of error of $p < 0.05$.

5.7 OUTCOME OF THE TREATMENT AND INTERPRETATION OF THE DATA

Four rats died within the first hour after the TBI modeling. No other exclusion criteria were applied due to the unsuccessful operation. Assessments on days 3 and 10 after TBI modeling revealed no differences between the animals. All rats were assessed as visually healthy. All rats survived for 14 days until the experiment's endpoint. In this case, no survival analysis was performed. This TBI model exhibited no mortality or neurological symptoms in rats over two weeks.

MRI images of the brains of 18 rats were obtained. Not all animals were studied due to logistical difficulties. Of these, 7 animals belonged to the treatment group, 7 to the control group, 3 were sham-operated, and 1 remained intact. As expected, MR imaging revealed several brain tissue defects, primarily in animals from the control group. These include hippocampal sclerosis, intracranial hematoma, cortical defects, and edema (Figure 5.3).

The table below presents the results of the MRI image analysis for rats from the experimental and control groups (Table 5.1). The assessment was conducted in a randomized and blinded manner. The data obtained were used for statistical processing.

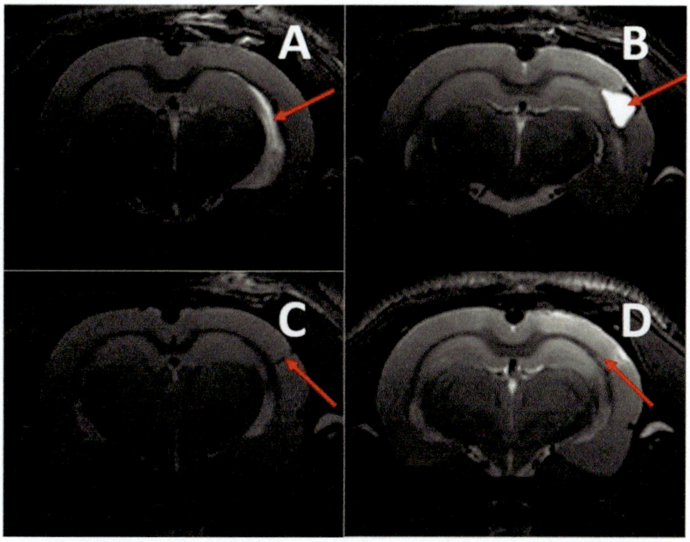

FIGURE 5.3 Samples of pathological signs in the rat brain after TBI modeling (marked with red arrows). A—hippocampal sclerosis, B—intracranial hematoma, C—linear defect in the cortical zone, D—focal edema in the gray matter.

TABLE 5.1
Results of the MRI Image Animal-by-Animal Evaluation

№	Group	Hippocampal Sclerosis	Brain Tissue Defect	Edema	Hemorrhage	Sum
14	iMSC-derived secretome	1	0	0	0	1
16	iMSC-derived secretome	0	0	0	0	0
17	DMEM/F12 (Control)	1	2	1	0	4
18	DMEM/F12 (Control)	2	2	0	0	4
20	iMSC-derived secretome	0	1	1	0	2
21	DMEM/F12 (Control)	1	1	1	0	3
22	Sham-operated	0	0	0	0	0
23	iMSC-derived secretome	0	1	1	0	2
24	DMEM/F12 (Control)	1	2	2	1	6
26	iMSC-derived secretome	0	1	0	0	1
27	Sham-operated	0	0	0	0	0
28	iMSC-derived secretome	0	0	0	0	0
30	DMEM/F12 (Control)	1	1	0	0	2
31	Sham-operated	0	0	0	0	0
32	iMSC-derived secretome	0	1	1	0	2
34	DMEM/F12 (Control)	1	2	0	0	3
35	DMEM/F12 (Control)	2	2	0	0	4
36	Intact	0	0	0	0	0

To assess the drug's (iMSC-derived secretome) effectiveness, two groups were compared: the treatment group and the control group. First, we determined whether the distribution was normal. The Shapiro-Wilk test P-value was 0.512, indicating normality. Levene's test: $p = 0.640$, indicating equal variances. Student's t-test: $p = 0.00085$, which is significantly less than 0.05, suggesting that the treatment is effective (Figure 5.4). The average score for the control group was 3.714 ± 1.254, and for the treatment group, 1.143 ± 0.900.

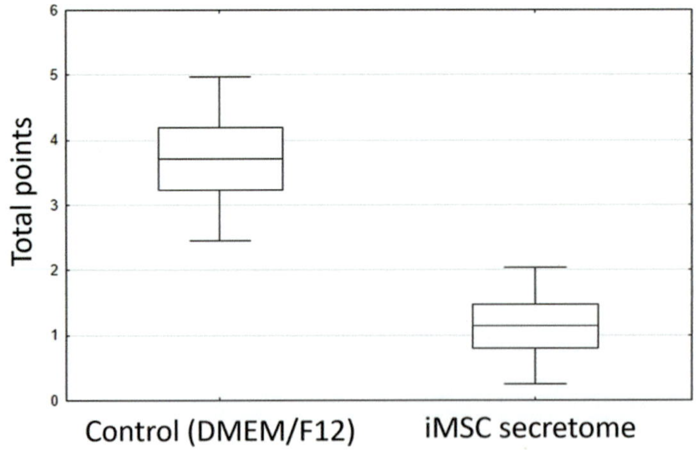

FIGURE 5.4 The total points for the incidence of the pathological signs in MRI image assessment in the control and experimental groups.

To avoid inaccuracies in applying this statistical criterion, a nonparametric criterion was also employed, given that the number of observations in each group is 7. The Mann-Whitney U test also showed a statistically significant difference between the groups ($p = 0.004$). Median and quartile values: control—4 [3.4], treatment—1 [0.2]. Additionally, it is worth noting that a score of 0 in all intact and sham-operated observations indicates that the scoring system does not produce false negative results—the model does indeed cause damage. We also evaluated each feature separately and came to the following conclusion. The incidence of hippocampal sclerosis ($p = 0.00037$; $p = 0.00279$) and cortical defect ($p = 0.00128$; $p = 0.01060$) changed most significantly with the administration of iMSC-derived secretome. Edema and hematoma did not show statistically significant differences between the groups. The above p-values for Student's t-tests and Mann-Whitney U tests are similar to those calculated for the total score.

Hematoxylin/eosin staining allowed assessment of possible foci of necrosis, leukocyte infiltration, and gliosis in the brain sample slices. Visual evaluation of the brain slices did not reveal any significant changes in tissue structure in any of the micro-preparations. No significant difference was found between the experimental groups in this study. Immunohistochemical staining enabled visualization of activated microglia and cell nuclei. A key finding was the activation of microglia at the border between the cortex and hippocampus on the affected side. Photographs of the hippocampus from the treatment and control groups are shown in Figure 5.5A.

To evaluate the neuroprotective effect of the MSC secretome, images were quantified to assess the number of CD68-positive cells. During the count, several preparations had to be excluded due to the impossibility of objectively evaluating the presence or absence of foci. The cell count results are shown in Table 5.2. The evaluation was conducted in a randomized and blinded manner. We used these data for statistical processing.

To analyze the ability of iMSC-derived secretome to suppress neuroinflammation in the rat hippocampus, two groups were compared: the treatment and control groups. First, it was determined whether the distribution was normal. The Shapiro-Wilk test P-value was 0.092, confirming normality. Levene's test: $p = 0.233$, indicating equal variances. Student's t-test: $p = 0.01178$, which is less than 0.05, suggesting that the treatment is effective. The average number of CD68-positive cells for the control group was 403.6 ± 226.6, and for the treatment group, 111.4 ± 126.9 (Figure 5.5B). As in the previous analysis, the data formally passed the normality criteria, but the sample size remains debatable. For a more accurate analysis, we will use a nonparametric criterion, as there are 5 and 8 observations in each group. According to the Mann-Whitney U test, a statistically significant difference is also observed between the groups ($p = 0.0214$). Median and quartile values: control—331 [227.617], treatment—72 [0.222].

5.8 INTERPRETATION OF THE DATA

This study demonstrated that the secretome of immortalized MSCs possesses neuroprotective properties and can suppress hippocampal neuroinflammation induced by traumatic brain injury in a lateral hydrodynamic impact (LHI) model in rats. The data obtained show that administration of the iMSC-derived secretome reduces the incidence of in vivo morphological changes in brain tissue caused by trauma and neuroinflammation, consistent with previous studies in this field (Cozene et al., 2021; Pischiutta et al., 2022).

Early mortality (within 1 hour after TBI modeling) in 4 of 26 rats is consistent with previous studies in this model (Komoltsev et al., 2022; Komoltsev et al., 2020). Mortality was associated with both the impact itself and flaws in hole drilling (1 animal). No pathological neurological symptoms were observed in the animals after TBI compared with control animals. In future studies, it may be necessary to administer additional behavioral or cognitive tests to identify significant differences between the experimental groups.

Analyses of brain tissue damage on MR images indicate the influence of iMSC-derived secretome on the incidence of brain tissue damage signs (hippocampal sclerosis, brain tissue defect,

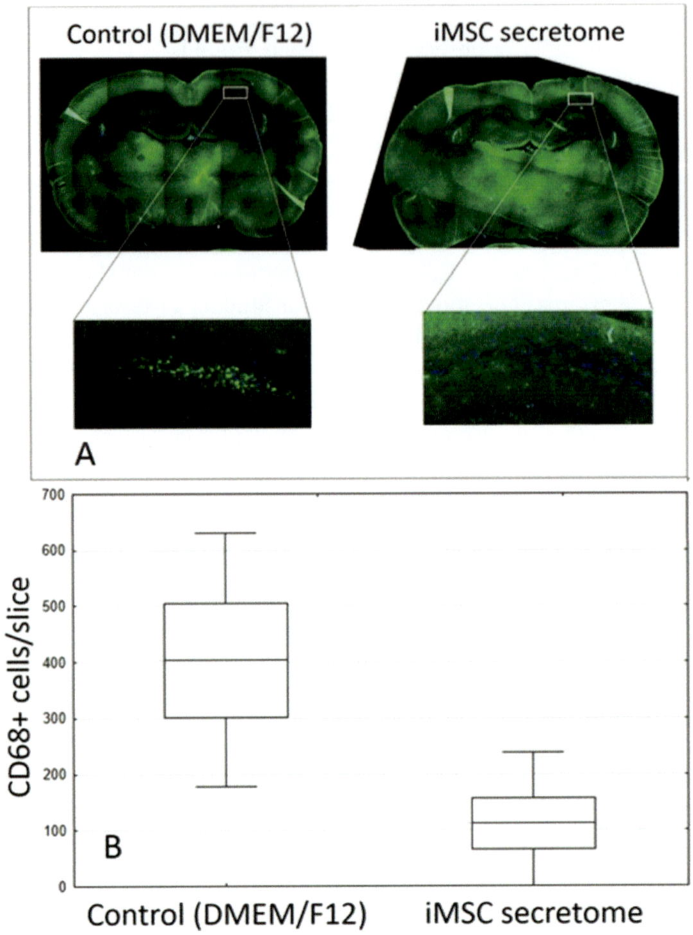

FIGURE 5.5 Neuroinflammation in the ipsilateral hippocampus of rats with TBI. A—samples of photos of hippocampus from treatment and control groups, green—CD68 positive cells, blue—cell nuclei (DAPI staining). B—Comparison of the number of CD68-positive cells between control and iMSC groups.

edema, and hemorrhage). Of particular interest is the effect of the iMSC-derived secretome on reducing the incidence of hippocampal sclerosis, which warrants further study. Hippocampal sclerosis is characterized by the loss of neurons (particularly pyramidal neurons) and gliosis (scarring) in the hippocampus. Hippocampal sclerosis is frequently found in patients with temporal epilepsy (Thom et al., 2014) and may be associated with cognitive deficits and mood disorders (depression). Our data suggest that the iMSC-derived secretome may prevent hippocampal sclerosis.

Histochemical evaluation did not reveal any signs of brain tissue damage, including necrosis, infiltration by foreign cells, or gliosis, in the damaged area. However, immunohistochemical staining of brain sections containing the lesion site showed the presence of microglial activation in the CA1 region of the ipsilateral hippocampus. Even a single administration of iMSC-derived secretome statistically significantly reduced microglial activation. This suggests that the potential to modify the course of TBI-induced neuroinflammation exists. In the long-term treatment of neuroinflammation, the possibility of administering the iMSC-derived secretome should be considered. It can be suggested that the course of treatment with the iMSC secretome will have a more pronounced and stable neuroprotective and anti-neuroinflammatory effect.

There are apparent similarities in the neuroprotective effects of the MSC secretome across various TBI models when our data are compared with previously reported results. For example, other

TABLE 5.2
Results of the Counting of CD68-Positive Cells in Animal-by-Animal Mode

№	Group	CD68-Positive Cells
8	iMSC-derived secretome	0
10	DMEM/F12 (Control)	331
11	DMEM/F12 (Control)	737
12	iMSC-derived secretome	0
14	iMSC-derived secretome	307
16	iMSC-derived secretome	0
17	DMEM/F12 (Control)	496
19	Sham-operated	0
20	iMSC-derived secretome	217
21	DMEM/F12 (Control)	322
22	Sham-operated	0
23	iMSC-derived secretome	143
24	DMEM/F12 (Control)	132
26	iMSC-derived secretome	224
28	iMSC-derived secretome	0

studies observed reduced inflammatory responses and improved neuronal regeneration after administration of the MSC secretome (Mugenyi et al., 2023), confirming the hypothesis that the MSC secretome possesses powerful anti-inflammatory and neuroprotective properties, potentially helpful in treating TBI.

The mechanism underlying the neuroprotective activity of intravenously administered iMSC-derived secretome was not established in our experiments (another limitation of this study). However, given the secretome's composition, we can assume its positive effect results from a combination of a direct neuroprotective effect on neuronal cells (via the neurotrophic factors present in the secretome) and the suppression of neuroinflammation (via anti-inflammatory molecules). Several studies confirmed this direct neuroprotection hypothesis: it has been shown that the MSC secretome includes specific proteins (Marshall et al., 1992; Striggow et al., 2000; Hayes et al., 2021; Farshbaf et al., 2021) and molecular complexes (microvesicles) (Davis et al., 2021; Rubtsov et al., 2017) able to cross the blood-brain barrier, providing neuroprotection and improving neurological outcomes. Moreover, it has been demonstrated that under conditions of hemorrhage or local ischemia, the permeability of the blood-brain barrier increases (Bernardo-Castro et al., 2020; Muhammad et al., 2022), potentially leading to a rise in the influx of neuroprotective molecules to the lesion site. MSC secretomes suppress the activation of T lymphocytes and monocyte-macrophage cells, as well as their extravasation into the lesion site (Roozenbeek et al., 2013; Liu et al., 2020; Bruno et al., 2015). Presumably, MSC secretome administered intravenously can increase the activation threshold of T cells and monocytes/macrophages, thereby suppressing their recruitment to the site of hemorrhage and preventing the progression of neuroinflammation and secondary damage (Ceserani et al., 2016). The anti-inflammatory activity of the MSC secretome may also mediate indirect neuroprotection of brain tissue.

The mechanisms of neuroprotective activity of endovascularly circulating growth factors in acute cerebrovascular accidents remain obscure. The results obtained are crucial for clinical practice, as they demonstrate the potential of the secretome from immortalized MSCs in treating TBI. The prospects for using iMSC-derived secretome in clinical practice include its application to reduce both primary and secondary brain damage and improve functional outcomes in patients with TBI. Our study used the Lateral hydrodynamic impact (LHI) model, which is well-validated for TBI research in animals and is regarded as the gold standard for rodent TBI models. This model reproduces many aspects of TBI, including inflammatory responses, apoptosis, and neuronal regeneration. Analysis of brain tissue damage and recovery was performed using modern methods, including MRI, histology, immunohistochemistry, and neurological tests, thereby ensuring high accuracy and reliability of the data obtained.

Given the promising data obtained with a limited number of experimental animals, it is clear that further studies are needed to evaluate the long-term effects of iMSC-derived secretome and its safety with respect to different routes, time, and frequency of administration. The drug was administered once, which may also affect treatment effectiveness. These aspects are necessary for the further development and optimization of therapeutic strategies.

Future studies should focus on further investigation of the molecular mechanisms of action of iMSC-derived secretome in TBI, as well as on standardizing its composition and methods of production. This may enable translation of the experimental data into clinical trials to evaluate the efficacy and safety of iMSC-derived secretome in patients with TBI. These steps will help confirm the clinical significance of the data obtained and accelerate the introduction of new therapeutic approaches into practice.

5.9 CONCLUSION AND FUTURE PERSPECTIVE

The results obtained allow us to conclude that the iMSC-derived secretome exhibits high neuroprotective activity in the TBI model, even under conditions closely resembling those in clinical settings, and is a promising candidate with high translation potential for the development of drugs to treat acute brain injuries, including traumatic brain injuries and intracerebral hemorrhages. The next step in this work is to standardize the composition of the iMSC secretome and identify its active components, which may enable replacing the total secretome with a combination of synthesized active molecules in the future.

REFERENCES

Alder J, Fujioka W, Lifshitz J, Crockett DP, Thakker-Varia S. Lateral fluid percussion: model of traumatic brain injury in mice. *J Vis Exp*. 2011 Aug 22;(54):3063. doi: 10.3791/3063

Alvarez-Viejo M, Haider KH. Mesenchymal stem cells. In: KH Haider, editor. *Handbook of Stem Cell Therapy*. Singapore: Springer; 2022. https://doi.org/10.1007/978-981-16-6016-0_6-1.

Balami JS, Buchan AM. Complications of intracerebral haemorrhage. *Lancet Neurol*. 2012;11(1):101–18.

Bernardo-Castro S, Sousa JA, Brás A, Cecília C, Rodrigues B, Almendra L, Machado C, et al. Pathophysiology of blood-brain barrier permeability throughout the different stages of ischemic stroke and its implications on hemorrhagic transformation and recovery. *Front Neurol*. 2020 Dec 9;11:594672. doi: 10.3389 / fneur 2020.594672

Bruno S, Deregibus MC, Camussi G. The secretome of mesenchymal stromal cells: Role of extracellular vesicles in immunomodulation. *Immunol Lett*. 2015 Dec;168(2):154–8. doi: 10.1016/j.imlet.2015.06.007

Butcher KS, Baird T, MacGregor L, Desmond P, Tress B, Davis S. Perihematomal edema in primary intracerebral hemorrhage is plasma-derived. Stroke. 2004 Aug;35(8):1879–85. doi: 10.1161/01.STR.0000131807.54742.1a

Ceserani V, Ferri A, Berenzi A, Benetti A, Ciusani E, Pascucci L, Bazzucchi C, et al. Angiogenic and anti-inflammatory properties of micro-fragmented fat tissue and its derived mesenchymal stromal cells. *Vasc Cell*. 2016 Aug 18;8:3. doi: 10.1186/s13221-016-0037-3.

Chang CF, Goods BA, Askenase MH, Hammond MD, Renfroe SC, Steinschneider AF, Landreneau MJ, Ai Y, Beatty HE, da Costa LHA, Mack M, Sheth KN, Greer DM, et al. Erythrocyte efferocytosis modulates macrophages towards recovery after intracerebral hemorrhage. *J Clin Invest*. 2018 Feb 1;128(2):607–24. doi: 10.1172/JCI95612

ClinicalTrials.Gov, A. https://clinicaltrials.gov/study/NCT04063215

ClinicalTrials.Gov, B. https://clinicaltrials.gov/study/NCT02742857

ClinicalTrials.Gov, C. https://clinicaltrials.gov/study/NCT03629275

Coppé JP, Desprez PY, Krtolica A, Campisi J. The senescence-associated secretory phenotype: the dark side of tumor suppression. *Annu Rev Pathol.* 2010;5:99–118. doi: 10.1146/annurev-pathol-121808-102144.

Cozene B, Sadanandan N, Farooq J, Kingsbury C, Park YJ, Wang ZJ, Moscatello A, et al. Mesenchymal stem cell-induced anti-neuroinflammation against traumatic brain injury. *Cell Transplant.* 2021 Jan-Dec;30:9636897211035715. doi: 10.1177/09636897211035715

Cui L, Saeed Y, Li H, Yang J. Regenerative medicine and traumatic brain injury: from stem cell to cell-free therapeutic strategies. *Regen Med.* 2022 Jan;17(1):37–53. doi: 10.2217/rme-2021-0069

Davis C, Savitz SI, Satani N. Mesenchymal stem cell derived extracellular vesicles for repairing the neurovascular unit after ischemic stroke. *Cells.* 2021 Mar 31;10(4):767. doi: 10.3390/cells10040767

Dewan MC, Rattani A, Gupta S, Baticulon RE, Hung YC, Punchak M, Agrawal A, et al. Estimating the global incidence of traumatic brain injury. *J Neurosurg.* 2018 Apr 27;130(4):1080–97. doi: 10.3171/2017.10.JNS17352

Dzhauari S, Basalova N, Primak A, Balabanyan V, Efimenko A, Skryabina M, Popov V, et al. The secretome of mesenchymal stromal cells in treating intracerebral hemorrhage: the first step to bedside. *Pharmaceutics.* 2023 May 29;15(6):1608. doi: 10.3390/pharmaceutics15061608.

Farshbaf JM, Alviña K. Multiple roles in neuroprotection for the exercise-derived Myokine Irisin. *Front Aging Neurosci.* 2021 Apr 16;13:649929. doi: 10.3389/fnagi.. 2021.649929

Gannushkina IV. Cerebral circulation in different types of circulatory hypoxia of the brain. *Bull Russ Acad Med Sci.* 2000;(9):22–27. [In Russian]

Godoy DA, Murillo-Cabezas F, Suarez JI, Badenes R, Pelosi P, Robba C. "THE MANTLE" bundle for minimizing cerebral hypoxia in severe traumatic brain injury. *Crit Care.* 2023 Jan 12;27(1):13. doi: 10.1186/s13054-022-04242-3

Guan B, Anderson DB, Chen L, Feng S, Zhou H. Global, regional and national burden of traumatic brain injury and spinal cord injury, 1990–2019: a systematic analysis for the global burden of disease study 2019. *BMJ Open.* 2023 Oct 6;13(10):e075049. doi: 10.1136/bmjopen-2023-075049

Gulyaeva NV. Glucocorticoids orchestrate adult hippocampal plasticity: growth points and translational aspects. *Biochemistry* (Mosc). 2023 May; 88(5):565–89. doi: 10.1134/S0006297923050012.

Gulyaeva NV, Onufriev MV, Moiseeva YV. Ischemic stroke, glucocorticoids, and remote hippocampal damage: a translational outlook and implications for modeling. *Front Neurosci.* 2021 Dec 9;15:781964. doi: 10.3389/fnins.2021.781964.

Haider KH, Aziz S. Paracrine hypothesis and cardiac repair. *Int J Stem Cell Res Transplant.* 2017; 5(1): 265–67.

Hasan A, Deeb G, Rahal R, Atwi K, Mondello S, Marei HE, Gali A, Sleiman E. Mesenchymal stem cells in the treatment of traumatic brain injury. *Front Neurol.* 2017 Feb 20;8:28. doi: 10.3389/fneur 2017.00028

Hayes CA, Valcarcel-Ares MN, Ashpole NM. Preclinical and clinical evidence of IGF-1 as a prognostic marker and acute intervention with ischemic stroke. *J Cereb Blood Flow Metab.* 2021 Oct;41(10):2475–91. doi: 10.1177/0271678X211000894

Hosseini SM, Sani M, Haider KH, Dorvash MR, Ziaee SM, Karimi A (2018) Concomitant use of mesenchymal stem cells and neural stem cells for treatment of spinal cord injury: a combo cell therapy approach. *Neurosc Lett.* 668: 138–46.

Jiang Y, Zhu W, Zhu J, Wu L, Xu G, Liu X. Feasibility of delivering mesenchymal stem cells via catheter to the proximal end of the lesion artery in patients with stroke in the territory of the middle cerebral artery. *Cell Transplant.* 2013;22(12):2291–8. doi: 10.3727/096368912X658818

Karagyaur M, Dzhauari S, Basalova N, Aleksandrushkina N, Sagaradze G, Danilova N, Malkov P, et al. MSC secretome as a promising tool for neuroprotection and neuroregeneration in a model of intracerebral hemorrhage. *Pharmaceutics.* 2021 Nov 29;13(12):2031. doi: 10.3390/pharmaceutics13122031.

Koliaraki V, Prados A, Armaka M, Kollias G. The mesenchymal context in inflammation, immunity and cancer. *Nat Immunol.* 2020 Sep;21(9):974–82. doi: 10.1038/s41590-020-0741-2

Komoltsev IG, Frankevich SO, Shirobokova NI, Volkova AA, Onufriev MV, Moiseeva JV, Novikova MR, Gulyaeva NV. Neuroinflammation and neuronal loss in the hippocampus are associated with immediate posttraumatic seizures and corticosterone elevation in rats. *Int. J. Mol. Sci.* 2021; 22(11):5883. doi: 10.3390/ijms22115883

Komoltsev IG, Gulyaeva NV. Brain trauma, glucocorticoids and neuroinflammation: dangerous liaisons for the hippocampus. *Biomedicines.* 2022 May 15;10(5):1139. doi: 10.3390/biomedicines10051139

Komoltsev IG, Sinkin MV, Volkova AA, Smirnova EA, Novikova MR, Kordonskaya OO, Talypov AE, Guekht AB, Krylov VV, Gulyaeva NV. A Translational study on acute traumatic brain injury: high incidence of epileptiform activity on human and rat electrocorticograms and histological correlates in rats. *Brain Sci.* 2020 Aug 19;10(9):570. doi: 10.3390/brainsci10090570

Komoltsev IG, Tret'yakova LV, Frankevich SO, Shirobokova NI, Volkova AA, Butuzov AV, Novikova MR, Kvichansky AA, Moiseeva YV, Onufriev MV, Bolshakov AP, Gulyaeva NV. Neuroinflammatory cytokine response, neuronal death, and microglial proliferation in the hippocampus of rats during the early period after lateral fluid percussion-induced traumatic injury of the neocortex. *Mol. Neurobiol.* 2022 Feb;59(2):1151–1167. doi: 10.1007/s12035-021-02668-4

Lee JS, Hong JM, Moon GJ, Lee PH, Ahn YH, Bang OY; STARTING collaborators. A long-term follow-up study of intravenous autologous mesenchymal stem cell transplantation in patients with ischemic stroke. *Stem Cells.* 2010 Jun;28(6):1099–106. doi: 10.1002/stem.430

Liu R, Zhang H, Cheng S, Sun Y, Li H, Xiao J, Huang Y. Association of brain iron overload with brain edema and brain atrophy after intracerebral hemorrhage. *Front Neurol.* 2020 Dec 18;11:602413. doi: 10.3389/fneur.2020.602413.

Lu D, Xu Y, Liu Q, Zhang Q. Mesenchymal stem cell-macrophage crosstalk and maintenance of inflammatory microenvironment homeostasis. *Front Cell Dev Biol.* 2021 Jun 25;9:681171. doi: 10.3389/fcell.2021.681171

Marshall LF, Marshall SB, Klauber MR, Van Berkum Clark M, Eisenberg H, Jane JA, Luerssen TG, et al. The diagnosis of head injury requires a classification based on computed axial tomography. *J Neurotrauma.* 1992 Mar;9 Suppl 1:S287–92

McGraw CP, Pashayan AG, Wendel OT. Cerebral infarction in the Mongolian gerbil exacerbated by phenoxybenzamine treatment. *Stroke.* 1976 Sep-Oct;7(5):485–8. doi: 10.1161/01.str.7.5.485

Menon D, Harrison D. Prognostic modelling in traumatic brain injury. *BMJ.* 2008 Feb 23;336(7641):397–8. doi: 10.1136/bmj.39461.616991.80

Mishra R, Ucros HEV, Florez-Perdomo WA, Suarez JR, Moscote-Salazar LR, Rahman MM, Agrawal A. Predictive value of Rotterdam score and Marshall score in traumatic brain injury: a contemporary review. *Indian J Neurotrauma.* 2022;19(2):69–77.

Moore EE, Moore HB, Kornblith LZ, Neal MD, Hoffman M, Mutch NJ, Schöchl H, et al. Trauma-induced coagulopathy. *Nat Rev Dis Primers.* 2021 Apr 29;7(1):30. doi: 10.1038/s41572-021-00264-3

MRC CRASH Trial Collaborators; Perel P, Arango M, Clayton T, Edwards P, Komolafe E, Poccock S, et al. Predicting outcomes after traumatic brain injury: practical prognostic models based on a large cohort of international patients. *BMJ.* 2008 Feb 23;336(7641):425–9. doi: 10.1136/bmj.39461.643438.25

Mugenyi N, Sakaiwa N, Darko K, Shituluka M, Tango T, Tunde O, Lordstrong A, et al. Assessing the neuroprotective efficacy of atorvastatin in traumatic brain injury: a systematic review protocol. *J Surg Protoc Res Methodol.* 2023;2023(3):1–4.

Muhammad SA, Abbas AY, Imam MU, Saidu Y, Bilbis LS. Efficacy of stem cell secretome in the treatment of traumatic brain injury: a systematic review and meta-analysis of preclinical studies. *Mol Neurobiol.* 2022 May;59(5):2894–909. doi: 10.1007/s12035-022-02759-w

NeurologyLive.Com. https://www.neurologylive.com/view/adipose-derived-mesenchymal-stem-cell-therapy-meets-primary-end-point-phase-2-trial-relapsing-remitting-ms

Orr TJ, Lesha E, Kramer AH, Cecia A, Dugan JE, Schwartz B, Einhaus SL. Traumatic brain injury: a comprehensive review of biomechanics and molecular pathophysiology. *World Neurosurg.* 2024 May;185:74–88. doi: 10.1016 /j.wneu.2024.01.084

Pischiutta F, Caruso E, Cavaleiro H, Salgado AJ, Loane DJ, Zanier ER. Mesenchymal stromal cell secretome for traumatic brain injury: Focus on immunomodulatory action. *Exp Neurol.* 2022 Nov;357:114199. doi: 10.1016/j.expneurol.2022.114199.

Podgorny OV, Gulyaeva NV. Glucocorticoid-mediated mechanisms of hippocampal damage: Contribution of subgranular neurogenesis. *J. Neurochem.* 2021 May;157(3):370–392. doi: 10.1111/jnc.15265

Primak A, Kalinina N, Skryabina M, Usachev V, Chechekhin V, Vigovskiy M, Chechekhina E, et al. Novel immortalized human multipotent mesenchymal stromal cell line for studying hormonal signaling. *Int J Mol Sci.* 2024a Feb 19;25(4):2421. doi: 10.3390/ijms25042421

Primak AL, Kalinina NI, Skryabina MN, Usachev VA, Chechekhin VI, Vigovsky MA, Chechekhina ES, et al. Creation and characterization of human mesenchymal stromal cell cultures with prolonged proliferative potential for regenerative medicine tasks. *Regener Organs Tissues.* 2024b;2:24–45 [In Russian]

Qiao LY, Huang FJ, Zhao M, Xie JH, Shi J, Wang J, Lin XZ, Zuo H, Wang YL, Geng TC. A two-year follow-up study of cotransplantation with neural stem/progenitor cells and mesenchymal stromal cells in ischemic stroke patients. *Cell Transplant.* 2014;23 Suppl 1:S65–72. doi: 10.3727/096368914X684961

Rao SM, Harrington DL, Haaland KY, Bobholz JA, Cox RW, Binder JR. Distributed neural systems underlying the timing of movements. *J Neurosci.* 1997 Jul 15;17(14):5528–35. doi: 10.1523/JNEUROSCI.17-14-05528.1997

Roozenbeek B, Maas AI, Menon DK. Changing patterns in the epidemiology of traumatic brain injury. *Nat Rev Neurol.* 2013 Apr;9(4):231–6. doi: 10.1038/nrneurol.2013.22

Rubtsov Y, Goryunov K, Romanov A, Suzdaltseva Y, Sharonov G, Tkachuk V. Molecular mechanisms of immunomodulation properties of mesenchymal stromal cells: a new insight into the role of ICAM-1. *Stem Cells Int.* 2017;2017:6516854. doi: 10.1155/2017/6516854

Sagaradze G, Grigorieva O, Nimiritsky P, Basalova N, Kalinina N, Akopyan Z, Efimenko A. Conditioned medium from human mesenchymal stromal cells: towards the clinical translation. *Int J Mol Sci.* 2019 Apr 3;20(7):1656. doi: 10.3390/ijms20071656

Samsonraj RM, Law SF, Chandra A, Pignolo RJ. An unbiased proteomics approach to identify the senescence-associated secretory phenotype of human bone marrow-derived mesenchymal stem cells. *Bone Rep.* 2023 Mar 21;18:101674. doi: 10.1016/j.bonr.2023.101674

Sengupta V, Sengupta S, Lazo A, Woods P, Nolan A, Bremer N. Exosomes derived from bone marrow mesenchymal stem cells as treatment for severe COVID-19. *Stem Cells Dev.* 2020 Jun 15;29(12):747–54. doi: 10.1089/scd.2020.0080

Shin MS, Park SY, Park SR, Seol SH, Kwon JS. Clinical and empirical applications of the rey-osterrieth complex figure test. *Nat Protoc.* 2006;1(2):892–9. doi: 10.1038/nprot.2006.115

Steinberg GK, Kondziolka D, Wechsler LR, Lunsford LD, Coburn ML, Billigen JB, Kim AS, Johnson JN, Bates D, King B, Case C, McGrogan M, Yankee EW, Schwartz NE. Clinical outcomes of transplanted modified bone marrow-derived mesenchymal stem cells in stroke: a phase 1/2a study. *Stroke.* 2016 Jul;47(7):1817–24. doi: 10.1161/STROKEAHA.116.012995

Striggow F, Riek M, Breder J, Henrich-Noack P, Reymann KG, Reiser G. The protease thrombin is an endogenous mediator of hippocampal neuroprotection against ischemia at low concentrations but causes degeneration at high concentrations. *Proc Natl Acad Sci U S A.* 2000 Feb 29;97(5):2264–9. doi: 10.1073/pnas.040552897

Taran S, Pelosi P, Robba C. Optimizing oxygen delivery to the injured brain. *Curr Opin Crit Care.* 2022 Apr 1;28(2):145–56. doi: 10.1097/MCC.0000000000000913

Thom M. Review: Hippocampal sclerosis in epilepsy: a neuropathology review. *Neuropathol Appl Neurobiol.* 2014 Aug;40(5):520–43. doi: 10.1111/nan.12150

Vishnubhatla I, Corteling R, Stevanato L, Hicks C, Sinden J. The development of stem cell-derived exosomes as a cell-free regenerative medicine. *J Circ Biomark.* 2014;3(1):1–14.

Wafaisade A, Lefering R, Tjardes T, Wutzler S, Simanski C, Paffrath T, Fischer P, Bouillon B, Maegele M; Trauma registry of DGU. Acute coagulopathy in isolated blunt traumatic brain injury. *Neurocrit Care.* 2010 Apr;12(2):211–19. doi: 10.1007/s12028-009-9281-1

Xi G, Reiser G, Keep RF. The role of thrombin and thrombin receptors in ischemic, hemorrhagic and traumatic brain injury: deleterious or protective? *J Neurochem.* 2003 Jan;84(1):3–9. doi: 10.1046/j.1471-4159.2003.01268.x

Xia J, Arai S, Haider KH. Researches and applications of stem cell secretome. In: Haider KH. (ed.) *Stem Cells: From Potential to Promise.* Singapore: Springer, 2021. https://doi.org/10.1007/978-981-16-0301-3_7.

Zehtabchi S, Soghoian S, Liu Y, Carmody K, Shah L, Whittaker B, Sinert R. The association of coagulopathy and traumatic brain injury in patients with isolated head injury. *Resuscitation.* 2008 Jan;76(1):52–6. doi: 10.1016/j.resuscitation.2007.06.024

Zhu X, Tao L, Tejima-Mandeville E, Qiu J, Park J, Garber K, Ericsson M, et al. Plasmalemma permeability and necrotic cell death phenotypes after intracerebral hemorrhage in mice. *Stroke.* 2012 Feb;43(2):524–31. doi: 10.1161/STROKEAHA.111.635672

6 Therapeutic Potential of Secretome in Ophthalmic Disorders

Kevin Y. Wu[1], Amelia T. Yuan[2], Syeda T. Rizvi[2], Marjorie Carbonneau[1], and Simon D. Tran[3]

[1]Department of Surgery, Division of Ophthalmology, University of Sherbrooke, Canada

[2]Faculty of Medicine, Western University, London, Canada

[3]Faculty of Dental Medicine and Oral Health Sciences, McGill University, Canada

6.1 ABBREVIATIONS

a-SMA	Alpha-smooth muscle actin
ACG	Angle-closure glaucoma
AION	Arteritic ischemic optic neuropathy
AKT	Protein kinase B
AMD	Age-related macular degeneration
APC	Antigen-presenting cell
ATP	Adenosine triphosphate
Bcl-2	B-cell lymphoma 2
Bax	Bcl-2-associated X protein
CCL2	C-C motif chemokine ligand 2
CCL21	C-C motif chemokine ligand 21
CD44	Cluster of differentiation 44
CD68	Cluster of differentiation 68
COX2	Cyclooxygenase-2
CRISPR/Cas9	Clustered regularly interspaced short palindromic repeats/CRISPR-associated protein 9
CSC	Corneal stromal cells
CS	Chondroitin sulfate
CTGF	Connective tissue growth factor
CXCL	C-X-C motif chemokine ligand
DED	Dry eye disease
DMEK	Descemet membrane endothelial keratoplasty
DSAEK	Descemet stripping automated endothelial keratoplasty
EAU	Experimental autoimmune uveoretinitis

DOI: 10.1201/9781003653387-6

ECM	Extracellular matrix
ERK	Extracellular signal-regulated kinase
FGF	Fibroblast growth factor
HA	Hyaluronic acid
HCECs	Human corneal ECs
HGF	Hepatocyte growth factor
HMGB1	High mobility group box 1
ICAM-1	Intercellular adhesion molecule 1
IFN-γ	Interferon-gamma
IL	Interleukin
IOP	Intraocular pressure
MCP-1	Monocyte chemotactic protein-1
MHC-II	Major histocompatibility complex class II
miR	microRNA
MIGS	Microinvasive glaucoma surgery
MMP	Matrix metalloproteinase
MSC	Mesenchymal stem cell
MSC-CM	Mesenchymal stem cell-conditioned medium
MSC-EVs	Mesenchymal stem cell-derived EVs.
MSC-S	Mesenchymal stem cell-secretome
NAION	Non-arteritic ischemic optic neuropathy
NEDD4	Neural precursor cell expressed developmentally downregulated protein 4
NF-κB	Nuclear factor kappa-light-chain-enhancer of activated B cells
NM	Nitrogen mustard
NPDR	Non-proliferative diabetic retinopathy
PACAP38	Pituitary adenylate cyclase-activating polypeptide 38
PDR	Proliferative diabetic retinopathy
PD-L1	Programmed death-ligand 1
PEDF	Pigment epithelium-derived factor
PI3K	Phosphoinositide 3-kinase
POAG	Primary open-angle glaucoma
PTEN	Phosphatase and Tensin Homolog
RPE	Retinal pigment epithelium
sFLT-1	Soluble fms-like tyrosine kinase-1
siRNA	Small interfering RNA
SPARC	Secreted protein acidic and rich in cysteine
SS	Sjogren's syndrome
T1D	Type 1 diabetes
TAB2	TGF-beta activated kinase 1/MAP3K7 binding protein 2
TAT	Transactivator of transcription
Tfh	T follicular helper
Th	T helper
TIMP	Tissue inhibitor of metalloproteinases
TLR4	Toll-like receptor 4
TNF-α	Tumor necrosis factor-alpha
Treg	Regulatory T cell
TSP-1	Thrombospondin-1
uPAR	Urokinase plasminogen activator receptor
VEGF	Vascular endothelial growth factor
ZO-1	Zonula occludens-1

6.2 INTRODUCTION

Some ophthalmic disorders present significant challenges in clinical management due to their complex pathophysiology and the limitations of current therapeutic strategies. Traditional treatments, such as topical drugs, intravitreal injections, surgical interventions, and systemic immunosuppressive therapies, often face issues related to limited bioavailability, invasive procedures, and potential side effects (Appell et al., 2024). In recent years, the secretome—comprising EVs, exosomes, cytokines, and growth factors—has emerged as a promising area of research in regenerative medicine, with its potential applications in ophthalmology now actively being explored (Seong et al., 2023).

This chapter provides an overview of the therapeutic potential of secretome-based strategies in the management of ophthalmic diseases. In retinal diseases such as age-related macular degeneration, diabetic retinopathy, and retinitis pigmentosa, secretome-derived interventions may facilitate neuroprotection, targeted drug delivery, and enhanced vascular repair, potentially reducing the need for frequent intravitreal injections (Van Lookeren Campagne et al., 2014). In glaucoma, secretome-based therapies could contribute to trabecular meshwork regeneration and optic nerve protection, offering new approaches for intraocular pressure regulation and neurodegeneration mitigation (Seong et al., 2023). For corneal and anterior segment disorders, secretome components have been investigated for their ability to promote epithelial wound healing, modulate extracellular matrix remodeling, and suppress inflammation, potentially reducing corneal fibrosis and neovascularization. In uveitis, the immunomodulatory properties of the secretome may help regulate inflammatory responses, providing a potential adjunct or alternative to conventional corticosteroid and immunosuppressive therapies.

While pre-clinical studies indicate promising effects (Table 6.1), significant challenges remain in translating secretome-based therapies into clinical practice (Aslam and Haider, 2018). Further research is needed to refine delivery methods, optimize dosing regimens, and ensure long-term safety and efficacy. This chapter aims to provide a comprehensive review of the current understanding of secretome-based approaches in ophthalmology, outlining their potential advantages and the obstacles that must be addressed before they can be translated into clinical practice.

6.3 RETINA

6.3.1 Current Challenges in Managing Retinal Diseases

Retinal diseases range from age-related macular degeneration to diabetic retinopathy. In retinal damage management, the primary issue at hand is the physiological barriers that protect the retina, thereby impeding the effective delivery of treatment.

Current therapies used in the management of retinal diseases include topical treatments, intravitreal injections, and intravitreal implants. In the case of topical administration, the primary concern is the low bioavailability of the drug to the posterior aspect of the globe, and therefore, the requirement for multiple administrations per day (Appell et al., 2024). The primary reason for the limited bioavailability is the tear film that rests above the cornea, as well as induced lacrimal fluid production and subsequent drainage. Eye drops can then be absorbed by the conjunctiva and distributed systemically.

To bypass the barriers at the eye's surface, intravitreal injections can be used. This includes intravitreal injections of anti-vascular endothelial growth factor (VEGF) to treat retinal and choroidal neovascularization caused by advanced conditions, such as wet age-related macular degeneration and proliferative diabetic retinopathy. These injections are typically administered once a month. However, with intravitreal injections, complications can arise, including infectious endophthalmitis and retinal detachment (Appell et al., 2024). To increase the injection frequency, surgical implants were developed. These devices improve patient adherence; however, accurately predicting the released medication dosage for all patients can be challenging. Additionally, the materials from

TABLE 6.1
Summary of Key Findings of Exosome or Secretome-Based Studies on Retinal Diseases

Disease	Stage	Treatment	Key Points	References
Diabetic retinopathy	pre-clinical	MSC-derived Exo-loaded with miRNA	• Reversed hyperglycemia-induced retinal inflammation. • Can delay inflammatory response and angiogenesis, attenuate neovascularization in the diabetic retinopathy pathway of ERK1/2 and HIF-1. • Counteracts retinal apoptosis, maintains retina structural integrity. • Retinal repair through miRNA-222 expression. • Can promote Müller cell proliferation.	(Zhang et al., 2019a; Guo et al., 2021; Cai et al., 2021; Gu et al., 2020; Sun et al., 2022; Safwat et al., 2018; Li W et al., 2021; Wang et al., 2024; Fu et al., 2024; Bai and Wang, 2024; Liang et al., 2022)
Diabetic retinopathy	pre-clinical	Exo-rich medium from amniotic MSC	• Vascular attenuation, delayed progression of DR, and cataract formation.	(Kim et al., 2023)
AMD	pre-clinical	Exo as biomarkers	• Expression of new miRNA and phosphoproteins different than that of the control.	(Biasutto et al., 2013; Grassmann et al., 2014)
AMD	pre-clinical	T-cell-derived Exo	• Reduction of choroidal neovascularization, loaded with anti-VEGF antibody, can also do this with the target of glutathione synthetase.	(Tian et al., 2021; Zhang et al., 2024)
AMD	pre-clinical	MSC-derived Exo	• Reduces oxidative stress, inhibits mitochondrial-mediated apoptosis of retinal pigment epithelium, suppresses subretinal fibrosis with miR-27b.	(Peng et al., 2024; Li D et al., 2021)
Retinitis pigmentosa	Pilot case series	MSC Secretome	• Minimal effects on late-stage RP patients (4 total), peribulbar injection.	(Mangungson et al., 2021)
Retinal ischemia	pre-clinical	MSC-derived Exo	• Inappropriate neovascularization reduced. • No immune response. • Increased angiogenesis in avascular regions. • Reduced inflammation.	(Moisseiev et al., 2017; Dellett et al., 2017; Mathew et al., 2023; Sanghani et al., 2022; Mathew et al., 2021)
Retinal injury	pre-clinical	MSC-derived Exo	• Reduced inflammation, monocyte chemotactic protein in the retinal injury model.	(Yu et al., 2016)
Retinal degeneration	pre-clinical	MSC-derived Exo	• Reduced photoreceptor loss, protected retinal function in a mouse model. • Similar effects for the overall secretome.	(Deng et al., 2021; Huang et al., 2024; Usategui-Martín et al., 2020)

Note: MSC: mesenchymal stem cell; RP: retinitis pigmentosa; AMD: age-related macular degeneration.

which the devices are made may need to be removed periodically, reintroducing the risks associated with procedures into the environment.

Therefore, there appears to be a need for a medication delivery system that reduces the risk of repeated injections, prolongs the duration of the administered medication's effect, and promotes safe, localized delivery of the therapeutic agent, particularly for the treatment of retinal diseases (Figure 6.1).

6.3.2 Age-Related Macular Degeneration

Age-related macular degeneration (AMD) is a progressive disease in which the macula, the area of the retina responsible for central vision, deteriorates. It is the third leading cause of blindness worldwide (Appell et al., 2024). Early-stage AMD, also known as dry AMD, is characterized by drusen (lipid deposits) beneath the retinal pigment epithelium and macular atrophy (Campagne et al., 2014). Early-stage AMD is often managed with observation. Advanced AMD, on the other hand, is neovascular or wet AMD, in which choroidal angiogenesis occurs. This is the primary reason why anti-VEGF is used to control the formation of these new vessels, which alter the macula's landscape, rendering it non-functional. Newer therapies being investigated include subretinal gene therapy and nanoparticle injection, both of which address the frequency by which current anti-VEGF injections must be administered and aim to continue to provide therapeutic benefit to the macula in a targeted manner (AbbVie, 2021; Huang and Chau, 2021; Mu et al., 2018; Anand et al., 2023).

Exosome-based therapies are currently in pre-clinical development. Firstly, exosomes can be used as biomarkers to identify which proteins or RNA subtypes are most valuable for ameliorating AMD, and which biomarkers and targets should be considered when loading exosomes. Some of the targets identified include cathepsin D and cytokeratin 8, both involved in the autophagy-lysosomal

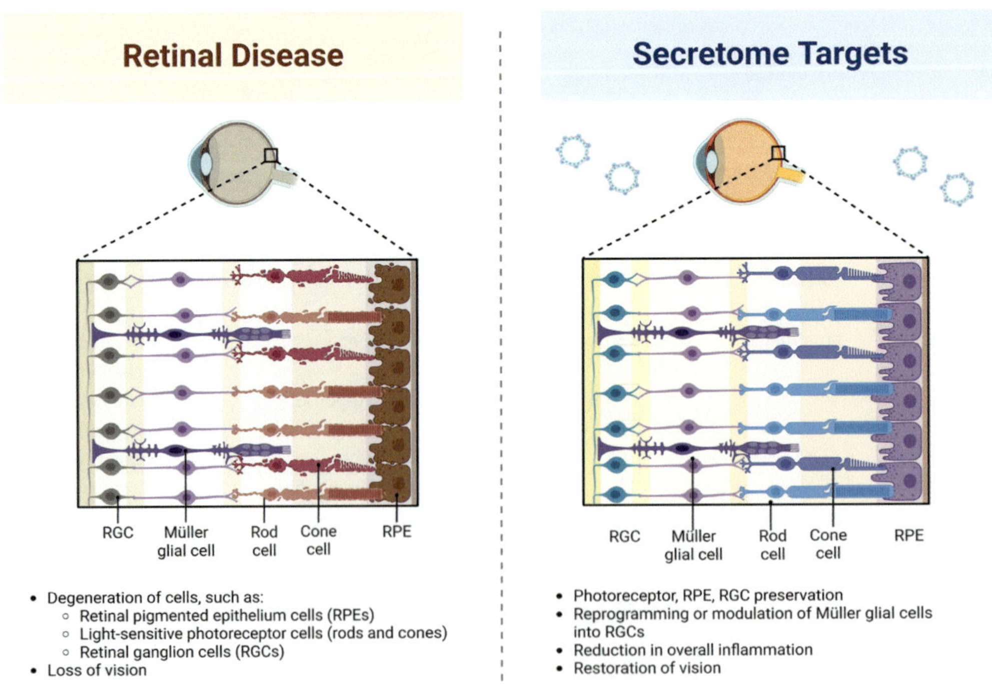

FIGURE 6.1 Retinal disease degeneration and therapeutic secretome targets. This figure illustrates that cell degeneration, including retinal pigment epithelium, photoreceptor, and retinal ganglion cells, is the primary cause of vision loss in the retina.

pathway, as well as miRNAs, including miR-301, 361, 424, 486, and 626, which regulate pathways involving proteins such as Akt, Src, and ERK (Van Lookerenet al., 2014; Grassmann et al., 2014).

In the context of exosome-based therapeutics, Tian et al. developed an anti-VEGF antibody-coated exosome therapy derived from regulatory T cells (Tian et al., 2021). These anti-VEGF particles would be cleaved by matrix metalloproteinase *in vivo*, effectively suppressing angiogenesis in the choroid of mouse and non-human primary models. This model therefore reduces the number of anti-VEGF injections needed by lowering the dosage and increasing the potency of the anti-VEGF therapeutic effect, even though this study specifically introduced the exosome-based treatment via intravitreal injections, similar to the standard treatment. Similarly, Zhang et al. used exosome-liposome particles as carriers for a CRISPR/Cas9-based treatment targeting the glutathione system with the nucleic acid aptamer AS1411 (Zhang et al., 2024). The exosomes demonstrated effective targeting and specificity for cells of interest and significantly reduced intraocular neovascularization in a rabbit model of neovascular AMD. The primary appeal of this technology lies in its potential for a single injection and the long-lasting effects of gene editing, offering a potentially permanent solution to wet AMD. Using MSC-derived exosomes, Peng et al. derived both the exosomes themselves and the conditioned media from the stem cells to reduce oxidative stress-induced retinal pigment epithelium cell death, acting through the PI3K/Akt signaling pathway, similar to what was found in exosome biomarker studies (Peng et al., This was further supported *in vivo* in a mouse model system, where retinal degeneration was significantly reduced.

Likewise, Li et al. addressed the subretinal fibrosis that results from neovascular AMD by using the exosomes derived from human UC MSCs (D. Li D et al., 2021). The researchers found that subretinal fibrosis, as measured by collagen I levels, was significantly reduced with exosome therapy, which also reduced the migration of fibrosis-related cells. Additionally, they identified miR-27 as the leading player in this pathway (Figure 6.2).

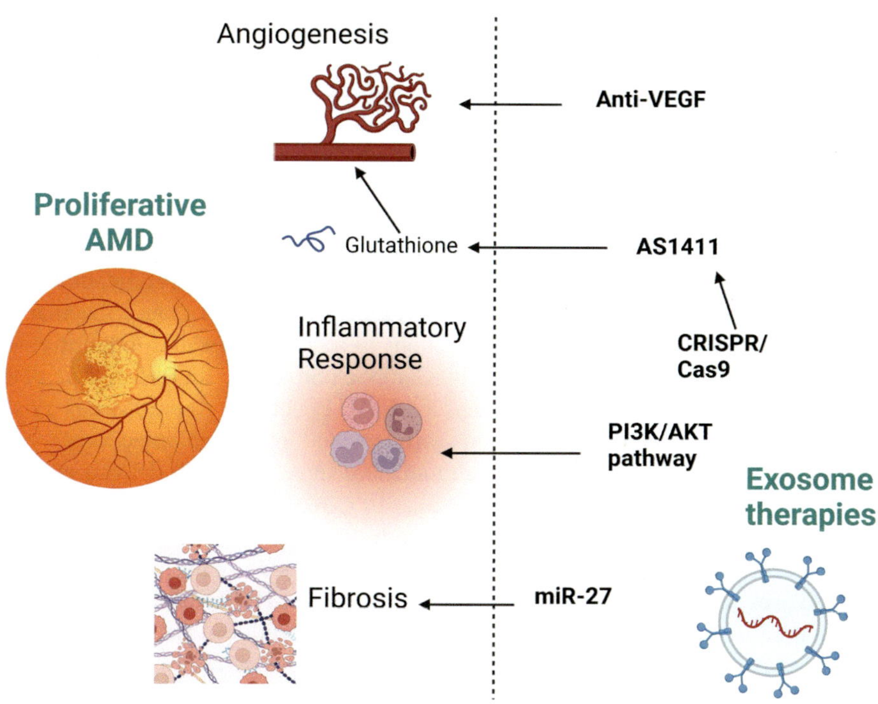

FIGURE 6.2 Pathways involved in proliferative age-related macular degeneration and addressed by exosome therapies.

6.3.3 Retinitis Pigmentosa

Retinitis Pigmentosa is a hereditary disease characterized by the degeneration of rod photoreceptor cells, followed by cone photoreceptor cells through an oxidative process. Currently established treatments for RPE65-mediated disease include gene therapy and the implantation of a retinal prosthesis, which stimulates retinal ganglion cells without activating photoreceptors (Russell et al., 2017; Schaffrath et al., 2019; Mathieson et al., 2012). Current research focuses on encapsulated gene therapy and on the delivery of genes and neurotrophic factors using the CRISPR/Cas9 system. These studies are mainly in the pre-clinical stage of development and have shown promising results, including a reduction in photoreceptor apoptosis, neuroprotective effects, preserved cone photoreceptor function, the development of cone features in rod photoreceptors, and overall rescue of retinitis pigmentosa (Wong et al., 2019; Zheng et al., 2020; Herrera-Barrera et al., 2023; Mashal et al., 2019; Rosa et al., 2022; Yu et al., 2017).

Interestingly, Mangungson et al. employed a secretome-based approach, administering a peribulbar injection of a secretome treatment derived from UC MSCs to four subjects with retinitis pigmentosa and severely damaged retinas, characterized by a visual field defect of 25–50% (Mangungson et al., 2021). Clinical visits were performed at pre-injection and at one, three, and six months post-injection. Electroretinography was performed, and the amplitude of the response, as well as the implicit times of the response, were recorded. There was little to no change in visual acuity, and electroretinography showed inconclusive patterns throughout the post-injection timeline. The researchers suggested that late-stage retinitis pigmentosa may not be amenable to visual rescue, and that more substantial and sustained paracrine factors, such as the direct injection of umbilical MSCs, may be needed in this case.

6.3.4 Diabetic Retinopathy

Diabetic retinopathy results from chronic hyperglycemia, leading to progressive retinal microvascular damage. It is classified into non-proliferative diabetic retinopathy (NPDR) and proliferative diabetic retinopathy (PDR). NPDR is characterized by microaneurysms, intraretinal hemorrhages, cotton-wool spots, and venous abnormalities due to capillary leakage and occlusion. PDR, the advanced stage, involves retinal ischemia-induced neovascularization, increasing the risk of vitreous hemorrhage and tractional retinal detachment (Appell et al., 2024). Management of diabetic retinopathy primarily involves controlling blood sugar levels, blood pressure, and lipid intake, especially in the non-proliferative and early stages of the disease. For proliferative diabetic retinopathy, pan-retinal photocoagulation—a laser treatment targeting the peripheral retina to reduce angiogenesis—is commonly used, along with strict control of diabetic markers (Evans et al., 2014). In addition, similar to AMD, diabetic retinopathy can be treated with anti-VEGF injections to reduce macular edema and disease progression, as well as corticosteroid injections to reduce intraocular inflammation (Nguyen et al., 2012; Wykoff et al., 2022; Boyer et al., 2014; Campochiaro et al., 2011. However, all these treatments carry a similar risk to the ones highlighted for AMD injections, where there is a risk of infection, retinal detachment, and other complications with repeated intravitreal injections.

New research now focuses on reducing the number of injections with the use of nanoparticles or intraocular implants (Erdal et al., 2023; Li et al., 2017; Won et al., 2020). However, in addition to these studies, EVs have been a key focus of research. In a pre-clinical study by Zhang et al., the authors derived exosomes from MSCs. Subsequently, they loaded it with miR-126 to promote vascular repair and provide anti-inflammatory effects (Zhang et al., 2019b). The levels of inflammatory proteins, such as IL-1, IL-18, and caspase-1, were significantly reduced with these exosomes, compared to fibroblast exosomes, which did not produce the same effect. Downstream proteins, such as HMGB1, were also analyzed and found to be significantly decreased in the presence of MSC-derived exosomes containing miR-126. The same group also investigated platelet-rich plasma-derived exosomes, finding that the treatment activated the TLR4 pathway to regulate inflammation in diabetic rats (Zhang et al., 2019b). This was in contrast to Sun et al., where the use of exosomes from MSCs was found to be a key player in regulating a different pathway, where it promoted the release of the neuronal precursor cell expressed developmentally downregulated 4 (NEDD4) compound to control retinal apoptosis in diabetic retinopathy (Sun et al., 2022). These exosomes were studied by Bai and Wang, who found that PI3K/AKT phosphorylation was suppressed and senescence in the retinal pigment epithelium was ameliorated (Bai and Wang, 2024). This finding is further supported by Liang et al., who demonstrated that MSC-exosomes exhibit similar effects on suppressing angiogenesis and oxidative stress via the miR-133 pathway (Liang et al., 2022).

In a slightly different approach, C. Gu et al. found that using adipose MSC-derived exosomes containing miR-192 ameliorated inflammation and angiogenesis in a diabetic rat model (Gu et al., 2021). In addition, they also reported that Müller cells, or retinal glial cells, which are present during the inflammatory process, were not induced by the introduction of these exosomes. These same exosomes were introduced into a rabbit model of diabetic retinopathy by Safwat et al., who observed the formation of regenerative retinal layers with increased miR-222 expression (Safwat et al., 2018). Interestingly, Dellett et al. found that even without stem cells, endothelial colony-forming cells have exosomes that can modulate angiogenesis (Dellett et al., 2017).

Gu et al. found that by using exosomes derived from retinal pigment epithelial cells, the endothelial to mesenchymal transition contributing to fibrosis in proliferative diabetic retinopathy was suppressed (Gu et al., 2020). Wang et al. used exosomes derived from M2 microglia instead, and found that pericyte apoptosis was reduced and vascular remodeling was improved in diabetic retinopathy-induced mouse models (Wang et al., 2024). Specifically, they addressed the vascular leakage that occurs during proliferative diabetic retinopathy, a key determinant in the degradation of visual acuity. Interestingly, unlike the study cited above by Gu et al., Fu et al. found that Müller cell-derived exosomes will regulate microglia polarization, and therefore, may not be particularly detrimental to the retinal reparative process (Gu et al., 2021; Fu et al., 2024).

Instead of using purified exosomes, Kim et al. opted to use exosome-rich conditioned medium derived from amniotic membrane stem cells, which only showed a delay in the progression of diabetic retinopathy through a more reactive electroretinogram and less severe vascular attenuation (Kim et al., 2023) (Figure 6.3).

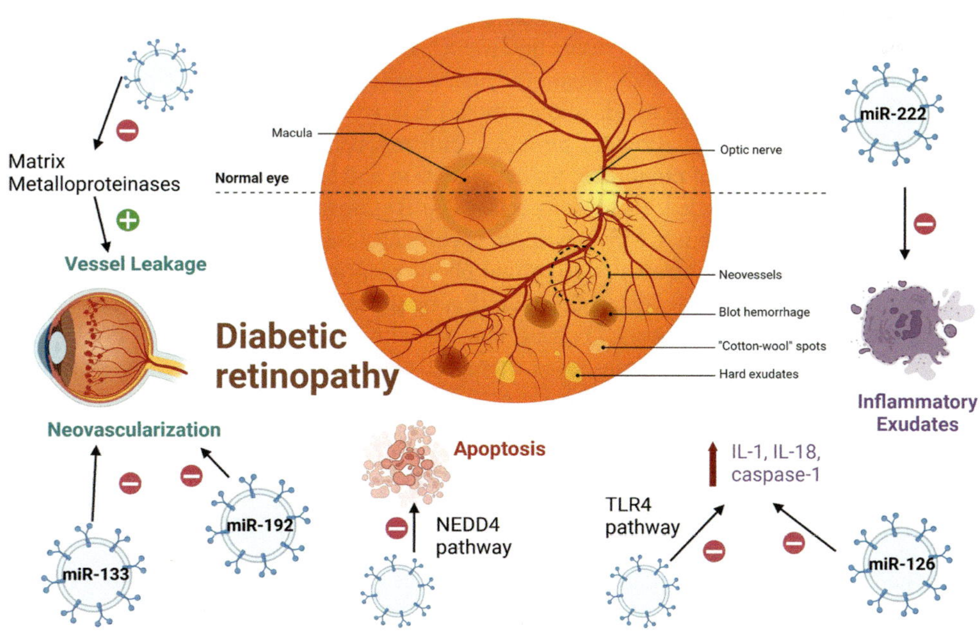

FIGURE 6.3 Exosome targets, pathways, and effectors in the pathogenesis and progression of diabetic retinopathy. The acronyms shown include neural precursor cell expressed, developmentally downregulated protein 4 (NEDD4) and toll-like receptor 4 (TLR4).

6.3.5 Retinal Ischemia

Retinal ischemia can be caused by central retinal artery occlusion, branch retinal artery occlusion, or central retinal venous occlusion. In terms of new studies, hypothermia, hyperglycemia, and IOP-lowering drugs have been studied to reduce ischemic retinal injury (Casson, 2004). However, exosome or secretome-based therapies are also highly researched in the case of retinal ischemia (Sanghani et al., 2022).

Moisseiev et al. found that by isolating exosomes from human MSCs and administering them to mice with oxygen-induced retinopathy via intravitreal injection, vascular flow was partially preserved and retinal thinning was reduced, although complete prevention or restoration was not achieved (Moisseiev et al., 2017). Mathew et al. employed a two-step approach to investigate the use of exosomes derived from MSCs in a retinal ischemia model (Mathew et al., 2021). This group first found that these exosomes can be injected intravitreally, are quickly cleared from the vitreous, and by 2 weeks post-administration, are concentrated in the retinal ganglion cells, as well as astrocytes, microglia, and Müller cells (Mathew et al., 2021). No toxicity was noted in the retinal ganglion cells. The next step they took was to identify key factors within these exosomes that resulted in a restorative effect in retinal ischemia models and reduced inflammatory cytokine release, leading to the identification of miR-424 (Mathew et al., 2021).

6.3.6 Retinal Degeneration Protection

More generally, research has also been conducted on reducing retinal degeneration using exosome strategies that address all forms of retinal disease, ranging from AMD to retinal ischemia. Yu et al. investigated the mechanism underlying the therapeutic effects of MSC-derived exosomes and found that monocyte chemotactic protein-1 (MCP-1) was downregulated, thereby protecting against further retinal injury (Yu et al., 2016). They explained that MCP-1 recruits monocytes and participates in photoreceptor degeneration; therefore, partial inhibition of its release via exosome-based therapy provides retinal protection. Deng et al. used miR-21-loaded exosomes derived from MSCs to mitigate photoreceptor apoptosis and retinal degeneration in a mouse model (Deng et al., 2021). They further analyzed the duration of exosomes' effect, noting that at 8 weeks post-retinal damage, the treatment continued to protect the retina and preserve cone photoreceptors.

Huang et al. take this a step further, where they used retinal progenitor cells from human embryonic stem cell-derived retinal organoids and extracted EVs to be introduced to Müller cells, which has previously been reported to allow for photoreceptor regeneration (S. Huang et al., 2024). They then observed the alleviation of photoreceptor degeneration and proposed a mechanism for miRNA-mediated regulation of the nuclear factor kappa B (NF-κB) transcription factor. Usategui-Martin et al. used the secretome from human MSCs without specifically isolating exosomes (Usategui-Martin et al.). They found that, when introduced to retinal photoreceptors *in vitro*, the secretome treatment provided neuroprotection and prevented degeneration by preserving a functional neuroretina organization and reducing reactive gliosis. Overall, this was similar to the effects observed with MSC introduction, indicating that secretome-based therapies were as effective as stem cell transplantation in neuroprotection for retinal diseases.

6.4 GLAUCOMA AND NON-GLAUCOMATOUS OPTIC NEUROPATHIES

6.4.1 Current Challenges in Managing Glaucoma

Current challenges in managing glaucoma will vary depending on the type of glaucoma. The most common type of glaucoma is open-angle glaucoma, in which the anterior chamber apparatus is standard on gonioscopy. The other diagnostic criteria include optic disc cupping and peripheral vision loss, which are characteristic of all types of glaucoma (Lee and Higginbotham, 2005). The primary target for primary open-angle glaucoma (POAG) is intraocular pressure (IOP). Another type of glaucoma is angle-closure glaucoma (ACG), characterized by a pupillary block of aqueous humor outflow, as determined by gonioscopy, which results in angle closure. ACG onset can be acute, with ocular pain and blindness occurring in an acute-to-subacute timeline (Sun X et al., 2017). The primary treatment for pupillary block ACG is laser iridotomy.

Lastly, there is normotensive primary glaucoma, in which the IOP is not elevated, but glaucomatous changes still occur. It is believed that vascular changes are one of the factors contributing to normotensive glaucoma. For lowering IOP, topical treatments are preferred. These medications include beta-adrenergic antagonists, alpha-adrenergic agonists, and prostaglandin analogs. Surgical approaches include laser procedures such as iridotomy for ACG and laser trabeculoplasty for POAG, as well as microinvasive glaucoma surgery (MIGS), in which small drainage devices can be implanted in the trabecular or subconjunctival spaces. New strategies for managing glaucoma include intraocular implants with prolonged medication release, such as travoprost and bimatoprost (prostaglandin analog) implants, which are currently in Phase 3 and Phase 2 trials, respectively (Glaukos Corporation, 2018; ForSight Vision5, Inc.,). Nanoparticle injections in the subconjunctival or intravitreal space as an option for prolonged release of topical drugs for treatment-resistant glaucoma are also an extensive research focus, with promising pre-clinical data on effective localization of coated drugs and significantly reduced IOP in pre-clinical models (Hu et al., 2022; Giannaccini et al., 2018).

6.4.2 Secretome Strategies for the Management of Glaucoma

pre-clinical studies on stem cell treatments for ocular diseases have proposed that the primary effectors of tissue regeneration, which are often impaired by disease, are growth and neurotrophic factors (Seong et al., 2023). These factors are often released as EVs, or exosomes, from stem cells and can also encapsulate other signal transmitters and cytokines, allowing for further regulation (Seong et al., 2023). The totality of substances released, including exosomes themselves, comprises the secretome. The use of exosomes and/or secretome formulations will reduce the number of stem cells required for an effective treatment, enabling cost-effective, transport-safe delivery of therapeutic compounds.

Here, we outline a typical pre-clinical study that describes secretome strategies for managing IOP. In a pre-clinical study reported by Seong et al., exosome-rich conditioned media were collected from human amniotic membrane MSCs and amniotic membrane epithelial stem cells (Seong et al., 2023). The media was analyzed for mean exosome size, despite exosome aggregation. Nevertheless, in a hyperosmotic glaucoma model in rats, normal IOP was obtained within 10 days of intravitreal injection, and retinal and ganglionic cell layer thickness was comparable to that of the standard control compared to the negative control. The proposed mechanism of IOP reduction involves induced trabecular meshwork regeneration, where platelet-derived growth factor plays a significant role (Manuguerra-Gagne et al., 2013).

Studies highlight the WNT/β-catenin signaling pathway as the underlying mechanism behind extracellular matrix expression and trabecular meshwork stiffening, which contributes to the decreased outflow of aqueous humor and, thus, increased IOP in POAG (Tabak et al., 2021; Lerner, Schreiber-Avissar, and Beit-Yannai. An exosome-based approach investigated by Lerner et al. used EVs from primary non-pigmented ciliary epithelium cells rather than stem cells (Pattabiraman et al., 2023). The microanalysis of secretome components from these cells revealed microRNAs (miRNAs) and proteins that regulate the WNT/β-catenin signaling pathway, thereby enabling IOP reduction. Similarly, Tabak et al. found that a specific siRNA can be injected into non-pigmented ciliary epithelium exosomes to silence a key component of the WNT/β-catenin signaling pathway, thereby preserving trabecular meshwork function more effectively (Tabak et al., 2021). Pattabiraman et al. offer some important insights into miRNA particles, including miR-27, 199, 23, 130b, and 200, which regulate genes involved in the WNT/β-catenin pathway and extracellular membrane regulation (Pattabiraman et al., 2023).

Li et al. highlight the protection of the trabecular meshwork as a key target (Li YC et al., 2021). Trabecular meshwork cells were subjected to oxidative stress, then treated with MSC-derived exosomes, and then exposed to oxidative stress again. The use of exosomes increased the viability of trabecular meshwork cells and reduced reactive oxygen species and pro-inflammatory factor production. Interestingly, Yu et al. addressed the mechanism of retinal ischemia-reperfusion injury in glaucoma, using TNFα-stimulated gingival MSC-derived exosomes to exert anti-inflammatory effects and reduce apoptosis via miR-21 (Yu et al., 2017). Similarly, Wang et al. identified ischemia-reperfusion injury as a consequence of increased IOP and optic nerve damage in glaucoma, targeting MSC stem cell-derived exosomes through the miR-150 functionality to decrease M1 microglial polarization and thus, inflammation (Wang Y et al., 2023) (Figure 6.4).

6.4.3 Optic Nerve Regeneration

The current treatment for glaucoma targets IOP reduction only (Kontiola, 2000). However, the pathophysiology of glaucoma extends beyond IOP; it also includes retinal ganglion cell damage, optic nerve degeneration, and optic disk hypoperfusion. The rationale for investigating exosomes is the potential to promote optic nerve regeneration or confer neuroprotective effects on the optic nerve (Hong, 2007). The current landscape of optic nerve regeneration is still in its infancy, with no approved drugs on the market and few pre-clinical studies. Nerve regeneration does not occur spontaneously in vivo; however, proteins such as nerve growth factor will promote neural growth and development. Recombinant factors have been developed in the past; however, they have been hindered by short half-lives, enzymatic degradation, and poor permeability through the blood-brain barrier, which is also present in the optic nerve head (Yang et al., 2020).

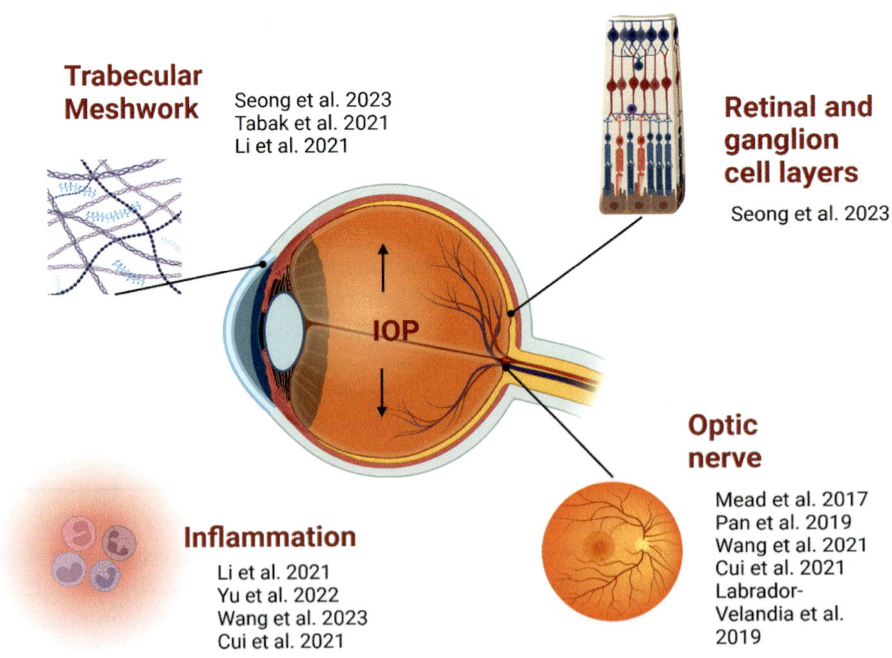

FIGURE 6.4 Secretome-based pre-clinical studies addressing different aspects of glaucoma and other optic neuropathies.

In neural diseases, nerve protection and regeneration have been addressed using exosome technology. In a study by Yang et al., nerve growth factor was found to be stable under storage, with high delivery efficiency and long-lasting therapeutic effects in stem cell-derived exosomes (Yang et al., 2020). This is further corroborated by Bucan et al., who found that exosomes containing neural growth factors promoted neurite growth and sciatic nerve regeneration after a crush injury (Bucan et al., 2019). Exosomes may be loaded with PTEN siRNA to enhance axonal growth in the spinal cord and promote neovascularization, or with mRNA to stimulate neurotrophins in peripheral nerve regeneration (Guo et al., 2019; Yang et al., 2021). Additionally, miRNA such as miR-214 can be utilized to stimulate the proliferation of supportive cells, including Schwann cells, thereby indirectly promoting axon regeneration (Zeng et al., 2023). Similarly, EVs derived from MSCs can facilitate microglial polarization and regulate mitochondrial function (Liu et al., 2021). The effect of exosomes was not found to be dose-dependent after a minimally effective dose was established (Zhao et al., 2020). To take a further step, neural grafts containing exosomes were used to promote nerve regeneration in a rat model of peripheral nerve crush (Hu et al., 2023).

In the context of optic nerve regeneration, several studies support the use of secretome-based technologies to promote the survival and regeneration of retinal ganglion cells, the primary cell type in the optic nerve. Mead and Tomarev found that isolated exosomes from MSCs promoted neuritogenesis, preserved retinal nerve fiber layer thickness, and facilitated axogenesis in retinal ganglion cells (Mead and Tomarev, 2017). They characterized this impact of the exosomes using the length of the longest neurite, the number of retinal ganglion cells bearing neurites, electroretinography as a function of visual input detection, retinal cell response, and the mean number of labeled axons. The hypothesis proposed by Mead et al. regarding the effectiveness of axogenesis was based on the presence of many miRNA particles that have yet to be identified as key players in nerve regeneration. The candidate miRNAs include miR-21, a regulator of PTEN expression, and miR-146a, which targets epidermal growth factor.

Pan et al. found that the use of UC MSC-derived exosomes promoted retinal ganglion cell survival following an experimentally induced optic nerve crush in a rat model, as well as enhanced glia activity in the retina; however, no axogenesis was reported (Pan et al., 2019). They characterized the miRNA makeup of human UC MSCs as miR-21, 125, 23, and 100, whereas other studies using MSC-derived MSCs identified miR-143, 10, 386, 22, and 21 as the most abundant. They attributed astrocyte activation to miR-21 and, therefore, suggested that this miRNA is a key player in the supportive environment that enables optic nerve protection and, if possible, eventual regeneration.

The loading of the endogenous neuroprotective peptide pituitary adenylate cyclase-activating polypeptide 38 (PACAP38) into retinal ganglion cell exosomes has been reported (Wang et al., 2021). PACAP has previously been found to be expressed in the ganglion cell layer and to change in response to injury to the optic nerve. Loaded exosomes administered through intravitreal injection promoted the survival of retinal ganglion cells and increased axon regeneration and visual function (measured through flash visual evoked potentials) in a rat model (Wang et al., 2021). The researchers also characterized the preserved retinal cell layer to demonstrate that it had no additional adverse effects on retinal functioning.

A more specific molecular study of the mechanism by which MSC-derived exosomes may impact optic nerve regeneration was reported by Cui et al. (2021). Exosomes were characterized by electron microscope and Western blot, and an intravitreal injection was administered to a rat model of optic nerve degeneration. The treatment promoted the survival of retinal ganglion cells and reduced pro-inflammatory factors, including TNF-α, IL-1, IL-6, IL-8, and MCP-1. The Bcl-2/Bax ratio was upregulated, along with downregulation of caspase-3, both of which are involved in the apoptotic pathway as negative regulators and positive markers, respectively. In addition, the PI3K/AKT pathway was downregulated, a key player in protein synthesis. However, improper activation of the PI3K/AKT pathway was found to promote apoptosis and may therefore not be a negative outcome of this study. A non-specific study involved physically separated co-cultures of neuroretina and human

MSCs, in which only paracrine effects were examined. However, exosomes were not explicitly mentioned as the main factor (Figure 6.5).

6.4.4 Other Optic Neuropathies

Other optic neuropathies that these secretome-based techniques may be applied to include ischemic optic neuropathy and traumatic optic neuropathy.

In ischemic optic neuropathy, the pathophysiology and management differ significantly between arteritic and non-arteritic forms. Arteritic ischemic optic neuropathy (AION) is most commonly caused by giant cell arteritis. This vasculitic condition leads to inflammation and occlusion of the short posterior ciliary arteries, which supply the optic nerve head. Prompt recognition and treatment with high-dose corticosteroids are essential to prevent further vision loss and involvement of the fellow eye. Non-arteritic ischemic optic neuropathy (NAION), on the other hand, results from vascular insufficiency due to systemic conditions such as hypertension, diabetes mellitus, or other factors contributing to a structurally crowded optic nerve head. Unlike AION, NAION is not associated with an inflammatory process, and there is no proven effective treatment to reverse the damage once it occurs. Management primarily focuses on controlling underlying risk factors to reduce the risk of further episodes. Ultimately, regardless of the cause, once the optic nerve is damaged, there is no established therapy to restore lost vision.

In glaucomatous optic neuropathy, no effective treatment currently exists to reverse or rescue optic nerve damage once it has occurred. Optic nerve regeneration, such as the pre-clinical techniques outlined above, is the only option that may be of value for this disease. In terms of treatment goals, traumatic optic neuropathy is closely related. The cause of traumatic optic neuropathy is often head trauma, which can be direct or indirect, resulting in visual field defects and optic atrophy. There is no consensus for the treatment of indirect traumatic optic neuropathy, but corticosteroids and optic canal decompression have been suggested with limited evidence (Wladis et al., 2021). For direct optic neuropathy, surgical management may be indicated, mainly if orbital compartment

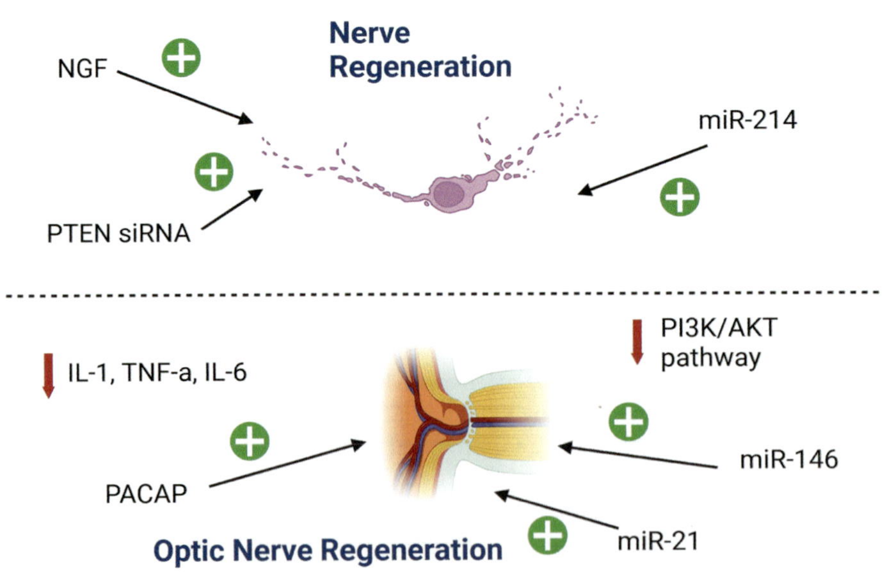

FIGURE 6.5 Comparison of effective proteins involved in peripheral nerve regeneration versus those involved in optic nerve regeneration. The acronyms are as follows: nerve growth factor (NGF), phosphatase and tensin homolog (PTEN), interleukin (IL), tumor necrosis factor (TNF), pituitary adenylate cyclase-activating peptide (PACAP), phosphoinositide 3-kinase (PI3K), and protein kinase B (Akt).

syndrome is caused (Blanch et al., 2024). Currently, there is no established treatment for optic nerve damage; however, potential future approaches may focus on regenerating the optic nerve. The studies outlined by both Wang et al. and Cui et al. specifically highlight its application in traumatic optic neuropathy as an isolated optic nerve defect, rather than a multitude of interplaying factors in the case of optic nerve degradation in glaucoma (Cui et al., 2021; Wang et al., 2021). Optic neuropathy treatments are summarized in Table 6.2.

6.5 CORNEAL AND ANTERIOR SEGMENT DISORDERS

6.5.1 Current Challenges in Corneal and Anterior Segment Disorders

The corneal epithelium, a non-keratinized stratified layer, serves as the eye's first line of defense, functioning as a barrier to the environment and renewing itself through re-epithelialization. However, exposure to toxic agents, mechanical injury, and/or infections that do not respond to treatment can lead to a non-healing corneal epithelial lesion, causing patients to experience discomfort and, in some cases, visual loss. The limited regenerative potential of the cornea, scar formation, perforation, and risk of vision loss highlight the importance of early diagnosis and management (Hazlett et al., 2016). However, several challenges exist for the successful management of corneal and anterior segment disorders. Topical treatments for conditions such as corneal ulcers also depend heavily on patient adherence. These challenges highlight the need for more effective, targeted therapies with fewer adverse effects (Hartley, 2010). Treatments such as allogeneic corneal transplantation are invasive intraocular surgical procedures that carry risks, including potential long-term complications such as graft rejection and the possible need for chronic immunosuppressive therapy. Recent studies have suggested that MSC-derived secretomes (MSC-secretomes) have been effective in treating corneal and anterior segment disorders.

6.5.2 Secretome in Corneal Wound Healing

The effects of secretomes and exosomes on corneal wounds, both *in vitro* using human corneal epithelial cells and in vivo in the corneal epithelium, have shown positive effects on corneal healing and improved corneal monolayer re-epithelialization (Samaeekia et al., 2018). It is important to note that secretome-enhanced corneal wound healing and closure occur in a dose-dependent manner. An et al. analyzed human corneal epithelial cell proliferation in exosome-depleted and exosome-present secretomes. The removal of exosomes from MSC-secretomes abolished wound-healing effects. However, most notably, exosome collection for 72 hours had greater wound-healing effects than 48-hour collection, indicating a dose-dependent relationship (An et al., 2023).

Several studies have focused on the modulatory effects of extracellular matrix (ECM) remodeling proteins, such as collagen proteins and matrix metalloproteinases (MMPs), on corneal wound healing, both in vitro and in vivo (Yu et al., 2022a; Shen et al., 2018; Tang et al., 2022). MMPs are ECM remodeling endopeptidases that contribute to abnormal ECM degradation. The suppression of MMPs helps regulate angiogenesis, tissue remodeling, and ECM turnover. Yu et al. utilized a microfluidic chip-based human corneal model to investigate wound healing in vitro after the application of exosomes derived from MSCs (MSC-EVs). Compared with the control group receiving serum-free cell culture medium, MSC-EVs significantly increased cell migration by 1.9-fold and decreased MMP-2 levels (Yu et al., 2022a). These results suggest that exosomes play a role in inhibiting corneal neovascularization and inflammation, while also promoting cell migration. Moreover, another study by Shen et al. complemented the findings on exosomes regulating MMPs, while also introducing cell apoptosis and ECM collagen regulation as key factors. Human Adipose-derived MSC-EVs were shown to significantly suppress apoptosis and MMP activity. Interestingly, the levels of ECM-related proteins, such as fibronectin and collagens, were increased in rabbit corneal stromal cells (CSCs) treated with MSC-EVs (Shen et al., 2018).

TABLE 6.2
Optic Neuropathy Treatments

Disease	Status	Treatment	Key Points	References
POAG	pre-clinical	Exo-rich conditioned media	• Facilitated retinal pigment epithelial cell proliferation. • Recovered IOP. • Involved in WNT/B-catenin signaling. • Promoted survival of retinal ganglion cells and regeneration of axons. • Inhibits expression of inflammatory cytokines.	Seong et al., 2023; Mead and Tomarev, 2017; Pan et al., 2019
POAG	pre-clinical	Exo-derived targets	• Includes miR27, 199, 23, 130b, 200. • Regulates Nrf2, Keap1, GSK3B, serine/threonine-protein phosphatase 2A. • Oxidative stress pathways are major targets.	Pattabiraman et al., 2023
POAG	pre-clinical	Exo-loaded with siRNA	• Targets mRNA knockdown in trabecular meshwork cells in vitro.	Tabak et al., 2021
Retinal ischemia-reperfusion	pre-clinical	Stem cell-derived Exo	• Significantly reduced inflammation, apoptosis. • Supported visual preservation.	Yu et al., 2022b; Wang Y et al., 2023
Nerve regeneration	pre-clinical	NGF mRNA via Exo	• Delivered NGF, reduced inflammation, was highly stable, tested on microglia and mouse models of stroke.	Yang et al., 2020
Nerve regeneration	pre-clinical	Melatonin-treated EVs	• Reduce production of ROS, facilitate M2-like polarization.	Liu et al., 2021
Nerve regeneration	pre-clinical	Intranasal delivery of Exo with siRNA	• Loaded with PTEN siRNA. • Enhanced axonal growth and neovascularization in spinal cord injury.	Guo et al., 2019
Nerve regeneration	pre-clinical	Stem cell-derived Exo	• Loaded with miRNA, promotes proliferation and migration of Schwann cells, promotes axon extension of dorsal root ganglion neurons, and reduces gliosis. • Increases myelin positive regulators and myelin proteins in Schwann cells. • Increases NGF expression.	Zeng et al., 2023; Bucan et al., 2019; Yang et al., 2021; Hu et al., 2023; Zhao et al., 2020.
Traumatic optic neuropathy	pre-clinical	Retinal ganglion cell-derived Exo	• Loaded with a neuroprotective peptide, found to support axon elongation and function preservation.	Wang et al., 2021; Cui et al., 2021; Pan et al., 2019

Note: POAG: primary open-angle glaucoma; IOP: intraocular pressure; NGF: nerve growth factor.

In addition to regulating ECM remodeling proteins, a study by Tang et al. highlighted that the presence of miR-432-5p in a thermosensitive hydrogel, combined with induced pluripotent stem cell-derived MSC-EVs, modulated the suppression of collagen biosynthesis and translocation-associated membrane protein 2 in CSCs. Surprisingly, *in vivo* reduction of mRNA expression of different types of collagens, alpha-smooth muscle actin (a-SMA), and vimentin was associated with decreased scar formation (Tang et al., 2022). The discrepancy in collagen expression across studies suggests that secretomes are involved in the differential downregulation of collagens associated with scar formation. In contrast, upregulation of collagen proteins is involved in ECM remodeling. Furthermore, the findings of these studies collectively support the notion that factors within secretomes govern corneal wound healing by increasing factors involved in ECM remodeling while suppressing matrix metalloproteinases that may be involved in angiogenesis and ECM remodeling.

Conversely, other studies investigate the role of secretomes in anti-fibrotic mechanisms and the expression of fibrotic markers during corneal wound healing. Kumar et al. (2018) analyzed the effects of secretomes from cryopreserved corneal stromal stem cells in corneal healing. The presence of secretomes enhanced corneal epithelial wound healing by promoting fibroblast proliferation and cell migration, while reducing fibrosis markers, including SPARC and fibronectin. Notably, the levels of connective tissue growth factor (CTGF) in corneal fibroblasts remained unchanged after treatment with secretomes. Interestingly, the regenerative potential of corneal stromal stem cells was retained even after 10 years of long-term cryopreservation (Kumar et al., 2018). Alternatively, levels of α-SMA, a marker of fibrosis, are suppressed in various studies (Carter et al., 2019; Shojaati et al., 2019; Tang et al., 2022). Carter et al. reported enhanced corneal wound closure associated with suppressed α-SMA. Interestingly, secretomes derived from MSCs co-cultured with electrospun fibers were more viable and led to less corneal opacification (Carter et al., 2019). This highlights the potential for future research to evaluate the impacts of MSC culture environments on the regulation of secreted factors in secretomes. Additionally, it is worth noting that other studies have investigated the role of microRNA (miRNA) in anti-fibrotic mechanisms that suppress α-SMA, a marker of fibrosis. Shojaati et al. and Tang et al. (2022) reported that knockdown of miRNAs in the secretome reverses its regenerative function, reducing corneal scarring and enhancing wound closure (Shojaati et al., 2019). Conversely, Liu et al. highlighted a different pathway in the context of miRNA-mediated corneal healing. They reported that activating the PTEN/PI3K/Akt pathway by suppressing PTEN enhanced corneal epithelial cell proliferation and migration, and improved corneal structural integrity, resulting in a more compact, regularly arranged structure. These effects were confirmed to be associated with miRNAs within secretomes, as miR-21 knockdown eliminated these positive effects (Liu et al., 2022).

In contrast, several other studies highlight the role of secretomes in regulating anti- and pro-inflammatory pathways to facilitate immunomodulation and successful corneal wound healing. Pro-inflammatory cytokines, including TNF-α, IL-6, IL-8, IL-1β, ICAM-1, and CXCL-2, were downregulated. At the same time, anti-inflammatory pathways, such as those associated with Regulatory T cell (Treg) phenotypes and IL-10, were upregulated across various studies where the application of MSC-secretomes resulted in enhanced corneal epithelial wound healing (Lai et al., 2018; Jabbehdari et al., 2020; Ma S et al., 2022). Furthermore, a study by Lai et al. found that tail vein injection of human MSC-derived exosomes in mouse models suppressed the cGVHD immune response via CD4 and TH17 cells. The use of tail vein injection in this study to deliver exosomes highlights exosomes' ability to treat corneal diseases when administered systemically (Lai et al., 2018). In addition to the immunomodulatory effects of secretomes, Ma S et al. (2022) reported a reduction in corneal haze and enhanced cell proliferation in the exosomes treatment group. Interestingly, the autophagy activator rapamycin (50 nM) combined with exosomes further enhanced the healing potential of secretomes, resulting in a more pronounced reduction in corneal haze and apoptosis of corneal epithelial cells. Conversely, in exosomes coupled with an autophagy inhibitor, the apoptosis was significantly increased (Ma S et al., 2022). Future research and clinical exploration of the use of autophagy activators coupled with MSC-exosomes could lead to a more effective treatment.

6.5.3 Secretome in Corneal Burns

Multiple studies using MSC secretomes for corneal epithelial regeneration after corneal burns in alkali-burn animal models have shown promising results. Fernandes-Cunha et al. administered a topical MSC-S within a viscoelastic gel consisting of hyaluronic acid (HA) and chondroitin sulfate (CS) to treat corneal burns. MSC-S in HA/CS was found to be the most effective in increasing epithelial wound closure and epithelial proliferation, as HA increased the regulation and activation of CD44 receptors. MSC-S successfully reduced corneal hemorrhage, scar formation, and neovascularization (Fernandes-Cunha et al., 2019). Tao et al., who used MSC-EVs with bioactive molecules to treat alkali-induced corneal burns, further support these findings. They reported an association between decreased neovascularization and inflammation-related molecules (e.g., IL-1β, TNF-α, NF-κB, MMP-2, MMP-9, and VEGF) and increased epithelial wound healing, as well as decreased cell apoptosis, angiogenesis, and inflammation (Tao et al., 2019). In a similar study, Sun et al. (2023) reported that MSC-EVs promoted corneal repair and epithelial cell proliferation by mediating Exos-miRNA 24-3p, thereby reducing corneal stromal fibrosis and macrophage activation (Sun et al., 2023). In another study, An et al. demonstrated severe ocular toxicity, increased markers (including COX-2, MMP-9, and VEGF), and separation of the stromal and epithelial layers following exposure to nitrogen mustard (NM) without treatment. However, treatment with MSC-secretomes had positive effects on corneal epithelial cell survival, primarily by enhancing mitochondrial function (via genes involved in glycolysis and ATP synthesis), reducing apoptosis and inflammation, and decreasing intracellular reactive oxygen species (An et al., 2022).

6.5.4 Secretome in Prevention of Corneal Neovascularization

Wound healing in corneal disorders is uncommonly associated with corneal neovascularization. However, the formation of new blood vessels can sometimes accompany new lymph vessel formation to aid in reducing corneal edema following corneal injury and is associated with a dysfunction in the balance between pro-angiogenic (e.g., VEGF, IL-1) and antiangiogenic factors (e.g., tissue inhibitor of metalloproteinase 3, soluble VEGF receptor). Notably, these new blood vessels can compromise corneal transparency, leading to changes in vision and/or vision loss. Therefore, treatments such as MSC-secretomes, which have been reported to regulate corneal neovascularization, hold important therapeutic potential for the effective management of corneal injury.

The application of MSC-secretomes to damaged corneas is associated with increased epithelial wound healing and a reduction in corneal neovascularization and inflammation in a dose-dependent manner. This process is reported to be influenced by decreased pro-inflammatory factors (including IL-2 and IFN-γ), decreased infiltration of CD4+ cells, and increased TSP-1 (Oh et al., 2008). In an important study, Eslani et al. confirmed the inhibition of corneal neovascularization development after treatment with Human Corneal-derived MSC-secretomes. This process was suggested to be mediated by the secretion of antiangiogenic factors, including PEDF and sFLT-1, as removing these factors from the secretome diminished its antiangiogenic effects. Additionally, angiogenic factor VEGF-A levels were reduced, affirming that a balance between pro-angiogenic and antiangiogenic factors governs neovascularization (Eslani et al., 2017).

6.5.5 Secretome in Corneal Ulcers and Infections

Corneal ulcers are defined as defects in the corneal epithelium and underlying stroma, characterized by visual changes, a red eye, and discomfort. The management of corneal ulcers is typically guided by bacterial sensitivity and ulcer severity, and commonly involves the use of topical antibiotics (Hartley, 2010). However, corneal healing often involves the formation of corneal scars, which can reduce visual acuity. This occurs through mechanisms that include light blocking or scattering, as well as the induction of irregular astigmatism (Hazlett et al., 2016). Pre-clinical animal studies using dog, rat, and mouse models with corneal ulcers have established the effectiveness of MSC-secretomes in treating corneal ulcers. They can therefore serve as an alternative to conventional topical antibiotic treatments. Casado-Santos et al. used topical adipose tissue-derived MSC-secretomes in dogs with complicated corneal ulcers. They suggested that treatment with MSC-secretomes improved outcomes in corneal ulcers without adverse reactions, reducing the mean time to ulcer healing and associated clinical signs (e.g., blepharospasm, photophobia, and conjunctival hyperemia) compared with other treatments. Interestingly, these findings were observed even in previously treated corneal ulcer cases (Casado-Santos et al., 2024). This suggests that MSC-secretomes may serve as a practical treatment option for complicated corneal ulcers.

Other research has suggested that the reversal of inflammation, neo-angiogenesis, and epithelial cell death, resulting in improvements in corneal ulcers, is governed by the downregulation of immunomodulatory and inflammatory markers (e.g., TNF-α, MIP-1α, PTGS2, and COX2) (Saccu et al., 2022). Importantly, Bermudez et al. (2015) also evaluated the bactericidal effects of human uterine-derived MSC-secretomes in rat models with NaOH-produced corneal ulcers. Researchers reported increased corneal epithelial cell proliferation, reduced apoptosis and macrophage inflammation, and a correlation with increased levels of bioactive compounds (e.g., TIMP-1, TIMP-2, FGF-6, FGF-7, uPAR, HGF). Interestingly, MSC-secretomes exerted bactericidal effects on E. coli and S. epidermidis by upregulating bactericidal proteins (e.g., chemokine ligands 1, 6, 8, 10, 15, and 20) (Bermudez et al., 2015). Thus, studies suggest that MSC-secretomes are effective in treating corneal ulcers by modulating immunomodulatory responses and exerting potent bactericidal effects.

6.5.6 Secretome in Increasing Corneal EC Density

The cornea, an avascular organ, is composed of an endothelial layer, the innermost cell layer, which plays a crucial role in maintaining corneal hydration and transparency. Any dysfunction of the corneal endothelium can cause stromal and epithelial edema, leading to loss of visual acuity (Tan et al., 2012; Tao et al., 2019). Furthermore, human corneal ECs gradually decline in number over time, as they are unable to proliferate in vivo. This degeneration is particularly significant in diseases such as Fuchs' endothelial dystrophy and posterior polymorphous corneal dystrophy, where progressive loss of ECs leads to corneal edema and vision impairment. As a result, surgical management, including Descemet membrane endothelial keratoplasty (DMEK) and Descemet stripping automated endothelial keratoplasty (DSAEK), remains the primary treatment option for restoring corneal clarity in these conditions (Tan et al., 2012). However, due to the shortage of donor corneas, this solution is not a viable long-term option. Additionally, procedures like DMEK and DSAEK carry operative and post-operative risks, including graft rejection, EC loss, visual disturbances, and the risk of graft displacement or detachment. This further underscores the need to explore other therapies, such as MSC secretomes, as regenerative treatments.

Some animal studies have found that HCECs treated with human MSCs-derived MSC secretomes exhibit enhanced proliferation, characterized by the same hexagonal cell morphology and functional phenotype as corneal ECs (Van den Bogerd et al., 2020). Nakahara et al. observed that MSC-CM led to the phosphorylation of the Akt and ERK1/2 pathways, which degraded p27, a potent inhibitor of the G1 phase of the cell cycle, as well as regulating the expression of cyclin D, which is crucial for G1/S progression and thus leading to cell proliferation (Nakahara et al., 2013). These results were further supported by Sun P et al. and Duan et al., who studied these effects using the secretomes of human orbital adipose-derived and human UC-derived MSCs, respectively. The researchers found that the proliferation ability of cultured HCECs was significantly enhanced, while maintaining the polygonal cell morphology of corneal ECs (Sun P et al., 2017; Duan et al., 2021). Interestingly, a key feature observed in the studies was that MSC-secretome-treated HCECs acquired stem-cell-like properties and expressed functional markers at intercellular junctions, including zona occludens-1 (ZO-1), Na+/K+-ATPase, and N-cadherin (Nakahara et al., 2013; Sun P et al., 2017). Moreover, Jung et al. found that Rabbit CECs in a combinatory culture comprised of MSC-CM and Rho-associated kinase inhibitor Y-27632 exhibited an increased passage number and cell proliferation with activation of genes such as MKI67/PCNA involved in cell proliferation, as well as low senescence, showcased by a significant decrease in senescence marker β-galactosidase-positive cells (Jung et al., 2021). These animal studies have demonstrated the potential of MSCs to treat corneal injuries by enhancing corneal EC proliferation, thereby increasing cell numbers.

6.5.7 Secretome in Dry Eye Disease

Dry eye disease (DED) is a multifactorial disorder of the ocular surface characterized by loss of tear film homeostasis, accompanied by ocular symptoms such as discomfort, visual disturbance, and dryness. It is associated with tear film instability, hyperosmolarity, inflammation, and neurosensory abnormalities, which can lead to corneal and conjunctival damage if left untreated (Rouen and White, 2018). Dry eyes, along with dry mouth, are often a key hallmark of Sjögren's syndrome (SS), an autoimmune disease characterized by lymphocyte infiltration of exocrine glands, such as the lacrimal and salivary glands (Zhao et al., 2022). The mainstay treatments for DED include artificial tears, topical anti-inflammatory agents, and autologous tear substitutes; however, certain limitations limit their optimal use (Rouen and White, 2018). For example, although artificial tear treatment is effective and safe, higher-viscosity formulations for those with severe disease can result in blurred vision by increasing contact time. On the other hand, biologic tear substitutes are often costly and have more complex storage and dosing requirements (Wu et al., 2022). Moreover, these treatments rely heavily on patient adherence, which can be inconsistent, further limiting their long-term effectiveness. Recent research suggests that MSC-derived secretomes have been effective in treating DED.

The use of secretomes across various studies has been associated with improved epithelial and goblet cell health, as well as the restoration of tear film layers (Lee et al., 2021; Wang L et al., 2023; Li et al., 2019; Ma F et al., 2022; Sendon-Lago et al., 2019). Additionally, MSC-exosomes have been beneficial for DED treatment, as they increase tear production and tear break-up time in DED animal models. Interestingly, enhancement of epithelium ultrastructure has been reported through increased corneal chondriosome and desmosomes, as well as reduced disorganization of corneal epithelial microvilli (Li et al., 2019). Lee et al. reaffirmed these findings by investigating the effects of adipose-derived MSC-secretomes in desiccating stress- and benzalkonium chloride-induced DED mouse models. As a result of topical secretome application, there was increased lacrimal tear secretion coupled with increased viability and goblet cell density of human corneal epithelial cells that were previously reduced by hyperosmolarity or desiccation. Interestingly, the corneal epithelium's tight junctions and microvilli were preserved in a process regulated by the increased expression of tight junction proteins, such as ZO-1, occludin, and keratin 12 (Lee et al., 2021). Similarly, in a more recent study, Wang L et al. (2023) suggested that the increased goblet cell density, tear volume, and corneal integrity, as well as the decreased apoptosis and aggregation of CD4+ T cells after MSC-EVs application, were regulated by immunomodulatory processes. As such, suppression of the IRAK1/TAB2/NF-κB pathway, which was activated in desiccation-induced DED mouse models, resulted in decreased CD4+ T cells and reduced levels of inflammatory cytokines, including IL-4, IL-8, IL-10, IL-17, and TNF-α (Wang L et al., 2023). The findings of these studies suggest that the effects of the secretome on epithelial and goblet cell health are multifactorial, including immunomodulation and increased expression of tight junction proteins.

Similarly, the mechanism of inflammation modulation in the treatment of DED with MSC-secretomes has been reported in various studies (Yu et al., 2020; Wang L et al., 2023; Guo et al., 2022; Zhou et al., 2022). The decrease in NLRP3 activation-mediated inflammation, caspase-1 activation, and IL-1β maturation emerges as a common pathway of immune modulation, combined with the upregulation of anti-inflammatory cytokines (including IL-10) and the downregulation of pro-inflammatory cytokines (including IL-1β, IL-6, IL-1α, IFN-γ, and TNF-α). This immunomodulation enables secretomes to effectively alleviate dry eye symptoms by increasing tear secretion, expanding conjunctival goblet cell density, mitigating corneal epithelial repair, and reducing apoptosis (Yu et al., 2020). In addition to regulating inflammatory cytokines, Guo et al. suggested that other inflammatory factors play a role in the successful treatment of DED with secretomes (Guo et al., 2022). They reported that mice receiving MSC-EVs, rather than PBS, exhibited lower DED disease severity, accompanied by decreased Th17 cell numbers, increased MHC-II expression, and reduced levels of inflammatory cytokines (including TNF-α, IL-6, and IL-1β). Interestingly, they found that MSC-EVs' suppression of inflammation in DED was due to reduced dendritic cell counts and maturation, thereby decreasing dendritic cell-mediated Th17 immune responses.

In several animal models, DED was regulated through the transition of macrophages into the M2 anti-inflammatory phenotype (Li N et al., 2022; Zhou et al., 2022; Sendon-Lago et al., 2019). Interestingly, this process was influenced by miR-204 targeting the interleukin-6 receptor (IL-6R), as miR-204 blockade in exosomes reduced the immunosuppressive effects. At the same time, those with miR-204 showed a markedly attenuated effect (Zhou et al., 2022). This transition of macrophages was associated with increased tear secretion, increased corneal and epithelial thickness, and decreased corneal edema and haze. Notably, Ma et al. combined ascorbic acid with exosome therapy and reported a similar pattern in a DED mouse model. The synergistic benefits of exosomes and ascorbic acid effectively reduced reactive oxygen species, corneal damage, and inflammation, while also enhancing stromal restoration (Ma F et al., 2022). This highlights the potential for future research to evaluate the effectiveness of combined therapeutic innovations in expanding the scope of MSC-secretomes. In line with secretome therapeutic innovations, Sendon-Lago et al. investigated the effects of MSC-secretomes containing TIMP-1 and TIMP-2 in the treatment of DED. They reported significantly increased corneal epithelial cell proliferation and reduced apoptosis in rabbits with atropine-induced dry eyes compared with controls (Sendon-Lago et al., 2019). Similarly, PD-L1-modulated exosomes were used to regulate T follicular helper (Tfh) cell responses and autoantibody production. It was observed that the disease progression, Tfh cell response, and Tfh differentiation from naïve T cells were significantly lower in DED mouse models treated with olfactory ecto-MSC-derived exosomes. Furthermore, to confirm the modulatory effects of PD-L1, PD-L1 was removed from the secretomes, resulting in decreased therapeutic effect, increased autoantibody production, and a sustained Tfh cell response (Rui et al., 2022). These studies demonstrate the effectiveness of MSC-secretomes in treating DED and identify avenues for exciting future research to enhance treatment outcomes further. Table 6.3 summarizes the potential of MSC-derived secretomes for treating corneal and anterior segment disorders.

6.6 UVEITIS AND INTRAOCULAR INFLAMMATION

6.6.1 Current Challenges in Managing Uveitis

Uveitis encompasses inflammation of the uvea (consisting of the iris, choroid, and ciliary body) due to non-infectious autoimmune or infectious causes. Untreated uveitis can be devastating and lead to reduced or even complete vision loss. Although many factors contribute to the pathophysiology of uveitis, it has been established that retinal antigen-specific T cells play a crucial role in the uveitis disease process. Notably, the use of experimental autoimmune uveoretinitis (EAU) animal models has significantly facilitated the understanding of how T cells, including natural T regulatory cells (Tregs), play a critical role in tissue damage through the inflammatory cascade following a microbial or traumatic stimulus.

Autoimmune uveitis is currently managed by various treatments based on etiology, severity, and other factors. Corticosteroids, immunosuppressive agents such as methotrexate, and various biologics are currently prescribed to manage uveitis. Although effective, these treatments have several

TABLE 6.3
MSC-Derived Secretomes in Corneal and Anterior Segment Disorders

Condition	Secretome/ Exosome Features	Route of Administration	Factor(s) Influenced	Overall Outcome(s)	Stage	Reference
Corneal burn	hBM-MSC-secretomes	Topical	-Increased: CD44 receptors, ALDH3A1 -Decreased: CD31	-Increased epithelial wound closure and epithelial cell proliferation in a dose-dependent manner. -Reduced corneal hemorrhage, scar formation, and neovascularization. -MSC-S in HA/CS improvements were greater than MSC-S, HA/CS, or saline alone.	Pre-clinical study (in vitro human corneal epithelial cells and in vivo rat and mouse models)	Fernandes-Cunha et al., 2019
Corneal burn	Human placenta-derived MSC-Exo	Topical	-Increased: IL-10 -Decreased: inflammatory markers (IL-1b rNA, IL-8 rNA, TNF-α rNA, NF-kB rNA, Cas-8), neovascularization markers (MMP-2, MMP-9, VEGF)	-Increased epithelial wound healing and corneal epithelial migration in a dose-dependent manner. -Reduced apoptosis and inflammation markers, corneal haze, and neovascularization.	Pre-clinical study (in vitro human corneal epithelial cells and in vivo mouse models)	Tao et al., 2019
Corneal burn	MSC-exosomes	Topical	-Partially increased: genes involved in inflammation (IL-1β, IL-10, TNF-α), ATP synthesis (ATP5S, ATP5D), glycolysis (HK2, PKM2, PFK), glutaminolysis (IDH2, IDH3G, GLS2) -Decreased: intracellular ROS	-Increased epithelial cell survival, corneal cell viability, restored morphology, epithelial wound healing, and reversed NM-induced cellular homeostatic damage through decreased ROS generation and increased mitochondrial function. -Maintenance of epithelial barrier and corneal stromal membrane integrity after NM exposure. -Reduced apoptosis and inflammation of corneal epithelium and stroma.	Pre-clinical study (in vitro corneal cells, ex vivo ocular surface of porcine, and in vivo mouse models)	An et al., 2022

Condition	Secretome/ Exosome Features	Route of Administration	Factor(s) Influenced	Overall Outcome(s)	Stage	Reference
Corneal burn	Rabbit AD-MSC-Exo	Topical	-Increased: CK3, ZO-1	-Increased corneal epithelial cell migration and corneal thickness, but a lack of tight junctions (poor morphology). -THH-3/Exos-miRNA 24-3p application increased epithelial maturation and corneal epithelial defect healing through decreased macrophage activation, corneal edema, and corneal stromal fibrosis.	Pre-clinical study (rabbit models)	Xiaomin Sun et al., 2023
Corneal ulcers	BM-MSC-secretomes	Topical	-Decreased: inflammatory markers (PTGS2, TNF-α, COX2)	-Improved corneal ulcer healing times and reduction of complications. -Reversal of corneal inflammation, neo-angiogenesis, and epithelial cell death with twice-daily application.	Pre-clinical study (in vitro human corneal epithelial cells and in vivo mouse models)	Saccu et al., 2022
Corneal ulcers	Dog AD-MSC-secretomes	Topical	N/R	-Enhanced corneal ulcer outcomes, even in cases that were previously treated with other treatments. -Decreased average ulcer healing time compared to other treatments.	Pre-clinical study (in vivo in 10 dogs with spontaneous chronic corneal epithelial defects)	Casado-Santos et al., 2024
Corneal ulcer and DED	Human uterine cervical-derived MSC-secretomes	Topical	-Increased: bioactive compounds (TIMP-1, TIMP-2, FGF-6, FGF-7, uPAR, HGF), bactericidal proteins (chemokine ligands 1, 6, 8, 10, 5, and 20) -Decreased: corneal MIP-1α, TNF-α mRNA expression	-Increased corneal epithelial cell proliferation, reduced apoptosis, and reduced corneal macrophage inflammation in corneal ulcers produced using NaOH. -Produced bactericidal effect on E. coli and S. epidermidis.	Pre-clinical study (in vitro rat corneal epithelial cells and in vivo rat models)	Bermudez et al., 2015

(Continued)

TABLE 6.3 CONTINUED
MSC-Derived Secretomes in Corneal and Anterior Segment Disorders

Condition	Secretome/ Exosome Features	Route of Administration	Factor(s) Influenced	Overall Outcome(s)	Stage	Reference
Corneal wound healing	BM-MSC-Exo	Topical	-Decreased: MMP-2	-Significantly increased cell migration (by ~1.9-fold). -Decreased corneal inflammation and neovascularization through a decrease in MMP-2.	Pre-clinical study (in vitro human corneal cells and porous membrane—"cornea on a chip")	Zitong Yu et al., 2022
Corneal wound healing	Cryopreserved corneal-derived MSC-secretomes	Topical	-Decreased: fibrotic markers expression (SPARC, fibronectin) -Unaltered: CTGF	-Enhanced corneal epithelial wound healing through fibroblast proliferation and cell migration, as well as a reduction in fibrosis markers. -Corneal stromal stem cells' regenerative potential was retained even after long-term cryopreservation (10 years).	Pre-clinical study (in vitro corneal fibroblasts)	Kumar et al., 2018
Corneal wound healing	hAD-MSC-Exo	Topical	-Increased: collagens, fibronectin -Decreased: expression of MMP1, MMP2, MMP3, MMP9	-Increased wound closure, increased corneal stromal cell growth, and reduced apoptosis in a dose-dependent manner. -Enhanced extracellular matrix (ECM) remodeling through upregulation of ECM-related proteins.	Pre-clinical study (in vitro corneal stromal cells)	Shen et al., 2018
Corneal wound healing	hBM-MSC-Exo	TVI	-Increased: IL-10-expressing cells -Decreased: pro-inflammatory cytokine relevant to Th17 (IL-17A, IL-21, IL-22, IL-2) -Suppressed: CD4 cells, TH17 cells	-Immunosuppressive effects on cGVHD through an increase in pro-inflammatory cytokine (IL-17A, IL-21). -Regulatory effect on GVHD effector T cells through enhanced Treg phenotype and decreased Th17 differentiation.	Pre-clinical study (in vivo mouse models)	Lai et al., 2018
Corneal wound healing	hBM-MSC-secretomes	Topical	-Decreased: α-SMA	-Enhanced corneal fibroblasts wound closure and viability in secretomes derived from MSCs co-culture with electrospun fiber. -Reduction in corneal opacification and scarring.	Pre-clinical study (in vitro corneal fibroblasts and ex vivo rabbit models)	Carter et al., 2019

Condition	Secretome/ Exosome Features	Route of Administration	Factor(s) Influenced	Overall Outcome(s)	Stage	Reference
Corneal wound healing	Human Corneal-derived MSC-exosomes	Topical	-Decreased: fibrotic genes (Col3a1), α-SMA2	-Reduced corneal scar formation and neutrophil infiltration. -Restoration of normal tissue morphology through reduction in fibrotic genes. -Exosomes reduced miRNA, lost their regenerative function, and were ineffective at reducing corneal scarring.	Pre-clinical study (in vitro human corneal epithelial cells and human corneal stromal fibroblasts)	Shojaati et al., 2019
Corneal wound healing	Human Corneal-derived MSC-exosomes or secretomes	Topical	-Expressed: CD9, CD63, CD81 -Not expressed: GM130, Calnexin, Cytochrome-C	-Significant decrease in remaining wound area and improved monolayer re-epithelialization in cMSC exosome-treated human corneal epithelial cells in vitro ($P < 0.005$). -Enhanced corneal wound healing in cMSC exosome-treated corneas in vivo ($P < 0.05$).	Pre-clinical study (in vitro human corneal MSCs, human corneal epithelial cells, human macrophages, and in vivo corneal epithelium)	Samaeekia et al., 2018
Corneal wound healing	Human Corneal-derived MSC-secretomes	Topical	N/R	-Increased wound healing and closure in a dose-dependent manner. -Removal of exosomes from MSC-conditioned medium terminated wound healing effects. -Collection of exosomes over 72 h had greater wound healing effects compared to 48 h collection.	Pre-clinical study (in vitro human corneal epithelial cell and in vivo mouse models)	An et al., 2023
Corneal wound healing	Human Corneal-derived MSC-secretomes	Topical	-Decreased: ICAM-1, TLR3, IL-6, IL-8, TNF-α	-Reduction in inflammatory gene expression ($P < 0.0001$) and enhanced corneal epithelial wound healing ($P < 0.05$) compared to control.	Pre-clinical study (in vitro human corneal limbal epithelial and in vivo mouse models)	Jabbehdari et al., 2020

(*Continued*)

TABLE 6.3 CONTINUED
MSC-Derived Secretomes in Corneal and Anterior Segment Disorders

Condition	Secretome/ Exosome Features	Route of Administration	Factor(s) Influenced	Overall Outcome(s)	Stage	Reference
Corneal wound healing	hUS-MSC-Exo	Topical	-Decreased: TNF-a, IL-1b, IL-6, CXCL-2	-Improved epithelial cell migration and viability with exosomes and autophagy activator (Rapamycin 50 nM). -Reduction in corneal hazing and enhanced cell proliferation in the exosome treatment group, which is increased further in the exosome + autophagy activator group. -Apoptosis of corneal epithelial cells is most significant in exosomes with an autophagy inhibitor. At the same time, it was lowest in exosomes with an autophagy activator.	Pre-clinical study (in vitro human corneal epithelial cells and in vivo mouse models)	Ma S et al., 2022
Corneal wound healing	hUS-MSC-Exo	Topical	-Increased: PTEN/ PI3K/Akt pathway -Decreased: PTEN	-Enhanced corneal epithelial cells proliferation and migration mediated through miRNAs, as miR-21 knockdown eliminated these effects. -Healing of defects in the cornea in vivo resulted in a more compact and regularly arranged structure.	Pre-clinical study (in vitro human corneal epithelial cells and in vivo rat models)	Liu et al., 2022
Corneal wound healing	IPSCs-derived MSC-Exo	Topical	-Increased: collagen expression (collagen type I alpha 1, collagen type V alpha 1, collagen type V alpha 2) -Decreased: α-SMA, vimentin, TRAM2	-Enhanced cell migration and wound closure in thermosensitive hydrogel exosomes. -Decreased expression of collagen and TRAM2 in corneal stromal stem cells, as well as a reduction in corneal scarring mediated through miR-432-5p, as the miR-432-5p inhibitor eliminated these effects.	Pre-clinical study (in vitro human corneal epithelial and corneal stromal stem cells and in vivo rat models)	Tang et al., 2022
Dry eye disease	hAD-MSC-Exo	Topical	-Decreased: NLRP3 activation-mediated inflammation (DS induced), caspase-1 activation, IL-1β maturation	-Increased tear production along with decreased cell apoptosis, goblet cell loss, inflammatory cytokine production, and corneal epithelial defects. -Potential protection against cell apoptosis that is induced by hyperosmotic stress.	Pre-clinical study (in vitro human corneal epithelial cells and in vivo mouse models)	Yu et al., 2020

Condition	Secretome/ Exosome Features	Route of Administration	Factor(s) Influenced	Overall Outcome(s)	Stage	Reference
Dry eye disease	hAD-MSC-secretomes	Topical	-Increased: expression of ZO-1, occludin, keratin 12, MUC16	-Increase in lacrimal tear secretion. -Increased viability and goblet cell density of human corneal epithelial cells that were previously reduced by hyperosmolarity or desiccation. -Corneal epithelium tight junction and microvilli were preserved.	Pre-clinical study (in vitro human corneal epithelial cells and in vivo mouse models)	Zhang, 2021
Dry eye disease	hUC-MSC-Exo	Topical	-Increased: IL-13 -Decreased: CD4+ T cells, inflammatory cytokines (IL-4, IL-8, IL-10, IL-17, TNF-α)	-Suppression of the IRAK1/TAB2/NF-κB pathway that was activated in desiccating-induced DED mouse models. -Increased goblet cell density, tear volume, and corneal integrity, as well as decreased apoptosis and CD4+ T cells aggregation. -Inflammatory cytokines were regulated in the ocular surface and tears.	Pre-clinical study (in vivo mouse models)	Wang L et al., 2023
Dry eye disease	hUC-MSC-Exo	Topical	-Decreased: expression of MHC-II, inflammatory cytokines (TNF-α, IL-6, IL-1β), Th17 cells	-Increased tear production and reduced disease severity in exosome-treated models compared to controls. -Reduction in dendritic cell count and maturation through downregulation of inflammatory cytokine expression.	Pre-clinical study (in vitro human corneal epithelial cells and in vivo mouse models)	Guo et al., 2022
Dry eye disease	hUC-MSC-Exo	Subconjunctival injection	-Increased: proportion of Tregs -Decreased: CD4+ T cell proliferation	-Rabbit autoimmune dacryoadenitis modulated through transition of macrophages into M2 anti-inflammatory phenotype -In vitro reduction of tissue damage and inflammation through inhibition of Treg generation and CD4+ T cell proliferation.	Pre-clinical study (in vivo rabbit models)	Li N et al., 2022

(*Continued*)

TABLE 6.3 CONTINUED
MSC-Derived Secretomes in Corneal and Anterior Segment Disorders

Condition	Secretome/ Exosome Features	Route of Administration	Factor(s) Influenced	Overall Outcome(s)	Stage	Reference
DED associated with GVHD	hUC-MSC-Exo (clinical study) and mice BM-MSC-Exo (in vivo study)	Topical	-Decreased: pro-inflammatory genes (Il-6, Il-1β, Il-17a, Cd86)	-Increased tear secretion, tear-film break-up time, and cornea and epithelium thickness, as well as decreased corneal edema, hazing, OSDI, and fluorescein scores. -Blockade of miR-204 in exosomes reduced immunosuppressive effects, while those with miR-204 showed a markedly attenuated effect. -Transition of macrophages into M2 anti-inflammatory phenotype through IL-6R targeted by miR-204.	Clinical study (14 patients with GVHD) and pre-clinical study (in vivo mouse models)	Zhou et al., 2022
DED	Human uterine cervical-derived MSC-secretomes	Topical	-Decreased: pro-inflammatory cytokines (TNF-α, MCP-1, MIP-1α, IL-6)	-Increased corneal epithelial cell proliferation ($P < 0.05$) and reduction in apoptosis compared to control in rabbits with atropine-induced dry eye. -These corneal regenerative effects were mediated by high levels of TIMP-1 and TIMP-2 found in Human uterine cervical-derived MSC-secretomes.	Pre-clinical study (in vitro rat corneal epithelial cells and in vivo rabbit models)	Sendon-Lago et al., 2019
DED	Mouse AD-MSC-Exo	Topical	-Increased: anti-inflammatory cytokine (IL-10) -Decreased: pro-inflammatory cytokine (IL-10-1β, IL-6, IL-1α, IFN-γ, TNF-α), NLRP3 inflammasome activation, caspase-1, IL-1β, and IL-18	-Increased tear secretion, conjunctival goblet cells density, and corneal epithelium repair, as well as reduction of apoptosis.	Pre-clinical study (in vivo mouse models)	Wang et al., 2022

Condition	Secretome/ Exosome Features	Route of Administration	Factor(s) Influenced	Overall Outcome(s)	Stage	Reference
DED	Mouse MSC-Exo coupled with Ascorbic acid	Topical	-Increased: CD206 associated with M2 macrophage -Decreased: ROS	-Increased tear secretion, central cornea thickness, and stroma layer restoration, as well as decreased corneal damage and inflammation. -Ascorbic acid coupled with exosome-mediated increase in M2 macrophages, which decreases hyperosmolarity of the tear film and inflammatory response through ROS removal.	Pre-clinical study (in vitro human corneal epithelial cells and in vivo mouse models)	Ma F et al., 2022
Dry eye disease	MSC-Exo	Topical	N/R	-Improved epithelium ultrastructure with increased corneal chondriosome/desmosomes. -Reduction in corneal fluorescein staining score, and disorganization of corneal epithelial microvilli in the exosome treatment group, similar to baseline without DED. -Increased tear production and tear break-up time.	Pre-clinical study (in vivo mouse models)	Long, 2019
DED	Olfactory ecto-derived MSC-Exo	Tail vein injection	-Increased: PD-L1 -Decreased: Tfh cell differentiation	-Reduction in disease progression, Tfh cell response, and Tfh differentiation from naïve T cells in mouse models. -Exosomes with PD-L1 knockdown showed decreased therapeutic effect, as well as increased autoantibody production and sustained Tfh cell response.	Pre-clinical study (in vivo mouse models)	Rui et al., 2022
Decreased corneal EC density	hBM-MSC-secretomes	Co-culture	-Increased: Ki67-positive cells, DNA incorporation of BrdU, phosphorylation of Akt and ERK1/2, cell cycle G1 proteins	-Increased epithelial cell proliferation in a dose-dependent manner. -Faster Scratch-induced wound closure (63% initial wound area coverage) in those treated with Human MSCs-derived MSC-secretomes. -Epithelial cells treated with secretomes assumed a hexagonal shape, which was similar to corneal ECs in vivo, compared to an irregular shape in controls.	Pre-clinical study (in vitro human corneal ECs)	Nakahara et al., 2013

(Continued)

TABLE 6.3 CONTINUED
MSC-Derived Secretomes in Corneal and Anterior Segment Disorders

Condition	Secretome/ Exosome Features	Route of Administration	Factor(s) Influenced	Overall Outcome(s)	Stage	Reference
Decreased corneal EC density	hBM-MSC-secretomes	Co-culture	-Increased: EGF, IGFBP2, IGFBP6	-Increased epithelial cell proliferation through the effect of secretomes, mediated by a combined effect of multiple growth factors (EGF, IGFBP2, IGFBP6). -Individual application of growth factors (EGF, IGFBP2, IGFBP6) did not reproduce epithelial cell proliferation.	Pre-clinical study (in vitro B4G12 human immortalized corneal ECs)	Van den Bogerd et al., 2020
Decreased corneal EC density	Human orbital AD-derived MSC-secretomes	Co-culture, then cell injection, transplant	-Increased: ZO-1, Na+/K+ ATPase, N-cadherin, Col8a2, SLC4A4	-Enhanced epithelial cell proliferation as well as maintained polygonal cell morphology. -Improved corneal transparency following in vivo injections.	Pre-clinical study (in vitro human corneal ECs and in vivo rabbit and monkey models)	Sun P et al., 2017
Decreased corneal EC density	hUC-MSC-secretomes	Co-culture	-Increased: CEC-relative functional markers, ZO-1, N-cadherin, Na+ /K+ -ATPase, mRNA expressions of CDK4 and CDK6 -Decreased p27 gene expression	-Increased epithelial cell proliferation as well as maintained characteristic phenotypes in vitro. -Epithelial cells grown in secretomes upregulated transcriptional levels of CEC-related functional markers.	Pre-clinical study (in vitro rabbit corneal ECs)	Duan et al., 2021
Decreased corneal EC density	hUC-MSC-secretomes combined + Rho-associated kinase inhibitor	Co-culture	-Increased: expression of corneal endothelial-specific marker genes (CLRN1, HTR1D, GRIP1, ZP4)	-Increase in proliferation and functional markers and reduction in senescence due to synergistic effects of secretomes and Y-27632. -Epithelial cells grown in secretomes upregulated expression of CEC-specific marker genes.	Pre-clinical study (in vitro rabbit corneal ECs)	Jung et al., 2021

Condition	Secretome/ Exosome Features	Route of Administration	Factor(s) Influenced	Overall Outcome(s)	Stage	Reference
Corneal neovascularization	Human Corneal-derived MSC-secretomes	Topical	-Increased: antiangiogenic factors (sFLT-1, PEDF), anti-inflammatory factors (TSG6) -Decreased: angiogenic factor (VEGF-A)	-Inhibition of human umbilical vein EC tube and sprout formation -Inhibition of corneal neovascularization development, mediated by secretion of PEDF and sFLT-1, as removal of these from the secretome diminished the secretome's antiangiogenic effects. -Secretome modulation of neovascularization indirectly mediated by anti-inflammatory factors (such as TSG6).	Pre-clinical study (in vivo mouse models)	Eslani et al., 2017
Corneal neovascularization	Rat primary MSC lines-derived MSC-secretomes	Topical	-Increased: TSP-1 -Decreased: infiltration of CD4+ cells, pro-inflammatory cytokines (IL-2, IFN-gamma, MMP-2)	-Increased epithelial wound healing, and reduction in corneal neovascularization and inflammation in a dose-dependent manner.	Pre-clinical study (in vivo rat models)	Oh et al., 2008

limitations, including side effects and variable efficacy, which make them less practical and appealing for managing uveitis. For example, prolonged topical corticosteroids can result in severe side effects, including cataracts and glaucoma. Similarly, immunosuppressive agents and biologics are commonly associated with an increased risk of infections, MSC suppression, gastrointestinal disturbances, and, in rare cases, demyelinating disorders and cancers. As such, investigating the roles of alternative therapies, such as MSC-secretomes (Figure 6.2), in managing autoimmune uveitis offers an opportunity to overcome some of these limitations and reduce the risk of complications.

6.6.2 Secretome in Autoimmune Uveitis

To understand the mechanisms underlying the successful treatment of autoimmune uveitis, several studies have analyzed the effects of MSC-derived secretomes in vivo EAU-induced mouse models. In 2017, Bai et al. first assessed the effects of MSC-derived exosomes (MSC-EVs) in rats with EAU following periocular injections of MSC-EVs, as well as in vitro. They reported that MSC-EVs were successful in reducing the severity of EAU through mechanisms that downregulated the infiltration of inflammatory cells (including T cell subsets). Interestingly, they reported that a significant decrease in the chemoattractive effects of CCL2 and CCL21 on inflammatory cells, in part, facilitated this process (Bai et al., 2017). This suggests that MSC-EVs help reduce the migration of inflammatory cells into the eyes, thereby mitigating the progression of EAU.

Conversely, while other future studies suggest a role of MSC-EVs in decreasing inflammatory cell migration, they also further clarify that MSC-EVs also act directly by suppressing the development of T helper cells (Th), including Th1 and Th17 cells (Shigemoto-Kuroda et al., 2017; Li N et al., 2022; Li H et al., 2022). Shigemoto-Kuroda et al. analyzed the effects of MSC-EVs on tail vein injections in established mouse models of EAU and type 1 diabetes (T1D). Researchers reported a significant reduction in damaged retinal photoreceptors, retinal pathology disease score, and CD3+ T cells and inflammatory infiltrates in EAU mice treated with exosomes, to levels that were similar to those of normal mouse retinas. Notably, MSC-EVs in this study prevented disease onset in both EAU and T1D by suppressing the development of Th1 and Th17 cells by inhibiting APC activation (Shigemoto-Kuroda et al., 2017). Similarly, Li et al. reported similar reductions in EAU disease severity, infiltrating inflammatory cells, and the development of Th1 and Th17 cells following MSC-EVs tail vein injections in EAU mouse models. Furthermore, they reported that MSC-EVs treatment decreased the infiltration of Th1 and Th17 cells in the eyes, spleen, and lymph nodes (Li Y et al., 2021). The findings of these studies highlight the potential systemic applications of MSC-EVs delivered via tail vein injection, beyond the management of ocular diseases. The study by Li et al. further supports these findings, which indicate decreased Th17 infiltration. More importantly, researchers in this study used subconjunctival injection of MSC-EVs loaded with Rapamycin to target EAU in mouse models. It was reported that Rapamycin-loaded exosomes reduced ocular inflammatory cell infiltration and EAU clinical scores, as well as the number and size of focal lesions near the optic disk (Li H et al., 2022). This finding underscores the synergistic role of Rapamycin in the treatment of EAU with MSC-EVs, highlighting a potential for future research to analyze the dose-dependent relationships between MSC-EVs and Rapamycin. The use of MSC-EVs in subconjunctival injection, which yielded the desired EAU amelioration in this study, also confirms that exosomes are effective at delivering drugs to the retina, making them a useful vector for drug transport in ocular diseases.

In a more recent study, Kaur et al. (2024) propose another distinct pathway through which MSC-EVs reduce EAU disease progression. The study analyzed levels of immunosuppressive molecules (including TGF-β1 and let-7b) to evaluate the immunomodulatory effect of tail vein injection of human MSC-derived MSC-EVs in EAU mouse models. They reported a reduction in EAU disease progression, accompanied by decreased apoptosis and chemotaxis of retina-reactive T cells, following treatment with MSC-EVs expressing high levels of TGF-β1. More importantly, they reported that the suppression of retina-reactive T cell infiltration of the eyes occurred in a TGF-β1-dependent manner and was facilitated by the inhibition of the MAPK/ERK pathway in activated T cells (Kaur et al., 2024). This study, therefore, highlights the therapeutic potential of MSC-EVs and identifies TGF-β1-dependent inhibition of the MAPK/ERK pathway as a target for future uveitis treatment.

Similarly, other studies have suggested that MSC-EVs ameliorate EAU in a dose-dependent manner. Ruotian et al. evaluated the manifestations of EAU in rats receiving periocular injections of MSC-EVs. The application of MSC-EVs with R16 peptides decreased EAU clinical and pathological manifestations, T cell proliferation, ocular macrophage infiltration, and ocular inflammatory cell infiltration, as evidenced by a lower retinal pathological score in the MSC-EVs treatment group. More notably, they reported a dose-dependent inhibition of the proliferation of CD68+, Th1, Th17, and Treg cell subsets in response to R16 (Ruotian et al., 2018). Similarly, Yang et al. suggested that MSC-EVs modified with TAT peptides effectively enhanced ocular barrier penetration, EAU prevention, and treatment compared with naive exosomes. In contrast, the control group receiving no exosomes showed no observable penetration. Researchers suggested that MSC-EVs TAT-peptide ameliorates EAU by reducing mRNA and protein expression of inflammatory cytokines (TNF-α, IL-1β, IL-6, IL-17A, IFN-γ, and IL-12) (Yang et al., 2024).

It is important to note that exosomes are the only secretome factors that have been investigated for their role in mediating autoimmune uveitis. It would therefore be valuable for future research to investigate the collective effects of exosomes and other secreted factors within the secretomes in the treatment of uveitis. The beneficial dose-dependent effects of MSC-EVs in reducing EAU severity highlight their significant potential and underscore the need for future studies to determine the optimal dose to optimize uveitis treatment protocols. Moreover, combining MSC-EVs with other proteins, such as the R16 and TAT peptides, provides insight into the role of combinative therapies and can enhance clinical applications in uveitis.

6.7 CONCLUSIONS

The exploration of secretome-based therapies in ophthalmology represents an evolving frontier with the potential to address key challenges in the management of retinal diseases, glaucoma, corneal disorders, and uveitis. Conventional treatments, while effective in many cases, often suffer from limitations such as the need for frequent administration, invasive procedures, and local and systemic side effects. Secretome-derived approaches, particularly those leveraging EVs and exosomes, offer promising avenues for targeted, sustained, and potentially regenerative therapies.

In retinal diseases, secretome-based interventions may facilitate neuroprotection, enhance drug delivery, and modulate angiogenesis, potentially improving outcomes while reducing the burden of frequent intravitreal injections. In glaucoma, secretome therapies have the potential to support trabecular meshwork repair and preserve the optic nerve, complementing existing intraocular pressure-lowering strategies. Corneal and anterior segment disorders, including non-healing wounds, neovascularization, and endothelial dysfunction, may benefit from the modulation of inflammation and extracellular matrix remodeling using secretome-based approaches. Similarly, in uveitis, secretome-derived immunomodulatory factors could provide alternative approaches to dampening ocular inflammation while reducing reliance on long-term corticosteroids and immunosuppressants.

While pre-clinical evidence supports the therapeutic potential of the secretome in ophthalmology, further research is required to optimize formulation, delivery methods, and long-term safety. Rigorous clinical trials will be essential to validate these findings and establish standardized treatment protocols. Crucially, continued collaboration between basic scientists and clinicians will be essential to translating these innovations into practical clinical applications, thereby bridging the gap between the bench and the bedside to improve patient outcomes. Table 6.4 summarizes the potential of MSC-derived secretomes for treating uveitis.

TABLE 6.4
MSC-Derived Secretomes in Uveitis

Condition	Secretome/ Exo Features	RoA	Factor(s) Influenced	Overall Outcome(s)	Stage	Reference
Experimental Autoimmune Uveitis	hBM-MSC-Exo	TVI	-Decreased: MAPK/ERK pathway in activated T cells	-Diminished EAU disease progression, decreased apoptosis, and chemotaxis of retina-reactive T cells through treatment with secretomes carrying high levels of TGF-β1. -Suppression of the MAPK/ERK pathway in activated T cells and retina-reactive T cells infiltration of the eyes, in a TGF-β1 levels-dependent manner.	Pre-clinical study (in vitro assays and in vivo mouse models)	Kaur et al., 2024
Experimental Autoimmune Uveitis and Type 1 Diabetes	hBM-MSC-Exo	TVI	-Decreased: Th1 (IFN-γ+CD4+ cells) and Th17 (IL-17+CD4+ cells)	-Effective prevention of EAU and T1D disease onset. -Reduction in damaged retinal photoreceptors, retinal pathology disease score, and CD3+ T cells and inflammatory infiltrates in EAU mice treated with secretome, which was similar to normal mouse retinas. -Decreased development of Th1 and Th17 cells.	Pre-clinical study (in vivo mouse models)	Shigemoto-Kuroda et al., 2017
Experimental Autoimmune Uveitis	hUS-MSC-Exo	TVI	-Increased: CD4+FOX3+cells in the spleen and LNs -Decreased: Th1 (IFN-γ+CD4+ cells) and Th17 (IL-17+CD4+ cells)	-Decreased EAU disease severity, mean clinical score ($P < 0.01$), infiltrating inflammatory cells, and photoreceptor layer damage in mice treated with secretomes containing recombinant IL-10. -Decreased infiltration of Th1 and Th17 cells in the eyes, spleen, and LNs.	Pre-clinical study (in vivo mouse models)	Li H et al., 2022
Experimental Autoimmune Uveitis	hUS-MSC-Exo	POI	-Decreased: CD68+ macrophages	-Exosome application ameliorates EAU through reduction in inflammatory cells (such as T cell subsets, CD3+ T cells, and CD68+ macrophages) migration in the eyes. -Significant decrease in CCL2 and CCL21 chemotactic effects on inflammatory cells.	Pre-clinical study (in vivo rat models)	Bai et al., 2017

Condition	Secretome/ Exo Features	RoA	Factor(s) Influenced	Overall Outcome(s)	Stage	Reference
Experimental Autoimmune Uveitis	Rat MSC-Exo	POI	-Decreased: CD68+, Th1, Th17 and Treg cell subsets	-Decreased EAU clinical and pathological manifestations, T cell proliferation, ocular macrophage infiltration, and inflammatory cells in the eye. -Lower retinal pathological score of the group treated with exosomes compared to the control ($P<0.01$). -Proliferation of T cells was inhibited in a R16 dose-dependent manner.	Pre-clinical study (in vivo rat models)	Ruotian et al., 2018
Experimental Autoimmune Uveitis	hUC-MSC-Exo—Rapamycin-loaded	Subconjunctival injection	-Increased: CD4+FOX3+cells in the spleen and LNs -Decreased: Th17 (IL-17+CD4+ cells)	-Rapamycin-loaded exosomes resulted in decreased ocular inflammatory cell infiltration and clinical scores. -Subconjunctival injection was successful in reaching the retinal foci, confirming the ability of exosomes to deliver drugs to the retina. -Lower number and size of focal lesions near the optic disk were detected in Rapamycin-loaded exosomes, compared to exosomes only. However, the exosomes-only group had fewer lesions compared to the control group, which exhibited multiple large, confluent lesions.	Pre-clinical study (in vivo mouse models)	Huan Li D et al., 2021
Experimental Autoimmune Uveitis	Mouse AD-MSC-Exo-TAT-peptide modified	Topical	-Decreased: mRNA and protein expression of inflammatory cytokines (TNF-α, IL-1β, IL-6, IL-17A, IFN-γ, and IL-12)	-MSCs exosomes-TAT-peptide ameliorate EAU in a dose-dependent manner. -Higher effectiveness of TAT-modified exosomes in ocular barrier penetration, EAU prevention, and treatment compared to naive exosomes, while the control group did not have observable penetration. -Decreased mean pathological score and inflammatory cytokines expression in the TAT-modified exosomes treatment group, compared to naïve exosomes only. However, the naïve exosomes had a lower mean pathological score and inflammatory cytokine expression compared to the control.	Pre-clinical study (ex vivo mouse models)	Yang et al., 2024

List of Abbreviations: hBM-MSC-Exo: Human UC-derived MSCs; hUC-MSC-Exo: Human UC-derived MSCs; MSCs: Mesenchymal stem cells; POI: Periocular Injection; TVI: Tail vein injection.

REFERENCES

AbbVie. A phase 2, open-label study to explore the pharmacodynamics of two doses in two formulations of RGX-314 gene therapy administered via subretinal delivery in participants with neovascular age-related macular degeneration. Identifier NCT04832724. 2021 Feb 22–2024 Mar 18. https://clinicaltrials.gov/study/NCT04832724

An S, Anwar K, Ashraf M, Lee H, Jung R, Koganti R, et al. Wound-healing effects of mesenchymal stromal cell secretome in the cornea and the role of exosomes. *Pharmaceutics*. 2023 May 13;15(5):1486.

An S, Shen X, Anwar K, Ashraf M, Lee H, Koganti R, et al. Therapeutic potential of mesenchymal stem cell-secreted factors on delay in corneal wound healing by nitrogen mustard. *Int. J. Mol. Sci.* 2022 Sep 29;23(19):11510.

Anand A, Jian HJ, Huang HH, Hean LE, Li YJ, Lai JY, et al. Anti-angiogenic carbon nanovesicles loaded with bevacizumab for the treatment of age-related macular degeneration. *Carbon*. 2023 Jan;201:362–70.

Appell MB, Pejavar J, Pasupathy A, Rompicharla SVK, Abbasi S, Malmberg K, et al. Next-generation therapeutics for retinal neurodegenerative diseases. *J Controlled Release*. 2024 Mar 1;367:708–36.

Bai L, Shao H, Wang H, Zhang Z, Su C, Dong L, et al. Effects of mesenchymal stem cell-derived exosomes on experimental autoimmune uveitis. *Sci Rep*. 2017 Jun 28;7(1):4323.

Bai L, Wang Y. Mesenchymal stem cells-derived exosomes alleviate senescence of retinal pigment epithelial cells by activating PI3K/AKT-Nrf2 signaling pathway in early diabetic retinopathy. *Exp Cell Res*. 2024 Aug;441(2):114170.

Bermudez MA, Sendon-Lago J, Eiro N, Trevino M, Gonzalez F, Yebra-Pimentel E, et al. Corneal epithelial wound healing and bactericidal effect of conditioned medium from human uterine cervical stem cells. *Invest Ophthalmol Vis Sci*. 2015 Feb 11;56(2):983–92.

Biasutto L, Chiechi A, Couch R, Liotta LA, Espina V. Retinal pigment epithelium (RPE) exosomes contain signaling phosphproteins affected by oxidative stress. *Exp Cell Res*. 2013 Aug 1;319(13):2113–2123.

Blanch RJ, Joseph IJ, Cockerham K. Traumatic optic neuropathy management: a systematic review. *Eye*. 2024 Aug;38(12):2312–8.

Boyer DS, Yoon YH, Belfort R, Bandello F, Maturi RK, Augustin AJ, et al. Three-year, randomized, sham-controlled trial of dexamethasone intravitreal implant in patients with diabetic macular edema. *Ophthalmology*. 2014 Oct;121(10):1904–14.

Bucan V, Vaslaitis D, Peck CT, Strauß S, Vogt PM, Radtke C. Effect of exosomes from rat adipose-derived mesenchymal stem cells on neurite outgrowth and sciatic nerve regeneration after crush injury. *Mol Neurobiol*. 2019 Mar;56(3):1812–24.

Cai C, Tahiri H, Fortin C, Ortiz C, Sintjago H, Yang C, et al. Lymphocytic microparticles suppress retinal angiogenesis via targeting Müller cells in the ischemic retinopathy mouse model. *Exp Cell Res*. 2021 Feb;399(2):112470.

Campochiaro PA, Brown DM, Pearson A, Ciulla T, Boyer D, Holz FG, et al. Long-term benefit of sustained-delivery fluocinolone acetonide vitreous inserts for diabetic macular edema. *Ophthalmology*. 2011 Apr;118(4):626–635.e2.

Carter K, Lee HJ, Na KS, Fernandes-Cunha GM, Blanco IJ, Djalilian A, et al. Characterizing the impact of 2D and 3D culture conditions on the therapeutic effects of human mesenchymal stem cell secretome on corneal wound healing in vitro and ex vivo. *Acta Biomater*. 2019 Nov;99:247–57.

Casado-Santos A, González-Cubero E, García-Rodríguez MB, Carrera-Serna Á, González-Fernández ML, Villar-Suárez V. The therapeutic potential of mesenchymal stem cell secretome in treating spontaneous chronic corneal epithelial defects in dogs [Internet]. 2024 [cited 2024 Dec 29]. Available from: https://www.ssrn.com/abstract=4995071

Casson RJ. The effect of hyperglycemia on experimental retinal ischemia. *Arch Ophthalmol*. 2004 Mar 1;122(3):361.

Cui Y, Liu C, Huang L, Chen J, Xu N. Protective effects of intravitreal administration of mesenchymal stem cell-derived exosomes in an experimental model of optic nerve injury. *Exp Cell Res*. 2021 Oct;407(1):112792.

Dellett M, Brown ED, Guduric-Fuchs J, O'Connor A, Stitt AW, Medina RJ, et al. MicroRNA-containing EVs released from endothelial colony-forming cells modulate angiogenesis during ischemic retinopathy. *J Cell Mol Med*. 2017 Dec;21(12):3405–19.

Deng CL, Hu CB, Ling ST, Zhao N, Bao LH, Zhou F, et al. Photoreceptor protection by mesenchymal stem cell transplantation identifies exosomal MiR-21 as a therapeutic for retinal degeneration. *Cell Death Differ*. 2021 Mar;28(3):1041–61.

Duan S, Zhang Y, Yuan F, Jiang A, Sang Y, Huang G. Corneal endothelial expansion using human umbilical cord mesenchymal stem cell-derived conditioned medium. *J Cell Physiol*. 2021 Apr;236(4):2606–15.

Erdal E, Bakici C, Arslan A, Batur B, Yaman ME, Alçığır ME, et al. Ocular pharmacokinetics and toxicity of nanoparticular acetazolamide: In vivo distribution and safety of PHBV-ACZ nanoparticle. *Int J Pharm.* 2023 Oct 15;645:123336.

Eslani M, Putra I, Shen X, Hamouie J, Afsharkhamseh N, Besharat S, et al. Corneal mesenchymal stromal cells are directly antiangiogenic via PEDF and sFLT-1. *Investig Ophthalmol Vis Sci.* 2017 Oct 26;58(12):5507.

Evans JR, Michelessi M, Virgili G. Laser photocoagulation for proliferative diabetic retinopathy. Cochrane Eyes and Vision Group, editor. *Cochrane Database Syst Rev* [Internet]. 2014 Nov 24 [cited 2024 Dec 27];2015(8).

Fernandes-Cunha GM, Na KS, Putra I, Lee HJ, Hull S, Cheng YC, et al. Corneal wound healing effects of mesenchymal stem cell secretome delivered within a viscoelastic gel carrier. *Stem Cells Transl Med.* 2019 May 1;8(5):478–89.

ForSight Vision5, Inc. A phase 2, multicenter, randomized, subject- and examiner-masked, controlled clinical trial designed to evaluate the safety and effectiveness of the bimatoprost ocular insert compared with topical timolol maleate ophthalmic solution (0.5%) in patients with glaucoma or ocular hypertension. 2013. Available from: https://clinicaltrials.gov/study/NCT01915940

Fu S, Sun W, Liu L, Xiao J, Xiong J, Hu Y, et al. Müller cells harboring exosomal lncRNA OGRU modulate microglia polarization in diabetic retinopathy by serving as miRNA sponges. *Diabetes.* 2024 Nov 1;73(11):1919–34.

Giannaccini M, Usai A, Chiellini F, Guadagni V, Andreazzoli M, Ori M, et al. Neurotrophin-conjugated nanoparticles prevent retina damage induced by oxidative stress. *Cell Mol Life Sci.* 2018 Apr 1;75(7):1255–67.

Glaukos Corporation. (2016, Mar 29 - 2020, Jul 21) Prospective, randomized phase III study comparing two models of a travoprost intraocular implant to timolol maleate ophthalmic solution. Identifier NCT02754596. https://clinicaltrials.gov/study/NCT02754596.

Grassmann F, Schoenberger PGA, Brandl C, Schick T, Hasler D, Meister G, et al. A Circulating MicroRNA profile is associated with late-stage neovascular age-related macular degeneration. Sørensen TL, editor. *PLoS ONE.* 2014 Sep 9;9(9):e107461.

Gu C, Zhang H, Gao Y. Adipose mesenchymal stem cells-secreted EVs containing microRNA-192 delay diabetic retinopathy by targeting ITGA1. *J Cell Physiol.* 2021 Jul;236(7):5036–51.

Gu S, Liu Y, Zou J, Wang W, Wei T, Wang X, et al. Retinal pigment epithelial cells secrete miR-202-5p-containing exosomes to protect against proliferative diabetic retinopathy. *Exp Eye Res.* 2020 Dec;201:108271.

Guo R, Liang Q, He Y, Wang C, Jiang J, Chen T, et al. Mesenchymal stromal cells-derived EVs regulate dendritic cell functions in dry eye disease. *Cells.* 2022 Dec 22;12(1):33.

Guo S, Perets N, Betzer O, Ben-Shaul S, Sheinin A, Michaelevski I, et al. Intranasal delivery of mesenchymal stem cell-derived exosomes loaded with phosphatase and tensin homolog siRNA repairs complete spinal cord injury. *ACS Nano.* 2019 Sep 24;13(9):10015–28.

Haider KH, Aslam M. Cell-free therapy with stem cell secretions: protection, repair and regeneration of the injured myocardium. In: KH Haider, S Aziz, editors. *Stem Cells: From Hype to Real Hope.* Berlin: De Gruyter; 2018.

Hartley C. Treatment of corneal ulcers: what are the medical options? *J Feline Med Surg.* 2010 May;12(5):384–97.

Hazlett L, Suvas S, McClellan S, Ekanayaka S. Challenges of corneal infections. *Expert Rev Ophthalmol.* 2016;11(4):285–97.

Herrera-Barrera M, Ryals RC, Gautam M, Jozic A, Landry M, Korzun T, et al. Peptide-guided lipid nanoparticles deliver mRNA to the neural retina of rodents and nonhuman primates. *Sci Adv.* 2023 Jan 13;9(2):eadd4623.

Hong S. Long-term intraocular pressure fluctuation and progressive visual field deterioration in patients with glaucoma and low intraocular pressures after a triple procedure. *Arch Ophthalmol.* 2007 Aug 1;125(8):1010.

Hu J, Li H, Zhao Y, Ke Y, Rupenthal ID, Liu H, et al. Critical evaluation of multifunctional betaxolol hydrochloride nanoformulations for effective sustained intraocular pressure reduction. *Int J Nanomedicine.* 2022 Dec; Volume 17:5915–31.

Hu T, Chang S, Qi F, Zhang Z, Chen J, Jiang L, et al. Neural grafts containing exosomes derived from Schwann cell-like cells promote peripheral nerve regeneration in rats. *Burns Trauma.* 2023 Jan 1;11:tkad013.

Huang S, Zeng Y, Guo Q, Zou T, Yin ZQ. Small EVs of organoid-derived human retinal stem cells remodel Müller cell fate via miRNA: A novel remedy for retinal degeneration. *J Controlled Release.* 2024 Jun;370:405–20.

Huang X, Chau Y. Enhanced delivery of siRNA to retinal ganglion cells by intravitreal lipid nanoparticles of positive charge. *Mol Pharm.* 2021 Jan 4;18(1):377–85.

Jabbehdari S, Yazdanpanah G, Kanu LN, Chen E, Kang K, Anwar KN, et al. Therapeutic effects of lyophilized conditioned-medium derived from corneal mesenchymal stromal cells on corneal epithelial wound healing. *Curr Eye Res.* 2020 Dec 1;45(12):1490–6.

Jung B, Lee H, Kim S, Tchah H, Hwang C. Effect of Rho-associated kinase inhibitor and mesenchymal stem cell-derived conditioned medium on corneal EC senescence and proliferation. *Cells.* 2021 Jun 11;10(6):1463.

Kaur G, Bae E, Zhang Y, Ciacciofera N, Jung KM, Barreda H, et al. Biopotency and surrogate assays to validate the immunomodulatory potency of EVs derived from mesenchymal stem/stromal cells for the treatment of experimental autoimmune uveitis. *J Extracell Vesicles.* 2024 Aug;13(8):e12497.

Kim H, Goh YS, Park SE, Hwang J, Kang N, Jung JS, et al. Preventive effects of exosome-rich conditioned medium from amniotic membrane-derived mesenchymal stem cells for diabetic retinopathy in rats. *Transl Vis Sci Technol.* 2023 Aug 23;12(8):18.

Kontiola AI. A new induction-based impact method for measuring intraocular pressure. *Acta Ophthalmol Scand.* 2000 Apr;78(2):142–5.

Kumar A, Xu Y, Yang E, Du Y. Stemness and regenerative potential of corneal stromal stem cells and their secretome after long-term storage: implications for ocular regeneration. *Investig Ophthalmol Vis Sci.* 2018 Jul 25;59(8):3728.

Lai P, Chen X, Guo L, Wang Y, Liu X, Liu Y, et al. A potent immunomodulatory role of exosomes derived from mesenchymal stromal cells in preventing cGVHD. *J Hematol OncolJ Hematol Oncol.* 2018 Dec;11(1):135.

Lee DA, Higginbotham EJ. Glaucoma and its treatment: a review. *Am J health Syst Pharm.* 2005 Apr;62(7):691–699.

Lee YC, Sun LY, Zhang JR. Protective effects of low-molecular-weight components of adipose stem cell-derived conditioned medium on dry eye syndrome in mice. *Sci Rep.* 2021 Nov 8;11(1):21874.

Li D, Zhang J, Liu Z, Gong Y, Zheng Z. Human umbilical cord mesenchymal stem cell-derived exosomal miR-27b attenuates subretinal fibrosis via suppressing epithelial–mesenchymal transition by targeting HOXC6. *Stem Cell Res Ther.* 2021 Dec;12(1):24.

Li H, Liu W, Sorenson CM, Sheibani N, Albert DM, Senanayake T, et al. Sustaining intravitreal residence with L-Arginine peptide-conjugated nanocarriers. *Investig Ophthalmol Vis Sci.* 2017 Oct 6;58(12):5142.

Li H, Zhang Z, Li Y, Su L, Duan Y, Zhang H, et al. Therapeutic effect of rapamycin-loaded small EVs derived from mesenchymal stem cells on experimental autoimmune uveitis. *Front Immunol.* 2022 Mar 29;13:864956.

Li J, Zhou Y, Long Q. Effects of mesenchymal stem cells derived exosomes on ultrastructure of corneal epithelium and function of the tear film in dry eye BALB/c mice. *Invest Ophthalmol Vis Sci.* 2019 Jul 22;60(9):4187–4187.

Li N, Gao Z, Zhao L, Du B, Ma B, Nian H, et al. MSC-derived small EVs attenuate autoimmune dacryoadenitis by promoting M2 macrophage polarization and inducing tregs via miR-100-5p. *Front Immunol.* 2022 Jul 6;13:888949.

Li W, Jin L, Cui Y, Nie A, Xie N, Liang G. Bone marrow mesenchymal stem cells-induced exosomal microRNA-486-3p protects against diabetic retinopathy through TLR4/NF-κB axis repression. *J Endocrinol Invest.* 2021 Jun;44(6):1193–207.

Li Y, Ren X, Zhang Z, Duan Y, Li H, Chen S, et al. Effect of small EVs derived from IL-10-overexpressing mesenchymal stem cells on experimental autoimmune uveitis. *Stem Cell Res Ther.* 2022 Mar 7;13(1):100.

Li Y, Zheng J, Wang XZ, Wang X, Liu WJ, Gao JL. Mesenchymal stem cell-derived exosomes protect trabecular meshwork from oxidative stress. *Sci Rep.* 2021 Jul 21;11(1):14863.

Liang G, Qin Z, Luo Y, Yin J, Shi Z, Wei R, et al. Exosomal microRNA-133b-3p from bone marrow mesenchymal stem cells inhibits angiogenesis and oxidative stress via FBN1 repression in diabetic retinopathy. *Gene Ther.* 2022 Dec;29(12):710–9.

Liu W, Tang P, Wang J, Ye W, Ge X, Rong Y, et al. EVs derived from melatonin-preconditioned mesenchymal stem cells containing USP29 repair traumatic spinal cord injury by stabilizing NRF2. *J Pineal Res.* 2021 Dec;71(4):e12769.

Liu X, Li X, Wu G, Qi P, Zhang Y, Liu Z, et al. Umbilical cord mesenchymal stem cell-derived small EVs deliver miR-21 to promote corneal epithelial wound healing through PTEN/PI3K/Akt pathway. Arnhold S, editor. *Stem Cells Int.* 2022 Jul 14;2022:1–15.

Ma F, Feng J, Liu X, Tian Y, Wang WJ, Luan FX, et al. Ascorbic acid-coupled mesenchymal stem cell-derived exosomes ameliorate dry eye disease. *SSRN Electron J* [Internet]. 2022 [cited 2024 Dec 29]; Available from: https://www.ssrn.com/abstract=4151353

Ma S, Yin J, Hao L, Liu X, Shi Q, Diao Y, et al. Exosomes from human umbilical cord mesenchymal stem cells treat corneal injury via autophagy activation. *Front Bioeng Biotechnol.* 2022 Apr 11;10:879192.

Mangungson C, Wirohadidjojo Y, Djaja M, Sartika C, Putera B, Sasongko M. Multifocal electroretinography results before and after peribulbar injection of allogeneic umbilical mesenchymal stem cell secretome for late-stage retinitis pigmentosa. *Indian J Public Health Res Dev* [Internet]. 2021;12(3):322–329.

Manuguerra-Gagne R, Boulos PR, Ammar A, Leblond FA, Krosl G, Pichette V, et al. Transplantation of mesenchymal stem cells promotes tissue regeneration in a glaucoma model through laser-induced paracrine factor secretion and progenitor cell recruitment. *Stem Cells.* 2013 Jun 1;31(6):1136–48.

Mashal M, Attia N, Martínez-Navarrete G, Soto-Sánchez C, Fernández E, Grijalvo S, et al. Gene delivery to the rat retina by non-viral vectors based on chloroquine-containing cationic niosomes. *J Controlled Release.* 2019 Jun;304:181–90.

Mathew B, Acha LG, Torres LA, Huang CC, Liu A, Kalinin S, et al. MicroRNA-based engineering of mesenchymal stem cell EVs for treatment of retinal ischemic disorders. *Acta Biomater.* 2023 Mar;158:782–97.

Mathew B, Torres LA, Gamboa Acha L, Tran S, Liu A, Patel R, et al. Uptake and distribution of administered bone marrow mesenchymal stem cell EVs in retina. *Cells.* 2021 Mar 25;10(4):730.

Mathieson K, Loudin J, Goetz G, Huie P, Wang L, Kamins TI, et al. Photovoltaic retinal prosthesis with high pixel density. *Nat Photonics.* 2012 Jun;6(6):391–7.

Mead B, Tomarev S. Bone marrow-derived mesenchymal stem cells-derived exosomes promote survival of retinal ganglion cells through miRNA-dependent mechanisms. *Stem Cells Transl Med.* 2017 Apr 1;6(4):1273–85.

Moisseiev E, Anderson JD, Oltjen S, Goswami M, Zawadzki RJ, Nolta JA, et al. Protective effect of intravitreal administration of exosomes derived from mesenchymal stem cells on retinal ischemia. *Curr Eye Res.* 2017 Oct 3;42(10):1358–67.

Mu H, Wang Y, Chu Y, Jiang Y, Hua H, Chu L, et al. Multivesicular liposomes for sustained release of bevacizumab in treating laser-induced choroidal neovascularization. *Drug Deliv.* 2018 Jan 1;25(1):1372–83.

Nakahara M, Okumura N, Kay EP, Hagiya M, Imagawa K, Hosoda Y, et al. Corneal endothelial expansion promoted by human bone marrow mesenchymal stem cell-derived conditioned medium. Covas DT, editor. *PLoS One.* 2013 Jul 23;8(7):e69009.

Nguyen QD, Brown DM, Marcus DM, Boyer DS, Patel S, Feiner L, et al. Ranibizumab for diabetic macular edema. *Ophthalmology.* 2012 Apr;119(4):789–801.

Oh JY, Kim MK, Shin MS, Lee HJ, Ko JH, Wee WR, et al. The anti-inflammatory and anti-angiogenic role of mesenchymal stem cells in corneal wound healing following chemical injury. *Stem Cells.* 2008 Apr 1;26(4):1047–55.

Pan D, Chang X, Xu M, Zhang M, Zhang S, Wang Y, et al. UMSC-derived exosomes promote retinal ganglion cell survival in a rat model of optic nerve crush. *J Chem Neuroanat.* 2019 Mar;96:134–9.

Pattabiraman PP, Feinstein V, Beit-Yannai E. Profiling the miRNA from exosomes of non-pigmented ciliary epithelium-derived identifies key gene targets relevant to primary open-angle glaucoma. *Antioxidants.* 2023 Feb 7;12(2):405.

Peng ZQ, Guan XH, Yu ZP, Wu J, Han XH, Li MH, et al. Human amniotic mesenchymal stem cells-derived conditioned medium and exosomes alleviate oxidative stress-induced retinal degeneration by activating the PI3K/Akt/FoxO3 pathway. *Exp Eye Res.* 2024 Jul;244:109919.

Rosa RH, Xie W, Zhao M, Tsai SH, Roddy GW, Su MG, et al. Intravitreal administration of stanniocalcin-1 rescues photoreceptor degeneration with reduced oxidative stress and inflammation in a porcine model of retinitis pigmentosa. *Am J Ophthalmol.* 2022 Jul;239:230–43.

Rouen PA, White ML. Dry eye disease: prevalence, assessment, and management. *Home Healthc Now.* 2018 Mar;36(2):74–83.

Rui K, Shen Z, Peng N, Zhao F, Tang Y, Liu S, et al. Olfactory ecto-mesenchymal stem cell-derived exosomes ameliorate murine Sjögren's syndrome via suppressing Tfh cell response. *Rheumatol Immunol Res.* 2022 Dec 31;3(4):198–207.

Ruotian X, Lingling B, Jing Y, Yongtao L, Lijie D, Feifei M, et al. Effects of rat mesenchymal stem cell-derived exosomes on rat experimental autoimmune uveitis. *Chin J Ophthalmol.* 2018 Dec 26;34(6):562–7.

Russell S, Bennett J, Wellman JA, Chung DC, Yu ZF, Tillman A, et al. Efficacy and safety of voretigene neparvovec (AAV2-hRPE65v2) in patients with RPE65-mediated inherited retinal dystrophy: a randomised, controlled, open-label, phase 3 trial. *The Lancet.* 2017 Aug;390(10097):849–60.

Saccu G, Menchise V, Gai C, Bertolin M, Ferrari S, Giordano C, et al. Bone marrow mesenchymal stromal/stem cell-derived EVs promote corneal wound repair by regulating inflammation and angiogenesis. *Cells.* 2022 Dec 2;11(23):3892.

Safwat A, Sabry D, Ragiae A, Amer E, Mahmoud R, Shamardan R. Adipose mesenchymal stem cells–derived exosomes attenuate retina degeneration of streptozotocin-induced diabetes in rabbits. *J Circ Biomark.* 2018 Jan;7:184945441880782.

Samaeekia R, Rabiee B, Putra I, Shen X, Park YJ, Hematti P, et al. Effect of human corneal mesenchymal stromal cell-derived exosomes on corneal epithelial wound healing. *Investig Ophthalmol Vis Sci.* 2018 Oct 25;59(12):5194.

Sanghani A, Andriesei P, Kafetzis KN, Tagalakis AD, Yu-Wai-Man C. Advances in exosome therapies in ophthalmology–From bench to clinical trial. *Acta Ophthalmol* (Copenh). 2022 May;100(3):243–52.

Schaffrath K, Schellhase H, Walter P, Augustin A, Chizzolini M, Kirchhof B, et al. One-year safety and performance assessment of the argus II retinal prosthesis: a postapproval study. *JAMA Ophthalmol.* 2019 Aug 1;137(8):896.

Sendon-Lago J, Seoane S, Martinez-Ordoñez A, Eiro N, Saa J, Vizoso FJ, et al. TIMP-1 and TIMP-2 mediate corneal regeneration by conditioned medium of human uterine cervical stem cells. *Exp Eye Res.* 2019 Mar;180:110–21.

Seong HR, Noh CH, Park S, Cho S, Hong SJ, Lee A young, et al. Intraocular pressure-lowering and retina-protective effects of exosome-rich conditioned media from human amniotic membrane stem cells in a rat model of glaucoma. *Int. J. Mol. Sci.* 2023 Apr 29;24(9):8073.

Shen T, Zheng QQ, Shen J, Li QS, Song XH, Luo HB, et al. Effects of adipose-derived mesenchymal stem cell exosomes on corneal stromal fibroblast viability and extracellular matrix synthesis. *Chin Med J* (Engl). 2018 Mar 20;131(6):704–12.

Shigemoto-Kuroda T, Oh JY, Kim D ki, Jeong HJ, Park SY, Lee HJ, et al. MSC-derived EVs attenuate immune responses in two autoimmune murine models: type 1 diabetes and uveoretinitis. *Stem Cell Rep.* 2017 May;8(5):1214–25.

Shojaati G, Khandaker I, Funderburgh ML, Mann MM, Basu R, Stolz DB, et al. Mesenchymal stem cells reduce corneal fibrosis and inflammation via extracellular vesicle-mediated delivery of miRNA. *Stem Cells Transl Med.* 2019 Nov 1;8(11):1192–201.

Sun F, Sun Y, Zhu J, Wang X, Ji C, Zhang J, et al. Mesenchymal stem cell-derived small EVs alleviate diabetic retinopathy by delivering NEDD4. *Stem Cell Res Ther.* 2022 Dec;13(1):293.

Sun P, Shen L, Zhang C, Du L, Wu X. Promoting the expansion and function of human corneal ECs with an orbital adipose-derived stem cell-conditioned medium. *Stem Cell Res Ther.* 2017 Dec;8(1):287.

Sun X, Dai Y, Chen Y, Yu DY, Cringle SJ, Chen J, et al. Primary angle closure glaucoma: what we know and what we do not know. *Prog Retin Eye Res.* 2017 Mar;57:26–45.

Sun X, Song W, Teng L, Huang Y, Liu J, Peng Y, et al. MiRNA 24-3p-rich exosomes functionalized DEGMA-modified hyaluronic acid hydrogels for corneal epithelial healing. *Bioact Mater.* 2023 Jul;25:640–56.

Tabak S, Feinshtein V, Schreiber-Avissar S, Beit-Yannai E. Non-pigmented ciliary epithelium-derived EVs loaded with SMAD7 siRNA attenuate Wnt signaling in trabecular meshwork cells in vitro. *Pharmaceuticals.* 2021 Aug 27;14(9):858.

Tan DTH, Dart JKG, Holland EJ, Kinoshita S. Corneal transplantation. *Lancet Lond Engl.* 2012 May 5;379(9827):1749–61.

Tang Q, Lu B, He J, Chen X, Fu Q, Han H, et al. Exosomes-loaded thermosensitive hydrogels for corneal epithelium and stroma regeneration. *Biomaterials.* 2022 Jan;280:121320.

Tao H, Chen X, Cao H, Zheng L, Li Q, Zhang K, et al. Mesenchymal stem cell-derived EVs for corneal wound repair. *Stem Cells Int.* 2019 Dec 9;2019:1–9.

Tian Y, Zhang F, Qiu Y, Wang S, Li F, Zhao J, et al. Reduction of choroidal neovascularization via cleavable VEGF antibodies conjugated to exosomes derived from regulatory T cells. *Nat Biomed Eng.* 2021 Jul 26;5(9):968–82.

Usategui-Martín R, Puertas-Neyra K, García-Gutiérrez MT, Fuentes M, Pastor JC, Fernández-Bueno I. Human mesenchymal stem cell secretome exhibits a neuroprotective effect on in vitro retinal photoreceptor degeneration. *Mol Ther - Methods Clin Dev.* 2020 Jun;17:1155–66.

Van den Bogerd B, Zakaria N, Matthyssen S, Koppen C, Ní Dhubhghaill S. Exploring the mesenchymal stem cell secretome for corneal endothelial proliferation. *Stem Cells Int.* 2020 Feb 5;2020:1–10.

Van Lookeren Campagne M, LeCouter J, Yaspan BL, Ye W. Mechanisms of age-related macular degeneration and therapeutic opportunities. *J Pathol.* 2014 Jan;232(2):151–64.

Wang G, Li H, Long H, Gong X, Hu S, Gong C. Exosomes derived from mouse adipose-derived mesenchymal stem cells alleviate benzalkonium chloride-induced dry eye model via inhibiting NLRP3 inflammasome. *Ophthalmic Res.* 2022;65(1):40–51.

Wang L, Wang X, Chen Q, Wei Z, Xu X, Han D, et al. MicroRNAs of EVs derived from mesenchymal stromal cells alleviate inflammation in dry eye disease by targeting the IRAK1/TAB2/NF-κB pathway. *Ocul Surf.* 2023 Apr;28:131–40.

Wang T, Li Y, Guo M, Dong X, Liao M, Du M, et al. Exosome-mediated delivery of the neuroprotective peptide PACAP38 promotes retinal ganglion cell survival and axon regeneration in rats with traumatic optic neuropathy. *Front Cell Dev Biol.* 2021 Apr 6;9:659783.

Wang X, Xu C, Bian C, Ge P, Laei J, Wang J, et al. M2 microglia-derived exosomes promote vascular remodeling in diabetic retinopathy. *J Nanobiotechnology.* 2024 Feb 9;22(1):56.

Wang Y, Qin W yun, Wang Q, Liu X na, Li X hui, Ye X qi, et al. Young Sca-1+ bone marrow stem cell-derived exosomes preserve visual function via the miR-150-5p/MEKK3/JNK/c-Jun pathway to reduce M1 microglial polarization. *J Nanobiotechnology.* 2023 Jun 15;21(1):194.

Wladis EJ, Aakalu VK, Sobel RK, McCulley TJ, Foster JA, Tao JP, et al. Interventions for indirect traumatic optic neuropathy. *Ophthalmology.* 2021 Jun;128(6):928–37.

Won JY, Kim J, Gao G, Kim J, Jang J, Park YH, et al. 3D printing of drug-loaded multi-shell rods for local delivery of bevacizumab and dexamethasone: A synergetic therapy for retinal vascular diseases. *Acta Biomater.* 2020 Oct;116:174–85.

Wong FSY, Tsang KK, Chu AMW, Chan BP, Yao KM, Lo ACY. Injectable cell-encapsulating composite alginate-collagen platform with inducible termination switch for safer ocular drug delivery. *Biomaterials.* 2019 May;201:53–67.

Wu KY, Chen WT, Chu-Bédard YK, Patel G, Tran SD. Management of Sjogren's dry eye disease—advances in ocular drug delivery offering a new hope. *Pharmaceutics.* 2022 Dec 31;15(1):147.

Wykoff CC, Abreu F, Adamis AP, Basu K, Eichenbaum DA, Haskova Z, et al. Efficacy, durability, and safety of intravitreal faricimab with extended dosing up to every 16 weeks in patients with diabetic macular oedema (YOSEMITE and RHINE): two randomised, double-masked, phase 3 trials. *The Lancet.* 2022 Feb;399(10326):741–55.

Yang H, Wu P, Wang T, Yu Y, Li J, Liu R, et al. Topical ophthalmic instillation of engineered hMSCs-derived exosomes: A novel non-invasive therapeutic strategy for ocular posterior-segment disorder. *Biochem Biophys Res Commun.* 2024 Sep;723:150212.

Yang J, Wu S, Hou L, Zhu D, Yin S, Yang G, et al. Therapeutic effects of simultaneous delivery of nerve growth factor mRNA and protein via exosomes on cerebral ischemia. *Mol Ther - Nucleic Acids.* 2020 Sep;21:512–22.

Yang Z, Yang Y, Xu Y, Jiang W, Shao Y, Xing J, et al. Biomimetic nerve guidance conduit containing engineered exosomes of adipose-derived stem cells promotes peripheral nerve regeneration. *Stem Cell Res Ther.* 2021 Dec;12(1):442.

Yu B, Shao H, Su C, Jiang Y, Chen X, Bai L, et al. Exosomes derived from MSCs ameliorate retinal laser injury partially by inhibition of MCP-1. *Sci Rep.* 2016 Sep 30;6(1):34562.

Yu C, Chen P, Xu J, Liu Y, Li H, Wang L, et al. hADSCs-derived EVs inhibit NLRP3 inflammasome activation and dry eye. *Sci Rep.* 2020 Sep 3;10(1):14521.

Yu W, Mookherjee S, Chaitankar V, Hiriyanna S, Kim JW, Brooks M, et al. Nrl knockdown by AAV-delivered CRISPR/Cas9 prevents retinal degeneration in mice. *Nat Commun.* 2017 Mar 14;8(1):14716.

Yu Z, Hao R, Du J, Wu X, Chen X, Zhang Y, et al. A human cornea-on-a-chip for the study of epithelial wound healing by EVs. *iScience.* 2022a May;25(5):104200.

Yu Z, Wen Y, Jiang N, Li Z, Guan J, Zhang Y, et al. TNF-α stimulation enhances the neuroprotective effects of gingival MSCs-derived exosomes in retinal ischemia-reperfusion injury via the MEG3/miR-21a-5p axis. *Biomaterials.* 2022b May;284:121484.

Zeng X, Bian W, Liu Z, Li J, Ren S, Zhang J, et al. Muscle-derived stem cell exosomes with overexpressed miR-214 promote the regeneration and repair of rat sciatic nerve after crush injury to activate the JAK2/STAT3 pathway by targeting PTEN. *Front Mol Neurosci.* 2023 May 26;16:1146329.

Zhang M, Lu X, Luo L, Dou J, Zhang J, Li G, et al. Targeting glutamine synthetase with AS1411-modified exosome-liposome hybrid nanoparticles for inhibition of choroidal neovascularization. *J Nanobiotechnology.* 2024 Nov 13;22(1):703.

Zhang W, Dong X, Wang T, Kong Y. Exosomes derived from platelet-rich plasma mediate hyperglycemia-induced retinal endothelial injury via targeting the TLR4 signaling pathway. *Exp Eye Res.* 2019a Dec;189:107813.

Zhang W, Wang Y, Kong Y. Exosomes derived from mesenchymal stem cells modulate miR-126 to ameliorate hyperglycemia-induced retinal inflammation via targeting HMGB1. *Investig Ophthalmol Vis Sci.* 2019b Jan 18;60(1):294.

Zhao J, An Q, Zhu X, Yang B, Gao X, Niu Y, et al. Research status and prospects of EVs in primary Sjögren's syndrome. *Stem Cell Res Ther.* 2022 Dec;13(1):230.

Zhao J, Ding Y, He R, Huang K, Liu L, Jiang C, et al. Dose-effect relationship and molecular mechanism by which BMSC-derived exosomes promote peripheral nerve regeneration after crush injury. *Stem Cell Res Ther.* 2020 Dec;11(1):360.

Zheng M, Mitra RN, Weiss ER, Han Z. Rhodopsin genomic loci DNA nanoparticles improve expression and rescue of retinal degeneration in a model for retinitis pigmentosa. *Mol Ther.* 2020 Feb;28(2):523–35.

Zhou T, He C, Lai P, Yang Z, Liu Y, Xu H, et al. miR-204–containing exosomes ameliorate GVHD-associated dry eye disease. *Sci Adv.* 2022 Jan 14;8(2):eabj9617.

7 The Potential of Extracellular Vesicles in Regenerative Medicine

Yongtao Wang[1,2], Shihui Xu[1,2], Yuxin Hu[2], Dragos Cretoiu[3,4], Michail Spanos[5,6], and Junjie Xiao[1,2]

[1]Institute of Geriatrics (Shanghai University), Affiliated Nantong Hospital of Shanghai University (The Sixth People's Hospital of Nantong), School of Medicine, Shanghai University, Nantong, China

[2]Cardiac Regeneration and Ageing Lab, Institute of Cardiovascular Sciences, Shanghai Engineering Research Center of Organ Repair, Joint International Research Laboratory of Biomaterials and Biotechnology in Organ Repair (Ministry of Education), School of Life Science, Shanghai University, Shanghai, China

[3]Department of Medical Genetics, Carol Davila University of Medicine and Pharmacy, Bucharest, Romania

[4]Materno-Fetal Assistance Excellence Unit, Alessandrescu-Rusescu National Institute for Mother and Child Health, Bucharest, Romania

[5]Cardiovascular Division of the Massachusetts General Hospital and Harvard Medical School, Boston, MA, USA

[6]Albert Einstein College of Medicine, Department of Internal Medicine, NCB, Bronx, NY, USA

7.1 LIST OF ABBREVIATIONS

β-TCP	β-tricalcium phosphate
AD	Alzheimer's disease
AI	Artificial intelligence
ASOs	Antisense oligonucleotides
BBB	Blood-brain barrier
BBTB	Blood-brain tumor barrier
BMP-2	Bone morphogenetic protein-2
BNDF	Brain-derived neurotrophic factor
BP	Black phosphorus
DPSC	Dental pulp stem cell
DRG	Dorsal root ganglion
ECM	Extracellular matrix
EGFR	Epidermal growth factor receptor

DOI: 10.1201/9781003653387-7

EVs	Extracellular vesicles
FDA	Food and Drug Administration
FGL1	Fibrinogen-like protein 1
GMP	Good manufacturing practice
GVDH	Graft-versus-host disease
HA	Hyaluronic acid
HF	Heart failure
HIF-1α	Hypoxia-inducible factor 1α
I/RI	Ischemia/reperfusion injury
IFN-γ	Interferon-γ
IL6	Interleukin-6
LOXL2	Lysyl oxidase-like protein 2
MET	Mesenchymal to epithelial transition
MHC	Major histocompatibility complex
MI	Myocardial infarction
MMP1	Matrix metalloproteinase-1
MS	Multiple sclerosis
MSCs	Mesenchymal stem cells
PCNA	Proliferating cell nuclear antigen
PD	Parkinson's disease
PD-L1	Programmed death-ligand 1
SCI	Spinal cord injury
TAMs	Tumor-associated macrophages
TGF-β	Transforming growth factor-β
TNF-α	Tumor necrosis factor-α
VEGF	Vascular endothelial growth factor

7.2 INTRODUCTION

Regenerative medicine is an interdisciplinary research field dedicated to repairing, replacing, and regenerating damaged tissues and organs to restore their physiological function (Zhu et al., 2023). It encompasses multiple branches such as stem cell biology, tissue engineering, and cell therapy, and offers promising novel therapies for many diseases and injuries (Tenreiro et al., 2021; Chen et al., 2024). With the growing global elderly population and the increasing public health burden caused by chronic diseases, such as cardiovascular diseases, neurodegenerative diseases, and diabetes, the limitations of current medical therapies, such as organ transplantation and drug therapies, are becoming increasingly apparent (Wang et al., 2025a; Jiang et al., 2025). These therapies are unable to reverse the underlying causes of tissue degeneration, which are associated with limited organ donors, immunologic rejection, and long-term side effects (Petrus-Reurer et al., 2021). Fortunately, regenerative medicine, a revolutionary technology, activates the body's self-repair mechanism or artificially rebuilds functional tissues to achieve tissue repair and organ regeneration. However, regenerative medicine also encounters many challenges (Seifert et al., 2023). Cell-based therapy, the primary approach to tissue regeneration, is limited by poor cell survival after transplantation, potential immunologic rejection, and tumorigenesis induced by uncontrolled cell proliferation (Hoang et al., 2022; Zhang et al., 2025). Some synthetic biomaterials designed to mimic the extracellular matrix (ECM) can serve as scaffolds for cell culture and for repairing or rebuilding damaged tissue. Still, they cannot recapitulate the ECM's biochemical and mechanical properties, which limits their regenerative performance (Liu et al., 2025). Therefore, it is necessary to find new, effective, biosafe, and scalable strategies to promote the clinical application of regenerative medicine.

Extracellular vesicles (EVs), membrane-bound nanoscale particles secreted by nearly all cell types, may help to solve this dilemma (Xiang et al., 2025). EVs can be classified into exosomes, microvesicles, and apoptotic bodies based on their mode of biogenesis, size, and cargo (Miron et al., 2024). EVs have been reported to encapsulate a variety of bioactive factors, including nucleic acids (mRNA and miRNA), peptides (e.g., growth factors and enzymes), and lipids (Cecchin et al., 2023). These abundant signaling molecules make EVs important regulators of cell-cell communication, and they can transfer their cargoes to physiologically switch the phenotype of target cells. EVs modulate cell division, differentiation, immunity, and homeostasis of recipient cells (Buzas et al., 2022; Mushtaq et al., 2024). EVs are attractive for cell-free therapy in tissue regeneration (Liang et al., 2025a).

Due to their low immunogenicity, good biocompatibility, and relatively stable circulation, EVs are potential candidates for tissue regeneration (Mondal et al., 2022). Recent advances in EVs have enabled even greater translational potential. The preclinical studies demonstrate that EVs from mesenchymal stem cells (MSCs) can present the curative outcomes consistent with their parent cells for modulating inflammation and attenuating fibrosis, as well as promoting angiogenesis in myocardial infarction, neural injury, and bone repair (Chuah et al., 2021; Kundu et al., 2024; Shen et al., 2025). It has been reported that EVs derived from MSCs can inhibit apoptosis in hypoxic cardiomyocytes, promote angiogenesis in the border zone of the infarct myocardium, and reduce fibrosis of the myocardium (Li et al., 2025a; Habib et al., 2025). The improvements in EV isolation methods, such as size-exclusion chromatography, ultracentrifugation, and microfluidics, have enhanced the yield, purity, and reproducibility of EVs, which have been limited by concerns about high heterogeneity (Zhang et al., 2023). In addition, surface modification can be used to engineer EVs or load other therapeutic cargoes onto them to enhance delivery precision and efficiency (Zhang et al., 2022). When combined with biomaterial scaffolds and delivery systems, EVs can further expand their therapeutic potential through localized, spatiotemporally controlled delivery (Zhang et al., 2024a). Thus, EVs have been positioned as a highly potent next-generation agent for regenerative medicine, offering minimally invasive, scalable, cell-free therapies and promising clinical translational potential.

7.3 HETEROGENEITY OF EVS

EVs display a high degree of diversity in their biogenesis, physical properties, and molecular composition, offering promise but also posing challenges for the clinical translation of these particles in regenerative medicine (Wakker et al., 2023) (Figure 7.1A).

All cell types appear capable of secreting EVs via distinct mechanisms. Exosomes, with a diameter ranging from 30 to 100 nm, are produced via the endosomal pathway and delivered by the fusion of MVBs with the plasma membrane (Carney et al., 2024). In contrast, microvesicles, sized between 100 and 1000 nm, are produced through outward budding from the plasma membrane (Figure 7.1B). Apoptotic bodies (ABs, >1000 nm) can be used to program cell apoptosis and death (Manno et al., 2025). In addition to varying widely in size and origin, the three main classes of EVs also exhibit considerable diversity in their cargo, comprising unique combinations of nucleic acids, lipids, polypeptides, and bioactive enzymes (Dechantsreiter et al., 2022). This diversity is a double-edged sword for standard EVs used in basic and translational studies, as the properties of vesicles produced by different cells or culture conditions can lead to highly variable biological activities.

Heterogeneity of EVs is mainly determined by their cell source and the microenvironment in which they are produced (Dixson et al., 2023). MSC-derived EVs typically contain high levels of regenerative molecules, including vascular endothelial growth factor (VEGF) and transforming growth factor-β (TGF-β) (Ahmed et al., 2023). In contrast, EVs produced by immune cells (T lymphocytes or DCs) typically contain immunomodulatory factors, including the major histocompatibility complex (MHC) and cytokines (Li et al., 20224a; Rajeev et al., 2024). EVs originating from cancer cells contain oncogenic proteins (e.g., epidermal growth factor receptor (EGFR) and mesenchymal-to-epithelial transition factor (MET)) and miRNAs that promote tumor growth by evading immune surveillance and promoting angiogenesis (Dai et al., 2024). Furthermore, different

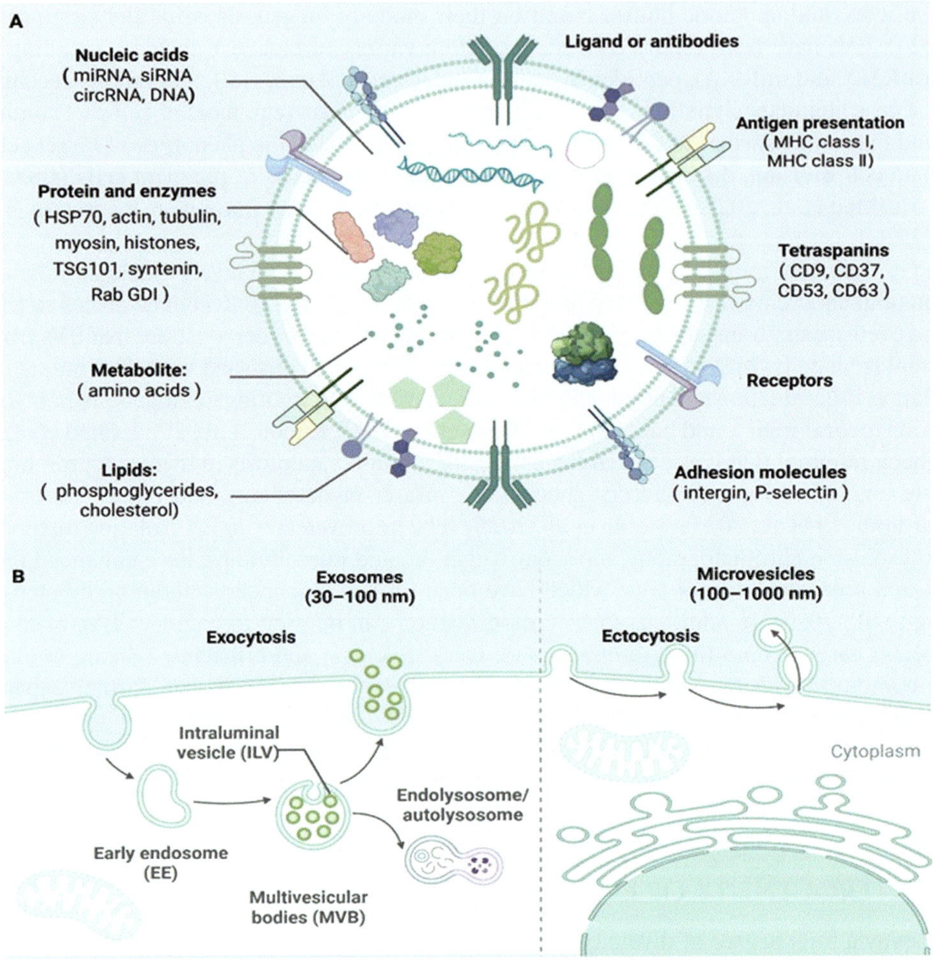

FIGURE 7.1 Overview of EV heterogeneity. (A) Schematic representation of EV molecular diversity, highlighting the range of internal cargo, including nucleic acids, proteins and enzymes, metabolites, and lipids, and surface modifications such as ligands, antibodies, antigen-presenting molecules, tetraspanins, receptors, and adhesion molecules. (B) Genesis of EVs in cells, such as exosomes and microvesicles.

environmental stimuli will further modulate the composition of EVs. For example, upon hypoxic exposure, exosome composition changes, and hypoxia-inducible factor 1α (HIF-1α) is enriched in exosomes, thereby enhancing their angiogenic properties (Li et al., 2025b). Even within the same cell type, different EV subpopulations can change dramatically in response to distinct metabolic conditions, mechanical strain, or pharmacologic agents, further complicating their molecular characterization.

Due to current technical limitations in isolation, purification, and characterization, the clinical potential of EVs is constrained by their biological significance. Differential ultracentrifugation is considered a common isolation technology due to its relatively convenient operation and safety. However, due to the heterogeneity of extracellular fluids, other substances are often co-isolated with EVs, and additional purification steps are required to eliminate these contaminants (Jia et al., 2022). Size-exclusion chromatography and density gradient centrifugation can increase purity but decrease yield, which limits their application in clinical settings (Foers et al., 2018). Microfluidic sorting and antibody-based capture can enhance the specific enrichment of EVs by targeting surface markers expressed on EVs; however, this may also lead to the exclusion of potentially biologically relevant EV subpopulations that lack these markers (Wu et al., 2025). In addition, owing to the lack of consensus protocols for EV quantification by nanoparticle tracking analysis (NTA), there remain variations in size distributions and concentration reports (Su et al., 2025).

EV heterogeneity plays a crucial role in biomedical applications. Immunomodulatory EVs, originating from immune cells, are used for cancer therapy, or regenerative EVs, originating from MSCs, are developed for regenerative medicine. Standardization of isolation protocols and analytical protocols is strictly needed to define EV subtypes and discriminate their physiological or non-functional relevance (Jiang et al., 2024). The solution to these problems will originate from transdisciplinary efforts focusing on developing reproducible, scalable platforms to standardize isolation protocols and analytical protocols, while standardizing the isolation protocols and analytical protocols will not come at the expense of EV homogeneity, and clinical-grade quality control will be ensured to turn EVs into precision tools for regenerative medicine and diagnosis.

7.4 PHYSIOLOGICAL FUNCTIONS OF EVS IN REGENERATIVE MEDICINE

Due to their capacity to participate in the repair of damaged tissues, the inflammatory response, and disease development, EVs have shown a significant impact on tissue regeneration (Jarrige et al., 2021). These characteristics of EVs are due to their function as natural mediators of bioactive factors, including microRNAs (miRNAs), messenger RNAs (mRNAs), peptides, and cytokines, which control the phenotype of target cells and close the difference between the behavior of living cells and synthetic biomaterials (Herrmann et al., 2021). Therefore, EVs deliver molecular cargoes from donors to recipient cells, transmitting the regenerative properties of their parent cells while avoiding potential side effects of cell therapy, such as immunological rejection. Furthermore, the high biocompatibility and stability of the nanovesicles, along with their versatility, make EVs an interesting vector for overcoming the current limitations of regenerative strategies.

Intersomatic communication mediated by EVs is a key event during the regenerative physiology of tissues (Racchetti et al., 2023). EVs modulate autocrine, paracrine, and endocrine signaling pathways and can cross the pericellular matrix and ECM to transfer molecular messages. Several groups have proposed the involvement of EVs in interactions with the ECM. Still, the effects of these interactions on both EV function and ECM homeostasis remain poorly defined (Niel et al., 2022). There is debate about the uptake of EVs by recipient cells, which has been described using models such as receptor-mediated binding, endocytosis, and membrane fusion. However, the mechanisms underlying the recognition and internalization of EVs remain under investigation. Specific interactions between EV surface ligands and cellular receptors may explain the preferential targeting and uptake by particular cell types (Isaac et al., 2021). Upon binding, EVs may activate signaling pathways at the cell surface or deliver their cargo into the cytoplasm via yet-to-be-characterized mechanisms (Haider and Aramini, 2020). While some contents of EVs may perform functions within the cytoplasm, others may be targeted for lysosomal degradation, providing metabolic substrates to host cells (Isaac et al., 2021).

In addition to their ability to communicate with other cells, EVs exhibit remarkable immunomodulatory activities and may help prevent or treat inflammatory and immune diseases. The promoting or inhibiting effect of EVs on the immune response depends on their origin and molecular cargo (Li et al., 2024b). In models of autoimmune diseases and chronic inflammation, MSC-EVs exert an anti-inflammatory effect by transferring miR-146a and miR-21 to macrophages, thereby suppressing pro-inflammatory cytokines, such as interleukin-6 (IL-6) and tumor necrosis factor-α (TNF-α) (Yang et al., 2021). Indeed, MSC-EVs inhibit macrophage polarization toward a pro-inflammatory phenotype and reduce synovial inflammation and joint destruction in a rheumatoid arthritis model (Wang et al., 2023a). In addition, EVs can induce immune tolerance. This property is used in transplant medicine to minimize the risk of graft-versus-host disease (GVHD) or organ rejection. Murine models have demonstrated that exosomes targeting two putative targets, programmed death-ligand one (PD-L1) and fibrinogen-like protein one (FGL1), derived from MSCs and released from MSCs, can suppress cardiac transplant rejection (Tsai et al., 2021). In contrast, EVs from dendritic cells can induce T-cell polarization toward adaptive immunity by presenting tumor antigens on their surface, an approach used in cancer immunotherapy to improve antitumor immunity (Chen et al., 2023a).

EVs serve as biocompatible, low-immunogenicity carriers for targeted drug delivery, with enhanced therapeutic effects and reduced off-target effects (Liang et al., 2025b). Antifibrotic miRNAs contained in EVs released by cells could be transferred to myofibroblasts and inhibit activation, ECM deposition, and tissue architecture reconstruction in fibrotic diseases, such as liver cirrhosis or pulmonary fibrosis. When EVs containing miR-4665 from hepatocytes entered liver sinusoidal endothelial cells, the expression of lysyl oxidase-like protein two (LOXL2) in hepatic stellate cells was inhibited. Then the expression of collagen was further decreased. Similarly, when EVs containing enriched miR-4465 from MSCs enter liver sinusoidal endothelial cells, LOXL2 expression is downregulated, thereby alleviating hepatic fibrosis in preclinical models (Wang et al., 2024a). In addition, when EVs released from neural stem cells or astrocytes enter target cells, they could transfer anti-apoptotic proteins and miRNAs to downregulate the expression of anti-apoptotic genes and further suppress microglial activation and transfer growth factors, including brain-derived neurotrophic factor (BDNF), to upregulate the expression level of synaptic proteins and mediate synaptic plasticity (Liu et al., 2024a). Especially when EVs enter ischemic areas during neurological functional recovery after stroke, they can transfer anti-apoptotic genes and proteins to improve neuronal survival and enhance neurological functional recovery by modulating the inflammatory response and regenerative processes (Liu et al., 2024b). In the cardiovascular system, cardiomyocyte-EVs or MSC-EVs could also be used as therapeutic agents in myocardial infarction models. For example, when cardiomyocyte-EVs enter cardiomyocytes, miR-21 in cardiomyocyte-EVs transfers into target cardiomyocytes to improve mitochondrial respiration and decrease cardiomyocyte apoptosis (Jiang et al., 2023). The effects of EVs on cancer are complex. Tumor-EVs could enhance cancer progression and metastasis by remodeling the ECM and evading immune cell attack. In contrast, engineered EVs could act as a promising tool for targeted anticancer therapy. Immunogenic EVs (imEVs) generated by pulsed electric fields could be loaded with interferon-γ (IFN-γ) mRNA, engineered to express CD64, and then further bound to antibodies, such as anti-CD71 or PD-L1. The imEVs generated in this way selectively target GBM cells and exhibit strong antitumor effects *in vivo* (Dong et al., 2023).

Therefore, cells have been replaced by their cell-free derivatives as a paradigm in regenerative medicine. These carriers can mediate intercellular communication, modulate the immune response, and serve as carriers for targeted cargo delivery. They are expected to become the next generation of tools for repairing damaged tissues, resolving inflammation, and treating complex diseases. Therefore, the use of natural or engineered EVs will pave the way for precision regenerative therapy.

7.5 THERAPEUTIC APPLICATIONS OF EVS IN REGENERATIVE MEDICINE

EVs serve as vectors of bioactive factors (peptides and nucleic acids) and play a pivotal role in intercellular signaling and tissue repair (Fan et al., 2025; Lu et al., 2024). Due to their high biocompatibility and stability, EVs are resistant to *in vivo* degradation and can traverse various biological barriers. This allows them to deliver therapeutic cargo to sites of tissue injury effectively, facilitating regeneration (Moghassemi et al., 2024). Therefore, EVs can be widely applied in tissue regeneration, including bone repair and regeneration, nerve regeneration, cardiovascular repair, skin wound healing, and cancer therapy (Figure 7.2) (Xing et al., 2020; Tan et al., 2024).

7.5.1 Bone Repair and Regeneration

Due to their efficient repair of significant bone defects and ability to promote bone tissue regeneration, EVs have attracted considerable attention. EVs loaded with specific miRNAs, proteins, and

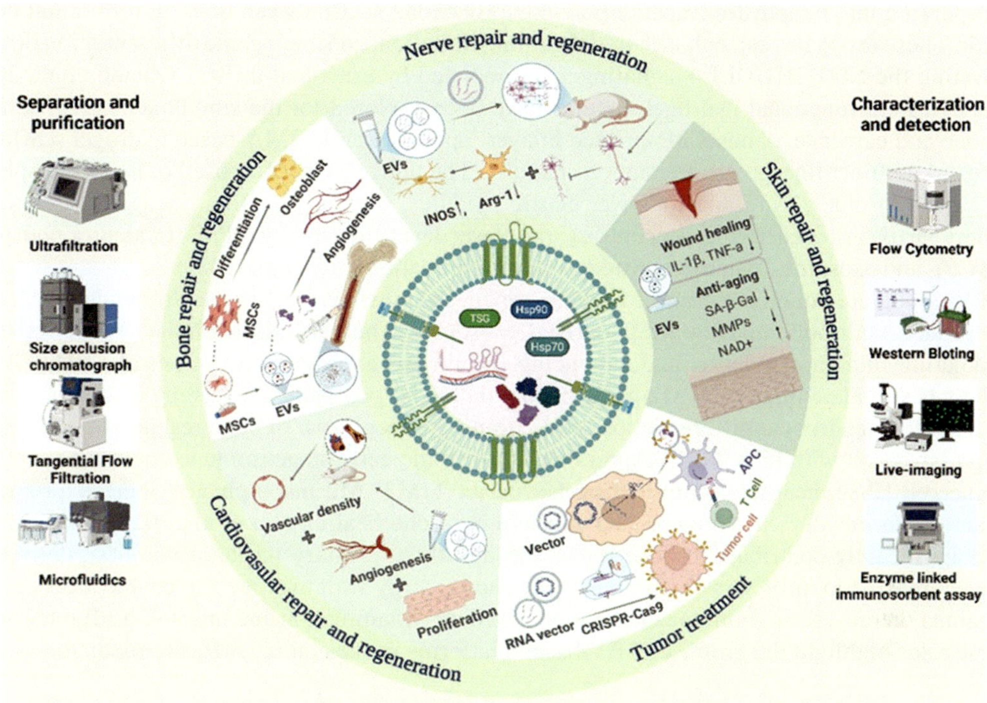

FIGURE 7.2 Therapeutic applications of EVs in regenerative medicine. EV-based technologies, including methods for EV separation, purification, characterization, and detection, are being advanced to explore their roles in tissue regeneration. Shown are five key areas where EVs show therapeutic potential: bone repair and regeneration, nerve repair and regeneration, cardiovascular repair and regeneration, skin repair and regeneration, and cancer therapy. Created by Biorender.com.

small molecules have been found to improve both osteogenesis and angiogenesis (Wang et al., 2022). The regenerative potential of EVs is further enhanced by engineering their parent cells to overexpress specific genes, producing EVs enriched in pro-osteogenic components. For example, EVs overexpressing proliferating cell nuclear antigen (PCNA) can rescue aged bone marrow-derived mesenchymal stem cells (BM-MSCs) and reverse the degenerative process, restoring regenerative potential (Lei et al., 2021). Neonatal umbilical cord-derived EVs can also be used to reactivate the regenerative capacity of BM-MSCs through angiogenesis and bone formation. Importantly, when hippocampal neurons are exposed to injury conditions, they can release small EVs enriched with osteogenic miRNAs (such as miR-328a-3p and miR-150-5p) that stimulate osteoprogenitors to promote bone regeneration (Xia et al., 2021). In addition, surface modifications of fibronectin-modified EVs can improve bone-targeted delivery and enhance healing effects. In addition, due to their high biocompatibility and low cost, cow milk-derived EVs can promote bone repair by regulating AP3B1-mediated chromatin remodeling in recipient cells (Dong et al., 2022).

Hydrogel systems serve as excellent matrices for EV delivery in bone tissue engineering. EVs encapsulated in biomaterial scaffolds not only facilitate sustained release but also enhance therapeutic effects. For instance, dental pulp stem cell (DPSC)-derived EVs encapsulated in multifunctional hydrogels have shown remarkable osteoinductive activity by upregulating bone morphogenetic protein-2 (BMP-2) expression and promoting mineralization and neovascularization (Wang et al., 2024b). Adhesion-modified hyaluronic acid (HA)-based hydrogels can enhance EV retention and delivery in inflammatory or hypoxic microenvironments, thereby promoting ECM remodeling and vascular infiltration. Umbilical MSC-derived exosomes loaded into HA hydrogels and further incorporated into nanohydroxyapatite/poly-ε-caprolactone scaffolds can be used for cranial defect repair. These exosomes can enhance proliferation, migration, and angiogenic differentiation through activating the NOTCH1/DLL4 signaling axis mediated by exosomal miR-21 (Zhang et al., 2021). Recently, multifunctional hydrogel systems have been explored for the simultaneous regeneration of bone and cartilage. A novel 3D-printed bilayer (upper-lower) GelMA-based hydrogel scaffold is designed to direct lineage differentiation spatially. The upper layer, composed of black phosphorus (BP) and exosomes, can impart a lower elastic modulus, thereby directing lineage differentiation toward cartilage regeneration. In contrast, the lower layer, formed by BP, β-tricalcium phosphate (β-TCP), and exosomes, can direct osteogenic differentiation (Sun et al., 2023).

EVs modulate immune responses for bone repair. In spinal cord injury, electroconductive hydrogels loaded with bone marrow MSC-derived exosomes synergistically stimulate tissue repair by modulating immunoregulation and enhancing myelinated axon growth (Fan et al., 2022). These hydrogels stimulate microglial M2 polarization via the NF-κB pathway, thereby enhancing neuronal and oligodendrocyte differentiation while limiting astrogliosis. In bone regeneration, immunosuppressive EVs with high TIM3 expression enhance the early-stage immune environment. These engineered EVs, encapsulated in hydrogels, recruit $TIM3^+$ M2 macrophages, thereby promoting an anti-inflammatory environment and accelerating bone healing (Lu et al., 2024). In summary, EVs significantly contribute to bone repair through three primary mechanisms: first, delivery of regenerative cargo (miRNAs and proteins); second, synergy with hydrogels to enable localized and sustained therapy; third, immune modulation to reduce inflammation and improve graft integration. These roles highlight the promise of EV-based platforms in skeletal regenerative medicine.

7.5.2 Nerve Repair and Regeneration

EVs exhibit significant therapeutic potential for nerve regeneration and repair by promoting axonal growth, modulating the immune environment, and enhancing neural stem cell function (Mushtaq et al., 2024). Using various molecular strategies, EVs have been reported to participate in neuroprotection, neuroregeneration, and functional rescue in peripheral and central nervous system injuries. Netrin-1 (NTN1)-engineered endothelial cell-derived EVs can establish a pre-regenerative niche and promote peripheral nerve regeneration (Huang et al., 2024a). Overexpression of NTN1 enhances angiogenesis, axonal regeneration, and repair-related phenotypes. Endothelial cell-derived EVs can induce the formation of adhesion spots and the phosphatidylinositol 3-kinase-AKT pathways, and also ameliorate downstream mTOR targets. By mimicking regenerative signals, these biofunctional EVs can stimulate neural stem cell division and differentiation. Platelet-rich plasma-derived exosomes (PRP-Exos) are also found to be able to promote peripheral nerve repair by modulating MSCs. PRP-Exos can improve the division, viability, motility, and stemness of MSCs while reducing apoptosis and senescence (Zhang et al., 2024b). When pretreated with PRP-Exos (pExo-MSCs), MSCs exhibit improved engraftment and therapeutic efficacy and can promote axonal regrowth, remyelination, and functional recovery. In coculture experiments, pExo-MSCs can promote Schwann cell proliferation and dorsal root ganglion (DRG) axonal extension. EVs can also modulate the extracellular microenvironment during neural regeneration (Coleman et al., 2023).

EVs modulate the immune microenvironment and exhibit anti-inflammatory, angiogenic, and neurostem cell-differentiation properties, all of which are conditions for the formation of a regenerative environment. Thus, the signaling information carried by EVs should be utilized to enhance nerve repair outcomes. As with spinal cord injury (SCI), engineered MSC-derived EVs also have great potential. The EVs embedded in the hydrogel can stimulate axon regeneration, synaptogenesis, and the expression of neurotrophic factors, e.g., synaptophysin (Sun et al., 2024), thereby dramatically enhancing axonal length and the number of branches in dorsal root ganglion neurons. Thus, these engineered exosomes can induce intrinsic neuronal repair programs. More interestingly, EVs can also downregulate the neuroinflammatory response, which is the original pathological phenotype of nerve injury. EVs can modify the immune microenvironment to protect neuronal tissue and create a more favorable regenerative microenvironment by regulating it. In particular, loaded EVs within a bionic peptide scaffold are designed to promote in situ polarization of M2-type macrophages, which is vital for long-distance peripheral nerve regeneration (Yang et al., 2023). In a pro-inflammatory injury model, impairment of M2 macrophage function is associated with a block in nerve regeneration. Thus, peptide-based hydrogel biomaterials modified with M2-derived cytokines and EVs may mimic the regenerative microenvironment, enhance Schwann cell division and migration, and promote axonal recovery and growth.

Besides injury repair, EVs have the emerging potential for neurodegenerative diseases, including Parkinson's disease (PD), multiple sclerosis (MS), and Alzheimer's disease (AD) (Yusuf et al., 2025; Ahmed et al., 2025). Neuron-derived EVs in AD can be isolated from peripheral blood and serve as a biomarker for assessing disease progression under the influence of exercise (Delgado-Peraza et al., 2023). Engineered MSC-derived EVs that overexpress the SHP2 phosphatase promote mitophagy (expulsion of damaged mitochondria). SHP2-EVs cross the BBB and reach target neurons, where they inhibit mitochondrial damage and downstream pathology, including NLRP3 inflammasome induction and neuronal apoptosis (Xu et al., 2022). SHP2-EVs maintain synaptic density and restore cognitive function in AD mice. For instance, MiR-124-3p-enriched EVs are engineered to promote neuron differentiation in cultures of subventricular zone stem cells (Esteves et al., 2022), while PD EVs *in vivo* shield the nigrostriatal pathway, i.e., protect the path above, which in turn spares neurons in the substantia nigra and the striatal projection. Similar observations have been made using engineered EVs in multiple sclerosis models, showing that, e.g., scaffold-engineered EVs delivered through hydrogel carriers enhance remyelination, decrease inflammation, and support neural regeneration. Hence, EVs harbor an array of bioactive molecules and genes, acting as nanocarriers that can readily cross the BBB and shuttle regenerative and/or immunomodulatory payloads directly to injured or diseased neural tissues. Their use as a dual therapeutic agent and delivery vectors renders them a promising agent for therapy against traumatic injuries and chronic neurodegenerative conditions alike.

7.5.3 Cardiovascular Repair and Regeneration

Native and bioengineered EVs have crucial therapeutic applications in the cardiovascular system, including ischemia/reperfusion injury (I/RI), cardiac hypertrophy, myocardial infarction (MI), heart failure (HF), atherosclerosis, and hypertension. Multi-omics approaches have identified tissue-derived EVs as key modulators in atherosclerosis and calcific aortic valve stenosis (Blaser et al., 2023).

Comparative proteomic analysis between human carotid plaques and calcified aortic valves reveals EVs' involvement in cardiovascular calcification and disease progression. Engineered platelet membrane-derived EVs (P-EVs) have been developed for immunomodulatory cardiac repair. These P-EVs, when administered intravenously, bind to peripheral monocytes and direct them to home toward the ischemic myocardium, where they differentiate into macrophages (Li et al., 2022). Upon infiltration, the P-EVs are preferentially endocytosed within the inflammatory milieu, where they escape lysosomal degradation and release their functional miRNAs into the cytoplasm. This process facilitates a phenotypic switch from pro-inflammatory M1 to reparative M2 macrophages, reprogramming the immune microenvironment to support cardiac healing. In a similar vein, M2 macrophage-derived EVs have been shown to suppress pro-inflammatory CCR2+ macrophages, thereby promoting post-MI cardiac repair. In ischemic stroke models, within this broader analytical framework, the therapeutic potential of encapsulating bacteria-derived outer membrane vesicles (OMVs) within neutrophils and delivering the neuroprotective agent pioglitazone (PGZ) is promising. This particular design appears to indicate substantially enhanced, targeted delivery to ischemic tissue, ostensibly improving therapeutic outcomes (Pan et al., 2023).

Given the complexity of these theoretical relationships, the bioengineered EVs are also frequently loaded onto biofunctional hydrogels, which appear to offer enhanced targeting, improved bioavailability, and controlled release for cardiovascular regeneration. For example, endothelial progenitor cell-derived EVs are often loaded with bioactive silicate ions, suggesting an augmentation of their regenerative function. In murine MI models, these EVs, delivered via hydrogel microspheres, appear to carry angiogenic mediators, including miR-126a-3p, VEGF, and SDF-1, thereby enhancing cardiac neovascularization and tissue recovery (Yu et al., 2023). Injectable GelMA hydrogel patches loaded with targeted miR-222-engineered EVs (TeEVs) are delivered in situ into the pericardial cavity, suggesting they ameliorate I/RI via force-sensing mechanotransduction in mouse models (Wang et al., 2025b). These TeEVs are predominantly loaded into force-sensing hydrogels to construct injectable TeEV-hydrogel cardiac patches and to facilitate prolonged EV release into the infarcted myocardium. Injectable TeEV-hydrogel cardiac patches mitigate cardiac remodeling post-I/RI. It is associated with focal adhesion formation, cytoskeletal force transmission, and nuclear mechanotransduction—all of which are thought to contribute to the attenuation of cardiac remodeling (Wang et al., 2024c). Preclinical porcine models have validated this technique, demonstrating feasibility and efficacy for translational cardiovascular therapies.

7.5.4 Skin Repair and Regeneration

EVs can deliver bioactive molecules directly to damaged skin tissue, thereby offering significant therapeutic potential for skin repair and regeneration. Metformin-engineered EVs have been proven to reinstate aging skin by remodeling cellular energy metabolism. These EVs mitigate cellular senescence and restore function in aging dermal cells by promoting mitophagy, reducing glycolysis, and enhancing mitochondrial ATP production (Zhuang et al., 2024). This metabolic shift supports activated wound repair in aged skin. In addition, RNA-enriched EVs are another promising therapeutic platform. For example, circCDK13-loaded EVs can enhance healing ability in diabetic wound models (Huang et al., 2024b). Upon delivery to wound sites, circCDK13 stimulates fibroblast and keratinocyte division and migration by enhancing CD44 and c-MYC, thereby contributing to chronic wound healing and re-epithelialization. Adipose-derived MSC-EVs, when embedded into ECM hydrogels, have shown superior outcomes in diabetic wound repair. These bioactive composites provide a sustained local release of EVs in the wound, facilitating angiogenesis, inhibiting inflammation, and promoting collagen deposition and cell proliferation (Song et al., 2023). The hydrogel-EV systems are highly biocompatible and biodegradable, making them suitable for clinical application.

EVs can modulate ECM remodeling to induce skin repair and regeneration. EVs from umbilical cord blood MSCs markedly enhanced type I collagen and elastic fibers synthesis in skin and downregulated the expression of matrix metalloproteinase-1 (MMP-1) by accelerating skin repair and regeneration (Narauskaite et al., 2021). Therefore, their ECM-modulatory properties make them promising candidates for use in aesthetic and clinical dermatology for anti-aging, scar and hypertrophic wound treatment, and others. Bioengineering strategies have been developed to use EVs for mRNA delivery to induce wound healing. Skin fibroblast-derived EVs encapsulating Gstm2 mRNA improve fibroblast-epidermal interactions and reduce oxidative stress in aging skin. The EVs maintain skin homeostasis and enhance epidermal regeneration in aged murine models (Wu et al., 2024). In parallel, epidermal organoids derived from pluripotent stem cells are used as biofactors for therapeutic EV production. These organoids, enriched in epidermal stem cells, produce EVs that express high levels of angiogenic VEGF and regenerative miRNAs and inhibit differentiation into other skin lineages (e.g., hair follicle cells), thereby providing a more stable EV profile for therapy. The results obtained with the produced EVs show that they can modulate *in vivo* wound healing by promoting cellular proliferation, migration, and neovascularization (Kwak et al., 2024).

7.5.5 Cancer Therapy

Bioengineered EVs are one type of cell-free cancer therapy (Sajid et al., 2025). Due to their good stability and biocompatibility, EVs could be ideal carriers for delivering proteins, nucleic acids, small-molecule drugs, and nanoparticles as therapeutic agents. Interesting work on fruit-derived EVs loaded with structurally engineered droplet-formulated chemotherapeutics to pass the BBB and blood-brain tumor barrier (BBTB) for glioblastoma chemotherapy. These EVs exhibit high tumor accumulation, deep tissue penetration, and efficient cellular internalization (Chen et al., 2023b). EVs are engineered to mediate immune reprogramming. Giving an example, exosomes have also been bioengineered with antisense oligonucleotides (ASOs), and targeting signal transducer and activator of transcription 6 (STAT6) has been used genetically to reprogram tumor-associated macrophages (TAMs), resulting in over 90% tumor inhibition in colorectal and hepatocellular carcinoma models (Kamerkar et al., 2022). In checkpoint blockade-based immunotherapy, CD47-functionalized hybrid EV nanoplatforms derived from M1 macrophages are designed to both evade immune clearance and deliver immunomodulatory signals. These EVs effectively reprogram TAMs to a pro-inflammatory phenotype, demonstrating potent antitumor activity in breast tumors and melanoma, suppressing both primary tumor growth and distant metastasis (Tang et al., 2023). These EVs serve as drug carriers, efficiently delivering and targeting tumor sites for cancer therapy.

Beyond serving as delivery platforms, EVs also mediate inter-organ communication that influences cancer progression and therapeutic response. Tumor-derived EVs, for example, have been implicated in hepatic metabolic reprogramming, promoting liver dysfunction and diminishing chemotherapy tolerance. This EV-mediated liver remodeling increases susceptibility to bone marrow suppression and cardiotoxicity, contributing to treatment-related side effects (Wang et al., 2023b). EVs secreted from failing or infarcted hearts, termed cardiac small EVs, have been shown to enhance tumor growth. These EVs, enriched in pro-tumorigenic factors, enhance cancer cell proliferation in models using lung, colon, melanoma, and breast cancer cell lines (Caller et al., 2024). Similarly, tissue fibrosis, through ECM stiffening, increases exosome secretion, thereby enhancing Akt pathway activation and tumor growth. Proteomic and transcriptomic analyses reveal that Notch signaling is a key driver in this mechanoresponsive EV secretion pathway, particularly in fibrotic liver tissues (Wu et al., 2023). Therefore, EVs can regulate the immune microenvironment to activate immune cells and enhance antitumor immune response, and interact with other organs, halting tumor progression, improving clinical outcomes, and contributing to personalized cancer therapy.

7.6 CONCLUSIONS AND FUTURE PERSPECTIVES

Compared with cell-based therapies, EVs offer several potential advantages, including lower immunogenicity and greater stability in circulation (Lou et al., 2023). Recently, the biological functions of EVs have been explored to regulate cellular communication, modulate the immune response, and act as cargo carriers in various regenerative scenarios, including bone, heart, nerve, and skin repair, as well as oncology. When the biology and function of EVs are understood, several challenges must be addressed to advance these promising preclinical results into standardized, safe, and effective clinical applications.

One significant limitation of EV therapy is its high heterogeneity. EVs originate from different cell types, and their composition changes with physiological or pathological conditions, leading to variations in size, molecular cargo, and biological activities (Zhao et al., 2021). This high diversity is one of EVs' strengths, as they can adapt to different conditions. Still, it also poses a challenge for developing uniform protocols for isolation, characterization, and quality control. Advances in universal standards for the classification and validation of EVs will be necessary to ensure reproducibility and meet regulatory requirements for clinical use. The functional complexity of EVs continues to present both opportunities and challenges in tissue engineering. EVs in drug delivery are truly innovative but also pose a challenge for bioengineering, as proper engineering is needed to avoid interference with EVs' natural function during therapeutic payload loading. Current strategies to engineer EVs for targeted delivery have focused on surface modification, electroporation, or sonication for cargo loading and the generation of hybrid EV-synthetic nanoparticle systems (Yang et al., 2024). Challenges for the field remain in EV stability, biodistribution, and the off-target effects, which will need to be addressed at the pharmacokinetic and toxicological levels.

Future work in EV-based regenerative medicine will rely on multidisciplinary innovation and cross-disciplinary work to bridge knowledge gaps and accelerate the clinical translation of these EVs. Emerging high-resolution single-EV evaluation technologies, such as high-resolution flow cytometry and Raman spectroscopy, will be developed to dissect EV heterogeneity and identify functionally distinct subpopulations. Machine learning will be employed to correlate EV molecular signatures to known therapeutic outcomes (Chen et al., 2025). Good manufacturing practice (GMP) regarding EV production, transportation, and preservation should be developed. This should also include the creation of reproducible metrics for purity, potency, and sterility assessment (Thery et al., 2018). Finally, EVs incorporated into biomaterials, such as hydrogels, three-dimensional (3D) printed scaffolds, or decellularized matrices, may provide a novel approach to augment EV retention and control release at injury sites. For example, EV-encapsulated hydrogels exhibit delayed regenerative effects on cutaneous wound healing by creating a moist microenvironment around the wound and preventing the early release of EVs (Zheng et al., 2024). Synthetic nanoparticles can replicate the structure and function of EVs. They may address limitations, such as scalability and consistency, allowing the preparation of patient-specific therapies based on patient-matched lentivirus populations, associated with CRISPR-Cas9 or RNA interference technology (Li et al., 2023). Nevertheless, in the meantime, for the administration, including the Food and Drug Administration (FDA), to determine the proper way of getting approval, a challenge will always remain. Besides intellectual property debates, isoelectric point isolation and engineering methods will always impede the commercialization of EV-based regenerative medicine.

EVs have been shown to proliferate in regenerative medicine and to address unmet clinical needs (Cheng et al., 2023). Although many challenges remain in standardization, controllability, and biosafety, the future of EV-based regenerative medicine remains optimistic, facilitated by biotechnology, materials science, and data-driven artificial intelligence (AI) innovations. As tissue regeneration moves from the laboratory to clinical application in the near future, tissue repair, immune modulation, and the production of personalized therapies will be realized in the next decade (Kumar et al., 2024).

7.7 ACKNOWLEDGMENTS

This work was supported by grants from the National Natural Science Foundation of China (82302401) and the Science and Technology Commission of Shanghai Municipality (25J22800800).

7.8 DISCLOSURES

The authors declare no competing interests.

REFERENCES

Ahmed LA, Al-Massri KF, Haider KH. Therapeutic efficacy of mesenchymal stem cells and their derived exosomes in different neurodegenerative disorders: current status and future perspectives. In: Haider KH, editor. *Handbook of Regenerative Medicine: Cell-Free Therapy*. Boca Raton (FL): CRC Press, 2025.

Ahmed ZT, Alkahlot MH, Haider KH. MSC-derived exosomes: advances in cell-free therapy. In: Haider KH, editor. *Handbook of Stem Cell Applications*. Singapore: Springer Nature Singapore, 2023, pp. 1–41.

Blaser MC, Buffolo F, Halu A, Turner ME, Schlotter F, Higashi H, et al. Multiomics of tissue extracellular vesicles identifies unique modulators of atherosclerosis and calcific aortic valve stenosis. *Circulation*. 2023;148(8):661–678.

Buzas EI. The roles of extracellular vesicles in the immune system. *Nature Reviews Immunology*. 2022;23(4):236–250.

Caller T, Rotem I, Shaihov-Teper O, Lendengolts D, Schary Y, Shai R, et al. Small extracellular vesicles from infarcted and failing hearts accelerate tumor growth. *Circulation*. 2024;149(22):1729–1748.

Carney RP, Mizenko RR, Bozkurt BT, Lowe N, Henson T, Arizzi A, et al. Harnessing extracellular vesicle heterogeneity for diagnostic and therapeutic applications. *Nature Nanotechnology*. 2024;20(1):14–25.

Cecchin R, Troyer Z, Witwer K, Morris K V. Extracellular vesicles: The next generation in gene therapy delivery. *Molecular Therapy*. 2023;31(5):1225–1230.

Chen J, Pan J, Liu S, Zhang Y, Sha S, Guo H, et al. Fruit-derived extracellular-vesicle-engineered structural droplet drugs for enhanced glioblastoma chemotherapy. *Advanced Materials*. 2023a;35(45):e2304187.

Chen J, Tian C, Xiong X, Yang Y, Zhang J. Extracellular vesicles: new horizons in neurodegeneration. *EBioMedicine*. 2025;113:105605.

Chen J, Zhang G, Wan Y, Xia B, Ni Q, Shan S, et al. Immune cell-derived exosomes as promising tools for cancer therapy. *Journal of Controlled Release*. 2023b;364:508–528.

Chen K, Yu W, Zheng G, Xu Z, Yang C, Wang Y, et al. Biomaterial-based regenerative therapeutic strategies for spinal cord injury. *NPG Asia Materials*. 2024;16(1):1–29.

Cheng K, Kalluri R. Guidelines for clinical translation and commercialization of extracellular vesicles and exosomes-based therapeutics. *Extracellular Vesicle*. 2023;2:100029.

Chuah SJ, Yong CW, Teo KYW, Chew JRJ, Cheow YA, Zhang S, et al. Mesenchymal stromal cell-derived small extracellular vesicles modulate macrophage polarization and enhance angio-osteogenesis to promote bone healing. *Genes & Diseases*. 2021;9(4):841–844.

Coleman LG. From the dust: Extracellular vesicles as regulators of development and neuroregeneration. *Neural Regeneration Research*. 2023;19(5):933–934.

Dai S, Liu Y, Liu Z, Li R, Luo F, Li Y, et al. Cancer-associated fibroblasts mediate resistance to anti-EGFR therapies in cancer. *Pharmacological Research*. 2024;206:107304.

Dechantsreiter S, Ambrose AR, Worboys JD, Lim JME, Liu S, Shah R, et al. Heterogeneity in extracellular vesicle secretion by single human macrophages revealed by super-resolution microscopy. *Journal of Extracellular Vesicles*. 2022,11(4):e12215.

Delgado-Peraza F, Nogueras-Ortiz C, Simonsen AH, Knight DD, Yao PJ, Goetzl EJ, et al. Neuron-derived extracellular vesicles in blood reveal effects of exercise in Alzheimer's disease. *Alzheimer's Research & Therapy*. 2023;15(1):156.

Dixson AC, Dawson TR, Di Vizio D, Weaver AM. Context-specific regulation of extracellular vesicle biogenesis and cargo selection. Nature Reviews *Molecular Cell Biology*. 2023;24(7):454–476.

Dong M, Shi C, Yu X, Yang Q, Wu S, Liu R, et al. Milk-derived small extracellular vesicles: Nanomaterials to promote bone formation. *Journal of Nanobiotechnology*. 2022;20(1):370.

Dong S, Liu X, Bi Y, Wang Y, Antony A, Lee D, et al. Adaptive design of mRNA-loaded extracellular vesicles for targeted immunotherapy of cancer. *Nature Communications*. 2023;14(1):6610.

Esteves M, Abreu R, Fernandes H, Serra-Almeida C, Martins PAT, Barão M, et al. MicroRNA-124-3p-enriched small extracellular vesicles as a therapeutic approach for Parkinson's disease. *Molecular Therapy*. 2022;30(10):3176–3192.

Fan H, Zhao H, Hou Y, Meng D, Jiang J, Lee EB, et al. Heterogeneous focal adhesion cytoskeleton nanoarchitectures from microengineered interfacial curvature to oversee nuclear remodeling and mechanotransduction of mesenchymal stem cells. *Cellular & Molecular Biology Letters*. 2025;30(1):10.

Fan L, Liu C, Chen X, Zheng L, Zou Y, Wen H, et al. Exosomes-loaded electroconductive hydrogel synergistically promotes tissue repair after spinal cord injury via immunoregulation and enhancement of myelinated axon growth. *Advanced Science*. 2022;9(13):e2105586.

Foers AD, Chatfield S, Dagley LF, Scicluna BJ, Webb AI, Cheng L, et al. Enrichment of extracellular vesicles from human synovial fluid using size exclusion chromatography. *Journal of Extracellular Vesicles*. 2018;7(1):1490145.

Habib SM, Haider KH, Martini MF. The potential of mesenchymal stem cell-derived exosomes in cell-free therapy for myocardial repair. In: Haider KH, editor. *Handbook of Regenerative Medicine: Cell-Free Therapy Approach* (1st ed.). CRC Press, 2025, Chapter 24. https://doi.org/10.1201/9781003611509.

Haider KH, Aramini B. Micrining the injured heart with stem cell-derived exosomes: an emerging strategy of cell-free therapy. *Stem Cell Research & Therapy*. 2020;11(1):23. doi: 10.1186/s13287-019-1548-7.

Herrmann IK, Wood MJA, Fuhrmann G. Extracellular vesicles as a next-generation drug delivery platform. *Nature Nanotechnology*. 2021;16(7):748–759.

Hoang DM, Pham PT, Bach TQ, Ngo ATL, Nguyen QT, Phan TTK, et al. Stem cell-based therapy for human diseases. *Signal Transduction and Targeted Therapy*. 2022;7(1):272.

Huang J, Li J, Li S, Yang X, Huo N, Chen Q, et al. Netrin-1-engineered endothelial cell exosomes induce the formation of a pre-regenerative niche to accelerate peripheral nerve repair. *Science Advances*. 2024a;10(26):eadm8454.

Huang Q, Chu Z, Wang Z, Li Q, Meng S, Lu Y, et al. circCDK13-loaded small extracellular vesicles accelerate healing in preclinical diabetic wound models. *Nature Communications*. 2024b;15(1):3904.

Isaac R, Reis FCG, Ying W, Olefsky JM. Exosomes as mediators of intercellular crosstalk in metabolism. *Cell Metabolism*. 2021;33(9):1744–1762.

Jarrige M, Frank E, Herardot E, Martineau S, Darle A, Benabides M, et al. The future of regenerative medicine: Cell therapy using pluripotent stem cells and acellular therapies based on extracellular vesicles. *Cells*. 2021;10(2):240.

Jia Y, Yu L, Ma T, Xu W, Qian H, Sun Y, et al. Isolation and separation of small extracellular vesicles: Current techniques, pending questions, and clinical applications. *Theranostics*. 2022;12(15):6548–6575.

Jiang J, Zhang X, Wang H, Spanos M, Jiang F, Ni L, et al. Closer to the heart: Harnessing the power of targeted extracellular vesicle therapies. *Advanced Biology*. 2023;8(2):2300141.

Jiang Q, Liu J, Huang S, Wang XY, Chen X, Liu GH, et al. Antiageing strategy for neurodegenerative diseases: from mechanisms to clinical advances. *Signal Transduction and Targeted Therapy*. 2025;10(1):76.

Jiang Y, Wang Y, Chen G, Sun F, Wu Q, Huang Q, et al. The nicotinamide metabolism face-off between macrophages and fibroblasts manipulates the microenvironment in gastric cancer. *Cell Metabolism*. 2024;36(8):1806–1822.

Kamerkar S, Leng C, Burenkova O, Jang SC, McCoy C, Zhang K, et al. Exosome-mediated genetic reprogramming of tumor-associated macrophages by exoASO-STAT6 leads to potent monotherapy antitumor activity. *Science Advances*. 2022;8(7):eabj7002.

Kumar MA, Baba SK, Sadida HQ, Marzooqi SAl, Jerobin J, Altemani FH, et al. Extracellular vesicles as tools and targets in disease therapy. *Signal Transduction and Targeted Therapy*. 2024;9(1):27.

Kundu D, Shin SY, Chilian WM, Dong F. The potential of mesenchymal stem cell-derived exosomes in cardiac repair. *International Journal of Molecular Sciences*. 2024;25(24):13494.

Kwak S, Song CL, Lee J, Kim S, Nam S, Park YJ, et al. Development of pluripotent stem cell-derived epidermal organoids that generate effective extracellular vesicles in skin regeneration. *Biomaterials*. 2024;307:122522.

Lei Q, Gao F, Liu T, Ren W, Chen L, Cao Y, et al. Extracellular vesicles deposit PCNA to rejuvenate aged bone marrow-derived mesenchymal stem cells and slow age-related degeneration. *Science Translational Medicine*. 2021;13(578):eaaz8697.

Li H, Liu Y, Lin Y, Li S, Liu C, Cai A, et al. Cardiac repair using regenerating neonatal heart tissue-derived extracellular vesicles. *Nature Communications*. 2025a,16(1):1292.

Li Q, Huang Z, Wang Q, Gao J, Chen J, Tan H, et al. Targeted immunomodulation therapy for cardiac repair by platelet membrane engineering extracellular vesicles via hitching peripheral monocytes. *Biomaterials*. 2022;284:121529.

Li S, Li W, Wu X, Zhang B, Liu L, Yin L. Immune cell-derived extracellular vesicles for precision therapy of inflammatory-related diseases. *Journal of Controlled Release*. 2024a;368:533–547.

Li S, Li Y, Zhu K, He W, Guo X, Wang T, et al. Exosomes from mesenchymal stem cells: Potential applications in wound healing. *Life Sciences*. 2024b;357:123066.

Li T, Zhang L, Lu T, Zhu T, Feng C, Gao N, et al. Engineered extracellular vesicle-delivered CRISPR/CasRx as a novel RNA editing tool. *Advanced Science*. 2023;10(10):2206517.

Li Z, Bai Y, Wu H, Feng Y, Wang X, Zhao C, et al. PTEN/PI3K/AKT pathway activation with hypoxia-induced human umbilical vein endothelial cell exosomes for angiogenesis-based diabetic skin reconstruction. *Materials Today Bio*. 2025b;32:101651.

Liang X, Gupta D, Xie J, Van Wonterghem E, Van Hoecke L, Hean J, et al. Engineering of extracellular vesicles for efficient intracellular delivery of multimodal therapeutics, including genome editors. *Nature Communications*. 2025a;16(1):4028.

Liang Y, Xia J, Xu L, Zhang Y, Xu X, Haider KH. Cell-free and cell-based therapy in regenerative medicine. Haider KH, editor. *Handbook of Regenerative Medicine: Cell-Free Therapy Approach* (1st ed.). CRC Press, 2025b. https://doi.org/10.1201/9781003611509.

Liu L, Liu W, Han Z, Shan Y, Xie Y, Wang J, et al. Extracellular vesicles-in-hydrogel (EViH) targeting pathophysiology for tissue repair. *Bioactive Materials*. 2024a;44:283–318.

Liu S, Chen L, Guo M, Li Y, Liu Q, Cheng Y. Targeted delivery of engineered RVG-BDNF-exosomes: A novel neurobiological approach for ameliorating depression and regulating neurogenesis. *Research*. 2024b;7:0402.

Liu Y, Bai Y, Heng BC, Wang Y, Zhang X, Nguyen TD, et al. Biomimetic electroactive materials and devices for regenerative engineering. *Nature Reviews Electrical Engineering*. 2025;2:188–204.

Lou P, Liu S, Wang Y, Lv K, Zhou X, Li L, et al. Neonatal-tissue-derived extracellular vesicle therapy (NEXT): a potent strategy for precision regenerative medicine. *Advanced Materials*. 2023;35(33):e2300602.

Lu T, Liu Y, Huang X, Sun S, Xu H, Jin A, et al. Early-responsive immunoregulation therapy improved the microenvironment for bone regeneration via engineered extracellular vesicles. *Advanced Healthcare Materials*. 2024;13(11):e2303681.

Manno M, Bongiovanni A, Margolis L, Bergese P, Arosio P. The physico-chemical landscape of extracellular vesicles. *Periodontology* . 2025;3:68–82.

Miron RJ, Estrin NE, Sculean A, Zhang Y. Understanding exosomes: Part 2-Emerging leaders in regenerative medicine. *Periodontology* 2000. 2024;94(1):257–414.

Moghassemi S, Dadashzadeh A, Sousa MJ, Vlieghe H, Yang J, Leon-Felix CM, et al. Extracellular vesicles in nanomedicine and regenerative medicine: A review over the last decade. *Bioactive Materials*. 2024;36:126–156.

Mondal J, Pillarisetti S, Junnuthula V, Saha M, Hwang SR, Park IK, et al. Hybrid exosomes, exosome-like nanovesicles, and engineered exosomes for therapeutic applications. *Journal of Controlled Release*. 2022;353:1127–1149.

Mushtaq M, Zineldeen DH, Mateen MA, Haider KH. Mesenchymal stem cells' "garbage bags" at work: Treating radial nerve injury with mesenchymal stem cell-derived exosomes. *World Journal of Stem Cells* 2024; 16(5):467–478.

Narauskaite D, Vydmantaite G, Rusteikaite J, Sampath R, Rudaityte A, Stasyte G, et al. Extracellular vesicles in skin wound healing. *Pharmaceuticals*. 2021;14(8):811.

Niel G, Carter DRF, Clayton A, Lambert DW, Raposo G, Vader P. Challenges and directions in studying cell-cell communication by extracellular vesicles. Nature Reviews. *Molecular Cell Biology*. 2022;23(5):369–382.

Pan J, Wang Z, Huang X, Xue J, Zhang S, Guo X, et al. Bacteria-derived outer-membrane vesicles hitchhike neutrophils to enhance ischemic stroke therapy. *Advanced Materials*. 2023;35(38):e2301779.

Petrus-Reurer S, Romano M, Howlett S, Jones JL, Lombardi G, Saeb-Parsy K. Immunological considerations and challenges for regenerative cellular therapies. *Communications Biology*. 2021;4(1):798.

Racchetti G, Meldolesi J. Four distinct cytoplasmic structures generate and release specific vesicles, thus opening the way to intercellular communication. *Extracellular Vesicles and Circulating Nucleic Acids*. 2023;4(1):44–58.

Rajeev KS, Sakthiswary R, Lokanathan Y. Potential therapeutic application and mechanism of action of stem cell-derived extracellular vesicles (EVs) in systemic lupus erythematosus (SLE). *International Journal of Molecular Sciences*. 2024;25(4):2444.

Sajid MI, Bukhari S, Ilyas H, Malik R, Fatima M, Tiwari RK, Nauli SM, Haider KH. Exosomes and encapsulated ExomiRs in breast cancer: theragnostic applications and clinical implications. In: Haider KH, editor. *Handbook of Regenerative Medicine: Cell-Free Therapy*. Taylor & Francis Group and CRC Press; 2025, Chapter-23, 357–370.

Seifert AW, Duncan EM, Zayas RM. Enduring questions in regenerative biology and the search for answers. *Communications Biology*. 2023;6(1):1139.

Shen M, Hou Y, Xu S, Tan J, Zhou H, Miao Q, et al. Biofunctionalized patterned platform as microarray biochip to supervise delivery and expression of pDNA nanolipoplexes in stem cells via mechanotransduction. *Journal of Nanobiotechnology*. 2025;23(1):22.

Song Y, You Y, Xu X, Lu J, Huang X, Zhang J, et al. Adipose-derived mesenchymal stem cell-derived exosomes biopotentiate extracellular matrix hydrogels, accelerating diabetic wound healing and skin regeneration. *Advanced Science*. 2023;10(30):e2304023.

Su Y, He W, Zheng L, Fan X, Hu TY. Toward clarity in single extracellular vesicle research: Defining the field and correcting missteps. *ACS Nano*. 2025;19(17):16193–16203.

Sun T, Feng Z, He W, Li C, Han S, Li Z, et al. Novel 3D-printing bilayer GelMA-based hydrogel containing BP, beta-TCP, and exosomes for cartilage-bone integrated repair. *Biofabrication*. 2023;16(1):015008.

Sun Y, Liu Q, Qin Y, Xu Y, Zhao J, Xie Y, et al. Exosomes derived from CD271(+)CD56(+) bone marrow mesenchymal stem cell subpopulation identified by single-cell RNA sequencing promote axon regeneration after spinal cord injury. *Theranostics*. 2024;14(2):510–527.

Tan F, Li X, Wang Z, Li J, Shahzad K, Zheng J. Clinical applications of stem cell-derived exosomes. *Signal Transduction Targeted Therapy*. 2024;9(1):17.

Tang L, Yin Y, Cao Y, Fu C, Liu H, Feng J, et al. Extracellular vesicle-derived hybrid nanoplatforms for amplified CD47 blockade-based cancer immunotherapy. *Advanced Materials*. 2023;35(35):e2303835.

Tenreiro MF, Louro AF, Alves PM, Serra M. Next generation of heart regenerative therapies: progress and promise of cardiac tissue engineering. *NPJ Regenerative Medicine*. 2021;6(1):30.

Thery C, Witwer KW, Aikawa E, Alcaraz MJ, Anderson JD, Andriantsitohaina R, et al. Minimal information for studies of extracellular vesicles 2018 (MISEV2018): A position statement of the International Society for Extracellular Vesicles and update of the MISEV2014 guidelines. *Journal of Extracellular Vesicles*. 2018;7(1):1535750.

Tsai H, Wu Y, Liu X, Xu Z, Liu L, Wang C, et al. Engineered small extracellular vesicles as a FGL1/PD-L1 dual-targeting delivery system for alleviating immune rejection. *Advanced Science*. 2021;9(3):2102634.

Wakker SI, Meijers FM, Sluijter JPG, Vader P. Extracellular vesicle heterogeneity and its impact for regenerative medicine applications. *Pharmacological Reviews*. 2023;75(5):1043–1061.

Wang D, Cao H, Hua W, Gao L, Yuan Y, Zhou X, et al. Mesenchymal stem cell-derived extracellular vesicles for bone defect repair. *Membranes*. 2022;12(7):716.

Wang G, Li J, Bojmar L, Chen H, Li Z, Tobias GC, et al. Tumour extracellular vesicles and particles induce liver metabolic dysfunction. *Nature*. 2023b;618(7964):374–382.

Wang L, Wei X, He X, Xiao S, Shi Q, Chen P, et al. Osteoinductive dental pulp stem cell-derived extracellular vesicle-loaded multifunctional hydrogel for bone regeneration. *ACS Nano*. 2024b;18(12):8777–8797.

Wang N, Hou Y, Lin L, Xu S, Lee K, Yang Y, et al. Advanced microarrays as a heterogeneous force-remodeling coordinator to orchestrate nuclear configuration and force-sensing mechanotransduction in stem cells. *Advanced Science*. 2025a;12(14):2416482.

Wang Y, Chatterjee E, Li G, Xu J, Xiao J. Force-sensing protein expression in response to cardiovascular mechanotransduction. *EBioMedicine*. 2024c;110:105412.

Wang Y, Chen Y, Yang F, Yu X, Chu Y, Zhou J, et al. MiR-4465-modified mesenchymal stem cell-derived small extracellular vesicles inhibit liver fibrosis development via targeting LOXL2 expression. *Journal of Zhejiang University Science B*. 2024a;25(7):594–604.

Wang Y, Meng D, Shi X, Hou Y, Zang S, Chen L, et al. Injectable hydrogel with miR-222-engineered extracellular vesicles ameliorates myocardial ischemic reperfusion injury via mechanotransduction. *Cell Reports Medicine*. 2025b;6(3):101987.

Wang Z, Zhang C, Meng J, Jiao Z, Bao W, Tian H, et al. A targeted exosome therapeutic confers both CfDNA scavenging and macrophage polarization for ameliorating rheumatoid arthritis. *Advanced Materials*. 2023a,35(48):e2302503.

Wu B, Liu DA, Guan L, Myint PK, Chin L, Dang H, et al. Stiff matrix induces exosome secretion to promote tumour growth. *Nature Cell Biology*. 2023;25(3):415–424.

Wu H, Yao Z, Li H, Zhang L, Zhao Y, Li Y, et al. Improving dermal fibroblast-to-epidermis communications and aging wound repair through extracellular vesicle-mediated delivery of Gstm2 mRNA. *Journal of Nanobiotechnology*. 2024;22(1):307.

Wu Q, Zeng Y, Wang W, Liu S, Huang Y, Zhang Y, et al. Profiling nascent tumor extracellular vesicles via metabolic timestamping and aptamer-driven specific click chemistry. *Journal of the American Chemical Society*. 2025;147(12):10737–10749.

Xia W, Xie J, Cai Z, Liu X, Wen J, Cui ZK, et al. Damaged brain accelerates bone healing by releasing small extracellular vesicles that target osteoprogenitors. *Nature Communications*. 2021;12(1):6043.

Xiang B, Zhang S, Zhao IS, Gan X, Zhang Y. Microenvironmental modulation for therapeutic efficacy of extracellular vesicles. *Advancement of Science*. 2025;12:2503027.

Xing Z, Zhao C, Liu H, Fan Y. Endothelial progenitor cell-derived extracellular vesicles: a nvel candidate for regenerative medicine and disease treatment. *Advanced Healthcare Materials*. 2020;9(12):e2000255.

Xu F, Wu Y, Yang Q, Cheng Y, Xu J, Zhang Y, et al. Engineered extracellular vesicles with SHP2 high expression promote mitophagy for Alzheimer's disease treatment. *Advanced Materials*. 2022;34(49):e2207107.

Yang C, Sun J, Tian Y, Li H, Zhang L, Yang J, et al. Immunomodulatory effect of MSCs and MSCs-derived extracellular vesicles in systemic lupus erythematosus. *Frontiers in Immunology*. 2021;12:714832.

Yang C, Xue Y, Duan Y, Mao C, Wan M. Extracellular vesicles and their engineering strategies, delivery systems, and biomedical applications. *Journal of Controlled Release*. 2024;365:1089–1123.

Yang P, Peng Y, Dai X, Jie J, Kong D, Gu X, et al. Bionic peptide scaffold in situ polarization and recruitment of M2 macrophages to promote peripheral nerve regeneration. *Bioactive Materials*. 2023;30:85–97.
Yu B, Li H, Zhang Z, Chen P, Wang L, Fan X, et al. Extracellular vesicles engineering by silicate-activated endothelial progenitor cells for myocardial infarction treatment in male mice. *Nature Communications*. 2023;14(1):2094.
Yusuf R, Haider KH. Theragnostic applications of exosomes in Alzheimer's disease. In: Haider KH, editor. *Handbook of Regenerative Medicine: Cell-Free Therapy*. Taylor & Francis Group and CRC Press; 2025, Chapter-255, pp. 390–409.
Zhang J, Ji C, Zhang H, Shi H, Mao F, Qian H, et al. Engineered neutrophil-derived exosome-like vesicles for targeted cancer therapy. *Science Advances*. 2022;8(2):eabj8207.
Zhang Q, Jeppesen DK, Higginbotham JN, Franklin JL, Coffey RJ. Comprehensive isolation of extracellular vesicles and nanoparticles. *Nature Protocols*. 2023;18(5):1462–1487.
Zhang W, Hou Y, Yin S, Miao Q, Lee K, Zhou X, et al. Advanced gene nanocarriers/scaffolds in nonviral-mediated delivery systems for tissue regeneration and repair. *Journal of Nanobiotechnology*. 2024a;22(1):376.
Zhang W, Liu S, Hou Y, Xu S, An J, Lee K, et al. Functional nanoplatform for modulating cellular forces to enhance antitumor immunity via mechanotransduction. *Journal of Controlled Release*. 2025;379:850–865.
Zhang Y, Xie Y, Hao Z, Zhou P, Wang P, Fang S, et al. Umbilical mesenchymal stem cell-derived exosome-encapsulated hydrogels accelerate bone repair by enhancing angiogenesis. *ACS Applied Materials & Interfaces*. 2021;13(16):18472–18487.
Zhang Y, Yi D, Hong Q, Cao J, Geng X, Liu J, et al. Platelet-rich plasma-derived exosomes boost mesenchymal stem cells to promote peripheral nerve regeneration. *Journal of Controlled Release*. 2024b;367:265–282.
Zhao Z, Wijerathne H, Godwin AK, Soper SA. Isolation and analysis methods of extracellular vesicles (EVs). *Extracellular Vesicles and Circulating Nucleic Acids*. 2021;2(1):80–103.
Zheng Y, Pan C, Xu P, Liu K. Hydrogel-mediated extracellular vesicles for enhanced wound healing: the latest progress, and their prospects for 3D bioprinting. *Journal of Nanobiotechnology*. 2024;22(1):57.
Zhu F, Nie G, Liu C. Engineered biomaterials in stem cell-based regenerative medicine. *Life Medicine*. 2023;2(4):Inad027.
Zhuang Y, Jiang S, Deng X, Lao A, Hua X, Xie Y, et al. Energy metabolism as a therapeutic target for aged wound repair by engineered extracellular vesicles. *Science Advances*. 2024;10(15):eadl0372.

8 Vesicle-Based Cell-Free Therapy in Liver Diseases

A Promising Therapeutic Frontier

Mansoor Bolideei[1], Rambod Barzigar[2], Sayan Paul[3], and Mohammad Javad Mehran[2]

[1]Department of Respiratory and Critical Care Medicine, the Center for Biomedical Research, NHC Key Laboratory of Respiratory Diseases, Tongji Hospital, Tongji Medical College, Huazhong University of Science and Technology, Wuhan, China

[2]Department of Biotechnology, JSS Research Foundation, SJCE Technical Campus, University of Mysore, Mysore, India

[3]Department of Biochemistry and Molecular Biology, The University of Texas Medical Branch at Galveston, Galveston, Texas, USA

8.1 LIST OF ABBREVIATIONS

3D	Three-dimensional
α-SMA	α-smooth muscle actin
AIH	Autoimmune hepatitis
ALF	Acute liver failure
ALT	Alanine aminotransferase
APCs	Antigen-presenting cells
AST	Aspartate aminotransferase
BM-MSC-Ex	Bone marrow MSC-derived exosomes
Cas9	CRISPR-associated protein 9
CRISPR	Clustered regularly interspaced short palindromic repeats
DAMPs	Damage-associated molecular patterns
DCs	Dendritic cells
ECM	Extracellular matrix
EVs	Extracellular vesicles
FXR	Farnesoid X receptor
GPI	Glycosylphosphatidylinositol
GSK3β	Glycogen synthase kinase 3-beta
HFD	High-fat diet
HGF	Hepatocyte growth factor
HIFs	Hypoxia-inducible factors
HSCs	Hepatic stellate cells
IFN-γ	Interferon gamma
IgG	Immunoglobulin G

DOI: 10.1201/9781003653387-8

IL-6	Interleukin-6
IRI	Ischemia–reperfusion injury
KC's	Kupffer cells
Lamp2B	Lysosome-associated membrane protein-2B
lncRNAs	Long non-coding RNAs
LPS	Lipopolysaccharide
MMPs	Matrix metalloproteinases
MPO	Myeloperoxidase
miRNAs	microRNAs
MSCs	Mesenchymal stem cells
MVEs	Multivesicular endosomes
MVs	Microvesicles
MPS	Mononuclear phagocyte system
MyD88	Myeloid differentiation primary response protein 88
NAFLD	Non-alcoholic fatty liver disease
NASH	Non-alcoholic steatohepatitis
NK	Natural killer
PCNA	Proliferating cell nuclear antigen
PRRs	Pattern recognition receptors
ROS	Reactive oxygen species
siRNAs	small interfering RNAs
TME	Tumor microenvironment
TNF-α	Tumor necrosis factor alpha
TLR4	Toll-like receptor 4
TGF-β	Transforming growth factor-beta
UCMSCs	Umbilical cord MSCs
UTR	Untranslated region

8.2 INTRODUCTION

Liver diseases represent a major global health challenge due to their high prevalence, significant morbidity, and associated mortality, which collectively impose a considerable economic burden (Devarbhavi et al., 2023). Various insults, including viral hepatitis, hepatotoxic drugs, and chronic alcohol consumption, can initiate acute or chronic liver injury, ultimately progressing to liver failure, cirrhosis, or hepatocellular carcinoma (Devarbhavi et al., 2023; Samant, Amiri, and Zibari 2021). Despite advances in clinical care, effective treatment options remain limited for many liver disorders, such as NAFLD and acute liver injury. In advanced cases of cirrhosis or liver failure, orthotopic liver transplantation remains the only definitive therapeutic option (Samant, Amiri, and Zibari 2021).

These limitations underscore the urgent need for novel and effective therapeutic strategies. Among emerging regenerative therapies, mesenchymal stromal cells (MSCs) have garnered considerable attention due to their ease of isolation and in vitro expansion, their multilineage differentiation potential, and their reparative paracrine effects, features that collectively position MSCs as near-ideal candidates for autologous cell-based therapies (Rahmani et al., 2023). Furthermore, allogenic MSCs can be cryopreserved with minimal functional loss, facilitating their off-the-shelf, ready-to-use availability at the point of care (Mahla 2016). Substantial preclinical evidence supports the therapeutic potential of MSCs in restoring damaged tissues across a range of organ systems, including experimental models of liver injury such as hepatic resection, acute liver failure, chemically induced hepatotoxicity, and cirrhosis. However, the mechanisms underlying the reparative effects of MSCs in liver pathology remain incompletely understood (Kouroupis et al., 2019; Parekkadan and Milwid, 2010).

Extracellular vesicles (EVs), which are membrane-bound particles secreted by nearly all cell types, including stem cells, play a crucial role in intercellular communication (Zhao et al., 2020). EVs are broadly categorized into exosomes, microvesicles, and apoptotic bodies, and mediate their biological effects through ligand-receptor interactions, membrane fusion, or endocytosis (Haider and Najimi, 2022). They serve as vehicles for transferring surface proteins, lipids, messenger RNAs, and microRNAs (miRNAs) to recipient cells (Zhao et al., 2020; Liu et al., 2022). Recent studies have shown that MSC-derived EVs can confer therapeutic benefits in multiple disease models (Yusuf and Haider, 2025; Najimi and Haider, 2024; Haider and Aramini, 2020). For instance, systemic delivery of MSC-derived extracellular vesicles (MSC-EVs) reduced infarct size in a rat model of acute myocardial infarction (Rostom et al., 2020). Similar protective effects have been observed in models of stroke, kidney injury, pulmonary inflammation, and skin wounds. In the context of liver injury, MSC-EVs have been shown to enhance hepatic regeneration and attenuate fibrosis and inflammation in models of drug-induced damage. These effects are primarily attributed to their anti-inflammatory and anti-fibrotic properties, supporting the potential of MSC-derived secretome and EVs as a novel cell-free therapeutic modality for a broad range of diseases (Rostom et al., 2020; Bruno et al., 2012; Haider and Aslam, 2018). MSCs from adipose tissue, bone marrow, umbilical cord, and muscle can secrete EVs, including exosomes and microvesicles, which have demonstrated therapeutic effects in preclinical models of liver, kidney, and cardiovascular diseases (Zhao et al., 2020; Bruno et al., 2012; Psaraki et al., 2022) (Figure 8.1).

8.3 MSC-DERIVED EXTRACELLULAR VESICLES

EVs encompass two major subtypes: microvesicles (MVs) and exosomes. Microvesicles are relatively large vesicles ranging from 100 to 1000 nm in diameter, generated through the outward budding of the plasma membrane (Haider and Najimi et al., 2022). These vesicles carry a diverse array of bioactive components, including proteins, lipids, and RNA molecules (Psaraki et al., 2022). In contrast, exosomes are smaller (typically 40–150 nm in diameter), bilayered lipid vesicles that play a pivotal role in intercellular communication and immunomodulation. Exosomes originate from the endosomal system, where they are formed through the inward budding of the limiting membrane of early endosomes, resulting in the formation of multivesicular endosomes (MVEs) (Ahmed et al., 2023).

These MVEs subsequently fuse with the plasma membrane, releasing exosomes into the extracellular space (Harrell et al., 2019). Exosomes are enriched with a broad and heterogeneous cargo, including proteins, lipids, messenger RNAs (mRNAs), microRNAs (miRNAs), small interfering RNAs (siRNAs), and long non-coding RNAs (lncRNAs) (Mateescu et al., 2017). Among these, non-coding RNAs (ncRNAs) constitute a relatively small but biologically significant portion of the EV cargo. The composition of ncRNAs can vary considerably between MVs and exosomes. Although some studies have reported that miRNAs may be present in relatively low quantities in EVs (Liu et al., 2021; Ferguson et al., 2018), numerous reports have documented the presence of miRNAs in EVs derived from various cell types, including MSCs, with their abundance ranging from minimal to substantial (Spinelli et al., 2018; Herwijnen et al., 2018). Notably, MSC-EVs are enriched with ncRNAs implicated in the regulation of key molecular pathways involved in liver disease pathogenesis. Importantly, MSC-derived exosomes exhibit an enhanced safety profile compared to their parent cells, as they are non-replicative, do not pose a risk of microvascular embolism, and retain their bioactivity during storage (Stenger et al., 2022; Xiong et al., 2017). These features render MSC-exosomes an attractive candidate for the development of cell-free therapies (Habib et al., 2025). Accordingly, ongoing research is actively exploring their therapeutic potential in diverse pathological conditions, including kidney injury, cardiovascular diseases, cutaneous wound healing, and liver failure (Psaraki et al., 2022).

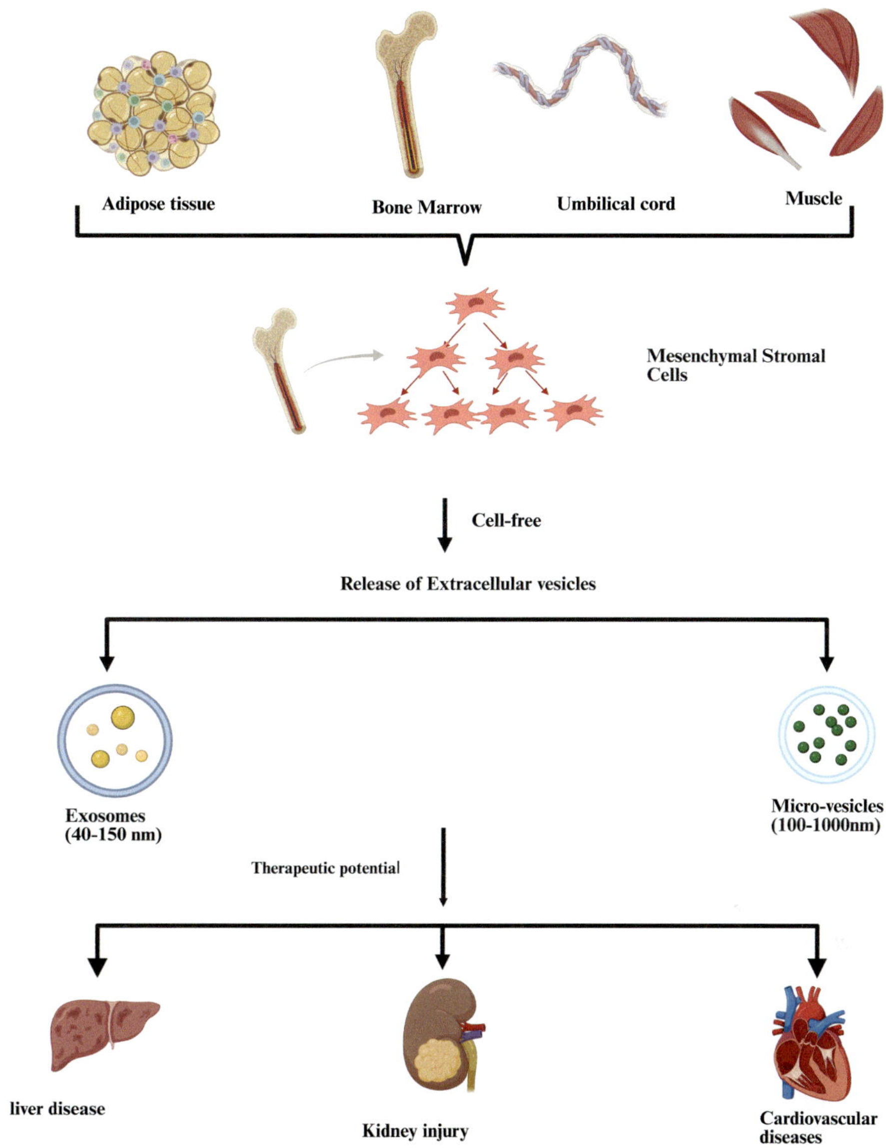

FIGURE 8.1 Generation of MSC-EV and therapeutic applications. Abbreviations: EV—extracellular vesicle; MSC—mesenchymal stromal cell.

8.4 APPLICATION OF MSC-EVS IN LIVER DISEASES

A growing body of research has demonstrated the therapeutic potential of mesenchymal stem cell-derived extracellular vesicles (MSC-EVs) in various pathological conditions, including pulmonary fibrosis, osteoporosis, dermatological disorders, neurological pathologies, oncology, cardiovascular diseases, and a broad spectrum of liver diseases (Haider and Aramini, 2020; Ahmed et al., 2025; Sajid et al., 2025; Yusuf and Haider, 2025; Najimi and Haider, 2022). Emerging evidence suggests that MSC-EVs can replicate many of the therapeutic benefits observed with MSC-based therapies (Lai et al., 2019). As a form of cell-free therapy, MSC-EVs retain the benefits of stem cell treatment while circumventing many of its inherent risks (Liang et al., 2025). Key benefits of MSC-EVs include (1) low immunogenicity, (2) ease of storage and handling (Rahmani et al., 2023), suitability as drug delivery vehicles with enhanced loading capacity, and (3) potential to serve as off-the-shelf therapeutics due to their amenability to cryopreservation (Zhang, Li et al., 2022). Although MSCs have been shown in vitro and in animal models to home to the tumor microenvironment (TME), supporting their potential utility in tumor-targeted therapies (Shojaei et al., 2019), clinical evidence for MSC recruitment in hepatocellular carcinoma remains limited. Additionally, the proliferative and multipotent nature of MSCs raises concerns regarding their potential to undergo malignant transformation or inadvertently promote tumor progression. As a result, most studies in this area remain confined to the preclinical stage (Shojaei et al., 2019; Zhang, Guo, et al., 2020). To mitigate these risks while leveraging their therapeutic potential, researchers have turned to MSC-derived EVs, which do not carry the same oncogenic concerns associated with other EVs. Recent investigations have revealed promising anti-tumor properties of MSC-EVs in liver cancer. For instance, exosomes derived from adipose tissue MSCs have been shown to enhance chemosensitivity in liver cancer cells by modulating the mTOR signaling pathway and upregulating genes associated with drug responsiveness. These effects were primarily attributed to the delivery of miR-199a into tumor cells (Lou et al., 2020; Jahangiri et al., 2022). Similarly, bone marrow MSC-derived exosomes (BM-MSC-Ex) enriched with miR-222-3p were found to suppress cancer cell proliferation and induce apoptosis (Zhang et al., 2020). Another study demonstrated that exosomal delivery of miR-302a from umbilical cord MSCs (UCMSC-Ex) inhibited cyclin D1 expression and downregulated the AKT signaling cascade, resulting in suppressed tumor growth (Li et al., 2019). Moreover, MSC-EVs have been explored as carriers for targeted drug delivery. They can be loaded with various anti-cancer agents and engineered to target tumor sites, bind to cancer cells, and release their therapeutic cargo (Bagheri et al., 2020; Pan et al., 2022). While these applications hold considerable promise, MSC-Ex–based interventions are still in the early stages of development, and further research is necessary to optimize their clinical utility. Notably, due to their large size, MSCs administered intravenously often become trapped in the pulmonary vasculature. In contrast, EVs—being substantially smaller—can more readily traverse biological barriers and reach target tissues, including injured liver parenchyma. MSC-EVs exert their therapeutic effects through three principal mechanisms, as evidenced by in vitro studies, preclinical models, and early clinical trials (1) stimulation of cellular proliferation and inhibition of apoptosis, (2) modulation of immune responses, and (3) attenuation of fibrosis (Hazrati et al., 2022) (Table 8.1).

8.4.1 Non-alcoholic Fatty Liver Disease

Non-Alcoholic Fatty Liver Disease (NAFLD) encompasses a spectrum of hepatic disorders ranging from simple steatosis (NAFL) to its progressive inflammatory form, non-alcoholic steatohepatitis (NASH). NAFLD has emerged as a global health concern, with an estimated prevalence of approximately 24%, contributing substantially to healthcare costs and the burden of chronic liver disease (Xiao et al., 2023; Younossi et al., 2018). The pathogenesis of NAFLD is intricately linked to persistent low-grade inflammation, driven by both hepatic and extrahepatic insults, with immune cells such as macrophages and neutrophils playing central roles (Schuster et al., 2018). Kupffer cells (KCs), the liver-resident macrophages, play a key role in mediating inflammation in NAFLD. They secrete proinflammatory cytokines and chemokines, including TNF, CCL2, and CCL5, which facilitate monocyte recruitment, hepatic steatosis, and fibrogenesis (Baeck et al., 2012). Furthermore, the accumulation of toxic lipids within KCs has been implicated in initiating early-stage hepatic inflammation (Leroux et al., 2012). Neutrophils, commonly observed in inflamed hepatic tissue, exert proinflammatory effects largely through myeloperoxidase (MPO) activity. Conversely, dendritic cells (DCs) appear to exert a protective role in the pathophysiology of NAFLD. Despite its high prevalence, therapeutic options for NAFLD remain limited, with lifestyle modifications—primarily weight loss—being the mainstay of management. While numerous pharmacological agents, such as farnesoid X receptor (FXR) agonists, are undergoing clinical trials, no drug has yet demonstrated definitive efficacy (Henning et al., 2013; Friedman et al., 2018).

TABLE 8.1
Mechanisms of MSC-EV Action in Liver Disease Models

Disease Model	Key Therapeutic Mechanisms	MSC-EV Cargo/Pathway	Preclinical Evidence	References
NAFLD/NASH	Anti-inflammation, macrophage modulation, apoptosis, and autophagy regulation.	miR-223, miR-96-5p, p-STAT3, HIFs	Reduced ALT/AST, steatosis, and fibrosis in the mouse.	(Zhao et al., 2018; El-Derany and AbdelHamid 2021; Chen et al., 2018; Pulido-Escribano et al., 2022)
Autoimmune hepatitis	Immune regulation, Treg/Th17 balance, cytokine modulation.	miR-223-3p, TGF-β, HGF	Decreased liver necrosis and transaminase levels; increased Treg cells.	(Khosrojerdi et al., 2023; Tamura, Uemoto, and Tabata, 2016)
Liver fibrosis	HSC inhibition, ECM degradation, and anti-inflammatory effects.	miR-150-5p, miR-486-5p, MMP9, MMP13	Reduced collagen deposition, suppressed HSC activation in vivo and in vitro.	(Li et al., 2013; Huang et al., 2018; Kim et al., 2021; Du et al., 2021)
Acute liver failure	Inhibition of proinflammatory macrophages, reduction of hepatocyte apoptosis, stimulation of regeneration.	miR-21, miR-199a, PCNA, cyclin D1	Improved survival and function in ALF models.	(Jiang et al., 2019; Eshghi et al., 2022; Tan et al., 2014; Li et al., 2020)
Ischemia–reperfusion injury	Immune modulation, Th17/Treg balance, suppression of inflammation.	miR-1246, CCT2, TGF-β1, IL-10	Lowered liver enzyme levels, reduced inflammatory cytokines, and decreased cell death.	(Zheng et al., 2020; Xie et al., 2019a, 2019b; Yang et al., 2020)

Consequently, interest is growing in alternative therapeutic strategies, including cell-free approaches. MSC-derived extracellular vesicles (MSC-EVs) have shown promising potential in attenuating hepatic inflammation and fibrosis. Preclinical studies have consistently reported that MSC-EVs can reduce serum levels of alanine aminotransferase (ALT) and aspartate aminotransferase (AST), mitigate liver fibrosis, and improve overall liver function in NAFLD models (Nickel et al., 2022; Watanabe et al., 2020). The immunomodulatory properties of MSC-EVs in NAFLD appear to be primarily mediated through interactions with macrophages. In NASH, lipopolysaccharide (LPS) activates KCs via the Toll-like receptor 4 (TLR4) pathway, initiating a cascade that promotes inflammation (Betrapally, Gillevet, and Bajaj 2017).

MSC-EVs have been shown to inhibit this cascade by suppressing M1 macrophage polarization and downregulating key proinflammatory cytokines, including TNF-α, IL-1β, and IL-6. Mechanistically, activation of the LPS/TLR4 pathway recruits myeloid differentiation primary response protein 88 (MyD88), leading to the activation of TNF receptor-associated factor 6 (TRAF6), which subsequently induces the phosphorylation of NF-κB subunits (p65 and IκB-α), promoting their nuclear translocation and transcriptional activation. MSC-EVs inhibited the phosphorylation of p65 and IκB-α, thereby reducing NF-κB activation. However, they did not impede NF-κB activity in cells overexpressing TRAF6, suggesting that MSC-EVs act at upstream points in the TLR4 signaling axis (Kataru et al., 2009; Ohara et al., 2018).

Additionally, MSC-EV treatment increased hepatic populations of $CD11b^+F4/80^+Ly6c$ low anti-inflammatory macrophages in NASH models (Nickel et al., 2022). These $F4/80^+$ macrophages exhibit high phagocytic activity, clearing apoptotic cells, microbial agents, and cellular debris, thereby promoting the resolution of inflammation (Kataru et al., 2009). Further, MSC-EVs contain phosphorylated STAT3 (p-STAT3), which binds to STAT3-responsive DNA elements and activates transcription of arginase-1 (ARG-1). This promotes the polarization of macrophages toward the M2 phenotype, characterized by the enhanced production of anti-inflammatory mediators, such as ARG-1 and IL-10 (Zhao et al., 2018). Beyond immunomodulation, MSC-EVs influence metabolic and cellular homeostasis in NAFLD. Studies have demonstrated their capacity to reduce hepatocyte apoptosis and enhance autophagic activity. El-Derany et al. reported that MSC-EVs reduce apoptosis in hepatocytes while promoting mitochondrial autophagy via miR-96-5p upregulation and caspase-2 suppression in a high-fat diet (HFD)-induced NASH model (El-Derany and AbdelHamid 2021; Heo and Kim 2022). Similarly, He et al. found that MSC-EVs activate the AMP-activated protein kinase (AMPK) signaling pathway in L-O2 hepatocytes, thereby enhancing autophagy and improving glucose and lipid metabolism (He et al., 2020).

8.4.2 Autoimmune Hepatitis

Autoimmune hepatitis (AIH) is a chronic, immune-mediated liver disorder characterized by persistent hepatic inflammation, the presence of circulating autoantibodies, and elevated serum immunoglobulin G (IgG) levels, particularly hypergammaglobulinemia (Schmeltzer and Russo 2018). Diagnosis relies primarily on liver biopsy, with hallmark histological features including interface hepatitis, hepatocyte rosette formation, and lymphocytic infiltration. Current first-line treatment consists of immunosuppressive therapy with corticosteroids, with or without azathioprine (Mack et al., 2020). In the absence of adequate control, AIH can rapidly progress to advanced fibrosis, cirrhosis, or hepatocellular carcinoma. The pathogenesis of AIH involves a breakdown of self-tolerance, triggered through mechanisms such as molecular mimicry, aberrant immune activation, and failure of regulatory T cell (Treg) function (Terziroli Beretta-Piccoli, Mieli-Vergani, and Vergani 2022). $CD4^+$ T cells are central to disease development. During immune responses to foreign antigens, structurally similar self-antigens—such as cytochrome P450 2D6 (CYP2D6)—may be mistakenly recognized, initiating autoreactive responses through molecular mimicry (Flikshteyn et al., 2024; Christen and Hintermann, 2018). These autoantigens are processed and presented by hepatic antigen-presenting cells (APCs), including dendritic cells (DCs), macrophages, and B cells, to naïve

$CD4^+$ T helper (Th0) cells. Under the influence of specific cytokine milieus, Th0 cells differentiate into Th1, Th2, and Th17 subsets, which subsequently secrete proinflammatory cytokines that sustain liver inflammation. Importantly, Tregs are functionally and numerically deficient in AIH, particularly during active disease phases (Liberal et al., 2015). This imbalance between effector T cells and Tregs disrupts immune tolerance and underpins AIH pathophysiology.

MSC-EVs have emerged as promising immunomodulatory agents capable of restoring hepatic immune homeostasis and mitigating inflammation in AIH. In a murine model of Concanavalin A (ConA)-induced AIH, Tamura et al. demonstrated that MSC-EV administration significantly reduced hepatic necrosis and serum ALT levels. These effects were associated with decreased expression of the proinflammatory cytokine IL-2 and elevated expression of anti-inflammatory mediators such as transforming growth factor-beta (TGF-β) and hepatocyte growth factor (HGF), in a dose-dependent manner. Furthermore, the number of Treg cells was markedly increased following treatment. The authors proposed that Kupffer cells (KCs) internalized MSC-EVs, subsequently releasing TGF-β, which facilitated the expansion of Treg cells and suppressed hepatic inflammation (Khosrojerdi et al., 2023; Tamura, Uemoto, and Tabata 2016).

In a separate study, Lu et al. reported that MSC-EVs reduced liver inflammation and serum ALT/AST levels in ConA-induced AIH mice, although no increase in Treg cell numbers was observed. Instead, the therapeutic effect was associated with an elevated Treg/Th17 ratio. The authors identified miR-223-3p in MSC-EVs as a key regulator, downregulating LPS-induced STAT3 and phosphorylated STAT3 (p-STAT3) expression in macrophages. STAT3 is a pivotal upstream modulator of IL-1β and IL-6, both of which drive Th17 differentiation. IL-6 induces naïve T cells to differentiate into Th17 cells, while IL-1β facilitates the conversion of Tregs into Th17 cells. Additionally, STAT3 directly activates RORγt, thereby enhancing IL-17 production. Thus, MSC-EV-mediated suppression of this axis shifted the immune balance toward a regulatory phenotype, alleviating AIH symptoms (Yang et al., 2017).

Additionally, miR-223 in MSC-EVs has been shown to suppress NLRP3 and caspase-1 expression in hepatocytes, thereby reducing pyroptosis, a form of programmed cell death that contributes to hepatic injury in autoimmune and inflammatory contexts (Chen et al., 2018).

8.4.3 Liver Fibrosis

Liver fibrosis is a common pathological consequence of most chronic liver diseases, including those caused by viral infections, NAFLD, and chronic alcohol consumption. It is characterized by the progressive accumulation of extracellular matrix (ECM) components, primarily cross-linked type I and type III collagen, leading to the replacement of functional parenchyma with fibrotic scar tissue (Friedman 2003). Myofibroblasts represent the principal cellular source of ECM during fibrogenesis. These cells can originate from hepatic stellate cells (HSCs) or portal fibroblasts that undergo activation and transdifferentiation in response to liver injury. Upon activation, HSCs express α-smooth muscle actin (α-SMA), secrete collagen type I, and migrate to injury sites where they contribute to fibrotic remodeling by depositing ECM components. Activation signals for HSCs include β-catenin and key profibrotic cytokines such as transforming growth factor-beta (TGF-β), both of which promote the transcription of fibrosis-associated genes (Iwaisako et al., 2014; Kim et al., 2024).

Macrophages exhibit a dual role in liver fibrosis. On one hand, they facilitate fibrogenesis by producing TGF-β and other cytokines that activate HSCs. On the other hand, macrophages can contribute to fibrosis resolution by secreting anti-inflammatory mediators and matrix-degrading enzymes, particularly matrix metalloproteinases (MMPs) like MMP-9 and MMP-12, which participate in the degradation and clearance of ECM (Ramachandran et al., 2012). Traditionally, M1-type macrophages are considered proinflammatory and pro-fibrogenic, while M2-type macrophages are viewed as anti-fibrotic due to their role in tissue remodeling and hepatocyte survival. However, recent evidence suggests that M1 macrophages may also participate in fibrosis resolution under certain conditions (Ma et al., 2017; Wang, Du, et al., 2023). Recent preclinical studies have increasingly highlighted the anti-fibrotic potential of mesenchymal stem cell-derived MSC-EVs across various organs, including the lungs, liver, kidneys, and heart.

In the context of liver fibrosis, MSC-EVs contribute to tissue repair by mitigating inflammation, thereby slowing its progression. These vesicles enhance the expression of matrix metalloproteinases (MMPs), such as MMP-9 and MMP-13, and anti-inflammatory cytokines, including TGF-β1 and IL-10, while simultaneously suppressing proinflammatory mediators, such as TNF-α, IL-1β, and IL-6. Additionally, MSC-EVs have been shown to attenuate carbon tetrachloride (CCl_4)-induced hepatic fibrosis by inhibiting the epithelial-to-mesenchymal transition (EMT) (Li et al., 2013; Huang et al., 2018). Furthermore, MSC-EVs have been shown to suppress hepatic stellate cell (HSC) activation. For instance, Rong et al. found that MSC-EVs modulate the Wnt/β-catenin signaling pathway to inhibit HSC activation both in vitro and in vivo (Rong et al., 2019). It has been reported that miR-486-5p within MSC-EVs binds to the 3′ untranslated region (UTR) of SMO mRNA, suppressing its expression and thereby deactivating HSCs (Kim et al., 2021). Similarly, Du et al. observed that miR-150-5p is underexpressed, while CXCL1 is overexpressed, in fibrotic liver tissue. However, MSC-EVs can deliver miR-150-5p to HSCs, thereby inhibiting their activation by downregulating CXCL1 expression (Du et al., 2021).

8.4.4 Acute Liver Failure

Acute liver failure (ALF) is a rapidly progressing clinical syndrome characterized by sudden and severe hepatocellular injury, leading to coagulopathy and hepatic encephalopathy in individuals without preexisting chronic liver disease (Stravitz and Lee 2019). The primary etiologies of ALF include viral hepatitis, drug-induced hepatotoxicity, excessive alcohol intake, and autoimmune liver disorders. Clinically, ALF presents with a spectrum of symptoms such as jaundice, coagulopathy, and neurological dysfunction. Although advances in intensive care and liver transplantation have improved survival rates, the need for effective alternative or adjunctive therapies remains urgent (Maiwall et al., 2024). Macrophages play a central role in the immunopathogenesis of ALF (Ni et al., 2024). Damage-associated molecular patterns (DAMPs), released from necrotic hepatocytes, interact with pattern recognition receptors (PRRs) on macrophages, triggering the release of proinflammatory cytokines and chemokines. This cascade activates inflammasomes, notably NLRP3, and drives the recruitment of neutrophils and monocyte-derived macrophages into the liver. In contrast, resident Kupffer cells (KCs) can exert a protective role by secreting IL-10 and clearing apoptotic cells and debris (You et al., 2013). Numerous studies have demonstrated the therapeutic potential of MSC-EVs in ALF. MSC-EVs have been shown to inhibit the activation and proliferation of proinflammatory macrophages, thereby reducing the secretion of key cytokines, including IL-1β, IL-6, IL-18, and TNF-α, which are critically involved in the pathophysiology of ALF (Jiang et al., 2019; Eshghi et al., 2022). These anti-inflammatory effects are thought to be mediated, at least in part, through suppression of the NLRP3 inflammasome signaling pathway (Jiang et al., 2019). Beyond their immunomodulatory functions, MSC-EVs contribute to hepatoprotection by attenuating hepatocyte apoptosis, a hallmark of early-stage ALF (Eshghi et al., 2022).

Several mechanisms have been proposed, including the induction of autophagy within hepatocytes, which enhances cell survival (Zhao, Liu, and Pu 2019) and mitigates oxidative stress-related injuries. Additionally, MSC-EVs play a key role in promoting liver regeneration. They have been reported to stimulate the dedifferentiation of hepatocytes into hepatic progenitor (oval) cells, capable of extensive proliferation (Chiabotto et al., 2025; Sitbon et al., 2024). Tan et al. observed increased expression of proliferative markers, such as proliferating cell nuclear antigen (PCNA) and cyclin D1, in MSC-EV-treated livers, suggesting that MSC-EVs enhance the priming phase of hepatocyte regeneration (Tan et al., 2014). Similarly, Damania et al. demonstrated that MSC-EV administration increased the residual liver-to-estimated liver weight ratio in murine models of liver failure induced by partial hepatectomy or ischemia–reperfusion injury (Damania et al., 2018).

8.4.5 Ischemia–Reperfusion Injury

Ischemia–reperfusion injury (IRI) is a significant contributor to hepatic damage during surgical procedures such as hepatectomy and liver transplantation. IRI encompasses both ischemic insult due to oxygen deprivation and subsequent inflammation-mediated injury upon reperfusion. The pathophysiological process involves a complex interplay of inflammatory responses, oxidative stress mediated by reactive oxygen species (ROS), hepatocyte apoptosis, and dysregulated autophagy (Peralta et al, 2013). Immune activation is central to IRI, involving Kupffer cells (KCs), dendritic cells (DCs), natural killer (NK) cells, and T lymphocytes, as well as a cascade of proinflammatory cytokines and inflammasome signaling pathways (Zhai, Busuttil, and Kupiec-Weglinski 2011). MSC-EVs have emerged as promising modulators of liver IRI. Zheng et al. demonstrated that CCT2, a component enriched in MSC-EVs, suppresses the expression of CD154 on $CD4^+$ T cells in the liver by regulating the Ca^{2+}–calcineurin–NFAT1 pathway (Zheng et al., 2020). Given that the CD154–CD40 interaction is a key driver of both innate and adaptive immune responses and is known to exacerbate liver IRI, this finding underscores the immunomodulatory potential of MSC-EVs (Shen et al., 2009). Additionally, MSC-EVs carrying miR-1246 have been shown to target glycogen synthase kinase 3β (GSK3β) in hepatocytes, thereby inhibiting Wnt1, Wnt3a, and β-catenin expression. This downregulation attenuates the Wnt/β-catenin signaling pathway, resulting in a significant reduction in IRI-associated inflammatory cytokines, including TNF-α, IL-6, and IL-1β (Xie et al., 2019a, 2019b). Furthermore, miR-1246 has been reported to upregulate anti-inflammatory cytokines, such as TGF-β1 and IL-10, while concurrently suppressing IL-6 and IL-17 levels (Xie et al., 2019a, 2019b). The cytokine milieu plays a crucial role in determining the fate of CD4+ T cell subsets. TGF-β alone favors the differentiation of naïve $CD4^+$ T cells into regulatory T cells (Tregs), whereas its combination with IL-6 promotes differentiation into proinflammatory Th17 cells via the IL-6–gp130–STAT3 axis. By modulating this balance, MSC-EVs enriched with miR-1246 help shift the immune response from a proinflammatory to a regulatory phenotype, thereby mitigating liver IRI. In addition to immune modulation, MSC-EVs have been shown to reduce hepatic infiltration by neutrophils and macrophages, thereby further limiting tissue damage (Piao et al., 2022; Nong et al., 2016). Their role in cellular stress responses has also been explored. Bo et al. reported that MSC-EVs attenuate hepatocyte apoptosis during IRI, a process likely mediated through enhanced autophagic activity (Yang et al., 2020).

8.5 ENGINEERED MSC-EVS AND THEIR APPLICATION AS A DRUG DELIVERY SYSTEM

Recent clinical research has increasingly explored the potential of engineered MSC-EVs as targeted delivery systems for therapeutic agents (Panda et al., 2021). Modifications can be made to either the EV cargo or membrane to enhance their drug-carrying and targeting capabilities. Cargo engineering of MSC-EVs can be achieved via two main strategies: passive and active loading. In passive loading, therapeutic compounds are incorporated into EVs post-isolation using techniques such as incubation under concentration gradients, electroporation, or lipofection—methods that exploit the physical and chemical properties of the therapeutic molecule relative to the vesicle size. Active loading, on the other hand, involves manipulating EV biogenesis by fusing therapeutic proteins or RNA molecules to endogenous EV-associated components, thereby incorporating the desired cargo during vesicle formation (Kim et al., 2016; Kim et al., 2020). Surface engineering further enhances the targeting specificity of MSC-EVs. This is primarily achieved by modifying surface proteins, such as lysosome-associated membrane protein-2B (Lamp2B) and glycosylphosphatidylinositol

(GPI)-anchored proteins, thereby enabling preferential interaction with receptors on target tissues or cells. Such engineering approaches enhance delivery efficiency and reduce nonspecific uptake by the mononuclear phagocyte system (MPS), thereby increasing therapeutic efficacy (Kooijmans et al., 2016; Yang et al., 2017). MSC-EVs offer notable advantages over traditional nanocarriers such as liposomes or synthetic polymeric nanoparticles. Their biocompatibility, intrinsic targeting ability, and minimal immunogenicity make them particularly attractive as natural nanocarriers (Lou et al., 2017). Engineered EVs have demonstrated promising outcomes in preclinical models of brain and bone tumors, as well as cardiac ischemia. While the application of engineered MSC-EVs for liver disease remains relatively underexplored, largely because unmodified MSC secretomes and native EVs have already demonstrated efficacy, this approach holds substantial promise. Rational design, surface engineering, and cargo modification of MSC-EVs could represent a next-generation strategy for the targeted treatment of both acute and chronic liver diseases, potentially enhancing therapeutic precision and minimizing off-target effects (Lou et al., 2017; Ciullo et al., 2019).

8.6 STRATEGIES TO ENHANCE THE THERAPEUTIC POTENTIAL OF MSC-EVS

MSC-EVs hold considerable promise in regenerative medicine and immune modulation. However, their clinical translation is currently hindered by limitations such as low production yield, limited targeting specificity, and variable therapeutic efficacy. To address these challenges, a range of bioengineering and culture optimization strategies is being explored. These include modifying MSC culture conditions, applying exogenous stimuli, genetically engineering EV cargo, and refining delivery platforms to enhance therapeutic outcomes (Zheng et al., 2023). One of the primary obstacles is the inherently low yield of EVs under conventional MSC culture conditions. The cellular microenvironment critically influences MSC proliferation, differentiation, and secretory profiles, thereby shaping the quality and therapeutic capacity of their EVs (Powsner et al., 2025). Mimicking physiological conditions—such as through hypoxia or three-dimensional (3D) culture systems—has been shown to enhance both the quantity and bioactivity of MSC-EVs. Furthermore, exogenous stimulation or genetic modifications can enhance the release of therapeutic factors, thereby augmenting the regenerative and immunomodulatory potential of MSC-EVs (Powsner et al., 2025; Zheng et al., 2023). Hypoxic preconditioning, in particular, has emerged as a potent method to enhance MSC-EV production. Culturing MSCs under low-oxygen conditions (2–5% O_2) mimics their native niche, leading to increased EV secretion and enrichment of therapeutic proteins. Hypoxia activates hypoxia-inducible factors (HIFs), which regulate key pathways involved in inflammation, cell proliferation, and tissue repair (Haider and Ashraf, 2012). Hypoxia-conditioned MSC-EVs have demonstrated superior regenerative effects in preclinical models of liver injury, promoting tissue repair and modulating immune responses more effectively than normoxic MSC-EVs (Pulido-Escribano et al., 2022).

Three-dimensional culture systems further amplify EV yield and functionality by replicating in vivo cellular architecture. Unlike conventional two-dimensional (2D) monolayer cultures, 3D scaffolds or bioreactors provide physiologically relevant mechanical and biochemical cues, enhancing cell–cell and cell–matrix interactions. This results in the production of EVs with enriched miRNA and protein cargo, improved therapeutic potency, and sustained support for hepatocyte regeneration, fibrosis attenuation, and maintenance of stemness (Haraszti et al., 2018). Exogenous stimulation, through exposure to inflammatory cytokines (e.g., TNF-α, IL-6, IFN-γ), pharmacologic agents, or bioactive compounds, can further enhance MSC-EV function. Preconditioning MSCs with such factors increases the anti-inflammatory, anti-fibrotic, and hepatoprotective properties of their EVs. For example, hydrogen sulfide donors and certain traditional Chinese medicines have been reported to augment EV-mediated liver regeneration. However, these approaches require precise optimization to preserve MSC viability and prevent unwanted alterations in EV composition (Wang, Yao, et al., 2023; Zheng et al., 2023).

Genetic engineering offers a powerful means to tailor the content of MSC-EVs. By introducing genetic modifications into parental MSCs through plasmid transfection, viral vectors, or electroporation, researchers can induce the overexpression of specific therapeutic miRNAs or proteins (Haider, 2024). This enables the customization of EVs to enhance their anti-inflammatory, anti-apoptotic, and tissue-repairing functions. Genetically engineered MSCs may also replace the need for exogenous stimulation by directly activating key intracellular pathways involved in EV biogenesis and function (Phinney and Pittenger 2017). Efforts are also directed at optimizing in vivo biodistribution and targeting. Enhancing MSC-EVs' ability to reach specific organs, such as the liver, improves therapeutic efficacy while minimizing off-target effects. Strategies include surface modification to evade immune clearance, encapsulation in biomaterials to extend circulation time, and the use of targeting ligands to direct EVs to specific cell types (Phinney and Pittenger 2017). To overcome rapid uptake by the mononuclear phagocyte system following systemic administration, researchers are developing targeted delivery systems to enhance the therapeutic efficacy of these compounds. These involve engineering the surface of MSC-EVs, either by modifying the parent MSCs or directly altering isolated EVs, to increase their affinity for diseased tissues. Surface functionalization with peptides, antibodies, or receptor-specific ligands facilitates homing, adhesion, and uptake by target cells, thereby enhancing the therapeutic index of MSC-EV-based interventions (Li et al., 2020; Mardpour et al., 2019). In summary, integrating advanced culture methodologies, exogenous priming, genetic engineering, and targeted delivery approaches offers a promising path to unlock the full therapeutic potential of MSC-EVs. These strategies aim to enhance EV yield, specificity, efficacy, and safety, thereby accelerating their clinical application in the treatment of complex liver disorders, including acute injury and chronic fibrosis (Zheng et al., 2023; Wang et al., 2024).

Multiple strategies are being investigated to overcome the limitations of MSC-EV therapy, including low production yield, restricted tissue targeting, and variable therapeutic outcomes. Key approaches include optimizing MSC culture conditions, bioengineering EV content, and refining administration methods (Figure 8.2).

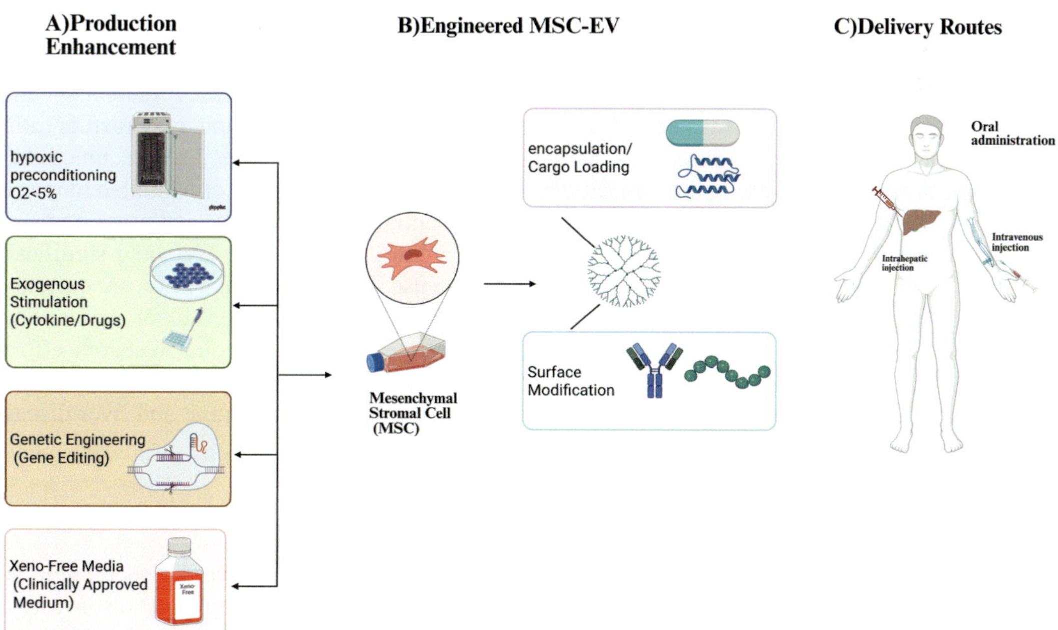

FIGURE 8.2 Strategies to improve the therapeutic efficacy of MSC-EV in treating liver disease. Abbreviations: MSC—mesenchymal stromal cell; EV—extracellular vesicle.

8.7 LIMITATIONS IN THE USE OF MSC-DERIVED EVS

MSC-EVs have garnered substantial attention for their regenerative and immunomodulatory properties. However, despite their promising therapeutic profile, several critical limitations continue to hinder their widespread clinical application. One of the foremost challenges is the intrinsic heterogeneity of MSC-EVs, stemming from multiple factors, including the tissue origin of MSCs, inter-donor variability, and even heterogeneity within cells from a single donor. This variability significantly influences the molecular composition and functional potency of the resulting EVs, complicating reproducibility and consistency in therapeutic outcomes. Contemporary isolation techniques and available optimized protocols, such as ultracentrifugation, generally fail to selectively separate EV subpopulations, limiting our ability to obtain homogeneous, functionally well-defined vesicle preparations. Although recent advances in high-resolution flow cytometry, nanoparticle tracking analysis, and Raman spectroscopy have improved EV characterization, no universally accepted method has yet resolved the issue of heterogeneity (Yang et al., 2025).

A significant challenge with MSC-derived EVs is their short biological half-life, typically ranging from a few minutes to less than an hour. Rapid clearance by the mononuclear phagocyte system significantly limits their bioavailability and narrows the therapeutic window. However, strategies such as encapsulating EVs in biocompatible hydrogels have shown potential for prolonging EV stability and modulating release kinetics. These hydrogels facilitate sustained delivery and enhance regenerative potential, as demonstrated by preclinical data from experimental models of chronic liver injury. However, optimizing controlled-release systems and ensuring efficient targeting remain active areas of investigation (Mei et al., 2024; Yoo et al., 2025). Standardization and quality control also pose significant challenges to the clinical translation of MSC-EV therapies. The efficacy, safety, and consistency of MSC-EV preparations are influenced by numerous variables, including donor characteristics (e.g., age, sex, health status), tissue source, culture conditions, passage number, and collection protocols (Safwan et al., 2024). Currently, there is no consensus on defining the biological potency or identity of MSC-EVs, resulting in substantial batch-to-batch variability and limiting the scalability of clinical-grade manufacturing processes (Mei et al., 2024; Shams et al., 2023).

Lastly, the clinical safety and efficacy profiles of MSC-EVs remain incompletely defined. Although early-phase clinical studies have reported a favorable safety profile with minimal adverse events, critical parameters—such as optimal dosing, administration frequency, and long-term safety—remain poorly established, particularly in the context of liver diseases. Patient-specific factors, including immunological status, comorbidities, and genetic background, may further influence therapeutic outcomes. Addressing these uncertainties requires comprehensive preclinical studies and rigorously designed clinical trials. Additionally, the development of validated biomarkers to monitor therapeutic efficacy and safety will be essential for advancing regulatory approval and clinical adoption (Huai et al., 2023). In summary, while MSC-derived EVs hold significant therapeutic potential, their clinical translation and routine use as a therapeutic option are currently hindered by multiple factors, including biological variability, rapid clearance, lack of standardization, and limited clinical data. Overcoming these challenges through ongoing research efforts in EV engineering, scalable production, delivery strategies, and rigorous clinical validation will be crucial for realizing the full potential of MSC-EVs in regenerative medicine and liver disease therapy (Table 8.2).

8.8 FUTURE AND PERSPECTIVES

Despite significant advancements in understanding and harnessing MSC-derived secretome and EVs for therapeutic applications in liver diseases, several critical areas require further investigation to ensure their successful clinical translation (Varderidou-Minasian and Lorenowicz 2020; Xia et al., 2021). Addressing these areas will be crucial to unlocking the full therapeutic potential of MSC-EVs. First, enhancing the consistency and reproducibility of MSC-EVs remains a paramount

TABLE 8.2
Current Limitations and Future Strategies for MSC-EV-Based Liver Therapy

Limitations	Impact on Therapy	Representative Solution(s)	Supporting Evidence/ Approach	References
Heterogeneity of MSC-EVs	Inconsistent efficacy	Standardized donor selection; advanced isolation (flow cytometry, Raman spectroscopy).	Enhanced characterization and consistency	(Yoo et al., 2025)
Short in vivo half-life	Reduced bioavailability	Encapsulation in hydrogels; surface engineering.	Prolonged release, improved targeting	(Yoo et al., 2025; Shams et al., 2023)
Lack of standardization	Batch variability, regulatory hurdles	Defined protocols for culture, isolation, and potency assessment.	Reproducible, scalable manufacturing	(Huai et al., 2023)
Incomplete safety/efficacy data	Uncertainty in clinical application	Comprehensive preclinical and clinical trials.	Safety and efficacy profiling	(Varderidou-Minasian and Lorenowicz 2020)
Limited targeting	Off-target effects, lower efficiency	Surface modification with ligands, peptides, or biomaterials.	Increased organ/tissue specificity	(Mardpour et al., 2019; Wang et al., 2024; Yang et al., 2025)
Low yield	Insufficient for clinical use	Hypoxic/3D culture, exogenous stimulation, genetic engineering.	Greater EV output, enriched cargo	(Powsner et al., 2025; Pulido-Escribano et al., 2022; Haraszti et al., 2018; Wang, Yao, et al., 2023)

priority in the field. It is vital to develop innovative isolation techniques that can effectively distinguish among EV subpopulations by size, density, or molecular composition. To achieve this, we should delve deeper into advanced methods, such as microfluidics, high-resolution flow cytometry, and affinity-based purification techniques, to fully exploit their potential. Furthermore, standardizing MSC-EV production through optimized protocols for donor selection, MSC cultivation, EV extraction, characterization, and storage is essential for enhancing batch-to-batch consistency and ensuring predictable therapeutic outcomes (Yáñez-Mó et al., 2015; Théry et al., 2018). Second, strategies to prolong the bioavailability and target the delivery of MSC-EVs must be optimized. Enhanced bioengineering approaches, such as hydrogels, biomaterials, and surface-functionalization techniques, can extend the in vivo half-life of MSC-EVs, facilitating sustained therapeutic effects. Further refinement of targeting strategies, including ligand-based surface modifications or engineering MSCs for targeted EV secretion, can enhance specificity and reduce off-target accumulation, thereby improving safety and therapeutic efficacy (Esmaeili et al., 2022).

Third, a comprehensive exploration of the mechanisms of action underlying MSC-EV-mediated therapeutic effects is vital. Understanding the precise molecular pathways by which MSC-EVs exert anti-inflammatory, anti-apoptotic, regenerative, and immunomodulatory effects—particularly in various forms of liver injury—will facilitate targeted therapeutic interventions (Lou et al., 2017; Maji et al., 2017). Advances in multi-omics technologies, such as proteomics, transcriptomics, and metabolomics, provide powerful tools for identifying key bioactive components and pathways involved, thereby enabling more tailored therapeutic strategies (Governini et al., 2025).

Fourth, enhancing the therapeutic potency of MSC-EVs via genetic and biochemical engineering represents a promising frontier. Gene-editing technologies, such as CRISPR/Cas9 or RNA interference, can enable precise control over the composition of EV cargo, thereby increasing their effectiveness in treating liver diseases (Bolideei et al., 2025). Moreover, the exploration of combinational approaches, such as integrating MSC-EVs with existing therapeutic agents or leveraging synergistic effects with pharmacological interventions, may significantly amplify clinical efficacy (Ramasubramanian, Kumar, and Wang 2019). Lastly, robust, carefully designed clinical trials are indispensable for assessing the safety, dosage regimens, therapeutic efficacy, and long-term outcomes of MSC-EVs in liver diseases. Developing standardized biomarkers for real-time monitoring of therapeutic responses will enhance clinical evaluation and treatment outcomes. Collaborative research efforts involving academic institutions, industry partners, and regulatory agencies will be crucial in accelerating clinical translation, ensuring compliance with regulatory requirements, and ultimately facilitating broader patient access to MSC-EV-based therapies (Claridge et al., 2021; Lou et al., 2017).

8.9 CONCLUSION

MSC-derived EVs have emerged as a promising cell-free therapeutic modality for a broad spectrum of liver diseases, including NAFLD, autoimmune hepatitis, liver fibrosis, acute liver failure, and ischemia–reperfusion injury. Their therapeutic benefits stem from potent immunomodulatory properties, anti-inflammatory effects, attenuation of fibrosis, reduction of hepatocyte apoptosis, and promotion of hepatic regeneration. Additionally, engineered MSC-EVs provide an innovative platform for targeted drug delivery, offering considerable advantages over conventional therapies. Nevertheless, several challenges remain that need to be addressed before MSC-EVs can achieve widespread clinical implementation. These challenges include biological heterogeneity, limited in vivo stability, insufficient standardization of production methods, and an incomplete understanding of their clinical safety and efficacy profiles. To address these challenges, it is essential to implement advanced isolation technologies, bioengineering strategies, and optimized delivery systems, with rigorous clinical validation being critical.

Moving forward, multidisciplinary efforts are needed to refine MSC-EV-based therapies, enhancing their specificity, efficacy, safety, and scalability. Continued research in bioengineering, omics analyses, targeted delivery, and clinical trials will not only improve our understanding but also accelerate the translation of MSC-EVs from the bench to the bedside, transforming the therapeutic landscape for liver diseases and potentially extending benefits to other complex medical conditions.

REFERENCES

Ahmed OTF, Ahmed ZT, Dairi AW, Zain Al-Abeden MS, Alkahlot MH, Alkahlot RH, Al Jowf GI, et al., The inconclusive superiority debate of allogeneic versus autologous MSCs in treating patients with HFrEF: a systematic review and meta-analysis of RCTs. *Stem Cell Research & Therapy*. 2025 Apr 12;16(1):175. https//doi.org/10.1186/s13287-025-04209-5.

Ahmed ZT, Alkahlot MH, Haider KH. MSC-derived exosomes: advances in cell-free therapy. In: KH Haider, editor. *Handbook of Stem Cell Applications*. Singapore: Springer; 2023. https//doi.org/10.1007/978-981-99-0846-2_60-1

Baeck C, Wehr A, Karlmark KR, Heymann F, Vucur M, Gassler N, et al., Pharmacological inhibition of the chemokine CCL2 (MCP-1) diminishes liver macrophage infiltration and steatohepatitis in chronic hepatic injury. *Gut*. 2012;61(3):416–26.

Bagheri E, Abnous K, Farzad SA, Taghdisi SM, Ramezani M, Alibolandi M. Targeted doxorubicin-loaded mesenchymal stem cells-derived exosomes as a versatile platform for fighting against colorectal cancer. *Life Sciences*. 2020;261:118369.

Betrapally NS, Gillevet PM, Bajaj JS. Gut microbiome and liver disease. *Translational Research*. 2017;17949–59.

Bolideei M, Barzigar R, Gahrouei RB, Mohebbi E, Haider KH, Paul S, Paul MK, et al., Applications of gene editing and nanotechnology in stem cell-based therapies for human diseases. *Stem Cell Reviews and Reports*. 2025 Feb 27. doi 10.1007/s12015-025-10857-0.

Bruno S, Grange C, Collino F, Deregibus MC, Cantaluppi V, Biancone L, et al., Microvesicles derived from mesenchymal stem cells enhance survival in a lethal model of acute kidney injury. *PLoS One*. 2012;7(3):33115.

Chen L, Lu FB, Chen DZ, Wu JL, Hu ED, Xu LM, et al., BMSCs-derived miR-223-containing exosomes contribute to liver protection in experimental autoimmune hepatitis. *Molecular Immunology*. 2018;9338–46.

Chiabotto G, Semnani A, Ceccotti E, Bruno S. Extracellular vesicles: emerging therapeutic agents for liver fibrosis. *Extracellular Vesicles and Circulating Nucleic Acids*. 2025;6(2):216–44.

Christen U, Hintermann E. Can autoantibodies in autoimmune hepatitis reveal epitopes that tell us about the etiology of the disease? *Frontiers in Immunology*. 2018;9:163.

Ciullo A, Biemmi V, Milano G, Bolis S, Cervio E, Fertig ET, et al., Exosomal expression of CXCR4 targets cardioprotective vesicles to the myocardium after myocardial infarction and improves outcome after systemic administration. *International Journal of Molecular Sciences*. 2019;20(3):468.

Claridge B, Lozano J, Poh QH, Greening DW. Development of extracellular vesicle therapeutics challenges, considerations, and opportunities. *Frontiers in Cell and Developmental Biology*. 2021;9:734720.

Damania A, Jaiman D, Teotia AK, Kumar A. Mesenchymal stromal cell-derived exosome-rich fractionated secretome confers a hepatoprotective effect in liver injury. *Stem Cell Research & Therapy*. 2018;9:1–12.

Devarbhavi H, Asrani SK, Arab JP, Nartey YA, Pose E, Kamath PS. Global burden of liver disease 2023 update. *Journal of Hepatology*. 2023;79(2):516–37.

Du Z, Wu T, Liu L, Luo B, Wei C. Extracellular vesicle-derived miR-150-5p secreted by adipose-derived mesenchymal stem cells inhibits CXCL1 expression to attenuate hepatic fibrosis. *Journal of Cellular and Molecular Medicine*. 2021;25(2):701–15.

El-Derany MO, AbdelHamid SG. Upregulation of miR-96-5p by bone marrow mesenchymal stem cells and their exosomes alleviates non-alcoholic steatohepatitis, with a focus on inhibiting caspase-2 signaling. *Biochemical Pharmacology*. 2021;190:114624.

Eshghi F, Tahmasebi S, Alimohammadi M, Soudi S, Khaligh SG, Khosrojerdi A, et al., Study of the immunomodulatory effects of mesenchymal stem cell-derived exosomes in a mouse model of LPS-induced systemic inflammation. *Life Sciences*. 2022;310:120938.

Esmaeili A, Alini M, Baghaban Eslaminejad M, Hosseini S. Engineering strategies for customizing extracellular vesicle uptake in a therapeutic context. *Stem Cell Research & Therapy*. 2022;13(1):129.

Ferguson SW, Wang J, Lee CJ, Liu M, Neelamegham S, Canty JM, et al., The microRNA regulatory landscape of MSC-derived exosomes: a systems view. *Scientific Reports*. 2018;8(1):1419.

Flikshteyn B, Amer K, Tafesh Z, Pyrsopoulos NT. Diagnosis of autoimmune hepatitis. *Clinics in Liver Disease*. 2024;28(1):37–50.

Friedman SL, Neuschwander-Tetri BA, Rinella M, Sanyal AJ. Mechanisms of NAFLD development and therapeutic strategies. *Nature Medicine*. 2018;24(7):908–22.

Friedman SL. Liver fibrosis–from bench to bedside. *Journal of Hepatology*. 2003;3838–53.

Governini L, Haxhiu A, Shaba E, Vantaggiato L, Mori A, Bruttini M, et al., Unraveling the multi-omic landscape of extracellular vesicles in human seminal plasma. *Biomolecules*. 2025;15(6):836.

Habib SM, Martini MF, Al Ahmed O, Qassab AQ, Haider KH. The potential of mesenchymal stem cell–derived exosomes in cell-free therapy for myocardial repair. In: KH Haider, editor. *Handbook of Regenerative Medicine: Cell-Free Therapy*. Boca Raton (FL): CRC Press; 2025. p. 372–390.

Haider KH, Aramini B. Micrining the injured heart with stem cell-derived exosomes: an emerging strategy of cell-free therapy. *Stem Cell Research & Therapy*. 2020;11(1):23. https://doi.org/10.1186/s13287-019-1548-7.

Haider KH, Ashraf M. A preconditioning approach in stem cell therapy for the treatment of infarcted hearts. *Progress in Molecular Biology and Translational Science*. 2012 Jan 1;111323–56.

Haider KH, Najimi M. Exosome-based cell-free therapy in regenerative medicine for myocardial repair. In: KH Haider, editor. *Handbook of Stem Cell Therapy*. Singapore: Springer; 2022. https//doi.org/10.1007/978-981-16-6016-0_9-1.

Haider KH. Priming mesenchymal stem cells to develop "super stem cells". *World Journal of Stem Cells* 2024;16(6): 623–640. https://doi.org/10.4252/wjsc.v16.i6.623.

Haraszti RA, Miller R, Stoppato M, Sere YY, Coles A, Didiot MC, et al., Exosomes produced from 3D MSC cultures by tangential flow filtration exhibit higher yields and improved activity. *Molecular Therapy*. 2018;26(12):2838–47.

Harrell CR, Jovicic N, Djonov V, Arsenijevic N, Volarevic V. Mesenchymal stem cell-derived exosomes and other extracellular vesicles as new remedies in the therapy of inflammatory diseases. *Cells*. 2019;8(12):1605.

Hazrati A, Malekpour K, Soudi S, Hashemi SM. Mesenchymal stromal/stem cells and their extracellular vesicles in the application of acute and chronic inflammatory liver diseases. Emphasizing the anti-fibrotic and immunomodulatory mechanisms. *Frontiers in Immunology*. 2022;13:865888.

He Q, Wang L, Zhao R, Yan F, Sha S, Cui C, et al., Mesenchymal stem cell-derived exosomes exert ameliorative effects in type 2 diabetes by improving hepatic glucose and lipid metabolism via enhancing autophagy. *Stem Cell Research & Therapy*. 2020;11:1–14.

Henning JR, Graffeo CS, Rehman A, Fallon NC, Zambirinis CP, Ochi A, et al., Dendritic cells limit fibroinflammatory injury in nonalcoholic steatohepatitis in mice. *Hepatology*. 2013;58(2):589–602.

Heo JS, Kim S. Human adipose mesenchymal stem cells modulate inflammation and angiogenesis through exosomes. *Scientific Reports*. 2022;12(1):2776.

Herwijnen MJV, Driedonks TA, Snoek BL, Kroon AT, Kleinjan M, Jorritsma R, et al., Abundantly present miRNAs in milk-derived extracellular vesicles are conserved between mammals. *Frontiers in Nutrition*. 2018;5:81.

Huai Q, Zhu C, Zhang X, Dai H, Li X, Wang H. Mesenchymal stromal/stem cells and their extracellular vesicles in liver diseases: insights on their immunomodulatory roles and clinical applications. *Cell & Bioscience*. 2023;13(1):162.

Huang B, Lu J, Ding C, Zou Q, Wang W, Li H. Exosomes derived from human adipose mesenchymal stem cells improve ovarian function of premature ovarian insufficiency by targeting SMAD. *Stem Cell Research & Therapy*. 2018;9:1–12.

Iwaisako K, Jiang C, Zhang M, Cong M, Moore-Morris TJ, Park TJ, et al., Origin of myofibroblasts in the fibrotic liver in mice. *Proceedings of the National Academy of Sciences*. 2014;111(32):3297–305.

Jahangiri B, Khalaj-Kondori M, Asadollahi E, Dizaj LP, Sadeghizadeh M. MSC-Derived exosomes suppress colorectal cancer cell proliferation and metastasis via miR-100/mTOR/miR-143 pathway. *International Journal of Pharmaceutics*. 2022;627:122214.

Jiang L, Zhang S, Hu H, Yang J, Wang X, Ma Y, et al., Exosomes derived from human umbilical cord mesenchymal stem cells alleviate acute liver failure by reducing the activity of the NLRP3 inflammasome in macrophages. *Biochemical and Biophysical Research Communications*. 2019;508(3):735–41.

Kataru RP, Jung K, Jang C, Yang H, Schwendener RA, Baik JE, et al., Critical role of CD11b+ macrophages and VEGF in inflammatory lymphangiogenesis, antigen clearance, and inflammation resolution. *Blood, The Journal of the American Society of Hematology*. 2009;113(22):5650–9.

Khosrojerdi A, Soudi S, Hosseini AZ, Khaligh SG, Hashemi SM. The combination of mesenchymal stem cell-and hepatocyte-derived exosomes, along with imipenem, ameliorates inflammatory responses and liver damage in a sepsis mouse model. *Life Sciences*. 2023;326:121813.

Kim H, Kim EH, Kwak G, Chi SG, Kim SH, Yang Y. Exosomes: cell-derived nanoplatforms for the delivery of cancer therapeutics. *International Journal of Molecular Sciences*. 2020;22(1):14.

Kim HY, Sakane S, Eguileor A, Weber RCG, Lee W, Liu X, et al., The origin and fate of liver myofibroblasts. *Cellular and Molecular Gastroenterology and Hepatology*. 2024;17(1):93–106.

Kim J, Lee C, Shin Y, Wang S, Han J, Kim M, et al., sEVs from tonsil-derived mesenchymal stromal cells alleviate activation of hepatic stellate cells and liver fibrosis through miR-486-5p. *Molecular Therapy*. 2021;29(4):1471–86.

Kim MS, Haney MJ, Zhao Y, Mahajan V, Deygen I, Klyachko NL, et al., Development of exosome-encapsulated paclitaxel to overcome MDR in cancer cells. *Nanomedicine Nanotechnology, Biology and Medicine*. 2016;12(3):655–64.

Kooijmans SA, Aleza CG, Roffler SR, Solinge WW, Vader P, Schiffelers RM. Display of GPI-anchored anti-EGFR nanobodies on extracellular vesicles promotes tumour cell targeting. *Journal of Extracellular Vesicles*. 2016;5(1):31053.

Kouroupis D, Sanjurjo-Rodriguez C, Jones E, Correa D. Mesenchymal stem cell functionalization for enhanced therapeutic applications. Tissue Engineering Part B Reviews. 2019;25(1):55–77.

Lai P, Weng J, Guo L, Chen X, Du X. Novel insights into MSC-EVs therapy for immune diseases. *Biomarker Research*. 2019;7:1–10.

Leroux A, Ferrere G, Godie V, Cailleux F, Renoud ML, Gaudin F, et al., Toxic lipids stored by Kupffer cells correlate with their pro-inflammatory phenotype at an early stage of steatohepatitis. *Journal of Hepatology*. 2012;57(1):141–9.

Li T, Yan Y, Wang B, Qian H, Zhang X, Shen L, et al., Exosomes derived from human umbilical cord mesenchymal stem cells alleviate liver fibrosis. Stem Cells and Development. 2013;22(6):845–54.

Li X, Liu LL, Yao JL, Wang K, Ai H. Human umbilical cord mesenchymal stem cell-derived extracellular vesicles inhibit endometrial cancer cell proliferation and migration through delivery of exogenous miR-302a. *Stem Cells International* 2019(1):810:8576.

Li Z, Zhou Y, Zhang L, Jia K, Wang S, Wang M, et al., microRNA-199a-3p inhibits hepatic apoptosis and hepatocarcinogenesis by targeting PDCD4. *Oncogenesis*. 2020;9(10):95.

Liang Y, Xia J, Xu L, Zhang Y, Xu X, Haider KH. Cell-free and cell-based therapy in regenerative medicine. In: KH Haider, editor. *Handbook of Regenerative Medicine: Cell-Free Therapy Approach*. Boca Raton (FL): CRC Press; 2025, Chapter 24. doi.org/10.1201/9781003611509

Liberal R, Grant CR, Holder BS, Cardone J, Martinez-Llordella M, Ma Y, et al., In autoimmune hepatitis type 1 or the autoimmune hepatitis–sclerosing cholangitis variant, defective regulatory T-cell responsiveness to IL-2 results in low IL-10 production and impaired suppression. *Hepatology*. 2015;62(3):863–75.

Liu H, Chen Y, Yin G, Xie Q. Therapeutic prospects of MicroRNAs carried by mesenchymal stem cells-derived extracellular vesicles in autoimmune diseases. *Life Sciences*. 2021;277:119458.

Liu S, Wu X, Chandra S, Lyon C, Ning B, Fan J, et al., Extracellular vesicles: emerging tools as therapeutic agent carriers. *Acta Pharmaceutica Sinica B*. 2022;12(10):3822–42.

Lou G, Chen L, Xia C, Wang W, Qi J, Li A, et al., MiR-199a-modified exosomes from adipose tissue-derived mesenchymal stem cells improve hepatocellular carcinoma chemosensitivity through the mTOR pathway. *Journal of Experimental & Clinical Cancer Research*. 2017:39:1–9.

Lou G, Chen Z, Zheng M, Liu Y. Mesenchymal stem cell-derived exosomes as a new therapeutic strategy for liver diseases. *Experimental & Molecular Medicine*. 2020;49(6):e346.

Ma PF, Gao CC, Yi J, Zhao JL, Liang SQ, Zhao Y, et al., Cytotherapy with M1-polarized macrophages ameliorates liver fibrosis by modulating the immune microenvironment in mice. *Journal of Hepatology*. 2017;67(4):770–9.

Mack CL, Adams D, Assis DN, Kerkar N, Manns MP, Mayo MJ, et al., Diagnosis and Management of Autoimmune Hepatitis in Adults and Children: 2019 Practice Guidance and Guidelines from the American Association for the Study of Liver Diseases. *Hepatology*. 2020;72(2):671–722.

Mahla RS. Stem cells applications in regenerative medicine and disease therapeutics. *International Journal of Cell Biology*. 2016(1):6940283.

Maiwall R, Kulkarni AV, Arab JP, Piano S. Acute liver failure. *The Lancet*. 2024;404(10454):789–802.

Maji S, Matsuda A, Yan IK, Parasramka M, Patel T. Extracellular vesicles in liver diseases. *American Journal of Physiology–Gastrointestinal and Liver Physiology*. 2017;312(3):G194–G200.

Mardpour S, Ghanian MH, Sadeghi-Abandansari H, Mardpour S, Nazari A, Shekari F, et al., Hydrogel-mediated sustained systemic delivery of mesenchymal stem cell-derived extracellular vesicles improves hepatic regeneration in chronic liver failure. *ACS Applied Materials & Interfaces*. 2019;11(41):37421–33.

Mateescu B, Kowal EJ, Balkom BW, Bartel S, Bhattacharyya SN, Buzás EI, et al., Obstacles and opportunities in the functional analysis of extracellular vesicle RNA–an ISEV position paper. *Journal of Extracellular Vesicles*. 2017;6(1):1286095.

Mei R, Wan Z, Yang C, Shen X, Wang R, Zhang H, et al., Advances and clinical challenges of mesenchymal stem cell therapy. *Frontiers in Immunology*. 2024;15:1421854.

Najimi M, Haider KH. Extracellular vesicle-based cell-free therapy for liver regeneration. In: KH Haider, editor. *Handbook of Stem Cell Therapy*. Singapore: Springer; 2022. https//doi.org/10.1007/978-981-16-6016-0_9-1.

Ni L, Chen D, Zhao Y, Ye R, Fang P. Unveiling the macrophage pyroptosis and its crucial role in liver diseases. *Frontiers in Immunology*. 2024;15:1338125.

Nickel S, Christ M, Schmidt S, Kosacka J, Kühne H, Roderfeld M, et al., Human mesenchymal stromal cells resolve lipid load in high-fat diet-induced non-alcoholic steatohepatitis in mice by mitochondria donation. *Cells*. 2022;11(11):1829.

Nong K, Wang W, Niu X, Hu B, Ma C, Bai Y, et al., Hepatoprotective effect of exosomes from human-induced pluripotent stem cell–derived mesenchymal stromal cells against hepatic ischemia-reperfusion injury in rats. *Cytotherapy*. 2016;18(12):1548–59.

Ohara M, Ohnishi S, Hosono H, Yamamoto K, Yuyama K, Nakamura H, et al., Extracellular vesicles from amniotic-derived mesenchymal stem cells ameliorate hepatic inflammation and fibrosis in rats. *Stem Cells International* 2018(1):3212643.

Pan Y, Wang X, Li Y, Yan P, Zhang H. Human umbilical cord blood mesenchymal stem cell-derived exosomal microRNA-503-3p inhibits the progression of human endometrial cancer cells through downregulation of MEST. *Cancer Gene Therapy*. 2022;29(8):1130–9.

Panda B, Sharma Y, Gupta S, Mohanty S. Mesenchymal stem cell-derived exosomes as an emerging paradigm for regenerative therapy and nano-medicine: a comprehensive review. *Life*. 2021;11(8):784.

Parekkadan B, Milwid JM. Mesenchymal stem cells as therapeutics. *Annual Review of Biomedical Engineering*. 2010;12(1):87–117.

Peralta C, Jiménez-Castro MB, Gracia-Sancho J. Hepatic ischemia and reperfusion injury effects on the liver sinusoidal milieu. *Journal of Hepatology*. 2013;59(5):1094–106.

Phinney DG, Pittenger MF. Concise review: MSC-derived exosomes for cell-free therapy. *Stem Cells*. 2017;35(4):851–8.

Piao C, Sang J, Kou Z, Wang Y, Liu T, Lu X, et al., Effects of exosomes derived from adipose-derived mesenchymal stem cells on pyroptosis and regeneration of injured liver. *International Journal of Molecular Sciences*. 2022;23(20):12065.

Powsner EH, Kronstadt SM, Nikolov K, Aranda A, Jay SM. Mesenchymal stem cell extracellular vesicle vascularization, bioactivity, and production yield are responsive to cell culture substrate stiffness. *Bioengineering & Translational Medicine*. 2025;e10743.

Psaraki A, Ntari L, Karakostas C, Korrou-Karava D, Roubelakis MG. Extracellular vesicles derived from mesenchymal stem/stromal cells: the regenerative impact in liver diseases. *Hepatology*. 2022;75(6):1590–603.

Pulido-Escribano V, Torrecillas-Baena B, Camacho-Cardenosa M, Dorado G, Gálvez-Moreno MÁ, Casado-Díaz A. Role of hypoxia preconditioning in therapeutic potential of mesenchymal stem-cell-derived extracellular vesicles. *World Journal of Stem Cells*. 2022;14(7):453.

Rahmani E, Rezaei M, Tavakoli R, Ghadirzadeh E, Sarnaghy FJ, Khorram R, et al., Role of regenerative medicine in the treatment of diseases. *Kindle*. 2023;3(1):1–184.

Ramachandran P, Pellicoro A, Vernon MA, Boulter L, Aucott RL, Ali A, et al., Differential Ly-6C expression identifies the recruited macrophage phenotype, which orchestrates the regression of murine liver fibrosis. *Proceedings of the National Academy of Sciences*. 2012;109(46):3186–95.

Ramasubramanian L, Kumar P, Wang A. Engineering extracellular vesicles as nanotherapeutics for regenerative medicine. *Biomolecules*. 2019;10(1):48.

Rong X, Liu J, Yao X, Jiang T, Wang Y, Xie F. Human bone marrow mesenchymal stem cells-derived exosomes alleviate liver fibrosis through the Wnt/β-catenin pathway. *Stem Cell Research & Therapy*. 2019;10:1–11.

Rostom DM, Attia N, Khalifa HM, Abou Nazel MW, El Sabaawy EA. The therapeutic potential of extracellular vesicles versus mesenchymal stem cells in liver damage. *Tissue Engineering and Regenerative Medicine*. 2020;17:537–52.

Safwan M, Bourgleh MS, Alshakaki H, Molhem A, Haider KH (2024). Morbid cell status and donor age significantly alter mesenchymal stem cell functionality and reparability. In: KH Haider, editor. *Handbook of Stem Cell Applications*. Singapore: Springer Nature Singapore; 2024. p. 1359–1387. https//doi.org/10.1007/978-981-99-7119-0_62

Sajid MI, Bukhari S, Ilyas H, Malik R, Fatima M, Tiwari RK, Nauli SM, Haider KH. Exosomes and encapsulated exomiRs in breast cancer theragnostic applications and clinical implications. In: KH Haider, editor. *Handbook of Regenerative Medicine: Cell-Free Therapy*. Boca Raton (FL): CRC Press; 2025. p. 357–370.

Samant H, Amiri HS, Zibari GB. Addressing the worldwide hepatocellular carcinoma epidemiology, prevention and management. *Journal of Gastrointestinal Oncology*. 2021;12(Suppl 2):361.

Schmeltzer PA, Russo MW. Clinical narrative autoimmune hepatitis. *Official Journal of the American College of Gastroenterology| ACG*. 2018;113(7):951–8.

Schuster S, Cabrera D, Arrese M, Feldstein AE. Triggering and resolution of inflammation in NASH. *Nature Reviews Gastroenterology & Hepatology*. 2018;15(6):349–64.

Shams F, Pourjabbar B, Hashemi N, Farahmandian N, Golchin A, Nuoroozi G, et al., Current progress in engineered and nano-engineered mesenchymal stem cells for cancer From mechanisms to therapy. *Biomedicine & Pharmacotherapy*. 2023;167:115505.

Shen X, Wang Y, Gao F, Ren F, Busuttil RW, Kupiec-Weglinski JW, et al., CD4 T cells promote tissue inflammation via CD40 signaling without de novo activation in a murine model of liver ischemia/reperfusion injury. *Hepatology*. 2009;50(5):1537–46.

Shojaei S, Hashemi SM, Ghanbarian H, Salehi M, Mohammadi-Yeganeh S. Effect of mesenchymal stem cells-derived exosomes on tumor microenvironment tumor progression versus tumor suppression. *Journal of Cellular Physiology*. 2019;234(4):3394–409.

Sitbon A, Delmotte PR, Pistorio V, Halter S, Gallet J, Gautheron J, et al., Mesenchymal stromal cell-derived extracellular vesicles therapy openings new translational challenges in immunomodulating acute liver inflammation. *Journal of Translational Medicine*. 2024;22(1):480.

Spinelli C, Adnani L, Choi D, Rak J. Extracellular vesicles as conduits of non-coding RNA emission and intercellular transfer in brain tumors. *Non-coding RNA*. 2018;5(1):1.

Stenger E, Giver CR, Langston A, Kota D, Das PK, Chinnadurai R, et al., Safety of autologous freshly expanded mesenchymal stromal cells for the treatment of graft-versus-host disease. *Frontiers in Immunology*. 2022;139:59658.

Stravitz RT, Lee WM. Acute liver failure. *The Lancet*. 2019;394(10201):869–81.

Tamura R, Uemoto S, Tabata Y. Immunosuppressive effect of mesenchymal stem cell-derived exosomes on a concanavalin A-induced liver injury model. *Inflammation and Regeneration*. 2016;36:1–11.

Tan CY, Lai RC, Wong W, Dan YY, Lim SK, Ho HK. Mesenchymal stem cell-derived exosomes promote hepatic regeneration in drug-induced liver injury models. *Stem Cell Research & Therapy*. 2014;5:1–14.

Terziroli Beretta-Piccoli B, Mieli-Vergani G, Vergani D. Autoimmmune hepatitis. *Cellular & Molecular Immunology*. 2022;19(2):158–176. https://doi.org/10.1038/s41423-021-00768-8

Théry C, Witwer KW, Aikawa E, Alcaraz MJ, Anderson JD, Andriantsitohaina R, et al., Minimal information for studies of extracellular vesicles 2018 (MISEV2018) a position statement of the International Society for Extracellular Vesicles and update of the MISEV2014 guidelines. *Journal of Extracellular Vesicles*. 2018;7(1):1535750.

Varderidou-Minasian S, Lorenowicz MJ. Mesenchymal stromal/stem cell-derived extracellular vesicles in tissue repair challenges and opportunities. *Theranostics*. 2020;10(13):5979–97.

Wang Y, Wen J, Lu T, Han W, Jiao K, Li H. Mesenchymal stem cell-derived extracellular vesicles in bone-related diseases intercellular communication, Messengers, and therapeutic engineering Protagonists. *International Journal of Nanomedicine*. 2024;32:33–57.

Wang Z, Du K, Jin N, Tang B, Zhang W. Macrophage in liver Fibrosis Identities and mechanisms. *International Immunopharmacology* . 2023;120:110357.

Wang Z, Yao L, Hu X, Yuan M, Chen P, Liu P, et al., Advancements in mesenchymal stem cell therapy for liver cirrhosis unveiling origins, treatment mechanisms, and current research frontiers. *Tissue and Cell*. 2023;84:102198.

Watanabe T, Tsuchiya A, Takeuchi S, Nojiri S, Yoshida T, Ogawa M, et al., Development of a non-alcoholic steatohepatitis model with rapid accumulation of fibrosis, and its treatment using mesenchymal stem cells and their small extracellular vesicles. *Regenerative Therapy*. 2020;14:252–61.

Xia J, Arai S, Haider KH. Researches and applications of stem cell secretome. In: KH Haider, editor. *Stem Cells: From Potential to Promise*. https://doi.org/10.1007/978-981-16-0301-3_7.

Xiao J, Ng CH, Chan KE, Fu C, Tay P, Yong JN, et al., Hepatic, extra-hepatic outcomes and causes of mortality in NAFLD–an umbrella overview of systematic review of meta-analysis. *Journal of Clinical and Experimental Hepatology*. 2023;13(4):656–65.

Xie K, Liu L, Chen J, Liu F. Exosomal miR-1246 derived from human umbilical cord blood mesenchymal stem cells attenuates hepatic ischemia reperfusion injury by modulating T helper 17/regulatory T balance. *IUBMB Life*. 2019a;71(12):2020–30.

Xie K, Liu L, Chen J, Liu F. Exosomes derived from human umbilical cord blood mesenchymal stem cells improve hepatic ischemia reperfusion injury via delivering miR-1246. *Cell Cycle*. 2019b;18(24):3491–501.

Xiong Y, Mahmood A, Chopp M. Emerging potential of exosomes for treatment of traumatic brain injury. *Neural Regeneration Research*. 2017;12(1):19–22.

Yáñez-Mó M, Siljander PR, Andreu Z, Zavec AB, Borràs FE, Buzas EI, et al., Biological properties of extracellular vesicles and their physiological functions. *Journal of Extracellular Vesicles*. 2015;427066.

Yang BO, Duan WU, Wei L, Zhao Y, Han Z, Wang J, et al., Bone marrow mesenchymal stem cell-derived hepatocyte-like cell exosomes reduce hepatic ischemia/reperfusion injury by enhancing autophagy. Stem cells and development. 2020;29(6):372–9.

Yang H, Yang H, Wang Q, Ji H, Qian T, Qiao Y, et al., Mesenchymal stem cells and their extracellular vesicles new therapies for cartilage repair. *Frontiers in Bioengineering and Biotechnology*. 2025;131:591400.

Yang Y, Hong Y, Nam GH, Chung JH, Koh E, Kim IS. Virus-mimetic fusogenic exosomes for direct delivery of integral membrane proteins to target cell membranes. *Advanced Materials (Deerfield Beach, Fla.)*. 2017;29(13):1605604.

Yoo D, Jung SY, Go D, Park JY, You DG, Jung WK, et al., Functionalized extracellular vesicles of mesenchymal stem cells for regenerative medicine. *Journal of Nanobiotechnology*. 2025;23(1):1–20.

You Q, Holt M, Yin H, Li G, Hu CJ, Ju C. Role of hepatic resident and infiltrating macrophages in liver repair after acute injury. *Biochemical Pharmacology*. 2013;86(6): 836–43.

Younossi Z, Anstee QM, Marietti M, Hardy T, Henry L, Eslam M, et al., Global burden of NAFLD and NASH trends, predictions, risk factors, and prevention. *Nature Reviews Gastroenterology & Hepatology*. 2018;15(1):11–20.

Yusuf R, Haider KH. Theragnostic applications of exosomes in Alzheimer's disease. In: KH Haider, editor. *Handbook of Regenerative Medicine: Cell-Free Therapy*. Boca Raton (FL): CRC Press; 2025. p. 390–409.

Zhai Y, Busuttil RW, Kupiec-Weglinski JW. Liver ischemia and reperfusion injury: new insights into mechanisms of innate—adaptive immune-mediated tissue inflammation. *American Journal of Transplantation*. 2011;11(8):1563–9.

Zhang F, Guo J, Zhang Z, Qian Y, Wang G, Duan M, et al., Mesenchymal stem cell-derived exosome: a tumor regulator and carrier for targeted tumor therapy. *Cancer Letters*. 2020;52:629–40.

Zhang X, Li N, Zhu Y, Wen W. The role of mesenchymal stem cells in the occurrence, development, and therapy of hepatocellular carcinoma. *Cancer Medicine*. 2022;11(4):931–43.

Zhao AG, Shah K, Cromer B, Sumer H. Mesenchymal stem cell-derived extracellular vesicles and their therapeutic potential. *Stem Cells International* 2020(1):88:25771.

Zhao H, Shang Q, Pan Z, Bai Y, Li Z, Zhang H, et al., Exosomes from adipose-derived stem cells attenuate adipose inflammation and obesity by polarizing M2 macrophages and promoting beige adipocyte formation in white adipose tissue. *Diabetes*. 2018;67(2):235–47.

Zhao S, Liu Y, Pu Z. Bone marrow mesenchymal stem cell-derived exosomes attenuate D-GaIN/LPS-induced hepatocyte apoptosis by activating autophagy in vitro. *Drug Design, Development, and Therapy*. 2019;28:87–97.

Zheng J, Lu T, Zhou C, Cai J, Zhang X, Liang J, et al., Extracellular vesicles derived from human umbilical cord mesenchymal stem cells protect liver ischemia/reperfusion injury by reducing CD154 expression on CD4+ T cells via CCT2. *Advanced Science*. 2020;7(18):1903746.

Zheng L, Gong H, Zhang J, Guo L, Zhai Z, Xia S, et al., Strategies to improve the therapeutic efficacy of mesenchymal stem cell-derived extracellular vesicle (MSC-EV) a promising cell-free therapy for liver disease. *Frontiers in Bioengineering and Biotechnology*. 2023;11:1322514.

9 Potential Therapies Using Engineered Mesenchymal Stem Cell-Derived Exosomes in Gastrointestinal Diseases

Lamiaa A. Ahmed[1], *Manar A. Didamoony*[2], *and Khaled F. Al-Massri*[3]

[1]Pharmacology and Toxicology Department, Faculty of Pharmacy, Cairo University, Cairo, Egypt

[2]Pharmacology and Toxicology Department, Faculty of Pharmacy, Egyptian Russian University, Cairo, Egypt

[3]Pharmacy and Biotechnology Department, Faculty of Medicine and Health Sciences, University of Palestine, Gaza, Palestine

9.1 LIST OF ABBREVIATIONS

ABs	Apoptotic bodies
AD	Adipose tissue
AD-MSCs	Adipose tissue-derived mesenchymal stem cells
ALI	Acute liver injury
AM-MSC-EVs	Extracellular vesicles from amnion-derived mesenchymal stem cells
AKT	Protein kinase B
BAX	Bcl2-associated X protein
Bcl2	B-cell lymphoma-2
Ber	Berberine
BM	Bone marrow
BM-MSC-EXOs	Bone marrow MSC-derived exosomes
CCl4	Carbon tetrachloride
DSS	Dextran sulfate sodium
ECM	Extracellular matrix
EVs	Extracellular vesicles
EXOs	Exosomes
GI	Gastrointestinal
GPX	Glutathione peroxidase
GSH	Glutathione
HGF	Hepatocyte growth factor
HIF-1α	Hypoxia-inducible factor 1alpha
HMGB1	High mobility group box1 protein

DOI: 10.1201/9781003653387-9

HSCs	Hepatic stellate cells
HUC	Human umbilical cord
HUC-MSC-EXOs	Human umbilical cord MSC-derived EXOs
HUC-MSCs	Human umbilical cord-derived mesenchymal stem cells
IBD	Inflammatory bowel disease
IGF	Insulin-like growth factor
IL	Interleukin
Lamp2b	Lysosomal-associated membrane protein
LPS	Lipopolysaccharide
LUT	Luteolin
miRs	Micro ribonucleic acids
mRNAs	Messenger RNAs
MSC-EVs	MSC-derived extracellular vesicles
MSC-EXOs	Mesenchymal stem cell-derived exosomes
MSCs	Mesenchymal stem cells
mTOR	Mammalian target of rapamycin
MVs	Microvesicles
NAFLD	Non-alcoholic fatty liver disease
NASH	Non-alcoholic steatohepatitis
NF-κB	Nuclear factor-kappa B
NLRP3	NOD-like receptor pyrin domain containing 3
OPN	Osteopontin
PEG	Polyethylene glycol
PI3K	Phosphoinositide 3-kinase
PLK-1	Polo-like kinase 1
PPARα	Peroxisome proliferator-activated receptor alpha
SEC	Size-exclusion chromatography
sEVs	Small extracellular vesicles
siRNA	Small interfering RNA
siRNA-OPN	siRNA-targeting osteopontin
STAT	Signal transducer and activator of transcription
TGF-β1	Transforming growth factor-beta1
Th17	T-helper 17 cells
TNF-α	Tumor necrosis factor-α
Treg	T regulatory cell
UC	Ulcerative colitis

9.2 INTRODUCTION

Gastrointestinal (GI) diseases encompass a range of inflammatory conditions that affect any segment of the GI tract—from the esophagus to the rectum—and the accessory digestive organs, including the liver, gallbladder, and pancreas. Motility disorders, visceral hypersensitivity, modified mucosal and immunological functions, and altered intestinal microbiota are characteristic features of these diseases (Drossman, 2016). Inflammatory bowel diseases (IBD), gastroesophageal reflux disease, hepatic disorders, peptic ulcers, pancreatitis, and GI malignancies represent a subset of the numerous conditions classified as GI disorders, which rank among the most prevalent ailments impacting patients globally. Numerous GI disorders adversely affect patients' quality of life and productivity (Wang et al., 2023). Notably, there are no discernible symptoms in the initial phases of GI diseases; consequently, these diseases are primarily identified in the intermediate and advanced stages, where the prognosis is unfavorable and is inadequately treated with existing pharmacological treatments

(Cheng et al., 2021). Consequently, it is imperative to develop innovative and effective ways for managing GI problems.

In recent decades, stem cell therapy has gained interest as a feasible treatment for various clinical disorders (Bhatti and Haider, 2025). Mesenchymal stem cells (MSCs) are effective therapies due to their capacity for self-renewal and the ability to differentiate into multiple cell types (Kou et al., 2022; Álvarez-Viejo and Haider, 2022). They are often sourced from bone marrow (BM), amniotic fluid, adipose tissue (AD), tooth pulp, and umbilical cord (Musiał-Wysocka et al., 2019; Kamal et al., 2022). Numerous preclinical and clinical investigations have demonstrated the potential effects of MSCs in GI protection and repair (Kubo et al., 2015; Trounson and McDonald, 2015). It was previously believed that the therapeutic efficacy of MSCs is attributed to their capacity to move to and integrate into target tissues. Subsequently, it was demonstrated that the biological effects observed after MSC administration are likely attributable to their soluble secreted factors, including cytokines, chemokines, growth factors, and extracellular vesicles (EVs) (Haider and Aziz, 2017; Gowen et al., 2020). These biological factors influence MSCs through autocrine functions that sustain self-renewal, differentiation, and proliferation, or through predominant paracrine functions that regulate the immune system, the inflammatory response, and apoptosis, and that promote neoangiogenesis (Habib et al., 2025a, 2025b). Moreover, EVs are the primary mediators of MSC paracrine effects (Ahmed et al., 2025; Rahimi et al., 2021).

EVs are membrane-enclosed nanovesicles, ranging 30–1000 nm, that convey essential biomaterials such as growth factors, cytokines, signaling lipids, micro ribonucleic acids (miRs), and messenger RNAs (mRNA) between cells, thereby modulating various cellular processes in both physiological and pathological contexts (Ahmed and Al-Massri, 2022). EVs originating from MSCs predominantly consist of exosomes (EXOs), microvesicles (MVs), and apoptotic bodies (ABs). MSC-EVs have demonstrated regenerative, anti-oxidant, anti-inflammatory, anti-necroptotic, and anti-fibrotic properties in various experimental models of GI diseases, including hepatic fibrosis, IBD, severe acute pancreatitis, non-alcoholic fatty liver disease (NAFLD), and acute liver injury (ALI) (Didamoony et al., 2023a; Dong et al., 2020; Mao et al., 2017; Niu et al., 2022).

Growing evidence suggests that MSC-EVs, rather than MSCs themselves, are primarily responsible for the therapeutic effects attributed to MSCs (Ahmed and Al-Massri, 2024; Zhao et al., 2020). Consequently, MSC-EVs may serve as an alternative to MSCs due to their superior safety profile, reduced immunogenicity, and ability to cross biological barriers (Natasha et al., 2014; Haider and Aslam, 2018). Furthermore, issues associated with MSC use, including the potential for ectopic tumor development, trapping within the lung microvasculature, and immunological rejection, may be mitigated by using MSC-derived EVs (Fennema et al., 2018). Notwithstanding their evident potential, the therapeutic use of MSC-EVs faces several challenges, including the inability of EVs to maintain their efficacy and stability over time after in vivo transplantation (Wiklander et al., 2015). Consequently, novel techniques, such as the preconditioning of MSC-EVs or their progenitor cells, may be investigated to enhance their efficacy and stability upon application (Lee and Kang, 2020).

In this chapter, we first summarize different approaches for the isolation, characterization, purification, and preservation of MSC-EVs. We also demonstrated their use in several GI models and the mechanisms behind their efficacy. Furthermore, we examined recent research and methodologies aimed at enhancing the therapeutic efficacy of MSC-EVs through various biological or pharmacological preconditioning strategies. Moreover, we identified problems and limitations that need to be addressed to advance the clinical application of MSC-EVs in various GI disorders.

9.3 BIOLOGICAL CHARACTERISTICS OF EVS

9.3.1 Definition and Origin of EVs

EVs are heterogeneous nanoparticles encased in a phospholipid membrane. They contain a variety of molecules that transduce signals, including transmembrane proteins, cytosolic proteins, organelles, transcription factors, mRNAs, miRs, and many more (Mushtaq et al., 2024). Typically, EVs are found in MSCs, fibroblasts, tumor cells, neurons, epithelial cells, and endothelial cells. They facilitate a wide range of physiological and pathological processes by acting as versatile messengers between nearby or distant cells (Didamoony et al., 2023a). In the 1980s, EVs were identified in the reticulocytes of sheep (Raposo and Stoorvogel, 2013) and were believed to be released to expel undesirable substances from the cell (Johnstone et al., 1987).

Secreted EVs primarily influence target cells by transferring information through multiple mechanisms, including internalization, ligand-receptor interactions, secreted factors, and fusion-mediated transfer of surface receptors (Rezaie et al., 2022). EVs are typically categorized by release mechanism and size into three primary types: EXOs, MVs, and ABs (He et al., 2018). Two distinct secretory processes are associated with EXOs and MVs, where EXOs are generated through the endocytic pathway and subsequently fuse with either lysosomes or the plasma membrane (Raposo and Stoorvogel, 2013) in response to increased intracellular calcium or other downstream effects of stimuli, such as stress (Wei et al., 2021). MVs form when the cell membrane extends outward, creating a closed sphere that encapsulates the cytoplasm (Van Niel et al., 2018). Various stress situations, such as oxidative or shear stress, hypoxia, or tissue damage, can trigger their release. ABs are formed during cell apoptosis, as elevated hydrostatic pressure during cell contraction detaches the plasma membrane from the cytoskeleton, leading to their formation (Doyle and Wang, 2019). ABs encompass items pertinent to cellular components, including organelles and DNA fragments (Monsel et al., 2016).

9.3.2 Isolation of EVs

Given the significant potential of EVs in preclinical and clinical research, there is an urgent necessity to prepare and obtain a considerable quantity of pure EVs. Various methods for EV isolation have been utilized, including centrifugation, polymer precipitation, size-based separation, immunoaffinity capture, and microfluidic techniques. Centrifugation technology is acknowledged as the benchmark for EV separation due to its simplicity and superior efficiency. However, this approach is time-consuming and yields low EVs purity (Tiwari et al., 2021). Polyethylene glycol (PEG) is commonly used in polymer precipitation methods for EV separation to displace water molecules, thereby decreasing solubility and precipitating EV particles. This method improves the quality of EVs, but it unavoidably results in the incorporation of polymer contaminants (Sidhom et al., 2020).

Size-dependent EV separation technologies primarily encompasses ultrafiltration and size-exclusion chromatography (SEC). Ultrafiltration approach utilizes membranes with designated pore diameters to remove proteins and nucleic acids that are smaller than the intended EV sizes. This method generates a pure EV without the necessity for specialized equipment or chemicals (Lobb et al., 2015). SEC is a method for separating EVs with varying molecular weights based on their elution times in a chromatographic column. However, SEC lacks specificity on EV subgroups of differing sizes (Stranska et al., 2018). Also, Immunocapture is an effective method for EV separation that utilizes the interactions between EV surface proteins and antibodies or ligands bound to magnetic beads (Taylor and Shah, 2015). This method can specifically separate particular EV subtypes based on their surface markers, including CD9, CD63, and CD81. Nonetheless, immunocapture is inappropriate for substantial sample volumes and is restricted to cell-free materials pathways (Jia et al., 2022).

Microfluidic technology has evolved as a novel approach for the separation of EVs by utilizing their biological, physical, and chemical characteristics, including size, density, and immunoreactivity (Contreras-Naranjo et al., 2017). Microfluidic technology offers several advantages, including the ability to handle small sample volumes, enhanced sensitivity, rapid detection, increased productivity, and cost-effective detection. Moreover, it may be effortlessly incorporated with traditional EV extraction methods and integrated with acoustic wave and microfluidic viscoelasticity techniques (Lee et al., 2015). Collectively, no singular separation method can attain both high purity and yield of EVs. Consequently, ongoing research is crucial to developing novel and reliable EV separation techniques that enable efficient, cost-effective isolation of pure EVs. Integrating an effective separation technique, such as SEC, with other procedures like ultracentrifugation, ultrafiltration, or PEG-based retrieval may yield optimal results (Sidhom et al., 2020).

9.3.3 Characterization of EVs

Accurate characterization of EVs is crucial, as stated in the International Society for EVs' basic specification report, to validate their separation processes and molecular and biological properties. A comprehensive characterization of EVs includes assessing their dimensions, morphology, composition, and surface markers (Casado-Díaz et al., 2020). The general characterization can be performed using Western blotting or an enzyme-linked immunosorbent assay to identify at least three positive and one negative EV protein markers. The positive markers should include at least one of the transmembrane/lipid-bound proteins (e.g., CD63, CD9, CD81) in addition to one cytosolic protein (e.g., TSG101, ALIX) (Abraham and Krasnodembskaya, 2020). Moreover, the characterization of individual vesicles employs imaging techniques, including transmission electron microscopy, atomic force microscopy, and scanning electron microscopy, to get high-resolution images of EV shape. Biophysical characterization can also be employed to investigate single vesicles, including nanoparticle tracking analysis, dynamic light scattering, and flow cytometry (Shao et al., 2018). While electron microscopy is accepted as the most effective method for analyzing EV structure, no single technology exists that can concurrently assess both the structural and biological characteristics of EVs (Gurunathan et al., 2019). Alternative quantification and characterization methods have been established for analyzing EVs, including nanoparticle tracking analysis and several optical flow-based techniques that can accurately quantify these vesicles. Nonetheless, these techniques cannot differentiate between particulate and membrane-bound vesicles, an issue that can be addressed by electron microscopy (Gimona et al., 2017).

9.3.4 Storage and Stability of EVs

Numerous studies have evaluated the effects of varying storage temperatures (4°C, 20°C, and −80°C) and freeze-thaw cycles on the dimensions, composition, and functionality of isolated EVs (Jeyaram and Jay, 2017). It was demonstrated that −80°C is the best temperature for preserving the stability and contents of EVs for subsequent molecular profiling (Sun et al., 2022). Freeze-thaw cycles, conversely, can cause aggregation or lysis of EVs, leading to cargo loss after application (Gandham et al., 2020).

9.4 UTILIZATION OF MSC-EVS IN VARIOUS GI MODELS

There is growing evidence that EVs are solely accountable for the therapeutic effects of MSCs in several GI disorders, including ALI, hepatic fibrosis, NAFLD, and ulcerative colitis (UC) (Didamoony et al., 2023a). Moreover, prior research indicates that MSC-EVs can be localized in damaged tissues, attenuating inflammation, apoptosis, and fibrogenesis, while concurrently regulating immune cells (Zhao et al., 2019). Consequently, research has concentrated on utilizing MSC-EVs as a substitute for MSCs in the treatment of diseases (Tamura et al., 2016).

9.4.1 Acute Liver Injury

ALI is recognized as a prominent life-threatening condition characterized by a rapid decline in normal liver functions, unfavorable clinical outcomes, and elevated death rates (Didamoony et al., 2022). The progression of the disease typically triggers a range of clinical symptoms, including jaundice, coagulation abnormalities, hepatic encephalopathy, and ascites (Wendon et al., 2017). In

ALI, many mechanisms concurrently induce liver injury by eliciting oxidative stress, inflammation, and apoptosis in response to infections, pharmaceuticals, and chemical toxins (Basir et al., 2022; Didamoony et al., 2022). Increasing evidence suggests the effective use of MSC-derived EXOs (MSC-EXOs) in the treatment of ALI due to their anti-inflammatory, antioxidant, and anti-apoptotic properties (Wu et al., 2021a; Zhao et al., 2019). BM-MSC-EXOs significantly mitigated concanavalin A-induced hepatic damage by enhancing tissue regeneration and promoting the expression of anti-inflammatory cytokines and the activity of T regulatory cells (Treg) (Tamura et al., 2016; Najimi and Haider, 2022). BM-MSC-EXOs may potentially mitigate ALI by inhibiting apoptosis. Its anti-apoptotic potential may result from reduced pro-apoptotic proteins, specifically B-cell lymphoma-2 (Bcl-2)-associated X protein (BAX) and cleaved caspase-3, while enhancing the expression of autophagy markers, such as LC3-II and Beclin-1, and the anti-apoptotic marker Bcl-2 (Zhao et al., 2019). Human umbilical cord (HUC)-MSC-EXOs demonstrated beneficial therapeutic effects on acetaminophen-induced ALI both in vivo and in vitro by reducing oxidative stress-induced inflammation and apoptosis through the activation of extracellular regulated protein kinases 1/2 and phosphoinositide 3-kinase/protein kinase B (PI3K/AKT) pathways (Wu et al., 2021a). In this context, HUC-MSC-EXOs loaded with glutathione peroxidase 1 (GPX1) decreased oxidative stress and apoptosis in hepatocytes, while enhancing EXOs protection both in vivo and in vitro (Yan et al., 2017). Subsequent research has demonstrated that HUC-MSC-EXOs mitigate hepatic inflammation and apoptosis in an ischemia/reperfusion model by transporting miR-1246, which, in turn, inhibits glycogen synthase kinase 3β-mediated Wnt/β-catenin signaling (Xie et al., 2019a). A separate hepatic ischemia/reperfusion study demonstrated that exosomal miR-1246, derived from HUC-MSCs, could suppress inflammation and regulate the balance between Treg and T-helper 17 cells (Th17) through the interleukin-6 (IL-6)/glycoprotein 130/signal transducer and activator of transcription 3 (STAT3) pathway (Xie et al., 2019b). The capacity of HUC-MSC-EXOs to reduce aminotransferase levels in ALI might also be facilitated by the downregulation of NOD-like receptor pyrin domain containing 3 (NLRP3), caspase-1, IL-1 beta (IL-1β), and IL-6 (Jiang et al., 2019). Liu et al. (2018) found that AD-MSC-EXOs could mitigate D-galactosamine and lipopolysaccharide (LPS)-induced ALI in a miR-17-dependent manner, diminishing NLRP3 inflammasome activation and thioredoxin-interacting protein in macrophages. Specifically, long-chain non-coding RNA H19 in AD-MSC-EVs mitigated hepatic necrosis, associated inflammatory cytokines, inflammatory cell infiltration, and hepatocyte proliferation in ALI in rats by activating the hepatocyte growth factor (HGF) receptor signaling pathway (Jin et al., 2018). Regarding BM-MSC-EXOs, it has been demonstrated that miR-223 inhibits NLRP3/caspase-1 signaling and reduces the levels of inflammation-related cytokines in antigen S100-induced liver injury, thereby mitigating hepatitis (Chen et al., 2018). A recent study demonstrated that MSC-EXOs therapy markedly inhibited liver isoenzymes and proinflammatory cytokines, such as IL-6 and tumor necrosis factor-alpha (TNF-α), in mice with carbon tetrachloride (CCl4)-induced acute liver damage (Kao et al., 2024). These results were correlated with improved histological examination alongside the overexpression of transcription factor E26 transformation-specific 1 and heme oxygenase-1.

9.4.2 Hepatic Fibrosis

Hepatic fibrosis is a common clinical condition characterized by the activation of immune cells and the release of inflammatory cytokines, resulting in the activation of hepatic stellate cells (HSCs) and the subsequent accumulation of extracellular matrix (ECM) proteins (Acharya et al., 2021; Didamoony et al., 2023b). The advancement of hepatic fibrosis leads to permanent cirrhosis, hepatocellular carcinoma, and finally liver failure (Didamoony et al., 2023a) . MSC-EVs have improved heart function in various experimental models by mitigating hepatic oxidative damage, inflammatory cytokine production, collagen deposition, and HSC activation, as demonstrated by AD-MSC-EVs, which inhibited HSC activation via the transfer of miR-150-5p, leading to the downregulation of chemokine (C-X-C motif) ligand 1 (Du et al., 2021). Furthermore, AD-MSC-EXOs that express miR-122 inhibited HSC activation in the hepatic failure paradigm (Lou et al., 2017). Moreover, Rong et al. (2019) showed that rat BM-MSC-EXOs could alleviate CCl4-induced liver failure by inhibiting hepatic stellate cell activation through the Wnt/β-catenin pathway, both in vivo and in vitro. Furthermore, miR-181-5p released by AD-MSC-EXOs has been demonstrated to inhibit the STAT3/Bcl-2/Beclin1 pathway, thereby enhancing autophagy and reducing HSC activation induced by transforming growth factor-beta1 (TGF-β1), thereby impeding liver failure (Qu et al., 2017). Notably, MSC-EXOs derived from HUC mitigated hepatic fibrosis by inhibiting the TGF-β1/Smad signaling pathway and the epithelial-to-mesenchymal transition (Li et al., 2013). MiR-148a, derived from HUC-MSC-EXOs, was demonstrated to modulate intrahepatic macrophages and regulate Kruppel-like factor 6/STAT3 activity, thereby inhibiting the development of hepatic failure (Tian et al., 2022). Tan et al. (2022) demonstrated that Beclin1 derived from HUC-MSC-EXOs induced ferroptosis in HSCs and decreased GPX4 levels. Moreover, HUC-MSC-EVs demonstrated their inhibitory effect on liver failure caused by Schistosoma japonicum through the downregulation of alpha-smooth muscle actin, collagen I, and collagen III, as well as inflammatory mediators such as interferon-gamma, TNF-α, and IL-1β (Dong et al., 2020). Similarly, MSC-EXOs carrying miR-125b and miR-486-5p were shown to successfully mitigate CCl4-induced liver failure by reducing smoothened expression and subsequently preventing hedgehog pathway activation (Hyun et al., 2015; Kim et al., 2021). Furthermore, Ohara et al. (2018) reported that amnion-derived MSC-EVs (AM-MSC-EVs) mitigated inflammation and hepatic failure by inhibiting the activation of HSCs and Kupffer cells (KCs). Ma et al. (2023) demonstrated that exosomal circular RNA circCDK13 from BM-MSCs suppressed hepatic fibrosis by regulating milk fat globulin-EGF factor 8 expression through the miR-17-5p/lysine acetyltransferase 2B axis. Didamoony et al. (2023c) proved that BM-MSC-EXOs alleviated hepatic inflammation, necroptosis, and fibrosis by overexpressing miR-200a and modulating platelet activating factor/receptor-interacting protein kinase 3/mixed lineage kinase domain-like protein/high mobility group box-1 protein (HMGB1), and TGF-β1/hedgehog trajectories. EXOs originating from human AD-MSCs improved hepatic fibrosis by suppressing the PI3K/Akt/mammalian target of rapamycin (mTOR) pathway and regulating choline metabolism. Enhancements in phosphatidylcholine metabolism and the regulation of PI3K/AKT/mTOR signaling appear to be the fundamental mechanisms that restore cell membrane integrity, reduce stellate cell activation, and limit the progression of hepatic fibrosis (Zhang et al., 2023). Moreover, MSC-EXOs could mitigate liver failure associated with primary sclerosing cholangitis by inhibiting Th17 differentiation, suggesting their promising therapeutic potential for hepatic fibrosis or Th17-related disorders (Chen et al., 2024).

9.4.3 Non-alcoholic Fatty Liver Disease

NAFLD is characterized by the accumulation of triglycerides within hepatocytes and the simultaneous activation of the immune system, leading to histological changes, tissue damage, and clinical manifestations, all of which are attributed to a sedentary lifestyle and calorically dense diets (Moayedfard et al., 2022). It includes a spectrum of diseases from mild steatosis (pure NAFLD) to non-alcoholic steatohepatitis (NASH), culminating in cirrhosis and hepatocellular carcinoma (Mahmoudi et al., 2022). The progression of NASH is often associated with metabolic disorders, including obesity, insulin resistance, and dysregulation of glucose and lipid metabolism. Furthermore, cellular and molecular alterations include oxidative stress, inflammation, compromised immune function, and dysfunction of the microvasculature and energy metabolism (Du et al., 2022). MSC-EVs have demonstrated protective benefits in NAFLD by improving insulin resistance caused by fat deposition, oxidative stress, dysregulated lipid metabolism, and inflammatory responses (Niu et al., 2022). Niu et al. (2022) demonstrated that miR-223-3p, abundant in AD-MSC-EVs, mitigates NAFLD by inhibiting the expression of E2F transcription factor 1, thereby decreasing lipid accumulation and hepatic fibrosis. Significantly, exosomal miR-627-5p produced from HUC-MSCs mitigated liver damage in NAFLD by improving glucose and lipid metabolism and reducing fat mass. The metabolic consequences result from the inhibition of fatty acid oxidation, facilitated by the obesity-related gene and peroxisome proliferator-activated receptor alpha (PPARα) (Cheng et al., 2021). A separate study indicated that miR-96-5p-shuttled BM-MSC-EXOs stimulated mitochondrial autophagy by inhibiting downstream caspase-2, a key factor in high-fat diet-induced NASH (El-Derany and AbdelHamid, 2021). Kang et al. (2022) and Du et al. (2022) demonstrated that NASH was significantly alleviated following the administration of HUC-MSC-EXOs, clarifying the implication of miR-24-3p/Kelch-like ECH-associated protein 1 /PPARα and nuclear factor erythroid 2-related factor 2 /NADPH quinone oxidoreductase 1 pathways. Furthermore, HUC-MSC-EXOs mitigated NASH induced by a methionine- and choline-deficient diet in mice by enhancing macrophage anti-inflammatory activity and increasing PPARα expression (Shi et al., 2022). Moreover, AM-MSC-EVs significantly inhibited the activation of HSCs and KCs. They mitigated hepatocyte inflammation and fibrogenesis in NASH by modulating the LPS/Toll-like receptor 4 pathway (Ohara et al., 2018). Notably, a prior investigation by Yang et al. (2023) demonstrated that calcium/calmodulin-dependent protein kinase 1, overexpressed in HUC-MSC-EXOs, effectively improves NAFLD in vitro and in vivo by enhancing fatty acid oxidation and decreasing fatty acid synthesis through the activation of adenosine monophosphate-activated protein kinase. EXOs derived from curcumin-preconditioned MSCs effectively mitigated NASH. They prevented its recurrence by modulating inflammatory responses, oxidative stress, and mitochondrial-dependent apoptosis through regulation of apoptosis signal-regulating kinase/c-Jun N-terminal kinase/BAX gene expression (Tawfeek and Kasem, 2023). Recently, HUC-MSC-EXOs successfully diminished liver steatosis and fibrosis in NASH by modifying macrophage polarization and promoting an anti-inflammatory environment. The observed outcomes have been associated with elevated levels of miR-24-3p in MSC-EXOs, a crucial regulator of macrophage polarization that suppresses interferon-stimulated genes expression (Jiang et al., 2025).

9.4.4 Ulcerative Colitis

UC is a prevalent kind of IBD characterized by recurrent inflammation and ulceration of the colonic mucosa, extending variably from the rectum to the cecum. Untreated UC may elevate the chance of colorectal cancer development (Olén et al., 2020). Inflammation and oxidative stress are critical elements in the etiology of UC (Soubh et al., 2015) and are regarded as primary targets of MSC-EVs treatment (Yang et al., 2015). The therapeutic efficacy of MSC-EVs has been attributed to their autonomous targeting abilities, which enable them to access injured colonic tissues, diminish inflammatory cell infiltration, and thereby preserve the integrity of the colonic mucosa while alleviating the severity of UC symptoms (Cai et al., 2021; Yang et al., 2015). Mao et al. (2017) documented that HUC-MSC-EXOs improved dextran sulfate sodium (DSS)-induced UC, resulting in a significant reduction in the recruitment of inflammatory M1 macrophages to the injured colon and a decrease in the release of proinflammatory cytokines, including TNF-α, IL-1β, and IL-6. The research conducted by Wu et al. (2018) indicated that HUC-MSC-EXOs modulated UC and suppressed ubiquitin-associated molecule (K48, K63, and FK2)-mediated inflammation by inhibiting the nuclear factor-kappa B (NF-κB) and mTOR pathways. HUC-MSC-EXOs enriched with miR-326 alleviated DSS-induced IBD in mice by inhibiting the neddylation of cullin 1 and NF-κB activation (Wang et al., 2020a). HUC-MSC-EXOs alleviated DSS-induced IBD by reducing macrophage pyroptosis through the regulation of the miR-378a-5p/NLRP3 pathway (Cai et al., 2021). Furthermore, Yang et al. (2021) highlighted the anti-inflammatory properties of TNF-α-stimulated gene 6 in HUC-MSC-EXOs, which effectively alleviated UC while enhancing intestinal function and immunological homeostasis. Furthermore, olfactory ecto-MSCs-EXOs mitigated experimental colitis by inhibiting the differentiation of pro-inflammatory Th1/Th17 cells and promoting the differentiation of anti-inflammatory Treg cells (Tian et al., 2020). Additionally, BM-MSC-EXOs were found to mitigate DSS-induced UC by promoting M2 macrophage polarization and regulating the Janus kinase 1/STAT1/STAT6 pathway (Cao et al., 2019).

A prior study indicated that HUC-MSC-EXOs alleviated IBD by inhibiting lipid peroxidation and ferroptosis, hence diminishing intestinal inflammation by targeting ACSL4 with miR-129-5p (Wei et al., 2023). A recent study indicated that MSCs, in conjunction with EXOs, may represent a potential therapeutic strategy alongside conventional medications for IBD. This study found that EXOs alone did not significantly alleviate the symptoms of experimentally induced colitis; however, when combined with umbilical cord MSCs and mesalazine, they efficiently reduced disease symptoms (Kheradmand et al., 2025). Engineered MSC-EXOs loaded with berberine (Ber) improved the solubility and oral bioavailability of Ber for the treatment of UC, leveraging the well-documented antioxidant and anti-inflammatory effects of these modified EXOs via regulation of the mitogen-activated protein kinase signaling pathway (Deng et al., 2024).

To summarize, studies comparing EVs with their parent cells have revealed either superior or equivalent efficacy for the generated EVs, accompanied by an improved safety profile (Liang et al., 2025). Del Fattore et al. (2015) demonstrated that MSC-EVs have superior immunomodulatory effects compared to their progenitor cells, promoting the expansion of CD4+, CD25+, and CD127low Tregs and the production of anti-inflammatory cytokines. Another study has shown that MSC-EVs can cross the pulmonary vasculature, ameliorate heart failure, and restore function, with efficacy comparable to that of their progenitor MSCs, by targeting diverse cell types in the liver (Gupta et al., 2022). The lack of MHC class I-II on both allogeneic and autologous MSC-EVs enhances their safety, as demonstrated by Sengupta et al. (2020) in a prospective, non-randomized cohort study involving allogeneic BM-MSC-EVs administered to COVID-19 patients, reporting no adverse effects and favorable safety endpoints (Safwan et al., 2025). Conversely, MSCs present numerous safety concerns, including the risk of aberrant differentiation or spontaneous malignancy, which has prompted consideration of EVs as an alternative (Lalu et al., 2012). Notably, Maji et al. (2017) and Sun et al. (2016) demonstrated that MSC-EVs lack genotoxic or adverse effects on the liver and kidneys, respectively, both in vitro and in vivo.

9.5 MSC-EVS ENGINEERING METHODS AND THEIR PRACTICAL IMPLEMENTATION

According to Cheng et al. (2017) and Didamoony et al. (2025) MSC-EXOs can be engineered to improve their therapeutic efficacy and regenerative potential. Engineering of MSC-EXOs, as shown in Table 9.1, involves several methods.

An overview of emerging engineering techniques highlights efforts to enhance the therapeutic efficacy of MSC-EXOs in GI diseases. As illustrated in Figure 9.1, these strategies are broadly divided into direct and indirect methods. Direct approaches involve modifying EXOs by loading them with cargo (drugs, proteins, or nucleic acids) or by surface modification. At the same time, indirect methods focus on engineering/priming the parent cells to produce more effective EXOs (Haider, 2024). Both strategies aim to enhance the targeting, stability, and overall therapeutic performance of MSC-EXOs within the GI microenvironment.

9.5.1 Cargo Loading into EXOs

9.5.1.1 Cell-Based Loading (Endogenous Cargo Loading)

It is feasible to genetically alter MSCs using various biological tools, such as plasmids or viral vectors, as the MSC condition determines the characteristics of the produced EXOs (Wang et al., 2020b; Haider, 2024). These methods enable cells to generate endogenous substances, which can subsequently be encapsulated into MSC-EXOs using the cells' biomolecular synthesis machinery (Noronha et al., 2019). According to Didamoony et al. (2025), parent MSCs could be genetically engineered to enhance the immunomodulatory, reparative, and regenerative properties of their EXOs and to minimize batch-to-batch variability.

TABLE 9.1

Examples of the Therapeutic Efficacy of Engineered MSC-EXOs in Gastrointestinal Diseases

No.	Disease	Title of Study	Method	Outcomes	References
1	ALI	Protective effects of bone marrow mesenchymal stem cell-derived exosomes loaded with cerium dioxide nanoparticles against deoxynivalenol-induced liver damage.	Using extrusion techniques, BM-MSC-EXOs were loaded with CeO2 NPs.	The modified EXOs demonstrated improved protective effects against liver injury through enhanced anti-oxidant and anti-inflammatory actions.	(Meng et al., 2025)
		Sodium hydrosulfide modification of mesenchymal stem cell-exosomes improves liver function in CCl4-induced hepatic injury in mice.	MSC-EXOs were preconditioned with sodium hydrosulfide.	H2S-exhibited reduced levels of inflammatory cytokines (IL-6, TNF-α) and improved liver aminotransferases, total oxidant levels, and cellular apoptosis.	(Sameri et al., 2023)
		Exosomes derived from human umbilical cord mesenchymal stem cells ameliorate IL-6-induced acute liver injury through miR-455-3p.	EXOs were used with overexpressed miR-455-3p.	Enhanced expression of miR-455-3p could inhibit macrophage infiltration, reduce local liver damage, and lower levels of inflammatory cytokines by targeting the PI3K signaling pathway.	(Shao et al., 2020)
		One-step engineered mesenchymal stem cell-derived exosomes against hepatic ischemia–reperfusion injury.	MSC-EXOs were modified with OPDEA-PCL or liver-targeting DSPE-PEG$_{2000}$-Galactose.	OP-EXOs and GPEG-EXOs significantly inhibited ALT, AST, and LDH levels after hepatic injury, thereby alleviating liver injury through a decrease in the expression of S100A8, S100A9, SELP, and ANXA2 in the liver.	(Lu et al., 2025)
		Fully bioactive nanodrugs: Stem cell-derived exosomes engineered with biomacromolecules to treat CCl4- and extreme hepatectomy-induced acute liver failure.	Harnessed MSC-EXOs were used as a natural carrier for GSH and HGF (HG@EXOs).	HG@EXOs revealed potential therapeutic effects through several mechanisms, including inhibition of oxidative stress, improvement of inflammation, and enhanced hepatocyte regeneration.	(Sun et al., 2024)
2	Hepatic fibrosis	BMP7-loaded human umbilical cord MSCs-derived small EVs. Ameliorate liver fibrosis by targeting the activated hepatic stellate cells.	Parental HUC-MSCs were transfected using lentivirus to obtain BMP7-loaded MSC-sEVs.	Overexpressing the anti-fibrotic BMP7 gene in parental MSCs and improving the anti-fibrotic potential of MSC-sEVs through promotion of aHSC phenotypic reversion and reduction of its proliferation.	(Zhu et al., 2024)
		MiR-122 modification enhances the therapeutic efficacy of adipose tissue-derived mesenchymal stem cells against liver fibrosis.	MSCs were transduced with lentivirus to enhance miR-122 expression.	Enhancement of the anti-fibrotic potentials of AD-MSCs through EXOs-mediated miR-122 communication between donor AMSCs and host HSCs, thus reducing HSCs' proliferation and collagen maturation.	(Lou et al., 2017)

No.	Disease	Title of Study	Method	Outcomes	References
		HU-MSC-derived exosomes modified with the targeting peptide of aHSCs for liver fibrosis therapy.	HSTP1 peptide was fused with exosomal-enriched membrane protein (Lamp2b) on the EXO surface.	Enhanced EXOs' capacity to reduce liver fibrosis, inhibit collagen deposition, and modulate M2 macrophage polarization.	(Lin et al., 2022)
		Effective delivery of osteopontin minor-interference RNA using exosomes suppresses liver fibrosis via TGF-β1 signaling.	EXOs were electroporated to load them with OPN siRNA.	Alleviation of fibrosis and deposition of ECM via prevention of the OPN/HMGB1/TGF-β1 axis.	(Tang et al., 2022)
		Obeticholic acid-loaded exosomes attenuate liver fibrosis through dual targeting of the FXR signaling pathway and ECM remodeling.	MSC-EXOs were loaded with obeticholic acid using ultrasound.	Enhancement of the FXR axis, thereby increasing ECM breakdown through elevation of MMP-13 levels and reduction of TIMP-1 levels, and subsequently inhibiting fibrosis.	(Azizsoltani et al., 2023)
		Luteolin-loaded exosomes derived from bone marrow mesenchymal stem cells: a promising therapy for liver fibrosis.	MSC-EXOs were loaded with LUT using sonication.	Enhanced LUT uptake into hepatocytes with further reduction of fibrosis using LUT-EXOs when compared to blank EXOs or LUT suspension.	(Ashour et al., 2022)
3	NAFLD	Curcumin-preconditioned mesenchymal stem cells-derived exosomes transplantation ameliorates and protects against non-alcoholic steatohepatitis by regulating the expression of key genes involved in inflammation and oxidative stress.	MSCs/EXOs were preconditioned with curcumin (MSCs/ EXOs-Cur).	MSCs/EXOs-Cur revealed enhanced protection against the recurrence of NASH through the inhibition of inflammatory, oxidative stress, and mitochondrial-dependent apoptosis.	(Tawfeek and Kasem, 2023)
		Exosomes derived from diabetic microenvironment-preconditioned mesenchymal stem cells ameliorate non-alcoholic fatty liver disease and inhibit pyroptosis of hepatocytes.	PMSCs–EXOs were collected from MSCs preconditioned in a diabetic microenvironment.	PMSC-EXOs significantly improved lipid metabolism, hepatic inflammation, steatosis, and fibrosis, and inhibited the progression of NAFLD by blocking pyroptosis.	(Wang et al., 2024b)
4	UC	Exosomes derived from mesenchymal stem cells containing berberine for ulcerative colitis therapy.	MSC-EXOs were loaded with Ber (EXOs-Ber).	EXOs-Ber enhance anti-inflammatory and antioxidant activities, contributing to the mitigation of UC, possibly through modulation of MAPK signaling.	(Deng et al., 2024)
		Hypoxic mesenchymal stem cell-derived exosomes alleviate ulcerative colitis injury by limiting reactive oxygen species accumulation in intestinal epithelial cells and DNA damage through the HIF-1α pathway.	EXOs were derived from bone MSCs under hypoxic conditions.	Hypoxic-EXOs inhibited ROS accumulation, immune homeostasis, and DNA damage in intestinal epithelial cells by overexpressing HIF-1α.	(Zhu et al., 2022)

(Continued)

TABLE 9.1 CONTINUED
Examples of the Therapeutic Efficacy of Engineered MSC-EXOs in Gastrointestinal Diseases

No.	Disease	Title of Study	Method	Outcomes	References
		Orally-delivered, cytokine-engineered extracellular vesicles for targeted treatment of inflammatory bowel disease.	Gal-IL10-EVs were prepared by modifying IL-10-containing EVs with galactose, thereby enabling targeted delivery of IL-10 to inflammatory macrophages. Chitosan/alginate hydrogel coating on Gal-IL10-EVs, with removal of C/A hydrogel coating at the diseased sites.	Gal-IL10-EVs reduced the production of ROSand inhibited the expression of pro-inflammatory cytokines, thereby improving IBD symptoms and disrupting the colonic barrier.	(Liu et al., 2023)

Abbreviations: AD-MSCs: adipose tissue-derived mesenchymal stem cells; aHSCs: activated hepatic stellate cells; ALI: acute liver injury; ALT: alanine aminotransferase; ANXA2: annexin A2; AST: aspartate aminotransferase; Ber: Berberine; BM-MSC-EXOs: bone marrow mesenchymal stem cell-derived exosomes; BMP7: bone morphogenic protein 7; CCl4: carbon tetrachloride; CeO_2: cerium dioxide; Cur: curcumin; DSPE-PEG_{2000}-Galactose: 1,2-distearoyl-sn-glycero-3-phosphoethanolamine-polyethylene glycol $_{2000}$-galactose; ECM: extracellular matrix; EXOs: exosomes; FXR: farnesoid X receptor; Gal-IL10-EVs: Galactose-modified interleukin-10-loaded extracellular vesicles; GPEG-EXOs: exosomes modified with DSPE-PEG2000-Ggalactose; GSH: glutathione; H2S: Hydrogen sulfide; HG@Exo: hepatocyte growth factor and glutathione-loaded exosomes; HGF: hepatocyte growth factor; HIF-1α: hypoxia-inducible factor 1alpha; HMGB1: high mobility group protein box1; HUC-MSCs: human umbilical cord-derived mesenchymal stem cells; IBD: Inflammatory bowel disease; IL: interleukin; Lamp2b: lysosomal-associated membrane protein 2; LDH: lactate dehydrogenase; LUT: luteolin; MAPK: mitogen-activated protein kinase; miR: micro-ribonucleic acid; MMP-13: matrix metalloproteinase-13; MSC-EXOs: mesenchymal stem cells-derived exosomes; NAFLD: non-alcoholic fatty liver disease; NASH: non-alcoholic steatohepatitis; OPDEA-PCL:Poly[2-(N-oxide-N, N-diethylamino)ethyl methacrylate]-polycaprolactone; OP-EXOs: exosomes modified with OPDEA-PCL; OPN: osteopoqntin; PI3K: phosphoinositide 3-kinase; PMSCs: preconditioned mesenchymal stem cells; ROS: reactive oxygen species; S100A8: S100 calcium-binding protein A8; S100A9: S100 calcium-binding protein A9; SELP: P-Selectin; sEVs: small extracellular vesicles; siRNA: small interfering RNAs; TGF-β1: transforming growth factor-beta 1; TIMP1: tissue inhibitor matrix metalloproteinase 1; TNF-α: tumor necrosis factor-α; UC: ulcerative colitis.

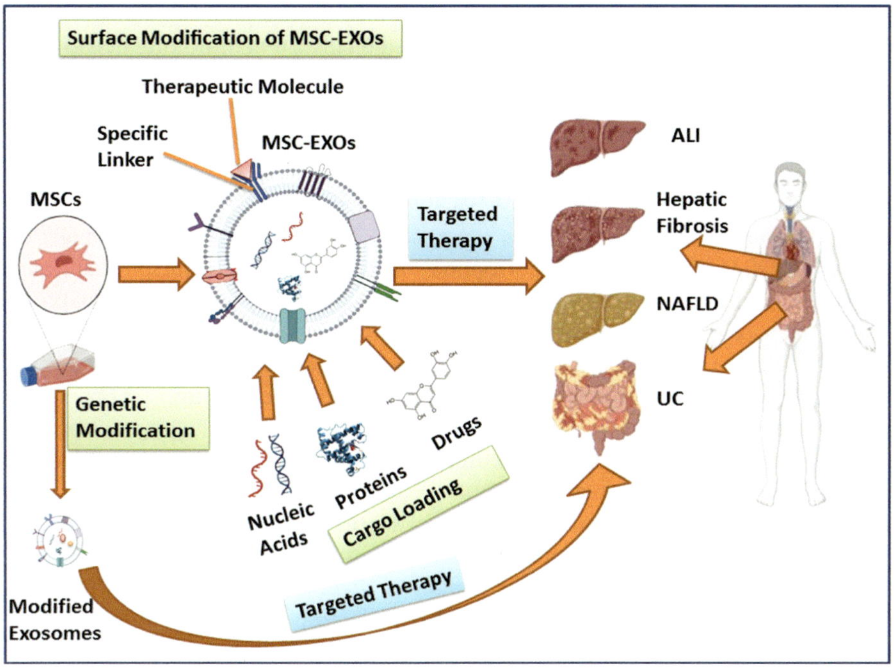

FIGURE 9.1 Overview of novel engineering techniques for enhancing the therapeutic efficacy of MSC-EXOs in GI diseases. MSC-EXO engineering strategies are broadly categorized into direct and indirect methods. Direct methods involve modifications, such as cargo loading (drugs, proteins, or nucleic acids) or surface alteration of exosomes, while indirect methods focus on engineering the exosome-producing MSCs. These approaches aim to improve the targeting capability, stability, and overall therapeutic potential of MSC-EXOs in the GI microenvironment. Abbreviations: MSC-EXOs—MSC-derived exosomes; GI—gastrointestinal; MSCs—mesenchymal stem cells.

The effectiveness of EXOs may be influenced by miRs overexpression when MSCs are genetically modified (Kim et al., 2012). In their study, Lou et al. (2017) found that EXOs with overexpressed miR-122 could target HSCs in vitro, inhibiting their activation and improving fibrosis. This was achieved by reducing the expression of insulin-like growth factor (IGF) receptor 1, prolyl-4-hydroxylase α1, and cyclin G. As a result, the researchers prevented HSC proliferation and collagen maturation, in contrast to native EXOs. This was achieved by transfecting MSCs with lentivirus encoding miR-122. In a similar vein, EXOs derived from AD-MSCs overexpressing miR-181b increased autophagy. They decreased TGF-β1-induced hepatic fibrosis in mouse HSCs in vitro and CCl4-induced hepatic fibrosis in vivo by inhibiting the STAT3/Bcl-2 and Beclin 1 axis (Qu et al., 2017). Additionally, Wang et al. (2024a) confirmed that EXOs generated by transfecting AD-MSCs with HGF alleviated hepatic fibrosis by blocking the Rho pathway and, consequently, reducing collagen formation. Significant anti-fibrotic effects were achieved by Xu et al. (2024) using EVs loaded with miR-200a derived from human AM-MSCs, which increased miR-200a expression in hepatocytes. This, in turn, inhibited the intracellular zinc finger E-box binding homeobox 1/PI3K regulatory subunit 3. Similarly, Cao et al. (2024) found that miR-26a-enriched MSC-EXOs can regulate SLC7A11 to improve hepatic fibrosis through inducing ferroptosis of HSCs.

In addition to miRs, the endogenous protein piggybacking has increased the therapeutic efficacy of MSC-EXOs in several disorders. The anti-fibrotic effects of human IGF-I-MSC-EVs were more pronounced in thioacetamide-induced hepatic fibrosis compared to EVs-depleted conditioned medium, according to research by Fiore et al. (2020), who utilized an adenoviral transfection system to transfect IGF-I into HUC perivascular cells . In addition, compared to blank or negatively controlled-transfected EXOs, HUM-MSC-EXOs loaded with bone morphogenic protein 7 had a far more pronounced therapeutic effect in reversing activated HSCs and inhibiting hepatic fibrosis (Zhu et al., 2024). Additionally, via the miR-141-3p/phosphatase and tensin homolog/AKT axis, EXOs were generated by human BM-MSCs overexpressing circDIDO1, a circular RNA derived from the DIDO1 gene, which inhibited HSC activation and hepatic fibrosis (Ma et al., 2022).

9.5.1.2 Non-cell-Based EXOs Loading (Exogenous Cargo Loading)

Direct exogenous cargo loading introduces therapeutic cargo into EXOs following their separation (Liang et al., 2025). Sonication, electroporation, and a specified buffer agent can facilitate loading of EXOs with various cargos, including anticancer drugs, hydrophobic natural products, proteins, small interfering RNAs (siRNAs), and transfection agents. Loading methods can be classified as passive or active, with both potentially combined to improve loading efficiency (Haney et al., 2015). Passive loading involves the diffusion of therapeutic cargo into EXOs, whereas active loading requires exosomal membrane disruption via methods such as electroporation or sonication to facilitate cargo entry. After the process is completed, the membrane is preserved by incubating at room temperature or 37°C (Walker et al., 2019).

Most exogenous cargo is introduced into EXOs via electroporation (Woo et al., 2020). This method utilizes an external electric field to generate minuscule, reparable pores in the phospholipid bilayer. The pores facilitate the incorporation of diverse small molecules into EXOs under an applied electric field. Liang et al. (2021) utilized electroporation to integrate the exogenous anticancer agent norcantharidin into MSC-EXOs. Due to the intrinsic liver-targeting properties of EXOs, these modified EXOs were employed to treat animal models of hepatic carcinoma. In alternative experimental models of colon and breast cancer , anticancer agents, such as doxorubicin (El-Nablaway et al., 2025), were effectively delivered into MSC-EXOs via electroporation, resulting in a significant reduction in tumor growth rates (Gomari et al., 2019; Xing et al., 2022). Furthermore, MSC-EXOs can be infused with exogenous substances using methods such as co-incubation, sonication, and dialysis (Wu et al., 2021b). For instance, researchers co-incubated MSC-EXOs with tensin homologs and siRNA-targeting phosphatases to enhance the safety and efficacy of MSC-EXOs in spinal cord injury, resulting in improved axonal regeneration and restored motor function in rats (Guo et al., 2019).

There are numerous advantages to targeting GI diseases with miRs and siRNAs. Electroporation of MSC-EXOs loaded with antisense oligonucleotide and siRNA targeting STAT3 improved liver functions, prevented fibrosis, and lowered STAT3 levels and ECM deposition in an experimental model of CCl_4-induced hepatic fibrosis (Tang et al., 2021). In a similar vein, Tang et al. (2022) found that engineered EXOs through electroporating MSC-EXOs with siRNA-targeting osteopontin (siRNA-OPN) were more effective than naked siRNA-OPN in reducing hepatic fibrosis and improving ECM deposition through blocking the OPN/HMGB1/TGF-β1 axis. Galectin-9 siRNA-treated MSC-EXOs also improved pancreatic ductal adenocarcinoma, downregulated Treg cells, and prevented macrophage polarization (Zhou et al., 2021). To silence polo-like kinase 1 (PLK-1) in bladder cancer cells, Greco et al. (2016) electroporated MSC-EXOs with PLK-1 siRNA. To further halt the migration of triple-negative breast cancer cells, Shojaei et al. (2019) employed electroporation to transfer miR-381-3p mimic into MSC-EXOs. These results highlight the practicality of applying these methods in GI disorders.

Indeed, the research into liver diseases has also examined methods for loading different medications into MSC-EXOs using an exogenous loading strategy. In a study conducted by Azizsoltani et al. (2023), the researchers used water bath ultrasound to inject obeticholic acid into human Warton's Jelly MSC-EXOs. This process activated the farnesoid X receptor axis, which, in turn, upregulated matrix metalloproteinase-13 and downregulated tissue inhibitor of matrix metalloproteinase 1. As a result, ECM degradation improved, and hepatic fibrosis was alleviated. Additionally, Fang et al. (2021) successfully treated ALI in mice by using AD-MSC-EXOs loaded with quercetin and vitamin A. The quercetin enhanced the therapeutic activity of the EXOs, while vitamin A enhanced their targeting to the liver. Fang and Liang (2021) found that MSC-EXOs supplemented with quercetin and vitamin A reduced the accelerated senescence response caused by acute liver damage. Moreover, EXOs loaded with the plant flavonoid luteolin (LUT) had superior anti-fibrotic potential compared to LUT suspension. The presence of biological activity in blank EXOs may account for this observation, increasing the drug's effectiveness. In addition, EXOs may facilitate the penetration of LUT into liver cells, thereby extending the drug's duration of action (Ashour et al., 2022).

9.5.2 Modification of EXO Surfaces

Prior studies have demonstrated that surface engineering of EXOs can modify the homing peptides or ligands on their surface, thereby extending their half-life in systemic circulation and enhancing their targeting capability and therapeutic efficacy (Ahmed and Al-Massri, 2022; Xu et al., 2021). The engineering alterations applied to the surface of MSC-EXOs can be categorized as either indirect or direct. Genetic engineering enables the expression of functional peptides or proteins on the membranes of parental cells, yielding EXOs containing analogous membrane proteins. Conversely, altered EXOs can be achieved directly by modifying their surface functional groups via ligand-receptor interactions, chemical coupling reactions, lipid fusion, and multivalent electrostatic interactions (Villata et al., 2020). Lysosomal-associated membrane protein 2 (Lamp2b) is the most commonly employed location for fusion with targeting moieties for adhesion (Salunkhe et al., 2020). Using genetic engineering, the HSTP1 peptide can be conjugated to Lamp2 on the surface of EXOs. HSTP1-EXOs can selectively target the HSC region after intravenous administration and enhance therapeutic efficacy against hepatic fibrosis in an in vivo study (Lin et al., 2022). The glycosylated sequence GNSTM, conjugated to Lamp2b, increased Lamp2b fusion protein expression in cells and EXOs, thereby preventing degradation of Lamp2b protein-linked targeting peptides and enhancing the efficacy of exosome-linked targeting peptides (Hung and Leonard, 2015). Enhancing the hepatic targeting of MSC-EXOs can also be achieved by modifying their surface with cationized pullulan. This biomaterial exhibits a strong affinity for the liver through multivalent electrostatic interactions. In comparison to unmodified EXOs, pullulan-modified EXOs exhibited greater accumulation in the liver following acute injury induced by concanavalin in mice (Tamura et al., 2017). EVs may be targeted using targeting ligands attached to PEG to selectively bind tumor cells that overexpress the epidermal growth factor receptor, enhancing cargo delivery and prolonging systemic circulation (Kooijmans et al., 2016). Additionally, anionic EXOs, generated by conjugating surface proteins to near-infrared fluorophores, showed significant liver uptake, with the hepatobiliary pathway as the primary route of EXOs excretion (Hwang et al., 2019). Research indicates that modifying the arginine-rich cell-penetrating peptide on the surface of EXOs may enhance their cellular uptake by promoting macropinocytosis (Nakase et al., 2017).

9.6 CHALLENGES AND FUTURE PERSPECTIVES

Addressing existing obstacles in the deployment of MSC-EXOs could improve their clinical translation. One potential constraint is MSC-EXOs heterogeneity, which can result from different cell passages and culture conditions and can be addressed by standardizing MSC-EXOs isolation techniques. To minimize variability in EXOs size, composition, and function, researchers should use standardized protocols for EXOs separation. Furthermore, MSCs and their EXOs should be produced more consistently under standardized culture conditions (Lotfy et al., 2023). Other problems to be resolved a preventing MSCs from senescence after several passages and keeping MSC-EXOs alive for an extended period. Lyophilizing MSC-EXOs or adding stabilizing compounds such as sugars or PEG could extend the lifetime of EXOs, according to numerous research studies (Driscoll et al., 2022; Guo et al., 2022; Liu et al., 2021).

Yet another challenge in the application of MSC-EXOs is their rapid removal from the body following systemic delivery, which may limit their capacity to provide long-term therapeutic benefits

(Harrell et al., 2019). This limitation could be mitigated by using modified EXOs with enhanced targeting and degradation resistance, as demonstrated herein, and by combining them with pharmacological drugs. Importantly, creating improved animal models in preclinical investigations is critical for bridging the gap between preclinical and clinical research. Humanized and genetically engineered systems, for example, could more accurately replicate human physiological conditions, particularly in complex diseases such as hepatic fibrosis (Benten et al., 2018). These models could help improve understanding of EXOs' safety profile and biodistribution, as well as identify reliable biomarkers for monitoring EXOs effectiveness and disease progression. Collectively, the use of these improved preclinical models could facilitate clinical translation and reduce the likelihood of failure in human trials.

9.7 CONCLUSIONS

MSC-EXOs are among the promising therapies being studied for the treatment of GI diseases. However, their clinical utility is limited by poor targeting and limited durability following in vivo transplantation. One of the most recent techniques to improve the therapeutic efficacy and molecular targeting features of MSC-EXOs is engineering. This method modulates EXO components by modifying their parent cells or directly altering the generated EXOs using specialized therapeutic chemical loading or surface modification. As a result, these approaches can control the activation or inhibition of the primary signaling pathways involved in GI diseases or their progression. Despite their promising results, engineering approaches require additional research and testing to optimize the techniques used, including physical loading, chemical alteration, and transfection methods. As a result, further research and trials are required to overcome the hurdles and problems that prevent the successful clinical applications of EXOs in GI diseases.

9.8 AUTHORS CONTRIBUTIONS

Manar A. Didamoony prepared the original draft. Khaled F. Al-Massri contributed to writing and revising the draft, and to designing the table and figure. Lamiaa A. Ahmed conceived the original idea and developed it in detail, providing valuable feedback and assisting in writing and revising the draft. All authors participated in drafting, writing, and editing the chapter and approving it for submission.

Acknowledgments: All of the persons who contributed to the chapter are listed as authors in the chapter.

Funding: This chapter received no specific grant from the public, commercial, or not-for-profit sectors.

Conflict of interest: The authors declare that they have no conflict of interest.

Ethical statement: No animal experiments were carried out for this chapter.

REFERENCES

Abraham A and Krasnodembskaya A. Mesenchymal stem cell-derived extracellular vesicles for the treatment of acute respiratory distress syndrome. *Stem Cells Translational Medicine*. 2020; 9(1):28–38.

Acharya P, Chouhan K, Weiskirchen S, and Weiskirchen R. Cellular mechanisms of liver fibrosis. *Frontiers in Pharmacology*. 2021; 12:671640.

Ahmed L and Al-Massri K. New approaches for enhancement of the efficacy of mesenchymal stem cell-derived exosomes in cardiovascular diseases. *Tissue Engineering & Regenerative Medicine*. 2022; 19(6):1129–1146.

Ahmed LA and Al-Massri KF. Exploring the role of mesenchymal stem cell-derived exosomes in diabetic and chemotherapy-induced peripheral neuropathy. *Molecular Neurobiology*. 2024; 61(8):5916–5927.

Ahmed LA, Al-Massri KF, and Haider KH. Therapeutic efficacy of mesenchymal stem cells and their derived exosomes in different neurodegenerative disorders: Current status and future perspectives. In: Haider KH, editor. *Handbook of Regenerative Medicine: Cell-Free Therapy Approach*. 1st ed. Boca Raton (FL): CRC Press; 2025. pp. 55–67.

Alvarez-Viejo M, and Haider KH. Mesenchymal stem cells: from identification and characterization to clinical applications. In: Haider KH, editor. *Handbook of Stem Cell Therapy*. Singapore: Springer; 2022. pp. 127–162.

Ashour AA, El-Kamel AH, Mehanna RA, Mourad G, and Heikal LA. Luteolin-loaded exosomes derived from bone marrow mesenchymal stem cells: a promising therapy for liver fibrosis. *Drug Delivery*. 2022; 29(1):3270–3280.

Azizsoltani A, Hatami B, Zali MR, Mahdavi V, Baghaei K, and Alizadeh E. Obeticholic acid-loaded exosomes attenuate liver fibrosis through dual targeting of the FXR signaling pathway and ECM remodeling. *Biomedicine & Pharmacotherapy*. 2023; 168:115777.

Basir HRG, Karbasi A, Ravan AP, Abbasalipourkabir R, and Bahmani M. Is human umbilical cord mesenchymal stem cell-derived conditioned medium effective against oxidative and inflammatory status in CCl4-induced acute liver injury? *Life Sciences*. 2022; 305:120730.

Benten D, Kluwe J, Wirth JW, Thiele ND, Follenzi A, Bhargava KK, et al. A humanized mouse model of liver fibrosis following expansion of transplanted hepatic stellate cells. *Laboratory Investigation*. 2018; 98(4):525–536.

Bhatti MA, Haider KH. Mesenchymal stem cells as living bio-drugs: Adding a novel dimension to the cardiology and cardiovascular sciences. *Journal of Cardiology and Cardiovascular Sciences*. 2025;6(1):30–35.

Cai X, Zhang ZY, Yuan JT, Ocansey DKW, Tu Q, Zhang X, et al. hucMSC-derived exosomes attenuate colitis by regulating macrophage pyroptosis via the miR-378a-5p/NLRP3 axis. *Stem Cell Research & Therapy*. 2021; 12(1):416.

Cao L, Xu H, Wang G, Liu M, Tian D, and Yuan Z. Extracellular vesicles derived from bone marrow mesenchymal stem cells attenuate dextran sodium sulfate-induced ulcerative colitis by promoting M2 macrophage polarization. *International Immunopharmacology*. 2019; 72:264–274.

Cao Y, Yang H, Huang Y, Lu J, Du H, and Wang B. Mesenchymal stem cell-derived exosomal miR-26a induces ferroptosis, suppresses hepatic stellate cell activation, and ameliorates liver fibrosis by modulating SLC7A11. *Open Medicine*. 2024; 19(1):20240945.

Casado-Díaz A, Quesada-Gómez JM, and Dorado G. Extracellular vesicles derived from mesenchymal stem cells (MSC) in regenerative medicine: applications in skin wound healing. *Frontiers in Bioengineering & Biotechnology*. 2020; 8:146.

Chen L, Lu F Bin, Chen D Zhi, et al. BMSCs-derived miR-223-containining exosomes contribute to liver protection in experimental autoimmune hepatitis. *Molecular Immunology*. 2018; 93:38-46.

Chen W, Lin F, Feng X, Yao Q, Yu Y, Gao F, et al. MSC-derived exosomes attenuate hepatic fibrosis in primary sclerosing cholangitis through inhibition of Th17 differentiation. *Asian Journal of Pharmaceutical Sciences*. 2024; 19(1):100889.

Cheng L, Yu P, Li F, Jiang X, Jiao X, Shen Y, et al. Human umbilical cord-derived mesenchymal stem cell-exosomal miR-627-5p ameliorates non-alcoholic fatty liver disease by repressing FTO expression. *Human Cell*. 2021; 34(6):1697–1708.

Cheng L, Zhang K, Wu S, Cui M, and Xu T. Focus on mesenchymal stem cell-derived exosomes: opportunities and challenges in cell-free therapy. *Stem Cells International*. 2017; 2017(1):6305295.

Contreras-Naranjo JC, Wu H-J, and Ugaz VM. Microfluidics for exosome isolation and analysis: enabling liquid biopsy for personalized medicine. *Lab on a Chip*. 2017; 17(21):3558–3577.

Del Fattore A, Luciano R, Pascucci L, Goffredo BM, Giorda E, Scapaticci M, et al. Immunoregulatory effects of mesenchymal stem cell-derived extracellular vesicles on T lymphocytes. *Cell Transplantation*. 2015; 24(12):2615–2627.

Deng C, Zhang H, Li Y, Cheng X, Liu Y, Huang S, et al. Exosomes derived from mesenchymal stem cells containing berberine for ulcerative colitis therapy. *Journal of Colloid & Interface Science*. 2024; 671:354–373.

Didamoony MA, Atwa AM, Abd El-Haleim EA, and Ahmed LA. Bromelain ameliorates D-galactosamine-induced acute liver injury: role of SIRT1/LKB1/AMPK, GSK3β/Nrf2 and NF-κB p65/TNF-α/caspase-8,-9 signalling pathways. *Journal of Pharmacy and Pharmacology*. 2022; 74(12):1765–1775.

Didamoony MA, Atwa AM, and Ahmed LA. Modulatory effect of rupatadine on mesenchymal stem cell-derived exosomes in hepatic fibrosis in rats: A potential role for miR-200a. *Life Sciences*. 2023c; 324:121710.

Didamoony MA, Atwa AM, and Ahmed LA. A novel mechanistic approach for the anti-fibrotic potential of rupatadine in rat liver via amendment of PAF/NF-κB p65/TGF-β1 and hedgehog/HIF-1α/VEGF trajectories. *Inflammopharmacology*. 2023b; 31(2):845–858.

Didamoony, MA, Soubh, AA, and Ahmed, LA. Cutting-edge insights into liver fibrosis: advanced therapeutic strategies and future perspectives using engineered mesenchymal stem cell-derived exosomes. *Drug Delivery & Translational Research*. 2025; 15(8):2608–2623.

Didamoony MA, Soubh AA, Atwa AM, and Ahmed LA. Innovative preconditioning strategies for improving the therapeutic efficacy of extracellular vesicles derived from mesenchymal stem cells in gastrointestinal diseases. *Inflammopharmacology*. 2023a; 31(6):2973–2993.

Dong L, Pu Y, Chen X, Qi X, Zhang L, Xu L, et al. hUCMSC-extracellular vesicles downregulated hepatic stellate cell activation and reduced liver injury in S. japonicum-infected mice. *Stem Cell Research & Therapy*. 2020; 11(1):21

Doyle LM and Wang MZ. Overview of extracellular vesicles, their origin, composition, purpose, and methods for exosome isolation and analysis. *Cells*. 2019; 8(7):727.

Driscoll J, Yan IK, and Patel T. Development of a lyophilized off-the-shelf mesenchymal stem cell-derived acellular therapeutic. *Pharmaceutics*. 2022; 14(4):849.

Drossman DA. Functional gastrointestinal disorders: history, pathophysiology, clinical features, and Rome IV. *Gastroenterology*. 2016; 150(6):1262–1279.

Du X, Li H, Han X, and Ma W. Mesenchymal stem cells-derived exosomal miR-24-3p ameliorates non-alcoholic fatty liver disease by targeting Keap-1. *Biochemical & Biophysical Research Communications*. 2022; 637:331–340.

Du Z, Wu T, Liu L, Luo B, and Wei C. Extracellular vesicle-derived miR-150-5p secreted by adipose-derived mesenchymal stem cells inhibits CXCL1 expression to attenuate hepatic fibrosis. *Journal of Cellular and Molecular Medicine*. 2021; 25(2):701–715.

El-Derany MO and AbdelHamid SG. Upregulation of miR-96-5p by bone marrow mesenchymal stem cells and their exosomes alleviates non-alcoholic steatohepatitis, with a focus on inhibiting caspase-2 signaling. *Biochemical Pharmacology*. 2021; 190:114624.

El-Nablaway M, Sonpol HMA, Elewa YHA, Ali MAM, Adel M, Zayed ES, Alhelf M, Didamoony MA, Dawood AF, Embaby EM, et al. Copper Chelation by Penicillamine Protects Against Doxorubicin-Induced Cardiomyopathy by Suppressing FDX1-Mediated Cuproptosis. *Biomolecules*. 2025; 15(9):1320.

Fang J and Liang W. ASCs-derived exosomes loaded with vitamin A and quercetin inhibit rapid senescence-like response after acute liver injury. *Biochemical & Biophysical Research Communications*. 2021; 572:125–130.

Fennema EM, Tchang LA, Yuan H, van Blitterswijk CA, Martin I, Scherberich A, et al. Ectopic bone formation by aggregated mesenchymal stem cells from bone marrow and adipose tissue: a comparative study. *Journal of Tissue Engineering and Regenerative Medicine*. 2018; 12(1):e150-e158.

Fiore E, Dominguez LM, Bayo J, Malvicini M, Atorrasagasti C, Rodriguez M, et al. Human umbilical cord perivascular cells-derived extracellular vesicles mediate IGF-I transfer to the liver and ameliorate hepatic fibrogenesis in mice. *Gene Therapy*. 2020; 27(1):62–73.

Gandham S, Su X, Wood J, Nocera AL, Alli SC, Milane L, et al. Technologies and standardization in research on extracellular vesicles. *Trends in Biotechnology*. 2020; 38(10):1066–1098.

Gimona M, Pachler K, Laner-Plamberger S, Schallmoser K, and Rohde E. Manufacturing of human extracellular vesicle-based therapeutics for clinical use. *International Journal of Molecular Sciences*. 2017; 18(6):1190.

Gomari H, Forouzandeh Moghadam M, Soleimani M, Ghavami M, and Khodashenas S. Targeted delivery of doxorubicin to HER2-positive tumor models. *International Journal of Nanomedicine*. 2019;14:5679–5690.

Gowen A, Shahjin F, Chand S, Odegaard KE, and Yelamanchili SV. Mesenchymal stem cell-derived extracellular vesicles: challenges in clinical applications. *Frontiers in Cell & Developmental Biology*. 2020; 8:149.

Greco KA, Franzen CA, Foreman KE, Flanigan RC, Kuo PC, and Gupta GN. PLK-1 silencing in bladder cancer by siRNA delivered with exosomes. *Urology*. 2016; 91:241. e1–7.

Guo S, Perets N, Betzer O, Ben-Shaul S, Sheinin A, Michaelevski I, et al. Intranasal delivery of mesenchymal stem cell-derived exosomes loaded with phosphatase and tensin homolog siRNA repairs complete spinal cord injury. *ACS Nano*. 2019; 13(9):10015–10028.

Guo Y, Zhai Y, Wu L, Wang Y, Wu P, and Xiong L. Mesenchymal stem cell-derived extracellular vesicles: pleiotropic impacts on breast cancer occurrence, development, and therapy. *International Journal of Molecular Sciences*. 2022; 23(6):2927.

Gupta S, Pinky, Vishal, Sharma H, Soni N, Rao EP, et al. Comparative evaluation of anti-fibrotic effect of tissue-specific mesenchymal stem cells-derived extracellular vesicles for the amelioration of CCl4-induced chronic liver injury. *Stem Cell Reviews & Reports*. 2022; 18(3):1097–1112.

Gurunathan S, Kang M-H, Jeyaraj M, Qasim M, and Kim J-H. Review of the isolation, characterization, biological function, and multifarious therapeutic approaches of exosomes. *Cells.* 2019; 8(4):307.

Habib SM, Martini MF, Al Ahmed O, Qassab AR, Haider KH. Mesenchymal stem cell secretomes for cell-free therapy of the broken heart. In: Haider KH, editor. *Handbook of Regenerative Medicine: Cell-Free Therapy.* Boca Raton (FL): CRC Press; 2025a. pp. 274–293.

Habib SM, Martini MF, Al Ahmed O, Qassab AR, Haider KH. The potential of mesenchymal stem cell–derived exosomes in cell-free therapy for myocardial repair. In: Haider KH, editor. *Handbook of Regenerative Medicine: Cell-Free Therapy.* Boca Raton (FL): CRC Press; 2025b. pp. 372–390.

Haider KH. Melatonin-based priming of stem cells to alleviate oxidative stress. *World Journal of Stem Cells.* 2024; 16(11):985–989.

Haider KH and Aziz S. Paracrine hypothesis and cardiac repair. *Journal of Stem Cell Research and Transplantation.* 2017; 5(1):265–267.

Haider KH and Aslam M. Cell-free therapy with stem cell secretions: protection, repair, and regeneration of the injured myocardium. In: Haider KH and Aziz S, editors. *Stem Cells: From Hype to Real Hope.* Berlin: De Gruyter; 2018. pp. 34–70.

Haney MJ, Klyachko NL, Zhao Y, Gupta R, Plotnikova EG, He Z, et al. Exosomes as drug delivery vehicles for Parkinson's disease therapy. *Journal of Controlled Release.* 2015; 207:18–30.

Harrell CR, Jankovic MG, Fellabaum C, Volarevic A, Djonov V, Arsenijevic A, et al. (2019) Molecular mechanisms responsible for anti-inflammatory and immunosuppressive effects of mesenchymal stem cell-derived factors. *Advances in Experimental Medicine & Biology.* 2019;1084:187–206

He C, Zheng S, Luo Y, and Wang B. Exosome theranostics: biology and translational medicine. *Theranostics.* 2018; 8(1):237–255.

Hung ME and Leonard JN. Stabilization of exosome-targeting peptides via engineered glycosylation. *Journal of Biological Chemistry.* 2015; 290(13):8166–8172.

Hwang DW, Jo MJ, Lee JH, Kang H, Bao K, Hu S, et al. Chemical modulation of bioengineered exosomes for tissue-specific biodistribution. *Advanced Therapeutics.* 2019; 2(11):1900111.

Hyun J, Wang S, Kim J, Kim GJ, and Jung Y. MicroRNA125b-mediated Hedgehog signaling influences liver regeneration by chorionic plate-derived mesenchymal stem cells. *Scientific Reports.* 2015; 5(1):14135.

Jeyaram A and Jay SM. Preservation and storage stability of extracellular vesicles for therapeutic applications. *The AAPS Journal.* 2017;20(1):1

Jia Y, Yu L, Ma T, Xu W, Qian H, Sun Y, et al. Small extracellular vesicles isolation and separation: Current techniques, pending questions and clinical applications. *Theranostics.* 2022; 12(15):6548–6575.

Jiang L, Zhang S, Hu H, Yang J, Wang X, Ma Y, et al. Exosomes derived from human umbilical cord mesenchymal stem cells alleviate acute liver failure by reducing the activity of the NLRP3 inflammasome in macrophages. *Biochemical & Biophysical Research Communications.* 2019; 508(3):735–741.

Jiang W, Zeng Q, Liu C-H, Wang Y, Wang S, Chen E, et al. Huc-MSCs-derived exosomes alleviate non-alcoholic steatohepatitis by regulating macrophage polarization through the miR-24-3p/STING axis. *Stem Cell Research & Therapy.* 2025; 16(1):74.

Jin Y, Wang J, Li H, Gao S, Shi R, Yang D, et al. Extracellular vesicles secreted by human adipose-derived stem cells (hASCs) improve the survival rate of rats with acute liver failure by releasing lncRNA H19. *EBioMedicine.* 2018; 34:231–242.

Johnstone RM, Adam M, Hammond J, Orr L, and Turbide C. Vesicle formation during reticulocyte maturation. Association of plasma membrane activities with released vesicles (exosomes). *Journal of Biological Chemistry.* 1987; 262(19):9412–9420.

Kamal M, Kassem D, Haider KH. Sources and therapeutic strategies of mesenchymal stem cells in regenerative medicine. In: Haider KH, editor. *Handbook of Stem Cell Therapy.* Singapore: Springer; 2022. pp. 1–28.

Kang Y, Song Y, Luo Y, Song J, Li C, Yang S, et al. Exosomes derived from human umbilical cord mesenchymal stem cells ameliorate experimental non-alcoholic steatohepatitis via Nrf2/NQO-1 pathway. *Free Radical Biology and Medicine.* 2022; 192:25–36.

Kao Y-H, Chang C-Y, Lin Y-C, Chen P-H, Lee P-H, Chang H-R, et al. Mesenchymal stem cell-derived exosomes mitigate acute murine liver injury by upregulating Ets-1 and heme oxygenase-1. *Current Stem Cell Research & Therapy.* 2024; 19(6):906–918.

Kheradmand F, Yasaman Rahimzadeh SF, Esmaeili S-A, Negah SS, Farkhad NK, Nazari SE, et al. Efficacy of umbilical cord-derived mesenchymal stem cells and exosomes in conjunction with standard IBD drug on immune responses in an IBD mouse model. *Stem Cell Research & Therapy.* 2025; 16(1):5.

Kim HW, Jiang S, Ashraf M, Haider KH. Stem cell-based delivery of Hypoxamir-210 to the infarcted heart: implications on stem cell survival and preservation of infarcted heart function. *Journal of Molecular Medicine*. 2012; 90(9):997–1010.

Kim J, Lee C, Shin Y, Wang S, Han J, Kim M, et al. sEVs from tonsil-derived mesenchymal stromal cells alleviate activation of hepatic stellate cells and liver fibrosis through miR-486-5p. *Molecular Therapy*. 2021; 29(4):1471–1486.

Kooijmans S, Fliervoet L, Van Der Meel R, Fens M, Heijnen H, en Henegouwen PvB, et al. PEGylated and targeted extracellular vesicles display enhanced cell specificity and circulation time. *Journal of Controlled Release*. 2016; 224:77–85.

Kou M, Huang L, Yang J, Chiang Z, Chen S, Liu J, et al. Mesenchymal stem cell-derived extracellular vesicles for immunomodulation and regeneration: a next-generation therapeutic tool? *Cell Death & Disease*. 2022; 13(7):580.

Kubo K, Ohnishi S, Hosono H, Fukai M, Kameya A, Higashi R, et al. Human amniotic-derived mesenchymal stem cell transplantation ameliorates liver fibrosis in rats. *Transplantation Direct*. 2015; 1(4): e16.

Lalu MM, McIntyre L, Pugliese C, Fergusson D, Winston BW, Marshall JC, et al. Safety of cell therapy with mesenchymal stromal cells (safecell): a systematic review and meta-analysis of clinical trials. *PLoS One*. 2012; 7(10):e47559.

Lee B-C and Kang K-S. Functional enhancement strategies for immunomodulation of mesenchymal stem cells and their therapeutic application. *Stem Cell Research & Therapy*. 2020; 11(1): 397.

Lee K, Shao H, Weissleder R, and Lee H. Acoustic purification of extracellular microvesicles. *ACS Nano*. 2015; 9(3):2321–2327.

Li T, Yan Y, Wang B, Qian H, Zhang X, Shen L, et al. Exosomes derived from human umbilical cord mesenchymal stem cells alleviate liver fibrosis. *Stem Cells & Development*. 2013; 22(6):845–854.

Liang L, Zhao L, Wang Y, and Wang Y. Treatment for hepatocellular carcinoma is enhanced when norcantharidin is encapsulated in exosomes derived from bone marrow mesenchymal stem cells. *Molecular Pharmaceutics*. 2021; 18(3):1003–1013.

Liang Y, Xia J, Xu L, Zhang Y, Xu X, Haider KH. Cell-free and cell-based therapy in regenerative medicine. In: Haider KH, editor. *Handbook of Regenerative Medicine: Cell-Free Therapy Approach*. 1st ed. Boca Raton (FL): CRC Press; 2025. pp. 134–152.

Lin Y, Yan M, Bai Z, Xie Y, Ren L, Wei J, et al. Huc-MSC-derived exosomes modified with the targeting peptide of aHSCs for liver fibrosis therapy. *Journal of Nanobiotechnology*. 2022; 20(1):432.

Liu A, Lin D, Zhao H, Chen L, Cai B, Lin K, et al. Optimized BMSC-derived osteoinductive exosomes immobilized in hierarchical scaffold via lyophilization for bone repair through Bmpr2/Acvr2b competitive receptor-activated Smad pathway. *Biomaterials*. 2021; 272:120718.

Liu J, Ren H, Zhang C, Li J, Qiu Q, Zhang N, et al. Orally-delivered, cytokine-engineered extracellular vesicles for targeted treatment of inflammatory bowel disease. *Small*. 2023; 19(50):2304023.

Liu Y, Lou G, Li A, Zhang T, Qi J, Ye D, et al. AMSC-derived exosomes alleviate lipopolysaccharide/d-gala ctosamine-induced acute liver failure by miR-17-mediated reduction of TXNIP/NLRP3 inflammasome activation in macrophages. *EBioMedicine*. 2018; 36:140–150.

Lobb RJ, Becker M, Wen Wen S, Wong CS, Wiegmans AP, Leimgruber A, et al. Optimized exosome isolation protocol for cell culture supernatant and human plasma. *Journal of Extracellular Vesicles*. 2015; 4(1):27031.

Lotfy A, AboQuella NM, and Wang H. Mesenchymal stromal/stem cell (MSC)-derived exosomes in clinical trials. *Stem Cell Research & Therapy*. 2023; 14(1):66.

Lou G, Yang Y, Liu F, Ye B, Chen Z, Zheng M, et al. MiR-122 modification enhances the therapeutic efficacy of adipose tissue-derived mesenchymal stem cells against liver fibrosis. *Journal of Cellular and Molecular Medicine*. 2017; 21(11):2963–2973.

Lu X, Hu H, Zhou Y, Zhang H, Xie C, Sun Y, et al. One-step engineered mesenchymal stem cell-derived exosomes against hepatic ischemia–reperfusion injury. *International Journal of Pharmaceutics*. 2025; 672:125292.

Ma J, Li Y, Chen M, Wang W, Zhao Q, He B, et al. hMSCs-derived exosome circCDK13 inhibits liver fibrosis by regulating the expression of MFGE8 through miR-17-5p/KAT2B. *Cell Biology and Toxicology*. 2023; 39(2):1–22.

Ma L, Wei J, Zeng Y, Liu J, Xiao E, Kang Y, et al. Mesenchymal stem cell-originated exosomal circDIDO1 suppresses hepatic stellate cell activation by miR-141-3p/PTEN/AKT pathway in human liver fibrosis. *Drug Delivery*. 2022; 29(1):440–453.

Mahmoudi A, Butler AE, Jamialahmadi T, and Sahebkar A. The role of exosomal miRNA in nonalcoholic fatty liver disease. *Journal of Cellular Physiology*. 2022; 237(4):2078–2094.

Maji S, Yan IK, Parasramka M, Mohankumar S, Matsuda A, and Patel T. In vitro toxicology studies of extracellular vesicles. *Journal of Applied Toxicology*. 2017; 37(3):310–318.

Mao F, Wu Y, Tang X, Kang J, Zhang B, Yan Y, et al. Exosomes derived from human umbilical cord mesenchymal stem cells relieve inflammatory bowel disease in mice. *Biomed Research International*. 2017; 2017(1):5356760.

Meng Z, Tang M, Xu S, Zhou X, Zhang Z, Yang L, et al. Protective effects of bone marrow mesenchymal stem cell-derived exosomes loaded with cerium dioxide nanoparticles against deoxynivalenol-induced liver damage. *Journal of Nanobiotechnology*. 2025; 23(1):215.

Moayedfard Z, Sani F, Alizadeh A, Bagheri Lankarani K, Zarei M, and Azarpira N. The role of the immune system in the pathogenesis of NAFLD and potential therapeutic impacts of mesenchymal stem cell-derived extracellular vesicles. *Stem Cell Research & Therapy*. 2022; 13(1):242.

Monsel A, Zhu Y-g, Gudapati V, Lim H, and Lee JW. Mesenchymal stem cell-derived secretome and extracellular vesicles for acute lung injury and other inflammatory lung diseases. *Expert Opinion on Biological Therapy*. 2016; 16(7):859–871.

Musiał-Wysocka A, Kot M, and Majka M. The pros and cons of mesenchymal stem cell-based therapies. *Cell Transplantation*. 2019; 28(7):801–812.

Mushtaq M, Zineldeen DH, Mateen MA, and Haider KH. Mesenchymal stem cells' "garbage bags" at work: Treating radial nerve injury with mesenchymal stem cell-derived exosomes. *World Journal of Stem Cells* 2024; 16(5):467–478.

Nakase I, Noguchi K, Aoki A, Takatani-Nakase T, Fujii I, and Futaki S. Arginine-rich cell-penetrating peptide-modified extracellular vesicles for active macropinocytosis induction and efficient intracellular delivery. *Scientific Reports*. 2017; 7(1):1991.

Najimi M, Haider KH. Extracellular vesicle-based cell-free therapy for liver regeneration. In: Haider KH, editor. *Handbook of Stem Cell Therapy*. Singapore: Springer; 2022. https://doi.org/10.1007/978-981-16-6016-0_9-1.

Natasha G, Gundogan B, Tan A, Farhatnia Y, Wu W, Rajadas J, et al. Exosomes as immunotheranostic nanoparticles. *Clinical Therapeutics*. 2014; 36(6):820–829.

Niu Q, Wang T, Wang Z, Wang F, Huang D, Sun H, et al. Adipose-derived mesenchymal stem cell-secreted extracellular vesicles alleviate non-alcoholic fatty liver disease via delivering miR-223-3p. *Adipocyte*. 2022; 11(1):572–587.

Noronha NdC, Mizukami A, Caliári-Oliveira C, Cominal JG, Rocha JLM, Covas DT, et al. Priming approaches to improve the efficacy of mesenchymal stromal cell-based therapies. *Stem Cell Research & Therapy*. 2019; 10(1):131.

Ohara M, Ohnishi S, Hosono H, Yamamoto K, Yuyama K, Nakamura H, et al. Extracellular vesicles from amniotic-derived mesenchymal stem cells ameliorate hepatic inflammation and fibrosis in rats. *Stem Cells International*. 2018; 2018(1):3212643.

Olén O, Erichsen R, Sachs MC, Pedersen L, Halfvarson J, Askling J, et al. Colorectal cancer in ulcerative colitis: a Scandinavian population-based cohort study. *The Lancet*. 2020; 395(10218):123–131.

Qu Y, Zhang Q, Cai X, Li F, Ma Z, Xu M, et al. Exosomes derived from miR-181-5p-modified adipose-derived mesenchymal stem cells prevent liver fibrosis via autophagy activation. *Journal of Cellular and Molecular Medicine*. 2017; 21(10):2491–2502.

Rahimi B, Panahi M, Saraygord-Afshari N, Taheri N, Bilici M, Jafari D, et al. The secretome of mesenchymal stem cells and oxidative stress: challenges and opportunities in cell-free regenerative medicine. *Molecular Biology Reports*. 2021; 48(7):5607–5619.

Raposo G and Stoorvogel W. Extracellular vesicles: exosomes, microvesicles, and friends. *Journal of Cell Biology*. 2013; 200(4):373–383.

Rezaie J, Nejati V, Mahmoodi M, and Ahmadi M. Mesenchymal stem cells-derived extracellular vesicles: a promising nanomedicine for drug delivery system. *Biochemical Pharmacology*. 2022; 203:115167.

Rong X, Liu J, Yao X, Jiang T, Wang Y, and Xie F. Human bone marrow mesenchymal stem cells-derived exosomes alleviate liver fibrosis through the Wnt/β-catenin pathway. *Stem Cell Research & Therapy*. 2019; 10(1):98.

Safwan M, Bourgleh M, Al-Ruqi A, Shrebaty O, Almujaydil F, Alothaim B, Alrashdi N, Haider KH. Living bio-drugs MSCs and their derivative exosomes to rescue ventilated patients with ARDS: a systematic review and meta-analysis. *World Journal of Stem Cells*. 2025; 17(10):111978

Salunkhe S, Basak M, Chitkara D, and Mittal A. Surface functionalization of exosomes for target-specific delivery and in vivo imaging & tracking: Strategies and significance. *Journal of Controlled Release*. 2020; 326:599–614.

Sameri MJ, Belali R, Neisi N, Razliqi RN, Mard SA, Savari F, et al. Sodium Hydrosulfide modification of mesenchymal stem cell-exosomes improves liver function in CCL4-induced hepatic injury in mice. *Reports of Biochemistry & Molecular Biology*. 2023; 11(4):644–655.

Sengupta V, Sengupta S, Lazo A, Woods P, Nolan A, and Bremer N. Exosomes derived from bone marrow mesenchymal stem cells as treatment for severe COVID-19. *Stem Cells and Development*. 2020; 29(12):747–754.

Shao H, Im H, Castro CM, Breakefield X, Weissleder R, and Lee H. New technologies for analysis of extracellular vesicles. *Chemical Reviews*. 2018; 118(4):1917–1950.

Shao M, Xu Q, Wu Z, Chen Y, Shu Y, Cao X, et al. Exosomes derived from human umbilical cord mesenchymal stem cells ameliorate IL-6-induced acute liver injury through miR-455-3p. *Stem Cell Research & Therapy*. 2020; 11 (1): 37.

Shi Y, Yang X, Wang S, Wu Y, Zheng L, Tang Y, et al. Human umbilical cord mesenchymal stromal cell-derived exosomes protect against MCD-induced NASH in a mouse model. *Stem Cell Research & Therapy*. 2022; 13(1):517.

Shojaei S, Hashemi SM, Ghanbarian H, Salehi M, and Mohammadi-Yeganeh S. Effect of mesenchymal stem cells-derived exosomes on tumor microenvironment: tumor progression versus tumor suppression. *Journal of Cellular Physiology*. 2019; 234(4):3394–3409.

Sidhom K, Obi PO, and Saleem A. A review of exosomal isolation methods: is size exclusion chromatography the best option? *International Journal of Molecular Sciences*. 2020; 21(18):6466.

Soubh AA, Abdallah DM, and El-Abhar HS. Geraniol ameliorates TNBS-induced colitis: Involvement of Wnt/β-catenin, p38MAPK, NFκB, and PPARγ signaling pathways. *Life Sciences*. 2015; 136:142–150.

Stranska R, Gysbrechts L, Wouters J, Vermeersch P, Bloch K, Dierickx D, et al. Comparison of membrane affinity-based method with size-exclusion chromatography for isolation of exosome-like vesicles from human plasma. *Journal of Translational Medicine*. 2018; 16(1):1.

Sun H, Zhang T, and Gao J. Extracellular vesicles derived from mesenchymal stem cells: a potential biodrug for acute respiratory distress syndrome treatment. *BioDrugs*. 2022; 36(6):701–715.

Sun L, Xu R, Sun X, Duan Y, Han Y, Zhao Y, et al. Safety evaluation of exosomes derived from human umbilical cord mesenchymal stromal cells. *Cytotherapy*. 2016; 18(3):413–422.

Sun M, Li M, Hu M, Fan Y, Liu Y, Sun J, et al. Fully Bioactive nanodrugs: Stem cell-derived exosomes engineered with biomacromolecules to treat CCl4-and extreme hepatectomy-induced acute liver failure. *ACS Nano*. 2024; 18(50):33907–33921.

Tamura R, Uemoto S, and Tabata Y. Immunosuppressive effect of mesenchymal stem cell-derived exosomes on a concanavalin A-induced liver injury model. *Inflammation and Regeneration*. 2016; 36: 26.

Tamura R, Uemoto S, and Tabata Y. Augmented liver targeting of exosomes by surface modification with cationized pullulan. *Acta Biomaterialia*. 2017; 57:274–284.

Tan Y, Huang Y, Mei R, Mao F, Yang D, Liu J, et al. HucMSC-derived exosomes that deliver BECN1 induce ferroptosis of hepatic stellate cells via regulating the xCT/GPX4 axis. *Cell Death & Disease*. 2022; 13(4):319.

Tang M, Chen Y, Li B, Sugimoto H, Yang S, Yang C, et al. Therapeutic targeting of STAT3 with small-interference RNAs and antisense oligonucleotides embedded in exosomes in liver fibrosis. *The FASEB Journal*. 2021; 35(5):e21557.

Tang M, Guo C, Sun M, Zhou H, Peng X, Dai J, et al. Effective delivery of osteopontin small-interfering RNA using exosomes suppresses liver fibrosis via the TGF-β1 signaling pathway. *Frontiers in Pharmacology*. 2022; 13:882243.

Tawfeek GA-E and Kasem HA. Curcumin preconditioned mesenchymal stem cells-derived exosomes transplantation ameliorates and protects against non-alcoholic steatohepatitis by regulating the expression of key genes of inflammation and oxidative stress. *Transplant Immunology*. 2023; 78:101837.

Taylor DD, Shah S. Methods of isolating extracellular vesicles impact downstream analyses of their cargoes. *Methods*. 2015; 87:3–10.

Tian J, Zhu Q, Zhang Y, Bian Q, Hong Y, Shen Z, et al. Olfactory ecto-mesenchymal stem cell-derived exosomes ameliorate experimental colitis via modulating Th1/Th17 and Treg cell responses. *Frontiers in Immunology*. 2020; 11:598322.

Tian S, Zhou X, Zhang M, Cui L, Li B, Liu Y, et al. Mesenchymal stem cell-derived exosomes protect against liver fibrosis by delivering miR-148a to target the KLF6/STAT3 pathway in macrophages. *Stem Cell Research & Therapy*. 2022; 13(1):330.

Tiwari S, Kumar V, Randhawa S, and Verma SK. Preparation and characterization of extracellular vesicles. *American Journal of Reproductive Immunology.* 2021; 85(2):e13367.

Trounson A and McDonald C. Stem cell therapies in clinical trials: progress and challenges. *Cell Stem Cell.* 2015; 17(1):11–22.

Van Niel G, d'Angelo G, and Raposo G. Shedding light on the cell biology of extracellular vesicles. *Nature Reviews Molecular Cell Biology.* 2018; 19(4):213–228.

Villata S, Canta M, and Cauda V. EVs and bioengineering: from cellular products to engineered nanomachines. *International Journal of Molecular Sciences.* 2020; 21(17):6048.

Walker S, Busatto S, Pham A, Tian M, Suh A, Carson K, et al. Extracellular vesicle-based drug delivery systems for cancer treatment. *Theranostics.* 2019; 9(26):8001–8017.

Wang A, Li B, Su W, Zhang H, Hu R, Zhang Y, et al. Exosomes derived from diabetic microenvironment-preconditioned mesenchymal stem cells ameliorate nonalcoholic fatty liver disease and inhibit pyroptosis of hepatocytes. *Experimental Cell Research.* 2024b; 443(2):114325.

Wang G, Yuan J, Cai X, Xu Z, Wang J, Ocansey DKW, et al. HucMSC-exosomes carrying miR-326 inhibit neddylation to relieve inflammatory bowel disease in mice. *Clinical and Translational Medicine.* 2020a; 10(2):e113.

Wang J, Bonacquisti EE, Brown AD, and Nguyen J. Boosting the biogenesis and secretion of mesenchymal stem cell-derived exosomes. *Cells.* 2020b; 9(3):660.

Wang J, Ye W, Jiang M, Zhou Y, and Zheng J. Therapeutic potential of exosomes derived from hepatocyte growth factor-overexpressing adipose mesenchymal stem cells in TGFβ1-stimulated hepatic stellate cells. *Cytotechnology.* 2024a; 76(2):217–229.

Wang R, Li Z, Liu S, and Zhang D. Global, regional, and national burden of 10 digestive diseases in 204 countries and territories from 1990 to 2019. *Frontiers in Public Health.* 2023; 11:1061453.

Wei H, Chen Q, Lin L, Sha C, Li T, Liu Y, et al. Regulation of exosome production and cargo sorting. *International Journal of Biological Sciences.* 2021; 17(1):163–177.

Wei Z, Hang S, Wiredu Ocansey DK, Zhang Z, Wang B, Zhang X, et al. Human umbilical cord mesenchymal stem cell-derived exosome shuttling miR-129-5p attenuates inflammatory bowel disease by inhibiting ferroptosis. *Journal of Nanobiotechnology.* 2023; 21(1):188.

Wendon J, Cordoba J, Dhawan A, Larsen FS, Manns M, Samuel D, et al. EASL Clinical Practical Guidelines on the management of acute (fulminant) liver failure. *Journal of Hepatology.* 2017; 66(5):1047–1081.

Wiklander OP, Nordin JZ, O'Loughlin A, Gustafsson Y, Corso G, Mäger I, et al. Extracellular vesicle in vivo biodistribution is determined by cell source, route of administration, and targeting. *Journal of Extracellular Vesicles.* 2015; 4(1):26316.

Woo CH, Kim HK, Jung GY, Jung YJ, Lee KS, Yun YE, et al. Small extracellular vesicles from human adipose-derived stem cells attenuate cartilage degeneration. *Journal of Extracellular Vesicles.* 2020; 9(1):1735249.

Wu H-Y, Zhang X-C, Jia B-B, Cao Y, Yan K, Li J-Y, et al. Exosomes derived from human umbilical cord mesenchymal stem cells alleviate acetaminophen-induced acute liver failure through activating ERK and IGF-1R/PI3K/AKT signaling pathway. *Journal of Pharmacological Sciences.* 2021a; 147(1):143–155.

Wu P, Zhang B, Ocansey DKW, Xu W, and Qian H. Extracellular vesicles: A bright star of nanomedicine. *Biomaterials.* 2021b; 269:120467.

Wu Y, Qiu W, Xu X, Kang J, Wang J, Wen Y, et al. Exosomes derived from human umbilical cord mesenchymal stem cells alleviate inflammatory bowel disease in mice through ubiquitination. *American Journal of Translational Research.* 2018; 10(7):2026–2036.

Xie K, Liu L, Chen J, and Liu F. Exosomal miR-1246 derived from human umbilical cord blood mesenchymal stem cells attenuates hepatic ischemia reperfusion injury by modulating T helper 17/regulatory T balance. *IUBMB Life.* 2019b; 71(12):2020–2030.

Xie K, Liu L, Chen J, and Liu F. Exosomes derived from human umbilical cord blood mesenchymal stem cells improve hepatic ischemia reperfusion injury via delivering miR-1246. *Cell Cycle.* 2019a; 18(24):3491–3501.

Xing H, Liang C, Xu X, Sun H, Ma X, and Jiang Z. Mesenchymal stroma/stem-like cells of GARP knockdown inhibit cell proliferation and invasion of mouse colon cancer cells (MC38) through exosomes. *Journal of Cellular & Molecular Medicine.* 2020; 24(23):13984–13990.

Xu A-L, Han L, Yan J, Liu D, and Wang W. Effects of mesenchymal stem cells-derived extracellular vesicles on inhibition of hepatic fibrosis by delivering miR-200a. *Tissue Engineering and Regenerative Medicine.* 2024; 21(4):609–624.

Xu M, Feng T, Liu B, Qiu F, Xu Y, Zhao Y, et al. Engineered exosomes: desirable target-tracking characteristics for cerebrovascular and neurodegenerative disease therapies. *Theranostics*. 2021; 11(18):8926–8944.

Yan Y, Jiang W, Tan Y, Zou S, Zhang H, Mao F, et al. hucMSC exosome-derived GPX1 is required for the recovery of hepatic oxidant injury. *Molecular Therapy*. 2017; 25(2):465–479.

Yang F, Wu Y, Chen Y, Xi J, Chu Y, Jin J, et al. Human umbilical cord mesenchymal stem cell-derived exosomes ameliorate liver steatosis by promoting fatty acid oxidation and reducing fatty acid synthesis. *JHEP Reports*. 2023; 5(7):100746.

Yang J, Liu X-X, Fan H, Tang Q, Shou Z-X, Zuo D-M, et al. Extracellular vesicles derived from bone marrow mesenchymal stem cells protect against experimental colitis via attenuating colon inflammation, oxidative stress, and apoptosis. *PLoS One*. 2015; 10(10):e0140551.

Yang S, Liang X, Song J, Li C, Liu A, Luo Y, et al. A novel therapeutic approach for inflammatory bowel disease by exosomes derived from human umbilical cord mesenchymal stem cells to repair intestinal barrier via TSG-6. *Stem Cell Research & Therapy*. 2021; 12(1):315.

Zhang Z, Shang J, Yang Q, Dai Z, Liang Y, Lai C, et al. Exosomes derived from human adipose mesenchymal stem cells ameliorate hepatic fibrosis by inhibiting the PI3K/Akt/mTOR pathway and remodeling choline metabolism. *Journal of Nanobiotechnology*. 2023; 21(1):29.

Zhao S, Liu Y, and Pu Z. Bone marrow mesenchymal stem cell-derived exosomes attenuate D-GaIN/LPS-induced hepatocyte apoptosis by activating autophagy in vitro. *Drug Design, Development & Therapy*. 2019; 13:2887–2897.

Zhao Y, Zhu X, Zhang L, Ferguson CM, Song T, Jiang K, et al. Mesenchymal stem/stromal cells and their extracellular vesicle progeny reduce injury in poststenotic swine kidneys via distinct mechanisms. *Stem Cells & Development*. 2020; 29(18):1190–1200.

Zhou W, Zhou Y, Chen X, Ning T, Chen H, Guo Q, et al. Pancreatic cancer-targeting exosomes for enhancing immunotherapy and reprogramming the tumor microenvironment. *Biomaterials*. 2021; 268:120546.

Zhu D, Sun Z, Wei J, Zhang Y, An W, Lin Y, et al. BMP7-Loaded human umbilical cord mesenchymal stem cell-derived small extracellular vesicles ameliorate liver fibrosis by targeting activated hepatic stellate cells. *International Journal of Nanomedicine*. 2024;19:3475–3495.

Zhu F, Wei C, Wu H, Shuai B, Yu T, Gao F, et al. Hypoxic mesenchymal stem cell-derived exosomes alleviate ulcerative colitis injury by limiting reactive oxygen species accumulation and DNA damage in intestinal epithelial cells through HIF-1α. *International Immunopharmacology*. 2022; 113:109426.

10 Stem Cell Exosomes
Nature's Nanoparticles for Respiratory Disorders

Avinash Kharat[1], *Avinash Sanap*[1], *and Ramesh Bhonde*[1]

[1]Regenerative Medicine Laboratory, Dr. D. Y. Patil Dental College and Hospital, Dr. D. Y. Patil Vidyapeeth, Pune, India

10.1 LIST OF ABBREVIATIONS

AECs	Alveolar epithelial cells
AMs	Alveolar macrophages
ARDS	Acute respiratory distress syndrome
COPD	Chronic obstructive pulmonary disease
CCL2	C-C motif ligand 2 (CCL2)
ECM	Extracellular matrix
EMT	Epithelial-mesenchymal transition
FEV1	Forced expiratory volume in one second
FVC	Forced vital capacity
IL	Interleukin
ICS	Inhaled corticosteroid
LABAs	Long-acting beta-adrenoceptor agonists
LAMA	Long-acting muscarinic antagonist
LABA	Long-acting beta-agonist
MSCs	Mesenchymal stem cells
MSC-Exos	Mesenchymal stem cells exosome
MCP-1	Monocytic chemotactic protein 1
mRNA	Messenger ribonucleic acid
miRNA	microRNA
PF	Pulmonary fibrosis
RV	Residual volume
SABAs	Short-acting beta agonists
Th2	T helper type 2
TGF-β	Transforming growth factor-beta
TLC	Total lung capacity
TNF-α	Tumor necrosis factor alpha
VEGF	Vascular endothelial growth factor

10.2 INTRODUCTION

Cells release exosomes, which are nanoscale extracellular vesicles (30–150 nm), into the extracellular environment. A lipid bilayer surrounds them and contains a cargo of bioactive molecules,

DOI: 10.1201/9781003653387-10

including proteins, mRNA, microRNA (miRNA), and lipids.(Műzes G, 2022) Exosomes play a vital role in cell-to-cell communication and have been implicated in various physiological and pathological processes (Qiu et al., 2020). They are taken up by recipient cells, where their cargo molecules can influence cellular behavior and function (Mushtaq et al., 2024). The potential of stem cell-derived exosomes to support lung health and repair tissue damaged by conditions such as acute respiratory distress syndrome (ARDS), pulmonary fibrosis, and chronic obstructive pulmonary disease (COPD) has garnered increasing attention in recent years (Safwan et al., 2025). One of the most remarkable features of stem cell exosomes is their ability to transfer bioactive molecules to recipient cells. This transfer can influence various cellular processes, including inflammation, apoptosis, and tissue regeneration. The unique composition of these exosomes allows them to modulate the immune response and promote healing in damaged tissues, making them a valuable resource in the treatment of lung diseases.

Research has shown that mesenchymal stem cell (MSC) exosomes possess anti-inflammatory properties (Ahmed et al., 2023; Lo Sisso et al., 2017). In models of lung injury, the exosomes from lung spheroid cells have been shown to reduce inflammation and promote tissue repair (Dinh et al., 2020). This is particularly important in conditions like COPD, where chronic inflammation leads to progressive lung damage. By delivering anti-inflammatory signals directly to lung cells, these exosomes can help restore normal lung function and alleviate symptoms. Stem cell exosomes possess anti-inflammatory properties, yet they are also crucial for angiogenesis, the process by which new blood vessels form. This process is vital for tissue repair and regeneration, particularly in the lungs, where a sufficient blood supply is essential for proper function. Research has demonstrated that exosomes can stimulate endothelial cell growth and proliferation, which are pivotal in the formation of new blood vessels (Ateeq et al., 2024). By facilitating angiogenesis, exosomes support the healing of lung tissue affected by diseases such as pulmonary fibrosis. These nanoparticles have been found to carry various growth factors and cytokines that promote cell survival and proliferation. For instance, exosomes containing transforming growth factor-beta (TGF-β) and vascular endothelial growth factor (VEGF) have been shown to support the survival of lung epithelial cells under stressful conditions. This property is critical for preventing cell death associated with lung injury, thereby preserving lung function.

10.2.1 Therapeutic Potential of Stem Cell-Derived Exosomes

The therapeutic applications of stem cell exosomes are gaining momentum in clinical settings for various pathological conditions (Haider and Najimi, 2022; Yusuf and Haider, 2025; Lamiaa et al., 2025). Several studies have explored exosome-based therapies for the treatment of lung diseases (Zargar et al., 2022). In animal models, MSC-derived exosomes have been shown to improve lung function, reduce inflammation, and enhance tissue repair (Feng et al., 2024). These promising results have spurred interest in translating this research into human clinical trials. One of the key advantages of using stem cell exosomes as a therapeutic approach is their safety profile. As naturally occurring nanoparticles, exosomes are generally well-tolerated by the body, reducing the risk of adverse effects often associated with traditional pharmaceutical interventions. Additionally, their small size enables efficient delivery to target tissues, thereby enhancing their therapeutic efficacy. Nonetheless, obstacles still exist in the clinical use of exosome-based treatments. It is essential to standardize methods for isolating and characterizing exosomes to ensure consistent therapeutic outcomes. Furthermore, understanding the mechanisms underlying exosome action in the lungs will be necessary for optimizing their use. Ongoing research aims to address these issues and pave the way for effective exosome-based treatments for lung diseases.

Exosomes are released by various cell types, including stem cells, into the extracellular environment (Al-Omar et al., 2022; Ahmed et al., 2023). A lipid bilayer surrounds them and carries a diverse array of bioactive molecules, such as proteins, lipids, mRNA, and microRNA (miRNA)

(Haider and Aramini, 2020). These molecules facilitate communication between cells by delivering their cargo to target cells, thereby affecting their biological activities. Exosomal cargo can target immune cells, dampening excessive inflammatory responses. MSC-derived exosomes have been shown to reduce pro-inflammatory cytokines and induce a shift toward anti-inflammatory phenotypes in macrophages and other immune cells. It can also modulate fibroblast differentiation and activation, thereby reducing their excessive proliferation and extracellular matrix (ECM) production. miRNAs can directly target profibrotic genes, thereby inhibiting the expression of collagen and other extracellular matrix (ECM) components. Exosomes can directly stimulate the proliferation and migration of alveolar epithelial cells (AECs) (Akram et al., 2013). It also promotes lung regeneration and improves lung microenvironments by reducing excessive extracellular matrix (ECM) and inflammation.(Sang et al., 2022) They can also promote the reduction of cellular senescence in damaged lung tissues. Exosomes contain anti-inflammatory cytokines and factors that can reduce the hyperinflammatory response in asthmatic airways (Cañas et al., 2021; Trubiani et al., 2019). They can suppress the activation of inflammatory cells, such as mast cells, eosinophils, and T helper type 2 (Th2) cells, which are central to the pathogenesis of asthma (Cheema et al., 2025). Exosomes contain factors that facilitate tissue repair and remodeling by promoting angiogenesis and extracellular matrix remodeling (Zhao et al., 2023). They can also stimulate the regeneration of damaged epithelial cells, restoring normal airway function (Hynds, 2022). While less well understood, stem cell exosomes may influence smooth muscle contraction in the airways, thereby contributing to bronchodilation. This book chapter focuses on the therapeutic potential of stem cell-derived exosomes in the context of various respiratory disorders including ARDS, highlighting the mechanisms of action, preclinical evidence, and future challenges for clinical translation (Figure 10.1).

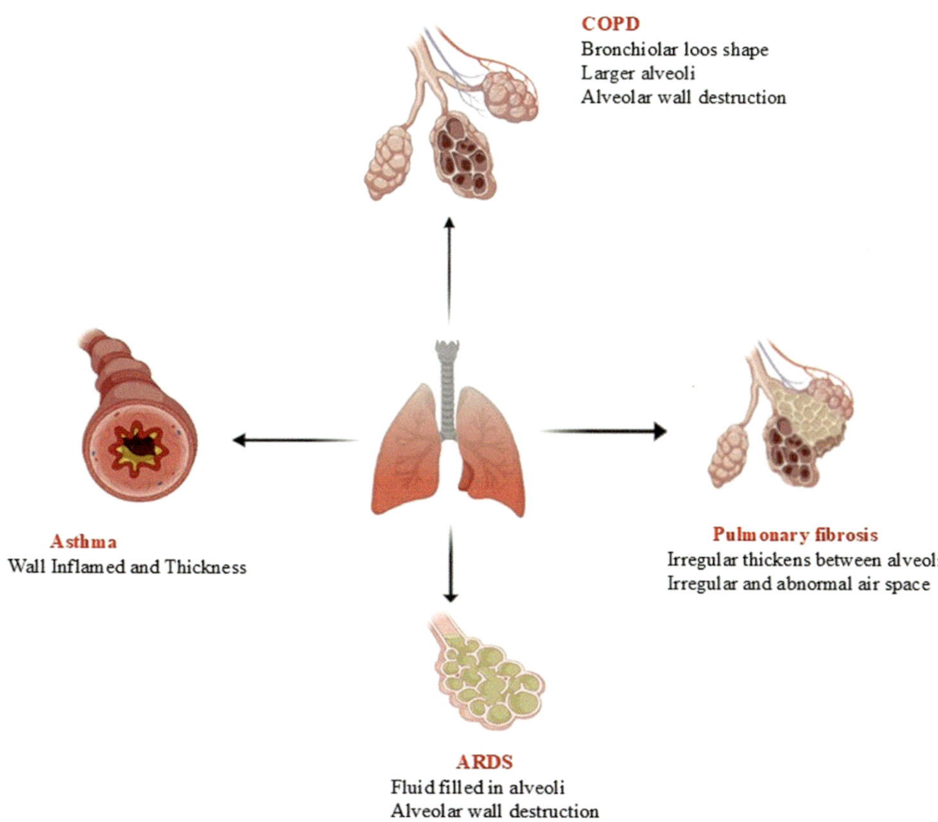

FIGURE 10.1 Common lung conditions and their effects on the respiratory system.

10.3 THE STRUCTURE OF THE LUNGS

The lungs are the primary organs of the respiratory system, and a remarkable example of biological engineering, designed for efficient gas exchange, vital for survival (Powell et al., 2015). Their complex structure, from the macroscopic lobes to the microscopic alveoli, is meticulously crafted to maximize surface area and minimize diffusion distance, making them exquisitely adapted for their crucial role in oxygenating the blood and removing carbon dioxide. Lung structure includes the trachea, bronchi, bronchioles, alveoli, and alveolar capillaries. The trachea, a rigid tube reinforced by C-shaped cartilaginous rings, ensures an open airway (Figure 10.2). At the sternal angle, the trachea splits into the right and left main bronchi, which each enter their corresponding lung at the hilum (Hakim & Usmani, 2014).

The alveolar-capillary membrane, also known as the respiratory membrane, is an astonishingly delicate structure, typically measuring less than 0.5 micrometers in thickness. It is composed of a surfactant layer that lowers surface tension, the alveolar epithelium, and the fused basement membranes of both the alveolar epithelium and the capillary endothelium (Connie et al., 2016). Bronchioles, unlike bronchi, lack cartilaginous support and are primarily composed of smooth muscle. This smooth muscle is critical for regulating airflow resistance through bronchodilation and bronchoconstriction. The bronchioles ultimately lead to terminal bronchioles, marking the end of the conducting zone, where no gas exchange occurs. The primary purpose of the conducting zone is to warm, humidify, and filter the incoming air, preparing it for gas exchange.

Beyond the terminal bronchioles lies the respiratory zone, the site where gas exchange occurs. Terminal bronchioles branch into respiratory bronchioles, which have occasional alveolar outpouchings, the functional units of the lung. Respiratory bronchioles lead into alveolar ducts, elongated airways lined almost entirely by alveoli. Alveolar ducts terminate in alveolar sacs, clusters of alveoli resembling bunches of grapes.

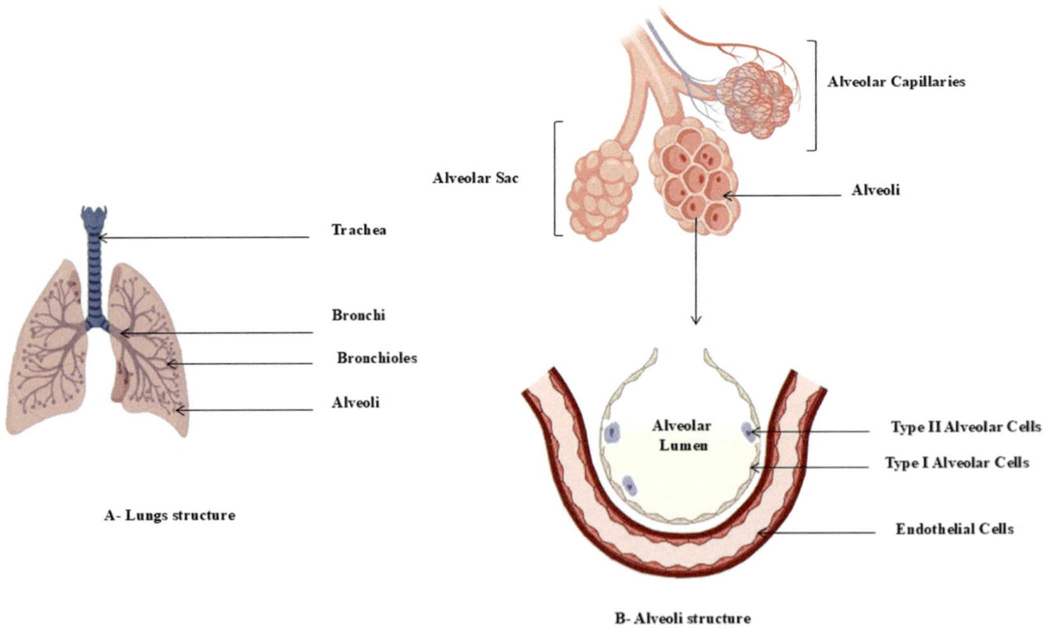

FIGURE 10.2 A. Overview of lung anatomy: B. Internal structure of alveoli: alveolar wall with alveolar epithelium composed of two main types of epithelial cells (Type I alveolar cells and Type II alveolar cells).

Each alveolus is a tiny, thin-walled sac surrounded by a dense network of capillaries. This close association between the alveoli and the capillaries forms the alveolar-capillary membrane, the critical interface for gas exchange.

The alveolar wall is composed primarily of two types of alveolar cells: Type I Alveolar Cells: these are thin, squamous epithelial cells that form the structural lining of the alveoli. Their thinness minimizes the diffusion distance for gases. Type II Alveolar Cells: these cells are more cuboidal and are responsible for producing surfactant, a phospholipid-protein mixture that reduces surface tension in the alveoli. Surfactant prevents alveolar collapse, making it easier to inflate the lungs (Song et al., 2024).

In addition to these cells, alveolar macrophages, also known as dust cells, patrol the alveolar surface, engulfing inhaled particles and pathogens, providing a critical defense mechanism against infection (Pervizaj-Oruqaj et al., 2024; Song et al., 2024).

The alveolar-capillary membrane, also known as the respiratory membrane, is a delicate structure that typically measures less than 0.5 micrometers in thickness. It consists of a layer of surfactant that reduces surface tension, alveolar epithelium, fused basement membranes of the alveolar epithelium, and capillary endothelium (Meegan et al., 2024). Oxygen can quickly diffuse from the alveoli into the blood, and carbon dioxide from the blood into the alveoli, via this very thin membrane. The large surface area of the alveoli, estimated at around 70 square meters in humans, further enhances the efficiency of gas exchange.

10.4 PATHOPHYSIOLOGY OF RESPIRATORY DISORDER

Respiratory disorders encompass a broad spectrum of conditions that affect the structures and function of the respiratory system, ultimately impairing gas exchange and leading to various systemic consequences (Fish, 2020). The respiratory system is crucial for oxygenating the blood and removing carbon dioxide, both of which are essential for life. Disruptions to this intricate system, caused by a multitude of factors like genetic predisposition, environmental exposures, and infectious agents, can lead to a range of respiratory disorders (Figure 10.3).

These disorders frequently manifest as shortness of breath, wheezing, coughing, chest pain, and decreased exercise tolerance. Understanding the underlying pathophysiology of these conditions

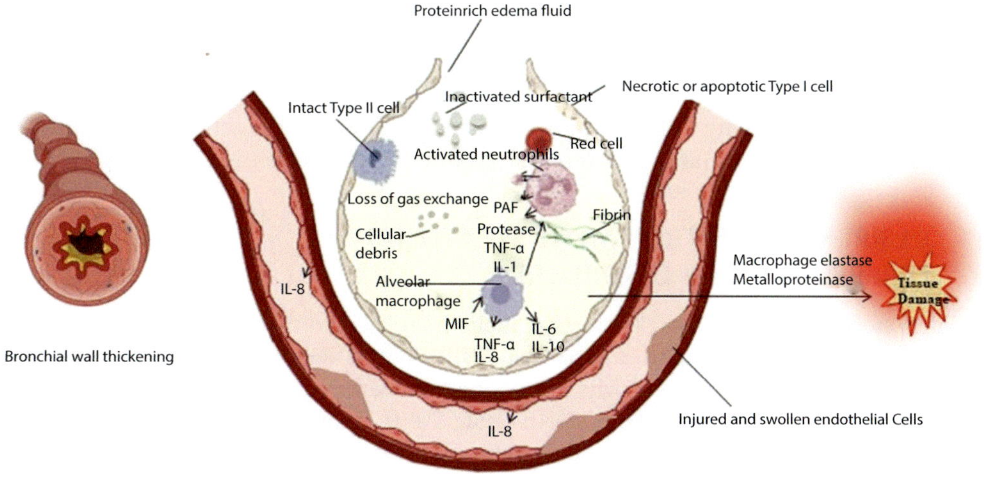

FIGURE 10.3 Respiratory disorder pathophysiology involves disrupted airway function, impaired gas exchange, inflammation, and structural damage caused by factors such as infection, chronic irritation, immune responses, or genetic abnormalities.

is paramount for developing effective therapeutic strategies and improving patient outcomes. Obstructive lung diseases are characterized by airflow limitation caused by narrowing or obstruction of the airways. This results in greater resistance in the airways, air trapping, and excessive lung inflation. Asthma is a long-lasting inflammatory condition characterized by temporary airflow obstruction, increased bronchial responsiveness, and airway inflammation. The way asthma develops is influenced by a mix of inherited traits, outside conditions, and how the immune system reacts. Exposure to allergens or irritants triggers an inflammatory cascade involving mast cells, eosinophils, T helper 2 (Th2) cells, and other immune cells. These cells release inflammatory mediators such as histamine, leukotrienes, prostaglandins, and cytokines (e.g., IL-4, IL-5, IL-13), leading to contraction of airway smooth muscle due to direct stimulation by mediators and increased parasympathetic tone, increased permeability of the bronchial microvasculature, leading to fluid leakage and airway wall thickening, increased production of mucus by goblet cells, contributing to airway obstruction. Long-lasting inflammation can lead to physical alterations in the airways, including thickening of submucosal tissue, increased smooth muscle growth, and the formation of new blood vessels (Hough et al., 2020). This remodeling contributes to irreversible airflow limitation. These pathological changes result in increased airway resistance and reduced airflow rates (e.g., a decreased FEV1/FVC ratio), air trapping and lung hyperinflation, as well as hypoxemia and hypercapnia during severe exacerbations.

COPD includes chronic bronchitis and emphysema, both of which are marked by irreversible damage. Limitations in airflow usually get worse as time goes on and are connected to an unusual inflammatory reaction. This is how the lungs respond to harmful substances or gases. Exposure to irritants like cigarette smoke causes chronic inflammation of the bronchial walls, leading to: hyperplasia of goblet cells (leading to excessive mucus production, contributing to airway obstruction), hypertrophy of bronchial glands (further exacerbating mucus production), ciliary dysfunction (impaired mucociliary clearance, leading to accumulation of mucus and increased susceptibility to infection), bronchial wall thickening and fibrosis (resulting in narrowing of the airways and increased airway resistance), and emphysema (defined pathologically as destruction of alveolar walls, leading to enlargement of air spaces distal to the terminal bronchioles).

Smoking triggers an inflammatory response in the lungs, leading to an imbalance between proteases and antiproteases. Increased activity of proteases (e.g., elastase) released by neutrophils and macrophages, coupled with decreased activity of antiproteases (e.g., α1-antitrypsin), leads to destruction of elastin in the alveolar walls. This is particularly prominent in individuals with α1-antitrypsin deficiency. Oxidative stress: cigarette smoke contains free radicals that damage lung tissue and inactivate antiproteases—alveolar wall destruction: reduction in the surface area for gas exchange and loss of elastic recoil.

In COPD, these processes cause airflow limitation and air trapping, increased residual volume (RV) and total lung capacity (TLC), decreased diffusion capacity for carbon monoxide (DLCO), hypoxemia and hypercapnia, and, in severe cases, pulmonary hypertension.

10.5 CURRENT DRUG TREATMENT STRATEGIES FOR RESPIRATORY DISORDERS

Bronchodilators, including beta-2 agonists, anticholinergics, and theophylline, have been used to treat respiratory disorders (Spahn & Israelsen, 2018). Beta-2 agonists activate beta-2 adrenergic receptors on airway smooth muscle cells, triggering a sequence of intracellular events that leads to the relaxation of smooth muscle (Hallsworth et al., 2001). Short-acting beta-2 agonists (SABAs), such as albuterol and levalbuterol, provide quick relief from sudden bronchospasms. On the other hand, long-acting beta-2 agonists (LABAs), such as salmeterol and formoterol, provide ongoing airway dilation and are often combined with inhaled corticosteroids for the long-term treatment of asthma and chronic obstructive pulmonary disease (COPD) (Aziz et al., 2018). Antagonist medications block the activity of acetylcholine on muscarinic receptors in the airways, leading to bronchodilation (Buels & Fryer, 2012). Inhaled corticosteroids (ICS) are potent anti-inflammatory agents that reduce airway inflammation and hyperreactivity in asthma and chronic obstructive pulmonary disease (COPD) (Adcock et al., 2022). They function by inhibiting the activation of inflammatory cells and decreasing the generation of inflammatory mediators. Common examples include beclomethasone, budesonide, fluticasone, and mometasone (Matera et al., 2019). ICSs are primarily used for long-term asthma control and are often combined with LABAs in fixed-dose inhalers to improve adherence and efficacy. Combining bronchodilators and ICS in a single inhaler is a common strategy for managing asthma and COPD (Ricciardolo et al., 2015). These fixed-dose combinations offer synergistic benefits, improving both bronchodilation and anti-inflammatory control. Examples include LABA/ICS combinations (e.g., salmeterol/fluticasone) and LAMA/LABA combinations (e.g., tiotropium/olodaterol). Emerging triple therapies combining ICS/LABA/LAMA are also gaining traction for patients with severe COPD.

Antibiotics are essential for treating bacterial respiratory infections, such as pneumonia, bronchitis, and exacerbations of chronic respiratory conditions. The choice of antibiotic depends on the suspected or confirmed pathogen and local patterns of antibiotic resistance. Common antibiotics used in respiratory infections include macrolides, beta-lactams, fluoroquinolones, and tetracyclines (Suresh Babu et al., 2013).

Mucus in the lungs is a common symptom in chronic lung diseases such as COPD. Mucolytics are drugs that help to break down and clear mucus from the airways, facilitating expectoration and improving lung function (Gipsman et al., 2024). Examples include: N-acetylcysteine (NAC): This agent breaks disulfide bonds in mucus, reducing its viscosity. Dornase alfa (recombinant human deoxyribonuclease I): This enzyme cleaves extracellular DNA in mucus, reducing its viscosity and improving mucociliary clearance. Hypertonic saline: Inhalation of hypertonic saline draws water into the airways, thinning mucus and stimulating a cough. They are often used in conditions characterized by excessive mucus production, such as cystic fibrosis, chronic bronchitis, and COPD (Roe et al., 2025).

Anti-inflammatory agents are used as add-on therapy for asthma control. Leukotriene Receptor Antagonists (LTRAs) and Phosphodiesterase-4 (PDE4) Inhibitors have been used as add-on therapy. LTRAs, such as montelukast and zafirlukast, block the actions of leukotrienes, inflammatory mediators involved in the pathogenesis of asthma. PDE4 inhibitors, such as roflumilast, are used to reduce inflammation in COPD by increasing intracellular cAMP levels. It is primarily used to reduce the frequency of COPD exacerbations.

Biologic agents are revolutionizing the treatment of severe asthma and other inflammatory respiratory disorders. These therapies target specific inflammatory pathways or immune cells involved in disease pathogenesis (Xiao et al., 2021). Examples include: Anti-IgE antibodies (Omalizumab). These antibodies bind to IgE, preventing it from binding to mast cells and basophils, thereby reducing allergic inflammation in asthma (Gon et al., 2022). Anti-IL-5 antibodies (Mepolizumab, Reslizumab): These antibodies inhibit IL-5, a cytokine that promotes eosinophil growth and survival, thereby lowering eosinophilic inflammation in asthma. Benralizumab, an antibody that binds to the IL-5 receptor on eosinophils, induces antibody-dependent cell-mediated cytotoxicity (ADCC), resulting in the depletion of these cells (Al et al., 2025; Eger et al., 2021). Dupilumab, an anti-IL-4 receptor alpha antibody, inhibits the signaling of IL-4 and IL-13, which are implicated in type 2 inflammatory responses in atopic dermatitis and asthma (Le Floc'h et al., 2020). Anti-TSLP antibodies (Tezepelumab): This antibody blocks TSLP, an epithelial-derived cytokine that plays a crucial role in initiating allergic airway inflammation (Shinkai & Yabuta, 2023).

10.6 ROLE OF GROWTH FACTORS IN RESPIRATORY DISORDERS

Numerous growth factors, each with unique functions in promoting tissue remodeling, fibrosis, and airway hyperactivity, have been linked to the pathophysiology of respiratory illnesses.

10.6.1 Transforming Growth Factor-Beta

TGF-β is a potent profibrotic cytokine playing a central role in the development of pulmonary fibrosis, including idiopathic pulmonary fibrosis (IPF). It stimulates fibroblast proliferation and differentiation into myofibroblasts, which produce excessive extracellular matrix (ECM) components, such as collagen. TGF-β also inhibits matrix degradation and promotes epithelial-to-mesenchymal transition (EMT), further contributing to fibrosis. Studies have shown elevated levels of TGF-β in the bronchoalveolar lavage fluid (BALF) and lung tissue of patients with idiopathic pulmonary fibrosis (IPF). Blocking TGF-β signaling has shown promise in preventing fibrosis in preclinical models (Chong et al., 2023; Saito et al., 2018).

10.6.2 Platelet-Derived Growth Factor

Platelet-Derived Growth Factor (PDGF) is another potent mitogen for fibroblasts and smooth muscle cells. It promotes cell migration, proliferation, and the production of the extracellular matrix (ECM). Elevated levels of PDGF have been detected in BALF from patients with asthma and COPD. In asthma, PDGF contributes to airway smooth muscle hypertrophy and hyperplasia, which leads to airway hyperreactivity. In COPD, airway remodeling and emphysema contribute (Kardas & Daszyn, 2020).

10.6.3 Vascular Endothelial Growth Factor

Although primarily known for its role in angiogenesis, VEGF also plays a role in airway remodeling and inflammation in respiratory disorders (Shi et al., 2025). In asthma, VEGF contributes to increased vascular permeability and airway edema. In COPD, VEGF can promote angiogenesis in response to hypoxia, potentially contributing to pulmonary hypertension (Yun et al., 2016). Conversely, in emphysema, VEGF deficiency may contribute to alveolar destruction (Shi et al., 2025).

10.6.4 Epidermal Growth Factor

Epidermal Growth Factor (EGF) and its receptor EGFR are involved in epithelial cell proliferation and repair. In some contexts, EGF signaling can be protective by promoting wound healing. However, chronic EGF signaling can contribute to airway remodeling and fibrosis in asthma and COPD. Increased EGFR expression has been observed in the airways of patients with these diseases (Y. Liu et al., 2023).

10.6.5 Cytokines in Respiratory Disorders

Cell cytokines (T helper (Th)) are critical mediators of inflammation and immune responses in the lung. Their dysregulation contributes to the pathogenesis of various respiratory disorders.

10.6.5.1 Th1 Cytokines

Traditionally associated with the response to intracellular pathogens, Th1 cytokines can contribute to airway inflammation and tissue damage in specific contexts. IFN-γ can promote epithelial cell apoptosis and contribute to COPD pathogenesis (Tang et al., 2023; Yu et al., 2020).

10.6.5.2 Th2 Cytokines

These cytokines are central to the pathogenesis of allergic asthma. IL-4 promotes the production of IgE by B cells, thereby sensitizing mast cells to allergens. IL-5 is employed and activates eosinophils, which contribute to airway inflammation and hyperactivity. IL-13 induces mucus hypersecretion, airway fibrosis, and airway hyperactivity. Therapies targeting IL-5 and IL-4/IL-13 signaling have proven effective in treating severe asthma (Khan et al., 2025; Mi & Guo, 2022; Pellicano et al., 2023).

10.6.5.3 Th17 Cytokines

Th17 cells and their related cytokines, IL-17 and IL-22, are implicated in neutrophilic inflammation and airway remodeling in chronic obstructive pulmonary disease (COPD), cystic fibrosis, and severe asthma. IL-17 promotes neutrophil recruitment and activation and contributes to mucus production and airway hyperactivity (Aksakal & Gorgun, 2024; Manni et al., 2014).

10.6.5.4 Pro-inflammatory Cytokines

These cytokines are produced by various immune cells and structural cells in response to injury, infection, and inflammation. They contribute to the acute inflammatory response in ARDS and other lung injuries. TNF-α promotes leukocyte recruitment and activation, while IL-1β and IL-6 contribute to systemic inflammation and fever. Chronic elevation of these cytokines contributes to the pathogenesis of chronic obstructive pulmonary disease (COPD) and pulmonary fibrosis (Schultheiß et al., 2022; Sordillo et al., 2022).

10.6.5.5 Chemokines

Chemokines are a family of chemoattractant cytokines that recruit immune cells to the site of inflammation (Cambier et al., 2023). CXCL8/IL-8 is a potent chemoattractant for neutrophils, playing a crucial role in neutrophilic inflammation in conditions such as chronic obstructive pulmonary disease (COPD) and cystic fibrosis. CCL2/MCP-1 recruits monocytes and macrophages, contributing to chronic inflammation and fibrosis in pulmonary fibrosis (Li, 2023).

10.7 STEM CELL EXOSOMES: A PROMISING AVENUE

Mesenchymal stem cells (MSCs), exosomes (MSCs-Ex), are extracellular vesicles (EVs) released by stem cells. They contain a complex cargo of proteins, lipids, and nucleic acids (DNA, mRNA, microRNA, lncRNA) that can be transferred to recipient cells, modulating their function and influencing various cellular processes. MSCs-Ex have been demonstrated to have therapeutic potential in preclinical models of multiple diseases, including cardiovascular disease, neurodegenerative disorders, and autoimmune diseases (Xu et al., 2020; Lamiaa et al., 2025; Yusuf and Haider, 2025; Najimi and Haider, 2022). The composition of stem cell exosomes is complex and reflects the specific stem cell type, culture conditions, and the stem cells' physiological state.

Stem cell exosomes mediate their therapeutic effects through several mechanisms, including direct delivery of cargo, modulation of gene expression, activation of signaling pathways, and immunomodulation (Abdulmalek et al., 2024). Stem cell exosomes have the potential to regulate the immune response by promoting tissue repair and suppressing inflammation (Wei et al., 2025). They can interact with immune cells, such as macrophages and T cells, and alter their activation state and cytokine production (Yi et al., 2025; Zubair et al., 2025).

Stem cell exosomes offer a promising opportunity for regenerative medicine, providing a cell-free therapeutic approach that circumvents some of the constraints of cell-based therapies (Choudhery et al., 2024; Liang et al., 2025). Their capacity to transport a multifaceted array of bioactive molecules to recipient cells and modulate their function holds excellent potential for treating a wide range of diseases and injuries (Kumar et al., 2024). However, additional studies are required to comprehensively elucidate their modes of action, enhance their manufacture and distribution, and assess their safety and efficacy in clinical studies (Figure 10.4).

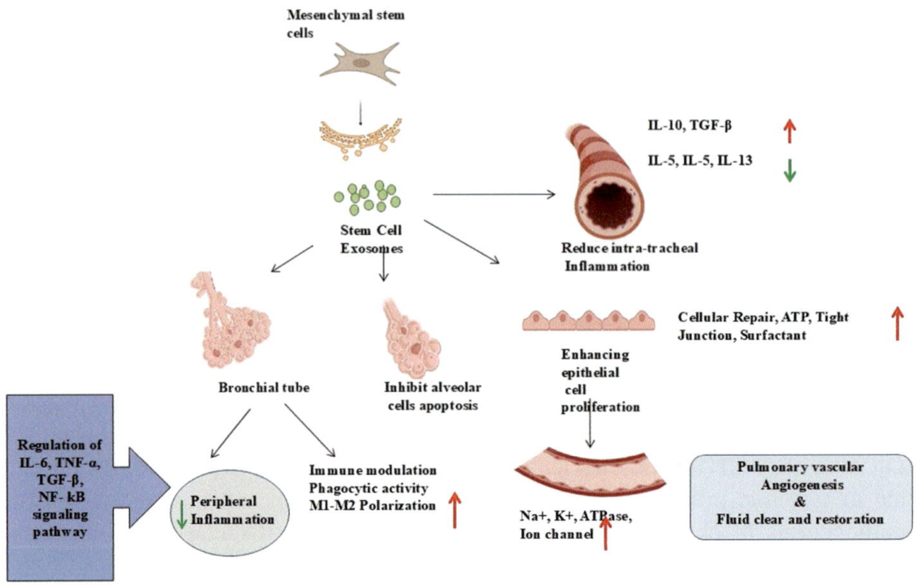

FIGURE 10.4 Stem cell exosome therapy for respiratory disorders functions by delivering bioactive molecules that reduce inflammation, promote tissue repair, modulate immune responses, and enhance regeneration of damaged lung tissue.

10.7.1 Stem Cell Exosome Inhalation: A Novel Therapeutic Approach

Respiratory disorders, such as COPD, asthma, ARDS, and pulmonary fibrosis, pose a significant global health burden. Current treatments often fall short of providing effective long-term solutions. Stem cell therapy holds promise for regenerative medicine in these conditions; however, direct cell transplantation faces significant challenges. Stem cell-derived exosomes, nanoscale extracellular vesicles mediating intercellular communication, have emerged as a promising alternative. The inhalation of stem cell exosomes represents a novel therapeutic approach for respiratory disorders, with a focus on their mechanisms of action, preclinical and clinical evidence, potential challenges, and future directions.

Inhalation provides a direct, efficient route for therapeutics to reach the lungs, bypassing systemic circulation and minimizing off-target effects (Figure 10.5). The advantages of using inhaled stem cell exosomes in respiratory disorders include targeted delivery, non-invasive administration, reduced risk of rejection, and enhanced tissue repair (Borghardt et al., 2018).

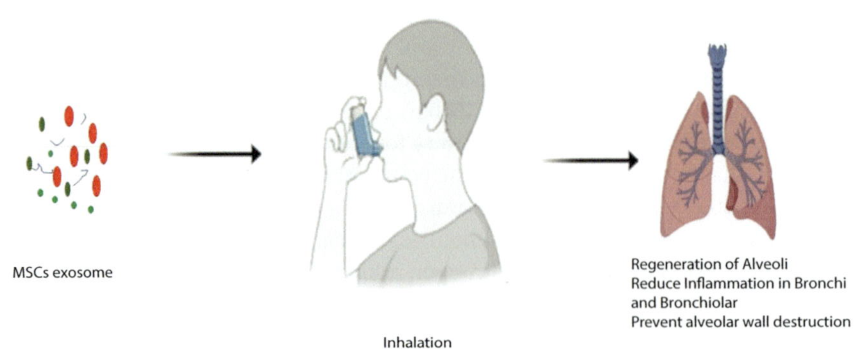

FIGURE 10.5 Therapeutic inhalation of exosomes from stem cells.

10.7.2 Exosome for Chronic Obstructive Pulmonary Disease

COPD encompasses a spectrum of chronic respiratory disorders, primarily chronic bronchitis and emphysema, characterized by persistent airflow limitation that is not fully reversible. The main risk factor for COPD is cigarette smoke (CS) exposure, although other environmental pollutants and genetic predispositions also contribute. The disease is characterized by chronic inflammation in the airways and lung parenchyma, resulting in alveolar destruction, mucus hypersecretion, and ultimately, irreversible airflow obstruction. Present treatment methods primarily focus on controlling symptoms and slowing disease progression, but they do not effectively repair lung damage.

The understanding of COPD pathogenesis centers on direct lung tissue injury from inhaled irritants, which leads to cellular dysfunction and inflammatory responses. However, recent advances have highlighted the importance of cell-to-cell communication in perpetuating and amplifying the inflammatory cascade and driving tissue remodeling. Exosomes secreted by virtually all cell types have emerged as crucial mediators of this intercellular communication.

COPD is a debilitating respiratory illness characterized by progressive airflow limitation and chronic inflammation (Adhikari et al., 2020). The pathophysiology of COPD is significantly influenced by the dysfunction of alveolar macrophages (AMs), which are crucial for tissue regeneration, host defense, and lung homeostasis (Vlahos & Bozinovski, 2014). Recent research suggests that multiple apoptotic (AM) death pathways significantly contribute to the development and progression of lung inflammation, which, in turn, plays a crucial role in the onset and exacerbation of various airway disorders (Xia et al., 2024).

Currently, existing treatments primarily focus on symptom management, underscoring the pressing need for new approaches. Stem cell-based treatments have emerged as promising options, and recently, the healing potential of stem cell-derived exosomes has garnered considerable interest. These nanoscale vesicles, carrying a cargo of bioactive molecules, offer a cell-free approach to deliver regenerative and immunomodulatory effects to the damaged lung tissue (Zhu et al., 2022).

10.7.3 Stem Cell Exosome Inhalation for Pulmonary Fibrosis

Pulmonary fibrosis (PF) is a long-lasting, progressive, and frequently deadly lung condition characterized by an overproduction of extracellular matrix and reduced lung function (Liu et al., 2024). MSC-exosomes represent a novel therapeutic strategy for PF, with a focus on their mechanisms of action, preclinical evidence, and translational challenges.

PF comprises a diverse collection of long-term lung diseases characterized by the gradual scarring of lung tissue, which impairs gas exchange and leads to respiratory failure (J. Wang et al., 2024). Idiopathic pulmonary fibrosis (IPF), the most common type, is marked by a continuous worsening, with few treatment choices and an unfavorable outlook (Glass et al., 2022). The underlying pathophysiology of PF involves a complex interplay of inflammatory, profibrotic, and remodeling processes, including dysregulated fibroblast activation, excessive extracellular matrix (ECM) deposition, and aberrant epithelial-mesenchymal transition (EMT) (Luo et al., 2024).

Current therapies, such as anti-fibrotic drugs (pirfenidone and nintedanib), primarily slow down disease progression but do not reverse established fibrosis or improve long-term survival significantly (Aimo et al., 2022; Fuster-Martínez & Calatayud, 2024; Man et al., 2024). Lung transplantation remains the only curative option, but its applicability is limited by donor organ availability and post-transplant complications. These limitations underscore the urgent need for the development of novel therapeutic strategies that can effectively target the underlying mechanisms of PF and promote lung tissue regeneration.

Stem cell-based therapies, particularly those involving MSCs, have demonstrated encouraging preclinical results in PF models by exhibiting antifibrotic, anti-inflammatory, and regenerative properties (Rojo & Conget, 2018). However, challenges such as poor cell engraftment, potential immune rejection, and ethical concerns have hampered the clinical translation of cell transplantation. This has led to the investigation of cell-free therapies, with a focus on the therapeutic potential of exosomes derived from stem cells (Habib et al., 2025b).

MSC-derived exosomes, in particular, have shown promise for therapeutic applications due to their potent immunomodulatory, anti-inflammatory, and regenerative properties (An et al., 2025). The cargo of MSC exosomes is highly influenced by the cellular environment and the method of cell culture, which can alter their biological activity.

Well-designed preclinical and clinical trials are crucial for determining the true potential of inhaled stem cell exosomes in treating pulmonary fibrosis. Despite these challenges, the field is rapidly advancing, and stem cell exosome inhalation holds significant promise for improving the lives of patients with this devastating disease.

10.7.4 Stem Cell Exosome Inhalation for Asthma

Asthma is characterized by airway inflammation, heightened airway sensitivity, and temporary airflow obstruction. The development of the disease involves a complicated interaction of genetic factors, environmental influences, and issues with the immune system. This results in ongoing inflammation caused by Th2 cytokines, the presence of eosinophils, and changes in the structure of the airways. It is characterized by airway inflammation and hyper-responsiveness (Bradding et al., 2024). Conventional medications used to treat asthma patients include short- and long-acting beta-2 agonists, anticholinergics, leukotriene antagonists, and inhaled corticosteroids (Vähätalo et al., 2022; Van Ganse et al., 2020). However, even with standard treatment, 5–10% of asthma symptoms do not go away. Therefore, there is a pressing need for novel therapies that can target the root causes of asthma and potentially offer long-term disease modification. In recent years, stem cell-derived exosomes have emerged as a promising cell-free therapy due to their ability to convey therapeutic signals and modulate immune responses (Abdulmalek et al., 2024). MSCs' exosomes offer a novel therapeutic approach to treat asthma patients. Stem cell exosomes possess immunomodulatory, anti-inflammatory, and tissue-regenerative properties, making them a compelling candidate for asthma management (Abbaszadeh et al., 2022; Lv et al., 2025). In asthma, MSCs have shown promise due to their immunomodulatory properties (Yuan, 2024). MSCs can suppress inflammatory responses, promote tissue repair, and modulate the immune system, all of which are essential for asthma management. However, the mechanism of action of MSCs has moved beyond the traditional concept of cell engraftment and direct differentiation, highlighting the importance of paracrine signaling, particularly through exosomes. However, further research is needed to optimize exosome isolation, dosage, and delivery methods for clinical translation.

10.7.5 Stem Cell Exosome Inhalation for Acute Respiratory Distress Syndrome

ARDS is a life-threatening inflammatory lung state characterized by severe hypoxemia and high mortality rates (Matthay et al., 2018). Conventional therapies often fail to adequately address the underlying pathophysiology of ARDS, emphasizing the urgent need for novel treatment approaches (Safwan et al., 2025). Recent advancements have focused on the therapeutic potential of stem cell-derived exosomes, nanoscale extracellular vesicles that contain bioactive molecules. These exosomes have demonstrated immunomodulatory, anti-inflammatory, and regenerative properties, making them attractive candidates for ARDS therapy (Wang et al., 2021).

ARDS is a devastating clinical syndrome characterized by the acute onset of bilateral pulmonary edema and hypoxemia, resulting from a severe inflammatory response within the lung. A variety of clinical insults, including pneumonia, sepsis, trauma, and aspiration, can trigger this complex condition (Serazin et al., 2021). ARDS is linked to considerable health complications and death, even with improvements in supportive treatments and mechanical breathing assistance. The underlying pathophysiology involves a dysregulated inflammatory cascade leading to alveolar damage, increased vascular permeability, and impaired gas exchange. Current therapies, such as mechanical ventilation and corticosteroids, aim to support respiratory function and reduce inflammation, but often fall short in effectively reversing the lung pathology (Al-rasheedi et al., 2021). Hence, there is a pressing need for innovative therapies that can directly target the mechanisms driving ARDS and promote lung repair (Quint et al., 2023).

Stem cell therapy has emerged as a promising approach in regenerative medicine for treating various conditions, including lung diseases. However, concerns about the safety and practicality of cell transplantation have prompted investigations into cell-free therapies. Stem cells secrete paracrine factors, including soluble factors contained in the secretome and extracellular vesicles, such as exosomes, which play an active role in intercellular communication (Habib et al., 2025a & 2025b). These exosomes contain essential molecules, such as proteins, lipids, and nucleic acids, that can influence the activities of the cells they are received by.

10.8 PRECLINICAL AND CLINICAL STUDIES

Multiple preclinical studies in animal models of respiratory disorders such as COPD, lung fibrosis, and ARDS have demonstrated. These studies commonly induce ARDS in mice, rats, and other large animals using LPS, acid aspiration, or other inflammatory stimuli (Table 10.1).

Administered via intravenous or intratracheal routes, the exosomes have often been shown to improve gas exchange (increase PaO2/FiO2 ratio), reduce pulmonary edema, decrease lung inflammation (reduced immune cell infiltration and pro-inflammatory cytokine levels), promote alveolar repair (increased epithelial cell proliferation and reduced fibrosis), and improve survival rates. Several preclinical studies have demonstrated the therapeutic potential of stem cell-derived exosomes in various models of pulmonary fibrosis. Stem cells have shown efficacy in reducing fibrotic changes in animal models of pulmonary fibrosis (PF) induced by bleomycin or other fibrotic stimuli. Studies have reported decreased lung collagen deposition, improved lung function, and reduced inflammation following exosome administration. Several routes of administration, including intratracheal, intravenous, and aerosolized delivery, have shown promising results in animal models.

10.9 CHALLENGES

Despite the promising preclinical results, several challenges hinder the clinical translation of MSC-Exos for respiratory disorders. The production of MSC-Exos is highly variable, influenced by factors such as cell source, culture conditions, and the isolation techniques used. This variability can lead to inconsistent therapeutic efficacy. Establishing standardized protocols for MSC-Exo production and characterization is crucial for ensuring reproducibility and comparability across different studies. Current methods for MSC-Exo isolation, such as ultracentrifugation, are often inefficient and difficult to scale up for clinical applications. Developing more efficient and scalable manufacturing methods, such as tangential flow filtration or microfluidic devices, is crucial for meeting the demands of clinical trials. A comprehensive characterization of MSC-Exo composition and function is necessary to ensure product quality and safety. Standardized methods for quantifying MSC-Exo concentration, size distribution, protein and nucleic acid cargo, and biological activity are needed. Systemic administration of MSC-Exos may lead to off-target effects and limited accumulation in the lungs. Developing strategies to improve MSC-Exo targeting to the lungs, such as surface modification with targeting ligands or encapsulation in biocompatible carriers, is essential to maximize therapeutic efficacy and minimize side effects. While MSC-Exos are generally considered less immunogenic than stem cells, potential immunogenicity and safety concerns remain to be addressed. Thorough preclinical safety studies are necessary to evaluate the potential for adverse effects, including immune reactions and tumorigenesis. A deeper understanding of the mechanisms by which MSC-Exos exert their therapeutic effects in respiratory disorders is crucial for optimizing their clinical application. Identifying the key signaling pathways and target cells involved in MSC-Exo-mediated lung repair will enable the development of more targeted and effective therapies. The rules and guidelines for MSC-Exo-related treatments

TABLE 10.1
Summary of preclinical and clinical studies of Stem Cell Therapy or stem cell exosomes for Respiratory Disorders

Preclinical Studies					
Disorder	**Cell Type**	**Model/Study Type**	**Delivery Route**	**Key Findings**	**Reference**
COPD/ Emphysema	UC-MSCs exosome	Mice (Papain induced)	Intravenous	Reduced lung inflammation, improved lung function, and decreased emphysema development.	(Chen et al., 2022)
Asthma	Gingival and Adipose MSCs	Mice (Allergic)	Intravenous	Lowered allergic asthma. It primarily focuses on CD11b+CD11c+ pro-inflammatory dendritic cells, thereby restoring a healthy balance between Th2 and Treg cells.	(Fang et al., 2024; Trzil et al., 2016)
Fibrosis	MSCs (Adipose-derived), iPSCs,	Mouse (Bleomycin-induced)	Intravenous	Reduced fibrosis, improved lung function, and decreased collagen deposition.	(Chakraborty et al., 2024; Lee et al., 2024; Zhao, Wu, et al., 2023)
ARDS/ALI	MSCs (Bone Marrow)	Pig (Oleic acid-induced injury)	Intravenous	Reduced lung edema, decreased inflammation.	(Farkhad et al., 2022; Moodley et al., 2016)
Clinical Studies					
Disorder	**Cell Type**	**Model/Study Type**	**Delivery Route**	**Key Findings**	**Reference**
COPD	MSCs (Bone Marrow)	Phase I/II	Intravenous/ Intrabronchial	There are some improvements in quality of life, limited improvements in lung function, and it is generally safe.	(Hoang et al., 2021; Stessuk et al., 2013)
Asthma	MSCs (Umbilical Cord)	Phase I/II	Intravenous	Rescue inhaler use decreased to once a month, representing a more than 90% drop. Also, he used the nebulizer 70% less often. This progress continued over the next six months.	(Sharan et al., 2022)
ARDS	Multipotent adult progenitor cells, Bone marrow MSCs	Phase I/II	Intravenous	Reduced mortality, improved oxygenation, decreased inflammation (in some studies), and is generally safe.	(Ichikado et al., 2023; Wilson et al., 2015)

are still changing. There is a need for well-defined standards and regulations to guarantee that these therapies are safe and effective for use in medical settings.

10.10 FUTURE DIRECTIONS

To address the challenges outlined above and accelerate the clinical translation of MSC-Exos for respiratory disorders, several future directions should be prioritized. There is a need to develop standardized protocols for MSC-Exo production, including selection of optimal cell sources, culture conditions, and isolation techniques. Implement scalable manufacturing methods, such as tangential flow filtration or microfluidic devices, to meet the demands of clinical trials and ensure efficient production. Explore the use of bioreactors for the large-scale production of mesenchymal stem cell-derived exosomes (MSC-Exo). Explore local delivery methods, such as inhalation or intratracheal administration, to minimize systemic exposure and maximize therapeutic efficacy in the lungs. Investigate the immunogenicity of MSC-Exos in preclinical models and in vitro assays. Develop strategies to reduce the immunogenicity of MSC-Exos, such as surface modification with immunosuppressive molecules. Conduct thorough preclinical safety studies to assess potential adverse effects, including immune reactions and tumorigenesis. Utilize omics technologies to comprehensively analyze the cellular and molecular changes induced by MSC-Exo treatment. Design well-controlled clinical trials with clearly defined endpoints to assess the safety and efficacy of MSC-Exo-based therapies for respiratory disorders.

10.11 CONCLUSION

Stem cell-derived exosomes represent a promising therapeutic approach for respiratory disorders that overcomes some of the limitations of traditional therapies and stem cell transplantation. Their ability to modulate inflammation, promote tissue repair, and induce immunomodulation makes them a compelling candidate for future clinical therapies. Further research is needed to standardize production protocols, assess safety, and optimize dosing strategies to fully realize the potential of stem cell exosome therapy in alleviating the burden of respiratory disorders.

Stem cell-derived exosomes offer significant potential as a new treatment method for respiratory conditions. Their ability to adjust the immune system, reduce inflammation, promote lung healing, and improve oxygen levels makes them appealing options for addressing the ongoing medical challenges in respiratory disorders. Preclinical evidence has demonstrated their efficacy in various respiratory disorder models, including COPD. The field is advancing toward clinical translation. However, challenges remain in terms of standardization, safety, and clinical trial design. Overcoming these hurdles will help to unlock the full potential of stem cell exosomes as a revolutionary therapy for ARDS and other complex lung diseases.

REFERENCES

Abbaszadeh H, Ghorbani F, Abbaspour-Aghdam S, Kamrani A, Valizadeh H, Nadiri M, et al. Chronic obstructive pulmonary disease and asthma: mesenchymal stem cells and their extracellular vesicles as potential therapeutic tools. *Stem Cell Res Ther* [Internet]. 2022;13(1):1–15. https://doi.org/10.1186/s13287-022-02938-5.

Abdulmalek OAAY, Al-Khalifa KHHKAA, Alturani MMAB, Alexandra E. Butler. Therapeutic applications of stem cell-derived exosomes. *Int J Mol Sci*. 2024;25(6):3562.

Adcock IM, Bhatt SP, Balkissoon R, Wise RA. The use of inhaled corticosteroids for patients with COPD who continue to smoke cigarettes: an evaluation of current practice. *Am J Med* [Internet]. 2022;135(3):302–12. https://doi.org/10.1016/j.amjmed.2021.09.006.

Adhikari TB, Acharya P, Högman M, Neupane D, Drews A, Cooper BG, et al. Prevalence of chronic obstructive pulmonary disease and its associated factors in Nepal: findings from a community-based household survey. *Int J Chron Obstruct Pulmon Dis*. 2020;15:2319–31.

Ahmed ZT, Alkahlot MH, Haider KH. MSC-derived exosomes: advances in cell-free therapy. In: KH Haider, editor. *Handbook of Stem Cell Applications*. Singapore: Springer Nature Singapore; 2023. p. 1–41.

Aimo A, Spitaleri G, Nieri D, Tavanti LM, Meschi C, Panichella G, et al. Pirfenidone for idiopathic pulmonary fibrosis and beyond idiopathic pulmonary fibrosis: pirfenidone for lung and cardiac disorders clinical trials on pirfenidone for the treatment of idiopathic pulmonary fibrosis pirfenidone in the treatment of idiopathic pulmonary fibrosis pirfenidone in the treatment of idiopathi. *Card Fail Rev*. 2022;8(12):1–5.

Akram KM, Samad S, Spiteri MA, Forsyth NR. Mesenchymal stem cells promote alveolar epithelial cell wound repair in vitro through distinct migratory and paracrine mechanisms. *Respir Res*. 2013;14(1):1–16.

Aksakal S, Gorgun S. The protective role of IL-17 and IL-22 in COVID-19 infection. *Iran J Immunol*. 2024;21(3):225–33.

Al CKET, Kyriakopoulos C, Papadopoulou E, Potonos D, Exarchos K, Beris E, et al. Effectiveness of anti-IL-5 / 5R α biologics in severe asthma in real-world studies: a systematic review and meta-analysis effectiveness of anti-IL-5/5R / 5R α biologics in severe asthma in real-world studies: a systematic review and meta-analysis. *ERJ Open Res* [Internet]. 2025;1(11). https://doi.org/10.1183/23120541.00625-2024.

Al-Omar MT, Alnajjar MT, Ahmed ZT, Salaas FMI, Alrefaei TSM, Haider KH. Endothelial progenitor cell-derived small extracellular vesicles for myocardial angiogenesis and revascularization. *J Clin Transl Res*. 2022;8(6):1–12.

Al-Rasheedi M, Alhazmi Y, Almaqwashi N, Alreshidi MA, Kardam A, Sharaf M, Haider KH. Corticosteroid therapy for 2019-nCoV-infected patients: a case series of eight mechanically ventilated patients. *Clin Case Rep*. 2021;9(5): e04066. https://doi.org/10.1002/ccr3.4066.

An T, Li D, Wang N, Ma F, Li T, He N, et al. Extracellular vesicle mechanisms and potential of mesenchymal stem cell-derived exosomes for treating radiation-induced intestinal injury. *Extracell Vesicle*. 2025:100078. https://doi.org/10.1016/j.vesic.2025.100078.

Ateeq M, Broadwin M, Sellke FW, Abid MR. Extracellular vesicles' role in angiogenesis and altering angiogenic signaling. *Med Sci* (Basel, Switzerland). 2024;12(1):1–21.

Aziz MIA, Tan LE, Wu DBC, Pearce F, Chua GSW, Lin L, et al. Comparative efficacy of inhaled medications (ICS/LABA, LAMA, LAMA/LABA, and SAMA) for COPD: a systematic review and network meta-analysis. *Int J COPD*. 2018;13:3203–31.

Borghardt JM, Kloft C, Sharma A. Inhaled therapy in respiratory disease: the complex interplay of pulmonary kinetic processes. *Can Respir J*. 2018;2018:1–9.

Bradding P, Porsbjerg C, Côté A, Dahlén SE, Hallstrand TS, Brightling CE. Airway hyperresponsiveness in asthma: the role of the epithelium. *J Allergy Clin Immunol*. 2024;153(5):1181–93.

Buels KS, Fryer AD. Muscarinic receptor antagonists: effects on pulmonary function. *Handb Exp Pharmacol*. 2012;208(208):317–41.

Cambier S, Gouwy M, Proost P. The chemokines CXCL8 and CXCL12: molecular and functional properties, role in disease, and efforts towards pharmacological intervention. *Cell Mol Immunol*. 2023;20(3):217–51.

Cañas JA, Rodrigo-Muñoz JM, Gil-Martínez M, Sastre B, Del Pozo V. Exosomes: A key piece in asthmatic inflammation. *Int J Mol Sci*. 2021;22(2):127.

Chakraborty A, Wang C, Hodgson-Garms M, Broughton BRS, Frith JE, Kelly K, et al. Induced pluripotent stem cell-derived mesenchymal stem cells reverse bleomycin-induced pulmonary fibrosis and related lung stiffness. *Biomed Pharmacother*. 2024;178(June):117259. https://doi.org/10.1016/j.biopha.2024.117259.

Cheema NA, Castagna A, Ambrosani F, Argentino G, Friso S, Zurlo M, et al. Extracellular vesicles in Asthma: intercellular cross-talk in TH2 inflammation. *Cells*. 2025;14(7):1–25.

Chen Q, Lin J, Deng Z, Qian W. Exosomes derived from human umbilical cord mesenchymal stem cells protect against papain-induced emphysema by preventing apoptosis through activating VEGF-VEGFR2-mediated AKT and MEK/ERK pathways in rats. *Regen Ther*. 2022;21:216–24. https://doi.org/10.1016/j.reth.2022.07.002.

Chong DLW, Mikolasch TA, Sahota J, Rebeyrol C, Garthwaite HS, Booth HL, et al. Investigating the role of platelets and platelet-derived transforming growth factor-β in idiopathic pulmonary fibrosis. *Am J Physiol - Lung Cell Mol Physiol*. 2023;325(4):L487–99.

Choudhery MS, Arif T, Mahmood R, Harris DT. Stem cell-based acellular therapy: insight into biogenesis, bioengineering, and therapeutic applications of exosomes. *Biomolecules*. 2024;14(7):1–26.

Connie C.W. Hsia, Hyde DM, Weibel ER. Lung structure and gas exchange. *Physiol Behav*. 2016;176(1):100–6.

Dinh Puyen C, Paudel D, Brochu H, Popowski KD, Gracieux MC, Cores J, et al. Inhalation of lung spheroid cell secretome and exosomes promotes lung repair in pulmonary fibrosis. *Nat Commun*. 2020;11. https://doi.org/10.1038/s41467-020-14344-7.

Eger K, Kroes JA, ten Brinke A, Bel EH. Long-term therapy response to Anti–IL–5 biologics in severe asthma—a real-life evaluation. *J Allergy Clin Immunol Pract.* 2021;9(3):1194–200. https://doi.org/10.1016/j.jaip.2020.10.010.

Fang Q, Wu W, Xiao Z, Zeng D, Liang R, Wang J, et al. Gingival-derived mesenchymal stem cells alleviate allergic asthma inflammation via HGF in animal models. *iScience.* 2024;27(5):109818. https://doi.org/10.1016/j.isci.2024.109818.

Farkhad NK, Sedaghat A, Reihani H, Moghadam AA, Moghadam AB, Ghaebi NK, et al. Specific clinical and immunological changes following mesenchymal stem cell transplantation in COVID–19–induced acute respiratory distress syndrome patients: a phase-I clinical trial. *Iran J Allergy, Asthma Immunol.* 2022;21(6):687–703.

Feng Y, Guo K, Jiang J, Lin S. Mesenchymal stem cell-derived exosomes as delivery vehicles for non-coding RNAs in lung diseases. *Biomed Pharmacotherapy.* 2024;170:116008. https://doi.org/10.1016/j.biopha.2023.116008.

Fish B. Chronic respiratory diseases: An introduction and need for novel drug delivery approaches. In: KDua et al., editor. *Targeting Chronic Inflammatory Lung Diseases Using Advanced Drug Delivery Systems.* LondonLondon: Academic Press; 2020. p. 1–31.

Fuster-Martínez I, Calatayud S. The current landscape of antifibrotic therapy across different organs: a systematic approach. *Pharmacol Res.* 2024;205(May):107245. https://doi.org/10.1016/j.phrs.2024.107245.

Gipsman AI, Bhandari A, Bhandari V. Use of mucolytics and inhaled antibiotics in the NICU. *J Perinatol* [Internet]. 2024;45:5–12. https://doi.org/10.1038/s41372-024-02178-w.

Glass DS, Grossfeld D, Renna HA, Agarwala P, Spiegler P, DeLeon J, et al. Idiopathic pulmonary fibrosis: current and future treatment. *Clin Respir J.* 2022;16(2):84–96.

Gon Y, Maruoka S, Mizumura K. Omalizumab and IgE in the control of severe allergic asthma. *Front Pharmacol.* 2022;13(March):1–8.

Habib SM, Martini MF, Al Ahmed O, Qassab AR, Haider KH. Mesenchymal stem cell secretomes for cell-free therapy of the broken heart. In: KH Haider, editor. *Handbook of Regenerative Medicine: Cell-Free Therapy.* Boca Raton (FL): CRC Press; 2025a. p. 274–293.

Habib SM, Martini MF, Al Ahmed O, Qassab AR, Haider KH. The potential of mesenchymal stem cell–derived exosomes in cell-free therapy for myocardial repair. In: KH Haider, editor. *Handbook of Regenerative Medicine: Cell-Free Therapy.* Boca Raton (FL): CRC Press; 2025b. p. 372–390.

Haider KH, Aramini B. Mircrining the injured heart with stem cell-derived exosomes: an emerging strategy of cell-free therapy. *Stem Cell Res Ther.* 2020;11(1):23. https://doi.org/10.1186/s13287-019-1548-7.

Haider KH, Najimi M. Exosome-based cell-free therapy in regenerative medicine for myocardial repair. In: KH Haider, editor. *Handbook of Stem Cell Therapy.* Singapore: Springer; 2022. https://doi.org/10.1007/978-981-16-6016-0_9-1.

Hakim A, Usmani OS. Structure of the lower respiratory tract. In: *Reference Module in Biomedical Sciences.* 3rd ed. Amsterdam: Elsevier; 2014. p. 1–8. https://doi.org/10.1016/B978-0-12-801238-3.00215-4.

Hallsworth MP, Twort CHC, Lee TH, Hirst SJ. β2-Adrenoceptor agonists inhibit the release of eosinophil-activating cytokines from human airway smooth muscle cells. *Br J Pharmacol.* 2001;132(3):729–41.

Hoang DM, Nguyen KT, Nguyen AH, Nguyen BN, Nguyen LT. Allogeneic human umbilical cord-derived mesenchymal stem/stromal cells for chronic obstructive pulmonary disease (COPD): study protocol for a matched case-control, phase I/II trial. *BMJ Open.* 2021;11(5):1–16.

Hough KP, Curtiss ML, Blain TJ, Liu RM, Trevor J, Deshane JS, et al. Airway remodeling in asthma. *Front Med.* 2020;7(May):191.

Hynds RE. Exploiting the potential of lung stem cells to develop pro-regenerative therapies. *Biol Open.* 2022;11(10):1–11.

Ichikado K, Kotani T, Kondoh Y, Imanaka H, Johkoh T, Fujimoto K, et al. Clinical efficacy and safety of multipotent adult progenitor cells (invimestrocel) for acute respiratory distress syndrome (ARDS) caused by pneumonia: a randomized, open-label, standard therapy–controlled, phase 2 multicenter study (ONE-BRIDGE). *Stem Cell Res Ther.* 2023;14(1):1–13. https://doi.org/10.1186/s13287-023-03451-z.

Kardas G, Daszyn A. Role of platelet-derived growth factor (PDGF) in asthma as an immunoregulatory factor mediating airway remodelling and possible pharmacological target. *Front Pharmacol.* 2020;11(February):1–9.

Khan M, Alteneder M, Reiter W, Krausgruber T, Dobnikar L, Madern M, et al. Single-cell and chromatin accessibility profiling reveals regulatory programs of pathogenic Th2 cells in allergic asthma. *Nat Commun* [Internet]. 2025;16(1). https://doi.org/10.1038/s41467-025-57590-3.

Kumar MA, Baba SK, Sadida HQ, Marzooqi S Al, Jerobin J, Altemani FH, et al. Extracellular vesicles as tools and therapeutic targets for diseases. *Signal Transduct Target Ther.* 2024;9(1).

Lamiaa AA, Al-Massri KF, Haider KH. Therapeutic efficacy of mesenchymal stem cells and their derived exosomes in different neurodegenerative disorders: current status and future perspectives. In: KH Haider, editor. *Handbook of Regenerative Medicine: Cell-Free Therapy.* Boca Raton (FL): CRC Press; 2025.

Le Floc'h A, Allinne J, Nagashima K, Scott G, Birchard D, Asrat S, et al. Dual blockade of IL-4 and IL-13 with dupilumab, an IL-4Rα antibody, is necessary to broadly inhibit type 2 inflammation. *Allergy Eur J Allergy Clin Immunol.* 2020;75(5):1188–204.

Lee KS, Yeom SH, Kim MK, Woo CH, Choi YC, Choi JS, et al. Therapeutic potential of human stem cell-derived extracellular vesicles in idiopathic pulmonary fibrosis. *Extracellular Vesicle.* 2024;4:100045. https://doi.org/10.1016/j.vesic.2024.100045.

Li H. Update on the role of chemokines and chemokine receptors in liver fibrosis. *J Gastrointest Liver Dis.* 2023;32(2):241–56.

Liang Y, Xia J, Xu L, Zhang Y, Xu X, Haider KH. Cell-free and cell-based therapy in regenerative medicine. In: KH Haider, editor. *Handbook of Regenerative Medicine: Cell-Free Therapy.* Boca Raton (FL): CRC Press; 2025.

Liu H, Heenan KM, Coyle L, Chaudhuri N. Progressive pulmonary fibrosis: a need for real-world data to solve real-world clinical problems. *BMJ Med.* 2024;3(1):e000911.

Liu Y, Li P, Jiang T, Li Y, Wang Y, Cheng Z. Epidermal growth factor receptor in asthma: a promising therapeutic target? *Respir Med.* 2023;207:107117. https://doi.org/10.1016/j.rmed.2023.107117.

Lo Sicco C, Reverberi D, Balbi C, Ulivi V, Principi E, Pascucci L, Becherini P, et al. Mesenchymal stem cell-derived extracellular vesicles as mediators of anti-inflammatory effects: endorsement of macrophage polarization. *Stem Cell Transl Med.* 2017;6:1018–28.

Luo L, Zhang W, You S, Cui X, Tu H, Yi Q, et al. The role of epithelial cells in fibrosis: mechanisms and treatment. *Pharmacol Res.* 2024;202:107144. https://doi.org/10.1016/j.phrs.2024.107144.

Lv K, Gao J, Yang L, Yuan X. The role of mesenchymal stem cell-derived exosomes in asthma (Review). *Mol Med Rep.* 2025;31(6):1–9.

Man RK, Gogikar A, Nanda A, Janga LSN, Sambe HG, Yasir M, et al. A comparison of the effectiveness of nintedanib and pirfenidone in treating idiopathic pulmonary fibrosis: a systematic review. *Cureus.* 2024;16(2):e54268.

Manni ML, Robinson KM, Alcorn JF. A tale of two cytokines: IL-17 and IL-22 in asthma and infection. *Expert Rev Respir Med.* 2014;8(1):25–42.

Matera MG, Rinaldi B, Calzetta L, Rogliani P, Cazzola M. Pharmacokinetics and pharmacodynamics of inhaled corticosteroids for asthma treatment. *Pulm Pharmacol Ther.* 2019;58(June 2019):101828. https://doi.org/10.1016/j.pupt.2019.101828.

Matthay MA, Zemans RL, Zimmerman GA, Arabi YM, Beitler JR, Mercat A, et al. Acute respiratory distress syndrome. *Nat Rev Dis Prim.* 2018;5(1). https://doi.org/10.1038/s41572-019-0069-0.

Meegan JE, Rizzo AN, Schmidt EP, Bastarache JA. Cellular mechanisms of lung injury: current perspectives. *Clin Chest Med* [Internet]. 2024;45(4):821–33. https://doi.org/10.1016/j.ccm.2024.08.004.

Mi LL, Guo WW. Crosstalk between ILC2s and Th2 CD4+T cells in lung disease. *J Immunol Res.* 2022;2022:1–12.

Moodley Y, Sturm M, Shaw K, Shimbori C, Tan DBA, Kolb M, et al. Human mesenchymal stem cells attenuate early damage in a ventilated pig model of acute lung injury. *Stem Cell Res.* 2016;17(1):25–31.

Mushtaq M, Zineldeen DH, Mateen MA, Haider KH. Mesenchymal stem cells' "garbage bags" at work: treating radial nerve injury with mesenchymal stem cell-derived exosomes. *World J Stem Cells* 2024;16(5). https://dx.https://doi.org/10.4252/wjsc. v16.i5.0000.

Műzes G SF. Mesenchymal stem cell-derived secretome: a potential therapeutic option for autoimmune and immune-mediated. *Cells.* 2022;11(15):2300.

Najimi M, Haider KH. Extracellular vesicle-based cell-free therapy for liver regeneration. In: KH Haider, editor. *Handbook of Stem Cell Therapy.* Singapore: Springer; 2022. https://doi.org/10.1007/978-981-16-6016-0_9-1.

Pellicano C, Vantaggio L, Colalillo A, Pocino K, Basile V, Marino M, et al. Type 2 cytokines and interstitial lung disease in scleroderma. *Clin Exp Med.* 2023;23(7):3517–25. https://doi.org/10.1007/s10238-023-01125-x.

Pervizaj-Oruqaj L, Ferrero MR, Matt U, Herold S. The guardians of pulmonary harmony: alveolar macrophages orchestrating the symphony of lung inflammation and tissue homeostasis. *Eur Respir Rev* [Internet]. 2024;33(172). https://doi.org/10.1183/16000617.0263-2023.

Powell FL, Wagner PD, West JB. Ventilation, blood flow, and gas exchange. In: *Murray & Nadel's textbook of respiratory medicine*. 6th ed. Philadelphia (PA): Elsevier; 2015. p. 44–75.e3.

Qiu C, Ge Z, Cui W, Yu L, Li J. Human amniotic epithelial stem cells: A promising seed cell for clinical applications. *Int J Mol Sci*. 2020;21(20):1–26.

Quint JK, Ariel A, Barnes PJ. Rational use of inhaled corticosteroids for the treatment of COPD. *NPJ Prim Care Respir Med*. 2023;33(1):24–31.

Ricciardolo FLM, Blasi F, Centanni S, Rogliani P. Therapeutic novelties of inhaled corticosteroids and bronchodilators in asthma. *Pulm Pharmacol Ther*. 2015;33(May):1–10. https://doi.org/10.1016/j.pupt.2015.05.006.

Roe T, Talbot T, Terrington I, Johal J, Kemp I, Saeed K, et al. Physiology and pathophysiology of mucus and mucolytic use in critically ill patients. *Crit Care*. 2025;29(1). https://doi.org/10.1186/s13054-025-05286-x.

Rojo D, Conget P. Acellular derivatives of mesenchymal stem cells prevent peritoneal adhesions in an animal model. *J Surg Res* [Internet]. 2018;223:198–206. https://doi.org/10.1016/j.jss.2017.11.018.

Safwan M, Bourgleh MS, Al-Ruqi A, Shrebaty O, Almujaydil F, Alothaim B, Alrashdi N, Haider KH. Living bio-drugs MSCs and their derivative exosomes to rescue ventilated patients with ARDS: a systematic review and meta-analysis. *World J Stem Cells*. 2025;17(10):1–17.

Saito A, Horie M, Nagase T. TGF-β signaling in lung health and disease. *Int J Mol Sci*. 2018;19(8):1–18.

Sang L, Guo X, Fan H, Shi J, Hou S, Lv Q. Mesenchymal stem cell-derived extracellular vesicles as idiopathic pulmonary fibrosis microenvironment targeted delivery. *Cells*. 2022;11(15):2322.

Schultheiß C, Willscher E, Paschold L, Gottschick C, Klee B, Henkes SS, et al. The IL-1β, IL-6, and TNF cytokine triad is associated with post-acute sequelae of COVID-19. *Cell Reports Med*. 2022;3(6):100663.

Serazin NA, Edem B, Williams SR, Ortiz JR, Kawade A, Das MK, et al. Acute respiratory distress syndrome (ARDS) as an adverse event following immunization: case definition & guidelines for data collection, analysis, and presentation of immunization safety data. *Vaccine* [Internet]. 2021;39(22):3028–36. https://doi.org/10.1016/j.vaccine.2021.01.053.

Sharan J, Barmada A, Band N, Liebman E, Prodromos C. First report in a human of successful treatment of asthma with mesenchymal stem cells: A case report with review of literature. *Curr Stem Cell Res Ther*. 2022;18(7):1026–9.

Shi X, Ma X, Wang H, He X, Fan H, Mao Y. Elevated serum levels of HIF-1α and VEGF as potential biomarkers in connective tissue disease-associated pulmonary arterial hypertension. *Sci Rep*. 2025;15(1):1–11.

Shinkai M, Yabuta T. Tezepelumab: An anti-thymic stromal lymphopoietin monoclonal antibody for the treatment of asthma. *Immunotherapy*. 2023;15(17):1435–47.

Song L, Li K, Chen H, Xie L. Cell cross-talk in alveolar microenvironment: From lung injury to fibrosis. *Am J Respir Cell Mol Biol*. 2024;71(1):30–42.

Sordillo PP, Allaire A, Bouchard A, Salvail D, Labbe SM. The complex lipid SPPCT-800 reduces lung damage, improves pulmonary function, and decreases pro-inflammatory cytokines in a murine model of LPS-induced acute respiratory distress syndrome (ARDS). *Pharm Biol*. 2022;60(1):1255–63. https://doi.org/10.1080/13880209.2022.2087689.

Spahn JD, Israelsen R. Bronchodilator therapy for asthma. In: Craig T, Ledford DK, editors. *Allergy and Asthma*. Cham: Springer; 2018. p. 1–31.

Stessuk T, Ruiz MA, Greco OT, Bilaqui A, Ribeiro-Paes MJ de O, Ribeiro-Paes JT. Phase I clinical trial of cell therapy in patients with advanced chronic obstructive pulmonary disease: Follow-up of up to 3 years. *Rev Bras Hematol Hemoter*. 2013;35(5):352–7.

Suresh Babu K, Kastelik J, Morjaria JB. The role of long-term antibiotics in chronic respiratory diseases. *Respir Med*. 2013;107(6):800–15. https://doi.org/10.1016/j.rmed.2013.02.009.

Tang W, Rong Y, Zhang H, Zhan Z, Yuan L, Ning Y, et al. The correlation between a Th1/Th2 cytokines imbalance and vitamin D level in patients with early chronic obstructive pulmonary disease (COPD), based on screening results. *Front Physiol*. 2023;14(March):1–8.

Trubiani O, Pizzicannella J, Caputi S, Marchisio M, Mazzon E, Paganelli R, et al. Periodontal ligament stem cells: Current knowledge and future perspectives. *Stem Cells Dev*. 2019;28(15):995–1003.

Trzil JE, Masseau I, Webb TL, Chang CH, Dodam JR, Liu H, et al. Intravenous adipose-derived mesenchymal stem cell therapy for the treatment of feline asthma: a pilot study. *J Feline Med Surg*. 2016;18(12):981–90.

Vähätalo I, Lehtimäki L, Tuomisto LE, Karjalainen J, Niemelä O, Ilmarinen P, et al. Long-term use of short-acting β2-Agonists in patients with adult-onset asthma. *J Allergy Clin Immunol Pract*. 2022;10(8):2074–83.e7. https://doi.org/10.1016/j.jaip.2022.03.027.

Van Ganse E, Texier N, Dima AL, Belhassen M, Laforest L, Herbage S, et al. Effects of short- and long-acting beta-agonists on asthma exacerbations: a prospective cohort. *Ann Allergy, Asthma Immunol.* 2020;124(3):254–60. https://doi.org/10.1016/j.anai.2019.12.012.

Vlahos R, Bozinovski S. Role of alveolar macrophages in chronic obstructive pulmonary disease. *Front Immunol.* 2014;5(September):1–7.

Wang J, Li K, Hao D, Li X, Zhu Y, Yu H, et al. Pulmonary fibrosis: pathogenesis and therapeutic strategies. *MedComm.* 2024;5(10):1–27.

Wang M, Zhou T, Zhang Z, Liu H, Zheng Z, Xie H. Current therapeutic strategies for respiratory diseases using mesenchymal stem cells. *MedComm.* 2021;(November 2020):1–30.

Wei X, Wang Q, Wen W, Yang L, Chen H, Xu G, et al. Stem cell-derived exosomes as nanotherapeutics for inflammatory diseases. *MedComm - Futur Med.* 2025;4(1):e70016.

Wilson JG, Liu KD, Zhuo H, Caballero L, McMillan M, Fang X, et al. Mesenchymal stem (stromal) cells for treatment of ARDS: A phase 1 clinical trial. *Lancet Respir Med* [Internet]. 2015;3(1):24–32. https://doi.org/10.1016/S2213-2600(14)70291-7.

Xia S, Gu X, Wang G, Zhong Y, Ma F, Liu Q, et al. Regulated cell death of alveolar macrophages in acute lung inflammation: Current knowledge and perspectives. *J Inflamm Res.* 2024;17(December):11419–36.

Xiao Q, Li X, Li Y, Wu Z, Xu C, Chen Z, et al. Biological drug and drug delivery-mediated immunotherapy. *Acta Pharm Sin B.* 2021;11(4):941–60.

Xu B, Chen S si, Liu M zhuo, xiaGan C , Li J qi, Guo G hua. Stem cell-derived exosomes-based therapy for acute lung injury and acute respiratory distress syndrome: A novel therapeutic strategy. *Life Sci.* 2020;254(April):117766. https://doi.org/10.1016/j.lfs.2020.117766.

Yi YF, Fan ZQ, Liu C, Ding YT, Chen Y, Wen J, et al. Immunomodulatory effects and clinical application of exosomes derived from mesenchymal stem cells. *World J Stem Cells.* 2025;17(3):1–16.

Yu Y, Zhao L, Xie Y, Xu Y, Jiao W, Wu J, et al. Th1/Th17 cytokine profiles are associated with disease severity and exacerbation frequency in COPD patients. *Int J COPD.* 2020;15:1287–99.

Yuan D. Mesenchymal stromal cell-based therapy in lung diseases; from research to clinic. *Am J Stem Cells.* 2024;13(2):37–58.

Yun EJ, Lorizio W, Seedorf G, Abman SH, Vu TH. VEGF and endothelium-derived retinoic acid regulate the development of both the lung's vascular and alveolar systems. *Am J Physiol - Lung Cell Mol Physiol.* 2016;310(4): L287–98.

Yusuf R, Haider KH. Theragnostic applications of exosomes in Alzheimer's disease. In: KH Haider, editor. *Handbook of Regenerative Medicine: Cell-Free Therapy.* Boca Raton (FL): CRC Press; 2025. p. 390–409.

Zargar MJ, Kaviani S, Vasei M, Soufi Zomorrod M, Heidari Keshel S, Soleimani M. Therapeutic role of mesenchymal stem cell-derived exosomes in respiratory disease. *Stem Cell Res Ther.* 2022;13(1):1–11. https://doi.org/10.1186/s13287-022-02866-4.

Zhao X, Wu J, Yuan R, Li Y, Yang Q, Wu B, et al. Adipose-derived mesenchymal stem cell therapy for reversing bleomycin-induced experimental pulmonary fibrosis. *Sci Rep.* 2023;13(1):1–15. https://doi.org/10.1038/s41598-023-40531-9.

Zhao X, Zhang W, Fan J, Chen X, Wang X. Application of mesenchymal stem cell exosomes in the treatment of skin wounds. *Smart Mater Med.* 2023;4:578–89. https://doi.org/10.1016/j.smaim.2023.04.006.

Zhu Z, Lian X, Su X, Wu W, Zeng Y, Chen X. Exosomes derived from adipose-derived stem cells alleviate cigarette smoke-induced lung inflammation and injury by inhibiting alveolar macrophages' pyroptosis. *Respir Res* [Internet]. 2022;23(1):1–15. https://doi.org/10.1186/s12931-022-01926-w.

Zubair M, Abouelnazar FA, Iqbal MA, Pan J, Zheng X, Chen T, et al. Mesenchymal stem cell-derived exosomes as a plausible immunomodulatory therapeutic tool for inflammatory diseases. *Front Cell Dev Biol.* 2025;13(March):1–23.

11 The Role of Exosomes in Primary Sjögren's Syndrome

Pathogenesis, Diagnostics, and Therapeutics

Kevin Y. Wu[1], Fahad Butt[2], Alaina Mahalanabis[2], Mazen Y. Choulakian[1], and Simon D. Tran[3]

[1]Department of Surgery, Division of Ophthalmology, University of Sherbrooke, Canada

[2]Faculty of Medicine, Western University, London, Canada

[3]Faculty of Dental Medicine and Oral Health Sciences, McGill University, Canada

11.1 LIST OF ABBREVIATIONS

AC9	Adenylate cyclase 9
AIDs	Autoimmune diseases
ALIX	ALG-2 Interacting protein X (an exosome marker)
APCs	Antigen-presenting cells
AQP5	Aquaporin 5
AUC	Area under the curve
BAFF	B cell-activating factor
BCL-6	B cell lymphoma 6 protein
BLIMP1	B Lymphocyte-Induced Maturation Protein 1 (also known as PRDM1)
Ca2+	Calcium ions
cAMP	Cyclic adenosine monophosphate
CD	Cluster of differentiation (e.g., CD14, CD15, CD20, CD31, CD36, CD146)
CEA	Carcinoembryonic antigen
CP	Ceruloplasmin
CREB	cAMP response element-binding protein
CRTAC1	Cartilage acidic protein 1
CTLA4	Cytotoxic T-lymphocyte-associated protein 4
DEPs	Differentially expressed proteins
DPSC-Exos	Dental pulp stem cell-derived exosomes
EBV	Epstein-Barr virus
ECM	Extracellular matrix
ECs	Endothelial cells
EMT	Epithelial-to-mesenchymal transition

DOI: 10.1201/9781003653387-11

ERK	Extracellular signal-regulated kinase
ESSPRI	EULAR Sjögren's Syndrome Patient Reporting Index
ESR	Erythrocyte sedimentation rate
EVs	Extracellular vesicles
FGF	Fibroblast growth factor
FOXP3	Forkhead box P3
GC	Germinal centers
GPER	G Protein-coupled estrogen receptor
GSK-3	Glycogen synthase kinase-3
GVHD	Graft-versus-host disease
HCC	Hepatocellular carcinoma
HLA	Human leukocyte antigen
IFN	Interferon (e.g., IFN-α, IFN-γ)
IgG	Immunoglobulin G
IgG4	Immunoglobulin G4
IL	Interleukin (e.g., IL-1β, IL-6, IL-7, IL-10, IL-12, IL-17, IL-21, IL-23)
iMEP	Immunomagnetic exosomal PCR
iMSC	induced pluripotent stem cell-derived mesenchymal stem cells
iPSCs	induced pluripotent stem cells
IP3	Inositol 1,4,5-trisphosphate
IRF5	Interferon regulatory factor 5
ITGAM/CD11b	Integrin alpha M
JAK-STAT	Janus kinase/signal transducer and activator of transcription pathway
LG	Lacrimal gland
LGMSC-Exos	Labial gland mesenchymal stem cell-derived exosomes
LSGB	Labial salivary gland biopsy
M3Rs	Muscarinic acetylcholine type 3 receptors
MAPK	Mitogen-activated protein kinase
MALT	Mucosa-associated lymphoid tissue
MHC	Major histocompatibility complex
miR	microRNA (e.g., miR-17, miR-19b, miR-142-3p, miR-146)
MIF	Macrophage migration inhibitory factor
MSCs	Mesenchymal stem cells
MSG-MSCs	Minor salivary gland-derived mesenchymal stem cells
NSAIDS	Non-steroidal anti-inflammatory drugs
NSCLC	Non-small cell lung cancer
OE-MSC-Exos	Olfactory ecto-mesenchymal stem cell-derived exosomes
OLFM4	Olfactomedin-4
PAMPs	Pathogen-associated molecular patterns
PEE	Punctate epithelial erosions
PI3K/AKT	Phosphoinositide 3-kinase/protein kinase b pathway
PKA	Protein kinase A
PRDM1	PR/SET domain 1 (also known as BLIMP1)
PSA	Prostate-specific antigen
PTPN22	Protein tyrosine phosphatase non-receptor type 22
pSS	primary Sjögren's syndrome
PZP	Pregnancy zone protein
RAB10	Ras-related protein RAB10
RA	Rheumatoid arthritis
RF	Rheumatoid factor
Rho/Ro	A common abbreviation for the SSA autoantigen (see SSA)

RyR2	Ryanodine receptor 2
SERS	Surface-enhanced raman spectroscopy
SERPINA4	Serpin family a member 4
SGs	Salivary glands
sSS	secondary Sjögren's syndrome
SS	Sjögren's syndrome (general term)
SSA/Ro	Sjögren's syndrome-related antigen A
SSB/La	Sjögren's syndrome-related antigen B
STAT5	Signal transducer and activator of transcription 5
TBUT	Tear breakup time
TGF-β	Transforming growth factor beta
TF	Transferrin
Th	T helper cell (e.g., Th1, Th2, Th17)
THBS4	Thrombospondin 4
TLRs	Toll-like receptors
Treg	T regulatory cell
TREX1	Three prime repair exonuclease 1
TTR	Transthyretin
TSG101	Tumor susceptibility gene 101 (an exosome marker)
UCMSC-Exos	Umbilical cord mesenchymal stem cell-derived exosomes
UTR	Untranslated region (e.g., 3' UTR)
VEGF	Vascular endothelial growth factor

11.2 INTRODUCTION

Primary Sjögren's Syndrome (pSS) is a chronic autoimmune disease characterized by lymphocytic infiltration of exocrine glands, resulting in dry eyes and oral cavity symptoms, as well as systemic manifestations. While the precise etiology of pSS remains not fully elucidated, genetic predisposition, environmental triggers, and immune dysregulation are believed to contribute to its pathogenesis. Autoantibody production, aberrant T cell and B cell activation, and persistent inflammation play central roles in disease progression (Amft et al., 2001). Despite advancements in understanding pSS, diagnosis remains challenging due to its heterogeneous clinical presentation and the absence of a single definitive biomarker.

Exosomes, a subset of EVs, have gained increasing attention in autoimmune research due to their potential roles in immune regulation, intercellular communication, and disease progression (Ahmed et al., 2023). These nanosized vesicles carry proteins, lipids, and nucleic acids, which may influence immune responses and contribute to the pathophysiology of primary pSS. The potential of exosomes in immune modulation offers exciting possibilities for pSS treatment, instilling hope in the audience. For instance, exosomes can modulate the activity of immune cells, such as T cells and B cells, and regulate the production of inflammatory cytokines, thereby potentially mitigating the autoimmune response in pSS. Recent studies suggest that exosomes may also be potential biomarkers for early diagnosis and disease monitoring. Furthermore, exosome-based approaches are being explored for their possible therapeutic applications, including immune modulation, targeted drug delivery, and tissue regeneration (Jeong et al., 2021).

This chapter provides a comprehensive review of the role of exosomes in pSS, discussing their possible contributions to disease mechanisms, their potential as diagnostic and prognostic markers, and their emerging therapeutic implications. The potential of exosomes in the management of pSS is vast, ranging from early diagnosis to targeted drug delivery. As exosome-based applications in pSS remain largely investigational, further research is crucial to validate their clinical relevance, optimize their isolation and characterization, and assess their therapeutic efficacy. The need for continued research in this area is paramount to unlocking the full potential of exosomes in the management of pSS, evoking curiosity and eagerness for future developments.

11.3 OVERVIEW OF PRIMARY SJÖGREN'S SYNDROME

11.3.1 Definition and Epidemiology

pSS is an autoimmune disease characterized by lymphocyte infiltration of the exocrine glands, particularly the lacrimal and SGs. The exocrine disorder can occur alone, pSS, or secondary to another autoimmune disease (secondary Sjögren's Syndrome, sSS), such as rheumatoid arthritis. pSS predominantly affects women, with a female-to-male ratio of 9:1, and typically presents after the age of 50 (Alamanos et al., 2006). The overall prevalence of pSS is 0.29–0.77% (Zhang et al., 1995). pSS appears to be more prevalent in Caucasians and the Asian population. However, further research is needed to confirm this finding. There is often a diagnostic delay for pSS, with diagnosis taking several years from disease onset (Huang et al., 2021).

11.3.2 Clinical Landscapes

pSS typically presents with sicca symptoms, primarily dry eyes and dry mouth. Patients experience aqueous tear deficiency, also known as dry eye, which worsens throughout the day. Ocular symptoms for pSS include discomfort, such as burning, foreign body sensation, and itching; these symptoms are nonspecific. Visual disturbances, including blurred or fluctuating vision, are also common, along with sticky or mucous discharge that may resemble infectious or allergic conjunctivitis (Al-Hashimi et al., 2001; Leonardi et al., 2015). Clinically, tear film instability is evident, with a Tear Breakup Time (TBUT) of <10 seconds in moderate-to-severe cases ("Methodologies to Diagnose and Monitor Dry Eye Disease: *Report of the Diagnostic Methodology Subcommittee of the International Dry Eye Workshop (2007)*" 2007). Corneal and conjunctival findings vary in severity. Mild cases present with punctate epithelial erosions (PEE) on both the conjunctiva and cornea, often accompanied by conjunctival injection. As the disease progresses, mucus debris accumulates in the tear film, and filamentary keratitis may develop. In advanced stages, neovascularization and pannus formation occur, along with corneal and conjunctival keratinization, scarring, thinning, and, in severe cases, perforation (Vivino et al., 2001).

Patients with pSS have decreased saliva production, which can make swallowing difficult and alter taste sensation. They may also have parotid and submandibular gland swelling, difficulty talking, halitosis, and dental caries ("Sjögren's Syndrome and Halitosis: A Case Report," n.d.). Patients with pSS have a higher risk of developing MALT lymphoma (Ekström Smedby et al., 2008). As pSS is a chronic inflammatory condition, patients have elevated ESR and C-reactive protein, polyclonal hypergammaglobulinemia, and are often RF+ (García et al., 2012; Sène et al., 2011).

11.3.3 Current Understanding of Pathogenesis

The pathophysiology of pSS involves a complex interplay of genetic predisposition, environmental factors, and dysregulation of both the innate and adaptive immune systems. Viral infections are considered a key trigger, initiating the disease process by induction of apoptosis in SG epithelial cells. This exposes intracellular proteins, which are typically hidden from the immune system, to the extracellular environment. Viruses activate the innate immune system, triggering the production of cytokines and chemokines, which in turn activate the adaptive immune system. Once the adaptive immune system encounters these intracytoplasmic proteins, autoantibody production begins, leading to chronic inflammation of the SGs. A signal transduction cascade is initiated when pathogen-associated molecular patterns (PAMPs) on the surfaces of viral cells bind to Toll-like receptors (TLRs) on innate immune cells. This cascade produces cytokine production, which recruits and activates the adaptive system. Dendritic cells are the primary innate immune cell type involved in type 1 IFN production, which acts through the JAK-STAT signaling pathway to induce cytokine production, including IL-6, IL-7, IL-21, and IL-23. These cytokines promote B cell activation and proliferation. Meanwhile, IFN-γ promotes the recruitment and activation of T cells by inducing the production of chemokines (Gottenberg et al., 2006). Overproduction of T cells and reduced levels of T regulatory cells (Treg) play a key role in pSS. Aberrant signaling, such as increased IL-17, leads to decreased Treg cell numbers. B cells and T cells (Th) differentiate into the Th1 and Th2 subtypes. Th1 cells increase IFN-gamma and IL-12 production, further recruiting T cells and dysregulating the Th/Treg ratio (Cha et al., 2004). In addition to activating T cells, interferons promote the production of B cell-activating factor (BAFF), which drives B cell activation, proliferation, survival, and hypergammaglobulinemia. T and B cells form ectopic germinal centers (GCs). These GCs are the central component of an immune microenvironment in which more antibody-secreting B cells are produced (Amft et al., 2001).

In addition to activating the immune system, viruses cause SG cell apoptosis. The Epstein-Barr virus can destroy the SGs by transfecting them with EBV-miR-BART13-3p, disrupting the Ca^{2+} entry pathway, and leading to SG apoptosis (Gallo et al., 2016). La and Rho are intracellular protein complexes in SGs that are exposed to B cells upon SG apoptosis. B cells then start producing autoantibodies against Ro/SSA and La/SSB, two diagnostic antibodies for pSS.

Genetics also plays a role in the formation of pSS. HLA-DRB1 and HLA-DQA1 are strongly associated with primary Sjögren's syndrome (pSS). These HLA variants effectively present antigens to B cells and T cells, thereby promoting a heightened immune response. Furthermore, genetic variants in genes such as IRF5, STAT5, and BAFF predispose individuals to pSS by increasing interferon and B cell-activating factor levels, thereby promoting B cell survival and proliferation (Teos & Alevizos 2017). Variants in genes involved in apoptotic cell clearance are also associated with primary pSS. For example, mutations in TREX1, which encodes for an exonuclease, can impair immune cells' ability to effectively clear out apoptotic SGs, thus exposing intracellular proteins to B cells (Barizzone et al., 2013). Genes that maintain T cell and B cell tolerance are also implicated in pSS. Usually, autoreactive B cells are destroyed in the bone marrow during development. However, genes like PTPN22 allow autoreactive B cells to survive, leading to unregulated autoantibody production (Cadena-Sandoval et al., 2024). Similarly, variants in FOXP3 and CTLA4 can result in T cell overactivation, indirectly contributing to increased autoreactive B cell production and antibody production (Ríos-Ríos et al., 2020).

It is essential to acknowledge that the complete pathogenesis of pSS remains unclear (Figure 11.1). There is considerable heterogeneity in the clinical representation of pSS, with symptom presentation and disease progression varying between individuals (Both et al., 2017). This variability likely reflects the complex and multifactorial nature of pSS. While many of the individual factors are known to an extent, the sheer number of factors makes reproducing the exact conditions in a given patient's immune system complicated and not necessarily directly applicable to others.

11.4 EXOSOMES IN THE PATHOGENESIS OF PRIMARY SJÖGREN'S SYNDROME

11.4.1 Exosome Biology

Exosomes are the smallest class of EVs. They are formed by the fusion of multivesicular bodies with the plasma membrane, facilitating the release of intraluminal vesicles into the ECM. Structurally, exosomes are characterized by their lipid bilayer-delimited discoid shape with diameters between 30 and 100 nanometers (Yurtsever et al., 2021). The composition of exosomal membranes is similar to that of cells, due to their endosomal origin and nanoscopic size, which allows exosomes to readily traverse many biological membranes. As such, exosomes can be studied in and extracted from samples of bodily fluids, including blood, urine, and saliva (Zhou et al., 2023).

Exosomal cargo comprises major cellular macromolecules, including proteins, lipids, and nucleic acids, as well as peptides and various metabolites (Duréndez-Sáez et al., 2022). Studies have described their functional involvement in cellular waste disposal and local and systemic cellular communication (Mushtaq et al., 2024). Under normal physiological conditions, exosomes can

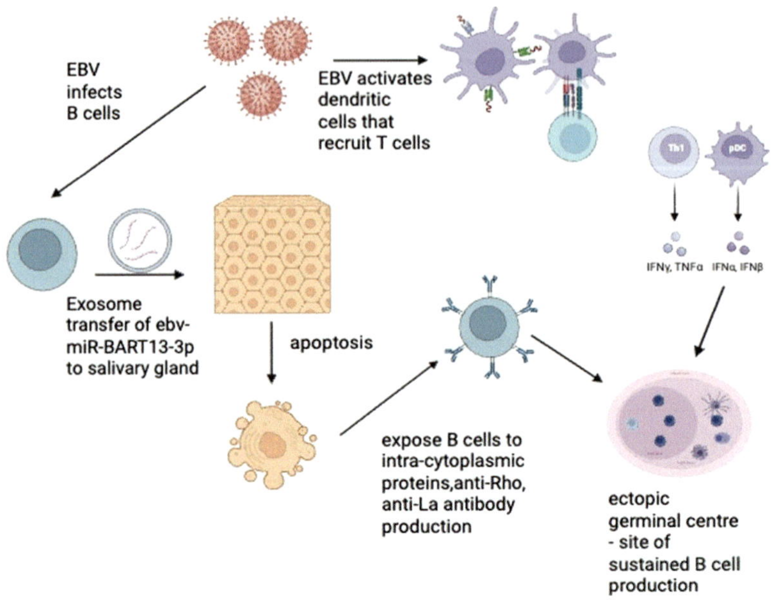

FIGURE 11.1 Pathophysiology of Primary Sjogren's Syndrome.

perform numerous crucial functions, including modulating the immune system and regulating cell differentiation (Habib et al., 2026; Losurdo et al., 2020; Narayanan et al., 2016). These functions are achieved through the secretion of exosomal cytokines and chemokines, which are transferred to T cells, thereby promoting cell recruitment and altering tissue phenotype (Ferguson et al., 2018). However, studies have also implicated exosomes in pathological processes, such as carcinogenesis and autoimmunity (Adem et al., 2024; Ji et al., 2024). In such cases, exosomes can facilitate the excessive transfer of bioactive molecules, including TGF-β and macrophage MIF, promoting a microenvironment conducive to metastasis (Costa-Silva et al., 2015). Several key marker proteins have been identified, enabling the continued study of exosome involvement in both physiological and pathophysiological processes. The most popular of these markers include tetraspanins, heat shock proteins, ALIX, and TSG101 due to their consistent relative abundances (Willms et al., 2016).

11.4.2 Physiological Significance of Exosomes

11.4.2.1 Exosome-Mediated Immune Dysregulation

Exosomes are implicated in the pathogenesis of various autoimmune diseases, including rheumatoid arthritis, systemic lupus erythematosus, and autoimmune thyroid diseases (see Table 11.1) (Zhou et al., 2024). In autoimmune diseases (AIDs), exosomes produced in antigen-presenting cells (APCs), such as dendritic cells, can enter the lymphatic circulation and expose antigen-specific T and B cells to MHC/polypeptide complexes.

This interaction can trigger the activation of the adaptive immune system and exacerbate AIDS (Hazrati et al., 2022). Additionally, the miR content of exosomes, such as miR-19b, can promote the imbalance in Th17/Treg differentiation, exacerbating autoimmune intolerance by disrupting immune homeostasis (Jiang et al., 2023). In rheumatoid arthritis, synoviocyte-derived exosomes can promote the secretion of inflammatory cytokines and chemokines from articular tissue (Li et al., 2024). These exosomes may also contain degradative enzymes and citrullinated autoantigens. Ultimately, the secretion of these contents from exosomes correlates with increased inflammation and disease progression (Figure 11.2). Moreover, several miRs contained within exosomes

TABLE 11.1
Summary of Immune System Interactions of Exosomal Cargo Biomarkers Extracted from Biofluids

Exosomal Cargo Type	Biosample Source	Specific Biomarker	Up or down-regulated	Involvement in Autoimmunity	Reference
miR	Mouth Rinse	let-7b-5p	Upregulated	Increases B cell survival Overproduction of autoantibodies.	(Cheng et al., 2023)
miR	Mouth Rinse	miR-34a-5p	Upregulated	Contributes to Treg/Th17 imbalance Increases the expression of proteins associated with severe autoimmune indices.	(Ismail et al., 2021)
miR	Plasma and Salivary	miR-146	Upregulated	Impaired regulation of pro-inflammatory cytokines Chronic inflammation and autoimmunity.	(Talotta et al., 2019)
Protein	Plasma	Complements (i.e., C5)	Upregulated	Increased complement complex activation and inflammation.	(Peng et al., 2023)
Protein	Plasma	Ceruloplasmin	Downregulated	Iron homeostasis dysregulation and increased ferroptosis.	(Shang et al., 2020)
Protein	Plasma	Transferrin	Downregulated	Iron homeostasis dysregulation and increased ferroptosis.	(Yan et al., 2021)
Protein	Unstimulated Whole Saliva	Pro-inflammatory cytokines (i.e., IL-7)	Upregulated	Excess leukocyte recruitment and inflammation.	(Lundström et al., 2013)
Protein	Unstimulated Whole Saliva	Anti-histone, anti-transglutaminase, anti-SSA, and anti-SSB	Upregulated	Autoantigenicity.	(Toker et al., 2004)

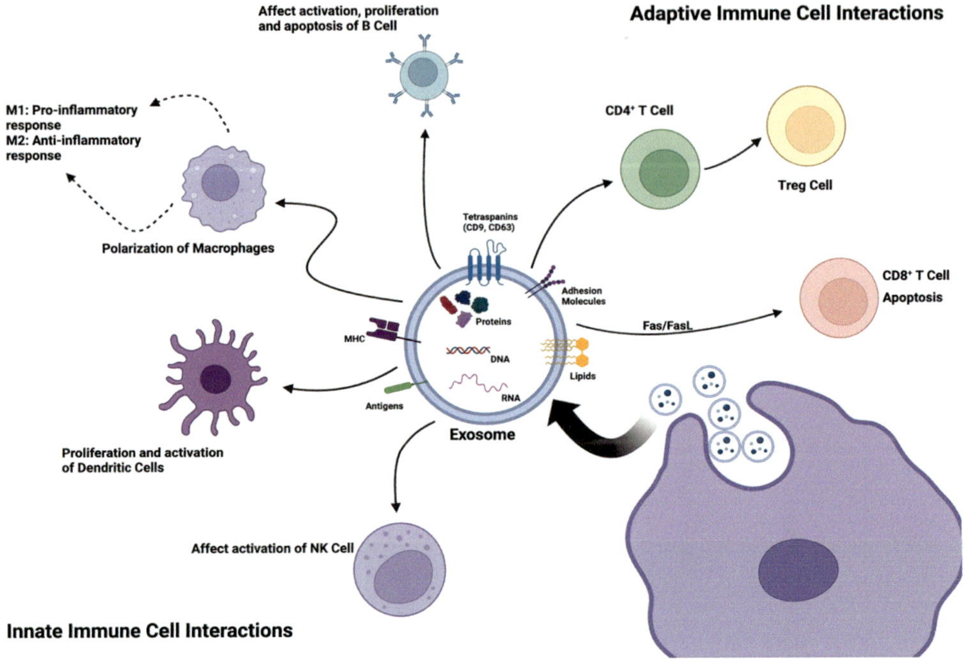

FIGURE 11.2 Exosomal mediation of chronic autoimmune conditions.

are hypothesized to be associated with the pathogenesis of rheumatoid arthritis. For example, serum exosomal miR-17 is upregulated and inhibits Treg induction by targeting TGFBR2, thereby disrupting immune cell regulation (Wang et al., 2018). In patients with systemic lupus erythematosus, exosomes have been associated with increased cytokine secretion, including IFN-α, IL-1β, and IL-6. Similar to rheumatoid arthritis, increased miR secretion and imbalances in miR relative abundances have been detected in exosomes and are associated with disease progression (Duan et al., 2019). In patients with pSS, plasma and salivary exosomes are secreted in significant abundance and, notably, are found to contain various autoantigenic proteins such as epithelial-specific cytoskeletal proteins, Ro/SSA, La/SSB, and Sm ribonucleoproteins ("SG Epithelial Cell Exosomes: A Source of Autoantigenic Ribonucleoproteins," n.d.). Upon presentation of these autoantigens to plasma B cells, anti-Ro/SSA and anti-La/SSB immunoglobulins are produced. Anti-Ro/SSA and anti-La/SSB promote local inflammation and lymphocyte infiltration into SG epithelial tissue. This response is then responsible for the dysfunction and tissue lesioning (Toker et al., 2004).

11.4.2.2 Altered Exosome Production in Primary Sjögren's Syndrome

Patients with pSS demonstrate distinct differences in the quantity and quality of plasma exosomes compared to healthy individuals. For example, an increase in the abundance of EV subpopulations has been observed in patients with pSS compared to their healthy donor counterparts in the literature. A study by Ferrant et al. using flow cytometry reported increased concentrations of CD31 and CD146 endothelial EVs, as well as cytokeratin epithelial and CD15 neutrophil-derived EVs, among Annexin V-positive EVs in pSS samples. Most of these EVs were believed to be upregulated in response to injury to surrounding tissue, such as the endothelium. This study also observed a marked decrease in protein count per EV in pSS patients (752, standard deviation 56) compared with healthy donors (908, standard deviation 83). Furthermore, differential expression protein (DEP) analysis identified 22 proteins that were upregulated, including 12 immunoglobulins, TTR, THBS4, SERPINA4, CRTAC1, TGFBI, PZP, RAB10, CD36, OLFM4, and ITGAM/CD11b. While the functions of the various DEP-identified proteins are unique, many are suspected to be present due to augmented immune function and tissue damage. Hence, the proteome of the plasma pSS EVs is distinct from that of healthy donor counterparts (Ferrant et al., 2023). Another study used DEP analyses to identify 17 upregulated and seven downregulated proteins in the plasma exosomes of pSS patients compared to healthy controls (Peng et al., 2023). These proteins have been linked to ferroptosis, a novel cell death pathway triggered by iron-dependent lipid peroxidation (Gao et al., 2016). This process can be coupled with apoptosis in pathological tissue (Hong et al., 2017). Hence, the exosomal cargo of diseased and non-diseased samples is measurably divergent.

11.4.2.3 Interaction with Salivary and Lacrimal Gland Epithelial Cells

Xerostomia (dry mouth) and xerophthalmia (dry eyes) are hallmark symptoms of primary pSS (Kachaner et al., 2023). Research has revealed a potential role for exosomes in the onset of these sicca symptoms. In a study by Troncoso et al., the function of T cell exosomal miR-142-3p was investigated. It was found that T cell exosomal miR-142-3p binds to the 3' UTRs of the SERCA2B, RyR2, and AC9 genes and downregulates their gene expression. When comparing biopsied epithelial cells from SGs of pSS patients with those of healthy donors, the pSS samples showed lower abundances of all three proteins, accompanied by increased miR-142-3p (Cortes-Troncoso et al., 2020). The SERCA2B pump and RyR2 channel are both involved in maintaining Ca^{2+} signaling pathways, while the AC9 membrane-bound enzyme is responsible for the catalysis of ATP into cAMP (Inoue et al., 2019; Yano et al., 2009; Qi et al., 2022). These two signaling pathways are involved in most exocrine functions in secretory glands, including amylase secretion. Hence, the transfer of miR-142-3p from activated T cells to SG epithelial cells via exosomes is believed to be associated with sicca symptoms in pSS patients.

Similarly, the autoantigenic cargo within exosomes has also been positively correlated with impaired salivary secretion. Specifically, pSS IgGs formed against Ro/SSA and La/SSB autoantigens have been found to promote internalization of Muscarinic Acetylcholine Type 3 Receptors (M3Rs) (Jin et al., 2012). M3Rs have been studied as parasympathetic regulators of SG function. M3Rs are associated with increased intracellular free-Ca^{2+} concentration through upregulation of inositol 1,4,5-trisphosphate (IP3) production. This is induced by the coupling of M3R with $G_{q/11}$ and results in the activation of Ca^{2+} gated K^+ and Cl^- channels, which in turn allows for the generation and secretion of saliva (Ham et al., 2024). The binding of M3Rs by pSS IgGs facilitates the phosphorylation of M3Rs, which in turn was confirmed to induce clathrin coat-dependent internalization and endocytosis of the receptor. Notably, this process takes place without the activation of Gq signaling pathways (Jin et al., 2012). Ultimately, exosomal Ro/SSA and La/SSB increase the concentration of pSS IgG and decrease M3R expression, significantly diminishing saliva production.

11.4.3 Contribution to Systemic Manifestations

Although xerostomia and xerophthalmia are the most common presenting symptoms of pSS, with 98% of patients exhibiting one and 89% exhibiting both, the clinical manifestations of pSS onset and progression can occur pervasively (Baldini et al., 2014). Among these manifestations, SS-related annular erythema and leukocytoclastic vasculitis are particularly interesting due to their association with anti-Ro and anti-La autoantibodies. The formation of raised, ring-shaped lesions with wide, well-defined boundaries, central clearing, and flattening characterizes SS-related annular erythema (Katayama et al., 2010). One study demonstrated that 91% of pSS patients presenting with annular erythema had elevated anti-Ro/SSA and anti-La/SSB autoantibodies (Ramos-Casals et al., 2015). Similarly, a study of pSS patients with leukocytoclastic vasculitis found abundant anti-Ro and anti-La antibodies in biopsied tissue in 70% of cases (Scofield, 2011). Although there is a lack of literature investigating the exosomal mechanisms underlying these findings, the prevalence of Ro/SSA and La/SSB autoantigens in the exosomes of SG epithelial cells suggests that similar observations may be made upon analyzing exosomes from affected skin biopsies. Many other systemic manifestations of pSS have been investigated, such as arthralgia, which afflicts up to 75% of patients with pSS, and nonspecific interstitial pneumonia. However, such symptoms have been described as downstream effects of xerostomia and xerophthalmia, or as tissue-specific mechanisms (Baldini et al., 2014).

11.5 EXOSOMES AS DIAGNOSTIC AND PROGNOSTIC MARKERS

11.5.1 Diagnostic Challenges in pSS

In current clinical practice, the diagnosis of pSS presents as a challenge due to its complex clinical presentation, disease mimicry, slow disease progression, and lack of strong evidence of early disease staging. The disease is known to have a heterogeneous and pernicious clinical presentation, which can significantly increase the difficulty of a diagnosis by traditional disease pattern recognition. Patients with pSS often present with a wide array of clinical symptoms and potential systemic manifestations. The presentation ranges from local sicca symptoms to extraglandular features, including involvement of the joints, skin, lungs, and kidneys (Ramos-Casals et al., 2015; Nguyen et al., 2024). Additionally, patients may first present with non-sicca systemic manifestations, which increases the complexity of disease recognition and diagnosis. Other systemic diseases, such as pSS, can also increase diagnostic difficulty by affecting the salivary and lacrimal glands. The symptom overlap and disease mimicry are broad, including diseases such as sarcoidosis, IgG4-related disease, chronic viral infections such as HIV or hepatitis C virus, all of which increase the potential for missed or delayed diagnosis (Sutcliffe, 2021). pSS is also known to have an indolent clinical presentation, with gradual deterioration of salivary and lacrimal function, which makes early disease detection more difficult. A population-based study in Taiwan by Huang et al. reported an average lag of 115 weeks between the onset of initial sicca symptoms and a diagnosis of pSS (Huang et al., 2021). Currently, there is no single definitive diagnostic test for pSS. Diagnosis relies on clinical findings, serological markers, and histological assessments. Furthermore, only 70% of pSS patients are positive for anti-SSA antibodies, the standard serological marker for pSS (Jonsson et al., 2018). With these diagnostic challenges come opportunities to utilize exosomes and their cargo as non-invasive biomarkers for the diagnosis of pSS. The immune system interactions of exosomal cargo biomarkers extracted from biofluids are summarized in Table 11.1.

11.5.2 Exosomal Content as Biomarkers

11.5.2.1 Exosomal miRs

Exosomal miRs play a significant role in intercellular communication and biological function, and have recently been presented as potential biomarkers for early disease diagnosis. There is evolving evidence supporting the notion that miR packaging and sorting into exosomes is not an incidental process, but rather that these miRs have implications in immune modulation, cell migration, apoptosis, proliferation, and several other cellular and biological functions (Rahimian et al., 2023). A substantial body of evidence supports the use of miRs as biomarkers for cardiovascular disease. Markers such as miR-1, miR-208, and miR-499 are detectable in circulation after acute myocardial infarction and serve as early markers of cardiac injury (Cheng et al., 2019). Outside the detection of these markers, variability in their levels also provides an opportunity to monitor disease progression; for instance, exo-miR-29a has been noted to increase over the course of heart failure.

The application of exosomal miRs for pSS is an evolving field of study that presents opportunities for early disease diagnosis, progression, and mapping. Recent studies have identified several miRs with altered expression profiles in patients with pSS. This includes human studies with healthy controls and bench studies with animal models that advance this field of research. In a survey conducted by Yamashiro et al., 12 potential miR biomarkers were identified from exosomes in mouth rinse samples, comparing patients with Sjögren's syndrome (SS) to controls (Yamashiro et al., 2022). Among the 12 markers initially identified by microarrays, let-7b-5p, miR-1290, miR-34a-5p, and miR-3648 showed higher expression in SS patients than in controls. The researchers conducted logistical regression analysis to develop a model that accurately classified SS patients from control samples, using the combination of miR-1290 and let-7b-5p (AUC: 0.856; sensitivity: 91.7%; specificity: 83.3%) (Yamashiro et al., 2022). Another study by Talotta et al., which used both plasma and salivary samples from patients with pSS and healthy controls, reported higher salivary exosomal miR-146 expression (Talotta et al., 2019). The expression level was also associated with worsening of the EULAR Sjogren's Syndrome Patient-Reported Index (ESSPRI), suggesting that miR-146 could be used as a biomarker to monitor disease progression (Talotta et al., 2019). An older study by Pauley et al. found that miR-155a expression was significantly higher in patients with pSS than in controls in peripheral blood mononuclear cells and in the salivary and lacrimal glands (Pauley et al., 2011). The study further investigated miR-146a expression in murine models and demonstrated its role in enhancing phagocytic activity and decreasing the production of inflammatory cytokines, leading to a dysregulated innate immune response (Pauley et al., 2011).

In an animal study completed by Kakan et al., using a murine model in which serum exosomal RNA was analyzed, dysregulated miRs in SS mice were identified as compared to controls (Kakan et al., 2020). The study identified five miRs—miR-127-3p, miR-409-3p, miR-410-3p, miR-540-3p, and miR-541-5p—as being significantly upregulated in SS mice (Kakan et al., 2020). Pathway analysis on the identified miR revealed an association with immune dysregulation. Downregulation of miR-127-3p has been demonstrated to contribute to the overstimulation of Type I Interferon signaling in autoimmune lupus nephritis (Wu et al., 2021). Similarly, miR-541-5p has been linked to the regulation of oxidative stress in acute lung injury, with decreased levels associated with pro-inflammatory effects (Shen et al., 2022). The miR-409-3p was linked to islet inflammation in type 1 diabetes, with levels of expression directly correlating to insulitis severity (Ventriglia et al., 2020). Additionally, miR-410-3p has been noted in the literature to regulate the cellular proliferation and apoptosis of fibroblast-like synoviocytes in rheumatoid arthritis, with significantly decreased levels identified in RA patients compared to controls (Wang et al., 2019).

11.5.2.2 Exosomal Protein Signatures

Isolated exosomes also serve as a vehicle for protein cargo, which, under proteomic analysis, can provide specific signatures and markers for the diagnosis of pSS. A study by Peng et al., analyzing plasma exosomes from patients with pSS, identified 24 differentially expressed proteins (DEPs) between patients and healthy controls (Peng et al., 2023). Among the 24 identified DEPs, 17 were upregulated, and seven were downregulated. These proteins were associated with various biological processes; notably, the DEPs were linked to ferroptosis pathways. Key DEPs that were downregulated included ceruloplasmin (CP) and transferrin (TF), both of which play crucial roles in iron metabolism (Peng et al., 2023). CP's primary role is to catalyze the oxidation of ferrous iron into ferric iron, a crucial step in iron loading onto transferrin and iron efflux from storage sites (Roeser et al., 1970). The TF role in iron metabolism is the primary iron transport protein, responsible for binding and delivering iron throughout the body and playing a crucial role in iron homeostasis (Camaschella et al., 2020). Reduced levels of CP and TF may disrupt iron homeostasis, leading to iron accumulation and subsequent production of reactive oxygen species. The differentiated expression of these proteins makes them potential biomarkers for early-stage pSS and may be associated with disease activity and progression. Similarly, the study found an increase in the production of complement C5 and C9, both of which are associated with the innate immune inflammatory response (Peng et al., 2023). These markers suggest increased activation of the complement system and inflammation in pSS, making these proteins potential biomarkers of disease activity and possibly enabling measurement of treatment response in pSS patients. Notably, the study also reported that plasma- and salivary-derived pSS exosomes contained increased levels of autoantigens SSA (Ro) and SSB (La) (Peng et al., 2023). These autoantibodies are frequently detected in patients with pSS. However, they are not specific to the disease and can be found in other autoimmune conditions

(Veenbergen et al., 2021). This further enhances the potential for protein analysis in pSS diagnosis, as detecting a combination of protein markers may help overcome the limitations in specificity.

While Peng et al.'s study provides strong evidence for the potential of exosomal proteomics in pSS analysis, Ferrant et al.'s findings, which analyzed a broader range of protein expression in EVs in pSS patients, are essential (Ferrant et al., 2023). The study by Ferrant et al. also analyzed pSS protein-specific signatures, which differed from the exosome-specific signature and extended to a broader category of EVs (Ferrant et al., 2023). The study found that EV concentrations in the plasma of pSS patients were significantly higher than those in healthy controls. This presents the option of using EV concentration as a biomarker for pSS. More specifically, specific subpopulations of plasma EVs were found to be increased, particularly those derived from neutrophils, endothelial cells (ECs), and epithelial cells (Ferrant et al., 2023). A specific proteomic signature was also identified in pSS plasma EVs, including proteins involved in neutrophil, epithelial, and endothelial functions, such as integrin alpha M, olfactomedin-4, Ras-related protein RAB10, and CD36 (Ferrant et al., 2023). This study provides initial evidence supporting the use of plasma EVs and protein content as biomarkers for the diagnosis and prognosis of pSS.

Furthermore, in the analysis of salivary exosomal cargo conducted by Finamore et al., protein profiles were compared among healthy individuals and pSS patients (Finamore et al., 2021). The researchers identified 121 differentially expressed proteins, among which 82% were upregulated in pSS patients. This upregulation indicated a significant deviation from normal protein production in the SGs. Among the DEPs, several proteins have been identified as playing a role in autoimmunity by modulating immune cells and mediating inflammatory responses (Finamore et al., 2021). This included proteins from the S100 protein family, annexin A2, resistin, serpin peptidase inhibitors, azurocidin, and the CD14 monocyte differentiation antigen. S100 proteins are strongly associated with pro-inflammatory states in autoimmune conditions such as rheumatoid arthritis, lupus, and psoriasis (Holzinger et al., 2019). Often presenting with elevated levels in these autoimmune conditions, and contributing to chronic disease conditions. Similarly, annexin A2 is present in conditions such as systemic lupus erythematosus. However, it can become an immune target, potentially contributing to the disease pathogenesis (Pianta et al., 2015). Resistin is an adipokine that acts as a pro-inflammatory molecule, exhibiting similar properties to S100 proteins. Increased resistin levels have been observed in chronic autoimmune conditions (Bokarewa et al., 2005). Specific serpin proteins, such as SERPINB3/B4, have been reported in the literature to be overexpressed in psoriasis and other inflammatory cutaneous conditions (Gatto et al., 2013). Azurocidin, a neutrophil granule protein, is associated with increased stimulation of pro-inflammatory cytokine production (Soehnlein & Lindbom, 2009). CD14, a monocyte differentiation antigen, is a pattern recognition receptor that contributes to autoimmune reactions by promoting monocyte activation and the differentiation of inflammatory macrophages (Sharygin et al., 2023). The results of this study highlight the potential of salivary protein biomarkers for disease diagnosis and the development of targeted therapies.

11.5.3 Advantages of Exosomal-Based Biomarkers

11.5.3.1 Non-invasive Sampling

Currently, few non-invasive methods exist for diagnosing pSS. The current golden standard for pSS diagnosis and classification is a labial SG biopsy (LSGB), particularly in patients with anti-SSA and anti-SSB seronegativity (Shiboski et al., 2012, 2017; Vitali et al., 2002). A retrospective analysis by Giovelli et al. demonstrated that LSGB had high specificity (97.43%), positive predictive value (95%), and negative predictive value (92.6%) for diagnosing pSS (Giovelli et al., 2015). However, the invasive nature of the procedure, which involves an incisional extraction of five or more SGs for analysis of lymphocytic infiltration, is a significant drawback. Other invasive methods of pSS diagnosis include parotid gland biopsy and sialography. However, both are less frequently employed than LSGB (Deroo et al., 2023). These invasive methods offer high specificity and sensitivity benefits, providing histopathological confirmation of pSS. However, they have limitations, including patient discomfort, interpreter variability, and unsuitability for patients with contraindications.

One of the key benefits of exosome-based biomarkers for pSS is the potential for non-invasive sampling. A sample for exosomal analysis can be acquired from non-invasive methods. Exosomes are readily accessible and can be isolated from various biological fluids, including saliva, blood (serum and plasma), tears, and urine (Gao et al., 2023). The accessibility of these biofluids makes them valuable in the development of non-invasive diagnostic tools for pSS. The extraction site and biofluid type also present an opportunity to study a specific subset of diseases in a greater context. Salivary exosomes are being investigated as biomarkers for oral and systemic metabolic-related diseases, head and neck cancer screening, and other chronic diseases (Hofmann et al., 2022; Deng et al., 2021; Song et al., 2023). Blood exosomes present the opportunity to investigate neurodegenerative diseases such as Alzheimer's (Shun Hu et al., 2024). Similarly, urine exosomes present the opportunity to identify biomarkers for kidney disease and urological cancers such as kidney, bladder, and prostate cancer (Lee et al., 2022). Tear-extracted exosomes, in particular, are gaining interest as potential biomarkers of ocular disorders, such as dry eye disease and diabetic retinopathy (Hu et al., 2022). Exosome-based biomarkers can decrease the need for biopsy procedures and improve patient comfort and compliance with diagnostic and screening procedures. These methods enhance the generalizability of diagnostic techniques to patient populations, including testing for patients who may have contraindications such as bleeding disorders that prevent them from getting LSGB. In the future, if the accessibility and ease of collection of these biofluids for exosome testing improve, there is potential for more frequent testing, which is essential for monitoring disease progression in chronic conditions such as pSS. This may also enhance the clinical understanding of the disease, enabling long-term follow-up. Furthermore, technological advancements may also allow the development of point-of-care diagnostic tools, thereby improving clinical diagnostic capabilities and monitoring (Figure 11.3).

11.5.3.2 Stability

Exosomes provide significant benefits as potential biomarkers for pSS due to their mechanical stability. Exosomes have a lipid bilayer and provide a mechanically strong, protective environment for intracellular cargo. This may ensure the integrity of miR and protein biomarkers during sample collection, preservation, and analysis (Sanz-Rubio et al., 2018). Exosomes also demonstrate stability for long-term storage, exhibiting stability at −80°C for conservation of the collected biofluid and exosomal content (Ge et al., 2014). Additionally, exosomes exposed to multiple freeze-thaw cycles do

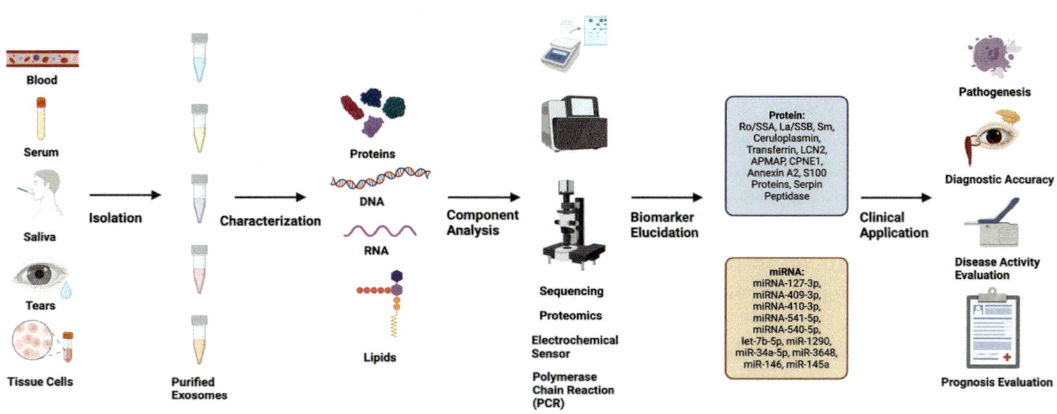

FIGURE 11.3 Exosome isolation from biofluids for characterization of biomarkers.

not exhibit a significant change in size or structural integrity (Sokolova et al., 2011). Cryoprotective proteins, such as trehalose and mannitol, can be added to the exosomal preparation before lyophilization to enhance this ability. These cryoprotectant agents prevent the development of ice crystals, thereby preserving the structural and functional integrity of exosomes upon rehydration (Frank et al., 2018). These agents also ensure the stability and activity of cargo molecules of interest, such as miRs and proteins. This ability may enable sample preservation and facilitate supplemental biomarker research and analysis.

11.5.3.3 Sensitivity and Specificity

Exosomal markers and disease detection are emerging fields of research. Several studies provide evidence of exosomes' high sensitivity and specificity, as well as their associated cargo, in identifying various conditions. Regarding pSS, the aforementioned study by Yamashiro et al. demonstrates the early, promising diagnostic ability of their combined miR model, including miR-1290 and let-7b-5p, achieving 91.7% sensitivity and 83.3% specificity (Yamashiro et al., 2022). There continues to be investigations into pSS diagnosis, with the development of innovative biomarker panels, such as Sjögren's diagnostic test, which integrates conventional serological markers with three proprietary biomarkers—salivary protein-1, carbonic anhydrase-6, and parotid secretory protein (Shen et al., 2016). Collectively, the proprietary and conventional biomarkers demonstrated compelling diagnostic performance, with 91.4% cumulative sensitivity and 79.8% cumulative specificity. These values fall within the corresponding ranges for specificity (63.5% to 93.7%) and sensitivity (61.2% to 100%) when using labial minor SG biopsy to diagnose pSS. Therefore, it suggests that the Sjo test is comparable to the labial minor SG biopsy for pSS diagnostic efficacy (Guellec et al., 2013).

Beyond pSS diagnosis, exosomes continue to attract research interest due to their potential for early detection of other pathologies. In a study by Niu et al., serum exosomes were utilized to improve the diagnosis of non-small cell lung cancer (NSCLC) (Niu et al., 2019). During the investigation, the exosomal protein biomarkers alpha-2-HS-glycoprotein and ECM protein 1 were combined with carcinoembryonic antigen (CEA)—a standard NSCLC biomarker—to produce an AUC of 0.938 for NSCLC. Furthermore, the identified biomarkers are readily detectable even in early-stage NSCLC, offering avenues for diagnosis and treatment. Similarly, exosomal miRs have the potential to serve as sensitive, non-invasive biomarkers for the diagnosis of hepatocellular carcinoma (HCC). In one study, exosomal miRs were used to detect HCC non-invasively, demonstrating high diagnostic accuracy with a pooled sensitivity of 86%. This finding demonstrates the applicability of leveraging exosomes and their cargo for non-invasive diagnosis of HCC (Tiyuri et al., 2023). Another study by Hu et al. highlighted the ability of exosome-based assays to detect disease-specific biomarkers with high sensitivity (Hu et al., 2024). The researchers employed immunomagnetic exosomal PCR (iMEP), which detected key proteins associated with Alzheimer's disease, including amyloid-β and phosphorylated tau, despite their presence at femtogram-per-milliliter levels within blood samples. When differentiating between samples from patients with Alzheimer's disease and those from healthy controls, the exosome-based assay demonstrated 95% sensitivity and specificity (Hu et al., 2024). Similarly, a study by Salciccia et al. demonstrates that exosome-based diagnostic tools can outperform existing standards. The developed immunocapture-based ELISA system, which identifies PSA-expressing exosomes, differentiated between prostate cancer and benign prostatic hyperplasia with 98.57% sensitivity and 80.28% specificity. These results exceed comparable metrics for traditional serum PSA tests (Salciccia et al., 2023). While limited data exist on exosome sensitivity and specificity in pSS diagnosis, the growing body of evidence supports investigating exosomes and their cargo for biomarker-based systems in disease detection.

11.5.3.4 Technological Advancements in Exosome Isolation and Analysis

Advancements in technology also offer novel opportunities to integrate exosomes into disease diagnosis. In addition to the systems above, exosomal cargo can be integrated with other biomarkers to develop more comprehensive diagnostic panels. Integrating exosomal cargo into panels is already under investigation for rheumatoid arthritis diagnostics. The inclusion of exomiR-451a and exomiR-25-3p, alongside soluble TWEAK, in a diagnostic panel for rheumatoid arthritis demonstrated the ability to classify 95.6% of patients with 100% specificity and 85.7% sensitivity (Rodríguez-Muguruza et al., 2021). The results of this study indicate the potential for similar research investigating exosomes and exosomal cargo integration into existing diagnostic panels for pSS.

There have been continued advancements in exosomal isolation and analysis methods, enhancing the clinical applicability of exosome-based diagnostic tools. Microfluidic technologies offer several benefits for isolating and analyzing exosomes from clinical specimens, including greater efficiency, higher specificity, and preservation of exosome integrity. One of the key benefits of microfluidic devices is the small sample volume required, which may provide clinical benefit in situations of limited sample availability (Wu et al., 2022). The precision control afforded by microfluidic systems enables more effective separation of exosomes from other cellular components, thereby enhancing their purity and yield. This precision also allows the use of specific capture agents and surface biomarkers, thereby improving the devices' sensitivity and specificity. Microfluidic devices are adept at integrating multiple functions into a single platform. This capability combines processes for isolating, detecting, and analyzing exosomes while reducing the risk of sample loss and contamination. Consolidating these processes into a single device reduces sample processing time, which is critical for real-time monitoring of disease progression (Wu et al., 2022). Similarly, researchers have investigated the use of Surface-Enhanced Raman Spectroscopy (SERS) for the identification of exosomes in early cancer detection (Yang et al., 2022). This method offers several benefits, including high sensitivity, good selectivity, and rapid analysis. This method provides the opportunity for unlabeled SERS deception, enabling the improved maintenance of exosome content integrity. This method allows researchers to isolate and detect exosomes in bodily fluids, including blood and urine. Further research has been conducted on isolation methods for more efficient exosome extraction, with advancements in field-flow fractionation, label-free magnetic isolation, and functional micro- and nanoscale isolation (Lim et al., 2022; Zeng et al., 2021; Kim et al., 2022). Advancements in exosome isolation techniques continue to enhance the clinical applicability of exosome-based diagnostic methods, with improvements in isolation, more accurate profiling of exosomal markers, and characterization of cargo content. Advancements in the speed of isolation and analysis also offer opportunities to monitor disease progression and assess potential treatment responses, thereby improving clinicians' ability to provide patient-specific care.

11.6 THERAPEUTIC POTENTIAL OF EXOSOMES IN PSS

11.6.1 Limitations of Current pSS Treatments and Potential of Exosomes

Current treatments for pSS include NSAIDs, artificial tear substitutes, corticosteroids, punctal plugs, and biologics like monoclonal antibodies (Fox et al., 2021). Current therapies have limited efficacy and primarily focus on symptomatic relief rather than addressing the underlying pathophysiology. Many of these treatments are not specific to pSS but are general dry-eye treatments. There are sometimes severe long-term side effects with the existing treatment methods for pSS. For instance, monoclonal antibodies can increase patients' susceptibility to infection and present significant cost barriers (Vikse et al., 2019). Punctal plugs exhibit limited efficacy and carry risks such as discomfort, canaliculitis, and plug migration ("Silicone Punctal Plug Migration Resulting in Dacryocystitis and Canaliculitis," n.d.). Similarly, corticosteroids have broad systemic side effects and can lead to a long-term increase in the production of autoantibodies (Reksten et al., 2011). Exosomes can present a potential avenue for addressing these limitations by offering targeted and personalized treatment, with the opportunity for more efficient scalability and cost-effectiveness (Zhang et al., 2022). An overview of exosome therapy in primary Sjögren's syndrome is presented in Table 11.2.

TABLE 11.2
. Summary of Exosome Therapy in Primary Sjögren's Syndrome

Exosome Role	Origin of Exosome	Therapy Mechanism	Therapy outcome	References
Immune modulation	Olfactory ecto-MSC-derived exosomes (OE-MSC-Exos)	OE-MSC-Exos enhances the suppression function of myeloid-derived suppressor cells (MDSCs) by secreting IL-6 and activating the JAK2/Stats3 pathway in MDSCs.	Improved salivary flow rate and reduced tissue damage in mouse SGs. Decreased serum antibodies.	(Rui et al., 2021)
Immune modulation	iPSCs-derived MSCs (iMSC)	Suppress the development of Th1 and Th17 cells, likely by inhibiting the activation of antigen-presenting cells (APCs) and T cells and by inducing the regulatory cytokine TGF-β1.	P5 EVs reduced mRNA levels of markers for Th1 and Th17 cells and increased those of regulatory cytokines IL-10 and TGF-β1. Early-stage iMSCs were more effective than late-stage iMSCs.	(Kim et al., 2021)
Immune modulation	UCMSC-Exos	Inhibited the excessive proliferation of CD4+ T cells in pSS patients.	Decrease in $CD4^+$ T cells.	(Ma et al., 2023)
Immune modulation	Stem cells from a human exfoliated deciduous tooth	Inhibiting the Akt/GSK-3/Slug pathway and increasing salivary secretion in pSS patients.	Saliva secretion increased and was accompanied by decreased levels of p-Akt/Akt and p-GSK-3β/GSK-3β.	(Du et al., 2023)
Immune modulation	iPSCs from Dental Pulp Stem Cells (DPSCs)	Exosomes inhibit pro-inflammatory cytokines by suppressing TLR4 expression.	Increased saliva volume, decreased anti-SS-A and anti-SS-B bodies, and reduced inflammatory cell infiltration in SG tissue.	(Ogata et al., 2024)
Immune Modulation	Labial gland MSCs	miR-125b In the exosome, it binds to BLIMP1 in plasma cells to suppress plasma cell differentiation.	miR125b Directly Targets the 3'-UTR of PRDM1, restraining its expression. LGMSC-Exos reduced the area and number of lymphocyte infiltration foci and increased the saliva flow rate.	(Xing et al., 2022)
Immune Modulation	MSCs	Targeting Bcl–6 with miR–10a–5p in EVs generated from MDSC reduces germinal centre B cells.	MDSC-EVs inhibit B cell proliferation, activation, and GC B cell differentiation. Intravenous administration of MDSC-EVs significantly increased the saliva flow rate and reduced SG tissue damage.	(Zhou et al., 2023)

Exosome Role	Origin of Exosome	Therapy Mechanism	Therapy outcome	References
Immune Modulation	Human umbilical cord MSC	miR-100-5p converted macrophages into an anti-inflammatory M2 phenotype and elevated Tregs.	Preventive and therapeutic effects of dry eyes in rabbits.	(Li et al., 2022)
Regenerative Medicine	MSCs	Promoted ECM formation by upregulating the BMP7 gene, preventing fibrosis by downregulating TGF-β1, and downregulating pro-inflammatory cytokines.	Higher salivary and tear flow rate compared to controls, less lymphocyte infiltration, and inhibited anti-Rho/SSa antibodies.	(Abughanam et al., 2019)
Regenerative Medicine	DPSC-Derived Exosomes	Revitalize SG epithelial cell function during SS via the GPER-mediated cAMP/PKA/CREB pathway.	IFN-γ+DPSC-Exos-treated SGEC showed higher expression of AQP5, p-PKA, cAMP, and intracellular Ca2+ levels compared with IFN-γ-treated SGEC.	(Hu et al., 2023)
Regenerative medicine	Lacrimal gland-derived MSCs	Lipocalin-2, Prosaposin, Rac1, STAT1 upregulated.	MSG-MSCs exhibit similar wound-healing and angiogenesis-promoting functions to those of ADSC-exos, with enhanced anti-scar and anti-inflammatory effects.	(Xiang et al., 2024)
Regenerative Medicine	Human ADSC	Inhibiting Epithelial-Mesenchymal Transition through miR-199a-3p.	96% increase in salivary excretion, upregulation of repair genes, and preserved functional cells in treated mice.	(Guo et al., 2025)
Regenerative Medicine	Minor SG MSCs	Promote angiogenesis through VEGF, FGF, and ITGB1.	70% of the identified LG organoid proteins were shown to be exosomal. Organoids developed from patients with SS had histological features similar to those of SS-derived LG tissue.	(Jeong et al., 2021)

Abbreviations: ADSCs: adipose-derived stem cells; DPSC: Dental pulp-derived MSCs; MSCs: Mesenchymal stem cells; SGs: Salivary glands.

11.6.2 Exosome Therapy for Autoimmune and Inflammatory Eye Conditions with Implications for pSS

The therapeutic potential of exosome therapy lies primarily in its ability to modulate the immune system and promote tissue regeneration. Given their demonstrated efficacy in treating ocular diseases with pathophysiological mechanisms similar to those of primary Sjögren's syndrome (pSS), these modalities offer a promising avenue for exosome therapy in pSS.

11.6.2.1 Immune Modulation in Autoimmune-Mediated Ocular Inflammation

pSS pathophysiology is characterized by T and B cell dysregulation, leading to chronic inflammation and increased pro-inflammatory cytokines. These inflammatory mechanisms are also central to other autoimmune ocular conditions, such as autoimmune uveitis and graft-versus-host disease (GVHD)-associated eye disease. A 2017 study by Bai et al. found that periocular injection of mesenchymal stem cell (MSC)-derived exosomes inhibits the autoimmune response in experimental rat models of autoimmune uveitis. The study then confirmed that the exosomes acted as the primary immunosuppressant by administering allogenic skin grafts to mice, followed by exosome injections. The exosome injection delayed GVHD for 2 days by inhibiting the migration of inflammatory cells (Bai et al., 2017). This immunosuppressive action of exosomes has also been proven in humans (Kordelas et al., 2014). Given the shared pathophysiological features and overlapping treatment strategies between autoimmune uveitis and pSS, the successful application of exosome therapy for immunosuppression in uveitis could serve as a proof of concept for its utilization in the treatment of pSS (Marsh & Pflugfelder, 1999; Wakefield et al., 2000).

A clinical study investigated the therapeutic potential of exosomes in treating GVHD-associated dry eye disease following hematopoietic stem cell transplants. This is an inflammatory condition caused by allogeneic grafts. Patients who received the exosome treatment had increased tear production and a reduced immune response (Shigemoto-Kuroda et al., 2017). One advantage of this treatment is that it showed regenerative potential and was not limited to symptom reduction. The exosome therapy also showed long-term symptom relief. Dry eyes are the hallmark symptom of pSS, present in 90% of patients, so the long-term relief provided in the studies above can be directly applied to pSS.

11.6.2.2 Regenerative Potential of Exosomes in Ocular Damage

Beyond immune modulation, exosomes play a role in regenerating damaged tissue. Injection with bone marrow-derived exosomes has successfully treated glaucoma in mouse models (Mead & Tomarev 2017). The exosomes promoted the survival and axonal regeneration of retinal ganglion cells. This regenerative capacity is relevant to pSS, where neuroinflammatory mechanisms contribute to peripheral neuropathy (Sène et al., 2011) Given exosomes' ability to promote axon regeneration in the retina, similar mechanisms may apply to nerve repair in pSS-associated neuropathy.

11.6.3 Immunomodulation Using Stem Cells Derived Exosomes in pSS

11.6.3.1 Mechanism of Immune Modulation Using Stem Cell-Derived Exosomes

Stem cell-derived exosomes modulate the immune system through various mechanisms to treat pSS. One therapeutic mechanism is to suppress the innate immune system, as demonstrated by a 2021 study by Rui et al., which showed that exosomes from olfactory ecto-MSCs increased the activity of myeloid-derived suppressor cells (MDSC) by secreting IL-6 and activating the Jak2/Stat3 pathway in MDSC (Rui et al., 2021). Although the activation of suppressor cells is attributed to IL-6, IL-6 expression is increased in pSS and is a key inflammatory cytokine that drives disease progression. Conversely, Ma et al. (2023) reported that human umbilical cord MSC-derived exosomes inhibited CD4+ T cell proliferation by suppressing the production of IL-6 and other cytokines (Ma et al., 2023). This indicates that IL-6 acts in a context-dependent manner. IL-6 enhances the activity of

B and T cells in autoimmune conditions and downregulates regulatory T cell activity. In contrast, IL-6 has the opposite effect in cancer, promoting immune tolerance by suppressing T cell activity. Further research is needed to elucidate the precise mechanisms by which IL-6 contributes to the pathogenesis of pSS and serves as a therapeutic target.

Supporting the role of MDSCs, Rui et al. (2021) also found that injecting MDSCs from tumor-induced mice into the pSS mouse model attenuated disease progression. Mechanistically, another group found that MDSCs contain miR–10a–5p, which interacts with BCL-6 receptors to reduce B cell germinal center formation (Zhou et al., 2023). However, directly inhibiting miR–10a–5p also slowed pSS progression, suggesting that miR–10a–5p may act in a context-dependent manner. These studies' contradictory findings further highlight our incomplete understanding of the pathophysiology of pSS.

In addition to modulating the innate immune system, stem cell-derived exosomes influence the adaptive immune system, thereby slowing the progression of disease. A hallmark of pSS is a decrease in T regulatory cells (Tregs) and an increase in T cells and B cells. Several studies have identified a key exosomal cargo, TGF-β1, that regulates the Th/Treg ratio by promoting Treg cell differentiation (Kim et al., 2021; Ma et al., 2023; N. Li et al., 2022; Hai et al., 2018). TGF-β1 may also play a role in exosome cargo sorting, as demonstrated by a study that found that source iPSCs with higher TGF-β1 levels produced exosomes enriched in let-7 family miRs, leading to increased production of anti-inflammatory cytokines through downstream pathways (Ogata et al., 2024). In addition to TGF-β1, several exosomal miRs have been identified as suppressors of lymphocyte activation in pSS, including miR-100-5p, which induces Tregs, miR-125b, which inhibits the expression of pro-inflammatory cytokines, and miR-21, a negative regulator of inflammation (Li et al., 2022; Kim et al., 2021). Another study identified that miR-29a-3p suppresses T-bet, a transcription factor that promotes the differentiation of CD4+ cells into Th1 cells and IFN-γ production (Du et al.). In addition to modulating T cells and cytokines, exosomes have been shown to modulate B cells. Several studies have identified exosomal miRs that inhibit B cells, including miR-10a-5p, which decreases the number of germinal B cells in the SG, and miR-125b, which suppresses plasma cells (Zhou et al., 2023; Xing et al., 2022).

The immunomodulatory effects of exosomes in these studies have been observed using exosomes derived from various sources. A 2021 study by Kim et al. showed that exosomes from newly differentiated iPSCs, known as early passage iPSCs, have greater therapeutic effects than late passage iPSCs (Kim et al., 2021). This shows that the secretome of early passage iPSCs contains more immunosuppressive molecules, which decrease as iPSCs replicate.

11.6.3.2 Advantages over Stem Cell Therapy

As research into the therapeutic applications of exosomes in pSS develops, it is essential to consider their advantages over traditional stem cell therapies. Exosomes isolated from stem cells offer several therapeutic advantages over direct stem cell transplantation (Liang et al., 2025). One key advantage of exosomes over stem cells is their safety profile (Ahmed et al., 2025). Exosomes are not self-replicating because they lack their own DNA and replication machinery, thereby eliminating the risk of tumorigenesis (Salunkhe et al., 2020). Furthermore, when stem cells are injected, they may differentiate into non-desired cell types, influenced by the microenvironment into which they are injected. Local signaling modules in the host tissue can cause stem cells to differentiate into unintended cell types with potentially devastating consequences (Breitbach et al., 2007). This is especially relevant to pSS, where the inflammatory microenvironment could cause unpredictable stem cell differentiation with potentially harmful outcomes. Exosomes are also safer because they are less likely to trigger an immune response. They typically have fewer cell-surface proteins than stem cells and lack an MHC class I molecule (Quah & O'Neill, 2005).

Exosomes also provide enhanced targeting capabilities for specific cell types compared to stem cells. This is primarily due to their surface protein composition, a subset of surface proteins found on parent stem cells, allowing for more controlled cellular targeting (Habib et al., 2025). Exosomes are approximately 400 times smaller than stem cells; this feature enables them to penetrate inflammatory microenvironments (Radley et al., 1999; Ge et al., 2014). In a condition like pSS, where germinal B centers form around the exocrine gland, exosomes may penetrate more readily than stem cells, allowing them to deliver their therapeutic cargo more efficiently.

From a logistical standpoint, exosomes are easier and more cost-effective to produce and store, as they require fewer cryoprotectants. They are also less likely to get damaged during thaw cycles (Ge et al., 2014). Exosomes are less sensitive to oxygen levels, temperature, and pH. Their membranes are more stable due to a higher cholesterol content (Jeyaram & Jay, 2017).

Given these advantages, exosomes present an up-and-coming alternative to stem cell therapy for treating pSS. Their enhanced safety, immune compatibility, precise targeting, and ease of storage make them ideal candidates for addressing the immune dysregulation seen in pSS.

11.6.3.3 Exosome Isolation from Stem Cells

Exosomes are typically isolated from stem cell culture supernatants of various sources by centrifugation. Immunological separation is an emerging third mechanism of exosome isolation (Boriachek et al., 2018). This method uses antibodies to selectively isolate exosomes bearing specific surface markers. In the context of pSS, exosomes with surface proteins allowing for directed migration to the affected tissue can be selected. Advances in this exosome isolation method could facilitate the specificity required for the real-world application of exosome-based therapy for pSS.

11.6.3.4 Advantages over Monoclonal Antibodies

Rituximab, a monoclonal antibody used in the treatment of pSS, targets all CD20-positive B cells, leading to nonspecific systemic immune suppression and an increased risk of infections (Vikse et al., 2019). In contrast, exosomes can be engineered to carry diverse biological molecules, such as miRs and cytokines, thereby minimizing this risk. The versatility of exosome loading means they can be loaded to target multiple pathways involved in a complex condition like pSS and be individualized to address the genetic heterogeneity driving pSS pathogenesis. While monoclonal antibodies primarily function through immune suppression, exosomes can suppress the immune system and promote tissue regeneration, providing long-term benefits for symptom management in patients with pSS. Given these advantages, exosome treatment plans are likely to be preferred for candidates at a later stage of the disease, where tissue regeneration is needed alongside immune suppression for effective treatment once the disease has progressed sufficiently. In contrast to the costly repeat injections required for monoclonal antibody therapies, exosomes can potentially modify recipient T cells, leading to longer-lasting therapeutic effects (Hauser et al., 2020).

11.6.4 Exosome-Based Drug Delivery Systems

Immune modulation using stem cell-derived exosomes involves isolating exosomes from MSCs, which naturally contain immunomodulatory molecules. However, this approach faces challenges with standardization, scalability, and cargo specificity. A novel therapeutic strategy, discussed in further detail below, involves directly loading exosomes with targeted RNA and protein cargo, thereby bypassing the need for stem cell-derived exosomes. By leveraging exosomes' inherent advantages as drug carriers, this approach has the potential to revolutionize treatment for pSS, offering a more personalized, efficient, and targeted intervention.

11.6.4.1 Advantages of Exosomes as Drug Carriers

Exosomes offer several advantages over traditional drug delivery systems. These include personalized biocompatibility, reduced immunogenicity, and reduced patient toxicity.

Exosomes have been proven to traverse biological barriers within the body. It has been shown that exosome-associated adeno-associated virus (AAV) can penetrate the retina and deliver a dye-based marker through the eye following intravitreal injection (Wassmer et al., 2017). A key finding from this paper is that exosome-based systems can even show positive results in nerve fiber penetration. They have also been shown to cross the blood-brain barrier, a significant challenge in traditional drug delivery systems. Both play key roles here, given the aforementioned low immunogenicity and the tiny exosome size (30–150 nm). This is significant in the treatment of pSS, as approximately 20% of patients have shown optic neuropathy, and up to 5% have shown neurological issues with the central nervous system (Bak et al., 2017). Exosome-based treatment is one of the few methods for effectively delivering therapeutic cargo to affected nerve cells.

Exosomes can carry a diverse range of biomolecules as cargo, including proteins, lipids, miRNA, RNA, and DNA (Haider & Aramini, 2020). Once isolated by ultracentrifugation, exosomes can be selected for a specific drug or target, making them highly versatile. MiR loading potential is particularly relevant for treating pSS, as both miR-21 and miR-125b have been shown to positively affect the Th/Treg ratio. Exosomes, as a method of delivering these miRs directly to affected areas, could increase the cargo load successfully delivered compared to traditional methods.

Exosomes can also be modified to provide enhanced targeted delivery. A 2020 study by Bagheri et al. used exosomes to deliver doxorubicin to mouse colon adenocarcinoma cells. The exosomes were loaded with doxorubicin and 5TR1 via electroporation. 5TR1 causes exosomes to covalently bind to MUC1 in cancer cells, delivering treatment directly to tumor sites (Bagheri et al., 2020). Similarly, exosomes could be engineered with specific aptamers or proteins to target the immune cells or inflamed glandular tissues in pSS.

11.6.5 Exosomes for Lacrimal and SG Regeneration

Exosome-based regenerative medicine approaches promote the growth of SG epithelial cells, increase angiogenesis and neurogenesis, remodel the ECM, and generate transplantable organoids and spheroids. These processes often occur concurrently, with exosomes simultaneously acting on multiple regenerative pathways to enhance tissue repair and functional recovery.

Exosomes promote cell proliferation and survival, making them a powerful tool in regenerative medicine. A 2023 study treating the SGs of NOD/ltj mouse models with pSS using dental pulp stem cell-derived exosomes (DPSC-Exos) demonstrated the therapeutic potential of these exosomes. DPSC-Exos restored SGEC function via the GPER-mediated cAMP/PKA/CREB pathway (Hu et al., 2023). Similarly, in a mouse model of ethanol-induced lacrimal gland injury, lacrimal gland MSC exosomes enhanced the expression of four proteins: Lipocalin-2, Prosaposin, Rac1, and STAT1 (Xiang et al., 2024). The study focused primarily on the role of proteins in promoting cell survival, but also found that these proteins are involved in cell migration and proliferation.

In addition to promoting cell survival and proliferation, exosomes can also mitigate pathological processes that disrupt tissue homeostasis, such as the epithelial-to-mesenchymal transition (EMT) observed in diseases like pSS. Recent studies have shown that, in many inflammatory conditions, such as pSS, acinar epithelial cells lose their epithelial characteristics and acquire mesenchymal properties (Sarrand & Soyfoo, 2023). This is known as the EMT. In the early stages of pSS, EMT drives fibroblast production for healing, but over time, this process leads to fibrosis and hypofunction of SG cells. In a mouse model of radiation-induced SG injury, injection with human adipose-derived stem cell exosomes increased saliva secretion by 96%. The exosome contained miR-133b-3p, which inhibited EMT by interacting with Twist1, thereby targeting the TGFB1/Smad3 pathway. Suppressing this pathway decreased fibrosis and caused epithelial cell proliferation (Guo et al., 2025).

In addition to restoring epithelial function, exosomes are also critical in promoting angiogenesis, the formation of new blood vessels, particularly in the context of chronic inflammation and fibrosis seen in pSS. A 2024 study found that exosomes derived from minor SG-derived MSCs (MSG-MSCs) significantly promoted EC proliferation, migration, and tube formation. These exosomes carry pro-angiogenic factors, such as vascular endothelial growth factor (VEGF) and fibroblast growth factor (FGF), as well as other growth factors that stimulate EC proliferation and migration

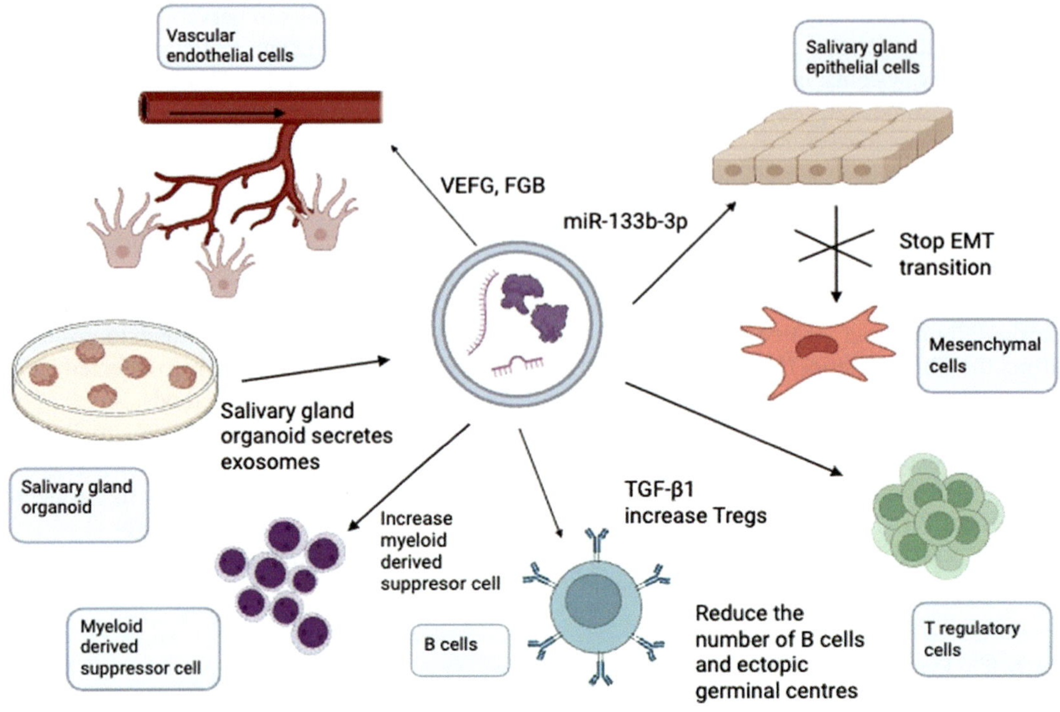

FIGURE 11.4 Exosome treatment modalities for pSS.

(Xiang et al., 2024). As ECs migrate, they form cadherin junctions with other ECs. Exosomes carry proteins that promote cadherin binding, ultimately leading to tube formation. The existing ECM must be broken down and remodeled for new blood vessels to form. ITGB1 in the exosome regulates EC adhesion to the ECM through the PI3K/AKT and MAPK/ERK pathways. The ability of MSG-MSC to promote angiogenesis could be utilized in pSS, where blood vessels are damaged by chronic inflammation and fibrosis. Guo et al. replicated the finding that exosomes encourage angiogenesis. They found that, in addition to inhibiting EMT in pSS, the exosomes promoted lacrimal gland regeneration through angiogenesis and neurogenesis (Guo et al., 2025).

The regenerative potential of exosomes extends to more complex processes, such as organogenesis. One study generated lacrimal gland organoids by harvesting healthy lacrimal gland cells from patients and allowing them to self-organize within a hydrogel. Pilocarpine was used to stimulate saliva production in the lacrimal gland organoid. Proteomic analysis of the organoids revealed that 70% of the proteins were associated with exosomes (Jeong et al., 2021). This highlights the crucial role exosomes play in SG formation and function. A 2017 study by Hayashi et al. also showed that exosomes play a key role in SG organogenesis. The paper identified that miR-133b-3p is transferred from the SG mesenchyme to epithelial cells via exosomes, promoting epithelial cell proliferation and expansion (Hayashi et al., 2017) (Figure 11.4).

11.7 CONCLUSION

In conclusion, this chapter has outlined the multifactorial pathogenesis of pSS and examined the emerging role of exosomes as potential diagnostic and therapeutic tools. By highlighting exosomes' roles in immune modulation, glandular injury, and systemic manifestations, we underscore their promise as non-invasive biomarkers and targeted treatment agents for pSS. Although these applications remain largely investigational, the potential benefits are substantial.

Bridging the gap between basic science and clinical practice will be crucial in translating these findings into meaningful patient outcomes. Collaboration between ophthalmologists, immunologists, and biomedical researchers will accelerate the development of exosome-based strategies, ultimately enhancing the diagnosis and management of pSS. Continued interdisciplinary efforts hold the potential to bring innovative, patient-centered solutions to clinical practice in the future.

REFERENCES

Abughanam G, Elkashty OA, Liu Y, Bakkar MO, Tran SD. Mesenchymal stem cells extract (MSCsE)-based therapy alleviates xerostomia and keratoconjunctivitis sicca in Sjögren's syndrome-like disease. *Int. J. Mol. Sci.* 2019 Jan;20(1):4750.

Adem B, Bastos N, Ruivo CF, Sousa-Alves S, Dias C, Vieira PF, et al. Exosomes define a local and systemic communication network in the healthy pancreas and pancreatic ductal adenocarcinoma. *Nat Commun.* 2024 Feb 21;15(1):1496.

Ahmed LA, Al-Massri KF, Haider KH. Therapeutic efficacy of mesenchymal stem cells and their derived exosomes in different neurodegenerative disorders: Current status and future perspectives. In: *Handbook of Regenerative Medicine: Cell-Free Therapy Approach.* 2025. p. 55–67.

Ahmed ZT, Alkahlot MH, Haider KH. MSC-derived exosomes: advances in cell-free therapy. In: KH Haider, editor. *Handbook of Stem Cell Applications.* Singapore: Springer Nature Singapore; 2023. p. 1–41.

Alamanos Y, Tsifetaki N, Voulgari PV, Venetsanopoulou AI, Siozos C, Drosos AA. Epidemiology of primary Sjögren's syndrome in north-west Greece, 1982–2003. *Rheumatology.* 2006 Feb 1;45(2):187–91.

Al-Hashimi I, Khuder S, Haghighat N, Zipp M. Frequency and predictive value of the clinical manifestations in Sjögren's syndrome. *J Oral Pathol Med.* 2001;30(1):1–6.

Amft N, Curnow SJ, Scheel-Toellner D, Devadas A, Oates J, Crocker J, et al. Ectopic expression of the B cell–attracting chemokine BCA-1 (CXCL13) on endothelial cells and within lymphoid follicles contributes to establishing germinal center–like structures in Sjögren's syndrome. *Arthritis Rheum.* 2001;44(11):2633–41.

Haider KH, Aramini B. Mircrining the injured heart with stem cell-derived exosomes: an emerging strategy of cell-free therapy. *Stem Cell Res Ther.* 2020 Jan 9;11(1):23. https://doi.org/10.1186/s13287-019-1548-7.

Bagheri E, Abnous K, Farzad SA, Taghdisi SM, Ramezani M, Alibolandi M. Targeted doxorubicin-loaded mesenchymal stem cells-derived exosomes as a versatile platform for fighting against colorectal cancer. *Life Sci.* 2020 Nov 15;261:118369.

Bai L, Shao H, Wang H, Zhang Z, Su C, Dong L, et al. Effects of mesenchymal stem cell-derived exosomes on experimental autoimmune uveitis. *Sci Rep.* 2017 Jun 28;7:4323.

Bak E, Yang HK, Hwang JM. Optic neuropathy associated with primary sjögren's syndrome: a case series. *Optom Vis Sci.* 2017 Apr;94(4):519–26.

Baldini C, Pepe P, Quartuccio L, Priori R, Bartoloni E, Alunno A, et al. Primary Sjögren's syndrome as a multi-organ disease: impact of the serological profile on the clinical presentation of the disease in a large cohort of Italian patients. *Rheumatology.* 2014 May 1;53(5):839–44.

Barizzone N, Monti S, Mellone S, Godi M, Marchini M, Scorza R, et al. Rare variants in the TREX1 gene and susceptibility to autoimmune diseases. *BioMed Res Int.* 2013;2013:471703.

Bokarewa M, Nagaev I, Dahlberg L, Smith U, Tarkowski A. Resistin, an adipokine with potent proinflammatory properties1. *J Immunol.* 2005 May 1;174(9):5789–95.

Boriachek K, Islam MdN, Möller A, Salomon C, Nguyen NT, Hossain MdSA, et al. Biological functions and current advances in isolation and detection strategies for exosome nanovesicles. *Small.* 2018;14(6):1702153.

Both T, Dalm VASH, van Hagen PM, van Daele PLA. Reviewing primary Sjögren's syndrome: beyond the dryness - From pathophysiology to diagnosis and treatment. *Int J Med Sci.* 2017 Feb 23;14(3):191–200.

Breitbach M, Bostani T, Roell W, Xia Y, Dewald O, Nygren JM, et al. Potential risks of bone marrow cell transplantation into infarcted hearts. *Blood.* 2007 Aug 15;110(4):1362–9.

Cadena-Sandoval D, Montúfar-Robles I, Barbosa-Cobos RE, Hernández-Molina G, Salas-García AK, Sánchez-Zauco N, et al. Interactions between TNFAIP3, PTPN22, and TRAF1-C5 gene polymorphisms in patients with primary Sjögren's syndrome. *Arch Rheumatol.* 2024;39(1):060–70.

Camaschella C, Nai A, Silvestri L. Iron metabolism and iron disorders revisited in the hepcidin era. *Haematologica.* 2020 Feb 1;105(2):260–72.

Cha S, Brayer J, Gao J, Brown V, Killedar S, Yasunari U, et al. A dual role for interferon-γ in the pathogenesis of sjögren's syndrome-like autoimmune exocrinopathy in the nonobese diabetic mouse. *Scand J Immunol.* 2004;60(6):552–65.

Cheng M, Yang J, Zhao X, Zhang E, Zeng Q, Yu Y, et al. Circulating myocardial miRs from infarcted hearts are carried in exosomes and mobilise bone marrow progenitor cells. *Nat Commun.* 2019 Feb 27;10(1):959.

Cheng Z, Hou G, Shen N. Evolving understandings for the roles of non-coding RNAs in autoimmunity and autoimmune disease. *J Autoimmun.* 2023 May 1;137:102948.

Cortes-Troncoso J, Jang SI, Perez P, Hidalgo J, Ikeuchi T, Greenwell-Wild T, et al. T cell exosome–derived miR-142-3p impairs glandular cell function in Sjögren's syndrome. *JCI Insight* [Internet]. 2020 May 7 [cited 2025 Feb 6];5(9). Available from: https://insight.jci.org/articles/view/133497

Costa-Silva B, Aiello NM, Ocean AJ, Singh S, Zhang H, Thakur BK, et al. Pancreatic cancer exosomes initiate the formation of a pre-metastatic niche in the liver. *Nat Cell Biol.* 2015 Jun;17(6):816–26.

Deng Y, Cao Y, Wang L, Ye D. The role and application of salivary exosomes in malignant neoplasms. *Cancer Manag Res.* 2021 Jul 22;13:5813–20.

Deroo L, Genbrugge E, Dochy F, Creytens D, Achten H, De Boeck K, et al. Ultrasound-guided core needle biopsy and incisional biopsy of the parotid gland are comparable in diagnosing primary Sjögren's syndrome. *Rheumatol Oxf Engl.* 2023 Aug 1;62(8):2765–72.

Du Z, Wei P, Jiang N, Wu L, Ding C, Yu G. SHED-derived exosomes ameliorate hyposalivation caused by Sjögren's syndrome via Akt/GSK-3β/Slug-mediated ZO-1 expression. *Chin Med J* (Engl). 2023 Nov 5;136(21):2596–608.

Duan W, Zhang W, Jia J, Lu Q, Eric Gershwin M. Exosomal miR in autoimmunity. *Cell Mol Immunol.* 2019 Dec;16(12):932–4.

Duréndez-Sáez E, Calabuig-Fariñas S, Torres-Martínez S, Moreno-Manuel A, Herreros-Pomares A, Escorihuela E, et al. Analysis of exosomal cargo provides accurate clinical, histologic, and mutational information in non-small cell lung cancer. *Cancers.* 2022 Jan;14(13):3216.

Ekström Smedby K, Vajdic CM, Falster M, Engels EA, Martínez-Maza O, Turner J, et al. Autoimmune disorders and risk of non-Hodgkin lymphoma subtypes: a pooled analysis within the InterLymph Consortium. *Blood.* 2008 Apr 15;111(8):4029–38.

Ferguson S, Kim S, Lee C, Deci M, Nguyen J. The phenotypic effects of exosomes secreted from distinct cellular sources: a comparative study based on miR composition. *AAPS J.* 2018 Apr 30;20(4):67.

Ferrant J, Pontis A, Zimmermann F, Dingli F, Poullet P, Loew D, et al. Phenotypic and proteomic analysis of plasma extracellular vesicles highlights them as potential biomarkers of primary Sjögren syndrome. *Front Immunol* [Internet]. 2023 Jul 17 [cited 2025 Feb 6];14. Available from: https://www.frontiersin.org/journals/immunology/articles/10.3389/fimmu.2023.1207545/full

Finamore F, Cecchettini A, Ceccherini E, Signore G, Ferro F, Rocchiccioli S, et al. Characterization of extracellular vesicle cargo in Sjögren's Syndrome through a SWATH-MS proteomics approach. *Int J Mol Sci.* 2021 Jan;22(9):4864.

Fox RI, Fox CM, Gottenberg JE, Dörner T. Treatment of Sjögren's syndrome: current therapy and future directions. *Rheumatology.* 2021 May 1;60(5):2066–74.

Frank J, Richter M, de Rossi C, Lehr CM, Fuhrmann K, Fuhrmann G. Extracellular vesicles protect glucuronidase model enzymes during freeze-drying. *Sci Rep.* 2018 Aug 17;8(1):12377.

Gallo A, Jang SI, Ong HL, Perez P, Tandon M, Ambudkar I, et al. Targeting the Ca2+ sensor STIM1 by exosomal transfer of Ebv-miR-BART13-3p is associated with Sjögren's syndrome. *eBioMedicine.* 2016 Aug 1;10:216–26.

Gao J, Li A, Hu J, Feng L, Liu L, Shen Z. Recent developments in isolating exosome methods. *Front Bioeng Biotechnol.* 2023 Jan 13;10:1100892.

Gao M, Monian P, Pan Q, Zhang W, Xiang J, Jiang X. Ferroptosis is an autophagic form of cell death. *Cell Res.* 2016 Sep;26(9):1021–32.

García M, a-Carrasco M, Mendoza-Pinto C, Cé J, Jimé S, Néz-Hernández et al. Serologic features of primary Sjogren's syndrome: clinical and prognostic correlation. *Int J Clin Rheumatol.* 2012 Dec 1;7(6):651–60.

Gatto M, Iaccarino L, Ghirardello A, Bassi N, Pontisso P, Punzi L, et al. Serpins, immunity and autoimmunity: old molecules, new functions. *Clin Rev Allergy Immunol.* 2013 Oct;45(2):267–80.

Ge Q, Zhou Y, Lu J, Bai Y, Xie X, Lu Z. miR in plasma exosome is stable under different storage conditions. *Mol Basel, Switzerland.* 2014 Jan 27;19(2):1568–75.

Giovelli RA, Santos MC, Serrano ÉV, Valim V. Clinical characteristics and biopsy accuracy in suspected Sjögren's syndrome referred to labial salivary gland biopsy. *BMC Musculoskelet Disord.* 2015 Feb 15;16(1):30.

Gottenberg JE, Cagnard N, Lucchesi C, Letourneur F, Mistou S, Lazure T, et al. Activation of IFN pathways and plasmacytoid dendritic cell recruitment in target organs of primary Sjögren's syndrome. *Proc Natl Acad Sci.* 2006 Feb 21;103(8):2770–5.

Guellec D, Cornec D, Jousse-Joulin S, Marhadour T, Marcorelles P, Pers JO, et al. Diagnostic value of labial minor salivary gland biopsy for Sjögren's syndrome: a systematic review. *Autoimmun Rev.* 2013 Jan;12(3):416–20.

Guo X, Huang Z, Wu F, Jiang W, Li Y, Wang T, et al. Exosomes from human adipose stem cells mitigate irradiation-induced injury to salivary glands by inhibiting epithelial-mesenchymal transition through miR-199a-3p targeting of Twist1 and regulating the TGFβ1/Smad3 pathway. *Theranostics.* 2025 Jan 2;15(5):1622–41.

Habib SM, Martini MF, Al Ahmed O, Qassab AR, Haider KH. The potential of mesenchymal stem cell–derived exosomes in cell-free therapy for myocardial repair. In: Haider KH. (ed.) *Handbook of Regenerative Medicine: Cell-Free Therapy.* Taylor & Francis Group CRC Press; 2025), Chapter-24, p. 372–390.

Habib SY, Martini MF, Hamdan YNA, Shrebaty OMM, Haider KH. Mesenchymal stem cell secretome and exosomes as potential advanced therapy medicinal products for treating a failing heart. *World J Stem Cells.* 2026:55–67.

Hai B, Shigemoto-Kuroda T, Zhao Q, Lee RH, Liu F. Inhibitory effects of iPSC-MSCs and their extracellular vesicles on the onset of sialadenitis in a mouse model of Sjögren's syndrome. *Stem Cells Int.* 2018;2018(1):2092315.

Ham D, Inoue A, Xu J, Du Y, Chung KY. Molecular mechanism of muscarinic acetylcholine receptor M3 interaction with Gq. *Commun Biol.* 2024 Mar 23;7(1):1–14.

Hauser SL, Bar-Or A, Cohen JA, Comi G, Correale J, Coyle PK, et al. Ofatumumab versus teriflunomide in multiple sclerosis. *N Engl J Med.* 2020 Aug 6;383(6):546–57.

Hayashi T, Lombaert IMA, Hauser BR, Patel VN, Hoffman MP. Exosomal MiR transport from salivary mesenchyme regulates epithelial progenitor expansion during organogenesis. *Dev Cell.* 2017 Jan 9;40(1):95–103.

Hazrati A, Soudi S, Malekpour K, Mahmoudi M, Rahimi A, Hashemi SM, et al. Immune cell-derived exosomes function as a double-edged sword, with roles in both disease progression and therapeutic applications. *Biomark Res.* 2022 May 12;10(1):30.

Hofmann L, Medyany V, Ezić J, Lotfi R, Niesler B, Röth R, et al. Cargo and functional profile of saliva-derived exosomes reveal biomarkers specific for head and neck cancer. *Front Med.* 2022;9:904295.

Holzinger D, Tenbrock K, Roth J. Alarmins of the S100-family in juvenile autoimmune and auto-inflammatory diseases. *Front Immunol* [Internet]. 2019 Feb 15 [cited 2025 Feb 6];10. Available from: https://www.frontiersin.org/journals/immunology/articles/10.3389/fimmu.2019.00182/full

Hong SH, Lee DH, Lee YS, Jo MJ, Jeong YA, Kwon WT, et al. Molecular crosstalk between ferroptosis and apoptosis: emerging role of ER stress-induced p53-independent PUMA expression. *Oncotarget.* 2017 Dec 8;8(70):115164–78.

Hu L, Zhang T, Ma H, Pan Y, Wang S, Liu X, et al. Discovering the secret of diseases by incorporated tear exosomes analysis via rapid-isolation system: iTEARS. *ACS Nano.* 2022 Aug 23;16(8):11720–32.

Hu S, Chen B, Zhou J, Liu F, Mao T, Pathak JL, et al. Dental pulp stem cell-derived exosomes revitalize salivary gland epithelial cell function in NOD mice via the GPER-mediated cAMP/PKA/CREB signaling pathway. *J Transl Med.* 2023 Jun 3;21(1):361.

Hu S, Zhang L, Su Y, Liang X, Yang J, Luo Q, et al. Sensitive detection of multiple blood biomarkers via immunomagnetic exosomal PCR for the diagnosis of Alzheimer's disease. *Sci Adv.* 2024 Mar 27;10(13):eabm3088.

Huang YT, Lu TH, Chou PL, Weng MY. Diagnostic delay in patients with primary sjögren's syndrome: a population-based cohort study in Taiwan. *Healthcare.* 2021 Mar 23;9(3):363.

Inoue M, Sakuta N, Watanabe S, Zhang Y, Yoshikaie K, Tanaka Y, et al. Structural basis of sarco/endoplasmic reticulum Ca2+-ATPase 2b regulation via transmembrane helix interplay. *Cell Rep.* 2019 Apr 23;27(4):1221–30.e3.

Ismail NM, Toraih EA, Mohammad MHS, Alshammari EM, Fawzy MS. Association of miR-34a rs2666433 (A/G) variant with systemic lupus erythematosus in female patients: a case-control study. *J Clin Med.* 2021 Jan;10(21):5095.

Jeong SY, Choi WH, Jeon SG, Lee S, Park JM, Park M, et al. Establishment of functional epithelial organoids from human lacrimal glands. *Stem Cell Res Ther.* 2021 Apr 21;12(1):247.

Jeyaram A, Jay SM. Preservation and storage stability of extracellular vesicles for therapeutic applications. AAPS J. 2017 Nov 27;20(1):1.

Ji J, He Q, Xia Y, Sha X, Liang Q, Xu Y, et al. Circulating plasma-derived exosomes from systemic lupus erythematosus aggravate lupus nephritis through miR-122-5p/FOXO3-mediated macrophage activation. *J Nanobiotechnology.* 2024 Dec 19;22(1):779.

Jiang Q, Wang Q, Tan S, Cai J, Ye X, Su G, et al. Effects of plasma-derived exosomal miR-19 b-3p on Treg/T B-cells 17 cell imbalance in Behçet's Uveitis. *Invest Ophthalmol Vis Sci.* 2023 Apr 24;64(4):28.

Jin M, Hwang SM, Davies AJ, Shin Y, Bae JS, Lee JH, et al. Autoantibodies in patients with primary Sjögren's syndrome induce the internalization of muscarinic type 3 receptors. *Biochim Biophys Acta BBA - Mol Basis Dis.* 2012 Feb 1;1822(2):161–7.

Jonsson R, Brokstad KA, Jonsson MV, Delaleu N, Skarstein K. Current concepts on Sjögren's syndrome – classification criteria and biomarkers. *Eur J Oral Sci.* 2018 Oct;126(Suppl 1):37–48.

Kachaner A, Bergé E, Desmoulins F, Pajolec CL, Rousseau A, Labetoulle M, et al. Comparison between primary Sjögren's disease patients with high or low levels of dryness. *RMD Open.* 2023 Oct 1;9(4):e003291.

Kakan SS, Janga SR, Cooperman B, Craig DW, Edman MC, Okamoto CT, et al. Small RNA deep sequencing identifies a unique miR signature released in serum exosomes in a mouse model of Sjögren's syndrome. *Front Immunol* [Internet]. 2020 Jul 17 [cited 2025 Feb 6];11. Available from: https://www.frontiersin.org/journals/immunology/articles/10.3389/fimmu.2020.01475/full

Katayama I, Kotobuki Y, Kiyohara E, Murota H. Annular erythema associated with Sjögren's syndrome: review of the literature on the management and clinical analysis of skin lesions. *Mod Rheumatol.* 2010 Apr 1;20(2):123–9.

Kim H, Zhao Q, Barreda H, Kaur G, Hai B, Choi JM, et al. Identification of molecules responsible for therapeutic effects of extracellular vesicles produced from iPSC-derived MSCs on Sjögren's syndrome. *Aging Dis.* 2021 Sep 1;12(6):1409–22.

Kim YB, Lee GB, Moon MH. Size separation of exosomes and microvesicles using flow field-flow fractionation/multiangle light scattering and lipidomic comparison. *Anal Chem.* 2022 Jun 28;94(25):8958–65.

Kordelas L, Rebmann V, Ludwig AK, Radtke S, Ruesing J, Doeppner TR, et al. MSC-derived exosomes: a novel tool to treat therapy-refractory graft-versus-host disease. *Leukemia.* 2014 Apr;28(4):970–3.

Liang Y, Xia J, Xu L, Zhang Y, Xu X, Haider KH. Cell-free and cell-based therapy in regenerative medicine. Haider, KH. (ed.) *Handbook of Regenerative Medicine: Cell-Free Therapy Approach* (1st ed.). CRC Press; 2025, Chapter 24. https://doi.org/10.1201/9781003611509.

Lee N, Canagasingham A, Bajaj M, Shanmugasundaram R, Hutton A, Bucci J, et al. Urine exosomes as biomarkers in bladder cancer diagnosis and prognosis: From functional roles to clinical significance. *Front Oncol.* 2022 Sep 20;12:1019391.

Leonardi A, Piliego F, Castegnaro A, Lazzarini D, La Gloria Valerio A, Mattana P, et al. Allergic conjunctivitis: a cross-sectional study. *Clin Exp Allergy.* 2015;45(6):1118–25.

Li G, Fang Y, Xu N, Ding Y, Liu D. Fibroblast-like synoviocytes-derived exosomal circFTO deteriorates rheumatoid arthritis by enhancing N6-methyladenosine modification of SOX9 in chondrocytes. *Arthritis Res Ther.* 2024 Feb 22;26(1):56.

Li N, Gao Z, Zhao L, Du B, Ma B, Nian H, et al. MSC-derived small extracellular vesicles attenuate autoimmune dacryoadenitis by promoting M2 macrophage polarization and inducing tregs via miR-100-5p. *Front Immunol* [Internet]. 2022 Jul 6 [cited 2025 Jan 4];13. Available from: https://www.frontiersin.org/journals/immunology/articles/10.3389/fimmu.2022.888949/full

Lim J, Kang B, Son HY, Mun B, Huh YM, Rho HW, et al. Microfluidic device for one-step detection of breast cancer-derived exosomal mRNA in blood using signal-amplifiable 3D nanostructure. *Biosens Bioelectron.* 2022 Feb 1;197:113753.

Losurdo M, Pedrazzoli M, D'Agostino C, Elia CA, Massenzio F, Lonati E, et al. Intranasal delivery of mesenchymal stem cell-derived extracellular vesicles exerts immunomodulatory and neuroprotective effects in a 3xTg model of Alzheimer's disease. *Stem Cells Transl Med.* 2020 Sep 1;9(9):1068–84.

Lundström W, Highfill S, Walsh STR, Beq S, Morse E, Kockum I, et al. Soluble IL7Rα potentiates IL-7 bioactivity and promotes autoimmunity. *Proc Natl Acad Sci.* 2013 May 7;110(19): E1761–70.

Ma D, Wu Z, Zhao X, Zhu X, An Q, Wang Y, et al. Immunomodulatory effects of umbilical mesenchymal stem cell-derived exosomes on CD4+ T-cells in patients with primary Sjögren's syndrome. *Inflammopharmacology.* 2023 Aug 1;31(4):1823–38.

Marsh P, Pflugfelder SC. Topical nonpreserved methylprednisolone therapy for keratoconjunctivitis sicca in Sjögren syndrome1. *Ophthalmology.* 1999 Apr 1;106(4):811–16.

Mead B, Tomarev S. Bone marrow-derived mesenchymal stem cells-derived exosomes promote survival of retinal ganglion cells through miR-dependent mechanisms. *Stem Cells Transl Med.* 2017 Apr 1;6(4):1273–85.

Methodologies to diagnose and monitor dry eye disease: report of the diagnostic methodology subcommittee of the international dry eye workshop (2007). *Ocul Surf.* 2007 Apr 1;5(2):108–52.

Mushtaq M, Zineldeen DH, Mateen MA, Haider KH. Mesenchymal stem cells' "garbage bags" at work: Treating radial nerve injury with mesenchymal stem cell-derived exosomes. *World J Stem Cells* 2024;16(5). https://doi.org/10.4252/wjsc. v16.i5.0000.

Narayanan R, Huang CC, Ravindran S. Hijacking the cellular mail: exosome mediated differentiation of mesenchymal stem cells. *Stem Cells Int.* 2016;2016(1):3808674.

Nguyen Y, Nocturne G, Henry J, Ng WF, Belkhir R, Desmoulins F, et al. Identification of distinct subgroups of Sjögren's disease by cluster analysis based on clinical and biological manifestations: data from the cross-sectional Paris-Saclay and the prospective ASSESS cohorts. *Lancet Rheumatol.* 2024 Apr 1;6(4):e216–25.

Niu L, Song X, Wang N, Xue L, Song X, Xie L. Tumor-derived exosomal proteins as diagnostic biomarkers in non-small cell lung cancer. *Cancer Sci.* 2019 Jan;110(1):433–42.

Ogata K, Moriyama M, Kawado T, Yoshioka H, Yano A, Matsumura-Kawashima M, et al. Extracellular vesicles from induced pluripotent stem cells (iPS cells), which are highly capable of producing HGF and TGF-β1, can attenuate Sjögren's syndrome by regulating innate immunity. *Cell Signal.* 2024 Jan 1;113:110980.

Pauley KM, Stewart CM, Gauna AE, Dupre LC, Kuklani R, Chan AL, et al. Altered miR-146a expression in Sjögren's syndrome and its functional role in innate immunity. *Eur J Immunol.* 2011 Jul;41(7):2029–39.

Peng X, Hou L, Wu X, Liu Z, Wang Y, Zeng P, et al. The plasma exosomes from patients with primary Sjögren's syndrome contain epithelial cell–derived proteins involved in ferroptosis. *J Mol Med.* 2023 Oct 1;101(10):1289–304.

Pianta A, Drouin EE, Crowley JT, Arvikar S, Strle K, Costello CE, et al. Annexin A2 is a target of autoimmune T and B cell responses associated with synovial fibroblast proliferation in patients with antibiotic-refractory Lyme arthritis. *Clin Immunol Orlando, Fla.* 2015 Oct;160(2):336–41.

Qi C, Lavriha P, Mehta V, Khanppnavar B, Mohammed I, Li Y, et al. Structural basis of adenylyl cyclase 9 activation. *Nat Commun.* 2022 Feb 24;13(1):1045.

Quah BJC, O'Neill HC. The immunogenicity of dendritic cell-derived exosomes. *Blood Cells Mol Dis.* 2005 Sep;35(2):94–110.

Radley JM, Ellis S, Palatsides M, Williams B, Bertoncello I. Ultrastructure of primitive hematopoietic stem cells isolated using probes of functional status. *Exp Hematol.* 1999 Feb;27(2):365–9.

Rahimian N, Nahand JS, Hamblin MR, Mirzaei H. Exosomal MiR profiling. *Methods Mol Biol Clifton NJ.* 2023;2595:13 47.

Ramos-Casals M, Brito-Zerón P, Seror R, Bootsma H, Bowman SJ, Dörner T, et al. Characterization of systemic disease in primary Sjögren's syndrome: EULAR-SS task force recommendations for articular, cutaneous, pulmonary, and renal involvements. *Rheumatology.* 2015 Dec 1;54(12):2230–8.

Reksten TR, Brokstad KA, Jonsson R, Brun JG, Jonsson MV. Implications of long-term medication of oral steroids and antimalarial drugs in primary Sjögren's syndrome. *Int Immunopharmacol.* 2011 Dec 1;11(12):2125–9.

Ríos-Ríos W de J, Sosa-Luis SA, Torres-Aguilar H. T-cells subsets in the immunopathology and treatment of Sjogren's syndrome. *Biomolecules.* 2020 Nov;10(11):1539.

Rodríguez-Muguruza S, Altuna-Coy A, Castro-Oreiro S, Poveda-Elices MJ, Fontova-Garrofé R, Chacón MR. A serum biomarker panel of exomiR-451a, exomiR-25-3p, and soluble TWEAK for early diagnosis of rheumatoid arthritis. *Front Immunol* [Internet]. 2021 Nov 15 [cited 2025 Feb 6];12. Available from: https://www.frontiersin.org/journals/immunology/articles/10.3389/fimmu.2021.790880/full

Roeser HP, Lee GR, Nacht S, Cartwright GE. The role of ceruloplasmin in iron metabolism. *J Clin Invest.* 1970 Dec;49(12):2408–17.

Rui K, Hong Y, Zhu Q, Shi X, Xiao F, Fu H, et al. Olfactory ecto-mesenchymal stem cell-derived exosomes ameliorate murine Sjögren's syndrome by modulating the function of myeloid-derived suppressor cells. *Cell Mol Immunol.* 2021 Feb;18(2):440–51.

Salciccia S, Frisenda M, Bevilacqua G, Gobbi L, Bucca B, Moriconi M, et al. Exosome analysis in prostate cancer: how they can improve biomarkers' performance. *Curr Issues Mol Biol.* 2023 Jul 21;45(7):6085–96.

Salivary gland epithelial cell exosomes: A source of autoantigenic ribonucleoproteins. [cited 2025 Feb 6]; Available from: https://acrjournals.onlinelibrary.wiley.com/doi/10.1002/art.21005

Salunkhe S, Dheeraj, Basak M, Chitkara D, Mittal A. Surface functionalization of exosomes for target-specific delivery and in vivo imaging & tracking: Strategies and significance. *J Controlled Release.* 2020 Oct 10;326:599–614.

Sanz-Rubio D, Martin-Burriel I, Gil A, Cubero P, Forner M, Khalyfa A, et al. Stability of circulating exosomal miRs in healthy subjects. *Sci Rep.* 2018 Jul 9;8(1):10306.

Sarrand J, Soyfoo MS. Involvement of Epithelial-Mesenchymal Transition (EMT) in autoimmune diseases. *Int J Mol Sci*. 2023 Jan;24(19):14481.

Scofield RH. Vasculitis in Sjögren's syndrome. *Curr Rheumatol Rep*. 2011 Dec 1;13(6):482–8.

Sène D, Jallouli M, Lefaucheur JP, Saadoun D, Costedoat-Chalumeau N, Maisonobe T, et al. Peripheral neuropathies associated with primary sjögren syndrome: immunologic profiles of nonataxic sensory neuropathy and sensorimotor neuropathy. *Medicine (Baltimore)*. 2011 Mar;90(2):133.

Shang Y, Luo M, Yao F, Wang S, Yuan Z, Yang Y. Ceruloplasmin suppresses ferroptosis by regulating iron homeostasis in hepatocellular carcinoma cells. *Cell Signal*. 2020 Aug 1;72:109633.

Sharygin D, Koniaris LG, Wells C, Zimmers TA, Hamidi T. Role of CD14 in human disease. *Immunology*. 2023 Jul;169(3):260–70.

Shen J, Yan J, Wang Q, Zhuang L, Luo Y. MiR-541-5p regulates type II alveolar epithelial cell proliferation and activity by modulating the HMGB1 expression. *Shock Augusta Ga*. 2022 Apr;57(4):536–43.

Shen L, Malyavantham K, Suresh L, Kapsogeorgou EK, Tzioufas AG, Sun Y, et al. Early detection of Sjögren's syndrome; Sensitivity and specificity of the Sjö® diagnostic test. *Invest Ophthalmol Vis Sci*. 2016 Sep 26;57(12):5681.

Shiboski CH, Shiboski SC, Seror R, Criswell LA, Labetoulle M, Lietman TM, et al. 2016 American College of Rheumatology/European League against rheumatism classification Criteria for Primary Sjögren's Syndrome: a consensus and data-driven methodology involving three international patient cohorts. *Arthritis Rheumatol Hoboken NJ*. 2017 Jan;69(1):35–45.

Shiboski S, Shiboski C, Criswell L, Baer A, Challacombe S, Lanfranchi H, et al. American college of rheumatology classification criteria for Sjögren's Syndrome: a data-driven, expert consensus approach in the SICCA Cohort. *Arthritis Care Res*. 2012 Apr;64(4):475–87.

Shigemoto-Kuroda T, Oh JY, Kim D ki, Jeong HJ, Park SY, Lee HJ, et al. MSC-derived extracellular vesicles attenuate immune responses in two autoimmune murine models: type 1 diabetes and uveoretinitis. *Stem Cell Rep*. 2017 May 9;8(5):1214–25.

Silicone punctal plug migration resulting in dacryocystitis and canaliculitis [Internet]. [cited 2025 Feb 18]. Available from: https://oce-ovid-com.proxy1.lib.uwo.ca/article/00003226-199705000-00022/HTML

Soehnlein O, Lindbom L. Neutrophil-derived azurocidin alarms the immune system. *J Leukoc Biol*. 2009 Mar;85(3):344–51.

Sokolova V, Ludwig AK, Hornung S, Rotan O, Horn PA, Epple M, et al. Characterisation of exosomes derived from human cells by nanoparticle tracking analysis and scanning electron microscopy. *Colloids Surf B Biointerfaces*. 2011 Oct 1;87(1):146–50.

Song M, Bai H, Zhang P, Zhou X, Ying B. Promising applications of human-derived saliva biomarker testing in clinical diagnostics. *Int J Oral Sci*. 2023 Jan 4;15(1):1–17.

Sutcliffe N. Mimics of Sjögren's syndrome. In: Price EJ, Tappuni AR, Price EJ, Tappuni AR, (eds.) *Oxford Textbook of Sjögren's Syndrome* [Internet]. Oxford University Press; 2021 [cited 2025 Feb 6]. https://doi.org/10.1093/med/9780198806684.003.0027

Talotta R, Mercurio V, Bongiovanni S, Vittori C, Boccassini L, Rigamonti F, et al. Evaluation of salivary and plasma miR expression in patients with Sjögren's syndrome, and correlations with clinical and ultrasonographic outcomes. *Clin Exp Rheumatol*. 2019;37 Suppl 118(3):70–7.

Teos LY, Alevizos I. Genetics of Sjögren's syndrome. *Clin Immunol*. 2017 Sep 1;182:41 7.

Tiyuri A, Baghermanesh SS, Davatgaran-Taghipour Y, Eslami SS, Shaygan N, Parsaie H, et al. Diagnostic accuracy of serum-derived exosomes for hepatocellular carcinoma: a systematic review and meta-analysis. *Expert Rev Mol Diagn*. 2023;23(11):971–83.

Toker E, Yavuz Ş, Direskeneli H. Anti-Ro/SSA and anti-La/SSB autoantibodies in the tear fluid of patients with Sjögren's syndrome. *Br J Ophthalmol*. 2004 Mar 1;88(3):384–7.

Veenbergen S, Kozmar A, van Daele PLA, Schreurs MWJ. Autoantibodies in Sjögren's syndrome and its classification criteria. *J Transl Autoimmun*. 2021 Dec 27;5:100138.

Ventriglia G, Mancarella F, Sebastiani G, Cook DP, Mallone R, Mathieu C, et al. miR-409-3p is reduced in plasma and islet immune infiltrates of NOD diabetic mice and is differentially expressed in people with type 1 diabetes. *Diabetologia*. 2020 Jan;63(1):124–36.

Vikse J, Jonsdottir K, Kvaløy JT, Wildhagen K, Omdal R. Tolerability and safety of long-term rituximab treatment in systemic inflammatory and autoimmune diseases. *Rheumatol Int*. 2019 Jun 1;39(6):1083–90.

Vitali C, Bombardieri S, Jonsson R, Moutsopoulos H, Alexander E, Carsons S, et al. Classification criteria for Sjögren's syndrome: a revised version of the European criteria proposed by the American-European Consensus Group. *Ann Rheum Dis*. 2002 Jun;61(6):554–8.

Vivino FB, Minerva P, Huang CH, Orlin SE. Corneal melt as the initial presentation of primary Sjögren's syndrome. *J Rheumatol.* 2001 Feb 1;28(2):379–82.

Wakefield D, Jennings A, McCluskey PJ. Intravenous pulse methylprednisolone in the treatment of uveitis associated with multiple sclerosis. *Clin Experiment Ophthalmol.* 2000;28(2):103–6.

Wang L, Wang C, Jia X, Yu J. Circulating exosomal miR-17 inhibits the induction of regulatory T-cells via suppressing TGFBR II expression in rheumatoid arthritis. *Cell Physiol Biochem.* 2018 Nov 1;50(5):1754–63.

Wang Y, Jiao T, Fu W, Zhao S, Yang L, Xu N, et al. miR-410-3p regulates proliferation and apoptosis of fibroblast-like synoviocytes by targeting YY1 in rheumatoid arthritis. *Biomed Pharmacother.* 2019 Nov 1;119:109426.

Wassmer SJ, Carvalho LS, György B, Vandenberghe LH, Maguire CA. Exosome-associated AAV2 vector mediates robust gene delivery into the murine retina upon intravitreal injection. *Sci Rep.* 2017 Mar 31;7(1):45329.

Willms E, Johansson HJ, Mäger I, Lee Y, Blomberg KEM, Sadik M, et al. Cells release subpopulations of exosomes with distinct molecular and biological properties. *Sci Rep.* 2016 Mar 2;6(1):22519.

Wu L, Han X, Jiang X, Ding H, Qi C, Yin Z, et al. Downregulation of renal Hsa-miR-127-3p contributes to the overactivation of type I interferon signaling pathway in the kidney of lupus nephritis. *Front Immunol* [Internet]. 2021 Oct 21 [cited 2025 Feb 6];12. Available from: https://www.frontiersin.org/journals/immunology/articles/10.3389/fimmu.2021.747616/full

Wu Y, Wang Y, Lu Y, Luo X, Huang Y, Xie T, et al. Microfluidic technology for the isolation and analysis of exosomes. *Micromachines.* 2022 Sep 22;13(10):1571.

Xiang H, Ding P, Qian J, Lu E, Sun Y, Lee S, et al. Exosomes derived from minor salivary gland mesenchymal stem cells: a promising novel exosome exhibiting pro-angiogenic and wound healing effects similar to those of adipose-derived stem cell exosomes. *Stem Cell Res Ther.* 2024 Dec 3;15(1):462.

Xing Y, Li B, He J, Hua H. Labial gland mesenchymal stem cell derived exosomes-mediated miR-125b attenuates experimental sjogren's syndrome by targeting PRDM1 and suppressing plasma cells. *Front Immunol* [Internet]. 2022;13. https://www.frontiersin.org/journals/immunology/articles/10.3389/fimmu.2022.871096/full

Yamashiro K, Hamada T, Mori K, Nishi K, Nakamura M, Beppu M, et al. Exosome-derived miRs from mouthrinse have the potential to be novel biomarkers for Sjögren syndrome. *J Pers Med.* 2022 Sep 10;12(9):1483.

Yan H fa, Zou T, Tuo Q zhang, Xu S, Li H, Belaidi AA, et al. Ferroptosis: mechanisms and links with diseases. *Signal Transduct Target Ther.* 2021 Feb 3;6(1):1–16.

Yang L, Jia J, Li S. Advances in the application of exosomes identification using surface-enhanced raman spectroscopy for the early detection of cancers. *Front Bioeng Biotechnol.* 2022 Jan 11;9:808933.

Yano M, Yamamoto T, Kobayashi S, Matsuzaki M. Role of ryanodine receptor as a Ca2+ regulatory center in normal and failing hearts. *J Cardiol.* 2009 Feb 1;53(1):1–7.

Yurtsever A, Yoshida T, Behjat AB, Araki Y, Hanayama R, Fukuma T. Structural and mechanical characteristics of exosomes from osteosarcoma cells explored by 3D-atomic force microscopy. *Nanoscale.* 2021 Apr 9;13(13):6661–77.

Zeng L, Chen X, Du J, Yu Z, Zhang R, Zhang Y, et al. Label-free separation of nanoscale particles by an ultrahigh gradient magnetic field in a microfluidic device. *Nanoscale.* 2021 Feb 25;13(7):4029–37.

Zhang NZ, Shi CS, Yao QP, Pan GX, Wang LL, Wen ZX, et al. Prevalence of primary Sjögren's syndrome in China. *J Rheumatol.* 1995 Apr;22(4):659–61.

Zhang R, Bu T, Cao R, Li Z, Wang C, Huang B, et al. An optimized exosome production strategy for enhanced yield without sacrificing cargo loading efficiency. *J Nanobiotechnology.* 2022 Oct 29;20(1):463.

Zhou H, Zhu Q, Mao Z, Li M, Zhang Y, Yang J, et al. Extracellular vesicle–encapsulated miR-10a-5p derived from MDSCs restrains germinal center B-cells in experimental Sjögren's syndrome. *Immunol Res.* 2023 Oct 1;71(5):760–70.

Zhou S, Huang J, Zhang Y, Yu H, Wang X. Exosomes in action: unraveling their role in autoimmune diseases and exploring potential therapeutic applications. *Immune Netw* [Internet]. 2024 Feb 20 [cited 2025 Feb 6];24(2). https://doi.org/10.4110/in.2024.24.e12

12 MSC-Derived Exosomes for the Therapy of Diabetes Mellitus and Associated Complications

M.A. Konoplyannikov[1,2], A.V. Proshkin[3], and P.S. Timashev[2]

[1]Federal Research Clinical Center of FMBA of Russia, Moscow, Russia

[2]Sechenov First Moscow State Medical University, Moscow, Russia

[3]Dokar Vita Cells, Bangkok, Thailand

12.1 LIST OF ABBREVIATIONS

AMPK	Adenosine monophosphate-activated protein kinase
ALT	Alanine aminotransferase
AP	Alkaline phosphatase
AT-MSC	Adipose tissue-derived mesenchymal stem cells
BAD	BCL2-associated agonist of cell death
BAX	BCL2-associated X protein
BDNF	Brain-derived neurotrophic factor
BG	Bioactive glass
BM	Bone marrow
BM-MSC	Bone marrow-derived mesenchymal stem cells
CCSMC	Corpora cavernosa's smooth muscle cells
cGMP	Cyclic guanosine monophosphate
CHOP	C/EBP homologous protein
CKD	Chronic kidney disease
CMCS	Carboxymethylchitosan
DFU	Diabetic foot ulcer
DNA	Deoxyribonucleic acid
DPN	Diabetic peripheral neuropathy
DPP-4	Dipeptidyl peptidase-4
ED	Erectile dysfunction
eNOS	Endothelial nitric oxide synthase
ERK	Extracellular signal-regulated kinase
ESCs	Embryonic stem cells
ET-1	Endothelin 1
FGF	Fibroblast growth factor
GLP-1	Glucagon-like peptide-1

DOI: 10.1201/9781003653387-12

HGF	Hepatocyte growth factor
HOMA-IR	Homeostasis model assessment of insulin resistance
IENF	Intraepidermal nerve fibers
IGF1	Insulin-like growth factor 1
iNOS	Inducible nitric oxide synthase
iPSC	Induced pluripotent stem cells
IV	Intravenous
lncRNA	Long noncoding RNA
MAPK	Mitogen-activated protein kinase
MenMSC	Menstrual blood-derived mesenchymal stem cells
miRNA	Micro-ribonucleic acid
MMP	Matrix metalloproteinase
MODY	Maturity-onset diabetes of the young
MSC	Mesenchymal stem cells
mTOR	Mammalian target of rapamycin
NCT	National clinical trial
NF-κB	Nuclear factor kappa-light-chain-enhancer of activated B cells
NGF	Nerve growth factor
nNOS	Neuronal nitric oxide synthase
OGD	Oxygen-glucose deprivation
PDCD4	Programmed cell death protein 4
PI3K	Phosphoinositide 3-kinase
PTEN	Phosphatase and tensin homolog
PTP1B	Protein-tyrosine phosphatase 1B
RNA	Ribonucleic acid
SDF1	Stromal cell-derived factor 1
SGLT2	Sodium-glucose cotransporter 2
SMA	Smooth muscle actin
STAT	Signal transducer and activator of transcription
T1DM	Type 1 diabetes mellitus
T2DM	Type 2 diabetes mellitus
TLR4	Toll-like receptor 4
TNFα	Tumor necrosis factor-alpha
UC-MSC	Umbilical cord-derived mesenchymal stem cells
VEGF	Vascular endothelial growth factor
WJ-MSC	Wharton's jelly-derived mesenchymal stem cells

12.2 INTRODUCTION

Diabetes mellitus is a chronic disease that occurs when the body either cannot produce enough insulin or cannot effectively use the insulin it makes. The lack of insulin or the cells' inability to respond to it leads to high blood glucose levels (hyperglycemia), a clinical sign of diabetes mellitus. The high blood glucose level, if not treated for a prolonged time, may result in damage to many systems in the body, which, in turn, leads to disabling and life-threatening complications. An estimated number of 589 million adults aged 20–79 live with diabetes worldwide, which constitutes 11.1% of the world's total population in this age group. The whole number of adults with diabetes is predicted to reach 853 million (13%) by 2050. More than $1 trillion was spent on diabetes treatment in 2024. This sum comprises 12% of the world's healthcare expenses. More than 3.4 million persons died as a result of diabetes in 2024, which corresponds to 9.3% of the all-cause mortality rate (https://diabetesatlas.org/resources/idf-diabetes-atlas-2025). Diabetes is mainly divided into four categories: Type 1 diabetes mellitus (T1DM), Type 2 diabetes mellitus (T2DM), gestational diabetes, and MODY (Maturity-onset diabetes of the young).

T1DM is the primary type of diabetes in children and young adults, but it can occur at any age. People with T1DM need insulin injections to survive. T2DM is associated with the majority of diabetes cases in the world (more than 90%). The inability of body cells to adequately respond to insulin in T2DM is known as insulin resistance. Insulin resistance leads to increased insulin production, which may become inadequate in the future, as pancreatic cells produce insulin with a delayed response to demand.

Patients with T2DM are susceptible to the increased risk of cardiovascular diseases, including a 72% higher risk of a heart attack, a 52% higher risk of stroke, and an 84% higher risk of heart failure. Diabetic patients have a 56% higher risk of dementia. The earlier the onset of T2DM, the higher the probability of dementia development at an older age. Almost every fourth diabetic patient has retinopathy, and every tenth of those with diabetic retinopathy risks losing their eyesight. Diabetes may also lead to nerve damage (neuropathy), kidney damage (nephropathy), lower limb amputation, liver disease, cancer, and asthenia.

Gestational diabetes (diabetes in pregnancy) is characterized by elevated blood glucose levels that occur during pregnancy. The excess glucose in the blood in gestational diabetes can have a toxic effect on the walls of blood vessels, nerve fibers, and internal organs. Insulin resistance grows during pregnancy. Such diabetes may resolve on its own after delivery. However, it happens rather often that several years after the pregnancy, chronic type 1 or 2 diabetes develops, as well. Besides, the child also has an increased risk of diabetes in the future. MODY (Maturity-Onset Diabetes of the Young) refers to a group of diseases characterized by mutations in one of the genes responsible for glucose regulation. This term was first introduced in 1974 by American scientists as a definition of an atypical, slowly progressing form of diabetes in young patients with a genetic predisposition.

The treatment of T1DM includes insulin injections or an insulin pump, as well as frequent blood glucose testing and carbohydrate calculations (https://www.mayoclinic.org/diseases-conditions/diabetes/diagnosis-treatment/drc-20371451). Some patients with T1DM may be treated with pancreatic or islet cell transplantation. The treatment of T2DM primarily involves lifestyle changes, blood sugar control, and the intake of oral antidiabetic drugs or insulin injections. Some antidiabetic drugs help the pancreas to secrete more insulin. Other preparations prevent glucose production and release from the liver, so less insulin is needed to transport sugar into cells. Additionally, some modern drugs can block the action of gastric or intestinal enzymes that break down carbohydrates, slowing their absorption, or make the patient's tissues more sensitive to insulin. Metformin is usually the first-line drug prescribed for T2DM. Besides other pharmacological options, diabetes is also treated with modern medications, including glucagon-like peptide-1 (GLP-1) agonists, dipeptidyl peptidase-4 (DPP-4) inhibitors, and sodium-glucose cotransporter 2 (SGLT2) inhibitors (Yang et al., 2025). Novel surgical procedures are increasingly available and are encouraged as efficient methods of diabetes treatment. In recent years, the following surgical procedures have been incorporated into clinical protocols: sleeve gastrectomy, gastric bypass, simultaneous pancreas-kidney transplantation, and pancreas and islet transplantation.

Unfortunately, extraneous insulin-based therapy is sometimes insufficient and cannot serve as an adequate replacement for functioning β-cells (Godfrey et al., 2012). In such cases, a whole pancreas transplant may be performed (Kochar et al., 2021). Unfortunately, this surgical procedure is associated with a lack of donors, some risk of mortality (1–3%), as well as the necessity of lifelong administration of immunosuppressants, increasing the risk of infections. Other complications include thrombosis, hemorrhage, and pancreatitis of the allotransplant (Martin et al., 2023).

Transplantation of isolated islets harvested from a cadaver pancreas is a less traumatic method than whole pancreas transplantation. In 2000, Shapiro et al. published the famous Edmonton Protocol, which involves islet administration to the portal vein (Shapiro et al., 2000). Later, an international, multicenter trial (ClinicalTrials.gov number, NCT00014911) demonstrated that 44% of patients (16 out of 36) did not require insulin injections within 1 year after islet transplantation

(Shapiro et al., 2006). Unfortunately, the condition of the majority of patients worsened in 2 years, and insulin injections were needed again. Only five subjects remained insulin-independent for 2 years.

In general, islet transplantation appears less efficient than whole pancreas transplantation. This procedure is also associated with limited donor availability and a need for immunosuppressants (Shapiro et al., 2014).

12.3 STEM CELLS FOR DIABETES THERAPY

Since there is always a shortage of donor pancreases and islets, the transplantation of stem cells to replace β-cells in the pancreas is a promising approach for treating diabetes (Sordi et al., 2023). Several different stem cell sources exist, including embryonic stem cells (ESCs), mesenchymal stem cells (MSCs), and induced pluripotent stem cells (iPSCs), among others, each with its advantages and disadvantages (Rufaihah et al., 2010; Omole et al., 2022; Rajab et al., 2019; Al-Khani et al., 2022).

In 2022, researchers at Peking University made a significant scientific breakthrough in the treatment of Type 1 Diabetes Mellitus (T1DM). For the first time in the history of medicine, a 25-year-old female patient successfully restored her ability to secrete insulin through the transplantation of insulin-secreting cells obtained from her own body (Mallapaty et al., 2024). Two and a half months after the surgery, her cells started to produce an amount of insulin sufficient for the patient to stop the exogenous hormonal therapy. The result was sustained for more than a year, with the blood glucose level remaining mainly within the normal range, without the dramatic oscillations characteristic of patients with diabetes.

The researchers' methods rely on reprogramming the patient's cells into induced pluripotent stem cells (iPSCs) (Ibrahim et al., 2016). iPS cells were subsequently differentiated into insulin-secreting cells, which were then implanted into the patient's abdominal wall muscles (Saima and Haider, 2021). In contrast to conventional liver islet transplantation techniques, the new procedure enables more efficient monitoring of cells and their removal when necessary. It is noteworthy, though, that the patient is still on immunosuppression since earlier she underwent liver transplantation, requiring the administration of immunosuppressants. This fact does not allow researchers to thoroughly assess the potential of transplanted iPS cells to reduce the risk of "graft rejection" (Albalawi et al., 2025). Additionally, immunosuppression is a necessary step to prevent the autoimmune destruction of islet cells in patients with T1DM. Such advances in diabetes treatment are not limited to T1DM. In April 2023, another research group from Shanghai successfully transplanted insulin-secreting cells into the liver of a 59-year-old male patient with T2DM (Wu et al., 2024). The insulin-secreting cells were obtained from reprogrammed stem cells from his organism. Following transplantation, the man ceased insulin injections, demonstrating the potential effectiveness of cell therapy in treating T2DM.

In June 2025, Vertex Pharmaceuticals (Boston, MA, USA) presented the updated results of their studies, in which donor ESC-derived islets were transplanted into the liver of 12 participants with T1DM (https://news.vrtx.com/news-releases/news-release-details/vertex-presents-positive-data-zimislecel-type-1-diabetes). One year post-transplantation, all the patients began secreting insulin in a blood glucose-responsive manner, and 10 patients out of the 12 (83%) no longer required exogenous insulin at month 12 of follow-up. To prevent the rejection of new cells, the participants received immunosuppressants.

On April 14, 2025, Daisuke Yabe from Kyoto University Hospital, Japan, reported the successful transplantation of iPS-derived pancreatic cell sheets into a diabetic patient, marking the first clinical application of such cell sheets in Japan (https://www.asahi.com/ajw/articles/15711029). These cells were protected from the immune response, minimizing the risk of rejection and enabling long-term success with this therapy. The patient was diagnosed with T1DM in 2001, underwent surgery in February 2025, and was dismissed with a favorable outcome in April 2025. Currently, Kyoto

University Hospital is conducting a clinical trial involving three patients with T1DM, aged 20–64, including the patient mentioned above.

The reported studies demonstrate that cell therapy for diabetes is a promising approach, capable of transforming the way the disease is treated, affecting millions of people worldwide. The application of one's own or donor stem cells, combined with new technologies that protect against the immune response, paves the way for a cure for diabetes and a significant improvement in patients' quality of life.

12.4 MSC-EXOSOMES IN DIABETES AND RELATED COMPLICATIONS

MSCs exhibit several characteristic properties (self-renewal, trilineage differentiation, low immunogenicity, etc.) (Álvarez-Viejo and Haider, 2022). Over the last few decades, they have demonstrated their efficacy in diabetes therapy in both preclinical and clinical studies (Mishra et al., 2020). MSCs may be obtained from various sources, including adipose tissue, bone marrow, umbilical cord, placenta, dental pulp, menstrual blood, and other sources (Kamal et al., 2022). MSCs can differentiate into insulin-secreting cells at a low rate, and their therapeutic effect is based on paracrine action (Shigemoto-Kuroda et al., 2017). Besides secreting a secretome containing soluble factors (Habib et al., 2025a), MSCs also secrete exosomes that carry numerous biologically active factors (Mushtaq et al., 2024).

Exosomes are extracellular vesicles characterized by bilayer lipid membranes and a diameter ranging from 30 to 150 nm. Exosomes are absorbed by recipient cells and participate in intercellular communication through the following mechanisms: ligand-receptor interaction, direct membrane fusion, and endocytosis/phagocytosis (He et al., 2018; Liang et al., 2025). The nanodimensions of exosomes facilitate their easy absorption, their ability to cross the blood-brain barrier, and their avoidance of endosomal and lysosomal degradation (Ahmed et al., 2023). The bilayer lipid membrane protects the cargo in exosomes from degradation in the medium, providing prolonged release. Exosomes contain DNA, RNA, lipids, and proteins from the parent cells and perform critical biological functions (Haider and Najimi, 2022). Under normal physiological conditions, exosomes play roles in cell proliferation, immune regulation, neuronal communication, and homeostasis (Li et al., 2022).

MSC-derived exosomes, like MSCs themselves, possess potent physiological properties that influence the repair of damaged tissues (Zhao et al., 2019; Habib et al., 2025b). At the same time, exosomes are more stable than MSCs and safer, as they do not trigger the host's immune response or provoke other reactions to the administration of cell preparations (Habib et al., 2026). Their payload composition may be engineered by priming the cells using various approaches (Konoplyannikov et al., 2013; Haider, 2024). MSC-derived exosomes have demonstrated pronounced efficacy in preclinical studies for the treatment of diabetes and related complications, comparable to or even higher than that of MSCs themselves (Haider and Najimi, 2022; Liang et al., 2025) (Figure 12.1).

Meligy et al. compared the effects of bone marrow-derived MSCs and MSC exosomes using a model of streptozotocin-induced diabetes in rats (Meligy et al., 2025). Three days after the streptozotocin administration to rats, MSC (3×10^6) or MSC-exosomes (80 g per rat) were applied in a single injection. Both MSCs and MSC-exosomes have been shown to promote the histological restoration of pancreatic tissue. In both cases, the amount of secreted insulin was increased, and the levels of autophagy markers, P62 and LC3, were significantly reduced. The results indicate that MSC-exosomes are an excellent alternative to MSCs for diabetes therapy.

12.4.1 MSC-Exosomes for the Therapy of T1DM

In T1DM, there is a deficiency of insulin and elevated blood glucose levels, induced by autoimmune-reactive cells (Pang et al., 2020). Suppressing inflammation in islets and maintaining a balance between autoreactive effector T cells and regulatory T cells reduces β-cell damage and death.

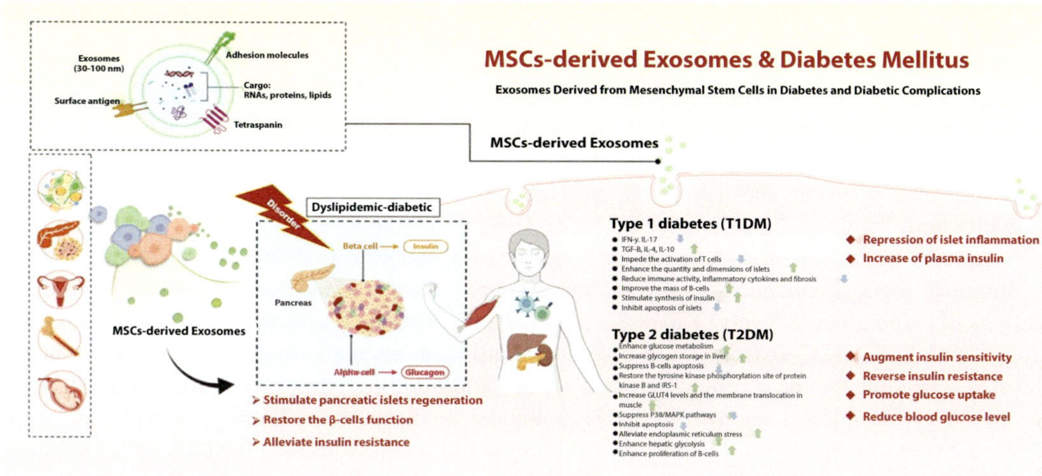

FIGURE 12.1 MSC-derived exosomes for diabetes mellitus treatment. Reproduced from (Jiao et al., 2024) under a Creative Commons Attribution (CC BY) license.

MSC-exosomes have been shown to have a potent immunomodulatory effect. For example, MSC-exosomes from adipose tissue-derived MSC (AT-MSC) have an immunomodulatory effect on T cells. MSC-exosomes alleviate hyperglycemia symptoms in mice with T1DM. At the same time, the number of regulatory T cells increases without affecting lymphocyte proliferation (Nojehdehi et al., 2018). MSC-derived exosomes from the menstrual blood (MenMSC) promote the regeneration of islets, improve the mass of β-cells, and enhance insulin secretion in rats with T1DM (Mahdipour et al., 2019). Similarly, four weeks after treating T1DM rats with exosomes from bone marrow-derived MSC (BM-MSC), a significant decrease in blood glucose levels was observed, along with an increase in blood insulin levels. Besides, the cells of Langerhans islets regenerated, and the number and size of the islets grew, while fibrosis and inflammation were reduced. It is noteworthy that these indicators improved significantly after MSC-exosome therapy, to a greater extent than after MSC therapy alone (Sabry et al., 2020).

After islet transplantation, it is essential to ensure their survival and prevent islet cell apoptosis. The key survival factor, expressed in intact islets but significantly attenuated in isolated islets, is vascular endothelial growth factor (VEGF), which is needed for the long-term viability of the transplant and provides revascularization of the islets (Cross et al., 2007). Exosomes from Wharton jelly-derived MSC (WJ-MSC) boost the VEGF expression in co-cultured islets. Exosomes from WJ-MSC also suppress the expression of key apoptotic genes, such as BAD and BAX, while conversely enhancing the expression of BCL-2, an antiapoptotic gene, and PI3K, a survival-promoting gene. PI3K activates the Akt signaling pathway and inhibits BAD and BAX (Downward et al., 2004), thereby increasing islet cell viability. BM-MSC-derived exosomes may deliver siFAS and anti-miR-375, attenuating the expression of genes stimulated by pro-inflammatory cytokines and associated with apoptosis and hypoxia. The expression of these genes leads to the death of implanted islet cells (Fas, miR-375, caspase-3, iNOS) (Wen et al., 2016). In the future, MSC-exosome therapy is expected to help T1DM patients reduce or discontinue insulin use.

12.4.2 MSC-Exosomes for the Therapy of T2DM

T2DM (90% of diabetic patients) is characterized by insulin resistance in hepatic and peripheral tissues, dysfunction, and loss of β-cell mass (Nyenwe et al., 2011). Using a rat model of T2DM induced by a high-fat and streptozotocin diet, Sun et al. demonstrated that intravenous injection of exosomes isolated from human umbilical cord-derived MSC reduced blood glucose levels and partially relieved insulin resistance (Sun et al., 2018). hUC-MSC-exosomes were shown to restore phosphorylation (the tyrosine fragment) of insulin receptor substrate 1 and protein kinase B, to promote the expression and membrane translocation of glucose transporter type 4, and to expand the resource of glycogen in the liver for sustaining glucose homeostasis. It was also demonstrated that hUC-MSC-exosomes inhibited streptozotocin-induced apoptosis of β-cells, thus restoring insulin secretion. In conclusion, the authors suggest that hUC-MSC-exosomes may alleviate T2DM by reversing peripheral insulin resistance and mitigating β-cell damage.

In another study, Yap et al. demonstrated that treating T2DM rats with a low dose of hUC-MSC-exosomes (1 mg/kg) enhanced insulin tolerance by reducing HOMA-IR and suppressing the fluctuations in blood glucose levels (Yap et al., 2022). The biochemical tests performed in the treatment groups with hUC-MSC-exosomes did not show significant changes in creatinine levels; however, elevated levels of alanine aminotransferase (ALT) and alkaline phosphatase (AP) were observed compared to the normal control group. Histological analysis showed that hUC-MSC-exosomes reduced structural damage in the pancreas, kidneys, and liver. Moreover, the blood profile showed that treatment with hUC-MSC-exosomes may be beneficial for diabetic rats by alleviating anemia-like symptoms and enhancing immunity, as evidenced by increased erythrocyte and hemoglobin levels, as well as supported leukocyte levels. The results demonstrate that hUC-MSC-exosomes may improve insulin resistance and exert a protective effect in T2DM rats.

In diabetes, increased oxygen uptake due to elevated glucose levels causes β-cells to become hypoxic. In these conditions, the selective downregulation of Mafa, Pdx1, Slc2a2, Ndufa5, Kcnj11, Ins1, Wfs1, Foxa2, and Neurod1 genes, which play crucial roles in β-cells, was observed in both cultured cells and isolated pancreatic islands. As a result, hypoxia leads to β-cell dysfunction and apoptosis in T2DM mice (Sato et al., 2014). Chen et al. demonstrated that hUC-MSC-exosomes could protect β-cells from hypoxia-induced apoptosis, primarily due to the transfer of miR-21, which suppresses endoplasmic reticulum stress (via the inhibition of GRP78, GRP94, p-eIF2α, and CHOP proteins) and inhibits p38 MAPK signaling. Thus, the authors have demonstrated that MSC-exosomes may enhance the viability of encapsulated pancreatic islets (Chen et al., 2020).

It is well established that the incidence of T2DM is significantly higher in older individuals (Bellary et al., 2021), who are characterized by insulin resistance. Su et al. demonstrated that nano-sized exosomes released by BM-MSCs of aged mice can be captured by adipocytes, myocytes, and hepatocytes, leading to insulin resistance both in vivo and in vitro (Su et al., 2019). The authors also found that miR-29b-3p levels were drastically increased in exosomes released by BM-MSCs from aged mice; this overexpression of miR-29b-3p caused insulin resistance in young mice. The results indicate that exosomal miR-29b-3p derived from BM-MSCs may modulate age-related insulin resistance, making it a potential therapeutic target (Su et al., 2019).

In summary, preclinical studies have demonstrated the efficacy of MSC-exosomes in treating T1DM and T2DM, offering hope for their potential clinical application in the near future (Kamaraj et al., 2025).

12.4.3 MSC-Exosomes for the Therapy of Diabetes Complications

The long-standing insufficient control of hyperglycemia in diabetic patients leads to a plethora of complications, related mainly to the damage to small and/or large blood vessels (micro- and macroangiopathy). The most frequent diabetes complications include: diabetic foot, diabetic nephropathy, diabetic retinopathy, erectile dysfunction, diabetic neuropathy, diabetic encephalopathy, diabetic osteoporosis, and diabetic cardiomyopathy.

MSC-exosomes have been widely used in recent decades for the treatment of various diabetes complications (Figure 12.2).

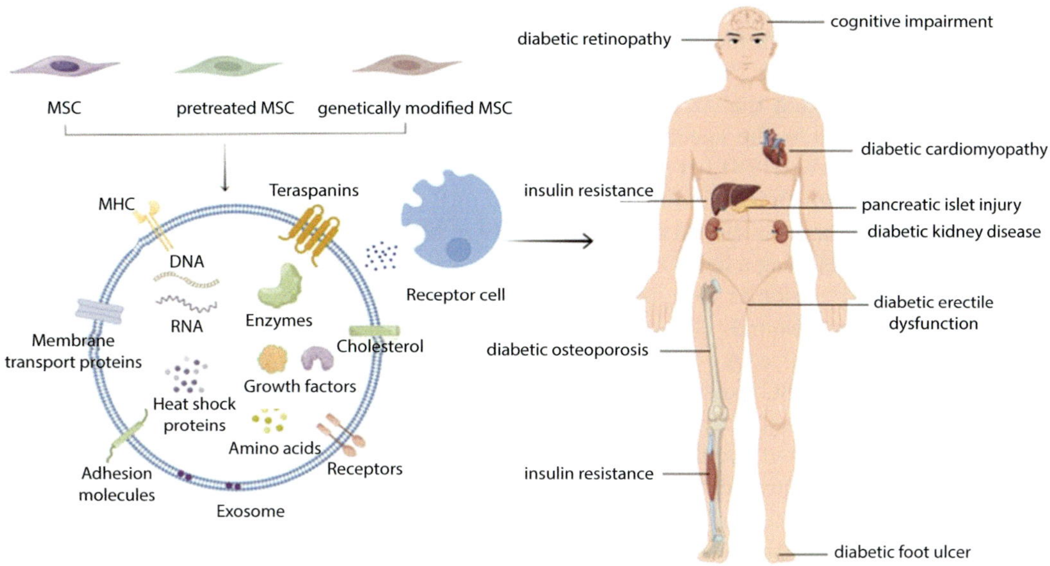

FIGURE 12.2 Application of exosomes derived from mesenchymal stem cells in diabetes and related complications. Reproduced from (Yang et al., 2022) under a Creative Commons Attribution (CC BY) license.

12.4.4 MSC-Exosomes for the Diabetic Foot Therapy

Diabetic foot is one of the most serious complications of diabetes. It manifests as pathological foot lesions, ulcers, bone and joint damage, and loss of skin sensitivity. The disease has a high incidence among diabetic patients, and if neglected, without timely consultation with a physician, it may lead to amputation. The cause of diabetic foot lies in the general impairment of blood circulation against the diabetic background. When an excessive concentration of glucose is present in the blood for a long time, carbohydrate and lipid metabolism is inevitably affected. The alterations in exchange processes can either cause or act as a catalyst for atherosclerosis, i.e., the narrowing of arteries and a worsening of the body's blood supply. Due to physiological factors, the lower limb tissues, especially nerve endings, are the first to suffer: cells do not receive the nutrients they need, and they gradually die or undergo irreversible changes. After that, minor foot wounds heal more slowly, and the pain syndrome is weakened due to the improper functioning of nerve endings. As a result, wounds may expand, become infected with bacteria and viruses, leading to purulent wounds that rapidly transform into trophic ulcers without treatment. Thus, the initial cause of diabetic foot development is the impairment of blood circulation due to hyperglycemia. The following worsening of the condition occurs under the negative influence of injuries, fungal lesions, narrow shoes, and hypertension.

Approximately 6.3% of diabetic patients worldwide develop diabetic foot, and a large number of patients finally undergo a limb amputation within a year after being diagnosed with diabetic foot (minor amputation in 17% and major amputation in 5% of cases) (Zhang et al., 2017). Diabetic neuropathy and peripheral vascular disease represent the leading causes of diabetic foot and may act together or separately. The treatment of diabetic foot includes drug therapy and surgical intervention, including percutaneous transluminal angioplasty of the damaged lower limb arteries. Angiogenesis plays a vital role in the healing of diabetic wounds (Sinwar et al., 2015).

Numerous studies have demonstrated the efficacy of MSC-exosomes in treating ulcers (wounds) in diabetic foot. Shabbir et al. found that MSC-exosomes induce proliferation and migration of normal and chronic fibroblasts in wounds, and also enhance angiogenesis in vitro (Shabbir et al., 2015). MSC-exosomes were shown to activate such key signal pathways for wound healing as Akt, ERK, and STAT3, as well as to induce the expression of several growth factors [hepatocyte growth factor (HGF), insulin-like growth factor 1 (IGF1), nerve growth factor (NGF), and stromal cell-derived factor 1 (SDF1)].

Li et al. demonstrated that wound healing in mice with streptozotocin-induced diabetic foot (DFU) was facilitated by the injection of exosomes derived from MSCs overexpressing long non-coding RNA (lncRNA) H19 (Li et al., 2020). Exosomal lncRNA H19 from those MSCs prevented apoptosis and inflammation by inhibiting PTEN via miR-152-3p, thereby stimulating wound healing in DFU. The authors concluded that lncRNA H19 in MSC-exosomes may be a promising therapeutic target for developing a new DFU treatment method.

Zhao et al. used exosomes from human adipose tissue-derived MSC (AT-MSC-exosomes) for the therapy of dermal wounds (Zhao et al., 2021). They created a model of hard-to-heal wounds using female diabetic mice. A ventral skin wound with a size of 1.5 cm^2 was inflicted in 6-week-old db/db mice. AT-MSC-exosomes accelerated wound closure and epithelialization. Besides, AT-MSC-exosomes enhanced collagen production in the skin, promoted angiogenesis, stimulated cell proliferation, inhibited apoptosis, restored skin barrier function, and diminished inflammation in skin lesions. The downregulation of MMP1 and MMP3 expression upon the action of AT-MSC-exosomes boosted collagen synthesis, thus stimulating wound healing.

Gondaliya et al. suggested that a combination of an miR-155 inhibitor and MSC-exosomes would efficiently promote healing of diabetic wounds (Gondaliya et al., 2022). Indeed, MSC-exosomes loaded with an miR-155 inhibitor demonstrated synergistic effects on keratinocyte migration, restoration of FGF-7 levels, and anti-inflammatory effects, thereby accelerating wound healing through miR-155 downregulation. Moreover, treatment with MSC-exosomes loaded with an miR-155 inhibitor enhanced collagen deposition, angiogenesis, and re-epithelialization in diabetic wounds. The authors assume that the application of miRNA encapsulation, combined with antibiotics, in MSC-exosomes will be an efficient future strategy for improving the treatment of chronic, hard-to-heal diabetic wounds.

Shang et al. applied a bioactive composite hydrogel dressing containing MSC-exosomes for the treatment of acute and chronic diabetic skin wounds (Shang et al., 2024). In this study, a CMCS-CEBT hydrogel was synthesized by encapsulating exosomes in a carboxymethylchitosan matrix, with chitosan nanoparticles, BG, and TiO_2 nanoparticles. The in vitro analysis revealed that the hydrogel demonstrated excellent cell compatibility, stimulating endothelial cell adhesion and proliferation, and exhibited anti-inflammatory, angiogenic, and antibacterial properties. In in vivo studies, the composite hydrogel dressing accelerated wound healing, stimulated angiogenesis, enhanced collagen deposition, and increased the expression of anti-inflammatory factors.

Weng et al. created a hydrogel loaded with MSC-exosomes, in combination with exosomes of plant origin from Momordica charantia (MC-exosomes), a plant known for its glucose-reducing effect (Weng et al., 2025). The hydrogel was obtained by photopolymerization of methacrylated gelatin (GelMA) and dopamine (MEMC-Gel). The prepared MEMC-Gel demonstrated favorable mechanical properties, adhesion, degradability, absorbability, and biocompatibility. In in vitro studies, MEMC-Gel has shown the ability to withstand inflammation and oxidative stress, facilitate fibroblast migration, support endothelial cell angiogenesis, and regulate macrophage polarization. In a murine model of diabetic wound healing, MEMC-Gel accelerated wound healing by suppressing inflammation and oxidative stress, modulating macrophage immune responses, and mitigating hyperglycemia in the microenvironment, thereby promoting angiogenesis and enhancing epithelialization. This multifunctional hydrogel, derived from exosomes from two distinct sources, represents an innovative therapeutic strategy for treating diabetic wounds and has significant clinical potential.

12.4.5 MSC-Exosomes for the Therapy of Diabetic Nephropathy

Diabetic nephropathy is a specific pathological alteration of the renal blood vessels that occurs in diabetes. The complication leads to glomerulosclerosis, suppresses the kidneys' filtration function, and promotes the development of chronic kidney disease (CKD). The clinical signs of diabetic nephropathy are proteinuria and microalbuminuria, nephrotic syndrome, arterial hypertension, uremia, and CKD (Samsu et al., 2021).

Diabetic nephropathy is also one of the most severe complications of T1DM and T2DM, and it manifests itself at the later stages of the disease. It presents one of the leading causes of mortality. Diabetic nephropathy is diagnosed in 10–20% of diabetic patients. This disease is observed most frequently in male patients or in persons with T1DM who developed during puberty. The peak of diabetic nephropathy development is found 15–20 years after the onset of diabetes. The prognosis for diabetic nephropathy depends on the stage and timely application of treatment. The only reversible stage of the disease is microalbuminuria. At the stage of proteinuria, one can only slow the progression of diabetic nephropathy into CKD. If a patient reaches the terminal stage of diabetic nephropathy, the prognosis is most unfavorable, since this stage is incompatible with life. Diabetic nephropathy and related CKD are indications for the replacement therapy—kidney transplantation or hemodialysis. In such a case, CKD is the cause of 15% of lethal cases for T1DM patients younger than 50.

An essential pathology found in the kidneys of diabetic patients is defective autophagy. Autophagy is a highly conserved lysosomal degradation pathway that removes protein aggregates and damaged organelles to maintain cellular homeostasis (Yang et al., 2018). Hyperglycemia associated with diabetes causes kidney damage, and restoring autophagy may protect against this damage.

Ebrahim et al. investigated the potential role of exosomes in enhancing autophagy activity for the treatment of diabetic nephropathy in a rat model of streptozotocin-induced diabetes (Ebrahim et al., 2018). The study results showed that MSC-exosomes significantly improved kidney function, with histological restoration of renal tissues, accompanied by a substantial increase in LC3 and Beclin-1, and a marked reduction in the expression of mTOR and the fibrotic marker in renal tissue. The autophagy inhibitors, chloroquine and 3-MA, partially canceled all the previous effects. The authors concluded that MSC-exosome-induced autophagy might diminish diabetic nephropathy.

Jin et al. demonstrated that AT-MSC-secreted exosomes ease diabetic nephropathy, kidney fibrosis, and T2DM in mice (Jin et al., 2019). The protein level in the urine, creatinine level in the serum, and the blood urea nitrogen level were reduced in diabetic mice upon the action of AT-MSC-exosomes. In the in vitro experiment, AT-MSC-exosomes also enhanced the viability of MPC5 cells, thereby suppressing the apoptosis induced by high glucose levels. The following mechanism was proposed: AT-MSC exosomes enhance autophagy and mitigate podocyte damage by inhibiting mTOR signaling in MPC5 cells derived from diabetic mice. It was found that miR-486 mediated the effect of AT-MSC-exosomes in easing diabetic nephropathy in vivo and in vitro. miR-486 reduces Smad1 expression, potentially inhibiting mTOR activation and thereby enhancing autophagy and suppressing apoptosis in podocytes.

Hao et al. demonstrated that exosomes obtained from adipose tissue-derived MSCs carried miRNA-125a, which protected against diabetic nephropathy by targeting histone deacetylase 1 and downregulating endothelin-1 (Hao et al., 2021). The results obtained using a rat model of diabetic nephropathy showed that AT-MSC-exosomes reduced blood glucose levels, serum creatinine, 24-hour urinary protein, UACR, and kidney weight-to-body weight ratio, and also suppressed mesangial hyperplasia and kidney fibrosis. miR-125a directly bound to histone deacetylase 1 (HDAC1), while HDAC1 additionally regulated the endothelin 1 (ET-1) activation. This study demonstrated that AT-MSC-exosomes inhibited the progression of diabetic nephropathy and alleviated symptoms by transferring miR-125a, which in turn inhibited HDAC1 and ET-1.

Zhang et al. demonstrated that both MSCs from the human umbilical cord (UC-MSCs) and exosomes derived from UC-MSCs (UC-MSC-exosomes) reduced kidney damage and inhibited epithelial-mesenchymal transition and renal fibrosis in rats with streptozotocin-induced diabetic nephropathy (Zhang et al., 2024). The action of UC-MSC and UC-MSC-exosomes leads to the suppressed expression of Smoothened (SMO), the key G protein-coupled receptor in the Hedgehog signaling pathway. The level of the latter is substantially elevated in renal tissue from patients with diabetic nephropathy and positively correlates with epithelial-mesenchymal transition and renal fibrosis. UC-MSC-exosomes inhibit epithelial-mesenchymal transition in cells of renal tubules and renal fibrosis by acting on the Hedgehog/SMO signaling pathway. Thus, the authors consider UC-MSC and UC-MSC-exosomes to be a new, efficient therapy for diabetic nephropathy.

12.4.6 MSC-Exosomes for the Therapy of Diabetic Retinopathy

Diabetic retinopathy is damage to the retina caused by diabetes. Prolonged hyperglycemia alters the walls of retinal blood vessels, making them more permeable and allowing fluid to enter the intraretinal space. The development of diabetic retinopathy most frequently occurs 5–10 years after the onset of diabetes.

The course of diabetic retinopathy varies depending on the type of diabetes. In T1DM, it is a rapidly progressing disease; in this case, proliferative diabetic retinopathy (the late stage of the disease) develops quickly. In T2DM, changes and developments occur in the central region of the retina. Diabetic retinopathy is a disease that can lead to blindness, primarily affecting individuals aged 20 to 65. The retina's regenerative capacity is minimal.

Mathew et al. tested the effect of MSC-exosomes using a model of oxygen-glucose deprivation (OGD) of retinal ischemia in vitro (Mathew et al., 2019)—the treatment of retinal R28 cells with MSC-exosomes substantially reduced cell death. The endocytosis of MSC-exosomes depended on dose and temperature, was saturable, and proceeded via heparin sulfate proteoglycans on the cell surface, mediated by the caveolar endocytic pathway. Injecting MSC-exosomes into the vitreous body 24 hours after retinal ischemia in a rat model enhanced functional recovery and reduced neuroinflammation and apoptosis. The retinal neurons and microglia absorbed MSC-exosomes. They were present in the vitreous body for 4 weeks after the intravitreal injection with saturable binding to the vitreous body's components. In general, the study highlighted the potential of MSC-exosomes in regenerative therapy for diabetic retinopathy.

Fu et al. estimated the protective effect of human umbilical cord-derived MSC-exosomes (hUC-MSC-exosomes) in a rat model of streptozotocin-induced diabetes with the use of different retinal biotests (Fu et al., 2021). Fundus fluorescein angiography and optical coherence tomography confirmed that hUC-MSC-exosomes efficiently prevented the early damage of the retinal blood vessels and the retina's thickening. Hematoxylin-eosin staining of the rat retina's cross-sections showed that MSC-exosomes effectively suppressed the structural damage of the retina caused by diabetes.

Gao et al. demonstrated the protective effect of hUC-MSC-exosomes on the rat retinal neurons in conditions of high glucose levels via the activation of the BDNF-TrkB pathway (Gao et al., 2021). The study results showed that hUC-MSC-exosomes enhanced the viability of retinal neurons, accompanied by increased TrkB levels in these neurons. At the same time, when hUC-MSC-exosomes are treated with BDNF inhibitors, the mentioned protective effect is reduced. As a result, hUC-MSC-exosomes may deliver BDNF to rat retinal neurons, inhibiting their apoptosis by activating the BDNF-TrkB pathway.

In a similar way to other diabetes complications, various microRNAs (miRNAs) play a key role in diabetic retinopathy.

Indeed, miR-17-3p, which targets signal transducer and activator of transcription 1 (STAT1) in hUC-MSC-exosomes, suppresses the inflammatory response and antioxidant damage in mice with diabetic retinopathy (Li et al., 2021a). The study results showed that mice with diabetic retinopathy had reduced miR-17-3p levels and increased STAT1 levels in retinal tissue. Injecting MSC-exosomes containing miR-17-3p reduces blood glucose and glycosylated hemoglobin levels, decreases the concentrations of inflammatory factors and VEGF, and mitigates oxidative damage and inhibits retinal cell apoptosis in mice with diabetic retinopathy by inhibiting STAT1.

Another study by the same authors demonstrated that miR-486-3p, present in exosomes from bone marrow-derived MSCs, suppressed oxidative stress, inflammation, and apoptosis, and promoted Müller cell proliferation in high-glucose medium by inhibiting the TLR4/NF-κB signaling pathway (Li et al., 2021b).

Liu et al. performed intravitreal injections of hUC-MSC-exosomes to rats with streptozotocin-induced diabetes and diabetic retinopathy (Liu et al., 2025). The intravitreal injections of hUC-MSC-exosomes improved histological morphology and blood-retina barrier function, suppressed Müller cell reactive gliosis, reduced PTP1B expression, oxidative-reductive stress, and apoptosis in the retina of diabetic rats. Conversely, the addition of an miR-125a-5p inhibitor canceled the protective effects of hUC-MSC-exosomes. Elevated PTP1B expression in Müller cells also attenuated the effects of hUC-MSC exosomes. The authors concluded that miR-125a-5p in MSC-exosomes facilitated Müller cell damage in diabetic retinopathy, modulating PINK1/Parkin-mediated mitophagy via the PTP1B pathway.

12.4.7 MSC-Exosomes for the Therapy of Erectile Dysfunction

Long-term exposure to high blood glucose levels can damage nerves, blood vessels, and muscles. This fact explains several diabetes complications, including erectile dysfunction (ED), or difficulty reaching and sustaining an erection. The diabetic damage to the penile nerves and vasculature causes a reduction in the blood flow to the penis. About 70% of diabetic patients experience symptoms of erectile dysfunction in the course of the disease, with the degree of the disorder depending on age and disease duration. Among male diabetic patients older than 60, 50–60% experience erectile dysfunction.

Hence, the ED treatment necessitates an approach that promotes blood vessel growth, protects nerves, and stimulates regeneration. Numerous drugs are used in the treatment of this condition, with Type 5 phosphodiesterase inhibitors usually prescribed. Nevertheless, the use of these inhibitors is limited by adverse effects, including smooth muscle relaxation and collagen accumulation. Besides, they cannot work for all patients and provide only temporary relief without addressing the fundamental cause.

In the last few decades, MSCs have been successfully applied in ED treatment (Patel et al., 2025). With their proangiogenic, antifibrotic, and antiapoptotic properties, MSCs may enhance the therapy of ED and other urological diseases that are poorly responsive to surgical or pharmacological treatment. Preclinical animal studies have confirmed this. The meta-analysis by Hou et al. summarized 20 studies involving 248 rats and found that adipose tissue-derived MSC therapy was effective in restoring cavernous tissue (Hou et al., 2017). This study also demonstrated that the use of adipose tissue-derived MSC modified to enhance growth factor expression (NGF, VEGF, HGF, and BDNF) resulted in significantly improved restoration of erectile function and corpora cavernosa compared to native MSC. Additional improvement was achieved when using hypoxic preconditioning of AT-MSC.

Since MSCs act basically through tissue-specific paracrine mechanisms (Xia et al., 2021; Haider and Aslam, 2018), it is reasonable to presume that the same action is expected from MSC-derived exosomes. Studies on animal models have shown that MSC-exosomes may enhance the expression of specific biomarkers (CD31, α-SMA, eNOS, and nNOS) in rats with ED. These markers indicate the presence of endothelium and smooth muscle in the corpora cavernosa. These findings testify that MSC-exosomes are capable of improving the corpora cavernosa's structure, enhancing erectile function. Besides, the expression of TGF-β1 and caspase-3 is reduced upon injecting MSC-exosomes, which points to their ability to prevent fibrosis and apoptosis and preserve the functioning corpora cavernosa's endothelium and content of smooth muscles in ED (Li et al., 2013). ED is characterized by the downregulation of the NO/cyclic guanosine monophosphate (cGMP) pathway, which is needed for the control of penile erection (Ángulo et al., 2010). NO, which is produced by eNOS in cells of the cavernous endothelium and by nNOS in cavernous neurons, induces erection by increasing the cGMP concentration in the smooth muscles of the corpora cavernosa (Yafi et al., 2016). According to the study by Song et al., application of exosomes from smooth muscle cells improves ED by regulating the NO/cGMP pathway (Song et al., 2020). It is assumed that MSC-exosomes, through regulation of this signaling pathway, confer functional improvements in the corpora cavernosa by increasing the expression of nNOS and eNOS.

Zhu et al. used exosomes derived from AT-MSCs (AT-MSC-exosomes) to treat ED in streptozotocin-induced diabetic rats (Zhu et al., 2018). AT-MSC-exosomes demonstrated proangiogenic properties in vitro, induced proliferation of endothelial cells, restored erectile function in vivo, and also decreased the corpora cavernosa's fibrosis. Besides, AT-MSC-exosomes were found to contain specific proangiogenic microRNA (miR-126, miR-130a, and miR-132) and antifibrotic microRNA (miR-let7b and miR-let7c). The researchers suggested that AT-MSC-exosomes transport key microRNA to target cells, thereby enhancing the functional restoration or activation of endogenous repair mechanisms.

Huo et al. studied the functional role of exosomal miR-21-5p in a rat model of diabetes-induced ED (Huo et al., 2020). It was demonstrated that miR-21-5p-containing MSC-exosomes suppressed PDCD4 (Programmed cell death protein 4) expression, improving erectile function and smooth muscle density in rats with ED. They reduced the apoptosis and enhanced the proliferation of cells of the corpora cavernosa's smooth muscles (CCSMC).

12.4.8 MSC-Exosomes for the Therapy of Diabetic Neuropathy

Diabetic neuropathy is damage to the peripheral nerves that occurs in the course of diabetes and is caused by dysmetabolic processes. Diabetic neuropathy represents one of the most frequent diabetes complications found in approximately 30–50% of patients. It develops within 5–15 years after the onset of diabetes. Diabetic peripheral neuropathy (DPN) is characterized by a complex of sensory and motor disorders, most pronounced in the extremities. Lower limb DPN manifests as numbness, burning, or stinging sensations, pain in the feet, fingers, and toes, and short muscle cramps. In general, DPN contributes significantly to mortality, since currently there exists no efficient treatment for this severe diabetes complication.

Exosomes derived from BM-MSCs can alleviate neurovascular dysfunction and promote functional recovery in mice with DPN by suppressing pro-inflammatory genes (Fan et al., 2020). Injecting MSC-exosomes intravenously significantly reduced the threshold of thermal and mechanical stimuli and increased the nerve conduction velocity in mice. It was also shown that MSC-exosomes substantially increased the density of blood vessels and the number of intraepidermal nerve fibers (IENF), myelin thickness, and the sizes of the sciatic nerve's axons. Furthermore, MSC-exosomes significantly suppressed the production of pro-inflammatory cytokines. A thorough analysis had also shown that MSC-exosomes contained numerous miRNAs, including miR-17, miR-23a, and miR-125b, that target the Toll-like receptor (TLR)4/NF-κB signaling pathway.

The same authors later created MSC-exosomes loaded with miR-146a (exo-146a) and compared their therapeutic effects with those of native MSC-exosomes in mice with DPN (Fan et al., 2021). Exo-146a exhibited a highly pronounced ability to accumulate in peripheral nerve tissues upon systemic administration. The exo-146a therapy of mice with DPN for two weeks significantly increased the nerve conduction velocity and, conversely, reduced the threshold of thermal and mechanical irritation, while the same effect was observed only 4 weeks after the application of native MSC-exosomes. Besides, exo-146a more efficiently suppressed the infiltration of macrophages and inflammation compared to native MSC-exosomes by inhibiting the synthesis of inflammatory TNFα and IL-1β cytokines via the TLR4/NF-κB pathway. As a result, exo-146a provided better neurovascular remodeling and functional restoration in DPN.

12.4.9 MSC-Exosomes for the Therapy of Diabetic Encephalopathy

Diabetic encephalopathy has a multifactorial mechanism of development, which includes vascular and metabolic components. Vascular disorders caused by macro- and microangiopathy impair cerebral hemodynamics and lead to hypoxia in brain cells. The pathobiochemical reactions that develop under hyperglycemic conditions induce the activation of anaerobic rather than aerobic glycolysis, resulting in the energetic deprivation of neurons. The emerging free radicals have a damaging effect on the cerebral tissue. The formation of glycosylated hemoglobin, which weakly binds oxygen, aggravates the hypoxia of neurons caused by vascular disorders. Hypoxia and dysmetabolism lead to neuronal death, resulting in diffuse or minor focal organic alterations in the cerebral substance, i.e., encephalopathy. The destruction of interneuronal connections results in the gradual progression of impairment of cognitive functions, which, in turn, leads to a decrease in the level of intellect and dementia. The latter is a cause of significant disability in diabetic patients, limiting their self-care abilities. Unfortunately, no efficient treatment of this severe diabetes complication has been developed to date.

Nakano et al. demonstrated the efficacy of BM-MSC-exosomes in treating diabetes-related cognitive impairment (Nakano et al., 2016). MSC-exosomes were administered by an intracerebroventricular injection to mice with streptozotocin-induced diabetes, which exhibited cognitive disorders 12 weeks after the streptozotocin administration. Following treatment, the restoration of mental functions and histological anomalies was demonstrated, comparable to that observed after BM-MSC injection. MSC-exosomes labeled with a fluorescent dye were found to be internalized by astrocytes and neurons, thereby facilitating their repair. There is a reason to believe that MSC-exosomes mediate the transfer of microRNA, in particular, miR-133b, into astrocytes and neurons. miR-133b suppresses the expression of RhoA (Ras homolog family member A) and is capable of activating the PI3K/Akt and ERK1/2 signaling pathways, thereby promoting neurite growth and restoring function (Lu et al., 2015; Xin et al., 2013).

When studying the therapeutic mechanism of MSC-exosomes' effect in the treatment of diabetic cognitive disorders, the same authors revealed an essential role for miR-146a (Kubota et al., 2018). It was found that the level of miR-146a in the blood of rats with streptozotocin-induced diabetes was reduced in comparison with that of non-diabetic rats. Administration of MSC-exosomes to diabetic rats stimulates the elevated secretion of exosomal miR-146a by endogenous MSCs in the bone marrow, which has an anti-inflammatory effect on damaged astrocytes and prevents diabetes-related cognitive impairment.

12.4.10 MSC-Exosomes in the Therapy of Diabetic Osteoporosis

Osteoporosis in diabetic patients is a serious, although somewhat underestimated, complication. The overall incidence of osteoporosis among all diabetes complications is 6–10% (Schwait et al., 2003). The metabolic anomalies, vascular and neurological diabetes complications, lead to pathological changes in the bone tissue. Such changes are found, on average, in half of diabetic patients, which determines their vulnerability to osteoporotic fractures. The insulin deficit, characterized by poor metabolic control, creates the conditions that lead to osteopenia and osteoporosis. In the condition of insulin deficiency, osteoblast activity decreases, while osteoclast activity increases, leading to enhanced bone resorption (Janghorbani et al., 2007). Due to elevated glucose levels and high concentrations of final glycosylation products in bone collagen, osteoclastic bone resorption may be stimulated (Pittas et al., 2007). Among the various mechanisms contributing to the accelerated development of osteoporosis in diabetes, diabetic microangiopathies, which worsen the blood supply to the bone, are of primary importance (Schwait et al., 2003).

As mentioned above, MSCs, including those from the BM-MSC and adipose tissue (AT-MSCs), play a vital role in the regeneration of damaged tissues, particularly in bone tissue regeneration, due to the secretion of various factors by MSC-exosomes, which are responsible for the MSCs' paracrine effect. Using a diabetic mouse model, Han et al. demonstrated that exosomes from BM-MSC of normal mice enhanced osteogenesis and suppressed adipogenesis; however, these effects were weaker with exosomes from BM-MSC of diabetic mice (Han et al., 2023). It was found that miR-221, highly expressed in diabetic BM-MSC-exosomes, suppressed osteogenesis and enhanced adipogenesis in vitro and in vivo. Conversely, elevating miR-221 levels in normal BM-MSC-exosomes diminished their ability to regulate both osteogenesis and adipogenesis. Thus, it was demonstrated that exosomal miR-221 obtained from BM-MSCs is a key regulator of diabetic osteoporosis and, correspondingly, a potential therapeutic target.

Li et al. demonstrated that miR-17 was suppressed in exosomes derived from BM-MSC of T2DM rats (Li et al., 2023). It was further shown that exosomes obtained from non-diabetic BM-MSCs essentially facilitated bone regeneration in rats with T2DM by upregulating miR-17. At the same time, exosomes from diabetic BM-MSC destroy the stimulating effect of osteogenesis by regulating the miR-17/SMAD7 axis. In another study, Wang et al. demonstrated that miR-140-3p levels were significantly different in exosomes from BM-MSC and AT-MSC and in serum from diabetic rats (Wang et al., 2022). When transplanted, exosomes from non-diabetic BM-MSCs deliver miR-140-3p, which promotes osteogenesis by inhibiting the plexinB1/RhoA/ROCK signaling pathway, thereby accelerating bone defect regeneration in diabetic rats.

Zhang et al. studied the mechanism of the anti-inflammatory effect of AT-MSC-exosomes in diabetic osteoporosis in both in vitro and in vivo models (Zhang et al., 2021). MSC-exosomes inhibited the secretion of the pro-inflammatory cytokines IL-1β and IL-18 in diabetic osteoclasts in vitro and restored the bone mass in rats with streptozotocin-induced diabetes. Exosomes from AT-MSCs alleviated diabetic osteoporosis by suppressing NLRP3 inflammasome activation in osteoclasts. The NLRP3 inflammasome is a key multiprotein complex involved in the innate immune response and inflammation, activating the caspase-1-dependent release of pro-inflammatory cytokines.

12.4.11 MSC-Exosomes in the Therapy of Diabetic Cardiomyopathy

Diabetic cardiomyopathy represents a pathological condition of the cardiac muscle caused by the diabetes-related electrolyte, metabolic, and vascular disorders. This disease emerges in any type of diabetes. In adults with T2DM, it may develop gradually and proceed without symptoms, manifesting only 10–15 years after the onset of hyperglycemia. According to the statistics, the mortality of diabetic patients from cardiomyopathy is 20–25%. Lin et al. demonstrated that exosomes derived from BM-MSCs alleviated myocardial damage and fibrosis in streptozotocin-induced diabetic rats by inhibiting the TGF-β1/Smad2 signaling pathway (Lin et al., 2019).

The onset of diabetic cardiomyopathy is characterized by autophagy, inflammation, and mitochondrial damage. One potential target for treating diabetic cardiomyopathy may be the regulation of autophagy. Zhang et al. used exosomes from UC-MSCs to treat diabetic cardiomyopathy in rats induced by a high-fat, high-sugar diet combined with streptozotocin (Zhang et al., 2022). The study results showed that UC-MSC-exosomes controlled the cardiac dysfunction in diabetic cardiomyopathy and protected autophagy-damaged cardiomyocytes. UC-MSC-exosomes activated the AMPK/ULK1 signaling pathway and suppressed excessive autophagy in cardiomyocytes, resulting in a protective effect against myocardial damage in diabetic cardiomyopathy. This study confirms that inhibiting excessive autophagy is a practical approach to alleviating heart failure and ventricular remodeling in vivo, suggesting a novel therapeutic strategy for diabetic cardiomyopathy.

12.5 CLINICAL TRIALS BASED ON THE USE OF MSC EXOSOMES

Over the last decade, extensive clinical studies on exosomes in the treatment of various diseases have been initiated. As of June 2025, 27 clinical trials on exosome research in diabetes and its complications are registered on clinicaltrials.gov. At the same time, it is worth noting that only 5 of them are directly related to the therapy of these diseases using MSC-exosomes (Table 12.1). The other studies either involve exosomes from different cell types (e.g., red blood cells) or focus on mechanisms for enhancing diabetes treatment with exosomes, with the miRNAs contained therein considered prognostic biomarkers for studying the pathogenesis of diabetes and its complications.

TABLE 12.1
Major Completed and Active Clinical Trials on the MSC-Exosomes Therapy for Diabetes and Related Complications

No.	Disease	Patients	Cell source	Intervention/Delivery	Study Location
1	T1DM	20	Allogeneic UC-MSC	Exosomes produced from 1.22–1.51 × 10^6/kg/IV (2 doses, 7 days between).	Sahel Teaching Hospital, Cairo, Egypt (NCT02138331, unknown status).
2	Diabetic foot ulcers	110 (n = 40 treated group, n = 35 control group, n = 35 placebo group)	Allogeneic WJ-MSC	WJ-MSC exosomes are administered once weekly for 4 weeks, followed by a follow-up for 16 weeks/topical use.	Kafrelsheikh University, Kafrelsheikh, Egypt (NCT06812637, completed).
3	Diabetic foot ulcers	30	Allogeneic BM-MSC	Personalized Nutritional Intervention Together With the Application of MSC-derived Exosomes, for/topical use.	Maimónides Biomedical Research Institute of Córdoba, Córdoba, Andalucía, Spain (NCT05243368, recruiting).
4	Diabetic foot ulcers	38	Allogeneic WJ-MSC	MSC Conditioned Medium containing exosomes for gel/topical use.	Tarumanagara University, Jakarta, Indonesia (NCT04134676, completed).
5	Erectile dysfunction is associated with T2DM	30	Autologous AT-MSC	AT-MSC or AT-MSC-derived exosome injections/Intra-cavernosally.	Jumeirah American Clinic, Dubai, United Arab Emirates (NCT06605508, recruiting).

Abbreviations: AT-MSC: Adipose tissue-derived MSC; BM-MSC: Bone marrow-derived mesenchymal stem cells; MSC: Mesenchymal stem cells; T1DM: Type 1 diabetes mellitus; T2DM: Type 2 diabetes mellitus; WJ-MSC: Wharton's Jelly-derived mesenchymal stem cells; IV: Intravenous; NCT Number: The National Clinical Trial number, also called the ClinicalTrials.gov Identifier.

One of the five mentioned clinical trials is dedicated to the T1DM therapy, three are related to diabetic foot ulcers, and the last one is to T2DM-associated erectile dysfunction (see Table 12.1).

The clinical trial conducted at Tarumanagara University (Jakarta, Indonesia) showed that local application of the human UC-MSC secretome in a 10% gel may efficiently facilitate healing of diabetic and trophic ulcers (Tan et al., 2023). The significant reduction in ulcer length, width, and area confirmed this finding.

In Phase 2 of the clinical trial, the therapy demonstrated its safety, with no local or systemic effects reported. The authors plan a broader clinical study (Phase 3) with a larger patient cohort and the use of control groups receiving either antibiotics or a placebo. The latter study will aim to expand the database on the local application of the UC-MSC secretome in the treatment of chronic ulcers.

Considering that the majority of registered clinical trials were initiated in the last several years, no results from those studies have been published yet. At the same time, the researchers from the Sahel Teaching Hospital (Cairo, Egypt), who have not published the results of the T1DM exosome therapy yet, already published the results of their study on the treatment of chronic kidney disease (grade III-IV) by UC-MSC-derived extracellular vesicles (Nassar et al., 2016). In the latter study, 50% of the included patients (10 out of 20) were T1DM patients with chronic kidney disease as a direct diabetes complication. The study results demonstrated that the applied therapy was "safe and can ameliorate the inflammatory immune reaction and improve the overall kidney function in grade III-IV CKD patients." This conclusion is supported by significantly enhanced indices of eGFR, Scr, Bun, and UACR, as well as increases in TGF-β1 and IL-10 levels and a decrease in TNF-α levels in the serum of treated patients.

12.6 CONCLUSIONS

In the last decade, MSC-exosome therapy has been explored for various diseases, including diabetes and its complications. There are already 47 companies worldwide developing MSC-exosome therapies (https://biotech-careers.org/business-area/exosomes). As mentioned above, MSC-exosomes have several advantages over native MSCs, including their lack of immunogenicity and their high selectivity for gene delivery to cells.

It is pertinent to note that further studies are necessary to expand the clinical applications of MSC-exosomes. Different conditions of MSC culture and varying cell passages lead to heterogeneity in MSC-exosomes and, hence, to differing efficiency in similar diabetic models. This, in turn, results in varying efficiency in clinical trials, even when patients have a similar clinical picture of diabetic damage. It is necessary to standardize protocols for isolating MSC-exosomes, establish optimal sources for MSC collection, and define optimal protocols for lyophilizing MSC-exosomes for long-term storage. It is also required to determine the optimal dosage and administration method for MSC-exosomes.

In general, the existing technologies for MSC-exosomes' application in the therapy of diabetes and its complications, as well as the strategy for additional studies in this field, hold promise for broad clinical application of MSC-exosomes in the near future.

REFERENCES

Ahmed ZT, Alkahlot MH, Haider KH. MSC-derived exosomes: advances in cell-free therapy. In: Haider KH. (ed.) *Handbook of Stem Cell Applications*. Singapore: Springer Nature Singapore; 2023. p. 1–41.

Albalawi RYN, Alchalabi Z, Ljajah L, Almasri S, Haider KH. Pluripotent stem cells: a hope to move forward for cardiovascular applications. In: Haider KH. (ed.) *Handbook of Regenerative Medicine: Cell-Based Therapy Approach*. Boca Raton (FL): CRC Press; 2025. p. 418–436.

Al-Khani AM, Khalifa MA, Haider KH. Mesenchymal stem cells for cardiac repair. In: Haider KH. (ed.) *Handbook of Stem Cell Therapy*. Springer; 2022. https://doi.org/10.1007/978-981-16-6016-0_11-1

Álvarez-Viejo M, Haider KH. Mesenchymal stem cells. In: Haider KH. (ed.) *Handbook of Stem Cell Therapy.* Springer; 2022. https://doi.org/10.1007/978-981-16-6016-0_6-1.

Ángulo J., González-Corrochano R., Cuevas P., Fernández A., Fuente J.M.L., Rolo F., et al. Diabetes exacerbates the functional deficiency of the NO/cGMP pathway associated with erectile dysfunction in the human corpus cavernosum and penile arteries. *J Sex Med.* 2010;7:758–768. doi: 10.1111/j.1743-6109.2009.01587.x.

Bellary S, Kyrou I, Brown JE, Bailey CJ. Type 2 diabetes mellitus in older adults: clinical considerations and management. *Nat Rev Endocrinol.* 2021 Sep;17(9):534–548. doi: 10.1038/s41574-021-00512-2.

Chen J, Chen J, Cheng Y, Fu Y, Zhao H, Tang M, Zhao H, et al. Mesenchymal stem cell-derived exosomes protect beta cells against hypoxia-induced apoptosis via miR-21 by alleviating ER stress and inhibiting p38 MAPK phosphorylation. *Stem Cell Res Ther.* 2020 Mar 4;11(1):97. doi: 10.1186/s13287-020-01610-0.

Cross SE, Richards SK, Clark A, Benest AV, Bates DO, Mathieson PW, Johnson PR, et al. Vascular endothelial growth factor as a survival factor for human islets: effect of immunosuppressive drugs. *Diabetologia.* 2007 Jul;50(7):1423–32. doi: 10.1007/s00125-007-0670-8.

Downward J. PI 3-kinase, Akt, and cell survival. *Semin Cell Dev Biol.* 2004 Apr;15(2):177–182. doi: 10.1016/j.semcdb.2004.01.002.

Ebrahim N, Ahmed IA, Hussien NI, Dessouky AA, Farid AS, Elshazly AM, Mostafa O, et al. Mesenchymal stem cell-derived exosomes ameliorated diabetic nephropathy by autophagy induction through the mTOR signaling pathway. *Cells.* 2018 Nov 22;7(12):226. doi: 10.3390/cells7120226.

Fan B, Chopp M, Zhang ZG, Liu XS. Treatment of diabetic peripheral neuropathy with engineered mesenchymal stromal cell-derived exosomes enriched in microRNA-146a results in enhanced therapeutic efficacy. *Exp Neurol.* 2021;341:113694. doi: 10.1016/j.expneurol.2021.113694.

Fan B, Li C, Szalad A, Wang L, Pan W, Zhang R, Chopp M, et al. Mesenchymal stromal cell-derived exosomes ameliorate peripheral neuropathy in a mouse model of diabetes. *Diabetologia.* 2020 Feb;63(2):431–443. doi: 10.1007/s00125-019-05043-0.

Fu Y, Gao X, He GH, Chen S, Gu ZH, Zhang YL, Li LY. Protective effects of umbilical cord mesenchymal stem cell exosomes in a diabetic rat model through live retinal imaging. *Int J Ophthalmol.* 2021 Dec 18;14(12):1828–1833. doi: 10.18240/ijo.2021.12.04.

Gao X, He GH, Zhang XT, Chen S. Protective effect of human umbilical cord mesenchymal stem cell-derived exosomes on rat retinal neurons in hyperglycemia through the brain-derived neurotrophic factor/TrkB pathway. *Int J Ophthalmol.* 2021 Nov 18;14(11):1683–1689. doi: 10.18240/ijo.2021.11.06.

Godfrey KJ, Mathew B, Bulman JC, Shah O, Clement S, Gallicano GI. Stem cell-based treatments for Type 1 diabetes mellitus: bone marrow, embryonic, hepatic, pancreatic, and induced pluripotent stem cells. *Diabet Med.* 2012;29:14–23. doi: 10.1111/j.1464–5491.2011.03433.x.

Gondaliya P, Sayyed AA, Bhat P, Mali M, Arya N, Khairnar A, Kalia K. Mesenchymal stem cell-derived exosomes loaded with miR-155 inhibitor ameliorate diabetic wound healing. *Mol Pharm.* 2022 May 2;19(5):1294–1308. doi: 10.1021/acs.molpharmaceut.1c00669.

Habib SM, Martini MF, Al-Ahmed O, Qassab AR, Haider KH. Mesenchymal stem cell secretomes for cell-free therapy of the broken heart. In: Haider KH. (ed.) *Handbook of Regenerative Medicine: Cell-Free Therapy.* Boca Raton (FL): CRC Press; 2025a. p. 274–293.

Habib SM, Haider KH, Martini MF. The potential of mesenchymal stem cell-derived exosomes in cell-free therapy for myocardial repair. In: Haider KH. (ed.) *Handbook of Regenerative Medicine: Cell-Free Therapy Approach.* 1st ed. Boca Raton (FL): CRC Press; 2025b, Chapter 24. https://doi.org/10.1201/9781003611509.

Habib SY, Martini MF, Hamdan YNA, Shrebaty OMM, Haider KH. Mesenchymal stem cell secretome and exosomes as potential advanced therapy medicinal products for treating a failing heart. *World J Stem Cells.* 2026 March 26; 18(3): 113871. doi: 10.4252/wjsc.v18.i3.113871.

Haider KH. Priming mesenchymal stem cells to develop "super stem cells". *World J Stem Cells* 2024; 16(6): 623–640. doi: 10.4252/wjsc.v16.i6.623

Haider KH, Aslam M. Cell-free therapy with stem cell secretions: protection, repair, and regeneration of the injured myocardium. In: Haider KH., Aziz S. (eds.) *Stem Cells: From Hype to Real Hope.* Berlin: De Gruyter; 2018.

Haider KH, Najimi M. Exosome-based cell-free therapy in regenerative medicine for myocardial repair. In: Haider KH. (ed.) *Handbook of Stem Cell Therapy.* Springer; 2022. https://doi.org/10.1007/978-981-16-6016-0_9-1.

Han F, Wang C, Cheng P, Liu T, Wang WS. Bone marrow mesenchymal stem cell-derived exosomal miRNAs can modulate diabetic bone-fat imbalance. *Front Endocrinol* (Lausanne). 2023 Apr 14;14:1149168. doi: 10.3389/fendo.2023.1149168.

Hao Y, Miao J, Liu W, Cai K, Huang X, Peng L. Mesenchymal stem cell-derived exosomes carry microRNA-125a to protect against diabetic nephropathy by targeting histone deacetylase 1 and downregulating endothelin-1. *Diabetes Metab Syndr Obes.* 2021 Mar 25;14:1405–1418. doi: 10.2147/DMSO.S286191.

He C, Zheng S, Luo Y, Wang B. Exosome theranostics: Biology and translational medicine. *Theranostics.* 2018;8(1):237–255. doi: 10.7150/thno.21945.

Hou QL, Ge MY, Zhang CD, Tian DD, Wang LK, Tian HZ, Wang WH, et al. Adipose tissue-derived stem cell therapy for erectile dysfunction in rats: a systematic review and meta-analysis. *Int Urol Nephrol.* 2017 Jul;49(7):1127–1137. doi: 10.1007/s11255-017-1590-2.

Huo W, Li Y, Zhang Y, Li H. Mesenchymal stem cells-derived exosomal microRNA-21-5p downregulates PDCD4 and ameliorates erectile dysfunction in a rat model of diabetes mellitus. *FASEB J.* 2020 Oct;34(10):13345–13360. doi: 10.1096/fj. 202000102RR.

Ibrahim AY, Mehdi Q, Abbas AO, Alashkar A, Haider KH. Induced pluripotent stem cells: next-generation cells for tissue regeneration. *J Biomed Sci Eng.* 2016;9(4):226–244.

Janghorbani M., Van Dam R.M., Willett W.C., F.B. Hu. Systematic review of type 1 and type 2 diabetes mellitus and risk of fracture. *Amer J Epidem.* 2007;166(5):495–505.

Jiao, YR., Chen, KX., Tang, X. *et al.* Exosomes derived from mesenchymal stem cells in diabetes and diabetic complications. *Cell Death Dis* 2024;15:271. doi: 10.1038/s41419-024-06659-w.

Jin J, Shi Y, Gong J, Zhao L, Li Y, He Q, Huang H. Exosomes secreted from adipose-derived stem cells attenuate diabetic nephropathy by promoting autophagy flux and inhibiting apoptosis in podocyte. *Stem Cell Res Ther.* 2019 Mar 15;10(1):95. doi: 10.1186/s13287-019-1177-1.

Kamal M, Kassem D, Haider KH. Sources and therapeutic strategies of mesenchymal stem cells in regenerative medicine. In: Haider KH. (ed.) *Handbook of Stem Cell Therapy.* Singapore: Springer; 2022. https://doi.org/10.1007/978-981-19-2655-6_2

Kamaraj R, Mehran MJ, Tiwari H, Shukla MK, Cetiz MV, Basharat Z, Barzigar R, et al. The role of stem cells in diabetes, liver disease, and cancer treatment. In: Haider KH. (ed.) *Handbook of Regenerative Medicine: Cell-Based Therapy Approach.* Boca Raton (FL): CRC Press; 2025, Chapter 26, pp. 402–417.

Kochar IS, Jain R. Pancreas transplant in type 1 diabetes mellitus: the emerging role of islet cell transplant. *Ann Pediatr Endocrinol Metab.* 2021 Jun;26(2):86–91. doi: 10.6065/apem.2142012.006.

Konoplyannikov M, Haider KH, Ahmed RPH, Jiang S, Vien KL, Ashraf M. Activation of diverse signaling pathways by delivery of multiple cytokines for myocardial repair. *Stem Cell Dev.* 2013; 22(2): 204–215.

Kubota K, Nakano M, Kobayashi E, Mizue Y, Chikenji T, Otani M, Nagaishi K, et al. An enriched environment prevents diabetes-induced cognitive impairment in rats by enhancing exosomal miR-146a secretion from endogenous bone marrow-derived mesenchymal stem cells. *PLoS One.* 2018 Sep 21;13(9):e0204252. doi: 10.1371/journal.pone.0204252.

Li B, Luan S, Chen J, Zhou Y, Wang T, Li Z, Fu Y, et al. The MSC-derived exosomal lncRNA H19 promotes wound healing in diabetic foot ulcers by upregulating PTEN via MicroRNA-152-3p. *Mol Ther Nucleic Acids.* 2020 Mar 6;19:814–826. doi: 10.1016/j.omtn.2019.11.034.

Li DF, Yang MF, Xu J, Xu HM, Zhu MZ, Liang YJ, Zhang Y, et al. Extracellular vesicles: the next generation theranostic nanomedicine for inflammatory bowel disease. *Int J Nanomedicine.* 2022 Sep 5;17:3893–3911. doi: 10.2147/IJN.S370784.

Li T., Yan Y., Wang B., Qian H., Zhang X., Shen L., et al. Exosomes derived from human umbilical cord mesenchymal stem cells alleviate liver fibrosis. *Stem Cells Dev.* 2013;22:845–854. doi: 10.1089/scd.2012.0395.

Li W, Jin LY, Cui YB, Xie N. Human umbilical cord mesenchymal stem cells-derived exosomal microRNA-17-3p ameliorates inflammatory reaction and antioxidant injury of mice with diabetic retinopathy via targeting STAT1. *Int Immunopharmacol.* 2021a Jan;90:107010. doi: 10.1016/j.intimp.2020.107010.

Li W, Jin L, Cui Y, Nie A, Xie N, Liang G. Bone marrow mesenchymal stem cells-induced exosomal microRNA-486-3p protects against diabetic retinopathy through TLR4/NF-κB axis repression. *J Endocrinol Invest.* 2021b Jun;44(6):1193–1207. doi: 10.1007/s40618-020-01405-3.

Li Z, Zhang B, Shang J, Wang Y, Jia L, She X, Xu X, et al. Diabetic and nondiabetic BMSC-derived exosomes affect bone regeneration via regulating the miR-17-5p/SMAD7 axis. *Int Immunopharmacol.* 2023 Dec;125(Pt B):111190. doi: 10.1016/j.intimp.2023.111190.

Liang Y, Xia J, Xu L, Zhang Y, Xu X, Haider KH. Cell-free and cell-based therapy in regenerative medicine. In: KH Haider, editor. *Handbook of Regenerative Medicine: Cell-Free Therapy Approach.* 1st ed. Boca Raton (FL): CRC Press; 2025, Chapter 24. doi: 10.1201/9781003611509.

Lin Y, Zhang F, Lian XF, Peng WQ, Yin CY. Mesenchymal stem cell-derived exosomes improve diabetes mellitus-induced myocardial injury and fibrosis via inhibition of TGF-β1/Smad2 signaling pathway. *Cell Mol Biol* (Noisy-le-grand). 2019 Sep 30;65(7):123–126.

Liu, C., Xiang, J., Chen, Y. et al. MiR-125a-5p in MSC-derived small extracellular vesicles alleviates Müller cells injury in diabetic retinopathy by modulating mitophagy via the PTP1B pathway. *Cell Death Discov.* 11, 226 (2025). doi: 10.1038/s41420-025-02439-3.

Lu XC, Zheng JY, Tang LJ, Huang BS, Li K, Tao Y, Yu W, et al. MiR-133b promotes neurite outgrowth by targeting the expression of RhoA. *Cell Physiol Biochem.* 2015;35(1):246–58. doi: 10.1159/000369692.

Mahdipour E, Salmasi Z, Sabeti N. Potential of stem cell-derived exosomes to regenerate β islets through Pdx-1 dependent mechanism in a rat model of type 1 diabetes. *J Cell Physiol.* 2019 Nov;234(11):20310–20321. doi: 10.1002/jcp. 28631.

Mallapaty S. Stem cells reverse woman's diabetes - a world first. *Nature.* 2024 Oct;634(8033):271–272. doi: 10.1038/d41586-024-03129-3.

Martin D, Alberti P, Demartines N, Phillips M, Casey J, Sutherland A. Whole-organ pancreas and islets transplantations in the UK: an overview and future directions. *J Clin Med.* 2023 May 1;12(9):3245. doi: 10.3390/jcm12093245.

Mathew B, Ravindran S, Liu X, Torres L, Chennakesavalu M, Huang CC, Feng L, et al. Mesenchymal stem cell-derived extracellular vesicles and retinal ischemia-reperfusion. *Biomaterials.* 2019 Mar;197:146–160. doi: 10.1016/j.biomaterials.2019.01.016.

Meligy FY, Mohammed HSE, Abou Elghait AT, Mohamed HK, Ashry IEM, Abdel-Rahman Sayed A, Hussein OA, et al. Mesenchymal stem cells versus mesenchymal stem cell-derived exosomes as potential autophagy pathway modulators in a diabetic model. *Adv Med Sci.* 2025 Mar;70(1):152–165. doi: 10.1016/j.advms.2025.02.004.

Mishra VK, Shih HH, Parveen F, Lenzen D, Ito E, Chan TF, Ke LY. Identifying the therapeutic significance of mesenchymal stem cells. *Cells.* 2020 May 6;9(5):1145. doi: 10.3390/cells9051145.

Mushtaq M, Zineldeen DH, Mateen MA, Haider KH. Mesenchymal stem cells' "garbage bags" at work: Treating radial nerve injury with mesenchymal stem cell-derived exosomes. *World J Stem Cells.* 2024 May 26;16(5):467–478. doi: 10.4252/wjsc.v16.i5.467.

Nakano M, Nagaishi K, Konari N, Saito Y, Chikenji T, Mizue Y, Fujimiya M. Bone marrow-derived mesenchymal stem cells improve diabetes-induced cognitive impairment by exosome transfer into damaged neurons and astrocytes. *Sci Rep.* 2016 Apr 22;6:24805. doi: 10.1038/srep24805.

Nassar W, El-Ansary M, Sabry D, Mostafa MA, Fayad T, Kotb E, Temraz M, et al. Umbilical cord mesenchymal stem cells-derived extracellular vesicles can safely ameliorate the progression of chronic kidney disease. *Biomater Res.* 2016 Aug 5;20:21. doi: 10.1186/s40824-016-0068-0. Erratum in: *Biomater Res.* 2017 Apr 6;21:3. doi: 10.1186/s40824-017-0089-3.

Nojehdehi S, Soudi S, Hesampour A, Rasouli S, Soleimani M, Hashemi SM. Immunomodulatory effects of mesenchymal stem cell-derived exosomes on experimental type-1 autoimmune diabetes. *J Cell Biochem.* 2018. 119(11):9433–9443. doi: 10.1002/jcb. 27260.

Nyenwe EA, Jerkins TW, Umpierrez GE, Kitabchi AE. Management of type 2 diabetes: evolving strategies for the treatment of patients with type 2 diabetes. *Metabolism.* 2011 Jan;60(1):1–23. doi: 10.1016/j.metabol.2010.09.010.

Omole AE, Fakoya AOJ, Nnawuba KC, Haider KH. Common ethical considerations of human-induced pluripotent stem cell research. In: Haider KH. (ed.) *Handbook of Stem Cell Therapy.* Springer; 2022. https://doi.org/10.1007/978-981-16-6016-0_40-1.

Pang H, Luo S, Huang G, Xia Y, Xie Z, Zhou Z. Advances in knowledge of candidate genes acting at the beta-cell level in the pathogenesis of T1DM. *Front Endocrinol (Lausanne).* 2020 Mar 12;11:119. doi: 10.3389/fendo.2020.00119.

Patel AA, Shafie A, Mohamed AH, Ali SA, Tayeb FJ, Waggiallah HA, Ahmad I, et al. The promise of mesenchymal stromal/stem cells in erectile dysfunction treatment: a review of current insights and future directions. *Stem Cell Res Ther.* 2025 Feb 26;16(1):98. doi: 10.1186/s13287-025-04221-9.

Pittas A.G., Lau J., F. Hu, Dawson-Hughes B. The role of vitamin D and calcium in type 2 diabetes. A systematic review and meta-analysis. *J Clin Endocrinol Metab.* 2007;92(6):2017–2029.

Rajab AM, Rajab TM, AlJundi S, Haider KH. Bone stem cell therapy in the clinical perspective: a focus on nonrandomized and randomized trials. In: Haider KH. (ed.) *Stem Cells: From Myth to Reality and Evolving.* Berlin: Medicine & Life Sciences, De Gruyter; 2019, Geithner Straße 13- 10785.

Rufaihah AJ, Haider KH, Heng BC, Ye L, Tan RS, Toh WS, Tian XF, Sim EK, Cao T. Therapeutic angiogenesis by transplantation of human embryonic stem cell-derived CD133+ endothelial progenitor cells for cardiac repair. *Regenerative Med.* 2010;5:231–244.

Sabry D, Marzouk S, Zakaria R, Ibrahim HA, Samir M. The effect of exosomes derived from mesenchymal stem cells in the treatment of induced type 1 diabetes mellitus in rats. *Biotechnol Lett.* 2020 Aug;42(8):1597–1610. doi: 10.1007/s10529-020-02908-y. Erratum in: *Biotechnol Lett.* 2020 Dec;42(12):2761–2762. doi: 10.1007/s10529-020-02975-1.

Saima K, Haider KH. Stem cells: a renewable source of pancreatic β-cells and future for diabetes treatment. In: Haider KH. (ed.) *Stem Cells.* Cham: Springer; 2021, Chapter 12, p. 185–202. https://doi.org/10.1007/978-3-030-77052-5_12.

Samsu N. Diabetic nephropathy: challenges in pathogenesis, diagnosis, and treatment. *Biomed Res Int.* 2021 Jul 8;2021:1497449. doi: 10.1155/2021/1497449.

Sato Y, Inoue M, Yoshizawa T, Yamagata K. Moderate hypoxia induces β-cell dysfunction with HIF-1-independent gene expression changes. *PLoS One.* 2014 Dec 12;9(12):e114868. doi: 10.1371/journal.pone.0114868.

Schwait A.V. Diabetes mellitus: does it affect bone? *Calcif Tissue Int.* 2003; 73(6): 515–519.

Shabbir A, Cox A, Rodriguez-Menocal L, Salgado M, Van Badiavas E. Mesenchymal stem cell exosomes induce proliferation and migration of normal and chronic wound fibroblasts, and enhance angiogenesis in vitro. *Stem Cells Dev.* 2015 Jul 15;24(14):1635–1647. doi: 10.1089/scd.2014.0316.

Shang S, Zhuang K, Chen J, Zhang M, Jiang S, Li W. A bioactive composite hydrogel dressing that promotes healing of both acute and chronic diabetic skin wounds. *Bioact Mater.* 2024 Jan 3;34:298–310. doi: 10.1016/j.bioactmat.2023.12.026.

Shapiro AM, Lakey JR, Ryan EA, Korbutt GS, Toth E, Warnock GL, Kneteman NM, et al. Islet transplantation in seven patients with type 1 diabetes mellitus using a glucocorticoid-free immunosuppressive regimen. *N Engl J Med.* 2000 Jul 27;343(4):230–238. doi: 10.1056/NEJM200007273430401.

Shapiro AM, Ricordi C. Long-term outcomes after islet transplantation. In: *Textbook of Organ Transplantation.* Hoboken (NJ): Wiley-Blackwell; 2014. p. 1314–1333. doi:10.1002/9781118873434.ch108.

Shapiro AM, Ricordi C, Hering BJ, Auchincloss H, Lindblad R, Robertson RP, Secchi A, et al. International trial of the Edmonton protocol for islet transplantation. *N Engl J Med.* 2006 Sep 28;355(13):1318–1330. doi: 10.1056/NEJMoa061267.

Shigemoto-Kuroda T, Oh JY, Kim DK, Jeong HJ, Park SY, Lee HJ, Park JW, et al. MSC-derived extracellular vesicles attenuate immune responses in two autoimmune murine models: type 1 diabetes and uveoretinitis. *Stem Cell Reports.* 2017 May 9;8(5):1214–1225. doi: 10.1016/j.stemcr.2017.04.008.

Sinwar PD. The diabetic foot management: recent advances. *Int J Surg.* 2015 Mar;15:27–30. doi: 10.1016/j.ijsu.2015.01.023.

Song J., Sun T., Tang Z., Ruan Y., Liu K., Rao K., et al. Exosomes derived from smooth muscle cells ameliorate diabetes-induced erectile dysfunction by inhibiting fibrosis and modulating the NO/cGMP pathway. *J Cell Mol Med.* 2020;24:13289–13302. doi: 10.1111/jcmm.15946.

Sordi V, Monaco L, Piemonti L. Cell therapy for type 1 diabetes: from islet transplantation to stem cells. *Horm Res Paediatr.* 2023;96(6):658–669. doi: 10.1159/000526618.

Su T, Xiao Y, Xiao Y, Guo Q, Li C, Huang Y, Deng Q, et al. Bone marrow mesenchymal stem cells–derived exosomal miR-29b-3p regulates aging-associated insulin resistance. *ACS Nano.* 2019 Feb 26;13(2):2450–2462. doi: 10.1021/acsnano.8b09375.

Sun Y, Shi H, Yin S, Ji C, Zhang X, Zhang B, Wu P, et al. Human mesenchymal stem cell-derived exosomes alleviate type 2 diabetes mellitus by reversing peripheral insulin resistance and relieving β-cell destruction. *ACS Nano.* 2018 Aug 28;12(8):7613–7628. doi: 10.1021/acsnano.7b07643.

Tan, S. T., Aisyah, P. B., Firmansyah, Y., Nathasia, N., Budi, E., & Hendrawan, S. Effectiveness of secretome from human umbilical cord mesenchymal stem cells in gel (10% SM-hUCMSC gel) for chronic wounds (diabetic and trophic ulcer): phase 2 clinical trial. *J Multidiscip Healthc.* 2023;16:1763–1777. doi: 10.2147/JMDH.S408162.

Wang N, Liu X, Tang Z, Wei X, Dong H, Liu Y, Wu H, et al. Increased BMSC exosomal miR-140-3p alleviates bone degradation. It promotes bone restoration by targeting Plxnb1 in diabetic rats. *J Nanobiotechnology.* 2022 Mar 2;20(1):97. doi: 10.1186/s12951-022-01267-2.

Wen D, Peng Y, Liu D, Weizmann Y, Mahato RI. Mesenchymal stem cells and derived exosomes as small RNA carriers and immunomodulators to improve islet transplantation. *J Control Release.* 2016 Sep 28;238:166–175. doi: 10.1016/j.jconrel.2016.07.044.

Weng J, Chen Y, Zeng Y, Jin W, Ji Y, Zhang W, Wang S, et al. A novel hydrogel loaded with plant exosomes and stem cell exosomes as a new strategy for treating diabetic wounds. *Mater Today Bio.* 2025 Apr 27;32:101810. doi: 10.1016/j.mtbio.2025.101810.

Wu J, Li T, Guo M, Ji J, Meng X, Fu T, Nie T, et al. Clinical Group. Treating a type 2 diabetic patient with impaired pancreatic islet function by personalized endoderm stem cell-derived islet tissue. *Cell Discov.* 2024 Apr 30;10(1):45. doi: 10.1038/s41421-024-00662-3.

Xia J, Arai S, Haider KH. Research and applications of stem cell secretome. In: Haider KH. (ed.) *Stem Cells: From Potential to Promise.* Springer; 2021. https://doi.org/10.1007/978-981-16-0301-3_7.

Xin H, Li Y, Liu Z, Wang X, Shang X, Cui Y, Zhang ZG, Chopp M. MiR-133b promotes neural plasticity and functional recovery after stroke treatment with multipotent mesenchymal stromal cells in rats via transfer of exosome-enriched extracellular particles. *Stem Cells.* 2013 Dec;31(12):2737–2746. doi: 10.1002/stem. 1409].

Yafi F.A., Jenkins L., Albersen M., Corona G., Isidori A.M., Goldfarb S., et al. Erectile dysfunction. *Nat Rev Dis Primers.* 2016;2 doi: 10.1038/nrdp. 2016.3.

Yang D, Livingston MJ, Liu Z, Dong G, Zhang M, Chen JK, Dong Z. Autophagy in diabetic kidney disease: regulation, pathological role and therapeutic potential. *Cell Mol Life Sci.* 2018 Feb;75(4):669–688. doi: 10.1007/s00018-017-2639-1.

Yang J, Kim G, Shim JH, An J. Second-line antidiabetic drugs: Friend or foe of the liver. *J Liver Cancer.* 2025 Jun 26. doi: 10.17998/jlc. 2025.06.25.

Yang M, Chen J, Chen L. The roles of mesenchymal stem cell-derived exosomes in diabetes mellitus and its related complications. *Front Endocrinol (Lausanne).* 2022 Oct 21;13:1027686. doi: 10.3389/fendo.2022.1027686.

Yap SK, Tan KL, Abd Rahaman NY, Saulol Hamid NF, Ooi J, Tor YS, Daniel Looi QH, et al. Umbilical cord mesenchymal stem cell-derived small extracellular vesicles ameliorated insulin resistance in type 2 diabetes mellitus rats. *Pharmaceutics.* 2022 Mar 16;14(3):649. doi: 10.3390/pharmaceutics14030649.

Zhang K, Zheng S, Wu J, He J, Ouyang Y, Ao C, Lang R, et al. Human umbilical cord mesenchymal stem cell-derived exosomes ameliorate renal fibrosis in diabetic nephropathy by targeting the Hedgehog/SMO signaling pathway. *FASEB J.* 2024 Apr 15;38(7):e23599. doi: 10.1096/fj.202302324R.

Zhang L, Wang Q, Su H, Cheng J. Exosomes from adipose-derived mesenchymal stem cells alleviate diabetic osteoporosis in rats through suppressing NLRP3 inflammasome activation in osteoclasts. *J Biosci Bioeng.* 2021 Jun;131(6):671–678. doi: 10.1016/j.jbiosc.2021.02.007.

Zhang P, Lu J, Jing Y, Tang S, Zhu D, Bi Y. Global epidemiology of diabetic foot ulceration: a systematic review and meta-analysis. *Ann Med.* 2017 Mar;49(2):106–116. doi: 10.1080/07853890.2016.1231932.

Zhang Z, Chen L, Chen X, Qin Y, Tian C, Dai X, Meng R, et al. Exosomes derived from human umbilical cord mesenchymal stem cells (HUCMSC-EXO) regulate autophagy through the AMPK-ULK1 signaling pathway to ameliorate diabetic cardiomyopathy. *Biochem Biophys Res Commun.* 2022 Dec 3;632:195–203. doi: 10.1016/j.bbrc.2022.10.001.

Zhao B, Zhang X, Zhang Y, Lu Y, Zhang W, Lu S, Fu Y, et al. Human exosomes accelerate cutaneous wound healing by promoting collagen synthesis in a diabetic mouse model. *Stem Cells Dev.* 2021 Sep 15;30(18):922–933. doi: 10.1089/scd.2021.0100.

Zhao T, Sun F, Liu J, et al. Emerging role of mesenchymal stem cell-derived exosomes in regenerative medicine. *Curr Stem Cell Res Ther.* 2019;14(6):482–494. doi: 10.2174/1574888X14666190228103230.

Zhu LL, Huang X, Yu W, Chen H, Chen Y, Dai YT. Transplantation of adipose tissue-derived stem cell-derived exosomes ameliorates erectile function in diabetic rats. *Andrologia.* 2018 Mar;50(2). doi: 10.1111/and.12871.

13 The Use of Mesenchymal Stem Cell-Sourced Exosomes in the Treatment of Neurocognitive Disease

From Bench to Bedside

Ana Volarevic[1], *Carl Randall Harrell*[2], *and Vladislav Volarevic*[3]

[1]Department of Medicine and Society, Faculty of Medicine, University of Nis, Nis, Serbia

[2]Regenerative Processing Plant, Florida, USA

[3]Department of Immunology, Faculty of Medicine, University of Nis, Nis, Serbia

13.1 LIST OF ABBREVIATIONS

6-OHDA	6-hydroxydopamine
AD	Alzheimer's disease
Aβ	Amyloid-β
BBB	Blood-brain barrier
BNDF	Brain-derived neurotropic factor
CNS	Central nervous system
DCs	Dendritic cells
EVs	Extracellular vesicles
FTD	Frontotemporal dementia
GMP	Good Manufacturing Practice
HGF	Hepatocyte growth factor
IDO	Indoleamine 2,3-dioxygenase
IL	Interleukin
IL-1β	Interleukin-1β
iPSCs	Induced pluripotent stem cells
MAO-B	Monoamine oxidase-B
miRNAs	microRNAs
MPP	1-methyl-4-phenylpyridinium
MSC-Exos	Exosomes derived from MSCs
MSCs	Mesenchymal stem cells
NGF	Nerve growth factor
PD	Parkinson's disease

DOI: 10.1201/9781003653387-13

ROS	Reactive oxygen species
SOD	Superoxide dismutase
TGF-β	Transforming growth factor-beta
T-MSCs-Exos	Trophoblast stage-derived MSCs
TNF-alpha	Tumor necrosis factor-alpha
TSG-6	Tumor necrosis factor-stimulated gene-6
VCI	Vascular cognitive impairment

13.2 INTRODUCTION

Neurocognitive disorders encompass a broad and increasingly prevalent class of neurological conditions characterized by progressive deterioration in cognitive domains, including memory, attention, executive functioning, language, and visuospatial abilities (Restrepo-Martinez et al., 2025). These disorders represent a major global health challenge due to their profound impact on patient autonomy, caregiver burden, and healthcare infrastructure. Among the most prominent and debilitating of these conditions are Alzheimer's disease (AD), Parkinson's disease (PD), frontotemporal dementia (FTD), and vascular cognitive impairment (VCI), all of which share overlapping clinical features and neuropathological substrates (Restrepo-Martinez et al., 2025; Ma et al., 2025). Despite decades of intense biomedical research and the development of numerous pharmacological agents, the current therapeutic landscape for these disorders remains predominantly palliative. Existing treatments primarily target symptom management, offering only modest and transient relief without fundamentally altering the trajectory of neuronal degeneration or disease progression. The failure to develop disease-modifying therapies has underscored the urgent need for innovative approaches that can intervene at earlier stages of pathogenesis and promote neural repair or regeneration.

In recent years, regenerative medicine, particularly the application of stem cell technologies, has emerged as a promising frontier in the search for such transformative therapies (Harrell et al., 2021). Initially, much of the focus centered on the transplantation of mesenchymal stem cells (MSCs) derived from bone marrow, adipose tissue, or umbilical cord, to replace lost neuronal populations or support endogenous repair processes (Kamal et al., 2022). However, accumulating evidence suggests that the therapeutic efficacy of MSCs is not primarily due to their engraftment or direct differentiation into neural lineages, but instead to their paracrine activity, specifically the secretion of extracellular vesicles (EVs), including exosomes (Harrell et al., 2021). These nano-sized, lipid bilayer-bound vesicles, once regarded as cellular waste, are now recognized as potent conveyors of intercellular signals (Mushtaq et al., 2024). Exosomes derived from MSCs (MSC-Exos) have garnered significant interest for their ability to carry bioactive molecular cargo, including proteins, lipids, mRNAs, and microRNAs (miRNAs), that can modulate the recipient cell phenotype and promote neuroprotective and neurodegenerative processes (Ahmed et al., 2023).

A particularly advantageous feature of MSC-Exos is their inherent capacity to cross the blood-brain barrier (BBB), a significant obstacle to delivering therapeutic agents to the central nervous system (CNS) (Lamiaa et al., 2025). Once within the CNS milieu, MSC-Exos can influence a wide range of biological activities. Preclinical studies have demonstrated that these vesicles can attenuate neuroinflammation by modulating microglial activation, inhibit neuronal apoptosis, stimulate endogenous neurogenesis, and enhance synaptic plasticity and connectivity (Harrell et al., 2021). In models of AD, MSC-Exos have been shown to facilitate the clearance of amyloid-beta aggregates and reduce tau hyperphosphorylation, two pathological hallmarks of the disease (Abdi et al., 2025). In PD models, they have exhibited the capacity to inhibit alpha-synuclein aggregation and support the survival of dopaminergic neurons within the substantia nigra (Abrishamdar et al., 2023). These pleiotropic effects are primarily attributed to the diverse cargo of regulatory miRNAs and signaling proteins packaged within exosomes, including miR-21, miR-29, and BDNF (Abrishamdar et al., 2023; Abdi et al., 2025).

The appeal of MSC-Exos as a therapeutic platform lies not only in their biological efficacy but also in their practical and safety-related advantages over whole-cell therapies (Haider and Aramini, 2020). Unlike transplanted stem cells, exosomes do not replicate, eliminating concerns related to uncontrolled proliferation, ectopic tissue formation, or tumorigenicity (Harrell et al., 2021). They exhibit lower immunogenicity and greater stability during storage and systemic circulation, facilitating their formulation into off-the-shelf, standardized therapeutic products. Moreover, MSC-Exos can be engineered to enhance their targeting capabilities or loaded with specific therapeutic agents, thereby expanding their utility across diverse neurological indications (Haider, 2024). These attributes render them especially suitable for clinical application in neurocognitive disorders, where treatment requires precision, long-term safety, and effective CNS delivery.

Nonetheless, translating MSC-Exos from bench to bedside remains an intricate and multifaceted challenge (Habib et al., 2025a). While preclinical data are promising, several key scientific and technical hurdles must be overcome before clinical implementation becomes feasible. Standardization of exosome isolation, purification, and characterization protocols is critical, as variability in source MSC populations, culture conditions, and processing methods can significantly influence exosome composition and therapeutic efficacy. Determining optimal dosing regimens, administration routes (e.g., intravenous, intranasal, or intracerebral), and treatment frequency is another crucial area that requires systematic investigation. Furthermore, robust long-term safety studies are essential for evaluating potential off-target effects, biodistribution, and the persistence of therapeutic outcomes (Harrell et al., 2021).

From a regulatory perspective, MSC-Exos represent a novel and complex class of biologics that do not neatly fit into existing frameworks for small molecules, biologics, or cell-based therapies (Habib et al., 2026). The heterogeneity of exosome populations, combined with their multifaceted mechanisms of action, poses significant obstacles to standardization, quality control, and reproducibility—factors essential for regulatory approval and clinical translation (Habib et al., 2025a). Moreover, while several studies have elucidated key molecular pathways influenced by MSC-Exos, the precise mechanisms by which they exert their neuroprotective and neurorestorative effects remain incompletely understood (Xiao et al., 2025). Elucidating these mechanisms will not only enhance the rational design of exosome-based therapeutics but also enable the identification of predictive biomarkers for treatment responsiveness and patient stratification.

In light of these challenges and opportunities, the translational development of MSC-Exos requires a multidisciplinary, collaborative approach (Xiao et al., 2025). Bridging the gap between fundamental neuroscience, bioengineering, clinical neurology, and regulatory science will be essential to advance MSC-Exos along the clinical development pipeline (Harrell et al., 2021). This chapter will explore the current state of research on MSC-Exos in neurocognitive disease models, highlighting both the scientific progress achieved and the strategic hurdles yet to be addressed. It will also discuss emerging innovations, such as engineering exosomes to enhance targeting specificity, loading them with therapeutic nucleic acids or proteins, and integrating them into combination regimens with existing pharmacotherapies. Such strategies may significantly augment their therapeutic potential and broaden their applicability to a broader spectrum of CNS pathologies, including not only AD and PD but also cognitive deficits secondary to traumatic brain injury, stroke, multiple sclerosis, and chronic neuroinflammatory states. As the field evolves, MSC-derived exosomes may well constitute a transformative paradigm in the treatment of neurocognitive disorders, offering not only symptomatic relief but also genuine disease modification and neurorestoration.

13.3 MSC-EXOS AS NOVEL REMEDIES FOR NEUROCOGNITIVE DISORDER TREATMENT

In recent years, MSCs have emerged as a promising and versatile candidate for treating neurocognitive disorders, owing to their distinctive biological properties and multimodal mechanisms of action

(Soares et al., 2022). These adult multipotent stromal cells, originally isolated from bone marrow by Friedenstein and colleagues, have since been derived from a wide array of tissue sources, including adipose tissue, the umbilical cord (particularly Wharton's jelly), dental pulp, the placenta, synovial fluid, and the amniotic membrane (Kamal et al., 2022). Irrespective of their source, MSCs are commonly defined by three key criteria established by the International Society for Cellular Therapy (ISCT): adherence to plastic under standard culture conditions, the ability to differentiate into mesodermal lineages—namely osteoblasts, chondrocytes, and adipocytes—and the expression of a characteristic surface marker profile (typically CD73+, CD90+, and CD105+), alongside the absence of hematopoietic and endothelial markers such as CD34, CD45, CD14, CD11b, CD79α, and HLA-DR (Álvarez-Viejo and Haider, 2022).

What makes MSCs uniquely attractive for neurological applications, and particularly for the treatment of neurocognitive disorders, is their inherent capacity to interact dynamically with the surrounding microenvironment and exert context-dependent therapeutic effects (Miceli et al., 2021). Unlike terminally differentiated somatic cells, MSCs possess environmental sensing capabilities and respond to injury or inflammation by secreting a broad spectrum of bioactive molecules (Al-khani et al., 2022). This secretome includes, but is not limited to, anti-inflammatory cytokines (e.g., IL-10, TGF-β), chemokines (e.g., CXCL12), neurotrophic factors (e.g., BDNF, GDNF, NGF, HGF), antioxidative enzymes (e.g., superoxide dismutase, catalase), and EVs, including exosomes (Habib et al., 2025a, 2025b). These molecules act synergistically to modulate innate and adaptive immune responses, attenuate oxidative stress, suppress pro-inflammatory glial activation, and enhance intrinsic regenerative mechanisms within the CNS. In contrast to strategies focused on direct neuronal replacement, MSC-based therapies primarily operate via paracrine signaling, inducing neuroprotective, angiogenic, and neurogenic effects that collectively restore neural homeostasis (Miceli et al., 2021).

Preclinical studies employing diverse models of neurocognitive disorders have demonstrated the therapeutic efficacy of MSCs delivered via multiple administration routes—including intravenous, intranasal, intrathecal, and intracerebral. In murine and rodent models of AD, MSC administration has been associated with reduced amyloid-β (Aβ) plaque burden, decreased hyperphosphorylation of tau protein, mitigation of oxidative and inflammatory cascades, and measurable improvements in spatial learning and memory performance. These effects appear to be mediated not only by the downregulation of neurotoxic pathways but also by enhanced clearance of pathological aggregates via microglial phagocytosis and improved synaptic function. In models of PD, MSCs have demonstrated the capacity to improve dopaminergic neuronal survival within the substantia nigra pars compacta, increase dopamine concentrations in the striatum, and alleviate motor dysfunction. These beneficial outcomes have been mainly attributed to the secretion of neurotrophic factors, notably BDNF and GDNF, which promote neuronal survival, axonal integrity, and synaptic maintenance (Miceli et al., 2021).

A particularly noteworthy aspect of MSC therapy in the context of neurodegeneration is their ability to modulate glial phenotypes (Laroni et al., 2015). Chronic activation of microglia and astrocytes into pro-inflammatory states (M1 and A1 phenotypes, respectively) is a key contributor to sustained neuroinflammation and progressive neuronal damage in both AD and PD. MSCs have been shown to induce a phenotypic switch toward anti-inflammatory (M2 and A2) states, thereby reducing the release of cytotoxic mediators, such as TNF-α, IL-1β, nitric oxide, and reactive oxygen species (ROS), while increasing the production of reparative molecules, including IL-4, IL-13, and neurotrophic factors. Furthermore, MSCs have been implicated in stimulating endogenous neural stem/progenitor cells, thereby promoting neurogenesis in neurogenic niches, including the subventricular zone and the hippocampal dentate gyrus. They also contribute to neurovascular remodeling by promoting angiogenesis and stabilizing the neurovascular unit, effects crucial for restoring cerebral perfusion and meeting the metabolic demands of recovering neurons.

Despite the breadth of preclinical evidence supporting the therapeutic potential of MSCs, translating these cells for the treatment of neurocognitive disorders faces several significant limitations and unresolved challenges (Miceli et al., 2021). One of the most critical barriers is the limited survival, engraftment, and functional integration of MSCs following transplantation into the hostile CNS environment. The diseased brain is often characterized by a pro-inflammatory milieu, high oxidative stress, impaired vascularization, and disrupted extracellular matrix architecture, all of which compromise MSC viability and diminish their capacity to exert sustained therapeutic effects. Although some beneficial outcomes can be achieved through transient paracrine signaling, the short-lived presence of transplanted MSCs raises questions regarding the durability and reproducibility of clinical responses (Miceli et al., 2021; Soares et al., 2022).

Moreover, the ability of systemically administered MSCs to effectively home to sites of CNS pathology remains controversial (Miceli et al., 2021). While specific studies suggest that chemokine gradients and adhesion molecules facilitate targeted migration across the BBB, others have reported that the majority of intravenously delivered MSCs are sequestered in peripheral organs such as the lungs, spleen, and liver—a phenomenon known as the "first-pass effect." This inefficiency often necessitates direct CNS administration (e.g., intracerebral or intrathecal injection), which, although more effective in delivering cells to the target tissue, carries greater procedural risk, higher cost, and reduced patient acceptability, particularly in chronic diseases requiring repeated treatments.

Another critical issue pertains to the intrinsic heterogeneity of MSC populations. The biological activity and therapeutic profile of MSCs can vary markedly based on numerous factors, including their tissue source, donor age and health status, isolation and expansion protocols, passage number, and cryopreservation methods. Such variability poses significant challenges to standardization, reproducibility, and quality control—parameters essential for Good Manufacturing Practice (GMP) compliance and regulatory approval. Furthermore, prolonged in vitro culture can induce cellular senescence, telomere shortening, and loss of immunomodulatory function, which negatively impact their therapeutic efficacy and scalability for clinical use (Miceli et al., 2021; Soares et al., 2022).

Additional safety concerns, although relatively limited for MSCs compared to other stem cell types such as induced pluripotent stem cells (iPSCs), must also be taken into account (Omole et al., 2022). These include the potential for unwanted differentiation, ectopic tissue formation, and rare but possible immune reactions in allogeneic transplantation settings (Ahmed et al., 2025). While autologous MSCs circumvent some of these risks, they are less practical in elderly patients with neurodegenerative diseases, where stem cell function may already be compromised (Volarevic et al., 2018). Finally, the regulatory landscape for cell-based therapies remains complex and evolving. MSC products must meet rigorous safety, identity, potency, and purity criteria. Developers must demonstrate consistent manufacturing processes, robust preclinical safety data, and long-term monitoring plans for clinical trials. The high cost, lengthy development timelines, and uncertain reimbursement frameworks further hinder the broad adoption of MSC-based treatments.

Given these limitations, there is growing interest in developing alternative or complementary approaches that harness the therapeutic potential of MSCs while circumventing their inherent drawbacks (Harrell et al., 2021). Among the most promising of these are cell-free therapies based on the MSC secretome, particularly the use of EVs, such as MSC-Exos, which retain many of the beneficial properties of MSCs without the associated risks of live cell transplantation. This emerging strategy offers the potential for safer, more controllable, and scalable interventions for neurocognitive diseases, opening new avenues for innovation in regenerative neurology. In recent years, MSC-Exos have gained considerable attention as a cell-free therapeutic platform. These nanoscale EVs, typically ranging from 30 to 150 nanometers in diameter, originate from the endosomal compartment of MSCs and are released upon the fusion of multivesicular bodies (MVBs) with the plasma membrane. MSC-Exos are morphologically and biogenetically distinct from other EV subsets, such as microvesicles, which bud directly from the plasma membrane, and apoptotic bodies, which result from programmed cell death. Structurally, MSC-Exos are spherical vesicles surrounded by a lipid bilayer that protects their molecular cargo and facilitates interaction with recipient cells. Under transmission electron microscopy (TEM), exosomes often appear cup-shaped—an artifact introduced by the sample preparation process. In contrast, size distribution and concentration can be quantitatively assessed using nanoparticle tracking analysis (NTA) and dynamic light scattering (DLS) (Harrell et al., 2021).

Phenotypically, MSC-Exos express a set of canonical exosomal markers, including the tetraspanins CD9, CD63, and CD81, as well as components of the ESCRT (endosomal sorting complexes required for transport) machinery, such as TSG101 and Alix. Additionally, they contain chaperone proteins, such as HSP70 and HSP90, which play roles in vesicle stability and cargo loading (Harrell et al., 2021). Importantly, MSC-Exos also retain surface markers representative of their parent MSCs, including CD73, CD90, and CD105, enabling both lineage-specific identification and potential functional interactions with target tissues. This dual signature is advantageous for tracing and characterizing MSC-Exos both in vitro and in vivo, facilitating their study in therapeutic models and eventual clinical translation.

Functionally, MSC-Exos recapitulate many of the paracrine-mediated therapeutic effects of their parent MSCs. These vesicles actively participate in intercellular communication and exert broad immunomodulatory, anti-inflammatory, and neuroprotective effects. In models of neurocognitive disorders, MSC-Exos have been shown to attenuate neuroinflammation, suppress neuronal apoptosis, and enhance synaptic plasticity and cognitive function. They achieve these effects in part by influencing the phenotype and activity of microglia and macrophages—shifting them from pro-inflammatory (M1-like) states toward anti-inflammatory (M2-like) phenotypes—and by promoting the secretion of anti-inflammatory cytokines such as interleukin-10 (IL-10) and transforming growth factor-beta (TGF-β), while simultaneously suppressing pro-inflammatory mediators including tumor necrosis factor-alpha (TNF-α), interleukin (IL)-1 beta (IL-1β), and IL-6 (Harrell et al., 2021).

The molecular cargo contained within MSC-Exos is highly heterogeneous and functionally rich, comprising proteins, lipids, messenger RNAs (mRNAs), miRNAs, long non-coding RNAs (lncRNAs), and even fragments of genomic or mitochondrial DNA (Harrell et al., 2021). Of particular interest is the miRNA content, which has been implicated in regulating key cellular pathways involved in inflammation, apoptosis, neurogenesis, and synaptic function. Notably, miRNAs such as miR-21, miR-146a, and miR-124 have been identified within MSC-Exos. They play crucial roles in modulating neuroinflammation, enhancing neuronal survival, and promoting neural progenitor differentiation. Proteomic profiling of MSC-Exos has further revealed the presence of growth factors, adhesion molecules, metabolic enzymes, and matrix remodeling proteins, all of which contribute to their regenerative and neuroprotective properties. Critically, MSC-Exos possess the ability to cross the BBB. This feature enables them to deliver their therapeutic cargo directly to CNS cells, including neurons, astrocytes, microglia, and endothelial cells, through mechanisms such as receptor-mediated endocytosis, transcytosis, or fusion with the target cell membrane.

From a translational standpoint, MSC-Exos offer several strategic and practical advantages over whole-cell therapies. As acellular products, they circumvent key safety concerns associated with live MSC transplantation, such as uncontrolled proliferation, differentiation into unintended lineages, immune rejection, and tumorigenicity. Being non-replicative and non-integrative, exosomes present a substantially lower risk profile while maintaining therapeutic efficacy through the delivery of bioactive cargo. Additionally, MSC-Exos demonstrate superior physicochemical stability compared to live cells. They can be cryopreserved at −80°C or lyophilized for extended storage with minimal functional degradation, thereby facilitating large-scale production, standardized quality control, and integration into existing pharmaceutical frameworks. Their small size and lipid-based structure allow for efficient penetration of physiological barriers, including the BBB, and enable less invasive administration routes (e.g., intravenous or intranasal), improving patient compliance and expanding clinical applicability (Harrell et al., 2022).

Furthermore, advances in nanomedicine and bioengineering have significantly enhanced the therapeutic potential of MSC-Exos (Harrell et al., 2024a, 2025). Techniques such as electroporation, sonication, or passive incubation can be employed to load exosomes with exogenous therapeutic agents, including small-molecule drugs, siRNAs, antisense oligonucleotides, and synthetic miRNAs (Harrell et al., 2024b). Additionally, genetic modification of parent MSCs can yield exosomes enriched with specific therapeutic payloads or surface ligands engineered for tissue-specific targeting. These innovations not only enhance the precision and potency of MSC-Exos but also facilitate the development of customizable exosome-based therapies tailored to the unique pathophysiological profiles of individual CNS diseases.

Given their capacity to deliver a wide range of therapeutic molecules, cross biological barriers, modulate immune and inflammatory pathways, and avoid many of the safety and logistical concerns associated with cell therapies, MSC-Exos represent a paradigm shift in the treatment of neurocognitive disorders. By combining the regenerative potential of MSCs with the scalability and safety of acellular therapies, MSC-Exos hold the potential to bridge the gap between experimental findings and clinical translation, offering a next-generation strategy for addressing complex and currently intractable neurodegenerative diseases such as Alzheimer's disease, Parkinson's disease, and other forms of dementia (Harrell et al., 2021, 2025).

13.4 THERAPEUTIC POTENTIAL OF MSC-EXOS IN THE TREATMENT OF PARKINSON'S DISEASE

13.4.1 Parkinson's Disease: Pathological Mechanisms, Immune Involvement, and Current Therapies

PD is a chronic and progressively debilitating neurodegenerative condition marked by both cellular and molecular hallmarks (Morris et al., 2024). A key pathological feature of PD is the selective and progressive degeneration of dopaminergic neurons located in the substantia nigra pars compacta (SNpc), a midbrain region essential for motor control (Bloem et al., 2021). One of the signature findings in the brains of individuals with PD is the accumulation of Lewy bodies, intracellular inclusions composed primarily of aggregated alpha-synuclein (α-syn), a presynaptic protein prone to misfolding and pathological accumulation within neurons (Morris et al., 2024). These aggregates interfere with normal cellular functions, including protein degradation systems and mitochondrial activity, ultimately impairing neuronal viability. The resulting loss of dopamine-producing neurons leads to a profound reduction in striatal dopamine levels, a key component of the basal ganglia involved in movement regulation (Bloem et al., 2021). The disruption of dopaminergic neurotransmission gives rise to the classical motor symptoms of PD, including resting tremors, muscular rigidity, bradykinesia (slowness of movement), and postural instability (Morris et al., 2024). Beyond motor deficits, many patients also experience a variety of non-motor symptoms, such as cognitive impairment, depression, anxiety, and disturbances in sleep and autonomic functions. These non-motor manifestations are believed to result from widespread neurodegeneration and reactive gliosis, which exacerbate neuronal loss and disrupt neurotransmitter homeostasis (Bloem et al., 2021).

The etiology of PD is multifactorial, arising from a complex interplay between genetic predisposition and environmental factors. Familial forms of PD, though less common than sporadic cases, have been linked to mutations in several key genes involved in neuronal maintenance and protein processing. Among the most studied are SNCA (which encodes alpha-synuclein), LRRK2 (Leucine-rich repeat kinase 2), and PARK7 (DJ-1). Mutations in SNCA are typically associated with early-onset PD and lead to overproduction or misfolding of α-synuclein, accelerating its aggregation (Morris et al., 2024). Meanwhile, LRRK2 mutations disrupt kinase activity that is crucial for synaptic vesicle trafficking and dopamine release, resulting in toxic gain-of-function effects and neuronal damage (Bloem et al., 2021). In addition, autosomal recessive forms of PD have been linked to mutations in genes PINK1 (PTEN-induced kinase 1), PRKN (Parkin), and DJ-1, all of which play critical roles in mitochondrial dynamics and cellular stress responses. Dysfunction in these genes compromises the cell's ability to manage oxidative stress and maintain mitochondrial integrity.

Environmental factors also contribute significantly to PD pathogenesis. Long-term exposure to neurotoxic substances, such as pesticides, heavy metals, and industrial chemicals, has been linked to an increased risk of PD. These toxins may trigger oxidative damage, mitochondrial dysfunction, or direct dopaminergic toxicity. Aging itself stands as the most significant risk factor for PD, given that the likelihood of disease onset rises sharply with age. This suggests that age-related declines in proteostasis, mitochondrial efficiency, and immune regulation may converge to promote neurodegeneration over time (Morris et al., 2024).

A major pathological contributor to disease progression is chronic neuroinflammation, which is driven by both innate and adaptive immune responses. Microglia, the resident immune cells of the central nervous system, play a central role in this process. Upon sensing misfolded α-synuclein and other neuronal damage-associated molecules, microglia become activated and adopt a pro-inflammatory phenotype (Morris et al., 2024). In this activated state, they release a cascade of pro-inflammatory cytokines—including TNF-α, IL-1β, and IL-6—that amplify oxidative stress and disrupt the local microenvironment, promoting further injury to already vulnerable dopaminergic neurons. This inflammatory milieu extends beyond microglia. Activated microglia also influence the integrity of the BBB by increasing the expression of adhesion molecules, such as E- and P-selectins, on endothelial cells, thereby facilitating the infiltration of peripheral immune cells. Consequently, circulating T lymphocytes, particularly IFN-γ–producing Th1 cells and IL-17–secreting Th17 cells, can enter the brain and further escalate neuroinflammatory responses. This crosstalk between CNS-resident immune cells and infiltrating peripheral immune populations creates a vicious cycle that exacerbates neuronal degeneration and impairs the brain's ability to initiate neurorestorative processes such as neurogenesis, synaptic remodeling, and plasticity (Bloem et al., 2021). Neuroinflammation not only perpetuates cell death but also disrupts neurotransmitter systems critical to PD pathology. Inflammatory cytokines can directly impair dopamine synthesis and release, thereby exacerbating motor deficits. Furthermore, these inflammatory mediators have been associated with mood disorders, cognitive decline, and other non-motor symptoms frequently observed in PD, linking immune dysregulation with clinical heterogeneity. Given the central role of inflammation in disease progression, therapeutic strategies that modulate neuroinflammation are gaining momentum as potentially disease-modifying approaches for PD management (Morris et al., 2024).

Currently, therapeutic interventions for PD are symptomatic, targeting motor impairments rather than halting the underlying neurodegenerative process (Ellis et al., 2021). The gold standard remains levodopa, a dopamine precursor that significantly alleviates symptoms such as bradykinesia and rigidity once it crosses the BBB and is converted into dopamine in the brain (Ellis et al., 2021; Vijiaratnam et al., 2021). However, chronic levodopa use is often associated with complications such as motor fluctuations and levodopa-induced dyskinesia. To mitigate these issues, levodopa is frequently co-administered with carbidopa, which inhibits peripheral conversion of levodopa, thereby increasing central availability. Additional pharmacological agents include dopamine agonists (e.g., pramipexole, ropinirole), monoamine oxidase-B (MAO-B) inhibitors (e.g., selegiline, rasagiline), and COMT inhibitors, which prolong levodopa's effects by modulating its metabolism. In parallel, non-pharmacological therapies, such as physical, occupational, and speech therapy, are vital for preserving mobility, independence, and quality of life in patients with PD (Ellis et al., 2021). These approaches address specific functional impairments and contribute to overall well-being.

Despite the symptomatic benefits of current therapies, there is growing recognition of the need for neuroprotective and disease-modifying treatments. MAO-B inhibitors, such as selegiline and rasagiline, have shown some promise in reducing oxidative stress and preserving dopaminergic neurons in early-stage PD. More recently, experimental strategies targeting the underlying mechanisms of neurodegeneration, such as gene therapy to enhance dopamine synthesis or mitochondrial function, and immunotherapies aimed at modulating α-synuclein accumulation, have been under investigation in both preclinical models and early-phase clinical trials (Ellis et al., 2021). An exciting frontier involves regulating neuroinflammation and immune activity using advanced delivery platforms, such as viral vectors, exosomes, or nanoparticles, engineered to deliver anti-inflammatory cytokines, neurotrophic factors, or gene-editing tools directly to affected brain regions. These approaches aim to protect vulnerable neuronal populations, restore immune homeostasis, and support long-term brain resilience. The ongoing exploration of such innovative strategies holds promise for fundamentally altering the trajectory of PD, offering hope for more effective, personalized, and disease-modifying therapies in the near future.

13.4.2 MSC-Exos-Derived Immunoregulatory Cytokines and Growth Factors for PD Attenuation

MSC-Exos have emerged as a promising therapeutic approach for PD, leveraging their ability to modulate neuroinflammation, enhance neuronal survival, and restore dopaminergic function (Volarevic et al., 2025). These vesicles encapsulate a diverse array of bioactive molecules, including proteins, lipids, and nucleic acids, which collectively contribute to their therapeutic efficacy. Neuroinflammation increases BBB permeability, facilitating the passage of MSC-Exos into the CNS. MSC-Exos express various chemokine receptors, including CCR1, CCR2, CCR4, and CXCR4, which interact with corresponding ligands such as CCL2 and CXCL12, which are upregulated in the inflamed endothelial cells of the PD brain (Morris et al., 2024). This chemotactic response enables MSC-Exos to home in on areas of neuronal injury and inflammation, delivering therapeutic agents directly to affected regions (Volarevic et al., 2025).

MSC-Exos are enriched with neuroprotective factors, including BDNF, GDNF, and NGF, which support the growth, survival, and differentiation of dopaminergic neurons. Additionally, MSC-Exos-derived enzymes, such as superoxide dismutase (SOD) and catalase, mitigate oxidative stress by neutralizing ROS, thereby protecting neurons from oxidative damage (Volarevic et al., 2025). MSC-Exos-sourced miR-133b, miR-21, and miR-124 regulate gene expression to promote neuronal resilience and inhibit apoptosis. Beyond neuroprotection, MSC-Exos possess immunomodulatory properties. They contain cytokines such as interleukin-1 receptor antagonist (IL-1Ra), interleukin (IL)-10, transforming growth factor-beta (TGF-β), and HGF, which attenuate neuroinflammation by suppressing pro-inflammatory cytokine production and promoting anti-inflammatory responses. These immunomodulatory effects help restore immune homeostasis within the CNS, further supporting neuronal health. MSC-Exos-derived IL-1Ra plays a critical role in mitigating neuroinflammation. IL-1Ra, an endogenous inhibitor of the pro-inflammatory cytokine interleukin-1β (IL-1β), is notably upregulated in PD and is closely linked to disease progression. By competitively binding to IL-1 receptors on target cells without activating them, MSC-Exos-sourced IL-1Ra prevents IL-1β from initiating its pro-inflammatory signaling cascade. This blockade is significant because IL-1β signaling amplifies neuroinflammatory responses, leading to increased neuronal injury, glial activation, and further recruitment of peripheral immune cells to the brain. By disrupting this pathway, IL-1Ra helps preserve neural tissue and maintain a less reactive, more stable CNS environment.

Moreover, MSC-Exos contain MSC-sourced IL-10, a potent immunoregulatory cytokine with broad anti-inflammatory effects (Porro et al., 2020). IL-10 suppresses the synthesis and release of several primary inflammatory mediators, including TNF-α, IL-1β, and IL-6, which are typically produced by activated macrophages and neutrophils during neuroinflammatory episodes in PD. By inhibiting these cytokines, IL-10 not only reduces the intensity of the inflammatory response but also supports neuronal resilience by curbing the neurotoxic effects associated with chronic inflammation. Beyond cytokine inhibition, MSC-Exos-derived IL-10 exerts profound modulatory effects on the brain-infiltrated immune cells. Notably, IL-10 promotes a phenotypic shift in microglia, from a pro-inflammatory M1 state to a neuroprotective, anti-inflammatory M2 phenotype. This polarization is associated with enhanced phagocytosis of cellular debris, increased secretion of trophic factors, and decreased levels of harmful pro-inflammatory molecules. Through these actions, IL-10 contributes to neuroprotection and tissue homeostasis, fostering an environment favorable for neuronal repair and survival (Porro et al., 2020).

In synergy with IL-10, TGF-β, another anti-inflammatory cytokine secreted by MSCs, further shapes the immune microenvironment. Both IL-10 and TGF-β induce a tolerogenic state in dendritic cells (DCs), promoting their transformation into a phenotype that limits T cell activation and supports immune tolerance. Additionally, these cytokines promote an alternative, M2-like macrophage activation, reducing tissue-destructive inflammation and facilitating regeneration (Roussa et al., 2009; Huang et al., 2022). Together, these effects establish an immunosuppressive milieu within the inflamed brain that effectively shields vulnerable dopaminergic neurons from immune-mediated damage. Another key immunoregulatory component of MSC-Exos is indoleamine 2,3-dioxygenase (IDO). This enzyme plays a crucial role in modulating adaptive immune responses, particularly by shaping T cell phenotypes. IDO catalyzes the degradation of tryptophan, an amino acid essential for T cell proliferation, thereby indirectly suppressing the expansion of pro-inflammatory T helper 1 (Th1) and T helper 17 (Th17) lymphocytes, which are implicated in chronic neuroinflammation in PD. Furthermore, the metabolic changes induced by IDO activity favor the differentiation and expansion of regulatory T cells (Tregs). These Tregs exert potent suppressive functions on autoreactive and inflammatory T cell populations, both through direct cell-cell contact and by secreting immunosuppressive cytokines. Thus, IDO contributes to the establishment of long-lasting immunotolerance and protection of neural tissues from T cell-driven injury (Roussa et al., 2009; Huang et al., 2022).

MSC-Exos are also enriched with tumor necrosis factor-stimulated gene-6 (TSG-6) and HGF, two additional molecules with demonstrated neuroprotective and anti-inflammatory properties. TSG-6 is particularly effective in mitigating oxidative stress and has been shown to counteract neurotoxic insults, such as those induced by 1-methyl-4-phenylpyridinium (MPP+), a mitochondrial toxin used to model PD in experimental systems. TSG-6 modulates intracellular signaling pathways, including the STAT3/miR-7/LRRK2 axis, to suppress oxidative damage, inhibit apoptosis, and prevent α-SN aggregation. This signaling cascade is critical in PD pathology, as dysregulation of LRRK2 is linked to both familial and sporadic forms of the disease. Similarly, HGF released via MSC-Exos promotes survival and functional recovery of dopaminergic neurons by exerting anti-apoptotic, anti-inflammatory, and regenerative effects. One of its key mechanisms is the suppression of microglial activation. By dampening microglial-mediated inflammatory responses, HGF indirectly protects neurons from the cytotoxic consequences of chronic glial activation. HGF also supports axonal growth, enhances synaptic plasticity, and stimulates angiogenesis, further contributing to the restoration of neural circuits disrupted by PD-related degeneration (Volarevic et al., 2025).

Taken together, the complex cargo of MSC-Exos, comprising cytokines, enzymes, and growth factors, interacts with various components of the brain's immune and neuronal networks to orchestrate a multifaceted therapeutic response (Huang et al., 2022; Volarevic et al., 2025). By modulating innate and adaptive immunity, reducing oxidative damage, promoting neuronal repair, and restoring homeostasis in the inflamed microenvironment of the PD brain, MSC-Exos present a promising, biologically sophisticated alternative to traditional mono-target pharmacological interventions. These features highlight their potential to alter the trajectory of neurodegenerative diseases like PD fundamentally.

13.4.3 The Beneficial Effects of MSC-Exos-Derived miRNAs in the Alleviation of PD

Recent research highlights the crucial role of MSC-exosome-derived miRNAs in slowing the progression of PD. These miRNAs play a vital role in modulating key cellular pathways, including the oxidative stress response and autophagy, thereby offering therapeutic potential for neuroprotection. In a murine model of PD induced by 1-methyl-4-phenyl-1,2,3,6-tetrahydropyridine (MPTP), Exos derived from trophoblast stage-derived MSCs (T-MSCs-Exos) demonstrated significant therapeutic effects (He et al., 2023). Upon intravenous administration, these Exos traversed the BBB and localized within the SNpc, where dopaminergic neurons internalized them through endocytosis. This uptake enhanced neuronal viability and function, thereby improving motor performance in MPTP-treated mice. Sequencing analyses identified several miRNAs enriched in T-MSCs-Exos, with miR-100-5p being particularly abundant. Bioinformatics predictions suggest that miR-100-5p targets the 3′ untranslated region (UTR) of the NOX4 gene, resulting in its downregulation. NOX4 is a NADPH oxidase that, when activated, contributes to the generation of ROS, thereby inducing oxidative stress and neuronal damage in PD. By suppressing NOX4 expression, miR-100-5p alleviates ROS accumulation, mitigates mitochondrial dysfunction, and reduces neuronal apoptosis. Furthermore, this modulation promotes the dissociation of the nuclear factor erythroid 2-related factor 2 (Nrf2) from its inhibitor, Keap1, facilitating Nrf2 nuclear translocation and the subsequent activation of antioxidant genes such as superoxide dismutase 1 (SOD1) and SOD2. Collectively, these actions diminish oxidative stress, protect dopaminergic neurons, and ameliorate motor deficits in PD mice (He et al., 2023).

In a 6-hydroxydopamine (6-OHDA)-induced rat model of PD, Exos derived from adipose tissue-derived MSCs (AT-MSCs-Exos) were shown to harbor miR-23b-3p, which plays a crucial role in autophagy induction within dopaminergic neurons (Geng et al., 2023a). Autophagy is a cellular process responsible for the degradation and recycling of damaged proteins and organelles, including α-SN. The activation of the Wnt/β-catenin signaling pathway is known to promote autophagy. In this context, miR-23b-3p enhances β-catenin phosphorylation, leading to its stabilization and nuclear translocation, and thereby activating transcription of autophagy-related genes such as Beclin-1 and LC3B. This cascade facilitates the clearance of α-SN aggregates and dysfunctional mitochondria, thereby reducing oxidative stress and preserving neuronal integrity. The therapeutic efficacy of miR-23b-3p was further confirmed by the administration of ICG-001, an inhibitor of the Wnt/β-catenin pathway, which abrogated the beneficial effects of AT-MSCs-EVs, underscoring the critical role of Wnt/β-catenin signaling in miR-23b-3p-mediated neuroprotection (Geng et al., 2023b).

Using a mouse model of MPTP-induced PD, Li and colleagues demonstrated that the neuroprotective effects of AT-MSC-Exos were primarily mediated through miR-188-3p (Li et al., 2021). This microRNA exerted its therapeutic influence by simultaneously targeting two critical pathological mechanisms: the NLRP3 inflammasome, which drives neuroinflammation, and cell division protein kinase 5 (CDK5), which contributes to pyroptotic cell death of dopaminergic neurons. In neurodegenerative conditions, both CDK5 and the NLRP3 inflammasome are aberrantly activated in response to mitochondrial dysfunction and the accumulation of neurotoxic proteins, such as α-synuclein (Restrepo-Martinez et al., 2025). CDK5 activation leads to abnormal tau phosphorylation, disrupting cytoskeletal stability and compromising neuronal structure, thereby increasing susceptibility to pyroptosis (Restrepo-Martinez et al., 2025). Concurrently, the activation of the NLRP3 inflammasome in microglia leads to the heightened production of pro-inflammatory cytokines, particularly IL-1β and IL-18, which exacerbate neuroinflammatory damage within the substantia nigra pars compacta (SNpc). Together, these two molecular cascades establish a highly toxic neural environment that accelerates the demise of dopaminergic neurons and worsens disease progression (Restrepo-Martinez et al., 2025). Li and colleagues found that miR-188-3p, present in AT-MSC-Exos, bound directly to the mRNAs of both CDK5 and NLRP3, leading to their translation repression and subsequent transcript degradation (Li et al., 2021). Through this dual targeting, miR-188-3p effectively suppressed CDK5-driven tau hyperphosphorylation and halted NLRP3 inflammasome activation in microglia. As a result, inflammation and pyroptosis were attenuated, helping preserve the viability and function of dopaminergic neurons and slowing disease progression in the animal model of PD.

While levodopa remains a cornerstone in PD therapy due to its ability to replenish dopamine and its inherent antioxidant properties, it is not without limitations. It fails to address non-dopaminergic symptoms and can cause side effects with long-term use (Ellis et al., 2021). In light of these shortcomings, Mohamed and colleagues conducted a comparative study in a rat model of PD to evaluate the efficacy of MSC-Exos compared with levodopa (Mohamed et al., 2023). The MSC-Exos were administered intravenously once weekly for 3 weeks. Remarkably, the treatment not only inhibited α-SN aggregation but also outperformed levodopa in restoring both motor and cognitive functions in PD-afflicted rats. The therapeutic impact of MSC-Exos in this context was attributed to their ability to modulate gene expression via non-coding RNAs. Specifically, MSC-Exos downregulated the expression of circular RNA (circRNA).2837 while simultaneously upregulating miR-34b in the brains of treated animals. MiR-34b binds to the 3'-untranslated region (3'-UTR) of α-SN mRNA, suppressing its translation and thereby reducing the accumulation of α-SN protein. In contrast, circRNA.2837 acts as a competing endogenous RNA, or molecular "sponge," that sequesters miR-34b, thereby neutralizing its inhibitory effect on α-SN expression (Mohamed et al., 2023). When circRNA.2837 levels rise, and miR-34b levels fall, this imbalance promotes α-SN aggregation in the SNpc, leading to motor symptoms such as bradykinesia and postural instability, which are commonly observed in PD patients (Restrepo-Martinez et al., 2025). By restoring the balance, reducing circRNA.2837 expression and boosting miR-34b levels, MSC-Exos effectively interrupted this pathological feedback loop. This resulted in diminished α-SN accumulation and significant improvement in motor performance in PD rats, demonstrating the promise of MSC-Exos as a potent therapeutic alternative to traditional PD treatments (Venugopal et al., 2017).

13.4.4 Experimental Evidence of the Therapeutic Efficacy of MSC-Exos in PD Treatment

Recent experimental studies have provided compelling evidence supporting the efficacy of MSC-Exos in PD treatment (Volarevic et al., 2025), and in a study conducted by Huang et al., intravenous administration of human umbilical cord MSC-derived Exos EVs (UC-MSC-Exos) in a mouse model of PD resulted in the vesicles crossing the BBB and accumulating in SNpc (Huang et al., 2024). These Exos were internalized by dopaminergic neurons, leading to enhanced neuronal activity and improved motor functions (Huang et al., 2024). Additionally, UC-MSC-Exos modulated the phenotype and function of astrocytes and microglia, inhibiting their activation and reducing the production of pro-inflammatory cytokines. This suppression of neuroinflammation protected dopaminergic neurons and prevented PD progression in the experimental mice. However, the study by Huang et al. primarily focused on the effects of UC-MSC-Exos on microglia-driven neuroinflammation, without addressing their impact on adaptive immunity, particularly T cell responses in the inflamed brains of PD mice. Given the crucial role of T cell-microglia interactions in PD pathogenesis (Soares et al., 2022), future studies should investigate the effects of UC-MSC-Exos on both innate and adaptive immune responses to fully understand their therapeutic potential in PD treatment.

Supporting these findings, Zhang et al. demonstrated that intravenously administered UC-MSC-Exos successfully crossed the BBB and accumulated in dopaminergic neurons and microglia of 6-OHDA-treated PD rats. MSC-Exos suppressed microglial activation and reduced the production of inflammatory cytokines, including IL-1β and IL-18, thereby alleviating inflammation-induced injury to dopaminergic neurons. Consequently, the treated rats exhibited improved neuronal viability, increased levels of dopamine and serotonin, and enhanced recovery of motor function (Zhang et al., 2023). Despite these promising results, the study by Zhang et al. did not assess the impact of UC-MSC-Exos on the pathological accumulation of α-SN. Accordingly, future research should explore whether UC-MSC-Exos can modulate α-SN aggregation and clearance, as this could further elucidate their therapeutic mechanisms and potential in treating PD.

Aliakbari and colleagues engineered UC-MSC-Exos to deliver hydrophobic compounds, baicalein and oleuropein, via zwitterionic nanoliposomes (NLPs), enhancing MSC-Exos-based neuroprotection in PD treatment (Aliakbari et al., 2024). Baicalein, a flavonoid derived from the root of Scutellaria baicalensis, has demonstrated the ability to inhibit α-SN fibrillation and aggregation, processes implicated in PD pathogenesis. Oleuropein, a polyphenolic compound found in olive leaves and oil, exhibits potent antioxidant properties, scavenging ROS and activating neuroprotective pathways such as the BDNF/CREB/Akt signaling cascade. To enhance the delivery of these hydrophobic compounds to the brain, Aliakbari et al. developed hybrid nanoparticles by encapsulating baicalein and oleuropein within NLPs, which were then loaded into UC-MSC-Exos. These UC-MSC-Exos-NLP hybrids effectively crossed the BBB and targeted dopaminergic neurons. Notably, the incorporation of NLPs stabilized baicalein and oleuropein, preserving their antioxidant and neuroprotective activities. The UC-MSC-Exos-NLP-baicalein and UC-MSC-Exos-NLP-oleuropein hybrids successfully prevented αSN fibrillation and aggregation in dopaminergic neurons, thereby mitigating neurotoxicity associated with αSN accumulation. These findings underscore the potential of MSC-Exos as vehicles for targeted delivery of therapeutic agents to the brain. The combination of UC-MSC-Exos with NLPs and bioactive compounds offers a promising strategy for enhancing the efficacy of PD treatments. However, further experimental and clinical studies are necessary to confirm the long-term neuroprotective effects of these hybrid formulations and to evaluate their safety and efficacy in human populations.

Using the 6-OHDA model of PD, Chen and colleagues investigated the therapeutic potential of intravenously administered UC-MSC-Exos to restore dopamine synthesis in damaged dopaminergic neurons in the SNpc (Chen et al., 2020). Their findings revealed that UC-MSC-Ex significantly reduced apomorphine-induced rotational asymmetry in treated rats and decreased the number of apoptotic dopaminergic neurons, suggesting a neuroprotective effect. Histopathological analysis of SNpc tissues from UC-MSC-Exos-treated rats revealed the absence of classical markers of neuronal death, such as cytoplasmic eosinophilia, pyknosis, and karyolysis, which were prevalent in untreated PD rats. The SNpc regions in MSC-Exos-treated animals contained well-preserved multipolar neurons characterized by distinct nucleoli and basophilic, granular cytoplasm, indicating restored neuronal integrity. Biochemical assessments further supported these observations, showing that UC-MSC-Exos markedly elevated levels of dopamine and its primary metabolites, 3,4-dihydroxyphenylacetic acid (DOPAC) and homovanillic acid (HVA), in the striatum of rats with 6-OHDA-induced injury (Chen et al., 2020). These changes indicate a substantial restoration of dopaminergic function following UC-MSC-exosome-based therapy. While these results are promising and suggest that UC-MSC-Exos can reverse neurochemical deficits and promote recovery of dopaminergic neurons in PD models, the specific intracellular signaling pathways and molecular mechanisms underlying these effects remain unclear. Future experimental studies should aim to clarify the exact molecular targets and signaling cascades activated in dopaminergic neurons by UC-MSC-Exos, thereby optimizing and refining this regenerative strategy for clinical application in PD treatment.

In a study using the α-SN A53T transgenic mouse model of PD, Xu and colleagues investigated the therapeutic potential of bone marrow-derived MSC-exosomes (BM-MSC-Exos) in restoring dopaminergic function (Xu et al., 2022a). Their findings revealed that BM-MSC-Exos significantly enhanced motor, learning, and memory abilities in these mice by modulating phospholipid composition and cholesterol metabolism in hippocampal neurons. Cholesterol plays a pivotal role in the CNS, accounting for approximately 25% of the body's total cholesterol. It is essential for lipid raft formation. However, dysregulated cholesterol metabolism can lead to the accumulation of neurotoxic metabolites such as 24S-hydroxycholesterol and 27-hydroxycholesterol (Tong et al., 2024). These oxidized derivatives can disrupt neuronal function by inducing oxidative stress, promoting inflammation, and impairing mitochondrial function, all of which are detrimental to dopaminergic neurons. Intrastriatal injection of BM-MSC-Exos in α-SN A53T transgenic mice resulted in a decrease in the levels of 24S-hydroxycholesterol and an increase in total cholesterol levels in the brain. Lipidomic analysis using liquid chromatography-mass spectrometry revealed that BM-MSC-Exos modulated brain cholesterol metabolism by downregulating the relative percentages of phosphatidylglycerol and phosphatidylethanolamine, and upregulating phosphatidylinositol, phosphatidylserine, and phosphatidylcholine. These changes in phospholipid composition are critical for maintaining membrane integrity and supporting neuronal function. These findings underscore the importance of lipid metabolism in neurodegenerative diseases and highlight the potential of BM-MSC-Exos as a therapeutic strategy for PD. By modulating lipid composition and cholesterol metabolism, BM-MSC-Exos may restore neuronal function and slow disease progression, offering a promising avenue for future PD treatments (Xu et al., 2022a).

The therapeutic efficacy of BM-MSC-Exos in PD treatment was compellingly demonstrated by Cai and colleagues, who showed that these vesicles significantly suppressed microglia-mediated neuroinflammation and decelerated disease progression in animal models (Cai et al., 2022). A key mechanism underlying this effect was the delivery of Glioma-associated oncogene homolog 1 (Gli-1) to activated microglia. Through targeted delivery, BM-MSC-Exos reprogrammed pro-inflammatory microglia into a more immunosuppressive and neuroprotective phenotype. This reprogramming process involved upregulating anti-inflammatory cytokines IL-10 and TGF-β while simultaneously suppressing TNF-α and IL-6, both of which are known to contribute to neurotoxicity and dopaminergic neuron injury. By shifting the microglial response from pro-inflammatory to anti-inflammatory, BM-MSC-Exos helped alleviate neuroinflammatory damage in the SNpc. Beyond their immunomodulatory role, BM-MSC-Exos-derived Gli-1 also exerted direct neuroprotective effects on dopaminergic neurons. Specifically, Gli-1 was found to inhibit the activation of specificity protein 1 (SP-1), a transcription factor that is typically upregulated in response to oxidative stress and inflammation. SP-1 plays a critical role in promoting neuronal apoptosis by inducing the expression of the pro-apoptotic Bax gene. In this context, Gli-1-mediated suppression of SP-1 effectively reduced Bax expression, thereby preventing programmed cell death in dopaminergic neurons and preserving neuronal viability (Cai et al., 2022).

MSC-Exos have also shown notable potential in alleviating sleep disturbances, one of the most prevalent and debilitating non-motor symptoms associated with PD (Li et al., 2023). In a 6-OHDA rat model of PD, Li and colleagues demonstrated that systemically administered MSC-Exos were capable of crossing the BBB and accumulating in affected brain regions, where they enhanced the synthesis of key neurotransmitters, including dopamine and serotonin (5-HT), both of which play essential roles in sleep regulation and circadian rhythm maintenance. A particularly striking finding was that MSC-Exos promoted the expression of Brain and Muscle ARNT-Like 1 (BMAL1) and Period Circadian Regulator 2 (PER2), two core proteins of the circadian clock machinery. These molecules regulate not only sleep-wake cycles but also neuronal metabolism and the survival of damaged neurons. By upregulating BMAL1 and PER2, MSC-Exos appeared to reestablish disrupted circadian rhythms, which are commonly observed in PD and closely linked to sleep abnormalities. Furthermore, MSC-Exos restored mitochondrial membrane potential in dopaminergic neurons, suggesting improved mitochondrial function and energy homeostasis, which are critical for maintaining proper neuronal activity and survival in PD. MSC-Exos also activated peroxisome proliferator-activated receptor gamma (PPARγ). This nuclear receptor modulates the expression of genes involved in lipid metabolism, insulin sensitivity, and the cellular response to oxidative stress. This activation likely contributed to the observed neuroprotective effects and metabolic stabilization in the damaged neurons. Behaviorally, rats treated with MSC-Exos exhibited significant improvements in sleep architecture. Measures of total sleep duration, slow-wave (deep) sleep, and fast-wave (REM-like) sleep were all prolonged, suggesting meaningful recovery of sleep function in these PD models. These improvements underscore the potential of MSC-Exos to treat non-motor symptoms of PD by targeting circadian dysregulation and neurochemical imbalances. However, it is essential to acknowledge a limitation of this study. While the short-term effects of MSC-EVs on sleep quality were beneficial, Li and colleagues did not investigate whether these effects are sustainable over the long term. The durability of circadian rhythm restoration and sleep improvement remains unaddressed, highlighting the need for follow-up studies to evaluate the clinical relevance and therapeutic longevity of MSC-EVs in managing sleep disorders in PD (Li et al., 2023).

Emerging evidence highlights the therapeutic potential of MSC-exosomes to promote neoangiogenesis and counteract neurodegeneration in PD. The formation of new blood vessels is pivotal for enhancing cerebral perfusion, delivering essential nutrients and oxygen to compromised brain regions, and facilitating the removal of metabolic waste products. These processes are crucial for maintaining neuronal health and function, particularly in the SNpc, where dopaminergic neurons are progressively lost in PD. In a study conducted by Xue and colleagues (Xue et al., 2021), MSC-Exos were shown to stimulate angiogenesis in the SNpc of PD mice by activating the ICAM1-SMAD3 and p38/MAPK signaling pathways in endothelial cells (ECs). The activation of SMAD3 promoted EC proliferation and migration, facilitating the formation of vascular networks that improved blood flow to degenerated neurons. Simultaneously, the p38/MAPK pathway induced the secretion of pro-angiogenic factors and enhanced EC survival under neurotoxic conditions, thereby mitigating PD progression in experimental animals. Despite these promising findings, the study did not assess potential adverse effects associated with MSC-Exos-induced enhanced angiogenesis. Uncontrolled or excessive angiogenesis can lead to abnormal blood vessel formation, increased inflammation, and possible disruption of the BBB, potentially counteracting the intended neuroprotective benefits in PD. Therefore, while MSC-Exos hold promise in promoting angiogenesis and supporting neuronal survival in PD, further research is essential to evaluate the long-term effects and safety of this therapeutic approach. Future studies should focus on optimizing MSC-Exos-based therapies to ensure controlled angiogenesis, minimize potential adverse effects, and maximize their neuroprotective efficacy in PD treatment.

13.4.5 MSC-Exos as Nanoscale Vehicles for Targeted Delivery of Bioactive Compounds to Dopaminergic Neurons

In addition to their immunomodulatory and neuroprotective properties, MSC-Exos can also serve as vehicles for targeted delivery of bioactive molecules to dopaminergic neurons, offering new and promising therapeutic approaches for PD (Volarevic et al., 2025). Their ability to deliver therapeutic agents directly to dopaminergic neurons, modulate pathogenic mechanisms, and enhance cellular resilience makes them desirable candidates for the treatment of PD.

BDNF has long been recognized for its neuroprotective properties, particularly in PD (Huang et al., 2022). BDNF plays a pivotal role in safeguarding dopaminergic neurons from oxidative stress-induced damage, fostering synaptic plasticity, enhancing learning and memory, and boosting dopamine production by promoting the synthesis of tyrosine hydroxylase (TH). This enzyme converts tyrosine into L-DOPA. Building on these insights, Wang and colleagues developed a novel therapeutic approach by encapsulating BDNF within MSC-exosomes, yielding MSC-exosome-BDNF hybrids (Wang et al., 2024). These hybrid vesicles were designed to deliver BDNF directly to injured dopaminergic neurons. Upon intravenous administration, MSC-Exos-BDNF successfully crossed the BBB and accumulated within the targeted neurons, significantly enhancing their survival. Moreover, MSC-Exos-BDNF facilitated synaptic remodeling and bolstered the brain's capacity to adapt and maintain function despite substantial neuronal loss. This was achieved by enhancing microtubule-associated protein 2 (MAP2) synthesis and regulating phosphorylated tau (p-tau) levels, thereby stabilizing the neuronal cytoskeleton. Additionally, MSC-Exos-BDNF activated the nuclear factor erythroid 2-related factor 2 (Nrf2) signaling pathway, strengthening dopaminergic neurons' antioxidant defenses and promoting mitochondrial function, thereby contributing to their overall resilience in PD models (Wang et al., 2024).

In a related study, Geng and colleagues addressed the role of N6-methyladenosine (m6A) RNA modification in PD pathogenesis (Geng et al., 2023a). M6A modification is a critical post-transcriptional process that influences RNA stability, translation, and splicing. Alterations in m6A levels can affect the translation of mRNA encoding αSN, potentially altering its production and aggregation, both of which are implicated in PD. Specifically, a decrease in m6A levels in the striatum leads to a notable reduction in dopamine production. To counteract this, Geng et al. encapsulated small interfering RNA (siRNA) targeting the Fat mass and obesity-related (FTO) gene, an RNA demethylase involved in m6A modification, within MSC-Exos, creating MSC-Exos-si-FTO (Geng et al., 2023a). This approach effectively delivered si-FTO into dopaminergic neurons, inhibiting FTO expression and thereby restoring m6A levels. Consequently, MSC-Exos-si-FTO reduced αSN levels, prevented the decrease of TH, and increased dopamine production. Additionally, this intervention protected dopaminergic neurons from programmed cell death, thereby slowing the progression of PD in the experimental model.

13.5 THERAPEUTIC POTENTIAL OF MSC-EXOS IN THE TREATMENT OF ALZHEIMER'S DISEASE

13.5.1 Pathogenesis of Alzheimer's Disease

Alzheimer's disease (AD) is a complex neurodegenerative disorder characterized by progressive cognitive decline and profound neuronal loss, particularly in the hippocampus and cortex (Khan et al., 2020). The molecular pathogenesis of AD is multifactorial, involving the accumulation of β-amyloid (Aβ) plaques, hyperphosphorylated tau protein aggregates (neurofibrillary tangles), synaptic dysfunction, oxidative stress, and widespread neuroinflammation. In recent years, growing evidence has highlighted the pivotal role of immune cells, particularly microglia and astrocytes, and the chronic neuroinflammatory response in the initiation and progression of AD (Twarowski et al., 2023).

At the core of AD pathology is the accumulation of extracellular Aβ peptides, generated by the sequential cleavage of the amyloid precursor protein (APP) by β-secretase (BACE1) and γ-secretase (Bondi et al., 2017). Aggregated Aβ oligomers are neurotoxic and disrupt synaptic function even before the formation of plaques. These oligomers activate microglia through pattern recognition receptors such as toll-like receptors (TLRs) and the NOD-like receptor protein 3 (NLRP3) inflammasome (Twarowski et al., 2023). Activated microglia attempt to clear Aβ through phagocytosis, but with chronic exposure and aging, their phagocytic capacity becomes impaired. Instead, they shift to a pro-inflammatory phenotype, releasing cytokines such as IL-1β, TNF-α, and IL-6, which contribute to neuronal dysfunction and promote tau hyperphosphorylation by activating GSK-3β and CDK5. The NLRP3 inflammasome plays a particularly critical role in mediating this inflammatory response. Upon Aβ stimulation, the NLRP3 complex becomes activated in microglia, leading to the cleavage and maturation of pro-inflammatory cytokines such as IL-1β and IL-18 via caspase-1. This not only exacerbates neuroinflammation but also enhances the aggregation of both Aβ and tau proteins, creating a vicious cycle of immune activation and proteinopathy. Additionally, inflammasome activation can induce a form of programmed cell death known as pyroptosis, which further damages neuronal and glial cells.

Astrocytes also contribute significantly to the inflammatory landscape of AD, acting not only as responders to pathological stimuli but also as amplifiers and sustainers of neuroinflammation (Twarowski et al., 2023). Under normal physiological conditions, astrocytes play a vital role in CNS homeostasis. They regulate ion balance, neurotransmitter clearance, support the BBB, provide metabolic support to neurons, and modulate synaptic activity. In this homeostatic state, astrocytes facilitate normal neuronal communication and plasticity, contributing to learning and memory. However, as AD progresses, astrocytes undergo a reactive state, often referred to as astrogliosis, characterized by hypertrophy, proliferation, and altered transcriptional profiles (Twarowski et al., 2023). These reactive astrocytes lose many of their neuroprotective functions and instead become sources of pro-inflammatory and neurotoxic factors that exacerbate neurodegeneration. In the brains of AD patients, astrocyte reactivity is primarily triggered by the accumulation of Aβ plaques and hyperphosphorylated tau aggregates, which directly and indirectly activate these glial cells. Upon activation, astrocytes begin secreting a wide range of pro-inflammatory cytokines (IL-1β, TNF-α, IL-6) and chemokines that recruit additional immune cells into the CNS.

This pro-inflammatory milieu not only contributes to the activation of neighboring microglia and perpetuation of neuroinflammation but also directly impairs neuronal function and survival. In parallel, reactive astrocytes produce increased levels of ROS and NO, further intensifying oxidative stress, damaging cellular components, and exacerbating mitochondrial dysfunction in neurons

(Twarowski et al., 2023). A fundamental and emerging mechanism by which astrocytes contribute to neurodegeneration in AD is through the dysregulation of the complement cascade, a key component of the innate immune system. In AD, astrocytes upregulate the production of complement proteins such as C3 and C1q, which tag synapses for elimination. While synaptic pruning via the complement system is essential during development for refining neural circuits, its reactivation in the adult brain is pathological. Complement-tagged synapses are recognized and engulfed by microglia, leading to widespread and inappropriate synapse loss. This aberrant complement-mediated synaptic pruning is now recognized as a major contributor to the cognitive decline observed in AD, as synapse density is a strong correlate of mental function. Additionally, astrocyte-derived complement activation is not merely a secondary consequence of microglial inflammation, but can occur independently and precede microglial activation in specific contexts. Astrocytes in AD have been shown to adopt distinct transcriptional profiles, sometimes categorized as "A1" neurotoxic reactive astrocytes, which are induced by microglial-secreted factors but sustain their pro-inflammatory phenotype autonomously. These A1 astrocytes not only secrete complement proteins but also lose their ability to promote neuronal survival and synaptogenesis, further compounding the neurodegenerative environment.

Recent genome-wide association studies (GWAS) have significantly advanced our understanding of the genetic underpinnings of AD, revealing that many of the strongest risk variants are located in genes involved in immune regulation and inflammation, particularly those expressed in microglia (Zheng et al., 2025). Among the most prominent of these genes are TREM2 (Triggering Receptor Expressed on Myeloid cells 2), CD33, CR1 (Complement Receptor 1), and ABI3 (Abelson interactor family member 3). These genes are not associated with the classical amyloid or tau pathways but instead highlight the central role of the innate immune system, particularly microglial function, in the pathogenesis and progression of AD. TREM2 has emerged as one of the most critical immune-related genes implicated in AD. It encodes a transmembrane receptor found almost exclusively on microglia in the brain, playing a pivotal role in regulating microglial activation, survival, chemotaxis, and phagocytosis. Under physiological conditions, TREM2 facilitates microglial responses to injury and promotes the clearance of cellular debris, including Aβ aggregates. When TREM2 is activated, it signals through the DAP12 (TYROBP) adaptor protein, which initiates downstream pathways, including the PI3K-Akt and MAPK pathways, leading to cytoskeletal remodeling, metabolic reprogramming, and the upregulation of genes necessary for phagocytic function. However, rare missense mutations in TREM2, such as the R47H variant, are associated with a substantially increased risk of late-onset AD, comparable to that conferred by a single copy of the *APOE4* allele (Zheng et al., 2025). These mutations result in partial loss-of-function (hypomorphic) alleles that impair TREM2 signaling, thereby reducing microglial ability to respond effectively to Aβ plaque accumulation. Impaired TREM2 function leads to failure to transition microglia into a "disease-associated microglia" phenotype, which is essential for containing and clearing amyloid deposits. Instead, microglia remain in a less activated, homeostatic state or shift toward a maladaptive, pro-inflammatory phenotype. This dysfunctional state is characterized by reduced phagocytic activity, impaired lipid metabolism, and increased secretion of inflammatory cytokines, all of which contribute to chronic neuroinflammation and neuronal damage. Beyond TREM2, other GWAS-identified genes, such as CD33 and CR1, further underscore the involvement of the complement and phagocytic pathways in AD. *CD33*, another microglia-expressed receptor, acts antagonistically to TREM2 by inhibiting phagocytosis and promoting an anti-clearance, pro-inflammatory state when overexpressed or hyperactive. Variants in CD33 that enhance its expression have been correlated with increased Aβ burden and poorer cognitive outcomes (Twarowski et al., 2023). Similarly, CR1 is involved in complement-mediated synapse elimination and Aβ clearance, and risk variants are believed to alter the efficiency of these processes, potentially accelerating synaptic loss.

ABI3, though less well understood, has also been implicated in regulating microglial cytoskeletal dynamics and motility, both of which are essential for effective plaque surveillance and debris clearance (Zheng et al., 2025). Variants in ABI3 may disrupt these functions, contributing to the inefficient immune response seen in AD. Taken together, these genetic findings strongly support the view that impaired microglial function, driven by inherited variants in immune-related genes, plays a critical role in the onset and progression of AD. Dysfunctional microglia are unable to clear pathological protein aggregates effectively and instead perpetuate a toxic, inflammatory environment that accelerates neurodegeneration. These insights have shifted therapeutic efforts toward targeting microglial receptors and signaling pathways to restore immune balance in the AD brain.

13.5.2 Therapeutic Potential of MSC-Exos in the Treatment of AD

Several recently published studies have demonstrated the therapeutic potential of MSC-Exos in treating AD, highlighting their ability to enhance neurogenesis and improve cognitive function (Yusuf and Haider, 2025). Cone and colleagues used the 5XFAD mouse model, which carries five familial AD mutations and exhibits early-onset Aβ accumulation, to investigate the therapeutic potential of MSC-Exos in ameliorating AD-related cognitive impairment (Cone et al., 2021). Over four months, 5XFAD mice received MSC-Exos intranasally every four days. Behavioral assessments, including the novel object recognition test, Y-maze, and Barnes maze, demonstrated that MSC-Exos-treated 5XFAD mice exhibited cognitive performance comparable to that of control animals and significantly better than saline-treated 5XFAD mice. Histological analyses revealed a reduction in Aβ plaque load in the hippocampus, accompanied by a decrease in the colocalization of Aβ plaques with glial fibrillary acidic protein (GFAP)-positive astrocytes, suggesting a reduction in neuroinflammation. Additionally, GFAP levels were lower in the hippocampus of EV-treated 5XFAD mice compared with their saline-treated counterparts, indicating decreased astrocyte activation. These findings suggest that multiple intranasal administrations of MSC-Exos can efficiently attenuate AD-like phenotypes in the 5XFAD mouse model by enhancing cognitive function, reducing Aβ plaque burden, and modulating neuroinflammation. The successful delivery of MSC-Exos via the intranasal route underscores the potential of this non-invasive method for brain-targeted therapies.

Similar conclusions were drawn by Losurdo and colleagues, who confirmed the efficacy of intranasally delivered MSC-Exos in alleviating AD (Losurdo et al., 2020). They used a 3xTg mouse model of AD that expresses three familial AD mutations—PS1M146V, APPSwe, and tauP301L—leading to the development of amyloid plaques, tau tangles, and neurodegeneration. MSC-Exos were isolated from mouse BM-derived MSCs that were preconditioned with pro-inflammatory cytokines TNFα and IFNγ for 48 hours to enhance their immunomodulatory properties. For therapeutic administration, MSC-Exos were resuspended in phosphate-buffered saline at 300 μg/mL and delivered intranasally to 7-month-old female 3xTg mice. Each mouse received two intranasal injections, 18 hours apart, over 21 days. Immunohistochemical analysis revealed a significant reduction in the number of Iba-1-positive microglial cells in MSC-Exos-treated AD mice compared to controls. Additionally, MSC-Exos-based therapy decreased CD68 expression, a marker of microglial activation, in the hippocampus, entorhinal cortex, and prefrontal cortex. Interestingly, no significant changes in CD206 expression, an anti-inflammatory marker, were observed, suggesting that MSC-Exos primarily modulate microglial activation rather than induce a complete phenotypic shift. To further investigate the mechanisms underlying MSC-Exos-based immunomodulation, Losurdo et al. examined the impact of MSC-Exos on microglial polarization *in vitro*. Primary murine microglia were treated with TNF-α and IFN-γ to induce a pro-inflammatory phenotype and then exposed to MSC-Exos. Cytokine release assays revealed that MSC-Exos treatment reduced the secretion of pro-inflammatory cytokines IL-6 and IL-1β, while increasing the release of the anti-inflammatory cytokine IL-10 (Losurdo et al., 2020).

Additionally, Losurdo and colleagues employed Golgi-Cox staining, a classical yet highly effective neuroanatomical technique, to visualize and quantify dendritic spines in the hippocampal and entorhinal cortical neurons of the 3xTg-AD mouse model. Dendritic spines are tiny, actin-rich protrusions extending from neuronal dendrites that serve as the primary sites of excitatory synaptic input. These structures are crucial for synaptic transmission, plasticity, and the overall integration of neural signals, making them essential components of learning and memory processes. In AD, the loss or structural remodeling of dendritic spines is closely linked to cognitive decline and synaptic dysfunction. Numerous studies have shown that reductions in spine density correlate with impaired memory performance and the progression of AD pathology, particularly in brain regions like the hippocampus and entorhinal cortex, which are critical for memory consolidation and spatial navigation (Losurdo et al., 2020). Following intranasal administration of MSC-Exos, treated mice showed a significant increase in dendritic spine density in both brain regions compared with saline-treated controls. This observation suggests a restorative effect of MSC-Exos on synaptic architecture. The increase in dendritic spine density following MSC-Exos treatment indicates that these vesicles may exert neuroprotective and neuroregenerative effects, possibly by modulating the local microenvironment to support synaptogenesis and structural plasticity. MSC-Exos-derived neuroprotective factors could stimulate intracellular signaling cascades that promote the growth and maturation of new spines or prevent the degeneration of existing ones. Consequently, the observed spine recovery may reflect improved neuronal connectivity and enhanced synaptic efficacy, which are essential for counteracting the synaptic deficits characteristic of AD.

By utilizing a sporadic AD mouse model induced by intracerebroventricular injection of streptozotocin (STZ), Liu and colleagues aimed to explore the effects of MSC-Exos on neuroinflammation, Aβ accumulation, tau hyperphosphorylation, and brain-derived neurotrophic factor (BDNF)-related neuropathology (Liu et al., 2022). MSC-Exos were isolated from the supernatant of mice MSCs, which were derived from the femur and tibia of adult C57BL/6 mice. MSC-Exos were characterized using transmission electron microscopy, flow cytometry, and Western blotting to confirm their size, morphology, and the presence of exosomal markers, including CD63, CD9, and TSG101. For therapeutic administration, MSC-Exos were injected into STZ-induced AD mice via two routes: the lateral ventricle (LV) and the caudal vein (CV). The LV route was chosen for its direct access to the brain, while the CV route served as a systemic delivery method. Behavioral assessments were conducted to evaluate the impact of MSC-Exos treatment on cognitive function. The open field test (OFT), elevated plus maze (EPM) test, novel object recognition (NOR) test, Y-maze test, and tail suspension test (TST) were employed to assess locomotor activity, anxiety-like behavior, memory, and depressive-like behavior. Results indicated that the LV administration of MSC-Exos significantly improved performance in the NOR and Y-maze tests, suggesting enhanced memory and cognitive function.

In contrast, CV administration did not yield significant improvements, highlighting the importance of direct brain delivery for the therapeutic efficacy of MSC-Exos in AD treatment. Immunofluorescence staining for glial fibrillary acidic protein (GFAP) and ionized calcium-binding adapter molecule one (IBA1) revealed increased activation of astrocytes and microglia in the hippocampus of STZ-induced AD mice. As observed during the evaluation of cognitive impairment, MSC-Exos administered via the LV route significantly reduced the number of activated glial cells, whereas those administered via the CV route had no effect. MSC-Exos-based inhibition of glial cells was accompanied by attenuated neuroinflammation and by reduced accumulation of Aβ plaques (Liu et al., 2022). Western blot analysis demonstrated elevated levels of Aβ1–42, β-site amyloid precursor protein cleaving enzyme one (BACE1), and phosphorylated tau (p-Tau Ser396) in the hippocampus of STZ-induced AD mice. LV administration of MSC-Exos significantly decreased the expression of these proteins, while CV administration did not produce notable changes. Immunofluorescence staining further confirmed the reduction in Aβ and p-Tau deposition in the hippocampus following MSC-Exos administration via LV. In addition to the analysis of neuroinflammatory and neurodegenerative changes in the brains of MSC-Exos-treated AD mice, Liu and colleagues also assessed the expression of neurogenesis marker doublecortin (DCX). They evaluated the expression of synaptic proteins, including synaptotagmin-1 (Syt-1), synapsin-1, and BDNF, in AD mice treated with MSC-Exos (Liu et al., 2022). Immunofluorescence staining revealed a reduction in DCX-positive cells in the hippocampus of STZ-induced AD mice. LV administration of MSC-Exos significantly increased the number of DCX-positive cells, indicating enhanced neurogenesis. Similarly, Western blot analysis showed decreased levels of Syt-1, synapsin-1, and BDNF in the hippocampus of AD mice, while MSC-Exos-based therapy, given via the LV route, significantly elevated the expression of BDNF and ameliorated AD-related symptoms. The hippocampal expression of BDNF was positively correlated with the expression of synaptic proteins Syt-1 and synapsin-1. It was negatively correlated with levels of glial activation markers IBA1 and GFAP, as well as with inflammatory cytokines IL-1β, IL-6, and TNF-α, and with Aβ1–42 in the hippocampus. These findings provided compelling evidence that MSC-Exos can alleviate cognitive decline in AD-like mice by modulating neuroinflammation, reducing amyloid and tau pathologies, and enhancing BDNF-related neuropathology.

Kim and colleagues recently conducted a Phase I clinical trial to evaluate the feasibility, safety, and tolerability of repeated intracerebroventricular injections of human umbilical cord blood-derived MSCs (hUCB-MSCs) in patients with mild-to-moderate AD-related dementia (Kim et al., 2021). Nine patients were divided into two groups: three received a low dose of 1.0×10^7 hUCB-MSCs/2 mL, and six received a high dose of 3.0×10^7 hUCB-MSCs/2 mL. Each patient underwent three repeated intracerebroventricular injections at four-week intervals, administered via an Ommaya reservoir implanted into the right lateral ventricle. The primary objective of this trial was to evaluate the safety of hUCB-MSC injections. Post-injection, all participants experienced fever, the most common adverse event, followed by headache, nausea, and vomiting. These symptoms were transient and resolved within 36 hours without the need for specific treatments. Notably, there were three serious adverse events in two participants, including extended hospitalization due to fever and nausea with vomiting, which were considered related to the investigational product. Importantly, no dose-limiting toxicities were observed, and the Ommaya reservoir implantation procedure was well-tolerated by all patients. Brain imaging studies, including CT and MRI scans, confirmed the absence of structural abnormalities such as hemorrhages or tumors. Cerebrospinal fluid analyses revealed an increase in white blood cell count one day after each hUCB-MSC injection, which normalized within four weeks. This transient elevation suggests an acute inflammatory response, likely triggered by the introduction of stem cells.

Additionally, levels of AD-related biomarkers, including total tau, phosphorylated tau, and amyloid-beta 42 (Aβ42), decreased 1 day post-injection but returned to baseline after 4 weeks, indicating a transient modulation of these biomarkers (Kim et al., 2021). Markers associated with MSC activity, such as galectin-3, soluble intercellular adhesion molecule-1 (sICAM-1), progranulin, and growth differentiation factor 15 (GDF-15), also exhibited transient increases following each injection, further supporting the biological activity of the administered cells. Positron emission tomography (PET) imaging with Pittsburgh compound B (PiB) and florbetaben showed a reduction in amyloid-beta burden in some participants, with changes in standardized uptake value ratios (SUVRs) observed at 12-week follow-up. These findings suggest potential therapeutic effects of hUCB-MSCs on amyloid deposition, a hallmark of AD pathology. Cognitive and functional assessments, including the Alzheimer's Disease Assessment Scale-Cognitive Subscale (ADAS-Cog), Mini-Mental State Examination (MMSE), Seoul Instrumental Activities of Daily Living (S-IADL), and Neuropsychiatric Inventory (NPI), were conducted at baseline and follow-up intervals. While the study primarily focused on safety, these assessments provide a foundation for evaluating the potential efficacy of hUCB-MSCs in future trials. Following the initial 12-week observation period, an extended follow-up study was conducted over 36 months. Five participants completed this extended phase, during which no additional serious adverse events were reported. This prolonged observation period underscores the long-term safety profile of hUCB-MSC injections. The encouraging safety data from this Phase I trial have paved the way for subsequent clinical studies, including a placebo-controlled, randomized Phase IIa trial, which will further assess the therapeutic potential of hUCB-MSCs in AD (Kim et al., 2021).

A significant challenge in MSC-based therapy of AD is the low retention of administered cells within the brain. To address this issue, Jung and colleagues incorporated iron oxide nanoparticles

(IONPs) in human Wharton's jelly-derived MSCs (hWJ-MSCs), obtaining hWJ-MSCs-IONPs (Jung et al., 2023). IONPs were incorporated into hWJ-MSCs using the nanoparticles' magnetic properties to facilitate their internalization by the stem cells. This strategy aimed to enhance the therapeutic potential of hWJ-MSCs by improving their engraftment in the brain (Jung et al., 2023). Following intracerebroventricular injection, the hWJ-MSC-IONPs demonstrated enhanced brain retention compared to non-incorporated IONPs. Results from the murine AD model indicated that hWJ-MSC-IONPs exhibited greater therapeutic efficacy than non-incorporated MSCs. Significant improvements in AD-related pathologies, including reduced amyloid-β accumulation and attenuated neuroinflammation, were observed in mice treated with hWJ-MSC-IONPs, confirming their therapeutic potential for AD. Additionally, Jung et al. compared the therapeutic efficacy of allogeneic hWJ-MSCs with that of autologous MSCs from aged AD patients. The findings indicated that allogeneic MSCs had superior therapeutic potential than autologous MSCs. This observation highlights the potential advantages of using allogeneic MSCs, which may possess greater proliferative and regenerative capacities than autologous MSCs, which may be compromised by the patient's age and disease status (Jung et al., 2023). These results demonstrated that magnetic guidance could enhance the retention of MSC-IONPs in the brains of AD patients, and that the use of allogeneic MSCs may offer additional benefits for MSC-based therapeutic efficacy.

Hong and colleagues investigated the therapeutic potential of nasal olfactory mucosa-derived (OM-MSCs) in the treatment of AD (Hong et al., 2022). OM-MSCs possess unique properties, such as the secretion of neurotrophic factors and the ability to transdifferentiate into neuronal-like cells, making them promising candidates for AD therapy. OM-MSCs secreted significant levels of various neurotrophic factors, including BDNF, NGF, and GDNF, which support neuronal survival, promote neurogenesis, and enhance synaptic plasticity. Hong et al. used a transgenic mouse model of AD that overexpresses human amyloid precursor protein (APP) and presenilin-1 (PS1), leading to the development of Aβ plaques and cognitive deficits. OM-MSCs were transplanted into the hippocampus of these mice, and the effects were assessed at multiple time points post-transplantation. Behavioral assessments, including the Morris water maze and Y-maze tests, demonstrated significant improvements in spatial learning and memory in OM-MSC-treated mice compared to controls, indicating that OM-MSC transplantation can ameliorate AD-related cognitive deficits. Histological examination of brain tissue revealed a reduction in Aβ plaque burden in the hippocampus and cortex of OM-MSC-treated mice. Immunohistochemical staining for GFAP and ionized calcium-binding adapter molecule 1 (Iba1), markers of astrocyte and microglial activation, respectively, showed decreased glial cell activation in the treated mice. These results indicate that OM-MSC transplantation not only reduces Aβ accumulation but also mitigates neuroinflammation, a key contributor to AD pathogenesis. To elucidate the mechanisms underlying the neuroprotective effects of OM-MSC transplantation, the researchers performed gene expression analyses. The results showed upregulation of genes associated with synaptic plasticity and neuroprotection, including BDNF, NGF, and GDNF, in the brains of mice treated with OM-MSCs. Additionally, there was increased expression of genes involved in mitochondrial function and antioxidant defense, suggesting that OM-MSCs may enhance cellular resilience against oxidative stress and mitochondrial dysfunction, both of which are implicated in AD (Hong et al., 2022). These findings suggested that the neuroprotective effects of OM-MSCs were likely mediated through the secretion of neurotrophic factors, reduction of Aβ plaque burden, attenuation of neuroinflammation, and enhancement of synaptic plasticity and mitochondrial function, supporting the potential of OM-MSCs-derived factors as a promising therapeutic strategy for AD and warrant further investigation into their clinical applicability. Therefore, further experimental and clinical studies should evaluate the therapeutic efficacy of OM-MSC-Exos in alleviating AD, as a potentially novel approach to treating this neurocognitive disorder.

Chen and colleagues recently provided compelling evidence that MSC-Exos could ameliorate AD pathology and improve cognitive deficits through multiple mechanisms, including reductions in Aβ levels, restoration of synaptic plasticity-related genes, enhancement of brain glucose metabolism, and inhibition of astrocyte activation (Chen et al., 2021). The researchers utilized a human neural cell culture model with familial AD (FAD) mutations and AD transgenic mice to evaluate the effects of MSC-Exos-based treatment. Chen et al. first established a human neural cell culture model by overexpressing human amyloid precursor protein (APP) carrying both the Swedish (K670N/M671L) and London (V717I) FAD mutations in a human neuroblastoma cell line. These FAD cells exhibited elevated levels of Aβ and downregulated expression of neuronal memory and synaptic plasticity-related genes, including BDNF, synaptophysin (SYP), glutamate receptor subunits GluR1 and GluR2, and NMDA receptor subunit NR2B (Grin2b). Upon co-culturing these FAD cells with purified MSC-Exos for 1 week, Chen et al. observed a significant reduction in Aβ levels and a restoration of gene expression to levels comparable to those in wild-type (WT) cells. These findings suggest that MSC-Exos can modulate the molecular pathways associated with AD pathology *in vitro*. Results from in vivo experiments conducted in AD transgenic mice confirmed the impact of MSC-Exos-based therapy on brain glucose metabolism and cognitive function. Notably, MSC-Exos treatment did not impair memory in WT mice, suggesting that the therapy specifically benefits AD pathology without causing detrimental effects in healthy individuals.

They also investigated the effects of MSC-Exos treatment on Aβ plaque burden and astrocyte activation in the brains of AD transgenic mice. Western blot analysis demonstrated a significant reduction in soluble Aβ species up to 90 kDa in the brains of MSC-Exos-treated mice. Immunofluorescence staining for glial fibrillary acidic protein (GFAP), a marker of reactive astrocytes, revealed a decrease in GFAP-positive cells in the hippocampus of treated mice, indicating a reduction in astrocyte activation. Additionally, quantitative real-time reverse transcription polymerase chain reaction (RT-PCR) analysis showed an upregulation of synaptic plasticity-related genes, including BDNF, SYP, and GluR1, in the brains of MSC-Exos-treated mice. These results suggest that MSC-exosomes can mitigate neuroinflammation and promote synaptic health in the AD brain. To elucidate the underlying mechanisms, Chen et al. identified miR-29a as a key microRNA enriched in MSC-Exos. MiR-29a was found to target histone deacetylase 4 (HDAC4), a histone-modifying enzyme implicated in synaptic plasticity and memory. *In vitro* experiments demonstrated that MSC-Exos treatment decreased HDAC4 expression and upregulated its target genes, including Homerl, Synapsin II (Syn2), and leucine-rich glioma inactivated 1 (Lgil), in FAD cells. Similarly, *in vivo* analysis revealed reduced nuclear HDAC4 levels and increased expression of these target genes in the brains of MSC-exosome-treated AD transgenic mice. These findings support the potential of MSC-Exos as a novel, cell-free therapeutic strategy for AD. However, further research is needed to optimize the delivery and efficacy of MSC-Exos and to evaluate their long-term safety and clinical applicability.

Having in mind that dysfunction is a hallmark of AD, Xu and colleagues engineered MSC-derived extracellular vesicles (MSC-EVs) to overexpress Src homology region 2 domain-containing phosphatase-2 (SHP2), a tyrosine phosphatase involved in various cellular processes, including mitophagy (Xu et al., 2022b). This newly engineered MSC-EV platform, termed MSC-EVs-SHP2, modulated mitophagy and successfully addressed mitochondrial dysfunction, alleviating neuroinflammation, neuronal apoptosis, and cognitive decline in AD mice. SHP2-overexpressing MSCs were cultured to produce EVs, which were then isolated and characterized. Flow cytometry analysis revealed that the engineered MSC-EVs-SHP2 exhibited enhanced uptake by neuronal cells compared to unmodified MSC-EVs, indicating their potential for targeted delivery to the brain. To assess the functional efficacy of MSC-EVs-SHP2, the researchers treated neuronal cell lines with these EVs and evaluated mitophagy induction. Confocal microscopy with MitoTracker and LysoTracker dyes revealed that MSC-EVs-SHP2 treatment increased colocalization of mitochondria with lysosomes, a hallmark of mitophagy. Western blot analysis showed upregulation of mitophagy-related proteins, including Parkin and PINK1, and downregulation of mitochondrial fission and fusion proteins, suggesting enhanced mitophagic activity. Furthermore, MSC-EVs-SHP2 treatment reduced mitochondrial ROS levels and restored mitochondrial membrane potential, indicating improved mitochondrial function. Intravenous administration of MSC-EVs-SHP2 resulted in the accumulation of SHP2 in the brains of AD mice, as confirmed by immunofluorescence staining (Xu et al.,

2022b). Behavioral assessments, including the Morris water maze and Y-maze tests, demonstrated significant improvements in spatial learning and memory in MSC-EVs-SHP2-treated mice compared to controls. Histological analysis revealed a reduction in Aβ plaque burden and decreased levels of neuroinflammatory markers, such as GFAP and ionized calcium-binding adapter molecule 1 (Iba1), indicating alleviation of neuroinflammation. Moreover, transmission electron microscopy (TEM) analysis of brain sections revealed increased mitophagy in neurons, corroborating the *in vitro* findings. To elucidate the underlying mechanisms, the researchers investigated the signaling pathways modulated by SHP2 in MSC-EVs. Western blot analysis revealed that SHP2 overexpression activated the AMP-activated protein kinase (AMPK) pathway, a known regulator of mitophagy. Inhibition of AMPK with compound C abrogated the mitophagic effects of MSC-EVs-SHP2, confirming the involvement of this pathway.

Additionally, SHP2 overexpression suppressed NLRP3 inflammasome activation, as evidenced by decreased IL-1β and caspase-1 levels, suggesting reduced neuroinflammation (Xu et al., 2022b). The findings of this study highlight the potential of MSC-EVs-SHP2 as a therapeutic strategy for AD. By enhancing mitophagy, MSC-EVs-SHP2 not only improve mitochondrial function but also mitigate neuroinflammation and neuronal apoptosis, leading to cognitive improvements in AD mouse models. The use of EVs as delivery vehicles offers several advantages, including biocompatibility, low immunogenicity, and the ability to cross the BBB, making them promising candidates for clinical translation. However, further studies are needed to optimize the production and delivery of MSC-EVs-SHP2, assess their long-term safety and efficacy, and evaluate their potential in clinical settings.

13.6 CONCLUSIONS

MSC-Exos have emerged as a promising therapeutic modality in the treatment of neurocognitive diseases. Their ability to cross the blood-brain barrier, deliver bioactive molecules, and modulate both immune responses and neural microenvironments makes them particularly suited for addressing the multifactorial pathology of neurodegenerative conditions. Current preclinical studies consistently demonstrate that MSC-Exos could efficiently reduce neuroinflammation, promote neuronal survival, and enhance synaptic plasticity, all of which are critical in slowing or even reversing neurodegenerative processes. Despite encouraging results, several challenges remain before MSC-Exos can be translated into routine clinical practice. Standardization of isolation and purification protocols, understanding optimal dosing and delivery routes, and rigorous assessment of long-term safety are essential steps that must be addressed. Furthermore, the heterogeneity of MSCs from different tissue sources complicates the reproducibility of therapeutic outcomes. In summary, MSC-Exos offer a novel, cell-free therapeutic strategy for neurodegenerative diseases, with the potential to overcome many limitations associated with conventional treatments and cell-based therapies. While clinical translation is still in its early stages, ongoing research continues to elucidate the mechanisms of action and therapeutic scope of these therapies, underscoring their potential to transform the future landscape of neurocognitive disease treatment.

REFERENCES

Abdi SMY, Al-Bakri SSM, Nordin N. Insights on the characteristics and therapeutic potential of mesenchymal stem cell-derived exosomes for mitigation of Alzheimer's disease's pathogenicity: a systematic review. *Cell Biochem Biophys.* 2025; 83:1399–1414.

Abrishamdar M, Jalali MS, Yazdanfar N. The role of exosomes in pathogenesis and the therapeutic efficacy of mesenchymal stem cell-derived exosomes against Parkinson's disease. *Neurol Sci.* 2023; 44:2277–2289.

Ahmed LA, Al-Massri KF, Haider KH. Therapeutic efficacy of mesenchymal stem cells and their derived exosomes in different neurodegenerative disorders: current status and future perspectives. In: Haider KH. (ed.) *Handbook of Regenerative Medicine: Cell-Free Therapy.* Boca Raton (FL): CRC Press; 2025. Chapter 32

Ahmed ZT, Alkahlot MH, Haider KH. MSC-derived exosomes: advances in cell-free therapy. In: *Handbook of Stem Cell Applications*. Singapore: Springer Nature Singapore; 2023 Aug 18. p. 1–41.

Ahmed OTF, Ahmed ZT, Dairi AW, Zain Al-Abeden MS, Alkahlot MH, Alkahlot RH, Al Jowf GI, et al. The inconclusive superiority debate of allogeneic versus autologous MSCs in treating patients with HFrEF: a systematic review and meta-analysis of RCTs. *Stem Cell Res Ther.* 2025; 16(1):175. doi: 10.1186/s13287-025-04209-5.

Aliakbari F, Marzookian K, Parsafar S, Hourfar H, Nayeri Z, Fattahi A, Raeiji M, et al. The impact of hUC MSC-derived Exosome-nanoliposome hybrids on α-synuclein fibrillation and neurotoxicity. *Sci Adv.* 2024; 10(14):eadl3406.

Álvarez-Viejo M, Haider KH. (2022) Mesenchymal stem cells. In: Haider KH, editor. *Handbook of Stem Cell Therapy*. Springer, Singapore. doi: 10.1007/978-981-16-6016-0_6-1.

Bloem BR, Okun MS, Klein C. Parkinson's disease. *Lancet.* 2021; 397:2284–2303.

Bondi MW, Edmonds EC, Salmon DP. Alzheimer's disease: Past, present, and future. *J Int Neuropsychol Soc.* 2017; 23:818–831.

Cai Y, Zhang MM, Wang M, Jiang ZH, Tan ZG. Bone marrow-derived mesenchymal stem cell-derived exosomes containing Gli1 alleviate microglial activation and neuronal apoptosis in vitro and in a mouse Parkinson's disease model by direct inhibition of Sp1 signaling. *Journal of Neuropathology & Experimental Neurology.* 2022; 81(7):522–534.

Chen HX, Liang FC, Gu P, Xu BL, Xu HJ, Wang WT, Hou JY, et al. Exosomes derived from mesenchymal stem cells repair a Parkinson's disease model by inducing autophagy. *Cell Death Dis.* 2020; 11(4):288.

Chen YA, Lu CH, Ke CC, Chiu SJ, Jeng FS, Chang CW, Yang BH, et al. Mesenchymal stem cell-derived exosomes ameliorate Alzheimer's disease pathology and improve cognitive deficits. *Biomedicines.* 2021; 9:594.

Cone AS, Yuan X, Sun L, Duke LC, Vreones MP, Carrier AN, Kenyon SM, et al. Mesenchymal stem cell-derived extracellular vesicles ameliorate Alzheimer's disease-like phenotypes in a preclinical mouse model. *Theranostics.* 2021; 11:8129–8142.

Ellis TD, Colón-Semenza C, DeAngelis TR, Thomas CA, Hilaire MS, Earhart GM, Dibble LE. Evidence for early and regular physical therapy and exercise in Parkinson's disease. *Semin Neurol.* 2021; 41:189–205.

Geng Y, Long X, Zhang Y, Wang Y, You G, Guo W, Zhuang G, et al. FTO-targeted siRNA delivery by MSC-derived Exosomes synergistically alleviates dopaminergic neuronal death in Parkinson's disease via m6A-dependent regulation of ATM mRNA. *J Transl Med.* 2023a; 21(1):652.

Geng X, Zou Y, Li J, Li S, Qi R, Zhong L, Yu H. Mesenchymal stem cell exosomes rich in miR-23b-3p affect the Wnt signaling pathway and promote neuronal autophagy to alleviate PD symptoms. *Neurosci Lett.* 2023b; 814:137437.

Habib SM, Martini MF, Al Ahmed O, Qassab AR, Haider KH. The potential of mesenchymal stem cell–derived exosomes in cell-free therapy for myocardial repair. In: Haider KH. (ed.) *Handbook of Regenerative Medicine: Cell-Free Therapy.* Boca Raton (FL): CRC Press; 2025a. Chapter 24, p. 372–390. doi: 10.1201/9781003611509.

Habib SM, Martini MF, Al-Ahmed O, Qassab AR, Haider KH. Mesenchymal stem cell secretomes for cell-free therapy of the broken heart. In: Haider KH. (ed.) *Handbook of Regenerative Medicine: Cell-Free Therapy.* Boca Raton (FL): CRC Press; 2025b. Chapter 18, p. 274–293.

Habib SY, Martini MF, Hamdan YNA, Shrebaty OMM, Haider KH. Mesenchymal stem cell secretome and exosomes as potential advanced therapy medicinal products for treating a failing heart. *World J Stem Cells.* 2026; 27:209.

Haider KH. Priming mesenchymal stem cells to develop "super stem cells". *World J Stem Cells* 2024; 16(6):623–640. doi: 10.4252/wjsc.v16.i6.623.

Haider KH, Aramini B. Mircrining the injured heart with stem cell-derived exosomes: an emerging strategy of cell-free therapy. *Stem Cell Res Ther.* 2020 Jan 9; 11(1):23. doi: 10.1186/s13287-019-1548-7.

Harrell CR, Djonov V, Volarevic A, Arsenijevic A, Volarevic V. Molecular mechanisms responsible for the therapeutic potential of mesenchymal stem cell-derived exosomes in the treatment of lung fibrosis. *Int J Mol Sci.* 2024a; 25:4378.

Harrell CR, Volarevic A, Djonov V, Arsenijevic A, Volarevic V. The role of MicroRNAs in mesenchymal stem cell-based modulation of pulmonary fibrosis. *Cell Transplant.* 2024b; 33:9636897241281026.

Harrell CR, Djonov V, Volarevic A, Arsenijevic A, Volarevic V. Mesenchymal stem cell-sourced exosomes as potentially novel remedies for severe dry eye disease. *J Ophthalmol.* 2025; 2025:5552374.

Harrell CR, Volarevic A, Djonov V, Volarevic V. Mesenchymal stem cell-derived exosomes as new remedy for the treatment of neurocognitive disorders. *Int J Mol Sci.* 2021; 22:1433.

Harrell CR, Volarevic V, Djonov V, Volarevic A. Therapeutic potential of exosomes derived from adipose tissue-sourced mesenchymal stem cells in the treatment of neural and retinal diseases. *Int J Mol Sci.* 2022; 23:4487.

He S, Wang Q, Chen L, He YJ, Wang X, Qu S. miR-100a-5p-enriched exosomes derived from mesenchymal stem cells enhance the anti-oxidant effect in a Parkinson's disease model via regulation of Nox4/ROS/Nrf2 signaling. *J Transl Med.* 2023; 21(1):747.

Hong CG, Chen ML, Duan R, Wang X, Pang ZL, Ge LT, Lu M, et al. Transplantation of nasal olfactory mucosa mesenchymal stem cells benefits Alzheimer's disease. *Mol Neurobiol.* 2022; 59:7323–7336.

Huang Y, Wu Q, Tam PKH. Immunomodulatory mechanisms of mesenchymal stem cells and their potential clinical applications. *Int J Mol Sci.* 2022; 23:10023.

Huang W, Zhang T, Li X, Gong L, Zhang Y, Luan C, Shan Q, Gu X, Zhao L. Intranasal administration of umbilical cord mesenchymal stem cell exosomes alleviates Parkinson's disease. *Neuroscience.* 2024; 549:1–12.

Jung M, Kim H, Hwang JW, Choi Y, Kang M, Kim C, Hong J, et al. Iron oxide nanoparticle-incorporated mesenchymal stem cells for alzheimer's disease treatment. *Nano Lett.* 2023; 23:476–490.

Kamal M, Kassem D, Haider KH. Sources and therapeutic strategies of mesenchymal stem cells in regenerative medicine. In: Haider KH. (ed.) *Handbook of Stem Cell Therapy.* Springer; 2022. doi: 10.1007/978-981-19-2655-6_2

Khan S, Barve KH, Kumar MS. Recent advancements in pathogenesis, diagnostics, and treatment of Alzheimer's disease. *Curr Neuropharmacol.* 2020; 18:1106–1125.

Al-Khani AM, Khalifa MA, Haider KH. Mesenchymal stem cells: how close we are to their routine clinical use? In: Haider KH. (ed.) *Handbook of Stem Cell Therapy.* Springer; 2022. doi: 10.1007/978-981-16-6016-0_11-1.

Kim HJ, Cho KR, Jang H, Lee NK, Jung YH, Kim JP, Lee JI, et al. Intracerebroventricular injection of human umbilical cord blood mesenchymal stem cells in patients with Alzheimer's disease dementia: a phase I clinical trial. *Alzheimer's Res Ther.* 2021; 13:154.

Laroni A, de Rosbo NK, Uccelli A. Mesenchymal stem cells for the treatment of neurological diseases: Immunoregulation beyond neuroprotection. *Immunol Lett.* 2015; 168:183–190.

Li Q, Wang Z, Xing H, Wang Y, Guo Y. Exosomes derived from miR-188-3p-modified adipose-derived mesenchymal stem cells protect against Parkinson's disease. *Mol Ther Nucleic Acids.* 2021; 23:1334–1344.

Li Z, Li Y, Xu X, Gu J, Chen H, Gui Y. Exosomes rich in Wnt5 improved circadian rhythm dysfunction via enhanced PPARγ activity in the 6-hydroxydopamine model of Parkinson's disease. *Neurosci Lett.* 2023; 802:137139.

Liu S, Fan M, Xu JX, Yang LJ, Qi CC, Xia QR, Ge JF. Exosomes derived from bone-marrow mesenchymal stem cells alleviate cognitive decline in AD-like mice by improving BDNF-related neuropathology. *J Neuroinflammation.* 2022; 19:35.

Losurdo M, Pedrazzoli M, D'Agostino C, Elia CA, Massenzio F, Lonati E, Mauri M, et al. Intranasal delivery of mesenchymal stem cell-derived extracellular vesicles exerts immunomodulatory and neuroprotective effects in a 3xTg model of Alzheimer's disease. *Stem Cells Transl Med.* 2020; 9:1068–1084.

Ma L, Jasem HJ, Gu WJ, Zeng Q, Wang X, Liu XD. Postoperative neurocognitive disorders in the elderly: how can we stop the harm? A literature review. *Front Med* (Lausanne). 2025; 12:1525639.

Miceli V, Bulati M, Iannolo G, Zito G, Gallo A, Conaldi PG. Therapeutic properties of mesenchymal stromal/stem cells: the need for cell priming for cell-free therapies in regenerative medicine. *Int J Mol Sci.* 2021; 22:763.

Mohamed AS, Abdel-Fattah DS, Abdel-Aleem GA, El-Sheikh TF, Elbatch MM. Biochemical study of the effect of mesenchymal stem cells-derived exosomes versus L-DOPA in experimentally induced Parkinson's disease in rats. *Mol Cell Biochem.* 2023; 478(12):2795–2811.

Morris HR, Spillantini MG, Sue CM, Williams-Gray CH. The pathogenesis of Parkinson's disease. *Lancet.* 2024; 403:293–304.

Mushtaq M, Zineldeen DH, Mateen MA, Haider KH. Mesenchymal stem cells' "garbage bags" at work: Treating radial nerve injury with mesenchymal stem cell-derived exosomes. *World J Stem Cells* 2024; 16(5). doi: 10.4252/wjsc. v16.i5.0000.

Omole AE, Fakoya AOJ, Nnawuba KC, Haider KH. (2022). Common ethical considerations of human-induced pluripotent stem cell research. In: Haider KH. (ed.) *Handbook of Stem Cell Therapy.* Springer. doi: 10.1007/978-981-16-6016-0_40-1.

Porro C, Cianciulli A, Panaro MA. The regulatory role of IL-10 in neurodegenerative diseases. *Biomolecules.* 2020; 10:1017.

Restrepo-Martinez M, Bell V, Ramirez-Bermudez J. Cognitive disorders in patients with neuroimmunological disease. *Curr Opin Psychiatry*. 2025; 38:126–133.

Roussa E, von Bohlen und Halbach O, Krieglstein K. TGF-beta in dopamine neuron development, maintenance and neuroprotection. *Adv Exp Med Biol*. 2009; 651:81–90.

Soares MBP, Gonçalves RGJ, Vasques JF, da Silva-Junior AJ, Gubert F, Santos GC, de Santana TA, et al. Current status of mesenchymal stem/stromal cells for treatment of neurological diseases. *Front Mol Neurosci*. 2022; 15:883378.

Tong B, Ba Y, Li Z, Yang C, Su K, Qi H, Zhang D, et al. Targeting dysregulated lipid metabolism for the treatment of Alzheimer's disease and Parkinson's disease: Current advancements and future prospects. *Neurobiol Dis*. 2024; 196:106505.

Twarowski B, Herbet M. Inflammatory processes in Alzheimer's disease-pathomechanism, diagnosis and treatment: a review. *Int. J. Mol. Sci*. 2023; 24:6518.

Venugopal C, Shamir C, Senthilkumar S, Babu JV, Sonu PK, Nishtha KJ, Rai KS, et al. Dosage and passage dependent neuroprotective effects of exosomes derived from rat bone marrow mesenchymal stem cells: an in vitro analysis. *Curr Gene Ther*. 2017; 17:379–390.

Vijiaratnam N, Simuni T, Bandmann O, Morris HR, Foltynie T. Progress towards therapies for disease modification in Parkinson's disease. *Lancet Neurol*. 2021; 20:559–572.

Volarevic A, Harrell CR, Arsenijevic A, Djonov V, Volarevic V. Therapeutic potential of mesenchymal stem cell-derived extracellular vesicles in the treatment of Parkinson's disease. *Cells*. 2025; 14:600.

Volarevic V, Markovic BS, Gazdic M, Volarevic A, Jovicic N, Arsenijevic N, Armstrong L, et al. Ethical and safety issues of stem cell-based therapy. *Int J Med Sci*. 2018; 15:36–45.

Wang CC, Hu XM, Long YF, Huang HR, He Y, Xu ZR, Qi ZQ. Treatment of Parkinson's disease model with human umbilical cord mesenchymal stem cell-derived Exosomes loaded with BDNF. *Life Sci*. 2024; 356:123014.

Xiao H, Yu X, Liu Y, Jiang W, Meng X, Dong Z, Wang F. Mesenchymal stem cell-derived exosomes: a promising therapeutic approach to improve neurocognitive disorders in chronic obstructive pulmonary disease. *Stem Cell Res Ther*. 2025; 16:314.

Xu X, Li Z, Zuo H, Chen H, Gui Y. Mesenchymal stem cell-derived exosomes alter neuron cholesterol metabolism via the Wnt5a-LRP1 axis and alleviate cognitive impairment in a progressive Parkinson's disease model. *Neurosci Lett*. 2022a; 787:136810.

Xu F, Wu Y, Yang Q, Cheng Y, Xu J, Zhang Y, Dai H, et al. Engineered extracellular vesicles with SHP2 high expression promote mitophagy for Alzheimer's disease treatment. *Adv Mater*. 2022b; 34:e2207107.

Xue C, Li X, Ba L, Zhang M, Yang Y, Gao Y, Sun Z, et al. MSC-derived exosomes can enhance the angiogenesis of human brain MECs and show therapeutic potential in a mouse model of Parkinson's disease. *Aging Dis*. 2021; 12(5):1211–1222.

Yusuf R, Haider KH. Theragnostic applications of exosomes in Alzheimer's disease. In: Haider KH. (ed.) *Handbook of Regenerative Medicine: Cell-Free Therapy*. Taylor & Francis Group CRC Press; 2025, Chapter-255, pp. 390–409.

Zhang ZX, Zhou YJ, Gu P, Zhao W, Chen HX, Wu RY, Zhou LY, et al. Exosomes derived from human umbilical cord mesenchymal stem cells alleviate Parkinson's disease and neuronal damage through inhibition of microglia. *Neural Regen Res*. 2023; 18(10):2291–2300.

Zheng Q, Wang X. Alzheimer's disease: insights into pathology, molecular mechanisms, and therapy. *Protein Cell*. 2025; 16:83–120.

14 The Utilization of Exosome-Derived Secretome in Cardiac Repair in Heart Failure

Alexander E. Berezin[1] *and Tetiana A. Berezina*[2]

[1]Department of Internal Medicine II, Division of Cardiology, Paracelsus Medical University, Salzburg, Austria

[2]Department of Internal Medicine & Nephrology, Vita Center, Zaporozhye, Ukraine

14.1 LIST OF ABBREVIATIONS

AMI	Acute myocardial infarction
CVD	Cardiovascular disease
DNA	Deoxyribonucleic acid
EGF	Epidermal growth factor
EVs	Extracellular vesicles
FasL	Fas ligand
FGF	Fibroblast growth factor
HF	Heart failure
HFpEF	HF with preserved ejection fraction
HFrEF	HF with reduced ejection fraction
HMGB-1	High-mobility group box protein 1
HSP	Heat shock proteins
MHC	Major Histocompatibility Complex
miRNA	microRNA
MSC	Mesenchymal stem cells
MSC-EVs	Mesenchymal stem cell-derived extracellular vesicles
ncRNA	Non-coding RNA
RNA	Ribonucleic acid
TfR	Transferrin receptor
TGF	Transforming growth factor
TNF	Tumor necrosis factor
TRAIL	TNF-related apoptosis-inducing ligand
VEGF	Vascular endothelial growth factor

14.2 INTRODUCTION

Heart failure (HF) remains a potentially fatal consequence of enormous cardiovascular diseases (CVD) and confers a significant economic burden to patients, their families, and health care resources (Shahim et al., 2023). HF affects more than 64 million people worldwide, and its prevalence is

DOI: 10.1201/9781003653387-14

increasing. However, the incidence of new cases of HF with reduced ejection fraction (HFrEF) is stabilizing, particularly in developed countries, but not in developing countries (Savarese et al., 2023). Although HF is a heterogeneous condition presenting certain variability in clinical signs and symptoms depending on its phenotype, especially at the early stage, at moderate-to-severe stages of HF, patients do not differ from each other in their mortality rate. In contrast, quality of life and the risk of hospital admission may vary significantly (Bragazzi et al., 2021). However, the prevalence, incidence, and prognosis of HF show significant geographical, social, age, and gender variations, depending on the different aetiologies of HF, clinical characteristics, concomitant comorbidities, affordability of health care system resources, and the implementation of effective management (Wintrich et al., 2022; Li et al., 2024).

The natural history of this syndrome is closely associated with adverse cardiac remodeling, which is characterized by significant myocardial injury, irreversible cardiomyocyte death, and extensive cardiac fibrosis (Bhatt et al., 2017). Changes in cardiac cavity dimensions, cardiac architecture, and myocardial function involve complex interactions among cellular and extracellular components, regulated by neurohormones, as well as by inflammatory and immune responses and by remote organ dysfunction. The interaction of these pathogenetic factors leads to the transition from a hemodynamically stable HF to an advanced HF with an inappropriately high risk of death (Chaudhry et al., 2016).

Over the past two decades, numerous breakthroughs have been made in conventional approaches to cardiovascular medicine and cardiac surgery, resulting in a significant reduction in heart failure-related mortality, particularly among patients with HFrEF (McDonagh et al., 2021; Heidenreich et al., 2022). The mortality due to HF with preserved ejection fraction (HFpEF), but not HFpEF hospitalization, exerted a less prominent benefit than HFrEF, probably due to the higher variability in the concomitant comorbidity profile. Despite significant advances in modern HF management, guideline-recommended medical therapy for HF only delays the progression of adverse cardiac remodeling (Beghini et al., 2025). Patients at stage C are considered candidates for life-prolonging therapies, such as cardiac transplantation or long-term mechanical circulatory support, or require palliative therapies (Crespo-Leiro et al., 2021; Blum et al., 2023).

The concept of utilizing stem cells or progenitor cells has emerged in the last decade as a promising approach for cardiac regenerative therapy, aiming to address the reversibility of adverse cardiac remodeling and prevent the progression of HF (Poglajen et al., 2020; AlKhani et al., 2023; Kalou et al., 2023; Muqresh et al., 2024). Although autologous and/or allogeneic cell therapy of advanced HF regardless of ischemic and non-ischemic etiology has revealed incredible potency for promoting myocardial reverse remodeling and recovery (Ahmed et al., 2025), it remains unclear whether the benefits in the reversibility of cardiac remodeling and in improving revascularization are maintained through the biochemical microenvironment which could influence the efficacy of cell therapy (Nair et al., 2020; Ahmed et al., 2022). In line with it, previous clinical studies and meta-analyses have shown the greater variability in response to stem cell therapy in HF patients, which could also be secondary to the comorbidity profile (Fu & Chen, 2020; Kavousi et al., 2024; Qayyum et al., 2023a; Bolli et al., 2022; Al-khani et al., 2023). Indeed, allogeneic cell therapies, including intramyocardial injections of adipose tissue-derived mesenchymal stromal cells (MSCs), have demonstrated safety in symptomatic HFrEF patients (Ahmed et al., 2022); however, they have not provided sufficient restoration of clinical symptoms and cardiac function in both short- and long-term follow-up (Kavousi et al., 2024; Qayyum et al., 2023b). Unfortunately, the application of MSCs in a clinical setting is sufficiently restricted due to the marked risks of genetic instability, undesired differentiation, pulmonary embolism, immune reactions, including graft-versus-host disease, and malignancy (Rosland et al., 2009; Kunter et al., 2007; Breitbach et al., 2007; Zangi et al., 2009; Cui et al., 2015; Nauta et al., 2006; Rajab et al., 2019). Previous studies have shown that MSCs implanted and differentiated at the injury site have a very short lifespan (Eggenhofer et al., 2012; Jiang et al., 2008). Cell-free therapy based on extracellular vesicles (EVs) may offer a promising alternative to MSC-based therapy for advanced HF (Haider & Najimi, 2022; Ahmed et al., 2023; Liang et al., 2025).

Emerging studies have provided strong evidence that MSC-derived exosomes can mimic the effects of MSC secretomes on cardiac repair by directly and indirectly regulating MSC paracrine protection (Shao et al., 2017; Zhang et al., 2016; Habib et al., 2025a, & 2025b). The secretome of MSCs refers to the complex mixture of bioactive molecules, including proteins, lipids, and nucleic acids, that these cells secrete, and it plays a crucial role in their paracrine function (Habib et al., 2026). This is achieved by transporting microRNAs (miRNAs), messenger RNAs, regulatory proteins, growth factors, and active molecules to target cells (Haider & Aramini, 2020). MSC-exosomes have been shown to mediate cardiac repair and regeneration after acute myocardial infarction (AMI) by promoting angiogenesis and cell proliferation, reducing apoptosis, and attenuating ischemia-induced adverse cardiac remodeling (Viola et al., 2021). In non-ischemic HF, MSC secretome is likely to be effective in suppressing cardiac myocyte apoptosis, immune responses, and myocardial fibrosis (Karnas et al., 2023). Therefore, the secretome of MSCs offers a potential therapeutic alternative that avoids the risks and disadvantages associated with stem cell therapies (Habib et al., 2025b). In this context, exosomal secretomes from different allogenic cell lines appear promising for cardiac reparative treatment in individuals with severe or advanced HF (Rosca et al., 2017). In this chapter, we summarize the key characteristics of the exosomal secretomes and explore the potential mechanisms underlying cell-free cardiac repair through the delivery of biologically active regenerative substances.

14.3 DEFINITION AND CHARACTERISTICS OF MSC-DERIVED EXTRACELLULAR VESICLES

MSC-derived extracellular vesicles (MSC-EVs), which package various signaling molecules for intercellular communication within a phospholipid bilayer membrane, include exosomes, microvesicles, and apoptotic bodies (Table 14.1). The different categories are distinguished by size, content, immune labeling, and secretion method (Börger et al., 2020). Exosomes (40–150 nm) are the smallest EVs, apoptotic bodies (1000–5000 nm) are the largest, and microvesicles (150–1000 nm) are intermediate in size (Théry et al., 2018). Exosomes are produced and released using a more complex pathway. Lysosomes may initiate degradation or release the material by fusing with the cell membrane (van Niel et al., 2018). The formation of microvesicles is triggered by budding, which occurs when the plasma membrane undergoes a series of alterations. These microvesicles contain a proportion of the intracellular cytoplasm (van Niel et al., 2018). The term "apoptotic bodies" is employed to denote the remnants of apoptotic cells that have undergone membrane blebbing, and which contain cytoplasmic contents that are utilized for the following processes: disposal, downstream signaling, material recycling, and control of immune responses via the transmission of autoantigens (van Niel et al., 2018).

Exosomes are known to contain a variety of cargo, including but not limited to structural (beta-actin, Actrlb protein, tubulin), transmembrane (CD9, CD63, coagulation factor II and V, fibulin-2) and regulatory (integral membrane protein TAPA-1, TCP-1 chaperone family subunits, heat shock proteins) proteins, adhesive molecules (intercellular adhesion molecule 1), autophagy-related proteins, transcription and growth factors, enzymes (matrix metalloproteinases, NADP-dependent malic enzyme, malate dehydrogenase, citrate synthase, phosphoglycerate kinase 1, fatty acid binding protein-3 and -4), hormones (adiponectin), cytokines, chemokines, lipids (ceramides, phosphatidylcholine, phosphatidylinositol, phosphatidylserine, and cholesterol), receptors, deoxyribonucleic acid (DNA), micro ribonucleic acid (miRNA), and non-coding ribonucleic acids (ncRNAs) (Sgaglione et al., 2025; Barile et al., 2014; Wu et al., 2020). They are involved in various cellular communication functions, including the enrichment of signaling pathways, immunomodulation and immunological responses, tissue repair, and protection against oxidative stress, ischemia/hypoxia, inflammation, and mitochondrial dysfunction (Figure 14.1).

TABLE 14.1
The Characteristics of MSC-Derived Extracellular Vesicles

Characteristics	Exosomes	Microvesicles	Apoptotic Bodies
Dimension	40–150 nm	150–1000 nm	1000–5000 nm
Density	1.13–1.19 g/ml	1.04–1.07 g/ml	1.16–1.28 g/ml
Releasing pathway	Lysosomes degrade or fuse with the cell membrane.	Budding and blebbing from the plasma membrane.	Through blebbing by programmed cell death and signal cell engulfment.
Intracellular cytoplasm components	Cargo from intracellular contents.	A small proportion of the cytoplasm.	A large number of cytoplasmic contents.
Shape	Spheroid	Irregular	High variable
Biochemical cargo composition	Structural (actin, tubulin, cofilin) and regulatory proteins (caveolin-1, 2 and 3) including MHC Class I and II, tetraspanins (CD9, CD53, CD63), endosomal sorting complex required for transport proteins, adhesion molecules (integrin alpha and beta, P-selectin), HSP (HSP20, HSP60, HSP70, HSP90), DNA, RNAs (mRNAs, small non-coding RNAs, long non-coding RNA, ribosomal RNA, transfer RNA, small nuclear RNA, small nucleolar RNA), lipids (ceramides, phosphatidylinostol, phosphatidylserine, phosphatidylcholine, cholesterol), growth factors (TRAIL, TGF, FGF, EGF, HGF), transcription factors, hormones, receptors (TfR, FasL), cytokines (TNF-alpha), enzymes (MMPs, malate dehydrogenase, citrate synthase, phosphoglycerate kinase 1), LAMP-1,-2 and-3, and metabolites (amino acids, adenosine triphosphate, amides, sugars, etc.).	Structural (actin, tubulin, cofilin) and regulatory proteins, integrins, selectins, CD40 ligand, flotillin-2, mRNAs, lipids, and metabolites (adenosine diphosphate ribosylation factor 6, phosphatidylserine).	DNA, histones, Annexin V, phosphatidylserine, cytosol portions.

Abbreviations: DNA: deoxyribonucleic acid; EGF: epidermal growth factor; MHC: Major Histocompatibility Complex; HSP: heat shock proteins; HGF: hepatocyte growth factor; FGF: fibroblast growth factor; LAMP: lysosomal-associated membrane protein; MMPs: metalloproteinases; RNA: ribonucleic acid; TNF: tumor necrosis factor; TGF: transforming growth factor; TRAIL: TNF-related apoptosis-inducing ligand; FasL: Fas ligand; TfR: Transferrin receptor.

For example, the secretome of cardiac progenitor and MSCs may contain miRNAs, such as miR-210, miR-132, miR-146a-3p, miR-4732-3p, and miR-497, which play a role in regulating cell death, proliferation, migration, and angiogenesis (Barile et al., 2014; Wu et al., 2020; Wang et al., 2017; Sánchez-Sánchez et al., 2021). Indeed, miR-210 has been shown to downregulate its known targets (Wang et al., 2017; Kim et al., 2012). Interestingly, the miR-4732-3p mimic was more effective than EVs at stimulating angiogenesis in vitro and in vivo, as well as reducing fibroblast differentiation when treated with transforming growth factor beta (Sánchez-Sánchez et al., 2021). Similarly, miR-132 downregulates its target RasGAP-p120, enhancing tube formation in vascular endothelial cells.

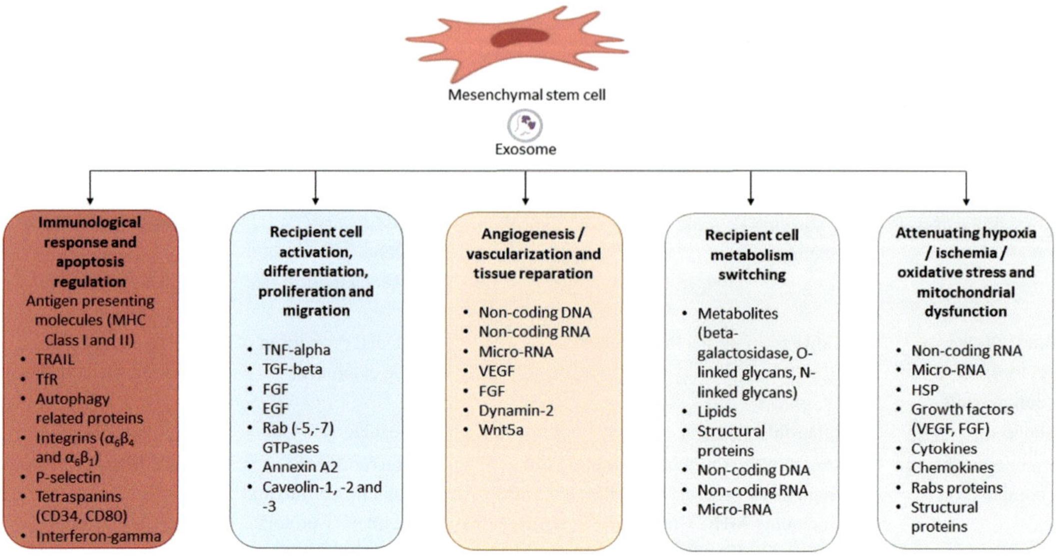

FIGURE 14.1 Biological effects of exosomes depending on their cargo components. Abbreviations: DNA: deoxyribonucleic acid; EGF: epidermal growth factor; MHC: Major Histocompatibility Complex; HSP: heat shock proteins; FGF: fibroblast growth factor; RNA: ribonucleic acid; TNF: tumor necrosis factor; TGF: Transforming growth factor; TRAIL: TNF-related apoptosis-inducing ligand; FasL: Fas ligand; TfR: Transferrin receptor; VEGF: vascular endothelial growth factor.

Nevertheless, miR-132 is likely to regulate the expression of the Ephrin A3 and Human Protein Tyrosine Phosphatase 1B (PTP1B), which, via the EGFR and JAK/STAT cascades, are also involved in the inhibition of apoptosis in cardiomyocytes and promotion of cardiomyocyte survival (Liu et al., 2022). Ren et al. (2021) reported that miR-29a-3p carried by MSCs-derived EVs downregulated the expression of inflammatory markers and improved cardiac function to attenuate myocardial injury via downregulation of the high-mobility group protein B1 (HMGB1)/Toll-like receptor-4/Akt/Akt/Akt axis. Thus, the cardiac protective effects of human pluripotent stem cell-derived exosomes can be enhanced by conditioning the cells under hypoxic conditions, and MALAT1 partially mediates this effect via targeting miRNAs (Anderson et al., 2016).

Another aspect of the regulation of recipient cell metabolism is an ability of exosomes to decrease the inflammatory and fibrotic responses of immune reactions via indoleamine 2,3-dioxygenase, prostaglandin E2, cytokines (interleukin [IL]-4, IL-10, tumor necrosis factor-alpha), nitric oxide (NO) and growth factors (hepatocyte growth factor, fibroblast growth factor, vascular endothelial growth factor) (van Balkom et al., 2019; Berezin, 2016). MSC-derived exosomes also encapsulate a number of proteins associated with the beneficial effects of implanted MSCs, such as cell adhesion molecules (CD29, CD44, CD47, CD73, fibronectin-1, ezrin, IQ motif containing GTPase activating protein 1, integrins and galectin LGALS1/LGALS3), lysosome-associated molecules (lysosomal-associated membrane protein-1, -2, and -3), surface receptors (epidermal growth factor receptor, platelet-derived growth factor receptor beta, toll-like receptor-2 and plasminogen activator, urokinase receptor), signaling molecules (G Protein Subunit Alpha 13/G Protein Subunit Gamma 12, CDC42, RRAS/NRAS, mitogen-activated protein kinase-1, VAV Guanine Nucleotide Exchange Factor-2), enzymes (matrix metalloproteinase-3,-9) and MSC-associated antigens (CD9, CD63, CD81, CD109, CD151, CD248 and CD276) (Gomzikova et al., 2019; Qiu et al., 2019; Berezin & Berezin, 2020; Ruvolo et al., 2020). Therefore, lipids, enzymes, and metabolites can regulate blood coagulation and matrix remodeling (Hade et al., 2021). Our knowledge of the biological cargo carried by EVs is continually developing. Integrated proteome, transcriptome, and lipidome databases, including Vesiclepedia (http://www.microvesicles.org), EVpedia (http://evpedia.info), and Exocarta (http://www.exocarta.org), now provide up-to-date information on exosomes from a variety of sources. Finally, due to the favorable biological properties of exosomal cargos, as well as the biocompatibility, stability, low toxicity, and immunogenicity of exosomes, they are promising for the paracrine modulation of recipient cell activities, and potentially for cardiac regenerative medicine (Pan et al., 2023).

14.4 IMPACT OF EXOSOMAL CARGOS ON DEVELOPING AND PROGRESSING HEART FAILURE

Experimental investigation has demonstrated that MSC-exosomes can accumulate in the heart and blood vessels. This, in turn, has been shown to have a significant inhibitory effect on extracellular matrix accumulation, cardiac fibrosis, adverse cardiac remodeling, inflammation, and apoptosis. Additionally, it has been discovered that they can stimulate angiogenesis and tissue repair (Figure 14.2).

14.4.1 Attenuating Adverse Cardiac Remodeling

Increasing evidence suggests that MSC-derived exosomes can significantly improve adverse cardiac remodeling and enhance cardiac function (Table 14.2).

Indeed, Zhu et al. (2022) have reported that intrapericardially injected MSC-derived exosomes provided an immunosurveillance effect on myocardium through forkhead box O3 (Foxo3) activation via the protein phosphatase (PP)-2A/p-Akt/forkhead box O3 (Foxo3) pathway. The authors concluded that MSC-derived exosomes are likely to present major histocompatibility complex (MHC)-II+ antigens for antigen-presenting cells and to induce Foxo3 activation, which subsequently

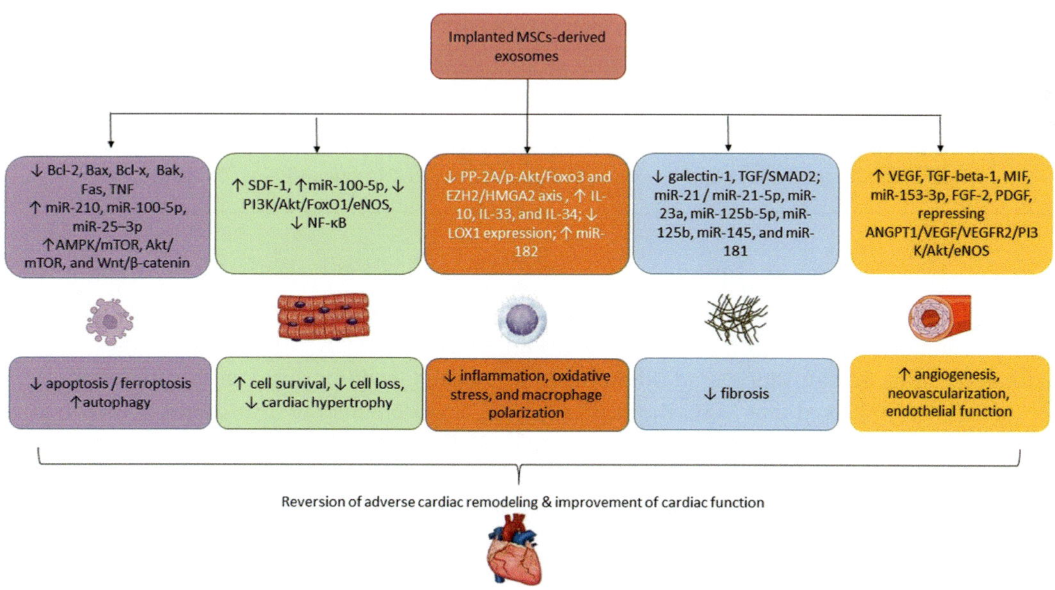

FIGURE 14.2 Underlying pathogenetic mechanisms for enhancing cardiac structure and function of MSC-derived exosomes. The figure was partly generated using free tools of BioRender.comAbbreviations: Foxo3: forkhead box O3; PP-2A/p-Akt/Foxo3: protein phosphatase (PP)-2A/p-Akt/forkhead box O3; SDF-1: stromal-derived factor-1; FGF-2: fibroblast growth factor-2; PDGF: platelet-derived growth factor; ↑: increase; ↓: decrease.

TABLE 14.2
The Effects of MSC-Derived Exosomes on Adverse Cardiac Remodeling

Model of HF	Route of Implantation	Underlying Biological Mechanisms	Biological Effects	References
Rat model of AMI	Intrapericardial injection	Decreases PP-2A/p-Akt/ Foxo3, MHC donation, increased IL-10, IL-33, and IL-34.	Improves cardiac function, reduces inflammatory reaction.	Zhu et al., 2022
Rat model of AMI	Intrapericardial injection	Decreases EZH2 expression.	Improves cardiac function, reduces cardiac fibrosis.	Jiao et al., 2022
Rat model of AMI	I/v injection	Decreases expression of FASL, PTEN, and increases HMGA2.	Improves cardiac function, reduces cell loss rate, and provides cardiac protection against ischemia.	Peng et al., 2020
Rat model of pressure overload HF	Cell culture	Decreases the senescence of myofibroblasts and apoptosis of cardiac myocytes.	Reduces cardiac hypertrophy and fibrosis.	Chen et al., 2020
Rat model of AMI	Intramuscular injection of umbilical cord-derived MSCs' exosomes	Transfer of miR-21, miR-23a, miR-125b, and miR-145 to recipient cells.	Reduces cardiac fibrosis.	Fang et al., 2016
Rat model of AMI	Intramuscular injection of umbilical cord-derived MSCs' exosomes	Transfer of miR-21-5p and miR-125b-5p.	Reduces cardiac fibrosis.	Zhang et al., 2021
—	Cell culture	Horizontal transfer of miR-21, miR-146a, and miR-181.	Reduces cardiac fibrosis.	Ti et al., 2016

Abbreviations: AMI: acute myocardial infarction; PP-2A/p-Akt/Foxo3: protein phosphatase (PP)-2A/p-Akt/forkhead box O3; HF: heart failure; MHC: major histocompatibility complex; miRNA: micro-RNAs; I/v: intravenous.

leads to the upregulated expression of anti-inflammatory cytokines (IL-10, IL-33, and IL-34) and the expansion of Treg cells in ischemic myocardium in a model of AMI. Ultimately, these factors facilitated cardiac repair, suppressed inflammatory reactions, and improved cardiac function.

Another study showed that MSC-derived exosomes have demonstrated sufficient ability to alleviate cardiac fibrosis and enhance cardiac function in AMI rats by inhibiting EZH2 expression (Jiao et al., 2022). It has been suggested that EZH2 intervened in HMGA2 expression by increasing the H3K27me3 modification, and PI3K/AKT pathway activity could be potentiated by the influence of MSC-derived exosomes on the EZH2/HMGA2 axis (Jiao et al., 2022). Interestingly, the cardioprotective effects of MSC-derived exosomes, observed in vitro and in vivo, could be related to miR-25-3p. It has directly targeted the pro-apoptotic genes FASL and PTEN, decreased the levels of EZH2 and H3K27me3, and consequently repressed the cardioprotective gene eNOS and the anti-inflammatory gene SOCS3 (Peng et al., 2020).

There is strong evidence that MSC-derived exosomes can significantly protect the myocardium against excessive cardiac hypertrophy, attenuate fibrosis, promote premature senescence of myofibroblasts, and reduce myocardial apoptosis when pressure overload occurs (Chen et al., 2020). Aligned with it, exosomal MSC microRNAs, such as miR-21 / miR-21-5p, miR-23a, miR-125b-5p, miR-125b, miR-145, and miR-181, are key players in preventing cardiac fibrosis and excessive accumulation of extracellular matrix (Fang et al., 2016; Zhang et al., 2021; Ti et al., 2016). Indeed, Fang et al. (2016) reported that a specific group of microRNAs (miR-21, miR-23a, miR-125b, and miR-145) transferred by exosomes from umbilical cord-derived MSCs can mediate myofibroblast formation by inhibiting excessive α-smooth muscle actin and collagen deposition, and this effect is associated with the transforming growth factor-β/SMAD2 signaling pathway. Additionally, umbilical cord blood MSC-derived exosomes, which contain miR-21-5p and miR-125b-5p, regulate collagen fiber distribution by inhibiting TGF-β receptor type II (TGFBR2) and TGF-β receptor type I (TGFBR1) (Zhang et al., 2021). Lastly, the cardioprotective properties of intravenously injected MSC-derived exosomes in a mouse model of pressure-overload HF depended markedly on circulating adiponectin (Nakamura et al., 2020).

14.4.2 Promoting Angiogenesis and Cell Viability

Accumulating evidence has shown that MSC-derived exosomes can mediate cardiac repair following AMI and non-ischemic HF by promoting angiogenesis and enhancing cardiac cell viability (Table 14.3).

To note, in a rat model of AMI, intramuscular injection into the peri-infarct region of MSC-derived exosomes, embossed with a plasmid of cardioprotective cytokine—macrophage migration inhibitory factor (MIF)—significantly suppressed mitochondrial fragmentation, reactive oxygen species generation, and cell apoptosis under hypoxia via activating adenosine 5'-monophosphate-activated protein kinase signaling (Liu et al., 2020). Moreover, this injection of MIF-MSC-exosomes

TABLE 14.3
The Effects of MSC-Derived Exosomes on Angiogenesis and Cardiac Cell Survival

Model of HF	Route of Implantation	Underlying Biological Mechanisms	Biological Effects	References
Rat model of AMI	Intramuscular injection	Transfer of MIF plasmid	Decreases mitochondrial fragmentation, ROS generation, and cell apoptosis.	Liu et al., 2020
Rat model of AMI	Intramuscular injection	Transfer of miR-153-3p	Improves cardiac cell survival, angiogenesis, and decreases apoptosis.	Ning et al., 2021
Mice model of ischemia/ reperfusion	Intramuscular injection	Transfer of SDF-1α	Improves cardiac function, decreases apoptosis and autophagy, and increases the regeneration of endothelial cells.	Gong et al., 2019
Mice model of ischemia/ reperfusion	Intramuscular injection	Transfer of SDF-1α	Increases angiogenesis.	Zhang et al., 2019
Rat model of AMI	Intramuscular injection	Transfer of HIF-1α	Increases angiogenesis, reduces fibrosis.	Sun et al., 2020

Abbreviations: MIF: macrophage migration inhibitory factor; HIF-1α: hypoxia-inducible factor-1α; SDF-1: stromal-derived factor-1; ROS: reactive oxygen species; ↑: increase; ↓: decrease.

was associated with sufficient enhancement of myocardial function and reduced adverse cardiac remodeling (Liu et al., 2020). The ANGPT1/VEGF/VEGFR2/PI3K/ Akt/eNOS pathway will likely be mediated by miR-153-3p captured in MSC-derived exosomes (Ning et al., 2021). Consequently, MSC-derived exosomes with low miR-153-3p expression can notably prevent endothelial cell and cardiomyocyte damage under hypoxia, reduce their apoptosis, and promote their viability and angiogenesis (Ning et al., 2021). Perhaps these facts explain the roles of the PI3K/Akt/FoxO1/ eNOS pathways in exosome-derived paracrine signaling and the preconditioning status of cardiac myocytes (Peng et al., 2024). In another study, stromal-derived factor-1-overexpressing MSC-derived exosomes suppressed apoptosis and autophagy in myocardial cells in mice with ischemia/ reperfusion injury. Also, they promoted microvascular endothelial cell regeneration in ischemic myocardium (Gong et al., 2019). The authors believe that MSCs-derived exosomes, which secrete stromal-derived factor-1, interfere with the PI3K signaling pathway and promote protective effects on the heart and blood vessels. Finally, exosomes from human umbilical cord mesenchymal stem cells (MSCs) can induce angiogenesis not only by increasing the expression of VEGF and hypoxia-inducible factor-1α (HIF-1α), but also by enhancing the proliferation, migration, and tube formation of endothelial progenitor cells (Zhang et al., 2019). In the rat model of AMI, HIF-1α-overexpressed MSC-derived exosomes preserved heart function by promoting neovascularization and inhibiting cardiac fibrosis (Sun et al., 2020). Additionally, the expression levels of proangiogenic factor mRNA and proteins were elevated after exosomes overexpressing HIF-1α were applied, both in vitro and in vivo (Sun et al., 2020). Indeed, there is mounting evidence that non-coding RNAs, particularly long non-coding RNAs (lncRNAs) and microRNAs, which are transferred by MSC-derived exosomes, play essential roles in regulating angiogenesis in damaged myocardium (Zhao et al., 2020).

14.4.3 Inhibiting Apoptosis of Cardiac Myocytes

The suppression of key regulators of apoptosis in ischemia/reperfusion myocardium is a promising approach to enhance cellular viability and prevent myocardial cell loss (Cheng et al., 2020a, 2020b). Table 14.4 illustrates the possible impact of MSC-derived exosomes on cardiac cell apoptosis.

TABLE 14.4
The Effects of MSC-Derived Exosomes on Cardiac Cell Apoptosis

Model of HF	Route of Implantation	Underlying Biological Mechanisms	Biological Effects	References
Rat model of doxorubicin-induced HF	Intramuscular injection	Transfer of miR-100-5p	Reduced cardiac myocytes, apoptosis, and oxidative stress, and increased cardiac cell viability.	Zhong et al., 2021
Rat model of AMI	Intramuscular injection	miR-210	Increased cardiac cell survival, angiogenesis, and decreased apoptosis.	Cheng et al., 2020a, 2020b
Cell culture H9c2	—	miR-21-5p	Reduced apoptosis.	Ji & Wang, 2024
Rat model of AMI with advanced HF	Intramuscular injection	miR-30e	Reduced apoptosis and autophagy.	Pu et al., 2021
Rat model of AMI	Intramuscular injection	miR-143-3p	Reduced apoptosis and autophagy.	Chen et al., 2021

Abbreviations: AMI: acute myocardial infarction; HF: heart failure.

Indeed, in the doxorubicin-induced HF umbilical cord model, MSC-derived exosomes suppressed oxidative stress and cardiac myocyte apoptosis, thereby improving their viability via the miR-100-5p/NOX4 pathway (Zhong et al., 2021). A recent study showed that MSC-exosomes, particularly those containing miR-210, can significantly reduce the regulatory effects of the apoptotic markers cleaved caspase-3, Bad, and Bax, while increasing the levels of the anti-apoptotic protein BCL-2 in rats with acute myocardial infarction (AMI) (Cheng et al., 2020a, 2020b). Furthermore, it has been demonstrated that exosomal miR–25–3p from MSC-derived exosomes can alleviate adverse cardiac remodeling by targeting key pro-apoptotic proteins (FASL and PTEN) and EZH2 (Peng et al., 2020).

In addition, it has been suggested that these exosomes could significantly upregulate miR-21-5p levels in H9c2 cells and abrogate their function, thereby reducing their protective effects on cardiac myocytes. Moreover, the authors have revealed that miR-21-5p can exert its effect through a binding site in the 3' UTR of YAP1, which directly inhibits YAP1 expression (Ji & Wang, 2024). Additionally, exosomal miR-30e from rat bone marrow-derived MSCs significantly reduced LOX1 expression, thereby decreasing NF-κB p65/Caspase-9 signaling activity and improving advanced HF after AMI in rats (Pu et al., 2021). Interestingly, MSC-derived exosomal miR-143-3p is likely effective against myocardial cell apoptosis by regulating autophagy via the CHK2-Beclin2 pathway (Chen et al., 2021). Thus, MSC-exosomes are an excellent and promising cell-free therapy for advanced HF because they inhibit apoptosis, regulate inflammation, ameliorate cardiac remodeling, and promote angiogenesis.

14.4.4 Ameliorating Systemic and Microvascular Inflammation

Several previous studies have demonstrated protective effects against inflammation and promoted the recovery of cardiac and vascular function by regulating complementary systems via MSC-derived exosomes (Table 14.5). Ischemia/reperfusion injury typically triggers local and systemic inflammatory responses, which can lead to adverse outcomes (Bhaskara et al., 2023; Binek et al., 2017).

TABLE 14.5
The Effect of MSC-Derived Exosomes on Inflammation

Model of HF	Route of Implantation	Underlying Biological Mechanisms	Biological Effects	References
Mouse model of myocardial ischemia/reperfusion injury	Intramyocardial injection	Transfer of miRNA-181a	Decreased oxidative stress and increased Treg polarization.	Wei et al., 2019
Mouse model of myocardial ischemia/reperfusion injury	Intramuscular injection	Transfer of miR-182	Decreased inflammation and macrophage polarization.	Zhao et al., 2019
Mouse model of myocardial ischemia/reperfusion injury	Intramuscular injection	Transfer of miR-1246	Decreased hypoxia-induced myocardial tissue damage.	Wang et al., 2021
Rat model of myocardial ischemia-reperfusion injury	Intramuscular injection	Transfer of miR-146a	Decreased apoptosis and autophagy.	Meng et al., 2024
Rat model of HF/cellular model of hypoxia	Intramuscular injection	Regulating the UL3/Hippo pathway	Decreased apoptosis, ferroptosis, cellular damage, and cardiac fibrosis.	Ren & Zhao, 2024

Abbreviations: AMI: acute myocardial infarction; HF: heart failure

Wei et al. (2019) investigated whether MSC-derived exosomes carrying miRNA-181a-overexpressing lentiviruses intramyocardially injected into a mouse myocardial ischemia/reperfusion injury model could produce anti-inflammatory and cardiac protective effects. The authors reported that overexpression of miR-181a in MSC-exosomes can target the c-Fos gene and inhibit the immune-inflammatory response by increasing Treg polarization (Wei Z. et al., 2019). In addition, MSC-derived exosomes, which contain miR-182 as a potent candidate mediator of macrophage polarization and toll-like receptor 4 (TLR4), were found to reduce inflammation in a mouse model of myocardial ischemia/reperfusion injury (Zhao et al., 2019). In another study, Wang Z. et al. (2021) found that exosomal miR-1246 released from human umbilical cord MSCs attenuated hypoxia-induced myocardial tissue damage by targeting serine protease 23 and suppressing the activation of the Snail/alpha-smooth muscle actin signaling pathway. However, injection of exosomal miR-146a derived from MSCs improved cardiac function in a rat model of myocardial ischemia-reperfusion injury (Meng et al., 2024). The cardioprotective effect was associated with reducing myocardial tissue apoptosis and suppressing inflammation. Therefore, in a rat model of HF and a cellular model of hypoxia, bone marrow MSC-derived exosomes sufficiently restored cardiac function and inhibited oxidative stress, apoptosis, cellular damage, cardiac fibrosis, and iron deposition in myocardial tissues by regulating the UL3/Hippo pathway (Ren & Zhao, 2024). Taken together, the use of exosomal cargo, including miRNAs with cardioprotective properties, may be a promising strategy for treating HF following ischemia/reperfusion damage (Wang et al., 2018).

14.5 POSSIBLE PERSPECTIVES FOR CLINICAL IMPLICATIONS OF MSC-DERIVED EXOSOMES IN THE MANAGEMENT OF HF IN THE FUTURE

The therapeutic benefits of exosome-based cell-free therapy for HF could be associated with improvement of cardiac function, including restoring left ventricular ejection fraction, attenuating cardiac fibrosis, preventing ischemia/reperfusion damage of myocardium, reducing microvascular inflammation and obstruction, increasing survival probability, improving quality of life, and reducing mortality rate (Daneshi et al., 2022). In this context, MSC-derived secretomes are often considered an off-the-shelf therapeutic approach that mitigates safety risks and the risk of microvascular occlusion (Habib et al., 2025b). They also prevent unregulated growth, particularly when administering large amounts of viable cells. In clinical settings, promising results have been obtained with MSC-derived exosomes, similar to those observed with MSC transplantation. Although the application of exosomes in clinical therapy is still some way off due to several challenges (e.g., maximizing cardiac regenerative effects, eliminating off-target effects, and generating a highly reproducible and consistent population), they could certainly offer a new perspective approach for the future treatment of ischemia-induced advanced HF and perhaps non-ischemic HF. Extensive clinical studies are necessary to determine whether the therapeutic benefits of MSC-derived exosomes are likely to be superior to those of MSC cell therapy.

14.6 CONCLUSIONS

In in vitro and in vivo studies, MSC-derived exosomes demonstrate potential comparable to that of MSCs in reversing cardiac remodeling, improving inflammation, reducing cardiomyocyte apoptosis, and enhancing myocardial function. It remains challenging to control the immunomodulatory properties and tissue repair activity of MSC-derived exosomes in specific microenvironments, thereby developing the most optimal cocktail of specific MSC components, such as miRNAs/non-coding RNAs, regulatory peptides, metabolites, and growth factors. Although the precise mechanisms underlying cardioprotection and the improvement in cardiac function associated with the implantation of MSC-derived exosomes remain unclear, the lack of high-quality clinical studies precludes discussion of their therapeutic benefits in clinical settings. Future clinical studies will likely overcome this knowledge gap and provide new evidence for MSC-derived exosome implantation as an alternative to stem cell therapy in HF patients.

14.7 DECLARATION OF GENERATIVE AI IN SCIENTIFIC WRITING

The authors did not use generative artificial intelligence and artificial intelligence-assisted technologies in the writing process.

14.8 DECLARATION OF INTEREST

The authors have no relevant affiliations or financial involvement with any organization or entity with a financial interest in or financial conflict with the subject matter or materials discussed in the manuscript. This includes employment, consultancies, honoraria, stock ownership or options, expert testimony, grants or patents received or pending, or royalties.

14.9 FUNDING SOURCE

This research received no external funding.

REFERENCES

Ahmed OTF, Ahmed ZT, Dairi AW, Zain Al-Abeden MS, Alkahlot MH, Alkahlot RH, Al Jowf GI, et al. The inconclusive superiority debate of allogeneic versus autologous MSCs in treating patients with HFrEF: a systematic review and meta-analysis of RCTs. *Stem Cell Res Ther.* 2025;16(1):175. doi: 10.1186/s13287-025-04209-5.

Ahmed ZT, Alkahlot MH, Haider KH. MSC-derived exosomes: advances in cell-free therapy. In: *Handbook of Stem Cell Applications.* Singapore: Springer Nature Singapore; 2023 Aug 18. p. 1–41.

Ahmed ZT, Zain Al-Abeden MS, Al Abdin MG, Muqresh MA, Al Jowf GI, Eijssen LMT, Haider KH. The dose-response relationship of MSCs as living Bio-drugs in HFrEF patients is a systematic review and meta-analysis of RCTs. *Stem Cell Res Ther.* 2024 Jun 13;15(1):165. doi: 10.1186/s13287-024-03713-4.

Al-Khani AM, Kalou Y, Haider KH. Response to letter to the editor: comment on bone marrow mesenchymal stem cells for heart failure treatment: a systematic review and meta-analysis. *Heart, Lung and Circulation.* 2023 Sep;32(9):e59–e60. doi: 10.1016/j.hlc.2023.07.001.

Al-Khani AM, Khalifa MA, Haider KH. Mesenchymal stem cells for cardiac repair. In: KH Haider, editor. *Handbook of Stem Cell Therapy.* Singapore: Springer Nature Singapore; 2022. https://doi.org/10.1007/978-981-16-6016-0_11-1

Anderson JD, Johansson HJ, Graham CS, Vesterlund M, Pham MT, Bramlett CS, et al. Comprehensive proteomic analysis of mesenchymal stem cell exosomes reveals modulation of angiogenesis via nuclear factor-kappaB signaling. *Stem Cells.* 2016;34(3):601–613. doi: 10.1002/stem.2298

Barile L, Lionetti V, Cervio E, Matteucci M, Gherghiceanu M, Popescu LM, et al. Extracellular vesicles from human cardiac progenitor cells inhibit cardiomyocyte apoptosis and improve cardiac function after myocardial infarction. *Cardiovasc Res.* 2014;103(4):530–41. doi: 10.1093/cvr/cvu167.

Beghini A, Sammartino AM, Papp Z, von Haehling S, Biegus J, Ponikowski P, Adamo M, et al. 2024 update in heart failure. *ESC Heart Fail.* 2025 Feb;12(1):8–42. doi: 10.1002/ehf2.14857.

Berezin A. Epigenetics in heart failure phenotypes. *BBA Clin.* 2016;6:31–7. doi: 10.1016/j.bbacli.2016.05.005.

Berezin AE, Berezin AA. Extracellular endothelial cell-derived vesicles: emerging role in cardiac and vascular remodeling in heart failure. *Front Cardiovasc Med.* 2020; 7:47. doi: 10.3389/fcvm. 2020.00047.

Bhaskara M, Anjorin O, Wang M. Mesenchymal stem cell-derived exosomal microRNAs in cardiac regeneration. *Cells.* 2023;12(24):2815. doi: 10.3390/cells12242815.

Bhatt AS, Ambrosy AP, Velazquez EJ. Adverse remodeling and reverse remodeling after myocardial infarction. *Curr Cardiol Rep.* 2017;19(8):71. doi: 10.1007/s11886-017-0876-4.

Bian S, Zhang L, Duan L, Wang X, Min Y, Yu H. Extracellular vesicles derived from human bone marrow mesenchymal stem cells promote angiogenesis in a rat myocardial infarction model. *J Mol Med (Berl).* 2014;92(4):387–97. doi: 10.1007/s00109-013-1110-5.

Binek A, Fernández-Jiménez R, Jorge I, Camafeita E, López JA, Bagwan N, Galán-Arriola C, et al. Proteomic footprint of myocardial ischemia/reperfusion injury: longitudinal study of the at-risk and remote regions in the pig model. *Sci Rep.* 2017;7(1):12343. doi: 10.1038/s41598-017-11985-5.

Blum M, Goldstein NE, Jaarsma T, Allen LA, Gelfman LP. Palliative care in heart failure guidelines: a comparison of the 2021 ESC and the 2022 AHA/ACC/HFSA guidelines on heart failure. *Eur J Heart Fail.* 2023;25(10):1849–1855. doi: 10.1002/ejhf.2981.

Bolli R, Solankhi M, Tang XL, Kahlon A. Cell therapy in patients with heart failure: a comprehensive review and emerging concepts. *Cardiovasc Res.* 2022;118(4):951–976. doi: 10.1093/cvr/cvab135.

Börger V, Weiss DJ, Anderson JD, Borràs FE, Bussolati B, Carter DRF, Dominici M, et al. International society for extracellular vesicles and international society for cell and gene therapy statement on extracellular vesicles from mesenchymal stromal cells and other cells: considerations for potential therapeutic agents to suppress coronavirus disease-19. *Cytotherapy.* 2020;22(9):482–485. doi: 10.1016/j.jcyt.2020.05.002.

Bragazzi NL, Zhong W, Shu J, Abu Much A, Lotan D, Grupper A, Younis A, et al. Burden of heart failure and underlying causes in 195 countries and territories from 1990 to 2017. *Eur J Prev Cardiol.* 2021;28(15):1682–1690. doi: 10.1093/eurjpc/zwaa147.

Breitbach M, Bostani T, Roell W, Xia Y, Dewald O, Nygren JM, et al. Potential risks of bone marrow cell transplantation into infarcted hearts. *Blood.* 2007; 110:1362–9. doi: 10.1182/blood-2006-12-063412

Chaudhry SP, Stewart GC. Advanced heart failure: prevalence, natural history, and prognosis. *Heart Fail Clin.* 2016;12(3):323–33. doi: 10.1016/j.hfc.2016.03.001.

Chen F, Li X, Zhao J, Geng J, Xie J, Xu B. Bone marrow mesenchymal stem cell-derived exosomes attenuate cardiac hypertrophy and fibrosis in pressure overload-induced remodeling. *In Vitro Cell Dev Biol Anim.* 2020;56(7):567–576. doi: 10.1007/s11626-020-00481-2.

Chen G, Wang M, Ruan Z, Zhu L, Tang C. Mesenchymal stem cell-derived exosomal miR-143-3p suppresses myocardial ischemia-reperfusion injury by regulating autophagy. *Life Sci.* 2021;280:119742. doi: 10.1016/j.lfs.2021.119742.

Cheng H, Chang S, Xu R, Chen L, Song X, Wu J, Qian J, et al. Hypoxia-challenged MSC-derived exosomes deliver miR-210 to attenuate post-infarction cardiac apoptosis. *Stem Cell Res Ther.* 2020a;11(1):224. doi: 10.1186/s13287-020-01737-0.

Cheng H, Ge X, Wang, L, Xu L. Four-quadrant sensorless operation of switched reluctance machine over the wide speed range. *Electric Power Components and Systems* 2020b;48(1–2):224–40. doi: 10.1080/15325008.2020.1731874

Crespo-Leiro MG, Barge-Caballero E. Advanced heart failure: definition, epidemiology, and clinical course. *Heart Fail Clin.* 2021;17(4):533–545. doi: 10.1016/j.hfc.2021.06.002.

Cui LL, Kerkelä E, Bakreen A, Nitzsche F, Andrzejewska A, Nowakowski A, Janowski M, et al. The cerebral embolism induced by intra-arterial delivery of allogeneic bone marrow mesenchymal stem cells in rats is associated with cell dose and infusion velocity. *Stem Cell Res Ther.* 2015;6(1):11. doi: 10.1186/scrt544.

Daneshi N, Bahmaie N, Esmaeilzadeh A. Cell-free treatments: a new generation of targeted therapies for treatment of ischemic heart disease. *Cell J.* 2022;24(7):353–363. doi: 10.22074/cellj.2022.7643.

Eggenhofer E, Benseler V, Kroemer A, Popp FC, Geissler EK, Schlitt HJ, Baan CC, et al. Mesenchymal stem cells are short-lived and do not migrate beyond the lungs after intravenous infusion. *Front Immunol.* 2012;3:297. doi: 10.3389/fimmu.2012.00297.

Fang S, Xu C, Zhang Y, Xue C, Yang C, Bi H, Qian X, et al. Umbilical cord-derived mesenchymal stem cell-derived exosomal MicroRNAs suppress myofibroblast differentiation by inhibiting the transforming growth factor-β/SMAD2 pathway during wound healing. *Stem Cells Transl Med.* 2016;5(10):1425–1439. doi: 10.5966/sctm.2015-0367.

Fu H, Chen Q. Mesenchymal stem cell therapy for heart failure: a meta-analysis. *Herz.* 2020;45(6):557–563. doi: 10.1007/s00059-018-4762-7.

Gomzikova MO, James V, Rizvanov AA. Therapeutic application of mesenchymal stem cells derived extracellular vesicles for immunomodulation. *Front Immunol.* 2019; 10:2663. doi: 10.3389/fimmu.2019.02663.

Gong XH, Liu H, Wang SJ, Liang SW, Wang GG. Exosomes derived from SDF1-overexpressing mesenchymal stem cells inhibit ischemic myocardial cell apoptosis and promote cardiac endothelial microvascular regeneration in mice with myocardial infarction. *J Cell Physiol.* 2019;234(8):13878–13893. doi: 10.1002/jcp.28070.

Habib SM, Martini MF, Al Ahmed O, Qassab AR, Haider KH. Mesenchymal stem cell secretomes for cell-free therapy of the broken heart. In: KH Haider, editor. *Handbook of Regenerative Medicine: Cell-Free Therapy.* Boca Raton (FL): CRC Press; 2025. Chapter 18, p. 274–293.

Habib SM, Martini MF, Al Ahmed O, Qassab AR, Haider KH. The potential of mesenchymal stem cell–derived exosomes in cell-free therapy for myocardial repair. In: KH Haider, editor. *Handbook of Regenerative Medicine: Cell-Free Therapy.* Boca Raton (FL): CRC Press; 2025. Chapter 24, p. 372–390.

Habib SY, Martini MF, Hamdan YNA, Shrebaty OMM, Haider KH. Mesenchymal stem cell secretome and exosomes as potential advanced therapy medicinal products for treating a failing heart. *World J Stem Cells.* 2026; 18(3): doi: 10.4252/wjsc.v18.i3.113871.

Hade MD, Suire CN, Suo Z. Mesenchymal stem cell-derived exosomes: applications in regenerative medicine. *Cells.* 2021;10(8):1959. doi: 10.3390/cells10081959.

Haider KH, Najimi M (2022) Exosome-based cell-free therapy in regenerative medicine for myocardial repair. In: KH Haider, editor. *Handbook of Stem Cell Therapy.* Singapore: Springer Nature Singapore; 2022. https://doi.org/10.1007/978-981-16-6016-0_9-1.

Haider KH, Aramini B. Micrining the injured heart with stem cell-derived exosomes: an emerging strategy of cell-free therapy. *Stem Cell Res Ther.* 2020;11(1):23. doi: 10.1186/s13287-019-1548-7.

Heidenreich PA, Bozkurt B, Aguilar D, Allen LA, Byun JJ, Colvin MM, et al; ACC/AHA guideline for the management of heart failure: a report of the American College of Cardiology/American Heart Association Joint Committee on Clinical Practice Guidelines. *Circulation.* 2022;145(18):e895–e1032. doi: 10.1161/CIR.0000000000001063.

Ji Z, Wang C. Mesenchymal stem cell-derived exosomal miR-21-5p inhibits YAP1 expression and improves outcomes in myocardial infarction. *BMC Cardiovasc Disord.* 2024;24(1):547. doi: 10.1186/s12872-024-04197-z.

Jiang S, Haider KH, Niagara MI, Lu G, Ashraf M (2008) Stable therapeutic effects of mesenchymal stem cell-based multiple gene delivery for cardiac repair. *Cardiovasc Res.* 77(3): 525–33.

Jiao W, Hao J, Xie Y, Meng M, Gao W. EZH2 mitigates the cardioprotective effects of mesenchymal stem cell-secreted exosomes against infarction via HMGA2-mediated PI3K/AKT signaling. *BMC Cardiovasc Disord.* 2022;22(1):95. doi: 10.1186/s12872-022-02533-9.

Kalou Y, Al-Khani AM, Haider KH. Bone marrow mesenchymal stem cells for heart failure treatment: a systematic review and meta-analysis. *Heart Lung Circ.* 2023 S1443-9506(23)00068–9. doi: 10.1016/j.hlc.2023.01.012. Mar 26; 16(3): 237–244. doi: 10.4252/wjsc.v16.i3.237.

Karnas E, Dudek P, Zuba-Surma EK. Stem cell-derived extracellular vesicles as new tools in regenerative medicine: immunomodulatory role and future perspectives. *Front Immunol.* 2023;14:1120175. doi: 10.3389/fimmu.2023.1120175.

Kavousi S, Hosseinpour A, Bahmanzadegan Jahromi F, Attar A. Efficacy of mesenchymal stem cell transplantation on major adverse cardiovascular events and cardiac function indices in patients with chronic heart failure: a meta-analysis of randomized controlled trials. *J Transl Med.* 2024;22(1):786. doi: 10.1186/s12967-024-05352-y.

Kim HW, Mallick F, Durrani S, Ashraf M, Jiang S, Haider KH. Concomitant activation of miR-107/PDCD10 and hypoxamir-210/Casp8ap2 and their role in cytoprotection during ischemic preconditioning of stem cells. *Antioxid Redox Signal.* 2012 Oct 15;17(8):1053–65. doi: 10.1089/ars.2012.4518.

Kunter U, Rong S, Boor P, Eitner F, Muller-Newen G, Djuric Z, et al. Mesenchymal stem cells prevent the progression of experimental renal failure but maldifferentiate into glomerular adipocytes. *J Am Soc Nephrol.* 2007; 18:1754–64. doi: 10.1681/ASN.2007010044

Li J, Lei L, Wang W, Ding W, Yu Y, Pu B, et al. Social Risk Profile And Cardiovascular-kidney-metabolic Syndrome in US adults. *J Am Heart Assoc.* 2024;13(16):e034996. doi: 10.1161/JAHA.124.034996.

Liang Y, Xia J, Xu L, Zhang Y, Xu X, Haider KH. Cell-free and cell-based therapy in regenerative medicine. In: KH Haider, editor. *Handbook of Regenerative Medicine: Cell-Free Therapy.* Boca Raton (FL): CRC Press; 2025. Chapter 10, p. 134–153.

Liu R, Mathieu C, Berthelet J, Zhang W, Dupret JM, Rodrigues Lima F. Human protein tyrosine phosphatase 1B (PTP1B): from structure to clinical inhibitor perspectives. *Int J Mol Sci.* 2022;23(13):7027. doi: 10.3390/ijms23137027.

Liu X, Li X, Zhu W, Zhang Y, Hong Y, Liang X, Fan B, et al. Exosomes from mesenchymal stem cells overexpressing MIF enhance myocardial repair. *J Cell Physio*l. 2020;235(11):8010–8022. doi: 10.1002/jcp.29456.

McDonagh TA, Metra M, Adamo M, Gardner RS, Baumbach A, Böhm M, et al. ESC scientific document group. 2021 ESC Guidelines for the diagnosis and treatment of acute and chronic heart failure: Developed by the Task Force for the diagnosis and treatment of acute and chronic heart failure of the European Society of Cardiology (ESC). With the special contribution of the Heart Failure Association (HFA) of the *ESC. Eur J Heart Fail.* 2022;24(1):4–131. doi: 10.1002/ejhf.2333.

Meng WT, Zhu J, Wang YC, Shao CL, Li XY, Lu PP, Huang MY, et al. Targeted delivery of miR-146a via IMTP-modified milk exosomes exerted cardioprotective effects by inhibiting NF-κB signaling after myocardial ischemia-reperfusion injury. *J Nanobiotechnology.* 2024;22(1):382. doi: 10.1186/s12951-024-02631-0.

Muqresh MA, Harbieh I, Haider KH. Myogenic cardiac regeneration: clinical studies using skeletal myoblasts. In: KH Haider, editor. *Handbook of Stem Cell Applications.* Singapore: Springer Nature Singapore; 2024. p. 849–885. doi: 10.1007/978-981-99-7119-0_5

Nair N, Gongora E. Stem cell therapy in heart failure: Where do we stand today? *Biochim Biophys Acta Mol Basis Dis.* 2020;1866(4):165489. doi: 10.1016/j.bbadis.2019.06.003.

Nakamura Y, Kita S, Tanaka Y, Fukuda S, Obata Y, Okita T, Nishida H, et al. Adiponectin stimulates exosome release to enhance mesenchymal stem-cell-driven therapy of heart failure in mice. *Mol Ther.* 2020; 28(10):2203-2219. doi: 10.1016/j.ymthe.2020.06.026

Nauta AJ, Westerhuis G, Kruisselbrink AB, Lurvink EG, Willemze R, Fibbe WE. Donor-derived mesenchymal stem cells are immunogenic in an allogeneic host and stimulate donor graft rejection in a nonmyeloablative setting. *Blood.* 2006;108(6):2114–20. doi: 10.1182/blood-2005-11-011650.

Ning W, Li S, Yang W, Yang B, Xin C, Ping X, et al. Blocking exosomal miRNA-153-3p derived from bone marrow mesenchymal stem cells ameliorates hypoxia-induced myocardial and microvascular damage by targeting the ANGPT1-mediated VEGF/PI3k/Akt/eNOS pathway. *Cell Signal.* 2021;77:109812. doi: 10.1016/j.cellsig.2020.109812.

Pan Y, Wu W, Jiang X, Liu Y. Mesenchymal stem cell-derived exosomes in cardiovascular and cerebrovascular diseases: From mechanisms to therapy. *Biomed Pharmacother.* 2023;163:114817. doi: 10.1016/j.biopha.2023.114817.

Peng C, Yan J, Jiang Y, Wu L, Li M, Fan X. Exploring cutting-edge approaches to potentiate mesenchymal stem cell and exosome therapy for myocardial infarction. *J Cardiovasc Transl Res.* 2024;17(2):356–375. doi: 10.1007/s12265-023-10438-x.

Peng Y, Zhao JL, Peng ZY, Xu WF, Yu GL. Exosomal miR-25-3p from mesenchymal stem cells alleviates myocardial infarction by targeting pro-apoptotic proteins and the epigenetic regulator EZH2. *Cell Death Dis.* 2020;11(5):317. doi: 10.1038/s41419-020-2545-6.

Poglajen G, Frljak S, Zemljič G, Cerar A, Okrajšek R, Šebeštjen M, et al. Stem cell therapy for chronic and advanced heart failure. *Curr Heart Fail Rep.* 2020;17(5):261–270. doi: 10.1007/s11897-020-00477-9.

Pu L, Kong X, Li H, He X. Exosomes released from mesenchymal stem cells overexpressing microRNA-30e ameliorate heart failure in rats with myocardial infarction. *Am J Transl Res.* 2021;13(5):4007–4025.

Qayyum AA, Mouridsen M, Nilsson B, Gustafsson I, Schou M, Nielsen OW, et al. Danish phase II trial using adipose tissue-derived mesenchymal stromal cells for patients with ischaemic heart failure. *ESC Heart Fail.* 2023a;10(2):1170–1183. doi: 10.1002/ehf2.14281.

Qiu G, Zheng G, Ge M, Wang J, Huang R, Shu Q, Xu J. Functional proteins of mesenchymal stem cell-derived extracellular vesicles. *Stem Cell Res Ther.* 2019; 10(1):359. doi: 10.1186/s13287-019-1484-6.

Rajab AM, Rajab TM, AlJundi S, Haider KH. Bone stem cell therapy in the clinical perspective: a focus on nonrandomized and randomized trials. In: KH Haider, editor. *Stem Cells: From Myth to Reality and Evolving.* Berlin (Germany): De Gruyter; 2019.

Ren S, Pan L, Yang L, Niu Z, Wang L, Feng H, Yuan M. miR-29a-3p transferred by mesenchymal stem cells-derived extracellular vesicles protects against myocardial injury after severe acute pancreatitis. *Life Sci.* 2021;272:119189. doi: 10.1016/j.lfs.2021.119189.

Ren Y, Zhao X. Bone marrow mesenchymal stem cells-derived exosomal lncRNA GAS5 mitigates heart failure by inhibiting UL3/Hippo pathway-mediated ferroptosis. *Eur J Med Res.* 2024;29(1):303. doi: 10.1186/s40001-024-01880-x.

Rosca AM, Rayia DM, Tutuianu R. Emerging role of stem cells - derived exosomes as valuable tools for cardiovascular therapy. *Curr Stem Cell Res Ther.* 2017;12(2):134–138. doi: 0.2174/1574888X10666151026115320.

Rosland GV, Svendsen A, Torsvik A, Sobala E, McCormack E, Immervoll H, et al. Long-term cultures of bone marrow-derived human mesenchymal stem cells frequently undergo spontaneous malignant transformation. *Cancer Res.* 2009; 69:5331 9. doi: 10.1158/0008-5472.CAN-08-4630

Ruvolo PP, Ma H, Ruvolo VR, Zhang X, Post SM, Andreeff M. LGALS1 acts as a pro-survival molecule in AML. *Biochim Biophys Acta Mol Cell Res.* 2020;1867(10):118785. doi: 10.1016/j.bbamcr.2020.118785.

Sánchez-Sánchez R, Gómez-Ferrer M, Reinal I, Buigues M, Villanueva-Bádenas E, Ontoria-Oviedo I, Hernándiz A, et al. miR-4732-3p in Extracellular vesicles from mesenchymal stromal cells is cardioprotective during myocardial Ischemia. *Front Cell Dev Biol.* 2021;9:734143. doi: 10.3389/fcell.2021.734143.

Savarese G, Becher PM, Lund LH, Seferovic P, Rosano GMC, Coats AJS. Global burden of heart failure: a comprehensive and updated review of epidemiology. *Cardiovasc Res.* 2023 Jan 18;118(17):3272–3287. doi: 10.1093/cvr/cvac013. *Erratum Cardiovasc Res.* 2023 Jun 13;119(6):1453. doi: 10.1093/cvr/cvad026.

Sgaglione J, Neufeld EV, Swami P, Grande DA. Clinical status of exosomes: A review. *HSS J.* 2025: 15566316251362179. doi: 10.1177/15566316251362179.

Shahim B, Kapelios CJ, Savarese G, Lund LH. Global public health burden of heart failure: an updated review. *Card Fail Rev.* 2023;9:e11. doi: 10.15420/cfr.2023.05.

Shao L, Zhang Y, Lan B, Wang J, Zhang Z, Zhang L, et al. MiRNA-sequence indicates that mesenchymal stem cells and exosomes have a similar mechanism to enhance cardiac repair. *Biomed Res Int.* 2017;2017:4150705. doi: 10.1155/2017/4150705.

Sun J, Shen H, Shao L, Teng X, Chen Y, Liu X, Yang Z, et al. HIF-1α overexpression in mesenchymal stem cell-derived exosomes mediates cardioprotection in myocardial infarction by enhanced angiogenesis. *Stem Cell Res Ther.* 2020;11(1):373. doi: 10.1186/s13287-020-01881-7.

Théry C, Witwer KW, Aikawa E, Alcaraz MJ, Anderson JD, Andriantsitohaina R, et al. Minimal information for studies of extracellular vesicles 2018 (MISEV2018): a position statement of the International Society for Extracellular Vesicles and update of the MISEV2014 guidelines. *J Extracell Vesicles.* 2018;7(1):1535750. doi: 10.1080/20013078.2018.1535750.

Ti D, Hao H, Fu X, Han W. Mesenchymal stem cells-derived exosomal microRNAs contribute to wound inflammation. *Sci China Life Sci.* 2016;59(12):1305–1312. doi: 10.1007/s11427-016-0240-4.

Van Balkom BWM, Gremmels H, Giebel B, Lim SK. Proteomic signature of mesenchymal stromal cell-derived small extracellular vesicles. *Proteomics.* 2019; 19(1–2):e1800163. doi: 10.1002/pmic.201800163.

van Niel G, D'Angelo G, Raposo G. Shedding light on the cell biology of extracellular vesicles. *Nat Rev Mol Cell Biol.* 2018;19(4):213–228. doi: 10.1038/nrm.2017.125.

Viola M, de Jager SCA, Sluijter JPG. Targeting inflammation after myocardial infarction: a therapeutic opportunity for extracellular vesicles? *Int. J. Mol. Sci.* 2021 Jul 22;22(15):7831. doi: 10.3390/ijms22157831.

Wang N, Chen C, Yang D, Liao Q, Luo H, Wang X,. Mesenchymal stem cell-derived extracellular vesicles, via miR-210, improve infarcted cardiac function by promoting angiogenesis. *Biochim Biophys Acta Mol Basis Dis.* 2017;1863(8):2085–2092. doi: 10.1016/j.bbadis.2017.02.023.

Wang X, Chen Y, Zhao Z, Meng Q, Yu Y, Sun J, Yang Z, et al. Engineered exosomes with ischemic myocardium-targeting peptide for targeted therapy in myocardial infarction. *J Am Heart Assoc.* 2018; 7(15):e008737. doi: 0.1161/JAHA.118.008737.

Wang Z, Gao D, Wang S, Lin H, Wang Y, Xu W. Exosomal microRNA-1246 from human umbilical cord mesenchymal stem cells potentiates myocardial angiogenesis in chronic heart failure. *Cell Biol Int.* 2021;45(11):2211–2225. doi: 10.1002/cbin.11664.

Wei Z, Qiao S, Zhao J, Liu Y, Li Q, Wei Z, Dai Q, et al. miRNA-181a over-expression in mesenchymal stem cell-derived exosomes influenced inflammatory response after myocardial ischemia-reperfusion injury. *Life Sci.* 2019;232:116632. doi: 10.1016/j.lfs.2019.116632.

Wintrich J, Berger AK, Bewarder Y, Emrich I, Slawik J, Böhm M. Neues zur Diagnostik und Therapie der Herzinsuffizienz [Update on diagnostics and treatment of heart failure]. *Herz.* 2022;47(4):340–353. German. doi: 10.1007/s00059-021-05062-x.

Wu Q, Wang J, Tan WLW, Jiang Y, Wang S, Li Q, Yu X, et al. Extracellular vesicles from human embryonic stem cell-derived cardiovascular progenitor cells promote cardiac infarct healing by reducing cardiomyocyte death and promoting angiogenesis. *Cell Death Dis.* 2020;11(5):354. doi: 10.1038/s41419-020-2508-y

Zangi L, Margalit R, Reich-Zeliger S, Bachar-Lustig E, Beilhack A, Negrin R, Reisner Y. Direct imaging of immune rejection and memory induction by allogeneic mesenchymal stromal cells. *Stem Cells.* 2009 Nov;27(11):2865–2874. doi: 10.1002/stem.217.

Zhang Y, Hao Z, Wang P, Xia Y, Wu J, Xia D, Fang S, et al. Exosomes from human umbilical cord mesenchymal stem cells enhance fracture healing through HIF-1α-mediated promotion of angiogenesis in a rat model of stabilized fracture. *Cell Prolif.* 2019;52(2):e12570. doi: 10.1111/cpr.12570.

Zhang Y, Pan Y, Liu Y, Li X, Tang L, Duan M, Li J, et al. Exosomes derived from human umbilical cord blood mesenchymal stem cells stimulate regenerative wound healing via transforming growth factor-β receptor inhibition. *Stem Cell Res Ther.* 2021;12(1):434. doi: 10.1186/s13287-021-02517-0.

Zhang Z, Yang J, Yan W, Li Y, Shen Z, Asahara T. Pretreatment of cardiac stem cells with exosomes derived from mesenchymal stem cells enhances myocardial repair. *J Am Heart Assoc.* 2016;5(1):e002856. doi: 10.1161/JAHA.115.002856.

Zhao J, Li X, Hu J, Chen F, Qiao S, Sun X, Gao L, et al. Mesenchymal stromal cell-derived exosomes attenuate myocardial ischaemia-reperfusion injury through miR-182-regulated macrophage polarization. *Cardiovasc Res*. 2019;115(7):1205–1216. doi: 10.1093/cvr/cvz040.

Zhao Z, Sun W, Guo Z, Zhang J, Yu H, Liu B. Mechanisms of lncRNA/microRNA interactions in angiogenesis. *Life Sci*. 2020;254:116900. doi: 10.1016/j.lfs.2019.116900.

Zhong Z, Tian Y, Luo X, Zou J, Wu L, Tian J. Extracellular vesicles derived from human umbilical cord mesenchymal stem cells protect against DOX-induced heart failure through the miR-100-5p/NOX4 pathway. *Front Bioeng Biotechnol*. 2021; 9:703241. doi: 10.3389/fbioe.2021.703241.

Zhu D, Liu S, Huang K, Wang Z, Hu S, Li J, et al. Intrapericardial exosome therapy dampens cardiac injury via activating foxo3. *Circ Res*. 2022;131(10):e135–e150. doi: 10.1161/CIRCRESAHA.122.321384.

15 Living Biodrugs and Their Derived Advanced Therapy Medicinal Products to Treat Autoimmune Pathologies

Mohammad Ammar Aloudah[1], Abdulrahman Bashar Aldura[1], Rawnaq Osama Alzoubi[1], Rahaf Osama Alzoubi[1], Hussam Ghassub Alatae[1], Ahmad Al Mohamad Almustafa[1], and Khawaja H. Haider[1]

[1]College of Medicine, Sulaiman Al Rajhi University, Al-Bukairiyah, Saudi Arabia

15.1 LIST OF ABBREVIATIONS

ACPA	Anti-citrullinated protein antibodies
ADA	American Diabetes Association
AD-MSCs	Adipose tissue-derived MSCs
ADs	Autoimmune diseases
ADMIRE-CD	AD-MSCs for induction of remission in Crohn's disease
Anti-GAD	Anti-glutamic acid decarboxylase (antibody)
Anti-ICA	Anti-islet cell cytoplasmic antibody
APCs	Antigen-presenting cells
ATF2	Activating transcription factor 2
BD	Behçet's disease (disease)
BDNF	Brain-derived neurotrophic factor
BM-MSCs	BM-derived mesenchymal stem cells
BMP	Bone morphogenetic protein
BU	Behçet's uveitis
CCL2	C-C Motif chemokine ligand 2
CCL21	C-C Motif chemokine ligand 21
CCR5	CC-Chemokine receptor 5
CCP	Cyclic Citrullinated peptide
CD (disease)	Crohn's disease
CD4+	Cluster of differentiation 4
COX-2	Cyclooxygenase-2
CXCR3	CXC-chemokine receptor 3
DIP	Distal interphalangeal (joint)
DKA	Diabetic ketoacidosis
DM1/T1DM	Diabetes mellitus type 1/type 1 diabetes mellitus

DOI: 10.1201/9781003653387-15

DMARD	Disease-modifying anti-rheumatic drug
DSS	Dextran sodium sulfate
EAE	Experimental autoimmune encephalomyelitis
EAU	Experimental autoimmune uveitis
EDSS	Expanded disability status scale
ERK	Extracellular signal-regulated kinase
ESCs	Embryonic stem cells
ESR	Erythrocyte sedimentation rate
EVs	Extracellular vesicles
Exos	Exosomes
FGF-2	Fibroblast growth factor-2
FLS	Fibroblast-like synoviocyte
FDA	Food and drug administration
Gal1	Galectin 1
GBD	Global burden of disease
GI/GIT	Gastrointestinal/gastrointestinal tract
HbA1c	Hemoglobin A1c
HGF	Hepatocyte growth factor
HLA	Human leukocyte antigen
HLA-B51	Human leukocyte antigen B51
HO-1	Heme oxygenase-1
HOTTIP	HOXA Transcript at the distal tip
IBD	Inflammatory bowel disease
IDO	Indoleamine 2,3-dioxygenase
IFN-γ	Interferon-gamma
IFN+Exo	Interferon-γ–primed exosomes
IFN-Exo	Non-primed exosomes
IGF	Insulin-like growth factor
IL	Interleukin
INO	Internuclear ophthalmoplegia
iNOS	Inducible nitric oxide synthase
JAM	Junctional adhesion molecule
lncRNA	Long non-coding RNA
MiR	MicroRNA
MLF	Medial longitudinal fasciculus
MMP14	Matrix metallopeptidase 14
MS	Multiple sclerosis
MSCs	Mesenchymal stem cells
MSC-Exos	MSC-derived exosomes
mTOR	Mechanistic target of rapamycin
MVB	Multivesicular body
NFAT	Nuclear factor of activated T cells
NF-κB	Nuclear factor kappa B
NGF	Nerve growth factor
NK cells	Natural killer cells
NLRP3	NOD-like receptor protein 3
NO	Nitric oxide
NOD2	Nucleotide-binding oligomerization domain-containing protein 2
NTA	Nanoparticle tracking analysis
PBMC	Peripheral blood mononuclear cells
PD1	Programmed cell death protein 1

PDL1	Programmed death-ligand 1
PGE_2	Prostaglandin E_2
PI3K/Akt	Phosphoinositide 3-kinase/protein kinase B
PIP	Proximal interphalangeal (joint)
RA	Rheumatoid arthritis
RCT	Randomized controlled trials
RT-PCR	Real-time polymerase chain reaction
SEGRA	Selective glucocorticoid receptor agonists
SHEs	Severe hypoglycemic events
SLE	Systemic lupus erythematosus
STAT3	Signal transducer and activator of transcription 3
TB	Tuberculosis
TGF-β	Transforming growth factor-beta
Th1	T helper 1
Th2	T helper 2
Th17	T helper 17
TNBS	2,4,6-Trinitrobenzenesulfonic acid
TNF-α	Tumor necrosis factor-alpha
Tob1	Transducer of ERBB2, 1
Treg	Regulatory T cell
TSG-6	TNF-stimulated gene-6
UC	Ulcerative colitis
UPB	Ultrasmall Prussian blue
VEGF	Vascular endothelial growth factor
ZO-1	Zonula occludens-1

15.2 INTRODUCTION

Autoimmune diseases constitute a heterogeneous group of disorders in which the immune system fails to discriminate between self and non-self, thus giving an aberrant response to the body's own proteins (Song et al., 2025). The problem is multifactorial and involves intricate mechanisms arising from interactions among environmental, genetic, and immunological factors that may lead to a breakdown of central and peripheral tolerance. As a result, dysregulated tolerance mechanisms activate autoreactive T and B lymphocytes, triggering an inflammatory cascade of events that cause irreversible tissue damage. In simple words, this misdirected immunity will cause a range of clinical symptoms, corresponding to a non-uniform set of signs and symptoms for each autoimmune condition (Pisetsky et al., 2023). For example, one of the many causes of autoimmune response has the underlying mechanism linking environmental pathogens to autoimmune conditions, which is molecular mimicry, wherein microbial antigens resemble a host tissue antigen (Rojas et al., 2018). This similarity between tissue proteins and microbial antigens remains undiscriminated by the host immune system, thereby triggering a cross-reactive immune response to the self-antigens. When foreign antigens activate autoreactive T or B cells, the activated cells mistakenly target the host tissues with similar epitopes (Rose et al., 2000). Rheumatic fever is one of the most common examples of this mechanism of autoimmune dysregulation, wherein antibodies and T cells targeting the M protein, an antigen of the bacterium *Streptococcus pyogenes*, cross-react with cardiac myosin and heart valve components (Blank et al., 2013). There are more than 100 autoimmune diseases, some of which are listed in Table 15.1.

This book chapter delves into discussing pathogenesis, symptoms, and the fast-emerging stem cell-based as well as cell-free therapy approach for the treatment of Behçet's Disease (BD), Systemic lupus erythematosus (SLE), Rheumatoid Arthritis (RA), Multiple Sclerosis (MS), Inflammatory bowel disease (IBD), and Diabetes mellitus type 1 (DM1), primarily using mesenchymal stem cells

TABLE 15.1
Pathophysiology and Clinical Features of Autoimmune Diseases.

Autoimmune Disease	Pathophysiology	Clinical Features
Behçet's disease (BD)	Autoimmune vasculitis affects both veins and arteries of different sizes. It is usually due to genetic factors in individuals. prone, associated with the presence of immune complexes that trigger CD4+ T cell expansion (Adil et a., 2025). This, in turn, releases pro-inflammatory cytokines, like TNF-α, IL-6, and IL-17, leading to tissue damage and vascular inflammation (Leccese et al., 2019).	Recurrent sores in the genitals, eye disorders (retinal vasculitis, uveitis), Hemoptysis (pulmonary artery aneurysm), thrombosis (venous and arterial), and oral aphthous ulcers (Adil et al., 2025).
Systemic lupus erythematosus (SLE)	SLE causes inflammation in various organs by producing antibodies that target nuclear and cytoplasmic antigens (Bartels, 2019). Immune complexes accumulate in the microvasculature during Type III hypersensitivity, triggering complement activation and tissue damage in the kidneys, joints, skin, and other organs. Type II hypersensitivity is another mechanism by which antibodies can attack cell-surface antigens, leading to cytopenia. Women of childbearing age are more often affected by SLE with a relapsing and remitting course (Aringer et al., 2019).	Weight loss, fever, and fatigue. 90% of cases involve joint involvement with distal symmetric polyarthritis, damage to the fingers, wrists, and knees, and skin symptoms such as discoid rash, Raynaud phenomenon, and malar rash (Aringer et al., 2019).
Rheumatoid arthritis (RA)	Persistent inflammation is associated with synovial membrane inflammation. It usually affects the joints, especially those in the hands and feet, although it can also affect other organs. The cause is a mixture of environmental factors, such as smoking, and genetic predispositions, like the presence of HLA-DR4 or HLA-DR1 (Jahid et al., 2023). These factors trigger an autoimmune response in which autoantibodies, i.e., anti-citrullinated protein antibodies (ACPA), target modified self-epitopes, ending in chronic inflammation and joint damage.	Symmetrical joint pain that lasts more than 30 minutes and improves with exercise, morning stiffness, and swelling. Boutonnière (PIP flexion, DIP hyperextension) and swan neck (PIP hyperextension, DIP flexion) (Chauhan et al., 2023).

(*Continued*)

TABLE 15.1 CONTINUED
Pathophysiology and Clinical Features of Autoimmune Diseases.

Autoimmune Disease	Pathophysiology	Clinical Features
Multiple sclerosis (MS)	Chronic autoimmune disease that affects the CNS by causing inflammation, demyelination, gliosis, and brain cell death. The pathogenesis of MS involves the activation of autoreactive T lymphocytes, resulting in demyelination and inflammation, and leading to the formation of acute plaques. With time, axonal loss and oligodendrocyte atrophy lead to chronic plaques, gliosis, and incomplete demyelination (Tafti et al., 2024).	The earliest MS symptom is unilateral optic neuritis. While spinal cord myelin loss causes ataxia, balance problems, and gait impairment, as well as cranial nerve lesions and autonomic dysfunction, MLF lesions cause INO.
Inflammatory bowel disease (IBD)	In CD, there is immune dysregulation, especially of the IL-23-Th17 signaling pathway, which is involved in IBD-induced inflammation. Overactive Th17 cells produce IL-17, which disrupts immune responses and induces the release of inflammatory cytokines, leading to edema, ulcers, and necrosis. This results in bowel obstruction, scarring, and stricture formation. Genomic investigations have shown that mutations in the NOD2 gene are also associated with CD (Boyapati et al., 2015).	The terminal ileum and colon are the sites affected by CD, which causes GIT problems, weight loss, and a mild fever. It can result in malabsorption, chronic diarrhea, sporadic lower GI bleeding, and complications such as fistulas and abscesses (Zajec et al., 2022).
Diabetes mellitus type 1 (DM1)	Insulitis in T1DM initiates an immune response that causes β-cell destruction and insulin deficiency, resulting in impaired glucose uptake and elevated blood sugar levels. Environmental factors, such as viral infections and autoantibodies (anti-ICA or anti-GAD), facilitate the autoimmune process against the pancreas (Zajec et al., 2022).	• Diabetic ketoacidosis (DKA) • Polyuria • Polydipsia • Polyphagia

Abbreviations: anti-GAD: anti-glutamic acid decarboxylase antibody; CD: Crohn's disease; T1DM: type 1 diabetes mellitus; TNF-α: tumor necrosis factor-alpha; IL: interleukin; HLA: human leukocyte antigen; and ACPA: anti-citrullinated protein antibodies. Anatomical and clinical terms are also abbreviated, PIP and DIP: proximal and distal interphalangeal joints; CNS: central nervous system; MLF: medial longitudinal fasciculus; INO: internuclear ophthalmoplegia; GIT: gastrointestinal tract; and DKA: diabetic ketoacidosis.

TABLE 15.2
MSC-Based Drug Products Approved by Drug Agencies of Different Countries for Anti-inflammatory and Immunomodulatory Applications in Clinical Settings

Product and Manufacturer	Approved in (Country)	Potential Uses
Alofisel® (Tigenix)	European Union	Perianal Fistula in Patients with Crohn's disease
Cupistem® (Anterogen)	South Korea	Crohn's fistula
Neuronata® (Corstem)	South Korea	Amyotrophic lateral sclerosis
TEMCELL® (JCR Pharma Ltd)	Japan, Canada, and New Zealand	COPD, Crohn's disease, MI, Acute GvHD, Type 1 diabetes.
Cymerus™ (Cynata Therapeutics Ltd)	UK, Medicines and Healthcare Products Regulatory Agency (MHRA)	GvDH
Ryoncil (Parenchymal, Remestemcel-Lrknd; Mesoblast Inc.)	Canada and the USA FDA	Steroid-refractory GvDH

and their derived advanced medicinal therapy products (AMTP). Among these, RA and MS are the most common autoimmune diseases, with a prevalence of 20.89 and 35.9 per 100,000 population, respectively (GBD et al., 2021; Walton et al., 2020). Extensive population data indicate that autoimmune diseases affect nearly 10% of the population and are more common in females than in males, with a 2:1 ratio (Angum et al., 2020). In some cases, this may reach 11:1 in favor of females, as observed in SLE (Albrecht et al., 2025).

It is pertinent to note that both innate and adaptive immune cells play a pivotal role in the pathogenesis of autoimmune diseases, underscoring the link between the two forms of immunity. Simply put, antigen-presenting cells (APCs), such as dendritic cells and macrophages, initiate the immune response by presenting antigens to T lymphocytes. CD4+ T helper cells, especially Th1 and Th17, release cytokines such as interleukin-1 (IL-1), interleukin-6 (IL-6), interleukin-17 (IL-17), interferon-gamma (IFN-γ), and tumor necrosis factor-alpha (TNF-α) (Shlomchik et al., 2001). This will trigger the recruitment and activation of immune cells, leading to a localized inflammatory response (Kuwabara et al., 2017). At the same time, B lymphocytes contribute to the autoimmune response primarily by producing autoantibodies and by participating in antigen presentation and cytokine release (Lundy, 2009). The cytokine-mediated inflammatory loop is a primary feature that is a common contributory mechanism across a diverse array of autoimmune pathologies. On the positive note, the inflammatory response offers an excellent therapeutic target for the fast-emerging novel anti-inflammatory and immunomodulatory interventions based on mesenchymal stem cells (MSCs) and their derivative AMTP-based treatment approaches (Zubair et al., 2025). Derived from diverse tissue sources and with regenerative potential, MSCs have well-established anti-inflammatory and immunomodulatory effects either through direct cell-to-cell contact or by secreting paracrine factors (especially via the release of secretome and exosomes) (Álvarez-Viejo & Haider, 2022; Kamal et al., 2022). Given their anti-inflammatory and immunomodulatory properties, drug agencies in various countries have already approved MSC-based drug products for routine clinical use (Bhatti & Haider, 2025). Table 15.2 lists some MSC-based products currently available in the market, used for anti-inflammatory and immunomodulatory effects.

15.2.1 Cell-Free Therapy Approach

In addition to cell-based therapy, MSC-derived exosomes (MSC-Exos) have attracted significant attention as a promising cell-free therapeutic approach due to their low immunogenicity and high permeability, without safety concerns (Mushtaq et al., 2024). Exosomes are nano-sized spherical extracellular vesicles (EVs) released from MSCs as part of intercellular communication (Habib et al., 2025b). They have a lipid bilayer membrane with a diameter of 40–100 nm and carry a specific cargo of bioactive molecules, including proteins, lipids, and nucleic acids (Liang et al., 2025). Besides other functions, exosomes carry many anti-inflammatory and immunomodulatory molecules as part of their cargo that regulate the immune response and, hence, provide a potential treatment for autoimmune disorders (Arun et al., 2025). Furthermore, they can be engineered, i.e., genetically engineered for their cargo, or surface-modified, to enhance therapeutic activity and targeted delivery of their cargo. They have also shown therapeutic potential in other diseases such as myocardial infarction, ischemic stroke, acute kidney injury, hepatic injury, and spinal cord injury (Haider & Najimi, 2022; Najim & Haider, 2022; Zhu et al., 2023; Cao et al., 2020; Habib et al., 2026).

15.3 CURRENT TREATMENT OPTIONS FOR AUTOIMMUNE DISEASES

Among the pharmacological options, corticosteroids such as prednisone and methylprednisolone remain the mainstay of immunosuppression in the treatment of autoimmune diseases due to their potent and rapid anti-inflammatory effects via both non-genomic pathways and genomic modulation of pro-inflammatory cytokine expression. Their use alleviates symptoms and induces remission by limiting damage to target tissues. They are more effective at managing acute flare-ups of diseases such as RA, vasculitis, and SLE. On the contrary, they are less effective against some of the autoimmune disorders, such as type 1 diabetes, Hashimoto's thyroiditis, Graves' disease, and advanced primary biliary cholangitis (Elkoshi, 2024). Chronotherapy, in which dosages are timed to match circadian peaks in inflammatory activity, showed even better results and tolerability. Although newer biologic agents offer targeted treatment alternatives, glucocorticoids remain important for immediate control and as bridging agents. However, long-term use of corticosteroids is not without risks, such as osteoporosis, muscle atrophy, metabolic disturbances, hypertension, increased risk of infection, psychological consequences, adrenal suppression, and diabetes. These side effects can be minimized by using the lowest effective dose and timed-release formulations. Some more recent approaches, such as targeted delivery systems and selective glucocorticoid receptor agonists (SEGRAs), aim to preserve therapeutic efficacy while reducing systemic toxicity (Babaoğlu et al., 2024).

Besides corticosteroids, antimetabolites such as methotrexate, azathioprine, and mycophenolate mofetil suppress autoimmunity by inhibiting DNA synthesis and cell proliferation. Calcineurin inhibitors (cyclosporine & tacrolimus) prevent NFAT transcription, thereby blocking T cell activation. The use of mTOR inhibitors such as sirolimus disrupts mTORC1/mTORC2 signaling to regulate cellular metabolism, growth, and proliferation. These conventional agents are effective in autoimmune diseases with partial tissue damage and sufficient regenerative capacity, such as RA, SLE, psoriasis, and early-stage type 1 DM, as well as in primary biliary cholangitis. In these diseases, preserving the remaining functional cells may delay disease progression. However, their effect is significantly reduced in conditions with poor regenerative capacity, such as advanced type 1 DM, Hashimoto's thyroiditis, and late-stage primary biliary cholangitis. In these diseases, extensive tissue loss makes recovery unlikely even with immune suppression. Possible side effects may include BM suppression, hepatotoxicity, nephrotoxicity, hypertension, hyperlipidemia, GI disturbances, improper wound healing, and increased infection risk. This highlights the need for careful patient selection and attentive monitoring, depending on the disease stage and the organ's capacity for repair or regeneration (Elkoshi et al., 2024).

Biologic therapies target specific molecules or pathways involved in dysregulated immune responses and have demonstrated significant efficacy in conditions such as RA, SLE, psoriasis, psoriatic arthritis, and IBD. The biologics include anti-TNF agents (i.e., infliximab, etanercept), IL-6 receptor inhibitors (tocilizumab), B cell-targeted agents (rituximab), and T cell costimulation blockers (abatacept). They reduce inflammation, improve physical function, and slow structural damage, even in patients who are unresponsive to conventional disease-modifying therapies. In some cases, biologics have also allowed for steroid-sparing regimens, which reduce the long-term side effects associated with corticosteroid use. On the other hand, biologics have many limitations, such as the need for parenteral administration, high cost, and adverse effects, including injection-site reactions, risk of infections (e.g., TB & opportunistic pathogens), cytopenias, elevated liver enzymes, and lipid abnormalities. Rare but severe conditions, such as malignancy, demyelinating disorders, and progressive multifocal leukoencephalopathy, have also been reported. Also, these agents show decreased efficacy in patients with advanced irreversible tissue damage, so pre-treatment screening is important. Furthermore, the need for ongoing safety monitoring and the variability in individual responses underscore the importance of tailoring treatment and administering it early in suitable candidates (Huashan et al., 2019).

15.4 CELL-BASED THERAPY AS A TREATMENT OPTION

15.4.1 OVERVIEW OF MESENCHYMAL STEM CELLS

MSCs are non-hematopoietic, spindle-shaped, multipotent stem cells derived from diverse mesenchymal tissues (Al-khani et al., 2022). Besides other near-ideal sterling features, they are known for their anti-inflammatory and immunomodulatory properties (Weiss et al., 2019). They have been extensively studied in experimental animal models of autoimmune disease (Table 15.3) and in clinical trials (Table 15.4).

TABLE 15.3
Preclinical Studies in Experimental Animal Models of Autoimmune Conditions Using Mesenchymal Stem Cells from Different Tissue Sources

Author/Year	Disease	Animal Model	MSC Source	Key Findings
Liu et al., 2024	RA	CIA mice	UC	↓ TNF-α/IL-17; 70% decrease in joint erosion.
Wei Park et al., 2019	SLE	MRL/lpr mice	AT	↓ Anti-dsDNA antibodies; 80% survival at 8 weeks.
Duan et al., 2020	IBD	DSS-colitis mice	Placenta	50% ↑ mucosal healing; ↓ IL-1β/IL-6.
Madec et al., 2009	T1D	NOD mice	Wharton's jelly	40% ↓ insulitis; delayed diabetes onset.
Sultan et al., 2025	AIH	Wister albino rats	BM	↓ Liver inflammation/injury vs control. Short-term improvement in the AIH model.
Liu et al., 2022	IBD	C57BL/6 mice	BM	↑ intestinal mucosal repair, ↓ inflammation.
Duan et al., 2020	IBD	BALB/c mice	Placenta	↑ clinical symptoms by inhibiting inflammation and oxidative stress to promote mucosal healing.
Jiang et al.,2017	MS	Lewis rats	Placenta	↓ inflammatory infiltration, preserved axons, ameliorated demyelination, ↑ and neurological functions.
Dan et al., 2019	RA	CIA mice	UC	↑ arthritis, delaying radiological progression, and inhibiting synovial hyperplasia.

TABLE 15.4
Clinical Studies using MSCs from Different Tissue Sources for Treating Autoimmune Diseases

Trial ID and Disease	Study Type and Phase	Sample Size	Cell source	ROA	Mechanism of action	Outcome measures	Conclusion
NCT00550498; Behçet's Disease (ocular/retinal vasculitis)	Interventional; Phase I	5	Autologous BM-MSCs	Intravitreal/ intraocular injection into one eye per patient (~1.8 million MSCs).	immunomodulation	Visual acuity (VA); Disease Activity Index (DAI) for anterior uveitis, posterior uveitis, and retinal vasculitis.	The procedure failed to rescue vision in these advanced cases.
NCT03219801; SLE	Interventional; Phase I	10	hUC-MSC	IV infusion	immunomodulation	Safety and efficacy of the MSC transplantation.	Study ongoing
NCT02633163; SLE	Interventional; Phase II	81	Allogeneic UC-MSCs	IV infusion, 1 x 10^6 cells	Immunomodulation/ anti-inflammatory effects of MSCs.	Safety and efficacy	Study ongoing
NCT03673748; Lupus nephritis	Interventional; Phase II	20	Allogeneic BM-MSCs	IV injection, 2 x 10^6 cells	Immunomodulation/ anti-inflammatory/ regenerative effects of MSCs.	Safety and efficacy	Study ongoing
NCT04184258; SLE	Interventional; Phase I/II	7	Allogenic olfactory mucosa-MSCs	NA	Immunomodulation/ regenerative effects of MSCs	Safety and efficacy	Study ongoing
NCT03562065; SLE	Interventional; Phase I/II	8	Allogeneic UC-MSCs	Single IV infusion 2 × 10^6 or 4 × 10^6 MSCs/kg	immunomodulation	Safety and efficacy	
NCT04835883; SLE	Interventional; Phase II	10	Allogeneic BM-MSCs	IV infusion of CS20AT04	Immunomodulation/ anti-inflammatory/ regenerative effects of MSCs	Safety and efficacy	Study ongoing
NCT05631717; Lupus nephritis	Interventional; Phase III	40	hUC-MSCs	IV drip, 1 × 10^6 cells/ kg	immunomodulation	Safety and efficacy	Study ongoing

(Continued)

TABLE 15.4 CONTINUED
Clinical Studies using MSCs from Different Tissue Sources for Treating Autoimmune Diseases

Trial ID and Disease	Study Type and Phase	Sample Size	Cell source	ROA	Mechanism of action	Outcome measures	Conclusion
NCT03174587; Lupus nephritis	Interventional; Phase I	7	Allogeneic BM-MSCs (CS20AT04 product)	Intravenous infusion, 2.0×10^6 cells/kg—3.0×10^6 cells/kg	CS20AT04 has anti-inflammatory effects, controls the activity of immune cells, and reduces autoantibodies.	Primary: safety, dose-limiting toxicity, tolerability, adverse events.	The BM-MSC therapy (CS20AT04) was well-tolerated, without dose-limiting toxicity at doses up to 3.0×10^6 cells/kg. The maximum tolerated dose was defined as 3.0×10^6 cells/kg.
NCT04318600; Lupus nephritis	Interventional; Phae I	16	Allogeneic amniotic MSCs	IV infusion, 1×10^6 cells/kg, monthly, ×3 infusions	Immunomodulation/anti-inflammatory/regenerative effects of MSCs	Safety/adverse events; renal function/proteinuria/SLEDAI	No results posted
NCT03171194; SLE	Interventional; Phase I	6	UC-MSCs	IV drip, 1×10^6 cells/kg	immunomodulation	Safety/adverse	MSC infusions in refractory SLE appear very safe
NCT02166021; MS	Interventional; Phae II	48	Autologous BM-MSCs	Intrathecal (IT) or IV (dose: 1×10^6 MSCs per kg	Immunomodulation, neuroprotection, and neuroregeneration; balancing the CNS inflammatory milieu.	Treatment failure rate/Safety/Tolerability	MSC therapy was well-tolerated and showed short-term effects; the IT route had advantages over the IV route.
NCT02239393; MS	Interventional; Phae II	31	Autologous BM-MSCs	IV infusion, 1–2 × 10^6 MSCs per kg	Immunomodulation and neuroprotection	Number of gadolinium-enhancing lesions on brain MRI at 6 months/EDSS/(MSFC)/safety/adverse events.	MSC therapy was safe and feasible, but did not meet the primary MRI efficacy endpoint.
NCT04823000; MS (Relapsing-Remitting and Secondary Progressive)	Interventional; Phase I/II	24	Autologous AD-MSCs	(IV) infusion, 1×10^6 MSCs per kg	Immunomodulation, reduction of autoreactive T cell activity, and neuroprotection	Safety/MRI lesion activity/EDSS/(MSQoL-54	Treatment was well-tolerated without SAE. Some patients had reduced inflammatory biomarkers.
NCT01730547; MS	Interventional; Phase II	2	Autologous, BM-MSCs	IV infusion	Immunomodulatory and neuroprotective	Safety and exploratory efficacy	Study ongoing

Trial ID and Disease	Study Type and Phase	Sample Size	Cell source	ROA	Mechanism of action	Outcome measures	Conclusion
NCT02034188; MS	Interventional; Phase II	20	UC-MSC	IV infusion: 20×10^6 UC-MSC per day for 7 days (total ~140×10^6)	Immunomodulatory/ neuroprotective	safety/EDSS/PASAT/ SNRS/PASAT	Treatment was safe; improvements were observed in EDSS, upper extremity function, and walking.
NCT00395200; MS	Open-label	18	Autologous BM-MSCs	IV infusion, single dose, mean 1.6×10^6 cells/kg)	Neuroprotection and possible remyelination	safety/feasibility/visual acuity/VEP latency amplitude/optic nerve area/OCT/EDSS	MSC infusion was safe and feasible in patients with SPMS, resulting in functional and structural improvements and neuroprotection.
NCT00813969; MS	Interventional; open-label; Phase I	24	BM-MSCs	IV infusion, 1.9×10^6 cells/kg	Immunomodulatory/ neuroprotective	safety/tolerability.	MSC transplantation was feasible, safe, and well-tolerated
NCT01377870; MS	Interventional; double-blind, placebo-controlled; Phase I/II	30	Autologous BM-MSCs	IT (1 million cells per Kg) and IV infusion (20 million cells)	Immunomodulatory/ neuroprotective	Primary: safety/feasibility; Secondary: clinical stabilization/ immunological effects, migration of labeled cells on MRI.	
NCT01745783; MS	Double-blind, placebo-controlled; Phase I/II	26	(BM-MSCs)	(IV) infusion, $1–2 \times 10^6$ cells/kg	Immunomodulatory/ neuroprotective	Safety/MRI lesion reduction/efficacy/(EDSS, MSFC, QoL scales).	IV administration of BM-MSCs is feasible and safe,
NCT03326505; MS	Interventional; RCT double-blind, placebo-controlled; Phase I/II	60	WJ-MSCs	Three administrations over 3 months: 1) Month 0: 100×10^6 IT + 50×10^6 IV; 2) Month 1: 100×10^6 IT + 50×10^6 IV; 3) Month 3: 8–10 mL IT.	Neuro-regenerative/ immunomodulatory	EDSS/inflammatory and neuro-regeneration biomarkers (blood/CSF), VEP, OCT.	The protocol was aimed to test whether PTE + WJ-MSCs improve motor and non-motor outcomes vs either alone; results pending.

(Continued)

TABLE 15.4 CONTINUED
Clinical Studies using MSCs from Different Tissue Sources for Treating Autoimmune Diseases

Trial ID and Disease	Study Type and Phase	Sample Size	Cell source	ROA	Mechanism of action	Outcome measures	Conclusion
NCT02035514; MS	Interventional; Phase I/II	9	BM-MSCs	IV infusion 1x10^6 Cells/kg	Immunomodulatory/ neuroprotective	Safety/change in MRI gadolinium-enhancing lesions/EDSS, MSFC, relapse rate, T2 lesion count, OCT, QoL, immunology markers.	
NCT05116540; Relapsing-Remitting MS	Randomized, double-blind, placebo-controlled; Phase II	24	Autologous AD-MSCs	Six IV infusions over ~32 weeks (200 million cells per infusion)	immunomodulation + neuro-repair/ regeneration	MSQOL-54 Physical and Mental Health Composite Scores. Secondary: EDSS, depression (PHQ-9), safety (AEs/ deaths).	Treatment group showed significant improvements in QoL scores vs placebo, and a favorable safety profile, with no treatment-related deaths reported in the news release.
NCT03799718; Progressive MS	Open-label, multicenter; Phase II	18	Autologous BM-MSCs	3 dose Intrathecal	Immunomodulation + neuroprotection	Functional improvements (T25FW, 9-HPT, SDMT, LCLA), biomarker changes, safety outcomes.	The treatment was well-tolerated and showed biological and functional signals of efficacy in progressive MS. No treatment-related deaths occurred.
NCT03355365; Secondary-Progressive MS	RCT double-blind, placebo-controlled; Phase II	54	Autologous BM-MSCs	Six intrathecal injections (up to 10 × 10^6 MSC-NP cells) every 2 months in year one.	immunomodulation + neuro-repair/ regeneration	Primary: "EDSS Plus" improvement (EDSS, T25FW, or 9-HPT)	IT MSC-NP therapy was safe, well-tolerated, and produced functional and biological improvements.

Trial ID and Disease	Study Type and Phase	Sample Size	Cell source	ROA	Mechanism of action	Outcome measures	Conclusion
NCT01056471; Secondary-Progressive MS	RCT Triple-blinded, placebo-controlled, Phase I/II	30	Autologous AD-MSCs	IV infusion	Immunomodulation + neuroprotection	Safety and some efficacy in SPMS	no finished
NCT01933802; Progressive MS	Open-label, single-arm; Phase I	20	Autologous BM-MSC-derived neural progenitors	Intrathecal injections of up to 1×10^7 cells/dose, three doses.	Neuro-regenerative/ immunomodulatory	Safety/tolerability (primary); EDSS, T25FW, muscle strength, bladder function (secondary).	Treatment was safe and well-tolerated; possible signals of efficacy (improved EDSS in some, improved muscle strength/ bladder function).
NCT02221258; RA	Interventional; Phase I	9	hUCB-MSC	Single IV infusion of FURESTEM-RA Injection	immunomodulation	Safety profile—frequency and severity of SAE within four weeks	NCT01663116; RA
Single-masked; Phase I/II	53	Allogeneic AD-MSCs	IV infusions on Days 1, 8, 15 of Cx611 at doses of 1×10^6/kg (cohort A), 2×10^6/kg (cohort B), or 4×10^6/kg (cohort C).	Immunomodulatory MSCs: anti-inflammatory, immune-suppressive	Number of adverse events and severe adverse events at 6 months/ ACR20/50/70 response rates, DAS28-ESR	IV Infusion of Cx611 was well-tolerated, with no evidence of dose-related toxicity.	NCT03618784; RA
RCT double-blind, placebo-controlled; Phase I/II	33	Allogeneic UC-MSCs	IV infusion of UC-MSCs; details: three IV infusions (5×10^7 or 1×10^8 cells per patient) spaced 4 weeks apart.	Immunomodulatory and anti-inflammatory effects.	safety/efficacy	results not publicly published.	NCT01157650; IBD
Interventional; Phase I/II	15	Autologous AD-MSCs	Local implantation	immunomodulatory and pro-repair effects	safety/tolerability/ fistula healing efficacy		NCT00294112 CD
Interventional (open-label); Phase II	10	Allogeneic BM-MSCs	Two intravenous infusions	immunomodulation	Reduction in Crohn's Disease Activity Index (CDAI)		NCT01233960 CD

(Continued)

TABLE 15.4 CONTINUED
Clinical Studies using MSCs from Different Tissue Sources for Treating Autoimmune Diseases

Trial ID and Disease	Study Type and Phase	Sample Size	Cell source	ROA	Mechanism of action	Outcome measures	Conclusion
Interventional (open-label); Phase III	73	Allogeneic BM-derived MSCs (PROCHYMAL®/ remestemcel-L)	Infusions of PROCHYMAL on Day 42, Day 84, and Day 126. Each infusion contains 200 million cells.	immunomodulation	Disease remission at 180 days (CDAI ≤150 + IBDQ increase).		NCT00482092 Crohn's disease
Intervention. RCT; Phase III	330	Allogeneic BM-MSCs (product: PROCHYMAL®)	Four intravenous infusions over two weeks	immunosuppressive and healing ability, migrate to sites of inflammation.	(CDAI between 250–450 at baseline)		NCT00999115 CD (Rectovaginal fistula)
Interventional (single-group); Phase I/II	10	Allogeneic expanded AD-MSCs (eAD-MSCs)	Intralesional injection of 20 million cells at baseline; if no healing by 12 weeks, a second dose of 40 million cells is administered.	immunomodulatory/ anti-inflammatory	safety and feasibility; Efficacy: complete fistula healing	Allogeneic AD-MSCs appear safe and feasible, with a ~60% healing rate in non-excluded patients.	NCT01541579 CD (perianal fistulas)
Intervention. RCT double-blind, placebo-controlled; Phase III	278(40)	Allogeneic AD-MSCs (darvadstrocel)	Single local administration, 120 million cells	immunomodulatory/ anti-inflammatory	clinical remission	The treatment was well-tolerated; remission at 104 weeks: 14/25 (56%) in the treatment arm vs 6/15 (40%) in the control arm.	NCT03279081 Crohn's disease (perianal fistulas)
Interventional; Phase III	568	allogeneic expanded (eAD-MSCs; product: Cx601)	Single intralesional injection of Cx601, 120 million cells	immunomodulation	Combined remission at Week 24: closure of all treated external openings + absence of collections >2 cm on MRI.		NCT02445547 Crohn's disease

Trial ID and Disease	Study Type and Phase	Sample Size	Cell source	ROA	Mechanism of action	Outcome measures	Conclusion
Open-label RCT; Phase I/II	83	UC-MSCs	1×10^6 cells/kg via IV infusion once weekly for 4 weeks	immunomodulatory/ anti-inflammatory	Changes in Crohn's Disease Activity Index (CDAI), Harvey–Bradshaw Index (HBI).	UC-MSC therapy significantly improved disease status and reduced steroid dose; it was safe in this setting.	NCT01144962 Crohn's disease
Intervention, placebo-controlled; Phase I/II	21	BM-MSCs	Local injection	immunomodulatory/ anti-inflammatory	Fistula healing: absence of discharge and MRI fluid collection <2 cm.	No SAE; injection of 3×10^7 MSCs appeared to promote fistula healing.	NCT00475410 Crohn's disease (perianal fistulas)
Intervention, single-masked RCT; Phase III	200	Autologous eAD-MSCs	Local injection20 million MSCs	immunomodulatory/ anti-inflammatory	Fistula healing: re-epithelization of external opening + absence of collection >2 cm by MRI.	Treatment with ASCs was safe and achieved ~40% healing at 6 months and >50% at 1 year; no significant difference vs fibrin-glue alone.	NCT01068951 T1DM
Interventional, Randomized, Controlled; N/A	20	Autologous BM-MSCs	IV infusion of MSCs shortly after diagnosis of type 1 DM.	Preservation of β-cell function by immunomodulation and support of residual islet cells	Change in C-peptide response (mixed-meal tolerance test) to assess β-cell function.	MSC treatment was safe and appeared to preserve endogenous insulin secretion compared to control.	NCT03920397 T1DM
Interventional, open-label; Phase II	30	Allogeneic AD-MSCs	Single intravenous infusion of ASCs (1×10^6 cells/kg)	Immunomodulation and preservation of β-cell function through MSC.	Safety, C-peptide (β-cell function), insulin dose, HbA1c.	Safe and feasible; the treatment group showed stability of C-peptide, lower insulin requirement.	NCT05308836 T1DM
Interventional; Phase I	10	Allogeneic AD-MSCs	Intravenous infusion of AD-MSCs	Immunomodulation	Safety of AD-MSC transplantation in T1D patients.		NCT00690066 T1DM

(Continued)

TABLE 15.4 CONTINUED

Clinical Studies using MSCs from Different Tissue Sources for Treating Autoimmune Diseases

Trial ID and Disease	Study Type and Phase	Sample Size	Cell source	ROA	Mechanism of action	Outcome measures	Conclusion
Intervention Double-Blind, RCT; Phase II	63	Allogeneic hMSCs (PROCHYMAL®)	Multiple infusions of PROCHYMAL®	Immunomodulation	Safety and efficacy (β-cell function/C-peptide).		

Abbreviations: ACR20/50/70: American College of Rheumatology improvement criteria, AD-MSCs: adipose-derived MSCs; ASCs/eASCs: adipose-derived stem cells; BM-MSC: bone marrow-derived MSCs; CDAI: Crohn's disease activity index; CNS: central nervous system; CS20AT04: specific BM-MSC product code; DAS28-ESR: disease activity score (28 joints, ESR-based); DAI: disease activity index; EDSS: expanded disability status;scale HbA1c: glycated hemoglobin; HBI: Harvey–Bradshaw index; hUC-MSC/UC-MSC/Huc-MSC: human umbilical cord-derived MSCs; IBD: inflammatory bowel disease; IBDQ: Inflammatory Bowel Disease Questionnaire; IT: intrathecal; IV: intravenous; LCLA: low-contrast letter acuity; MSC: mesenchymal stem cells; MSC-NP: MSC–derived neural progenitors; MRI: magnetic resonance imaging; MSFC: Multiple Sclerosis Functional Composite; MSQoL-54: multiple sclerosis quality-of-life 54-item scale; OCT: optical coherence tomography; QoL: quality of life; RA: rheumatoid arthritis; SAE: severe adverse events; SDMT: Symbol Digit Modalities Test; SLE: systemic lupus erythematosus; SLEDAI: SLE Disease Activity Index; T1D/T1DM: Type 1 diabetes mellitus; T25FW: timed 25-foot walk; VA: visual acuity; VEP: visual evoked potential.

Source: Clinicaltrials.gov.

Although the underlying mechanism remains less well-defined, both preclinical and clinical research data support their immunomodulatory potential. There is mounting evidence to show that MSCs affect almost every immune cell type, i.e., NK cells, DCs, macrophages, as well as B and T lymphocytes, and prevent their activation, proliferation, and differentiation (Ma et al., 2016). This has primarily been attributed to their paracrine activity (Figure 15.1).

Some of the important pathways involved in their immunomodulatory activity are:

- The chemokine–iNOS–IDO axis: MSCs can acquire an immunosuppressive phenotype when exposed to inflammatory cytokines (Shi et al., 2018). MSCs are activated by IFNγ combined with either tumor necrosis factor (TNF) or IL-1 to produce CC-chemokine receptor 5 (CCR5) and CXC-chemokine receptor 3 (CXCR3) ligands, which recruit T cells close to MSCs. Subsequently, MSCs will suppress T cell activity and proliferation by expressing inducible nitric oxide synthase (iNOS) or Indoleamine 2,3-dioxygenase (Figure 15.2A and B).
- The MSC secretome: It contains growth factors, cytokines, anti-inflammatory proteins, TNF-stimulated gene six protein (TSG-6), heme oxygenase 1 (HO-1), and exosomes. It also contains a large number of mRNAs and microRNAs (miRNAs) that have been researched for their therapeutic effects (Shi et al., 2018; Tingfen et al., 2013). All of these molecules have Immunomodulatory effects.
- MSCs also use cell-to-cell contact to modulate immunity via two substances: the costimulatory molecule PDL1 and the TNF ligand superfamily member 6. The interaction of these molecules with receptors on the cell surface modulates downstream signaling pathways in immune cells, thereby affecting cell proliferation and survival. The binding of PDL1 to programmed cell death protein 1 (PD1) or CD80 expressed on immune cells yields an inhibitory signal to the activated T cells (Sheng et al., 2008).

It is important to note that the immunomodulatory properties of MSCs are affected by various factors, which include but are not limited to donor variability, for example, the age or tissue source, that

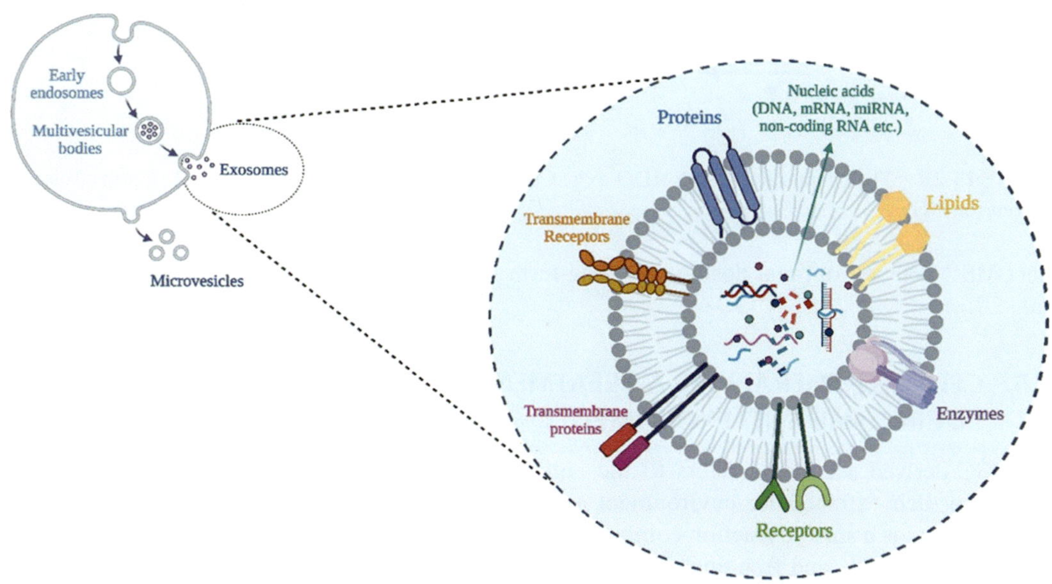

FIGURE 15.1 Exosomes' composition shows their origin from multivesicular bodies and their cargo of proteins, lipids, nucleic acids, enzymes, and receptors. Created in BioRender. Oudah M. (2025), https://biorender.com/yipbasy.

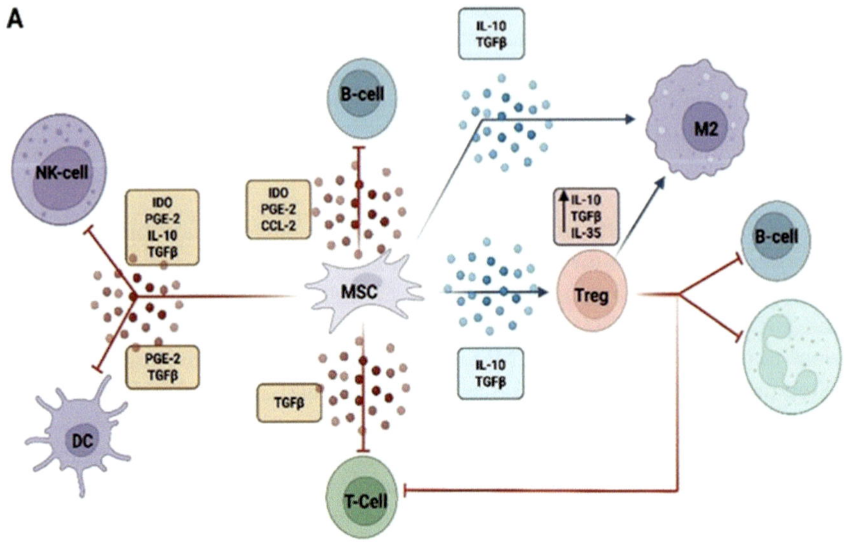

FIGURE 15.2A The general immunomodulatory effects of MSCs on different immune cells. Created in BioRender. Oudah, M. (2025), https://biorender.com/i6ofb11.

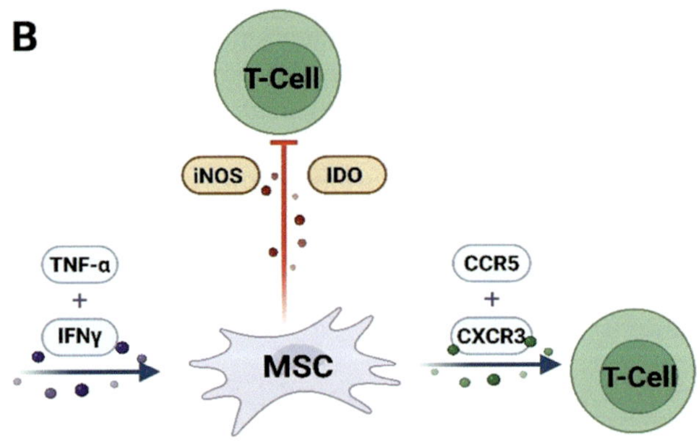

FIGURE 15.2B The chemokine–iNOS–IDO axis. Created in BioRender. Oudah, M. (2025), https://biorender.com/i6ofb11.

affect MSC potency, optimal dosing, and long-term safety, and require further research (Galipeau et al., 2018).

15.5 CELL-FREE THERAPY AS A TREATMENT OPTION (SECRETOME AND EXOSOME)

The MSC-derived secretome refers to the entire collection of bioactive molecules released by MSCs into their extracellular environment as part of their paracrine activity (Haider & Aslam, 2018). It contains a soluble fraction composed of proteins (growth factors, cytokines, chemokines, and enzymes), lipids, and free nucleic acids, and an insoluble fraction containing EVs, including exosomes and microvesicles, which also contain their protein, lipid, and RNA cargo (Xia et al., 2021). The secretome composition differs with MSC source (BM, AD, UC/perinatal tissues), culture conditions (oxygen tension, inflammatory priming), and isolation/collection methods, as well as the cells' priming status, which together determine potency and functionality (Trigo et al., 2025;

Advani et al., 2025). Similarly, the composition of the payload carried by EVs varies between cells, driven by multiple factors, including tissue source and cell priming status (Habib et al., 2025a). Some of the well-documented key effects of treatment with secretome or exosomes include:

- modulation of innate immune cells—changing macrophage phenotype from pro-inflammatory M1 to reparative M2.
- lowering neutrophil infiltration.
- suppression of adaptive immune system activation—preventing dendritic cell maturation and T cell proliferation.
- triggering regulatory T cell (Treg) induction.
- trophic and cytoprotective effects—enhancing epithelial/MSCs survival, proliferation.
- angiogenesis through growth factors (e.g., VEGF, HGF, IGF).
- anti-apoptotic and anti-oxidative effects mediated by secreted enzymes and miRNA delivered inside EVs.

Many of these functions are mediated by a combination of soluble mediators and EV cargo, such as proteins and small RNAs, working in concert (Trigo et al., 2025; Kavaldzhieva et al., 2025). Although the use of secretome avoids several practical and safety concerns of live-cell-based therapy (e.g., cell viability during transport and the potential for ectopic engraftment, tumorigenicity, and immunogenicity), its therapeutic translation requires standardization of MSC source, optimization of cell culture conditions and priming protocols, as well as secretome collection, purification, and characterization techniques, etc., while ensuring to maintain both reproducibility and efficacy. Recent papers focus on batch characterization (proteomics, miRNA profiling, functional bioassays) as important steps in the clinical development of the secretome as an AMTP (Singh et al., 2025).

15.5.1 Exosome

Exosomes are nature's nanotechnology at work and originate from the endosomal pathway in the cells of their origin (Ahmed et al., 2023). They are produced when early endosomes mature into multivesicular bodies (MVBs), which later fuse with the plasma membrane, releasing intraluminal vesicles into the extracellular space (Sengul et al., 2025). They are rich in proteins (e.g., tetraspanins CD9, CD63, CD81; adhesion molecules; cytokines; immunoregulatory proteins such as TGF-β, PD-L1, IDO; lipids (i.e., sphingomyelin, cholesterol, and nucleic acids such as mRNAs, microRNAs, other non-coding RNAs, DNA fragments) (Janockova et al., 2021; Liang et al., 2025). The produced cargo reflects the tissue source and the physiological/primed state of the parental MSCs. There are many isolation methods, such as ultracentrifugation, size-exclusion chromatography, polymer-based precipitation, and immunoaffinity capture. Exosome identity is typically confirmed using nanoparticle tracking analysis (NTA), electron microscopy, and western blotting for EV markers such as CD63 and CD81 (Habib et al., 2026). When used for their immunomodulatory potential, purified MSC-Exos act on distinct immune and stromal cell targets, modulating immune and tissue responses by delivering cargo proteins, lipids, and RNAs to target cells and initiating the desired response and activity of those cells, as discussed in the following sections and summarized in Table 15.5.

Additionally, exosomes downregulate IL-6, TNF-α, IL-12, and IFN-γ, and upregulate IL-10 and TGF-β1 by inhibiting NF-κB signaling. They also modulate different pathways via the transfer of miRNAs, i.e., miR-17, miR-21, and miR-223, which are angiogenesis modulators, Toll-like receptor pathways, and inflammasome activation, all of which help to reduce inflammation (Liang et al., 2025). Exosome-based treatment also enhances angiogenesis and the repair of the epithelial barrier by transporting growth factors (i.e., VEGF, HGF) and regulatory RNAs to the endothelial and epithelial cells (Sengul et al., 2025).

TABLE 15.5
Key Immunomodulatory Actions of MSC-Exos

Target Cell	Mechanism	Representative Cargo	Functional Effect
Macrophages	M1→M2 polarization; metabolic shift	let-7b, miR-146a, PGE_2, TGF-β	Anti-inflammatory phenotype
T cells	Th1/Th2 rebalance; Treg induction; apoptosis of activated T cells	TGF-β, IDO, miR-125a-3p	Immune tolerance
B cells	Inhibit proliferation and antibody secretion	miR-155-5p, IL-10	Reduced autoantibody damage
Dendritic cells	Inhibit maturation and antigen presentation	TGF-β, PGE_2	Decreased T/B activation
All immune cells	NF-κB pathway suppression	TGF-β, PGE_2	Reduced pro-inflammatory cytokines
Stromal cells	Anti-apoptotic, pro-angiogenic	VEGF, miR-21, HGF	Tissue repair and regeneration

15.5.2 Innate Immunity Modulation

One of the primary mechanisms of immunomodulation underlying exosomal treatment is the transition of macrophage polarization from the pro-inflammatory M1 phenotype (TNF-α, IL-1β, IL-6-producing) to the anti-inflammatory M2 phenotype (IL-10-producing). This is driven by a diverse group of molecules, including let-7b, miR-146a, miR-181c, PGE_2, and TGF-β, which also change macrophage metabolism from glycolysis to oxidative phosphorylation. Additionally, through proteins such as COX-2 and TNF-stimulated gene 6, they prevent mast cells and neutrophils from migrating into inflammatory tissue and from degranulating. They also reduce the activation of T and B lymphocytes by preventing DC growth and antigen presentation (Sengul et al., 2025).

15.5.3 Adaptive Immunity Regulation

Exosomes are potent suppressors of effector T cell proliferation by inducing cell cycle arrest via p27kip1/Cdk2 signaling, restoring the Th1/Th2 balance, promoting Treg cell differentiation, and inducing apoptosis of activated T cells (Lee et al., 2020). The mediators of this activity include TGF-β, IDO, miR-125a-3p, and adenosine. Kahrizi et al. have implied negative regulation of PI3K/Akt and MAPK/ERK signaling in T cell suppression (Kahrizi et al., 2022). Their effects on B cells include decreased proliferation, suppression of immunoglobulin secretion, and promotion of IL-10-producing regulatory B cells via miR-155-5p–mediated inhibition of the PI3K/Akt pathway (Yi et al., 2025). When tested preclinically in murine colitis and experimental autoimmune encephalomyelitis (EAE), MSC-Exos decreased IL-17 and IFN-γ, increased TGF-β and IL-10, enhanced Treg expansion, and supported mucosal/neuronal repair (Sengul et al., 2025; Huashan et al., 2019).

Similarly, experimental studies in chemically induced murine colitis models showed that MSC-Exos reduced clinical and histologic inflammation, modulated cytokine profiles, and enhanced mucosal healing, demonstrating the effectiveness of exosome-based therapy in IBD. Earlier studies in some autoimmune diseases (e.g., Behçet's disease, SLE, and RA) have demonstrated improvements in symptoms, reduced levels of inflammatory cytokines, and increased Treg levels. However, larger studies are needed to confirm the efficacy of the treatment approach (Swain et al., 2024). These findings support that MSC-Exos may be a promising cell-free therapeutic approach

TABLE 15.6
Comparison between Cell-Based Therapy and Cell-Free Therapy

Cell-Based Therapy	Cell-Free Therapy	Definition
Transplantation of living cells into a patient to repair or replace diseased or damaged tissues or cells (Mount et al., 2015).	Using nano-molecules released from cells instead of whole cells (Sengul et al., 2025).	Mechanism of action
Cells integrate and interact with host tissue (Csöbönyeiová et al., 2022).	Effects are mediated by interaction with signaling, not integration (Csöbönyeiová et al., 2022).	Examples
Stem cell therapy (MSC) and Hematopoietic stem cell therapy (Csöbönyeiová et al., 2022).	Secretomes and exosomes (Sengul et al., 2025).	Safety
Risk of tumor formation (insertional mutagenesis) and immune rejection.	Lower risk of immune rejection and no risk of tumor formation (Csöbönyeiová et al., 2022).	Manufacturing and storage
Manufacturing is complex and sensitive to storage (Mount et al., 2015).	Easier to manufacture and standardized (Rahgoshay et al., 2025).	

for autoimmune diseases, with potential advantages in safety, stability, and barrier penetration compared with whole-cell MSC therapy. A direct comparison of cell-based therapy using MSCs and their derived exosomes, highlighting their respective advantages and limitations, is summarized in Table 15.6.

15.6 THERAPEUTIC POTENTIAL OF MSCS/MSC-EXOS IN AUTOIMMUNE DISEASE

A summary of the clinical studies using MSC-Exos for the treatment of autoimmune diseases is presented in Table 15.7.

15.6.1 Behçet's Disease

Behcet's disease (BD) is an immune-mediated, chronic, relapsing disorder involving multiple systems due to vasculitis (Direskeneli et al., 2006). Mostly observed in youngsters, it affects both genders equally (Davatchi et al., 2013). A combination of genetic predisposition, mainly HLA-B51, epigenetic alterations, environmental factors, and prolonged infections can activate the immune system and increase pro-inflammatory cytokines, contributing to the pathogenesis of the disease (Mattioli et al., 2021; Houwen et al., 2022. It is generally treated with immunosuppressants and biological agents, which can cause systemic side effects. However, current therapy at best remains only symptomatic, with a high possibility of treatment resistance. This necessitates a search for novel therapeutic options. One of the treatment options being studied is the use of MSCs and their derivative exosomes to suppress the autoimmune response by restoring the immunosuppressive functions of Treg cells, suppressing the activation of Th1 and Th17 cells, and decreasing inflammation (Mazaheri et al., 2012). In an experimental study in mice, herpes simplex virus type 1 was used to develop a model of BD. Intraperitoneal MSCs were administered at different times (prior to, during, and after infection with the virus). MSCs suppressed TNF-α, IFN-γ, and IL-17 expression, thereby reducing pro-inflammatory cytokine levels. DC maturation and the activity of T cells, B cells, and NK cells were also suppressed as compared to. These findings demonstrated that, by reducing inflammation, MSCs have therapeutic potential to delay the progression of illness.

TABLE 15.7
Summary of the Clinical Studies Involving the Cell-Free Therapy Approach Using MSC-Derived Exosomes

Trial ID and Disease	Study Type and Phase	Sample Size	MSC Source	RoA	Mechanism of Action	Outcome Measures	Status
NCT07146087; MS	Interventional; Phase I	20	Allogenic MSC-derived exosomes	IV infusion	Neuroprotection, immunomodulation, and anti-inflammatory effects to slow disability progression	EDSS score, MRI changes, cognitive tests, biomarkers (neuroinflammatory and neurodegenerative)	Ongoing
NCT06853522; UC	Interventional; Phase I	40	hucMSC-derived Exos	IV infusion	Immunomodulatory and anti-inflammatory effects promoting mucosal repair	Primary: Safety and efficacy—Secondary: Preliminary clinical activity and biomarker changes	Ongoing
NCT05499156; CD (Resistant perianal fistula)	Interventional; Phase II	80	Placenta MSC-derived exosomes	Injection into the local and surrounding tissue of the fistula	Tissue repair, anti-inflammatory, and regenerative effects of EVs/exosomes from MSCs	Safety, Efficacy, and Inflammatory Markers	Ongoing
IRCT20130812014333N182; SLE	Interventional; Phase II/III	20	Placenta-derived MSC exosomes	IV infusion of incremental four units over 2 weeks	Immunomodulation and anti-inflammatory activity from MSC-EVs	Glucose, Antiphospholipid antibodies, Triglyceride, ANA	Ongoing
NCT05843799; NF-κB–related Inflammatory and Autoimmune Diseases	Interventional; Phase I	18	Engineered HEK293 cell-derived Exos (ILB-202)	IV Single ascending dose	immunomodulation, NF-κB inhibition via IκBα super-repressor cargo	Safety, tolerability, cytokine profile, NK cell count, and immune markers	ILB-202 was well-tolerated without serious or dose-limiting toxicities

Abbreviations: ANA: antinuclear antibodies; EDSS: Expanded Disability Status Scale; EVs: extracellular vesicles; Exos: exosomes; hucMSC-Exos: human umbilical cord MSC-derived exosomes; ILB-202: engineered EV derived from HEK293 cells; IV: intravenous; MSC: mesenchymal stem cells; MSC-EVs: MSC–derived extracellular vesicles; MRI: magnetic resonance imaging; NF-κB: nuclear factor kappa B; NK: natural killer cells; ROA: Route of administration; SLE: systemic lupus erythematosus; UC-MSC: Umbilical cord-derived MSCs.

Li et al. reported a case demonstrating the effectiveness of MSC-based treatment for leg ulcers complicated by BD (Li et al., 2018). Given the abnormal innate and adaptive immune responses in BD pathogenesis, MSC-Exos can inhibit the autoinflammatory process by modulating immune cell phenotypes, thereby improving clinical symptoms (Zhang et al., 2021). In a study involving three patients with advanced retinal vasculitis due to BU and unresponsive to immunosuppressive therapy, intravitreal delivery of BM-MSCs (1.8×10^6 cells) into one eye per patient successfully managed inflammation and the autoimmune reaction. However, the procedure primarily failed because vasculitis was in an advanced and severe stage (Davatchi et al., 2013). A subsequent clinical case report of a 55-year-old woman with BD manifestations of recurrent oral, genital, and leg ulcers that affected walking and were resistant to conventional immunosuppressive therapy and anti-TNF inhibitors successfully responded to MSCs combined with low-dose prednisone and thalidomide. The combined treatment strategy led to complete improvement of the ulcers on one leg and significant improvement on the other, allowing the patient to walk normally again (Yanhong et al., 2018).

The following section describes several important studies on exosome-based treatment for BD.

- Bai et al. developed an experimental autoimmune uveitis (EAU) model in rats and treated them with periocular MSC-Exos injections for 7 days at the onset of the disease. They were immunized with the peptide interphotoreceptor retinol-binding protein 1177–1191 to examine the therapeutic efficacy of MSC-Exos on EAU. A significant reduction in EAU severity was observed, driven by reduced infiltration of T cell subsets and other inflammatory cells into ocular tissues (Bai et al., 2017). Elucidating the underlying mechanism, the authors reported that MSC-Exos treatment regulated the migration of immune cells into the eye by inhibiting the chemotactic effects of C-C motif chemokine ligands 2 (CCL2) and 21 (CCL21) on inflammatory cells.
- A recent study aimed to determine whether exosomes from primed (IFN+Exo) and non-primed (IFN-Exo) AD-derived MSCs (AD-MSCs) could be used to treat uveitis in BD (Gozel et al., 2024). Peripheral blood mononuclear cells (PBMCs) from patients with BU were treated with exosomes and assessed by flow cytometry to evaluate the effects of the treatment on lymphocyte proliferation, survival, and apoptosis. The results showed that IFN+Exo triggered lymphocyte apoptosis in cells from BU, suppressing T lymphocyte survival (Gozel et al., 2024).
- Elajami et al. have reported a case of a 38-year-old woman suffering from BD with oral, genital, and ocular lesions, in addition to problems with wound healing that resulted in keloid formation and joint pain (Elajami et al., 2024). The patient experienced keloid growth and delayed wound healing following an abdominoplasty two years ago. She was using colchicine and corticosteroids. Exosome treatment during surgery led to a good postoperative outcome without complications, suggesting that the combined approach of preoperative inflammation control and regenerative medicine may be an effective treatment for high-risk BD patients.

15.6.2 Systemic Lupus Erythematosus

SLE is a chronic autoimmune disorder characterized by the production of autoantibodies against nuclear antigens that cause chronic inflammation in multiple systems, i.e., the gastrointestinal, cardiovascular, hematologic, musculoskeletal, pulmonary, and renal systems (Hoi et al., 2024; Siegel et al., 2024). However, lupus nephritis (LN) is one of the most severe presentations of SLE, occurring in about 40-60% of patients, leading to end-stage renal disease (Anders et al., 2020).

Hydroxychloroquine remains the drug of choice in almost all SLE patients, besides the use of glucocorticoids to control flares in moderate-to-severe cases. Immunosuppressants are used based on severity, including azathioprine, mycophenolate mofetil, cyclophosphamide, and methotrexate. Targeted biologic therapies are approved and used in treatment as a BBAF inhibitor (belimumab) and a type 1 interferon receptor blocker (anifrolumab) (Pawlak et al., 2023). Despite the improvements in some patients with the existing treatment, some SLE patients show resistance and experience disease relapse, necessitating safer and more effective treatment options (Barbado et al., 2024).

MSCs have shown significant promise in SLE treatment because of their immunomodulatory effect by inhibiting B cells, restricting the differentiation of CD4+ cells into autoreactive T cells, and suppressing DC function as APCs (Barbado et al., 2024), besides elevating regulatory cytokines, i.e., IL-10 and TGF-β, and by diminishing the pro-inflammatory cytokines such as IFN-γ, TNF-α, and IL-12 (Radmanesh et al., 2020). A recent meta-analysis of 12 clinical studies involving 586 LN patients revealed a notable decline in disease activity during a 12-month follow-up. There was a clear improvement in renal function markers, including eGFR, BUN, creatinine, albumin, and protein. A rate of 28.1% during the 12-month follow-up was observed, with a low incidence of adverse events (Yang et al., 2021). Another study involving 81 drug-resistant SLE patients treated with allogeneic MSCs showed that 27% of patients had complete remission, and 7% were in partial remission. Laboratory markers, including SLEDAI scores, serum albumin levels, and proteinuria, showed significant improvement, demonstrating that allogeneic MSC therapy is safe and effective in SLE patients (Wang et al., 2018).

As with other autoimmune disorders, MSC-Exos are used to treat SLE because of their potential to modulate the immune response through multifactorial mechanisms, as discussed earlier. This treatment will suppress glycolysis and mitochondrial metabolism, leading to T cell activation and helping to reduce inflammation (Cunha et al., 2020). MSC-Exos can also modulate B cell activity, which is crucial in the pathophysiology of SLE, where B cells promote autoantibody production. It has been reported that miRNA-155-5p and miRNA-497-5p in MSC-Exos can regulate the PI3K-AKT signaling pathway, thereby suppressing B cell activation and proliferation (Adamo et al., 2019). MSC-derived secretomes can also modulate B cell immunity (Yordanova et al., 2024). However, no completed RCTs using MSC-Exos or MSC secretome for SLE treatment have been reported to date.

15.6.3 Rheumatoid Arthritis

RA involves dysregulated innate and adaptive immune systems that severely affect the synovium of joints and cause synovitis, swelling, and destruction of bony and cartilaginous components, leading to pain and disability. RA can also affect other body systems; for instance, it increases the risk of cardiovascular diseases (Gibofsky et al., 2014). Although the existing pharmacological options provide symptomatic relief, they can only slow down the disease process. MSCs are being studied for their potential in treating RA, with encouraging results (Sarsenova et al., 2021). On the mechanistic note, MSCs increase the number and activity of Treg and IL-10-expressing CD4 T cells, while decreasing Th1 and Th17 cells, thereby modulating the immune response and inhibiting autoimmunity in RA (Lopez et al., 2021). Moreover, MSC-based therapy also decreases pro-inflammatory cytokines such as TNFα, IL-4, IL-17, and INFγ, while increasing anti-inflammatory cytokines such as IL-10 and TGFβ. Furthermore, MSCs secrete numerous immunosuppressive factors as part of their paracrine activity, including IDO, NO, TGFβ, PGE2, COX2, IL-6, IL-10, and HGF, which inhibit T cell proliferation and activity and promote Treg cells. A study by Liang et al. showed that infusion of allogenic MSCs in patients with refractory RA decreased erythrocyte sedimentation rate (ESR) and anti-CCP levels with a concomitant improvement in the 28-joint disease activity score (Liang et al., 2012). An RCT including 105 patients, of whom 52 received MSC therapy, and 53 were in the control group (Yang et al., 2018). Twenty-eight patients in the MSC treatment group showed slowed disease progression. The remaining 24 patients in the MSC treatment group and patients in the control group did not show improvement.

Exosome-based therapy has shown promise in treating RA via a multifactorial mechanism. For example, a study by Chen et al. showed that miR-486-5p in fibroblast-like synoviocytes (FLS) promoted osteoblast differentiation by activating the BMP signaling pathway and reducing Tob1 expression, a target of miR-486-5p, thereby alleviating bone damage in RA (Chen et al., 2020). Similarly, exosomal miR-150-5p was found to suppress VEGF and MMP14, thereby inhibiting angiogenesis and synovial hyperplasia (Chen et al., 2018). Liangyu et al. used miR-451a-loaded exosomes for treatment targeting ATF2 to inhibit synovial fibroblast proliferation and migration (Liangyu et al., 2024). These data demonstrated the effectiveness of microRNA manipulation via exosomes in treating RA. Zhang et al. used ultrasmall Prussian blue (uPB) exosomes to treat RA, which successfully reduced oxidative stress in FLS and inhibited pro-inflammatory factors, thereby helping regulate Th17/Treg cells (Zhang et al., 2022). Moreover, suppression of lncRNA HOTTIP can slow the progression of RA via the miR-1908/STAT3 axis (Yao et al., 2021). In another study, the use of exosomes containing PD-L1 and GAL-1 has been demonstrated (Shen et al., 2021). Molecular studies revealed that PD-L1 exosomes regulated Treg cell development, while GAL-1 exosomes inhibited T cell growth and induced T cell apoptosis, and regulated immune responses through Treg cells.

15.6.4 Multiple Sclerosis

MS is a chronic immune-mediated disorder of the CNS. The pathophysiology of MS involves a complex interaction between environmental factors and genetic susceptibility. The blood-brain barrier is damaged by abnormal autoreactive lymphocytes, thereby facilitating immune cell infiltration and leading to inflammatory demyelination, axonal loss, and progressive neurodegeneration. The conventional disease-modifying therapies merely modify the immune response and reduce its relapse, but fail to prevent the disease progression. The emergence of regenerative and cell-based therapies provides hope for RA patients for a possible cure. These emerging strategies can potentially modulate neuroinflammation, promote remyelination, and enhance neuronal survival (Sengul et al., 2025). Preclinical and early clinical studies using MSCs and their derived exosomes have already shown significant promise (Hauser et al., 2020).

MSCs have attracted attention in MS treatment for their strong neuroprotective and regenerative properties, as well as their immunomodulatory potential. They do this primarily through paracrine actions rather than engraftment and differentiation (Kråkenes et al., 2024). The released bioactive paracrine factors suppress DC maturation, suppress autoreactive T and B lymphocytes, and support an anti-inflammatory environment by increasing M2-polarized microglia/macrophages and regulatory T cells (Sheikhi et al., 2025). The neurotrophic and angiogenic factors in the secretome promote oligodendrocyte survival and remyelination while preventing glial scar formation and apoptosis. Some of the important factors include BDNF, NGF, HGF, IGF-1, VEGF, TGF-β, and FGF-2 (Sengul et al., 2025).

The effectiveness of cell-based and cell-free therapy approaches has been extensively studied in experimental animal models of MS, especially experimental autoimmune encephalomyelitis (EAE). MSC-based therapy has demonstrated a reduction in clinical disability scores, decreased central nervous system inflammation, and promoted remyelination (Hamidi et al., 2024; Sheikhi et al., 2025). A meta-analysis of MSC-derived extracellular vesicles (MSC-EVs) in rodent MS models showed a substantial effect on functional recovery (standardized mean difference ≈ -2.17). However, variability across studies and the lack of standardized dosing protocols remain challenging (Xun et al., 2022).

MSC-based therapy has also progressed to the advanced clinical phases, demonstrating that the treatment approach is safe and effective in MS patients, particularly when autologous BM-MSCs or UC-MSCs are administered intravenously or intrathecally (Connick et al., 2012; Harris et al., 2018; Nabavi et al., 2019). Preliminary indications of potential improvements in visual outcomes, disability scores (e.g., EDSS), and MRI lesion burden were observed in small, uncontrolled cohorts. Extensive placebo-controlled randomized trials with long-term follow-up are still needed to confirm the efficacy and optimize specific factors such as cell source, dosage, route, and frequency of administration (Islam et al., 2023; Jamali et al., 2024).

Going beyond cell-based therapy, MSC-Exos offer multiple benefits, including improved blood-brain barrier permeability without tumorigenicity.

Both in vitro and in vivo studies have shown that MSC-Exos enhance remyelination, reduce neuroinflammation, and convert activated microglia to an anti-inflammatory phenotype (Shen et al., 2021). A recent meta-analysis of experimental studies showed that administration of MSC-EVs or exosomes in rodent EAE models improved symptoms, reduced demyelination, and increased regulatory T cell populations (Xun et al., 2022). The authors also discussed some unsettled issues, such as variation in dosing metrics and the lack of standardized protocols required for translation. A direct comparison of stem cell therapy and exosome-based treatment in a canine model of MS revealed that although MSCs yielded better outcomes in glial cell differentiation and remyelination, MSC-Exos were nonetheless effective and safe, suggesting their practical usefulness as alternatives (Abdallah et al., 2024). Based on a recent narrative review, MSC-Exos could be a promising cell-free therapeutic option to prevent the progression of MS and promote neuroregeneration (Kråkenes et al., 2024).

15.6.5 Inflammatory Bowel Disease

IBD is a chronic, recurrent autoimmune disorder that primarily affects the gastrointestinal system. The two main manifestations of IBD are Crohn's disease and ulcerative colitis. The underlying pathophysiology is multifactorial and has a strong genetic component, including NOD2 gene mutations, combined with immunological dysregulation, particularly activation of the IL-23/Th17 pathway, leading to excessive cytokine release and chronic intestinal inflammation. Dysregulated immune activity results in mucosal damage, strictures, fistulas, and progressive tissue injury, leading to a severely compromised quality of life. Most pharmacological options are aimed at alleviating inflammatory response, but they frequently fail to achieve sustained effects and mucosal repair.

MSCs have shown great potential for IBD treatment due to their immunomodulatory effects and their ability to enhance tissue repair (Saadh et al., 2023). Based on preclinical research, MSCs release trophic factors and paracrine signaling that increase Treg cells, inhibit pathogenic Th1/Th17 responses, and promote mucosal and epithelial repair (Wei et al., 2024). A systematic review and meta-analysis of both animal and clinical studies on UC confirmed that MSC treatment significantly improves disease activity indices compared with conventional therapies (Shi et al., 2019). A recently published meta-analysis of local MSC injections for CD, primarily refractory perianal fistulizing disease, reported combined remission rates of about 62% at 6 months and 58% at 12 months post-treatment, suggesting sustained effects (Guillo et al., 2025). The effectiveness of MSC in complex clinical cases was highlighted by phase III trials (such as ADMIRE-CD, which led to the EU's approval of darvadstrocel) that also showed ~50–60% remission in complex perianal fistulas (Tian et al., 2023; Wang et al., 2023). Intravenous infusion of MSCs for luminal IBD was safe but showed variable efficacy. However, outcomes across studies remain heterogeneous, indicating the need for optimized cell source, dose, and administration protocols (Garcia-Olmo et al., 2022).

Compared with MSCs, MSC-Exos provide a safe and effective alternative. Intraperitoneal delivery of human UC-MSC exosomes has been shown to reduce GI inflammation in experimental mouse models of colitis (DSS and TNBS). They improved barrier integrity by upregulating tight junction proteins (ZO-1, Occludin, and Claudin-1) and increasing the number of goblet cells (Yang et al., 2021). Molecular studies revealed that TSG-6 mediated mucosal healing. Furthermore, combining MSCs with exosomes and standard IBD therapy (mesalazine) resulted in synergistic effects in colitis models, alleviating histological damage (Kheradmand et al., 2025). Treatment with MSC-Exos involved multiple signaling pathways, including the epigenetic inhibition of NF-κB activation, the promotion of neutrophil polarization toward an anti-inflammatory M2 phenotype with ERK pathway modulation, and the suppression of inflammasome activation (NLRP3) through the delivery of miR-378a-5p (Zhao et al., 2023). In conclusion, preclinical evidence suggests that MSC-Exos are potent modulators of inflammation and healing in IBD; however, further clinical evidence is needed to support their routine use.

15.6.6 Diabetes Mellitus Type 1

T1DM is a genetic disorder due to the autoimmune destruction of insulin-producing β-cells in the pancreas. Given the limited pharmacological options, cell-based and cell-free approaches offer hope for T1DM patients. In this regard, Latindra (Doniclecel) is the first cell-based therapy product approved by the FDA for T1DM patients (Erbasan et al., 2025). The product contains allogenic insulin-producing beta cells from deceased pancreatic donors and is infused into the hepatic portal vein of the recipient. Although approved, further research is needed to enhance graft survival after delivery and reduce its dependence on immunosuppression (Thite et al., 2025). Vertex has initiated phase III trials of VX-880 (Zimislecel), a stem cell-based therapy derived from embryonic stem cells (ESCs), for type 1 diabetes (Rech et al., 2025). This investigational allogeneic stem cell-derived treatment is designed to produce functional insulin-producing islet cells (Thite et al., 2025). The outcome is favorable, with data from 12 patients who received a full dose of Zimislecel in a single infusion and were followed for 1 year through October 2024. Importantly, all 12 patients (100%) show glucose-responsive C-peptide production and met the American Diabetes Association (ADA) targets for HbA1c $<7\%$ and $>70\%$ time in range, while getting an average 92% decrease in daily insulin use. In month 12, 83% of patients were free of exogenous insulin. Additionally, the trial met its Phase 1/2 primary endpoint and reduced severe hypoglycemic events (SHEs) in patients with HbA1c$<7\%$. Fortunately, there were no severe adverse events, and most were mild or moderate, suggesting the possibility of a new treatment (Reichman et al., 2025). Another pluripotent stem cell option is induced pluripotent stem cells (Ibrahim et al., 2016), which also offer the option of using autologous patient-derived cells for treatment and may overcome the problems of graft acceptance and the need for immunosuppression (Saima & Haider, 2021).

Among adult stem cell options, BM-MSCs are a promising approach for treating T1DM. Their lack of HLA class II antigens and costimulatory molecules, such as CD40, CD80, and CD86, helps them avoid immune recognition, reducing the risk of immune rejection (Ghoneim et al., 2024). MSCs offer a scalable, immune-evasive therapeutic option that can avoid these problems (Myshko et al., 2025). As an alternative to BM-MSCs, skeletal myoblasts have also been used to treat diabetes, with significant data (Law et al., 2019). In experimental settings, treatment with skeletal myoblasts has shown antihyperglycemic effects, reduced hyperinsulinemia, and attenuation of glucose intolerance (Lei et al., 2009).

MSC-Exos have emerged as an alternative for the treatment of T1DM. Preclinical studies in experimental animal models have shown that they can protect pancreatic beta cells from autoimmune attacks, promote beta cell regeneration, and enhance insulin secretion. Additionally, MSC-Exos help in glucose homeostasis, increase plasma insulin levels, improve islet function by increasing beta cell size and number, and reduce inflammation (Pang et al., 2020). According to one estimate, 3×10^6 exosomes could be produced by a single MSC, making it more practical than using millions of MSCs for treatment (Meligy et al., 2025). Although exosome research is still in its infancy, several limitations must be overcome before the procedure can be effectively used to treat T1DM, despite these encouraging findings (Rasha et al., 2024; Zhang et al., 2025).

15.7 CONCLUSION AND FUTURE PERSPECTIVES

MSCs and MSC-Exos have shown promise as therapeutic options in the treatment of autoimmune diseases. Despite significant data supporting their contribution to maintaining glucose homeostasis, their use still has many limitations. Engineering MSCs and their exosomes, such as genetic engineering and surface modification, is associated with high production costs and immune-related reactions (Zhu et al., 2023). Non-viral gene transfer for MSC functional improvement does not appear to have a significant effect, and surface modification does not consistently improve mobilization or the therapeutic efficacy of the cells, since nanomaterials in tissue engineering may cause cytotoxic effects and diminish stem cell differentiation (Rayat et al., 2025). Furthermore, there are no clearly defined protocols for generating, purifying, and dosing MSC-Exos, making it challenging to interpret results and translate them into clinical use. Long-term safety assessment has yet to be established, including tumorigenesis, immune response, and characterization and quality of MSCs and their exosomes, which may significantly influence treatment success. The solutions to these barriers include improving production strategies, producing MSC-Exos with greater specificity and a defined cargo profile, and tailoring treatment to the patient's needs (Chengmei et al., 2021; Zhu et al., 2023).

REFERENCES

Abdallah AN, Shamaa AA, El-Tookhy OS, Bahr MM. Comparison between stem cell therapy and stem cell-derived exosomes on induced multiple sclerosis in dogs. *BMC Vet Res.* 2024;8;20(1):90.

Adamo A, Brandi J, Caligola S, Delfino P, Bazzoni R, Carusone R. Extracellular vesicles mediate mesenchymal stromal cell-dependent regulation of B cell PI3K-AKT signaling pathway and actin cytoskeleton. *Front Immunol.* 2019 Mar 12.

Adil A, Goyal A, Quint JM. *Behcet Disease.* Treasure Island, FL: StatPearls Publishing; 2025.

Advani D, Farid N, Tariq MH, Kohli N. A systematic review of mesenchymal stem cell secretome: Functional annotations, gene clusters, and proteomics analyses for bone formation. *Bone.* 2025;190.

Ahmed ZT, Alkahlot MH, Haider KH. MSC-derived exosomes: Advances in cell-free therapy. In: *Handbook of Stem Cell Applications* (pp. 1–41). Singapore: Springer Nature; 2023 Aug 18.

Albrecht K, Troll W, Callhoff J, Strangfeld A, Ohrndorf S, Mucke J. Sex- and gender-related differences in systemic lupus erythematosus: A scoping review. *Rheumatol Int.* 2025 June 27;

Al-Khani AM, Khalifa MA, Haider KH. Mesenchymal stem cells: How close are we to their routine clinical use? In: Haider K.H. (eds) *Handbook of Stem Cell Therapy.* Singapore: Springer; 2022. https://doi.org/10.1007/978-981-16-6016-0_11-1.

Álvarez-Viejo M, Haider KH. Mesenchymal stem cells. In: Haider K.H. (eds) *Handbook of Stem Cell Therapy.* Springer. https://doi.org/10.1007/978-981-16-6016-0_6-1.

Anders HJ, Saxena R, Zhao M, Parodis I, Salmon JE, Mohan C. Lupus nephritis. *Nat Rev Dis Primers.* 2020 Jan;23;6(1):1–25.

Angum F, Khan T, Kaler J, Siddiqui L, Hussain A. The prevalence of autoimmune disorders in women: A narrative review. *Cureus.* 2020 May;13;12(5).

Aringer M, Costenbader K, Daikh D, Brinks R, Mosca M, Ramsey-Goldman R. European league against rheumatism/American College of Rheumatology classification criteria for systemic lupus erythematosus. *Arthritis Rheumatol.* 2019;6;71(9):1400–12.

Arun M, Rajasingh S, Madasamy P, Rajasingh J. Immunomodulatory and regenerative functions of MSC-derived exosomes in bone repair. *Bioengineering.* 2025; 12(8):844. https://doi.org/10.3390/bioengineering12080844.

Babaoğlu H, Karadağ Ö. Therapeutic mechanisms of glucocorticoids, their administration, and chronotherapy. *Acta Medica.* 2024;3:1–9.

Bai L, Shao H, Wang H, Zhang Z, Su C, Dong L, et al. Effects of mesenchymal stem cell-derived exosomes on experimental autoimmune uveitis. *Sci Rep.* 2017;7:4323.

Barbado J. Mesenchymal stem cell transplantation may induce immunological tolerance in systemic lupus erythematosus. *Biomed J.* 2024.

Bartels C. Systemic lupus erythematosus (SLE): Practice essentials, pathophysiology, etiology. Medscape.com. 2019 Feb 2.

Bhatti A, Haider KH. Mesenchymal stem cells as living bio-drugs: Adding a novel dimension to cardiology and cardiovascular sciences. *J Cardiol Cardiovasc Sci.* 2025;6(1):30–5. https://doi.org/10.29245/2578-3025/2025/1.1219.

Blank M, Israeli E, Gertel S, et al. Molecular mimicry in autoimmunity and vaccinations. In: Anaya JM, Shoenfeld Y, Rojas-Villarraga A, et al. (eds) *Autoimmunity: From Bench to Bedside.* Bogotá, Colombia: El Rosario University Press; 2013 Jul 18. Chapter 21.

Boyapati R, Satsangi J, Tzer Ho G. Pathogenesis of Crohn's disease. *F1000Prime Rep.* Apr. 2015;2;7:44.

Cao J, Wang B, Tang T, Lv L, Ding Z, Li Z, Hu R, et al. Three-dimensional culture of MSCs produces exosomes with improved yield and enhanced therapeutic efficacy for cisplatin-induced acute kidney injury. *Stem Cell Res Ther.* 2020 May 27;11(1):206. https://doi.org/10.1186/s13287-020-01719-2.

Chauhan K, Jandu JS, Brent LH, et al. *Rheumatoid Arthritis*. Treasure Island, FL: StatPearls Publishing; 2023 Jan. https://www.ncbi.nlm.nih.gov/books/NBK441999/.

Chen J, Liu M, Luo X, Peng L, Zhao Z, He C, He Y. Exosomal miRNA-486-5p derived from rheumatoid arthritis fibroblast-like synoviocytes induces osteoblast differentiation through the Tob1/BMP/Smad pathway. *Biomater Sci*. 2020 Jun 21;8(12):3430–42. https://doi.org/10.1039/c9bm01761e.

Chen Z. Therapeutic potential of mesenchymal cell-derived MIRNA-150-5p-expressing exosomes in rheumatoid arthritis mediated by the modulation of MMP14 and VEGF. *J Immunol*. 2018.

Chengmei He, Yang Y, Zheng K, Chen Y, Liu S, Li Y. Mesenchymal stem cell-based treatment in autoimmune liver diseases: underlying roles, advantages, and challenges. *Therap Adv Chron Dis*. 2021 Jan;12:204062232199344.

Connick P, Kolappan M, Crawley C, Webber DJ, Patani R, Michell AW, et al. Autologous mesenchymal stem cells for the treatment of secondary progressive multiple sclerosis: An open-label phase 2a proof-of-concept study. *Lancet Neurol*. 2012;11(2):150–6.

Csöbönyeiová M. Cell-based and selected cell-free therapies for myocardial infarction: How do they compare to the current treatment options? *Int J Mol Sci*. 2022.

Cunha FF, V AO, Almeida DC, Silva TB, Souza Breda CN, et al. Extracellular vesicles isolated from mesenchymal stromal cells modulate CD4+ T lymphocytes toward a regulatory profile. *Cells*. 2020;9(4):1059.

Dan Ma, Xu K, Zhang G, Liu Y, Gao J, Tian M, et al. Immunomodulatory effect of human umbilical cord mesenchymal stem cells on T lymphocytes in rheumatoid arthritis. *Int Immunopharmacol*. 2019 Sep.

Davatchi F, Nikbin B, Shams H, Sadeghi Abdollahi B, Mohyeddin M, Shahram F. Mesenchymal stem cell therapy unable to rescue the vision from advanced Behcet's disease retinal vasculitis: Report of three patients. *Int J Rheum Dis*. 2013;16:139–47.

Direskeneli H. Autoimmunity vs autoinflammation in Behcet's disease: Do we oversimplify a complex disorder? *Rheumatol*. 2006;45:1461–5.

Duan L, Huang H, Zhao X, et al. Extracellular vesicles derived from human placental mesenchymal stem cells alleviate experimental colitis in mice by inhibiting inflammation and oxidative stress. *Int J Mol Med*. 2020;46(4):1551–61. https://doi.org/10.3892/ijmm.2020.4679.

Elajami MH. The usefulness of exosomes in accelerating healing and preventing complications in Behçet's disease: A case report. *Cureus*. 2024;16(11):e74476. https://doi.org/10.7759/cureus 74476.

Elkoshi Z. Autoimmune diseases refractory to corticosteroids and immunosuppressants. *Front Immunol*. 2024 Sept;15:1447337.

Erbasan E, Aliciaslan M, Erendor F, Dandin O, Sanlioglu S. Lantidra (donislecel) in type 1 diabetes: An in-depth analysis of pharmacology, clinical effectiveness, safety, and the therapeutic role of the first FDA-approved allogeneic islet cell therapy. *Diabet Med*. 2025:e70168. https://doi.org/10.1111/dme.70168.

Galipeau J, Sensébé L. Mesenchymal stromal cells: Clinical challenges and therapeutic opportunities. *Cell Stem Cell*. 2018;22(6):824–33.

Garcia-Olmo D, Gilaberte I, Binek M, D Hoore AJL, Lindner D, Selvaggi F, Spinelli A, et al. Follow-up study to evaluate the long-term safety and efficacy of darvadstrocel (MSC treatment) in patients with perianal fistulizing Crohn's disease: ADMIRE-CD Phase 3 randomized controlled trial. *Dis Colon Rectum*. 2022;65(5):713–20. https://doi.org/10.1097/DCR.0000000000002325.

GBD 2021 Rheumatoid Arthritis Collaborators. Global, regional, and national burden of rheumatoid arthritis, 1990-2020, and projections to 2050: A systematic analysis of the Global Burden of Disease Study 2021. *Lancet Rheumatol*. 2023;5(10):e594–610. https://doi.org/10.1016/S2665-9913(23)00211-4.

Ghoneim MA, Gabr MM, El-Halawani SM, Refaie AF. Current status of stem cell therapy for type 1 diabetes: A critique and a prospective consideration. *Stem Cell Res Ther*. 2024 Jan 29.

Gibofsky A, M. Epidemiology, pathophysiology, and diagnosis of rheumatoid arthritis: A synopsis. *AJMC*, 2014.

Gozel M, Aydemir D, Kaleli HN, Cosar F, Guleser UY, Unlu EK, et al. Effects of primed adipose mesenchymal stem cell-derived exosomes on immunomodulation in behcet uveitis. bioRxiv. https://doi.org/10.1101/2024.11.24.625042.

Guillo L, Gravier Dumonceau R, Vélier M. Efficacy of mesenchymal stem cell-based therapies in the treatment of perianal fistulizing Crohn's disease: A systematic review and meta-analysis. *Stem Cell Res Ther*. 2025;16:152.

Habib SM, Martini MF, Al Ahmed O, Qassab AR, Haider KH. Mesenchymal stem cell secretomes for cell-free therapy of the broken heart. In: Haider K.H. (ed) *Handbook of Regenerative Medicine: Cell-Free Therapy*. Taylor & Francis Group/CRC Press, 2025a, Chapter 18, pp. 274–93.

Habib SM, Martini MF, Al Ahmed O, Qassab AR, Haider KH. The potential of mesenchymal stem cell–derived exosomes in cell-free therapy for myocardial repair. In: Haider K.H. (ed) *Handbook of Regenerative Medicine: Cell-Free Therapy*. Taylor & Francis Group/CRC Press, 2025b, Chapter 24, pp. 372–90.

Habib SM, Martini MF, Hamdan YNA, Shrebaty OMM, Haider KH. Mesenchymal stem cell secretome and exosomes as potential advanced therapy medicinal products for treating a failing heart. *World J Stem Cells*. 2026.

Haider KH, Aslam M. (2018). Cell-free therapy with stem cell secretions: Protection, repair, and regeneration of the injured myocardium. In: K.H. Haider and Aziz S (eds) *Stem Cells: From Hype to Real Hope*. Berlin: Medicine & Life Sciences, De Gruyter, 2018.

Haider KH, Najimi M. Exosome-based cell-free therapy in regenerative medicine for myocardial repair. In: Haider K.H. (ed) *Handbook of Stem Cell Therapy*. Singapore: Springer, 2022. https://doi.org/10.1007/978-981-16-6016-0_9-1.

Hamidi SH, Etebar N, Rahimzadegan M, Zali A, Roodsari NSR, S. Mesenchymal stem cells and their derived exosomes in multiple sclerosis disease: From paper to practice. *Mol Cell Biochem*. 2024;479(7):1643–71.

Harris VK, Stark J, Vyshkina T, Blackshear L, Joo G, Stefanova V, et al. Trial of intrathecal mesenchymal stem cell-derived neural progenitors in progressive multiple sclerosis. *EBioMedicine*. 2018. pp. 23–30.

Hauser SL, Cree BAC. Treatment of multiple sclerosis: A review. *Am J Med*. 2020;1390:e2.

Hoi A, Igel T, Mok CC, Arnaud L. Systemic lupus erythematosus. *Lancet*. 2024;1;403:10441.

Houwen TB, van Hagen PM, van Laar JAM. Immunopathogenesis of Behçet's disease and treatment modalities. *Semin Arthritis Rheum*. 2022;52:151956. https://doi.org/10.1016/j.semarthrit.2022.151956.

Huashan Liu, Liang Z, Wang F, Zhou C, Zheng X, Hu T, He X, Wu X, Lan P. Exosomes from mesenchymal stromal cells reduce murine colonic inflammation via a macrophage-dependent mechanism. *JCI Insight*. 2019;4:24. https://doi.org/10.1172/jci.insight.131273.

Ibrahim AY, Mehdi Q, Abbas AO, Alashkar A, Haider KH. Induced pluripotent stem cells: Next-generation cells for tissue regeneration. *J Biomedi Sci Eng*. 2016;9(4):226–44.

Islam MA, Alam SS, Kundu S, Ahmed S, Sultana S, Patar A, et al. Mesenchymal stem cell therapy in multiple sclerosis: A systematic review and meta-analysis. *J Clin Med*. 2023 Sept;30;12(19):6311.

Jahid M, Khan KU, Rehan-Ul-Haq, Ahmed RS. Overview of rheumatoid arthritis and scientific understanding of the disease. *Mediterr J Rheumatol*. 2023 Aug 1;34(3):284–91. https://doi.org/10.31138/mjr.20230801.oo.

Jamali F, Aldughmi M, Atiani S, Al-Radaideh A, Dahbour S, Alhattab D, et al. Human umbilical cord-derived mesenchymal stem cells in the treatment of multiple sclerosis patients: Phase I/II Dose-Finding Clinical Study. *Cell Transplant*. 2024;33.

Janockova J, Slovinska L, Harvanova D. New therapeutic approaches of mesenchymal stem cells-derived exosomes. *J Biomed Sci*. 2021;28:39.

Jiang JH, Zhang Y, Tian K, Wang B, Han S. Amelioration of experimental autoimmune encephalomyelitis through transplantation of placental-derived mesenchymal stem cells. *Sci Rep*. 2017;7(1).

Kahrizi MS, Maghsoodi S, Ghayoumi F, Etemadi H, Aminzadeh S, Safa K, Alesaeidi S, et al. Mesenchymal stem/stroma cell (MSC)-derived nanoscale exosomes inhibit T cell proliferation by negative regulation of PI3K/Akt and MAPK/ERK pathway. *Nanomed Res J*. 2022;7(2):189–94. https://doi.org/10.22034/nmrj.2022.02.008.

Kamal M, Kassem D, Haider KH. Sources and therapeutic strategies of mesenchymal stem cells in regenerative medicine. In: Haider, K.H. (ed), *Handbook of Stem Cell Therapy*. Singapore: Springer, 2022. https://doi.org/10.1007/978-981-19-2655-6_2.

Kavaldzhieva K, Mladenov N, Markova M, Belemezova K. Mesenchymal stem cell secretome: Potential applications in human infertility caused by hormonal imbalance, external damage, or immune factors. *Biomedicines*. 2025;13(3):586.

Kheradmand F, Yasaman Rahimzadeh SF, Esmaeili SA, Negah SS, Farkhad NK, Nazari SE, et al. Efficacy of umbilical cord-derived mesenchymal stem cells and exosomes in conjunction with standard IBD drug on immune responses in an IBD mouse model. *Stem Cell Res Ther*. 2025 Jan;7;16(1):5.

Kråkenes T, Sandvik CE, Ytterdal M, Gavasso S, Evjenth EC, Bø L, et al. The therapeutic potential of exosomes from mesenchymal stem cells in multiple sclerosis. *Int J Mol Sci*. 2024 Sept;24;25(19):10292.

Kuwabara T, Ishikawa F, Kondo M, Kakiuchi T. The role of IL-17 and related cytokines in inflammatory autoimmune diseases. *Mediators Inflamm*. 2017;2017:1–11.

Law PK, Law DM, Ye L, Haider KH, Song SJ, Lu P, Ma JH. Myoblasts provide safe and effective treatments for hereditary muscular dystrophies, cardiomyopathies, type 2 diabetes, solid tumors, and aging. In: *Stem Cells—From Hype to Real Hope*. Berlin: Walter de Gruyter GmbH; 2019, pp. 71–97.

Leccese P, Alpsoy E. Behçet's disease: An overview of etiopathogenesis. *Front Immunol.* 2019 May 10;10.

Lee S, Kim S, Chung H, Moon JH, Kang SJ, Park CG. Mesenchymal stem cell-derived exosomes suppress proliferation of T cells by inducing cell cycle arrest through p27kip1/Cdk2 signaling. *Immunol Lett.* 2020;225:16–22. https://doi.org/10.1016/j.imlet.2020.06.006.

Lei Y, Lee KO, Su LP, Toh WC, Haider KJ, Law PK, Zhang W, Chan SP, Sim EKW. Skeletal myoblast transplantation for attenuation of hyperglycemia, hyperinsulinemia, and glucose intolerance in a mouse model of type-2 diabetes mellitus. *Diabetologia.* 2009; 52(9):1925–34.

Li Y, Wang Z, Zhao Y, Luo Y, Xu W, Marion TN, Liu Y. Successful mesenchymal stem cell treatment of leg ulcers complicated by Behcet disease: A case report and literature review. *Medicine.* 2018;97(16):e0515. https://doi.org/10.1097/MD.0000000000010515.

Liang J, Li X, Zhang H, Wang D, Feng X, Wang H, Hua B, Liu B, Sun L. Allogeneic mesenchymal stem cells transplantation in patients with refractory RA. *Clin Rheumatol.* 2012;31(1):157–61. https://doi.org/10.1007/s10067-011-1816-0.

Liang Y, Xia J, Xu L, Zhang Y, Xu X, Haider KH. Cell-free and cell-based therapy in regenerative medicine. In: Haider, K.H. (ed), *Handbook of Regenerative Medicine: Cell-Free Therapy.* Taylor & Francis Group/CRC Press, 2025, Chapter 10, pp. 134–53.

Liangyu Mi, Gao J, Li N, Liu Y, Zhang N, Gao Y, et al. Human umbilical cord mesenchymal stem cell-derived exosomes loaded with miR-451a target ATF2 to improve rheumatoid arthritis. *Int Immunopharmacol.* 2024;127:111365.

Liu Di, Yu G, Yuan N, Nie D. The efficacy and safety of biologic or targeted synthetic DMARDs in rheumatoid arthritis treatment: One year of review 2024. *Allergol Immunopathol (Madr).* https://doi.org/10.15586/aei.v53i3.1278.

Liu P, Xie Xr, Wu H, et al. Conditioned medium of mesenchymal stem cells pretreated with H_2O_2 promotes intestinal mucosal repair in acute experimental colitis. *Sci Rep.* 2022;12:20772. https://doi.org/10.1038/s41598-022-24493-y.

Lopez-Santalla M, Bueren JA, Garin MI. Mesenchymal stem/stromal cell-based therapy for the treatment of rheumatoid arthritis: An update on Preclinical Studies, *EBioMed.* 2021.

Lundy SK. Killer B lymphocytes: The evidence and the potential. *Inflamm Res.* 2009 Mar;5;58(7):345–57.

Ma OK-F, Chan KH. Immunomodulation by mesenchymal stem cells: Interplay between mesenchymal stem cells and regulatory lymphocytes. *World J Stem Cells.* 2016 Sep 26;8(9):268–78. https://doi.org/10.4252/wjsc.v8.i9.268.

Madec AM, Mallone R, Afonso G, Abou Mrad E, Mesnier A, Eljaafari A, Thivolet C. Mesenchymal stem cells protect NOD mice from diabetes by inducing regulatory T cells. *Diabetologia.* 2009;52(7):1391–9. https://doi.org/10.1007/s00125-009-1374-z.

Mattioli I, Bettiol A, Saruhan-Direskeneli G, Direskeneli H, Emmi G. Pathogenesis of Behçet's syndrome: Genetic, environmental and immunological factors. *Front Med.* 2021;8:713052.

Mazaheri T, Esmaeilzadeh A, Mirzaei MH. Introducing the immunomodulatory effects of mesenchymal stem cells in an experimental model of Behçet's disease. *J Med Hypotheses Ideas.* 2012;6:23–7.

Meligy FY, Mohammed HSED, Abou Elghait AT, Mohamed HK. Mesenchymal stem cells versus mesenchymal stem cells-derived exosomes as potential autophagy pathway modulators in a diabetic model. 2025.

Mount NM. Cell-based therapy technology classifications and translational challenges. *Philos Trans R Soc Lond B Biol Sci.* 2015.

Mushtaq M, Zineldeen DH, Mateen MA, Haider KH. Mesenchymal stem cells' "garbage bags" at work: Treating radial nerve injury with mesenchymal stem cell-derived exosomes. *World J Stem Cells* 2024; 16(5). https://doi.org/10.4252/wjsc. v16.i5.0000.

Myshko D. Phase 3 trial of Vertex's islet cell therapy for type 1 diabetes under way. Managed Healthcare Executive. 2025.

Nabavi SM, Arab L, Jarooghi N, Bolurieh T, Abbasi F, Mardpour S, et al. Feasibility of intravenous and intrathecal injection of autologous bone marrow derived mesenchymal stromal cells in patients with amyotrophic lateral sclerosis: An open label phase I clinical trial. *Cell J.* 2019;20(4):592–8.

Najimi M, Haider KH. (2022). Extracellular vesicle-based cell-free therapy for liver regeneration. In: Haider, K.H. (eds) Handbook of Stem Cell Therapy. Springer, Singapore. https://doi.org/10.1007/978-981-16-6016-0_9-1.

Pang H, Luo S, Xiao Y, Xia Y, Li X, Huang G. Emerging roles of exosomes in T1DM. *Front Immunol.* 2020 Nov 26.

Pawlak-Buś K, Schmidt W, Ewa D, Piotr L. Current treatment of systemic lupus erythematosus: A clinician's perspective. *Rheumatol Int.* 2023;43(8):1395–407.

Pisetsky DS. Pathogenesis of autoimmune disease. *Nat Rev Nephrol.* 2023;10;19(8):1–16.

Radmanesh F, Mahmoudi M, Yazdanpanah E, Keyvani V, Kia N, Nikpoor AR. The immunomodulatory effects of mesenchymal stromal cell-based therapy in human and animal models of systemic lupus erythematosus. *IUBMB Life.* 2020;72(11):2366–81.

Rahgoshay M. Engineered exosomes: Advanced nanocarriers for targeted therapy and drug delivery in hematological malignancies - cancer nanotechnology, *BioMed Central.* 2025.

Rasha AM, M BH, Wadid NA, Akikwala MS. Exploring novel treatment modalities for type 1 diabetes mellitus: Potential and prospects. *Healthcare.* 2024;12(15):1485–5.

Rayat Pisheh H, Sani M. Mesenchymal stem cells derived exosomes: A new era in cardiac regeneration. *Stem Cell Res Ther.* 2025;16(1):16. https://doi.org/10.1186/s13287-024-04123-2.

Rech TA, Lanzoni G. Islet cell replacement and regeneration for type 1 diabetes: Current developments and future prospects. *BioDrugs.* 2025;39(2):261–80. https://doi.org/10.1007/s40259-025-00703-7.

Reichman TW, Markmann JF, Odorico J, Witkowski P, Fung JJ; VX-880-101 FORWARD Study Group. Stem cell-derived, fully differentiated islets for type 1 diabetes. *N Engl J Med.* 2025;393(9):858–68. https://doi.org/10.1056/NEJMoa2506549.

Rojas M, Restrepo-Jiménez P, Monsalve DM, Pacheco Y, Acosta-Ampudia Y, Ramírez-Santana C. Molecular mimicry and autoimmunity. *J Autoimmunity.* 2018;95:100–23.

Rose NR, Mackay IR. Molecular mimicry: a critical look at exemplary instances in human diseases. *Cell Mol Life Sci.* 2000;57(4):542–51.

Saadh MJ, Mikhailova MV, Rasoolzadegan S. Therapeutic potential of mesenchymal stem/stromal cells (MSCs)-based cell therapy for inflammatory bowel diseases (IBD) therapy. *Eur J Med Res.* 2023;28:47.

Saima Kh, Haider KH. Stem cells: A renewable source of pancreatic β-cells and future for diabetes treatment. In: Haider, K.H. (eds) *Stem Cells.* Cham: Springer, 2021, Chapter 12, pp. 185–202. https://doi.org/10.1007/978-3-030-77052-5_12.

Sarsenova M, Issabekova A, Abisheva S, Rutskaya-Moroshan K, Ogay V, Saparov A. Mesenchymal stem cell-based therapy for rheumatoid arthritis. *Int J Mol Sci.* 2021;22(21):11592. https://doi.org/10.3390/ijms222111592.

Sengul Bag F, Bag OF. Molecular and therapeutic effects of mesenchymal stem cell-derived exosomes on autoimmune diseases. *World J Stem Cells.* 2025;17(7):107202. https://doi.org/10.4252/wjsc.v17.i7.107202.

Sheikhi K, Ghaderi S, Firouzi H, Rahimibarghani S, Shabani E, Afkhami H, et al. Recent advances in mesenchymal stem cell therapy for multiple sclerosis: Clinical applications and challenges. *Front Cell Dev Biol.* 2025;13:1517369.

Shen Z, Huang W, Liu J, Tian J, Wang S, Rui K. Effects of mesenchymal stem cell-derived exosomes on autoimmune diseases. *Front Immunol.* 2021;12:749192.

Sheng H, et al. A critical role of IFNgamma in priming MSC-mediated suppression of T cell proliferation through up-regulation of B7-H1. *Cell Res.* 2008;18:846–57.

Shi X, Chen Q, Wang F. Mesenchymal stem cells for the treatment of ulcerative colitis: A systematic review and meta-analysis of experimental and clinical studies. *Stem Cell Res Ther.* 2019;23;10(1):266.

Shi Y, Wang Y, Li Q, Liu K, Hou J, Shao C, Wang Y. Immunoregulatory mechanisms of mesenchymal stem and stromal cells in inflammatory diseases. *Nat Rev Nephrol.* 2018;14(8):493–507. https://doi.org/10.1038/s41581-018-0023-5.

Shlomchik MJ, Craft JE, Mamula MJ. From T to B and back again: positive feedback in systemic autoimmune disease. *Nat Rev Immunol.* 2001;1(2):147–53.

Siegel CH, Sammaritano LR. Systemic lupus erythematosus: A review. *JAMA.* 2024;331(17):1480–91. https://doi.org/10.1001/jama.2024.2315.

Singh N, Choonara YE, Kumar P. Method standardization of secretome production, collection, and characterization: New insights and challenges. *Regen Ther.* 2025;29:466–73,.

Song X, Liang H, Nan F, Chen W, Li J, He L, Cun Y, Li Z, Zhang W, Zhang D. Autoimmune diseases: Molecular pathogenesis and therapeutic targets. 2025;6(7):e70262. https://doi.org/10.1002/mco2.70262.

Sultan N, Saadeldin Ramadhan W, Alkarim S. Bone marrow-derived mesenchymal stem cells: A potential therapy in an autoimmune hepatitis rat model. *Adv Life Sci.* 2025;31;12(1):105–113.

Swain HN, Boyce PD, Bromet BA, Barozinksy K, Hance L, Shields D, et al. Mesenchymal stem cells in autoimmune disease: A systematic review and meta-analysis of pre-clinical studies. *Biochimie.* 2024;223:54–73.

Tafti D. *Multiple Sclerosis.* Treasure Island, FL: StatPearls Publishing; 2024. 499849 p.

Thite T. Can stem cell therapy cure diabetes? Cas.org CAS. 2025 Aug 11.

Tian CM, Zhang Y, Yang MF, Xu HM, Zhu MZ, Yao J, et al. Stem cell therapy in inflammatory bowel disease: A review of achievements and challenges. *J Inflamm Res.* 2023;16:2089–119.

Tingfen Li, Yan Y, Wang B, Qian H, Zhang X, Shen L, Wang M, Zhou Y, Zhu W, Li W, Xu W. Exosomes derived from human umbilical cord mesenchymal stem cells alleviate liver fibrosis. *Stem Cells Dev.* 2013;22(6):845–54. https://doi.org/10.1089/scd.0395.

Trigo CM, Rodrigues JS, Camões SP, Solá S, Miranda JP. Mesenchymal stem cell secretome for regenerative medicine: Where do we stand? *J Adv Res.* 2025;70:103–24.

Walton C, King R, Rechtman L, et al. Rising prevalence of multiple sclerosis worldwide: Insights from the Atlas of MS, third edition. *Mult Scler.* 2020;26(14):1816–21. https://doi.org/10.1177/1352458520970841.

Wang D, Zhang H, Liang J, Wang H, Hua B, Feng X. A long-term follow-up study of allogeneic mesenchymal stem/stromal cell transplantation in patients with drug-resistant systemic lupus erythematosus. *Stem Cell Rep.* 2018;10(3):933–41.

Wang H, Jiang HY, Zhang YX, Jin HY, Fei BY, Jiang JL. Mesenchymal stem cells transplantation for perianal fistulas: A systematic review and meta-analysis of clinical trials. *Stem Cell Res Ther.* 2023;26;14(1):103.

Wei S, Li M, Wang Q, Zhao Y, Du F, Chen Y, et al. Mesenchymal stromal cells: New generation treatment of inflammatory bowel disease. *J Inflamm Res.* 2024;17:3307–34.

Wei S, Xie S, Yang Z, Peng X, Gong L, Zhao K, Zeng K, Lai K. Allogeneic adipose-derived stem cells suppress mTORC1 pathway in a murine model of systemic lupus erythematosus. *Lupus.* 2019;28(2):199–209. https://doi.org/10.1177/0961203318819131.

Weiss ARR, Dahlke MH. Immunomodulation by mesenchymal stem cells (MSCs): Mechanisms of action of living, apoptotic, and dead MSCs. *Front Immunol.* 2019;10:1191. https://doi.org/10.3389/fimmu.2019.01191.

Xia J, Arai S, Haider KH. Research and applications of stem cell secretome. In: Haider, K.H. (eds) *Stem Cells: From Potential to Promise.* Singapore: Springer, 2021. https://doi.org/10.1007/978-981-16-0301-3_7.

Xun C, Deng H, Zhao J, Ge L, Hu Z. Mesenchymal stromal cell extracellular vesicles for multiple sclerosis in preclinical rodent models: A meta-analysis. *Front Immunol.* 2022;13:972247.

Yang S, Liang X, Song J. A novel therapeutic approach for inflammatory bowel disease by exosomes derived from human umbilical cord mesenchymal stem cells to repair the intestinal barrier via TSG-6. *Stem Cell Res Ther.* 2021;12:315.

Yang Y. Serum IFN-γ levels predict the therapeutic effect of mesenchymal stem cell transplantation in active rheumatoid arthritis. *J Transl.* 2018.

Yanhong Li, Wang Z, Zhao Y, Luo Y, Xu W, Marion TN. Successful mesenchymal stem cell treatment of leg ulcers complicated by Behcet disease: A case report and literature review. *Medicine.* 2018;97(16):e0515.

Yao X, Wang Q, Zeng P, Hou L, Yang Y, Lu D, et al. LncRNA HOTTIP from synovial fibroblast-derived exosomes: A novel molecular target for rheumatoid arthritis through the miR-1908 5p/STAT3 axis. *Expe Cell Res.* 2021;409(2):112943.

Yi YF, Fan ZQ, Liu C, Ding YT, Chen Y, Wen J, Jian XH, Li YF. Immunomodulatory effects and clinical application of exosomes derived from mesenchymal stem cells. *World J Stem Cells.* 2025;17(3):103560. https://doi.org/10.4252/wjsc.v17.i3.103560.

Yordanova A, Ivanova M, Tumangelova-Yuzeir K, Angelov A, Kyurkchiev S, Belemezova K. Umbilical cord mesenchymal stem cell secretome: A potential regulator of B cells in systemic lupus erythematosus. *Int J Mol Sci.* 2024;25(23):12515.

Zajec A, Trebušak Podkrajšek K, Tesovnik T, Šket R, Čugalj Kern B, Jenko Bizjan B. Pathogenesis of type 1 diabetes: Established facts and new insights. *Genes.* 2022;16;13(4):706.

Zhang L. Nanoenzyme-engineered neutrophil-derived exosomes attenuate joint injury in advanced rheumatoid arthritis by regulating the inflammatory environment. *Bioact Mater.* 2022.

Zhang L. Progress in Exosomes in the treatment of diabetes mellitus. *Highl Sci Eng Technol.* 2025 June 27;144:112–8.

Zhang R, Miao J, Zhu P. Regulatory T cell heterogeneity and therapy in autoimmune diseases. *Autoimmun Rev.* 2021;20:102715.

Zhao Z, Zhang L, Ocansey DKW, Wang B, Mao F. The role of mesenchymal stem cell-derived exosomes in epigenetic modifications in inflammatory diseases. *Front Immunol.* 2023 May;14:1166536.

Zhu X, Ma D, Yang B, An Q, Zhao J, Gao X. Research progress of engineered mesenchymal stem cells and their derived exosomes and their application in autoimmune/inflammatory diseases. *Stem Cell Res Ther.* 2023 11;14(1).

Zubair M, Abouelnazar FA, Iqbal MA, Pan J, Zheng X, Chen T. Mesenchymal stem cell-derived exosomes as a plausible immunomodulatory therapeutic tool for inflammatory diseases. *Front Cell Dev Biol.* 2025;10;13.

16 Mesenchymal Stem Cells in Knee Osteoarthritis

Mechanisms, Evidence, and Future Directions

Moaz Safwan[1], Mariam S. Bourgleh[1], Batoul Suliman Aibour[1], Rawan Yousef Albalawi[1], Aseel Almsned[1], Mohammed Raed Hazaea[1], and Khawaja H. Haider[1]

[1]Sulaiman AlRajhi Medical School, Al Bukairiyah, Kingdom of Saudi Arabia

16.1 LIST OF ABBREVIATIONS

AD-MSCs	Adipose-derived mesenchymal stem cells
AE	Adverse event
BM-MSCs	Bone marrow-derived mesenchymal stem cells
EVs	Extracellular vesicles
MSC	Mesenchymal stem cell
MSCs	Mesenchymal stem cells
IL	Interleukin
TNF	Tumor necrosis factor
NF-κB	Nuclear factor kappa B
VEGF	Vascular endothelial growth factor
PGE2	Prostaglandin E2
TGF-β	Transforming growth factor-beta
OA	Osteoarthritis
UC-MSCs	Umbilical cord-derived mesenchymal stem cells
VAS	Visual analog scale
ISCT	International Society for Cellular Therapy
WOMAC	Western Ontario and McMaster Universities Osteoarthritis Index

16.2 INTRODUCTION

Osteoarthritis (OA) is a progressive degenerative disease of synovial joints, most commonly affecting the knee. It is characterized by synovial inflammation, cartilage loss, and subchondral bone remodeling, leading to varying degrees of functional impairment (Langworthy et al., 2024). It's considered the leading cause of disability among older populations, with increased prevalence in parallel with early onset due to joint injuries, obesity, and metabolic comorbidities (Dell'Isola et al., 2025). The typical clinical presentation includes chronic pain, stiffness, and limitations in functional mobility, resulting in reduced quality of life, increased unemployment, and a growing

DOI: 10.1201/9781003653387-16

socioeconomic burden (Sadosky et al., 2010). Data from the United States indicate that the direct and indirect costs of OA exceed $70 billion annually, with younger patients increasingly affected and expected to remain symptomatic for more extended periods of their lives (Leifer et al., 2022; Wang et al., 2024). Despite the high disease burden, the currently available therapeutic options are largely symptomatic and fail to halt or reverse disease progression. Non-pharmacological measures, pharmacological interventions, and intra-articular injections provide only symptomatic, temporary relief, with total knee replacement remaining the definitive treatment for end-stage or refractory disease (Jang et al., 2021). However, surgery is effective, but it carries significant risks, particularly for elderly and comorbid patients. These gaps in therapeutic options highlight the urgent need for disease-modifying strategies. Among emerging cell-based and cell-free therapy approaches, mesenchymal stem cell (MSC)-based treatments have attracted attention for their regenerative, anti-inflammatory, and immunomodulatory potential, making them a therapeutically promising candidate for modulating chronic disease progression (Safwan et al., 2025). In this chapter, we aim to provide an overview of the current evidence on MSC-based therapies for knee OA, integrating preclinical insights, translational advances, and clinical trial findings to clarify their therapeutic potential and future role in disease management.

16.3 BIOLOGICAL CHARACTERISTICS OF MSCS

MSCs are multipotent progenitor cells that can be isolated from various tissues, mainly bone marrow, adipose tissue, umbilical cord, and synovial fluid (Kamal et al., 2022; Safwan et al., 2024). The International Society for Cellular Therapy (ISCT) defines MSCs by three minimal criteria: (1) plastic adherence under standard culture conditions, (2) expression of surface markers CD73, CD90, and CD105, and absence of hematopoietic markers (CD14, CD34, CD45), and (3) capacity for trilineage differentiation into osteoblasts, adipocytes, and chondrocytes (Alvarez-Viejo and Haider 2022). Importantly, MSCs exhibit low immunogenicity through their expression of major histocompatibility complex (MHC) class I and lack of MHC class II, which reduces T cell activation and makes them favorable for allogeneic transplantation (Al-Khani et al., 2022; Rajab et al., 2019). Additionally, MSCs secrete a wide range of paracrine factors, including cytokines, chemokines, and extracellular matrix proteins, that modulate the immune response, promote angiogenesis, regulate inflammation, and support tissue regeneration (Ahmed et al., 2024). However, despite standardized criteria, MSC populations often display heterogeneity, leading to the broader use of the term "mesenchymal stromal cells" to reflect their phenotypic diversity (Dominici et al., 2006).

16.4 MSC SOURCES AND CHARACTERISTICS

16.4.1 Bone Marrow-Derived MSCs

Bone marrow-derived MSCs (BM-MSCs) are the most extensively studied cell type, but require invasive harvesting under anesthesia. They exhibit multipotent differentiation potential and predominantly differentiate into chondrocytes, adipocytes, and osteoblasts under specific, well-defined conditions. However, their yield, proliferation, and functional potency have been shown to decline with donor age (Safwan et al., 2024). While BM-MSCs secrete immunoregulatory cytokines (e.g., Interleukin (IL)-1α, IL-6, Tumor necrosis factor (TNF)-α), their immunosuppressive effect is considered weaker compared to MSCs from other sources. Still, they play an essential role in tissue repair via homing to injury sites and expressing regenerative surface markers (Barlow et al., 2008).

16.4.2 Umbilical Cord-Derived MSCs

UC-MSCs are isolated from Wharton's Jelly, cord lining, or perivascular regions. Although their isolation frequency from these tissue sources is lower than in bone marrow, they retain long-term multipotency and exhibit reduced expression of senescence markers (p53, p21, p16) (Chatzistamatiou et al., 2014). UC-MSCs also exhibit mild expression of embryonic markers, including Oct4, Sox2, Nanog, and KLF4, suggesting a more primitive phenotype than adult tissue-derived MSCs (Inoue et al., 2014).

16.4.3 Adipose-Tissue-Derived MSCs

Adipose-tissue-derived MSCs (ADMSCs) are the most abundant and easily harvested from subcutaneous fat without seriously invasive intervention, offering high cell yields with minimal invasiveness. They are identified by the presence of CD13, CD29, CD73, CD90, CD105, and CD106, which increase with in vitro expansion. Unique markers like CD34, CD36, and Stro-1 may indicate hypoxic survival and stem cell niche adaptation (Pachón-Peña et al., 2016; Gronthos et al., 2001). Moreover, ADMSCs exhibit high proliferation rates, especially in younger donors, and retain their differentiation potential with age (Shi et al., 2005). Their low expression of human leukocyte antigen (HLA) confers strong immunotolerance, making them suitable for allogeneic use (Strem et al., 2005). Previous studies have also demonstrated that ADMSCs respond to stimuli via pathways like beta-adrenergic signaling and release trophic factors that enhance their anti-inflammatory, angiogenic, and regenerative functions (Mazini et al., 2019). Table 16.1 summarizes the main characteristics of the three MSC types.

16.5 PATHOGENESIS OF OSTEOARTHRITIS

OA is a multifactorial joint pathology driven by diverse factors, including genetic, mechanical, and biochemical characteristics. It primarily affects synovial joints, where the articular cartilage, subchondral bone, and a thin layer of calcified cartilage function together as a load-distributing unit during movement and weight-bearing (Martel-Pelletier et al., 2016). Articular cartilage is mainly composed of water, type II collagen, and proteoglycans such as aggrecan, which are all maintained by the chondrocytes (Houard et al., 2013). These cells exist in a low-turnover state, supported by a specialized pericellular matrix and low-oxygen conditions, and are equipped with mechanosensors, such as primary cilia, that guide their responses to physical forces (Loeser et al., 2014; Xu et al., 2014). However, this ability to sense and respond to mechanical stress is impaired in OA. Simultaneously, the chondrocytes lose their capacity for matrix maintenance and adaptation, leading to early pathological changes such as surface fibrillation and disruption of the collagen meshwork. These structural changes progress to deep cartilage fissures, delamination, and cartilage fragmentation into the joint space, which contribute to inflammation and mechanical symptoms (Burr et al., 2012). As the disease progresses, chondrocytes become exposed to the interterritorial matrix, which alters their behavior and activates signaling pathways via specialized receptors, such as integrins and discoidin domain receptors (Xu et al., 2014). Additionally, OA is characterized by upregulation of matrix-degrading enzymes, particularly aggrecanases and metalloproteinases, which degrade key components of the cartilage matrix. This process is driven by catabolic signaling pathways, including nuclear factor kappa B (NF-κB) and mitogen-activated protein kinases (MAPKs), and is exacerbated by the loss of matrix integrity (Wang et al., 2013; Adam et al., 2024). Concurrently, structural changes occur in the subchondral bone, including increased volume and remodeling, formation of bone cysts and osteophytes, with the development of bone marrow lesions (Kaspiris et al., 2023). These bony structural alterations reflect an abnormal osteoblast and osteoclast activity, often triggered by mechanical overload and impaired tissue repair responses (Burr et al., 2012).

Inflammation has a central role in OA pathogenesis. Synovitis is frequently observed, characterized by infiltration of T and B lymphocytes and synovial membrane hyperplasia. Cartilage breakdown products act as damage-associated molecular patterns (DAMPs), which activate receptors such as toll-like receptors (TLRs) and integrins on chondrocytes and synovial cells. This abnormal activation induces further release of inflammatory mediators, creating a self-perpetuating vicious

TABLE 16.1
Characteristics of the Main MSC Types

Type	Source	Collection Method	Differentiation Potential	Immunological Features	Key Advantages/ Disadvantages
BM-MSCs	Bone Marrow	Highly invasive, painful; requires anesthesia; yield depends on patient condition and aspiration volume.	Adipocytes, chondrocytes, osteoblasts.	Weaker immunomodulatory effects; secretes IL-1α, IL-6, IL-8, TNF-α, TGF-β1.	Multipotent; potential decreases with age; crucial for tissue repair; homing ability to injury sites.
UC-MSCs	Wharton's Jelly, cord lining, perivascular area	Non-invasive; isolated from the umbilical cord after birth.	Multipotent; long-term potency.	Reduced senescence markers (p53, p21, p16); mild expression of Oct4, Nanog, Sox2, KLF4.	Primitive characteristics: lower MSC frequency than bone marrow, but longer multipotency.
AD-MSCs	Subcutaneous fat tissue	Minimally invasive; high yield from liposuction.	Adipocytes, others (maintained with age).	Low immunogenicity; <1% express HLA-DR; CD markers (e.g., CD34, CD10, CD36) vary with culture.	Fast growth; strong regenerative properties; low immune response; suitable for allogeneic transplant.

Abbreviations: MSCs: mesenchymal stem cells; CD: cluster of differentiation; MHC: major histocompatibility complex; ECM: extracellular matrix; BM-MSCs: bone marrow-derived mesenchymal stem cells; TGF-β1: transforming growth factor beta-1; IL: interleukin; UC-MSCs: umbilical cord-derived mesenchymal stem cells; Oct4: octamer-binding transcription factor 4; Sox2: SRY-box transcription factor 2; Nanog: Homeobox protein nanog; KLF4: Kruppel-like factor 4; ADMSCs: adipose-derived mesenchymal stem cells; HLA-DR: human leukocyte antigen—DR isotype.

cycle of tissue damage and inflammation (Knights et al., 2023). Moreover, inflammatory cytokines—especially IL-1 and TNF- are critical contributors. TNF sustains the inflammatory state and supports immune cell activation, while IL-1 promotes matrix degradation by upregulating catabolic enzymes, suppressing their inhibitors, and reducing the synthesis of matrix components (Molnar et al., 2021). Additional cytokines, such as IL-15, have been found to facilitate immune cell recruitment and synovial inflammation (Martel-Pelletier et al., 2016). Late-stage OA is marked by vascular invasion, tidemark duplication, and expansion of the calcified cartilaginous zone, driven by the release of angiogenic factors such as vascular endothelial growth factor (VEGF) (Hamilton et al., 2016). These changes further compromise cartilage integrity, promote hypertrophic chondrocyte differentiation, and facilitate matrix breakdown (Dreier et al., 2010). Altogether, the cumulative effect of these degenerative, inflammatory, and structural alterations leads to progressive joint deterioration, pain, stiffness, and loss of function, the classical characteristic observed in OA (Martel-Pelletier et al., 2004).

16.6 BIOLOGICAL BASIS OF MSC THERAPY IN KNEE OA

Given the degenerative and inflammatory state of OA, MSCs offer a promising therapeutic strategy through a combination of immunomodulatory and trophic mechanisms rather than direct tissue replacement (Habib et al., 2026). In inflammatory environments, MSCs release several immunosuppressive molecules such as prostaglandin E2 (PGE2), nitric oxide (NO), indoleamine 2,3-dioxygenase (IDO), and transforming growth factor-beta (TGF-β), which collectively inhibit the proliferation of CD4+ and CD8+ T cells and favor the induction of regulatory T cells (Tregs) (Gao et al., 2016). These factors also induce apoptosis of activated T cells via Fas–Fas ligand interactions and suppress dendritic cell maturation and natural killer (NK) cell cytotoxicity (Figure 16.1).

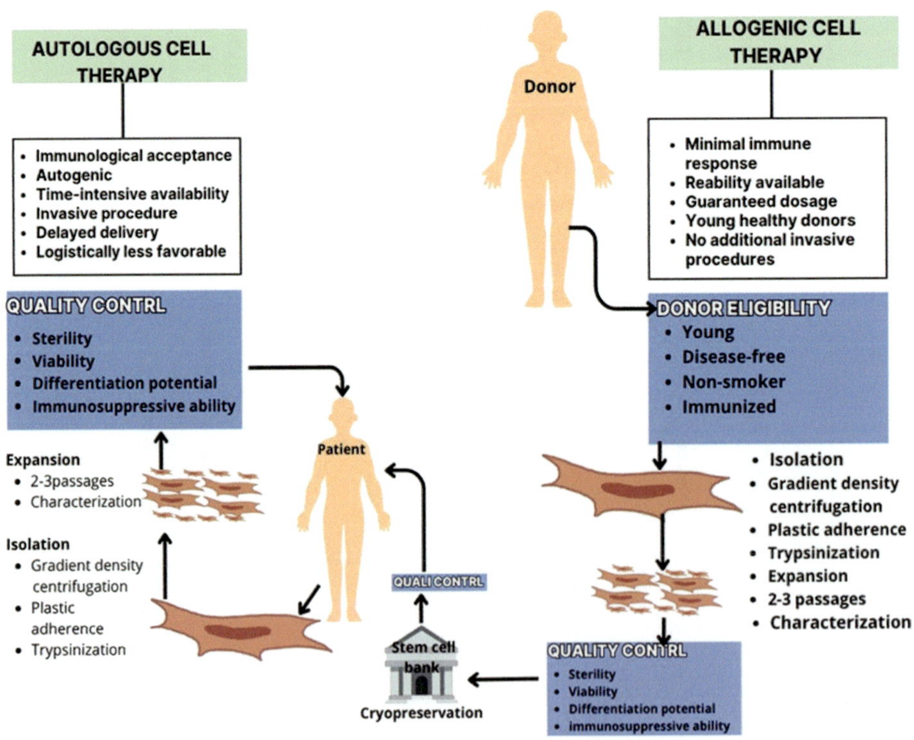

FIGURE 16.1 Autologous versus allogeneic mesenchymal stem cell (MSC)-based therapy: Workflow and clinical considerations.

MSCs further influence macrophage plasticity by shifting macrophages from the pro-inflammatory M1 to the anti-inflammatory M2 phenotype, mainly through PGE2, TGF-β, IL-10, and IDO (Wang et al., 2018). M2 macrophages play a significant role in tissue repair by enhancing phagocytosis and reducing levels of pro-inflammatory mediators, including interferon-γ (IFN-γ), IL-1β, and IL-12. Additionally, MSCs release factors like monocyte chemoattractant protein 1 (MCP-1), which function in recruiting and regulating macrophage activity at the injury site (Wang et al., 2019). Besides their effects on inflammatory cells, MSCs secrete cytokines such as IL-10 and PGE2, which suppress pro-inflammatory cytokines and reduce the activity of matrix-degrading enzymes, such as matrix metalloproteinases (MMPs), further preserving joint structure (Wang et al., 2018; Molnar et al., 2022).

The distinctive paracrine activity of MSCs—mediated by trophic factors such as TGF-β, insulin-like growth factor-1 (IGF-1), hepatocyte growth factor (HGF), VEGF, and fibroblast growth factor-2 (FGF2)—supports chondrocyte survival, inhibits apoptosis cascades, and promotes anabolic activity in cartilage tissue (Molnar et al., 2022). For instance, IGF-1 enhances the synthesis of type II collagen and aggrecan, while FGF2 and VEGF activate prosurvival signals via Bcl-2 and Akt pathways. MSCs contribute to chondroprotection by reducing catabolic enzyme activity and preventing chondrocyte apoptosis. TGF-β and FGF2 prevent hypertrophic degeneration, while VEGF improves cellular resilience under stress (Kong et al., 2025). They also promote extracellular matrix production, especially type II collagen and aggrecan, by stimulating resident chondrocytes and synoviocytes, rather than through direct differentiation. Mechanical cues and the microenvironment further enhance this regenerative activity (Wang et al., 2019). Additionally, MSCs exhibit anti-fibrotic effects by secreting stanniocalcin-1 (STC1), which enhances epithelial-mesenchymal communication and limits fibrotic remodeling, particularly under mechanical stress (Molnar et al., 2022).

MSCs also modulate angiogenesis and maintain synovial homeostasis by releasing VEGF, IL-6, and MCP-1. While excessive IL-6 release may aggravate OA, controlled angiogenesis is essential for nutrient supply and repair, and MSCs help maintain this balance (Molnar et al., 2022). During wound healing, MSCs secrete chemokines, such as MCP-1, MCP-2, and stromal cell-derived factor 1 (SDF-1a), to recruit immune cells essential for tissue repair, including monocytes, lymphocytes, and macrophages (Wang et al., 2019). Furthermore, MSCs are characterized by the release of exosomes—small vesicles containing microRNAs, proteins, and lipids—that recapitulate many of the immunomodulatory and regenerative effects of the parent cells. These exosomes hold promise as a cell-free therapeutic tool for OA by influencing matrix remodeling, immune regulation, and synovial stability (Wang et al., 2019) (Figure 16.2).

16.7 PRECLINICAL EVALUATION OF MSC THERAPY FOR KNEE OA: EVIDENCE FROM ANIMAL MODELS AND TRANSLATIONAL CHALLENGES

Experimental animal models have been widely used to evaluate MSC-based therapy for knee OA in vivo settings. Animal models in mice, rabbits, goats, and dogs have been developed to mimic various aspects of human OA pathology, providing a platform for assessing therapeutic outcomes under controlled conditions. Small animals such as mice and rabbits are widely used because of their availability, genetic similarity to humans, and the relative ease with which OA can be induced through surgical or chemical methods (Kraan et al., 2008). Rabbits are particularly useful for cartilage regeneration studies, as their larger joints allow for more accurate modeling of early degenerative changes that resemble human disease (Chiang et al., 2016). Larger animals, such as goats, offer insights into joint biomechanics, as their cartilage undergoes degeneration under mechanical stress in ways similar to those in humans. At the same time, dogs represent one of the few large-animal species that naturally develop OA, usually secondary to aging or trauma, making them well-suited for evaluating disease progression and the long-term outcomes of MSC therapy (McCoy et al., 2015;

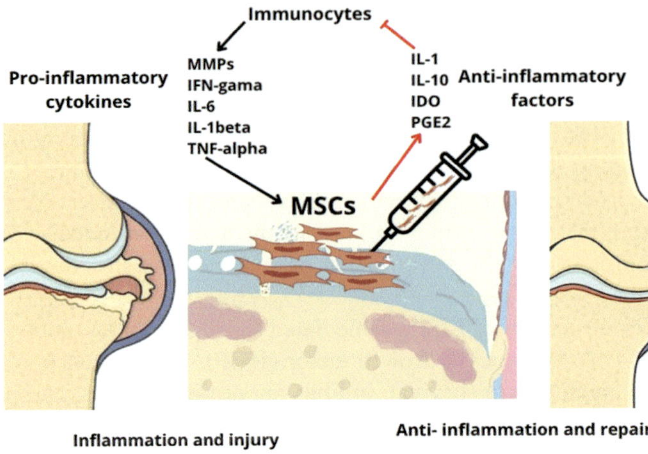

FIGURE 16.2 Therapeutic effects of mesenchymal stem cells (MSCs) in knee osteoarthritis (KOA). MSCs modulate inflammation by suppressing pro-inflammatory cytokines (MMPs, IFN-γ, IL-6, IL-1β, TNF-α) and promoting the release of anti-inflammatory mediators (IL-1 receptor antagonist, IL-10, IDO, PGE2), thereby reducing tissue damage and contributing to repair.

Vilar et al., 2014). The induction of OA in animal models is most commonly achieved through joint destabilization or mechanical overload. In small animals such as mice and rabbits, anterior cruciate ligament (ACL) transection or meniscal injury reliably produces progressive cartilage degeneration, while in goats and dogs, surgical destabilization or repetitive joint loading produces the mechanical stress typical of human knee osteoarthritis (McCoy et al., 2015; Chiang et al., 2016). These models enable systematic evaluation of MSC-based interventions, particularly their ability to repair cartilage, reduce pain, and restore joint function. Across multiple species, intra-articular administration of MSCs has been shown to exert both chondroprotective and regenerative effects. Following injection, MSCs migrate to the damaged cartilage, where they contribute to tissue repair, primarily through paracrine signaling (Zhang et al., 2022). In rabbits, MSCs have been associated with increased cartilage thickness and improved mobility. At the same time, in larger models such as goats and dogs, studies have demonstrated reductions in joint inflammation, enhanced cartilage repair, and measurable improvements in mobility and joint mechanics (Murphy et al., 2003; Black et al., 2008; Chiang et al., 2016). These therapeutic effects appear consistent across models, and they generally include pain relief through the downregulation of pro-inflammatory cytokines such as IL-1 and TNF-α, preservation of existing cartilage with stimulation of new matrix formation, and functional recovery reflected in better range of motion, weight-bearing, stiffness, and stability (Zhang et al., 2022).

The efficacy of MSCs in preclinical models has been assessed using several complementary methods. Histological scoring systems, most notably the Osteoarthritis Research Society International (OARSI) scale, are widely applied to quantify cartilage damage and have repeatedly shown reduced scores following MSC therapy, indicating less degeneration and improved tissue survival (Chiang et al., 2016). Magnetic resonance imaging (MRI) has enabled non-invasive monitoring of cartilage thickness, joint space narrowing, and synovial inflammation, with changes in animals paralleled by improvements later confirmed in clinical trials (Jiang et al., 2021). Biomechanical assessments further corroborate these findings by demonstrating superior joint stiffness, range of motion, and weight-bearing performance in MSC-treated animals compared to controls (Chiang et al., 2016). Despite the encouraging data, experimental animal models are inherently limited in their ability to replicate the multifactorial nature of human OA.

Most animal models reproduce cartilage degeneration but fail to incorporate comorbidities such as obesity or metabolic syndrome, which commonly influence disease progression in patients

(Gupta et al., 2016). Moreover, differences in immune responses between animals and humans may alter the mechanisms by which MSCs exert their effects (Zhang et al., 2022). The smaller joints of mice and rabbits do not capture the complex biomechanics of the human knee, while large animal models, although more representative, are costly and raise ethical concerns (Long et al., 2022). To optimize therapeutic outcomes and address the multifactorial nature of knee OA, preclinical studies have increasingly explored combinatorial approaches to leverage the best of multiple strategies. Biomaterial scaffolds, such as collagen, hyaluronic acid, and synthetic polymers, have been used to improve MSC survival and retention within cartilage defects, thereby enhancing regenerative outcomes in rabbit and goat models (Vinatier et al., 2016). Growth factors, including transforming bone morphogenetic proteins, TGF-β, and VEGF, have been incorporated to stimulate chondrogenesis, osteochondral repair, and angiogenesis, particularly when delivered through controlled-release systems (Zhang et al., 2022). In this regard, priming of the cells before use is gaining popularity (Haider et al., 2024a, 2024b). Genetically modified MSCs that overexpress growth factors have also been tested and shown to accelerate cartilage repair in several experimental settings, although questions remain regarding tumorigenicity, immune rejection, and scalability (Madry et al., 2016).

Collectively, preclinical evidence strongly supports the capacity of MSCs to alleviate pain, preserve cartilage, and restore joint function. However, their translation to human OA remains constrained by biological and mechanical differences between species, safety considerations, and the challenges of large-scale production and delivery. Addressing these barriers through refinement of delivery systems, standardization of outcome measures, and careful integration of combination therapies will be essential for enhancing MSC-based treatments toward routine clinical application.

16.8 MSCS FOR KNEE OA, CURRENT CLINICAL PERSPECTIVES

Given the favorable outcomes demonstrated in preclinical studies, the different MSC types discussed previously have been utilized in clinical settings, with BM-MSCs being the most extensively evaluated MSC source for knee OA (Table 16.2). Although the discussion about allogenic vs autologous MSCs for cell-based therapy continues (Ahmed et al., 2025a, 2025b), each has its advantages and limitations; autologous BM-MSCs remain an obvious choice due to their better immune compatibility, relatively safe profile, and avoidance of risks such as immune rejection or tumorigenesis (Safwan et al., 2025). Across the published trials, MSC-based therapies have been assessed using multiple clinical and imaging endpoints, primarily including the Visual Analogue Scale (VAS) for pain, the Western Ontario and McMaster Universities Osteoarthritis Index (WOMAC) for overall function, MRI for structural assessment, and the incidence of adverse events (AEs). Pain reduction, as measured by VAS, was consistently superior in BM-MSC groups compared with controls across several published randomized controlled trials. In a placebo-controlled study by Gupta et al., the VAS score declined by a mean of 32 points over 12 months in the BM-MSC group, compared to only 3.9 points in the placebo arm (Gupta et al., 2023). Similarly, Emadedin et al. reported a 20.8-point reduction in VAS scores at 6 months with 40 million BM-MSCs, which was significantly greater than the 15-point decrease observed with placebo (Emadedin et al., 2018). When compared to hyaluronic acid (HA)—an approved therapeutic for knee OA—Vega et al. demonstrated a 21-point VAS reduction over 12 months with BM-MSCs versus a 13-point reduction with HA treatment alone (Vega et al., 2015). Similar findings were reported in the Espinosa et al. study, in which both low- and high-dose BM-MSCs yielded significantly greater pain relief than HA at one year of follow-up (Espinosa et al., 2016). Additional favorable results were reported by Ho et al., who found that BM-MSCs led to a significant reduction in VAS scores at rest and during activity, compared with worsening pain in the HA group (Ho et al., 2022).

Furthermore, in a combinatorial approach, MSCs combined with plasma rich in growth factors have been used for knee OA (Aixirefu et al., 2024; He et al., 2025). BM-MSCs combined with plasma rich in growth factors (PRGF), another currently recommended injection therapy for knee OA, produced greater VAS improvements than PRGF alone. However, between-group differences

TABLE 16.2
Summary of Randomized Clinical Trial Evidence Evaluating Mesenchymal Stem Cell (MSC) Therapies in Knee Osteoarthritis (KOA), Stratified by Cell Source

MSC Source	Key Trials	Pain Outcomes (VAS)	Function (WOMAC/KOOS)	Imaging (MRI)	Safety
BM-MSCs	Gupta 2023, Emadedin 2018, Vega 2015, Espinosa 2016/2020, Ho 2022	Consistent, significant VAS reduction vs placebo/HA (e.g., −32 vs −3.9, Gupta 2023; ~−21 vs −13, Vega 2015).	Significant WOMAC improvements in pain, function, and stiffness (Emadedin 2018; Vega 2015; Espinosa 2016; Ho 2022).	Vega 2015: Reduced T2 relaxation, improved cartilage; Espinosa 2016: WORMS ↑ up to 22 points; Ho 2022: trend to cartilage improvement (not sig.).	Favorable; only mild, transient injection-site pain; no serious AEs.
UC-MSCs	Matas 2019, Mautner 2023	Significant VAS reduction with repeated dosing (Matas 2018: VAS 2.4 vs 22.1 HA; Mautner 2023: similar to steroid, not significant).	WOMAC improved with repeated UC-MSCs (Matas 2019); Mautner 2023: no sig. KOOS difference.	No sig. WORMS or MRI structural benefit in either study.	Favorable; mild synovitis/effusion in 33–44% (Matas 2018), more swelling vs steroid (Mautner 2023); no serious AEs.
AD-MSCs	Kuah et al., 2018, Chen 2021, Lu 2019, Sadri 2023, Kim 2023	Significant VAS reductions across studies (−32.7 mm PRG; sustained benefit with ELIXCYTE®, Re-Join®, etc.).	WOMAC consistently improved across trials (e.g., >70% reduction, Sadri 2023; early benefits Chen 2021).	Kuah et al., 2018: maintained tibial cartilage; Sadri 2023: ↑ tibial/femoral thickness; Lu 2019: ↑ total cartilage volume; Kim/Chen: no sig. WORMS change at six months.	Safe; mild transient swelling/pain, no serious AEs, no long-term complications.

Abbreviations: MSC: mesenchymal stem cell; BM: bone marrow; AD: adipose-derived; UC: umbilical cord-derived; VAS: visual analog scale; WOMAC: Western Ontario and McMaster Universities Osteoarthritis Index; KOOS: Knee Injury and Osteoarthritis Outcome Score; MRI: magnetic resonance imaging; WORMS: Whole-organ magnetic resonance imaging score; HA: hyaluronic acid; AE: adverse event; PRG: Progenza®.

were not statistically significant. When the WOMAC index assessed functional outcomes, the same studies frequently favored BM-MSC treatment. Emadedin et al. showed statistically significant improvements in the WOMAC pain and physical function subscales, indicating superiority to placebo. The WOMAC pain score decreased from 35 to 13.1 in the MSC group, and total WOMAC scores dropped below 60% at six months (Emadedin et al., 2018). Similarly, significant reductions in total WOMAC score at both 6 and 12 months of follow-up have been reported, with the control group showing minimal changes (Vega et al., 2015). Espinosa et al. observed significant improvements across all WOMAC subscales, particularly in patients receiving high-dose MSCs, with median symptom relief notably greater than in controls (Espinosa et al., 2016). Compared with HA, Ho et al. reported that the MSC-treated group showed statistically significant improvements in total WOMAC score, pain subscore, and percentage score, with favorable trends in stiffness and physical function (Ho et al., 2022). Structural changes in the knee after MSC-based treatment were evaluated through MRI, providing objective insight into cartilage preservation or regeneration. Vega et al. utilized T2 mapping to monitor cartilage quality, demonstrating significant decreases in T2 relaxation times and in the percentage of cartilage with abnormal signals, indicating improved cartilage composition after BM-MSC therapy (Vega et al., 2015). Espinosa et al. employed the Whole-Organ Magnetic Resonance Imaging Score (WORMS). They found a median improvement of 4 points in high-dose MSC recipients, with up to 22-point gains in a subset, while the control and low-dose groups showed no such benefit (Espinosa et al., 2016).

Larger reductions in T2 relaxation times among MSC-treated patients than in the HA group have been reported, with 6 of 10 subjects showing marked decreases (Ho et al., 2022). However, overall group differences did not reach significance due to high variability. Notably, the safety profile across all trials was favorable. Emadedin et al. (2018) reported no serious AEs, and similar findings were reported in the studies by Vega et al. (2015), Ho et al. (2022), Espinosa et al. (2016), and Espinosa et al. (2020). Minor joint pain post-injection was occasionally noted but was transient and self-limiting. No long-term complications or treatment-related toxicities were observed. Altogether, these findings from BM-MSC-based trials not only confirm the consistent safety of MSC intra-articular injections but also support their beneficial effects on pain relief, functional improvement, and potentially structural cartilage regeneration.

UC-MSCs have also been frequently evaluated in clinical settings, primarily due to their noninvasive and easy collection, high proliferative capacity, and immunoprivileged characteristics. Compared to HA, Matas et al. demonstrated a significant reduction in pain, particularly in patients receiving repeated intra-articular MSC injections. At 12-month follow-up, the VAS score decreased to 2.4 ± 2.1 in the repeated-dose UC-MSC group, significantly lower than the 22.1 ± 9.8 reported in the HA control group ($p = .03$) (Matas et al., 2019). Mautner et al. compared MSC treatment with corticosteroid injection and noted that the UC-MSC cohort showed a VAS decline of −20.1 points at 12 months, comparable to that with corticosteroid injection (−20.9). However, no statistically significant difference between groups was found (Mautner et al., 2023). The WOMAC outcomes in Matas et al. strongly supported the clinical utility of repeated UC-MSC dosing, with the MSC group achieving a total WOMAC score of 4.2 ± 3.9 compared to 15.2 ± 11 in the HA group ($p = .05$). Additionally, the MSC group experienced an 86% reduction in pain and 89% improvement in disability from baseline, in contrast to 38% and 50%, respectively, in patients receiving HA injection (Matas et al., 2019).

In comparison, Mautner et al. reported no statistically significant difference between MSCs and steroid injection in the Knee Injury and Osteoarthritis Outcome Score (KOOS) pain domain or in functional measures at 12 months, despite all treatments exceeding the minimally clinically significant difference (MCID) (Mautner et al., 2023). Regarding imaging-based structural outcomes, MRI evaluations in both trials showed no evidence of cartilage damage or adverse structural effects following UC-MSC treatment. Matas et al. found no significant changes from baseline in any of the 14 WORMS components at 24 or 48 weeks, while Mautner et al. similarly reported no significant MRI score differences across groups, with UC-MSCs yielding a mean change of −0.26, which was statistically and clinically comparable to corticosteroid injection (0.30) (Matas et al., 2019; Mautner et al., 2023). This suggests that while symptomatic improvements were notable—particularly with repeated dosing—morphological cartilage regeneration remains inconclusive in UC-MSC trials to date.

The safety profile of UC-MSCs was consistently favorable across both studies. Matas et al. reported no serious AEs, neoplasia, or infections (Matas et al., 2019). The most frequent AE was mild-to-moderate synovitis with transient knee effusion, noted in 33%–44% of UC-MSC-treated patients versus 22%–37.5% in the HA group. However, differences were not statistically significant ($p = .99$). Symptoms were self-limiting and responded well to rest and acetaminophen, with no patients requiring hospitalization or joint aspiration. Similarly, Mautner et al. found no serious procedure-related AEs across groups. However, UC-MSC recipients experienced a significantly higher incidence of joint swelling (24.1%) compared to the steroid injection group (7.4%, $p = .01$), though no post-procedural contusions or hematomas were reported in the UC-MSC group (Matas et al., 2019; Mautner et al., 2023). These trials highlight the clinical efficacy and reassuring safety profile of UC-MSCs compared to currently available injection therapies for knee OA, particularly when administered repeatedly, while emphasizing the need for further research into their structural and long-term disease-modifying effects in knee OA compared to other injections.

AD-MSCs are an alternative MSC type that have gained substantial attention for their minimally invasive harvesting process and high yield, making them attractive for clinical application. Across multiple randomized controlled trials, AD-MSCs consistently demonstrated a significant reduction in knee pain assessed by VAS. In the Kauh trial, both doses of Progenza (PRG) (3.9M and 6.7M) showed statistically significant within-group reductions in VAS scores at 12 months of follow-up, with mean decreases of –32.7 mm and –26.5 mm, respectively. Notably, up to 87.5% of PRG-treated patients were classified as pain responders, compared to 50% in the placebo group (Kuah et al., 2018). Similarly, Sadri et al. reported a significant decrease in VAS scores over 12 months in the AD-MSC group compared with patients receiving normal saline ($p < 0.001$) (Sadri et al., 2023). Kim et al. found significantly greater VAS improvement at 6 months of follow-up in the AD-MSC group compared to normal saline (mean difference of 10 points), with higher proportions of patients achieving MCID in both absolute (68.5% vs. 53.7%) and relative (75.0% vs. 60.5%) terms (Kim et al., 2023). A comparison between AD-MSCs with HA injection has reported early and sustained VAS reductions across all ELIXCYTE® dose groups, with significant differences from HA beginning as early as week 2 (Chen et al., 2021). Likewise, significantly lower VAS scores in the Re-Join® group have been observed as compared to HA at both 6 and 12 months of follow-up ($P < 0.05$) (Lu et al., 2019). Functional improvement, as assessed by WOMAC scores, was also consistently evaluated across these trials. Kauh et al. demonstrated statistically significant within-group reductions in the WOMAC pain, stiffness, and physical function subscales for PRG-treated patients at all measured time points, with no significant differences compared with placebo (Kuah et al., 2018). In Chen et al.'s study, all ELIXCYTE® dose groups exhibited substantial reductions in total WOMAC and all subscales compared with the HA group, with differences evident as early as week 2 and persisting through week 48 of follow-up (Chen et al., 2021).

Both Re-Join® and HA injections have been reported to lower the WOMAC scores over time, but a higher proportion of Re-Join® patients achieved ≥50% and ≥70% improvement at 12 months (Lu et al., 2019). In another study, WOMAC total scores declined by over 70% at 6 months post-injection in the AD-MSC group, with significant superiority over normal saline throughout the 12-month follow-up period ($p < 0.001$) (Sadri et al., 2023). MRI-based structural evaluations revealed cartilage preservation and, in some studies, even regeneration potential. While Kauh et al. found that the PRG 3.9M group maintained tibial cartilage volume over 12 months, in contrast to the significant loss observed in the placebo group (difference = 106.5 mm^3, $p = 0.028$), Sadri et al. showed increased thickness in several tibial and femoral regions at 1-year post-injection (Kuah et al., 2018; Sadri et al., 2023). Similarly, Lu et al. reported significant increases in total cartilage volume in both knees at 12 months in the Re-Join® group, with no similar benefit observed in the HA group (Lu et al.,

2019). While these improvements were observed after long-term follow-up, Kim et al. and Chen et al. reported no significant differences in WORMS score at 6 months, suggesting that structural changes following AD-MSC injection may take longer to manifest or require larger sample sizes to detect (Chen et al., 2021; Kim et al., 2023). Regarding safety, AD-MSC therapy was well tolerated across all studies, with no serious treatment-related AEs reported. Kauh et al. reported only mild, transient arthralgia and joint effusion, with no withdrawals due to AEs (Kuah et al., 2018). Chen et al. found that most treatment-related events occurred in the first 24 weeks, were mild to moderate (e.g., injection site pain, arthralgia), and resolved without sequelae (Chen et al., 2021). Sadri et al. also noted only brief local swelling and pain in two patients, with no systemic abnormalities on laboratory testing (Sadri et al., 2023). Lu et al. found a similar AE profile between Re-Join® and HA groups, with all events being mild and self-limited (Lu et al., 2019). Finally, Kim et al. observed no significant difference in AE rates between groups, with only mild procedure-related events from lipoaspiration (Kim et al., 2023).

16.9 MSC-DERIVED EXOSOMES AND EVS: CHARACTERISTICS, ISOLATION METHODS, MOLECULAR CONTENT, PRECLINICAL EFFICACY, AND CLINICAL TRANSLATION

MSC-derived exosomes and extracellular vesicles (EVs) have emerged as potent, cell-free therapeutic agents with significant regenerative, immunomodulatory, and anti-inflammatory properties (Ahmed et al., 2024). These nanosized vesicles, secreted by MSCs derived from diverse tissue sources, including BM, AD, and UC, facilitate intercellular communication and influence a wide range of biological responses (Mushtaq et al., 2024). EVs are primarily classified into exosomes (30–150 nm) and microvesicles (100–1000 nm), with both types enclosed by a lipid bilayer and carrying bioactive molecules such as proteins, lipids, messenger RNAs (mRNAs), microRNAs (miRNAs), and signaling molecules (Haider & Aramini, 2020; Théry et al., 2018). Compared to MSCs, EVs hypothetically offer several advantages, including reduced immunogenicity, enhanced safety, and greater stability in storage and transport, while retaining the beneficial paracrine effects of their parent cells (Kalluri et al., 2020; Tsiapalis et al., 2020).

The isolation of EVs is a critical step in their clinical translation, and various techniques have been utilized with comparable efficacy (Habib et al., 2025). Differential ultracentrifugation remains the most common method, although it is time-consuming and may result in the co-isolation of non-vesicular contaminants and in potential damage to vesicle integrity (Habib et al., 2026). Density gradient ultracentrifugation, using sucrose or iodixanol gradients, increases isolation purity by separating EVs based on their buoyant density, yet it is not easily scalable for clinical-grade manufacturing (Théry et al., 2018). Size-exclusion chromatography (SEC) is another method that offers a gentler, more reproducible alternative for separating particles based on size using porous beads. However, it may still yield co-eluted protein aggregates (Tsiapalis et al., 2020). Polymer-based precipitation methods, such as ExoQuick™ using polyethylene glycol (PEG), are rapid and scalable but co-precipitate unwanted proteins and lipoproteins, requiring additional isolate purification (Kalluri et al., 2020). Microfluidics is an emerging isolation platform that offers rapid, high-throughput extraction via size-, charge-, or marker-specific capture but still requires clinical validation (Habib et al., 2026).

The therapeutic potential of MSC-EVs is primarily attributed to their molecular cargo. Among these, miRNAs—small non-coding RNAs around 22 nucleotides in length—are key modulators of gene expression in recipient cells, often by silencing target mRNAs. MSC-EVs have been found to carry anti-inflammatory and pro-regenerative miRNAs such as miR-21, miR-146a, miR-155, and miR-210 (Kalluri et al., 2020; Haider & Aramini, 2020). For instance, miR-146a inhibits the NF-κB pathway, while miR-21 promotes anti-apoptotic pathways. Additionally, MSC-EVs are enriched in functional proteins, such as TGF-β, VEGF, IL-10, and IDO, which mediate immunomodulation and tissue repair (Théry et al., 2018). Structural proteins, including tetraspanins (CD63, CD81, CD9), are essential for vesicle formation and serve as exosomal markers. MSC-EVs also contain MMPs, facilitating extracellular matrix remodeling, and heat shock proteins (HSP60, HSP70, HSP90), which promote stress tolerance (Théry et al., 2018). These molecular constituents enable EVs to recapitulate many of the therapeutic effects of MSCs without the safety concerns of cell-based therapies, and MSC source and culture conditions can modulate their composition.

In OA models, MSC-derived EVs have shown significant preclinical efficacy. These vesicles have been shown to downregulate catabolic enzymes (e.g., MMP-13), upregulate anabolic markers (e.g., COL2A1, ACAN), and suppress inflammatory mediators in vitro and in vivo (Cosenza et al., 2018; He et al., 2020). He et al. (2020) demonstrated that intra-articular injections of BM-MSC-derived exosomes in a sodium iodoacetate-induced OA rat model led to improved histological outcomes and reduced MMP-13 expression. Pain reduction was confirmed by increased paw withdrawal thresholds and decreased dorsal root ganglion inducible nitric oxide synthase (iNOS) and Calcitonin gene-related peptide (CGRP) levels. Similarly, Jin et al. (2021) showed that exosomes from human BM-MSCs improved cartilage and subchondral bone structure in a rat model. In vitro, these exosomes attenuated IL-1β-induced chondrocyte senescence via the long non-coding RNA (lncRNA) maternally expressed gene 3 (MEG3). Engineered EVs, such as miR-199a-3p-enriched EVs (Zhao et al., 2023) and miR-140-5p-overexpressing dental pulp stem cell EVs (Lin et al., 2021), demonstrated enhanced chondroprotection, autophagy induction, and cartilage repair by modulating PI3K-Akt, Wnt, and mTOR signaling. Warmink et al. (2023) showed that MSC-EVs, unlike MSCs, did not exacerbate joint inflammation in diet-induced OA models.

In addition to therapeutic benefits, MSC-EVs may offer key logistical and clinical advantages over MSC-based cellular therapy. MSCs require cryopreservation at ultra-low temperatures (–150°C) with cryoprotectants such as dimethyl sulfoxide (DMSO), which can cause cell toxicity, compromise viability, and alter gene expression upon thawing (Mendt et al., 2018). In contrast, EVs remain stable at –80°C and are amenable to lyophilization for room-temperature storage without loss of bioactivity (Charoenviriyakul et al., 2018). Lyophilized EVs reconstituted with buffers maintain protein and miRNA contents, enabling batch consistency (Klyachko et al., 2014; Witwer et al., 2019). Additionally, EVs lack replicative capacity and a nucleus, thereby avoiding the risk of tumorigenic or ectopic growth associated with MSC infusion (Nakamura et al., 2015). MSC-EVs do not engraft but act transiently via biomolecule delivery, further improving their safety profile. Their lack of MHC class II expression further reduces the risk of alloimmunization (Kordelas et al., 2014; Keshtkar et al., 2018). EVs also offer superior scalability. While MSC expansion is hampered by senescence and donor variability, EV production can be standardized using bioreactors that continuously harvest vesicles from conditioned media (Whitford et al., 2019). More importantly, unlike exosomes, which are advanced medicinal therapeutic products, stem cells, as living biodrugs, require viability as a mandatory characteristic for their mechanistic participation in the repair process. Recent techniques, such as 3D cultures and hollow fiber bioreactors, enhance EV yield and quality, supporting reproducible large-scale manufacturing for broad clinical applications. Additionally, MSCs can be preconditioned or genetically engineered to enrich EV cargo for disease-specific therapy (Gurunathan et al., 2021).

Besides preclinical reports on the efficacy of MSC-EVs in OA, several early-phase human trials across different disease populations support their safety. Sengupta et al. administered BM-MSC-derived exosomes (ExoFlo) to 24 patients with severe COVID-19, observing improved oxygenation and reduced inflammatory markers with no adverse events (Sengupta et al., 2020). Kordelas et al. reported successful use of MSC-EVs in steroid-refractory graft-versus-host disease (GvHD) (Kordelas et al., 2014).

Despite the observed positive outcomes in various pathologies, regulatory pathways remain underdeveloped. In the United States, the Food and Drug Administration (FDA) may classify EVs as biologics or drugs. At the same time, the European Medicines Agency (EMA) considers them advanced therapy medicinal products (ATMPs), complicating global harmonization. Another reason is that regulatory agencies require a precise, explicit, and clear characterization of identity, potency, purity, and stability. However, EVs' heterogeneity and batch-to-batch variability complicate Good Manufacturing Practice (GMP) manufacturing. The International Society for Extracellular Vesicles (ISEV) guidelines recommend rigorous quality control, including flow cytometry, nanoparticle tracking analysis, and the presence of CD63, CD81, and TSG101 (Théry et al., 2018; Habib et al., 2025).

Understanding EV pharmacokinetics is also essential, as they function through transient cargo delivery rather than tissue integration. Thus, regulatory approval requires robust biodistribution and toxicity data from preclinical models (El Andaloussi et al., 2013). In summary, MSC-derived EVs are emerging as safe, stable, and scalable therapeutics with strong preclinical and initial clinical support. However, their successful clinical translation will depend on standardization, regulatory clarity, and quality-controlled production.

16.10 CLINICALLY AVAILABLE MSC-BASED THERAPIES FOR KNEE OA

By 2025, several MSC-based products will have been developed and will have undergone extensive preclinical and clinical testing. These products have received regulatory approval for clinical use under defined protocols designed to ensure safety, efficacy, and consistent quality (Bhatti and Haider, 2025) (Table 16.3). The world's first approved MSC-based therapeutic product was Cartistem, granted marketing authorization by the South Korean Ministry of Food and Drug Safety (MFDS) in January 2012 (Ancans et al., 2012; Bhatti and Haider, 2025). Cartistem is a cryopreserved, allogeneic, off-the-shelf product composed of UC-MSCs, GMP by Medipost Co., Ltd. Each dose contains 7.5×10^6 UC-MSCs suspended in 1.5 mL of 4% HA, which serves as a biological scaffold. The product is surgically implanted into drilled cartilage defects of the femoral condyle via mini-arthrotomy, offering a regenerative approach to articular cartilage repair in KOA (Park et al., 2017; Park et al., 2023). Following Cartistem's approval, several other MSC-based therapies for knee OA have emerged across Asia with varying formulations and delivery methods. In Japan, the Ministry of Health, Labour and Welfare (MHLW) approved JointStem for the treatment of knee OA in 2015 (Stem Cell Bio., n.d.). JointStem is developed by Biostar Stem Cell Research Institute, a division of Nature Cell Co., Ltd., based in South Korea. Unlike Cartistem, JointStem utilizes autologous AD-MSCs harvested from subcutaneous abdominal fat via liposuction. The product consists of 1×10^8 AD-MSCs suspended in 3 mL of saline and is administered via a single intra-articular injection (Jo et al., 2014; Lee et al., 2019). Notably, JointStem was granted Breakthrough

TABLE 16.3
Currently approved MSC-based products for knee osteoarthritis.

Product	Manufacturer	Approved by	Cell Source	Dosage
Cartistem	Medipost Co., Ltd.	South Korea MFDS	UCB-MSCs	7.5×10^6
JointStem	Biostar Stem Cell Research Institute	Japan MHLW	AD-MSCs	1×10^8
StemOne (Stempeucel®)	Stempeutics Research	India DCGI	BM-MSCs	2.5×10^7
MesestroCell	Celltech Pharmed	Iran FDA	BM-MSCs	2×10^7

Abbreviations: MFDS: Ministry of Food and Drug Safety; MHLW: Ministry of Health, Labour and Welfare; DCGI: Drug Controller General of India; FDA: Food and Drug Administration; UC-MSCs: umbilical cord-derived MSCs; AD-MSCs: adipose-derived MSCs; BM-MSCs: bone marrow-derived MSCs; MSCs: Mesenchymal stem cells.

Therapy Designation (BTD) by the United States FDA in March 2025, a process that expedites the development and review of therapies for severe conditions when early evidence suggests substantial improvement over existing treatments (FDA, 2025).

In India, regulatory approval was granted by the Drug Controller General of India (DCGI) for StemOne (Stempeucel®), developed by Stempeutics Research Pvt. Ltd. and licensed to Alkem Laboratories, a cryopreserved allogeneic BM-MSC product indicated for grade II and III knee OA. StemOne is delivered as a single intra-articular injection composed of 25 million MSCs in 1 mL of CryoStore medium containing 5% DMSO, combined with 1 mL of Plasmalyte-A and followed by 2 mL of HA (Gupta et al., 2023). In addition, Iran has also entered the MSC-based therapy landscape. In 2018, the Iranian FDA-approved MesestroCell, an autologous BM-MSC product for the intra-articular treatment of knee OA manufactured by Celltech Pharmed Company. A minimum dose of 2×10^7 cells is injected per knee, with bilateral treatment totaling 4×10^7 cells (Fernández-Garza et al., 2023). These diverse regional approvals reflect the growing international momentum toward MSC-based therapies for knee OA, despite variability in cell source, treatment methods, and regulatory frameworks.

While multiple MSC-based products for the treatment of knee OA have been approved in several Asian countries, including South Korea, Japan, and Iran, the regulatory landscape in the United States and Europe remains more restrictive. Despite encouraging clinical trial results, there are currently no FDA-approved MSC-based products for KOA in the United States (FDA, 2025). The only FDA-approved MSC product to date is remestemcel-L, indicated for pediatric steroid-refractory acute graft-versus-host disease (GvHD) (Kurtzberg et al., 2020; FDA, 2024). Similarly, as of mid-2025, no MSC-based therapies have been approved by the European Medicines Agency (EMA) for any indication (European Medicines Agency, 2025). The sole MSC product that previously held marketing authorization in Europe, Alofisel, an adipose-derived MSC therapy for complex perianal fistulas in Crohn's disease, was withdrawn in December 2024 due to failure of the post-approval confirmatory trial to meet the primary or secondary efficacy endpoints, thus failing to show that benefits outweigh risks (European Medicines Agency, 2024).

With MSC-based regenerative therapies demonstrating potential efficacy in treating various conditions for which existing treatments are ineffective, they have gained widespread popularity, inflated public expectations, and increased demand for rapid availability in clinical practice (Caulfield et al., 2016). However, the process of clinical validation and regulatory approvals is lengthy and complex, creating a gap that unapproved and unregulated stem cell clinics have exploited in several countries. These clinics often rely on aggressive advertising strategies, including social media campaigns and patient testimonials, to promote misleading efficacy claims while neglecting to disclose potential risks (Snyder et al., 2018; Lyons et al., 2022). In contrast, approved MSC-based products are manufactured in accordance with Good Clinical Practice (GCP) standards, ensuring product consistency and safety. They are administered by trained healthcare professionals using standardized protocols (Abou-El-Enein et al., 2013; Sensebé et al., 2013). Many unregulated clinics don't adhere to GCP and employ unqualified personnel, failing to ensure correct dosing and administration, and compromising product quality, which can compromise treatment safety and efficacy.

Patients with limited therapeutic options may resort to traveling abroad to access such unproven and unregulated stem cell treatments, a phenomenon known as "stem cell tourism," exposing patients to travel-related illness and therapy-related complications, including neurological, cardiovascular, infectious, and autoimmune complications, and may lead to death in some cases (Bauer et al, 2018). The leading countries in the international stem cell tourism market are the USA (27%), China (12%), India (12%), Thailand (11%), and Mexico (9%) (Lyons et al., 2022). In the US, orthopedic and sports medicine, and pain conditions, including knee OA, constitute the majority of the unapproved stem cell therapy market (Turner et al., 2021). The American Association of Hip and Knee Surgeons (AAHKS) has stated that biological therapies, including MSCs, cannot be recommended for routine treatment of advanced knee OA until their safety and efficacy are demonstrated through well-designed clinical trials (AAHKS, 2025). Moreover, the US FDA has issued public

FIGURE 16.3 Global regulatory approvals of mesenchymal stem cell (MSC)-based therapies for knee osteoarthritis (KOA). Countries shaded green represent locations where MSC-based products have received regulatory approval for clinical use.

alerts cautioning patients against seeking stem cell or exosome therapies for various conditions, including orthopedic disorders such as knee OA, as these treatments remain investigational and are illegal outside the context of approved clinical trials (FDA, 2024). Similarly, the International Society for Stem Cell Research (ISSCR) has condemned the use of unproven and unregulated stem cell therapies and has urged the FDA and EMA to enforce regulations on stem-cell-based therapies (ISSCR, 2025).

Given the variability in regulatory approval and clinical use of MSC-based therapies across countries, a coordinated global approach is essential (Figure 16.3). As Master et al. (2021) emphasized, the establishment of a World Health Organization (WHO) advisory committee on regenerative medicine could help develop a unified international regulatory framework with consistent standards, along with public educational campaigns to prevent misleading information (Master et al., 2021). These efforts would enhance patient safety and support regulated regenerative medical research.

16.11 CONCLUSION AND FUTURE PERSPECTIVE

Current data indicate that MSCs are a promising disease-modifying approach for knee OA, offering both structural and functional benefits beyond symptomatic relief. Evidence from preclinical models has consistently demonstrated their ability to modulate the underlying tissue inflammation, preserve cartilage, and improve joint function. Clinical studies across BM-MSCs, UC-MSCs, and AD-MSCs confirm their safety upon long-term injection and suggest meaningful reductions in pain and disability, with some trials showing signs of structural cartilage preservation on MRI. Nonetheless, the observed therapeutic outcomes vary across MSC sources, dosing strategies, and delivery protocols, highlighting the urgent demand for treatment standardization. Moreover, the efficacy of MSC treatment compared to the currently available therapy, especially steroid injection, remains to be fully established.

Future research should focus on optimizing cell sourcing, standardizing manufacturing protocols for isolation and bio-banking, and refining dosing regimens to reduce heterogeneity and improve reproducibility. Direct head-to-head comparisons among BM-MSCs, UC-MSCs, and AD-MSCs are still lacking, and such trials are needed to determine which cell source is most suitable for different patient populations. In addition, combinatorial strategies—such as MSCs delivered with scaffolds, growth factors, or gene modifications—warrant further evaluation to enhance chondrogenic and regenerative potential. The rise of MSC-derived EVs and exosomes represents a promising cell-free alternative, with potential advantages in safety, scalability, and regulatory approval. Long-term, multicenter randomized controlled trials with standardized imaging and functional endpoints are critical to defining the disease-modifying role of MSC-based therapies. Ultimately, the integration of MSCs or MSC-EV into cell-based and cell-free treatment approaches for clinical practice will depend on robust evidence of sustained benefit, cost-effectiveness, and streamlined regulatory frameworks.

REFERENCES

Abou-El-Enein M, Römhild A, Kaiser D, Beier C, Bauer G, Volk HD, Reinke P. Good Manufacturing Practices (GMP) manufacturing of advanced therapy medicinal products: a novel tailored model for optimizing performance and estimating costs. *Cytotherapy.* 2013 Mar;15(3):362–83. https://doi.org/10.1016/j.jcyt.2012.09.006.

Adam MS, Zhuang H, Ren X, Zhang Y, Zhou P. The metabolic characteristics and changes of chondrocytes in vivo and in vitro in osteoarthritis. *Front Endocrinol (Lausanne).* 2024 May 24;15:1393550. https://doi.org/10.3389/fendo.2024.1393550.

Al-Khani AM, Khalifa MA, Haider KH. Mesenchymal stem cells: How close are we to their routine clinical use? In: Haider KH (eds) *Handbook of Stem Cell Therapy.* Singapore: Springer, 2022. Vol. I, Chapter 10, pp. 269–322. https://doi.org/10.1007/978-981-16-6016-0_11-1.

Alofisel withdrawn from the EU market. European Medicines Agency (EMA). https://www.ema.europa.eu/en/news/alofisel-withdrawn-eu-market. 2024.

Alvarez-Viejo M, Haider KH. Mesenchymal stem cells. In: Haider KH (ed) *Handbook of Stem Cell Therapy.* Singapore: Springer, 2022. https://doi.org/10.1007/978-981-16-6016-0_6-1.

Ahmed OTF, Ahmed ZT, Dairi AW, Zain Al-Abeden MS, Alkahlot MH, Alkahlot RH, Al Jowf GI, Eijssen LMT, Haider KH. The inconclusive superiority debate of allogeneic versus autologous MSCs in treating patients with HFrEF: A systematic review and meta-analysis of RCTs. *Stem Cell Res Ther.* 2025a Apr 12;16(1):175. https://doi.org/10.1186/s13287-025-04209-5.

Ahmed OT, Ahmed ZT, Dairi AW, Alkahlot RH, Al Jowf GI, Haider KH. Autologous versus allogenic mesenchymal stem cells in heart disease: Revisiting preclinical and clinical trials. In: Haider KH (ed) *Handbook of Regenerative Medicine: Cell-Free Therapy.* Taylor & Francis Group/CRC Press, 2025b, Chapter 22, pp. 341–56. https://doi.org/10.1201/9781003611509-22.

Ahmed ZT, Alkahlot MH, Haider KH. MSC-derived exosomes: Advances in cell-free therapy. In: Haider KH (eds) *Handbook of Stem Cell Applications.* Singapore: Springer, 2024. Chapter 60, pp. 1295–326. https://doi.org/10.1007/978-981-99-0846-2_60-1.

Aixirefu A, Chen R, Wang H. Clinical efficacy of mesenchymal stem cells and platelet-rich plasma in the therapy of osteoarthritis: A meta-analysis. *Am J Transl Res.* 2024 Sep 15;16(9):4256–67. https://doi.org/10.62347/JUJV3321.

Ancans J. Cell therapy medicinal product regulatory framework in Europe and its application for MSC-based therapy development. *Front Immunol.* 2012 Aug 14;3:253. https://doi.org/10.3389/fimmu.2012.00253.

Approved Cellular and Gene Therapy Products | FDA [Internet] 2024. [cited 2025 Jul 5]. Available from: https://www.fda.gov/vaccines-blood-biologics/cellular-gene-therapy-products/approved-cellular-and-gene-therapy-products.

Barlow S, Brooke G, Chatterjee K, Price G, Pelekanos R, Rossetti T, Doody M, et al. Comparison of human placenta- and bone marrow-derived multipotent mesenchymal stem cells. *Stem Cells Dev.* 2008 Dec;17(6):1095–107. https://doi.org/10.1089/scd.2007.0154.

Bauer G, Elsallab M, Abou-El-Enein M. Concise Review: A comprehensive analysis of reported adverse events in patients receiving unproven stem cell-based interventions. *Stem Cells Transl Med.* 2018 Sep;7(9):676–85. https://doi.org/10.1002/sctm.17-0282.

Bhatti A, Haider KH. Mesenchymal stem cells as living bio-drugs: Adding a novel dimension to cardiology and cardiovascular sciences. *J Cardiol Cardiovasc Sci.* 2025;6(1):30–5. https://doi.org/10.29245/2578-3025/2025/1.1219.

Biologics. AAHKS. https://www.aahks.org/position-statements/biologics-for-advanced-hip-and-knee-arthritis/.
Breakthrough Therapy | FDA [Internet] 2025. [cited 2025 Jul 4]. Available from: https://www.fda.gov/patients/fast-track-breakthrough-therapy-accelerated-approval-.
Black LL, Gaynor J, Adams C, Dhupa S, Sams AE, Taylor R, Harman S, et al. Effect of intra-articular injection of autologous adipose-derived mesenchymal stem and regenerative cells on clinical signs of chronic osteoarthritis of the elbow joint in dogs. *Vet Ther.* 2008 Fall;9(3):192–200.
Burr DB, Gallant MA. Bone remodelling in osteoarthritis. *Nat Rev Rheumatol.* 2012 Nov;8(11):665–73. https://doi.org/10.1038/nrrheum.2012.130.
Caulfield T, Sipp D, Murray CE, Daley GQ, Kimmelman J. Scientific community. Confronting stem cell hype. *Science.* 2016 May 13;352(6287):776–7. https://doi.org/10.1126/science.aaf4620.
Charoenviriyakul C, Takahashi Y, Nishikawa M, Takakura Y. Preservation of exosomes at room temperature using lyophilization. Int J Pharm. 2018 Dec 20;553(1–2):1–7. https://doi.org/10.1016/j.ijpharm.2018.10.032. Epub 2018 Oct 11. Erratum in: *Int J Pharm.* 2019 Mar 25;559:427–8. https://doi.org/10.1016/j.ijpharm.2019.02.037.
Chatzistamatiou TK, Papassavas AC, Michalopoulos E, Gamaloutsos C, Mallis P, Gontika I, Panagouli E, et al. Optimizing isolation culture and freezing methods to preserve Wharton's jelly's mesenchymal stem cell (MSC) properties: An MSC banking protocol validation for the Hellenic Cord Blood Bank. *Transfusion.* 2014 Dec;54(12):3108–20. https://doi.org/10.1111/trf.12743.
Chen CF, Hu CC, Wu CT, Wu HH, Chang CS, Hung YP, Tsai CC, et al. Treatment of knee osteoarthritis with intra-articular injection of allogeneic adipose-derived stem cells (ADSCs) ELIXCYTE®: A phase I/II, randomized, active-control, single-masked, multiple-center clinical trial. *Stem Cell Res Ther.* 2021 Oct 30;12(1):562. https://doi.org/10.1186/s13287-021-02631-z.
Chiang ER, Ma HL, Wang JP, Liu CL, Chen TH, Hung SC. Allogeneic mesenchymal stem cells in combination with hyaluronic acid for the treatment of osteoarthritis in rabbits. *PLoS One.* 2016 Feb 25;11(2):e0149835. https://doi.org/10.1371/journal.pone.0149835.
Cosenza S, Toupet K, Maumus M, Luz-Crawford P, Blanc-Brude O, Jorgensen C, Noël D. Mesenchymal stem cells-derived exosomes are more immunosuppressive than microparticles in inflammatory arthritis. *Theranostics.* 2018 Feb 3;8(5):1399–410. https://doi.org/10.7150/thno.21072.
Dell'Isola A, Recenti F, Giardulli B, Lawford BJ, Kiadaliri A. Osteoarthritis year in review 2025: Epidemiology and therapy. *Osteoarthritis Cartilage.* 2025;33(11):1300–6. https://doi.org/10.1016/j.joca.2025.08.015.
Dominici M, Le Blanc K, Mueller I, Slaper-Cortenbach I, Marini F, Krause D, Deans R, et al. Minimal criteria for defining multipotent mesenchymal stromal cells. The International Society for Cellular Therapy position statement. *Cytotherapy.* 2006;8(4):315–7. https://doi.org/10.1080/14653240600855905.
Dreier R. Hypertrophic differentiation of chondrocytes in osteoarthritis: The developmental aspect of degenerative joint disorders. *Arthritis Res Ther.* 2010;12(5):216. https://doi.org/10.1186/ar3117.
EL Andaloussi S, Mäger I, Breakefield XO, Wood MJ. Extracellular vesicles: Biology and emerging therapeutic opportunities. *Nat Rev Drug Discov.* 2013 May;12(5):347–57. https://doi.org/10.1038/nrd3978.
Emadedin M, Labibzadeh N, Liastani MG, Karimi A, Jaroughi N, Bolurieh T, Hosseini SE, et al. Intra-articular implantation of autologous bone marrow-derived mesenchymal stromal cells to treat knee osteoarthritis: A randomized, triple-blind, placebo-controlled phase 1/2 clinical trial. *Cytotherapy.* 2018 Oct;20(10):1238–46. https://doi.org/10.1016/j.jcyt.2018.08.005.
European Medicines Agency. CAT quarterly highlights and approved ATMPs, May 2025. 2025 May 23 [cited 2025 Jul 5]. https://www.ema.europa.eu/en/documents/committee-report/cat-quarterly-highlights-approved-atmps-may-2025_en.pdf.
FDA approves first mesenchymal stromal cell therapy to treat steroid-refractory acute graft-versus-host disease. FDA. [cited 2025 Jul 5]. https://www.fda.gov/news-events/press-announcements/fda-approves-first-mesenchymal-stromal-cell-therapy-treat-steroid-refractory-acute-graft-versus-host.
Fernández-Garza LE, Barrera-Barrera SA, Barrera-Saldaña HA. Mesenchymal stem cell therapies approved by regulatory agencies around the world. *Pharmaceuticals (Basel).* 2023 Sep 21;16(9):1334. https://doi.org/10.3390/ph16091334.
Gao F, Chiu SM, Motan DA, Zhang Z, Chen L, Ji HL, Tse HF, et al., Mesenchymal stem cells and immunomodulation: current status and prospects. *Cell Death Dis.* 2016 Jan 21;7(1):e2062. https://doi.org/10.1038/cddis.2015.327.
Gronthos S, Franklin DM, Leddy HA, Robey PG, Storms RW, Gimble JM. Surface protein characterization of human adipose tissue-derived stromal cells. *J Cell Physiol.* 2001 Oct;189(1):54–63. https://doi.org/10.1002/jcp.1138.

Gupta A. StemOneTM/Stempeucel®: CDSCO approved, adult human bone marrow-derived, cultured, pooled, allogenic mesenchymal stem cells for knee osteoarthritis. *Biomedicines*. 2023 Oct 26;11(11):2894. https://doi.org/10.3390/biomedicines11112894.

Gupta PK, Chullikana A, Rengasamy M, Shetty N, Pandey V, Agarwal V, Wagh SY, et al. Efficacy and safety of adult human bone marrow-derived, cultured, pooled, allogeneic mesenchymal stromal cells (Stempeucel®): Preclinical and clinical trial in osteoarthritis of the knee joint. *Arthritis Res Ther.* 2016 Dec 20;18(1):301. https://doi.org/10.1186/s13075-016-1195-7.

Gupta PK, Maheshwari S, Cherian JJ, Goni V, Sharma AK, Tripathy SK, Talari K, et al. Efficacy and safety of stempeucel in osteoarthritis of the knee: A phase 3 randomized, double-blind, multicenter, placebo-controlled study. *Am J Sports Med.* 2023 Jul;51(9):2254–66. https://doi.org/10.1177/03635465231180323.

Gurunathan S, Kang MH, Jeyaraj M, Qasim M, Kim JH. Review of the isolation, characterization, biological function, and multifarious therapeutic approaches of exosomes. *Cells*. 2019 Apr 3;8(4):307. https://doi.org/10.3390/cells8040307. Erratum in: Cells. 2021 Feb 22;10(2):462. https://doi.org/10.3390/cells10020462.

Habib SM, Martini MF, Ahmed OA, Qassab AR, Haider KH. The potential of mesenchymal stem cell–derived exosomes in cell-free therapy for myocardial repair. In: Haider KH (ed) *Handbook of Regenerative Medicine: Cell-Free Therapy*. Taylor & Francis Group/CRC Press. 2025, Chapter 24, pp. 372–90. https://doi.org/10.1201/9781003611509-24.

Habib SM, Martini MF, Hamdan YNA, Shrebaty OMM, Haider KH. Mesenchymal stem cell secretome and exosomes as potential advanced therapy medicinal products for treating a failing heart. *WJSC* 2026.

Haider KH. Melatonin-based priming of stem cells to alleviate oxidative stress. *World J Stem Cells.* 2024a;16(11):985–89. https://doi.org/10.4252/wjsc.v16.i11.985.

Haider KH. Priming mesenchymal stem cells to develop "super stem cells." *World J Stem Cells*. 2024b Jun 26;16(6):623–40. https://doi.org/10.4252/wjsc.v16.i6.623.

Haider KH, Aramini B. Mircrining the injured heart with stem cell-derived exosomes: An emerging strategy of cell-free therapy. *Stem Cell Res Ther.* 2020 Jan 9;11(1):23. https://doi.org/10.1186/s13287-019-1548-7.

Hamilton JL, Nagao M, Levine BR, Chen D, Olsen BR, Im HJ. Targeting VEGF and its receptors for the treatment of osteoarthritis and associated pain. *J Bone Miner Res.* 2016 May;31(5):911–24. https://doi.org/10.1002/jbmr.2828.

He L, He T, Xing J, et al. Bone marrow mesenchymal stem cell-derived exosomes protect cartilage damage and relieve knee osteoarthritis pain in a rat model of osteoarthritis. *Stem Cell Res Ther.* 2020;11:276. https://doi.org/10.1186/s13287-020-01781-w.

He W, Zhao J, Liu J, Wang F, Xu Z. Adipose-derived mesenchymal stem cells combined with platelet-rich plasma are superior options for the treatment of osteoarthritis. *J Orthop Surg Res.* 2025;20:2. https://doi.org/10.1186/s13018-024-05396-2.

Ho KK, Lee WY, Griffith JF, Ong MT, Li G. Randomized control trial of mesenchymal stem cells versus hyaluronic acid in patients with knee osteoarthritis – A Hong Kong pilot study. *J Orthop Translat.* 2022 Oct 6;37:69–77. https://doi.org/10.1016/j.jot.2022.07.012.

Houard X, Goldring MB, Berenbaum F. Homeostatic mechanisms in articular cartilage and role of inflammation in osteoarthritis. *Curr Rheumatol Rep.* 2013 Nov;15(11):375. https://doi.org/10.1007/s11926-013-0375-6.

International Society for Stem Cell Research. ISSCR Urges FDA and EMA to Enforce Regulation of Clinics Offering Unproven and Unapproved Stem Cell-based Interventions [Internet] 2025. [cited 2025 Jul 5]. Available from: `https://www.isscr.org/isscr-news/isscr-urges-fda-and-ema-to-enforce-regulation-of-clinics-offering-unproven-and-unapproved-stem-cell-based-interventions.

Jang S, Lee K, Ju JH. Recent updates of diagnosis, pathophysiology, and treatment on osteoarthritis of the knee. *Int J Mol Sci.* 2021 Mar 5;22(5):2619. https://doi.org/10.3390/ijms22052619.

Jiang P, Mao L, Qiao L, Lei X, Zheng Q, Li D. Efficacy and safety of mesenchymal stem cell injections for patients with osteoarthritis: a meta-analysis and review of RCTs. *Arch Orthop Trauma Surg.* 2021. Available from: https://www.fda.gov/vacDoies-blood-biologics/consumers-biologics/consumer-alert-regenerative-medicine-products-including-stem-cells-and-exosomes.

Jin Y, Xu M, Zhu H, et al. Therapeutic effects of bone marrow mesenchymal stem cells-derived exosomes on osteoarthritis. *J Cell Mol Med.* 2021 Oct;25(19):9281–94. https://doi.org/10.1111/jcmm. 16860.

Jo CH, Lee YG, Shin WH, Kim H, Chai JW, Jeong EC, Kim JE, et al. Intra-articular injection of mesenchymal stem cells for the treatment of osteoarthritis of the knee: A proof-of-concept clinical trial. *Stem Cells.* 2014 May;32(5):1254–66. https://doi.org/10.1002/stem. 1634. Erratum in: Stem Cells. 2017 Jun;35(6):1651–2. https://doi.org/10.1002/stem.2589.

Kalluri R, LeBleu VS. The biology, function, and biomedical applications of exosomes. Science. 2020 Feb 7;367(6478):eaau6977. https://doi.org/10.1126/science.aau6977.

Kamal M, Kassem D, Haider KH. Sources and therapeutic strategies of mesenchymal stem cells in regenerative medicine. In: Haider KH (ed) *Handbook of Stem Cell Therapy*. Singapore: Springer, 2022. Vol. I, Chapter 2, pp. 23–52.

Kaspiris A, Hadjimichael AC, Lianou I, Iliopoulos ID, Ntourantonis D, Melissaridou D, Savvidou OD, et al. Subchondral bone cyst development in osteoarthritis: From pathophysiology to bone microarchitecture changes and clinical implementations. *J Clin Med.* 2023 Jan 19;12(3):815. https://doi.org/10.3390/jcm12030815.

Keshtkar S, Azarpira N, Ghahremani MH. Mesenchymal stem cell-derived extracellular vesicles: Novel frontiers in regenerative medicine. *Stem Cell Res Ther.* 2018 Mar 9;9(1):63. https://doi.org/10.1186/s13287-018-0791-7.

Kim KI, Lee MC, Lee JH, Moon YW, Lee WS, Lee HJ, Hwang SC, et al. Clinical efficacy and safety of the intra-articular injection of autologous adipose-derived mesenchymal stem cells for knee osteoarthritis: A phase III, randomized, double-blind, placebo-controlled trial. *Am J Sports Med.* 2023 Jul;51(9):2243–53. https://doi.org/10.1177/03635465231179223.

Klyachko NL, Haney MJ, Zhao Y, Manickam DS, Mahajan V, Suresh P, Hingtgen SD, et al. Macrophages offer a paradigm switch for CNS delivery of therapeutic proteins. *Nanomedicine* (Lond). 2014 Jul;9(9):1403–22. https://doi.org/10.2217/nnm.13.115.

Knights AJ, Redding SJ, Maerz T. Inflammation in osteoarthritis: the latest progress and ongoing challenges. *Curr Opin Rheumatol.* 2023 Mar 1;35(2):128–34. https://doi.org/10.1097/BOR.0000000000000923.

Kong, H., Zheng, J., Han, J. Meng X, Li X, Wang X, Zhang WX. The crosstalk among macrophages, chondrocytes and mesenchymal stem cells in osteoarthritis: The role of extracellular vesicles. *Stem Cell Res Ther.* 2025;16:553. https://doi.org/10.1186/s13287-025-04659-x.

Kordelas L, Rebmann V, Ludwig AK, Radtke S, Ruesing J, Doeppner TR, Epple M, et al. MSC-derived exosomes: A novel tool to treat therapy-refractory graft-versus-host disease. *Leukemia.* 2014 Apr;28(4):970–3. https://doi.org/10.1038/leu.2014.41.

Kuah D, Sivell S, Longworth T, James K, Guermazi A, Cicuttini F, Wang Y, et al. Safety, tolerability, and efficacy of intra-articular Progenza in knee osteoarthritis: A randomized double-blind placebo-controlled single ascending dose study. *J Transl Med.* 2018 Mar 6;16(1):49. https://doi.org/10.1186/s12967-018-1420-z.

Kurtzberg J, Abdel-Azim H, Carpenter P, Chaudhury S, Horn B, Mahadeo K, Nemecek E, et al. A phase 3, single-arm, prospective study of remestemcel-L, ex vivo culture-expanded adult human mesenchymal stromal cells for the treatment of pediatric patients who failed to respond to steroid treatment for acute graft-versus-host disease. *Biol Blood Marrow Transplant.* 2020 May;26(5):845–54. https://doi.org/10.1016/j.bbmt.2020.01.018.

Lamo-Espinosa JM, Blanco JF, Sánchez M, Moreno V, Granero-Moltó F, Sánchez-Guijo F, Crespo-Cullel Í, et al. Phase II multicenter randomized controlled clinical trial on the efficacy of intra-articular injection of autologous bone marrow mesenchymal stem cells with platelet-rich plasma for the treatment of knee osteoarthritis. *J Transl Med.* 2020 Sep 18;18(1):356. https://doi.org/10.1186/s12967-020-02530-6.

Lamo-Espinosa JM, Mora G, Blanco JF, Granero-Moltó F, Núñez-Córdoba JM, Sánchez-Echenique C, et al. Intra-articular injection of two different doses of autologous bone marrow mesenchymal stem cells versus hyaluronic acid in the treatment of knee osteoarthritis: Multicenter randomized controlled clinical trial (phase I/II). *J Transl Med.* 2016 Aug 26;14(1):246. https://doi.org/10.1186/s12967-016-0998-2.

Langworthy M, Dasa V, Spitzer AI. Knee osteoarthritis: Disease burden, available treatments, and emerging options. Ther Adv Musculoskelet Dis. 2024 Sep 15;16. https://doi.org/10.1177/1759720X241273009.

Lee WS, Kim HJ, Kim KI, Kim GB, Jin W. Intra-articular injection of autologous adipose tissue-derived mesenchymal stem cells for the treatment of knee osteoarthritis: A phase IIb, randomized, placebo-controlled clinical trial. *Stem Cells Transl Med.* 2019 Jun;8(6):504–11. https://doi.org/10.1002/sctm.18-0122.

Leifer VP, Katz JN, Losina E. The burden of OA-health services and economics. *Osteoarthritis Cartilage.* 2022 Jan;30(1):10–6. https://doi.org/10.1016/j.joca.2021.05.007.

Lin T, Wu N, Wang L, Zhang R, Pan R, Chen YF. Inhibition of chondrocyte apoptosis in a rat model of osteoarthritis by exosomes derived from miR 140 5p overexpressing human dental pulp stem cells. *Int J Mol Med.* 2021 Mar;47(3):7. https://doi.org/10.3892/ijmm.2020.4840.

Loeser RF. Integrins and chondrocyte-matrix interactions in articular cartilage. *Matrix Biol.* 2014 Oct;39:11–6. https://doi.org/10.1016/j.matbio.2014.08.007.

Long Z, Zhang M, Zhang T, Zeng L, Yang K, Yang T, Yu G, et al. The effectiveness and safety of mesenchymal stem cells in the treatment of osteoarthritis: A systematic review and meta-analysis of 28 randomized controlled trials. *Stem Cells Int.* 2022 Oct 12;2022:6151866. https://doi.org/10.1155/2022/6151866.

Lu L, Dai C, Zhang Z, Du H, Li S, Ye P, Fu Q, et al. Treatment of knee osteoarthritis with intra-articular injection of autologous adipose-derived mesenchymal progenitor cells: A prospective, randomized, double-masked, active-controlled, phase IIb clinical trial. *Stem Cell Res Ther.* 2019 May 21;10(1):143. https://doi.org/10.1186/s13287-019-1248-3.

Lyons S, Salgaonkar S, Flaherty GT. International stem cell tourism: A critical literature review and evidence-based recommendations. *Int Health.* 2022 Mar 2;14(2):132–41. https://doi.org/10.1093/inthealth/ihab050.

Madry H, Cucchiarini M. Gene therapy for human osteoarthritis: Principles and clinical translation. *Expert Opin Biol Ther.* 2016;16(3):331–46. https://doi.org/10.1517/14712598.2016.1124084.

Martel-Pelletier J, Barr AJ, Cicuttini FM, Conaghan PG, Cooper C, Goldring MB, Goldring SR, et al. Osteoarthritis. *Nat Rev Dis Primers.* 2016 Oct 13;2:16072. https://doi.org/10.1038/nrdp.2016.72.

Martel-Pelletier J. Pathophysiology of osteoarthritis. *Osteoarthritis Cartilage.* 2004;12(Suppl A):S31–3.

Pachón-Peña G, Serena C, Ejarque M, Petriz J, Duran X, Oliva-Olivera W, Simó R, et al. Obesity Determines the Immunophenotypic Profile and Functional Characteristics of Human Mesenchymal Stem Cells From Adipose Tissue. *Stem Cells Transl Med.* 2016 Apr;5(4):464–75. https://doi.org/10.5966/sctm.2015-0161.

Master Z, Matthews KRW, Abou-El-Enein M. Unproven stem cell interventions: A global public health problem requiring global deliberation. *Stem Cell Reports.* 2021 Jun 8;16(6):1435–45. https://doi.org/10.1016/j.stemcr.2021.05.004.

Matas J, Orrego M, Amenabar D, Infante C, Tapia-Limonchi R, Cadiz MI, Alcayaga-Miranda F, et al. Umbilical cord-derived mesenchymal stromal cells (MSCs) for knee osteoarthritis: Repeated MSC dosing is superior to a single MSC dose and to hyaluronic acid in a controlled randomized phase I/II trial. *Stem Cells Transl Med.* 2019 Mar;8(3):215–24. https://doi.org/10.1002/sctm.18-0053.

Mautner K, Gottschalk M, Boden SD, Akard A, Bae WC, Black L, Boggess B, et al. Cell-based versus corticosteroid injections for knee pain in osteoarthritis: A randomized phase 3 trial. Nat Med. 2023 Dec;29(12):3120–6. https://doi.org/10.1038/s41591-023-02632-w. Epub 2023 Nov 2. Erratum in: *Nat Med.* 2024 Mar;30(3):905. https://doi.org/10.1038/s41591-023-02776-9.

Mazini L, Rochette L, Amine M, Malka G. Regenerative capacity of adipose-derived stem cells (ADSCs): Comparison with mesenchymal stem cells (MSCs). *Int J Mol Sci.* 2019;20(10):2523.

McCoy AM. Animal models of osteoarthritis: comparisons and key considerations. *Vet Pathol.* 2015 Sep;52(5):803–18. https://doi.org/10.1177/0300985815588611.

Mendt M, Kamerkar S, Sugimoto H, McAndrews KM, Wu CC, Gagea M, Yang S, et al. Generation and testing of clinical-grade exosomes for pancreatic cancer. *JCI Insight.* 2018 Apr 19;3(8):e99263. https://doi.org/10.1172/jci.insight.99263.

Molnar V, Matišić V, Kodvanj I, Bjelica R, Jeleč Ž, Hudetz D, Rod E, et al. Cytokines and chemokines involved in osteoarthritis pathogenesis. *Int J Mol Sci.* 2021 Aug 26;22(17):9208. https://doi.org/10.3390/ijms22179208.

Molnar V, Pavelić E, Vrdoljak K, Čemerin M, Klarić E, Matišić V, Bjelica R, et al. Mesenchymal stem cell mechanisms of action and clinical effects in osteoarthritis: A narrative review. *Genes (Basel).* 2022 May 26;13(6):949. https://doi.org/10.3390/genes13060949.

Murphy JM, Fink DJ, Hunziker EB, Barry FP. Stem cell therapy in a caprine model of osteoarthritis. *Arthritis Rheum.* 2003 Dec;48(12):3464–74. https://doi.org/10.1002/art.11365.

Mushtaq M, Zineldeen DH, Mateen MA, Haider KH. Mesenchymal stem cells' "garbage bags" at work: Treating radial nerve injury with mesenchymal stem cell-derived exosomes. *World J Stem Cells.* 2024; 16(5). https://doi.org/https://dx.doi.org/10.4252/wjsc.v16.i5.0000.

Nagamura-Inoue T, He H. Umbilical cord-derived mesenchymal stem cells: Their advantages and potential clinical utility. *World J Stem Cells.* 2014 Apr 26;6(2):195–202. https://doi.org/10.4252/wjsc.v6.i2.195.

Nakamura Y, Miyaki S, Ishitobi H, Matsuyama S, Nakasa T, Kamei N, Akimoto T, et al. Mesenchymal-stem-cell-derived exosomes accelerate skeletal muscle regeneration. *FEBS Lett.* 2015 May 8;589(11):1257–65. https://doi.org/10.1016/j.febslet.2015.03.031.

Park YB, Ha CW, Lee CH, Yoon YC, Park YG. Cartilage regeneration in osteoarthritic patients by a composite of allogeneic umbilical cord blood-derived mesenchymal stem cells and hyaluronate hydrogel: Results from a clinical trial for safety and proof-of-concept with 7 years of extended follow-up. *Stem Cells Transl Med.* 2017 Feb;6(2):613–21. https://doi.org/10.5966/sctm.2016-0157.

Park YB, Lee HJ, Nam HC, Park JG. Allogeneic umbilical cord-blood-derived mesenchymal stem cells and hyaluronate composite combined with high tibial osteotomy for medial knee osteoarthritis with full-thickness cartilage defects. *Medicina (Kaunas).* 2023 Jan 11;59(1):148. https://doi.org/10.3390/medicina59010148.

Rajab AM, Rajab TM, AlJundi S, Haider KH. (2019). The clinical perspective of bone stem cell therapy encompasses both non-randomized and randomized trials. In: Haider KH (ed) *Stem Cells: From Myth to Reality and Evolving.* Berlin: Medicine & Life Sciences, De Gruyter.

Sadosky AB, Bushmakin AG, Cappelleri JC, Lionberger DR. Relationship between patient-reported disease severity in osteoarthritis and self-reported pain, function, and work productivity. *Arthritis Res Ther.* 2010;12(4):R162. https://doi.org/10.1186/ar3121.

Sadri B, Hassanzadeh M, Bagherifard A, Mohammadi J, Alikhani M, Moeinabadi-Bidgoli K, Madani H, et al., Cartilage regeneration and inflammation modulation in knee osteoarthritis following injection of allogeneic adipose-derived mesenchymal stromal cells: A phase II, triple-blinded, placebo-controlled, randomized trial. *Stem Cell Res Ther.* 2023 Jun 14;14(1):162. https://doi.org/10.1186/s13287-023-03359-8.

Safwan M, Bourgleh MS, Alshakaki H, Molhem A, Haider KH. Morbid cell status and donor age significantly alter mesenchymal stem cell functionality and reparability. In: Haider KH (ed) *Handbook of Stem Cell Applications.* Singapore: Springer, 2024.

Safwan M, Bourgleh MS, Baroudi L, Almsned A, AlBalawi R, Aibour B, AlFawaz S, Bourgleh SM, Haider KH. Entering the era of living biopharmaceuticals for treating knee osteoarthritis: A systematic review and network meta-analysis. *World J Stem Cells* 2025;17(8):107076. https://doi.org/10.4252/wjsc.v17.i8.107076.

Sengupta V, Sengupta S, Lazo A, Woods P, Nolan A, Bremer N. Exosomes derived from bone marrow mesenchymal stem cells as treatment for severe COVID-19. *Stem Cells Dev.* 2020 Jun 15;29(12):747–54. https://doi.org/10.1089/scd.2020.0080.

Sensebé L, Gadelorge M, Fleury-Cappellesso S. Production of mesenchymal stromal/stem cells according to good manufacturing practices: A review. *Stem Cell Res Ther.* 2013 Jun 7;4(3):66. https://doi.org/10.1186/scrt217.

Shi YY, Nacamuli RP, Salim A, Longaker MT. The osteogenic potential of adipose-derived mesenchymal cells is maintained with aging. *Plast Reconstr Surg.* 2005 Nov;116(6):1686–96. https://doi.org/10.1097/01.prs.0000185606.03222.a9.

Snyder J, Turner L, Crooks VA. Crowdfunding for unproven stem cell-based interventions. *JAMA.* 2018 May 8;319(18):1935–6. https://doi.org/10.1001/jama.2018.3057.

Strem BM, Hicok KC, Zhu M, Wulur I, Alfonso Z, Schreiber RE, Fraser JK, et al. Multipotential differentiation of adipose tissue-derived stem cells. *Keio J Med.* 2005 Sep;54(3):132–41. https://doi.org/10.2302/kjm.54.132.

Théry C, Witwer KW, Aikawa E, Alcaraz MJ, Anderson JD, Andriantsitohaina R, Antoniou A, et al., Minimal information for studies of extracellular vesicles 2018 (MISEV2018): A position statement of the International Society for Extracellular Vesicles and update of the MISEV2014 guidelines. *J Extracell Vesicles.* 2018 Nov 23;7(1):1535750. https://doi.org/10.1080/20013078.2018.1535750.

Tsiapalis D, O'Driscoll L. Mesenchymal stem cell derived extracellular vesicles for tissue engineering and regenerative medicine applications. *Cells.* 2020 Apr 16;9(4):991. https://doi.org/10.3390/cells9040991.

Turner L, Snyder J. Ethical issues concerning a pay-to-participate stem cell study. *Stem Cells Transl Med.* 2021 Jun;10(6):815–9. https://doi.org/10.1002/sctm. 20-0428.

van der Kraan PM, van den Berg WB. Osteoarthritis in the context of ageing and evolution. Loss of chondrocyte differentiation block during ageing. *Ageing Res Rev.* 2008 Apr;7(2):106–13. https://doi.org/10.1016/j.arr.2007.10.001.

Vega A, Martín-Ferrero MA, Del Canto F, Alberca M, García V, Munar A, Orozco L, et al. Treatment of knee osteoarthritis with allogeneic bone marrow mesenchymal stem cells: A randomized controlled trial. *Transplantation.* 2015 Aug;99(8):1681–90. https://doi.org/10.1097/TP.0000000000000678.

Vilar JM, Batista M, Morales M, Santana A, Cuervo B, Rubio M, Cugat R, et al. Assessment of the effect of intra-articular injection of autologous adipose-derived mesenchymal stem cells in osteoarthritic dogs using a double-blinded force platform analysis. *BMC Vet Res.* 2014 Jul 1;10:143. https://doi.org/10.1186/1746-6148-10-143.

Vinatier C, Guicheux J. Cartilage tissue engineering: From biomaterials and stem cells to osteoarthritis treatments. *Ann Phys Rehabil Med.* 2016 Jun;59(3):139–44. https://doi.org/10.1016/j.rehab.2016.03.002.

Wang AT, Feng Y, Jia HH, Zhao M, Yu H. Application of mesenchymal stem cell therapy for the treatment of osteoarthritis of the knee: A concise review. *World J Stem Cells.* 2019 Apr 26;11(4):222–35. https://doi.org/10.4252/wjsc.v11.i4.222.

Wang M, Sampson ER, Jin H, Li J, Ke QH, Im HJ, Chen D. MMP13 is a critical target gene during the progression of osteoarthritis. *Arthritis Res Ther.* 2013 Jan 8;15(1):R5. https://doi.org/10.1186/ar4133.

Wang M, Yuan Q, Xie L. Mesenchymal stem cell-based immunomodulation: Properties and clinical application. *Stem Cells Int.* 2018 Jun 14;2018:3057624. https://doi.org/10.1155/2018/3057624.

Wang Z, Xiao Z, Sun C, Xu G, He J. Global, regional and national burden of osteoarthritis in 1990-2021: A systematic analysis of the global burden of disease study 2021. *BMC Musculoskelet Disord.* 2024 Dec 19;25(1):1021. https://doi.org/10.1186/s12891-024-08122-5.

Warmink K, Rios JL, Varderidou-Minasian S, Torres-Torrillas M, van Valkengoed DR, Versteeg S, Eijkelkamp N, et al., Mesenchymal stem/stromal cells-derived extracellular vesicles as a potentially more beneficial therapeutic strategy than MSC-based treatment in a mild metabolic osteoarthritis model. *Stem Cell Res Ther.* 2023 May 24;14(1):137. https://doi.org/10.1186/s13287-023-03368-7.

Whitford W, Guterstam P. Exosome manufacturing status. *Future Med Chem.* 2019 May;11(10):1225–36. https://doi.org/10.4155/fmc-2018-0417.

Witwer KW, Théry C. Extracellular vesicles or exosomes? On primacy, precision, and popularity influencing a choice of nomenclature. *J Extracell Vesicles.* 2019 Aug 1;8(1):1648167. https://doi.org/10.1080/20013078.2019.1648167.

Xu L, Golshirazian I, Asbury BJ, Li Y. Induction of high temperature requirement A1, a serine protease, by TGF-beta1 in articular chondrocytes of mouse models of OA. *Histol Histopathol.* 2014 May;29(5):609-18. https://doi.org/10.14670/HH-29.10.609.

Zhang Y, Yang H, He F, Zhu X. Intra-articular injection choice for osteoarthritis: Making sense of cell source-an updated systematic review and dual network meta-analysis. *Arthritis Res Ther.* 2022 Nov 28;24(1):260. https://doi.org/10.1186/s13075-022-02953-0.

Zhao S, Xiu G, Wang J, Wen Y, Lu J, Wu B, Wang G, et al. Engineering exosomes derived from subcutaneous fat MSCs specially promote cartilage repair as miR-199a-3p delivery vehicles in osteoarthritis. *J Nanobiotechnology.* 2023 Sep 22;21(1):341. https://doi.org/10.1186/s12951-023-02086-9..

17 Stem Cells Extracellular Vesicles and Autoimmune Disease

Ranjeet Singh Mahla[1,2]

[1]Fred Hutch Cancer Centre, Seattle, WA, USA

[2]Kennedy Institute of Rheumatology, University of Oxford, Oxford, UK

17.1 LIST OF ABBREVIATIONS

ACPA	Anti-citrullinated protein antibody
AD-MSC-EVs	Adipose-derived mesenchymal stem cell extracellular vesicles
ANA	Antinuclear antibodies
APMAP	Adipocyte plasma membrane-associated protein
ApoVs/ApoV	Apoptotic vesicles
BAFF	B cell-activating factor
BAFF	B cell-activating factor
BBB	Blood-brain barrier
BMDSC-EVs	Bone marrow-derived stem cell EVs
BMSC/BM-MSC	Bone marrow-derived mesenchymal stem cell
BMSC-EVs	Bone marrow-derived MSC extracellular vesicles
C3/C3b	Complement component 3/activated C3 fragment
CD	Cluster of differentiation
Cdk2	Cyclin-dependent kinase 2
CNS	Central nervous system
CRISPR	Clustered regularly interspaced short palindromic repeats
CSF	Cerebrospinal fluid
CTLs	Cytotoxic T lymphocytes
DCs	Dendritic cells
dsDNA	Double-stranded DNA
EAE	Experimental autoimmune encephalomyelitis
EBV	Epstein–Barr virus
EGFR	Epidermal growth factor receptor
ESCRT	Endosomal sorting complex required for transport
FLS/FLSs	Fibroblast-like synoviocytes
GAD65	Glutamic acid decarboxylase 65
GM-CSF	Granulocyte-macrophage colony-stimulating factor
GTPase	Guanosine triphosphatase

DOI: 10.1201/9781003653387-17

HBMECs	Human brain microvascular endothelial cells
HSP70/HSP90	Heat shock protein 70/90
IA2	Islet antigen-2
IFNγ/IFNα	Interferon-gamma/interferon-alpha
IL	Interleukin
iNOS	Inducible nitric oxide synthase
iPSC	Induced pluripotent stem cell
iPSC-EVs	Induced pluripotent stem cell extracellular vesicles
MBP	Myelin basic protein
MDSCs	Myeloid-derived suppressor cells
MHC	Major Histocompatibility complex (contextually implied)
miRNA/miR	MicroRNA
MISEV	Minimal information for studies of extracellular vesicles
MMP1	Matrix metalloproteinase-1
MOG	Myelin oligodendrocyte glycoprotein
MS	Multiple sclerosis
MVBs	Multivesicular bodies
NF-κB	Nuclear factor kappa-light-chain-enhancer of activated B cells
NK cells	Natural killer cells
NOTCH1	Neurogenic locus notch homolog protein 1
OE-MSC-EVs	Olfactory epithelium-derived MSC EVs
pDCs	Plasmacytoid dendritic cells
PDLSCs	Periodontal ligament stem cells
PI3K/PI3K-Akt	Phosphoinositide 3-kinase pathway
PLP	Proteolipid protein
PMSC-EVs	Placenta-derived msc extracellular vesicles
PS	Phosphatidylserine
RA	Rheumatoid arthritis
RF	Rheumatoid factor
RORγt	Retinoic acid-related orphan receptor gamma t
ROS	Reactive oxygen species
SARDs	Systemic autoimmune rheumatic diseases
SEC	Size-exclusion chromatography
SGECs	Salivary gland epithelial cells
SIPRA	Signal-induced proliferation-associated protein
siRNA/shRNA	Small interfering RNA/short hairpin RNA
SLE	Systemic lupus erythematosus
SM	Sphingomyelin
SSc	Systemic sclerosis
SSc	Systemic sclerosis
T1D	Type 1 diabetes
TAB1	TGF-β activated kinase binding protein 1
TGFβ	Transforming growth factor-beta
Th1	T helper
TLR	Toll-like receptor
TSG101	Tumor susceptibility gene 101
UC-MSCs	Umbilical cord-derived mesenchymal stem cells
hUC-MSC	Human umbilical cord-derived mesenchymal stem cells
UCMSC-EVs	Umbilical cord-derived mesenchymal stem cells-derived EVs
ZnT8	Zinc transporter 8

17.2 INTRODUCTION

Autoimmune diseases encompass a diverse group of disorders in which the immune system mistakenly attacks the body's own tissues, resulting in chronic inflammation, tissue damage, and organ dysfunction. Collectively, autoimmune diseases affect millions of individuals worldwide and are associated with substantial morbidity and healthcare burden. These conditions may be organ-specific—such as type 1 diabetes (T1D), which affects the pancreas, or multiple sclerosis (Mohammadzadeh et al., 2024), which affects the central nervous system—or systemic, involving multiple organs simultaneously. Within this broad spectrum is a subset known as autoimmune rheumatic diseases, which primarily affect the joints, muscles, and connective tissues (Pisetsky, 2023). When these disorders involve widespread immune dysregulation across several organ systems, they are classified as systemic autoimmune rheumatic diseases (SARDs). Prominent examples include systemic lupus erythematosus (SLE), rheumatoid arthritis (RA), systemic sclerosis, and Sjögren's syndrome. Autoimmune diseases, whether organ-specific or systemic, often present with overlapping clinical features marked by persistent inflammation, autoantibody production and immune-mediated tissue injury. Their pathogenesis involves a complex interplay between genetic susceptibility, environmental triggers, epigenetic modifications, and immune dysregulation. Despite significant advances in understanding their underlying mechanisms, most autoimmune and rheumatic diseases remain incurable. Current therapies primarily focus on symptom management and immune suppression rather than achieving disease modification. In this context, EVs—particularly those derived from mesenchymal stem cells (MSCs)—have emerged as a promising cell-free therapeutic candidates for autoimmune and inflammatory diseases (Song et al., 2024; Ahmed et al., 2024). Importantly, EV-based therapies may overcome several limitations associated with direct MSC transplantation, including immune rejection, tumorigenic potential, and poor engraftment. MSC-derived EVs exhibit a robust immunomodulatory and therapeutic approach. MSCs are multipotent stromal cells the can be isolated from various tissues and organs, including bone marrow, umbilical cord, umbilical cord blood, peripheral blood, fetal liver, placenta, adipose tissue, and dental pulp (Yang et al., 2021; Kamal et al., 2022).

EVs are small, membrane-bound particles naturally released by cells. Although they lack a nucleus, they are encapsulated by a lipid bilayer and vary widely in size from about 30 nm to 10 μM. EVs are highly diverse, with classification based on size, cellular origin, and molecular content (Haider and Aramini, 2020). Their core role is to shuttle biologically active molecules—such as DNA, RNA, proteins, and lipids—making them essential players in cell-to-cell communication. Several intrinsic characteristics make EVs highly attractive for therapeutic use (Y. Chen et al., 2025). EVs carry molecular cargo that mirrors the physiological state of their parent cells, offering insights into disease presence and progression. Thanks to their small size, EVs can traverse various tissues. At the same time, their lipid membranes and surface markers enable efficient uptake by target cells, positioning them as effective delivery vehicles for therapeutic agents. In immunotherapy, EVs have already shown promise in modulating and enhancing immune responses (Hallal et al., 2022). For instance, EVs derived from endothelial cells have been shown to cross the blood-brain barrier. They are detectable in the blood of patients with multiple sclerosis (MS) (Manna et al., 2024)—the functional properties of EVs are dependent on their source. EVs derived from mesenchymal stem cells (MSCs), for example, have shown strong regenerative potential, while those from dendritic cells exhibit immunomodulatory effects and can serve as carriers for vaccine components (Habib et al., 2026). The biological complexity of EVs enables them to perform a wide range of functions, offering novel therapeutic opportunities for the treatment of autoimmune diseases beyond conventional therapies (Van Hoecke & Lucci, 2025). Despite their potential, several challenges persist, including EV heterogeneity in size, composition, cellular origin, and function; limited production yields; and difficulties achieving consistent purity during isolation. To address these issues, advanced bioengineering approaches are being developed to improve targeting precision, drug loading capacity, and scalable manufacturing processes. In recent years, research on EVs has attracted significant attention for cell-free therapy, as EVs are less immunogenic, easier to manage, and retain properties of parental cells. This chapter explores the current understanding of the biology of EVs and stem cell-derived EVs in the diagnosis and treatment of autoimmune diseases (Su et al., 2025). In particular, it examines how EV characterization and cargo profiling may serve as valuable biomarkers for disease diagnosis, prognosis, therapeutic monitoring, and the development of precision medicine approaches across a broad spectrum of autoimmune disorders.

17.3 EVS: CLASSIFICATION, BIOGENESIS, CARGO COMPOSITION ANALYSIS AND ISOLATION

17.3.1 EVs Classification and Biogenesis

The Minimal Information for Studies of Extracellular Vesicles (MISEV2023) guidelines define EVs as small particles released from all cells, delimited by a lipid bilayer, and incapable of replication (Welsh et al., 2024). MISEV2023 recommends classifying vesicles as extracellular vesicles, synthetic vesicles, or artificial cell-derived vesicles, and advises avoiding biogenesis-based terms such as "exosomes" and "microvesicles." Based on size and biogenesis pathways, native EVs are classified into exosomes (30–150 nm), microvesicles (approximately 100 nm to 1 μm), and apoptotic bodies (500 nm to a few μM). However, EV classification terminology should be used with caution, given the lack of evidence for the biogenesis pathway (Welsh et al., 2024). Exosomes were first discovered by Pan and Johnstone et al. in 1983. The term "exosomes" was coined to describe small membrane-bound vesicles that originate from the endosomal pathway and are secreted into the extracellular space by exocytosis (Johnstone et al., 1987; Pan et al., 1983). Exosome biosynthesis involves inward budding of the plasma membrane, the formation of early endosomes and multivesicular bodies (MVBs), cargo sorting and loading into intraluminal vesicles within MVBs, and subsequent release through fusion with the plasma membrane. The endosomal sorting complex required for transport (ESCRT; 4 multiprotein sub-complexes, namely ESCRT-0, I, II, and III) is essential for determining the fate of endocytosed cargo (Kalluri et al., 2020). However, exosomes can also be generated via an ESCRT-independent pathway, in which sphingolipid ceramide triggers inward membrane budding within MVBs (Trajkovic et al., 2008). Exosomes play a key role in cell-to-cell communication by transporting small biomolecules—such as miRNAs, proteins, mRNAs, long non-coding RNAs, and metabolic enzymes—to recipient cells (Mushtaq et al., 2024). They are also enriched with additional molecules like tetraspanins (e.g., CD63, CD9, and CD81), heat shock proteins (e.g., HSP70 and HSP90), endosomal associated proteins (e.g., TSG101 (tumor susceptibility gene 101), and Alix) and integrins, which are crucial for the assembly and intracellular trafficking of these biomolecules and are commonly used as markers to identify exosomes (Kalluri et al., 2020; Jeppesen et al., 2019). Microvesicles form via outward budding of the plasma membrane, driven by cytoskeletal reorganization, selective cargo sorting, and contractile machinery activation, regulated by GTPases (guanosine triphosphatases) like ARF1, ARF6, Rab22a, and RhoA (Tricarico et al., 2017). Apoptotic bodies are formed during programmed cell death, triggered by intrinsic or extrinsic pathways. The intrinsic pathway is activated by cellular stress, leading to increased mitochondrial permeability and the release of cytochrome c. In contrast, the extrinsic pathway involves Fas ligand binding to its receptor, which recruits the Fas-associated death domain protein. Both pathways converge on the activation of caspases 3, 6, and 7. During apoptosis, phosphatidylserine (PS) translocates to the outer membrane and binds annexin V. Simultaneously, oxidized surface molecules bind thrombospondin or complement protein C3b, thereby signaling phagocytes to engulf (M. Li et al., 2020).

In addition to exosomes, microvesicles, and apoptotic bodies, distinct types of EVs have been isolated that include oncosomes released from cancer cells, migrasomes (500 nm to 3 μm) released from retraction fibers during migracytosis, and spherosomes (40 to 125 nm) released as clusters enveloped within multivesicular spheres (Jiang et al., 2023; Junquera et al., 2016; Minciacchi et al., 2015), and exomers (<50 nm) and supermeres. As recommended by MISEV2023, the generic term "EV" should be used with precise operational definitions. EVs can be classified by size (small <200 nm, large >200 nm), density (low, medium, high), and cellular origin (e.g., MSC-EVs, endothelial EVs), based on their physical and biochemical properties (Welsh et al., 2024).

17.3.2 EVs Cargo Composition and Basic Features

EV cargo encompasses a diverse array of biomolecules, including protein, DNA, RNA, and lipids. The composition and relative abundance of cargo contents depend on the source cells and tissues of EVs, the biogenesis process, and the isolation techniques. EVs carry proteins involved in their biogenesis and secretion, including tetraspanins (CD9, CD63, CD81) and integrins. Studies using differential centrifugation and immunoprecipitation reveal size-dependent protein heterogeneity among EV subpopulations. Heterogeneity is also reflected in the tissue of origin and disease state (Hurwitz et al., 2016). EVs RNA cargo can have a wide range of sizes and types, such as short RNAs and long RNAs, and can be non-coding or protein-coding. Based on type, these include microRNA, piRNA, mRNA, circular RNA, tRNA, or small RNA-derived fragments. miRNAs play a central role in gene regulatory functions and are among the most widely studied RNA entities (Habib et al., 2026). RNA-binding proteins can be highly abundant in EVs, including autoantigens that are enriched in autoimmune conditions. EVs' RNA and cargo can be characterized using next-generation sequencing and proteomics (Amorim et al., 2017). Advanced transcriptomic and proteomic approaches enable comprehensive characterization of EV cargo profiles. Next-generation transcriptomics and proteomics analysis approaches require limited amounts of biological samples and can be performed on plasma, urine, and saliva specimens. Apart from RNA and protein cargo, EVs also contain DNA cargo and lipid cargo. DNA cargo includes genomic DNA, mitochondrial DNA, and cytoplasmic chromatin fragments. EVs' lipid bilayer also contains phosphatidylcholine (PC), sphingomyelin (SM), and cholesterol. EVs offer significant advantages over conventional drug delivery systems, including their natural ability to facilitate cell-to-cell communication and modulate immune responses, as well as high biocompatibility, excellent stability, and the capacity to cross biological barriers.

EVs are naturally occurring particles that exhibit remarkable stability, ensuring their integrity under various physiological conditions (T. Chen et al., 2025). EVs possess exceptional biocompatibility, facilitating their integration into biological system with minimal adverse responses. EVs containing cargo are protected from enzymatic degradation, enabling prolonged therapeutic efficacy. EVs' engineering can enable selective loading and targeted cargo delivery. Engineered EVs can be designed to carry therapeutic molecules such as siRNA/shRNA or CRISPR–Cas9 systems for targeted gene editing. EVs can carry specific siRNA/shRNA or CRISPR-Cas9 systems for gene editing. Even surface engineering can target EVs with particular cells of interest, such as autoreactive B cells.

17.3.3 EVs Isolation

EVs' isolation is essential for their application in cell-free therapies. The most common methods of EV isolation are ultracentrifugation, size-exclusion chromatography (SEC), precipitation-based kits, ultrafiltration, and immunoaffinity capture. Ultracentrifugation is widely used for its ability to isolate EVs with high yield, though it may co-purify contaminants. SEC and immunoaffinity capture offer higher purity, making them ideal for sensitive downstream applications like proteomics or RNA analysis. Overall, no single method is universally superior—rather, the optimal method depends on achieving an appropriate balance among yield, purity, scalability, and the intended biomarker discovery and therapeutic applications.

17.4 CORE PROPERTIES OF STEM CELL-DERIVED EVS

EVs derived from stem cells offer several advantages over direct stem cell transplantation. These include lower immunogenicity, greater stability, and easier storage compared to stem cells. Stem cell-derived exosomes are emerging as a promising alternative, as they retain many of the functional properties of their parent stem cells. In recent years, exosomes have attracted considerable interest as a cell-free therapeutic approach, offering a potential alternative to established stem cell-based therapies. Bone marrow-derived stem cells (BMSCs) are among the earliest and most extensively studied stem cell types. Exosomes derived from BMSCs have been evaluated for both safety and therapeutic efficacy in numerous studies. Their ease of isolation and rapid in vitro expansion make BMSCs an ideal model for exploring the therapeutic applications of EVs. BMSC EVs' biological activity can vary with donor age; however, umbilical cord stem cell (UCSC) EVs don't exhibit age-associated variability in exosome composition. In addition, menstrual blood stem cell-derived EVs (MenSC-EVs) possess superior proliferative properties and can be easily obtained non-invasively (Table 17.1).

17.5 EVS AND IMMUNE REGULATION

EVs have significant implications in cell-free therapy. Stem cell-derived EVs transfer bioactive compounds to recipient cells through endocytosis, membrane fusion, and specific receptor-ligand interactions. MSC-derived EVs can transform inflammatory phenotypes into immunosuppressive phenotypes, for example, M1 macrophages into M2 macrophages, DCs into tolerance DCs, and Th1 and Th17 cells into regulatory T cell phenotypes (Harrell et al., 2019). MSCs' EV-derived biomolecules are generally not used as diagnostic markers for autoimmunity, but rather as translational insights to advance therapeutics. MSC-EVs that encapsulate miR-23a and the TGFβ receptor two

TABLE 17.1
Key Features of Stem Cell-Derived EVs

Feature	Description
Immune modulation	Stem cell-derived EVs (e.g., from MSCs or iPSCs) carry immunoregulatory molecules (e.g., TGFβ and IL10) that suppress autoreactive immune responses.
Biocompatibility	Stem cell-derived EVs are naturally produced and well tolerated in the host, without risk of immune rejection in autoimmune conditions.
Stability in the inflamed tissue microenvironment	EVs remain structurally and functionally intact in inflamed tissues, including joints in RA, and the CNS in MS.
Cargo protection	EVs containing cargo (e.g., proteins, RNAs, and loaded drugs) are relatively stable from degradation in autoimmune environments.
Crossing biological barriers	Stem cell-derived EVs can cross the blood-brain barrier and are implicated in CNS autoimmune conditions, such as MS.
Engineering: Surface engineering	Surface receptor-engineered EVs can specifically target autoreactive immune cells (B cells/T cells, etc.) for cytotoxicity.
Engineering: Cargo loading	EVs can be preloaded with desired cargo (miRNAs, siRNAs, or CRISPR components targeting genes involved in autoimmunity (e.g., IL17, TNFα, NFκB).
High retention time	EVs are naturally derived and low in immunogenicity, enabling long-term systemic retention and improving therapeutic reach in autoimmunity.
Mimicking stem cell therapy	Carry the anti-inflammatory and regenerative factors of their parent stem cells, without risks of uncontrolled cell proliferation.
Tolerogenic antigen delivery	Can present self-antigens in a tolerogenic context, training the autologous immune system to become tolerant and preventing further attacks.

can promote the differentiation of $CD4^+$ T lymphocytes into Treg phenotypes. MSC-derived exosomes suppress T cell proliferation and induce cell cycle arrest via the p27^Kipl/Cdk2 signaling pathway. Additionally, MSC-EVs promote apoptosis of activated T cells, inhibit the proliferation of self-reactive lymphocytes, and enhance Treg production by increasing IL-10 and TGFβ secretion (Kim et al., 2020). BM-MSCs-EVs and hUC-MSCs-EVs are shown to decrease macrophage M1 polarization, indicated by downregulation of NOTCH1, IL1β, and iNOS, NOS2, and MCP1(Zhang et al., 2022; Sun et al., 2022). Moreover EVs profoundly downregulate proinflammatory cytokines GM-CSF (granulocyte-macrophage colony-stimulating factor), IL6, IL12, IL16, IL17, TNFα, IL1β, and IFNγ (Rajeev Kumar et al., 2024). UC-MSC administration also suppresses the differentiation of T cells into Th17 cells, as evidenced by decreased levels of IL-17, a cytokine commonly used as a biomarker for assessing disease activity in SLE (Tu et al., 2022). EVs derived from deciduous tooth pulp stem cells enhance the expression of $CD4^+$, $CD25^+$, and $Foxp3^+$ markers, suggesting a role in suppressing T cell activation. In contrast, apoptotic vesicles (ApoVs) promote the expansion of $CD4^+$ effector T cells while reducing the population of $CD4^+$ naïve T cells (30). Additionally, ApoVs inhibit the differentiation of Th1, Th2, and Th17 cells, as demonstrated by the downregulation of proinflammatory cytokines IL4, IL17A, and IFNγ (R. Wang et al., 2023) (Table 17.2).

17.6 STEM CELLS-DERIVED EVS IN AUTOIMMUNITY

Autoimmunity: refers to a condition in which the body's immune system mistakenly identifies its own healthy cells, tissues, and biomolecules (proteins, RNA, lipids, RNA-binding proteins) as foreign substances and attacks them. Unchecked anti-self-immune response leads to chronic inflammation, tissue damage, and disruption of normal physiological functions. Autoimmune diseases can be organ-specific (e.g., T1D affecting the pancreas) or systemic (e.g., systemic lupus erythematosus (SLE) and rheumatoid arthritis (RA) affecting multiple organs) (Bertolino et al., 2024). The causes are multifactorial, involving genetic predisposition, environmental triggers, and immune dysregulation. Treatments typically focus on reducing immune activity and managing symptoms, as most autoimmune diseases are currently incurable. EVs show great promise in the diagnosis, prognosis, and management of autoimmune diseases. Reflecting the molecular profile of their parent cells, EVs carry disease-specific biomarkers useful for early detection and monitoring disease progression. They also reveal key mechanisms underlying immune dysregulation by transferring inflammatory or regulatory signals between immune cells. In preclinical and in vitro studies, EVs—particularly from stem or regulatory cells—have demonstrated the ability to suppress inflammation and restore immune balance. Their non-invasive accessibility and functional relevance position EVs as valuable tools for advancing personalized treatment strategies in the management of autoimmune diseases. Recent advances further suggest that engineered stem cell-derived EVs may overcome limitations associated with conventional immunosuppressive therapies by enabling targeted delivery of immunomodulatory cargo, reducing systemic toxicity, and promoting tissue regeneration. In addition, EVs exhibit improved biocompatibility and lower immunogenicity compared with whole-cell therapies, thereby enhancing their translational potential in autoimmune disorders.

17.6.1 EVs in Rheumatoid Arthritis

Rheumatoid arthritis (RA) is a chronic, progressive, and systemic autoimmune disease characterized by the presence of anti-citrullinated protein antibodies (ACPAs) and rheumatoid factor (RF) autoantibodies. Clinically, RA manifests with joint inflammation, pannus formation, and progressive destruction of cartilage and bone, often accompanied by significant pain. Peptidyl arginine deiminase (PAD) enzymes play a key role in RA pathogenesis by catalyzing the citrullination of proteins such as fibrinogen, vimentin, enolase, and type II collagen—converting arginine residues into citrulline. These citrullinated proteins are recognized as autoantigens and targeted by ACPAs. In

TABLE 17.2
Immune Regulation and Stem Cell-Derived EVs

EVs Type	Target Immune Cells	Mechanism	Outcomes	Citation
MSC-EVs	Th1 cells	Regulation of glycolysis and cytokine signaling pathways.	Suppression of Th1 differentiation.	(Xu et al., 2020)
MSC-EVs	CD4+ T cells	miR-23a-3p mediated post-transcriptional regulation of TGFβ2 receptor.	Suppression of Th1 differentiation.	(Franco Da Cunha et al., 2020)
MSC-EVs	T cells	P27kip1/Cdk2 pathway.	Suppression of T cell activation and proliferation.	(Lee et al., 2020)
MSC-EVs	T cells	Increased production of IL-10 and TGFβ.	Suppression of T cell proliferation and induction of apoptosis.	(Kim et al., 2020)
MSC-EVs	B cells	miR-155-5p targets PI3K/AKT signaling pathway.	Suppression of B cell activation.	(Adamo et al., 2019)
MSC-Exosomes	DCs	Increased production of IL-10, TGFβ, and decreased production of IL-6, and lower expression of co-stimulatory markers.	Induction of tolerance DCs and promotion of Treg differentiation.	(Shahir et al., 2020)
MSC-EVs	DCs	Increased production of IL10, TGFβ and decreased production of IL6 and IL12p70.	Attenuation of DCs maturation and functions.	(Reis et al., 2018)
FL-MSC-Exosomes	NK cells	Regulation of TGFβ production.	Suppression of proliferation, activation, and cytotoxic activities of NK cells.	(Fan et al., 2019)
MSC-Exosomes	Macrophages	Decrease of IL22 and IL23.	Enhancement of anti-inflammatory phenotypes of macrophages.	(Hyvarinen et al., 2018)
MSC-Exosomes	Macrophages	Effects are mediated through the miR-223/pKNOX1 pathway.	Induction of anti-inflammatory M2 phenotypes.	(He et al., 2019)
MSC-EVs	Macrophages	Effects are executed via TLR4/NF-κB/PI3K/Akt signaling pathways.	Induction of anti-inflammatory M2 phenotypes.	(Xin et al., 2020)
UC-MSC-Exosomes	Macrophages	Induction of M2 phenotype.	Attenuation of diffuse alveolar inflammatory responses and alveolar hemorrhage.	(Chen et al., 2021)
FL-MSC-Exosomes	NK Cells	Induction of TGFβ.	Suppression of proliferation, activation, and cytotoxic activities of NK cells.	(Fan et al., 2019)

parallel, RF autoantibodies target the Fc portion of IgG antibodies, further contributing to immune complex formation and joint inflammation. Osteoblasts, osteoclasts, chondrocytes, and synovial fibroblasts (SFs), along with immune cells, are key joint-resident cells involved in RA pathogenesis.

17.6.1.1 Biomarker Discovery

EULAR-recognized biomarkers for RA diagnosis include serum levels of rheumatoid factor and anti-citrullinated peptide antibodies (ACPA). Emerging research highlights the potential of next-generation sequencing of EVs in identifying novel diagnostic markers. miRNA sequencing has revealed significant upregulation of miR-885-5p, miR-6894-3p, and miR-1268a in the serum exosomes of RA patients compared with healthy controls. Notably, miR-885-5p and miR-6894-3p also differentiate ACPA-negative RA patients (Gong et al., 2024). In addition to miRNAs, proteomic profiling of EVs has identified elevated levels of inflammatory mediators and matrix-degrading enzymes associated with synovial inflammation and cartilage destruction. These findings suggest that EV-derived molecular signatures may complement conventional serological markers for early RA diagnosis and disease stratification.

18.6.1.2 Therapeutic Interventions

MSCs and MSC-derived EVs modulate immune responses by inhibiting T cells, B cells, dendritic cells, and macrophages, while promoting Tregs and MDSCs. They suppress proinflammatory cytokines (TNFα, IL1β, IL6) and enhance anti-inflammatory cytokines (IL10, TGFβ (transforming growth factor-beta)). Additionally, they support chondrocyte and osteoblast differentiation, aiding bone and cartilage regeneration (Cosenza et al., 2018). EVs inhibit the activation, migration, and invasion of FLSs (fibroblast-like synoviocytes) by transferring microRNAs that modulate inflammatory, proliferative, and angiogenic pathways. BMSC-derived EVs also promote bone and cartilage regeneration and angiogenesis by delivering growth factors and bioactive molecules.(Cosenza et al., 2018; Chen et al., 2018; Meng et al., 2020). Bolandi et al. showed that AMSC-derived EVs modulate immune cells and FLSs by suppressing proinflammatory $CD4^+$ T cells and promoting anti-inflammatory subsets via specific miRNAs. They also support cartilage and bone regeneration by delivering growth factors and bioactive molecules (Li et al., 2021). Neutrophil-derived microvesicles can selectively modulate macrophage polarization and promote TGFβ release (Rhys et al., 2018). Furthermore, engineered EVs loaded with anti-inflammatory miRNAs or therapeutic agents have shown enhanced efficacy in experimental arthritis models by reducing synovial inflammation, inhibiting osteoclastogenesis, and preserving joint architecture. These findings support the potential of EV-based precision therapies for RA management.

17.6.2 EVs in Systemic Sclerosis (SSc)

SSc involves vasculopathy, fibrosis, and immune dysregulation. Key features include microvascular injury, endothelial dysfunction, and immune cell infiltration (lymphocytes and histiocytes) around vessels, leading to extracellular matrix accumulation and myofibroblast activation. Systemic Sclerosis (SSc) leads to progressive and widespread fibrosis, manifesting as skin thickening and severe internal organ involvement. Common complications include interstitial lung disease and scleroderma renal crisis, both of which significantly impair quality of life and contribute to increased morbidity and mortality.

17.6.2.1 Biomarker Discovery

Emerging evidence indicates that circulating EV-associated miRNAs and fibrosis-related proteins may serve as minimally invasive biomarkers for disease activity and progression in SSc. Elevated levels of EV-associated miR-21, miR-155, and pro-fibrotic signaling mediators have been associated with fibroblast activation and extracellular matrix deposition, highlighting their potential utility in monitoring fibrotic progression.

17.6.2.2 Translational Insights

MSC-EVs have been shown to immunomodulate the production and differentiation of immune cells, including B cells, T cells, and macrophages. Current EV-based approaches in SSc primarily aim to achieve anti-fibrotic effects by modulating key signaling pathways and extracellular matrix remodeling. Adipose-derived MSC-EVs, enriched in miR-29a, target TAB1, leading to reduced TIMP1 and increased MMP1 expression, collectively promoting anti-fibrotic effects (Rozier et al., 2021). Pre-activation of MSC-EVs with IFNγ enhances their anti-fibrotic effect in bleomycin-induced SSc mouse models. Bone marrow-derived MSC-EVs (BMDSC-EVs) significantly reduce TGFβ1-positive cells and α-SMA-positive myofibroblasts in mice, effects linked to the enrichment of let-7, miR-21a, miR-27b, miR-29a, and miR-143 (Rozier et al., 2021). Macrophage polarization is central to inflammation and fibrosis in autoimmune diseases. AKT phosphorylation drives proinflammatory M1 polarization, while its inhibition favors anti-inflammatory M2 polarization. BMSC-EVs suppress AKT phosphorylation, promoting M2 polarization, collagen synthesis, and angiogenesis. Additionally, BMSC-EVs deliver miR-34c-5p to macrophages via a CD81–EGFR (epidermal growth factor receptor) complex, inhibiting M1 activation (Hu et al., 2022). MSC-EVs modulate immune responses by influencing the Th17/Treg balance. They are internalized by $CD4^{+}$ T cells and inhibit Th17 differentiation by promoting RORγt degradation via K63 deubiquitination, reducing IL17 production, and shifting the response toward anti-inflammatory, immunosuppressive Treg activity in SSc (Jung et al., 2023).

17.6.3 Sjogren's Disease

SjD is the second most common autoimmune rheumatic disease. Anti-SSA (Ro52, Ro60) and anti-SSB (La) antibodies, as well as anti-ANA (antinuclear antibodies), are commonly used humoral diagnostic markers of active SjD. A large proportion of SjD patients also have rheumatoid factors, and occasionally some SjD patients also have anti-centromere and anti-dsDNA (double-stranded DNA) antibodies. Immune cell infiltration and ectopic germinal center formation occur in salivary glands, increasing the differentiation of CD4+ T cells to Th1 and Th17 phenotypes that support inflammation, and increased BAFF supports the survival and proliferation of autoreactive B cells. In addition to glandular dysfunction, systemic manifestations such as fatigue, arthritis, pulmonary involvement, and neuropathies are frequently observed in SjD patients, reflecting the multisystemic nature of the disease.

17.6.3.1 Biomarker Discovery

In SjD, the exosome biomarker may facilitate early diagnosis. pSS saliva EVs isolated by size-exclusion chromatography, liquid chromatography, and mass spectrometry identified biomarkers SIPRA and LSP1, essential for activation of the innate immune system, and APMAP (adipocyte plasma membrane-associated protein), essential for adipocyte differentiation (Aqrawi et al., 2017). Another study records EBV-specific microRNA, EBV-miR-bart13-3p, in salivary gland epithelial cells, suggesting that miRNA can be transferred from B cells to salivary gland epithelial cells via exosomes, affecting salivary gland epithelial cells' secretory functions (Gallo et al., 2016). Moreover, miR-142-3p derived from T cell exosomes modulates calcium signaling and cyclic AMP (cAMP) production in salivary gland epithelial cells (SGECs), thereby decreasing secretion and impairing SGEC function. These miRNAs, produced by specific salivary gland cells, not only indicate physiological and pathological processes in pSS but can also serve as diagnostic and prognostic markers. Recent evidence further indicates that salivary EV-associated proteins and nucleic acids may provide a non-invasive liquid biopsy platform for monitoring disease progression and therapeutic response in SjD patients.

17.8.3.2 Translational Insights

MSC-EVs have been shown to modulate B cell proliferation and function. Labial gland-derived MSC-EVs (LGMSC-EVs) alleviate Sjögren's disease (SjD) symptoms by suppressing B cell proliferation and differentiation, evidenced by reduced $CD19^+$ and $CD138^+$ plasma cells in the spleen of SjD mice. This effect is linked to miR-125b enrichment in LGMSC-EVs, which inhibits plasma cell differentiation (Xing et al., 2022). It is imperative to note that stem cell-derived exosomes hold great promise for modulating immune responses and restoring glandular function. Exosomes derived from mouse extra-olfactory mesenchymal stem cells (OE-MSCs) significantly enhance the immunosuppressive function of bone marrow-derived suppressor cells (MDSCs) by upregulating arginase expression and increasing ROS and NO levels. In a mouse model of Sjögren's syndrome, coadministration of OE-MSC exosomes delays disease progression and restores MDSC function. Labial gland MSC-derived exosomes (LGMSC-exosomes) treatment can reduce immune cell infiltration and restore salivary gland secretory function in NOD mice. Exosomes decrease Th17 differentiation and increase Treg differentiation in PBMCs from NOD mice and pSS patients, as evidenced by decreased IL17 and IFNγ and increased IL10 and TGFβ (Li et al., 2021). EVs from early-passage iPSCs exert more potent immunomodulatory effects than those from late-passage iPSCs. Kim et al. (Kim et al., 2021) have shown that EVs from early passage iPSCs are enriched in TGF-β1, TGFβ2, and miR-21, whereas late-passage iPSC-derived EVs are enriched in miR-125 b. The early passage BM-MSCs-EVs demonstrated significant modulation of the inhibitory effect of TGFβ1, miR-21, and miR-125b on Th17 production. Collectively, these findings support the therapeutic potential of stem cell-derived EVs as immunomodulatory agents capable of restoring glandular homeostasis and limiting autoimmune-mediated tissue injury in SjD.

17.6.4 Systemic Lupus Erythematosus

An imbalance among Th1, Th2, Th17, and Treg cells contributes to inflammatory responses and immune dysregulation in systemic lupus erythematosus (SLE). Impaired clearance of apoptotic cells contributes to the pathogenesis of SLE. Although activation and dysregulation of B cells play a central role in the pathogenesis of SLE, autoantibodies produced by autoreactive B cells mediate tissue damage and facilitate the presentation of autoantigens to self-reactive T cells. Activation of self-reactive B cells leads to the production of large quantities of autoantibodies, thereby driving disease progression, including BAFF. Immune complex deposition in tissues, particularly in the kidneys, skin, joints, and vasculature, contributes to chronic inflammation and multisystem organ damage characteristic of SLE.

17.6.4.1 Biomarkers Discovery

SLE pathogenesis modulates EV biosynthesis, reflected in changes in EV size, biogenesis, and cargo composition. In patients with SLE, serum-derived EV microRNAs—specifically miR-21, miR-152-3p, and miR-155—are upregulated, while miR-146a is downregulated. miR-152-3p results in increased BAFF expression by downregulating KLF5 (Luo et al., 2019). The expression of miR-21 negatively correlates with anti-SSA (anti-Ro60 or anti-Ro52) antibodies, and miR-146a negatively correlates with anti-dsDNA antibodies (W. Li et al., 2020). Additionally, the expression levels of miR-21 and miR-155 are associated with proteinuria in SLE patients. miR-21, let-7b, and miR-574 present in plasma EVs can activate plasmacytoid dendritic cells (pDCs) through the TLR7 pathway, enabling sustained production of IFNα and other proinflammatory cytokines, thereby contributing to a persistent trigger for SLE pathogenesis (Salvi et al., 2018). Circulating EVs enriched with miR-146a are internalized by mesenchymal stem cells (MSCs), where they activate the TRAF6/NF-κB signaling pathway and induce MSC senescence (Dong et al., 2019). Circulating exosomes that encapsulate miR-146a can serve as surrogate biomarkers for SLE diagnosis. Emerging studies also suggest that EV-associated nucleic acids and immune complexes may provide valuable prognostic information regarding lupus nephritis severity and treatment responsiveness.

17.6.4.2 Translational Insights

Adipose-derived MSC-EVs (AD-MSC-EVs) encapsulating miR-10a promote the differentiation of $CD4^+$ T cells into Th2 and Treg phenotypes (Bolandi et al., 2020). MiR-10a from AD-MSC-derived exosomes also regulates the balance between Th17 and Treg differentiation by modulating Foxp3 expression through the TGFβ signaling pathway (Wei et al., 2021). IgG, IgM, and C3 immune complexes are significantly decreased in the glomerular mesangial and endocapillary regions of the kidney following administration of EVs. EVs can also downregulate dsDNA, serum creatine levels, and urinary protein levels in patients with SLE (Wei et al., 2021). Chen et al. demonstrated that inhibiting miR-146a exacerbates lung injury, characterized by increased expression of NOTCH1, IL1β, and iNOS, and reduced levels of IL10 and TGFβ. In contrast, overexpression of miR-146a-5p via EVs in SLE patients upregulates IL10, CD206, Arg-1, and TGFβ, while downregulating NOTCH1, IL1β, and iNOS (Chen et al., 2022). These findings collectively highlight the therapeutic potential of EV-based immunomodulation in suppressing systemic inflammation, restoring immune tolerance, and mitigating organ damage in SLE.

17.6.5 Multiple Sclerosis

Multiple sclerosis (MS) is a chronic inflammatory demyelinating disease of the central nervous system (CNS), characterized by an immune response that progressively targets multiple myelin antigens—myelin basic protein (MBP), proteolipid protein (PLP), and myelin oligodendrocyte glycoprotein (MOG)—a phenomenon known as epitope spreading (Mohammadzadeh et al., 2024). MS involves dysregulation of both innate and adaptive immune responses. $CD4^+$ T cells predominantly differentiate into Th1 and Th17 subsets, cross the blood-brain barrier (BBB), and secrete IFNγ, promoting macrophage activation. This contributes to BBB disruption and chronic neuroinflammation. A proinflammatory cytokine milieu—IFNγ, IL1, IL6, IL17, and TNFα—overwhelms anti-inflammatory cytokines like IL10 and TGFβ. Macrophages infiltrate CNS lesions and degrade myelin, while activated microglia (resident macrophages) exacerbate inflammation and neurodegeneration. B cells act as antigen-presenting cells and produce pathogenic autoantibodies. In over 90% of MS patients, oligoclonal IgG bands in cerebrospinal fluid (CSF) reflect intrathecal antibody production. Axonal degeneration, mitochondrial dysfunction, and persistent neuroinflammation collectively contribute to progressive neurological disability and disease progression in MS patients.

17.6.5.1 Biomarker Discovery

Disruption of BBB permeability is a key aspect of MS pathogenesis. EVs facilitate cell-free communication by transferring signals, such as microRNAs (miRNAs), from source to target cells. EVs derived from monocytes can deliver specific miRNAs to human brain microvascular endothelial cells (HBMECs), inducing a proinflammatory phenotype that enhances leukocyte infiltration into the CNS. Recent studies have demonstrated that EV-associated miRNAs, cytokines, and myelin-related proteins in serum and CSF may serve as minimally invasive biomarkers for disease activity, neuroinflammation, and therapeutic responsiveness in MS. Elevated levels of EV-associated miR-155 and miR-326 have been linked to Th17-mediated neuroinflammatory responses and disease severity.

17.6.5.2 Translational Insights

Stem cell-derived EVs may provide a safer alternative due to their low immunogenicity and offer enhanced neuroprotective effects. Their small size allows them to cross the blood-brain barrier, enabling effective delivery to the CNS. Meta-analysis of EV treatment studies indicates that EVs derived from different stem cell sources (BM-MSCs, AD-MSCs, UC-MSCs, PDLSCs) reduced disease severity in mice/rats, humans, and monkeys. It is imperative to understand that EV treatment outcomes can vary significantly with recipient animal age, EAE (experimental autoimmune encephalomyelitis) scoring, treatment design, and EV isolation methods. EV treatment significantly reduces microglial activation and immune cell infiltration in the CNS, as measured by Iba-1 and H&E staining. Luxol fast blue staining indicated a profound reduction in demyelination in the CNS of EAE mice following EV administration. Autoreactive immune cells target MBP, whereas EV administration increases MBP expression in EAE mice. Moreover, EV administration promotes remyelination and reduces axonal damage. Administration of EVs before EAE induction results in a greater effect on restoring regulatory T cells. In the EAE mouse model of multiple sclerosis, UCMSC-EV treatment shapes immune activity by upregulating Lag-3 on $CD4^{+}/Foxp3^{+}$ T regulatory cells and limiting the entry of immune cells into the hypothalamus (Mohammadzadeh et al., 2024). Administration of high-dose PMSC-EVs improves motor deficits in the EAE mouse model of multiple sclerosis, an effect that appears to stem from their neuroprotective properties (Clark et al., 2019).

17.6.6 Type 1 Diabetes

T1D is a chronic autoimmune disease characterized by the immune system's aberrant attack on insulin-producing β-cells in the pancreatic islets of Langerhans, leading to insulin deficiency and hyperglycemia. T1D onset occurs with the breakdown of self-tolerance as a consequence of complex cell-to-cell interactions between immune cells and β-cells inside the islet microenvironment, where autoreactive T cells ($CD4^{+}$ T cells and $CD8^{+}$ T cells) are activated against pancreatic β-cell antigens [(such as insulin, GAD65 (glutamic acid decarboxylase) (Shi et al., 2018), IA2 (islet antigen-2), and ZnT8 (zinc transporter 8)]. To which macrophages and DCs play a pivotal role in antigen presentation to T cells, amplifying the autoimmune response by inducing the production of chemokines that subsequently facilitate cell-to-cell proapoptotic signaling (65). $CD8^{+}$ CTLs (cytotoxic T lymphocytes) function as primary effectors in β-cell killing through perforin/granzyme pathways, and $CD4^{+}$ Th1 cells amplify CTL response via inflammatory cytokines (IL1β, TNFα, IFNγ), leading to β-cell stress and apoptosis. B cells act as antigen-presenting cells and produce anti-insulin and anti-GAD65 antibodies. Persistent β-cell destruction ultimately results in severe insulin deficiency, metabolic dysregulation, and long-term microvascular and macrovascular complications.

17.6.6.1 Biomarker Discovery

Preclinical and clinical studies are underway. Evidently, autoreactive antibodies appear before the clinical onset of T1D, enabling early detection and intervention. Emerging evidence suggests that circulating EV-associated miRNAs and β-cell-specific proteins may serve as early indicators of β-cell stress, immune activation, and disease progression in individuals at risk of developing T1D.

17.6.6.2 Translational Insights

Stem cell-derived exosomes have shown multiple beneficial effects in T1D, including promoting pancreatic islet regeneration, restoring β-cell function, reducing insulin resistance, suppressing islet inflammation, and increasing plasma insulin levels. Exosomes derived from mesenchymal stem cells (MSCs) can inhibit the differentiation of Th1 and Th17 cells while enhancing IL-10 production. These actions collectively help suppress islet inflammation, significantly boost plasma insulin levels, and effectively delay the onset of T1D (Shigemoto-Kuroda et al., 2017). At four weeks post-administration, BMSC-derived exosomes significantly reduced blood glucose levels and increased insulin concentrations in diabetic rats. These effects were accompanied by pancreatic islet cell regeneration and reduced inflammation and fibrosis (Sabry et al., 2020). MSCs-derived exosomes can preserve islet cell survival and enhance insulin production in vitro by activating the phosphoinositide 3-kinase (PI3K) signaling pathway (Keshtkar et al., 2020). In addition, EV-mediated delivery of immunoregulatory miRNAs and regenerative growth factors may provide a promising strategy for preserving residual β-cell mass and restoring immune tolerance in T1D patients.

17.7 EVS ENGINEERING AND STORAGE

Stem cell-derived EVs, particularly those from MSCs, have emerged as potent cell-free therapeutics due to their ability to carry bioactive molecules, modulate immune responses, and mediate intercellular communication. In autoimmune diseases, where immune dysregulation leads to self-tissue damage, engineering these EVs to enhance their cargo, targeting, and therapeutic efficacy offers a promising therapeutic strategy. This section focuses on key engineering approaches used to optimize EVs for the treatment of autoimmune disorders.

17.7.1 Cargo Loading Strategies

Loading therapeutic cargo—such as RNA, proteins, or small molecules—is a key step in EV engineering to modulate immune responses. This can be achieved through direct (post-isolation) or indirect (pre-isolation) methods (Figure 17.1).

17.7.1.1 Passive Loading

In passive loading, EVs are incubated with cargo molecules, which then diffuse across the membrane. This method works best for small hydrophobic drugs that integrate into the lipid bilayer. However, it has low loading efficiency and is ineffective for large or hydrophilic molecules, such as RNA and proteins.

17.7.1.2 Active Loading

Active loading involves physical or chemical manipulation of EVs using methods like electroporation (Chen et al., 2024), sonication (Chen et al., 2024), extrusion, and freeze–thaw cycles (Hettich et al., 2022). Electroporation applies electric pulses to create temporary pores in the EV membrane, allowing the entry of negatively charged molecules, such as siRNAs and miRNAs. However, it may cause cargo aggregation and vesicle disruption (Chen et al., 2024).Sonication uses ultrasound waves to increase membrane permeability, allowing small molecules to enter (Chen et al., 2024). Extrusion forces EVs through nanopores under shear mechanical conditions at controlled temperature conditions, standardizing their size and aiding cargo loading—often alongside synthetic lipids (Luan et

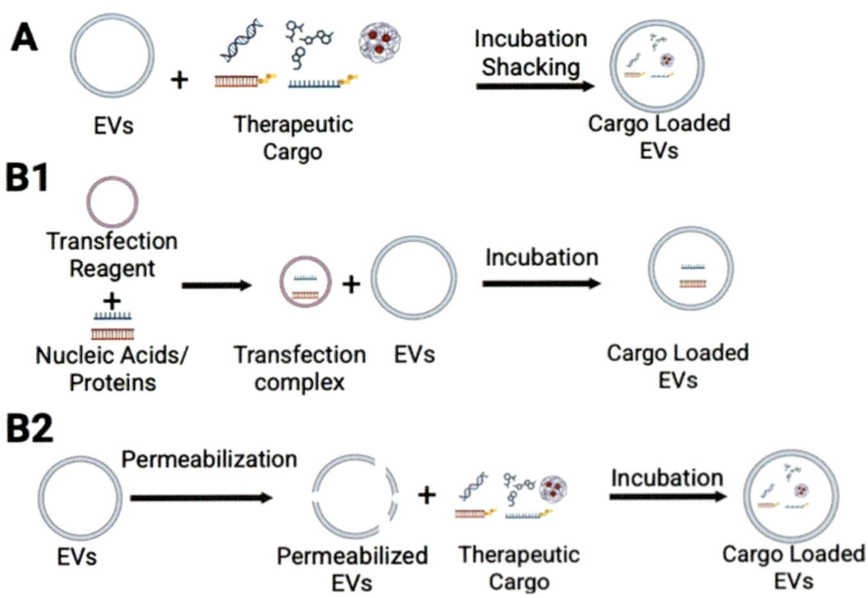

FIGURE 17.1 EV cargo loading strategies: (A) Passive loading. (B) Active loading.

al., 2017). Freeze–thawing, particularly through osmotic shock that repeatedly disrupts and reforms membranes, promotes fusion and incorporation of proteins into the vesicle membrane. It has been proven to be one of the most effective methods for EV loading, achieving high drug encapsulation while preserving structural and biological features (Hettich et al., 2022). Active loading not only requires loading exogenous cargo but also the depletion of endogenous cargo molecules. Chen et al. have shown that ultrasonication outperforms electroporation for depleting endogenous cargo. However, the choice of method can depend on the end goals of effective cargo loading and EV stability. Electroporation is favored over sonication due to ease of operation, speed, efficiency, and controllable parameters.

17.7.2 Endogenous Engineering

This is a parent cell-based loading approach enabling genetic and chemical manipulation of source cells, ensuring the loading of desired RNA, proteins, and small molecules loading in naturally packed EVs during the process of biogenesis (Mohammadzadeh et al., 2024; Walker et al., 2019). Donor cells can be transfected with plasmids or synthetic RNAs encoding miRNAs or regulatory proteins, which are then selectively incorporated into EVs (Pisetsky, 2023). RNA-binding proteins can be fused to EV marker proteins, such as CD63, to guide specific mRNA or miRNA into EVs via recognition sequences (Song et al., 2024). Preconditioning source cells with hypoxia, cytokines, or pharmacological agents modulates the EV cargo profile and chains therapeutic content for immunomodulation.

17.7.3 Surface Engineering of EVs for Targeting Autoimmune Tissues

Autoimmune diseases affect specific tissues and organs, such as T1D affecting the pancreas, RA affecting joints, MS affecting the CNS, and IBD affecting the gut mucosa. Surface engineering of EVs can improve EV homing and targeting to specific tissues and organs.

17.7.3.1 Genetic Engineering of Surface Proteins

Surface-engineered EVs can be generated by genetically modifying donor cells to express fusion proteins that link exosome membrane markers (e.g., Lamp2b, CD63, CD9) with targeting ligands. This approach enables tissue-specific delivery by directing EVs to disease-relevant sites. For instance, RGD peptides facilitate targeting of inflamed vasculature via integrin binding; RVG (rabies virus glycoprotein) peptides enable BBB crossing for CNS delivery in diseases like multiple sclerosis; and pancreatic β-cell–targeting peptides allow selective EV accumulation in islets for T1D therapy (Mohammadzadeh et al., 2024).

17.7.3.2 Incorporating Stealth and Functional Moieties

In autoimmune contexts, evading immune clearance while ensuring targeted delivery is essential. PEGylation of EVs with polyethylene glycol reduces opsonization and extends circulation time; using cleavable PEG linkers allows site-specific release at inflamed regions. Expression of CD47 on EV surfaces provides a "don't eat me" signal, minimizing phagocytic uptake (Zhang et al., 2024). Additionally, pH-sensitive moieties or cleavable peptides can be incorporated to enable controlled cargo release in the acidic microenvironment of inflamed tissues. Lipid dioleoyl phosphatidylethanolamine (DOPE) can serve as an anchor for isolated EVs, allowing conjugation with fluorophores, receptor-targeting peptides, or albumin-binding molecules to enable cellular uptake tracking (Zheng et al., 2023).

17.7.4 Engineering EV Mimetics and Hybrids

Due to limitations in natural EV yield, purity, and cargo control, engineered EV mimetics and hybrid systems provide scalable, customizable alternatives for therapeutic use.

17.7.4.1 EV–Liposome Hybrids

These combine the targeting capability, membrane proteins, and biocompatibility of exosomes with the high drug-loading efficiency and controlled release properties of synthetic liposomes. Fusion methods such as freeze–thaw cycles, extrusion (shear mechanical pressure under controlled temperature conditions) (Luan et al., 2017), or PEG-mediated fusion are commonly used. These hybrids can co-deliver nucleic acids and small molecules—such as miR-146a and methotrexate—for synergistic immunosuppression.

17.7.4.2 Cell-Mimetic Nanovesicles

Produced by physically disrupting cells using techniques such as extrusion or microfluidics, these artificial vesicles mimic the surface protein profile of natural EVs while offering improved scalability and consistency. They are particularly effective for delivering siRNAs, peptides, or anti-inflammatory drugs in autoimmune diseases such as rheumatoid arthritis.

17.7.5 Stimulus-Responsive EV Engineering

To enhance precision and therapeutic efficacy, EVs can be engineered to respond to specific physiological stimuli characteristic of autoimmune inflammation.

17.7.5.1 pH-Sensitive Systems

Inflammatory sites typically exhibit acidic microenvironments. EVs functionalized with pH-sensitive lipids or peptides can undergo conformational changes or degradation under low pH, enabling localized cargo release.

17.7.5.2 Enzyme-Cleavable Linkers

Elevated levels of matrix metalloproteinases (MMPs) or elastases at autoimmune sites can be exploited using surface modifications with enzyme-sensitive peptide linkers, allowing targeted release of therapeutic payloads in response to enzymatic activity.

17.7.5.3 Light or Magnetic Activation

Lysosomes are capable of metabolizing paramagnetic iron oxide nanoparticles (SPIONs). Therefore, EVs labeled with SPIONs can be externally guided to inflamed tissues, such as joints or the CNS, to deliver desired therapeutic cargo to the site of inflammation (Chen et al., 2023). Albeit SPION treatment of stem cells can trigger the assimilation of therapeutic cargo into EVs. Exosomes coated with magnetic nanoparticles can be enriched at the inflammation site using an external magnetic field. Due to slow in vivo ionization and assimilation of INOP, treatment of MSCs with iron oxide nanoparticles can increase the expression of anti-inflammatory tissue repair cytokines by activating the JNK pathway. Alternatively, the incorporation of photosensitive elements (Tong et al., 2022) such as titanium dioxide, 5-aminolevulinic acid, doxorubicin, hypericin, IR780, and paclitaxel, which enable light-triggered cargo release, can be particularly useful for autoimmune diseases affecting the skin or superficial joints.

17.7.6 Engineering for Immune Modulation in Autoimmunity

In autoimmune diseases, EVs must be engineered to modulate key immune cell subsets such as Tregs, Th17 cells, dendritic cells, and macrophages.

17.7.6.1 Targeting Specific Immune Cells

To target DCs, EVs can be functionalized with antibodies such as anti-DEC205 and anti-CD11c, enabling selective interaction with antigen-presenting cells (APCs)(Dang et al., 2024). The anti-DEC-205 antibody has demonstrated the potential to enhance antigen presentation by DCs and activate T cells. Similarly, controlled polarization of monocytes into macrophages M2 phenotype can be achieved by preloading EVs with microRNAs miR-223 and miR-146a (C. Wang et al., 2023), modulating their differentiation pathways.

17.7.6.2 miRNA and siRNA Cargo Engineering

MicroRNAs and siRNAs with immunomodulatory functions are promising candidates for EV-based therapy in autoimmune diseases. For example, miR-155 inhibitors reduce Th17 differentiation and suppress proinflammatory cytokine production, while miR-146a attenuates NF-κB signaling and supports regulatory T cell (Treg) stability (Testa et al., 2017). Additionally, siRNAs targeting inflammatory mediators such as TNFα and IL6 can be incorporated into EVs directly. These nucleic acids can be loaded into EVs either through donor cell transfection or post-isolation techniques such as electroporation, depending on the cargo type and therapeutic target.

17.8 EVS PRESERVATION AND STORAGE

The gold standard for EVs storage is −80 °C. Effective cryopreservation requires cryoprotective agents (CPAs) to modulate viscosity, control ice nucleation, and facilitate extracellular ice formation during slow cooling. Insufficient CPA concentrations can lead to chilling injury, while excessive levels can be cytotoxic. Thus, optimizing CPA concentration is critical. Alternative preservation methods, such as freeze-drying (lyophilization) and spray-drying, are being explored as viable non-frozen storage strategies for EVs (Kusuma et al., 2018). Multiple freeze–thaw cycles can result in cargo release and EV aggregation; therefore, EVs should be stored in their solutions and in small, one-time-use aliquots.

17.9 CLINICAL TRANSLATION AND CHALLENGES

Significant Challenges in the Therapeutic Applications of EVs While the development of EV-based therapies holds considerable promise for personalized medicine, several key challenges continue to hinder their widespread therapeutic application.

17.9.1 Production and Heterogeneity

One of the primary obstacles is the efficient production of EVs at a therapeutic scale. EV preparations are inherently heterogeneous, and their yield in standard cell culture systems is typically low. This necessitates large-scale production facilities capable of producing clinical-grade EVs that meet both quantitative and qualitative requirements. Achieving homogeneous and reproducible EV production remains a critical challenge.

17.9.2 Delivery Strategies

Another major limitation lies in determining the most effective route of administration. The optimal method for delivering EVs to specific tissues or cells is still under investigation, and current approaches often suffer from inefficiencies in targeting and uptake.

Bioavailability and Sustained Delivery: Ensuring the bioavailability of EVs in systemic circulation and maintaining sustained delivery of their cargo to target cells are additional hurdles. Rapid clearance, off-target effects, and variable uptake reduce the therapeutic efficacy of EV-based treatments.

17.9.3 Standardization

The lack of standardized protocols for EV isolation, characterization, and quantification significantly affects reproducibility across studies and hinders progress in both research and clinical translation. Scaling up the production of EVs for clinical use is technically demanding. Efficient and cost-effective methods for large-scale EV manufacturing are still in development.

Regulatory Challenges: Regulatory frameworks for EV-based therapeutics are still evolving. The complexity of EV compositions and their biological origin poses unique safety, quality control, and regulatory hurdles that must be addressed for clinical approval.

17.10 FUTURE DIRECTIONS AND CONCLUSION

Stem cell-derived EVs represent a convergence of cell-free regenerative medicine and immunotherapy. EVs can modulate immune responses, support tissue regeneration, and serve as both diagnostic tools and therapeutic agents, positioning them at the forefront of next-generation treatment options for autoimmune diseases. Future studies should define and optimize the most appropriate cell source for therapeutic EVs for disease-specific diagnosis and therapeutic EV production. Second, it is essential to elucidate the mechanisms underlying EV cargo selection and controlled release. Third, there is a need to enhance target precision and delivery efficiency. Fourth, well-designed clinical trials in animal models must be initiated to test their safety and efficacy. The potential of EVs in personalized medicine for autoimmune diseases is immense, offering hope for targeted, effective, and minimally invasive therapies. Although preclinical and early clinical results are promising, challenges related to standardization, scalability, and long-term safety must be resolved before clinical translation can be achieved.

REFERENCES

Adamo A, Brandi J, Caligola S, Delfino P, Bazzoni R, Carusone R, et al. Extracellular vesicles mediate mesenchymal stromal cell-dependent regulation of B cell PI3K-AKT signaling pathway and actin cytoskeleton. *Front Immunol.* 2019;10:446.

Ahmed ZT, Alkahlot MH, Haide KH. MSC-derived exosomes: Advances in cell-free therapy. In: Haider, K.H. (eds) *Handbook of Stem Cell Applications*. Singapore: Springer, 2024. Chapter 60, pp. 1295–326. https://doi.org/10.1007/978-981-99-0846-2_60-1

Amorim MG, Valieris R, Drummond RD, Pizzi MP, Freitas VM, Sinigaglia-Coimbra R, et al. A total transcriptome profiling method for plasma-derived extracellular vesicles: applications for liquid biopsies. *Sci Rep.* 2017;7(1):14395.

Aqrawi LA, Galtung HK, Vestad B, Ovstebo R, Thiede B, Rusthen S, et al. Identification of potential saliva and tear biomarkers in primary Sjogren's syndrome, utilising the extraction of extracellular vesicles and proteomics analysis. *Arthritis Res Ther.* 2017;19(1):14.

Bertolino GM, Maumus M, Jorgensen C, Noel D. Author correction: Therapeutic potential in rheumatic diseases of extracellular vesicles derived from mesenchymal stromal cells. *Nat Rev Rheumatol.* 2024;20(6):389.

Bolandi Z, Mokhberian N, Eftekhary M, Sharifi K, Soudi S, Ghanbarian H, et al. Adipose-derived mesenchymal stem cell exosomes loaded with miR-10a promote the differentiation of Th17 and Treg from naive CD4(+) T cells. *Life Sci.* 2020;259:118218.

Chen T, Chen D, Su W, Liang J, Liu X, Cai M. Extracellular vesicles as vital players in drug delivery: A focus on clinical disease treatment. *Front Bioeng Biotechnol.* 2025;13:1600227.

Chen X, Su C, Wei Q, Sun H, Xie J, Nong G. Exosomes derived from human umbilical cord mesenchymal stem cells alleviate diffuse alveolar hemorrhage associated with systemic lupus erythematosus in mice by promoting M2 macrophage polarization via the microRNA-146a-5p/NOTCH1 axis. *Immunol Invest.* 2022;51(7):1975–93.

Chen X, Wei Q, Sun H, Zhang X, Yang C, Tao Y, et al. Exosomes derived from human umbilical cord mesenchymal stem cells regulate macrophage polarization to attenuate systemic lupus erythematosus-associated diffuse alveolar hemorrhage in mice. *Int J Stem Cells.* 2021;14(3):331–40.

Chen Y, Hou S. Recent progress in the effect of magnetic iron oxide nanoparticles on cells and extracellular vesicles. *Cell Death Discov.* 2023;9(1):195.

Chen Y, Wang L, Yu X, Mao W, Wan Y. Ultrasonication outperforms electroporation for extracellular vesicle cargo depletion. *Extracell Vesicle.* 2024;4.

Chen Z, Wang H, Xia Y, Yan F, Lu Y. Therapeutic potential of mesenchymal cell-derived miRNA-150-5p-expressing exosomes in rheumatoid arthritis mediated by the modulation of MMP14 and VEGF. *J Immunol.* 2018;201(8):2472–82.

Clark K, Zhang S, Barthe S, Kumar P, Pivetti C, Kreutzberg N, et al. Placental mesenchymal stem cell-derived extracellular vesicles promote myelin regeneration in an animal model of multiple sclerosis. *Cells.* 2019;8(12).

Cosenza S, Toupet K, Maumus M, Luz-Crawford P, Blanc-Brude O, Jorgensen C, et al. Mesenchymal stem cell-derived exosomes are more immunosuppressive than microparticles in inflammatory arthritis. *Theranostics.* 2018;8(5):1399–410.

Dang XTT, Phung CD, Lim CMH, Jayasinghe MK, Ang J, Tran T, et al. Dendritic cell-targeted delivery of antigens using extracellular vesicles for anti-cancer immunotherapy. *Cell Prolif.* 2024;57(7):e13622.

Dong C, Zhou Q, Fu T, Zhao R, Yang J, Kong X, et al. Circulating exosomes derived-miR-146a from systemic lupus erythematosus patients regulates senescence of mesenchymal stem cells. *Biomed Res Int.* 2019;2019:6071308.

Fan Y, Herr F, Vernochet A, Mennesson B, Oberlin E, Durrbach A. Human fetal liver mesenchymal stem cell-derived exosomes impair natural killer cell function. *Stem Cells Dev.* 2019;28(1):44–55.

Franco da Cunha F, Andrade-Oliveira V, Candido de Almeida D, Borges da Silva T, Naffah de Souza Breda C, Costa Cruz M, et al. Extracellular vesicles isolated from mesenchymal stromal cells modulate CD4(+) T lymphocytes toward a regulatory profile. *Cells.* 2020;9(4).

Gallo A, Jang SI, Ong HL, Perez P, Tandon M, Ambudkar I, et al. Targeting the Ca(2+) sensor STIM1 by exosomal transfer of Ebv-miR-BART13-3p is associated with Sjogren's syndrome. *EBioMedicine.* 2016;10:216–26.

Gong J, Zhang X, Khan A, Liang J, Xiong T, Yang P, et al. Identification of serum exosomal miRNA biomarkers for diagnosis of Rheumatoid arthritis. *Int Immunopharmacol.* 2024;129:111604.

Habib SY, Martini MF, Hamdan YNA, Shrebaty OMM, Haider KH. Mesenchymal stem cell secretome and exosomes as potential advanced therapy medicinal products for treating a failing heart. *World J Stem Cells.* 2026;18.

Haider KH, Aramini B. Mircrining the injured heart with stem cell-derived exosomes: an emerging strategy of cell-free therapy. *Stem Cell Res Ther.* 2020 Jan 9;11(1):23. https://doi.org/10.1186/s13287-019-1548-7.

Hallal S, Tuzesi A, Grau GE, Buckland ME, Alexander KL. Understanding the extracellular vesicle surface for clinical molecular biology. *J Extracell Vesicles.* 2022;11(10):e12260.

Harrell CR, Jovicic N, Djonov V, Arsenijevic N, Volarevic V. Mesenchymal stem cell-derived exosomes and other extracellular vesicles as new remedies in the therapy of inflammatory diseases. *Cells.* 2019;8(12).

He X, Dong Z, Cao Y, Wang H, Liu S, Liao L, et al. MSC-derived exosome promotes M2 polarization and enhances cutaneous wound healing. *Stem Cells Int.* 2019;2019:7132708.

Hettich BF, Bader JJ, Leroux JC. Encapsulation of hydrophilic compounds in small extracellular vesicles: Loading capacity and impact on vesicle functions. *Adv Healthc Mater.* 2022;11(5):e2100047.

Hu X, Shen N, Liu A, Wang W, Zhang L, Sui Z, et al. Bone marrow mesenchymal stem cell-derived exosomal miR-34c-5p ameliorates RIF by inhibiting the core fucosylation of multiple proteins. *Mol Ther.* 2022;30(2):763–81.

Hurwitz SN, Rider MA, Bundy JL, Liu X, Singh RK, Meckes DG, Jr. Proteomic profiling of NCI-60 extracellular vesicles uncovers common protein cargo and cancer type-specific biomarkers. *Oncotarget.* 2016;7(52):86999–7015.

Hyvarinen K, Holopainen M, Skirdenko V, Ruhanen H, Lehenkari P, Korhonen M, et al. Mesenchymal stromal cells and their extracellular vesicles enhance the anti-inflammatory phenotype of regulatory macrophages by downregulating the production of interleukin (IL)-23 and IL-22. *Front Immunol.* 2018;9:771.

Jeppesen DK, Fenix AM, Franklin JL, Higginbotham JN, Zhang Q, Zimmerman LJ, et al. Reassessment of exosome composition. *Cell.* 2019;177(2):428–45.e18.

Jiang Y, Liu X, Ye J, Ma Y, Mao J, Feng D, et al. Migrasomes, a new mode of intercellular communication. *Cell Commun Signal.* 2023;21(1):105.

Johnstone RM, Adam M, Hammond JR, Orr L, Turbide C. Vesicle formation during reticulocyte maturation. Association of plasma membrane activities with released vesicles (exosomes). *J Biol Chem.* 1987;262(19):9412–20.

Jung S, Lee S, Kim HJ, Kim S, Moon JH, Chung H, et al. Mesenchymal stem cell-derived extracellular vesicles subvert Th17 cells by destabilizing RORgammat through posttranslational modification. *Exp Mol Med.* 2023;55(3):665–79.

Kalluri R, LeBleu VS. The biology, function, and biomedical applications of exosomes. *Science.* 2020;367(6478).

Kamal M, Kassem D, Haider, KH. Sources and therapeutic strategies of mesenchymal stem cells in regenerative medicine. In: Haider K.H. (eds) *Handbook of Stem Cell Therapy.* Singapore: Springer, 2022. Vol I, Chapter 2, pp. 23–52. https://doi.org/10.1007/978-981-16-6016-0_2-1.

Keshtkar S, Kaviani M, Sarvestani FS, Ghahremani MH, Aghdaei MH, Al-Abdullah IH, et al. Exosomes derived from human mesenchymal stem cells preserve mouse islet survival and insulin secretion function. *EXCLI J.* 2020;19:1064–80.

Kim H, Lee MJ, Bae EH, Ryu JS, Kaur G, Kim HJ, et al. Comprehensive molecular profiles of functionally effective MSC-derived extracellular vesicles in immunomodulation. *Mol Ther.* 2020;28(7):1628–44.

Kim H, Zhao Q, Barreda H, Kaur G, Hai B, Choi JM, et al. Identification of molecules responsible for therapeutic effects of extracellular vesicles produced from iPSC-derived MSCs on Sjögren's syndrome. *Aging Dis.* 2021;12(6):1409–22.

Kusuma GD, Barabadi M, Tan JL, Morton DAV, Frith JE, Lim R. To protect and to preserve: Novel preservation strategies for extracellular vesicles. *Front Pharmacol.* 2018;9:1199.

Lee S, Kim S, Chung H, Moon JH, Kang SJ, Park CG. Mesenchymal stem cell-derived exosomes suppress proliferation of T cells by inducing cell cycle arrest through p27kip1/Cdk2 signaling. *Immunol Lett.* 2020;225:16–22.

Li M, Liao L, Tian W. Extracellular vesicles derived from apoptotic cells: An essential link between death and regeneration. *Front Cell Dev Biol.* 2020;8:573511.

Li W, Liu S, Chen Y, Weng R, Zhang K, He X, et al. Circulating exosomal microRNAs as biomarkers of systemic lupus erythematosus. *Clinics (Sao Paulo).* 2020;75:e1528.

Luan X, Sansanaphongpricha K, Myers I, Chen H, Yuan H, Sun D. Engineering exosomes as refined biological nanoplatforms for drug delivery. *Acta Pharmacol Sin.* 2017;38(6):754–63.

Luo S, Ding S, Liao J, Zhang P, Liu Y, Zhao M, et al. Excessive miR-152-3p results in increased BAFF expression in SLE B-cells by inhibiting the KLF5 expression. *Front Immunol.* 2019;10:1127.

Manna I, De Benedittis S, Porro D. Extracellular vesicles in multiple sclerosis: Their significance in the development and possible applications as therapeutic agents and biomarkers. *Genes (Basel).* 2024;15(6).

Meng Q, Qiu B. Exosomal MicroRNA-320a derived from mesenchymal stem cells regulates rheumatoid arthritis fibroblast-like synoviocyte activation by suppressing CXCL9 expression. *Front Physiol.* 2020;11:441.

Minciacchi VR, You S, Spinelli C, Morley S, Zandian M, Aspuria PJ, et al. Large oncosomes contain distinct protein cargo and represent a separate functional class of tumor-derived extracellular vesicles. *Oncotarget.* 2015;6(13):11327–41.

Mohammadzadeh A, Lahouty M, Charkhian H, Ghafour AA, Moazzendizaji S, Rezaei J, et al. Human umbilical cord mesenchymal stem cell-derived exosomes alleviate the severity of experimental autoimmune encephalomyelitis and enhance LAG-3 expression on Foxp3+ CD4+ T cells. *Mol Biol Rep.* 2024;51(1):522.

Mushtaq M, Zineldeen DH, Mateen MA, Khawaja Husnain Haider. Mesenchymal stem cells' "garbage bags" at work: Treating radial nerve injury with mesenchymal stem cell-derived exosomes. *World J Stem Cells.* 2024; 16(5). https://dx.doi.org/10.4252/wjsc.v16.i5.0000.

Pan BT, Johnstone RM. Fate of the transferrin receptor during maturation of sheep reticulocytes in vitro: Selective externalization of the receptor. *Cell.* 1983;33(3):967–78.

Pisetsky DS. Pathogenesis of autoimmune disease. *Nat Rev Nephrol.* 2023;19(8):509–24.

Rajeev Kumar S, Sakthiswary R, Lokanathan Y. Potential therapeutic application and mechanism of action of stem cell-derived extracellular vesicles (EVs) in systemic lupus erythematosus (SLE). *Int J Mol Sci.* 2024;25(4).

Reis M, Mavin E, Nicholson L, Green K, Dickinson AM, Wang XN. Mesenchymal stromal cell-derived extracellular vesicles attenuate dendritic cell maturation and function. *Front Immunol.* 2018;9:2538.

Rhys HI, Dell'Accio F, Pitzalis C, Moore A, Norling LV, Perretti M. Neutrophil microvesicles from healthy control and rheumatoid arthritis patients prevent the inflammatory activation of macrophages. *EBioMedicine.* 2018;29:60–9.

Rozier P, Maumus M, Maria ATJ, Toupet K, Jorgensen C, Guilpain P, et al. Extracellular vesicles improve lung fibrosis from ifngamma-primed mesenchymal stromal cells in murine systemic sclerosis. *Cells.* 2021;10(10).

Sabry D, Marzouk S, Zakaria R, Ibrahim HA, Samir M. The effect of exosomes derived from mesenchymal stem cells in the treatment of induced type 1 diabetes mellitus in rats. *Biotechnol Lett.* 2020;42(8):1597–610.

Salvi V, Gianello V, Busatto S, Bergese P, Andreoli L, D'Oro U, et al. Exosome-delivered microRNAs promote IFN-alpha secretion by human plasmacytoid DCs via TLR7. *JCI Insight*. 2018;3(10).

Shahir M, Mahmoud Hashemi S, Asadirad A, Varahram M, Kazempour-Dizaji M, Folkerts G, et al. Effect of mesenchymal stem cell-derived exosomes on the induction of mouse tolerogenic dendritic cells. *J Cell Physiol*. 2020;235(10):7043–55.

Shi Y, Wang Y, Li Q, Liu K, Hou J, Shao C, et al. Immunoregulatory mechanisms of mesenchymal stem and stromal cells in inflammatory diseases. *Nat Rev Nephrol*. 2018;14(8):493–507.

Shigemoto-Kuroda T, Oh JY, Kim DK, Jeong HJ, Park SY, Lee HJ, et al. MSC-derived extracellular vesicles attenuate immune responses in two autoimmune murine models: Type 1 diabetes and uveoretinitis. *Stem Cell Reports*. 2017;8(5):1214–25.

Song Y, Li J, Wu Y. Evolving understanding of autoimmune mechanisms and new therapeutic strategies of autoimmune disorders. *Signal Transduct Target Ther*. 2024;9(1):263.

Su X, Wang H, Li Q, Chen Z. Extracellular vesicles: A review of their therapeutic potentials, sources, biodistribution, and administration routes. *Int J Nanomedicine*. 2025;20:3175–99.

Sun W, Yan S, Yang C, Yang J, Wang H, Li C, et al. Mesenchymal stem cells-derived exosomes ameliorate lupus by inducing M2 macrophage polarization and regulatory T cell expansion in MRL/lpr mice. *Immunol Invest*. 2022;51(6):1785–803.

Testa U, Pelosi E, Castelli G, Labbaye C. miR-146 and miR-155: Two key modulators of immune response and tumor development. *Noncoding RNA*. 2017;3(3).

Tong L, Zhang S, Huang R, Yi H, Wang JW. Extracellular vesicles as a novel photosensitive drug delivery system for enhanced photodynamic therapy. *Front Bioeng Biotechnol*. 2022;10:1032318.

Trajkovic K, Hsu C, Chiantia S, Rajendran L, Wenzel D, Wieland F, et al. Ceramide triggers budding of exosome vesicles into multivesicular endosomes. *Science*. 2008;319(5867):1244–7.

Tricarico C, Clancy J, D'Souza-Schorey C. Biology and biogenesis of shed microvesicles. *Small GTPases*. 2017;8(4):220–32.

Tu J, Zheng N, Mao C, Liu S, Zhang H, Sun L. UC-BSCs exosomes regulate Th17/Treg balance in patients with systemic lupus erythematosus via miR-19b/KLF13. *Cells*. 2022;11(24).

Van Hoecke L, Lucci C, Vandenbroucke RE. Mesenchymal stromal cell extracellular vesicles as immune modulators and drug carriers in neurodegenerative disorders. *Trends Neurosci*. 2025;48(11):919–34.

Walker S, Busatto S, Pham A, Tian M, Suh A, Carson K, et al. Extracellular vesicle-based drug delivery systems for cancer treatment. *Theranostics*. 2019;9(26):8001–17.

Wang C, Wang X, Zhang D, Sun X, Wu Y, Wang J, et al. The macrophage polarization by miRNAs and its potential role in the treatment of tumors and inflammation (Review). *Oncol Rep*. 2023;50(4).

Wei S, Zhang Z, Yan L, Mo Y, Qiu X, Mi X, et al. miR-20a overexpression in adipose-derived mesenchymal stem cells promotes therapeutic efficacy in murine lupus nephritis by regulating autophagy. *Stem Cells Int*. 2021;2021:3746335.

Welsh JA, Goberdhan DCI, O'Driscoll L, Buzas EI, Blenkiron C, Bussolati B, et al. Minimal information for studies of extracellular vesicles (MISEV2023): From basic to advanced approaches. *J Extracell Vesicles*. 2024;13(2):e12404.

Xin D, Li T, Chu X, Ke H, Yu Z, Cao L, et al. Mesenchymal stromal cell-derived extracellular vesicles modulate microglia/macrophage polarization and protect the brain against hypoxia-ischemic injury in neonatal mice by targeting delivery of miR-21a-5p. *Acta Biomater*. 2020;113:597–613.

Xing Y, Li B, He J, Hua H. Labial gland mesenchymal stem cell derived exosomes-mediated miRNA-125b attenuates experimental Sjogren's syndrome by targeting PRDM1 and suppressing plasma cells. *Front Immunol*. 2022;13:871096.

Xu H, Chen W, Zheng F, Tang D, Liu D, Wang G, et al. Reconstruction and analysis of the aberrant lncRNA-miRNA-mRNA network in systemic lupus erythematosus. *Lupus*. 2020;29(4):398–406.

Yang C, Sun J, Tian Y, Li H, Zhang L, Yang J, et al. Immunomodulatory effect of MSCs and MSCs-derived extracellular vesicles in systemic lupus erythematosus. *Front Immunol*. 2021;12:714832.

Zhang M, Johnson-Stephenson TK, Wang W, Wang Y, Li J, Li L, et al. Mesenchymal stem cell-derived exosome-educated macrophages alleviate systemic lupus erythematosus by promoting efferocytosis and recruitment of IL-17(+) regulatory T cells. *Stem Cell Res Ther*. 2022;13(1):484.

Zhang X, Li Y, Zhou Z. Lipid nanoparticle-based delivery system: A competing place for mRNA vaccines. *ACS Omega*. 2024;9(6):6219–34.

Zheng W, Schurz M, Wiklander RJ, Gustafsson O, Gupta D, Slovak R, et al. Surface display of functional moieties on extracellular vesicles using lipid anchors. *J Control Release*. 2023;357:630–40.

18 Theragnostic Uses of Exosomes

Sobia Razzaq[1], *Uzma Nisar*[1], *Waqar Aman*[1], *Sarah Rehman*[1], *and Abdul Qayyum Khan*[1]

[1]School of Pharmacy, University of Management and Technology, Lahore, Pakistan

18.1 LIST OF ABBREVIATIONS

αMβ2	Alpha M beta 2 integrin
α-syn	α-synuclein
111In-oxine	Indium-oxine
AM-MSCs	Amniotic-derived MSCs
Anti-HER2	Anti-human epidermal growth factor receptor 2
APP	Amyloid precursor protein
Aβ	Amyloid beta
BACE1	Beta-site amyloid precursor protein cleaving enzyme 1
CD36	Cluster of differentiation 36
CD8	Cluster of differentiation 8
CD9-HuR	CD9-tagged HuR (Human antigen R) fusion protein
CKAP4	Cytoskeleton-associated protein 4
CRC	Colorectal carcinoma
CRISPR/Cas9	Clustered regularly interspaced short palindromic repeats/CRISPR-associated protein 9
CSF	Cerebrospinal fluid
Cyr61	Cysteine-rich protein 61
DARP	Designed ankyrin repeated proteins
DM	Diabetes mellitus
EGF	Epidermal growth factor
EGFR	Epidermal growth factor receptor
EMNV	Exosome-mimetic nanovesicles
EPR	Enhanced permeability and retention
EVs	Extracellular vehicles
GALA-exosomes	Galactosylated exosomes
GBM	Glioblastoma multiforme
gLuc	Gaussian luciferase
HSP	Heat shock proteins
IBD	Inflammatory bowel disease
LA	Lactadherin
LAMP 3	Lysosome-associated membrane glycoprotein 3
lncRNAs	Long non-coding RNAs
MDSCs	Myeloid-derived suppressor cells

DOI: 10.1201/9781003653387-18

ME 491	Melanoma-associated antigen ME491
MHC	Major histocompatibility complex
miR	MicroRNA
MLA1	Major late antigen 1
MRSA	Methicillin-resistant *Staphylococcus aureus*
MSC	Mesenchymal stem cells
NDs	Neurodegenerative diseases
NFTs	Neurofibrillary tangles
PDAC	Pancreatic ductal adenocarcinoma
PDCD4	Programmed cell death protein 4
PD-L1	Programmed death-ligand 1
PEG	Polyethylene glycol
PH20	Hyaluronidase PH20
PKH67	Green fluorescent dye for cell labeling
PLK-1	Polo-like kinase 1
PSMA	Prostate-specific membrane antigen
RAD51	DNA repair protein RAD51 homolog 1
RAG2	Recombination activating gene 2
RBC	Red blood cells
RECK	Reversion-inducing cysteine-rich protein with Kazal motifs
RVG	Rabies viral glycoproteins
RVG-Lamp-2b	Rabies viral protein-lysosome-associated membrane protein-2b
SARS-CoV-2	Severe acute respiratory syndrome Coronavirus 2
SC-Exos	Stem cell-derived exosomes
siRNA	Small interfering RNA
SIRPα	Signal regulatory protein alpha
SMEs	Surface-modified exosomes
TEM	Tetraspanin-enriched microderms
TEXs	Tumor-derived exosomes
TM4SF	Transmembrane 4 superfamily
TRCs	Tumor-reactive cytotoxic T cells
TrKB	Tropomyosin receptor kinase B
TSG 101	Tumor susceptibility gene 101
WBC	White blood cells

18.2 INTRODUCTION

18.2.1 Exosomes and Their Origin

Exosomes are tiny vesicles produced by various cell types, including T cells and dendritic cells, typically measuring between 50 and 150 nanometers in diameter. They are formed within cells during endocytosis and are disseminated by almost all cell types (Ahmed et al., 2023). Vascular substances, such as blood, tears, saliva, and urine, contain large amounts of these vesicles (Matsuzaka et al., 2022). Exosomes were first discovered in the immature red blood cells (RBCs) of sheep in 1983 (Golubovskaya et al., 2020). Exosomes transport functional proteins, lipids, and nucleic acids, which are vital for cell-to-cell communication. Exosomes are released by cells and were initially thought to be irrelevant, considered waste products of cell damage or by-products of normal cellular processes. However, recent discoveries have shown that these extracellular vesicles (EVs) are essential and functional in carrying complex cargo, such as proteins, lipids, and nucleic acids

(Simpson et al., 2009). These cargos can be delivered to target cells, even at a distance from their release, by potentially reprogramming them. Exosomes represent a novel form of intercellular communication that may be crucial for various cellular functions, including antigen presentation, signal transduction, and immune responses (Lin et al., 2020). Exosomes can transport a variety of proteins, DNA, RNA, and miRs obtained from the cell. These physiologically active molecules can be delivered to distant cells via exosomes secreted by the exosome-generating cell (Mushtaq et al., 2024). Proteins, lipid rafts, phosphatidylserine, ceramide, sphingolipids, and cholesterol are all in the lipid bilayer that makes up the exosomal membrane, as shown in Figure 18.1. Animal exosomes are microscopic vesicles and are secreted by both healthy and diseased cells. Examples of animal exosomes include fibroblasts, fat cells, immunological cells, and tumor cells. Among blood cells, exosomes have been investigated most frequently in reticulocytes, RBCs, white blood cells (WBCs), and platelets. Modified exosomes are popular among researchers because they are easier to use, more efficient, and better at targeting specific cells than naive exosomes (Zhao et al., 2025).

18.2.2 Biological Contribution of Exosomes

Exosomes are gaining attention for their role in developing effective treatments for various diseases. Their unique properties, including biocompatibility, minimal toxicity, and high specificity, make them superior delivery systems compared to traditional methods such as liposomes and polymeric materials. They can efficiently transport biomolecules, drugs, and therapeutic compounds. In treating bacterial infections, exosomes loaded with antibiotics have shown greater efficacy than free antibiotics. Exosomes transport genetic material and protect it from degradation. They prefer DNA-based therapies for delivering small interfering RNA (siRNA) and non-coding RNAs (miRs) due to their precise delivery, zero or low immunogenicity, and reduced toxicity. The delivery of the anticancer RNA let-7a miR has been enhanced by conjugating a specific peptide to exosomes to target EGFR-expressing breast cancer cells. Administering these exosomes to tumor-induced RAG2-/- mice led to a localized reduction in tumor growth (Ohno et al., 2013). Exosomes can also contribute to the production of antigens for target cells, making them effective in vaccine development. Their pro-angiogenic, anti-inflammatory, and immunomodulatory properties are being utilized to create SARS-CoV-2 vaccines (Cacciottolo et al., 2023).

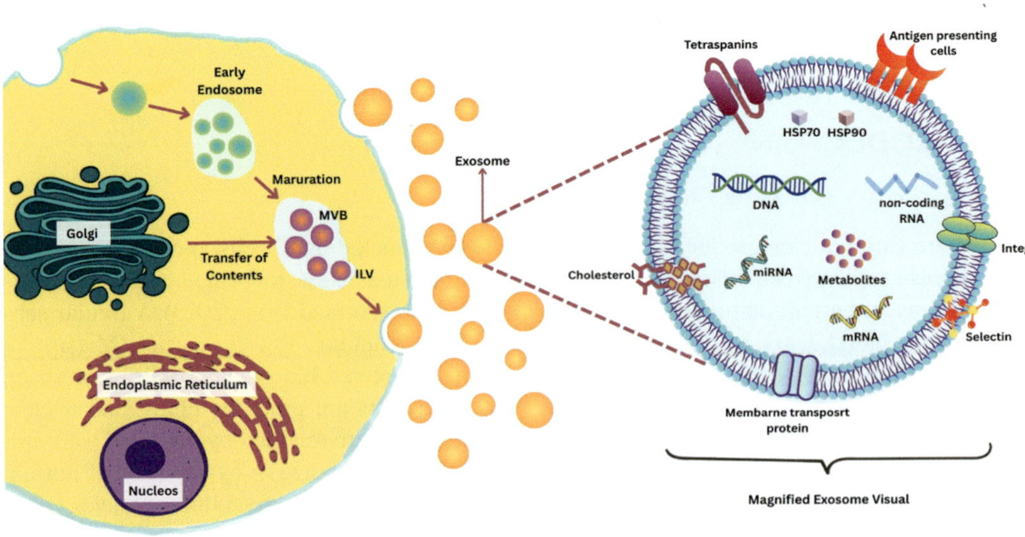

FIGURE 18.1 Illustration of exosome formation, showing endosomal membrane budding, MVB maturation, and exosome release with molecular details.

18.3 EXOSOMES IN MODERN THERAPEUTICS

Exosomes have emerged as a promising area of modern therapeutics due to their natural role in intercellular communication (Habib et al., 2026). Their ability to cross biological barriers, such as the blood-brain barrier, makes them particularly valuable for treating challenging conditions like neurodegenerative diseases and cancers (Mushtaq et al., 2024). Exosomes derived from stem cells or immune cells are also actively being studied for their potential in regeneration and immune modulation. This versatility makes exosomes an essential tool in personalized medicine and cell-free therapies (Sadeghi et al., 2023).

Recent research demonstrated that curcumin-encapsulated exosomes enhance curcumin's solubility, bioavailability, and stability in mice with Alzheimer's disease, outperforming free circulating curcumin. In mouse models, exosomes have been shown to improve cognitive function and exhibit strong anti-Alzheimer's effects (Ding et al., 2025). Moreover, when integrated into exosomes derived from mouse macrophages, linezolid, an antibiotic effective against methicillin-resistant Staphylococcus aureus (MRSA), significantly decreased infection rates compared with free linezolid (Wang et al., 2019). Significantly, this method did not adversely affect macrophage cytotoxicity. Exosomes deliver biomolecules, including proteins, nucleic acids, and growth factors, to targeted cells, thereby aiding in genetic modifications. Their unique properties are gaining attention in gene therapy and RNA interference.

Modified exosomes, such as GALA-exosomes, enhance the delivery of tumor cell antigens. For example, exosomes containing ovalbumin can stimulate adaptive immune responses without inducing immune cell anergy (Babaei et al., 2024). Exosomal delivery systems are groundbreaking systems designed to transport the powerful antioxidant catalase across the blood-brain barrier, opening up new possibilities for treating neurological disorders and enhancing brain health. This can be beneficial for treating neurodegenerative diseases (Liu, 2020). Additionally, PH20 hyaluronidase-loaded exosomes have been developed to improve tumor penetration (Almughem et al., 2020). Furthermore, loading the signal-regulating protein α (SIRPα) into exosomes is more effective in inhibiting tumor growth than using ferritin nanocages (Das et al., 2024). A poor delivery approach limits the clinical use of nucleic acid medications; however, exosomes offer a promising vehicle for therapeutic nucleic acids. For instance, exosomes from adipose stem cells delivered miR-122, contributing to chemoresistance in hepatocellular carcinoma (Sharma et al., 2024). Additionally, exosomes carrying siRNA targeting RAD51 reduced RAD51 expression and induced significant cell death in cervical and fibrosarcoma cells (Ruiz et al., 2018). Researchers have also developed tumor cell-derived vesicles loaded with chemotherapeutic drugs, which improved drug delivery to the nucleus and helped reverse treatment resistance in tumor-rejecting cells (TRCs). Furthermore, engineered exosomes containing miR-21 sponge constructs can downregulate miR-21 expression in glioma cells, upregulate target genes PDCD4 and RECK, and inhibit malignancy (Kar et al., 2023).

18.4 EXOSOMES AS GENE DELIVERY VEHICLE

Exosomes serve as effective carriers of genetic material for gene therapy, shielding nucleic acids from degradation. Small interfering RNA (siRNA) and non-coding RNAs, such as microRNAs (miRs), are favored over DNA-based therapies due to a reduced risk of mutations (Duan et al., 2021). However, these molecules are unstable in circulation and are vulnerable to RNase-mediated breakdown. Furthermore, they encounter obstacles in reaching their target cells, which may lead to being trapped and degraded in lysosomes. Due to their advantages for drug delivery, such as low toxicity and precise targeting, exosomes are regarded as ideal biocarriers for gene therapy (Lee et al., 2024).

Dendritic cell-derived exosomes in the targeted delivery of small interfering siRNAs, particularly in neurodegenerative diseases. This approach uses electroporation, which applies an electrical field to create temporary pores in the exosomal membrane. This enables efficient loading of siRNAs specifically designed to silence BACE1 expression, a key enzyme involved in the pathological processes of Alzheimer's disease (Liang et al., 2024). Once enriched with siRNA, these engineered exosomes are adept at delivering their therapeutic cargo to critical cell types within the brain, including neurons, microglia, and oligodendrocytes. By reducing BACE1 expression, the treatment aims to mitigate the formation of beta-amyloid plaques, a hallmark of Alzheimer's disease (Yusuf et al., 2025). This targeted delivery system not only enhances the bioavailability of siRNAs but also minimizes potential off-target effects, offering a novel strategy to address the complex challenges posed by Alzheimer's disease (Zhang et al., 2022).

Modified parent cells were developed to target a specific region of the epidermal growth factor receptor (EGFR), leveraging the elevated levels of epidermal growth factor (EGF) in cancer cells. These exosomes were then deployed therapeutically to deliver specific antitumoral microRNAs directly into tumor tissues, effectively targeting and killing malignant cells in vivo (Zhang et al., 2023). This innovative approach not only showcased the potential of bioengineering but also opened new frontiers in the fight against cancer. A recent study demonstrated the enhanced therapeutic effects of miR-675, a miR downregulated in senescent cells, aging muscles, and ischemic limbs (Sancho-Albero et al., 2020). Researchers found that incorporating miR-675 into exosomes retained within a silk fibroin hydrogel significantly improved its efficacy. The findings suggest that this encapsulation strategy not only enhances the anti-aging properties of miR-675 but also provides protective benefits against ischemic vascular dysfunction (Sancho-Albero et al., 2020).

Exosomes derived from stem cells are preferred for treating cardiovascular diseases as they protect their nucleic acid cargo from degradation in circulation and deliver medication effectively (Haider and Najimi, 2022). Recent studies have shown that the tumor suppressor miR-199a-3p is present in exosomes from ovarian cancer patients, indicating its potential to inhibit cancer cell growth and invasion (Lian et al., 2020). The gene-editing tool CRISPR/Cas9 has garnered significant attention, resulting in the development of new treatments (Bolideei et al., 2025). Scientists engineered exosomes to deliver CRISPR/Cas9 components by conjugating Cas9 with green fluorescent protein and the exosome membrane protein CD36, resulting in effective targeted gene delivery. Additionally, exosomes are ideal bio-vectors for treating familial hypercholesterolemia, as they help reduce liver lipid accumulation and LDL cholesterol synthesis (Gun Choi et al., 2025).

18.5 EXOSOMES IN IMMUNOTHERAPY

Exosomes influence inflammatory responses by regulating cytokine signaling; tumor-derived exosomes (TEX) release immunosuppressive factors that inhibit immune cell function. TEX can inhibit immune cell proliferation, induce apoptosis in activated CD8+ T cells, suppress NK cell function, disrupt monocyte differentiation, and promote the expansion of regulatory T cells and myeloid-derived suppressor cells (MDSCs). They also play a role in drug resistance and can affect tumor sensitivity to immunotherapy, presenting new treatment opportunities. TEX impairs T cell signaling and proliferation while promoting apoptosis in CD8+ T cells. In contrast, exosomes from immune cells stimulate T cell proliferation. Research shows that tumors can shift cytotoxic CD8+ T cells toward a more suppressive phenotype, characterized by loss of CD27/28 expression and triggered by TEX. These changes have been observed in both in vitro assays and in CD8+ T cells from head and neck cancer patients, with exosomal RNA inducing similar phenotypic changes (Olejarz et al., 2020).

Research into mesenchymal stromal/stem cell (MSC)-derived exosomes has investigated their potential as treatments for various diseases, including cardiovascular disease, graft-versus-host disease, and inflammatory bowel disease. Metastatic melanoma produces high levels of EVs, predominantly exosomes, which display PD-L1 on their surface (Cordonnier et al., 2020). Exosomal PD-L1 can directly bind to and inhibit T cell function like a drone. Additionally, exosomes consist of natural lipid bilayers and are rich in adhesion proteins, facilitating easy interaction with cell membranes, which makes them excellent candidates for drug delivery. Consequently, the potential clinical application of exosomes as biomarkers for disease diagnosis and as drug delivery carriers has attracted considerable attention in recent years. For instance, exosomal miRs have emerged as crucial regulators in the development of diabetes mellitus (DM), influencing the expression of genes related to pancreatic β-cell damage and insulin resistance. This suggests that DM may be treated by silencing or activating exosomal miRs in patients (Muskan et al., 2024).

Exosomes derived from dendritic cells (DC) can trigger anti-tumor immunity by activating T cells and NK cells, demonstrating promising results in clinical trials. Their effectiveness increases when DCs mature with Toll-like receptor agonists—in mouse models of hepatocellular carcinoma, exosomes from antigen-expressing DCs elicited robust immune responses, leading to extended survival. Furthermore, DCs travel to lymphoid tissues to activate alloreactive T cells in transplant settings, and donor DC exosomes have been associated with acute rejection in murine heart transplant studies. On the other hand, tumor cells produce exosomes that may inhibit immune responses, suggesting a potential use as customized cancer vaccines. Although exosomes from tumors generally promote immunosuppression, they can be specially modified to bolster immunity, triggering anti-tumor effects through CD8+ T cells and enhancing NK cell activity. However, these responses can be weak and may lead to tolerance. To improve their immunogenicity, strategies such as genetic modification and direct antigen fusion are being explored (Lu et al., 2024).

18.6 STEM CELL EXOSOMES IN REGENERATIVE MEDICINE

Released by various cell types, including stem cells, their composition is generally conserved but can vary depending on the origin and microenvironmental conditions. Stem cell-derived exosomes (SC-Exos) present a promising cell-free therapeutic option, addressing the challenges of direct stem cell delivery and minimizing tumorigenicity (Liang et al., 2025). Their ease of manipulation, storage, and availability makes them an attractive strategy, as SC-Exos transfer the molecular characteristics of their parent cells, influencing recipient cells like the original stem cells. Specific databases that play integral roles in exosome research are listed in Table 18.1.

18.6.1 Theragnostic Potential of Exosomes

Exosomes, which serve as theragnostic candidates, represent a promising frontier in personalized medicine, as they combine therapeutic and diagnostic capabilities. Their unique biological properties enable them to serve as both therapeutic agents and diagnostic biomarkers, particularly in the treatment of cancer, neurodegenerative diseases, and infectious diseases. Exosomes have shown great promise in the early detection and treatment of several diseases, including cancer, cardiovascular conditions, and neurological disorders, due to their ability to transport bioactive substances, such as nucleic acids, proteins, and lipids.

18.6.1.1 Cancer

Cancer remains one of the leading causes of disease and mortality worldwide. Exosomes are being utilized as biomarkers for cancer diagnosis (Sajid et al., 2025). The number of exosomes generated by cancer patients is higher than that of healthy individuals, and this number increases as the disease progresses. Tumors may disseminate more widely as they carry proteins and miRs, facilitating tumor growth (Aramini et al., 2020). Exosomes are beneficial for both cancer diagnosis and the treatment of numerous pathologies (Suh et al., 2017; Aramini et al., 2021). Anticancer medications, such as doxorubicin and paclitaxel, and cancer gene suppressors, including miR-134, miR-143, c-Myc siRNA, and PLK-1 siRNA, have been loaded into exosomes for cancer treatment (Wang et al., 2017).

18.6.1.2 Neurodegenerative Diseases

Exosomes are increasingly being associated with Alzheimer's disease (AD) (Lamiaa et al., 2025). Specific proteins, such as tau, Aβ, and APP, are carried by exosomes and may help disseminate

TABLE 18.1
Currently Available Databases for Exosome Research

Database	Description	Focus Areas	Website/Link	References
ExoCarta	A curated database of exosomal proteins, RNAs, and lipids.	Exosome contents: proteins, mRNA, miR, lipids.	http://www.exocarta.org	(Gummadi et al., 2024)
Vesiclepedia	A community-annotated database of EVs.	EVs from various cell types, not limited to exosomes.	http://www.microvesicles.org	(Hu and Huang, 2024)
EV-TRACK	A platform promoting transparency and standardization in EV research.	Experimental methods, standardization.	http://evtrack.org	(Consortium et al., 2017)
ExoRBase 2.0	Updated version with expanded RNA annotations from exosomes.	Cardiovascular, cancer research.	https://www.exorbase2.org	(Hu and Huang, 2024)
miRBase	Repository for microRNA sequences and annotations.	miR analysis, including exosomal miR.	https://www.mirbase.org	(Chen et al., 2021)
HMDB (Human Metabolome DB)	Database of small molecule metabolites found in human tissues and biofluids.	For metabolomic studies of exosome cargo.	https://hmdb.ca	(Wishart et al., 2022)
STRING	Protein–protein interaction database.	Interaction networks for exosomal proteins.	https://string-db.org	(Szklarczyk et al., 2025)
BioGRID	Database of protein and genetic interactions.	Functional annotation of exosome proteins.	https://thebiogrid.org	(Oughtred et al., 2019)
GeneCards	Comprehensive gene-centric database.	Exosomal gene and protein function search.	https://www.genecards.org	(Veneruso et al., 2023)

these proteins across cells, thereby accelerating the development of AD (D'Anca et al., 2019). In Alzheimer's disease, the endosomal-lysosomal system malfunctions, especially in those having one or two copies of the ApoE4 gene. As a result of this malfunction, exosome production rises (Caselli et al., 2019).

In the early stages of Alzheimer's disease, Aβ peptides are generated in a region of the neuron called the multivesicular bodies (MVBs). Studies reveal that exosomes are involved in the progression of AD, facilitating the spread of Aβ deposits and aiding early diagnosis and treatment (Yusuf and Haider, 2025). It has been proven that exosomes containing tau protein from CSF are very effective markers for early detection of AD. Exosomes play a crucial role in Parkinson's disease by facilitating the intercellular transmission of misfolded α-synuclein (α-syn) proteins, which behave like prions. Exosome machinery facilitates this distribution in PD (Yu et al., 2020).

18.6.1.3 Cardiovascular Diseases

Many of the CVDs, including ischemic heart disease, heart failure, hypertension, preeclampsia, and atherosclerosis, are associated with exosomal miR profile changes (Haider and Aramini, 2020). Exosomal cargo varies with disease severity and has been shown to carry elements that may serve as prognostic and diagnostic indicators for CVDs across various clinical contexts (Weber et al., 2020). Pigs with cardiac ischemia-reperfusion injury showed a considerable rise in exosome levels and circulating miRs, especially miR-133b, miR-208b, and miR-499 (Deddens et al., 2016). Therefore, these miR-containing exosomes can be utilized as biomarkers to identify and predict CVD (Habib et al., 2025). Exosomes containing isoforms of glycogen phosphate were detected in doxorubicin-treated cardiomyocytes, suggesting they mark the initial phases of cardiac injury (Yarana et al., 2018). Many cardiovascular conditions, including aneurysms, pulmonary arterial hypertension, chronic heart failure, and kidney disease, are protected by these exosomes by the transfer of miRs, which are critical regulators of various cellular activities, such as miR-144-5p, miR-1246, miR-181-5p, and miR-324-5p (Neves et al., 2023; Al-Omar et al., 2022).

18.6.2 Exosomes as Diagnostic Beacons

Exosomes represent a valuable biomedical tool due to their release into easily accessible body fluids, such as blood plasma, breast milk, saliva, serum, cerebrospinal fluid, and urine, enabling their utilization as potential biomarkers. The molecular content of exosomes, comprising DNA, RNA, lipids, and proteins, reflects the distinctive features of the cells that release them. Hence, they serve as a fingerprint in determining that cell's physiological or pathological state and as an essential diagnostic tool (Properzi et al., 2013). Therefore, these biological cargos are crucial targets for both diagnostic and therapeutic roles (Jayaseelan, 2020).

18.6.2.1 Tumor-Derived Exosome Markers

Due to their easy accessibility in biological fluids, exosomes serve as non-invasive biomedical tools for multiple types of liquid biopsies, including serum, plasma, saliva, and urine, thereby facilitating the diagnosis of cancer and other infectious diseases. Exosomes derived from tumor cells carry cancer-specific markers that help in the early detection of tumor cells. Moreover, they also give clues about the aggressiveness or complexity of the disease (McKiernan et al., 2016). Most of the FDA-approved cancer biomarkers are glycosylated proteins. CD63 and caveolin-1 represent non-invasive exosome markers for cancer detection (Zhao et al., 2019). CD 63, also known as LAMP 3, MLA1, or ME 491, belongs to the Transmembrane 4 superfamily (TM4SF). It is a glycoprotein that is found on cell membranes. TM4SF contains various proteins, such as CD9 (Sohn et al., 2015), which are involved in differentiation and motility. It exists in late endosomes and lysosome-related organelles. Late endosomes are membrane-bound compartments within the cell that facilitate the breakdown of substances. CD63 moves between these compartments and contributes to their diagnostic role by facilitating trafficking within the cell (Kwon et al., 2007). The level of CD63-positive exosomes isolated from tumor cells provides clues to the aggressiveness, age, and stage of cancer cells. Therefore, CD63 serves as an essential biomarker not only for diagnosis but also for evaluating progression and treatment response of various types of cancers, including lung, breast, and ovarian cancers (Nakamura et al., 2025).

Long non-coding RNAs (lncRNAs) and miR (miR-21) are found in cancer-derived exosomes, and they help predict distant metastasis and lymph node involvement. LncRNAs, either alone or in combination with miR-21, are specific in predicting tumor progression and staging (Peinado et al., 2012). Cytoskeleton-Associated Protein 4 (CKAP4) is a tumor-derived exosome (TEX), a type II transmembrane protein released by Pancreatic ductal adenocarcinoma (PDAC). CKAP4 isolated from a patient's blood sample can serve as a biomarker to identify PDAC stages and guide targeted treatment. Likewise, Tropomyosin receptor kinase B (TrKB) is used as a non-invasive biomarker for the diagnosis of glioblastoma. MicroRNAs (miR-21 and miR-155) from blood samples of patients with metastatic malignancies significantly predict prognosis and treatment outcomes (Sohn et al., 2015). The enhanced expression of tumor susceptibility gene 101 (TSG101) in colorectal carcinoma (CRC) makes it a promising biomarker for evaluating the aggressiveness, progression, and metastasis of CRC (Gheytanchi et al., 2021).

18.6.2.2 Biomarker for Neurodegenerative Diseases

Neurodegenerative diseases (NDs) pose a significant burden on healthcare systems worldwide, representing a major cause of mortality and lifelong disability. The most prevalent are Alzheimer's disease (AD) and Parkinson's disease (PD). The diseases are challenging as they cause progressive structural and functional loss of neurons, leading to motor dysfunction and cognitive decline, thereby compromising the quality of life of patients. The increasing burden of NDs demands rapid diagnosis and preventive strategies. Owing to the exosome's property of cell-cell communication and imperative role in the nerve regeneration process and synaptic plasticity, they serve as promising biomarkers for the diagnosis of NDs (Rastogi et al., 2021). Dementia is a broader term that represents loss of memory and cognitive dysfunction, and two-thirds of dementia patients develop AD. Two factors characterize the pathology of AD—first, the formation of extracellular plaques (senile plaques) containing amyloid beta (Aβ). Aβ is responsible for neuronal damage and death. Aβ derived from CSF exosomes is a promising diagnostic biomarker for AD (Bermejo-Pareja et al., 2020). The second factor is the formation of intracellular neurofibrillary tangles (NFTs). They cause malfunctioning of neuronal cells by disrupting the cell's internal transport system. Misfolded and hyperphosphorylated Tau proteins form NFTs. Hyperphosphorylation of Tau leads to the loss of the ability to stabilize microtubules. The diagnosis of AD is facilitated by the estimation of phosphorylated Tau and total Tau using exosomes as non-invasive tools (Soares Martins et al., 2021). The second most prevalent neurodegenerative disorder is Parkinson's disease (PD). Exosomes derived from CSF, saliva, blood, and other body fluids are used to diagnose PD. PD is characterized by mitochondrial dysfunction, which in turn is caused by oxidative stress, leading to the buildup of protein aggregates (amyloid β, huntingtin, tau, and α-synuclein). Altered levels of these exosomal proteins indicate the development of PD (Picca et al., 2019).

18.6.2.3 Biomarkers for Cardiovascular Disease

The heart is composed of various cell types, including vascular, neural, fibroblastic, and cardiomyocytic cells. Exosomes play a pivotal role in both normal physiology and the prognosis of cardiac diseases. Exosomal miRs offer greater specificity than conventional diagnostic biomarkers (Zamani et al., 2019). Specific molecular patterns characterize cardiovascular diseases, and these are well represented by exosomal cargo. To enhance test sensitivity and specificity, exosome contents or surface molecules are employed for the diagnosis of acute myocardial ischemia (AMI), coronary artery disease (CAD), arrhythmia, and congestive heart failure (CHF) (Chen and Luo, 2022). Serum exosomal miR-122-5p serves as an independent biomarker for the diagnosis of acute coronary syndrome (ACS). Likewise, plasma cysteine-rich protein 61 (Cyr61) has been used as an exosomal biomarker for accurate diagnosis of various cardiac diseases, including ACS (Li et al., 2021b).

18.6.2.4 Exosomes in Liquid Biopsy

Sampling and molecular analysis of biofluids derived from various diseased cells is referred to as a liquid biopsy. The exponential evolution of the liquid biopsy technique offers a more specific, non-invasive diagnostic procedure, thereby replacing conventional solid biopsies. It represents a more precise medical intervention for the diagnosis and prognosis of various life-threatening and debilitating medical conditions. Among multiple sources of liquid biopsies, exosomes offer significant superiority. This is because of their presence in almost all body fluids and provides higher stability due to the encapsulated lipid layer. This makes them clinically more acceptable, as high stability facilitates their handling and transportation and allows them to be stored for a more extended period (Cai et al., 2015). Moreover, exosomes are secreted by living cells; therefore, they contain significant information from the parent cell. Despite these factors, the use of exosomes poses challenges due to their intrinsic heterogeneity and nanoscale size. This is crucial for their isolation with high purity (Li et al., 2017). Additionally, cancer exosomes represent only a small fraction of the total exosome population in the body. Therefore, exosome-based cancer diagnostics require specific and ultrasensitive detection.

18.7 ENGINEERING EXOSOMES FOR THERAGNOSTIC

Exosomes are cell-derived extracellular nanovesicles that facilitate the transportation of numerous small molecule-based therapeutics, mRNA, and nucleic acids into the extracellular matrix. Exosomes are unique in that they are endogenously derived and lack immune responsiveness, thereby bypassing macrophage and neutrophil clearance. The avoidance of uptake by the reticuloendothelial system enables exosomes to be suitable for passive targeting, as they exhibit an enhanced permeability and retention (EPR) effect. Exosomes demonstrated physicochemical stability and a well-tolerated environment, promoting therapeutic efficacy and targeted drug delivery (Zeng et al., 2024). However, the application of exosomes in site-specific drug delivery necessitates modification of their surface by attaching a targeting ligand for therapeutic and diagnostic purposes, thereby determining their biodistribution. Engineered exosomes for theragnostic applications have gained significant importance in biomedical research (He et al., 2022). Numerous engineering strategies have been developed to modify exosomes, enhancing their ability to deliver drugs, diagnose diseases, and facilitate tissue regeneration.

18.7.1 Functionalization with Targeted Ligands

Exosomal membrane modification leads to the binding of various ligands. Proteomic studies have demonstrated that exosomal membranes contain multiple proteins, including actin, cofilin, tubulin, and annexins, which are involved in intracellular trafficking. Moreover, heat shock proteins (HSPs), such as HSP70 and HSP90, play a crucial role in the elimination of misfolded proteins. Exosomal surface membrane possesses major histocompatibility complex (MHC) class II, MHC class I, CD86, integrins ($\alpha M\beta 2$), tetraspanins, and lactadherin (LA). MHCs are involved in stimulating T cells, and integrins/tetraspanins are the source of adhesion for ligands (Salunkhe et al., 2020). Similarly, tetraspanin family components (CD37, CD53, CD63, CD81, and CD82) can interact with various protein factors to form tetraspanin-enriched microdomains (TEM). TEM is multifunctional, enhancing cellular adhesion, proliferation, and locomotion, while maintaining membrane integrity and generating binding sites for various ligands (Salunkhe et al., 2020). As far as bioluminescence labeling is concerned, Gaussia luciferase (gLuc) can be attached to exosomal protein lactadherin (LA), which allows the ligand to be expressed on the exosomal surface. LA has three domains at the N-terminus, including the secretion domain, the epidermal growth factor (EGF) domain, and a C-terminal lectin domain (Morishita et al., 2017). Imai et al. substituted the EGF-like domain with gLuc (gLuc-LA) to prevent reticuloendothelial uptake. Radiolabeling approaches include the fusion of LA with radiolabeled iodine (^{125}I) for studying the distribution pattern of exosomes (Takahashi et al., 2013). Smyth and coworkers incorporated indium-oxine (111In-oxine) directly into the exosomal membrane using a hydrophobic interaction (Smyth et al., 2015). Genetic engineering strategies use plasmid vectors encoding Lamp-2b fused peptides to achieve targeted modifications. Moreover, GPI-anchored proteins, such as decay-accelerating factor (CD55), enable the attachment of anti-EGFR nanobodies, facilitating the targeting of A431 cells (an epidermoid carcinoma cell line).

Various studies have examined the use of exosome sources in combination with ligands. Exosomes were separated from mouse fibroblasts, and LDL peptide was used as a therapeutic ligand to treat glioblastoma multiforme (GBM). The glioma volume was markedly reduced (Ye et al., 2018). Human embryonic stem cells were used to produce exosomes with a targeting moiety of PEG 2000, resulting in improved therapeutic activity for treating GBM. Exosomes derived from murine dendritic cells were conjugated with the ligand Rabies viral protein-Lysosome-associated membrane protein-2b (RVG-Lamp-2b) and used for Parkinson's disease. The results concluded a reduction in the expression of disease-overexpressed enzymes. Exosomes, when linked to an integrin ligand peptide, can prevent cerebral ischemia through targeted therapy and increased uptake (Ye et al., 2018).

Bone marrow-transfected exosomes interacting with RVG-Lamp-2b facilitated neurogenesis at the injured site after ischemic stroke, thereby enhancing the neurogenesis process (Guo et al., 2021). However, ankyrin protein ligands cross-linking with exosomes for breast cancer treatment led to increased uptake of modified exosomes (Zhang et al., 2023). The interaction of an anti-HER2 antibody with exosomes for breast cancer treatment reduced breast tumor volume (Wang et al., 2018). Designed Ankyrin Repeated Proteins (DARP) cross-linking with exosomes facilitated increased uptake in breast cancer cells (Limoni et al., 2019). The prostate-specific membrane antigen (PSMA) RNA aptamer (A9g) ligand, upon binding to the exosomal membrane, facilitated the knockdown of survivin (an apoptosis-inhibiting protein), thereby increasing the likelihood of survival in prostate cancer patients (Pi et al., 2018), as illustrated in Figure 18.2. Theragnostic exosomes were labeled with the fluorescent dye PKH67 to investigate the effect of temperature on exosome uptake in the peritoneal cavity. The results showed that fluorescence increased with increasing temperature

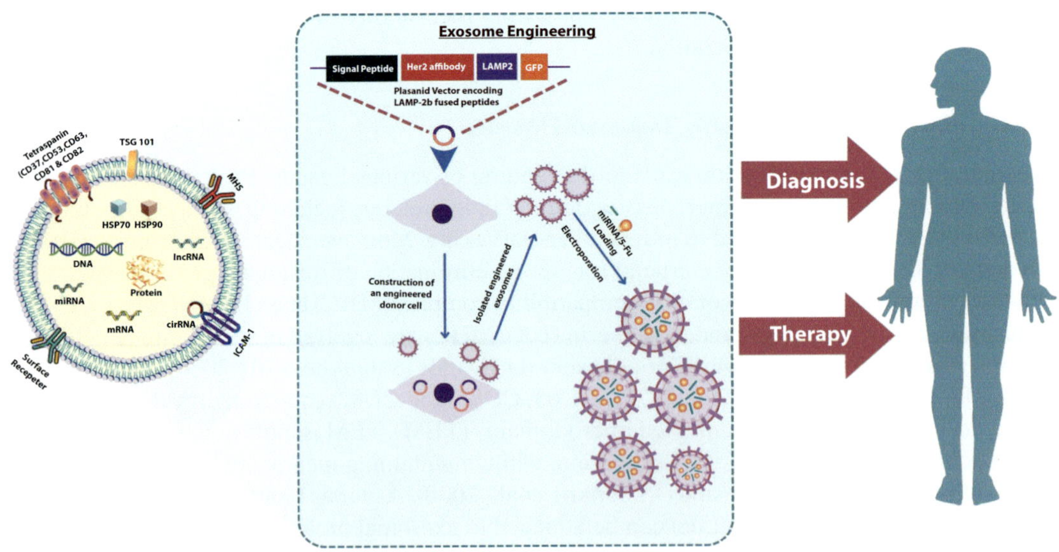

FIGURE 18.2 Engineering of exosomes and their theragnostic application.

(Matsumoto et al., 2017). PKH26 fluorescent dye-based ligand facilitates the in vitro internalization of labeled exosomes into normal urothelial cells (Greco et al., 2016).

18.7.2 Enhanced Theragnostic Specificity

Engineering exosomes for theragnostic applications has become increasingly important in biomedical research. Numerous engineering strategies have been developed to modify exosomes, enhancing their ability to deliver drugs, diagnose diseases, and even regenerate tissues. Some key strategies include genetic engineering, chemical modification, optimized delivery systems, and tumor-targeting exosomes.

18.7.2.1 Genetic Modification

Genetic engineering of exosomes via plasmid vectors is one of the most widely used approaches for producing surface-modified exosomes (SMEs). Many studies have been conducted in this regard. Kim and coworkers co-transfected HEK293 cells with a cardiac-targeting peptide (CTP) to generate cardiac-targeting exosomes (Kim et al., 2018). Erviti and coworkers developed rabies virus glycoprotein (RVG)-mediated exosomes through genetic modifications (Alvarez-Erviti et al., 2011). Zhelong Li et al. developed novel CD9-HuR functionalized exosomes via genetic engineering to enhance exosome drug-loading efficiency (Li et al., 2019). Peptides used in the transfection process during genetic engineering were non-immunogenic and stable. However, achieving the desired yield of peptide-functionalized exosomes is challenging due to proteolytic degradation of the peptide. Leonard and coworkers employed a genetic engineering approach to conjugate a targeting peptide with a glycosylation sequence, thereby preventing peptide degradation (Hung et al., 2015). However, genetic modification in exosome-producing cells is thought-provoking. The targeting moiety presented in the exosome membrane may affect the normal functions of exosome membrane proteins.

18.7.2.1.1 Covalent Modifications

Covalent modification of the exosomal surface can be achieved through click chemistry: an alkyne is attached to the isolated exosome via carbodiimide chemistry, followed by a condensation reaction and conjugation of an azido group to the targeting moiety. Click chemistry is an advanced technique that offers advantages over conventional chemical reactions, which often violate temperature and pressure constraints in osmotic environments due to fluctuating salt concentrations. Osmotic pressure fluctuations can damage or aggregate exosomes. However, click chemistry reactions are efficacious, as they can be carried out in both organic solvents and aqueous buffers for a limited duration. Moreover, the conjugation reaction does not affect exosome size or their uptake. Thereby, click chemistry is suitable for exosome surface modification (Salunkhe et al., 2020).

18.7.2.1.2 Non-Covalent Modification

Non-covalent modification methods for exosomes include multivalent electrostatic interactions, which involve coating the exosomal membrane with a cationic charge, thereby increasing its interaction and muco-penetration with biological membranes. Another approach to non-covalent modification is ligand-receptor interaction, in which ligands bind to receptors that are overexpressed. Moreover, the hydrophobic interaction between lipid nanoparticles and exosomes is an interesting approach: lipid nanoparticles are conjugated to a peptide, and exosomal fusion is achieved using the freeze-thaw method. All these non-covalent methods are effective in targeting without affecting the native functions of exosome membrane proteins (Salunkhe et al., 2020).

18.8 ADVANCING EXOSOMES FOR PRECISION MEDICINE

The field of exosome research is entering a transformative phase, driven by innovative technologies that enhance the isolation, characterization, and functional application of exosomes. Two key advancements—integrating artificial intelligence (AI) and developing exosome-mimetic systems—are driving cost-effective production and enhanced precision in medicine (Kang et al., 2024). The use of machine learning algorithms is becoming a significant breakthrough, particularly for detecting patterns in exosomal RNA, DNA, and protein content. This progress enables earlier detection of various diseases (Kalluri et al., 2020) and enhances clinical translation efforts while redefining strategies for detecting, monitoring, and treating complex diseases. By merging biological intricacies with engineering accuracy, exosomes and their synthetic alternatives are poised to become central to next-generation therapeutics. The application of AI in analyzing exosome content and nano-genomics ensures higher precision and reproducibility, and speeds up processes beyond the capabilities of traditional methods. This is promising for the future of medical science and patient care.

Personalized medicine seeks to tailor medical treatment to each patient's unique characteristics. Given their ability to transport patient-specific molecular cargo, exosomes play a crucial role in these strategies. Autologous exosomes can be engineered to deliver therapeutic agents, including CRISPR/Cas9, siRNA, miRNA, chemotherapeutics, and drugs targeting disease-specific pathways, which helps minimize immune rejection. For example, in oncology, tumor-derived exosomes may be modified to express anti-tumor agents and then reintroduced to the patient for personalized therapy (Patel et al., 2025). Exosomes reflect the physiological state of their originating cells, carrying distinct molecular signatures that are optimal for non-invasive diagnostics. They could facilitate liquid biopsies using circulating exosomes, enabling early disease detection, real-time monitoring, and prognosis prediction (Andre et al., 2024). MSC-derived exosomes encourage angiogenesis, regulate inflammation, and speed up tissue regeneration. These features position them as promising tools for autologous, cell-free therapies in personalized medicine (Zhang et al., 2025). The combination of exosomes and nanoparticles merges the biological specificity of exosomes with the customizable properties of synthetic particles. Well-established examples, such as gold or magnetic nanoparticles bonded with exosomes, enhance imaging, enable controlled drug release, and provide magnetic guidance (Vader, Mol et al., 2016). These hybrid systems pave the way for theragnostic approaches, enabling simultaneous diagnosis and therapy, improving efficiency, and reducing systemic toxicity.

18.8.1 Predictive Modeling and Treatment Optimization

AI platforms can model EV biodistribution and optimize dosage strategies for precision-targeted therapies, thereby enhancing the efficacy of these treatments. Predictive models enhance optimal patient dosing, identify drug-drug/disease interactions, support clinical decision-making, and inform treatment protocols. High-throughput omics datasets are now being more accurately analyzed using bioinformatics pipelines to identify disease-specific exosome signatures. These profiles are crucial for biomarker discovery and tracking therapeutic response (Amini et al., 2023). Bioinformatics has revolutionized exosome research by leveraging large databases (ExoCarta, Vesiclepedia, and EVpedia) for large-scale data mining, enabling effective exosome classification, disease progression prediction, and the development of personalized treatment approaches. It has enabled early disease diagnosis and effective treatments based on exosomal biomarkers. Overall, improvements can be anticipated in clinical decision-making and personalized medicine. Moreover, exosome-mimetic systems can also deliver chemotherapeutics (such as doxorubicin-loaded mimetics), siRNAs, miRs, proteins, and CRISPR components. Ongoing studies are validating their in vivo biodistribution, stability, and immunogenicity, paving the way for their use in complex neurodegenerative/CVS diseases (Patel et al., 2023).

18.9 CLINICAL EVIDENCE OF EXOSOMES

Exosomes deliver active biomolecules to target cells, play crucial roles in various physiological and pathological processes, and hold great therapeutic potential for disease.

In clinical trials, exosomes demonstrate potential as a platform for treating a wide range of diseases (Liu et al., 2019). Therapeutic approaches that utilize EVs, such as exosomes, have experienced rapid growth in recent years. Numerous clinical studies using exosomes were focused on treatments for skin infections, alopecia, and wound healing (Tienda-Vázquez et al., 2023). A new treatment for mild to moderate Alzheimer's disease was evaluated in a clinical trial involving exosomes derived from human fat-derived stem cells (Sharma et al., 2021). Exosomes were administered twice weekly via the nose for 12 weeks at three different doses. The treatment was considered safe, with no significant symptoms reported. Participants in the medium-dose group demonstrated enhanced memory and improved cognitive function. Although the treatment did not affect specific disease markers, the findings suggest that this method is promising for therapeutic efficacy. Exosomes derived from placental stem cells were also utilized to treat complex perianal fistulas that had not responded to conventional therapies. Patients with fistulas lasting more than a year received treatments during surgery and were monitored for six months (Xie et al., 2023). Most patients improved; five achieved complete healing, eight were discharged, and only one reported no improvement. No allergic reactions were noted, indicating that this treatment is effective and could provide a novel and efficient strategy for managing complex issues (Xie et al., 2023).

Currently, no effective antiviral drugs or vaccines are commercially available to treat coronavirus disease (COVID-19) (Al-Rasheedi et al., 2021; Hyder and Haider, 2020). The majority of COVID-19-related deaths are caused by respiratory failure, which is triggered by severe and uncontrolled inflammation, thus underscoring the critical need for effective treatment strategies. MSCs and their exosomes have been shown to possess anti-inflammatory properties (Safwan et al., 2025). Exosomes serve as natural carriers of bioactive molecules between cells, and clinical trials have explored their potential for COVID-19 therapy. The secretions of MSCs, as well as exosomes engineered with specific micro RNAs (miRs) or mRNAs, have shown promise as nanomedicines (Rezakhani et al., 2021).

Exosome technology in clinical applications has become a hallmark in the field of research over the last few years. The effective use of exosomes as diagnostic biomarkers and therapeutic carriers is garnering significant interest and financial investment from the pharmaceutical industry (Perocheau et al., 2021). In addition, several clinical trials have been conducted over the past 5 years in targeted delivery for cancer, neurodegeneration, inflammation, and immunology (Rezaie et al., 2022). According to the International Society of Extracellular Vesicles (ISEV), it is essential to characterize the exosomes based on their function and composition. Exosomes possess reduced immunogenicity compared to liposomes and other vectors for transporting therapeutic moieties and crossing the blood-brain barrier (Lee et al., 2024). The non-immunogenicity of this feature makes it an innovative and attractive option as both a biomarker and a therapeutic agent. Biomarker research on exosomes is based on the fact that they are released in numerous pathophysiological conditions, such as inflammation, neurodegeneration, and angiogenesis (Chen et al., 2019). Exosomes released in the diseases mentioned above contain transmembrane proteins that can be used for clinical diagnosis and staging. Tetraspanins present on the exosomal surface are directly involved in disease screening, as they correlate with disease progression and serve as diagnostic biomarkers (Sun et al., 2014). Viral hepatitis C causes an increase in CD81 levels, which then correlate with fibrosis progression to serve as a biomarker. Proteomics analysis of exosomes, which are secreted under both physiological and pathological conditions, can detect significant changes in protein expression. Protein expression in various cancers also resembles biological processes underlying adhesion, migration, and transport. Exosomes also contribute to angiogenesis, metastasis, tumor formation, and disease progression in lung, melanoma, and breast cancer (Chen et al., 2022).

Cathepsin D is a diagnostic marker for early-stage Parkinson's disease. Exosomes can be incorporated into nucleic acids, such as mRNA, to serve as non-invasive diagnostic markers for less invasive liquid biopsies (Batista et al., 2024). Exosomal mRNA was also used in the prognosis of prostate and pancreatic cancer. As therapeutic agents, exosomes can be used in cancer treatment by inhibiting oncogenes. The immune modulation therapy using exosomes was successfully employed to treat unresectable non-small cell lung cancer in 2015 (NCT01159288) by utilizing dendritic cell-derived exosomes loaded with targeted antigens. Tumor cells are used as ligands to modify exosomes for cancer therapeutics, owing to their tropism and their ability to induce an inflammatory response. Tumor cells employ a novel approach to neutralize the innate immune response. A Phase 1 trial is underway to inhibit tumorigenesis by targeting tumor tyrosine kinase cell-surface receptors in glioma cells. Numerous clinical studies using exosomes were focused on treatments for skin infections, alopecia, and wound healing (Tienda-Vázquez et al., 2023). MSCs and their exosomes have been shown to possess anti-inflammatory properties. Exosomes serve as natural carriers of bioactive molecules between cells, and clinical trials have explored their potential for COVID-19 therapy. The secretions of MSCs, as well as exosomes engineered with specific micro RNAs (miRs) or mRNAs, have shown promise as nanomedicines (Rezakhani et al., 2021). Moreover, anti-inflammatory and immunomodulatory techniques used in exosomal therapeutics are emerging due to their inherent anti-inflammatory properties (Bai et al., 2024). The pro-inflammatory cytokine-responsive behavior in patients was reduced after administration of exosomes derived from MSCs. However, because exosomes cross the blood-brain barrier, they can be used as a step toward brain targeting. A recent study was conducted on exosomes for treating ischemic stroke, where lesions of the stroke were targeted via intravenous injections. Reperfusion can be the only therapy for the reversal of brain damage in stroke, but it can be associated with potential inflammatory responses. Therefore, researchers suggested using neutrophil membrane-derived vesicles loaded with resolving D2 to prevent inflammation in the stroke lesion (Song et al., 2021). Exosomes exhibit great potential as a non-invasive approach in regenerative skin therapies. Currently, no effective antiviral drugs or vaccines are commercially available to treat coronavirus disease (COVID-19). Most COVID-19-related deaths are caused by respiratory failure, which is triggered by severe and uncontrolled inflammation, thus underscoring the critical need for effective treatment strategies (Jiang et al., 2022). Similarly, acute lung inflammation results from innate defensive mechanisms against microbes, whereas chronic inflammation occurs when the response fails to remove the stimulus. Studies have used mouse models of acute lung injury (ALI) to evaluate vesicle-based treatments, and results have shown that this approach suppresses the immune response and reduces the production of inflammatory mediators (Rahnama et al., 2024). Heart diseases can lead to an increase in the damage of myocardial cells, leading to blockage of coronary vessels, poor blood flow, and ischemia. Induction of exosome therapy can reduce infiltrated and inflamed cells after 7 days of treatment (Tschuschke et al., 2020). Kidney injury is accompanied by structural damage and loss of function, and causes a variety of symptoms such as sepsis, nephrotoxicity, and ischemia in patients. Acute kidney injury (AKI) is the most common renal disease, and exosomal treatment has been reported to re-protect mice in an acute injury model (Suh et al., 2021). Exosomes have also been used in the past to treat inflammatory bowel disease (IBD), resulting in reduced pain and inflammation and structural healing of the villi. Liver fibrosis and cirrhosis are fatal diseases resulting from chronic liver inflammation, and vesicles derived from human amniotic MSCs (AM-MSCs) significantly reduced the expression of inflammatory cytokines (Suh et al., 2021). Moreover, clinical outcomes of the exosomes are shown in Table 18.2.

18.10 CONCLUSION

Exosomes have evolved from simple waste products of cells to essential tools in new therapies. These tiny particles are compatible with the body and can carry active molecules, making them ideal for personalized medicine and targeted drug delivery. To utilize exosomes effectively, we need to transition from ideas to real-world applications. This involves creating efficient isolation methods, improving cargo loading, and ensuring precise targeting and safety. Advanced techniques, such as nanotechnology and gene editing, can enhance their treatment capabilities. It's also crucial to address the ethical, regulatory, and economic aspects of exosome therapies. Clear global rules and fair access strategies will support their safe use. With collaboration and research funding, exosomes

TABLE 18.2
Clinical Outcomes of Exosomes in Several Diseases

Diseases	Clinical Outcome	References
Crohn's disease	A Phase 1 study evaluated the safety of placenta-derived MSC exosome (MCS exosome) injections and their potential to improve therapeutic recovery in patients with perianal fistula who do not have Crohn's disease.	(Pak et al., 2023)
Alzheimer's disease	Clinical trial phase 1 or 2 discovered the effectiveness of intranasal MCS exosomes and improved cognitive function in Alzheimer's disease.	(Xie et al., 2023)
Diabetes	Extravascular vesicles enhance diabetic wound healing by miR-145-5p, indicating favorable clinical trial potential.	(Su et al., 2023)
Ischemic stroke	Small extravascular vesicles derived from cerebral endothelial cells enhance vascular repair and neurological recovery in ischemic stroke models, suggesting their potential for clinical applications.	(Li et al., 2021a)
Osteoarthritis	Exosomes-loaded hydrogel-mediated dual regulation of osteogenesis and macrophage polarization improves fracture repair.	(Mi et al., 2022)
Liver fibrosis	Hyaluronic acid-modified exosomes deliver forsythiaside to target CD44, inhibit NLRP3-mediated pyroptosis, and effectively treat liver fibrosis, thereby providing a promising therapeutic platform.	(Gong et al., 2023)
Tumor targeting	Low-pH extravascular vesicles, such as exosomes, enhance homologous tumor targeting and drug delivery by leveraging lipid-mediated targeting, offering an innovative and selective therapeutic approach.	(Gong et al., 2021)
Vein graft restenosis	Gelatin methacryloyl-loaded umbilical cord MSC exosomes promote endothelial healing and reduce restenosis after vein tissue surgery in preclinical models.	(Deng et al., 2023)

can enhance existing treatments and transform the way we prevent and treat diseases. Their integration into clinical guidelines could lead to a new era of personalized healthcare.

18.11 FUTURE PERSPECTIVE

Exosome-based drug delivery has emerged as a highly promising and innovative frontier in medical science, achieving notable experimental success and demonstrating significant potential to transform therapeutic practices. However, this advanced approach faces several challenges that offer numerous opportunities for further investigation and technological refinement. A particularly substantial concern is the inherent variability of exosomes, naturally occurring EVs that vary widely in origin, structure, and function, potentially complicating their biodistribution and overall therapeutic effectiveness. The biodistribution of exosomes significantly influences treatment outcomes, underscoring the need for a detailed understanding of their trafficking patterns and biological interactions. Notably, the incorporation of polyethylene glycol (PEG) into exosomal formulations has been shown to extend their detection time following injection from approximately 10 minutes to an impressive 60 minutes, marking a considerable advancement in the ability to monitor these innovative delivery vehicles within the body. To fully capitalize on the potential of exosomes as carriers for unconventional therapeutic agents, enhancing their intrinsic capabilities to penetrate diverse tissues while evading the body's immune responses (Alzhrani et al., 2021) (Table 18.3).

Presently, exosome production is constrained by low yields from cell cultures, typically yielding less than one gram of protein per milliliter. This limitation necessitates the extensive cultivation of cellular materials, a process that can be resource-intensive and time-consuming. The introduction

TABLE 18.3
Future Directions and Research Priorities

Research Priorities	Current Limitations	Future Opportunities
Isolation and purification	Low yield and purity	Scalable microfluidic and automated platforms.
Cargo loading	Poor encapsulation efficiency	AI-guided bioengineering and endogenous loading mechanisms.
Targeting and biodistribution	Variable effects and imprecise pharmacokinetics	Ligand engineering and *in vivo* tracking systems.
Safety and long-term effects	Limited clinical data	Family studies spanning generations and predictive toxicology models.
Regulatory and quality standards	No universal guidelines	International consortia for harmonize

of exosome-mimetic nanovesicles (EMNVs) offers a promising solution, as they can significantly increase production yields, thereby facilitating the practical application of exosomal technology for therapeutic purposes. Exosomes possess a remarkable capacity for transporting targeted therapeutic agents, making them suitable candidates for treating a wide range of diseases, including cancer and neurodegenerative disorders, and enhancing their efficacy as drug delivery vehicles necessitates strategic development aimed at optimizing several crucial characteristics: achieving high drug loading capacities, enhancing specificity to target cells, minimizing toxicity to healthy tissues, and reducing immunogenicity. Each of these factors is pivotal in determining the success of exosome-based therapeutic interventions and requires rigorous attention and research (Kimiz-Gebologlu et al., 2022).

Additionally, establishing standardized protocols for the isolation, classification, and purification of exosomes is crucial to ensuring that their clinical application is practical and reliable. Such standardization will provide a solid foundation for integrating exosome-based therapies into established medical treatments. While exosomes present considerable promise as natural nanocarriers for theragnostic applications, they face significant challenges that hinder their clinical implementation. A primary obstacle is the lack of standardized, scalable methods for isolating and purifying exosomes, leading to heterogeneous populations with varying purity and yield. This inconsistency directly affects reproducibility, dosing accuracy, and therapeutic efficacy. Moreover, the complex composition of exosomes, which includes a diverse mix of proteins, lipids, and nucleic acids, may produce unpredictable biological effects and raise concerns regarding potential immunogenicity and unintended interactions. These complexities complicate quality control measures and heighten safety concerns that must be thoroughly addressed before broad clinical deployment. The achievement of efficient and targeted delivery of therapeutic agents via exosomes remains a critical challenge. Natural exosomes often lack adequate targeting specificity and biodistribution, leading to insufficient accumulation in the intended tissues or cells and rapid clearance by the immune system. Current engineering approaches aimed at improving targeting accuracy and loading efficiency are still under development and often involve intricate modifications that may compromise exosome integrity or function. Consequently, significant advancements in exosome engineering are essential to enhance their stability, precision targeting, and controlled release of therapeutic cargo, while also deepening our understanding of their behavior in vivo. Addressing these multifaceted issues is fundamental to unlocking the full potential of exosomes as practical theragnostic tools and paving the way for innovative therapeutic strategies in modern medicine (Moghassemi et al., 2024).

REFERENCES

Ahmed LA, Al-Massri KF, Haider KH. Therapeutic efficacy of mesenchymal stem cells and their derived exosomes in different neurodegenerative disorders: Current status and future perspectives. In: *Handbook of Regenerative Medicine: Cell-Free Therapy.* London: Taylor & Francis Group/CRC Press, 2025. Chapter 32.

Ahmed ZT, Alkahlot MH, Haider KH. MSC-derived exosomes: Advances in cell-free therapy. In: *Handbook of Stem Cell Applications.* Singapore: Springer, 2023 Aug 18, pp. 1–41.

Almughem FA, Alsheri AA, Alomary MN. *Exosomes in Drug Delivery. Role of Exosomes in Biological Communication Systems.* 2020 Nov 21, pp. 337–60.

Al-Omar MT, Alnajjar MT, Ahmed ZT, Salaas FMI, Alrefaei TSM, Haider KH. Endothelial progenitor cell-derived small extracellular vesicles for myocardial angiogenesis and revascularization. *J. Clin. Transl. Res.* 2022;8(6):476–87. http://dx.doi.org/10.18053/jctres.08.202206.004.

AL-Rasheedi M, Alhazmi Y, Almaqwashi N, Alreshidi MA, Kardam A, Sharaf M, Haider KH. Corticosteroid therapy for 2019-nCoV-infected patients: A case series of eight mechanically ventilated patients. *Clin. Case Rep.* 2021;9(5):e04066. http://dx.doi.org/10.1002/ccr3.4066.

Alvarez-Erviti L, Seow Y, Yin H, Betts C, Lakhal S, Wood MJA. Delivery of siRNA to the mouse brain by systemic injection of targeted exosomes. *Nat. Biotechnol.* 2011;29:341–5.

Alzhrani GN, Alanazi ST, Sumayyah Y, Alshareef AM, Alshareef AA, Abdel-maksoud MS, Elsherbiny NM. Exosomes: Isolation, characterization, and biomedical applications. *Cell Biol. Int.* 2021;45:1807–31.

Amini H, Namjoo AR, Narmi MT, Mardi N, Narimani S, Naturi O, Khosrowshahi ND, et al. 2023. Exosome-bearing hydrogels and cardiac tissue regeneration. *Biomat. Res.* 2023;27(1):99.

Andre M, Caobi A, Miles JS, Vashist A, Ruiz MA, Raymond AD. Diagnostic potential of exosomal extracellular vesicles in oncology. *BMC Cancer.* 2024;24;322.

Aramini B, Masciale V, Grisendi G, Bertolini F, Mauer M, D'Amico R, Massimo DR, Haider KH. *Future Perspectives of Exosomal miR Payloads in Lung Cancer. Stem Cells: From Hype to Hope and More.* Singapore: Springer, 2021.

Aramini B, Masciale V, Haider KH. Defining lung cancer stem cells: Exosomal payload of miRs from a clinical perspective. *World J. Stem Cells.* 2020 Jun 26;12(6):406–21. http://dx.doi.org/10.4252/wjsc.v12.i6.406. PMID: 32742559; PMCID: PMC7360993.

Babaei S, Fadaee M, Abbasi-kenarsari H, Shanehbandi D, Kazemi T. Exosome-based immunotherapy as an innovative therapeutic approach in melanoma. *Cell Commun. Signal.* 2024;22:527.

Bai G, Truong TM, Pathak GN, Benoit L, Rao B. Clinical applications of exosomes in cosmetic dermatology. *Skin. Health Dis.* 2024;4348.

Batista IA, Machado JC, Melo SA. Advances in exosomes utilization for clinical applications in cancer. *Trends Cancer.* 2024;10.

Bermejo-Pareja F, Del Ser T, Valenti M, De La Fuente M, Bartolome F, Carro E. Salivary lactoferrin as biomarker for Alzheimer's disease: Brain-immunity interactions. *Alzheimer's Dementia.* 2020;16:1196–204.

Bolideei M, Barzigar R, Gahrouei RB, Mohebbi E, Haider KH, Paul S, Paul MK, Mehran MJ. Applications of gene editing and nanotechnology in stem cell-based therapies for human diseases. *Stem Cell Rev. Rep.* 2025 May;21(4):905–34. http://dx.doi.org/10.1007/s12015-025-10857-0.

Cacciottolo M, Nice JB, Li Y, Leclaire MJ, Twaddle R, Mora CL, Adachi SY, et al. Exosome-based multivalent vaccine: Achieving potent immunization, broadened reactivity, and strong T-cell responses with nanograms of proteins. *Microbiol. Spectrum.* 2023;1:503–23.

Cai X, Janku F, Zhan Q, Fan JB. Accessing genetic information with liquid biopsies. *Trends Genetics.* 2015;3:564–75.

Caselli RJ, Dueck AC, Osborne D, Sabbagh MN, Connor DJ, Ahern GL, Baxter LC, et al. Longitudinal modeling of age-related memory decline and the APOE ε4 effect. *N. Engl. J. Med.* 2019;36:255–63.

Chen L, Wang C, Sun H, Wang J, Liang Y, Wang Y, Wong G. The bioinformatics toolbox for circRNA discovery and analysis. *Brief. Bioinform.* 2021;22:1706–28.

Chen R, Xu X, Qian Z, Zhang C, Niu Y, Wang Z, Sun J, et al. The biological functions and clinical applications of exosomes in lung cancer. *Cell. Mol. Life Sci.* 2019;76:4613–33.

Chen X, Luo Q. Potential clinical applications of exosomes in the diagnosis, treatment, and prognosis of cardiovascular diseases: A narrative review. *Ann. Transl. Med.* 2022;10:372.

Consortium ET, Van Deun J, Mestdagh P, Agostinis P, Akay O, Anand S, Anckaert J, et al. EV-TRACK: transparent reporting and centralizing knowledge in extracellular vesicle research. *Nat. Methods.* 2017;14:228–32.

Cordonnier M, Nardin C, Chanteloup G, Derangere V, Algros MP, Arnould L, Garrido C, Aubin F, Gobbo J. Tracking the evolution of circulating exosomal-PD-L1 to monitor melanoma patients. *J. Extracell. Vesicles.* 2020;9:1710899.

D'anca M, Fenoglio C, Serpente M, Arosio B, Cesari M, Scarpini EA, Galimberti D. Exosome determinants of physiological aging and age-related neurodegenerative diseases. *Front. Aging Neurosci.* 2019;11:232.

Das S, Lyon CJ, Hu T. A panorama of extracellular vesicle applications: from biomarker detection to therapeutics. *ACS Nano.* 2024;18:9784–97.

Deddens JC, Vrijsen KR, Colijn JM, Oerlemans MI, Metz CH, Van Der Vlist EJ, Nolte-'t hoen EN, et al. Circulating extracellular vesicles contain miRNAs and can serve as early biomarkers of cardiac injury. *J. Cardiovasc. Transl. Res.* 2016;9:291–301.

Deng Y, Li Y, Chu Z, Dai C, Ge J. Exosomes from umbilical cord-derived mesenchymal stem cells combined with gelatin methacryloyl inhibit vein graft restenosis by enhancing endothelial functions. *J. Nanobiotechnol.* 2023;21:380.

Ding YN, Li MQ, Song JY, Guan PP, Wang P. Engineered exosomes improve the effects of curcumin on protecting mitochondria of neurons in Alzheimer's disease. *Materials Today Bio.* 2025;32:101738.

Duan L, Xu L, Xu X, Qin Z, Zhou X, Xiao Y, Liang Y, et al. Exosome-mediated delivery of gene vectors for gene therapy. *Nanoscale.* 2021;13:1387–97.

Gheytanchi E, Saeednejad ZL, Ghods R, Abolhasani M, Shahin M, Vafaei S, Naseri M, et al. High expression of tumor susceptibility gene 101 (TSG101) is associated with more aggressive behavior in colorectal carcinoma. *J. Cancer Res. Clin. Oncol.* 2021;147:1631–46.

Golubovskaya V, Bukrinsky MI, Grizzi F. Editorial on "exosomes, their biogenesis and role in inter-cellular communication, tumor microenvironment and cancer immunotherapy." *Vaccines.* 2020;8.

Gong C, Zhang X, Shi M, Li F, Wang S, Wang Y, Wei W, et al. Tumor exosomes reprogrammed by low pH are efficient targeting vehicles for smart drug delivery and personalized therapy against their homologous tumor. *Adv. Sci.* 2021;8:2002787.

Gong L, Zhou H, Zhang S, Wang C, Fu K, Ma C, Zhang Y, et al. CD44-targeting drug delivery system of exosomes loading forsythiaside, which combats liver fibrosis via regulating NLRP3-mediated pyroptosis. *Adv. Healthcare Mater.* 2023;12(2):202–28.

Greco KA, Franzen CA, Foreman KE, Flanigan RC, Kuo PC, Gupta GN. PLK-1 Silencing in bladder cancer by siRNA delivered with exosomes. *Urology.* 2016;91:241.e1–e7.

Gummadi S, Chitti SV, Kang T, Shahi S, Mathivanan S, Fonseka P. ExoCarta: A web-based repository of small extracellular vesicle cargo. *J. Mol. Biol.* 2024;437:169218.

Gun CB, Taehun YB. Exosome-based therapy in cardiovascular diseases: A new frontier in cardiovascular disease treatment. *Korean Circ. J.* 2025;55:e54.

Guo L, Huang Z, Huang L, Liang J, Wang P, Zhao L, Shi Y. Surface-modified engineered exosomes attenuated cerebral ischemia/reperfusion injury by targeting the delivery of quercetin towards impaired neurons. *J. Nanobiotechnol.* 2021;19:141.

Habib SM, Martini MF, Al Ahmed O, Qassab AR, Haider KH. The potential of mesenchymal stem cell–derived exosomes in cell-free therapy for myocardial repair. In: Haider, K.H. (eds) *Handbook of Regenerative Medicine: Cell-Free Therapy.* London: Taylor & Francis Group/CRC Press, 2025;24, pp. 372–90.

Habib SY, Martini MF, Hamdan YNA, Shrebaty OMM, Haider KH. Mesenchymal stem cell secretome and exosomes as potential advanced therapy medicinal products for treating a failing heart. *World J Stem Cells.* 2026;18.

Haider KH, Aramini B. Mircrining the injured heart with stem cell-derived exosomes: an emerging strategy of cell-free therapy. *Stem Cell Res. Ther.* 2020;11(1):23. http://dx.doi.org/10.1186/s13287-019-1548-7.

Haider KH, Najimi M. Exosome-based cell-free therapy in regenerative medicine for myocardial repair. In: Haider, K.H. (eds) *Handbook of Stem Cell Therapy.* Singapore: Springer, 2022. https://doi.org/10.1007/978-981-16-6016-0_9-1.

He J, Ren W, Wang W, Han W, Jiang L, Zhang D, Guo M. Exosomal targeting and its potential clinical application. *Drug Deliv. Transl. Res.* 2022;12:2385–402.

Hu J, Huang S. Introduction to online nucleic acid databases for EV research. In: *Extracellular Vesicles: From Bench to Bedside.* Springer, 2024.

Hung M E, Leonard JN. Stabilization of exosome-targeting peptides via engineered glycosylation. *J. Biol. Chem.* 2015;290:8166–72.

Hyder Q, Haider KH. The ongoing battle against COVID-19. *Iberoam J. Med.* 2020;2(4):360–6. http://dx.doi.org/10.5281/zenodo.3987277.

Jayaseelan VP. Emerging role of exosomes as a promising diagnostic tool for cancer. *Cancer Gene Therapy.* 2020;27:395–8.

Jiang Z, Wang H, Mou Y, Li L, Jin W. Functions and clinical applications of exosomes in pancreatic cancer. *Mol. Biol. Rep.* 2022;49:11037–48.

Kalluri R, Lebleu VS. The biology, function, and biomedical applications of exosomes. *Science.* 2020;367:69–77.

Kang W, Xu Z, Lu H, Liu S, Li J, Ding C, Lu Y. Advances in biomimetic nanomaterial delivery systems: Harnessing nature's inspiration for targeted drug delivery. *J. Mater. Chem.* B. 2024;12:7001–19.

Kar R, Dhar R, Mukherjee S, Nag S, Gorai S, Mukerjee N, Mukherjee D, et al. Exosome-based smart drug delivery tool for cancer theranostics. *ACS Biomater. Sci. Eng.* 2023;9:577–94.

Kim H, Yun N, Mun D, Kang JY, Lee SH, Park H, Joung B. Cardiac-specific delivery by cardiac tissue-targeting peptide-expressing exosomes. *Biochem. Biophys. Res. Commun.* 2018;499:803–8.

Kimiz-gebologlu I, Oncel SS. Exosomes: Large-scale production, isolation, drug loading efficiency, and biodistribution and uptake. *J. Control. Release.* 2022;347.

Kwon MS, Shin SH, Yim SH, Lee KY, Kang HM, Kim TM, Chung YJ. CD63 as a biomarker for predicting the clinical outcomes in adenocarcinoma of the lung. *Lung Cancer.* 2007;57:46–53.

Lee KWA, Chan LKW, Hung LC, Phoebe LKW, Park Y, Yi KH. Clinical applications of exosomes: A critical review. *Int. J. Mol. Sci.* 2024;25:77–94.

Li C, Wang C, Zhang Y, Alsrouji O K, Chebl A B, Ding G, Jiang Q, et al. Cerebral endothelial cell-derived small extracellular vesicles enhance neurovascular function and neurological recovery in rat models of acute ischemic stroke, including those treated with mechanical thrombectomy and tPA for embolic stroke. *J. Cereb. Blood Flow Metab.* 2021a;41:2090–104.

Li P, Kaslan M, Lee S H, Yao J, Gao Z. Progress in exosome isolation techniques. *Theranostics.* 2017;7(3):789–804.

Li W, Li Y, Zhi W, Liu C, Fan W, Miao Q, Gu X. Diagnostic value of using exosome-derived cysteine-rich protein 61 as a biomarker for acute coronary syndrome. *Exp. Ther. Med.* 2021b;22:1437.

Li Z, Zhou X, Wei M, Gao X, Zhao L, Shi R, Sun W, et al. In vitro and in vivo RNA inhibition by CD9-HuR functionalized exosomes encapsulated with miR or CRISPR/dCas9. *Nano Letters.* 2019;19:19–28.

Lian XY, Zhang H, Liu Q, Lu X, Zhou P, He SQ, Tang RX, et al. Ovarian cancer-excreted exosomal miR-199a-5p suppresses tumor metastasis by targeting hypoxia-inducible factor-2α in a hypoxic microenvironment. *Cancer Commun.* 2020;40:380–385.

Liang W, Luo Y, Xu A, Chu J, Ji W, Wang L, Gu Y, et al. Advances in carrier-delivered small interfering RNA-based therapeutics for the treatment of neurodegenerative diseases. *Biomater. Sci.* 2024;12:4927–45.

Liang Y, Xia J, Xu L, Zhang Y, Xu X, Haider KH. Cell-free and cell-based therapy in regenerative medicine. In: Haider, K.H. (eds) *Handbook of Regenerative Medicine: Cell-Free Therapy Approach* (1st ed.). CRC Press, 2025. Chapter 24. https://doi.org/10.1201/9781003611509.

Limoni SK, Moghadam MF, Moazzeni SM, Gomari H, Salimi F. Engineered exosomes for targeted transfer of siRNA to HER2 positive breast cancer cells. *Appl. Biochem. Biotechnol.* 2019;187:352–64.

Lin S, Yu Z, Chen D, Wang Z, Miao J, Li Q, Zhang D, et al. Progress in microfluidics-based exosome separation and detection technologies for diagnostic applications. *Small.* 2020;16(1):903–16.

Liu Q. Intranasal administration of exosome packaged catalase-SKL, a novel antioxidant delivery system across the blood-brain barrier in mice: Biodistribution and safety. The University of Western Ontario, Ontario, 2020.

Liu SL, Sun P, Li Y, Liu SS, Lu Y. Exosomes as critical mediators of cell-to-cell communication in cancer pathogenesis and their potential clinical application. *Transl. Cancer Res.* 2019;8:298–311.

Lu Y, Zheng J, Lin P, Lin Y, Zheng Y, Mai Z, Chen X, et al. Tumor microenvironment-derived exosomes: A double-edged sword for advanced T cell-based immunotherapy. *ACS Nano.* 2024;18:27230–60.

Matsumoto A, Takahashi Y, Nishikawa M, Sano K, Morishita M, Charoenviriyakul C, Saji H, et al. Role of phosphatidylserine-derived negative surface charges in the recognition and uptake of intravenously injected B16BL6-derived exosomes by macrophages. *J. Pharmaceut. Sci.* 2017;106:168–75.

Matsuzaka Y, Yashiro R. Regulation of extracellular vesicle-mediated immune responses against antigen-specific presentation. *Vaccines.* 2022;10(10):1691.

McKiernan J, Donovan M J, O'Neill V, Bentink S, Noerholm M, Belzer S, Skog J, et al. A novel urine exosome gene expression assay to predict high-grade prostate cancer at initial biopsy. *JAMA Oncol.* 2016;2:882–9.

Mi B, Chen L, Xiong Y, Yang Y, Panayi AC, Xue H, Hu Y, et al. Osteoblast/osteoclast and immune cocktail therapy using a multifunctional hydrogel for exosome and drug delivery accelerates fracture repair. *ACS Nano.* 2022;16:771–82.

Moghassemi S, Dadashzadeh A, Sousa MJ, Vlieghe H, Yang J, Leon-Felix CM, Amorim CA. Extracellular vesicles in nanomedicine and regenerative medicine: A review over the last decade. *Bioactive Mater.* 2024;36:126–56.

Morishita M, Takahashi Y, Nishikawa M, Takakura Y. Pharmacokinetics of exosomes: An important factor for elucidating the biological roles of exosomes and for the development of exosome-based therapeutics. *J. Pharmaceut. Sci.* 2017;106:2265–9.

Mushtaq M, Zineldeen DH, Mateen MA, Haider KH. Mesenchymal stem cells' "garbage bags" at work: Treating radial nerve injury with mesenchymal stem cell-derived exosomes. *World J. Stem Cells* 2024; 16(5). https://dx.doi.org/10.4252/wjsc.v16.i5.0000.

Muskan M, Abeysinghe P, Cecchin R, Branscome H, Morris KV, Kashanchi F. Therapeutic potential of RNA-enriched extracellular vesicles: The next generation in RNA delivery via biogenic nanoparticles. *Mol. Therapy.* 2024;32:2939–49.

Nakamura K, Ida N, Hirasawa A, Okamoto K, Vu TH, Ly DTH, Masuyama HJ. CD63 as a potential biomarker for patients with ovarian cancer. *Eur. J. Obstet. Gynecol. Reprod. Biol.* 2025;306:87–93.

Neves KB, Rios FJ, Sevilla-montero J, Montezano AC, Touyz RM. Exosomes and the cardiovascular system: role in cardiovascular health and disease. *J. Physiol.* 2023;60:4923–36.

Ohno SI, Takanashi M, Sudo K, Ueda S, Ishikawa A, Matsuyama N, Fujita K, et al. Systemically injected exosomes targeted to EGFR deliver antitumor microRNA to breast cancer cells. *Mol. Ther.* 2013;21:185–91.

Olejarz W, Dominiak A, Zolnierzak A, Kubiak-Tomaszewska G, Lorenc T. Tumor-derived exosomes in immunosuppression and immunotherapy. *J. Immunol. Res.* 2020;1:6272498.

Oughtred R, Stark C, Breitkreutz BJ, Rust J, Boucher L, Chang C, Kolas N, et al. The BioGRID interaction database: 2019 update. *Nucleic Acids Res.* 2019;47:529–41.

Pak H, Hadizadeh A, Heirani-tabasi A, Soleimani M, Asbagh RA, Fazeli MS, Kazemeini A, et al. Safety and efficacy of injection of human placenta mesenchymal stem cells-derived exosomes for treatment of complex perianal fistula in non-Crohn's cases: Clinical trial phase I. *J. Gastroenterol. Hepatol.* 2023;38:539–47.

Patel G, Agnihotri TG, Gitte M, Shinde T, Gomte SS, Goswami R, Jain A. Exosomes: A potential diagnostic and treatment modality in the quest for counteracting cancer. *Cell. Oncol.* 2023;46:1159–79.

Patel KK, Patel BS, Prajapati A, Narkhede SB, Luhar S. The exosome revolution: Bridging early detection and personalized treatment. Res. J. Pharmaceut. *Dosage Forms Technol.* 2025;17:26–30.

Peinado H, Aleckovic M, Lavotshkin S, Matei I, Costa-silva B, Moreno-bueno G, Hergueta-redondo M, et al. Melanoma exosomes instruct bone marrow progenitor cells to adopt a pro-metastatic phenotype via the MET pathway. *Nat Med.* 2012;18:883–91.

Perocheau D, Touramanidou L, Gurung S, Gissen P, Baruteau J. Clinical applications for exosomes: Are we there yet? Br. *J. Pharmacol.* 2021;178:2375–92.

Pi F, Binzel DW, Lee TJ, Li Z, Sun M, Rychahou P, Li H, et al. Nanoparticle orientation to control RNA loading and ligand display on extracellular vesicles for cancer regression. *Na. Nanotechnol.* 2018;13:82–9.

Picca A, Guerra F, Calvani R, Bucci C, Lo Monaco MR, Bentivoglio AR, Landi F, Bernabei R, et al. Mitochondrial-derived vesicles as candidate biomarkers in Parkinson's disease: Rationale, design and methods of the EXosomes in PArkiNson Disease (EXPAND) study. *Int. J. Mol. Sci.* 2019;20:2373.

Properzi F, Logozzi M, Fais S. Exosomes: The future of biomarkers in medicine. *Biomark. Med.* 2013;7:769–78.

Rahnama M, Heidari M, Poursalehi Z, Golchin A. Global trends of exosomes application in clinical trials: A scoping review. *Stem Cell Rev. Reports.* 2024;20:2165–93.

Rastogi S, Sharma V, Bharti PS, Rani K, Modi GP, Nikolajeff F, Kumar S. The evolving landscape of exosomes in neurodegenerative diseases: Exosomes characteristics and a promising role in early diagnosis. *Int. J. Mol. Sci.* 2021;22:440.

Rezaie J, Feghhi M, Etemadi T. A review on exosomes application in clinical trials: Perspective, questions, and challenges. *Cell Commun. Signal.* 2022;20:145.

Rezakhani L, Kelishadrokhi AF, Soleimanizadeh A, Rahmati S. Mesenchymal stem cell (MSC)-derived exosomes as a cell-free therapy for patients infected with COVID-19: Real opportunities and range of promises. *Chem. Phys. Lipids.* 2021;234:105009.

Ruiz G, Valencia-Gonzalez HA, Leon-Galicia I, Garcia-Villa E, Garcia-Carranca A, Garigolio P. Inhibition of RAD51 by siRNA and resveratrol sensitizes cancer stem cells derived from HeLa cell cultures to apoptosis. *Stem Cells Int.* 2018;2018:2493869.

Sadeghi S, Tehrani FR, Tahmasebi S, Shafiee A, Hashemi SM. Exosome engineering in cell therapy and drug delivery. *Inflammopharmacology.* 2023; 31:145–69.

Safwan M, Bourgleh MS, Al-Ruqi A, Shrebaty O, Almujaydil F, Alothaim B, Alrashdi N, Haider KH. Living bio-drugs MSCs and their derivative exosomes to rescue ventilated patients with ARDS: A systematic review and meta-analysis. *World J. Stem Cells.* 2025;17:111978.

Sajid MI, Bukhari S, Ilyas H, Malik R, Fatima M, Tiwari RK, Nauli SM, et al. Exosomes and encapsulated exomiRs in breast cancer: Theragnostic applications and clinical implications. In: Haider, K.H. (eds) *Handbook of Regenerative Medicine: Cell-Free Therapy.* London: Taylor & Francis Group/CRC Press, 2025, Chapter-23, pp. 357–70.

Salunkhe S, Dheeraj BM, Chitkara D, Mittal A. Surface functionalization of exosomes for target-specific delivery and in vivo imaging & tracking: Strategies and significance. *J. Control. Release.* 2020;326:599–614.

Sancho-Albero M, Medel-Martínez A, Martin-Duque P. Use of exosomes as vectors to carry advanced therapies. *RSC Adv.* 2020;10:23975–87.

Sharma S, Muthu S, Jeyaraman M, Ranjan R, Jha SK. Translational products of adipose tissue-derived mesenchymal stem cells: Bench to bedside applications. *World J. Stem Cells* 2021;13:1360.

Sharma V, Mukhopadhyay CD. Exosomes as drug delivery systems: Current advances. *Extracell. Vesicle.* 2024;3:100032.

Simpson RJ, Lim JW, Moritz RL, Mathivanan S. Exosomes: Proteomic insights and diagnostic potential. *Expert Rev. Proteomics.* 2009;6:267–83.

Smyth T, Kullberg M, Malik N, Smith-Jones P, Graner MW, Anchordoquy TJ. Biodistribution and delivery efficiency of unmodified tumor-derived exosomes. *J. Control. Release.* 2015;199:145–55.

Soares MT, Trindade D, Vaz M, Campelo I, Almeida M, Trigo G, Da Cruz e Silva OA, Henriques AG. Diagnostic and therapeutic potential of exosomes in Alzheimer's disease. *J. Neurochemistry.* 2021;156:162–81.

Sohn W, Kim J, Kang SH, Yang SR, Cho JY, Cho HC, Shim SG, et al. Serum exosomal microRNAs as novel biomarkers for hepatocellular carcinoma. *Exp Mol Med.* 2015;47:184.

Song Y, Kim Y, Ha S, Sheller-Miller S, Yoo J, Choi C, Park CH. The emerging role of exosomes as novel therapeutics: Biology, technologies, clinical applications, and the next. *Am. J. Reproduct. Immunol.* 2021;85:e13329.

Su J, Wei Q, Ma K, Wang Y, Hu W, Meng H, Li Q, et al. P-MSC-derived extracellular vesicles facilitate diabetic wound healing via miR-145-5p/CDKN1A-mediated functional improvements of high glucose-induced senescent fibroblasts. *Burns Trauma.* 2023;11:tkad010.

Suh JH, Joo HS, Hong EB, Lee HJ, Lee JM. Therapeutic application of exosomes in inflammatory diseases. *Int. J. Mol. Sci.* 2021;22(3):1144.

Suh N, Subramanyam D, Lee MY. 2017. Molecular signatures of secretomes from mesenchymal stem cells: Therapeutic benefits. *Mol. Cell. Toxicol.* 2017;13:133–41.

Sun Y, Liu J. Potential of cancer cell–derived exosomes in clinical application: A review of recent research advances. *Clin. Therap.* 2014;36:863–72.

Szklarczyk D, Nastou K, Koutrouli M, Kirsch R, Mehryary F, Hachilif R, Hu D, et al. The STRING database in 2025: Protein networks with directionality of regulation. *Nucleic Acids Res.* 2025;53:730–7.

Takahashi Y, Nishikawa M, Shinotsuka H, Matsui Y, Ohara S, Imai T, Takakura Y. Visualization and in vivo tracking of the exosomes of murine melanoma B16-BL6 cells in mice after intravenous injection. *J. Biotechnol.* 2013;165:77–84.

Tienda-Vázquez MA, Hanel JM, Márquez-Arteaga EM, Salgado-Álvarez AP, Scheckhuber CQ, Alanis-Gómez JR, Espinoza-silva JI, et al. Exosomes: A promising strategy for repair, regeneration, and treatment of skin disorders. *Cells.* 2023;12(12):1625.

Tschuschke M, Kocherova I, Bryja A, Mozdziak P, Angelova VA., Janowicz K, Sibiak R, et al. Inclusion biogenesis, methods of isolation, and clinical application of human cellular exosomes. *J. Clin. Med.* 2020;9:436.

Vader P, Mol EA, Pasterkamp G, Schiffelers RM. Extracellular vesicles for drug delivery. *Adv. Drug Deliv. Rev.* 2016;106:148–56.

Veneruso I, Ranieri A, Falcone N, Tripodi L, Scarano C, La Monica I, Pastore L, et al. The potential usefulness of the expanded carrier screening to identify hereditary genetic diseases: A case report from real-world data. *Genes.* 2023;14:1651.

Wang J, Xia L, Wang R, Cai Y. Linezolid and its immunomodulatory effect: In vitro and in vivo evidence. *Front. Pharmacol.* 2019;10:1389.

Wang J, Zheng Y, Zhao M. Exosome-based cancer therapy: Implication for targeting cancer stem cells. *Front. Pharmacol.* 2017;7:533.

Wang JH, Forterre AV, Zhao J, Frimannsson DO, Delcayre A, Antes TJ, Efron B, Jeffrey SS, et al. Anti-HER2 scFv-directed extracellular vesicle-mediated mRNA-based gene delivery inhibits growth of HER2-positive human breast tumor xenografts by prodrug activation. *Mol. Cancer Therap.* 2018;17:1133–42.

Weber B, Franz N, Marzi I, Henrich D, Leppik L. Extracellular vesicles as mediators and markers of acute organ injury: Current concepts. *Eur. J. Trauma Emerg. Surgery.* 2020;48:1525–44.

Wishart D S, Guo A, Oler E, Wang F, Anjum A, Peters H, Dizon R, et al. HMDB 5.0: The human metabolome database for 2022. *Nucleic Acids Res.* 2022;50:622–31.

Xie X, Song Q, Dai C, Cui S, Tang R, Li S, Chang J, et al. Clinical safety and efficacy of allogenic human adipose mesenchymal stromal cells-derived exosomes in patients with mild to moderate Alzheimer's disease: A phase I/II clinical trial. *Gen. Psychiatry.* 2023;36:e101143.

Yarana C, Carroll D, Chen J, Chaiswing L, Zhao Y, Noel T, Alstott M, et al. Extracellular vesicles released by cardiomyocytes in a doxorubicin-induced cardiac injury mouse model contain protein biomarkers of early cardiac injury. *Clin. Cancer Res.* 2018;24:1644–53.

Ye Z, Zhang T, He W, Jin H, Liu C, Yang Z, Ren J. Methotrexate-loaded extracellular vesicles functionalized with therapeutic and targeted peptides for the treatment of glioblastoma multiforme. *ACS Appl. Mater. Interface.* 2018;10:12341–50.

Yu H, Sun T, An J, Wen L, Liu F., Bu Z, Cui Y, et al. Potential roles of exosomes in Parkinson's disease: From pathogenesis, diagnosis, and treatment to prognosis. *Front Cell Develop Biol.* 2020;8:86.

Yusuf R, Haider KH. Theragnostic applications of exosomes in Alzheimer's disease. In: Haider, K.H. (eds) *Handbook of Regenerative Medicine: Cell-Free Therapy.* London: Taylor & Francis Group CRC Press, 2025, Chapter 255, pp. 390–409.

Zamani P, Fereydouni N, Butler AE, Navashenaq JG, Sahebkar A. The therapeutic and diagnostic role of exosomes in cardiovascular diseases. *Trends Cardiovasc. Med.* 2019;29:313–23.

Zeng M, Maozhu L, Xuelin T, Xi Y, Chao S, Wang X. Emerging trends in the application of extracellular vesicles as novel oral delivery vehicles for therapeutics in inflammatory diseases. *Int. J. Nanomed.* 2024;19:8573–601.

Zhang F, Zhang Z, Yang W, Peng Z, Sun J, Li G, Wei Y, et al. Engineering autologous cell-derived exosomes to boost melanoma-targeted radio-immunotherapy by cascade cGAS-STING pathway activation. *Small.* 2025;21:2408769.

Zhang M, Hu S, Liu L, Dang P, Liu Y, Sun Z, Qiao B, et al. Engineered exosomes from different sources for cancer-targeted therapy. *Signal Transduct. Target. Ther.* 2023;8:124.

Zhang X, Xu D, Song Y, He R, Wang T. Research progress in the application of exosomes in immunotherapy. *Front. Immunol.* 2022;13:731516.

Zhao B, Li Z, Li R. Exosomes in oral squamous cell carcinoma: Functions, challenges, and potential applications. *Front. Oncol.* 2025;14:1502283.

Zhao K, Wang Z, Li X, Liu J L, Tian L, Chen JQ. Exosome-mediated transfer of CLIC1 contributes to the vincristine-resistance in gastric cancer. *Biochemistry.* 2019; 462:97–105.

19 Exosome Engineering for Enhanced Therapeutic Activity

Mahmood S Choudhery[1], Taqdees Arif[1], and Ruhma Mahmood[2]

[1]Department of Human Genetics and Molecular Biology, University of Health Sciences, Lahore, Pakistan

[2]Allama Iqbal Medical College, Jinnah Hospital, Lahore, Pakistan

19.1 LIST OF ABBREVIATIONS

AA-PEG	Aminoethyl anisamide-polyethylene glycol
APCs	Antigen-presenting cells
CD	Cluster of differentiation
CRISPR	Clustered regularly interspaced short palindromic repeats
DMSO	Dimethyl sulfoxide
DCs	Dendritic cells
EFGR	Epidermal growth factor receptor
EGFP	Enhanced green fluorescent protein
ESCs	Embryonic stem cells
GALA	Glutamic acid-alanine-leucine-alanine
GAPDH	Glyceraldehyde-3-phosphate dehydrogenase
HCC	Hepatocellular carcinoma
ILVs	Intraluminal vesicles
MVB	Multivesicular body
MFHEs	Microfluidic fiber extrusions
MSCs	Mesenchymal stem cells
PEG	Polyethylene glycol
PTX	Paclitaxel
RNPs	Ribonucleoprotein complexes
SELEX	Systematic ligand evolution by exponential enrichment
SIRPα	Signal regulatory protein
sdAbs	Single-domain antibodies
scFvs,	Single-chain variable fragments
TAAs	Tumor-associated antigens
TNBC	Triple-negative breast cancer

19.2 INTRODUCTION

Exosomes are natural extracellular vesicles. They are spherical and are enclosed in a phospholipid bilayer. They range in size from 40 nm to 160 nm. Their molecular cargo consists of mitochondrial

DOI: 10.1201/9781003653387-19

DNA, various types of RNA (microRNAs, messenger RNAs, long non-coding RNAs, and other small RNAs), lipids, and proteins (Ahmed et al., 2023). They are released by multiple cell types, including epithelial cells, uterine macrophages, embryonic stem cells (ESCs), endometrial cells, oviductal epithelial cancer cells, stem cells (adult and neonatal), follicular cells, and dendritic cells. Exosomes are present in multiple body fluids, including amniotic fluid, saliva, semen, and plasma. Exosomes are released by parent cells through the endosomal route. Exosomes exhibit low immunogenicity, high physicochemical stability, the ability to penetrate cells and tissues, and an inherent capacity to facilitate long-distance intercellular communication. These features make them promising candidates for the delivery of therapeutic agents and for serving as therapeutic agents.

Exosomes transport metabolites, proteins, and diverse nucleic acids across cells, facilitating intracellular signaling (Mushtaq et al., 2024). Moreover, exosomes control biological processes and modify immunological responses. Exosomes also play a crucial role in tissue regeneration, underscoring their potential in regenerative medicine. Exosomes can replicate the properties of their donor cells in recipient cells by transferring their contents. These characteristics make the exosomes effective therapeutic agents. However, there are certain limitations in the use of natural exosomes. The inherent composition of exosomes often leads to off-target effects and decreased efficiency. Furthermore, the lack of standardized methods for characterizing and isolating exosomes may pose challenges for their clinical application (Choudhery et al., 2024; Li et al., 2021).

Exosome engineering, the process of modifying exosomes to bestow desired characteristics, plays a crucial role in addressing the limitations of natural exosomes (Liang et al., 2025). Various biophysical, biochemical, and molecular biology methods have been developed to manipulate exosomes, enhancing their specificity, stability, and therapeutic payload. This engineering process addresses current issues related to exosome use and enhances their therapeutic potential. The modified exosome may enhance exosome targeting efficiency, enabling them to recognize recipient cells more effectively and facilitating cellular uptake. The significance of exosome engineering in overcoming the limitations of natural exosomes will be a key focus of this chapter (Choudhery et al., 2024; Li et al., 2021).

This chapter provides a detailed overview of exosomes, their mechanism of action, and the limitations associated with native exosomes. It also explores the various methods of exosome engineering, offering a comprehensive understanding of how these vesicles can be manipulated to enhance their contents and targeting efficacy. Furthermore, the application and challenges of therapeutic exosome use are discussed, providing a holistic view of the topic.

19.3 OVERVIEW OF EXOSOMES: SIZE AND COMPOSITION

As extracellular vesicles, exosomes are released by various types of cells and play a crucial role in immune system modulation (Habib et al., 2026). Ranging in size from 40 nm to 160 nm in diameter, exosomes were once thought to be a means of cellular waste disposal. However, they have now been shown to play a key role in intercellular communication (Mushtaq et al., 2024). Exosomes contain surface markers, various proteins, nucleic acids, lipids, and other signaling molecules. The membranes of exosomes are coated with a variety of antigens that trigger an intracellular signaling pathway by binding to receptors on recipient cells. The synthesis, release, and uptake of exosomes involve tetraspanins, integrins, and Rab GTPases, which are components of the exosomal proteome. They are also used for diagnostics, therapeutics, and immune modulation. Lipid molecules, including cholesterol, sphingolipids, and lipid rafts, are present in the outer phospholipid bilayer of exosome membranes. These structures play roles in exosome stability, release, packaging, and interaction with the host cell. They can also play a role in the prognosis of different diseases, such as cancer and liver disease, highlighting their diverse functions and intriguing potential (Habib et al., 2025).

19.4 BIOGENESIS OF EXOSOMES

The biogenesis of exosomes, the process by which cells generate and release them, is a complex process that occurs in multiple steps. It begins with the formation of early endosomes, which arise through the inward budding of the plasma membrane (Figure 19.1). Early endosomes develop into multivesicular bodies (MVBs), also known as late endosomes. Intraluminal vesicles (ILVs) are produced inside the MVB by the ESCRT (endosomal sorting complex required for transport). These ILVs either fuse with lysosomes for degradation or fuse with the plasma membrane to release exosomes via exocytosis. In addition to exosome production, the release of other microvesicles, ectosomes, is facilitated by ectocytosis. Ectocytosis bypasses the endosomal process and involves the immediate budding of microvesicles from the plasma membrane.

Microvesicles are eventually released from the cell membrane. Both exosomes and microvesicles formation are essential for molecular transfer and intercellular communication (Choudhery et al., 2024).

19.5 CLASSIFICATION OF EXOSOMES

Exosomes are categorized into three types: natural, artificial, and engineered. This classification is based on the natural or modified contents of exosomes obtained through various engineering processes (Table 19.1).

19.5.1 Natural Exosomes

Both plant and animal cells produce and release natural exosomes. These exosomes facilitate intercellular communication and maintain nutritional balance. Natural exosomes are abundantly produced in body fluids by stem cells, epithelial cells, and immune cells such as macrophages and monocytes. The exosomes secreted from these cells play a significant role in the treatment of various diseases. The exosomes secreted by the immune cells are particularly important in developing adaptive immunity. Exosomes are also released by cancer cells (Aramini et al., 2020). These exosomes may be associated with immunosuppression, tumor growth, metastasis, and immune evasion. The tumor-derived exosomes stimulate the transformation of CD4+ T cells into CD39+ regulatory T cells. This leads to immunosuppression within the tumor microenvironment (Choudhery et al., 2024; Li et al., 2021).

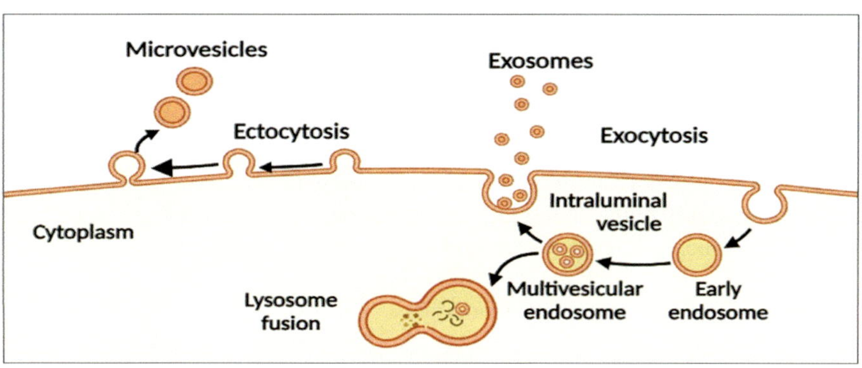

FIGURE 19.1 Exosome biogenesis. Macrovesicles bud directly from the plasma membrane via ectocytosis. Exosomes are formed within multivesicular endosomes through intraluminal vesicle formation and are released by exocytosis after fusion with the plasma membrane. Alternatively, lysosome fusion with the multivesicular endosome can occur, in which the multivesicular endosome fuses with a lysosome for degradation rather than exosome release.

TABLE 19.1
Comparison of Types of Exosomes

Classification	Description	Advantages	Disadvantages
Natural exosomes	Exosomes derived directly from cells facilitate cell-to-cell communication and maintain nutritional balance.	Biocompatible, natural, and non-toxic.	Heterogeneous, difficult to isolate and purify, and a limited understanding of the mechanism of exosome biogenesis and function.
Artificial exosomes	Engineered nanoparticles mimicking the characteristics of natural exosomes, designed to overcome the limitations of natural exosomes.	Improved targeting and cargo delivery, scalable production.	May lack natural biocompatibility and cellular uptake properties.
Modified/ engineered exosomes	Natural exosomes are intentionally altered to enhance therapeutic potential, targeting capabilities, or stability.	Retains natural biocompatibility, improved targeting, and therapeutic potential.	May require complex engineering processes, potential off-target effects.

However, selecting the stem cell source based on the treatment is necessary when treating a specific disease. The contents of the exosomes and the lipids and proteins on their surface are affected by the source from which they are harvested. Exosomes released by stem cells affect the ability of cells to self-renew, regulate the immune system, proliferate, and repair themselves. Additionally, these exosomes show great potential for cell-free rejuvenation (Choudhery et al., 2024; Liang et al., 2025). However, the size, composition, and function of natural exosomes are not uniform. Moreover, isolating and purifying clinical-grade exosomes for therapeutic use poses significant challenges. Furthermore, existing methods for delivering genes, medications, and other therapies are inadequate. Developing novel, safe, and effective modification procedures is essential to overcome these limitations. Therefore, technological advancements have led to new methods for synthesizing and engineering exosomes that mimic their natural components (Li et al., 2019).

19.5.2 Artificial Exosomes

Exosomes synthesized by manipulating natural exosomes are called artificial exosomes. They mimic the structural and functional characteristics of natural exosomes. The development of artificial exosomes can address the inherent limitations of natural exosomes, including poor yield, inefficient isolation procedures, undesired cargo, and complex structure. There are two types of artificial exosomes, namely hybrid exosomes and exosome mimics. Hybrid exosomes are generated when the natural exosomes (derived from parent cells) are combined with synthetic components (liposomes). Hybrid exosomes aim to utilize the characteristics of both natural and artificial exosomes. Exosome mimics are synthetic nanoparticles that can mimic the structure and function of natural exosomes. They typically comprise lipids, polymers, or other biomaterials assembled into vesicles with specific sizes, shapes, and surface properties (Choudhery et al., 2024; Li et al., 2021).

19.5.3 Hybrid Exosomes

Hybrid exosomes are created by fusing natural exosomes with synthetic nanoparticles. Various methods have been employed to develop hybrid exosomes. Direct incubation, extrusion, freeze-thaw cycles, and sonication are standard methods for generating hybrid exosomes. Hybrid exosomes are obtained by combining exosomes and liposomes in a specific ratio, followed by a process such as fusion, hybridization, or encapsulation. The targeting potential of hybrid exosomes can be further enhanced by adding diverse moieties, including antibodies, probes, peptides, PEGylated lipids, fluorescent tags, and pharmaceutical compounds.

19.5.3.1 Polyethylene Glycol Fusion

Polyethylene glycol (PEG) fusion is used as a fusogen to combine exosomes with liposomes, producing hybrid exosomes. PEG is routinely used to mediate cell-to-cell fusion (Haider and Stimson, 1994 &1996). PEG controls the close interaction of lipid bilayer membrane structures. It can combine exosomes and liposomes from different cell sources. The sizes, ratios, and concentrations of PEG molecules determine the efficacy of exosome-liposome fusion. It can prolong the circulation time of liposomes and exosomes in the bloodstream, facilitating their efficient fusion. Furthermore, this method is easy to use and yields consistent results. Hybrid exosomes may, however, lack sufficient PEG on their surface to achieve the undetectable qualities required to evade rapid uptake, thereby reducing cellular uptake (Steinberg et al., 2018).

19.5.3.2 Direct Incubation

Direct incubation of exosomes with liposomes at standard culture conditions is also used to create hybrid exosomes. This process relies on membrane fusion, a spontaneous process that utilizes the physicochemical surface properties of vesicles, such as hydrophobicity, electrostatic charge, and membrane curvature, to drive fusion. Hydrophobic or electrostatic interactions form hybrid exosomes without rupturing the lipid bilayer or permitting the vesicle's contents to escape. To create hybrid exosomes, for example, clustered regularly interspaced short palindromic repeats (CRISPR)/Cas9-expressing liposomes and HEK293FT cell-derived exosomes can be cultured at 37°C for 12 hours. This presents a novel method for delivering the CRISPR-Cas9 system safely and effectively. This approach does not cause significant damage to vesicles. However, the fusion efficiency is low (Lin et al., 2018).

19.5.3.3 Membrane Extrusion

Membrane extrusion enables the movement of large molecules or organelles out of the cell through the plasma membrane. To create hybrid exosomes, membrane extrusion simultaneously extrudes liposomes and exosomes through membrane pores under physical pressure, producing mixed-type vesicles. For example, hybrid nanovesicles designed to treat pulmonary fibrosis are made using fibroblast-derived exosomes and clodronate-loaded liposomes. This method involves pulsing and sonicating a suspension of synthetic liposomes with exosomes derived from L-929 fibroblasts, followed by extrusion through 400 nm and 200 nm polycarbonate ester films. Exosome-hybridized liposomes are effectively produced using this method. Hybrid vesicles produced via membrane extrusion have more homogeneous particle sizes than those produced by incubation or freeze-thaw methods. Despite the high fusion efficiency of membrane extrusion techniques, the shear stress generated during the process may compromise the integrity of natural exosomes (Poinsot et al., 2024).

19.5.3.4 Freeze-Thaw Methods

The freeze-thaw procedure is a widely used method for generating hybrid exosomes. In addition, this method is also frequently used for drug loading into liposomes. It is an easy-to-use method. The formation of ice crystals within and outside the plasma membrane can temporarily rupture it. It may allow water-soluble chemicals to enter the liposomes. In addition to affecting the drugs' biological activity within the vesicles, high-frequency freeze-thaw cycles may compromise the biological integrity of the exosome membrane. When employing this fusion strategy, it is crucial to carefully assess the types of drugs and the number of freeze-thaw cycles (Steinberg et al., 2018).

19.5.4 Exosomes Mimics

Exosome mimics are artificially synthetic or engineered vesicles. They are designed to mimic the functions and biological properties of natural exosomes. The limited exosome production by parent cells and their low loading efficiency pose significant challenges for clinical applications. This challenge can be reduced by using exosome mimetics. Not all components of exosomes are essential for their functionality. Exosome mimics only deliver the functional components and pharmaceutical agents. These exosome mimics can be produced using cell membrane-cloaked nanoparticles, serial extrusion, or liposomes containing only the essential elements of natural exosomes (Poinsot et al., 2024).

The plasma membrane of cells is forced through porous membranes with pores of 100–400 nm, breaking it down into smaller, spherical nanovesicles or polymer nanoparticles. Exosome mimetics are manufactured with a yield 100 times greater than natural exosomes, making them suitable for large-scale production for clinical applications. Exosome mimetics are less complex than native exosomes. They also exhibit more stability, dispersion, and immunocompatibility, identical to natural exosomes. These membrane-bound exosomes can be modified to enhance cellular uptake and targeting. Exosomes can be readily manipulated by fusing exosomes produced from transformed cells with liposomes containing antibodies, PEG, or peptides. Exosome mimetics could provide enhanced control and scalability for clinical applications. Furthermore, incorporating specific peptides onto cell membranes makes exosome mimics proper and safe vehicles for pharmaceutical delivery. However, the exosomal components essential for forming functional exosome mimetics remain poorly described (Poinsot et al., 2024).

19.5.5 Engineered Exosomes

Engineered exosomes refer to natural exosomes that have been modified to enhance their therapeutic potential, targeting capabilities, or stability, while retaining their natural biocompatibility and cellular uptake properties. For exosome engineering, the internal contents and/or surface (charge and composition) of natural exosomes can be modified for targeted therapeutic applications. It also includes incorporating therapeutic compounds and pharmaceutical agents into the exosomal cargo (Choudhery et al., 2024).

19.5.5.1 Modifications of Exosomal Payload

Electroporation, sonication, and gene-editing methods integrate drug(s) into natural exosomes. Electroporation is the process by which cargo enters the exosomes through transiently formed pores in the membrane's phospholipid bilayer. During this process, the exosomes and desired cargo are mixed, and a short, high-voltage electric current is applied. The electric pulse makes tiny pores in the cell's membrane. The cargo enters the exosomes through a ruptured membrane. Additionally, a suitable drug may be added to the parent cells by incubation, which causes the cells to take up the drug and encapsulate it within exosomes. The host body then naturally releases the drug-laden exosomes into the targeted cell population. Nevertheless, this process makes it challenging to control loading efficiency. Specific proteins and RNA molecules cannot be naturally loaded into exosomes. However, it is possible to incorporate these therapeutic molecules into exosomes using gene-editing tools, such as CRISPR-Cas9, to modify the parent cells that produce the exosomes (Choudhery et al., 2024).

19.5.5.2 Surface Modifications

The surface characteristics of exosomes are crucial for their biodistribution, cell-specific targeting, and potential therapeutic applications. The surface of exosomes can be modified to include necessary characteristics specifically to improve cell targeting. Either the membranes of the parent cells (i.e., exosome-generating cells) are altered, or the exosomes' membranes can be directly modified. Different types of vectors can be used to integrate the coding sequences of desired ligands into the cell membrane, thereby altering the cell's genetic makeup. Peptides are then expressed on exosome membranes. Cells secrete exosomes that contain peptides. Modifying the exosome membrane can also be attained through genetic engineering of parental cells (Haider, 2024a & 2024b). These surface modifications enable exosomes to target specific cells or tissues, increasing their therapeutic potential. For instance, exosomes can be engineered to target cancer cells and deliver antitumor drugs or RNA molecules that inhibit cancer growth. Similarly, exosomes can target immune cells, modulating the immune response to treat autoimmune diseases (Choudhery et al., 2024).

19.5.6 Methods of Exosome Engineering

Exosomes are derived from diverse natural sources, including various fruit and vegetable juices and mammalian biological fluids (such as blood, urine, saliva, and milk). They can be extensively manipulated with methods to assess their potential for biomedical applications. Exosome engineering entails the modification of exosomes to augment their functionality through the incorporation of specific cargo or alteration of their surface properties. Diverse direct methods, including covalent binding and non-covalent binding, as well as indirect methods of exosome engineering, are employed to generate modified exosomes (Bahadorani et al., 2024).

19.5.6.1 Covalent Binding

Several functional groups are covalently attached to exosomes. Covalent binding of functional groups modifies the surface of exosomes without altering their size or function through a cross-linking mechanism known as azide-alkyne cycloaddition, also referred to as click chemistry. For example, sulfhydryl groups are abundant on the surface of exosomes; they serve as binding sites for sulfhydryl groups and maleimides via the Michael addition reaction. The Michael addition reaction is a chemical reaction between a nucleophile (a molecule that donates electrons) and an α, β-unsaturated carbonyl compound (a molecule with a carbon-carbon double bond adjacent to a carbonyl group). This method involves adding an azide or an alkyl group to the exosome surface to create active chemical sites, enabling targeting moieties to bind to the exosome in aqueous buffers, including dimethyl sulfoxide (DMSO), alcohols, and water. This method is ideal for covalently attaching polymers, small molecules, and macromolecules to the exosome surface, thereby providing the required functionalities (Nan et al., 2022; Parada et al., 2021).

PEGylation is one of the most frequent chemical conjugation methods that uses covalent linkages. This process involves modifying exosome surfaces with polyethylene glycol (PEG) (Susa et al., 2019). Exosomes modified with aminoethyl anisamide-PEG (AA-PEG) can specifically target lung cancer cells overexpressing the sigma receptor. AA-PEG is a sigma receptor targeting the ligand (Kim et al., 2018). Furthermore, NRP-1 has been observed to be overexpressed in glioma cells and tumor vascular endothelium; therefore, labeling exosomal membranes with the NRP-1-targeting peptide (RGE peptide) can enhance glioma targeting and blood-brain barrier (BBB) penetration in orthotopic glioma mice. The accumulation of RGE-labeled exosomes in the glioma of U251 tumor-bearing mice was about 1.5 times higher 1 hour after exosome injection, and the exosomes remained retained in the tumor for an extended period (Jia et al., 2018). A peptide called c(RGDyK) was attached to the surface of exosomes generated from MSCs to treat stroke. This peptide has a strong affinity for integrin αvβ3, which is expressed in reactive cerebral vascular endothelial cells during ischemia. When contrasted with scrambled c(RGDyK) peptide-labeled exosomes, c(RGDyK)-labeled exosomes showed an 11-fold increase in tropism to the lesion region of the ischemic brain of mice (Tian et al., 2018). Although highly stable, hazardous substances typically form covalent bonds, making the use of covalent modification techniques in therapies problematic (Choi et al., 2021).

19.5.6.2 Non-Covalent Methods of Engineering Exosomes

Another crucial method for directly engineering the exosome surface is non-covalent binding, which primarily involves electrostatically engineered exosomes, aptamer-engineered exosomes, hydrophobic integration-engineered exosomes, and antibody-engineered exosomes to load functionalized agents on the exosome surface (Sadeghi et al., 2023). The non-covalent methods of exosome engineering are explained below.

19.5.6.2.1 Electrostatically Engineered Exosomes

Electrostatically engineered exosomes are modified exosomes that utilize electrostatic forces to load or attach therapeutic agents, such as nucleic acids, proteins, or small molecules, to their surfaces. This approach exploits the electrostatic interactions between the negatively charged exosome surface and positively charged therapeutic agents. The electrostatic attraction enables the loading of therapeutic agents onto the exosome surface (Sadeghi et al., 2023). A combination of the amphipathic glutamic acid-alanine-leucine-alanine (GALA) peptide and the electrostatic interaction between cationic lipids and the exosomal membrane was employed to improve the uptake of exosomal cargo. Sequences of glutamic acid, alanine, leucine, and alanine are repeated in the GALA peptide. Adding cationic lipids and the GALA peptide enhanced the payload's internalization. The GALA peptide mimics a viral protein at acidic pH by adopting a helical conformation. Substances can escape from endosomes with the help of this helical shape. The GALA peptide adopts this helical form to expose a hydrophobic leucine residue. Endosomal stability is compromised, and cargo is released into the cytoplasm due to this exposure, allowing the exosomal and endosomal membranes to fuse. Cellular absorption was enhanced by the interaction of the negatively charged glutamic acid residues in the GALA peptide with cationic lipids, leading to their accumulation on the exosome membrane (Nakase and Futaki, 2015).

One potential disadvantage of this technique is the cytotoxicity that some cationic nanomaterials can produce through hole formation and membrane weakening. The primary drawback of this strategy is that cationic nanoparticles are often internalized by cells via endocytosis, leading to lysosomal degradation of the payloads and thereby reducing their therapeutic efficacy (Sadeghi et al., 2023).

19.5.6.2.2 Aptamer-Engineered Exosome

Aptamers are short, single-stranded DNA or RNA molecules that bind specifically to a target molecule, such as a protein, peptide, or small molecule. They are often referred to as chemical antibodies due to their high affinity and specificity for their target molecules. Aptamers are combined with click chemistry to modify exosomes for particular uses. Click chemistry is a method for attaching molecules or surfaces via a simple, efficient chemical reaction. Click chemistry is used to attach aptamers to exosome surfaces, enabling targeted delivery of therapeutic payloads to specific cells or tissues. The target's optimal aptamer is determined through systematic ligand evolution by exponential enrichment (SELEX). SELEX is a laboratory technique for identifying and isolating specific nucleic acid sequences (such as aptamers) that bind to a particular target molecule with high affinity. DNA aptamers exhibit affinity levels comparable to those of monoclonal antibodies for non-covalent binding to macromolecules, such as proteins. Aptamer-exosome internalization via clathrin-mediated endocytosis results from the aptamers' modification of exosome uptake and cellular interaction. Angiopep-2 RNA and CD133 aptamers are coupled to exosomes to target the blood-brain barrier and glioblastoma. The altered carrier exhibited increased distribution in the brain, lungs, liver, spleen, heart, and kidneys. Furthermore, incorporating gold nanorods and aptamers on exosomes may enhance the efficacy of photothermal suppression of cancer cells. Aptamers offer numerous advantages, including enhanced immunological compatibility, specificity towards specific targets, and increased tissue penetration. The reduced size facilitates the binding of many aptamers to exosomes. However, challenges must be addressed for clinical applications, including aptamer aggregation, fast clearance, and limited in vivo longevity owing to the immune system's protein corona profile. Novel engineering methodologies are crucial for addressing these challenges (Sadeghi et al., 2023).

19.5.6.2.3 Hydrophobic Integration-Engineered Exosomes

Given the lipid bilayer surrounding exosomes, hydrophobic interactions are believed to result from targeting moieties being directly inserted into the exosome membrane. Exosomes can attach several functional groups to their surface, including amine- or carboxylic-ended phospholipid or transmembrane protein moieties (functional groups). Functionalized phospholipids modified with specific functional groups can be incorporated into the exosome membrane using a simple incubation method that employs a hydrophobic insertion technique. Hydrophobic sequestration is a technique that can load exosomes with small lipophilic medications, such as anti-inflammatory, chemotherapeutic, and photosensitizing agents. For example, exosome membrane stains prepared with the well-known dyes DiI, PKH-67, and BODIPY TR ceramide are commercially produced using this technique, requiring only simple co-incubation in environments with efficient loading that correlate positively with the hydrophobicity of the foreign species (Sadeghi et al., 2023).

19.5.6.2.4 Antibody-Engineered Exosomes

An alternate method for targeted drug delivery is the design of antibodies against antigens. For example, increased CD47 expression in cancer cells and its association with signal regulatory protein α (SIRPα) facilitate immune evasion; thus, blocking this interaction may regulate several malignancies. The exosome coupled to anti-CD47 or anti-SIRP demonstrated tumoricidal effects both in vitro and in vivo (Pham et al., 2021). Given the widespread distribution of CD47 throughout the body, therapeutic targeting of this molecule in cancer cells must be strategic and localized, minimizing effects on normal cells. The most straightforward approach for enhancing exosome targeting is to use antibodies. However, due to the limitations associated with their vast size, complicated structure, and ability to induce an immune response, simple antibody fragments, including single-domain antibodies (sdAbs) or single-chain variable fragments (scFvs), have been employed extensively (Pham et al., 2021).

19.5.6.3 Indirect Engineering of Exosomes

Indirect surface engineering of exosomes can be achieved by altering the genes of the cells that produce them (Haider, 2024b). This method has several advantages over directly manipulating exosomes. These benefits include increased expression and yield of the targeting moiety, as well as the stability of modified exosomes. Tetraspanins, ClC_2, and the Lamp2b domain of lactadherin are exosomal membrane components that can be coupled with targeting moieties (receptors, peptides, and antibodies) of engineered exosome-producing cells. The targeting moieties are expressed in surface-modified exosomes produced by the natural exosome biogenesis process in donor cells transfected with these vectors. The added targeting moiety on the surface of these exosomes, derived from genetically modified cells, is stable (Choi et al., 2021).

19.5.7 Cargo Loading of Engineered Exosomes

Passive and active methods can be used to deliver exosome cargo. In the passive technique, the cells or exosomes that release them are incubated with the intended payload. This method, while simple, could be improved to achieve more efficient cargo uptake. However, to increase membrane permeability, active exosome loading involves a range of techniques, including freeze-thaw cycles, sonication, electroporation, extrusion, and dialysis. These strategies enhance exosome loading with other molecules (Fu et al., 2020).

19.5.7.1 Active Exosome Loading

In this method, the cargo molecules are actively forced into exosomes rather than relying on passive diffusion or endocytosis. When exosomes are actively loaded, their membranes are disrupted, generating tiny pores. These tiny membrane pores can incorporate the cargo molecules into the exosomes. Various methods have been successfully employed to disrupt or manipulate the exosome membrane. These methods involve electroporation, incubation, heat shock, sonication, extrusion, freeze-thaw cycles, and hypotonic dialysis.

19.5.7.1.1 Electroporation

Electroporation uses brief electrical pulses to create transient openings in the membranes of cells, liposomes, or extracellular vesicles. An electric pulse alters the lipid composition of the cell membrane, creating reversible holes. These tiny openings facilitate the entry of molecules, such as DNA or RNA. The GAPDH and BACE1 genes have been successfully knocked down in vitro and in wild mice models, and RVG-engineered exosomes have been loaded with siRNA. Doxorubicin has been electroporated into exosomes for targeted tumor therapy in animal models (Fu et al., 2020). The doxorubicin-exosome combination in an electroporation buffer was maintained at 4°C. Loading was then performed at 150 mF and 350 V, followed by a 30-minute incubation at 37°C. The reported encapsulation efficiency with this method is 20%. However, it is crucial to optimize temperature, pulse duration, and buffer type and concentration for effective cargo loading via electroporation (Fu et al., 2020).

19.5.7.1.2 Selective Exosome Incubation with Membrane-Permeating Agents

Membrane-permeating agents, also known as membrane-permeabilizing agents or cell permeabilizers, are chemicals that temporarily or permanently alter the permeability of cell membranes. Saponin is an amphipathic glycoside. It is frequently used to form pores in exosomes and improve membrane permeability. It can specifically interact with cholesterol, extract it from the membrane, and create holes in the cell membrane. It has been used to package therapeutic compounds, such as catalase, into exosomes. Catalase and exosomes are incubated with 0.2% saponin. After 20 minutes of incubation, catalase is encapsulated into exosomes. These encapsulated exosomes have been demonstrated to exhibit neuroprotective properties in models of Parkinson's disease. Moreover, saponin has high loading efficiency. However, it can cause cytotoxicity, hemolysis, and rupture of erythrocyte exosome membranes, thereby limiting its therapeutic potential (Fu et al., 2020).

19.5.7.1.3 Heat Shock

Heat shock is a standard method employed for bacterial transformation. However, it can also induce membrane pores and facilitate exosome loading. Heat shock increases cellular fluidity and promotes membrane lipid remodeling, leading to the formation of membrane holes. Heat shock treatment in the presence of $CaCl_2$ can rupture the exosomal membrane, facilitating the loading of miRNA into exosomes. The loading of miRNA into exosomes cannot be induced by heat shock or $CaCl_2$ treatment alone. Combining both, i.e., heat shock and $CaCl_2$ treatment, is necessary to achieve efficient miRNA loading. The miRNA-15a level was significantly increased when macrophages were subjected to mild heat shock. This demonstrates the successful delivery of miRNA-15a to macrophage cells via exosomes. Furthermore, mice exposed to the exosome complex showed increased miRNA-15a levels in their alveolar macrophages. The exosome-miRNA complex was effectively delivered to their lungs. The miRNA-15a and exosome mixture was heated to 42°C for one minute during the loading procedure. It was subsequently refrigerated for 5 minutes on ice in a 0.1 M $CaCl_2$ solution to facilitate efficient miRNA loading and membrane permeabilization (Fu et al., 2020).

19.5.7.1.4 Freeze-Thaw

Freeze-thaw cycles can generate temporary holes in the cell membrane. This technique consists of three cycles of alternating phases at ambient temperature and −80°C. The freeze-thaw method has been widely employed to encapsulate drugs into exosomes. For example, this method has been successfully used to encapsulate aspirin into exosomes for the treatment of Parkinson's disease. This method achieved an 88% encapsulation effectiveness. Amphiphilic polymers were used to form an amorphous form of aspirin, aiming to enhance exosome-loading efficacy (Tran et al., 2019a). The freeze-thaw technique can be further enhanced by combining it with sonication before or after freeze-thaw cycles. In conjunction with sonication, the altered freeze-thaw technique enables the encapsulation of amorphous aspirin into the exosomes (Fu et al., 2020). Exosomes sourced from HT-29 and MDA-MB-231 cells were also used in conjunction with an EpCAM aptamer to enhance targeting. This method inhibited the growth of colorectal and breast cancer cells in cell cultures and animal models (Tran et al., 2019b).

19.5.7.1.5 Sonication

Sonication disrupts or fragments exosome membranes, thereby enhancing the uptake of molecules into cells. Ultrasound sonication can induce temporary cavitation, thereby enhancing membrane fluidity. Using high-frequency pulsed sounds, the exosome membrane can be rendered permeable to encapsulate certain compounds, such as siRNA. For instance, exosomes that were sonicated during loading were used to deliver HER-2-targeted siRNA (Lamichhane et al., 2016). Sonication at 35 kHz for 30 seconds increased siRNA exosome loading, thereby inhibiting HER2 mRNA and protein expression. In addition, M1-macrophage-derived exosomes were sonicated to load paclitaxel. The loaded exosomes exerted a significant anticancer effect in vivo and in vitro. Further, the loaded exosomes enhanced macrophage-mediated inflammation, which is hypothesized to improve the anticancer efficacy of PTX. Compared to electroporation, sonication does not cause exosome aggregation, making it a better technique (Fu et al., 2020).

19.5.7.1.6 Dialysis

Exosomes can be loaded with cargo using a dialysis membrane in a hypotonic solution, which facilitates the encapsulation of therapeutic molecules within the exosomal vesicles. In this method, the membrane is stretched, and transient holes form due to a hypotonic solution, leading to cell swelling and increased membrane permeability. Using a gradient created by combining exosomes with therapeutic cargo, the technique encapsulates the cargo by encouraging its transfer into the exosome. It is essential to note that while this method effectively contained porphyrin, it also led to exosome aggregation and reduced cell uptake (Fu et al., 2020).

19.5.7.1.7 Extrusion

Extrusion involves passing the target cells or exosomes through carbon filters with progressively decreasing pore sizes, ranging from 400 nm to 100 nm. The device is equipped with a heat blocker to reduce the temperature rise when applying force to rupture the cell or exosome membrane by passing samples through the holes approximately 31 times in both directions. This technique can generate artificial exosomes with consistent dimensions. Furthermore, this can produce vesicles laden with therapeutic cargo when loading agents are present. Extrusion was successfully used to integrate catalase into exosomes. They demonstrate significant effectiveness in preventing, releasing, and encapsulating protease degradation. Similarly, extrusion was used to produce doxorubicin-loaded exosome mimics. The exosome-mimicking encapsulation of doxorubicin demonstrated substantial antitumor efficacy both in vivo and in vitro (Fu et al., 2020).

19.5.7.2 Passive Exosome Loading

Passive loading is the process of loading chemicals into exosomes without the use of external force or energy. Passive loading methods include directly incubating exosomes with targeted chemicals or indirectly incubating the parent cells that produce the exosomes with those chemicals. The indirect approach involves loading the desired cargo into the parental cells, which spontaneously package it into the exosomes that the cells release. The direct technique allows cargo to be loaded into exosomes by incubating the exosomes directly with the required agents. This approach relies on concentration gradients for the internalization of cells or exosomes, enabling targeted medicines or agents to passively cross the cell or exosome membrane (Fu et al., 2020).

19.5.7.2.1 Exosome-Incubated Payloads

Exosomes have demonstrated the capacity to efficiently encapsulate drugs such as paclitaxel, doxorubicin, rhodamine, and curcumin through co-incubation with the target agent (Batrakova and Kim, 2015). A 5-minute incubation at 22°C is sufficient to successfully encapsulate the anti-inflammatory drugs curcumin and JS-124 (cucurbitacin I) in lymphocyte exosomes (EL-4). Intranasal administration of these exosomes, which are laden with these proteins, exhibits anti-inflammatory properties, indicating potential therapeutic uses for glioblastoma and autoimmune encephalomyelitis (EAE) (Zhuang et al., 2011). When considering its natural payloads, the exosome has a low loading efficiency. Nevertheless, exosomes can be loaded passively while preserving their membrane integrity (Fu et al., 2020).

19.5.7.2.2 Parental Cell Incubation Desired Payloads

Incorporating desired payloads with parental cells is crucial in loading exosomes with specific cargo for therapeutic or diagnostic applications. This has been applied in various applications. For example:

- When various anticancer drugs were incubated with hepatocellular carcinoma (HCC) cells, they naturally encapsulated into exosomes. These exosomes enhanced the cytotoxicity of natural killer cells by delivering anticancer drugs and encapsulating heat shock proteins (Lv et al., 2012).
- MSCs were treated with paclitaxel to show the anticancer effect of the released exosomes on pancreatic malignancy. After exposure to paclitaxel for a full day, the drug's anticancer effects on pancreatic carcinoma were evident in exosomes secreted by MSCs (Pascucci et al., 2014).
- Incorporating liposomes into donor cells was used to load a hydrophobic payload into exosomes. Zinc phthalocyanine (ZnPc), the hydrophobic component of photodynamic therapy, was encapsulated in a liposome. Cancer cells were subsequently exposed to ZnPc-loaded liposomes for 30 minutes. Following 48 hours, the naturally occurring cargo encapsulation within the exosome was identified. The liposome simplified the process of delivering a hydrophobic payload to spheroids. After irradiation, exosomes could infiltrate and exert anticancer effects on CT26 tumors and HeLa spheroids. The outer cells of the spheroids effectively encapsulated the cargo within exosomes. This technique facilitated the infiltration of cargo into the tumor microenvironment, including avascular areas (Lee et al., 2015).

19.6 LIMITATIONS OF NATIVE/NATURAL EXOSOMES

Native exosomes, despite their therapeutic potential, exhibit several intrinsic limitations that limit their efficacy and clinical applications. Key limitations of the naive exosomes include, but are not limited to, the following (Li et al., 2021; Corradetti et al., 2021):

- They have insufficient targeting specificity, which results in reduced therapeutic efficacy, high systemic toxicity, and off-target effects. This problem arises from the lack of particular targeting moieties on the surface of native exosomes (Koh et al., 2023).
- The restricted cargo capacity complicates the loading of therapeutic drugs. Moreover, the lack of standardized cargo loading and release protocols further intensifies this problem.
- It is challenging to isolate and purify exosomes in a substantial number. Additionally, large-scale exosome production requires consistent, high-quality isolation and purification procedures.

- There is variability in the sizes, shapes, contents, and biological origin of exosomes that further complicates the application of exosomes as therapeutic agents.
- The current regulation of native exosomes is unclear. There are concerns about the reliability and efficacy of exosomes.
- Standardized production of exosomes is another subject of debate.
- Regulatory guidelines regarding the use of exosomes are currently not available.

These limitations of naive exosomes underscore the need for exosome engineering to enhance their therapeutic efficacy, address intrinsic deficiencies, and facilitate clinical application.

19.7 THERAPEUTIC APPLICATIONS OF ENGINEERED EXOSOMES

Exosomes are emerging as one of the most promising tools in regenerative medicine and targeted therapy. They naturally facilitate intercellular communication, transport therapeutic drugs, and modulate disease-specific signaling pathways. Exosomes are the ideal candidates for clinical applications due to their endogenous origin, biocompatibility, ability to cross biological barriers, such as the blood-brain barrier, and relatively low immunogenicity. Therapeutically, exosomes play dual roles, i.e., as natural biomarkers and as bioengineered delivery vehicles. As biomarkers, exosomes offer a non-invasive and highly informative means to diagnose diseases and monitor therapeutic responses. They reflect the physiological or pathological state of their parent cells, making exosomes reliable indicators for cancer progression, neurodegenerative disorders, cardiovascular conditions, and metabolic syndromes. Additionally, exosome profiles are being investigated as a real-time method to assess organ-specific injury or treatment efficacy.

From a treatment perspective, engineered exosomes have gained significant traction as carriers for targeted delivery of drugs, RNA therapeutics (like siRNA and miRNA), and even gene-editing tools such as CRISPR-Cas9. Their ability to protect fragile therapeutic molecules from enzymatic degradation in circulation, target specific tissues via surface modifications, and induce specific immune responses enhances their utility across a wide range of medical disciplines. Current therapeutic strategies include the use of exosomes derived from MSCs, dendritic cells, or even tumor cells, engineered to deliver therapeutic payloads for diseases such as cancer, Parkinson's disease, Huntington's disease, and more. Ongoing advancements in exosome isolation, engineering, and cargo loading techniques are rapidly transforming exosome-based therapies from experimental tools into clinically viable treatment platforms.

Exosomes play a significant role in advancing personalized medicine by serving as valuable biomarkers and therapeutic agents (Sajid et al., 2025). Exosomes can serve as potential indicators for early disease diagnosis. Exosomal proteins play a pivotal role in diagnosing various diseases, including cancer, cardiovascular disorders, and liver disease. Exosome-based molecules, including proteins and nucleic acids, play a crucial role as biomarkers. Biomarkers include proteins such as TSG101, CD63, CD9, and CD81, as well as lipids such as cholesterol and sphingomyelin. Fetuin-A and proteins associated with the epidermal growth factor receptor (EGFR) pathway are biomarkers for acute renal disease. Membrane scaffold proteins used for exosome characterization exhibit differential expression across several cancer types, making them potential biomarkers for cancer diagnosis (Kul and Erbaş, 2022). Numerous microRNAs, which are frequently employed as disease indicators, are found in exosomes. miR-1246 and miR-21 are utilized in the diagnosis of breast cancer Zhang et al., 2020). Exosomes have been isolated from various biopsies to investigate disease progression, evaluate treatment efficacy, and improve therapeutic strategies. A clinical investigation analyzed the urine exosomes of 24 patients to identify hypertension. Angiotensin II and other angiotensin peptides were quantified in patient plasma to assess the renin–angiotensin–aldosterone system, which regulates the aberrant renal sodium concentration responsible for this medical disease. Urinary exosomes were selected as the primary outcome measure because they serve as indicators of renal function and provide a valid criterion for assessing the drug's therapeutic efficacy (Zhou et al., 2023).

Exosomes have been engineered in various ways to efficiently deliver active pharmaceutical ingredients across different clinical applications for the treatment of multiple diseases (Haider and Najimi, 2022; Najimi and Haider, 2022; Yusuf and Haider, 2025; Ahmed et al., 2025). For example, measuring α-synuclein levels in saliva and plasma can be a valuable indicator for diagnosing Parkinson's disease (Ahmed et al., 2025). Targeting α-synuclein mRNA using exosome-loaded antisense oligonucleotides improved mobility in the Parkinson's mouse model (α-synuclein A53T). Additionally, in Parkinson's disease mouse models, exosomes loaded with catalase showed impressive brain delivery following nasal administration. The brain of a Parkinson's animal model was subcutaneously implanted with modified cell lines that continuously released exosomes along with catalase mRNA. The exosomes carried a tiny reporter at the 3' end of CD63, RVG-Lamp2, and catalase mRNA to track them, target the brain, and counteract inflammation. The difficulty of quickly removing exosomes via intravenous injection was also addressed. Exosomes generated from dopamine-loaded blood cells exhibited over 15 times superior brain distribution, efficacy, and reduced toxicity compared to free dopamine (Qu et al., 2018).

This nanocarrier is more effective when administered intranasally for neurological diseases than intravenously, as the kidneys and liver excrete the former. Exosomes can serve as vehicles for siRNA to regulate gene expression. Modified exosomes significantly enhanced siRNA delivery and gene silencing in the brains of Huntington disease mouse models. An RNA-binding method was developed to strengthen siRNA exosome loading. This technique combined glyceraldehyde-3-phosphate dehydrogenase (GAPDH) with the RNA-binding protein and the phosphatidylserine-binding domain for siRNA binding and attachment to the exosome surface, respectively. Following injection, siRNA loading onto exosomes' surfaces resulted in a significant decrease in Huntington's gene activity. This study showed that binding siRNA to GAPDH prevented serum nucleases from degrading it (Dar et al., 2021).

The potential of exosomes as carriers for therapeutic chemicals, especially those involved in gene editing, has recently gained attention. Several carriers, including DNA, RNA, and ribonucleoprotein complexes (RNPs), have been utilized to deliver CRISPR-Cas9. Delivering CRISPR Cas9 plasmids via exosomes produced from cancer cells suppressed poly ADP-ribose polymerase-1 and caused apoptosis in ovarian cancer. Red blood cell exosomes were utilized as vehicles for the mRNA version of CRISPR-Cas9, both in vitro and in vivo. They were more effective than the plasmid form in leukemia and breast cancer cells (Kim et al., 2017; Usman et al., 2018).

Hybrid exosomes were developed by combining liposomes and exosomes to construct a potent vector for gene-editing systems, thereby enhancing exosomes' capacity to deliver large nucleic acids. First, the mRunx2 sgRNA plasmid was transfected into HEK293FT cells to produce sgRNA-loaded exosomes. The hybrid carrier was subsequently generated by incubating the sgRNA-loaded exosomes for 12 hours at 37°C with the dCas9 plasmid encoding enhanced green fluorescent protein (EGFP) and liposomes. Hybrid exosomes successfully delivered the gene-editing system to MSCs and significantly reduced the level of targeted gene RNA (mRunx2). This technique improved gene-editing efficiency in MSCs and addressed problems associated with exosome delivery. These findings suggest that exosomes, in combination with other delivery systems, could be a valuable tool for gene therapy applications (Lin et al., 2018).

Exosomes may be utilized as a form of cancer immunotherapy. They can transport different antigens, trigger immune pathways, and produce durable immune responses. They can transport tumor-associated antigens (TAAs) to antigen-presenting cells (APCs), such as dendritic cells (DCs). Exosomes can activate T cells and elicit an immune response against neoplastic cells. Exosomes were isolated from the patient's tumor cells and cultured with dendritic cells in vitro. This incubation facilitates the incorporation of tumor exosomes into the dendritic cells. Tumor-associated antigens are subsequently displayed on the dendritic cell surface. T cells are then activated to identify and respond to tumor-associated antigens following this presentation. Patients with advanced non-small cell lung cancer (NSCLC) showed an increase in natural killer cells after receiving immunotherapy. Given their low toxicity and good tolerance, exosomes offer cancer patients a promising therapeutic alternative. Drug delivery via modified exosomes shows promise for reducing tumor growth. Inhibiting CD47 and its association with SIRP-α on mast cells enhances cancer immunotherapy. The acidic tumor microenvironment activated exosomes engineered with anti-SIRP and anti-CD47, resulting in decreased tumor growth (Besse et al., 2016; Nie et al., 2020).

Cargo delivery was improved by evaluating the exosome travel step by step. The engineering of natural exosomes facilitates targeted delivery and evasion of clearance mechanisms. PEG may decrease circulatory clearance and enhance cellular targeting. PEG can reduce endosomal escape and cell membrane fusion despite its beneficial effects. However, endosomal escape, cargo transport, and cell fusion might all be enhanced by sheddable PEGylation. pH-sensitive linkers, including vinyl ether, hydrazone, phosphoramidate, diorthoester, and β-thiopropionate, induced PEG detachment in low-pH environments, such as lysosomes, endosomes, and tumor microenvironments, in liposome tests. The overexpression of CD47 on exosomes, in conjunction with PEG, has been shown to diminish macrophage phagocytosis and prolong exosome circulation. The integration of CD47 onto the modified exosome significantly influenced the cargo (siRNA) targeting KRAS G12D mRNA and reduced pancreatic tumor proliferation (Bahadorani et al., 2024).

Exosomes also demonstrated the ability to transport miRNA to specific cells. The co-delivery of a 5-fluorouracil and miRNA-21 inhibitor via modified exosomes demonstrated antitumor efficacy against colorectal cancer in a murine model. The exosomal co-delivery of miR-7-5p and Everolimus improved the anticancer efficacy of Everolimus in non-small cell lung cancer. A peptide (KSLSRHDHIHHHC) was used in a different investigation to modify exosomes and target the mesenchymal epithelial transition factor (c-MET). Doxorubicin was efficiently administered by smart exosomes that included overexpressed c-MET and were directed toward triple-negative breast cancer (TNBC). Site-specific, directed medication delivery and nanotechnology may overcome tumor cells' immunological evasion, multidrug resistance, and apoptosis. In addition to biological distribution, maximizing intracellular localization and cell targeting are essential for efficient drug delivery. Surface engineering and methods for loading cargo into exosomes can make them actively targeted carriers (Liang et al., 2020; Li et al., 2020).

Advances in exosome engineering have enabled the next generation of exosomes, transforming our daily lives in significant ways. Many efforts have been made to identify the multidisciplinary connections between the biological sciences and pharmacy, to develop practical clinical applications. Exosome-based research and therapeutics have significantly enhanced our understanding of these unique molecules due to their ability to cross the blood-brain barrier, stability across various body fluids, biocompatibility, and safety. Table 19.2 represents the clinical applications of engineered exosomes.

19.8 CONCLUSION

Exosomes are recognized as natural vectors in diagnostic and therapeutic applications. They are safer and more biocompatible alternatives for drug delivery. However, low yields, poor targeting, and limited encapsulation efficiency can hinder their use in various clinical applications. Modifying and designing exosomes for a specific target can resolve these problems. Engineered exosomes have numerous therapeutic applications, including drug delivery for the treatment of disorders such as Parkinson's disease, Huntington's disease, and cancer. They are highly adaptable and biocompatible because they can encapsulate and transport a wide range of molecular cargos, from RNAs to genome editing components (such as CRISPR-Cas9). As exosome engineering advances, these nanocarriers hold great potential to advance precision medicine, particularly in oncology, neurology, and regenerative therapies. Despite their potential, engineered exosomes are heterogeneous in terms of size, composition, biocompatibility, and activity, making consistency and clinical translation

TABLE 19.2
Clinical Applications of Engineered Exosomes

Sr No	Clinical Area	Application	Exosome Source/ Engineering Strategy	Cargo/Biomarkers	Mechanism/Outcome
1	Disease Diagnosis	Biomarker detection for cancer, cardiovascular diseases, liver, and renal disorders	Natural exosomes from body fluids	TSG101, CD63, CD9, CD81, Fettuin-A, EGFR proteins, lipids	Detection of disease-specific exosomal proteins and lipids for early diagnosis.
2	Breast Cancer	Diagnostic markers	Isolated plasma exosomes	miR-1246, miR-21	MicroRNA-based diagnosis of breast cancer.
3	Hypertension	Functional biomarker study	Urinary exosomes	Angiotensin II and other peptides	Evaluation of kidney function and drug efficacy via exosomal content.
4	Parkinson's Disease	Drug delivery: Catalase or antisense oligonucleotides to the brain	Modified blood or engineered exosomes	α-synuclein mRNA, catalase mRNA, dopamine	Improved brain targeting, reduced inflammation, and better mobility.
5	Huntington's Disease	Gene silencing via siRNA	GAPDH-modified exosomes with RNA-binding proteins	siRNA targeting the Huntingtin gene	Stable gene silencing in the brain via serum-protected delivery.
6	Ovarian and Breast Cancer	Gene editing using the CRISPR-Cas9 system	Cancer cell or RBC-derived exosomes	CRISPR-Cas9 plasmid or mRNA, PARP-1 target	Induction of apoptosis and targeted gene editing.
7	MSC Gene Editing	Enhanced CRISPR-Cas9 delivery	Hybrid exosome-liposome carrier system	mRunx2 sgRNA + dCas9-EGFP	Efficient gene transfer into MSCs with enhanced editing capability.
8	Cancer Immunotherapy	TAA delivery to dendritic cells	Tumor-cell-derived exosomes	Tumor antigens, anti-SIRP, anti-CD47	T cell activation, NK cell enhancement, tumor suppression.
9	Pancreatic Cancer	siRNA targeting KRAS G12D	PEGylated exosomes + CD47 expression	siRNA targeting KRAS G12D, PEG, CD47	Reduced phagocytosis, increased circulation, and tumor growth inhibition.
10	Colorectal Cancer	Co-delivery of drug and miRNA inhibitor	Modified tumor-targeting exosomes	5-Fluorouracil, miR-21 inhibitor	Synergistic antitumor effect via dual therapeutic delivery.
11	Non-Small Cell Lung Cancer (NSCLC)	Combination therapy delivery	Smart exosomes for targeted delivery	miR-7-5p, Everolimus	Enhanced drug sensitivity and anticancer response.
12	Triple-Negative Breast Cancer	Targeted doxorubicin delivery	c-MET peptide-decorated exosomes	Doxorubicin, c-MET targeting peptide	Site-specific delivery with improved targeting to TNBC cells.
13	General Drug Delivery Optimization	pH-sensitive and immune-evasive delivery systems	PEGylated exosomes with pH-cleavable linkers	siRNA, small molecules	Controlled release in acidic environments, reduced clearance, and better intracellular trafficking.

challenging. Addressing these pivotal elements is crucial to realizing the full potential of artificial exosomes in therapeutic applications.

REFERENCES

Ahmed LA, Al-Massri KF, Haider KH. Therapeutic efficacy of mesenchymal stem cells and their derived exosomes in different neurodegenerative disorders: Current status and future perspectives. In: *Handbook of Regenerative Medicine. Cell-free Therapy Approach*. 2025, pp. 55–67.

Ahmed ZT, Alkahlot MH, Haider KH. MSC-derived exosomes: advances in cell-free therapy. In: *Handbook of Stem Cell Applications*. Singapore: Springer, 2023 Aug 18, pp. 1–41.

Aramini B, Masciale V, Haider KH. Defining the exosomal payload of lung cancer stem cell miRNAs from a clinical perspective. *World J. Stem Cells*. 2020 Jun 26;12(6):406–21. https://doi.org/10.4252/wjsc.v12.i6.406.

Bahadorani M, Nasiri M, Dellinger K, Aravamudhan S, Zadegan R. Engineering exosomes for therapeutic applications: decoding biogenesis, content modification, and cargo loading strategies. *Int. J. Nanomedicine*. 2024;19:7137–64.

Batrakova EV, Kim MS. Using exosomes, naturally-equipped nanocarriers, for drug delivery. *J. Control. Release*. 2015;219:396–405.

Besse B, Charrier M, Lapierre V, Dansin E, Lantz O, Planchard D, et al. Dendritic cell-derived exosomes as maintenance immunotherapy after first-line chemotherapy in NSCLC. *Oncoimmunology*. 2016;5(4):e1071008.

Choi H, Choi Y, Yim HY, Mirzaaghasi A, Yoo J-K, Choi C. Biodistribution of exosomes and engineering strategies for targeted delivery of therapeutic exosomes. *Tissue Eng. Regen. Med*. 2021;18(4):499–511.

Choudhery MS, Arif T, Mahmood R, Harris DT. Stem cell-based acellular therapy: Insight into biogenesis, bioengineering, and therapeutic applications of exosomes. *Biomolecules*. 2024;14(7):792.

Corradetti B, González D, Mendes Pinto I, Conlan RS. Exosomes as therapeutic systems. *Front. Cell Dev. Biol*. 2021;9:714743.

Dar GH, Mendes CC, Kuan W-L, Speciale AA, Conceição M, Görgens A, et al. GAPDH controls extracellular vesicle biogenesis and enhances the therapeutic potential of EV-mediated siRNA delivery to the brain. *Nat. Commun*. 2021;12(1):6666.

Fu S, Wang Y, Xia X, Zheng JC. Exosome engineering: Current progress in cargo loading and targeted delivery. *NanoImpact*. 2020;20:100261.

Habib SM, Martini MF, Al Ahmed O, Qassab AR, Haider KH. The potential of mesenchymal stem cell–derived exosomes in cell-free therapy for myocardial repair. In: Haider, K.H. (eds) *Handbook of Regenerative Medicine: Cell-Free Therapy*. London: Taylor & Francis Group/CRC Press, 2025, Chapter 24, pp. 372–90.

Habib SY, Martini MF, Hamdan YNA, Shrebaty OMM, Haider KH. Mesenchymal stem cell secretome and exosomes as potential advanced therapy medicinal products for treating a failing heart. *World J Stem Cells*. 2026;18.

Haider KH. Melatonin-based priming of stem cells to alleviate oxidative stress. *World J. Stem Cells*. 2024b Nov 26;16(11):985–9. https://doi.org/10.4252/wjsc.v16.i11.985.

Haider KH. Priming mesenchymal stem cells to develop "super stem cells." *World J. Stem Cells*. 2024a;16(6):623–40. https://doi.org/10.4252/wjsc.v16.i6.623.

Haider KH, Stimson WH. Production and characterization of monoclonal antibodies specific for cardiac troponin-I. *Disease Markers*. 1994;12:187–91.

Haider KH, Najimi M. Exosome-based cell-free therapy in regenerative medicine for myocardial repair. In: Haider, K.H. (eds) *Handbook of Stem Cell Therapy*. Singapore: Springer, pp. 1007–34. https://doi.org/10.1007/978-981-16-6016-0_9-1.

Jia G, Han Y, An Y, Ding Y, He C, Wang X, et al. NRP-1-targeted and cargo-loaded exosomes facilitate simultaneous imaging and therapy of glioma in vitro and in vivo. *Biomaterials*. 2018;178:302–16.

Kim MS, Haney MJ, Zhao Y, Yuan D, Deygen I, Klyachko NL, et al. Engineering macrophage-derived exosomes for targeted paclitaxel delivery to pulmonary metastases: in vitro and in vivo evaluations. *Nanomedicine*. 2018;14(1):195–204.

Kim SM, Yang Y, Oh SJ, Hong Y, Seo M, Jang M. Cancer-derived exosomes as a delivery platform of CRISPR/Cas9 confer cancer cell tropism-dependent targeting. *J. Control. Release*. 2017;266:8–16.

Kul Y, Erbaş O. Exosomes: Classification, isolation, and therapeutic applications in various diseases. *J. Exp. Basic Med. Sci*. 2022;3(1):006–12.

Koh HB, Kim HJ, Kang S-W, Yoo T-H. Exosome-based drug delivery: Translation from bench to clinic. *Pharmaceutics*. 2023;15(8):2042.

Lamichhane TN, Jeyaram A, Patel DB, Parajuli B, Livingston NK, Arumugasaamy N, et al. Oncogene knockdown via active loading of small RNAs into extracellular vesicles by sonication. *Cell. Mol. Bioengineering*. 2016;9:315–24.

Lee J, Kim J, Jeong M, Lee H, Goh U, Kim H, et al. Liposome-based engineering of cells to package hydrophobic compounds in membrane vesicles for tumor penetration. *Nano Letters*. 2015;15(5):2938–44.

Li S, Wu Y, Ding F, Yang J, Li J, Gao X, et al. Engineering macrophage-derived exosomes for targeted chemotherapy of triple-negative breast cancer. *Nanoscale*. 2020;12(19):10854–62.

Li X, Corbett AL, Taatizadeh E, Tasnim N, Little JP, Garnis C, et al. Challenges and opportunities in exosome research—Perspectives from biology, engineering, and cancer therapy. *APL Bioengineering*. 2019;3(1).

Liang Y, Xia J, Xu L, Zhang Y, Xu X, Haider KH. Cell-free and cell-based therapy in regenerative medicine. In: Haider, K.H. (eds) *Handbook of Regenerative Medicine: Cell-Free Therapy*. London: Taylor & Francis Group/CRC Press, 2025, Chapter 24. https://doi.org/10.1201/9781003611509.

Li Y-J, Wu J-Y, Liu J, Xu W, Qiu X, Huang S, et al. Artificial exosomes for translational nanomedicine. *J Nanobiotechnology*. 2021;19:1–20.

Liang G, Zhu Y, Ali DJ, Tian T, Xu H, Si K, et al. Engineered exosomes for targeted co-delivery of miR-21 inhibitor and chemotherapeutics to reverse drug resistance in colon cancer. *J Nanobiotechnology*. 2020;18:1–15.

Lin Y, Wu J, Gu W, Huang Y, Tong Z, Huang L, et al. Exosome–liposome hybrid nanoparticles deliver the CRISPR/Cas9 system in MSCs. *Adv. Sci.* 2018;5(4):1700611.

Lv L-H, Wan Y-L, Lin Y, Zhang W, Yang M, Li G-L, et al. Anticancer drugs induce the release of exosomes containing heat shock proteins from human hepatocellular carcinoma cells, which elicit effective natural killer cell antitumor responses in vitro. *J. Biol. Chem.* 2012;287(19):15874–85.

Mushtaq M, Zineldeen DH, Mateen MA, Haider Kh. Mesenchymal stem cells' "garbage bags" at work: Treating radial nerve injury with mesenchymal stem cell-derived exosomes. *World J. Stem Cells* 2024;16(5). https://dx.doi.org/10.4252/wjsc.v16.i5.0000.

Najimi M, Haider KH. Extracellular vesicles-based Cell-free therapy in for liver regeneration. In: Haider, K.H. (eds) *Handbook of Stem Cell Therapy*. Singapore: Springer, 2022, pp. 221–38. https://doi.org/10.1007/978-981-16-6016-0_9-1.

Nan W, Zhang C, Wang H, Chen H, Ji S. Direct modification of extracellular vesicles and its applications for cancer therapy: A mini-review. Front. *Chemistry*. 2022;10:910341.

Nie W, Wu G, Zhang J, Huang LL, Ding J, Jiang A, et al. Responsive exosome nano-bioconjugates for synergistic cancer therapy. *Angew. Chem. Int. Ed.* 2020;59(5):2018–22.

Nakase I, Futaki S. Combined treatment with a pH-sensitive fusogenic peptide and cationic lipids enhances exosome cytosolic delivery. *Sci. Rep.* 2015;5(1):10112.

Parada N, Romero-Trujillo A, Georges N, Alcayaga-Miranda F. Camouflage strategies for therapeutic exosomes evasion from phagocytosis. *J. Adv. Res.* 2021;31:61-74.

Pascucci L, Coccè V, Bonomi A, Ami D, Ceccarelli P, Ciusani E, et al. Paclitaxel is incorporated into mesenchymal stromal cells and released via exosomes, inhibiting in vitro tumor growth: a new approach for drug delivery. *J. Control. Release*. 2014;192:262–70.

Pham TC, Jayasinghe MK, Pham TT, Yang Y, Wei L, Usman WM, et al. Covalent conjugation of extracellular vesicles with peptides and nanobodies for targeted therapeutic delivery. *J. Extracell. Vesicles*. 2021;10(4):e12057.

Poinsot V, Pizzinat N, Ong-Meang V. Engineered and mimicked extracellular nanovesicles for therapeutic delivery. *Nanomaterials*. 2024;14(7):639.

Qu M, Lin Q, Huang L, Fu Y, Wang L, He S, et al. Dopamine-loaded blood exosomes targeted to the brain for better treatment of Parkinson's disease. *J. Control. Release*. 2018;287:156–66.

Sadeghi S, Tehrani FR, Tahmasebi S, Shafiee A, Hashemi SM. Exosome engineering in cell therapy and drug delivery. *Inflammopharmacology*. 2023;31(1):145–69.

Sajid MI, Bukhari S, Ilyas H, Malik R, Fatima M, Tiwari RK, Nauli SM, et al. Exosomes and encapsulated exomiRs in breast cancer: Theragnostic applications and clinical implications. In: Haider, K.H. (eds) *Handbook of Regenerative Medicine: Cell-Free Therapy*. London: Taylor & Francis Group/CRC Press,2025, Chapter 23, pp. 357–70.

Steinberg GK, Kondziolka D, Wechsler LR, Lunsford LD, Kim AS, Johnson JN, et al. Two-year safety and clinical outcomes in chronic ischemic stroke patients after implantation of modified bone marrow–derived mesenchymal stem cells (SB623): A phase 1/2a study. *J. Neurosurgery*. 2018;131(5):1462–72.

Susa F, Limongi T, Dumontel B, Vighetto V, Cauda V. Engineered extracellular vesicles as a reliable tool in cancer nanomedicine. *Cancers*. 2019;11(12):1979.

Tian T, Zhang H-X, He C-P, Fan S, Zhu Y-L, Qi C, et al. Surface functionalized exosomes as targeted drug delivery vehicles for cerebral ischemia therapy. *Biomaterials*. 2018;150:137–49.

Tran PH, Wang T, Yin W, Tran TT, Barua HT, Zhang Y, et al. Development of a nanoamorphous exosomal delivery system as an effective biological platform for improved encapsulation of hydrophobic drugs. *Int. J. Pharmaceutics*. 2019a;566:697–707.

Tran PH, Wang T, Yin W, Tran TT, Nguyen TN, Lee B-J, et al. Aspirin-loaded nanoexosomes as cancer therapeutics. *Int. J. Pharmaceutics*. 2019b;572:118786.

Usman WM, Pham TC, Kwok YY, Vu LT, Ma V, Peng B, et al. Efficient RNA drug delivery using red blood cell extracellular vesicles. *Nat. Commun*. 2018;9(1):2359.

Yusuf R, Haider KH. Theragnostic applications of exosomes in Alzheimer's disease. In: Haider, K.H. (eds) *Handbook of Regenerative Medicine: Cell-Free Therapy*. London: Taylor & Francis Group/CRC Press, 2025, Chapter 255, pp. 390–409.

Zhang Y, Bi J, Huang J, Tang Y, Du S, Li P. Exosome: A review of its classification, isolation techniques, storage, diagnostic and targeted therapy applications. *Int. J Nanomed*. 2020;15:6917–34.

Zhuang X, Xiang X, Grizzle W, Sun D, Zhang S, Axtell RC, et al. Treatment of brain inflammatory diseases by delivering exosome-encapsulated anti-inflammatory drugs from the nasal region to the brain. *Mol. Therapy*. 2011;19(10):1769–79.

Zhou Z, Zhang D, Wang Y, Liu C, Wang L, Yuan Y, et al. Urinary exosomes: A promising biomarker of drug-induced nephrotoxicity. *Front. Med*. 2023;10:1251839.

20 Osteoimmunology and Intercellular Communication

The Role of The Secretome and Extracellular Vesicles in Bone Regeneration

Itziar Álvarez-Iglesias[1], Luis Gil-de-Gómez[1], Daniel García-Sánchez[1], Jennifer Cabello-Sanz[1], and Flor M. Pérez-Campo[1]

[1]Departamento de Biología Molecular, Grupo de Ingeniería de Tejidos, Universidad de Cantabria-IDIVAL, Santander, Cantabria, Spain

20.1 LIST OF ABBREVIATIONS

ALP	Alkaline phosphatase
ASM	Acid sphingomyelinase
BMP	Bone morphogenetic protein
BM-MSCs	Bone marrow-derived mesenchymal stem cells
BMPR	Bone morphogenetic protein receptor
BSP	Bone sialoprotein
CD	Cluster of differentiation
COL1	Collagen type I
CXCL	C-X-C motif chemokine ligand
CXCR	C-X-C motif chemokine receptor
DCs	Dendritic cells
DLX5	Distal-less homeobox 5
ECM	Extracellular matrix
EFNA3	Ephrin-A3
ESCRT	Endosomal sorting complex required for transport
EVs	Extracellular vesicles
FGF	Fibroblast growth factor
GVHD	Graft-versus-host disease
HGF	Hepatocyte growth factor
HLA	Human leukocyte antigen
HSP	Heat shock protein
HSPC	Hematopoietic stem and progenitor cells

DOI: 10.1201/9781003653387-20

IDO	Indoleamine 2,3-dioxygenase
IFN-γ	Interferon gamma
IGF-I	Insulin-like growth factor I
IL	Interleukin
ILVs	Intraluminal vesicles
iNOS	Inducible nitric oxide synthase
JAK	Janus kinase
lncRNA	Long non-coding RNA
M-CSF	Macrophage colony-stimulating factor
MAPK	Mitogen-activated protein kinase
miRNA:	miR: MicroRNA
MMP	Matrix metalloproteinase
MVB	Multivesicular body
MSCs	Mesenchymal stem cells
mTOR	Mechanistic target of rapamycin
NF-κB	Nuclear factor kappa-light-chain-enhancer of activated b cells
NFATc1	Nuclear factor of activated T cells, cytoplasmic 1
NK	Natural killer (Cells)
OPG	Osteoprotegerin
OP	Osteoporosis
OPN	Osteopontin
OSX	Osterix (Sp7)
PGE2	Prostaglandin E2
PI3K	Phosphoinositide 3-kinase
PlGF	Placental growth factor
RANK	Receptor activator of nuclear factor κB
RANKL	Receptor activator of nuclear factor κB ligand
RANTES	Regulated on activation, normal T cell expressed and secreted (also known as CCL5)
RUNX2	Runt-related transcription factor 2
SDF-1	Stromal cell-derived factor 1 (CXCL12)
SIBLING	Small integrin-binding ligand N-linked glycoprotein
SLRP	Small leucine-rich proteoglycan
SMAD	Homolog of drosophila mothers against decapentaplegic
SPP1	Secreted phosphoprotein 1 (Osteopontin)
STAT	Signal transducer and activator of transcription
TβRI:	TβRII: TGF-β receptor type I/II
TAB1	TAK1-binding protein 1
TAK1	TGF-β-activated kinase 1
TCA	Tricarboxylic acid (Cycle)
TGF-β	Transforming growth factor beta
Th	T helper (Cells)
TNF	Tumor necrosis factor
TLR	Toll-like receptor
Tr1	Type 1 regulatory T cell
Tregs	Regulatory T cells
VEGF	Vascular endothelial growth factor
VEGF-A	Vascular endothelial growth factor A
VEGFR	Vascular endothelial growth factor receptor

20.2 INTRODUCTION

Bone is a living, dynamic tissue with a remarkable capacity for regeneration. Its ability to heal after injury relies on the coordinated action of bone-resident cells, including mesenchymal stem cells (MSCs), immune cells, and a finely tuned microenvironment. However, this regenerative capacity can be compromised in conditions such as aging, chronic inflammation, or systemic disease, leading to delayed healing or structural fragility. Over the past two decades, the emerging field of osteoimmunology has reshaped our understanding of bone biology, revealing a close and bidirectional relationship between the immune system and skeletal homeostasis (Hu *et al.*, 2023; Okamoto, 2024).

Originally described as a dialog between osteoclasts and T cells, osteoimmunology now encompasses a much broader spectrum of cellular interactions and molecular signals. Immune cells do not simply infiltrate bone during pathology—they are essential participants in bone formation, resorption, and repair. Inflammatory cytokines, such as TNF-α, IL-1β, and IL-17, can drive osteoclastogenesis and bone loss, while regulatory signals like IL-10 and TGF-β support bone formation and immune resolution (Levescot *et al.*, 2021; Capobianco *et al.*, 2024).

At the center of this immune-skeletal dialog are MSCs, multipotent stromal cells that can differentiate into osteoblasts, chondrocytes, and adipocytes while exerting profound immunomodulatory effects through paracrine signaling (Álvarez and Haider, 2022). MSCs respond to inflammatory cues and, in turn, modulate the immune response by releasing a complex mixture of soluble factors and extracellular vesicles (EVs) (Gao *et al.*, 2016; Ferreira *et al.*, 2018; Ahmed *et al.*, 2023). This "secretome" has emerged as a key driver of MSC therapeutic actions (Habib et al., 2026).

Rather than relying on direct engraftment, the regenerative capacity of MSCs is now thought to depend mainly on paracrine signaling (Haider and Aziz, 2017). Their secretome comprises cytokines, chemokines, enzymes, growth factors, and nucleic acids, many of which are also packaged into EVs, such as exosomes and microvesicles (Habib et al., 2025a). These vesicles carry bioactive molecules in a protected, cell-targeted form, enabling long-range intercellular communication within the bone microenvironment (Boilard, 2018; Wan *et al.*, 2024). Significantly, both immune cells (e.g., macrophages and T cells) and bone cells (e.g., osteoclasts and osteoblasts) can be influenced by the content of MSC-derived EVs.

Bone homeostasis depends on a tightly regulated balance between osteoclast-mediated resorption and osteoblast-mediated formation. This balance is influenced not only by mechanical loading and hormonal signals but also by immune-mediated factors. Emerging evidence highlights the role of innate immune cells in bone homeostasis. Macrophages, traditionally classified into pro-inflammatory M1 and anti-inflammatory M2 subsets, display plasticity and can influence both bone resorption and formation depending on their activation status (Hu *et al.*, 2023). Bone-resident macrophages (osteomacs) seem to support osteoblast function and are crucial for fracture healing, while inflammatory macrophages can exacerbate bone loss (Capobianco *et al.*, 2024). Interestingly, MSC-derived EVs have been shown to promote M2 polarization and to inhibit key inflammatory pathways such as NF-κB and NLRP3 inflammasome activation (Wan *et al.*, 2024; Liu *et al.*, 2023).

Similarly, T cells orchestrate key immune–bone interactions, with subsets such as Th17 cells promoting osteoclastogenesis via cytokines, including IL-17. In contrast, regulatory T cells (Tregs) counterbalance this effect by secreting IL-10 and TGF-β, which suppress inflammation and osteoclast activity. Disruption of this immune balance has been implicated in bone loss associated with chronic inflammatory diseases, including rheumatoid arthritis and osteoporosis (Camacho *et al.*, 2020; Okamoto, 2024; Zhu *et al.*, 2024).

Another layer of complexity arises from EVs secreted by immune cells themselves. T cell-derived EVs can promote osteoclast formation, while macrophage-derived EVs influence MSC differentiation and matrix remodeling. These reciprocal interactions suggest that vesicle-mediated communication is a key mechanism in bone-immune crosstalk, relevant not only in pathological conditions but also in physiological bone turnover (Boilard, 2018).

The use of MSC secretomes and EVs as therapeutic tools has generated increasing interest (Habib et al., 2026). These cell-free therapies offer several advantages: they are less immunogenic, more stable, and easier to standardize and store than cell transplants (Liang *et al.*, 2025). Preclinical studies have demonstrated that EVs derived from MSCs can enhance fracture healing, reduce inflammation, and promote angiogenesis and mineralization (Liu *et al.*, 2023; Capobianco *et al.*, 2024; Wan *et al.*, 2024). However, several challenges remain to be addressed before secretome and EV-based therapies can be fully translated into clinical practice. These include standardizing production protocols, precisely characterizing secretome and EV contents, optimizing delivery methods, and elucidating the mechanisms of action *in vivo* (Habib et al., 2025b).

In this chapter, we first outline the fundamental principles of osteoimmunology and describe the cellular and molecular players involved. We then explore the composition and roles of the MSC secretome and EVs, with a special focus on their ability to modulate inflammation and stimulate bone regeneration. Finally, we discuss translational strategies and future perspectives in the field, considering how immune-aware biomaterials, engineered EVs, and advanced delivery systems could transform the management of skeletal disorders.

20.3 BONE BIOLOGY AND THE MICROENVIRONMENT

20.3.1 Structure and Function of Bone Tissue

The skeleton is a specialized form of connective tissue, composed of approximately 70% mineral content, 22% organic phase, and 8% water (Boskey and Posner, 1984). The inorganic matrix consists primarily of calcium and phosphate, which crystallize into hydroxyapatite structures ($Ca_5(PO_4)_3OH$). The dimensions of these crystals vary with age, typically ranging from 20 to 50 nm in length and 12 to 20 nm in width (Murshed, 2018). The adult human skeleton weighs approximately 10 kg, accounting for around 15% of body weight. Out of this, 1–1.5 kg is calcium (constituting about 99% of the body's total calcium) and 0.8–1 kg is phosphate (85% of the body's phosphate), both of which are held in the hydroxyapatite crystals. In addition, magnesium is the third most abundant mineral in the bones, containing 0.03–0.05 kg (50% of the body's magnesium). Other trace minerals contribute an additional 0.1 to 0.2 kg, dispersed throughout the skeleton (Yenchek *et al.*, 2012; Zofkova *et al.*, 2013; Skalny *et al.*, 2023).

The organic component of the bone, also known as osteoid, is predominantly composed of type I collagen fibers interspersed with non-collagenous proteins, such as osteocalcin and osteonectin. Type I collagen, the most abundant protein in the human body, accounts for approximately 90% of the osteoid. It is composed of two $\alpha 1$ chains and one $\alpha 2$ chain, which form a triple-helix structure, imparting tensile strength and structural organization to the osteoid (Shoulders and Raines, 2009). Collagen serves as a scaffold for mineral deposition and guides the alignment of hydroxyapatite crystals during mineralization. Mutations in the genes encoding collagen chains (e.g., *COL1A1* or *COL1A2*) result in disorders such as osteogenesis imperfecta or Ehlers-Danlos Syndrome (Van Dijk and Sillence, 2014). The remaining 10% of the osteoid consists of non-collagenous proteins, which are essential for regulating bone formation, remodeling, and mineralization. These include:

- **Proteoglycans:** Glycoproteins with glycosaminoglycan (GAG) chains that help hydrate the bone matrix (e.g., heparan sulfate, hyaluronan, small leucine-rich proteoglycans (SLRPs), and versican) (Hua and Jiang, 2021).
- **Glycoproteins:** Mediators of osteoblast and osteoclast activity involved in bone remodeling. Some of these proteins are alkaline phosphatase (ALP), osteonectin, and fibronectin (Carvalho *et al.*, 2021; Hamamura *et al.*, 2024).
- **SIBLING Proteins:** Small integrin-binding ligand N-linked glycoproteins that facilitate cell adhesion and signaling (e.g., osteopontin, sialoproteins) (Staines *et al.*, 2012).

- **Osteocalcin:** A γ-carboxyglutamic acid-containing protein that binds to hydroxyapatite crystals and supports mineral deposition (Neve *et al.*, 2013).

The bone matrix is organized into two distinct macroscopic structures: cortical (compact) bone and trabecular (spongy) bone. Cortical bone forms the dense outer layer of most bones and constitutes 80% of total bone mass. It is predominant in the diaphysis of long bones and the outer layers of short, flat, and irregular bones. Cortical bone displays central channels (Haversian and Volkmann's canals) that house blood and lymphatic vessels and nerves. Concentric lamellae surround these. Cortical bone serves to protect marrow, acts as a reservoir of calcium, and supports nutrient and waste exchange (Morgan et al., 2018). Trabecular bone consists of porous, lattice-like trabeculae (approximately 200 μm) and is found at the epiphyses of long bones and inside flat bones such as ribs, skull, and pelvis (Oftadeh et al., 2015). The trabecular network is filled with bone marrow and plays a key role in hematopoiesis. It is also responsible for distributing mechanical loads across bone structures.

Bone fulfills several essential physiological functions:

- Protection of internal organs (e.g., brain, heart, lungs, spinal cord).
- Support and movement through the musculoskeletal system.
- Hormones like parathyroid hormone and calcitonin regulate mineral homeostasis.
- Hematopoiesis in red marrow within trabecular bone.
- Endocrine regulation, e.g., secretion of osteocalcin and fibroblast growth factor 23 (FGF23) (DiGirolamo *et al.*, 2012).
- Energy storage via yellow bone marrow, which contains adipocytes.

To accomplish its functions, bone undergoes continuous remodeling every 4–6 months, allowing for adaptation to mechanical stress and repair of microdamage. This bone remodeling process involves two sequential stages: resorption by osteoclasts, followed by formation by osteoblasts (Martin and Seeman, 2008; Eriksen, 2010). The equilibrium between these two phases is tightly regulated, and its disruption can lead to bone pathologies, such as osteoporosis.

20.4 MAIN CELLULAR COMPONENTS OF THE BONE MICROENVIRONMENT

Bone-specific cells, including osteoblasts, osteoclasts, and osteocytes, each play key and interdependent roles in maintaining bone homeostasis. Osteoblasts (OBs) originate from MSCs, with their differentiation driven by the runt-related transcription factor 2 (RUNX2) (Ziros *et al.*, 2008). They represent 5% of all bone cells and are responsible for synthesizing collagen and mineralizing the matrix. During matrix deposition, osteoblasts become embedded in the newly formed bone and differentiate into osteocytes. In addition to their matrix-forming capacity, osteoblasts secrete several bioactive proteins that play key roles in bone homeostasis. Among them, the receptor activator of nuclear factor kB (RANKL) and osteoprotegerin (OPG) are needed for the bone remodeling process, vascular endothelial growth factor A (VEGF-A) is a major pro-angiogenic factor required for the vascularization of the bone tissue, and macrophage colony-stimulating factor 1 (M-CSF1) is essential for osteoclast differentiation, survival, and motility (Sims and Walsh, 2012).

Osteoclasts are multinucleated cells derived from the monocyte/macrophage lineage. M-CSF1 and RANKL stimulate their differentiation. Once differentiated, osteoclasts interact with the bone surface through their ruffled border membrane to mediate bone resorption by pumping protons and creating an acidic microenvironment. This microenvironment will dissolve hydroxyapatite and activate matrix metalloproteinases (MMPs) and cysteine proteinases to degrade the collagenous bone matrix (Peruzzi and Teti, 2011). To maintain the balance of the bone remodeling process, osteoblasts tightly regulate osteoclast activity through the RANKL/RANK/OPG axis, thus preserving skeletal integrity.

Lastly, osteocytes are the most abundant cells in bones, accounting for approximately 90% of all bone cells. They are embedded in the bone matrix and extend dendritic processes through canaliculi to communicate with other bone and surface cells (Kitaura *et al.*, 2020). Osteocytes serve as mechanosensors and key regulators of bone remodeling, orchestrating this process through the release of signaling molecules, including M-CSF, RANKL, OPG, and Wnt pathway inhibitors such as Dickkopf1 (DKK1) (Nakashima *et al.*, 2011; Kitaura *et al.*, 2020).

MSCs, although not terminally differentiated, play a pivotal role within the bone microenvironment. These multipotent progenitors are capable of differentiating into osteoblasts, chondrocytes, and adipocytes (Ding *et al.*, 2011). Importantly, MSCs also exert immunomodulatory functions. They lack HLA class II molecules and co-stimulatory surface proteins, such as CD40, which enable them to modulate the activity of T cells, B cells, natural killer (NK) cells, macrophages, and dendritic cells (Gao *et al.*, 2016). These immune-regulatory properties make MSCs essential contributors to both skeletal regeneration and the maintenance of immune balance during tissue repair.

20.4.1 Immune Cells in the Bone Microenvironment

In addition to classical bone-resident cells, a diverse population of immune cells, including macrophages, T cells, and B cells, resides in the bone marrow and plays crucial roles in maintaining skeletal homeostasis. These immune cells interact closely with osteoblasts, osteoclasts, and osteocytes, modulating their activity through paracrine signals, direct cell–cell contact, and the secretion of regulatory molecules, including cytokines and growth factors.

Macrophages, in particular, are highly adaptable innate immune cells that participate in both inflammatory responses and tissue regeneration. Within the bone niche, they regulate remodeling processes and influence the behavior of bone-forming and bone-resorbing cells, depending on physiological or pathological conditions. Their ability to alter their function in response to environmental stimuli makes them critical mediators of skeletal-immune crosstalk (Yunna *et al.*, 2020). T and B lymphocytes also contribute to the fine-tuning of bone metabolism. These cells respond to local and systemic signals and exert either stimulatory or inhibitory effects on bone turnover. Their involvement becomes especially relevant in the context of inflammation or immune imbalance, which can profoundly affect the bone microenvironment (El Khassawna *et al.*, 2017).

Together, the interplay between bone and immune cells creates a dynamic and tightly regulated niche that is essential for proper bone remodeling and repair. While bone cells serve as the direct effectors of structural changes—forming, resorbing, and sensing matrix dynamics—immune cells contribute by shaping the local environment and influencing the direction and intensity of the remodeling response. This complex interaction forms the basis of the field of osteoimmunology, which will be explored in greater detail in the following section.

20.4.2 Osteoimmunology: The Interplay between Bone and Immunity

Osteoimmunology is a cross-disciplinary field that describes the bidirectional interactions between immune cells and bone cells, in which immune responses influence bone formation and resorption (Adamopoulos *et al.*, 2025). Under physiological conditions, osteocytes and immune cells orchestrate the balance between osteoblast and osteoclast activity to maintain the equilibrium at the intraosseous microenvironment. Dysregulated immune function alters the bone microenvironment through various immune-derived signals, including cytokines and chemokines. Conversely, disruptions in the bone marrow niche's homeostasis can also affect immune cells, leading to immune dysfunction.

20.5 ROLES OF IMMUNE CELLS IN BONE REMODELING

Macrophages, T cells, and B cells play crucial roles in regulating bone turnover, both during development and in response to injury or disease.

20.5.1 Macrophages

Macrophages are highly plastic and dynamic innate immune cells that play a crucial role in bone immunity through their interaction with bone-resident cells (Hu *et al.*, 2023). These cells can phagocytose pathogens, cell debris, and apoptotic cells, as well as various exogenous and endogenous particles. In response to such stimuli, macrophages secrete a broad array of immune-regulatory molecules, particularly cytokines and chemokines, positioning them as key regulators of inflammation (J. Muñoz *et al.*, 2020). As immune effector cells, macrophages are closely linked to both bone formation and bone resorption, interacting directly with MSCs, osteoclasts, and osteoblasts. Through this extensive crosstalk, they contribute to the pathophysiology of skeletal tissues.

One of the hallmarks of macrophages is their phenotypic plasticity. In a process known as polarization, macrophages adapt their functional profile in response to environmental cues, modifying their phenotype to meet specific physiological or pathological demands (Razi *et al.*, 2023). Based on their activation state, they can be broadly classified into two subtypes: pro-inflammatory M1 and anti-inflammatory M2 macrophages.

M1 macrophages secrete a spectrum of inflammatory cytokines such as TNF-α, IL-1β, and IL-6, along with chemokines such as CXCL2, CXCL8, and CXCL10 (CXC motif chemokine Ligands 2, 8, and 10). These mediators enhance the differentiation and maturation of osteocytes, which are key regulators of bone metabolism and resorption. Notably, the transcription factor nuclear factor kappa B (NF-κB), which is constitutively active in M1 macrophages, regulates the expression of many of these pro-inflammatory genes and also enhances the osteoclastogenic potential of the bone microenvironment. Similarly, macrophage colony-stimulating factor (M-CSF), which is essential for macrophage survival and proliferation, also promotes the commitment of myeloid precursors toward the osteoclast lineage. It is also noteworthy that osteoclasts and macrophages share a common myeloid progenitor and, thus, may compete for lineage commitment. This competition can ultimately impact bone resorption dynamics by modulating osteoclastogenesis (Hu *et al.*, 2023).

When macrophage polarization is skewed toward the M1 phenotype and inflammatory factors are chronically overproduced, a persistent inflammatory state is established at the bone marrow, leading to disruption of bone homeostasis. Chronic inflammation of this nature underlies the pathophysiology of various bone disorders, including osteoporosis, rheumatoid arthritis, and steroid-induced avascular necrosis of the femoral head (SANFH), all of which are characterized by bone tissue degradation and impaired bone homeostasis (Molendijk *et al.*, 2018; J. Muñoz *et al.*, 2020; M. Muñoz et al., 2020; Wang *et al.*, 2018).

In contrast, M2 macrophages secrete anti-inflammatory cytokines such as IL-4 and IL-10, which inhibit osteoclast formation and play a crucial role in osteogenic activity and osteoblast differentiation. M2 macrophages exhibit a strong osteoinductive capacity, promoting bone and cartilage development and accelerating bone regeneration and repair. The phenotypic transition from M1 to M2 supports osteoblasts in upregulating osteogenic genes, such as *RUNX2*, *OPN*, osteocalcin, *COL1A1*, and *ALP*, thereby enhancing their capacity for matrix synthesis, secretion, and mineralization (You *et al.*, 2022). Moreover, M2 macrophages produce bone morphogenetic proteins 2 and 4 (BMP-2 and BMP-4), which further promote osteogenesis. Following BMP-2 recognition by the BMP receptor, BMP-2 induces the translocation of Runx2 to the nucleus by SMAD1 phosphorylation (Wu *et al.*, 2016). Consequently, the mineralization process is facilitated by the upregulation of alkaline phosphatase (ALP) and osteocalcin (Vimalraj, 2020).

20.5.2 T Lymphocytes

The bone microenvironment plays a fundamental role in maintaining physiological immune tolerance while limiting pathological immune responses. However, under inflammatory conditions, enhanced T cell activation and cytokine production disrupt the balance of bone remodeling, contributing to the pathogenesis of diseases such as OP and rheumatoid arthritis (De Filippo *et al.*,

2013; Song *et al.*, 2024). The functional heterogeneity and dynamic behavior of T cells support their role in regulating bone homeostasis. For instance, in a murine model of palatal expansion, an early increase in type 1 helper T (Th1) cells and type 17 helper T (Th17) cells was shown to promote osteoclastogenesis. This initial response was subsequently followed by the recruitment of regulatory T cells (Tregs), which ultimately suppressed Th1 and Th17 activity, thereby indirectly reducing osteoclast function (Li *et al.*, 2020).

Th1 cells are a pro-inflammatory T cell subset that plays a central role in orchestrating inflammatory responses. They have been shown to inhibit chondrocyte proliferation and induce apoptosis via interferon-mediated mechanisms, ultimately impairing both cartilage and bone repair capacity (Platzer *et al.*, 2020). Elevated numbers of activated Th1 cells are typically found in the synovial tissue of individuals with bone-related diseases, such as osteoarthritis (Yamada *et al.*, 2011). Th1 cells secrete inflammatory cytokines such as interferon-γ (IFN-γ), as well as chemokine receptors like CXCR3 and CCR5, which stimulate synovial macrophages to polarize toward the M1 phenotype. These M1 macrophages, in turn, release TNF-α, IL-1β, and IL-6, which foster synovial inflammation and the secretion of matrix-degrading enzymes such as MMP-3 and MMP-9 (Platzer *et al.*, 2020; Hu *et al.*, 2025). In addition, Th1 cells amplify inflammation through a self-sustaining feedback loop that prolongs the inflammatory state and drives chronic degradation of the skeletal system (Nabbe *et al.*, 2003).

Th17 cells, another significant pro-inflammatory T cell subset, play a crucial role in the progression of disease. Studies have demonstrated significantly higher frequencies of Th17 cells and elevated serum IL-17 levels in patients with osteoarthritis compared to healthy controls. Moreover, IL-17 levels have been shown to positively correlate with disease severity, highlighting its central role in joint inflammation and degeneration (Okuyan *et al.*, 2019). IL-17 not only promotes cartilage degradation via upregulation of matrix-degrading enzymes, including MMP and ADAMTS family proteins (Na *et al.*, 2020), but also induces senescence in chondrocytes and fibroblasts. Additionally, IL-17 acts synergistically with pro-inflammatory factors such as TNF-α and IL-1β to stimulate receptor activator of nuclear factor kappa B ligand (RANKL) expression, thereby enhancing osteoclastogenesis and contributing to inflammation-associated bone destruction (Lubberts *et al.*, 2003). In addition, IL-17 stimulation also enhances polarization of macrophages towards the pro-inflammatory M1 phenotype and promotes synovial angiogenesis and immune cell infiltration (Honorati *et al.*, 2006). Moreover, it suppresses tissue repair by inducing MCP-1 expression in synovial macrophages and enhancing JAK/STAT3-mediated monocyte recruitment, exacerbating the local inflammatory response (Faust *et al.*, 2020).

In contrast, Th2 cells help reduce inflammation and promote tissue regeneration. They exert an anti-inflammatory effect by releasing cytokines such as IL-4, IL-10, and IL-13, which help resolve inflammation and support repair mechanisms. IL-4 and IL-10 promote the polarization of synovial macrophages toward the M2 phenotype, thus reducing the release of inflammatory mediators and indirectly facilitating cartilage regeneration (Raphael et al., 2015). However, in persistent inflammatory conditions, the regulatory functions of Th2 cells may be insufficient to counteract tissue damage.

Unlike Th2 cells, Tregs have the capacity to limit excessive inflammatory responses. Tregs comprise approximately 30% of CD4+ T cells in the bone marrow and play a crucial role in maintaining immune homeostasis and preserving bone integrity (Zhao *et al.*, 2012). Bone marrow-resident Tregs exhibit distinct phenotypic features compared to their peripheral counterparts. They express markers such as CD150, which is typical of hematopoietic stem cells (HSCs), and are commonly found near HSCs in the bone marrow niche *(Hirata et al.*, 2018). The trafficking of bone marrow Tregs relies on the chemokine CXCL12, which is secreted primarily by LepR+ MSCs (Ding and Morrison, 2013). IL-10 secreted by Tregs can act on MSCs via IL-10 receptors (IL-10R), promoting phenotypic changes that ultimately support the expansion of hematopoietic stem and progenitor cells (HSPC) under stress conditions. This effect creates a regulatory feedback loop between Tregs and MSCs within the bone marrow (Camacho *et al.*, 2020).

In osteoarthritis, Tregs play a protective role by attenuating synovial inflammation and cartilage degradation. Mechanistically, they suppress the pro-inflammatory activity of Th1 and Th17 cells and secrete anti-inflammatory cytokines such as TGF-β and IL-10. Notably, the number and functionality of Tregs are often reduced in osteoarthritis patients, and an imbalance in the Th17/Treg ratio is strongly linked to uncontrolled inflammation and accelerated disease progression (Li *et al.*, 2022; Xia *et al.*, 2024). A promising strategy for joint preservation involves lipid nanoparticles designed to interact antigen-specifically with regulatory T cells (Tregs). This approach modulates cytokine expression, reduces immune cell infiltration into the joint, prevents chondrocyte apoptosis, and protects against cartilage matrix degradation in osteoarthritis models (McHugh, 2023). Nevertheless, chronic inflammation is often accompanied by insufficient immunosuppression, suggesting that enhancing Treg function may be necessary to improve the efficacy of current therapeutic approaches (Hu, Song *et al.*, 2025).

20.5.3 B Lymphocytes

B lymphocytes play a dual role in the pathophysiology of bone diseases by triggering both pro-inflammatory and anti-inflammatory responses. Under physiological conditions, B cells contribute to bone stability by secreting osteoprotegerin (OPG), a decoy receptor for RANK that inhibits its interaction with RANKL. The presence of OPG helps maintain the balance between bone formation and resorption (Horowitz et al., 2005). In contrast, under chronic inflammatory conditions, B cells upregulate RANKL expression while reducing OPG secretion. This imbalance between RANKL and OPG promotes osteoclastogenesis, leading to subchondral bone resorption and joint structural degradation (Guo *et al.*, 2020). In this pathological context, B cells participate in multiple immune mechanisms (Karampetsou *et al.*, 2019). Through direct interactions, B cells activate T cells that, in turn, promote the secretion of pro-inflammatory cytokines such as IFN-γ and IL-17. These cytokines drive synovial macrophage polarization toward the M1 phenotype, exacerbating local synovial inflammation and contributing to bone degradation by activating osteoclasts. This process results in irreversible structural lesions in osteoarthritis patients, characterized by extensive subchondral bone loss and joint degeneration (Hu *et al.*, 2025). Indeed, B cells are detected in the synovium in approximately 50% of patients with osteoarthritis, with a strong correlation observed between the degree of lymphocytic infiltration and disease progression (Da *et al.*, 2007).

Moreover, the function of B cells appears to be impaired under inflammatory conditions. For example, osteoarthritis patients exhibit accelerated differentiation of B cells into antibody-secreting plasma cells compared to healthy individuals (Xie *et al.*, 2023). This altered functional state may further influence the inflammatory milieu and the progression of bone pathology.

This complex interplay between innate and adaptive immune cells, as well as bone-resident cells, underscores the dynamic regulation of skeletal homeostasis by immune-derived factors. While direct cell–cell interactions are fundamental, secreted molecules such as cytokines, chemokines, and growth factors also serve as pivotal mediators of this crosstalk, influencing both physiological remodeling and disease-associated bone loss.

20.5.4 Cytokine Signaling in Osteoimmunology

Cytokines are key signaling molecules in osteoimmunology, functioning as critical mediators of the dialog between immune and skeletal cells. Produced by macrophages, T cells, B cells, and stromal cells, these molecules orchestrate inflammatory and regenerative responses in bone tissue. Pro-inflammatory cytokines, such as TNF-α, IL-1β, and IL-6, can enhance osteoclastogenesis and impair osteoblast activity, thereby contributing to bone loss in chronic inflammation. In contrast, anti-inflammatory cytokines like IL-10, IL-4, and TGF-β promote tissue repair and support osteogenic differentiation. These dual effects of cytokine signaling are highly context-dependent and play a decisive role in shaping the bone microenvironment. A deeper understanding of these signaling pathways is crucial for developing therapeutic strategies that rely on paracrine modulation. The following section explores in detail the role of the MSC-derived secretome—including cytokines, growth factors, and EVs—in bone regeneration.

20.6 THE MSC-DERIVED SECRETOME IN BONE REGENERATION

MSCs exert many of their regenerative and immunomodulatory effects through the secretion of a complex array of bioactive molecules collectively known as the "secretome" (Habib et al., 2025a). This mixture includes growth factors, cytokines, chemokines, enzymes, and EVs, which together influence bone formation, remodeling, and immune responses. In this section, we examine the primary molecular constituents of the MSC secretome, with a specific focus on the roles of pro- and anti-inflammatory cytokines, and discuss their translational implications for bone regeneration.

20.6.1 The Osteoimmunomodulatory Potential of the MSC Secretome

The secretome derived from MSCs plays a pivotal role in bone regeneration by modulating immune responses, promoting osteogenesis, and coordinating tissue remodeling. This mixture of soluble factors, including cytokines, growth factors, chemokines, and enzymes, mediates dynamic interactions between skeletal and immune cells, making it a central element of osteoimmunology.

20.6.2 Key Components of the MSC Secretome with Osteoimmunological Relevance

20.6.2.1 Bone Morphogenetic Proteins

The MSCs' secretome comprises a wide range of molecules that can influence bone cells and immune effectors. Among these, BMPs play a crucial role due to their effect on osteoblast differentiation. These factors, within the TGF-β superfamily (Urist, 1997; Rahman *et al.*, 2015), stimulate the expression of essential osteogenic transcription factors such as RUNX2, distal-less homeobox 5 (DLX5), and Sp7 (also known as Osterix), which orchestrate the transcriptional activation of a suite of osteoblastic genetic markers (COL1, osteocalcin, OPN, BSP, and ALP). Collectively, these molecular events not only govern osteoblastic differentiation but also regulate the synthesis and mineralization of the bone matrix by mature osteoblasts (Hyc *et al.*, 2022).

The osteogenic effects of BMPs are mediated through their binding to BMP type I (BMPR-I) and BMP type II (BMPR-II) receptors, which form heterotetrameric complexes that activate intracellular signaling via both canonical (Smad-dependent) and non-canonical (Smad-independent) pathways (Wan and Cao, 2005). In the canonical pathway, receptor activation leads to phosphorylation of Smad1/5/8, which form complexes with Smad4 and translocate into the nucleus to regulate gene expression involved in differentiation and matrix formation.

In parallel, BMPs also trigger non-canonical signaling. Specifically, BMPR-I activation can stimulate the MAPK pathway through TGF-β-activated kinase 1 and TAK1-binding protein 1 (TAB1), leading to p38 MAPK activation and subsequent upregulation of key transcription factors such as RUNX2, DLX5, and OSX, promoting osteogenesis independently of Smads (Hyc *et al.*, 2022).

20.6.2.2 Vascular Endothelial Growth Factor

Vascular endothelial growth factor (VEGF) is a key mediator of angiogenesis and vasculogenesis, playing a central role in both vascular development and skeletal tissue repair. VEGF belongs to a family of homodimeric proteins consisting of at least six members: VEGF-A (commonly referred to as VEGF), VEGF-B, VEGF-C, VEGF-D, VEGF-E, and placental growth factor (PlGF) (Hu and Olsen, 2016). Among these, VEGF-A is the most extensively studied in the context of bone biology.

VEGF expression is tightly regulated by signaling pathways involved in bone development and repair. Members of the TGF-β superfamily, including bone morphogenetic proteins BMP-2, BMP-4, and BMP-7, stimulate VEGF expression during osteogenic differentiation. Inflammatory mediators also modulate VEGF levels in early repair phases. IL-1, IL-6, IL-8, and prostaglandins E1 and E2, which increase after injury, are potent inducers of VEGF (Hu and Olsen, 2016). Following a fracture, hematoma formation establishes a microenvironment with VEGF levels up to 15 times those in plasma. This not only supports angiogenesis but also stimulates VEGF expression in local bone and inflammatory cells. Among these inflammatory cells, macrophages play a key role in secreting cytokines such as TNF-α, IL-1α, and IL-1β, which activate local endothelial cells and promote neovascularization at the injury site. VEGF acts as a chemotactic factor for macrophages, thereby facilitating their recruitment to the injury site. Given that macrophages themselves release angiogenic factors, VEGF contributes to angiogenesis both indirectly, by attracting macrophages, and directly, by acting on endothelial cells to stimulate new vessel formation (Hu and Olsen, 2016).

VEGF produced by osteoblast precursors is also essential during endochondral ossification, where it facilitates vascular invasion and recruitment of osteogenic and chondroclastic cells. OSX, expressed by hypertrophic chondrocytes within the developing cartilage, is a potent transcriptional regulator of VEGF expression. The upregulation of VEGF in these cells subsequently promotes the recruitment of chondroclasts to the hypertrophic cartilage, enhancing cartilage resorption and facilitating its replacement with mineralized bone tissue (Hu and Olsen, 2016). During endochondral ossification, VEGF produced by osteoblast precursors stimulates the migration of osteogenic cells along with blood vessels to the primary ossification center.

Finally, hypoxia, characteristic of the early inflammatory stages, further enhances VEGF production by osteoblasts and progenitors, ensuring the delivery of oxygen and osteogenic precursors. In turn, BMPs released from new vessels act synergistically to stimulate bone formation (Hu and Olsen, 2016).

20.6.2.3 Transforming Growth Factor Beta

The TGF-β superfamily comprises more than 40 multifunctional cytokines, which are subdivided into four subfamilies. Among them, the TGF-β subfamily, exclusive to mammals, is one of the most abundant cytokine families present in the bone matrix, where it plays pivotal roles in bone remodeling and immune regulation (Crane and Cao, 2014). TGF-β1, the predominant isoform, is highly concentrated in bone tissue and becomes locally activated during osteoclastic resorption. This activation leads to the formation of TGF-β gradients at bone resorption sites, which guide the recruitment of perivascular MSCs and promote their differentiation into osteoblasts, contributing to new bone formation (Crane and Cao, 2014).

The roles of TGF-β1 in osteoblastogenesis are stage-dependent. In the early phases, TGF-β1 promotes MSC proliferation and migration, supporting osteogenic progression. However, at later stages, it inhibits MSCs' differentiation into osteoblasts and suppresses matrix mineralization capacity in mature osteoblasts (Lamora *et al.*, 2016). TGF-β1 also exhibits dose-dependent effects on osteoclastogenesis. At low concentrations, it promotes the migration and maturation of osteoclast precursors, whereas at high concentrations, it inhibits these processes. This inhibition is mediated by TGF-β1-induced modulation of osteoblast RANKL and OPG expression. Notably, *in vivo* studies have shown that under certain conditions, TGF-βs may favor bone resorption, suggesting a dual role in skeletal homeostasis (Lamora *et al.*, 2016).

At the molecular level, TGF-β signals primarily through the canonical Smad-dependent pathway. TGF-β ligands bind to type II TGF-β receptors (TβRII), which transphosphorylate type I receptors (TβRI). The activated TβRI then phosphorylates receptor-regulated Smads (R-Smads), mainly SMAD2 and SMAD3. These receptor-regulated Smads form complexes with the common mediator SMAD4 and translocate into the nucleus to regulate genes involved in osteoblast and osteoclast differentiation, as well as immune signaling (Lamora *et al.*, 2016).

Beyond its role in bone, TGF-β1 has broad immunomodulatory effects. It can directly suppress immune effector cells, such as T cells, B cells, NK cells, and macrophages, or indirectly promote immune tolerance by enhancing Treg differentiation. Interestingly, TGF-β signaling can suppress MSC immunosuppression by inhibiting the expression of iNOS and indoleamine 2,3-dioxygenase (IDO) in response to inflammatory stimuli. In macrophages, this suppression occurs via iNOS protein degradation, while in MSCs, the TGF-β-induced suppression of iNOS expression is Smad3 dependent (Xu *et al.*, 2014).

TGF-β also affects the homeostasis of T helper (Th) cells. It inhibits Th1 differentiation by suppressing Th2 differentiation through the downregulation of GATA-3 and IL-4-mediated STAT6 signaling (Xu *et al.*, 2014). Moreover, TGF-β negatively regulates CD8+ T cell function. It negatively affects CD8+ T cell maintenance by antagonizing interleukin-15 (IL-15), which is essential for the survival and expansion of memory CD8+ T cells (Sheng *et al.*, 2015). Additional immunoregulatory roles include the suppression of B cell proliferation and the inhibition of NK cell activation. In the absence of TGF-β signaling, NK cells exhibit increased IFN-γ production and hyperproliferation, highlighting their essential role in controlling cytokine balance and immune cell expansion (Sheng *et al.*, 2015).

TGF-β also downregulates macrophage activation by reducing the expression of iNOS and MMP-12, further reinforcing its role as a key modulator of bone-immune homeostasis via Smad3-dependent mechanisms (Sheng *et al.*, 2015). These findings underscore the versatile and context-dependent functions of TGF-β in coordinating bone remodeling and immune homeostasis.

20.6.2.4 Cytokines and Chemokines

Cytokines and chemokines act as central mediators of bone remodeling, balancing bone formation and resorption by influencing osteoblasts, osteoclasts, and immune cells. This section focuses on key pro- and anti-inflammatory cytokines—namely TNF, IL-6, and IL-10—as well as chemokines such as CXCL12, which regulate the trafficking of immune cells and skeletal regeneration.

20.6.2.5 Tumor Necrosis Factor

TNF, including TNF-α and TNF-β, is a potent stimulator of osteoclastogenesis and a key contributor to pathological bone loss. TNF-α promotes osteoclast precursor cells by inducing RANKL and M-CSF expression in osteoclasts, stromal cells, and T cells, thereby enhancing osteoclast recruitment and activation (J. Xu *et al.*, 2023) In addition, TNF-α can act through RANKL-independent mechanisms, activating osteoclastogenesis via the NF-κB and PI3K/Akt signaling pathways (J. Xu *et al.*, 2023).

Clinically, elevated TNF-α levels are associated with conditions such as postmenopausal osteoporosis, where increased RANKL and decreased estrogen levels amplify its osteocatabolic effects. Notably, TNF-α displays a biphasic, concentration-dependent effect: at low levels, it may stimulate MSC osteogenic differentiation, whereas at high levels, it suppresses IGF-I expression and inhibits RUNX2 activity, impairing osteoblast maturation and function (J. Xu *et al.*, 2023).

20.6.2.6 Interleukin-6

Interleukin-6 (IL-6) belongs to a cytokine family that signals through the IL-6 receptor (IL-6R). This cytokine is produced by a variety of bone and immune cells, including osteoblasts, bone marrow-derived MSCs (BM-derived MSCs), osteoclasts, macrophages, T cells, and neutrophils. In bone, IL-6 primarily acts as a pro-osteoclastogenic cytokine, facilitating bone resorption through direct and indirect mechanisms (J. Xu *et al.*, 2023). It can directly stimulate osteoclastogenesis independently of RANKL and enhance this process by upregulating RANKL expression, a process mediated by the JAK/STAT3, SHP2/MEK2, and AHP2/AKT pathways. Beyond its pro-inflammatory role, IL-6 also plays a crucial role in the resolution of inflammation. It promotes M2 macrophage polarization by upregulating M2-associated markers and supports the production of IL-4 and IL-10 by naïve CD4+ T cells, contributing to a shift toward an anti-inflammatory Th2 phenotype (Newman *et al.*, 2021).

20.6.2.7 Interleukin-10

Interleukin-10 (IL-10) is a central Th2-type cytokine with broad anti-inflammatory and osteoprotective properties. Clinical studies indicate lower IL-10 levels in patients with osteoporosis compared to healthy individuals. Activated T and B lymphocytes secrete IL-10 and directly inhibit both osteoclast and osteoblast differentiation, primarily through modulation of the RANK/RANKL/OPG axis. Specifically, IL-10 upregulates OPG, reduces RANK and M-CSF expression in osteoclast precursors, and inhibits nuclear factor of activated T cells c1 (NFATc1) activation, a master transcription factor required for osteoclastogenesis (J. Xu *et al.*, 2023).

Moreover, IL-10 is secreted by various immune cells, including neutrophils, macrophages, dendritic cells, Th2 cells, and regulatory T cells, and plays a crucial role in resolving inflammation and facilitating bone repair. It supports macrophage polarization toward the M2 anti-inflammatory phenotype, thereby promoting a regenerative microenvironment (Newman *et al.*, 2021).

20.6.2.8 Chemokines and Their Receptors

Chemokines orchestrate the trafficking of immune cells and contribute to bone homeostasis by regulating the localization and recruitment of both bone-resident and inflammatory cells. One of these chemokines, CCL5 (RANTES), binds to multiple receptors, including CCR1, CCR3, CCR4, and CCR5, all of which are expressed in primary osteoblasts and might regulate their chemotaxis during bone repair. CXCL12 (stromal cell-derived factor-1, SDF-1) is one of the best-characterized chemokines in skeletal biology. It is indispensable for embryonic development, pro-angiogenic processes, and the recruitment of endothelial progenitors. Within the bone marrow, CXCL12 is produced by perivascular, endothelial, and skeletal progenitor cells. It maintains and supports hematopoietic stem cell niches, guiding the recruitment of progenitors and immune cells during regeneration. Finally, CXCL2, predominantly secreted by monocytes and macrophages, recruits neutrophils via CXCR2 and enhances osteoclastogenesis, especially when co-stimulated with RANKL. CXCL1, another CXCR2 ligand, exhibits similar effects (J. Xu *et al.*, 2023). These chemokines integrate immune cell trafficking with bone resorption and formation, reinforcing their role at the core of the osteoimmune axis.

20.6.2.9 Other Proteins and Enzymes

In the bone microenvironment, the interplay between extracellular matrix (ECM) components and signaling molecules is essential for coordinating skeletal remodeling and immune responses. Among the most relevant proteins secreted by MSCs are osteopontin and matrix MMPs, which function both structurally and as active participants in inflammation and repair processes.

20.6.2.9.1 Osteopontin

Osteopontin, also known as secreted phosphoprotein-1 (SPP1), is a glycosylated and highly phosphorylated ECM protein expressed by osteoblasts, osteoclasts, and osteocytes. It plays critical roles in cell adhesion, matrix organization, and immune cell recruitment. During bone formation, osteopontin is concentrated within the mineralized collagen matrix, where it enhances the adhesion of osteoblasts and osteoclasts, thereby facilitating their attachment to the bone surface. This adhesive property is primarily mediated by the integrin $\alpha v\beta 3$, which recognizes the RGD (arginine–glycine–aspartic acid) sequence within osteopontin (Icer and Gezmen-Karadag, 2018).

Osteopontin also regulates the structural integrity of bone tissue by organizing type I collagen fibrils and mediating matrix–matrix and cell–matrix interactions essential for maintaining mechanical strength (Depalle *et al.*, 2021). Beyond its structural and adhesive functions in bone, osteopontin acts as a pro-inflammatory cytokine-like molecule, serving as a potent chemoattractant for immune cells (macrophages and T lymphocytes) and facilitating migration to inflammatory sites. This dual role positions osteopontin as a critical mediator linking skeletal remodeling with immune system activity during injury, infection, and chronic inflammation (Icer and Gezmen-Karadag, 2018).

20.6.2.9.2 Matrix Metalloproteinases

MMPs represent a family of zinc- and calcium-dependent endopeptidases with critical roles in ECM remodeling and signaling regulation. In the bone context, MMPs contribute to bone turnover by enabling cell migration, modulating cell–matrix interactions, and facilitating communication among bone-resident cells (Paiva and Granjeiro, 2017). MMP-2, MMP-9, MMP-13, and MMP-14 are particularly relevant for bone physiology. MMP-9, predominantly expressed by osteoclasts, facilitates bone resorption by degrading collagen and other matrix proteins, contributing to the formation of osteoclastic resorptive lacunae. MMP-2 and MMP-13 are also involved in matrix digestion and the recruitment of osteoprogenitor and immune cells.

Importantly, several MMPs are responsible for releasing bioactive molecules that are sequestered within the ECM. MMP activity enables the liberation of transforming growth factor beta (TGF-β), which is tethered via latent TGF-β-binding proteins (LTBPs), VEGF, stored in association with proteoglycans, and BMP-2 and BMP-4, which are linked to fibrillin fibers (Liang *et al.*, 2016; Hardy and Fernandez-Patron, 2020). This proteolytic activity modulates the function of osteoblasts and osteoclasts, stimulates angiogenesis, and facilitates the deposition of new matrix components. Altogether, MMPs are essential modulators of the osteoimmune interface, orchestrating the balance between ECM degradation, growth factor release, and immune cell mobilization during bone regeneration.

20.7 BIOCHEMICAL PROPERTIES AND FUNCTIONAL MECHANISMS OF THE MSC SECRETOME

The MSC secretome comprises a rich repertoire of bioactive molecules with pleiotropic effects on skeletal and immune cells. However, many of these soluble proteins, such as growth factors, cytokines, and chemokines, have relatively short half-lives in circulation, typically ranging from minutes to a few hours, due to rapid enzymatic degradation and clearance by hepatic and renal mechanisms (Eleuteri and Fierabracci, 2019). This biochemical limitation presents a challenge for therapeutic applications, particularly in achieving sustained local concentrations at target sites.

To address this issue, several strategies have been developed to enhance the stability and potency and to favor the proper localization of secretome-derived factors. These include encapsulating secretome components within biocompatible biomaterials and preconditioning MSCs under specific conditions, such as hypoxia, which has been shown to promote the release of more stable, bioactive, and therapeutically potent factors (Chouw *et al.*, 2022).

The bioactive molecules in the MSC secretome exert their effects by binding to specific cell-surface receptors on target cells in the bone microenvironment. For instance, VEGF binds to VEGF receptors (VEGFRs) on endothelial cells, initiating angiogenesis signaling cascades essential for neovascularization and bone repair. TGF-β signals through its canonical Smad-dependent pathway by binding to TGF-β receptors, modulating MSCs differentiation, ECM synthesis, and immune responses. Similarly, IL-10 acts through its specific receptor (IL-10R) to downregulate pro-inflammatory cytokines, favoring immune tolerance. CXCL12, also known as stromal cell-derived factor-1 (SDF-1), a chemokine abundantly secreted by MSCs, binds to CXCR4 receptors on progenitor and immune cells, directing their homing and participation in repair processes (Cuesta-Gómez *et al.*, 2021).

The MSC secretome contributes to physiological bone remodeling and regeneration via two principal mechanisms: modulation of immune responses and enhancement of tissue repair. Firstly, MSC-derived factors actively suppress inflammation by downregulating pro-inflammatory mediators and promoting the expansion of regulatory T cells (Tregs), thereby fostering immune homeostasis (Eleuteri and Fierabracci, 2019). IDO, a key enzyme within the secretome, catalyzes the degradation of tryptophan into kynurenine and related metabolites, which inhibit T cell proliferation and effector function, thereby reinforcing the immunosuppressive microenvironment (Rahimi *et al.*, 2021). Secondly, the secretome promotes tissue regeneration by supporting ECM remodeling, neovascularization, and cellular proliferation. MMPs secreted by MSCs facilitate ECM degradation, enabling cell migration and angiogenesis while creating space for new tissue deposition (Thai *et al.*, 2023). Concurrently, growth factors and cytokines activate osteoblasts and endothelial cells, stimulating matrix synthesis and providing vascular support. Altogether, the biochemical versatility of the secretome enables it to function as a potent paracrine system integrating skeletal repair with immune regulation.

20.8 IMMUNOMODULATORY EFFECTS OF THE MSC-DERIVED SECRETOME IN BONE REPAIR

The MSC-derived secretome is notably rich in anti-inflammatory mediators, including IL-10, TGF-β, prostaglandin E2 (PGE2), and indoleamine 2,3-dioxygenase. These molecules collectively suppress the activation, proliferation, and differentiation of Th17 cells, reduce IL-2-induced NK cell proliferation, and prevent the maturation of monocytes into dendritic cells (DCs). IL-10 and TGF-β downregulate the production of pro-inflammatory cytokines, thereby contributing to the resolution of inflammation and promoting wound healing (Salgado and Gimble, 2013).

MSCs also regulate adaptive immunity by modulating both T and B lymphocyte responses. The MSCs' secretome inhibits T cell proliferation in a dose-dependent manner by secreting soluble factors, including hepatocyte growth factor (HGF), TGF-β1, IDO, PGE2, and IL-10. Moreover, it promotes the expansion and stabilization of Tregs and Tr1-like cells, which are key mediators of immune tolerance and protection against autoimmune responses (González-González *et al.*, 2020). Clinical studies have demonstrated that MSC-based therapies increase Treg and Tr1 populations in patients with graft-versus-host disease (GVHD), underscoring the immunosuppressive potential of the secretome.(González-González *et al.*, 2020).

In addition, MSCs modulate B cell activity by downregulating chemokine receptors, such as CXCR4, CXCR5, and CXCR7, and by suppressing STAT3 signaling, which is partially mediated by the chemokine CCL2. As a result, MSCs impair B cell activation and immunoglobulin production without interfering with antigen presentation, representing a selective regulatory mechanism for humoral immunity (González-González *et al.*, 2020).

Additionally, the MSC secretome downregulates key pro-inflammatory cytokines, including TNF-α, IL-12, and IFN-γ. This downregulation is crucial for controlling excessive inflammation and establishing a microenvironment favorable to tissue repair and regeneration (Planat-Benard *et al.*, 2021).

20.8.1 Macrophage Polarization and Inflammation Resolution

Macrophages play a crucial role in regulating inflammation and bone repair. The MSC secretome has been shown to exert potent effects on macrophage phenotype, favoring their polarization from the pro-inflammatory M1 state to the anti-inflammatory M2 phenotype (Gao et al., 2014; González-González *et al.*, 2020; Planat-Benard *et al.*, 2021). This transition is driven by key mediators present in the secretome, particularly IL-10, PGE2, and IDO (González-González *et al.*, 2020).

M2 macrophages are characterized by their immunoregulatory and tissue-remodeling properties, including enhanced phagocytic activity, expression of anti-inflammatory cytokines, and promotion of angiogenesis and ECM repair. Both in vitro and in vivo studies have demonstrated that MSC-derived factors can effectively induce M2 polarization, reduce the expression of pro-inflammatory markers, and accelerate the resolution of inflammation (Gao *et al.*, 2014; Song et al., 2020). This macrophage reprogramming is crucial for bone tissue regeneration, as persistent M1-driven inflammation hinders osteogenesis and delays healing. By facilitating the shift toward a reparative M2 phenotype, the MSC secretome directly contributes to orchestrating a bone-healing microenvironment.

20.9 EVS: OSTEOIMMUNOMODULATORY MEDIATORS WITHIN THE MSC SECRETOME

EVs are lipid bilayer-enclosed nanostructures actively secreted by virtually all cell types, including MSCs. They mediate intercellular communication by delivering functional cargo such as proteins, lipids, and nucleic acids to recipient cells (Silva et al., 2017). Within the bone microenvironment, EVs have emerged as critical regulators of both skeletal homeostasis and immune responses. By modulating osteoblast and osteoclast activity and influencing immune cell behavior, MSC-derived EVs significantly contribute to the coordination of bone remodeling and repair. This section examines the biogenesis and molecular cargo of EVs, followed by an in-depth discussion of their osteoinductive, immunomodulatory, and angiogenic functions in bone regeneration, highlighting their therapeutic relevance in osteoimmunology.

20.9.1 Biogenesis and Molecular Cargo of EVs

20.9.1.1 Formation and Types of EVs

EVs are commonly categorized into three primary types based on their origin and size: exosomes, microvesicles, and apoptotic bodies. Among these, exosomes and microvesicles are the most functionally characterized in regenerative medicine and osteoimmunology (O'Brien *et al.*, 2020).

Exosomes are typically 30–150 nm in diameter and originate from the endosomal compartment. Their biogenesis begins with the inward budding of the plasma membrane to form early endosomes, which subsequently mature into late endosomes. During this maturation process, the endosomal membrane invaginates to generate intraluminal vesicles (ILVs), which give rise to structures known as multivesicular bodies (MVBs) (Ren *et al.*, 2022). ILVs are released as exosomes upon fusion of the MVBs with the plasma membrane.

Exosome biogenesis varies across cell types and cargo. The endosomal sorting complex required for transport (ESCRT)-dependent mechanism involves over 30 proteins responsible for cargo sorting and membrane remodeling, while the ESCRT-independent mechanisms rely on ceramide and tetraspanins (e.g., CD9, CD63, CD81) to drive ILV formation (van Niel *et al.*, 2018; Fang et al., 2024). The fate of MVBs is tightly regulated: they may fuse with lysosomes for cargo degradation (mediated by RAB7) or with the plasma membrane for exosome release (via RAB11, RAB27A/B, and RAB35), with SNARE complexes facilitating the final fusion step (M. Xu *et al.*, 2023).

In contrast, microvesicles are larger (100–1000 nm) and are formed by direct outward budding of the plasma membrane. Their formation involves membrane lipid redistribution and cytoskeletal reorganization, often mediated by calcium-dependent enzymes such as flippases, scramblases, and calpains, as well as GTPases such as RhoA and ARF6 (Fang *et al.*, 2024; Wan *et al.*, 2024). Additional regulators such as acid sphingomyelinase (ASM) and ARRDC1 have been implicated in microvesicle formation, sometimes involving ceramide or partial ESCRT participation.

Distinguishing exosomes from microvesicles remains challenging, as their size ranges overlap and they can carry similar cargoes (Kim *et al.*, 2024). No definitive molecular markers exist for each; however, exosomes are typically enriched in tetraspanins (CD63, CD9), heat shock proteins (HSP70/90), and MHC-I, while microvesicles may contain cytoskeletal proteins, integrins, selectins, CD40, and are often rich in phosphatidylserine and cholesterol (Ratajczak and Ratajczak, 2020). The identification and characterization of EVs generally require a combination of analytical approaches, including electron microscopy, flow cytometry, western blotting, and multi-omics profiling.

20.9.1.2 Molecular Cargo of EVs

The molecular cargo of EVs reflects their cell of origin but is also dynamically shaped by the physiological state and microenvironmental conditions surrounding the producing cell. EVs can carry a broad spectrum of biomolecules, including proteins, lipids, RNAs—particularly microRNAs—and, to a lesser extent, DNA. This cargo does not always mirror the content of the parent cell, as selective loading mechanisms can diverge in response to stimuli such as hypoxia, metabolic stress, oncogenic signals, or cellular senescence (Liu and Wang, 2023; Kim *et al.*, 2024). These environmental and intracellular cues influence both the quantity and specificity of the molecules loaded into EVs, ultimately determining the functional effects of EVs on recipient cells. Despite advances in characterization, the intrinsic heterogeneity of EV populations continues to challenge efforts to define a universal cargo profile.

Among the proteins enriched in EVs are components of the ESCRT machinery (e.g., TSG101, Alix), GTPases (RAB27A, ARF6), and surface markers such as CD9, CD63, CD81, and MHC-I/II (Abels and Breakefield, 2016). Notably, cytokines, including IL-6 and TGF-β, can also be packaged into EVs, enhancing their immunomodulatory potential (Fitzgerald *et al.*, 2018). In addition, EVs may be engineered to carry therapeutic proteins for targeted delivery (Ilahibaks *et al.*, 2023).

Lipid content of EVs plays a fundamental role not only in maintaining structural integrity but also in signaling. Lipid species such as ceramide, phosphatidylserine, phosphatidic acid, and diacylglycerol are actively involved in vesicle formation, cell targeting, and immune modulation (Fyfe *et al.*, 2023). EVs may also contain bioactive lipids, such as prostaglandins and leukotrienes, which modulate inflammation and bone metabolism.

Although less abundant, DNA fragments—including mitochondrial and nuclear DNA—have been detected in EVs, particularly in apoptotic bodies (Jiao *et al.*, 2024). In pathological contexts such as cancer, tumor-derived EVs carrying genomic DNA fragments may serve as diagnostic biomarkers (Tsering *et al.*, 2024). However, the mechanisms underlying DNA loading remain incompletely understood.

EVs carry a variety of RNA molecules, including miRNAs, mRNAs, ribosomal RNAs (rRNAs), and long non-coding RNAs (lncRNAs). Their content varies depending on the type of vesicle and the cell origin, reflecting the selective packaging mechanisms. Exosomes are typically enriched in regulatory RNAs such as miRNAs and lncRNAs, while apoptotic bodies may contain abundant rRNA and even DNA (Fischer and Deindl, 2021). MiRNAs are the most extensively studied RNA species in EVs due to their stability and regulatory roles in gene expression. They have been associated with key biological functions, including inflammation, angiogenesis, cell proliferation, apoptosis, and immune modulation, and are considered promising tools for diagnostics and therapy (Xu *et al.*, 2022). However, some studies argue that their abundance in EVs is too low to exert functional effects, casting doubt on their biological relevance (Albanese *et al.*, 2021).

Taken together, the complexity of EV cargo enables these vesicles to act as potent mediators of intercellular signaling in bone and immune environments. Their ability to carry a cell-specific and bioactive payload makes them promising therapeutic tools in osteoimmunology, capable of orchestrating regenerative and immunomodulatory processes with high precision.

20.9.1.3 Promotion of Osteogenesis

MSC-derived EVs exhibit a robust capacity to promote osteogenesis both *in vitro* and *in vivo*. This osteo-inductive effect is primarily attributed to the bioactive molecules enclosed within EVs, particularly microRNAs (miRNAs), which regulate key transcriptional programs involved in osteoblast differentiation, matrix mineralization, and bone regeneration (Liu, Wu *et al.*, 2023).

Several studies have demonstrated that MSC-EVs can significantly enhance the expression of osteogenic markers, including RUNX2, alkaline phosphatase (ALP), osteocalcin, and OPN. For instance, Lin *et al.* demonstrated that miR-21-5p, delivered via MSC-EVs, promotes osteoblast differentiation by inhibiting SMAD7, a negative regulator of BMP signaling (Lin *et al.*, 2023).

Similarly, MSC-derived EVs seem to stimulate osteogenic differentiation without affecting proliferation, at least in part through miR-196a. When embedded in hydrogel scaffolds, these EVs accelerated calvarial bone regeneration in rat models (Qin *et al.*, 2016).

In another study, miR-148a-3p was identified and shown to promote both proliferation and osteogenic differentiation of BM-derived MSCs by targeting SMURF1, SMAD7, and BCL2, thereby shifting the balance toward bone formation (Huang *et al.*, 2020). Additional miRNAs found in MSC-EVs with confirmed osteogenic properties include miR-101 and miR-140-3p, which have been implicated in the regulation of osteoblast maturation and bone matrix deposition (Li *et al.*, 2021; Wang et al., 2022).

Transcriptomic analyses have further revealed that the miRNA content of EVs changes dynamically during osteoinduction. Yahao and Xinjia identified 221 differentially expressed miRNAs in exosomes derived from osteogenically induced MSCs, with 41 of them considered functionally relevant for bone regeneration (Yahao and Xinjia, 2021). Among these, miR-2110 and miR-328-3p showed progressive increases and were predicted to target genes involved in skeletal development. Preconditioning MSCs under osteogenic or hypoxic conditions has been shown to enhance the pro-osteogenic potential of their EVs. For example, miR-455-3p, miR-27a-3p, and miR-126 are enriched in EVs from preconditioned MSCs and promote osteoblast differentiation and vascularization (Ma *et al.*, 2022; Ren *et al.*, 2021; Liu *et al.*, 2020). Similarly, the use of strontium-containing biomaterials has been demonstrated to enrich EVs with miR-146a, thereby enhancing their capacity to stimulate both angiogenesis and bone formation (Liu *et al.*, 2021). In addition to their effects on osteoblasts, MSC-derived EVs influence other bone-related cell types. Otsuru *et al.* demonstrated that these EVs stimulate chondrocyte proliferation *in vitro* and promote growth plate expansion *in vivo.* The study showed that soluble proteins did not mediate these effects but rather depended on intact vesicular RNA, particularly miRNAs (Otsuru *et al.*, 2018).

MSC-derived EVs also regulate bone resorption by modulating osteoclastogenesis. For instance, hMSC-EVs carrying miR-122-5p have been shown to inhibit RANKL-induced osteoclast formation by downregulating key osteoclastogenic genes (Choi *et al.*, 2024). In another study, exosomes from Scx-overexpressing BM-derived MSCs delivered miR-6924-5p, which suppressed osteoclast differentiation and enhanced tendon–bone integration by reducing osteolytic activity (Feng *et al.*, 2021).

Collectively, these findings highlight the versatility of MSC-derived EVs in promoting bone regeneration. By delivering specific miRNAs and regulatory molecules, EVs influence both anabolic (osteoblast-driven) and catabolic (osteoclast-driven) pathways, offering a promising cell-free strategy for bone tissue engineering.

20.9.1.4 Immune Modulation and Macrophage Polarization

As key regulators of the bone microenvironment, MSCs also have an essential immunomodulatory role. Thus, MSC-derived EVs can modulate the inflammatory and immune responses in bone during healing and regeneration.

EVs derived from various MSC sources have been shown to exert immunomodulatory effects through diverse, but often convergent, molecular mechanisms. In experimental models of intestinal inflammation, EVs released by bone marrow-derived MSCs attenuated colitis symptoms by promoting macrophage polarization toward the anti-inflammatory M2 phenotype. This effect was associated with the activation of the JAK1/STAT1/STAT6 signaling cascade, resulting in reduced secretion of pro-inflammatory cytokines and increased production of IL-10 and TGF-β (Cao *et al.*, 2019).

Similar outcomes were reported using adipose-derived MSC-EVs, which protected intestinal epithelial integrity by enhancing the regeneration of $Lgr5^+$ intestinal stem cells and epithelial cells. This effect was attributed to activation of the Wnt/β-catenin pathway, likely mediated by the presence of Wnt2 within the EV cargo (Yu *et al.*, 2021). In another model, umbilical cord-derived MSC-EVs reduced inflammatory signaling by inhibiting ubiquitination machinery, leading to decreased activation of NF-κB and mTOR pathways (Wu *et al.*, 2018).

Although these findings derive from extrapulmonary and extraskeletal models, they demonstrate a consistent and conserved immunoregulatory potential of MSC-EVs. In bone-relevant contexts, specific miRNAs encapsulated within EVs have been identified as central regulators of immune function. For example, miR-23a-3p, which is enriched in BMSC-derived exosomes, has been shown to induce M2 macrophage polarization by targeting interferon regulatory factor 1 (IRF1). This process contributes to inflammation resolution and has been associated with improved outcomes in tendon–bone healing (Kou *et al.*, 2022).

Collectively, these observations support the concept that MSC-derived EVs actively participate in the control of inflammation through multiple mechanisms. These include direct cytokine suppression, inhibition of inflammasome activation, macrophage reprogramming, and possibly pyroptosis suppression. Through these coordinated effects, EVs contribute significantly to restoring immune balance and facilitating effective bone regeneration.

20.9.1.5 Angiogenesis and Vascularized Bone Formation

Bone regeneration is tightly dependent on the establishment of a functional vascular network that delivers oxygen, nutrients, immune mediators, and skeletal progenitor cells to the healing site (Hu and Olsen, 2017). Angiogenesis, the sprouting of new capillaries from pre-existing vessels, is not only essential for coupling bone formation and remodeling but also intimately regulated by immune responses. Within this context, MSC-derived EVs have emerged as critical paracrine regulators of angiogenesis.

EVs derived from MSCs have been shown to promote endothelial cell proliferation, migration, and tube formation by delivering angiogenic miRNAs. For example, miR-132, carried by MSC-derived EVs, enhances vascularization following myocardial infarction by targeting RASA1 in endothelial cells (Ma *et al.*, 2018). Similarly, miR-21-5p, delivered via bone marrow MSC-derived EVs (BM-MSC-derived EVs), has demonstrated potent pro-angiogenic effects in models of cerebral ischemia, where it improves neurological outcomes by enhancing capillary density (Hu *et al.*, 2022).

In bone regeneration, the interplay between osteogenesis and angiogenesis is tightly coordinated. Exosomal miR-29a, for instance, supports both processes by targeting vasohibin 1 (VASH1), a negative regulator of endothelial growth. Its delivery via BMSC-derived EVs has been shown to enhance vascular ingrowth and new bone formation in osteoporotic models, making it a promising candidate for therapeutic development (Lu *et al.*, 2020). Beyond the content of EVs themselves, strategies to enhance their therapeutic activity have shown promise.

Combining MSC-derived EVs with engineered biomaterials has demonstrated improved outcomes in bone repair. For example, Brennan *et al.* highlighted the value of integrating EVs into scaffolds to improve retention and sustain paracrine activity. Wu *et al.* developed a thermosensitive hydrogel encapsulating miR-21-enriched BM-MSC-derived EVs, which enhanced vascularized bone regeneration in vivo (Brennan *et al.*, 2020; Wu *et al.*, 2021b).

Further studies have aimed to enhance the pro-angiogenic content of EVs by modulating the parental MSCs. For instance, it has been shown that stimulating MSCs with magnetic nanoparticles and a static magnetic field increases miR-1260a expression in secreted EVs. This miRNA targets HDAC7 and COL4A2, two regulators of vascular and osteogenic differentiation (Wu et al., 2021a). Likewise, Liu *et al.* demonstrated that lithium-containing biomaterials stimulate the expression of exosomal miR-130a, which activates the PTEN/AKT signaling pathway in endothelial cells, thereby enhancing angiogenesis (Liu *et al.*, 2019). Hypoxic preconditioning is another effective strategy to improve the regenerative potential of EVs. Under low-oxygen conditions, MSCs increase miR-210-3p secretion, which targets EFNA3 and promotes angiogenesis via the PI3K/AKT pathway. This leads to enhanced vascularization and bone formation in models of critical-size defects *(Zhuang et al.*, 2022).

Altogether, these findings underscore the pivotal role of MSC-derived EVs in orchestrating vascular and skeletal regeneration. By acting on both endothelial and immune cells, EVs help establish a vascularized, immunomodulated environment that supports effective bone healing. Their ability to integrate osteogenic, angiogenic, and anti-inflammatory signals further reinforces their value as therapeutic tools in the osteoimmunological landscape.

20.10 CONCLUSION AND FUTURE PERSPECTIVES

The convergence of immunology and skeletal biology has prompted a paradigm shift in the established idea of bone regeneration. Osteoimmunology, a term initially coined to describe the bidirectional interactions between immune cells and bone cells (Rho *et al.*, 2004), has now evolved into a multidisciplinary field with significant implications for regenerative medicine, bridging immunology and bone biology to reveal how immune mechanisms influence bone remodeling in both physiological and pathological conditions. In this chapter, we have demonstrated that bone homeostasis and repair cannot be viewed solely from the perspective of osteoblast–osteoclast dynamics, but must be understood as the outcome of complex intercellular communication involving MSCs, immune mediators, EVs, and secretome factors. Far from being passive structural components, bone cells such as osteoblasts, osteoclasts, and osteocytes actively respond to immune signals and, in turn, regulate immune cell function. This reciprocal interaction is central not only to normal bone turnover but also to the pathogenesis of a wide range of bone disorders.

One key insight from this work is that the immune response is not merely a response to disease, but a crucial regulatory factor that influences the quality and extent of bone regeneration. This is particularly important in situations where inflammation becomes chronic, such as in osteoporosis, rheumatoid arthritis, and age-related bone loss. The polarization of macrophages, the activation state of T cells, and the cytokine milieu in the bone microenvironment can strongly influence the balance between bone formation and resorption. MSC-derived secretomes and EVs are natural mediators of this balance, capable of promoting M2 macrophage polarization, downregulating pro-inflammatory cytokines (e.g., TNF-α, IL-1β), and enhancing osteogenic signals that support bone formation, such as RUNX2 and BMP-2 (González-González *et al.*, 2020; Zohora *et al.*, 2023; Tang *et al.*, 2024).

The RANKL–RANK–OPG axis remains the central molecular pathway linking immune activity to bone resorption. Immune cells—including activated T cells and synovial fibroblasts—can upregulate RANKL expression under inflammatory conditions, tipping the balance toward enhanced osteoclastogenesis and bone destruction (Parolini, 2025). This is particularly relevant in diseases such as rheumatoid arthritis, inflammatory bowel disease, and postmenopausal osteoporosis, where chronic inflammation drives sustained bone loss.

Recent discoveries have identified specific subsets of immune and bone cells involved in inflammatory bone loss. For example, inflammatory osteoclasts, primed by cytokines such as TNF-α, IL-1β, and IL-17, exhibit enhanced bone-resorptive capacity and are a hallmark of chronic inflammatory conditions. These cytokines not only increase RANKL production but also directly stimulate osteoclast precursor differentiation and inhibit osteoblast function, thereby compounding bone destruction (Ibanez *et al.*, 2016).

Another promising direction is the use of MSC-derived EVs as immunomodulatory tools. These vesicles can modulate macrophage phenotype, shifting M1 macrophages toward an M2 reparative state, and also downregulate inflammasome activity, as demonstrated by EVs enriched in miR-210, which inhibit NLRP3 in vivo (Zhuang *et al.*, 2022). Moreover, miR-21-5p-enriched EVs can simultaneously suppress the TLR4/NF-κB pathway and enhance osteogenic SMAD signaling, revealing the dual targeting potential of these vesicles (Hu *et al.*, 2022). Significantly, many of these therapeutic effects are amplified when MSCs are primed/preconditioned under hypoxic conditions or genetically modified, enhancing the bioactivity of secreted EVs (Liu *et al.*, 2020; Chouw *et al.*, 2022; Haider, 2024).

Despite these advances, several limitations must be addressed before these strategies reach clinical practice. One challenge is standardizing EV production. Differences in MSC source, culture conditions, and isolation protocols lead to high variability in EV content and bioactivity (Safwan *et al.*, 2024a & 2024b). Furthermore, the stability and biodistribution of EVs *in vivo* remain poorly understood, although engineering strategies to modify EV surface molecules (e.g., adding bone-targeting ligands) are under investigation.

In parallel, the field of biomaterials is also evolving. Immunomodulatory scaffolds are being designed not only to provide structural support but also to actively guide the immune response toward tissue regeneration and repair. These materials can be embedded with EVs or engineered to release immunoregulatory cytokines in a spatiotemporally controlled manner. For instance, Schwann cell-derived exosomes embedded in hydrogels have shown promise in orchestrating the inflammatory, vascular, and osteogenic phases of bone healing (Hao *et al.*, 2023). This highlights the importance of spatiotemporal regulation in orchestrating multiple cellular processes during bone repair.

A fascinating area of development in osteoimmunology is the study of immune metabolism, specifically how the metabolic activity of immune cells, particularly macrophages, influences their function in bone repair. It's becoming clear that how macrophages generate energy determines whether they promote inflammation or support healing. Pro-inflammatory M1 macrophages mainly rely on glycolysis and an interrupted TCA cycle to produce molecules that drive inflammatory responses. On the other hand, M2 macrophages, which are associated with tissue repair, make use of a complete TCA cycle, fatty acid oxidation, and oxidative phosphorylation in the mitochondria. These metabolic differences help explain how macrophages adopt such distinct roles. Recent work has also highlighted the importance of glucose, lipid, and amino acid metabolism in regulating these immune cells. However, the specific contribution of these metabolic pathways to bone regeneration remains unclear. By targeting these metabolic programs, specifically encouraging energy profiles that favor healing over inflammation, we may be able to develop new strategies to promote bone repair in inflammatory conditions (Ma *et al.*, 2024).

Future research should also prioritize multi-omics approaches, including single-cell RNA sequencing and spatial transcriptomics, to dissect the complex cellular networks within the bone niche (Paul *et al.*, 2023). Such tools could unveil previously unrecognized interactions between immune, stromal, and vascular cells during bone formation. Additionally, the roles of systemic regulators, such as the gut microbiota and neuroendocrine factors, in shaping the immune-bone axis remain an emerging frontier.

In conclusion, osteoimmunology is advancing our understanding of bone biology and reshaping the therapeutic landscape of regenerative medicine. From manipulating cytokine networks and immune cell populations to deploying EVs and immunomodulatory biomaterials, the immune system is proving to be a key architect of skeletal repair. The integration of immunological insights into biomaterial design, EV therapy, and clinical strategies holds enormous promise for treating osteoporosis and other bone pathologies. Continued interdisciplinary collaboration, together with rigorous preclinical validation, will be essential to translate these breakthroughs into safe and effective clinical interventions.

REFERENCES

Abels ER, Breakefield XO. Introduction to extracellular vesicles: Biogenesis, RNA cargo selection, content, release, and uptake. *Cell Mol Neurobiol.* 2016;36(3):301–12.

Adamopoulos IE, Choi Y, Takayanagi H. Novel insights and recent progress in osteoimmunology. *Trends Immunol.* 2025;46(3):192–4.

Ahmed ZT, Alkahlot MH, Haider KH. MSC-derived exosomes: advances in cell-free therapy. In: Haider, K.H. (ed.) *Handbook of Stem Cell Applications*. Singapore: Springer Nature, 2023 Aug 18, pp. 1–41.

Albanese M, Chen YA, Huls C, Gartner K, Tagawa T, Mejias-Perez E, et al. MicroRNAs are minor constituents of extracellular vesicles that are rarely delivered to target cells. *PLoS Genet.* 2021;17(12):e1009951.

Álvarez-Viejo M, Haider KH. Mesenchymal stem cells. In: Haider, K.H. (ed.) *Handbook of Stem Cell Therapy*. Singapore: Springer, 2022. https://doi.org/10.1007/978-981-16-6016-0_6-1

Boilard E. Extracellular vesicles and their content in bioactive lipid mediators: More than a sack of microRNA. *J Lipid Res.* 2018;59(11):2037–46.

Boskey AL, Posner AS. Bone structure, composition, and mineralization. *Orthop Clin North Am.* 1984;15(4):597–612.

Brennan MA, Layrolle P, Mooney DJ. Biomaterials functionalized with MSC secreted extracellular vesicles and soluble factors for tissue regeneration. *Adv Funct Mater.* 2020;30(37).

Camacho V, Matkins VR, Patel SB, Lever JM, Yang Z, Ying L, et al. Bone marrow Tregs regulate stromal cell function and support hematopoiesis by producing IL-10. *JCI Insight.* 2020;5(22).

Cao L, Xu H, Wang G, Liu M, Tian D, Yuan Z. Extracellular vesicles derived from bone marrow mesenchymal stem cells attenuate dextran sodium sulfate-induced ulcerative colitis by promoting M2 macrophage polarization. *Int Immunopharmacol.* 2019;72:264–74.

Capobianco CA, Hankenson KD, Knights AJ. Temporal dynamics of immune-stromal cell interactions in fracture healing. *Front Immunol.* 2024;15:1352819.

Carvalho MS, Cabral JMS, da Silva CL, Vashishth D. Bone matrix non-collagenous proteins in tissue engineering: creating new bone by mimicking the extracellular matrix. *Polymers (Basel).* 2021;13(7).

Choi JH, Sung SE, Kang KK, Lee S, Sung M, Park WT, et al. Extracellular vesicles from human adipose tissue-derived mesenchymal stem cells suppress RANKL-induced osteoclast differentiation via miR122-5p. *Biochem Genet.* 2024;62(4):2830–52.

Chouw A, Facicilia G, Sartika CR, Faried A, Milanda T. Factors influencing the therapeutic potential of the MSC-derived secretome. *Regen Eng Transl Med.* 2022;8:384–93.

Crane JL, Cao X. Bone marrow mesenchymal stem cells and TGF-β signaling in bone remodeling. *J Clin Invest.* 2014;124:466–72.

Cuesta-Gómez N, Graham GJ, Campbell JDM. Chemokines and their receptors: Predictors of the therapeutic potential of mesenchymal stromal cells. *J Transl Med.* 2021;19.

Da RR, Qin Y, Baeten D, Zhang Y. B cell clonal expansion and somatic hypermutation of Ig variable heavy chain genes in the synovial membrane of patients with osteoarthritis. *J Immunol.* 2007;178(1):557–65.

De Filippo K, Dudeck A, Hasenberg M, Nye E, van Rooijen N, Hartmann K, et al. Mast cell and macrophage chemokines CXCL1 and CXCL2 control the early stage of neutrophil recruitment during tissue inflammation. *Blood.* 2013;121(24):4930–7.

Depalle B, McGilvery CM, Nobakhti S, Aldegaither N, Shefelbine SJ, Porter AE. Osteopontin regulates type I collagen fibril formation in bone tissue. *Acta Biomaterialia.* 2021;120:194–202.

DiGirolamo DJ, Clemens TL, Kousteni S. The skeleton as an endocrine organ. *Nat Rev Rheumatol.* 2012;8(11):674–83.

Ding DC, Shyu WC, Lin SZ. Mesenchymal stem cells. *Cell Transplant.* 2011;20(1):5–14.

Ding L, Morrison SJ. Haematopoietic stem cells and early lymphoid progenitors occupy distinct bone marrow niches. *Nature.* 2013;495(7440):231–5.

El Khassawna T, Serra A, Bucher CH, Petersen A, Schlundt C, Konnecke I, et al. T Lymphocytes influence the mineralization process of bone. *Front Immunol.* 2017;8:562.

Eleuteri S, Fierabracci A. Insights into the secretome of mesenchymal stem cells and its potential applications. *Int J Mol Sci.* 2019;20.

Eriksen EF. Cellular mechanisms of bone remodeling. *Rev Endocr Metab Disord.* 2010;11(4):219–27.

Fang F, Yang J, Wang J, Li T, Wang E, Zhang D, et al. The role and applications of extracellular vesicles in osteoporosis. *Bone Res.* 2024;12(1):4.

Faust HJ, Zhang H, Han J, Wolf MT, Jeon OH, Sadtler K, et al. IL-17 and immunologically induced senescence regulate response to injury in osteoarthritis. *J Clin Invest.* 2020;130(10):5493–507.

Feng W, Jin Q, Ming-Yu Y, Yang H, Xu T, You-Xing S, et al. MiR-6924-5p-rich exosomes derived from genetically modified Scleraxis-overexpressing PDGFRalpha(+) BMMSCs as novel nanotherapeutics for treating osteolysis during tendon-bone healing and improving healing strength. *Biomaterials.* 2021;279:121242.

Ferreira JR, Teixeira GQ, Santos SG, Barbosa MA, Almeida-Porada G, Gonçalves RM. Mesenchymal stromal cell secretome: Influencing therapeutic potential by cellular pre-conditioning. *Front Immunol.* 2018;9:2837.

Fischer S, Deindl E. Characterization of RNA in extracellular vesicles. *Appl Sci.* 2021;11(16).

Fitzgerald W, Freeman ML, Lederman MM, Vasilieva E, Romero R, Margolis L. A system of cytokines encapsulated in extracellular vesicles. *Sci Rep.* 2018;8(1):8973.

Fyfe J, Casari I, Manfredi M, Falasca M. Role of lipid signaling in extracellular vesicles-mediated cell-to-cell communication. *Cytokine Growth Factor Rev.* 2023;73:20–6.

Gao F, Chiu SM, Motan DA, Zhang Z, Chen L, Ji HL, et al. Mesenchymal stem cells and immunomodulation: Current status and prospects. *Cell Death Dis.* 2016;7(1):e2062.

Gao S, Mao F, Zhang B, Zhang L, Zhang X, Wang M, et al. Mouse bone marrow-derived mesenchymal stem cells induce macrophage M2 polarization through the nuclear factor-κB and signal transducer and activator of transcription 3 pathways. *Exp Biol Med.* 2014;239(3):366–75.

González-González A, García-Sánchez D, Dotta M, Rodríguez-Rey JC, Pérez-Campo FM. Mesenchymal stem cells secretome: The cornerstone of cell-free regenerative medicine. *World J Stem Cells.* 2020;12(12):1439–690.

Guo X, Xu T, Zheng J, Cui X, Li M, Wang K, et al. Accumulation of synovial fluid CD19(+)CD24(hi)CD27(+) B cells was associated with bone destruction in rheumatoid arthritis. *Sci Rep.* 2020;10(1):14386.

Habib SM, Martini MF, Al Ahmed O, Qassab AR, Haider KH. Mesenchymal stem cell secretomes for cell-free therapy of the broken heart. In: Haider, K.H. (ed.) *Handbook of Regenerative Medicine: Cell-Free Therapy.* London: Taylor & Francis Group/CRC Press, 2025a, Chapter 18, pp. 274–93.

Habib SM, Martini MF, Al Ahmed O, Qassab AR, Haider KH. The potential of mesenchymal stem cell–derived exosomes in cell-free therapy for myocardial repair. In: Haider, K.H. (ed.) *Handbook of Regenerative Medicine: Cell-Free Therapy.* London: Taylor & Francis Group/CRC Press, 2025b, Chapter 24, pp. 372–90.

Habib SY, Martini MF, Hamdan YNA, Shrebaty OMM, Haider KH. Mesenchymal stem cell secretome and exosomes as potential advanced therapy medicinal products for treating a failing heart. *World J Stem Cells.* 2026;18.

Haider KH. Priming mesenchymal stem cells to develop "super stem cells." *World J Stem Cells.* 2024;16(6):623–40. https://dx.doi.org/10.4252/wjsc.v16.i6.623.

Haider KH, Aziz S. Paracrine hypothesis and cardiac repair. *Int J Stem Cell Res Transplant.* 2017;5(1):265–7.

Hamamura K, Nagao M, Furukawa K. Regulation of glycosylation in bone metabolism. *Int J Mol Sci.* 2024;25(7).

Hao Z, Ren L, Zhang Z, Yang Z, Wu S, Liu G, et al. A multifunctional neuromodulation platform utilizing Schwann cell-derived exosomes orchestrates bone microenvironment via immunomodulation, angiogenesis, and osteogenesis. *Bioact Mater.* 2023;23:206–22.

Hardy E, Fernandez-Patron C. Destroy to rebuild: The connection between bone tissue remodeling and matrix metalloproteinases. *Front Physiol.* 2020;11.

Hirata Y, Furuhashi K, Ishii H, Li HW, Pinho S, Ding L, et al. CD150(high) bone marrow tregs maintain hematopoietic stem cell quiescence and immune privilege via adenosine. *Cell Stem Cell.* 2018;22(3):445–53.e5.

Honorati MC, Neri S, Cattini L, Facchini A. Interleukin-17, a regulator of angiogenic factor release by synovial fibroblasts. *Osteoarthritis Cartilage.* 2006;14(4):345–52.

Horowitz MC, Bothwell AL, Hesslein DG, Pflugh DL, Schatz DG. B cells, osteoblasts, and osteoclast development. *Immunol Rev.* 2005;208:141–53.

Hu H, Hu X, Li L, Fang Y, Yang Y, Gu J, et al. Exosomes derived from bone marrow mesenchymal stem cells promote angiogenesis in ischemic stroke mice via upregulation of MiR-21-5p. *Biomolecules.* 2022;12(7).

Hu K, Olsen BR. The roles of vascular endothelial growth factor in bone repair and regeneration. *Bone.* 2016;91:30–8.

Hu K, Olsen BR. Vascular endothelial growth factor control mechanisms in skeletal growth and repair. *Dev Dyn.* 2017;246.

Hu K, Shang Z, Yang X, Zhang Y, Cao L. Macrophage polarization and the regulation of bone immunity in bone homeostasis. *J Inflamm Res.* 2023;16:3563–80.

Hu K, Song M, Song T, Jia X, Song Y. Osteoimmunology in osteoarthritis: Unraveling the interplay of immunity, inflammation, and joint degeneration. *J Inflamm Res.* 2025;18:4121–42.

Hua R, Jiang JX. Small leucine-rich proteoglycans in the physiological and biomechanical function of bone. *Matrix Biol Plus.* 2021;11:100063.

Huang S, Li Y, Wu P, Xiao Y, Duan N, Quan J, et al. microRNA-148a-3p in extracellular vesicles derived from bone marrow mesenchymal stem cells suppresses SMURF1 to prevent osteonecrosis of the femoral head. *J Cell Mol Med.* 2020;24(19):11512–23.

Hyc A, Osiecka-Iwan A, Moskalewski S. Could BMPs therapy be improved if BMPs were used in composition acting during bone formation in endochondral ossification? *Int J Mol Sci.* 2022;23.

Ibanez L, Abou-Ezzi G, Ciucci T, Amiot V, Belaid N, Obino D, et al. Inflammatory osteoclasts prime TNFalpha-producing CD4(+) T cells and express CX(3) CR1. *J Bone Miner Res.* 2016;31(10):1899–908.

Icer MA, Gezmen-Karadag M. The multiple functions and mechanisms of osteopontin. *Clin Biochem.* 2018;59:17–24.

Ilahibaks NF, Ardisasmita AI, Xie S, Gunnarsson A, Brealey J, Vader P, et al. TOP-EVs: Technology of protein delivery through extracellular vesicles is a versatile platform for intracellular protein delivery. *J Control Release.* 2023;355:579–92.

Jiao Y, Gao L, Zhang T, He Z, Zheng SY, Liu W. Profiling DNA cargos in single extracellular vesicles via hydrogel-based droplet digital multiple displacement amplification. *Anal Chem.* 2024;96(3):1293–300.

Karampetsou MP, Comte D, Suárez-Fueyo A, Katsuyama E, Yoshida N, Kono M, et al. Signaling lymphocytic activation molecule family member 1 engagement inhibits T cell-B cell interaction and diminishes interleukin-6 production and plasmablast differentiation in systemic lupus erythematosus. *Arthritis Rheumatol.* 2019;71(1):99–108.

Kim HI, Park J, Zhu Y, Wang X, Han Y, Zhang D. Recent advances in extracellular vesicles for therapeutic cargo delivery. *Exp Mol Med.* 2024;56(4):836–49.

Kitaura H, Marahleh A, Ohori F, Noguchi T, Shen WR, Qi J, et al. Osteocyte-related cytokines regulate osteoclast formation and bone resorption. *Int J Mol Sci.* 2020;21(14).

Kou M, Huang L, Yang J, Chiang Z, Chen S, Liu J, et al. Mesenchymal stem cell-derived extracellular vesicles for immunomodulation and regeneration: A next-generation therapeutic tool? *Cell Death Dis.* 2022;13(7):580.

Lamora A, Talbot J, Mullard M, Brounais-Le Royer B, Redini F, Verrecchia F. TGF-β signaling in bone remodeling and osteosarcoma progression. *J. Clin. Med.* 2016;5.

Levescot A, Chang MH, Schnell J, Nelson-Maney N, Yan J, Martinez-Bonet M, et al. IL-1beta-driven osteoclastogenic Tregs accelerate bone erosion in arthritis. *J Clin Invest.* 2021;131(18).

Li J, Yu TT, Yan HC, Qiao YQ, Wang LC, Zhang T, et al. T cells participate in bone remodeling during the rapid palatal expansion. *Faseb J.* 2020;34(11):15327–37.

Li X, Xiao S, Li F, Fang K, Wen J, Gong H. Max interacting protein 1 induces IL-17-producing T helper/regulatory T imbalance in osteoarthritis by upregulating tectonic family member 2. *Tissue Cell.* 2022;78:101906.

Li Y, Wang J, Ma Y, Du W, Feng K, Wang S. miR-101-loaded exosomes secreted by bone marrow mesenchymal stem cells require the FBXW7/HIF1alpha/FOXP3 axis, facilitating osteogenic differentiation. *J Cell Physiol.* 2021;236(6):4258–72.

Liang HPH, Xu J, Xue M, Jackson C. Matrix metalloproteinases in bone development and pathology: Current knowledge and potential clinical utility. *Metalloproteinases Med.* 2016;3:93–102.

Liang Y, Xia J, Xu L, Zhang Y, Xu X, Haider KH. Cell-free and cell-based therapy in regenerative medicine. In: Haider, K.H. (ed.) *Handbook of Regenerative Medicine: Cell-Free Therapy Approach* (1st ed.). CRC Press, 2025. Chapter 24. https://doi.org/10.1201/9781003611509.

Lin Z, Xiong Y, Sun Y, Zeng R, Xue H, Hu Y, et al. Circulating MiRNA-21-enriched extracellular vesicles promote bone remodeling in patients with traumatic brain injury. *Exp Mol Med.* 2023;55(3):587–96.

Liu L, Liu Y, Feng C, Chang J, Fu R, Wu T, et al. Corrigendum to: "Lithium-containing biomaterials stimulate bone marrow stromal cell-derived exosomal miR-130a secretion to promote angiogenesis" [Biomaterials 192(2019) 523–536]. *Biomaterials.* 2019;204:82–4.

Liu L, Yu F, Li L, Zhou L, Zhou T, Xu Y, et al. Bone marrow stromal cells stimulated by strontium-substituted calcium silicate ceramics: Exosomal miR-146a release regulates osteogenesis and angiogenesis. *Acta Biomater.* 2021;119:444–57.

Liu R, Wu S, Liu W, Wang L, Dong M, Niu W. microRNAs delivered by small extracellular vesicles in MSCs as an emerging tool for bone regeneration. *Front Bioeng Biotechnol.* 2023;11:1249860.

Liu W, Li L, Rong Y, Qian D, Chen J, Zhou Z, et al. Hypoxic mesenchymal stem cell-derived exosomes promote bone fracture healing by the transfer of miR-126. *Acta Biomater.* 2020;103:196–212.

Liu YJ, Wang C. A review of the regulatory mechanisms of extracellular vesicle-mediated intercellular communication. *Cell Commun Signal.* 2023;21(1):77.

Lu GD, Cheng P, Liu T, Wang Z. BMSC-derived exosomal miR-29a promotes angiogenesis and osteogenesis. *Front Cell Dev Biol.* 2020;8:608521.

Lubberts E, van den Bersselaar L, Oppers-Walgreen B, Schwarzenberger P, Coenen-de Roo CJ, Kolls JK, et al. IL-17 promotes bone erosion in murine collagen-induced arthritis through loss of the receptor activator of NF-kappa B ligand/osteoprotegerin balance. *J Immunol.* 2003;170(5):2655–62.

Ma H, Gao L, Chang R, Zhai L, Zhao Y. Crosstalk between macrophages and immunometabolism and their potential roles in tissue repair and regeneration. *Heliyon.* 2024;10(18):e38018.

Ma H, Li M, Jia Z, Chen X, Bu N. MicroRNA-455-3p promotes osteoblast differentiation via targeting HDAC2. *Injury*. 2022;53(11):3636–41.

Ma T, Chen Y, Chen Y, Meng Q, Sun J, Shao L, et al. MicroRNA-132, delivered by mesenchymal stem cell-derived exosomes, promotes angiogenesis in myocardial infarction. *Stem Cells Int*. 2018;2018:3290372.

Martin TJ, Seeman E. Bone remodelling: its local regulation and the emergence of bone fragility. *Best Pract Res Clin Endocrinol Metab*. 2008;22(5):701–22.

McHugh J. T(reg) cell-inducing nanoparticles show promise for treating OA. *Nat Rev Rheumatol*. 2023;19(2):62.

Molendijk M, Hazes JM, Lubberts E. From patients with arthralgia, pre-RA and recently diagnosed RA: What is the current status of understanding RA pathogenesis? *RMD Open*. 2018;4(1):e000256.

Morgan EF, Unnikrishnan GU, Hussein AI. Bone mechanical properties in healthy and diseased states. *Annu Rev Biomed Eng*. 2018;20:119–43.

Muñoz J, Akhavan NS, Mullins AP, Arjmandi BH. Macrophage polarization and osteoporosis: A review. *Nutrients*. 2020;12(10).

Muñoz M, Robinson K, Shibli-Rahhal A. Bone health and osteoporosis prevention and treatment. *Clin Obstet Gynecol*. 2020;63(4):770–87.

Murshed M. Mechanism of bone mineralization. *Cold Spring Harb Perspect Med*. 2018;8(12).

Na HS, Park JS, Cho KH, Kwon JY, Choi J, Jhun J, et al. Interleukin-1-Interleukin-17 signaling axis induces cartilage destruction and promotes experimental osteoarthritis. *Front Immunol*. 2020;11:730.

Nabbe KC, van Lent PL, Holthuysen AE, Kolls JK, Verbeek S, van den Berg WB. FcgammaRI upregulation induced by local adenoviral-mediated interferon-gamma production aggravates chondrocyte death during immune-complex-mediated arthritis. *Am J Pathol*. 2003;163(2):743–52.

Nakashima T, Hayashi M, Fukunaga T, Kurata K, Oh-Hora M, Feng JQ, et al. Evidence for osteocyte regulation of bone homeostasis through RANKL expression. *Nat Med*. 2011;17(10):1231–4.

Neve A, Corrado A, Cantatore FP. Osteocalcin: skeletal and extra-skeletal effects. *J Cell Physiol*. 2013;228(6):1149–53.

Newman H, Shih YV, Varghese S. Resolution of inflammation in bone regeneration: From understandings to therapeutic applications. *Biomaterials*. 2021;277.

O'Brien K, Breyne K, Ughetto S, Laurent LC, Breakefield XO. RNA delivery by extracellular vesicles in mammalian cells and its applications. *Nat Rev Mol Cell Biol*. 2020;21(10):585–606.

Oftadeh R, Perez-Viloria M, Villa-Camacho JC, Vaziri A, Nazarian A. Biomechanics and mechanobiology of trabecular bone: A review. *J Biomech Eng*. 2015;137(1):0108021.

Okamoto K. Crosstalk between bone and the immune system. *J Bone Miner Metab*. 2024;42(4):470–80.

Okuyan HM, Terzi MY, Ozcan O, Kalaci A. Association of UCMA levels in serum and synovial fluid with severity of knee osteoarthritis. *Int J Rheum Dis*. 2019;22(10):1884–90.

Otsuru S, Desbourdes L, Guess AJ, Hofmann TJ, Relation T, Kaito T, et al. Extracellular vesicles released from mesenchymal stromal cells stimulate bone growth in osteogenesis imperfecta. *Cytotherapy*. 2018;20(1):62–73.

Paiva KBS, Granjeiro JM. Matrix metalloproteinases in bone resorption, remodeling, and repair. *Prog Mol Biol Transl Sci*. 2017;148:203–303.

Parolini C. Pathophysiology of bone remodelling cycle: Role of immune system and lipids. *Biochem Pharmacol*. 2025;235:116844.

Paul S, Umar PK, Paul MK, Tiwari H, Shukla MK, Kandasami L, Joliholi A, et al. Application of single-cell sequencing on stem cell research. In: Haider, K.H. (ed.) *Handbook of Stem Cell Applications*. Singapore: Springer Nature, 2023. https://doi.org/10.1007/978-981-99-0846-2_43.

Peruzzi B, Teti A. The physiology and pathophysiology of the osteoclast. *Clin Rev Bone Mineral Metab*. 2011;10(2):71–97.

Planat-Benard V, Varin A, Casteilla L. MSCs and inflammatory cells crosstalk in regenerative medicine: Concerted actions for optimized resolution driven by energy metabolism. *Front Immunol*. 2021;12.

Platzer H, Nees TA, Reiner T, Tripel E, Gantz S, Hagmann S, et al. Impact of mononuclear cell infiltration on chondrodestructive MMP/ADAMTS production in osteoarthritic knee joints - An ex vivo study. *J Clin Med*. 2020;9(5).

Qin Y, Wang L, Gao Z, Chen G, Zhang C. Bone marrow stromal/stem cell-derived extracellular vesicles regulate osteoblast activity and differentiation in vitro and promote bone regeneration in vivo. *Sci Rep*. 2016;6:21961.

Rahimi B, Panahi M, Saraygord-Afshari N, Taheri N, Bilici M, Jafari D, et al. The secretome of mesenchymal stem cells and oxidative stress: challenges and opportunities in cell-free regenerative medicine. *Mol Biol Rep*. 2021 Jul;48:5607–19.

Rahman MS, Akhtar N, Jamil HM, Banik RS, Asaduzzaman SM. TGF-β/BMP signaling and other molecular events: Regulation of osteoblastogenesis and bone formation. Bone Research: Sichuan University; 2015.

Raphael I, Nalawade S, Eagar TN, Forsthuber TG. T cell subsets and their signature cytokines in autoimmune and inflammatory diseases. *Cytokine.* 2015;74(1):5–17.

Ratajczak MZ, Ratajczak J. Extracellular microvesicles/exosomes: Discovery, disbelief, acceptance, and the future? *Leukemia.* 2020;34(12):3126–35.

Razi S, Yaghmoorian Khojini J, Kargarijam F, Panahi S, Tahershamsi Z, Tajbakhsh A, et al. Macrophage efferocytosis in health and disease. *Cell Biochem Funct.* 2023;41(2):152–65.

Ren J, Yu R, Xue J, Tang Y, Su S, Liao C, et al. How do extracellular vesicles play a key role in the maintenance of bone homeostasis and regeneration? A comprehensive review of literature. *Int J Nanomedicine.* 2022;17:5375–89.

Ren LR, Yao RB, Wang SY, Gong XD, Xu JT, Yang KS. MiR-27a-3p promotes osteogenic differentiation by activating the CRY2/ERK1/2 axis. *Mol Med.* 2021;27(1):43.

Rho J, Takami M, Choi Y. Osteoimmunology: interactions of the immune and skeletal systems. *Mol Cells.* 2004;17(1):1–9.

Safwan M, Bourgleh MS, Aldoush M, Haider KH. Tissue-source effect on mesenchymal stem cells as living biodrugs for heart failure: Systematic review and meta-analysis. *World J Cardiol.* 2024a;16(8):469–83. https://doi.org/10.4330/wjc.v16.i8.469.

Safwan M, Bourgleh MS, Alshakaki H, Molhem A, Haider KH. Morbid cell status and donor age significantly alter mesenchymal stem cell functionality and reparability. In: Haider, K.H. (ed.) *Handbook of Stem Cell Applications.* Singapore: Springer Nature, 2024b, pp.1359–1387. https://doi.org/10.1007/978-981-99-7119-0_62

Salgado AJ, Gimble JM. Secretome of mesenchymal stem/stromal cells in regenerative medicine. *Biochimie.* 2013;95:2195.

Sheng J, Chen W, Zhu HJ. The immunosuppressive function of transforming growth factor-β (TGF-β) in human diseases. *Growth Factors.* 2015;33(2):92–101.

Shoulders MD, Raines RT. Collagen structure and stability. *Annu Rev Biochem.* 2009;78:929–58.

Silva AM, Teixeira JH, Almeida MI, Gonçalves RM, Barbosa MA, Santos SG. Extracellular vesicles: Immunomodulatory messengers in the context of tissue repair/regeneration. *Eur J Pharm Sci.* 2017;98:86–95.

Sims NA, Walsh NC. Intercellular cross-talk among bone cells: new factors and pathways. *Curr Osteoporos Rep.* 2012;10(2):109–17.

Skalny AV, Aschner M, Silina EV, Stupin VA, Zaitsev ON, Sotnikova TI, et al. The role of trace elements and minerals in osteoporosis: A review of epidemiological and laboratory findings. *Biomolecules.* 2023;13(6).

Song C, Tong T, Dai B, Zhu Y, Chen E, Zhang M, et al. Osteoimmunology in bone malignancies: A symphony with evil. *J Natl Cancer Cent.* 2024;4(4):354–68.

Song N, Scholtemeijer M, Shah K. Mesenchymal stem cell immunomodulation: Mechanisms and therapeutic potential. Trends Pharmacol Sci. 2020;41:653–64.

Staines KA, MacRae VE, Farquharson C. The importance of the SIBLING family of proteins on skeletal mineralisation and bone remodelling. *J Endocrinol.* 2012;214(3):241–55.

Tang J, Wang X, Lin X, Wu C. Mesenchymal stem cell-derived extracellular vesicles: A regulator and carrier for targeting bone-related diseases. *Cell Death Discov.* 2024;10(1):212.

Thai VL, Ramos-Rodriguez DH, Mesfin M, Leach JK. Hydrogel degradation promotes angiogenic and regenerative potential of cell spheroids for wound healing. *Materials Today Bio.* 2023;22.

Tsering T, Nadeau A, Wu T, Dickinson K, Burnier JV. Extracellular vesicle-associated DNA: Ten years since its discovery in human blood. *Cell Death Dis.* 2024;15(9):668.

Urist MR. Bone morphogenetic protein: The molecularization of skeletal system development. *J Bone Mineral Res.* 1997;12(3):343–6.

Van Dijk FS, Sillence DO. Osteogenesis imperfecta: clinical diagnosis, nomenclature and severity assessment. *Am J Med Genet A.* 2014;164A(6):1470–81.

Van Niel G, D'Angelo G, Raposo G. Shedding light on the cell biology of extracellular vesicles. *Nat Rev Mol Cell Biol.* 2018;19(4):213–28.

Vimalraj S. Alkaline phosphatase: Structure, expression, and its function in bone mineralization. *Gene.* 2020;754:144855.

Wan M, Cao X. BMP signaling in skeletal development. *Biochem Biophys Res Commun.* 2005;328:651–7.

Wan X, Zhang W, Dai L, Chen L. The role of extracellular vesicles in bone regeneration and associated bone diseases. *Curr Issues Mol Biol.* 2024;46(9):9269–85.

Wang A, Ren M, Wang J. The pathogenesis of steroid-induced osteonecrosis of the femoral head: A systematic review of the literature. *Gene.* 2018;671:103–9.

Wang N, Liu X, Tang Z, Wei X, Dong H, Liu Y, et al. Increased BMSC exosomal miR-140-3p alleviates bone degradation and promotes bone restoration by targeting Plxnbl in diabetic rats. *J Nanobiotechnology.* 2022;20(1):97.

Wu D, Chang X, Tian J, Kang L, Wu Y, Liu J, et al. Stimulation of bone mesenchymal stem cells with magnetic nanoparticles and a static magnetic field: Exosomal miR-1260a release improves osteogenesis and angiogenesis. *J Nanobiotechnology.* 2021a;19(1):209.

Wu D, Qin H, Wang Z, Yu M, Liu Z, Peng H, et al. Bone Mesenchymal stem cell-derived sEV-encapsulated thermosensitive hydrogels accelerate osteogenesis and angiogenesis by release of exosomal miR-21. *Front Bioeng Biotechnol.* 2021b;9:829136.

Wu M, Chen G, Li YP. TGF-β and BMP signaling in osteoblasts, skeletal development, bone formation, homeostasis, and disease. *Bone Res.* 2016;4:16009.

Wu Y, Qiu W, Xu X, Kang J, Wang J, Wen Y, et al. Exosomes derived from human umbilical cord mesenchymal stem cells alleviate inflammatory bowel disease in mice through ubiquitination. *Am J Transl Res.* 2018;10(7):2026–36.

Xia Y, Yang Q, Li Q, Wen J, Li M, Wu Z, et al. Metallothionein-1 mitigates the advancement of osteoarthritis by regulating Th17/Treg balance. *Cell Immunol.* 2024;405–406:104877.

Xie X, Doody GM, Shuweihdi F, Conaghan PG, Ponchel F. B-cell capacity for expansion and differentiation into plasma cells is altered in osteoarthritis. *Osteoarthritis Cartilage.* 2023;31(9):1176–88.

Xu C, Yu P, Han X, Du L, Gan J, Wang Y, et al. TGF-β promotes immune responses in the presence of mesenchymal stem cells. *J Immunol.* 2014;192(1):103–9.

Xu D, Di K, Fan B, Wu J, Gu X, Sun Y, et al. MicroRNAs in extracellular vesicles: Sorting mechanisms, diagnostic value, isolation, and detection technology. *Front Bioeng Biotechnol.* 2022;10:948959.

Xu J, Yu L, Liu F, Wan L, Deng Z. The effect of cytokines on osteoblasts and osteoclasts in bone remodeling in osteoporosis: A review. *Front. Immunol.* 2023;14.

Xu M, Ji J, Jin D, Wu Y, Wu T, Lin R, et al. The biogenesis and secretion of exosomes and multivesicular bodies (MVBs): Intercellular shuttles and implications in human diseases. *Genes Dis.* 2023;10(5):1894–907.

Yahao G, Xinjia W. The role and mechanism of exosomes from umbilical cord mesenchymal stem cells in inducing osteogenesis and preventing osteoporosis. Cell Transplant. 2021;30:9636897211057465.

Yamada H, Nakashima Y, Okazaki K, Mawatari T, Fukushi J, Oyamada A, et al. Preferential accumulation of activated Th1 cells not only in rheumatoid arthritis but also in osteoarthritis joints. *J Rheumatol.* 2011;38(8):1569–75.

Yenchek RH, Ix JH, Shlipak MG, Bauer DC, Rianon NJ, Kritchevsky SB, et al. Bone mineral density and fracture risk in older individuals with chronic kidney disease. *Clin J Am Soc Nephrol.* 2012;7(7):1130–6.

You Y, Wang W, Li Y, Song Y, Jiao J, Wang Y, et al. Aspirin/PLGA-coated 3D-printed Ti-6Al-4V alloy modulates macrophage polarization to enhance osteoblast differentiation and osseointegration. *J Mater Sci Mater Med.* 2022;33(10):73.

Yu H, Yang X, Xiao X, Xu M, Yang Y, Xue C, et al. Human adipose mesenchymal stem cell-derived exosomes protect mice from DSS-induced inflammatory bowel disease by promoting intestinal-stem-cell and epithelial regeneration. *Aging Dis.* 2021;12(6):1423–37.

Yunna C, Mengru H, Lei W, Weidong C. Macrophage M1/M2 polarization. *Eur J Pharmacol.* 2020;877:173090.

Zhao E, Xu H, Wang L, Kryczek I, Wu K, Hu Y, et al. Bone marrow and the control of immunity. *Cell Mol Immunol.* 2012;9(1):11–9.

Zhu S, Zhou J, Xie Z. The balance between helper T17 and regulatory T cells in osteoimmunology and relevant research progress on bone tissue engineering. *Immun Inflamm Dis.* 2024;12(9):e70011.

Zhuang Y, Cheng M, Li M, Cui J, Huang J, Zhang C, et al. Small extracellular vesicles derived from hypoxic mesenchymal stem cells promote vascularized bone regeneration through the miR-210-3p/EFNA3/PI3K pathway. *Acta Biomater.* 2022;150:413–26.

Ziros PG, Basdra EK, Papavassiliou AG. Runx2: of bone and stretch. *Int J Biochem Cell Biol.* 2008;40(9):1659–63.

Zofkova I, Nemcikova P, Matucha P. Trace elements and bone health. *Clin Chem Lab Med.* 2013;51(8):1555–61.

Zohora FT, Aliyu M, Saboor-Yaraghi AA. *Secretome-based Acellular Therapy of Bone Marrow-derived Mesenchymal Stem Cells in Degenerative and Immunological Disorders: A Narrative Review. Heliyon*: Elsevier Ltd; 2023.

21 Advancing Regenerative Medicine

The Role of Secretome and Extracellular Vesicles in Bone and Cartilage Disorders

Wahyu Widowati[1], Bahareh Sadri[2] Aris Muhamad Nurjamil[3], Rizal Azis[4], Adilah Hafizha Nur Sabrina[3], Fadhilah Haifa Zahiroh[3], Teresa Liliana Wargasetia[1], and Massoud Vosough[2,5]

[1]Faculty of Medicine, Maranatha Christian University, Bandung, Indonesia

[2]Department of Regenerative Medicine, Cell Science Research Center, Royan Institute for Stem Cell Biology and Technology, ACECR, Tehran, Iran

[3]Aretha Medika Utama, Biomolecular and Biomedical Research Center, Bandung, Indonesia

[4]Biomedical Engineering, Department of Electrical Engineering, Faculty of Engineering, University of Indonesia, Depok, Indonesia

[5]Experimental Cancer Medicine, Institution for Laboratory Medicine, Karolinska Institute, Sweden

21.1 LIST OF ABBREVIATIONS

BMP2	Bone morphogenetic protein-2
BMSCs	Bone marrow stem cells
ECM	Extracellular matrix
EGF	Epidermal growth factor
EMA	European medicines agency
EVs	Extracellular vesicle
FGF	Fibroblast growth factor
GMP	Good Manufacturing Practice
IL12	Interleukin-12
ILVs	Intraluminal vesicles
IGF-1	Insulin-like growth factor-1
MZ	Marginal Zone
MSCs	Mesenchymal stromal cells

DOI: 10.1201/9781003653387-21

mRNAs	Messenger RNAs
MVBs	Multivesicular bodies
OA	Osteoarthritis
PDGF	Platelet-derived growth factor
rhBMP7	Recombinant human BMP-7
rAAV	Recombinant adeno-associated virus
TGF-β	Transforming growth factor-beta
UCSCs	Umbilical cord mesenchymal stem cells
FDA	US Food and Drug Administration
VEGF	Vascular endothelial growth factor

21.2 INTRODUCTION

Regenerative medicine is an emerging interdisciplinary field that endeavors to restore the structure and function of damaged tissues and organs. This domain focuses on the repair, restoration, and replacement of damaged body parts that contribute to the development of diseases (Aguilera et al., 2023). It encompasses therapeutic strategies that harness the body's endogenous repair mechanisms, drawing concepts from cell therapy, gene therapy, and tissue engineering. To promote tissue repair and regeneration, various approaches, including stem cell therapies and biomaterial-driven approaches, have been developed (Han et al., 2020).

One key principle of regenerative medicine is that numerous strategies are employed to address complex biological problems. In cell therapy, living cells (including stem cells), which have the unique characteristic of self-renewal and the ability to generate specialized cells, are implanted. Such property is a key element in therapy programs aimed at repairing or replacing damaged tissues (Akyash et al., 2016). On the one hand, gene therapy aims to alter gene expression in a patient's cells, either to repair genetic defects or to promote tissue regeneration (Jessop et al., 2016). Tissue engineering involves combining cells, biocompatible scaffolds, and growth factors to generate functional tissues in vitro, which can then be transplanted to restore compromised biological functions (Farshidfar et al., 2023).

Cell-based therapies have demonstrated considerable promise; however, researchers are increasingly turning to cell-free strategies, particularly those based on secretome and extracellular vesicle (EV) therapies (Ahmed et al., 2023). The secretome comprises diverse bioactive substances that cells release, such as growth factors and cytokines, as well as extracellular matrix components, for the repair and regeneration of tissues (Hellmund & Koksch, 2019; Habib et al., 2025). Similarly, EVs, small membrane-bound structures secreted by cells, play a crucial role in cell communication by carrying bioactive molecules that support various regenerative pathways (Correia et al., 2015).

Secretomes play a crucial role in addressing bone and cartilage issues by enhancing stem cell regenerative capacity and facilitating tissue repair (Chyzy & Płońska-Brzezińska, 2020). EVs, including exosomes and microvesicles, function as biological messengers, transporting proteins, lipids, and RNA molecules that regulate key cellular processes (Hellmund & Koksch, 2019). New evidence suggests that EVs can achieve many of the therapeutic effects of stem cells without the limitations of a cell-based paradigm. Due to their potential to reduce inflammation, stimulate angiogenesis, and promote tissue regeneration, they are being explored as a treatment for musculoskeletal injuries (Hellmund & Koksch, 2019).

The use of secretome- and EV-based therapies in regenerative medicine represents a paradigm shift in the treatment of bone and cartilage disorders. Utilizing the innate therapeutic capabilities of these biologics allows researchers and clinicians to develop more efficient, effective, novel, and safer therapeutic alternatives (Habib et al., 2025a & 2025b). Not only do these advances advance the field of regenerative medicine, but they also bring new hope to patients suffering from challenging musculoskeletal conditions. In this chapter, the composition, organization, therapeutic properties, future technological advances, and commercial prospects of secretome and EVs are systematically described to highlight their significant role in regenerative medicine.

21.3 COMPOSITION AND MECHANISMS OF SECRETOME AND EVS

21.3.1 Secretome Composition

The secretome, a complex repertoire of bioactive molecules, has become a linchpin in the development of tissue regeneration therapies. This complex mixture, comprising soluble proteins, growth factors, cytokines, and metabolites, acts synergistically to modulate cellular pathways essential for tissue repair (Ibrahim et al., 2022) (Figure 21.1). The secretome orchestrates the regenerative response by promoting proliferation, migration, differentiation, and angiogenesis, making it an attractive strategy for treating a wide range of clinical conditions (Wu et al., 2016).

The secretome comprises soluble proteins, cytokines, growth factors, and bioactive molecules. These components collectively contribute to five primary regenerative functions, including proliferation (supported by growth factors and soluble proteins), migration (driven by growth factors and bioactive molecules), differentiation (regulated by growth factors), angiogenesis (facilitated by soluble proteins and bioactive molecules), and tissue remodeling (modulated by cytokines such as TGF-β).

Soluble factors, such as vascular endothelial growth factor (VEGF), epitomize the strong angiogenic potential of the secretome. VEGF is an endothelial-specific mitogen that promotes the growth of new blood vessels by increasing endothelial cell proliferation and migration. This process ensures a sufficient nutrient and oxygen supply to the damaged tissue, providing a favorable microenvironment for tissue regeneration (Wei et al., 2022). VEGF has been incorporated into delivery systems, such as injectable hydrogels and growth factor-adsorbed scaffolds, to utilize its therapeutic potency. Such strategies have proven individually to be effective in enhancing tissue repair in various contexts, including liver, skeletal muscle, and periodontal regeneration (Griffeth et al., 2014).

Similarly, several growth factors (primarily EGF, VEGF, BMP-2, and PDGF) indicate that the secretome functions as an orchestrator of tissue regeneration. Such delivery technologies—specifically, custom-designed membranes and scaffolds that can localize activity and increase the

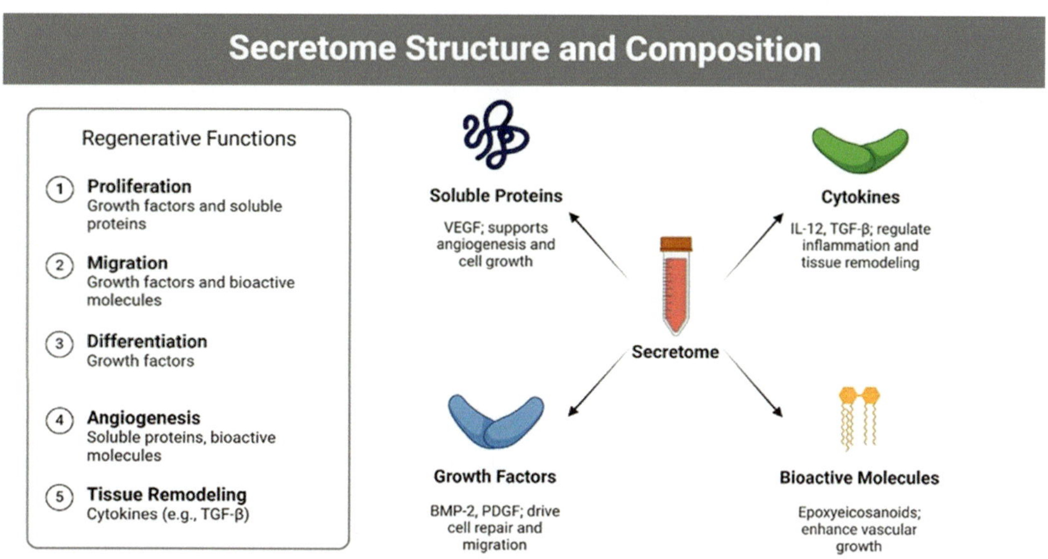

FIGURE 21.1 Secretome structure and composition.

bioavailability of these growth factors—will further enhance their therapeutic effects in the restoration of damaged tissues (Dong et al., 2020).

Cytokines in the secretome also play essential roles in shaping the inflammatory and regenerative microenvironments in tissue repair and remodeling. One example is interleukin-12 (IL-12), which has demonstrated an antiangiogenic effect that may play a role in specific contexts where vascular formation needs to be finely controlled (Kumar et al., 2015). Furthermore, transforming growth factor-beta (TGF-β) released from mesenchymal stromal cells (MSCs) promotes tissue restructuring and repair, for example, in cartilage regeneration (Sadri et al., 2023). TGF-β and encourages the restoration of functional tissue architecture by modulating cellular processes, highlighting the multifunctionality of the secretome (Winkler et al., 2016). Beyond proteins and cytokines, bioactive metabolites such as epoxyeicosanoids add another dimension to the secretome's regenerative potential. These metabolites promote angiogenesis, endothelial cell proliferation, and vascular remodeling, all of which are critical for effective tissue repair.

21.3.2 Extracellular Vesicles

EVs are membrane-bound structures that play a crucial role in cellular communication and the modulation of various cellular processes (Liu & Wang, 2023). Exosomes are a heterogeneous population of vesicles released by cells into the extracellular space, acting as carriers of bioactive molecules to mediate local and systemic signaling (Ke et al., 2022; Liang et al., 2025). Depending on their biogenesis and size, EVs are classified into exosomes and microvesicles (Bongiovanni et al., 2021), as shown in Figure 21.2. Endosomal-derived particles, also known as exosomes (30 to 150 nm) (Kim et al., 2024). Intriguingly, they originate from the inward budding of the membrane in multivesicular bodies (MVBs) to produce intraluminal vesicles (ILVs). Following plasma membrane fusion of the MVBs, these ILVs are secreted into the extracellular space as exosomes (Jeppesen et al., 2019). Microvesicles, on the other hand, are larger (100 to 1000 nm in diameter) and are formed from direct outward budding and fission from the plasma membrane (Bazzan et al., 2021). Although exosomes and microvesicles share many similar functions during intercellular communication, their distinct origins underscore the complexity of their biogenesis (Kreimer et al., 2015).

Exosomes (30–150 nm) are shed from multivesicular bodies (MVBs) inside early endosomes, while microvesicles (100–1000 nm) protrude from the plasma membrane. Exosomes are vesicles

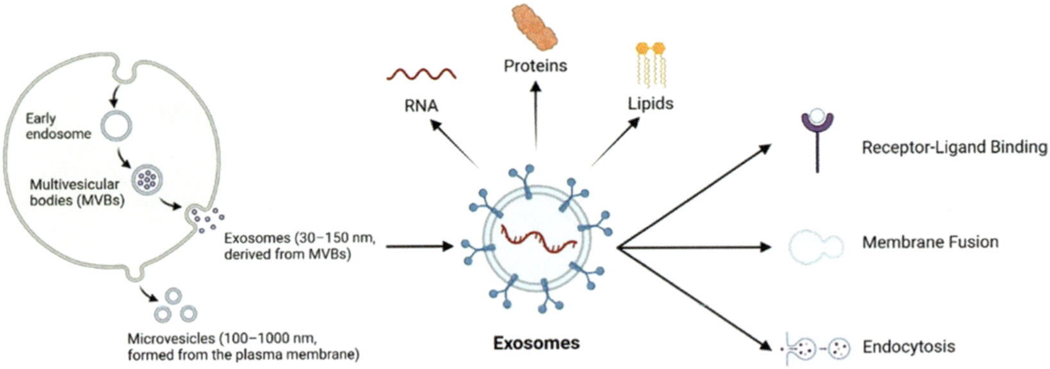

FIGURE 21.2 Schematic overview of the biogenesis and functions of extracellular vesicles (EVs). Exosomes (30–150 nm) are shed from multivesicular bodies (MVBs) inside early endosomes, while microvesicles (100–1000 nm) protrude from the plasma membrane. Exosomes are vesicles involved in the transfer of bioactive molecules (RNA, proteins, lipids) and mediate intercellular communication. Once released, exosomes can affect recipient cells through receptor-ligand binding, membrane fusion, or endocytosis, regulating various cellular processes.

that mediate intercellular communication by transferring bioactive molecules (RNA, proteins, lipids). Once released, exosomes can affect recipient cells through receptor-ligand binding, membrane fusion, or endocytosis, regulating various cellular processes.

EVs contain various bioactive malecules including proteins, lipids, and nucleic acids, such as DNA, messenger RNAs (mRNAs), and non-coding RNAs (e.g., microRNAs and long non-coding RNAs) (Shurtleff et al., 2017; Haider & Aramini, 2020). First, specific molecules are selectively packaged into EVs through a highly regulated process, making the EVs specialized vehicles for intercellular signaling. Proteomic analysis of EVs has demonstrated enrichment for proteins involved in pathways, including cell-surface receptors, ligands, adhesion molecules, signaling proteins, and cytoskeletal components, indicating their roles in cellular adhesion and signaling (Gerwing et al., 2020). Additionally, RNA cargo within these compartments comprises enriched mRNAs and non-coding RNAs that are believed to regulate gene expression in recipient cells (Statello et al., 2018).

The lipid profile of EVs also highlights their function-specificity. The membrane of EV is highly enriched with sphingolipids, phospholipids, and cholesterol, which maintain their structural integrity and stability in circulation (Hallal et al., 2022). Such a lipid composition not only protects the bioactive cargo but also enables exosome uptake by target cells through receptor-ligand interactions or direct membrane fusion (Wei et al., 2017). The selective packaging of bioactive molecules is one of the interesting properties of EVs (Raposo & Stahl, 2023). The molecular cargo in EVs frequently diverges from that in their parent cells, demonstrating the existence of a tightly controlled sorting mechanism. Such selective integration enables EVs to contain only the molecular signals necessary to modulate target cells, thereby simplifying the process. EVs play messenger roles by influencing biological processes such as immune response, tissue repair, and angiogenesis, explaining their potential in regenerative treatments (Wu et al., 2022). However, they are of interest for therapeutic applications, as they can mediate intercellular communication without the risks associated with cell-based therapies, such as immune rejection or tumorigenicity. Their ability to modulate inflammation, promote cell survival, and matrix remodeling in bone and cartilage disorders highlights the transformative role of EVs in regenerative strategies (Dayem et al., 2024).

21.3.3 Mechanisms of Action

EVs serve as key mediators of intercellular communication by transporting biological substances, including proteins, lipids, and nucleic acids, between cells (Rani et al., 2023). Now, EVs can also affect various cell types in multiple physiological and pathological contexts (Zhang et al., 2023). EVs, which are classified by biogenesis into exosomes and microvesicles, serve as essential means by which cells coordinate their responses and maintain tissue homeostasis. Exosomes are formed by the endosomal pathway and are released when multivesicular bodies merge with the plasma membrane, while microvesicles are released by outward budding of the plasma membrane (Abels & Breakefield, 2016).

A critical characteristic of EVs is their ability to modulate various cellular events to facilitate tissue repair and regeneration (Mushtaq et al., 2024). EVs play a central role in restoring the structural and functional integrity of damaged tissues through these mechanisms. Among these, one key mechanism by which EVs promote tissue restoration is the regulation of inflammatory signals. Inflammation is integral to the acute healing response; however, it must be carefully regulated to prevent potential chronic injury (Gao et al., 2024). EVs can deliver anti-inflammatory molecules, such as lipid mediators and cytokine antagonists, that suppress excessive inflammatory signaling and facilitate the resolution of inflammation. EV types derived from MSCs, for example, have been demonstrated to regulate the immune response by reducing the production of pro-inflammatory cytokines and increasing tissue repair processes (Kuriyama et al., 2020). EVs promote a microenvironmental transition from pro-inflammation to pro-healing, ultimately stimulating tissue regeneration.

EVs also play a key role in stimulating angiogenesis (the formation of new blood vessels) to ensure sufficient oxygen and nutrients for regenerating tissues. Tumor cell- or endothelial cell-derived EVs often contain proangiogenic factors, such as VEGF (Olejarz et al., 2020). These vesicles deliver VEGF and other signaling molecules to recipient endothelial cells, thereby activating angiogenic pathways and promoting increased capillary formation (Zhao et al., 2018). Boosted angiogenesis nourishes the healing of ischemic tissues and helps improve the entire regenerative process, contributing especially in avascular or under-vascularized areas (Sadri et al., 2024).

Besides their effects on inflammation and angiogenesis, EVs also stimulate cellular proliferation, a crucial mechanism in tissue repair and regeneration. Among their cargos are growth factors (e.g., fibroblast growth factor, FGF, and transforming growth factor beta, TGF-β) and cell-cycle-regulating microRNAs. These bioactive molecules induce the proliferation and recruitment of various cell types, including endothelial cells, fibroblasts, and stem/progenitor cells, to sites of injury (Potz et al., 2018). The increase in these cells contributes to the repopulation of damaged tissues and the regeneration of functional architectures.

21.4 THERAPEUTIC POTENTIAL OF SECRETOME AND EVS IN BONE AND CARTILAGE DISORDERS

21.4.1 Secretome-Mediated Chondrogenesis and Osteogenesis

A secretome of soluble proteins, growth factors, cytokines, and bioactive metabolites (known as the secretome) is critical in mediating tissue repair (Kearney et al. 2022). The secretome, a predominant component of cell-free regenerative therapies, shows promising potential to modulate cell behavior essential to tissue repair (Xia et al., 2021).

Multiple in vitro studies have demonstrated that the secretome induces MSC differentiation into chondrogenic and osteogenic lineages. Secretomes from chondrogenic-differentiated MSCs are, for example, enriched with extracellular proteins that promote cartilage matrix synthesis, such as proteoglycans and collagen II, both of which are important for cartilage integrity (Rocha et al., 2014). Like MSCs, exosomes derived from human embryonic MSCs have been effective in stimulating osteochondral regeneration by delivering bioactive molecules that activate signaling pathways involved in bone and cartilage repair (Zhang et al., 2016) (Figure 21.3).

MSC-derived secretome and EVs promote chondrogenesis and osteogenesis. TGF-β1 and BMP-2 induce cartilage matrix formation through chondrogenic differentiation, leading to the production of proteoglycans and collagen type II. BMP-2 promotes osteogenic differentiation, leading to the formation of osteoid matrix and mineralized bone tissue for bone regeneration. In addition to direct distinctions, the secretome impacts the broader cellular context. It also serves as a modulating role for different cell types (e.g., endothelial cells, fibroblasts, immune cells), all of which participate in the tissue-regenerative process (Görgün et al., 2021). Interestingly, MSC secretomes can prevent fibroblast differentiation into myofibroblasts, thereby limiting the fibrotic response and promoting optimal tissue repair (Kulebyakina et al., 2023).

In murine models of osteoarthritis, for instance, MSC secretome injections have been shown to reduce pain and prevent further cartilage damage by modulating inflammatory processes and promoting cartilage repair (Khatab et al., 2018). The MSC secretome has been shown to facilitate osteochondral regeneration in a rat critical-sized defect model, demonstrating its potential as a restorative factor for extensive tissue damage (Zhang et al., 2016).

The regenerative effects of the secretome are primarily due to its role in mediating inflammatory signals, angiogenesis, and cell proliferation (Chi et al., 2023). For instance, MSC-derived secretomes contain anti-inflammatory cytokines and lipid mediators that inhibit hyperimmunity, thereby promoting a tissue-friendly microenvironment for regeneration (Ahmadi & Rezaie, 2020). Furthermore, pro-angiogenic factors in the secretome, including vascular endothelial growth factor

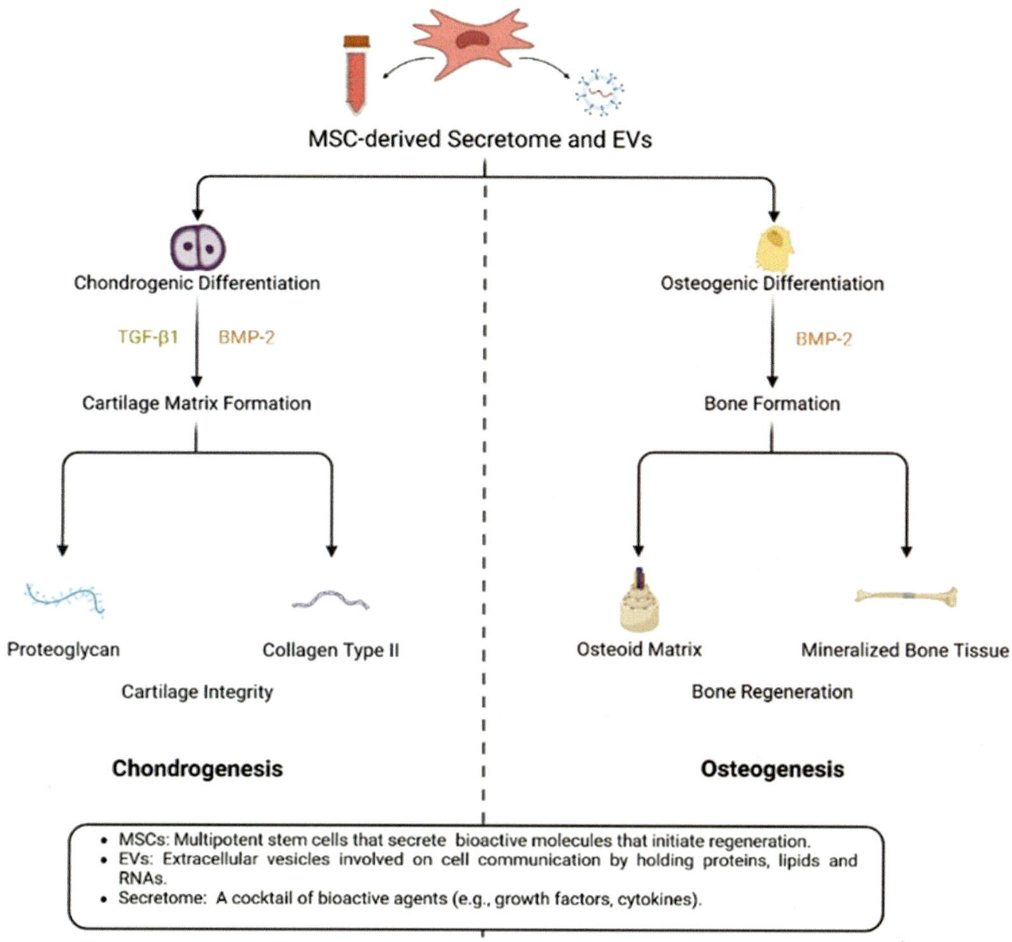

FIGURE 21.3 Mechanisms of chondrogenesis and osteogenesis mediated by secretome and EVs.

(VEGF), facilitate blood vessel development, which provides adequate oxygen and nutrient supply to the tissue in a regenerating microenvironment (Gao et al., 2022).

21.4.2 EVs in Bone and Cartilage Regeneration

EVs are important therapeutic mediators in regenerative medicine due to their ability to regulate the inflammatory microenvironment and improve cell survival (Wang & Pan, 2023). Both of these functions play a crucial role in effective tissue regeneration, particularly in diseases affecting bone and cartilage. EVs are naturally secreted by many cell types, including Marginal Zone (MZ)-MSCs, and they act as carriers of bioactive molecules that mediate various cellular and molecular pathways crucial for repair and recovery (Go et al., 2023).

One of the essential functions of EVs is to modulate inflammation. Although inflammation is necessary for repair, excessive inflammation tends to inhibit rather than promote tissue regeneration. MSC-derived EVs have been demonstrated to have significant anti-inflammatory effects in multiple preclinical models, thereby reducing fibrosis and promoting tissue repair. This is mediated, in part, by inhibiting the differentiation of fibroblasts into myofibroblasts, the cells responsible for excessive extracellular matrix (ECM) deposition that culminates in fibrosis (Itoh et al., 2023). Furthermore,

EVs are enriched in anti-inflammatory lipid mediators and cytokine antagonists, enabling them to modulate the inflammatory response. These molecules facilitate the shift of the surrounding environment from a pro-inflammatory state to one that promotes repair and regeneration (Yekula et al., 2020) (Figure 21.4).

MSC-derived secretome and EVs help resolve inflammation in injured tissue. MSC-derived secretome and EVs release anti-inflammatory molecules, including TGF-β and IL-10, which shift the environment of injured tissue toward healing. This leads to the presence of M2 anti-inflammatory macrophages, resulting in inflammation resolution, enhanced cell survival, and improved vascularization (Haider & Aslam, 2018).

EVs also promote the survival and proliferation of important cell types that contribute to tissue repair, including endothelial cells, fibroblasts, and stem/progenitor cells, in addition to their anti-inflammatory properties. These signals, which are mediated by the molecular cargo of EVs, including growth factors, microRNAs, and various bioactive molecules, stimulate pathways that enhance cell proliferation, migration, and differentiation (Ma et al., 2020). Such processes play a key role in replenishing injured tissues and restoring their functional integrity. Additionally, the promotion of angiogenesis by EVs enhances tissue regenerative capacity, as increased vascularization provides nutrients and oxygen to sustain repair processes (Li K et al., 2022).

EVs have demonstrated therapeutic potential in numerous preclinical models of bone and cartilage diseases. For example, their administration has been reported to suppress pain and prevent deteriorative cartilage changes in murine OA models (Khatab et al., 2018). Likewise, human embryonic MSC-derived EVs have been shown to repair critical-sized defects in rat models of osteochondral injury, further underscoring their potential for repairing complex tissue injuries (Zhang et al., 2016). Together, these findings highlight the multifunctional roles of EVs in regulating inflammatory responses, driving angiogenesis, and promoting cellular proliferation, which are essential components of effective tissue regeneration (Kearney et al., 2022).

Ongoing clinical trials are investigating the therapeutic potential of EVs in cartilage regeneration and wound healing. One such example is NCT06466850, a clinical trial investigating exosomes (EVs) for cartilage regeneration to evaluate their effectiveness in repairing damaged cartilage. In addition, another trial, NCT05060107, evaluates the safety and efficacy of EVs during musculoskeletal wound healing, whereas NCT06431152 investigates the potential of MSC EVs in treating osteoarthritis (Chen et al., 2025). These trials highlight a concerning trend in EVs and growing interest in their potential as a therapeutic option for joint diseases (Ding et al., 2024).

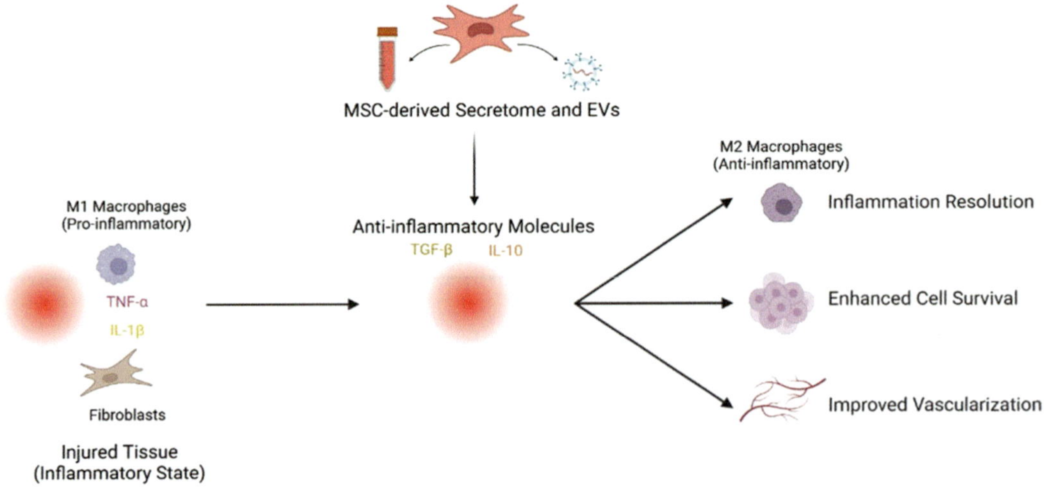

FIGURE 21.4 The role of EVs in reducing inflammation and supporting cell survival in injured tissues.

EVs are rich in proteins, lipids, and nucleic acids, which play key roles in cellular processes and thus possess regenerative properties. For instance, bone marrow MSC-derived EVs have been shown to promote the production of essential cartilage factors, including proteoglycans and type II collagen, thereby aiding in cartilage recovery (Yang et al., 2014). Considering their role in modulating immune responses, EVs have also been shown to induce a pro-chondrogenic microenvironment, which is required for successful cartilage regeneration (Li M et al., 2022). Additionally, human umbilical cord MSC-EVs have been shown to promote the polarization of M2 macrophages, thereby reducing inflammation and improving osteoarthritis (Jeyaraman et al., 2020).

21.4.3 Bone and Cartilage Regeneration with Secretome-Based Approaches

The use of growth factors, such as bone morphogenetic protein-2 (BMP-2) and transforming growth factor-beta 1 (TGF-β1), has been extensively investigated in both preclinical and clinical settings related to osteoarthritis and cartilage injuries. Such bioactive molecules are key regulators of the cellular milieu and participate in various molecular pathways that promote tissue regeneration, thereby demonstrating significant potential in regenerative medicine (Fortier et al., 2011; Diaz-Solano et al., 2024).

Bone morphogenetic protein–2 (BMP-2) has shown promising results in stimulating chondrogenesis while simultaneously downregulating chondrocyte hypertrophy, both pivotal processes in cartilage repair and regeneration. BMP-2 is also suggested to mediate the upregulation of Wnt/β-catenin signaling, a key pathway that promotes chondrocyte-specific gene upregulation while inhibiting hypertrophic differentiation, in preclinical studies (Papathanasiou et al., 2012). This pleiotropic property makes BMP-2 a promising option for regenerative therapy. Clinically, recombinant human BMP-7 (rhBMP-7) administered intra-articularly has been reported to reduce pain and cartilage degeneration in patients with osteoarthritis (Mainil-Varlet et al., 2012). Nevertheless, while these studies have shown promise, it cannot be concluded that BMP-2 alone can provide complete protection of cartilage from progressive degeneration; thus, additional combination therapies may be necessary to achieve optimal therapeutic efficacy (Chang et al., 2015).

Likewise, TGF-β1 has highlighted the importance of cartilage repair and regeneration. In preclinical studies, TGF-β1 was shown to promote the chondrogenic differentiation of MSCs and to increase extracellular matrix production (i.e., collagen and proteoglycans) to sustain cartilage integrity (Yang et al., 2014). A clinical trial (Frisch et al., 2016) demonstrated that TGF-β1 delivered via a recombinant adeno-associated virus (rAAV) vector exhibits chondrogenic activity in human bone marrow aspirates and can be used to stimulate cartilage repair in the clinic. Additionally, some studies have explored the synergistic effects of multiple growth factors to enhance regenerative effects. BMP-2, for example, has been shown to work in conjunction with insulin-like growth factor-1 (IGF-1) in preclinical models of complex fracture healing. Using a porous silk scaffold loaded with reverse gradients of BMP-2 and IGF-1, a study demonstrated that these gradients can stimulate human MSC differentiation, with high BMP-2 concentrations causing the most significant cellular changes (Lu et al., 2014). This strategy highlights opportunities to leverage spatially guided delivery systems to maximize cartilage regeneration by coordinating the actions of multiple growth factors.

Despite promising results from preclinical studies, the clinical application of growth factor-mediated therapies remains a considerable challenge. The complex and multifactorial nature of the disease makes interpreting clinical outcomes difficult, especially in heterogeneous osteoarthritis populations (Engen et al., 2010). Furthermore, the design of clinical trials, including the selection of patient cohorts, comparator groups, and clinical endpoints, has been recognized as crucial to their ability to yield robust and meaningful results (Mithoefer et al., 2011). The shortcomings highlighted must be remedied by representing more clinically relevant cohorts and standardizing the outcomes measured to facilitate an accurate assessment of therapeutic efficacy.

21.5 SUPPORTING TECHNOLOGIES FOR SECRETOME AND EV APPLICATIONS

21.5.1 Interaction between Secretome and the Extracellular Matrix

Although the secretome–extracellular matrix (ECM) interplay is pivotal for maintaining homeostasis and restoring bone and cartilage tissues, it remains largely unexplored in regenerative medicine. These secreted molecules, including soluble factors and EVs, released from cells in the tissue microenvironment, are collectively referred to as the secretome. They significantly participate in modulating ECM dynamics and cellular responses. In bone and cartilage disorders, the interaction between these two components is highly relevant, as the integrity of the ECM is disrupted.

Moreover, recent studies have focused on the protective roles of MSCs-derived secretomes on cartilage. For example, secretomes from bone marrow stem cells (BMSCs) and umbilical cord mesenchymal stem cells (UCSCs) have been reported to alleviate cartilage destruction and promote cartilage matrix deposition, thus ameliorating the microarchitecture of subchondral bone (El-Qashty, 2023). Indeed, these secretomes affect the joint environment by inhibiting chondrocyte apoptosis and diminishing ECM proteolysis, which is involved in osteoarthritis and other degenerative joint diseases (Bourmaud et al., 2021). Moreover, the MSC secretome contains pro-angiogenic factors that can also induce vascularization in the subchondral bone, which is crucial for supplying nutrients and removing waste products from cartilage, thereby enabling repair processes (Ragni et al., 2022).

EVs serve as secretome-editing centers that are also pivotal for ECM interactions. MSC-derived EVs, for instance, have shown potential to modulate inflammatory responses and stimulate tissue regeneration by delivering therapeutic cargoes to target cells (Murphy et al., 2019). These EVs can activate various signaling pathways that promote chondrocyte survival and proliferation, and inhibit catabolic pathways that degrade the ECM (D'Arrigo et al., 2019).

Furthermore, the relationship between secretome components and the ECM is reciprocal, with the ECM also influencing the composition and functions of the secretome. The ECM has mechanical and biochemical properties that can modulate a wide range of cell-secreted factors, thereby initiating tissue regeneration (Lin et al., 2020). For example, variations in ECM stiffness can modulate MSC behavior and the resulting secretome, which, in turn, can either facilitate or inhibit tissue repair (AlHalawani et al., 2022).

21.5.2 Scaffold Design to Optimize EV Delivery

EVs are nanoscale membrane-bound vesicles secreted by various types of cells that are effective in intercellular communication (Lazar et al., 2021). Designing scaffolds to deliver these vesicles efficiently would be crucial to maximizing their efficacy. The primary challenge in utilizing EVs for therapeutic purposes is maintaining their stability and bioavailability at the site of action. These scaffolds can be optimized to create a microenvironment that resembles the extracellular matrix in both prioritization and properties, thus promoting the retention and release of EVs. For example, bioactive scaffolds can be engineered to entrap EVs, enabling controlled, sustained in vitro and in vivo release, a process fundamental for promoting tissue regeneration (Lazar et al., 2021). Several studies on scaffolds modified with MSC-derived EVs have demonstrated significant improvements in vascularization and tissue integration, leading to enhanced regenerative outcomes (Hao et al., 2020).

Additionally, the physicochemical properties of scaffolds, including porosity, mechanical stability, and surface chemistry, can be tailored to optimize EV delivery. For instance, highly porous scaffolds can allow EVs to infiltrate the scaffold, thereby promoting cell migration, and specific surface modifications can increase the binding affinity of EVs to scaffold materials (Kang et al., 2021). The inclusion of hydrogels or nanofibrous structures can also facilitate a favorable arrangement for the collective of EVs, enabling sustained release and continuous cyclic treatment (Hao et al., 2020).

Beyond structural principles, functionalizing the scaffold with targeting ligands can increase the specificity of EV delivery to target cell types. This method can readily demonstrate the therapeutic effect of EVs by delivering them to the appropriate tissue and cellular targets (Kang et al., 2021). For instance, scaffolds can be designed to display specific peptides or antibodies (via non-covalent and covalent interactions) that promote the particular EV binding to target cells, thereby enhancing cellular uptake and biological response (Lazar et al., 2021). In addition, 3D bioprinting technologies have emerged as an innovative solution for providing complex scaffold architectures that accurately incorporate EVs. Using this technique, scaffold design can be tailored to meet specific therapeutic requirements, allowing the incorporation of different cell types and bioactive agents, such as EVs, into a single construct (Bari et al., 2021). Control over the spatial distribution of EVs in the scaffold will significantly improve their efficiency in stimulating tissue regeneration.

21.6 TISSUE ENGINEERING APPROACHES FOR ENHANCED TISSUE INTEGRATION

Tissue engineering is one of the most crucial approaches in regenerative medicine, particularly for treating bone and cartilage disorders. The incorporation of EVs and secretomes into tissue-engineering scaffolds would significantly accelerate tissue integration and regeneration. Herein, diverse strategies in scaffold design and fabrication are summarized, facilitating the delivery of EVs and their better incorporation into the tissue.

In scaffold design, one of the primary focuses is on mimicking the natural extracellular matrix to provide a biomimetic environment. This plays a crucial role in regulating cell adhesion, proliferation, and differentiation, which are essential for effective tissue regeneration. Innovations in scaffold manufacturing techniques, such as 3D printing or electrospinning, have enabled the production of customized scaffolds with defined characteristics that can accommodate EVs and secretomes (Xue et al., 2023). For example, 3D-printed scaffolds can be engineered to achieve specific porosity and mechanical strength, which are crucial for replicating the structural and functional features of native tissues (Turnbull et al., 2018).

Microfluidic technologies have also been used to enhance the scaffold's properties. Microfluidic foaming methods enable researchers to design scaffolds with specific pore sizes and morphologies for improving the permeability and diffusion rates of EVs within the scaffold. This structure provides uniform porosity, enabling better cell spreading and improved retention and release of bioactive molecules, which enhances tissue integration and regeneration (Costantini et al., 2015). Another attractive approach is to include dECM in scaffold designs. In addition, dECM scaffolds maintain both the biochemical signals of native tissues and their structural integrity, resulting in significant improvements in cellular responses and chondrogenic differentiation of stem cells (Collins et al., 2021). Such scaffolds derived from devitalized tissues have been shown to promote the chondrogenesis of adult stem cells, resulting in improved cartilage repair outcomes (Almeida et al., 2017). This alleviates the native ECM sourcing problem and the in vivo immune reactions of xenogeneic materials.

Moreover, hydrogels as scaffolding materials have emerged as a popular option due to their low immunogenicity and ability to encapsulate secretomes and EVs. Additionally, hydrogels provide a hydrated environment similar to that of the natural extracellular matrix (ECM), which supports cell migration and tissue ingrowth. Additionally, bioactive factors can be incorporated into the hydrogel to enhance its regenerative capacity, as their inclusion has been shown to induce cell proliferation and differentiation and to support the integration of engineered tissues with the host tissue (Silva et al., 2016). Alternatively, spheroids or stem cell aggregates in scaffolds have also led to innovative approaches. Its use has been shown to amplify EV paracrine signaling and increase cell-cell interactions, thereby accelerating cell integration into tissues. Additionally, the micro-molded technology has been shown to produce spheroids that are uniform in shape, and the ability to easily incorporate these spheroids into scaffolds improves the overall regenerative competence of the construct (Silva et al., 2016).

Furthermore, encapsulating growth factors and signaling molecules within scaffolds can promote matrix integration. As an example of factors released by scaffolds, VEGF can stimulate angiogenesis, a crucial process that provides nutrients and oxygen to regenerating tissue (Collins et al., 2021). This is especially relevant in bone and cartilage engineering, where vascularization is crucial for the proper incorporation and regeneration of these tissues.

21.7 CHALLENGES AND OPPORTUNITIES IN CELL-FREE THERAPIES

21.7.1 Bone and Cartilage Disorders: Osteoarthritis and Chronic Inflammation

Osteoarthritis (OA) is a degenerative joint disease characterized by progressive destruction of the articular cartilage, synovial inflammation, and remodeling of the subchondral bone. It's the leading cause of disability around the world, especially in the older population. OA pathogenesis is multifactorial, involving mechanical stress, metabolic factors, and chronic inflammation, all of which contribute to chondrocyte senescence and extracellular matrix dysregulation (De Liyis et al., 2022). In this regard, therapeutic applications of MSC-based and cell-free therapies using EVs offer an advanced approach (Safwan et al., 2024; AlBalawi et al., 2025). Recent theranostic modalities have further expanded the range of OA management approaches (Sadri et al., 2021).

The development of effective anti-inflammatory therapies to reduce the chronic inflammatory milieu characteristic of OA is one of the most noteworthy challenges in OA treatment. Studies have further shown that MSC-derived EVs inhibit the secretion of pro-inflammatory cytokines and promote tissue regeneration (Lou et al., 2024). For example, LPS-conditioned MSC-derived EVs have also been shown to prevent ECM degradation and inhibit inflammation in preclinical models of OA (Duan et al., 2021). Thus, EVs may serve as a non-invasive therapeutic strategy targeting the causal chronic inflammatory process, with potential advantages over the complications of cell-based therapies.

The greater therapeutic fidelity of EVs against OA can be directly correlated with their cargo, including microRNAs, proteins, and lipids, which govern critical cellular communication and tissue homeostasis (Vallabhaneni et al., 2015). Some of these microRNAs in MSC-derived EVs have been identified as those that enhance chondrocyte survival and function while inhibiting apoptosis and inflammation (Liu, 2023). This cargo can alter the behavior of receiving cells, thereby improving their regenerative capacity and facilitating cartilage repair (Yuan et al., 2023). Nevertheless, heterogeneity in EV composition, which depends on MSC source and the processing environment in which they are harvested, limits our ability to standardize clinical EV-based therapies (van de Wakker et al., 2023).

Optimizing the delivery method to ensure that vesicles reach the specific tissues involved in the desired interaction represents another significant barrier to the clinical translation of EVs in the treatment of OA (Jihwaprani et al., 2024). Currently, strategies for EV administration often involve intra-articular injections, which can provide localized treatment but may not achieve sustained release or distribution within the joint (Mohd Noor et al., 2021). Novel strategies, such as integrating EVs into biomimetic scaffolds or hydrogels, are being investigated to enhance retention and controlled delivery at the injury site (Zhang et al., 2023). These strategies focus on improving the microenvironment of EVs, thereby facilitating their integration with surrounding tissues and increasing therapeutic efficacy.

However, it is worth noting that EV-based therapies for OA present significant opportunities despite these challenges. Due to their low immunogenicity and potential to transport bioactive molecules, EVs serve as an appealing alternative to standard procedures, such as invasive operations or systemic therapy, which can lead to highly nonspecific adverse events (Wu et al., 2022). The ability to engineer EVs to increase their therapeutic efficacy or direct them toward specific alternative cell types signifies potential approaches to personalized medicine for the treatment of OA (Teng & Fussenegger, 2020; Zhang et al., 2024).

21.8 CLINICAL TRANSLATION BARRIERS: SECRETOME STABILITY, DOSING, AND DELIVERY METHODS

Several hurdles face MSC-derived secretomes and EVs as cell-free therapies for clinical development. Among the challenges are stabilizing the secretome and determining optimal dosing and delivery methods. A significant challenge for the clinical application of secretomes is their instability, which can be compromised by diverse storage conditions. For example, secretomes must be stored at low temperatures to preserve their functional integrity; however, prolonged storage can compromise key growth factors and cytokines essential for tissue regeneration (Yu et al., 2023). It has been demonstrated that the secretome can vary over time, potentially impacting its therapeutic potential (Rogulska et al., 2024). Therefore, creating standardized protocols for secretome collection, processing, and storage is essential to ensure these outcomes remain stable and effective during clinical delivery.

The right timing of secretomes is also a significant challenge. In contrast to traditional pharmaceuticals, for which the dose is typically standardized, the dosing of secretome will likely depend on factors such as PSA, the method of PS extraction, and/or the indication being treated (Wang et al., 2022). Although the biological impact of the secretome is dose-dependent (Liang et al., 2025), higher doses can thus enhance the therapeutic effects. Still, they may correspondingly increase risks of adverse effects (such as hyperinflammation or immune responses) (Sadri et al., 2025). Therefore, conducting dose-response studies is necessary to determine safe and effective doses for clinical use.

The delivery must be optimized to maximize the therapeutic potential of secretomes and EVs. Localized treatment methods for current delivery strategies, such as intra-articular injections, may, however, hinder the release duration or prevent sufficient distribution to the target tissues (Simader et al., 2017). New strategies, such as incorporating secretomes into hydrogels or scaffolds, are being developed to enhance retention and provide controlled release at the injury site (Wechsler et al., 2021). Hydrogels provide mechanical support while mimicking the natural extracellular environment, allowing for cell migration and stimulating tissue integration and development (Choi et al. 2024). However, the design of these delivery systems must also account for biocompatibility, degradation rates, and the retention of bioactive components in microbiomes over time.

In addition, the viscosity and particle size of the secretome are physicochemical properties that affect its delivery and distribution in tissues. For example, the aerosolized application of secretomes for lung regeneration would necessitate careful fine-tuning of both particle diameter and surface properties to facilitate deposition within target tissues (Bari et al., 2021). Continued investigation into the formulation and engineering of secretomes for various administration routes will be necessary to enhance their clinical translation and application (Maal et al., 2025).

21.9 FUTURE PERSPECTIVES: PERSONALIZED AND TARGETED EV-BASED THERAPIES

The role of EVs as therapeutic agents in regenerative medicine offers new opportunities for personalized, target-driven therapy, particularly for addressing bone and cartilage diseases, including osteoarthritis and critical-size bone defects (Sadri et al., 2022). The endosomal formation of bioactive molecules for EVs, along with their distinctiveness from other lipoprotein secretory vesicles, establishes an ideal platform for loading bioactive molecules via a wide range of biosynthesis-based methods, making EVs promising for precise therapy and future perspectives of personalized, targeted, evidence-based treatments.

Therapies based on exosomes represent one of the most promising paradigms, and they can be tailored to the characteristics of a given patient's disease. Recent investigations have focused on profiling cargo EVs, proteins, lipids, and microRNAs to discover potential biomarkers specific to the State (Ghafouri-Fard et al., 2021). By studying the content of EVs secreted from a patient's native cells, doctors can design treatment strategies that are much more likely to work for that patient. Specific microRNAs released from EVs associated with inflammatory or cartilage-degrading pathways can be targeted to develop efficacious therapeutic strategies underlying these processes (Rong et al., 2021). Such a strategy aligns with the philosophy of personalized medicine, which seeks to tailor therapeutic effects to a patient's unique biological signature to maximize the likelihood of success (Liu et al., 2023). EV profiling may soon be translated into clinical practice, influencing ongoing research as well.

The targeted delivery of EVs and their applications in personalized medicine are promising developments in this arena. To this end, multiple approaches are being investigated, including the functionalization of EVs with targeting ligands that bind to receptors overexpressed on target cells (Kim et al., 2024). In this context, EVs engineered to display ligands that bind specifically to chondrocytes or osteoblasts would enable targeted delivery to these cell types, thereby enhancing regenerative therapeutic effects (Kang et al., 2021). Moreover, the use of EVs in biomaterials or scaffolds can create a controlled-release system that ensures continuous delivery of therapeutic components to the injured site. As a result, this strategy not only enhances EV retention at the target tissue but also enables sustained release of EV cargo, thereby facilitating prolonged therapeutic effects. Such approaches are especially pertinent in the realm of bone and cartilage regeneration (Liu et al., 2023), where the maintenance-release of growth factors and anti-inflammatory agents can enhance healing efficacy.

Although there are great promises for the future of personalized and targeted EV-based therapies, numerous challenges remain. Methods for the isolation and characterization of EVs must be standardized to ensure their reproducibility and the safety of these therapies (Théry et al., 2018). Furthermore, relevant regulatory frameworks for the clinical application of EVs are still being developed, and their safety and efficacy will need to be established through large-scale clinical trials before these therapeutic agents can be adopted for clinical use (Petrosyan et al., 2022). In addition, variations in the biological content of EVs, depending on source and isolation methods, as well as the complexity of EV biology, pose challenges for establishing consistent, reproducible therapeutic products (Nagelkerke et al., 2021). Overcoming these challenges will require the collective efforts of researchers, clinicians, and regulatory agencies to develop guidelines and best practices for implementing EVs in regenerative medicine.

21.10 COMMERCIALIZATION OF SECRETOME AND EVS AS OFF-THE-SHELF PRODUCTS

Considering that secretomes and EVs can be commercialized, their use as off-the-shelf products represents a significant turning point in regenerative medicine for bone and cartilage diseases. These products promise standardized, scalable, and accessible therapeutics that can be easily applied in the clinic. Thereafter, in the realm of commercialization, the state-of-the-art limitations and prospects of secretome and EV-based therapies are examined.

Secretome and EV-based therapies have garnered increasing interest in recent years, following their demonstrated efficacy in preclinical research and early-phase clinical trials (Xia et al., 2021). Even MSC-derived secretomes hold potential for tissue repair and regeneration, particularly in osteoarthritis and other degenerative disease states (Ragni et al., 2022). Their ability to modulate inflammation, promote cell proliferation, and enhance tissue integration has made these products attractive for commercialization (Ragni et al., 2022).

Several entities are focusing on secretome- and exosome-based applications for development and commercialization. These products are often isolated from adipose tissue, bone marrow, or dental stem cells, each of which has distinct advantages in regenerative potential (Pulido-Escribano et al., 2022). Good Manufacturing Practice (GMP) is an essential requirement for the preparation of these products to achieve safety, efficacy, and regulatory compliance (Bari et al., 2018).

Secretome and EVs are promising therapeutic tool sets with supportive potential; however, to commercialize these products, several hurdles need to be overcome. The most significant issue is the need to standardize the production process. Source (allogeneic or autologous), culture conditions, and isolation methods variations can induce substantial differences in the secretome and extracellular vesicle (EV) composition and biological activity (Pulido-Escribano et al., 2022). Standardizing protocols for the preparation, processing, and characterization is a concerted effort necessary to ensure these products are consistently high-quality and effective.

The second challenge is in the regulatory space. Secretome and EV-based products currently reside in a regulatory gray area, as they do not neatly fit into existing classifications for cell-based therapies or biologics (D'Arrigo et al., 2019). Meanwhile, relevant regulatory agencies, including the U.S. Food and Drug Administration (FDA) and the European Medicines Agency (EMA), are developing frameworks to approve these products. Specific guidelines and pathways for regulatory approval will be key to assessing the success of secretome and EV therapies in advancing their commercialization (Bari et al., 2018).

Secretome- and extracellular vesicle (EV)-based therapies hold great promise, with numerous opportunities for development and commercialization. A significant opportunity lies in integrating these products with advanced delivery systems, such as 3D-printed scaffolds or hydrogels, which can enhance their therapeutic effects or provide localized treatment (Bari et al., 2021). Such combinations could maintain and control the secretomes and EVs released at injured tissues, enabling us to fully utilize their regenerative abilities.

Additionally, the growing demand for personalized medicine presents opportunities for developing tailored secretome and extracellular vesicle (EV) therapies. Clinicians could profile the specific cargo of EVs derived from a patient's cells to personalize therapies targeting the atypical properties characteristic of each condition (Beetler et al., 2022). Such methods potentiate therapeutic benefit while limiting the risk of adverse reactions often associated with allogeneic products.

These will also serve as a stimulus and encourage further understanding and investment in secretome and EV therapies, as confirmed by the ongoing influx of clinical experience demonstrating their safety and efficacy across various settings (González-González et al., 2020). With the consequence of more clinical trials and improved results, the acceptance and adoption of these at the clinical practice level should gradually expand, enabling commercialization as off-the-shelf products.

12.11 CONCLUSION AND FUTURE DIRECTIONS

Secretomes and EVs have recently become revolutionary tools in regenerative medicine, including the treatment of bone and cartilage diseases. Instead, these cell-free therapies employ bioactive molecules as a therapeutic cargo that can also help modulate inflammation, including promoting angiogenesis, enhancing cell survival, and facilitating tissue repair. They represent a paradigm shift from symptom management to utilizing decreased immunogenicity and simplified production, enabling the restoration of tissue function. Additional investigations are necessary to clarify mechanisms, establish standardized protocols, optimize delivery systems, and tailor administration to each patient's specific needs. Clinical trials will also be essential for verifying the safety and efficacy of secretome and EV therapies, thereby advancing them to a transformative role in regenerative medicine.

REFERENCE

Abdal Dayem A, Yan E, Do M, Kim Y, Lee Y, Cho SG, Kim DH. Engineering extracellular vesicles for ROS scavenging and tissue regeneration. *Nano Converg.* 2024 Jun 26;11(1):24. https://doi.org/10.1186/s40580-024-00430-9

Abels ER, Breakefield XO. Introduction to extracellular vesicles: Biogenesis, RNA cargo selection, content, release, and uptake. *Cell Mol Neurobiol.* 2016;36:301–12. https://doi.org/10.1007/s10571-016-0366-Z

Aguilera SB, McCarthy A, Khalifian S, Lorenc ZP, Goldie K, Chernoff WG. The role of calcium hydroxylapatite (Radiesse) as a regenerative aesthetic treatment: A narrative review. *Aesthet. Surg. J.* 2023;43(10):1063–1090.

Ahmadi M, Rezaie J. Tumor cells-derived exosomes as angiogenic agents: Possible therapeutic implications. *J Transl Med.* 2020;18(1). https://doi.org/10.1186/s12967-020-02426-5

Ahmed ZT, Alkahlot MH, Haider KH. MSC-derived exosomes: Advances in cell-free therapy. In KH Haider, editor. *Handbook of Stem Cell Applications.* Singapore: Springer Nature, 2023 Aug 18, pp. 1–41.

Akyash F, Javidpou M, Sadeghian Nodoushan F, Aflatoonian B. Human embryonic stem cell-derived mesenchymal stem/stromal cells and their use in regenerative medicine. *J Stem Cell Res Ther.* 2016;1(7):47. http://doi.org/10.15406/jsrt.2016.01.00047

AlBalawi R, Safwan M, Bourgleh MS, Baroudi L, Aybour B, Alfawaz S, Almesned A, et al. A systematic review and network meta-analysis of RCTs involving MSC-based therapy for knee osteoarthritis: Breaking barriers in regenerative medicine with living biodrugs. In: International Meeting on Inflammation and Regenerative Medicine, 12–13 June 2025, H10 Roma Citta, Rome, Italy.

Al Halawani A, Mithieux SM, Yeo GC, Hosseini-Beheshti E, Weiss AS. Extracellular vesicles: Interplay with the extracellular matrix and modulated cell responses. *Int. J. Mol. Sci.* 2022;23(6):3389. https://doi.org/10.3390/ijms23063389

Almeida HV, Liu Y, Cunniffe GM, Mulhall KJ, Matsiko A, Buckley CT, O'Brien FJ, Kelly DJ. Porous scaffolds derived from devitalized tissue engineered cartilaginous matrix support chondrogenesis of adult stem cells. *ACS Biomater. Sci. Eng.* 2017;3(6):1075–1082. https://doi.org/10.1021/acsbiomaterials.6b00654

Bari E, Ferrarotti I, Torre ML, Corsico AG, Perteghella S. Pilot production of mesenchymal stem/stromal freeze-dried secretome for cell-free regenerative nanomedicine: A validated GMP-compliant process. *Cells.* 2018;7(11):190. https://doi.org/10.3390/cells7110190

Bari E, Scocozza F, Perteghella S, Sorlini M, Auricchio F, Torre ML, Conti M. 3D bioprinted scaffolds containing mesenchymal stem/stromal lyosecretome as next generation controlled release device for bone regenerative medicine. *Pharmaceutics.* 2021;13(4):515. https://doi.org/10.3390/pharmaceutics13040515

Bazzan E, Tinè M, Casara A, Biondini D, Semenzato U, Cocconcelli E, Balestro E, et al. Critical review of the evolution of extracellular vesicles' knowledge: From 1946 to today. *Int. J. Mol. Sci.* 2021 Jun 15;22(12):6417. https://doi.org/10.3390/ijms22126417

Beetler DJ, Di Florio, DN, Bruno KA, Ikezu T, March KL. Extracellular vesicles as personalized medicine. *Mol. Asp. Med.* 2022;91:101155. https://doi.org/10.1016/j.mam.2022.101155

Bongiovanni L, Andriessen A, Wauben MHM, Nolte-'t Hoen ENM, de Bruin A. Extracellular vesicles: Novel opportunities to understand and detect neoplastic diseases. *Vet Pathol.* 2021 May;58(3):453–71. https://doi.org/10.1177/0300985821999328

Bourmaud M, Zarka M, Le Cozannet R, et al. Rubus idaeus reduces inflammation and prevents chondrocyte catabolism and cartilage loss. *Biomolecules.* 2021;11(2):245. https://doi.org/10.3390/biom11020245

Chang SF, Hsieh RZ, Huang KC, Chang CA, Chiu FY, Kuo HC, Chen CN, et al. Upregulation of bone morphogenetic protein-2 synthesis and consequent collagen II expression in leptin-stimulated human chondrocytes. *PLoS One.* 2015 Dec 4;10(12):e0144252. https://doi.org/10.1371/journal.pone.0144252

Chen X, Tian B, Wang Y, Zheng J, Kang X. Potential and challenges of utilizing exosomes in osteoarthritis therapy (Review). *Int J Mol Med.* 2025 Mar;55(3):43. https://doi.org/10.3892/ijmm.2025.5484

Chi B, Zou A, Mao L, Cai D, Xiao T, Wang Y, Wang Q, et al. Empagliflozin-pretreated mesenchymal stem cell-derived small extracellular vesicles attenuated heart injury. *Oxid. Med. Cell Longev.* 2023 Feb 18;2023:7747727. https://doi.org/10.1155/2023/7747727

Choi H, Kim J, Lee S. A review of advanced hydrogel applications for tissue engineering and drug delivery. *Gels.* 2024;10(11):693. https://doi.org/10.3390/gels10110693

Chyzy A, Plonska-Brzezinska ME. Hydrogel properties and their impact on regenerative medicine and tissue engineering. *Molecules.* 2020;25(24):5795. https://doi.org/10.3390/molecules25245795

Collins MN, Ren G, Young K, Pina S, Reis RL, Oliveira JM. Scaffold fabrication technologies and structure/function properties in bone tissue engineering. *Adv. Funct. Mater.* 2021;31(21):2010609. https://doi.org/10.1002/adfm.202010609

Correia CR, Reis RL, Mano JF. Nanostructured capsules for cartilage tissue engineering. *Cartilage Tissue Eng.* 2015;181–9. https://doi.org/10.1007/978-1-4939-2938-2_13

Costantini M, Colosi C, Jaroszewicz J, Tosato A, Święszkowski W, Dentini M, Garstecki P, Barbetta A. Microfluidic foaming: A powerful tool for tailoring the morphological and permeability properties of sponge-like biopolymeric scaffolds. *ACS Appl. Mater. Interfaces.* 2015;7(42):23660–23671. https://doi.org/10.1021/acsami.5b08221

D'Arrigo D, Roffi A, Cucchiarini M, Moretti M, Candrian C, Filardo G. Secretome and extracellular vesicles as new biological therapies for knee osteoarthritis: A systematic review. *J. Clin. Med.* 2019;8(11):1867. https://doi.org/10.3390/jcm8111867

De Liyis BG, Nolan J, Maharjana MA. Fibroblast growth factor receptor 1-bound extracellular vesicle as novel therapy for osteoarthritis. *Biomedicine.* 2022;12(2):1

Diaz-Solano D, Sadri B, Peshkova M, Shpichka A, Smirnova O, Shams R, et al. Advanced therapeutic medicinal products in bone and cartilage defects. *Curr. Rev. Clin. Exp. Pharmacol.* 2024 Jan 25;19(4):355–69. https://doi.org/10.2174/0127724328274436231207062008

Ding Y, Lu L, Guo Y, Zhu B. Bibliometric analysis of research on osteoarthritis and extracellular vesicles: trends and frontiers. *Heliyon.* 2024;10(16):e36127. https://doi.org/10.1016/j.heliyon.2024.e36127

Dong Z, Haines S, Coates D. Proteomic profiling of stem cell tissues during regeneration of deer antlers: A model of mammalian organ regeneration. *J. Proteome Res.* 2020;19(4):1760–75. https://doi.org/10.1021/acs.jproteome.0c00026

Duan A, Shen K, Li B, Li C, Zhou H, Kong R, et al. Extracellular vesicles derived from LPS-preconditioned human synovial mesenchymal stem cells inhibit extracellular matrix degradation and prevent osteoarthritis of the knee in a mouse model. *Stem Cell Res. Ther.* 2021;12:427. https://doi.org/10.1186/s13287-021-02507-2

El-Qashty R, Elkashty OA, Hany E. Photobiostimulation conjugated with stem cells or their secretome for temporomandibular joint arthritis in a rat model. *BMC Oral Health.* 2023;23:720. https://doi.org/10.1186/s12903-023-03434-5

Engen C, Engebretsen L, Årøen A. Patients with knee cartilage defects enrolled in randomized controlled trials are not representative of patients in orthopedic practice. *Cartilage.* 2010;1(4):312–9. https://doi.org/10.1177/1947603510373917

Farshidfar N, Fooladi S, Nematollahi MH, Iravani S.Carbon dots with tissue engineering and regenerative medicine applications. *RSC Adv.* 2023;13(21):14517–29. https://doi.org/10.1039/D3RA02336B

Fortier LA, Barker JU, Strauss EJ, McCarrel TM, Cole BJ. The role of growth factors in cartilage repair. *Clin. Orthop. Related Res.* 2011;469(10):2706–15. https://doi.org/10.1007/s11999-011-1857-3

Frisch J, Rey-Rico A, Venkatesan J, Schmitt G, Madry H, Cucchiarini M. TGF-β gene transfer and overexpression via AAV vectors stimulate chondrogenic events in human bone marrow aspirates. *J. Cell. Mol. Med.* 2016;20(3):430–40. https://doi.org/10.1111/jcmm.12774

Gao M, Guo H, Dong X, Wang Z, Yang Z, Shang Q, Wang Q. Regulation of inflammation during wound healing: The function of mesenchymal stem cells and strategies for therapeutic enhancement. *Front. Pharmacol.* 2024;15:1345779. https://doi.org/10.3389/fphar.2024.1345779

Gao T, Fan H, Wang J, Wang R. Extracellular vesicles derived from human umbilical cord mesenchymal stem cells stimulate angiogenesis in myocardial infarction via the microRNA-423-5p/efna3 axis. *Adv. Interv. Cardiol.* 2022;18(4):373–91. https://doi.org/10.5114/aic.2023.124797

Gerwing M, Kocman V, Stölting M, Helfen A, Masthoff M, Roth J, Barczyk-Kahlert K, et al. Tracking of tumor cell-derived extracellular vesicles in vivo reveals a specific distribution pattern with consecutive biological effects on target sites of metastasis. *Mol. Imaging. Biol.* 2020 Dec;22(6):1501–10. https://doi.org/10.1007/s11307-020-01521-9

Ghafouri-Fard S, Poulet C, Malaise M, Abak A, Mahmud Hussen B, Taheriazam A, Taheri M, Hallajnejad M. The emerging role of non-coding RNAs in osteoarthritis. *Front. Immunol.* 2021;12:773171. https://doi.org/10.3389/fimmu.2021.773171

Go Y, Lee C, Chae S, Song J. Regenerative capacity of trophoblast stem cell-derived extracellular vesicles on mesenchymal stem cells. *Biomat. Res.* 2023;27(1). https://doi.org/10.1186/s40824-023-00396-5

González-González A, García-Sánchez D, Dotta M, Rodríguez-Rey JC, Pérez-Campo FM. Mesenchymal stem cells secretome: the cornerstone of cell-free regenerative medicine. *World J. Stem Cells.* 2020;12(12):1529–1552. https://doi.org/10.4252/wjsc.v12.i12.1529

Görgün C, Ceresa D, Lesage R, Villa F, Reverberi D, Balbi C, et al. Dissecting the effects of preconditioning with inflammatory cytokines and hypoxia on the angiogenic potential of mesenchymal stromal cell (MSC)-derived soluble proteins and extracellular vesicles (EVs). *Biomaterials.* 2021;269:120633. https://doi.org/10.1016/j.biomaterials.2020.120633

Griffeth R, García-Párraga D, Mellado-López M, Crespo-Picazo J, Soriano-Navarro M, Martínez-Romero A, et al. Platelet-rich plasma and adipose-derived mesenchymal stem cells for regenerative medicine-associated treatments in bottlenose dolphins (Tursiops truncatus). *PLoS One*. 2014;9(9):e108439. https://doi.org/10.1371/journal.pone.0108439

Habib SM, Martini MF, Al Ahmed O, Qassab AR, Haider KH. Mesenchymal stem cell secretomes for cell-free therapy of the broken heart. In: KH Haider, editor. *Handbook of Regenerative Medicine: Cell-Free Therapy*. London: Taylor & Francis Group/CRC Press, 2025a, Chapter 18, pp. 274–93.

Habib SM, Martini MF, Al Ahmed O, Qassab AR, Haider KH. The potential of mesenchymal stem cell–derived exosomes in cell-free therapy for myocardial repair. In: KH Haider, editor. *Handbook of Regenerative Medicine: Cell-Free Therapy*. London: Taylor & Francis Group/CRC Press, 2025b, Chapter 24, pp. 372–90.

Haider KH, Aramini B. Mircrining the injured heart with stem cell-derived exosomes: An emerging strategy of cell-free therapy. *Stem Cell. Res. Ther.* 2020 Jan 9;11(1):23. https://doi.org/10.1186/s13287-019-1548-7.

Haider KH, Aslam M. Cell-free therapy with stem cell secretions: Protection, repair, and regeneration of the injured myocardium. In: KH Haider, S Aziz, editors. *Stem Cells: From Hype to Real Hope*. Berlin: Medicine & Life Sciences, De Gruyter, 2018, pp. 34–70.

Hallal S, Tűzesi Á, Grau GE, Buckland ME, Alexander KL. Understanding the extracellular vesicle surface for clinical molecular biology. *J. Extracell. Vesicles* 2022;11(10):e12260. https://doi.org/10.1002/jev2.12260

Han F, Wang J, Ding L, Hu Y, Li W, Yuan Z, et al. Tissue engineering and regenerative medicine: Achievements, future, and sustainability in Asia. *Fronti. Bioeng. Biotechnol.* 2020;8:83. https://doi.org/10.3389/fbioe.2020.00083

Hao D, Swindell HS, Ramasubramanian L, Liu R, Lam KS, Farmer DL, et al. Extracellular matrix mimicking nanofibrous scaffolds modified with mesenchymal stem cell-derived extracellular vesicles for improved vascularization. *Front. Bioeng. Biotechnol.* 2020;8:633. https://doi.org/10.3389/fbioe.2020.00633

Hellmund KS, Koksch B. Self-assembling peptides as extracellular matrix mimics to influence stem cell fate. *Front. Chem.* 2019;7:172. https://doi.org/10.3389/fchem.2019.00172

Ibrahim R, Mndlovu H, Kumar P, Adeyemi SA, Choonara YE. Cell secretome strategies for controlled drug delivery and wound-healing applications. *Polymers* 2022;14(14):2929. https://doi.org/10.3390/polym14142929

Itoh T, Degawa T, Ito Y, AkaoY, Okumura N. Role of royal jelly-treated adipose-derived stem cell-extracellular vesicles on fibroblast proliferation, migration, and collagen production. *Dermatol. Ther.* 2023;2023(1). https://doi.org/10.1155/2023/7950026

Jihwaprani MC, Sula I, Charbat MA, Haider KH. Establishing delivery route-dependent safety and efficacy of living biodrug mesenchymal stem cells in heart failure patients. *World J. Cardiol.* 2024;16(6):339–54. https://doi.org/10.4330/wjc.v16.i6.339

Jeppesen D, Fenix A, Franklin J, Higginbotham J, Zhang Q, Zimmerman L, et al. Reassessment of exosome composition. *Cell* 2019;177(2):428–45.e18. https://doi.org/10.1016/j.cell.2019.02.029

Jessop ZM, Al-Sabah A, Francis WR, Whitaker IS. Transforming healthcare through regenerative medicine. *BMC Med.* 2016;14:1–6. https://doi.org/10.1186/s12916-016-0669-4

Jeyaraman M, Muthu S, Gulati A, Jeyaraman N, Shivashankar P, Jain R. Mesenchymal stem cell–derived exosomes: A potential therapeutic avenue in knee osteoarthritis. *Cartilage* 2020;13(1_suppl):1572S–85S. https://doi.org/10.1177/1947603520962567

Kang M, Jordan V, Blenkiron C, Chamley LW. Bioactive scaffolds integrated with liposomal or extracellular vesicles for tissue regeneration. *Bioengineering*. 2021;8(10):137. https://doi.org/10.3390/bioengineering8100137

Ke Z, Wei X, Zhi Z, Zhang S, Cui L, Li Y, et al. Comparison of different isolation methods for plasma-derived extracellular vesicles in patients with hyperlipidemia. *Life*. 2022;12(11):1942. https://doi.org/10.3390/life12111942

Kearney C, Khatab S, Buul G, Plomp S, Korthagen N, Labberté M, et al. Treatment effects of intra-articular allogenic mesenchymal stem cell secretome in an equine model of joint inflammation. *Front. Vet. Sci.* 2022;9. https://doi.org/10.3389/fvets.2022.907616

Khatab S, Osch G, Kops N, Bastiaansen-Jenniskens Y, Bos P, Verhaar J, et al. Mesenchymal stem cell secretome reduces pain and prevents cartilage damage in a murine osteoarthritis model. *eCM* 2018;36:218–30. https://doi.org/10.22203/ecm.v036a16

Kim HI, Park J, Zhu Y, Wang X, Han Y, Zhang D. Recent advances in extracellular vesicles for therapeutic cargo delivery. *Exp. Mol. Med.* 2024;56(4):836–49. https://doi.org/10.1038/s12276-024-01201-6

Kreimer S, Belov A, Ghiran I, Murthy S, Frank D, Ivanov A. Mass-spectrometry-based molecular characterization of extracellular vesicles: Lipidomics and proteomics. *J. Proteome Res.* 2015;14(6):2367–84. https://doi.org/10.1021/pr501279t

Kulebyakina M, Basalova N, Butuzova D, Arbatsky M, Chechekhin V, Kalinina N, et al. Balance between pro- and antifibrotic proteins in mesenchymal stromal cell secretome fractions revealed by proteome and cell subpopulation analysis. *Int. J. Mol. Sci.* 2023;25(1):290.

Kumar AP, Vikram Reddy G, Raja Babu PJRG, Jagadish Reddy G. Platelet rich fibrin-A second regeneration platelet concentrate and advances in PRF. *Indian J. Dent.* 2015;7:251–4.

Kuriyama N, Yoshioka Y, Kikuchi S, Azuma N, OchiyaT. Extracellular vesicles are key regulators of tumor neovasculature. *Front. Cell Develop. Biol.* 2020:20208. https://doi.org/10.3389/fcell.2020.611039

Lazar S, Mor S, Chen J, Hao D, Zhang Z, Wu G. Bioengineered extracellular vesicle-loaded bioscaffolds for therapeutic applications in regenerative medicine. *Extracell Vesicles Circ. Nucleic Acids.* 2021;2(2):175–178. https://doi.org/10.20517/evcna.2021.10

Li K, Yan G, Huang H, Zheng M, Ma K, Cui X, Lu D, et al. Anti-inflammatory and immunomodulatory effects of the extracellular vesicles derived from human umbilical cord mesenchymal stem cells on osteoarthritis via M2 macrophages. *J. Nanobiotechnology.* 2022 Jan 20;20(1):38. https://doi.org/10.1186/s12951-021-01236-1

Li M, Fang F, Sun M, Zhang Y, Hu M, Zhang J. Extracellular vesicles as bioactive nanotherapeutics: An emerging paradigm for regenerative medicine. *Theranostics.* 2022;12(11):4879–903. https://doi.org/10.7150/thno.72812

Liang Y, Xia J, Xu L, Zhang Y, Xu X, Haider KH. Cell-free and cell-based therapy in regenerative medicine. In KH Haider, editor. *Handbook of Regenerative Medicine: Cell-Free Therapy Approach* (1st ed.). CRC Press, 2025. Chapter 24. https://doi.org/10.1201/9781003611509.

Lin X, Patil S, Gao YG, Qian A. The bone extracellular matrix in bone formation and regeneration. *Front. Pharmacol.* 2020;11:757. https://doi.org/10.3389/fphar.2020.00757

Liu YJ, Wang C. A review of the regulatory mechanisms of extracellular vesicles-mediated intercellular communication. *Cell Commun. Signal.* 2023 Apr 13;21(1):77. https://doi.org/10.1186/s12964-023-01103-6.

Lou K, Luo H, Jiang X, Feng S. Applications of emerging extracellular vesicles technologies in the treatment of inflammatory diseases. *Front. Immunol.* 2024;15:1364401. https://doi.org/10.3389/fimmu.2024.1364401

Lu S, Lam J, Trachtenberg JE, Lee EJ, Seyednejad H, van den Beucken JJJP, Tabata Y, et al. Dual growth factor delivery from bilayered, biodegradable hydrogel composites for spatially-guided osteochondral tissue repair. *Biomaterials.* 2014 Oct;35(31):8829–39. https://doi.org/10.1016/j.biomaterials.2014.07.006

Ma Z, Wang Y, Li H. Applications of extracellular vesicles in tissue regeneration. *Biomicrofluidics.* 2020;14(1). https://doi.org/10.1063/1.5127077

Maal AB, Kordi R, Madani H, Khadivi M, Moghadam N, Asnaashari A, et al. Intradiscal injection of allogeneic bone marrow-derived clonal mesenchymal stromal cells in discogenic low back pain: A phase I study on safety and feasibility (RELIEF: phase I). *Stem Cell Res. Therapy.* 2025 Sep 29;16(1). https://doi.org/10.1186/s13287-025-04606-w

Mainil-Varlet P, Schiavinato A, Ganster M. Efficacy evaluation of a new hyaluronan derivative, hyadd®4-g, to maintain cartilage integrity in a rabbit model of osteoarthritis. *Cartilage.* 2012;4(1):28–41. https://doi.org/10.1177/1947603512455193

Mithoefer K, Saris D, Farr J, Kon E, Zaslav K, Cole B, et al. Guidelines for the design and conduct of clinical studies in knee articular cartilage repair. *Cartilage* 2011;2(2):100–21. https://doi.org/10.1177/1947603510392913

Mohd Noor NA, Abdullah NA, Ahmad Mohd Zain MR, Wan Nor Aduni WK, Azlan M. Extracellular vesicles from mesenchymal stem cells as potential treatments for osteoarthritis. *Cells.* 2021;10(6):1287. https://doi.org/10.3390/cells10061287.

Murphy DE, de Jong OG, Brouwer M, Wood MJ, Lavieu G, Schiffelers RM, et al. Extracellular vesicle-based therapeutics: natural versus engineered targeting and trafficking. *Exp. Mol. Med.* 2019;51(3):1–12. https://doi.org/10.1038/s12276-019-0223-5

Mushtaq M, Zineldeen DH, Mateen MA, Haider KH. Mesenchymal stem cells' "garbage bags" at work: Treating radial nerve injury with mesenchymal stem cell-derived exosomes. *World J. Stem Cells* 2024;16(5). https://doi.org/10.4252/wjsc.v16.i5.0000

Nagelkerke A, Ojansivu M, van der Koog L, Whittaker TE, Cunnane EM, Silva AM, Dekker N, Stevens MM, Vader P. Extracellular vesicles for tissue repair and regeneration: evidence, challenges and opportunities. *Adv. Drug. Deliv. Rev.* 2021;175:113775. https://doi.org/10.1016/j.addr.2021.04.013

Olejarz W, Kubiak-Tomaszewska G, Chrzanowska A, Lorenc T. Exosomes in angiogenesis and anti-angiogenic therapy in cancers. *Int. J. Mol. Sci.* 2020;21(16):5840. https://doi.org/10.3390/ijms21165840

Papathanasiou I, Malizos K, Tsezou A. Bone morphogenetic protein-2-induced activation of the Wnt/β-catenin signaling pathway through enhanced catabolic activity of low-density lipoprotein receptor-related protein 5 contributes to hypertrophy in osteoarthritic chondrocytes. *Arth. Res. Ther.* 2012;14(2). https://doi.org/10.1186/ar3805

Petrosyan A, Martins PN, Solez K, Uygun BE, Gorantla VS, Orlando G. Regenerative medicine applications: an overview of clinical trials. *Front. Bioeng. Biotechnol.* 2022;10:942750. https://doi.org/10.3389/fbioe.2022.942750

Potz BA, Scrimgeour LA, Pavlov VI, Sodha NR, Abid MR, Sellke FW. Extracellular vesicle injection improves myocardial function and increases angiogenesis in a swine model of chronic ischemia. *J. Am. Heart Assoc.* 2018 Jun 12;7(12):e008344. https://doi.org/10.1161/JAHA.117.008344

Pulido-Escribano V, Torrecillas-Baena B, Camacho-Cardenosa M, Dorado G, Gálvez-Moreno MA, Casado-Díaz A. Role of hypoxia preconditioning in therapeutic potential of mesenchymal stem-cell-derived extracellular vesicles. *World J. Stem. Cells.* 2022;14(7):453–472. https://doi.org/10.4252/wjsc.v14.i7.453

Ragni E, Perucca Orfei C, de Girolamo L. Secreted factors and extracellular vesicles account for the immunomodulatory and tissue regenerative properties of bone-marrow-derived mesenchymal stromal cells for osteoarthritis. *Cells.* 2022;11(21):3501. https://doi.org/10.3390/cells11213501

Rani S, Lai A, Nair S, Sharma S, Handberg A, Carrion F, Möller A, Salomon C. Extracellular vesicles as mediators of cell-cell communication in ovarian cancer and beyond – A lipids focus. *Cytokine Growth Factor Rev.* 2023 Oct;73:52–68. https://doi.org/10.1016/j.cytogfr.2023.06.004

Raposo G, Stahl PD. Extracellular vesicles – On the cusp of a new language in the biological sciences. *Extracell. Vesicles Circ. Nucl. Acids.* 2023;4(2):240–54. https://doi.org/10.20517/evcna.2023.18. Epub 2023 May 31.

Rocha B, Calamia V, Casas V, Carrascal M, Blanco FJ, Ruiz-Romero C. Secretome analysis of human mesenchymal stem cells undergoing chondrogenic differentiation. *J Proteome Res.* 2014 Feb 7;13(2):1045–54. https://doi.org/10.1021/pr401030n

Rogulska O, et al. Storage conditions affect the composition of the lyophilized secretome of multipotent mesenchymal stromal cells. *Sci. Rep.* 2024;14:10110. https://doi.org/10.1038/s41598-024-60787-z

Rong Y, Zhang J, Jiang D, Ji C, Liu W, Wang J, et al. Hypoxic pretreatment of small extracellular vesicles mediates cartilage repair in osteoarthritis by delivering miR-216a-5p. *Acta Biomater.* 2021;122:325–342. https://doi.org/10.1016/j.actbio.2020.12.034

Sadri B, Hassanzadeh M, Bagherifard A, Mohammadi J, Alikhani M, Moeinabadi-Bidgoli K, et al. Cartilage regeneration and inflammation modulation in knee osteoarthritis following injection of allogeneic adipose-derived mesenchymal stromal cells: A phase II, triple-blinded, placebo-controlled, randomized trial. *Stem Cell Res. Ther.* 2023 Jun 14;14(1). https://doi.org/10.1186/s13287-023-03359-8

Sadri B, Labibzadeh N, Madani H, Maal AB, Niknejadi M, Hossein-Khannazer N, et al. Safety of autologous nucleus pulposus cell transplantation with fibrin glue following endoscopic discectomy in chronic low back pain: A preliminary phase I trial. *BMC Musculoskel. Dis.* 2025 Oct 1;26(1). https://doi.org/10.1186/s12891-025-09186-7

Sadri B, Labibzadeh N, Mirmorsali L, Ebrahimi M, Bagherifard A, Arab L, et al. Local transplantation of mesenchymal stromal cells is safe and could alleviate Kienböck disease's complications: A clinical trial study. *PubMed.* 2024 Sep 11;26(7):446–53. https://pubmed.ncbi.nlm.nih.gov/39290122

Sadri B, Nouraein Sh, Hossein-Khannazer N, Mohammadi J, Vosough M. Current and novel theranostic modalities for knee osteoarthritis. *Sechenov Med. J.* 2021 Sep 23;12(3):17–30. https://doi.org/10.47093/2218-7332.2021.293.03

Sadri B, Tamimi A, Nouraein S, Fard AB, Mohammadi J, Mohammadpour M, et al. Clinical and laboratory findings following transplantation of allogeneic adipose-derived mesenchymal stromal cells in knee osteoarthritis, a brief report. *Connect. Tissue Res.* 2022 Jul 20;63(6):663–74. https://doi.org/10.1080/03008207.2022.2074841

Safwan M, Bourgleh MS, Aldoush M, Haider KH. Tissue-source effect on mesenchymal stem cells as living biodrugs for heart failure: Systematic review and meta-analysis. *World J. Cardiol.* 2024;16(8):469–83. https://doi.org/10.4330/wjc.v16.i8.469

Shurtleff M, Yao J, Qin Y, Nottingham R, Temoche-Diaz M, Schekman R, Lambowitz AM. A broad role for ybx1 in defining the small non-coding RNA composition of exosomes. 2017. https://doi.org/10.1101/160556

Silva KR, Rezende RAF, Pereira FDAS, Gruber B, Stuart MP, Ovsianikov A, Brakke K, Kasyanov V, Bacakova L. Delivery of human adipose stem cells spheroids into lockyballs. *PLoS One*. 2016;11(11):e0166073. https://doi.org/10.1371/journal.pone.0166073

Simader E, Traxler D, Kasiri MM, Hofbauer H, Wolzt M, Glogner C, Storka A, Mildner M, Gouya G, Geusau A, Laggner M, Beer L, Gabriel C, Gruber F, Gschwandtner M, Pehamberger H, Dome B, Tschachler E, Grillari J, Ankersmit HJ. Safety and tolerability of topically administered autologous, apoptotic PBMC secretome (APOSEC) in dermal wounds: a randomized Phase 1 trial (MARSYAS I). *Sci Rep*. 2017;7:6216. https://doi.org/10.1038/s41598-017-06223-x

Statello L, Maugeri M, Garre E, Nawaz M, Wahlgren J, Papadimitriou A, Lundqvist C, et al. Identification of RNA-binding proteins in exosomes capable of interacting with different types of RNA: RBP-facilitated transport of RNAs into exosomes. *PLoS One*. 2018 Apr 24;13(4):e0195969. https://doi.org/10.1371/journal.pone.0195969

Teng F, Fussenegger M. Shedding light on extracellular vesicle biogenesis and bioengineering. *Adv. Sci.* 2020;8(1):2003505. https://doi.org/10.1002/advs.202003505

Théry C, Witwer KW, Aikawa E, Alcaraz MJ, Anderson JD, Andriantsitohaina R, et al. Minimal information for studies of extracellular vesicles 2018 (MISEV2018): a position statement of the International Society for Extracellular Vesicles and update of the MISEV2014 guidelines. *J. Extracell Vesicles*. 2018;7(1):1535750. https://doi.org/10.1080/20013078.2018.1535750

Turnbull G, Clarke J, Picard F, Riches P, Jia L, Han F, et al. 3D bioactive composite scaffolds for bone tissue engineering. *Bioact. Mater.* 2018;3(3):278–314. https://doi.org/10.1016/j.bioactmat.2017.10.001

Vallabhaneni KC, Penfornis P, Dhule S, Guillonneau F, Adams KV, Mo YY, Xu R, et al. Extracellular vesicles from bone marrow mesenchymal stem/stromal cells transport tumor regulatory microRNA, proteins, and metabolites. *Oncotarget*. 2015 Mar 10;6(7):4953–67. https://doi.org/10.18632/oncotarget 3211

van de Wakker SI, Meijers FM, Sluijter JPG, Vader P. Extracellular vesicle heterogeneity and its impact for regenerative medicine applications. *Pharmacol Rev*. 2023;75(5):1043–1061. https://doi.org/10.1124/pharmrev.123.000841.

Wang DR, Pan J. Extracellular vesicles: Emerged as a promising strategy for regenerative medicine. *World J. Stem Cells*. 2023 Apr 26;15(4):165–81. https://doi.org/10.4252/wjsc.v15.i4.165

Wang L, Liu T, Liang R. Delivery of stem cell secretome for therapeutic applications. *ACS Appl. Bio. Mater.* 2022;5(3):854–70. https://doi.org/10.1021/acsabm.1c01312

Wechsler ME, Rao VV, Borelli AN, Anseth KS. Engineering the MSC secretome: A hydrogel focused approach. *Adv. Healthc Mater.* 2021;10(7):2001948. https://doi.org/10.1002/adhm.202001948

Wei S, Li Z, Shi Q, Luan X, Yuan X, Li Y, Guo C, et al. Collagen-binding vascular endothelial growth factor (CBD-VEGF) promotes liver regeneration in murine partial hepatectomy. *Mol. Med. Rep.* 2022 Nov;26(5):326. https://doi.org/10.3892/mmr.2022.12842

Wei Z, Batagov AO, Schinelli S, Wang J, Wang Y, El Fatimy R, Rabinovsky R, et al. Coding and noncoding landscape of extracellular RNA released by human glioma stem cells. *Nat. Commun.* 2017 Oct 26;8(1):1145. https://doi.org/10.1038/s41467-017-01196-x

Winkler S, Hempel M, Brückner S, Tautenhahn HM, Kaufmann R, Christ B. Identification of pathways in liver repair potentially targeted by secretory proteins from human mesenchymal stem cells. *Int. J. Mol. Sci.* 2016 Jul 9;17(7):1099. https://doi.org/10.3390/ijms17071099.

Wu K, Huang R, Wu H, Liu Y, Yang C, Cao S, Wu C. Collagen-binding vascular endothelial growth factor attenuates CCL4-induced liver fibrosis in mice. *Mol. Med. Rep.* 2016;14(5):4680–6. https://doi.org/10.3892/mmr.2016.5826

Wu T, Zhang B, Lu J, Huang A, Wu H, Qiao J, Ruan H. Label-free relative quantitative proteomics reveals extracellular vesicles as a vehicle for Salmonella effector protein delivery. *Front. Microbiol.* 2022;13. https://doi.org/10.3389/fmicb.2022.1042111

Xia J, Arai S, Haider KH. Researches and applications of stem cell secretome. In: KH Haider, editor. *Stem Cells: From Potential to Promise*. Singapore: Springer, 2021. https://doi.org/10.1007/978-981-16-0301-3_7

Xue J, Xie J, Liu W, Xia Y. 3D printing of cell-delivery scaffolds for tissue regeneration. *Eng.* 2023;20:17–39.

Yang X, Trehan S, Guan Y, Sun C, Moore D, Jayasuriya C, Chen Q. Matrilin-3 inhibits chondrocyte hypertrophy as a bone morphogenetic protein-2 antagonist. *J. Biol. Chem.* 2014;289(50):34768–79. https://doi.org/10.1074/jbc.m114.583104

Yekula A, Yekula A, Muralidharan K, Kang K, Carter B, Balaj L. Extracellular vesicles in glioblastoma tumor microenvironment. *Front. Immunol.* 2020;10. https://doi.org/10.3389/fimmu.2019.03137

Yuan S, Li G, Zhang J, Chen X, Su J, Zhou F. Mesenchymal stromal cells-derived extracellular vesicles as potential treatments for osteoarthritis. *Pharmaceutics.* 2023;15(7):1814. https://doi.org/10.3390/pharmaceutics15071814

Zhang L, Feng B, Zhou Z, Huang H, Yu C, Wang X, et al. Extracellular vesicles-transmitted long non-coding RNA MTUS2-5 promotes proliferation and vascularization of human vascular endothelial cells in patients with Budd–Chiari syndrome. *J. Cell. Mol. Med.* 2023; 27(22):3431–42. https://doi.org/10.1111/jcmm.17911

Zhang S, Chu W, Lai R, Lim S, Hui J, Toh W. Exosomes derived from human embryonic mesenchymal stem cells promote osteochondral regeneration. *Osteoarthritis Cartilage.* 2016;24(12):2135–40. https://doi.org/10.1016/j.joca.2016.06.022

Zhang Y, Wu D, Zhou C, Bai M, Wan Y, Zheng Q, et al. Engineered extracellular vesicles for tissue repair and regeneration. *Burns Trauma.* 2024;12:tkae062. https://doi.org/10.1093/burnst/tkae062

Zhao H, Achreja A, Iessi E, Logozzi M, Mizzoni D, Raimo R, et al. The key role of extracellular vesicles in the metastatic process. *Biochim. Et Biophys. Acta (Bba).* 2018; 1869(1):64–77. https://doi.org/10.1016/j.bbcan.2017.11.005

22 Mesenchymal Stem Cell-Derived Secretome and Extracellular Vesicles

Therapeutic Innovations for Nervous System Disorders

Filipa Ferreira-Antunes[1,2], *João L. Afonso*[1,2], *Ana T. Palha*[1,2], *Carla Teixeira-Pereira*[1,2], *Belém Sampaio-Marques*[1,2], *and António J. Salgado*[1,2]

[1]Life and Health Sciences Research Institute (ICVS), Escola de Medicina, Universidadedo Minho, Campus Gualtar, Braga, Portugal

[2]ICVS/3B's—PT Government Associate Laboratory, Braga/Guimarães, Portugal

22.1 LIST OF ABBREVIATIONS

6-OHDA	6-hydroxidopamine
AD	Alzheimer's Disease
ALS	Amyotrophic lateral sclerosis
ASCs	Adipose stem cells
Asyn	alpha-synuclein
Aβ	Amyloid-beta plaques
BBB	Blood-brain barrier
BDNF	Brain-derived neurotrophic factor
BM	Bone marrow
BM-MSCs	Bone marrow mesenchymal stem cells
CD	Cluster of differentiation
CNS	Central nervous system
CysC	Cystatin C
DPSCs	Dental pulp stem cells
ECM	Extracellular matrix
EMA	European medicines agency
ERK	Extracellular signal-regulated protein kinase
EVs	Extracellular vesicles
FDA	US Food and Drug Administration
FGF2	Fibroblast growth factor 2
Gal-1	Galectin-1

DOI: 10.1201/9781003653387-22

GDNF	Glial cell line-derived neurotrophic factor
Hep	Glycosaminoglycan heparin
hESCs	Human embryonic stem cells
HGF	Nerve growth factor
hiPSCs	Human induced pluripotent stem cells
HUCPVCs	Human umbilical cord perivascular cells
ICAM-1	Intracellular adhesion molecule-1
IDO	Indoleamine 2,3-dioxygenase
IGF-1	Insulin-like growth factor 1
IL-6	Interleukin 6
IPSCs	Induced pluripotent stem cells
ISCT	International Society for Cellular Therapy
MAPK	Mitogen-activated protein kinase
MiRNAs	microRNAs
MMP2	Matrix metalloproteinase 2
MNs	Motor neurons
MSCs	Mesenchymal stem cells
MVBs	Multivesicular bodies
NDs	Neurodegenerative diseases
NEP	Neprilysin
NFT	Neurofibrillary tangles
NGF	Nerve growth factor
NSCs	Neural Stem Cells
NT-3	Neurotrophin 3
PD	Parkinson's Disease
PD-L1	Programmed death ligand 1
PEDF	Pigment epithelium-derived factor
PGA	Polyglycolic acid
PI3K	Phosphoinositide 3-kinase
PLA	Polylactic acid
RVG	Rabies viral glycoprotein
SCI	Spinal cord injury
SDF-1α	Stromal cell-derived factor 1
SNpc	Substantia nigra pars compacta
starPEG	Star-shaped poly(ethylene glycol)
STR	Striatum
TBI	Traumatic brain injury
TGF	Transforming growth factor
TrkB	Tropomyosin receptor kinase B
VEGF	Vascular endothelial growth factor
WJ-MSCs	Wharton's jelly MSCs

22.2 INTRODUCTION

Neurodegenerative diseases (NDs) encompass a range of acute and chronic conditions that affect the central nervous system (CNS) (Dugger & Dickson, 2017). Briefly, NDs are characterized by the gradual deterioration of neurons and glial cells in the brain and/or spinal cord, leading to declines in specific brain-associated functions, such as movement, memory, and cognition (Dugger & Dickson, 2017). Parkinson's disease (PD), Alzheimer's disease (AD), and amyotrophic lateral sclerosis (ALS) are the most common NDs, where the degeneration extends over a decade, and for which the underlying mechanisms are highly complex (Dugger & Dickson, 2017). Currently available therapies focus on mitigating the most prominent symptoms or targeting specific impaired pathways, but are insufficient to prevent degeneration. Given the complexity of NDs, disease-modifying treatments that halt the CNS degeneration remain an unmet need.

Over the past few decades, stem cell-based therapies have emerged as promising alternatives to modulate these disorders (Caplan et al., 2019; Teixeira & Salgado, 2020; Han et al., 2022). In particular, MSCs are a promising approach for regenerative therapies due to their unique features, including self-renewal, differentiation capacities, and immunomodulatory properties (Han et al., 2022; Caplan et al., 2019; Teixeira & Salgado, 2020). In recent years, new lines of evidence suggest that MSCs' potential in NDs primarily stems from their secretory capacity, collectively known as the secretome, which comprises a variety of soluble factors and EVs.

This chapter provides an overview of the most prominent characteristics of MSCs' secretome in regenerative medicine, drawing on evidence from comprehensive proteomics and functional analyses of MSC-derived secretomes, as well as preclinical models of NDs. We will also focus on the potential of specific bioactive molecules to shape therapeutic responses, as well as on strategies to shape MSC secretory profiles to target specific therapeutic needs (Figure 22.1).

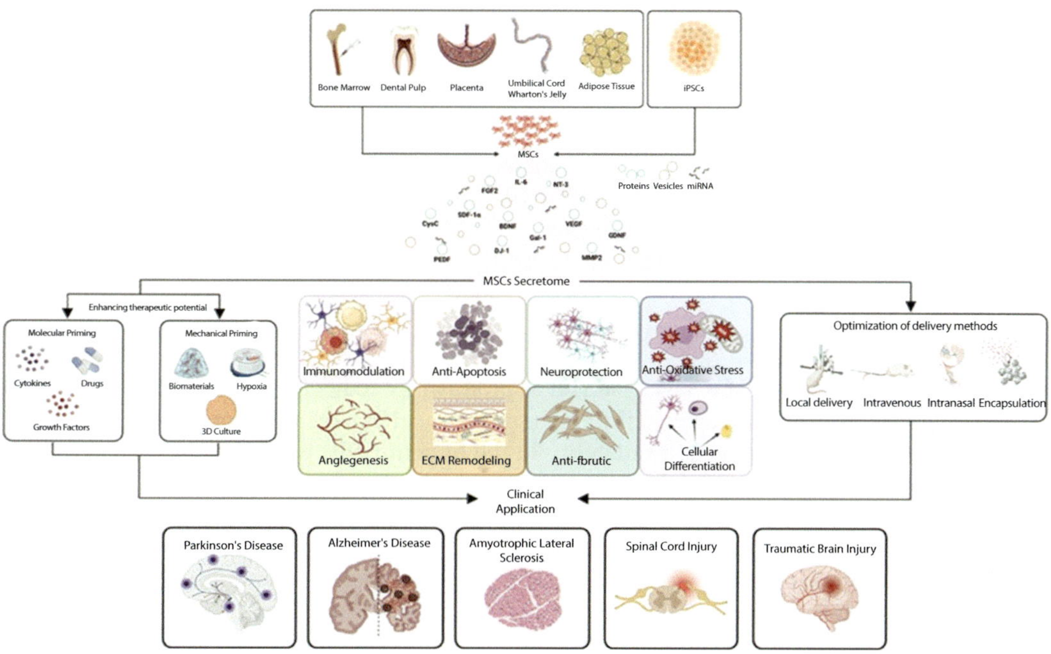

FIGURE 22.1 Mesenchymal Stem Cells Secretome: From Bench to Bedside. Mesenchymal Stem Cells (MSCs) can be obtained from a variety of tissues (e.g., Bone Marrow, Placenta, Dental Pulp) and from induced Pluripotent Stem Cells (iPSCs). MSCs secrete a variety of factors, collectively known as their secretome, such as interleukin 6 (IL-6), DJ-1, neurotrophin 3 (NT-3), fibroblast growth factor 2 (FGF2), Cystatin C (CysC), stromal cell-derived factor 1 (SDF-1α), Brain-derived neurotrophic factor (BDNF), Vascular Endothelial Growth Factor (VEGF), Galectin-1 (Gal-1), matrix metalloproteinase 2 (MMP2), Glial cell line-derived neurotrophic factor (GDNF), and Pigment Epithelium-Derived Factor (PEDF). MSC-derived secretome has shown important roles in disease-relevant pathways, including immunomodulation, anti-apoptosis, neuroprotection, anti-oxidative stress, angiogenesis, extracellular matrix (ECM) remodeling, anti-fibrosis, and cell differentiation. Enhancement of its therapeutic potential through mechanical and molecular priming, together with delivery routes optimization, has shaped the way to translate MSCs-derived secretome into clinics into a variety of neurodegenerative diseases, such as Parkinson's Disease, Alzheimer's Disease, amyotrophic lateral sclerosis, and traumatic injuries such as spinal cord injury (SCI) and traumatic brain injury (TBI). The Figure was generated with BioRender.

22.3 MESENCHYMAL STEM CELLS: AN OVERVIEW

MSCs were first identified in the bone marrow as clonogenic fibroblast precursor cells by Friedenstein et al. (1974) and later named MSCs by Arnold Caplan (Caplan, 1991). Since then, MSC-based therapies have been extensively studied over the past five decades, and their regenerative potential has been widely recognized in various disease contexts (Han et al., 2022).

22.3.1 MSCs' Identity and Sources

MSCs have a mesodermal origin and are defined as multipotent, non-hematopoietic progenitor cells with self-renewal and proliferative capacities. MSCs exhibit a spindle-shaped morphology upon adherence to tissue culture plastic (Álvarez-Viejo & Haider, 2022). They are further characterized by a unique expression profile of cluster of differentiation (CD) cell surface markers and trilineage differentiation potential, as described by the minimal criteria for designating a cell as an MSC established by the International Society for Cellular Therapy (ISCT; Dominici et al., 2006). According to these criteria, besides plastic adherence in standard culture conditions, MSCs must express CD105, CD73, and CD90, and lack expression of CD45, CD34, CD14, CD19, and HLA-DR surface molecules, as well as be able to differentiate into osteoblasts, adipocytes, and chondroblasts *in vitro* (Pittenger et al., 2019).

MSCs are widely distributed in the human body and can be isolated from various adult tissues, with bone marrow being the most common source (Pittenger et al., 2019). However, in the last few decades, cells with a MSC-like phenotype have been successfully isolated from fetal, neonatal, and adult tissues, including adipose tissue (ASCs), dental pulp, placenta, amniotic fluid, umbilical cord blood, umbilical cord Wharton's jelly (WJ-MSCs), and its perivascular region (HUCPVCs) (Kamal et al., 2022). More recently, it has also been shown that MSCs can be derived from human induced pluripotent stem cells (hiPSCs) (Pittenger et al., 2019).

22.3.2 Immunomodulatory and Regenerative Capabilities

Over the last few decades, MSCs have emerged as promising candidates for regenerative medicine and tissue engineering and have entered the biopharmaceutical market as living therapeutics (Bhatti & Haider, 2025). Their widespread availability in the human body, ease of isolation and *in vitro* expansion, and ability to maintain viability and regenerative function after long-term cryopreservation underscore their clinical utility (Han et al., 2019; Pittenger et al., 2019). Moreover, MSCs exhibit self-renewal and differentiation capacities, along with low tumorigenic and immunogenic properties—features that support their use in allogeneic transplantation approaches (Rajab et al., 2019). Based on these properties, initial therapeutic applications of MSCs focused on their differentiation potential in cell replacement strategies aimed at restoring compromised adult mesenchymal tissues (Bagher et al., 2016; Takeda & Xu, 2015). Despite the promising characteristics of MSCs for cell transplantation purposes, increasing evidence suggests that MSCs primarily exert their therapeutic effects through paracrine actions (Caplan & Dennis, 2006; Teixeira et al., 2013). This paracrine activity has been demonstrated through conditioned media, which comprises all soluble factors and vesicles secreted by these cells, collectively referred to as the secretome (Gnecchi et al., 2005; Teixeira et al., 2013; Habib et al., 2025a & 2025b).

22.4 MSC-DERIVED SECRETOME

The first evidence of the paracrine action of MSCs was described in 2005 by Gnecchi and collaborators in the context of ischemic heart injury (Gnecchi et al., 2005). Since then, several research groups, including our own, have demonstrated the benefits of the MSCs-derived secretome in various other pathological conditions (Volkman & Offen, 2017).

22.4.1 Composition and Mechanisms

The secretome of MSCs has been shown to exert significant trophic and immunomodulatory effects, supporting cellular homeostasis and tissue repair during regeneration (Han et al., 2022; Pinho et al., 2022; Teixeira et al., 2013). It contains diverse bioactive factors that modulate key processes under pathological conditions, including neuroprotection, anti-apoptotic activity, immune regulation, angiogenesis, cell proliferation, stem cell recruitment, re-epithelialization, ECM remodeling, and anti-fibrotic responses (Aziz & Haider, 2017; Pinho et al., 2020). These immunomodulatory properties have enabled the development of MSC-based therapies for graft-versus-host disease and autoimmune conditions, such as Crohn's disease (Fitzsimmons et al., 2018). The use of secretome as a cell-free alternative therapy offers advantages for clinical translation, overcoming several limitations of cell-based therapies (e.g., ethical issues and immune incompatibility) (Volarevic et al., 2018). However, the composition of MSCs' secretome is highly variable and influenced by factors such as donor characteristics, tissue source, and culture conditions (Safwan et al., 2024). As a result, comprehensive and consistent characterization of its molecular content remains a significant challenge (Xia et al., 2021). To address this, some studies have proposed fractionating the secretome into distinct components (i.e., soluble and vesicular fractions) to simplify and improve characterization. This approach may provide a more defined and reproducible composition, while potentially maintaining comparable therapeutic efficacy.

22.4.2 Protein-Soluble-Derived Fraction

The protein fraction comprises a variety of soluble proteins, including growth factors (e.g., VEGF, BDNF, FGF-9, EGF, HGF, LIF, NT-3, and TGF-β2), cytokines (e.g., IL-6, IL-8), chemokines (e.g., MCP-1, CCL4, RANTES, and IP-10), and other proteins (e.g., TIMP-1 and OPG) (Vilaça-Faria et al., 2021). The therapeutic potential of several of these proteins has been extensively studied in various contexts.

In the field of growth factors, BDNF (brain-derived neurotrophic factor) and VEGF (vascular endothelial growth factor) are actively studied for neuronal differentiation and protection (Numakawa et al., 2010; Calvo et al., 2024). This interest stems from several studies that report BDNF and its receptor, TrkB (tropomyosin receptor kinase B), are widely expressed in the developing and adult mammalian brain. BDNF/TrkB-stimulated intracellular signaling is critical for neuronal survival, morphogenesis, and plasticity (Numakawa et al., 2010). It is well known that binding of BDNF to TrkB elicits various intracellular signaling pathways, including mitogen-activated protein kinase/extracellular signal-regulated protein kinase (MAPK/ERK), phospholipase Cγ (PLCγ), and phosphoinositide 3-kinase (PI3K) pathways, and that BDNF exerts biological effects on neurons via activation of similar mechanisms (Numakawa et al., 2010). Other studies have also investigated VEGF, well known for its angiogenic activity, and recent evidence has revealed a neuroprotective effect of this factor on injured or diseased neurons (Calvo et al., 2024). Apart from growth factors, cytokines and chemokines also play a key role in CNS modulation, primarily through their interactions with glial cells. For instance, IL-6 is a multifunctional cytokine secreted by neurons and glial cells, playing a vital role in neuronal development and differentiation (Borsche et al., 2020). The combination of these therapeutically relevant proteins is undoubtedly of interest. However, some studies have shown that the protein alone is insufficient to elicit significant changes in neuronal differentiation or protection. This may be due to a low concentration of relevant proteins, or it may indicate that the most pertinent therapeutic potential observed after MSC-secretome treatment is associated with its vesicular component.

22.4.3 Extracellular Vesicles

MSCs' secretome contains a vesicular fraction, composed of exosomes and microvesicles packed with cell-type-specific combinations of proteins, lipids, coding, and non-coding RNAs (e.g., miRNA-106b, miRNA-124a, and miRNA-181a-2-3p) that are major mediators of cell–cell communication (Marote et al., 2016; Mushtaq et al., 2024). Similar to the secretome as a whole, the composition of MSC-EVs is dynamic and influenced by their cellular origin and surrounding environment. However, their protective lipid bilayer allows for stable and efficient delivery of cargo into recipient cells. In contrast, their low immunogenicity, ability to cross the blood-brain barrier (BBB), and targeted cell interactions enhance their therapeutic appeal (Marote et al., 2016).

EVs are typically classified by biogenesis and size into three major categories: exosomes, microvesicles, and apoptotic bodies. Exosomes (30–200 nm) are formed via inward budding of the endosomal membrane, leading to the development of multivesicular bodies (MVBs) (Ahmed et al., 2023). These MVBs either fuse with lysosomes for degradation or merge with the plasma membrane to release exosomes. Common markers include CD9, CD63, CD81, heat shock proteins (Hsp60, Hsp70, Hsp90), Flotillin, ESCRT-3, Alix, and tumor susceptibility gene 101 (TSG101). Microvesicles (100–1000 nm) are generated through outward budding of the plasma membrane. These vesicles may express integrins (e.g., macrophage-1 antigen), selectins (e.g., P-selectin), CD9, CD63, CD81, TSG101, Alix, Annexin V, and other regulatory molecules such as PD-L1 (programmed death ligand 1), Gal-1 (galectin-1), and TGF (transforming growth factor). Finally, apoptotic bodies (1000–5000 nm), released during regulated cell death (apoptosis), are larger vesicles typically marked by Annexin V, phosphatidylserine, and thrombospondin. All three EV subtypes have been identified in MSC secretomes. However, the precise characterization of these vesicles remains challenging due to overlapping size distributions and a lack of universally accepted molecular markers (Welsh et al., 2024).

Despite their heterogeneity, MSC-EVs possess several advantageous features that make them appealing for therapeutic applications, including a protective lipid bilayer that preserves cargo stability, low immunogenicity, the capacity to cross the BBB, and the ability to target specific cells (Tran & Damaser, 2015; Marote et al., 2016). Interestingly, once considered cellular waste, EVs are now recognized as natural carriers of diverse bioactive molecules, including proteins (e.g., growth factors, enzymes, cytokines), lipids, metabolites, genetic material (DNA and RNA), and even organelles, such as mitochondria, derived from their parental cells (Tran & Damaser, 2015; Mushtaq et al., 2024). EVs play essential roles in biological processes, including tissue remodeling, metabolic regulation, protein trafficking and removal, and intercellular communication (De Sousa et al., 2023). MSC-EVs are particularly rich in molecules that support immunomodulation and neural regeneration (Marote et al., 2016). These include neurotrophic factors (e.g., BDNF, NGF (nerve growth factor), IGF-1 (insulin-like growth factor-1)), synaptic proteins (e.g., nestin, synaptophysin), and anti-inflammatory cytokines (e.g., TGF-β, IL-10) (Marote et al., 2016; Pinho et al., 2020). Moreover, MSC-EVs transport antioxidants like catalase, protecting neurons from oxidative stress (Turano et al., 2023) and contain functionally relevant microRNAs (miRNAs), such as miR-22 (attenuates inflammation and supports nerve repair), miR-146a-5p & miR-155 (modulates immune responses), miR-126 (promotes angiogenesis and neurogenesis) and miR-210 (provides neuroprotection in hypoxic environments) (Kandeel et al., 2023; Habib et al., 2026).

The diverse repertoire of bioactive molecules carried by MSC-EVs holds considerable therapeutic potential, as demonstrated by several studies that show their capacity to interact with and modulate key neuronal pathways (Turano et al., 2023). The mechanisms by which EVs communicate with other cells have been studied. EVs interact with target cells through three main mechanisms: direct ligand–receptor interactions, internalization via endocytosis, and membrane fusion-based cargo delivery (Teng & Fussenegger, 2020). EVs' surface molecules—such as growth factor receptors and adhesion proteins—can bind to complementary receptors on neural cells, triggering intracellular signaling cascades without internalization. This can lead to phenotypic changes in neurons or glial cells (Kandeel et al., 2023). MSC-EVs can also enter cells via several endocytic pathways (e.g., clathrin-, caveolin, lipid raft-mediated endocytosis, phagocytosis, macropinocytosis) (Huo et al., 2021). In neurons, clathrin-mediated uptake is common in both dendrites and axons, whereas microglia efficiently phagocytose EVs (Huo et al., 2021). After internalization, EVs may release their cargo within endosomes or directly into the cytoplasm. Within neural cells, EVs can modulate gene expression via miRNAs (e.g., miR-29c-3p downregulates BACE1 in Alzheimer's models, activating Wnt/β-catenin signaling; Sha et al., 2021), stimulate neuroprotective pathways (e.g., pPI3K/Akt, Nrf2 in oxidative stress responses) and promote synaptic function and neuroplasticity via transferred proteins (e.g., nestin, GAP-43, synaptophysin) (Turano et al., 2023). MSC-EVs can also influence glial populations. EVs modulate astrocyte reactivity, shifting them toward a neuroprotective phenotype by delivering anti-inflammatory miRNAs (Provenzano et al., 2022). EV uptake can promote an anti-inflammatory state and reduce Aβ-induced inflammation (Kaniowska et al., 2022). They may also aid in remyelination and support oligodendrocyte maturation in diseases involving demyelination (e.g., Multiple sclerosis and ALS) (Turano et al., 2023). These findings highlight the multifaceted role of MSC-EVs in modulating neural and glial responses, underscoring their therapeutic potential across a range of neurodegenerative conditions (Ahmed et al., 2025). Nevertheless, the therapeutic potential of MSC-EVs can be further enhanced. For instance, EVs can be engineered to improve or modify their biological activity, further increasing their therapeutic potential as acellular, biocompatible agents for drug delivery, diagnostics, and regenerative medicine (De Sousa et al., 2023).

22.4.4 Exploring Therapeutic Potential: Whole Secretome or Fractionated Parts?

While the vesicular fraction contains several molecules (e.g., neurotrophic factors and miRNAs) with relevant therapeutic potential, the presence of specific proteins with beneficial effects in the protein-soluble fraction suggests that both fractions may have greater therapeutic value. Using the vesicular fraction alone could dilute potentially essential proteins for regeneration, suggesting a synergistic effect between the two fractions (Pinho et al., 2022). The fact is that, when comparing the impact of the whole secretome to its fractionated parts, the entire secretome yields better outcomes (Vilaça-Faria et al., 2021; González-Cubero et al., 2022). Although this does not dismiss the therapeutic value of MSC-EVs or the protein fraction of MSCs alone, in severe pathologies such as NDs, a more robust therapy is likely to yield better therapeutic outcomes.

22.5 THERAPEUTIC APPLICATIONS IN NERVOUS SYSTEM DISORDERS

Preclinical and early clinical data investigating the role of MSCs as a regenerative paradigm for CNS disorders are promising and a growing field of investigation. In this section, we will provide an overview of current advances in MSC secretome-based therapeutic approaches for nervous system disorders.

22.5.1 Neurodegenerative Diseases

22.5.1.1 Parkinson's Disease

Parkinson's Disease is characterized by the presence of misfolded and aggregated α-syn, degeneration of dopaminergic neurons in the substantia nigra pars compacta (SNpc), and dopamine depletion in the striatum (STR), leading to non-motor and motor impairments (e.g., bradykinesia, resting tremors, and rigidity) (Grayson, 2016). Moreover, several pathways, including oxidative stress, neuroinflammation, and impaired calcium and iron homeostasis, are implicated in Parkinson's Disease (Grayson, 2016). Numerous *in vivo* studies have demonstrated the effects of MSC secretome using different animal models of disease and different routes of administration, showing promising alterations in neurogenesis, neuroprotection, immunomodulation, and aggregate clearance (Teixeira et al., 2017; Mendes-Pinheiro et al., 2019; Sadan et al., 2009; Marques et al., 2024; Marques et al., 2021).

The secretome of MSCs has been shown to modulate neurogenic niches by increasing neural stem cell (NSC) proliferation, neuronal survival, and hippocampal neuronal differentiation in mice and rats. In Parkinson's Disease rat and mouse models, the same effect was observed in the subventricular zone, which may explain the reduced cell loss observed in the SNpc. This neurogenic effect of MSCs in Parkinson's Disease has been proposed to be mediated by BDNF secretion (Cova et al., 2010). However, other factors present in MSCs' secretome are known to play a role in neurogenesis, such as VEGF, GDNF, FGF2, and NT-3, thereby supporting the notion that MSCs' secretome can modulate neurogenic niches and restore neuronal function (Volkman & Offen, 2017).

Several *in vitro* studies have identified the MSC secretome's role in neuroprotection and the protective effects of some secreted factors against 6-hydroxydopamine (6-OHDA)-induced cytotoxicity (Cova et al., 2012; Parga et al., 2018). Among these factors, SDF-1α has been shown to mediate apoptotic cell death in PC12 cells, thereby increasing DA release. Moreover, prostaglandin E2 receptors have also been identified as the primary mediators of the observed neuroprotective and neurorescue activities of MSCs' secretome (Parga et al., 2018). Proteomic analysis of BM-MSCs secretome revealed the presence of critical neurotrophic factors, such as VEGF, BDNF, IL-6, and GDNF, as well as potential neuroregulatory molecules, namely DJ-1, Cys C, glial-derived nexin, Gal-1, and pigment epithelium-derived factor, which may mediate the behavioral recovery in 6-OHDA lesioned rats, supported by an increase of DA neurons and neuronal terminals in the SNpc and striatum, respectively (Teixeira et al., 2017; Mendes-Pinheiro et al., 2019; Pinho et al., 2020). Some studies have suggested that MSC secretion profiles can be modulated by exposure to neurotrophic factors or by overexpression of specific factors in MSCs. For instance, GDNF-transduced rat BM-MSCs have been shown to protect and promote sprouting of DA terminals in the striatum in neurotoxin- and inflammation-induced rat models of Parkinson's disease (Sadan et al., 2009; Hoban et al., 2015). On the other hand, HGF (nerve growth factor) overexpression in human umbilical cord-MSCs also protected SH-SY5Y cells exposed to MPP+, probably by regulating intracellular Ca2+ via modulation of CaBP-D28K, an intracellular calcium-binding protein (Liu et al., 2014).

The immunomodulatory properties of MSCs have also been suggested to play a significant role in the therapeutic potential for Parkinson's Disease (Kim et al., 2009; McCoy et al., 2008; Suzuki et al., 2015). Although both pro- and anti-inflammatory cytokines have been identified in the secretome of MSCs, anti-inflammatory actions, such as the suppression of microglia activation, have been described as accounting for the therapeutic effects of MSCs in Parkinson's Disease (Kim et al., 2009; Suzuki et al., 2015). Despite the immunomodulatory mediators of MSCs' paracrine activity on the peripheral immune system having been described, such as IL-6, IL-10, prostaglandin E2, and inducible indoleamine 2,3-dioxygenase (IDO) (Fan et al., 2020), the specific mechanism underlying suppression of microglia activation in Parkinson's Disease animal models has not been elucidated. Lastly, it was demonstrated that MSC secretome can reduce extracellular α-syn aggregates and protect dopaminergic neurons from α-syn toxicity, both *in vitro* and *in vivo*. This effect is thought to be partially mediated by metalloproteinase-2 (MMP-2) (Oh et al., 2017) and to block clathrin-mediated endocytosis, thereby inhibiting the transmission of α-synuclein (Oh et al., 2016).

Altogether, these findings strongly support the therapeutic potential of MSCs-derived secretome in addressing multiple pathological aspects of Parkinson's Disease. By simultaneously targeting various dysregulated mechanisms (e.g., oxidative stress, mitochondrial dysfunction, neuroinflammation, and proteostatic imbalance), the MSCs' secretome holds promise for mitigating disease progression. Its multifaceted mode of action not only enhances neuronal survival but also modulates key cellular processes implicated in Parkinson's Disease pathogenesis, reinforcing its potential as a novel and effective therapeutic strategy.

22.5.1.2 Alzheimer's Disease

Alzheimer's Disease (AD) has main pathological features, which are chronic accumulation of amyloid-beta (Aβ) plaques and neurofibrillary tangles (NFT) enriched by hyperphosphorylated protein tau, leading to a gradual loss of hippocampal neurons resulting in loss of memory and reduced cognitive function (Lane et al., 2018; Dugger & Dickson, 2017). This accumulation leads to the hyperactivation of microglia and astrocytes, and thus, neuroinflammation is known to have a significant role in the progression of Alzheimer's disease pathology (Dugger & Dickson, 2017). Given the complexity of Alzheimer's disease pathology, effective treatment requires a multitargeted approach that addresses both neuroinflammation and aggregate clearance to protect hippocampal neurons from degeneration (Yusuf & Haider, 2025). The secretory profile of MSCs has attracted significant attention in Alzheimer's disease therapies, particularly for its immunomodulatory properties and the presence of bioactive molecules with potential roles in aggregate clearance (Volkman & Offen, 2017).

Transplantation of hypoxia-preconditioned MSC-EVs into APP/PS1 mouse models has been shown to modulate the inflammatory response driven by reactive microglia, with mice exhibiting decreased expression of IL-1α, IL-1β, and TNF-α, and increased expression of IL-4 and IL-10 (Kaur et al., 2019). Moreover, MSC-EVs appear to reduce astrocytic activation and decrease AB plaque deposition, thereby improving memory and cognitive function. Previous studies support the notion that these effects can be mediated by MSC-derived CX3CL1, which is proposed to switch reactive microglia toward a protective state, thereby protecting hippocampal neurons against the toxic effects of inflammatory cytokine release (Giunti et al., 2021; Giunti et al., 2012). However, not all cytokines contribute to Alzheimer's disease pathology. A study by Yang et al. found that IL-6 secreted by MSCs reduces autophagy in hippocampal neurons by suppressing the AMPK/mTOR pathway via activation of the gp130-IL-6R receptor complex (Escrig et al., 2019). This modulation was accompanied by a significant downregulation of key autophagy-related proteins, Beclin 1 and LC3 II. Additionally, IL-6 was shown to inhibit the proliferation of activated astrocytes, also via interference with the AMPK/mTOR axis, suggesting a broader regulatory role for IL-6 in neural cell homeostasis under pathological conditions.

Another hypothesis is that MSCs-derived bioactive factors can recruit activated microglial cells to the inflamed area after exposure to Aβ deposition by secreting CCL5, a chemoattractant factor (Lee et al., 2012). Lee et al. injected BM-derived MSCs into the hippocampal region of Alzheimer's disease mice. They realized that CCL5 secretion recruited activated microglia to the injury site, leading to decreased Aβ plaques and attenuated memory impairments through IL-4 and neprilysin (NEP) secretion. In some studies, NEP, a protease enzyme, is inversely correlated with Aβ plaque levels and is thought to have a role in its degradation. *In vitro* studies have shown that MSC-derived soluble intracellular adhesion molecule-1 (ICAM-1) and CCL5 can induce Aβ plaque degradation by increasing neprilysin (NEP) levels via alternative microglial activation.

Another vital mechanism in Alzheimer's disease is the induction of neuronal apoptosis by Aβ plaques, either directly by upregulating the Bax protein or indirectly by increasing ROS-induced oxidative stress. MSCs-derived secretome significantly reduced neuronal apoptosis by altering the BAX/BCL-2 balance in favor of cell survival (Feng et al., 2018). The dental pulp stem cell (DPSC)-derived secretome can counteract Bax expression in neuroblastoma cell lines by upregulating the anti-apoptotic protein Bcl-2. Some groups suggest that this effect could be related to the expression of RANTES, Fractalkine, VEGF, and FGF2, which can promote Bcl-2 expression (Feng et al., 2018).

Nonetheless, it appears that MSC-derived secretome can be a powerful therapeutic tool to modulate inflammatory processes and proteostatic impairments during Alzheimer's disease progression, leading to reduced Aβ plaque deposition, improved neuronal survival, and ultimately enhanced symptom amelioration.

22.5.1.3 Amyotrophic Lateral Sclerosis

Amyotrophic lateral sclerosis (ALS) is a rapidly progressing and fatal neurodegenerative disease that attacks motor neurons (MNs) located in the anterior horn of the spinal cord, resulting in loss of muscle control (Sykova et al., 2021; Dugger & Dickson, 2017). It is marked by pronounced spinal cord atrophy and, in many cases, concurrent degeneration of both cerebral gray and white matter.

This widespread neurodegeneration impairs the ability to initiate and coordinate voluntary movements such as walking, talking, chewing, and breathing. A key genetic contributor to ALS is the *SOD1* gene, which encodes Cu/Zn superoxide dismutase 1, an enzyme that mitigates oxidative stress. Mutations in *SOD1* disrupt its standard folding, promoting the formation of toxic protein aggregates and persistent stress granules. These granules, rich in misfolded proteins, can propagate intercellularly, spreading dysfunction and driving neurodegeneration associated with the symptoms above (Riva et al., 2024).

Preclinical studies have demonstrated that the administration of MSCs (intraspinal, intravenous, or combined) can slow motor impairments, decrease inflammation, and stimulate the secretion of specific cytokines and growth factors that promote cell survival, thereby enabling symptomatic transgenic animals to survive longer (Sykova et al., 2021). However, these results were primarily due to their paracrine effect, which supports the resistance of neurons and glia to apoptosis by releasing anti-apoptotic and trophic factors, thereby maintaining a neuroprotective microenvironment (Vercelli et al., 2008). In particular, ASCs' secretome has received increased attention in the context of ALS, given its content of disease-relevant neuroprotective and neurotrophic factors, such as BDNF, NGF, VEGF, HGF, and IGF-1 (Wei et al., 2009). The secretome of ASCs has previously been shown to exert neuroprotective effects in both animal models and cell culture systems of neurodegeneration (Wei et al., 2009). Additionally, exosomes derived from ASCs have been shown to protect motor neurons harboring *SOD1* mutations by mitigating oxidative stress and enhancing cell viability (Wei et al., 2009). These EVs also restored complex I activity, mitochondrial membrane potential, and coupling efficiency (Calabria et al., 2019). Similarly, the secretome of MSCs has been shown to provide substantial protection against glutamate-induced toxicity, promote neurite outgrowth, and reduce LDH release and caspase-3/7 activity (Maguire et al., 2019). Several miRNAs enriched within MSC-derived EVs, including miR-467f, miR-466q, miR-466m-5p, miR-466i-3p, miR-466i-5p, miR-467g, miR-3082-5p, and miR-669c-3p, are known to exert therapeutic effects (Giunti et al., 2021). These miRNAs may contribute to the well-documented influence of MSCs on microglial activation and could be transferred to target cells via EV-mediated shuttling (Maguire et al., 2019).

Collectively, these findings underscore the potential of MSC-derived EVs and secretome as a promising therapeutic strategy for motor neuron diseases such as ALS.

22.6 TRAUMATIC INJURIES

Traumatic CNS injuries refer to damage to the brain and spinal cord, caused mainly by falls, traffic accidents, and sports (Silva et al., 2014). Neurotrauma conditions comprise traumatic brain injury (TBI) and spinal cord injury (SCI). They are often associated with a clinical manifestation of permanent functional changes, such as loss of motor, sensory, and autonomic functions (Thapa et al., 2021). The heterogeneous and complex pathophysiological profile of patients with both degenerative conditions is the main challenge for clinical management and therapeutic success, which is currently limited and inefficient (Silva et al., 2014). MSCs have been extensively assessed as a regenerative and neuroprotective approach in this research field, aiming to bridge the gap between immunomodulatory and neuro-restorative strategies (Hosseini et al., 2018). The safety and efficacy of this therapy in this specific disease context have been shown in several studies employing MSCs transplantation (intravenous, into the injury site, and intranasal for TBI), in both SCI and TBI animal models, and some conducted clinical trials (Cízková et al., 2006; Qu et al., 2011). As previously mentioned, MSCs' capacity to improve sensorimotor functions and cognitive deficits in neurotrauma conditions primarily derives from their paracrine capacity (Assunção Silva et al., 2022).

The TBI research field has focused on exploring the effects of MSC transplantation and their capacity to migrate and secrete neurotrophic factors *in situ*. In line with this, BM-MSCs have been the most commonly used source for transplantation, as they secrete VEGF and BDNF, which attenuate synapse protein loss and enhance synaptic plasticity, thereby promoting functional recovery

(Feng et al., 2017). The secretion of NGF has also been confirmed to ameliorate cholinergic neuronal loss. Together, these factors could promote internal repair processes by regenerating the BBB and neuronal and glial tissues (Ohta et al., 2011; Mahmood et al., 2006). Intravenous transplantation was also found to reduce glial reactivity and promote an anti-inflammatory microenvironment, mediated through TSG-6 expression (Zhang et al., 2013). Nonetheless, a growing field is focused on exploiting the effects of the secretome and its isolated vesicular fraction from MSCs. Notably, this secretome approach has been shown to offer benefits similar to, or even superior to, those of transplantation. Administration of ASCs-derived secretome could reduce hippocampal damage, thereby improving motor and cognitive functions and lowering contusion volume (Tajiri et al., 2014).

The cell-transplantation-free secretome-based approach has been widely explored in the SCI context, with extensive comparisons across different MSC sources. Several studies have confirmed that secretome delivery via other routes can promote functional recovery in various animal models, associated with attenuated inflammation and glial scar maturation, as well as increased tissue regeneration (Assunção-Silva et al., 2024). The mediation of neuronal survival and neurogenesis is strongly correlated with the secretion of BDNF and beta-NGF, which are detected in both BM and ASCs-derived secretomes (Martins et al., 2017). However, ASCs' conditioned medium has been shown to provide higher levels of axonal growth and increased viability of CNS cells (Assunção-Silva et al., 2018). This has been correlated to the results of an unbiased proteomic analysis, which revealed higher secretion levels of angiogenic and anti-apoptotic proteins from ASCs (Pires et al., 2016).

In general, the secretory properties of MSCs have similar protective and functional benefits in the two injury contexts, due to their similarity in pathophysiological events, showing powerful neurotrophic and immunomodulatory effects related to anti-inflammatory (TGF-β and IL-10), angiogenic (VEGF and HGF), and antioxidant factors (Teixeira et al., 2013). Gathering in vitro and in vivo studies on the effects of MSCs in neurotrauma degeneration, this potential therapeutic approach has demonstrated significant clinical relevance and requires further optimization and scalability.

22.7 DELIVERY METHODS FOR MSCS-DERIVED THERAPIES

Despite promising findings on the therapeutic potential of MSCs and their derivatives, developing efficient, disease-specific delivery methods remains a critical challenge to maximize clinical outcomes. The optimal route of administration is still under debate and varies significantly depending on the ND being targeted. Delivery strategies include intracerebral, intrathecal, and intraspinal injections, which enable targeted delivery, promote local differentiation, and support tissue grafting—essential requirements in diseases with focal degeneration (Caplan et al., 2019). For instance, nearly 70% of cell delivery strategies in SCI clinical trials utilize intraspinal or intrathecal injections due to their precision (Bartlett et al., 2020). However, these invasive procedures often require neurosurgical intervention, limiting their feasibility for repeated or chronic administration. In contrast, systemic administration—particularly intravenous injection (IV)—is widely adopted in clinical trials for disorders such as Alzheimer's disease and Parkinson's Disease due to its minimally invasive nature and ease of protocol implementation (Vargas-Rodríguez et al., 2023; Kean et al., 2013). However, systemic routes present substantial limitations. A significant proportion of MSCs become sequestered in peripheral organs, such as the lungs, spleen, liver, and kidneys, reducing CNS bioavailability and raising concerns about off-target effects (Kean et al., 2013). Furthermore, the BBB and the blood-spinal cord barrier represent a substantial obstacle to the successful targeted delivery of drugs (Vargas-Rodríguez et al., 2023). While some degree of barrier disruption is observed in many NDs, these alterations are disease-specific, temporally dynamic, and mediated by diverse mechanisms, including tight junction breakdown or transporter dysfunction (Rosenberg, 2012). Given these complexities, there is growing interest in the use of MSC-derived secretomes and EVs, which offer a promising alternative for therapeutic delivery. Due to its smaller size, the secretome is expected to become less entrapped in unwanted tissues. Furthermore, EVs have demonstrated the capacity to cross the BBB and reach tissues, enabling the non-cell-based delivery of MSC-derived bioactive factors (Turano et al., 2023). While MSCs-derived secretome and EVs hold substantial therapeutic promise for CNS disorders, their non-specific biodistribution following systemic administration presents a significant hurdle, often leading to variable outcomes and reduced efficacy. Consequently, current research is increasingly focused on developing strategies to enhance the CNS-targeting capacity of secretome-based therapeutics. Due to the complexity and heterogeneous composition of the secretome, most targeting strategies have so far focused on EVs because of their well-defined structure, modifiability, and therapeutic relevance (Nieland et al., 2023). Typically, these vesicles are engineered to minimize off-target uptake, increase brain accumulation, and protect against enzymatic degradation (Nieland et al., 2023). For instance, Tian et al. conjugated cyclic RGD to the amine groups on the EV membrane, creating vesicles with high affinity for integrin $\alpha v\beta 3$, a molecule expressed by endothelial cells during angiogenesis and under CNS pathological conditions (Tian et al., 2018). Another method involves fusing the abundant EV surface protein Lamp2b with the rabies viral glycoprotein (RVG) peptide, enabling the selective delivery of EVs to neurons, oligodendrocytes, and microglia (Nieland et al., 2023).

The diversity of proteins that the different forms of vesicle engineering can target makes this field ever-evolving. However, unlike tumors, the upregulation of disease-specific proteins is not straightforward in other neuropathological conditions. Most CNS disorders lack consistent, disease-specific surface markers, necessitating personalized EV engineering and a deeper understanding of the molecular landscape at different disease stages. Furthermore, a non-vesicular fraction of the MSCs' secretome, including cytokines, growth factors, and soluble proteins, also significantly contributes to therapeutic efficacy. However, these components are often excluded from the EV-focused strategies. A comparative study in the SCI model showed that the whole secretome outperformed isolated fractions, underscoring the value of delivering the full spectrum of secreted bioactive molecules (Pinho et al., 2022). One promising strategy to enhance delivery while preserving the benefits of the entire secretome is to encapsulate it into systems with targeted mechanisms, such as nanoparticles. Lipid nanoparticles with diameters below 100 nm represent an interesting approach due to their ability to diffuse more readily across the BBB (Toader et al., 2024). Furthermore, these can be enhanced by functionalizing ligands for transcytosis receptors such as transferrin and integrins (Toader et al., 2024). Recently, Barata-Antunes and colleagues developed a lipid nanoparticle that could encapsulate the secretome for delivery. It was observed that different parts of the secretome were being incorporated into the particles, paving the way for an integrated matrix in which the secretome can be delivered (Barata-Antunes et al., 2025). Besides lipids, nanoparticles can be generated from various materials, such as PLGA, gold, chitosan, and PEG, among others, resulting in a wide array of properties. However, simpler molecules are often incorporated into particles lacking secretome-encapsulation techniques.

To enhance the delivery of either EVs or nanoparticles, some groups explore the use of focused ultrasound with microbubbles to temporarily disrupt the BBB, allowing the therapeutic component to enter. Along these lines, Haroon et al. employed low-intensity focused ultrasound to enhance EV concentrations in the brain without compromising the BBB (Haroon et al., 2023).

In parallel, localized delivery strategies are being explored to address interest, with various approaches under consideration. Specifically, for CNS applications, hydrogels and microneedles can serve as carriers for molecules, enabling the localized delivery of factors while shielding them from degradation (Arifka et al., 2022). These materials can also serve as instructive agents for cells and tissues, synergistically enhancing the therapeutic effect of the secretome. In a SCI rat model, Silva et al. encapsulated the entire secretome in star-shaped poly(ethylene glycol) (starPEG) and glycosaminoglycan heparin (Hep) hydrogels, which were then transplanted at the injury site (Silva et al., 2023). With a controlled release of factors over 10 days, this approach led to improved functional motor outcomes (Silva et al., 2023). Still in SCI, Ao Fang developed a microneedle patch with embedded MSCs that released secretome into the spinal cord, without direct invasion into the tissue

(Fang et al., 2023). While several hydrogel formulations have already been developed for secretome embedment and CNS disorders, the combination of the two is still in its early stages, presenting a window of opportunity for future research (Arifka et al., 2022).

More recently, to avoid the complexity of these technologies while simultaneously achieving a high delivery yield to the central nervous system (CNS), some groups are exploring intranasal delivery of secretome. This approach bypasses the BBB and skips peripheral metabolism. Furthermore, it allows for multi-dose administration (Turano et al., 2023). However, the mucosa also presents some degree of barrier, which needs to be addressed to guarantee absorption into the brain. In an Alzheimer's disease mouse model, weekly intranasal administration of MSCs' secretome induces memory recovery and improved histological outcomes (Santamaria et al., 2021). EVs derived from MSCs, when administered intranasally, have been shown in murine models of stroke, autism, Parkinson's Disease, and Alzheimer's disease to target disease-specific pathological regions in the brain (Perets et al., 2019). In an ALS mouse model, intranasal delivery of MSC-EVs produced an effect similar to that of intravenous injection (Bonafede et al., 2020). In an encephalomyelitis mouse model, EVs delivered by this method led to immunomodulation and functional improvements (Fathollahi et al., 2021).

Secretome delivery strategies have been evolving in parallel ways, with specific advantages for each technique. Refining these methods while developing new and improved ones will be a challenge that, if overcome, could significantly enhance the therapeutic effect of MSCs and their secretome. Furthermore, as synergism with other therapeutic agents may be of interest, careful consideration of individual or combined delivery routes for each or a synergistic approach is necessary. Still, the biggest challenge will remain in understanding the biomolecular mechanisms of disease and how the secretome modulates their pathological processes. Having this knowledge will allow for better target design to achieve maximum therapeutic efficacy.

22.8 ENHANCING MSCS-SECRETOME THERAPEUTIC EFFICACY: AN EYE TOWARD CLINICAL APPLICATION

Although MSC-based therapies have garnered significant attention in regenerative medicine, their clinical translation remains a complex and challenging endeavor. As previously discussed, the substantial heterogeneity of MSCs, arising from differences in tissue origin, donor variability, and culture conditions, poses a significant obstacle. Additionally, the broad and multifaceted nature of MSCs' secretome raises concerns about potential off-target effects, further complicating the safe and effective implementation of these cells in clinical settings.

22.8.1 Enhancing MSCs-Secretome Therapeutic Potential: Priming Approaches

Given MSCs' sensitivity to their microenvironment, various priming strategies have been developed to shape their secretory profile and enhance therapeutic efficacy (Haider, 2024a & 2024b). These strategies include the use of biomaterials, such as hydrogels, genetic modifications, and preconditioning methods (Bagno et al., 2018).

The incorporation of biomaterials into MSC research has significantly advanced their therapeutic applications, particularly in tissue engineering. Biomaterials can modulate the microenvironment of MSCs through both biochemical and biophysical cues, thereby enhancing their regenerative and immunomodulatory potential (Nava et al., 2012). By providing structural support and mimicking native ECM features, biomaterials guide MSCs' fate and improve tissue repair outcomes (Kusuma et al., 2017). Biomaterials can be broadly classified into natural and synthetic matrices (Noronha et al., 2019). Natural matrices include alginate, collagen, fibrin, chitosan, gelatin, and hyaluronic acid. Synthetic alternatives often comprise bioresorbable polymers such as polylactic acid (PLA), polyglycolic acid (PGA), polyethylene, and polypropylene (Zippel et al., 2010). These materials can be further functionalized with bioactive motifs, chemotactic factors, or ECM proteins to enhance cellular responses (Zippel et al., 2010). Hydrogels, in particular, have emerged as highly promising biomaterials due to their physicochemical similarity to the native ECM. Encapsulating MSCs in hydrogels not only supports cell survival and viability but also protects cells from shear forces during injection-based administration (Baldari et al., 2017). Additionally, hydrogel-based culture has been shown to enhance the osteogenic and chondrogenic differentiation of MSCs, accelerate wound healing, promote neovascularization, and improve cell homing, proliferation, and viability (Huang et al., 2022).

Preconditioning, also known as "priming," has also gained interest in MSCs research. Cell priming can be divided into biophysical and molecular. Biophysical priming relies on physical or mechanical stimuli (e.g., hypoxia, dynamic culture, or electrical stimulation). Hypoxic preconditioning, using either hypoxia chambers or hypoxia-mimetic agents, has been widely employed to explore the therapeutic potential of MSCs. This approach has been shown to upregulate chemoattractant properties, modulate inflammatory responses, and increase the secretion of angiogenic molecules (e.g., VEGF, HGF), thereby promoting tissue repair, remodeling, and regeneration (Lee et al., 2017; Xue et al., 2018; Page et al., 2014; Miceli et al., 2023). Interestingly, EVs derived from hypoxia-preconditioned MSCs also exhibit heightened angiogenic potential (Xue et al., 2018). Due to this, priming MSCs with hypoxia has been primarily explored in the context of cerebral ischemia, spinal cord injury (SCI), traumatic brain injury (TBI), and scar formation (Miceli et al., 2023).

Priming through 3D culture systems, most commonly in the form of MSCs spheroids, has emerged as another effective strategy (Miceli et al., 2023; Noronha et al., 2019). The conformation resembles the *in vivo* MSC niche and enriches the MSC phenotypic profile, increasing the secretion of functional factors with anti-inflammatory, angiogenic, anti-fibrotic, anti-apoptotic, and mitogenic properties (Noronha et al., 2019; Miceli et al., 2023). Curiously, EVs derived from 3D MSC spheroids have also shown higher yields and enhanced activity, stimulating angiogenesis, attenuating inflammation, and promoting tissue regeneration (Noronha et al., 2019; Miceli et al., 2023). Given this, 3D MSC spheroids-derived secretome has been studied in various in vitro models, including bone and cartilage repair, intervertebral disc degeneration, and acute myocardial infarction.

On the other hand, molecular priming involves exposing MSCs to specific bioactive compounds that shape cellular biomechanics, thereby altering secretory profiles (Haider, 2024). These priming agents can be cytokines and growth factors, which elicit subtle changes in cells, or pharmacological and natural compounds, which induce more robust alterations, often affecting entire pathways (Haider, 2024). For instance, priming MSCs with neurotrophic factors enhances their neuroprotective potential through an ER stress-induced mechanism, which is typically disrupted in NDs (Teli et al., 2023). The use of clinically approved drugs is also being explored, as they are safe for human use and can be easily applied for cell priming in clinics (Song et al., 2013). For instance, statins (e.g., lovastatin, an approved cholesterol-lowering drug) are commonly used as preconditioning mimetics (Alnasser et al., 2022). It has been successfully used to prime MSCs, accentuating their stemness and functionality, thereby reducing cardiomyocyte apoptosis and suppressing the inflammatory response in ischemic myocardium (Song et al., 2013).

Nonetheless, selecting priming agents or conditioning strategies requires careful consideration. As highlighted throughout this section, the optimal priming conditions can vary significantly depending on the disease context and individual patient needs. Therefore, a unique approach is unlikely to yield consistent therapeutic benefits. Instead, systematic evaluation of priming methods in disease-specific models is essential to unlock the full therapeutic potential of MSCs-derived secretome and to tailor interventions that address the unique pathophysiological features of each condition. For this purpose, stem cell-derived organoid models offer a promising platform for disease modeling and diagnosis (Panoutsopoulos, 2021). These systems provide a more physiologically relevant environment, facilitating the screening of therapeutic strategies and refining MSC-based approaches (Panoutsopoulos, 2021).

22.8.2 The Use of Organoids and Assembloids as Physiologically Relevant Models

Organoids are defined as self-organizing, 3D *in vitro* culture systems that contain a variety of specialized cell types and recapitulate specific structural and functional features of an organ or a region within an organ (Paşca et al., 2022; Paşca, 2018). These systems can be derived from multiple sources, including human induced pluripotent stem cells (iPSCs), human embryonic stem cells (hESCs), or primary tissues such as isolated adult stem cells, excised tumors with growth potential, or fetal tissue (Paşca, 2018). Building on organoid technology, assembloids are 3D cell culture systems formed by integrating distinct organoids or specialized cell types and exhibit self-organizing properties (Paşca et al., 2022; Kanton & Paşca, 2022). These additional components can arise from the direct differentiation of stem cells in 2D cultures or from isolated primary tissues (Kanton & Paşca, 2022). When organoids are combined with such specialized cell types, they are referred to as multi-lineage assembloids. Assembloids can also be classified based on their cellular origin. Those generated from organoids derived from different individuals are termed inter-individual assembloids, and those formed by combining organoids from other species (e.g., human and chimpanzee) are known as inter-species assembloids (Paşca et al., 2022; Kanton & Paşca, 2022).

In the context of NDs, various groups have already generated multiple organoids and assembloids, demonstrating proven pathophysiologic relevance and overcoming several limitations observed in 2D models (Wray, 2021). For instance, in the context of NDs with protein aggregates, such as Alzheimer's and Parkinson's, 3D systems provide an environment that allows these aggregation-prone species to form larger protein aggregates (Wray, 2021). Another significant advantage of cerebral organoids is their ability to generate models with specific mutations, thereby mimicking the natural progression observed in the human brain, which is often challenging to replicate in animal models. Additionally, cerebral organoids and assembloids are also promising approaches for studying early-onset and late-onset pathology as it progresses (Wray, 2021). The generation of such complex models enables the investigation of disease-modifying therapies, such as MSC-derived secretome, in human-based models. Our team has already explored the role of BM-MSCs' secretome in midbrain organoids, where the secretome elicited near-significant protection of the dopaminergic cell population, protecting midbrain organoids against 6-OHDA toxicity (Mendes-Pinheiro et al., 2023).

Therefore, advancing the use of organoids and assembloids as pathologically relevant models is essential for the preclinical evaluation of secretome-based therapies. These systems offer a powerful platform for disease- and patient-specific screening, enabling a more precise and physiologically meaningful assessment of therapeutic efficacy. By capturing the complexity of human neurodegenerative disorders *in vitro*, organoids represent a transformative step toward refining and personalizing secretome-based interventions, ultimately bridging the gap between experimental research and clinical application.

22.8.3 Integration of Secretome-Based Therapy into Clinical Practice

Decades have passed since MSCs were first described, and their therapeutic potential has been extensively studied across various disease contexts. The constant evolution of the field, along with promising pre-clinical results in animal-relevant models, has prompted clinicians to explore the use of MSC cell transplantation and/or derived factors in clinical settings (Maldonado et al., 2023).

The implantation of BM-MSCs bilaterally in patients with Parkinson's Disease led to promising results (Venkataramana et al., 2012). These patients reported improved speech clarity, reduced tremor, rigidity, and freezing attacks (Venkataramana et al., 2012). Moreover, patients transplanted in the early stages of the disease (less than 5 years after diagnosis) showed greater improvement and no further disease progression compared to those in the later stages (11–15 years after diagnosis) (Venkataramana et al., 2012). In Alzheimer's disease patients, stereotactic brain injections of hUCB-MSCs in the hippocampus of nine patients with mild-to-moderate pathology were proven to be safe. However, no conclusions regarding therapeutic effects were retrieved correctly (Kim et al., 2015). Autologous MSCs secreting neurotrophic factors have been transplanted in patients with ALS, demonstrating that this administration is also safe and may indicate possible clinical benefits, which need to be further explored (Petrou et al., 2016). This is also indicative that the secretory profile of these cells is therapeutically relevant and further supports the need to improve and tailor MSCs' EVs and secretome to meet disease requirements (Petrou et al., 2016). In SCI patients, administration of WJ-MSCs resulted in 13 of 16 patients with incomplete SCI responding to treatment, whereas patients with complete SCI showed no response (Liu et al., 2013). Another study also showed that 7 of 10 patients treated with an intraspinal injection of WJ-MSCs experienced improvement in movement, self-care, and muscular tension (Cheng et al., 2014).

Despite the promising results obtained so far, there are still some gaps to be filled before advancing effectively into clinics. Aside from the topics we have been discussing throughout this chapter, a lack of consensus remains regarding regulatory requirements, frameworks, and the general classification of MSC-derived products, which we will discuss in the next section.

22.9 CURRENT CHALLENGES AND LIMITATIONS

The rapid development of MSCs-derived secretome and EVs-based therapies has outpaced the establishment of clear regulatory pathways (Haider & Aslam, 2018). As a result, regulatory agencies worldwide continue to define standards to ensure safety, efficacy, and consistency in clinical applications. Unlike traditional cell-based therapies, MSC-derived secretome products are highly heterogeneous in composition and do not fit into a specific category, remaining in a grey area between laboratories. This complexity complicates their classification and the establishment of standard requirements for quality control, batch reproducibility, and safety (Sanchez-Guijo et al., 2024). As a result, one significant challenge in regulating MSC-derived products is their classification and categorization. For instance, the US Food and Drug Administration (FDA) considers these products as biologics. At the same time, the European Medicines Agency (EMA) may classify them as biological medicinal products or drugs, depending on their composition, mechanism of action, and intended use (Sanchez-Guijo et al., 2024; Wuschko et al., 2019). The absence of robust, GMP-compliant manufacturing protocols for secretome and EV products remains a significant obstacle. Critical aspects such as cell source, culture conditions, collection methods, purification techniques, and storage must be tightly controlled to ensure batch-to-batch reproducibility, safety, and therapeutic efficacy (Liang et al., 2025). Current GMP guidelines are primarily designed for traditional biologics or cell therapies, and adapting them to cell-free, vesicle-rich products requires the development of new standards for identity, potency, and purity testing. As MSCs-derived secretome therapies move closer to clinical translation, international efforts to harmonize regulatory standards and manufacturing practices are urgently needed to facilitate their safe and effective integration into clinical practice (Sanchez-Guijo et al., 2024). Importantly, clinical translation requires the production of large quantities of secretome, and in this regard, there are still limitations to overcome. For instance, MSC senescence represents an intrinsic limitation on secretome production capacity and large-scale application. A good alternative for tissue-derived MSCs is the use of iMSCs. iMSCs exhibit a more rejuvenated phenotype, characterized by superior proliferative capacity before the induction of replicative senescence, enabling the acquisition of a large number of cells with no alterations in their secretome profile (Marote et al., 2023). The use of bioreactors for large-scale production is also a viable option. Bioreactors enable cells to grow without continuous cell passage during production runs, making them more time- and cost-effective and reducing the need for cell-culture materials. Still, some investigation remains to be done, as bioreactors can alter the secretory profile; thus, standardization is also necessary with this approach (Teixeira et al., 2016). Adopting multiple strategies can be more effective and address multiple limitations simultaneously.

22.10 CONCLUSION

MSC-derived secretome holds immense promise as a cell-free therapeutic approach, particularly for neurological disorders (NDs), where its ability to modulate inflammation and oxidative stress, and to promote regeneration, is especially valuable. Advances in biomaterials and priming strategies have significantly enhanced the efficacy of the secretome, while organoids and assembloids now offer powerful, disease-relevant platforms for preclinical screening.

However, clinical translation remains challenging due to regulatory uncertainties, classification issues, and the need for standardized, Good Manufacturing Practice (GMP)-compliant manufacturing. Addressing these gaps will be essential for the safe and effective integration of secretome-based therapies into clinical practice. With continued progress, these therapies could revolutionize treatment strategies for complex neurodegenerative disorders.

Author Contributions: Conceptualization, FFA, BSM, and AJS; validation, AJS; investigation, FFA, JLA, ATP, CTP, and BSM; writing—original draft preparation, FFA, JLA, ATP, and CTP; writing—review and editing, BSM and AJS; supervision, BSM and AJS. All authors have read and agreed to the published version of the manuscript.

Funding: This work was supported by national funds through the Foundation for Science and Technology (FCT) under the following projects: UIDB/50026/2020 (DOI: 10.54499/UIDB/50026/2020), UIDP/50026/2020 (DOI: 10.54499/UIDP/50026/2020), LA/P/0050/2020 (DOI: 10.54499/LA/P/0050/2020), and 2023.13115.PEX. It was also funded by "la Caixa" Foundation and FCT, through Iniciativa Ibérica de Investigación y Innovación Biomedica, under the agreement LCF/PR/HP20/52300001. FCT provided additional support through a PhD Fellowship awarded to FFA (UI/BD/154989/2023, DOI: https://doi.org/10.54499/UI/BD/154989/2023) and JLA (UI/BD/08337/2021).

Institutional Review Board Statement: Not applicable.

Informed Consent Statement: Not applicable.

Data Availability Statement: Not applicable.

Conflicts of Interest: The authors declare that they have no conflicts of interest.

REFERENCES

Ahmed LA, Al-Massri KF, and Haider KH. Therapeutic efficacy of mesenchymal stem cells and their derived exosomes in different neurodegenerative disorders. In: *Current Status and Future Perspectives. Handbook of Regenerative Medicine. Cell-free Therapy Approach*. 2025, pp. 55–67.

Ahmed ZT, Alkahlot MH, Haider KH. MSC-derived exosomes: Advances in cell-free therapy. In: *Handbook of Stem Cell Applications*. Singapore: Springer Nature, 2023 Aug 18, pp. 1–41.

Alnasser MS, AL-Rasheedi, Alreshidi MA, Alqifari SF, Haider KH. Augmenting mesenchymal stem cell-based therapy of the infarcted myocardium with statins. In: Haider K.H. (ed.) *Handbook of Stem Cell Therapy*. Singapore: Springer, 2022. https://doi.org/10.1007/978-981-16-6016-0_20-1.

Álvarez-Viejo M, Haider KH. Mesenchymal stem cells. In: Haider K.H. (ed.) *Handbook of Stem Cell Therapy*. Singapore: Springer, 2022. https://doi.org/10.1007/978-981-16-6016-0_6-1

Arifka M, Wilar G, Elamin KM, Wathoni N. Polymeric hydrogels as mesenchymal stem cell secretome delivery system in biomedical applications. *Polymers (Basel)*. 2022;14(6).

Assunção-Silva RC, Mendes-Pinheiro B, Patrício P, Behie LA, Teixeira FG, Pinto L, Salgado AJ. Exploiting the impact of the secretome of MSCs isolated from different tissue sources on neuronal differentiation and axonal growth. *Biochimie*. 2018;155:83–91.

Assunção-Silva RC, Pinho A, Cibrão JR, Pereira IM, Monteiro S, Silva NA, Campos J, et al. Adipose tissue-derived stem cell secretome induces motor and histological gains after complete spinal cord injury. *Journal of Tissue Engineering*. 2024;15.

Assunção-Silva RC, Pinto L, Salgado AJ. Cell transplantation and secretome-based approaches in spinal cord injury regenerative. *Medicine Medicinal Research Reviews*. 2022;42(2):850–96.

Bagher Z, Azami M, Ebrahimi-Barough S, Mirzadeh H, Solouk A, Soleimani M, Ai J, Nourani MR, Joghataei MT. Differentiation of Wharton's jelly-derived mesenchymal stem cells into motor neuron-like cells on three-dimensional collagen-grafted nanofibers. *Molecular Neurobiology*. 2016; 53(4):2397–408.

Bagno L, Hatzistergos KE, Balkan W, Hare JM. Mesenchymal stem cell-based therapy for cardiovascular disease: Progress and challenges. *Molecular Therapy.* 2018;26(7):1610–1623.

Baldari S, Di Rocco G, Piccoli M, Pozzobon M, Muraca M, Toietta G. Challenges and strategies for improving the regenerative effects of mesenchymal stromal cell-based therapies. *International Journal of Molecular Sciences.* 2017;18(10).

Barata-Antunes S, Sousa RA, Salgado AJ, Silva BFB. The whole secretome of mesenchymal stem cells is fully incorporated in lipid bicontinuous cubic phases. *Frontiers in Medical Engineering.* 2025;3:1397406.

Bartlett RD, Burley S, Ip M, Phillips JB, Choi D. Cell therapies for spinal cord injury: Trends and challenges of current clinical trials. *Neurosurgery.* 2020;87(4):E456–472.

Bhatti MA, Haider KH. Mesenchymal stem cells as living bio-drugs: Adding a novel dimension to the cardiology and cardiovascular sciences. *Journal of Cardiology and Cardiovascular Sciences.* 2025;6(1):30–5.

Bonafede R, Turano E, Scambi I, Busato A, Bontempi P, Virla F, Schiaffino L, et al. ASC-exosomes ameliorate the disease progression in SOD1(G93A) murine model underlining their potential therapeutic use in human ALS. *International Journal of Molecular Sciences.* 2020;21(10).

Borsche M, König IR, Delcambre S, Petrucci S, Balck A, Brüggemann N, Zimprich A, et al. Mitochondrial damage-associated inflammation reveals biomarkers in PRKN/PINK1 Parkinson's disease. *Brain.* 2020;143(10):3041–51.

Calabria E, Scambi I, Bonafede R, Schiaffino L, Peroni D, Potrich V, Capelli C, et al. ASCs-exosomes recover coupling efficiency and mitochondrial membrane potential in an in vitro Model of ALS. *Frontiers in Neuroscience.* 2019;13:1070.

Calvo PM, Hernández RG, Pastor AM, de la Cruz RR. VEGF and neuronal survival. *Neuroscientist.* 2024;30(1):71–86.

Caplan AI. Mesenchymal stem cells. *Journal of Orthopaedic Research.* 1991;9(5):641–50.

Caplan AI, Dennis JE. Mesenchymal stem cells as trophic mediators. *Journal of Cellular Biochemistry.* 2006; 98(5):1076–84.

Caplan H, Olson SD, Kumar A, George M, Prabhakara KS, Wenzel P, Bedi S, et al. Mesenchymal stromal cell therapeutic delivery: Translational challenges to clinical application. *Frontiers in Immunology.* 2019;10:1645.

Cheng H, Liu X, Hua R, Dai G, Wang X, Gao J, An Y. Clinical observation of umbilical cord mesenchymal stem cell transplantation in treatment for sequelae of thoracolumbar spinal cord injury. *Journal of Translational Medicine.* 2014;12:253.

Cízková D, Rosocha J, Vanický I, Jergová S, Cízek M. Transplants of human mesenchymal stem cells improve functional recovery after spinal cord injury in the rat. *Cellular and Molecular Neurobiology.* 2006; 26(7–8):1167–80.

Cova L, Armentero MT, Zennaro E, Calzarossa C, Bossolasco P, Busca G, Lambertenghi-Deliliers G, Polli E, Nappi G, Silani V, Blandini F. Multiple neurogenic and neurorescue effects of human mesenchymal stem cells after transplantation in an experimental model of Parkinson's disease. *Brain Research.* 2010;1311:12–27.

Cova L, Bossolasco P, Armentero MT, Diana V, Zennaro E, Mellone M, Calzarossa C, et al. Neuroprotective effects of human mesenchymal stem cells on neural cultures exposed to 6-hydroxydopamine: implications for reparative therapy in Parkinson's disease. *Apoptosis.* 2012;17(3):289–304.

De Sousa KP, Rossi I, Abdullahi M, Ramirez MI, Stratton D, Inal JM. Isolation and characterization of extracellular vesicles and future directions in diagnosis and therapy. *Wiley Interdisciplinary Reviews: Nanomedicine and Nanobiotechnology.* 2023;15(1):e1835.

Dominici M, Le Blanc K, Mueller I, Slaper-Cortenbach I, Marini, F, Krause D, Deans R, et al. Minimal criteria for defining multipotent mesenchymal stromal cells. The International Society for Cellular Therapy position statement. *Cytotherapy.* 2006; 8(4):315–7.

Dugger BN, Dickson DW. Pathology of neurodegenerative diseases. *Cold Spring Harbor Perspectives in Biology.* 2017;9(7).

Escrig A, Canal C, Sanchis P, Fernández-Gayol O, Montilla A, Comes G, Molinero A, et al. IL-6 trans-signaling in the brain influences the behavioral and physio-pathological phenotype of the Tg2576 and 3xTgAD mouse models of Alzheimer's disease. *Brain, Behavior and Immunity.* 2019;82:145–59.

Fan XL, Zhang Y, Li X, Fu QL. Mechanisms underlying the protective effects of mesenchymal stem cell-based therapy. *Cellular and Molecular Life Sciences.* 2020;77(14): 2771–94.

Fang A, Wang Y, Guan N, Zuo Y, Lin L, Guo B, Mo A, et al. Porous microneedle patch with sustained delivery of extracellular vesicles mitigates severe spinal cord injury. *Nature Communications.* 2023;14(1):4011.

Fathollahi A, Hashemi SM, Haji Molla Hoseini M, Tavakoli S, Farahani E, Yeganeh F. Intranasal administration of small extracellular vesicles derived from mesenchymal stem cells ameliorated the experimental autoimmune encephalomyelitis. *International Immunopharmacology*. 2021;90:107207.

Feng MG, Liu CF, Chen L, Feng WB, Liu M, Hai H, Lu JM. MiR-21 attenuates apoptosis triggered by amyloid-β via modulating the PDCD4/PI3K/AKT/GSK-3β pathway in SH-SY5Y cells. *Biomedicine & Pharmacotherapy*. 2018;101:1003–7.

Feng Y, Ju Y, Cui J, Wang L. Bone marrow stromal cells promote neuromotor functional recovery, via upregulation of neurotrophic factors and synapse proteins following traumatic brain injury in rats. *Molecular Medicine Reports*. 2017;16(1):654–60.

Fitzsimmons REB, Mazurek MS, Soos A, Simmons CA. Mesenchymal stromal/stem cells in regenerative medicine and tissue engineering. *Stem Cells International*. 2018;2018:8031718.

Friedenstein AJ, Chailakhyan RK, Latsinik NV, Panasyuk AF, Keiliss-Borok IV. Stromal cells are responsible for transferring the microenvironment of the hemopoietic tissues. Cloning in vitro and retransplantation in vivo. *Transplantation*. 1974;17(4):331–40.

Giunti D, Marini C, Parodi B, Usai C, Milanese M, Bonanno G, Kerlero de Rosbo N, Uccelli A. Role of miRNAs shuttled by mesenchymal stem cell-derived small extracellular vesicles in modulating neuroinflammation. *Scientific Reports*. 2021;11(1):1740.

Giunti D, Parodi B, Usai C, Vergani L, Casazza S, Bruzzone S, Mancardi G, Uccelli A. Mesenchymal stem cells shape microglia effector functions through the release of CX3CL1. *Stem Cells*. 2012;30(9):2044–53.

Gnecchi M, He H, Liang OD, Melo LG, Morello F, Mu H, Noiseux N, et al. Paracrine action accounts for marked protection of the ischemic heart by Akt-modified mesenchymal stem cells. *Nature Medicine*. 2005;11(4):367–8.

González-Cubero E, González-Fernández ML, Olivera ER, Villar-Suárez V. Extracellular vesicle and soluble fractions of adipose tissue-derived mesenchymal stem cells' secretome modulate inflammatory cytokine levels in an in vitro model of discogenic pain. *The Spine Journal*. 2022;22(7):1222–34.

Grayson M. Parkinson's disease. *Nature*. 2016;538(7626):S1.

Habib SM, Martini MF, Al Ahmed O, Qassab AR, Haider KH. Mesenchymal stem cell secretomes for cell-free therapy of the broken heart. In: Haider, K.H. (ed.) *Handbook of Regenerative Medicine: Cell-Free Therapy*. London: Taylor & Francis Group/CRC Press, 2025a, Chapter 18, pp. 274–93.

Habib SM, Martini MF, Al Ahmed O, Qassab AR, Haider KH. The potential of mesenchymal stem cell–derived exosomes in cell-free therapy for myocardial repair. In: Haider, K.H. (ed.) *Handbook of Regenerative Medicine: Cell-Free Therapy*. London: Taylor & Francis Group/CRC Press, 2025b, Chapter 24, pp. 372–90.

Habib SY, Martini MF, Hamdan YNA, Shrebaty OMM, Haider KH. Mesenchymal stem cell secretome and exosomes as potential advanced therapy medicinal products for treating a failing heart. *World Journal of Stem Cells*. 2026;18.

Haider KH. Priming mesenchymal stem cells to develop "super stem cells." *World Journal of Stem Cells* 2024a;16(6):623–40. https://dx.doi.org/10.4252/wjsc.v16.i6.623

Haider KH. Melatonin-based priming of stem cells to alleviate oxidative stress. *World Journal of Stem Cells* 2024b;16(11):985–9. https://dx.doi.org/10.4252/wjsc.v16.i11.985

Haider KH, Aslam M. Cell-free therapy with stem cell secretions: Protection, repair, and regeneration of the injured myocardium. In: Haider K.H. and Aziz, S. (eds) *Stem Cells: From Hype to Real Hope*. Berlin: Medicine & Life Sciences, De Gruyter, 2018.

Haider KH, Aziz S. Paracrine hypothesis and cardiac repair. *International Journal of Stem Cell Research and Transplantation*. 2017;5(1):265–7.

Han Y, Li X, Zhang Y, Chang F, Ding J. Mesenchymal stem cells for regenerative medicine. *Cells*. 2019;8(8).

Han Y, Yang J, Fang J, Zhou Y, Candi E, Wang J, Hua D, et al. The secretion profile of mesenchymal stem cells and potential applications in treating human diseases. *Signal Transduction and Targeted Therapy*. 2022;7(1):92.

Haroon J, Aboody K, Flores L, McDonald M, Mahdavi K, Zielinski M, Jordan K, et al. Use of transcranial low-intensity focused ultrasound for targeted delivery of stem cell-derived exosomes to the brain. *Scientific Reports*. 2023;13(1):17707.

Hoban DB, Howard L, Dowd E. GDNF-secreting mesenchymal stem cells provide localized neuroprotection in an inflammation-driven rat model of Parkinson's disease. *Neuroscience*. 2015;303:402–11.

Hosseini SM, Sani M, Haider KH, Dorvash M, Ziaee SM, Karimi A, Namavar MR. Concomitant use of mesenchymal stem cells and neural stem cells for treatment of spinal cord injury: A combo cell therapy approach. *Neurosci Lett*. 2018 Mar 6;668:138–46. https://dx.doi.org/10.1016/j.neulet.2018.01.008

Huang Y, Li X, Yang L. Hydrogel encapsulation: Taking the therapy of mesenchymal stem cells and their derived secretome to the next level. *Frontiers in Bioengineering and Biotechnology.* 2022;10:859927.

Huo L, Du X, Li X, Liu S, Xu Y. The emerging role of neural cell-derived exosomes in intercellular communication in health and neurodegenerative diseases. *Frontiers in Neuroscience.* 2021;15:738442.

Kamal M, Kassem D, Haider KH. Sources and therapeutic strategies of mesenchymal stem cells in regenerative medicine. In: Haider, K.H. (ed.) *Handbook of Stem Cell Therapy.* Singapore: Springer, 2022. https://doi.org/10.1007/978-981-19-2655-6_2

Kandeel M, Morsy MA, Alkhodair KM, Alhojaily S. Mesenchymal stem cell-derived extracellular vesicles: An emerging diagnostic and therapeutic biomolecule for neurodegenerative disabilities. *Biomolecules.* 2023;13(8).

Kaniowska D, Wenk K, Rademacher P, Weiss R, Fabian C, Schulz I, Guthardt M, et al. Extracellular vesicles of mesenchymal stromal cells can be taken up by microglial cells and partially prevent the stimulation induced by β-amyloid. *Stem Cell Reviews and Reports.* 2022;18(3):1113–26.

Kanton S, Paşca SP. Human assembloids. *Development.* 2022;149(20).

Kaur D, Sharma V, Deshmukh R. Activation of microglia and astrocytes: A roadway to neuroinflammation and Alzheimer's disease. *Inflammopharmacology.* 2019;27(4):663–77.

Kean TJ, Lin P, Caplan AI, Dennis JE. MSCs: Delivery routes and engraftment, cell-targeting strategies, and immune modulation. *Stem Cells International.* 2013;2013:732742.

Kim HJ, Seo S W, Chang JW, Lee JI, Kim CH, Chin J, Choi SJ, et al. Stereotactic brain injection of human umbilical cord blood mesenchymal stem cells in patients with Alzheimer's disease dementia: A phase 1 clinical trial. *Alzheimer's & Dementia.* 2015;1(2):95–102.

Kim YJ, Park HJ, Lee G, Bang OY, Ahn YH, Joe E, Kim HO, Lee PH. Neuroprotective effects of human mesenchymal stem cells on dopaminergic neurons through anti-inflammatory action. *Glia.* 2009;57(1):13–23.

Kusuma GD, Carthew J, Lim R, Frith JE. Effect of the microenvironment on mesenchymal stem cell paracrine signaling: Opportunities to engineer the therapeutic effect. *Stem Cells & Development.* 2017;26(9):617–31.

Lane CA, Hardy J, Schott JM. Alzheimer's disease. *European Journal of Neurology.* 2018;25(1):59–70.

Lee JH, Yoon YM, Lee SH. Hypoxic preconditioning promotes the bioactivities of mesenchymal stem cells via the HIF-1α-GRP78-Akt axis. *International Journal of Molecular Sciences.* 2017;18(6).

Lee JK, Schuchman EH, Jin HK, Bae JS. Soluble CCL5 derived from bone marrow-derived mesenchymal stem cells and activated by amyloid β ameliorates Alzheimer's disease in mice by recruiting bone marrow-induced microglia immune responses. *Stem Cells.* 2012;30(7):1544–55.

Liang Y, Xia J, Xu L, Zhang Y, Xu X, Haider KH. Cell-free and cell-based therapy in regenerative medicine. In: Haider, K.H. (ed.) *Handbook of Regenerative Medicine: Cell-Free Therapy Approach* (1st ed.). CRC Press. Chapter 24, 2025. https://doi.org/10.1201/9781003611509

Liu J, Han D, Wang Z, Xue M, Zhu L, Yan H, Zheng X, et al. Clinical analysis of the treatment of spinal cord injury with umbilical cord mesenchymal stem cells. *Cytotherapy.* 2013;15(2):185–91.

Liu XS, Li JF, Wang SS, Wang YT, Zhang YZ, Yin HL, Geng S, et al. Human umbilical cord mesenchymal stem cells infected with an adenovirus expressing HGF promote the regeneration of damaged neurons in a Parkinson's disease model. *BioMed Research International.* 2014;2014:909657.

Maguire G, Paler L, Green L, Mella R, Valcarcel M, Villace P. Rescue of degenerating neurons and cells by stem cell-released molecules: using a physiological renormalization strategy. *Physiological Reports.* 2019;7(9):e14072.

Mahmood A, Lu D, Qu C, Goussev A, Chopp M. Long-term recovery after bone marrow stromal cell treatment of traumatic brain injury in rats. *Journal of Neurosurgery.* 2006;104(2):272–7.

Maldonado VV, Patel NH, Smith EE, Barnes CL, Gustafson MP, Rao RR, Samsonraj RM. Clinical utility of mesenchymal stem/stromal cells in regenerative medicine and cellular therapy. *Journal of Biological Engineering.* 2023;17(1):44.

Marote A, Santos D, Mendes-Pinheiro B, Serre-Miranda C, Anjo SI, Vieira J, Ferreira-Antunes F, et al. Cellular aging secretes: A comparison of bone-marrow-derived and induced mesenchymal stem cells and their secretome over long-term culture. *Stem Cell Reviews and Reports.* 2023;19(1):248–63.

Marote A, Teixeira FG, Mendes-Pinheiro B, Salgado AJ. MSCs-derived exosomes: Cell-secreted nanovesicles with regenerative potential. *Frontiers in Pharmacology.* 2016;7:231.

Marques CR, Campos J, Sampaio-Marques B, Antunes FF, Dos Santos Cunha RM, Silva D, Barata-Antunes S, et al. The secretome of bone marrow mesenchymal stromal cells cultured in a dynamic system induces neuroprotection and modulates microglial responsiveness in an α-synuclein-overexpression rat model. *Cytotherapy.* 2024:26(7):700–13.

Marques CR, Pereira-Sousa J, Teixeira FG, Sousa RA, Teixeira-Castro A, Salgado AJ. Mesenchymal stem cell secretome protects against alpha-synuclein-induced neurodegeneration in a Caenorhabditis elegans model of Parkinson's disease. *Cytotherapy.* 2021:23(10):894–901.

Martins LF, Costa RO, Pedro JR, Aguiar P, Serra SC, Teixeira FG, Sousa N, et al. BDNF mediates the secretome-induced axonal outgrowth of mesenchymal stem cells. *Scientific Reports.* 2017;7(1):4153.

McCoy MK, Martinez TN, Ruhn KA, Wrage PC, Keefer EW, Botterman BR, Tansey KE, Tansey MG. Autologous transplants of adipose-derived adult stromal (ADAS) cells afford dopaminergic neuroprotection in a model of Parkinson's disease. *Experimental Neurology.* 2008;210(1):14–29.

Mendes-Pinheiro B, Anjo SI, Manadas B, da Silva JD, Marote A, Behie LA, Teixeira FG, Salgado AJ. Bone marrow mesenchymal stem cells' secretome exerts neuroprotective effects in a Parkinson's disease rat model. *Frontiers in Bioengineering and Biotechnology.* 2019;7:294

Mendes-Pinheiro B, Campos J, Marote A, Soares-Cunha C, Nickels SL, Monzel AS, Cibrão JR, et al. Treating Parkinson's disease with human bone marrow mesenchymal stem cell secretome: A translational investigation using human brain organoids and different routes of in vivo administration. *Cells.* 2023;12(21).

Miceli V, Zito G, Bulati M, Gallo A, Busà R, Iannolo G, Conaldi PG. Different priming strategies enhance the distinct therapeutic capabilities of mesenchymal stromal/stem cells: Potential implications for their clinical application. *World Journal of Stem Cells.* 2023;15(5):400–20.

Mushtaq M, Zineldeen DH, Mateen MA, Haider KH. Mesenchymal stem cells' "garbage bags" at work: Treating radial nerve injury with mesenchymal stem cell-derived exosomes. *World Journal of Stem Cells* 2024;16(5). https://dx.doi.org/10.4252/wjsc.v16.i5.0000.

Nava MM, Raimondi MT, Pietrabissa R. Controlling self-renewal and differentiation of stem cells via mechanical cues. *Journal of Biomedicine and Biotechnology.* 2012;2012:797410.

Nieland L, Mahjoum S, Grandell E, Breyne K, Breakefield XO. Engineered EVs are designed to target CNS diseases. *Journal of Controlled Release.* 2023;356:493–506.

Noronha NC, Mizukami A, Caliári-Oliveira C, Cominal JG, Rocha JLM, Covas D T, Swiech K, Malmegrim KCR. Priming approaches to improve the efficacy of mesenchymal stromal cell-based therapies. *Stem Cell Research & Therapy.* 2019;10(1):131.

Numakawa T, Suzuki S, Kumamaru E, Adachi N, Richards M, Kunugi H. BDNF function and intracellular signaling in neurons. *Histology and Histopathology.* 2010; 25(2):237–58.

Oh SH, Kim HN, Park HJ, Shin JY, Bae EJ, Sunwoo MK, Lee SJ, Lee PH. Mesenchymal stem cells inhibit transmission of α-synuclein by modulating clathrin-mediated endocytosis in a Parkinsonian model. *Cell Reports.* 2016;14(4):835–49.

Oh SH, Kim HN, Park HJ, Shin JY, Kim DY, Lee PH. The cleavage effect of mesenchymal stem cell and its derived matrix metalloproteinase-2 on extracellular α-synuclein aggregates in Parkinsonian models. *Stem Cells Translational Medicine.* 2017; 6(3):949–61.

Ohta Y, Nagai M, Miyazaki K, Tanaka N, Kawai H, Mimoto T, Morimoto N, et al. Neuroprotective and angiogenic effects of bone marrow transplantation combined with granulocyte colony-stimulating factor in a mouse model of amyotrophic lateral sclerosis. *Cell Medicine.* 2011;2(2):69–83.

Page P, DeJong J, Bandstra A, Boomsma RA. Effect of serum and oxygen concentration on gene expression and secretion of paracrine factors by mesenchymal stem cells. *International Journal of Cell Biology.* 2014;2014:601063.

Panoutsopoulos AA. Organoids, assembloids, and novel biotechnology: Steps forward in developmental and disease-related neuroscience. *Neuroscientist.* 2021;27(5):463–72.

Parga JA, García-Garrote M, Martínez S, Raya Á, Labandeira-García JL, Rodríguez-Pallares J. Prostaglandin EP2 receptors mediate mesenchymal stromal cell-neuroprotective effects on dopaminergic neurons. *Molecular Neurobiology.* 2018;55(6):4763–76.

Paşca SP. The rise of three-dimensional human brain cultures. *Nature.* 2018;553(7689):437–45.

Paşca SP, Arlotta P, Bateup HS, Camp JG, Cappello S, Gage FH, Knoblich JA, et al. A nomenclature consensus for nervous system organoids and assembloids. *Nature.* 2022;609(7929):907–10.

Perets N, Betze O, Shapira R, Brenstein S, Angel A, Sadan T, Ashery U, Popovtzer R, Offen D. Golden exosomes selectively target brain pathologies in neurodegenerative and neurodevelopmental disorders. *Nano Letters Journal.* 2019;19(6):3422–31.

Petrou P, Gothelf Y, Argov Z, Gotkine M, Levy YS, Kassis I, Vaknin-Dembinsky A, et al. Safety and clinical effects of mesenchymal stem cells secreting neurotrophic factor transplantation in patients with amyotrophic lateral sclerosis: Results of phase 1/2 and 2a clinical trials. *JAMA Neurology.* 2016;73(3):337–44.

Pinho AG, Cibrão JR, Lima R, Gomes ED, Serra SC, Lentilhas-Graça J, Ribeiro C, et al. Immunomodulatory and regenerative effects of the full and fractioned adipose tissue-derived stem cells secretome in spinal cord injury. *Experimental Neurology.* 2022;351:113989.

Pinho AG, Cibrão JR, Silva NA, Monteiro S, Salgado AJ. Cell secretome: Basic insights and therapeutic opportunities for CNS disorders. *Pharmaceuticals (Basel)*. 2020;13(2).

Pires AO, Mendes-Pinheiro B, Teixeira FG, Anjo SI, Ribeiro-Samy S, Gomes ED, Serra SC, et al. Unveiling the differences of secretome of human bone marrow mesenchymal stem cells, adipose tissue-derived stem cells, and human umbilical cord perivascular cells: A proteomic analysis. *Stem Cells and Development*. 2016;25(14):1073–83.

Pittenger MF, Discher DE, Péault BM, Phinney DG, Hare JM, Caplan AI. Mesenchymal stem cell perspective: cell biology to clinical progress. *NPJ Regenerative Medicine*. 2019;4:22.

Provenzano F, Nyberg S, Giunti D, Torazza C, Parodi B, Bonifacino T, Usai C, et al. Micro-RNAs shuttled by extracellular vesicles secreted from mesenchymal stem cells dampen astrocyte pathological activation and support neuroprotection in in vitro models of ALS. *Cells*. 2022;11(23).

Qu C, Mahmood A, Liu XS, Xiong Y, Wang L, Wu H, Li B, et al. The treatment of TBI with human marrow stromal cells impregnated into a collagen scaffold: functional outcome and gene expression profile. *Brain Research*. 2011;1371:129–39.

Rajab AM, Rajab TM, AlJundi S, Haider KH. Bone stem cell therapy in the clinical perspective: A focus on nonrandomized and randomized trials. In: Haider, K.H. (ed.) *Stem Cells: From Myth to Reality and Evolving*. Berlin: Medicine & Life Sciences, De Gruyter, 2019.

Riva N, Domi T, Pozzi L, Lunetta C, Schito P, Spinelli EG, Cabras S, et al. Update on recent advances in amyotrophic lateral sclerosis. *Journal of Neurology*. 2024;271(7):4693–723.

Rosenberg GA. Neurological diseases related to the blood-brain barrier. *Journal of Cerebral Blood Flow & Metabolism*. 2012;32(7):1139–51.

Sadan O, Bahat-Stromza M, Barhum Y, Levy YS, Pisnevsky A, Peretz H, Ilan AB, et al. Protective effects of neurotrophic factor-secreting cells in a 6-OHDA rat model of Parkinson's disease. *Stem Cells and Development*. 2009;18(8):1179–90.

Safwan M, Bourgleh MS, Alshakaki H, Molhem A, Haider KH. (2024). Morbid cell status and donor age significantly alter mesenchymal stem cell functionality and reparability. In: Haider, K.H. (ed.) *Handbook of Stem Cell Applications*. Singapore: Springer Nature, pp.1359–87, 2024. https://doi.org/10.1007/978-981-99-7119-0_62

Sanchez-Guijo F, Vives J, Ruggeri A, Chabannon C, Corbacioglu S, Dolstra H, Farge D, et al. Current challenges in cell and gene therapy: A joint view from the European Committee of the International Society for Cell & Gene Therapy (ISCT) and the European Society for Blood and Marrow Transplantation (EBMT). *Cytotherapy*. 2024;26(7):681–5.

Santamaria G, Brandi E, Vitola P, Grandi F, Ferrara G, Pischiutta F, Vegliante G, et al. Intranasal delivery of mesenchymal stem cell secretome restores brain function in Alzheimer's mice. *Cell Death & Differentiation*. 2021;28(1):203–18.

Sha, S, Shen, X, Cao, Y, Qu, L. Mesenchymal stem cells-derived extracellular vesicles ameliorate Alzheimer's disease in rat models via the microRNA-29c-3p/BACE1 axis and the Wnt/β-catenin pathway. *Aging*. 2021;13(11):15285–306.

Silva D, Schirmer L, Pinho TS, Atallah P, Cibrão JR, Lima R, Afonso J, et al. Sustained release of human adipose tissue stem cell secretome from star-shaped poly(ethylene glycol) glycosaminoglycan hydrogels promotes motor improvements after complete transection in spinal cord injury rat model. *Advanced Healthcare Materials*. 2023;12(17):e2202803.

Silva NA, Sousa N, Reis RL, Salgado AJ. From basics to clinical: A comprehensive review on spinal cord injury. *Progress in Neurobiology*. 2014;114:25–57.

Song L, Yang Y J, Dong QT, Qian HY, Gao RL, Qiao SB, Shen R, et al. Atorvastatin enhances the efficacy of mesenchymal stem cell treatment for swine myocardial infarction via activation of nitric oxide synthase. *PLoS One*. 2013;8(5):e65702.

Suzuki S, Kawamata J, Iwahara N, Matsumura A, Hisahara S, Matsushita T, Sasaki M, et al. Intravenous mesenchymal stem cell administration exhibits therapeutic effects against 6-hydroxydopamine-induced dopaminergic neurodegeneration and glial activation in rats. *Neuroscience Letters*. 2015;584:276–81.

Sykova E, Cizkova D, Kubinova, S. Mesenchymal stem cells in treatment of spinal cord injury and amyotrophic lateral sclerosis. *Frontiers in Cell and Developmental Biology*. 2021;9:695900.

Tajiri N, Acosta SA, Shahaduzzaman M, Ishikawa H, Shinozuka K, Pabon M, Hernandez-Ontiveros D, et al. Intravenous transplants of human adipose-derived stem cells protect the brain from traumatic brain injury-induced neurodegeneration and motor and cognitive impairments: Cell graft biodistribution and soluble factors in young and aged rats. *Journal of Neuroscience*. 2014;34(1):313–26.

Takeda YS, Xu Q. Neuronal differentiation of human mesenchymal stem cells using exosomes derived from differentiating neuronal cells. *PLoS One.* 2015;10(8):e0135111.

Teixeira FG, Carvalho MM, Panchalingam KM, Rodrigues AJ, Mendes-Pinheiro B, Anjo S, Manadas B, et al. Impact of the secretome of human mesenchymal stem cells on brain structure and animal behavior in a rat model of Parkinson's disease. *Stem Cells Translational Medicine.* 2017;6(2):634–46.

Teixeira FG, Carvalho MM, Sousa N, Salgado AJ. Mesenchymal stem cells secretome: A new paradigm for central nervous system regeneration? *Cellular and Molecular Life Sciences.* 2013;70(20):3871–82.

Teixeira FG, Panchalingam KM, Assunção-Silva R, Serra SC, Mendes-Pinheiro B, Patrício P, Jung S, et al. Modulation of the mesenchymal stem cell secretome using computer-controlled bioreactors: Impact on neuronal cell proliferation, survival and differentiation. *Scientific Reports.* 2016;6:27791.

Teixeira FG, Salgado AJ. Mesenchymal stem cells secretome: Current trends and future challenges. *Neural Regeneration Research.* 2020;15(1):75–7.

Teli P, Nachanekar A, Kale V, Vaidya A. Priming mesenchymal stromal cells with neurotrophic factors boosts the neuro-regenerative potential of their secretome. *Regenerative Medicine.* 2023;18(4):329–46.

Teng F, Fussenegger M. Shedding light on extracellular vesicle biogenesis and bioengineering. *Advanced Science.* 2020;8(1):2003505.

Thapa K, Khan H, Singh TG, Kaur A. Traumatic brain injury: Mechanistic insight on pathophysiology and potential therapeutic targets. *Journal of Molecular Neuroscience.* 2021;71(9):1725–42.

Tian T, Zhang HX, He CP, Fan S, Zhu YL, Qi, C, Huang NP, et al. Surface functionalized exosomes as targeted drug delivery vehicles for cerebral ischemia therapy. *Biomaterials.* 2018;150:137–49.

Toader C, Dumitru AV, Eva L, Serban M, Covache-Busuioc RA, Ciurea AV. Nanoparticle strategies for treating CNS disorders: A comprehensive review of drug delivery and theranostic applications. *International Journal of Molecular Sciences.* 2024;25(24).

Tran C, Damaser MS. Stem cells as drug delivery methods: Application of stem cell secretome for regeneration. *Advanced Drug Delivery Reviews.* 2015;82–3:1–11.

Turano E, Scambi I, Virla F, Bonetti B, Mariotti, R. Extracellular Vesicles from mesenchymal stem cells: Towards novel therapeutic strategies for neurodegenerative diseases. *International Journal of Molecular Sciences.* 2023;24(3).

Vargas-Rodríguez P, Cuenca-Martagón A, Castillo-González J, Serrano-Martínez I, Luque RM, Delgado M, González-Rey E. Novel therapeutic opportunities for neurodegenerative diseases with mesenchymal stem cells: The focus on modulating the blood-brain barrier. *International Journal of Molecular Sciences.* 2023;24(18).

Venkataramana NK, Pal R, Rao SA, Naik AL, Jan M, Nair R, Sanjeev CC, et al. Bilateral transplantation of allogenic adult human bone marrow-derived mesenchymal stem cells into the subventricular zone of Parkinson's disease: A pilot clinical study. *Stem Cells International* 2012;2012:931902.

Vercelli A, Mereuta OM, Garbossa D, Muraca G, Mareschi K, Rustichelli D, Ferrero I, et al. Human mesenchymal stem cell transplantation extends survival, improves motor performance, and decreases neuroinflammation in a mouse model of amyotrophic lateral sclerosis. *Neurobiology of Disease.* 2008;31(3):395–405.

Vilaça-Faria H, Marote A, Lages I, Ribeiro C, Mendes-Pinheiro B, Domingues AV, Campos J, et al. Fractionating stem cells' secretome for Parkinson's disease modeling: Is it the whole better than the sum of its parts? *Biochimie.* 2021;189:87–98.

Volarevic V, Markovic BS, Gazdic M, Volarevic A, Jovicic N, Arsenijevic N, Armstrong L, et al. Ethical and safety issues of stem cell-based therapy. *International Journal of Medical Sciences.* 2018;15(1):36–45.

Volkman R, Offen D. Concise review: Mesenchymal stem cells in neurodegenerative diseases. *Stem Cells.* 2017;35(8):1867–80.

Wei X, Du Z, Zhao L, Feng D, Wei G, He Y, Tan J, et al. IFATS collection: The conditioned media of adipose stromal cells protect against hypoxia-ischemia-induced brain damage in neonatal rats. *Stem Cells.* 2009;27(2): 478–88.

Welsh JA, Goberdhan DCI, O'Driscoll L, Buzas EI, Blenkiron C, Bussolati B, Cai H, et al. Minimal information for studies of extracellular vesicles (MISEV2023): From basic to advanced approaches. *Journal of Extracellular Vesicles.* 2024;13(2):e12404.

Wray S. Modelling neurodegenerative disease using brain organoids. *Seminars in Cell and Developmental Biology.* 2021;111:60–6.

Wuschko S, Gugerell A, Chabicovsky M, Hofbauer H, Laggner M, Erb M, Ostler T, et al. Toxicological testing of allogeneic secretome derived from peripheral mononuclear cells (APOSEC): A novel cell-free therapeutic agent in skin disease. *Scientific Reports.* 2019;9(1):5598.

Xia J, Arai S, Haider KH. Research and applications of stem cell secretome. In: Haider, K.H. (ed.) *Stem Cells: From Potential to Promise*. Singapore: Springer, 2021. https://doi.org/10.1007/978-981-16-0301-3_7.

Xue C, Shen Y, Li X, Li B, Zhao S, Gu J, Chen Y, et al. Exosomes derived from hypoxia-treated human adipose mesenchymal stem cells enhance angiogenesis through the PKA signaling pathway. *Stem Cells and Development*. 2018;27(7):456–65.

Yusuf R, Haider KH. Theragnostic applications of exosomes in Alzheimer's disease. In: Haider, K.H. (ed.) *Handbook of Regenerative Medicine: Cell-Free Therapy*. London: Taylor & Francis Group/CRC Press, 2025, Chapter 255, pp. 390–409.

Zhang R, Liu Y, Yan K, Chen L, Chen XR, Li P, Chen FF, Jiang XD. Anti-inflammatory and immunomodulatory mechanisms of mesenchymal stem cell transplantation in experimental traumatic brain injury. *Journal of Neuroinflammation*. 2013;10:106.

Zippel N, Schulze M, Tobiasch E. Biomaterials and mesenchymal stem cells for regenerative medicine. *Recent Patents on Biotechnology*. 2010;4(1): 1–22.

23 Exosomes and Secretome in Urogenital Medicine

Beyond Cellular Therapies

Guihua Liu[1], Jintao Guo[1], Anthony Atala[1], and Yuanyuan Zhang[1]

[1]Wake Forest Institute for Regenerative Medicine, Wake Forest University, School of Medicine, Winston-Salem, NC, USA

23.1 LIST OF ABBREVIATIONS

ABU	Asymptomatic bacteriuria
AKI	Acute kidney injury
AR-V7	Androgen receptor splice variant 7
BMSC-Exo	Bone marrow MSC-derived exosomes
CKD	Chronic kidney disease
CP	Ceruloplasmin
DKD	Diabetic kidney disease
ECM	Extracellular matrix
ED	Erectile Dysfunction
EMT	Epithelial–mesenchymal transition
ESCs	Embryonic stem cells
EVs	Extracellular vesicles
Fas-L	Fas-lignad
FGF-2	Fibroblast-derived growth factor-2
HGF	Hepatocyte growth factor
hUCMSC-Exos	Human umbilical cord mesenchymal stem cells
IGF-1	Insulin like growth factor
ILVs	Intraluminal vesicles
iPSCs	Induced pluripotent stem cells
MHC	Major histocompatibility complex
miRNAs	microRNAs
MSC-Exo	MSC-derived exosomes
MSCs	Mesenchymal stem cells
MVBs	Multivesicular bodies
NOA	Non-obstructive azoospermia
OA	Obstructive azoospermia
ROS	Reactive oxygen species
RVG	Rabies viral glycoprotein
SA	Secretory azoospermia
SUI	Stress Urinary Incontinence

DOI: 10.1201/9781003653387-23

TNF	Tumor necrosis factor
TRAIL	Tumor necrosis factor-related apoptosis-inducing ligand
UTI	Urinary tract infection
VEGF	Vascular endothelial growth factor
α-SMA	Alpha smooth muscle actin

23.2 INTRODUCTION

23.2.1 The Shifting Paradigms in Regenerative Medicine

For decades, regenerative medicine has centered on cell-based therapies, particularly those utilizing mesenchymal stem cells (MSCs), embryonic stem cells (ESCs), and induced pluripotent stem cells (iPSCs), due to their potential for differentiation, immune modulation, and tissue repair. These approaches have demonstrated therapeutic promise in a wide array of conditions, including those affecting the urogenital system (Liu et al., 2023, Yin et al., 2024). However, accumulating evidence suggests that the majority of these therapeutic benefits are mediated not by direct cell engraftment or differentiation, but by paracrine mechanisms (Haider and Aziz, 2017; Haider & Aslam, 2018), primarily through the secretion of extracellular vesicles (EVs), including exosomes, and the broader spectrum of secreted soluble factors collectively referred to as the secretome (Foo et al., 2021, Ding et al., 2021).

Exosomes are nanosized (30–150 nm) membrane-bound vesicles actively secreted by almost all cell types (Mushtaq et al., 2024). They carry a rich cargo of bioactive molecules—proteins, lipids, mRNA, miRNA, and other nucleic acids—that can regulate recipient cell behavior, promote angiogenesis, reduce inflammation, inhibit oxidative stress, and prevent fibrosis (Moghassemi et al., 2024, Hu et al., 2024). The secretome refers to the total collection of substances released by cells under specific physiological or induced conditions. In addition to extracellular vesicles such as exosomes, microvesicles, and apoptotic bodies, it also includes soluble proteins, peptides, growth factors, chemokines, metabolites, and lipids (Ahmed et al., 2023). Together, these components form a bioactive microenvironment that supports cell proliferation, migration, and differentiation, thereby enhancing tissue regeneration and functional recovery (Li et al., 2024a; Habib et al., 2026).

Acellular therapies based on exosomes and the secretome offer several advantages over traditional cell therapies, including reduced risk of immune rejection, elimination of concerns regarding tumorigenicity, and the potential for engineered improvements in stability and targeted delivery (Moghassemi et al., 2024; Margiana et al., 2024; Liang et al., 2025). These features make them particularly attractive for urogenital conditions such as erectile dysfunction (ED), renal dysfunction, urinary incontinence, and male infertility—conditions that often involve inflammation, ischemia, or fibrotic tissue remodeling.

This chapter aims to comprehensively explore the therapeutic potential of acellular strategies—specifically exosomes and the secretome—in urogenital regenerative medicine. We will examine their biological mechanisms, current preclinical and clinical applications, and the challenges and prospects for their future translation and implementation.

23.2.2 Advantages of Acellular Therapies (Exosomes and Secretome) over Cell-Based Therapies

23.2.2.1 Reduced Immunogenicity

One advantage of acellular therapies, including exosomes and the secretome, over traditional cell-based therapies is their significantly lower immunogenicity (Table 23.1). Unlike transplanted cells that may express major histocompatibility complex (MHC) molecules and trigger immune rejection, exosomes and secretome components are less likely to elicit strong immune responses

TABLE 23.1
Comparative Advantages and Disadvantages of Cell-Based vs. Acellular (Exosome/Secretome) Therapies

Criteria	Cell-Based Therapies	Acellular Therapies (Exosome/Secretome)	References
Immunogenicity	Higher risk due to MHC expression and possible immune rejection.	Lower; exosomes and secretome components have minimal immunogenicity.	(Elahi et al., 2020) (Ha et al., 2020)
Stability	Requires cryopreservation and has a short shelf-life.	Stable at −80 °C or via lyophilization; longer shelf-life.	(Jeyaram et al., 2017) (Charoenviriyakul et al., 2018)
Delivery Control	Difficult to control cell homing and engraftment.	Better control via natural tropism and engineering of exosome surfaces.	(Zhang et al., 2019); (Álvarez-Erviti (Armstrong et al., 2017)
Risk of Off-Target Effects	High; potential for ectopic tissue formation or tumorigenesis.	Low; no replication or differentiation capability; effects mediated via signaling.	(Wang, 2023)
Scalability and Manufacturing Feasibility	Easy to isolate and possible for mass production.	Relatively low yield for large-scale manufacturing.	(Tan et al., 2024)
Regulatory Landscape	Well-developed regulatory guidelines.	Insufficient regulatory control.	(Tan et al., 2024)

(Elahi et al., 2020). Moreover, the absence of viable cells eliminates concerns related to uncontrolled proliferation, tumorigenicity, or long-term engraftment. Additionally, exosomes can carry immunomodulatory molecules that actively suppress local inflammation and promote immune tolerance within the target tissue microenvironment (Ha et al., 2020). For instance, mesenchymal stem cell–derived exosomes can increase regulatory T cell activity while simultaneously reducing pro-inflammatory Th17 responses, thereby creating a tolerogenic microenvironment that is favorable for tissue regeneration (Pianta et al., 2015). This makes acellular therapies particularly advantageous for urogenital applications, where minimizing host immune response is critical for successful regenerative outcomes.

23.2.2.2 Improved Stability and Shelf-Life

Another key advantage of acellular therapies, including exosomes and the secretome, is their superior stability and longer shelf life compared to cell-based therapies. Studies have shown that exosomes can maintain their integrity and biological activity when stored at −80°C for several months, and even at 4°C for short periods, without significant loss of function (Jeyaram et al., 2017). Moreover, methods such as lyophilization (freeze-drying) have been successfully applied to exosomes, enabling preservation at room temperature while maintaining structural integrity and bioactivity. Charoenviriyakul et al. demonstrated that exosomes derived from B16-BL6 murine melanoma cells retained their characteristic morphology, size distribution, and protein markers after lyophilization and storage at room temperature for several weeks. Importantly, these lyophilized exosomes maintained their cellular uptake capability, suggesting that their therapeutic functions remained intact (Charoenviriyakul et al., 2018). This enhanced stability facilitates not only storage and transportation but also scalable manufacturing, sterility assurance, and regulatory compliance, making acellular therapies more feasible for clinical translation.

23.2.2.2.1 Targeted Delivery Potential

One of the promising advantages of exosome-based therapies lies in their intrinsic and engineerable potential for targeted delivery (Habib et al., 2025a). Native exosomes express specific surface proteins, such as integrins and tetraspanins, that mediate tissue tropism and facilitate natural homing to particular cell types or injured tissues (Zhang et al., 2019). In addition to their natural targeting abilities, exosomes can be engineered to enhance their specificity by displaying targeting ligands or peptides on their membranes. For instance, exosomes modified to express neuron-specific rabies viral glycoprotein (RVG) peptides have been shown to cross the blood–brain barrier and deliver therapeutic cargo to neurons (Alvarez-Erviti et al., 2011). Furthermore, advancements in exosome surface modification have enabled the incorporation of targeting molecules, such as antibodies, aptamers, or homing peptides, thereby improving localization to pathological sites, including tumors and inflamed tissues (Armstrong et al., 2017). This capacity for targeted delivery not only enhances therapeutic efficacy but also reduces off-target effects, making exosomes an ideal vehicle for precision medicine applications in urogenital diseases and beyond.

23.2.2.2.2 Reduced Risk of Off-Target Effects

One of the safety concerns in cell-based therapies is the potential for uncontrolled proliferation or inappropriate differentiation of transplanted cells, which may lead to tumorigenesis or ectopic tissue formation (Wang, 2023). In contrast, exosomes are non-living nanoscale vesicles that lack the replicative or differentiative capacity of cells. They carry bioactive molecules—such as proteins, lipids, and RNAs—within a defined structure and exert their effects primarily through intercellular signaling, minimizing the risk of forming inappropriate tissues or promoting unregulated growth (Tan et al., 2024).

23.3 EXOSOMES: NANOSCALE MESSENGERS OF CELLULAR COMMUNICATION

23.3.1 Biogenesis and Characteristics of Exosomes

Exosomes are nanosized extracellular vesicles, typically 30–150 nanometers in diameter, that originate from the endosomal compartment of most eukaryotic cells (Haider and Najimi, 2022). Their biogenesis begins with the inward budding of the plasma membrane to form early endosomes, which subsequently mature into multivesicular bodies (MVBs) containing intraluminal vesicles (ILVs). Upon fusion of MVBs with the plasma membrane, these ILVs are released into the extracellular space as exosomes (Kalluri et al., 2020). Biogenesis is orchestrated by both ESCRT-dependent machinery—including ESCRT-0, I, II, III, and Vps4—and ESCRT-independent mechanisms involving tetraspanins and ceramide.(Gurunathan et al., 2019) The trafficking and secretion of exosomes depend on small Rab GTPases (Rab27a/b, Rab11, Rab35) and SNARE proteins, which facilitate the transport of MVBs to the plasma membrane and their subsequent exocytosis (Gurung et al., 2021). Exosomes encapsulate a diverse range of cargo, including proteins (tetraspanins, ALIX, TSG101), lipids (cholesterol, ceramide), and nucleic acids (mRNA, miRNA, dsDNA, ssDNA, mtDNA) (Lee et al., 2024).

23.3.2 Mechanisms of Action of Exosomes

Exosomes exert their biological effects on recipient cells through a series of well-coordinated mechanisms that begin with their binding to the cell surface. Their membranes are enriched with surface molecules such as integrins, tetraspanins, and glycoproteins, which mediate specific receptor–ligand interactions and enable initial recognition and adhesion to target cells (Isaac et al., 2021). Upon binding, exosomes are internalized via multiple pathways, including receptor-mediated endocytosis (via clathrin or caveolin), direct membrane fusion, macropinocytosis, and phagocytosis, depending on both the exosome's origin and the recipient cell type (Hu et al., 2020). While membrane fusion allows for the immediate release of cargo into the cytoplasm, endocytosed exosomes are typically processed within endosomal systems. They may subsequently fuse with endosomal membranes or escape into vesicles, releasing functional contents, such as microRNAs (miRNAs), messenger RNAs (mRNAs), and proteins, that modulate gene expression and cellular behavior (Joshi et al., 2020). Additionally, exosomes can trigger juxtacrine signaling by engaging surface receptors via Fas ligand (FasL), tumor necrosis factor (TNF), or TRAIL, thereby activating immune or apoptotic signaling cascades (Gurung et al., 2021). Together, these mechanisms enable exosomes to influence various aspects of recipient cell behavior, including promoting tissue repair, modulating immune responses, suppressing inflammation, and indirectly facilitating angiogenesis. A single pathway does not mediate these effects; instead, they occur through a complex interplay of membrane interactions, signal transduction, and functional cargo delivery (Shen et al., 2020). This mechanistic foundation underpins the growing interest in exosome-based therapeutic platforms in regenerative urogenital medicine.

23.3.3 Exosomes in Urogenital Medicine: Applications and Mechanisms

Exosomes derived from various cell types possess remarkable targeting capabilities and bioactive content, enabling them to function as modulators of tissue repair across complex pathological environments such as ischemia, fibrosis, and inflammation. In recent years, exosomes have been explored as "drug-like signaling vesicles" for the treatment of diverse urogenital conditions—including erectile dysfunction, renal impairment, male infertility, and urinary incontinence (summarized in Table 23.2).

Erectile Dysfunction (ED): Exosomes originating from corpus cavernosum smooth muscle cells or MSCs can restore endothelial and neural functions in penile tissue, attenuate fibrosis, and alleviate oxidative stress. In a rat model of arterial injury, MSC-Exos enhanced cavernous endothelium formation, increased eNOS/nNOS expression, reduced ROS, and improved ICP/MAP scores (Liu et al., 2019). Similarly, smooth muscle cell–derived exosomes inhibited corporal fibrosis and activated NO/cGMP signaling, thereby rescuing erectile function in diabetic ED rats (Song et al., 2020). A 2023 meta-analysis confirmed significant improvements in ICP/MAP, α-SMA, CD31, and other cavernosum structural markers following MSC-Exos treatment (Zhu et al., 2023).

Acute and chronic kidney injury: Stem cell-derived exosomes have demonstrated notable efficacy in ameliorating chronic kidney disease (CKD) and acute kidney injury (AKI). In CKD models, bone marrow MSC-derived exosomes (BMSC-Exos) reduced proteinuria and serum creatinine while upregulating the anti-aging protein Klotho, thus preserving glomerular structure and delaying disease progression (Wan et al., 2022). In diabetic kidney disease (DKD), exosomal miR-99b-5p modulated the mTOR/autophagy axis, thereby reducing fibrosis and glomerular hypertrophy (Li et al., 2025). For AKI, exosomes from endometrial regenerative cells promoted M2 macrophage polarization and reduced Th1/Th17 inflammation via ERK/p38 signaling, mitigating tubular necrosis and cytokine release (Shao et al., 2025). In ischemic models, engineered exosomes enhanced renal homing and delivered miR-125b-5p to suppress p53 signaling, improving tubular regeneration and function (Cao et al., 2021). Together, these findings underscore exosomes' capacity to mitigate inflammation, oxidative stress, apoptosis, and fibrosis, while promoting renal repair through immunomodulation and targeted molecular delivery.

Male infertility: Emerging evidence underscores the therapeutic potential of exosomes in male infertility. In a busulfan-induced non-obstructive azoospermia (NOA) model, exosomes derived from human umbilical cord mesenchymal stem cells (hUCMSC-Exos) restored spermatogenesis by promoting germ cell proliferation and reducing apoptosis. This was mediated by upregulation of germ cell-specific markers (VASA, DAZL, Stra8) and junctional proteins (β-catenin, Connexin 43),

TABLE 23.2
Summary of Studies Evaluating Exosome Therapy in Urogenital Conditions

Condition	Exosome Source	Route of Administration	Key Findings	References
Erectile dysfunction	Bone marrow MSCs	Intracavernosal injection	Enhanced cavernous endothelium formation, increased eNOS/nNOS expression, reduced ROS, and improved ICP/MAP scores.	(Liu et al., 2019) (Zhu et al., 2023)
	Corpus cavernosum SMCs	Intracavernosal injection	Inhibited corporal fibrosis and activated NO/cGMP signaling.	(Song et al., 2020)
Acute kidney injury	Endometrial regenerative cell	Intravenous injection	Promoted M2 macrophage polarization; Suppressed Th1/Th17 cytokines; Protected against ischemic AKI.	(Shao et al., 2025)
Chronic kidney disease	Bone marrow MSCs	Intravenous injection	Reduced proteinuria; restored Klotho expression; Delayed glomerulosclerosis in CKD mouse models.	(Wan et al., 2022)
Male infertility	Umbilical cord MSCs	Testicular injection	Restored spermatogenesis; Upregulated germ cell markers (VASA, DAZL); Inhibited Bax/Caspase-3.	(Yue et al., 2024)
	Amniotic fluid-derived	Testicular injection	Improved testicular architecture and sperm count; Reduced oxidative damage	(Mobarak et al., 2021)
Urinary incontinence	Urine-derived stem cells	Local injection to the pelvic floor	Activated satellite cells via ERK phosphorylation; Improved leak point pressure and bladder capacity.	(Wu et al., 2019)
	Adipose-Derived Stem Cell	Periurethral injection	Enhanced striated muscle regeneration; Modulated PI3K-Akt and Wnt pathways.	(Ni et al., 2018) (Dinescu et al., 2021) (Liu et al., 2018)

Abbreviations: CKD: chronic kidney disease; MSCs: mesenchymal stem cells; SMCs: smooth muscle cells.

as well as suppression of pro-apoptotic pathways (Bax/Caspase-3) (Yue et al., 2024). Amniotic fluid-derived exosomes (AF-Exos) significantly improved testicular architecture and partially restored sperm production by attenuating oxidative damage and enhancing cellular survival mechanisms (Mobarak et al., 2021). Furthermore, endogenous seminal exosomes—including epididymosomes and prostasomes—play essential roles in sperm maturation and motility. These vesicles transfer proteins, such as CRISPR1, CD9, and tetraspanins, that modulate capacitation, the acrosome reaction, and sperm-egg interaction, thereby improving fertilization potential (Baskaran et al., 2020). Collectively, these findings highlight exosomes' capacity to reduce apoptosis and oxidative stress, support germ cell proliferation, and enhance sperm maturation through paracrine signaling and targeted biomolecule delivery, offering a promising acellular strategy for treating male infertility.

Stress urinary incontinence: Preclinical studies have demonstrated that exosomes derived from urine-derived stem cells (USC-Exos) and adipose-derived stem cells (ADSC-Exos) significantly facilitate recovery from stress urinary incontinence (SUI). In a rat model of SUI induced by vaginal

balloon dilation and pudendal nerve transection, local USC-Exos injection enhanced activation, proliferation, and differentiation of pubococcygeus satellite cells via ERK phosphorylation, leading to improved leak point pressure and bladder capacity (Wu et al., 2019). Meanwhile, ADSC-Exos delivered non-invasively to the urethra increased bladder capacity and leak point pressure, restored striated muscle mass and peripheral nerve fibers, and outperformed stem cell therapy, primarily through the PI3K-Akt, JAK-STAT, and Wnt signaling pathways, which support muscle and nerve regeneration (Ni et al., 2018; Dinescu et al., 2021). Additionally, ADSC-Exos modulated collagen metabolism in vaginal fibroblasts by upregulating TIMP-1 and TIMP-3 and downregulating MMP-1 and MMP-2, thereby promoting extracellular matrix remodeling, which is critical for SUI recovery (Liu et al., 2018). Collectively, these findings highlight exosomes' ability to enhance muscle and nerve regeneration, remodel the extracellular matrix, and improve urinary function through ERK-mediated satellite cell activation, modulation of signaling pathways, and anti-fibrotic effects, establishing them as a potent cell-free therapy for SUI. Beyond therapeutic applications, exosomes are also emerging as critical non-invasive biomarkers in urogenital medicine. Their cargo, including RNAs, proteins, and lipids, can be detected in body fluids such as urine, plasma, and semen, providing valuable insights for disease diagnosis, prognosis, and stratification (summarized in Table 23.3).
Prostate cancer: Exosomes derived from prostate cancer are considered highly promising non-invasive biomarkers, particularly those isolated from urine and plasma. Nilsson et al. (2009) reported that urinary exosomes derived from prostate cancer cells contain tumor-associated RNA molecules, including PCA3 and TMPRSS2-ERG fusion transcripts. Compared with conventional serum PSA testing, exosome-based detection demonstrated greater specificity and stability, providing a novel and reliable platform for non-invasive diagnostics.(Nilsson et al., 2009) Furthermore, Del Re et al. (2017) detected androgen receptor splice variant 7 (AR-V7) in plasma-derived RNA from patients with metastatic prostate cancer. They confirmed that its presence was closely associated with resistance to androgen receptor–targeted therapies, including enzalutamide and abiraterone.

TABLE 23.3
Current and Potential Applications of Exosomes in Urogenital Diagnostics

Disease/Condition	Exosomal Biomarkers	Biological/ Clinical Significance	References
Prostate cancer	PCA3; TMPRSS2:ERG; AR-V7	Non-invasive urinary biomarkers; Improved specificity over PSA; AR-V7 predicts resistance to androgen receptor signaling inhibitors.	(Nilsson et al., 2009) (Del Re et al., 2017)
Bladder cancer	miRs (miR-93-5p, miR-21-5p, miR-200-family)	Distinct urinary exosomal miRNA signatures enable discrimination between bladder cancer and non-cancer controls; Associated with tumor grade and stage.	(Kural et al., 2024)
Kidney disease (AKI/CKD/DKD)	miR-16-5p, miR-24-3p, miR-200c-3p (early AKI) miR-9a, miR-141, miR-200a, miR-200c, miR-429 (early recovery from AKI); CP; Elf3	Non-invasive and sensitive indicators of AKI onset and recovery (miRNA temporal signatures); CP is elevated before overt proteinuria; Elf3 indicates podocyte injury, an early biomarker of DKD.	(Li et al., 2022) (Wang et al., 2024)
Testicular/male infertility	miR-205-5p, miR-31-5p; Epididymosomes	Distinguish the origin of azoospermia (obstructive vs. non-obstructive).	(Barceló et al., 2018) (Vickram et al., 2021)
Urinary tract infections	Akt/PKB; CD9	Akt and CD9 may distinguish UTI from asymptomatic bacteriuria.	(Mizutani et al., 2019)

These findings indicate that exosomes can serve not only as diagnostic tools but also as predictive biomarkers of therapeutic response and prognosis. Compared with traditional PSA testing, exosome biomarkers offer significant advantages in terms of specificity, functional relevance, and dynamic monitoring (Del Re et al., 2017). Collectively, urinary and plasma-derived exosomes provide a robust and accessible platform for prostate cancer diagnostics and monitoring. Exosome-derived biomarkers, including PCA3, TMPRSS2-ERG, and AR-V7, offer greater specificity and dynamic monitoring capabilities than conventional methods and hold great promise for advancing precision medicine in prostate cancer.

Bladder cancer: MicroRNAs have emerged as promising non-invasive biomarkers for bladder cancer, reflecting tumor biology and disease progression. Among them, miR-93-5p, miR-21-5p, and members of the miR-200 family have attracted particular attention due to their consistent alteration across studies. MiR-93-5p promotes proliferation, invasion, and angiogenesis in bladder cancer by targeting tumor suppressors such as BTG2 and modulating EMT-related pathways. Its expression in urinary exosomes correlates with muscle-invasive disease and TNM stage, with a diagnostic accuracy (AUC $\approx$ 0.84) superior to that of conventional urine cytology. Moreover, when combined with other miRNAs in a panel, its performance is further enhanced, underscoring its role as a robust biomarker for early detection and monitoring of disease progression. MiR-21-5p is another well-validated oncogenic miRNA that is frequently overexpressed in the urinary exosomes of patients with bladder cancer. It promotes migration, proliferation, and resistance to apoptosis, and exhibits high diagnostic performance (sensitivity ~75%, specificity ~96%). Importantly, it remains detectable even in low-grade or cytology-negative cases, suggesting value for early diagnosis and surveillance. However, the potential for false positives in inflammatory conditions, such as cystitis, highlights the need for validation in larger cohorts. The miR-200 family (miR-200a, miR-200b, miR-200c, miR-141, and miR-429) plays a central role in regulating epithelial–mesenchymal transition (EMT) by targeting ZEB1 and ZEB2.

Dysregulation of these miRNAs in urinary exosomes has been linked with tumor progression, invasion, and grade. For example, the downregulation of miR-200a distinguishes muscle-invasive from non-muscle-invasive tumors, while the altered expression of miR-200b/c correlates with early tumor progression. Collectively, the miR-200 family provides insight into bladder cancer pathogenesis and offers potential as stage-specific urinary biomarkers (Kural et al., 2024). Together, these findings demonstrate that miR-93-5p, miR-21-5p, and the miR-200 family not only capture the molecular signatures of bladder tumors but also outperform traditional diagnostic tools in sensitivity and specificity. Their integration into diagnostic panels may significantly improve non-invasive detection, disease monitoring, and patient stratification in bladder cancer management.

Kidney disease: Exosomal microRNAs have been recognized as critical mediators and biomarkers in the diagnosis and monitoring of acute kidney injury (AKI) and chronic kidney disease (CKD). Their expression patterns reflect both the etiology of renal injury and the dynamic changes that occur during injury and recovery. In infection-related AKI, urinary exosomal miR-21 is markedly elevated, showing a positive correlation with systemic inflammatory activity and an inverse association with renal function. This highlights its potential role as a non-invasive indicator for the early detection of AKI in infectious contexts. In experimental models of hypoxia-induced tubular injury, renal tubular epithelial cell–derived exosomal miR-20a-5p demonstrated protective functions by reducing serum creatinine and blood urea nitrogen, enhancing endothelial proliferation, and preventing renal cell apoptosis, thereby supporting renal recovery. Distinct temporal patterns of urinary exosomal microRNAs further characterize the progression of AKI. During the acute injury phase, elevated levels of miR-16-5p, miR-24-3p, and miR-200c-3p have been observed, whereas in the recovery stage, miR-9a, miR-141, miR-200a, miR-200c, and miR-429 become upregulated. These recovery-associated microRNAs converge on molecular targets, including ZEB1 and ZEB2, which are central regulators of TGF-β1 signaling and play pivotal roles in renal fibrosis.

Such differential expression patterns suggest that exosomal miRs may serve not only as biomarkers for the onset and resolution of AKI but also as mechanistic participants in the development of fibrosis and long-term kidney remodeling (Li et al., 2022). Beyond AKI, exosomal proteins have proven highly informative in CKD and diabetic kidney disease (DKD). Urinary exosomal ceruloplasmin (CP) is consistently elevated—up to 10–20 times higher in CKD patients than in healthy individuals—and its rise precedes the development of overt proteinuria, positioning it as a sensitive early biomarker of disease progression. Similarly, the transcription factor Elf3, uniquely detected in the urinary exosomes of patients with diabetic nephropathy, reflects irreversible podocyte injury and provides a valuable non-invasive indicator of early DKD pathology (Wang et al., 2024). Together, these findings illustrate that exosomes serve as a versatile diagnostic platform in kidney diseases, bridging early detection, mechanistic insight, and prognostic evaluation.

Testicular/male infertility: *Exosome-derived miRNAs have garnered increasing attention as non-invasive diagnostic biomarkers in male infertility, particularly in cases of* azoospermia. While traditional clinical indicators such as serum FSH levels and testicular volume remain insufficient to reliably distinguish obstructive azoospermia (OA) from secretory azoospermia (SA), seminal plasma exosomes provide a rich source of molecular signals that closely reflect testicular and epididymal physiology. Among these, miR-205-5p and miR-31-5p stand out for their diagnostic utility. Both exhibit distinct expression patterns in OA versus SA, with miR-31-5p demonstrating particularly excellent predictive performance. Its sensitivity and specificity for differentiating OA from SA have exceeded 90% in predictive models, and when integrated with conventional endocrine markers such as serum FSH, diagnostic accuracy is further improved. miR-205-5p, another key marker, adds complementary value by enhancing the discriminatory power between etiological subtypes of azoospermia. Together, these molecules form a strong basis for predictive models that can identify patients with realistic chances of sperm retrieval, thereby minimizing unnecessary testicular biopsies (Barceló et al., 2018). In addition to RNA cargo, epididymosomes—a specialized subpopulation of exosomes secreted by the epididymal epithelium—play a critical role in male fertility and may provide further diagnostic insights.

Epididymosomes are enriched in specific proteins such as CD9, CD26, and CD224, which mediate protein transfer to viable sperm, as well as ELSPBP1, which interacts with non-viable sperm to scavenge reactive oxygen species and protect functional cells. By delivering proteins and lipids, epididymosomes govern essential processes, including sperm motility, capacitation, and zona pellucida binding. Aberrant function or altered molecular content of epididymosomes may therefore serve as an additional layer of diagnostic information in male infertility (Vickram et al., 2021). Taken together, seminal plasma exosomal biomarkers—particularly miR-205-5p and miR-31-5p—combined with functional insights from epididymosomes, provide a powerful diagnostic platform. Their integration into clinical practice could improve patient stratification, reduce reliance on invasive procedures, and support precision medicine approaches in reproductive health.

Urinary tract infections: Accurate differentiation between urinary tract infection (UTI) and asymptomatic bacteriuria (ABU) remains a diagnostic challenge, as both conditions are commonly associated with pyuria and bacteriuria. Conventional markers such as urine WBC counts or bacterial culture lack specificity. Urinary exosomes have recently been explored as a promising alternative, reflecting host cellular responses to bacterial presence. Among exosomal candidates, Akt and CD9 have shown strong potential as diagnostic markers. Their levels are significantly elevated in the urinary exosomes of UTI patients compared with those of ABU, indicating the activation of host intracellular signaling pathways and enhanced exosome release from leukocytes and uroepithelial cells during infection. Importantly, exosomal Akt demonstrates excellent discriminative power, with high sensitivity and specificity. At the same time, CD9 provides additional supportive value, underscoring the diagnostic advantage of focusing on the exosomal fraction rather than whole urine (Mizutani et al., 2019). These findings suggest that urinary exosomal Akt and CD9 could serve as reliable, non-invasive markers for distinguishing UTI from ABU. Beyond diagnosis, they also provide a molecular window into host–pathogen interactions in the urinary tract, which may support a better understanding of infection dynamics. Larger clinical studies will be required to validate their utility and explore their potential in differentiating between cystitis and pyelonephritis, ultimately improving diagnostic precision and patient management.

23.4 THE SECRETOME: A SYMPHONY OF SOLUBLE BIOACTIVE FACTORS

23.4.1 Definition and Composition of the Secretome

The secretome comprises the full array of soluble bioactive molecules secreted by cells into their extracellular environment, including growth factors (e.g., VEGF, FGF, TGF-β), cytokines, chemokines, extracellular matrix (ECM) proteins, enzymes, lipids, nucleic acids, and extracellular vesicles such as microvesicles and exosomes (Li et al., 2024b). These components actively mediate paracrine signaling that orchestrates a range of regenerative processes (Xia et al., 2021).

23.4.2 Role of the Secretome in Tissue Regeneration

The secretome plays a fundamental role in orchestrating tissue regeneration by creating a pro-healing microenvironment through the coordinated action of various soluble factors (Habib et al., 2025b). These bioactive molecules—such as growth factors, cytokines, chemokines, and extracellular matrix regulators—act synergistically to promote cell proliferation, guide differentiation, facilitate migration, and remodel the ECM, all of which are critical to effective tissue repair. Growth factors, including FGF-2, HGF, IGF-1, and VEGF, stimulate the proliferation of both resident and recruited cells, thereby accelerating tissue regeneration (Hodge et al., 2023). For instance, conditioned media (CM) derived from mesenchymal stem cells (MSCs) have been shown to enhance keratinocyte and fibroblast proliferation, significantly improving wound closure in preclinical skin models (Kwon et al., 2023). Recent studies demonstrated that secretomes from pluripotent stem cell-derived smooth muscle progenitors can modulate extracellular matrix metabolism in the lower urinary tract and vagina, enhancing ECM turnover and providing a scaffold for tissue repair. Such secretomes are rich in growth factors, matrix metalloproteinases (MMPs), and ECM proteins, thereby promoting balanced matrix remodeling, which is essential for structural restoration (Zhuang et al., 2021). In addition to its roles in proliferation and ECM remodeling, the secretome also facilitates cell migration and homing through the chemokines and growth factors it contains. SDF-1 present in umbilical cord MSC secretomes promotes endothelial cell chemotaxis and neovessel formation (Burlacu et al., 2013). Collectively, by coupling enhanced proliferation, targeted migration, and ECM turnover, the secretome serves as a potent, cell-free strategy for regenerating urogenital tissues.

23.4.3 Conditioned Media as a Source of Secretome

While the therapeutic value of the secretome has been attributed to its rich repertoire of bioactive molecules, conditioned media (CM) serves as a practical and clinically translatable source of these components. CM is collected from the culture supernatant of stem or progenitor cells—such as mesenchymal stem cells (MSCs) or induced pluripotent stem cell-derived smooth muscle progenitors (pSMCs)—after serum-free incubation, and contains a broad spectrum of soluble regenerative factors, including growth factors, cytokines, chemokines, and ECM regulators. Building upon the mechanisms described above, CM has been applied directly in various preclinical models to promote structural and functional restoration in urogenital tissues. In models of stress urinary incontinence (SUI), peri-urethral delivery of pSMC-CM significantly improved tissue function and continence. This effect was demonstrated by increased leak point pressure and structural normalization in the urethra and vaginal wall.(Zhuang et al., 2021) Similarly, bone marrow MSC-CM (BMSC-CM) was shown to accelerate urethral repair in SUI rats by promoting fibroblast proliferation and matrix deposition (Jiang et al., 2021). Adipose-derived MSC-CM enhanced collagen synthesis in fibroblasts from women with SUI, reducing collagen degradation (Liu et al., 2018). These underscore CM as a practical, non-cell-based therapeutic tool that delivers a complex mixture of regenerative factors directly to urogenital injury sites.

23.4.4 Examples of Secretome Components and Their Urogenital Therapeutic Potential

Vascular endothelial growth factor (VEGF), a central pro-angiogenic molecule found in MSC-conditioned media, promotes endothelial proliferation, neovascularization, and improved tissue perfusion (Sun et al., 2019). In rodent pelvic floor injury models, MSC-CM-enhanced VEGF significantly increased vascular density, facilitating urethral functional recovery via improved blood supply and tissue remodeling (Jiang et al., 2021).

Insulin-like growth factor-1 IGF-1: IGF-1 is a multifunctional growth factor that plays a central role in promoting cell proliferation, differentiation, and angiogenesis (Jiang et al., 2010; Haider et al., 2008). In the context of stress urinary incontinence (SUI), studies have highlighted the therapeutic utility of muscle-derived stem cells (MDSCs) genetically modified to overexpress IGF-1. This approach not only enhances the myogenic potential of MDSCs but also improves sphincter muscle regeneration, angiogenesis, and extracellular matrix remodeling at the injection site.(Chen et al., 2022).

Neurotrophic factors, such as BDNF and NGF, are critical components of the mesenchymal stem cell (MSC) secretome and play a crucial role in supporting neuronal survival and regeneration. In a study by Kim et al., three-dimensionally cultured human MSC-conditioned medium (MSC-CM) was applied to a rat model of erectile dysfunction (ED) induced by bilateral cavernous nerve injury (Kim et al., 2022).

ECM modulators: MMPs and TIMPs are key secreted proteins that regulate ECM turnover and structural remodeling. The presence of matrix metalloproteinases (MMP-2, MMP-9) and their endogenous inhibitors, TIMP-1 and TIMP-2, enables controlled ECM degradation and renewal, which is particularly important for restoring tissue integrity in the urogenital tract. In secretomes derived from smooth muscle progenitor cells, MMPs and TIMPs work in concert to regulate the metabolism of collagen and elastin in urethral and vaginal tissues. Controlled MMP activity facilitates the removal of damaged matrix and the release of embedded growth factors, while TIMPs ensure that degradation remains balanced and does not compromise tissue structure.(Zhuang et al., 2021).

Together, these diverse bioactive components of the secretome act synergistically to orchestrate a multifaceted regenerative response in urogenital tissues. By promoting angiogenesis, neurogenesis, ECM remodeling, and cellular survival, secretome-based therapies offer a potent acellular alternative to traditional regenerative approaches. Understanding the specific roles and mechanisms of these factors is essential for optimizing therapeutic strategies and advancing clinical translation in the treatment of conditions such as erectile dysfunction, urinary incontinence, and pelvic floor injury.

23.5 CHALLENGES AND FUTURE DIRECTIONS IN EXOSOME AND SECRETOME-BASED THERAPIES

23.5.1 Standardization of Isolation and Characterization Techniques

One of the primary challenges in the clinical translation of exosome- and secretome-based therapies is the lack of standardized techniques for isolating and characterizing these biomarkers. Currently, various methods, including ultracentrifugation, size-exclusion chromatography, polymer-based precipitation, and microfluidic technologies, are used to isolate exosomes; however, these approaches often yield heterogeneous populations with varying purity and bioactivity (Konoshenko et al., 2018). This variability can significantly impact therapeutic outcomes and the reproducibility of results across studies. Moreover, robust and reproducible protocols are urgently needed to ensure the quality and consistency of therapeutic preparations. Parameters such as vesicle size, concentration, surface markers (e.g., CD9, CD63, CD81), and cargo content (RNAs, proteins) must be consistently characterized using techniques like nanoparticle tracking analysis (NTA), flow cytometry, transmission electron microscopy (TEM), and omics-based profiling (Margolis et al., 2019). Without these standards, comparison between studies and regulatory approval remains a significant obstacle. Establishing international consensus and quality control frameworks, such as those proposed by the International Society for Extracellular Vesicles (ISEV), is crucial for advancing the clinical translation of these acellular therapies (Théry et al., 2018). Only through standardized protocols can we ensure the safety, efficacy, and scalability of exosome- and secretome-based products for urogenital applications.

23.5.2 Understanding Mechanisms of Action

Exosomes deliver bioactive cargo, including miRNAs, proteins, and lipids, via receptor binding, membrane fusion, or endocytosis, thereby modulating multiple cellular pathways (Safwan et al., 2025). For instance, MSC-derived exosomal miR-21a-5p targets phosphofructokinase muscle isoform (PFKM) to suppress glycolysis and attenuate renal fibrosis in unilateral ureteral obstruction (UUO) models. However, it remains unknown how this metabolic modulation interacts with anti-inflammatory and anti-apoptotic effects, or how timing and expression levels of miRNAs coordinate functional outcomes (Xu et al., 2022a). Additionally, the dose–response relationship between active components and biological effects is largely unexplored. Urine-derived stem cell exosomes enriched in miR-146a-5p target IRAK1 and suppress NF-κB signaling, mitigating renal ischemia–reperfusion injury (IRI) in rats. While functionally effective, a quantitative understanding of optimal dosing is lacking, hindering safe and effective clinical translation (Li et al., 2020).

23.5.3 Optimization of Delivery Strategies

Efficient delivery of exosomes and secretome components to specific urogenital tissues is pivotal for therapeutic success. Various administration routes have been explored, each with unique advantages and limitations (Habib et al., 2025a). Local injections, such as intracavernous injection for erectile dysfunction or intraurethral delivery for urinary incontinence, allow for direct targeting of affected tissues, maximizing local concentration while minimizing systemic exposure. For instance, local injection of MSC-derived exosomes into the corpus cavernosum has significantly improved erectile function in diabetic rat models (Song et al., 2020). Systemic administration, including intravenous or intraperitoneal injection, offers a non-invasive alternative; however, it often suffers from off-target distribution and rapid clearance by the reticuloendothelial system. To overcome these limitations, researchers have engineered exosomes with surface ligands or peptides that enhance tissue-specific targeting. For example, exosomes modified with the RVG peptide have demonstrated targeted delivery to the nervous system, a concept that is adaptable to the pelvic nerve plexus in urology (Li et al., 2024b). Additionally, hydrogel-based delivery systems have gained increasing attention in urogenital regenerative therapy due to their biocompatibility, injectability, and ability to sustain the localized release of exosomes or secretome components. A recent study demonstrated that adipose-derived stem cell (ADSC) exosomes, when embedded in an injectable thermo-sensitive hydrogel, significantly improved erectile function in a rat model of cavernous nerve injury.

The hydrogel formulation enhanced exosome retention at the site of nerve damage, facilitated neuroregeneration and angiogenesis, and promoted functional recovery (Liu et al., 2023). Moreover, the integration of image-guided delivery technologies has emerged as a promising approach to enhance the spatial precision of exosome administration. A recent study introduced a photoacoustic imaging-guided system in which ADSC exosomes were embedded within a polydopamine-modified thermosensitive hydrogel. This platform enabled accurate intratunical injection into the corpus cavernosum under photoacoustic guidance, thereby improving exosome localization and retention in penile tissue. In a model of erectile dysfunction, this strategy significantly promoted angiogenesis and tissue regeneration, offering a feasible solution for targeted delivery to deep pelvic structures (Liang et al., 2022). Emerging delivery modalities such as exosome-loaded microneedles, nanoparticle carriers, and tissue-specific scaffolds further enhance spatial control and bioavailability. The combination of these approaches with advanced targeting ligands and imaging guidance holds great promise for improving therapeutic outcomes in urogenital regenerative medicine.

23.5.4 Scalability and Manufacturing

Scalable manufacturing remains a bottleneck for exosome- and secretome-based therapies. Standard 2D cultures yield minimal exosomes—often insufficient for clinical dosing. 3D bioreactors, microcarrier-based dynamic culture, and treatments such as hypoxia or mechanical stimulation can be employed to enhance exosome secretion efficiency, thereby increasing overall yield markedly (Ng et al., 2022). Isolation and purification techniques also present scalability bottlenecks. Ultracentrifugation, while widely used, is labor-intensive and time-consuming, with low recovery rates. In contrast, emerging methods such as tangential flow filtration (TFF) and size-exclusion chromatography (SEC) offer higher throughput, better reproducibility, and improved purity of exosome preparations (Busatto et al., 2018, Hua et al., 2023).

23.5.5 Regulatory Considerations

Acellular therapies—such as exosomes- and secretome-based products—are fundamentally distinct from traditional cell therapies, and their regulatory landscape is rapidly evolving. In the United States, EV-based therapies are considered investigational new drugs (INDs) and are regulated as biological medicinal products by the FDA's Center for Biologics Evaluation and Research (CBER). To initiate clinical trials, an IND application must be submitted that includes preclinical data, manufacturing protocols, the clinical study design, and the qualifications of the investigators. Once IND approval is granted and clinical trials are completed, a Biologics License Application (BLA) is required to obtain market authorization, with evidence demonstrating the product's quality, safety, purity, and potency (DiStefano et al., 2022). In Europe, regulatory bodies such as the European Medicines Agency (EMA) are also actively establishing frameworks to address challenges related to batch consistency, safety, potency, and quality control of EV-based therapies (Lener et al., 2015). The establishment of robust regulatory pathways is essential to ensuring patient safety and facilitating the clinical translation and global accessibility of these promising acellular therapies in urogenital medicine.

23.5.6 Future Research Opportunities

The future of exosome and secretome-based therapies in urogenital medicine holds significant promise, but key areas require further exploration to ensure safe and effective clinical translation (Haider & Aramini, 2020). One crucial direction is the systematic implementation of well-designed clinical trials to evaluate safety, dosage, and efficacy in conditions such as urinary incontinence and erectile dysfunction (Zhu et al., 2023; Yue et al., 2024). Despite the encouraging results from preclinical studies, human trials remain sparse and often limited in scale and design. Another pivotal research frontier lies in identifying reliable exosomal biomarkers (Xu et al., 2022b). Identifying predictive and diagnostic exosomal biomarkers, particularly those reflecting early-stage urogenital pathologies or therapeutic response, could enhance precision medicine and treatment monitoring. Additionally, the bioengineering of exosomes has emerged as a rapidly evolving field of research. Innovative strategies such as surface functionalization, encapsulation of therapeutic RNAs or peptides, and integration with biomaterial scaffolds have been developed to enhance the targeting specificity, drug-loading efficiency, and local retention of exosomes. These advancements aim to overcome the

TABLE 23.4
Challenges and Future Directions in Exosome and Secretome Research for Urogenital Medicine

Challenge/Research Area	Description/Key Points	Potential Strategies/Future Directions	References
Standardization	Heterogeneous exosome isolation, variable purity, and bioactivity.	Adopt standardized protocols; Robust characterization (size, markers, cargo).	(Konoshenko et al., 2018) (Margolis et al., 2019) (Théry et al., 2018)
Mechanisms of action	Incomplete understanding of cargo effects; Dose–response unclear.	Mechanistic studies; Optimize dosing and timing.	(Xu et al., 2022a); (Li et al., 2020)
Delivery optimization	Off-target distribution; Rapid clearance.	Surface engineering, hydrogel/ scaffold delivery, image-guided targeting.	(Song et al., 2020); (Li et al., 2024b)(Liu et al., 2023) (Liang et al., 2022)
Scalability	Low yield from 2D culture; Labor-intensive isolation.	3D culture, bioreactors; Scalable purification (TFF, SEC).	(Ng et al., 2022); (Busatto et al., 2018);(Hua et al., 2023)
Regulatory	Complex regulatory landscape; Quality control required.	Early regulatory engagement; Standardize manufacturing.	(DiStefano et al., 2022) (Lener et al., 2015)
Future research	Sparse clinical trials; Biomarker and ngineering gaps.	Conduct trials; identify biomarkers; Bioengineer exosomes; Precision medicine.	(Zhu et al., 2023);(Yue et al., 2024);(Xu et al., 2022b) (Huang et al., 2023)

inherent limitations of native exosomes and improve their therapeutic utility in clinical applications (Huang et al., 2023). In summary, integrating clinical validation, biomarker development, and bioengineering approaches will be pivotal to realizing the full therapeutic potential of exosomes and secretome products in urogenital regenerative medicine (Table 23.4).

23.6 CONCLUSION

23.6.1 The Promising Future of Acellular Therapies in Urogenital Medicine

Acellular therapies, particularly exosomes and the secretome, have emerged as transformative modalities in the treatment of urogenital diseases. These biologically active components, derived from stem or progenitor cells, have demonstrated remarkable potential to promote tissue regeneration, modulate inflammation, and restore organ function across a spectrum of conditions, including bladder dysfunction, erectile dysfunction, and prostate diseases. Compared to conventional cell-based therapies, exosomes and secretome-based treatments offer several key advantages, including reduced immunogenicity, greater storage and transport stability, and the elimination of risks associated with uncontrolled cell proliferation or differentiation. Their nano-scale size and bioactive cargo make them highly amenable to targeted engineering, potentially enabling precision therapies tailored to patient-specific pathologies. Despite these advantages, several challenges remain, including standardizing isolation methods, scaling up manufacturing, and developing a comprehensive understanding of pharmacokinetics and biodistribution. However, rapid advancements in extracellular vesicle (EV) biology, biomaterials science, and translational research continue to accelerate progress toward clinical application. In conclusion, exosome and secretome-based strategies represent a paradigm shift in regenerative medicine. While regulatory, technical, and biological hurdles persist, their unique therapeutic profile and growing body of supporting evidence underscore their promise as a future cornerstone in the management of urogenital disorders.

REFERENCES

Ahmed ZT, Alkahlot MH, Haider KH. MSC-derived exosomes: advances in cell-free therapy. In: Haider, K.H. (eds) *Handbook of Stem Cell Applications*. Singapore: Springer Nature, 2023 Aug 18, pp. 1–41.

Alvarez-Erviti L, Seow Y, Yin H, Betts C, Lakhal S, Wood MJ: Delivery of siRNA to the mouse brain by systemic injection of targeted exosomes. *Nature Biotechnology* 2011, 29:341–5.

Armstrong JP, Holme MN, Stevens MM: Re-engineering extracellular vesicles as smart nanoscale therapeutics. *ACS Nano* 2017, 11:69–83.

Barceló M, Mata A, Bassas L, Larriba S: Exosomal microRNAs in seminal plasma are markers of the origin of azoospermia and can predict the presence of sperm in testicular tissue. *Human Reproduction* 2018, 33:1087–98.

Baskaran S, Panner Selvam MK, Agarwal A: Exosomes of male reproduction. *Advances in Clinical Chemistry* 2020, 95:149–63.

Burlacu A, Grigorescu G, Rosca AM, Preda MB, Simionescu M: Factors secreted by mesenchymal stem cells and endothelial progenitor cells have complementary effects on angiogenesis in vitro. *Stem Cells and Development* 2013, 22:643–53.

Busatto S, Vilanilam G, Ticer T, Lin WL, Dickson DW, Shapiro S, Bergese P, et al.: Tangential flow filtration for highly efficient concentration of extracellular vesicles from large volumes of fluid. *Cells* 2018, 7(12): 273.

Cao JY, Wang B, Tang TT, Wen Y, Li ZL, Feng ST, Wu M, et al.: Exosomal miR-125b-5p deriving from mesenchymal stem cells promotes tubular repair by suppression of p53 in ischemic acute kidney injury. *Theranostics* 2021, 11:5248–66.

Charoenviriyakul C, Takahashi Y, Nishikawa M, Takakura Y: Preservation of exosomes at room temperature using lyophilization. *International Journal of Pharmaceutics* 2018, 553:1–7.

Chen C, Xu Y, Song Y: IGF-1-induced muscle-derived stem cells as a potential treatment of stress urinary incontinence in female rats. *Urology Journal* 2022, 19:451–9.

Del Re M, Biasco E, Crucitta S, Derosa L, Rofi E, Orlandini C, Miccoli M, et al.: The detection of androgen receptor splice variant 7 in plasma-derived exosomal RNA strongly predicts resistance to hormonal therapy in metastatic prostate cancer patients. *European Urology* 2017, 71:680–7.

Dinescu S, Dobranici A, Tecucianu R, Selaru A, Balahura R, Ignat S, Costache M: Exosomes as part of the human adipose-derived stem cells secretome- opening new perspectives for cell-free regenerative applications. *Advances in Experimental Medicine and Biology* 2021, 1312:139–63.

Ding Y, Li Y, Sun Z, Han X, Chen Y, Ge Y, Mao Z, et al.: Cell-derived extracellular vesicles and membranes for tissue repair. *Journal of Nanobiotechnology* 2021, 19(1):368.

DiStefano TJ, Vaso K, Danias G, Chionuma HN, Weiser JR, Iatridis JC: Extracellular vesicles as an emerging treatment option for intervertebral disc degeneration: Therapeutic potential, translational pathways, and regulatory considerations. *Advanced healthcare materials* 2022, 11:e2100596.

Elahi FM, Farwell DG, Nolta JA, Anderson JD: Preclinical translation of exosomes derived from mesenchymal stem/stromal cells. *Stem Cells* (Dayton, Ohio) 2020, 38:15–21.

Foo JB, Looi QH, Chong PP, Hassan NH, Yeo GEC, Ng CY, Koh B, et al.: Comparing the therapeutic potential of stem cells and their secretory products in regenerative medicine. *Stem Cells International* 2021, 2021:2616807.

Gurunathan S, Kang MH, Jeyaraj M, Qasim M, Kim JH: Review of the isolation, characterization, biological function, and multifarious therapeutic approaches of exosomes. *Cells* 2019, 8(4): 307.

Gurung S, Perocheau D, Touramanidou L, Baruteau J: The exosome journey: From biogenesis to uptake and intracellular signalling. *Cell Communication and Signaling* 2021, 19:47.

Ha DH, Kim HK, Lee J, Kwon HH, Park GH, Yang SH, Jung JY, et al.: Mesenchymal stem/stromal cell-derived exosomes for immunomodulatory therapeutics and skin regeneration. *Cells* 2020, 9(5): 1157.

Habib SM, Martini MF, Al Ahmed O, Qassab AR, Haider KH: The potential of mesenchymal stem cell–derived exosomes in cell-free therapy for myocardial repair. In: Haider, K.H. (eds) *Handbook of Regenerative Medicine: Cell-Free Therapy*. London: Taylor & Francis Group/CRC Press, 2025a, Chapter 24, pp. 372–90.

Habib SM, Martini MF, Al Ahmed O, Qassab AR, Haider KH: Mesenchymal stem cell secretomes for cell-free therapy of the broken heart. In: Haider, K.H. (eds) *Handbook of Regenerative Medicine: Cell-Free Therapy*. London: Taylor & Francis Group CRC Press, 2025b, Chapter 18, pp. 274–293.

Habib SY, Martini MF, Hamdan YNA, Shrebaty OMM, Haider KH: Mesenchymal stem cell secretome and exosomes as potential advanced therapy medicinal products for treating a failing heart. *World Journal of Stem Cells*. 2026; 18(3) (in Press).

Haider KH, Aramini B: Mircrining the injured heart with stem cell-derived exosomes: An emerging strategy of cell-free therapy. *Stem Cell Research & Therapy*. 2020 Jan 9;11(1):23. https://doi.org/10.1186/s13287-019-1548-7.

Haider KH, Aslam M: Cell-free therapy with stem cell secretions: Protection, repair and regeneration of the injured myocardium. In: Haider, K.H. and Aziz, S. (eds) *Stem Cells: From Hype to Real Hope*. Berlin: Medicine & Life Sciences, De Gruyter StraBe 13-10785 Berlin, Germany. 2018. https://doi.org/10.1515/9783110642438.

Haider KH, Aziz S: Paracrine hypothesis and cardiac repair. *International Journal of Stem Cells*. 2017; 5(1): 265–7.

Haider KH, Najimi M: Exosome-based cell-free therapy in regenerative medicine for myocardial repair. In: Haider, K.H. (eds) *Handbook of Stem Cell Therapy*. Singapore: Springer. (pp. 1007–1034; 2022) https://doi.org/10.1007/978-981-16-6016-0_9-1.

Haider KH, Jiang S, Niagara MI, Ashraf M: IGF-1 overexpressing mesenchymal stem cells accelerate bone marrow stem cell mobilization via paracrine activation of SDF-1α/CXCR4 signaling to promote myocardial repair. *Circulation Research*. 2008; 103: 1300–8.70.

Hodge JG, Robinson JL, Mellott AJ: Tailoring the secretome composition of mesenchymal stem cells to augment specific functions of epidermal regeneration: an in vitro diabetic model. *Frontiers in Medical Technology* 2023, 5:1194314.

Hu Q, Su H, Li J, Lyon C, Tang W, Wan M, Hu TY: Clinical applications of exosome membrane proteins. *Precision Clinical Medicine* 2020, 3:54–66.

Hu W, Wang W, Chen Z, Chen Y, Wang Z: Engineered exosomes and composite biomaterials for tissue regeneration. *Theranostics* 2024, 14:2099–126.

Hua X, Zhu Q, Liu Y, Zhou S, Huang P, Li Q, Liu S: A double tangential flow filtration-based microfluidic device for highly efficient separation and enrichment of exosomes. *Analytica Chimica Acta* 2023, 1258:341160.

Huang L, Wu E, Liao J, Wei Z, Wang J, Chen Z: Research advances of engineered exosomes as drug delivery carrier. *ACS Omega* 2023, 8:43374–87.

Isaac R, Reis FCG, Ying W, Olefsky JM: Exosomes as mediators of intercellular crosstalk in metabolism. *Cell Metabolism* 2021, 33:1744–62.

Jeyaram A, Jay SM: Preservation and storage stability of extracellular vesicles for therapeutic applications. *The AAPS Journal* 2017, 20:1.

Jiang M, Liu J, Liu W, Zhu X, Bano Y, Liao H, Li H, et al.: Bone marrow stem cells secretome accelerates simulated birth trauma-induced stress urinary incontinence recovery in rats. *Aging* 2021, 13:10517–34.

Jiang S, Ashraf M, Haider KH: Bone marrow stem cells mobilized in response to ex vivo transgenic overexpression of IGF-1 differentiate into functioning cardiomyocytes and contribute to myocardial regeneration. *Circulation* Nov. 2010, 122.

Joshi BS, de Beer MA, Giepmans BNG, Zuhorn IS: Endocytosis of extracellular vesicles and release of their cargo from endosomes. *ACS Nano* 2020, 14:4444–55.

Kalluri R, LeBleu VS: The biology, function, and biomedical applications of exosomes. *Science (New York, NY)* 2020, 367(6478): eaau6977.

Kim SG, You D, Kim K, Aum J, Kim YS, Jang MJ, Moon KH, et al.: Therapeutic effect of human mesenchymal stem cell-conditioned medium on erectile dysfunction. *The World Journal of Men's Health*, 2022, 40:653–62.

Konoshenko MY, Lekchnov EA, Vlassov AV, Laktionov PP: Isolation of extracellular vesicles: General methodologies and latest trends. *BioMed Research International* 2018, 2018:8545347.

Kural S, Jain G, Agarwal S, Das P, Kumar L: Urinary extracellular vesicles-encapsulated miRNA signatures: A new paradigm for urinary bladder cancer diagnosis and classification. *Urologic Oncology* 2024, 42:179–90.

Kwon JW, Savitri C, An B, Yang SW, Park K: Mesenchymal stem cell-derived secretomes-enriched alginate/extracellular matrix hydrogel patch accelerates skin wound healing. *Biomaterials Research* 2023, 27:107.

Lee YJ, Shin KJ, Chae YC: Regulation of cargo selection in exosome biogenesis and its biomedical applications in cancer. *Experimental & Molecular Medicine* 2024, 56:877–89.

Lener T, Gimona M, Aigner L, Börger V, Buzas E, Camussi G, Chaput N, et al.: Applying extracellular vesicles-based therapeutics in clinical trials – An ISEV position paper. *Journal of Extracellular Vesicles* 2015, 4:30087.

Li Q, Zhang Z, Yin M, Cui C, Zhang Y, Wang Y, Liu F: What do we actually know about exosomal microRNAs in kidney diseases? *Frontiers in Physiology* 2022, 13:941143.

Li R, Tao H, Pan K, Li R, Guo Z, Chen X, Li Z: Extracellular vesicles derived from mesenchymal stem cells alleviate renal fibrosis via the miR-99b-5p/mTOR/autophagy axis in diabetic kidney disease. *Stem Cell Research & Therapy* 2025, 16:142.

Li X, Liao J, Su X, Li W, Bi Z, Wang J, Su Q, et al.: Human urine-derived stem cells protect against renal ischemia/reperfusion injury in a rat model via exosomal miR-146a-5p, which targets IRAK1. *Theranostics* 2020, 10:9561–78.

Li X, Zhang D, Yu Y, Wang L, Zhao M: Umbilical cord-derived mesenchymal stem cell secretome promotes skin regeneration and rejuvenation: From mechanism to therapeutics. *Cell Proliferation* 2024a, 57:e13586.

Li Y, Chen Z, Guo J, Meng D, Pang X, Sun Z, Pu L, et al.: Enhanced brain-targeting and efficacy of cannabidiol via RVG-Exo/CBD nanodelivery system. *Biochemical and Biophysical Research Communications*, 2024b, 725: 150260.

Liang L, Shen Y, Dong Z, Gu X: Photoacoustic image-guided corpus cavernosum intratunical injection of adipose stem cell-derived exosomes loaded polydopamine thermosensitive hydrogel for erectile dysfunction treatment. *Bioactive Materials* 2022, 9:147–56.

Liang Y, Xia J, Xu L, Zhang Y, Xu X, Haider KH: Cell-free and cell-based therapy in regenerative medicine. In: Haider, K.H. (eds) *Handbook of Regenerative Medicine: Cell- Free Therapy Approach* (1st ed.). CRC Press, 2025. https://doi.org/10.1201/9781003611509.

Liu S, Li R, Dou K, Li K, Zhou Q, Fu Q: Injectable thermo-sensitive hydrogel containing ADSC-derived exosomes for the treatment of cavernous nerve injury. *Carbohydrate Polymers* 2023, 300:120226.

Liu X, Li T, Zhang J, Lin X, Wang W, Fan X, Wang L: Mesenchymal stem cell-based therapy for female stress urinary incontinence. *Frontiers in Cell and Developmental Biology* 2023, 11:1007703.

Liu X, Wang S, Wu S, Hao Q, Li Y, Guo Z, Wang W: Exosomes secreted by adipose-derived mesenchymal stem cells regulate type I collagen metabolism in fibroblasts from women with stress urinary incontinence. *Stem Cell Research & Therapy* 2018, 9:159.

Liu Y, Zhao S, Luo L, Wang J, Zhu Z, Xiang Q, Deng Y, et al.: Mesenchymal stem cell-derived exosomes ameliorate erection by reducing oxidative stress damage of corpus cavernosum in a rat model of artery injury. *Journal of Cellular and Molecular Medicine* 2019, 23:7462–73.

Margiana R, Pilehvar Y, Amalia FL, Lestari SW, Supardi S, I'Tishom R: Mesenchymal stem cell secretome: A promising therapeutic strategy for erectile dysfunction? *Asian Journal of Urology* 2024, 11:391–405.

Margolis L, Sadovsky Y: The biology of extracellular vesicles: The known unknowns. *PLoS Biology* 2019, 17:e3000363.

Mizutani K, Kawakami K, Horie K, Fujita Y, Kameyama K, Kato T, Nakane K, et al.: Urinary exosome as a potential biomarker for urinary tract infection. *Cellular Microbiology* 2019, 21:e13020.

Mobarak H, Heidarpour M, Rahbarghazi R, Nouri M, Mahdipour M: Amniotic fluid-derived exosomes improved spermatogenesis in a rat model of azoospermia. *Life Sciences* 2021, 274:119336.

Moghassemi S, Dadashzadeh A, Sousa MJ, Vlieghe H, Yang J, León-Félix CM, Amorim CA: Extracellular vesicles in nanomedicine and regenerative medicine: A review over the last decade. *Bioactive Materials* 2024, 36:126–56.

Mushtaq M, Zineldeen DH, Mateen MA, Haider KH. Mesenchymal stem cells' "garbage bags" at work: Treating radial nerve injury with mesenchymal stem cell-derived exosomes. *World Journal of Stem Cells* 2024, 16(5). https://dx.doi.org/10.4252/wjsc.v16.i5.0000.

Ng CY, Kee LT, Al-Masawa ME, Lee QH, Subramaniam T, Kok D, Ng MH, et al.: Scalable production of extracellular vesicles and its therapeutic values: A review. *International Journal of Molecular Sciences* 2022, 23(14):7986. https://doi.org/10.3390/ijms23147986.

Ni J, Li H, Zhou Y, Gu B, Xu Y, Fu Q, Peng X, et al.: Therapeutic potential of human adipose-derived stem cell exosomes in stress urinary incontinence – An in vitro and in vivo study. *Cellular Physiology and Biochemistry* 2018, 48:1710–22.

Nilsson J, Skog J, Nordstrand A, Baranov V, Mincheva-Nilsson L, Breakefield XO, Widmark A: Prostate cancer-derived urine exosomes: A novel approach to biomarkers for prostate cancer. *British Journal of Cancer* 2009, 100:1603–7.

Pianta S, Bonassi Signoroni P, Muradore I, Rodrigues MF, Rossi D, Silini A, Parolini O: Amniotic membrane mesenchymal cells-derived factors skew T cell polarization toward Treg and downregulate Th1 and Th17 cell subsets. *Stem Cell Reviews and Reports* 2015, 11:394–407.

Safwan M, Bourgleh MS, Al-Ruqi A, Shrebaty O, Almujaydil F, Alothaim B, Alrashdi N, Haider KH. Living bio-drugs MSCs and their derivative exosomes to rescue ventilated patients with ARDS: A systematic review and meta-analysis. *World Journal of Stem Cells*. 2025, 17(10):111978.

Shao B, Wang HD, Ren SH, Chen Q, Wang ZB, Xu YN, Liu T, et al.: Exosomes derived from a mesenchymal-like endometrial regenerative cells ameliorate renal ischemia reperfusion injury through delivery of CD73. *Stem Cell Research & Therapy* 2025, 16:148.

Shen AR, Zhong X, Tang TT, Wang C, Jing J, Liu BC, Lv LL: Integrin, exosome and kidney disease. *Frontiers in Physiology* 2020, 11:627800.

Song J, Sun T, Tang Z, Ruan Y, Liu K, Rao K, Lan R, et al.: Exosomes derived from smooth muscle cells ameliorate diabetes-induced erectile dysfunction by inhibiting fibrosis and modulating the NO/cGMP pathway. *Journal of Cellular and Molecular Medicine* 2020, 24:13289–302.

Sun DZ, Abelson B, Babbar P, Damaser MS: Harnessing the mesenchymal stem cell secretome for regenerative urology. *Nature Reviews Urology* 2019, 16:363–75.

Tan F, Li X, Wang Z, Li J, Shahzad K, Zheng J: Clinical applications of stem cell-derived exosomes. *Signal Transduction and Targeted Therapy* 2024, 9:17.

Théry C, Witwer KW, Aikawa E, Alcaraz MJ, Anderson JD, Andriantsitohaina R, Antoniou A, et al.: Minimal information for studies of extracellular vesicles 2018 (MISEV2018): A position statement of the International Society for Extracellular Vesicles and update of the MISEV2014 guidelines. *Journal of Extracellular Vesicles* 2018, 7:1535750.

Vickram AS, Srikumar PS, Srinivasan S, Jeyanthi P, Anbarasu K, Thanigaivel S, Nibedita D, et al.: Seminal exosomes – A crucial biological marker for various disorders and syndromes in human reproduction. *Saudi Journal of Biological Sciences* 2021, 28:3607–15.

Wan F, Yang RC, Tang YW, Tang XL, Ye T, Zheng J, Zhang HQ, et al.: BMSC-derived exosomes protect against kidney injury through regulating klotho in 5/6 nephrectomy rats. *European Journal of Medical Research* 2022, 27:118.

Wang L, Wang J, Xu A, Wei L, Pei M, Shen T, Xian X, et al.: Future embracing: exosomes driving a revolutionary approach to the diagnosis and treatment of idiopathic membranous nephropathy. *Journal of Nanobiotechnology* 2024, 22:472.

Wang Z: Assessing tumorigenicity in stem cell-derived therapeutic products: A critical step in safeguarding regenerative medicine. *Bioengineering (Basel, Switzerland)* 2023, 10.

Wu R, Huang C, Wu Q, Jia X, Liu M, Xue Z, Qiu Y, et al.: Exosomes secreted by urine-derived stem cells improve stress urinary incontinence by promoting repair of pubococcygeus muscle injury in rats. *Stem Cell Research & Therapy* 2019, 10:80.

Xia J, Arai S, Haider KH: Research and applications of stem cell secretome. In: Haider K.H. (eds) *Stem Cells: From Potential to Promise*. Singapore: Springer, 2021. https://doi.org/10.1007/978-981-16-0301-3_7.

Xu S, Cheuk YC, Jia Y, Chen T, Chen J, Luo Y, Cao Y, et al.: Bone marrow mesenchymal stem cell-derived exosomal miR-21a-5p alleviates renal fibrosis by attenuating glycolysis by targeting PFKM. *Cell Death & Disease* 2022a, 13:876.

Xu YX, Pu SD, Li X, Yu ZW, Zhang YT, Tong XW, Shan YY, et al.: Exosomal ncRNAs: Novel therapeutic target and biomarker for diabetic complications. *Pharmacological Research* 2022b, 178:106135.

Yin X, Li Q, Shu Y, Wang H, Thomas B, Maxwell JT, Zhang Y: Exploiting urine-derived induced pluripotent stem cells for advancing precision medicine in cell therapy, disease modeling, and drug testing. *Journal of Biomedical Science* 2024, 31:47.

Yue D, Wang F, Han Y, Xiong C, Yang R: Exosomes derived from umbilical cord mesenchymal stem cells ameliorate male infertility caused by busulfan in vivo and in vitro. *Ecotoxicology and Environmental Safety* 2024, 272:116063.

Zhang Y, Liu Y, Liu H, Tang WH: Exosomes: biogenesis, biologic function and clinical potential. *Cell & Bioscience* 2019, 9:19.

Zhu Y, Jiang T, Yao C, Zhang J, Sun C, Chen S, Chen M: Effects of stem cell-derived exosome therapy on erectile dysfunction: a systematic review and meta-analysis of preclinical studies. *Sexual Medicine* 2023, 11:qfac019.

Zhuang G, Wen Y, Briggs M, Shao Q, Tran D, Wang H, Chen B: Secretomes of human pluripotent stem cell-derived smooth muscle cell progenitors upregulate extracellular matrix metabolism in the lower urinary tract and vagina. *Stem Cell Research & Therapy* 2021, 12:228.

24 Advances in Cell-Derived Exosomes and Secretomes from Bench to Bedside in Oncology

Kazem Mashayekhi[1], Navid Ghasemzadeh[2], Gaida AlTamimi[3], Nouf Sufian[3], Maysam AlSkikhi[3], Aima Mustafa Bhatti[4], Mohammad Abdur Rahman[5], Maryam Rahnama[6], and Khawaja H. Haider[3]

[1]Department of Immunology, School of Medicine, Rafsanjan University of Medical Sciences, Rafsanjan, Iran

[2]Department of Clinical Biochemistry, School of Medicine, Urmia University of Medical Sciences, Urmia, Iran

[3]Department of Basic Sciences, Sulaiman AlRajhi Medical School, Kingdom of Saudi Arabia

[4]Sargent College, Boston University, Massachusetts, USA

[5]Withybush General Hospital, Fishguard Road, Haverfordwest, UK

[6]Immunology of Infectious Diseases Research Center, Research Institute of Basic Medical Sciences, Rafsanjan University of Medical Sciences, Rafsanjan, Iran

24.1 LIST OF ABBREVIATIONS

AFP	Alpha-fetoprotein
AI	Artificial intelligence
ATMPs	Advanced therapeutic medicinal products
AUC	Area under the receiver operating characteristic curve
BBB	Blood-brain barrier
bFGF	Basic fibroblast growth factor
CAFs	Cancer-associated fibroblasts
CAR	Chimeric antigen receptor
CE	Carcinoembryonic antigen
CLTC	Clathrin heavy chain
CPIs	Checkpoint inhibitors
CTLs	Cytotoxic T lymphocytes
DC	Dendritic cell

DOI: 10.1201/9781003653387-24

Del-1	Developmental endothelial locus-1
Dex	Dendritic cell-derived exosomes
EMA	European medicines agency
EMT	Epithelial-mesenchymal transition
EpCAM	Epithelial cell adhesion molecule
ER	Endoplasmic reticulum
EPR	Enhanced permeability and retention
ESCRT	Endosomal sorting complex required for transport
EVs	Extracellular vesicles
EXO-D	Exosomal DNA
EXO-IncRNA	Exosomal IncRNA
EXO-miR	Exosomal miRNAs
ECM	Extracellular matrix
EZR	Ezrin
FDA	Food and Drug Administration
FasL	Fas ligand
GKN1	EXO-Gastrokine 1
GMP	Good Manufacturing Practice
GPC-1	Glypican-1
HCC	Hepatocellular carcinoma
HER2	Human epidermal growth factor receptor 2
IFNγ	Interferon-gamma
IFNγ-Dex	Dendritic cell-derived exosomes treated with interferon-gamma
IL-10	Interleukin-10
ILVs	Intraluminal vesicles
IExosomes	Exosomes loaded with KRASg12dg-specific siRNA
IV	Intravenous
LncRNAs	Long non-coding RNAs
LUAD	Lung adenocarcinoma
MISEV	Minimal information for studies of extracellular vesicles
ML	Machine learning
MiRNAs	MicroRNAs
Mctx	Metronomic cyclophosphamide
mRNA	Messenger RNA
MSN	Moesin
MSCs	Mesenchymal stem cell
MVBs	Multivesicular bodies
MVs	Microvesicles
NIIC	Nanoplasmonic immunoassay integrated cell-on-a-chip
NK	Natural killer
NSCLC	Non-small cell lung cancer
OS	Overall survival
PEG	Polyethylene glycol
PFS	Progression-free survival
PLK	Polo-like kinases
PMDA	Pharmaceuticals and the medical devices agency
PSA	Prostate-specific antigen
TAAs	Tumor-associated antigens
TAM	Tumor-associated macrophages
TDEs	Tumor-derived exosomes
TFF	Tangential flow filtration

TGF-β	Tumor growth factor-beta
TLN1	Talin-1
TME	Tumor microenvironment
Tregs	T regulatory cells
VEGF	Vascular endothelial growth factor
siRNA	Small interfering RNA

24.2 INTRODUCTION

The worldwide impact of cancer remains a significant public health concern, necessitating ongoing efforts to deepen our understanding of its complex biology and to develop innovative diagnostic and therapeutic approaches (Li and Kuang 2024). Even with considerable advancements in cancer treatment—from surgical interventions and chemotherapy to radiotherapy—we still face persistent hurdles like drug resistance, disease recurrence, and metastasis (Kaur et al., 2023). These limitations have prompted the search for innovative therapeutic strategies to improve patient outcomes. In recent years, stem cell-derived exosomes and secretomes have become a focal point in cancer research, offering promising new possibilities for improving the detection, treatment, and monitoring of cancer progression (Habib et al., 2025a & 2025b). Their unique properties have led to innovative approaches in oncology, helping scientists and clinicians develop more effective strategies for patient care (Hanahan, 2022).

The cellular secretome represents the entire spectrum of biologically active molecules released by cells into their microenvironment, orchestrating complex intercellular signaling networks (Habib et al., 2025a). This dynamic repertoire comprises not only soluble proteins, lipids, and metabolites but also an extensive range of extracellular vesicles (EVs), including exosomes and nanovesicles such as exomeres (Ritchie et al., 2021).

Among these EVs, exosomes have garnered significant interest owing to their unique biogenesis, specialized cargo, and remarkable ability to enhance intercellular interaction. Once considered the cell's waste disposal system, they are now recognized as key players in intercellular communication (Mushtaq et al., 2024). They carry a heterogeneous payload of bioactive substances, including proteins (such as growth factors, enzymes, signaling proteins, and heat shock proteins), lipids (such as cholesterol, sphingomyelin, and ceramides), and nucleic acids, particularly messenger RNA (mRNAs), miRNAs, long non-coding RNAs (lncRNAs), and DNA fragments (Lee et al., 2024). Upon release, exosomes circulate through various physiological fluids, including blood, lymph, cerebrospinal fluid, and urine, and interact with recipient cells via receptor-mediated signaling, direct fusion, or endocytosis to deliver their payload (Waqas et al., 2022). Although exosomes constitute the insoluble subset of the secretome, the comparatively broader landscape of the secretome offers a more comprehensive understanding of the interaction between cells and their microenvironment. It includes microvesicles, apoptotic bodies, and soluble substances such as cytokines, chemokines, growth regulators, matrix-degrading enzymes, and extracellular matrix (ECM) components (López de Andrés et al., 2020).

24.3 THE ROLE OF SOLUBLE AND INSOLUBLE SECRETOME IN CANCER BIOLOGY

The secretome has a multifaceted role in cancer biology, influencing nearly all aspects of tumorigenesis. Cancer-derived exosomes and soluble factors carry oncogenic signals, promote tumor proliferation, remodel the ECM, and facilitate metastatic dissemination. Specific exosomal miRNAs have been shown to reduce cell proliferation, suppress apoptosis, and promote the epithelial-mesenchymal transition (EMT), a crucial mechanism for cancer cell migration and dissemination (Mashouri et al., 2019; Aramini et al., 2020b). The secretome produced by tumors plays a decisive role in reshaping the tumor microenvironment (TME) (Aramini et al., 2020a). This environment comprises a diverse mix of cells, including not only cancer cells but also cancer-associated fibroblasts (CAFs), endothelial and perivascular cells, various immune cells, and the ECM (Ritchie et al., 2021). Like normal cells, tumor cells release exosomes that can influence nearby stromal and immune cells, effectively "teaching" them to create conditions that promote tumor growth and help evade immune detection (Armani et al., 2022). For example, these exosomes can carry pro-angiogenic factors, such as vascular endothelial growth factor (VEGF) and basic fibroblast growth factor (bFGF), which stimulate the formation of new blood vessels. This process, known as neovascularization, ensures that the tumor receives the oxygen and nutrients it needs to thrive (Yang et al., 2025).

Additionally, tumor exosomes can inhibit immune surveillance by downregulating the activity of cytotoxic T lymphocytes and natural killer (NK) cells, polarizing macrophages toward the tumor-promoting M2 phenotype, and expanding regulatory T cell populations (Morini et al., 2025). Additionally, soluble immunomodulatory cytokines, such as tumor growth factor-beta (TGF-β) and interleukin-10 (IL-10), contribute to immune escape and stromal remodeling, thereby fostering a pro-inflammatory and pro-angiogenic niche (Yang et al., 2025). Exosomes and secretome components also play a central role in establishing pre-metastatic niches, a prerequisite for successful metastasis, the principal cause of cancer mortality. Tumor-derived exosomes (TDEs) can prime distant tissues by enhancing vascular permeability, inducing local inflammation, and modulating the ECM, thereby creating a receptive microenvironment before the arrival of circulating tumor cells (Guo et al., 2019). For instance, exosomes from pancreatic cancer cells expressing specific integrins preferentially fuse with Kupffer cells in the liver, triggering upregulation of fibronectin and immune activation, thereby supporting metastatic colonization (Casari et al., 2021).

A further challenge in oncology is the emergence of treatment resistance, wherein the secretome plays a detrimental role. Exosomes can transfer drug-resistance mediators, such as P-glycoprotein, between cells, enabling previously sensitive cells to survive chemotherapy. They also transport miRNAs and proteins that modulate apoptotic pathways or activate survival signaling cascades. Soluble components of the secretome can similarly interfere with drug efficacy by activating compensatory pathways, altering drug absorption, or restructuring the TME to prevent therapeutic penetration (Yang et al., 2022). Despite these challenges, the unique properties of exosomes and the components of the secretome present significant opportunities for innovative evaluation and treatment approaches. Their stability in biological fluids, coupled with molecular fingerprints reflective of their tumor origin, positions them as ideal tools for non-invasive liquid biopsy approaches (Yadav et al., 2024a).

Analyzing exosomal contents derived from biological fluids may provide prompt information on tumor presence, classification, stage, heterogeneity, and therapeutic response. Specific exosomal miRNAs, proteins (including Glypican-1 in pancreatic cancer, CD63, and TSG101), and DNA mutations have been identified as possible biomarkers for many tumors, including lung, breast, prostate, and colorectal cancers (Makler and Asghar, 2020; Yang et al., 2024). Such biomarkers could revolutionize early detection, guide personalized therapy, and monitor treatment response or recurrence. While exosome-based diagnostics are advancing, existing clinical tools such as circulating tumor DNA and protein markers (e.g., prostate-specific antigen (PSA), cancer antigen (CA-125)) remain in use, bolstered by innovations in proteomics and metabolomics (Ghosh et al., 2022). The integration of exosomal and soluble secretome markers may yield comprehensive, high-resolution diagnostic platforms (Yu et al., 2022; Zhou et al., 2024).

The integration of exosomal and soluble biomarkers could provide the most thorough and precise diagnostic and predictive insights (Aramini et al., 2020b). The therapeutic potential of exosomes and the secretome is being investigated using a range of new approaches (Wang et al., 2024c). Exosomes possess a natural ability to carry bioactive compounds, exhibit biocompatibility, demonstrate low immunogenicity, and inherently cross biological barriers, such as the blood-brain barrier (BBB), making them appealing nanocarriers for targeted drug delivery (Liang et al., 2021). Despite significant advances, achieving scalable exosome production that meets Good Manufacturing Practice (GMP) standards remains a critical barrier to their clinical translation. Consistent batch quality, reproducibility, and regulatory compliance are ongoing challenges that must be addressed to facilitate the widespread therapeutic application of these products (Rahnama et al., 2024).

Exosomes are not mere passive messengers; they are active substances that alter the tumor microenvironment, accelerate cancer spread, and make it harder for treatments to work. This complex interaction, on the other hand, also reveals numerous possibilities. The unique molecular markers found in exosomes hold significant potential to improve next-generation liquid biopsies for early cancer detection, prognosis, and treatment evaluation. Moreover, exosomes' inherent biocompatibility and cargo-transporting capacity position them as promising vectors for targeted drug delivery. In contrast, the therapeutic potential of native or manipulated exosomes and secretome components offers novel avenues for cancer treatment. This chapter aims to provide a comprehensive examination of this rapidly evolving field. As investigations clarify the complexities of these nanoscale effectors and the sophisticated biochemical language they use, we move closer to fully harnessing their potential in the diverse fight against cancer.

24.4 EXOSOME BIOGENESIS AND ITS CARGO

The biogenesis of the secretome involves the synthesis and secretion of a diverse array of proteins and bioactive molecules by cells into the extracellular space through both classical and non-classical secretory pathways, as well as EV-mediated mechanisms (Lorey, 2017). Critically, the secretome encompasses various classes of EVs: exosomes, which originate from multivesicular body fusion; microvesicles, which bud directly from the plasma membrane; and larger apoptotic bodies, released during programmed cell death (Haider & Najimi, 2022). These distinct biogenic pathways collectively determine the rich and dynamic composition of the secretome, which faithfully mirrors the physiological state and functional programming of the secreting cells (Rahnama et al., 2024). In the classical secretory route, nascent proteins are meticulously directed into the endoplasmic reticulum (ER) lumen. Here, they undergo crucial post-translational modifications, including the precise cleavage of signal peptides, meticulous folding into correct tertiary structures, and strategic glycosylation. This initial targeting is mediated by an N-terminal signal sequence that accurately guides ribosomes to the ER during polypeptide synthesis. Following this intricate ER processing, these modified proteins are systematically transported to the Golgi apparatus for further refinement before being efficiently secreted into the extracellular environment via exocytosis (Colombo et al., 2014).

In non-classical secretion mechanisms, specific molecules, such as growth factors and cytokines, which play key roles in immune responses, angiogenesis, cell growth, and differentiation, can be released without involving the ER–Golgi complex. This release can occur directly through the cell membrane or encapsulation in EVs (Gurunathan et al., 2021). They significantly contribute to the regulation of cellular signaling pathways and tissue repair. These diverse vesicles are broadly categorized based on their unique biogenesis and characteristic size (Najimi & Haider, 2022).

Microvesicles (MVs), the largest EV subtype, range in size from 100 to 1000 nm and are formed by the outward budding of the plasma membrane. They transport their cargo to target cells through interactions involving microtubules and SNARE proteins. Exosomes are small vesicles, typically 30–150 nm in diameter, whose complex biogenesis and unique molecular profiles closely mirror the physiological or pathological conditions of their parent cells. These vesicles originate from the endosomal system, where inward budding of the endosomal membrane generates intraluminal vesicles (ILVs) within multivesicular bodies (MVBs). The formation of ILVs within MVBs occurs through two main pathways: one dependent on the endosomal sorting complex required for transport (ESCRT) machinery and another that is independent of ESCRT (Han et al., 2022). The ESCRT system facilitates inward budding and effectively encapsulates cytoplasmic material. Meanwhile, specific lipids and tetraspanin proteins, such as CD9, CD63, and CD81, play essential roles in ESCRT-independent ILV formation. When MVBs fuse with the plasma membrane, these ILVs are released into the extracellular environment as exosomes. This endosomal origin imparts exosomes with a distinct molecular signature, notably enriched in specific tetraspanins and signaling molecules, which sets them apart from other EV types (Han et al., 2022). Consequently, sophisticated analytical methods are crucial for accurately identifying exosome subpopulations and differentiating them from other EVs, including apoptotic bodies and microvesicles (Youssef et al., 2025). EVs utilize various uptake mechanisms to enter recipient cells, such as micropinocytosis, caveolin-dependent pathways, clathrin-mediated endocytosis, and direct fusion with the plasma membrane. Exosomes, in particular, exhibit high specificity in delivering their cargo, a feature driven by their surface markers and lipid composition, which facilitate selective targeting and internalization by recipient cells. Conversely, microvesicles carry a more diverse cargo and exhibit less selective interactions with cells (Van De Wakker et al., 2023).

Functionally, exosomes serve as critical mediators of intercellular communication by transferring their cargo, which modulates numerous physiological and pathological processes in the recipient cells (Moon et al., 2022; Rahnama et al., 2024). In particular, exosomes influence cell signaling pathways by delivering bioactive molecules that can alter gene expression and cellular behavior in target cells (Van De Wakker et al., 2023).

Within the intricate context of the TME, exosomes play multifaceted and often detrimental roles that significantly contribute to cancer progression and metastasis (Gabaran et al., 2025; Li et al., 2021). TDEs facilitate complex communication between cancer cells and various stromal cells, thereby promoting processes vital to tumor survival, such as angiogenesis (the formation of new blood vessels), immune evasion (which helps cancer cells evade detection and destruction by the immune system), and ECM remodeling. They can carry oncogenic factors and immunosuppressive molecules that directly modulate immune cell behavior, thereby aiding tumor cells in escaping immune surveillance (Moeinzadeh et al., 2022). Furthermore, exosomes are active participants in preparing distant metastatic niches by conditioning target tissues, making them more receptive and hospitable to circulating tumor cells (Li et al., 2024a). The ability of exosomes to influence the TME highlights their potential as biomarkers for cancer diagnosis and prognosis, as well as therapeutic targets or delivery vehicles for anti-cancer agents (Basyoni et al., 2025; Rahnama et al., 2025; Wang et al., 2022). Current research focuses on understanding the regulatory mechanisms of exosome biogenesis and uptake to develop strategies for manipulating exosome-mediated communication in cancer treatment (Han et al., 2022).

24.5 ISOLATION AND CHARACTERIZATION OF EXOSOMES

One of the most critical steps in leveraging exosomes as biomarkers is isolating and purifying them from complex cell culture media in oncology research and from patients' biological fluids in clinical settings (Dilsiz et al., 2024). Given their extremely small size, complex membrane structure, and co-existence with a heterogeneous array of proteins, lipoproteins, and other extracellular vesicles, selecting an optimal isolation method remains a critical determinant of the purity and yield, and that too with intact payload, for theranostic applications of the recovered vesicles (Huang et al., 2019). Although there are a few standardized protocols, the presently available techniques, along with their respective advantages and limitations, have been summarized in Table 24.1.

Ultracentrifugation remains the most established and widely used method and serves as the gold standard for exosome isolation and purification. It primarily involves sequential centrifugation steps, starting with low-speed centrifugation to remove cells and their debris, followed by ultracentrifugation at forces greater than $100{,}000 \times g$ to pellet out the exosomes. A relatively clean exosome preparation is the primary strength of this technique. Still, its clinical utility is marred by its time-intensiveness and the mechanical stress on the vesicles, which may alter their structure or bioactivity. Building on ultracentrifugation, the density gradient technique yields a higher purity exosome product, but the steps involved are technically demanding and labor-intensive. Size-exclusion

TABLE 24.1
Summary of the Currently Available Isolation Techniques for Exosomes from Cell Culture Media and Body Fluids

Isolation Technique	Advantages	Limitations
Ultracentrifugation	• Gold standard method of isolation. • Well-documented and widely adopted. • Produces relatively clean preparations.	• Time-intensive • Needs access to costly ultracentrifuges. • Vesicles get subjected to mechanical stress that potentially alters their structure or bioactivity.
Density gradient centrifugation (DGC)	• DGC separates vesicles based on their buoyant density. Often using sucrose. • Yields vesicles of higher purity by removing protein aggregates and contaminating vesicles.	• Time-intensive and laborious. • Technically demanding • Better suited for laboratory-scale research than for high-throughput or clinical workflows.
Size exclusion chromatography (SEC)	• pass the samples through porous beads to fractionate particles by size. • SEC is a gentle method that preserves vesicle integrity. • Preserves functional surface markers that help in their use for biomarker analysis.	• There is a capacity limitation—large sample volumes are difficult to process. • Time-intensive exercise with large volume samples. • Potential for co-isolation of similarly sized particles. • Purity of the isolate remains a concern.
Precipitation-based methods	• Availability of Commercial kits such as ExoQuick to isolate exosomes from biological fluids. • Simple, straightforward, and user-friendly. • These methods are quick, inexpensive, and user-friendly.	• Purity of the final product remains an issue. • Chance of non-exosomal entities in the final product, such as serum proteins, which may interfere with use as biomarkers.
Microfluidics & Immunoaffinity Capture	• Most modern and sophisticated approach. • Potential to process tiny volume samples. • High precision, specificity, and selectivity. • Significant potential for clinical applications.	• They are expensive • Yet to be standardized for routine use. • Product yield is in smaller quantities of exosomes.

Abbreviations: SEC: Size exclusion chromatography; DGC: Density gradient centrifugation.

chromatography and precipitation-based approaches are gentle on exosomes' delicate structure, as they preserve the end-product's structural integrity. Still, their capacity to handle large volumes and co-isolate the similarly sized non-exosome contaminants remains an issue when the end product is intended for therapeutic use. The newer microfluidic-based platforms and immunoaffinity capture approaches are gaining popularity (Wu et al., 2022). They have the potential to process small-volume samples with high precision, while the immunocapture method offers high sensitivity and precision, which enhances their clinical translation potential (Xing et al., 2025).

Once isolated and purified, the exosome products are rigorously characterized using morphological, physical, and molecular methods, either alone or in combination. For example, transmission electron microscopy (TEM) remains the gold standard for visualizing exosomes and their morphological features, providing high-resolution images that reveal the hallmark "cup-shaped" morphology. Similarly, vesicle size is expected to range from 30 to 150 nm, as determined by Dynamic Light

Scattering (DLS). At the same time, surface topology is mapped using atomic force microscopy (AFM) without extensive sample preparation. Another essential feature of exosome characterization is molecular marker identification by Tracking Analysis, which is widely used to determine the size distribution and concentration of isolated vesicles. Together, these sophisticated and advanced isolation, purification, and characterization approaches ensure exosome preparations are pure and reproducible, enabling reliable downstream theragnostic use. Together, these strategies not only enhance our understanding of tumor biology but also pave the way for translating exosome science from bench to bedside, bridging laboratory discovery with real clinical impact (Wang et al., 2022).

24.6 THERAPEUTIC POTENTIAL OF EXOSOMES IN CANCER

Exosomes can transform the way cancer is treated. This section summarizes published data on exosome loading, customized administration, and the reduction of drug resistance. It also discusses their importance for individualized oncology for the treatment and management of cancer patients (Zemanek et al., 2025).

24.6.1 Dendritic Cell and NK Cell-Derived Exosomes

Dendritic cell-derived exosomes (Dex) display antigens well due to the presence of MHC class I and II, co-stimulatory molecules including CD80 and CD86, and tumor-associated antigens (TAAs). When pulsed with tumor-specific antigens, Dex helps the immune system activate naive T cells and cytotoxic T lymphocytes (CTLs), thereby promoting their ability to recognize and eliminate cancer cells (Pitt et al., 2014). The fast-emerging exosome-based cancer vaccines harness Dex's antigen-presenting machinery to establish systemic anti-tumor immunity without allogeneic rejection or excessive growth associated with whole-cell vaccines. In preclinical melanoma models, Dex pulsed with melanoma-associated antigens, including MAGE-A3, induced robust T-cell responses, resulting in tumor shrinkage (Escudier et al., 2005). Dex-based vaccines for non-small cell lung cancer (NSCLC) and colorectal cancer have shown safety and may engage the immune system by enhancing T-cell infiltration and cytokine production. Novel protocols are being developed to stimulate DCs with toll-like receptor agonists to boost Dex production.

This process improves exosome production and their immunogenicity, thus rendering pancreatic cancer vaccines more effective (Besse et al., 2016). On the same note, NK cells also release exosomes that carry cytotoxic chemicals, including perforin, granzymes, and Fas ligand (FasL), which can potentially induce the immediate death of cancer cells (Jiang et al., 2021). Exosomes derived from NK cells offer a cell-free alternative to NK cell therapy. They facilitate the delivery of cytotoxic payloads while reducing the likelihood of immunological rejection (Raue et al., 2021).

24.6.2 Mesenchymal Stem Cell-Derived Exosomes

Mesenchymal stem cells (MSCs) have emerged as one of the most well-studied adult stem cell types (Álvarez-Viejo & Haider, 2022), and the first to reach the market as living biodrugs for routine patient use across different diseases (Bhatti & Haider, 2025). They are among the very few cell types that have reached advanced phases of clinical assessment for diverse clinical applications (Kalou et al., 2023). MSCs-derived secretome and exosomes can modify the immune system responsiveness depending on their payload profile and the TME (Ahmed et al., 2023; Habib et al., 2026). They can have pro-tumor and anti-tumor effects. For example, the presence of anti-tumor microRNAs in the payload, such as miR-145, miR-143, or miR-100, suppresses tumor growth by inhibiting pro-oncogenic signaling, slowing tumor growth, and preventing angiogenesis (Jiang et al., 2021; Sheykhhasan et al., 2021). Recent research has demonstrated that MSC exosomes engineered to deliver IL-12 can convert "cold" tumors into "hot" ones, making T cells more likely to attack them and increasing their sensitivity to anti-PD-1 therapy in melanoma models. However, careful engineering is needed to mitigate any pro-tumor consequences, such as the administration of TGF-β, which can render the immune system less active in certain situations (Moreno 2021).

24.6.3 Engineered Exosomes for Immunotherapy

Exosome engineering is a rapidly evolving area of research that aims to modulate the immune system. For example, they can be engineered to deliver immune checkpoint inhibitors (e.g., anti-PD-1 or anti-PD-L1 antibodies) or stimulatory cytokines (e.g., IL-2 or IL-12) to support existing immunotherapies. For example, exosomes modified to express anti-PD-L1 antibodies have altered the PD-1-PD-L1 interaction in the TME, thereby enhancing T-cell function in experimental colorectal cancer models (Tian et al., 2022). Exosomes from chimeric antigen receptor (CAR)-T cells offer a safer and more scalable alternative to cellular therapies for leukemia treatment. These exosomes, carrying bioactive molecules such as CAR proteins, selectively target and eliminate leukemia cells while reducing the risk of cytokine release syndrome, providing a promising and precise therapeutic approach (Kang et al., 2024; Sani et al., 2024).

Exosomes can alter the TME, thereby enhancing their immunomodulatory abilities. For instance, exosomes carrying miR-155 convert macrophages into an anti-tumor M1 phenotype, which counteracts the immunosuppressive effects of M2 macrophages in the TME. Exosomes can also deliver RNA-based drugs, such as siRNAs that target immunosuppressive genes like STAT3, to enhance immune activation in glioblastoma models. Given their ability to perform various functions, exosomes are excellent platforms for immunotherapy (Guo et al., 2020; Liang et al., 2022). Researchers are working to enhance the effectiveness of their cargo and surface modifications for clinical applications. Their ongoing efforts aim to push the boundaries of what's possible in healthcare, making treatments more efficient and impactful (Gabaran et al., 2025).

24.6.4 Exosome Loading Strategies

Exosomes are well-suited to serve as delivery vehicles for therapeutic agents due to their biocompatibility, low immunogenicity, and protective lipid bilayer, which enables them to carry bioactive molecules (Wang et al., 2023a). To maximize the therapeutic potential of exosomes, novel strategies have been developed to modify their cargo profile to meet therapeutic needs (Tenchov et al., 2022). These cargo loading strategies are broadly classified as passive and active loading strategies. While passive loading strategies leverage the lipid bilayer's inherent permeability, this approach is flexible and straightforward, making it suitable for loading a wide range of entities, including small-molecule drugs, nucleic acids, and peptides. However, this may lead to problems achieving high loading efficiency or maintaining exosome structural integrity during the loading process (Tenchov et al., 2022). On the contrary, active loading strategies require the parent cell to be genetically modified or primed to produce the cargo molecules, which are then loaded during exosome biogenesis. Complex cargos, such as therapeutic proteins, tumor-suppressive miRNAs, or gene-editing tools like CRISPR/Cas9, can be delivered using this method (Kim et al., 2021; Haider, 2024). As the method guarantees cargo loading via the biogenesis pathway, it preserves exosome biology while enhancing treatment outcomes (Liu et al., 2024).

24.6.5 Targeted Delivery Strategies

To maximize therapeutic efficacy and minimize off-target effects, exosomes can be engineered to deliver their payload with precision to cancer cells or the TME. Targeted delivery reduces the required therapeutic dose, widens the therapeutic window, minimizes systemic toxicity, and enhances drug accumulation at the target site, thereby improving therapeutic efficacy and reducing the side effects. Two primary approaches for targeted delivery can be broadly classified as passive and active targeting, each employing the unique properties of exosomes and advanced engineering techniques to achieve site specificity (Song et al., 2021). In the passive targeting approach, exosomes accumulate in tumor tissues due to the leaky vasculature and impaired lymphatic drainage characteristic of solid tumors. The nanoscale size (30–150 nm) facilitates their deep penetration into the tumor interstitium, thereby overcoming the barriers posed by larger nanoparticles (Tenchov et al., 2022). Recent studies have shown that the size and membrane composition of exosomes significantly contribute toward their enhanced permeability and retention (EPR)-mediated accumulation, with smaller exosomes (<100 nm) demonstrating superior tumor penetration in aggressive cancers, such as pancreatic ductal adenocarcinoma and glioblastoma (Vagena et al., 2025). On the contrary, active targeting involves exosomal surface modification to enable specific ligand-receptor binding on cancer cells or TME components. In other words, this approach uses molecular recognition to direct exosomes to their intended targets, thereby improving therapeutic outcomes (Liang et al., 2021). To this end, exosomes are chemically modified to attach targeting ligands using techniques such as click chemistry, avidin-biotin interactions, or direct conjugation (Tang et al., 2023). Recent studies have used pH-sensitive linkers to trigger cargo release in the acidic TME, thereby enhancing specificity for cancers such as melanoma. Additionally, polyethylene glycol (PEG) conjugation has been employed to improve exosome stability and prolong circulation time while presenting targeting ligands, such as transferrin, for brain tumor targeting (Wang et al., 2015).

Another strategy involves the use of aptamers, short single-stranded oligonucleotides that offer high affinity and specificity but low immunogenicity. Exosomes modified with aptamers, such as AS1411 targeting nucleolin in nucleolin-overexpressing breast cancer cells or PSMA-specific aptamers for prostate cancer, achieve precise targeting (Han et al., 2020). Recent research has explored aptamer-functionalized exosomes for siRNA targeting of oncogenic pathways in ovarian cancer, demonstrating enhanced specificity and therapeutic efficacy (Wu et al., 2024a).

24.7 OVERCOMING DRUG RESISTANCE

In cancer treatment, resistance to treatment and even relapse remain unresolved issues. By delivering medications that alter cellular pathways, exosomes offer a novel option for treating innate and acquired drug resistance, as they can enter tumor cells and the TME and deliver their payload (Ji et al., 2025). For example, in one study aimed at overcoming chemoresistance in ovarian cancer models, exosomes containing siRNAs targeting RAB27A delayed the emergence of resistant exosomes (Ren et al., 2024a). Similarly, exosomes can be modulated to carry CRISPR/Cas9 components to correct mutations that render drugs less effective or to silence genes that reduce their effectiveness, such as MDR1, BCL2, or EGFR. For example, exosomes carrying CRISPR/Cas9 targeting oncogenic KRAS mutations in pancreatic cancer can render cells significantly more sensitive to gemcitabine with minimal off-target effects (McAndrews et al., 2021). Recent research has also explored the use of exosomes to deliver base-editing tools for single-nucleotide mutations in lung cancer animal models, thereby making them sensitive to tyrosine kinase inhibitors again. This method appears promising for malignancies with clear genetic causes of resistance (Jiang et al., 2025; Song et al., 2024).

Besides resistance, the spread of drug resistance in the nearby cells adds to the problem. TDEs contain mRNAs, miRNAs, or long non-coding RNAs that encode drug efflux pumps that can render the neighboring cells resistant. For example, in ovarian cancer models, exosomes carrying siRNAs targeting RAB27A (an exosome release regulator) delayed the emergence of resistant exosomes, thereby abrogating chemoresistance. Similarly, neutralizing antibodies against exosomal surface markers, such as CD63, can also reduce chemoresistance by preventing their uptake by the recipient cells. In recent years, GW4869 has been explored as a means to suppress exosomes that contribute to breast cancer cells' resistance to chemotherapy (He et al., 2025; Ji et al., 2025).

24.8 REGULATORY AND ETHICAL CONSIDERATIONS IN EXOSOME THERAPY

Exosome-based therapies are poised to transform oncology by transitioning it from bench to bedside (Wang et al., 2024c). However, the rapid advancement in this field is accompanied by intricate regulatory and ethical challenges that must be addressed to ensure patient safety. Therefore, there is an urgent need for a regulatory framework distinct from the conventional drug regulatory framework (Rahnama et al., 2024). Exosome-based products are diverse in origin (e.g., MSCs, immune cells, or tumor cells), differ in function (e.g., drug delivery, immunomodulation, or regenerative therapy), and vary in modifications for use (Welsh et al., 2024; Youssef et al., 2025). Depending on these factors, exosomes may be classified as biopharmaceuticals (similar to vaccines or blood derivatives), pharmaceuticals (if their primary action involves a specific pharmacological effect), or medical devices (when integrated into delivery or diagnostic systems) (Wang et al., 2024a). This diversity in their classification necessitates a specific regulatory framework across global jurisdictions.

In the United States, exosome products are regulated as pharmaceuticals and biopharmaceuticals under the Food and Drug Administration (FDA), the Cosmetic Act, and the Public Health Service Act, requiring rigorous premarket review and approval by the FDA to ensure safety and efficacy (Wang et al., 2024a; Youssef et al., 2025). At the time of this writing, no exosome-based drug products have received FDA approval. Instead, there are warnings about unproven exosome products (U.S. Food & Drug Administration). The Japanese Pharmaceuticals and Medical Devices Agency (PMDA) classifies exosomes as biopharmaceuticals, aligning them with standards for vaccines and blood products (Tsuchiya et al., 2022; Youssef et al., 2025). The European Medicines Agency (EMA) distinguishes between exosomes containing functionally translated RNA (classified as advanced therapeutic medicinal products (ATMPs)) and those directly derived from cells (naïve exosomes) (Europen Mecines Agency), which are classified with specifications (Wang et al., 2024a; Youssef et al., 2025). In the absence of a consensus global classification system, developers face ambiguity, inconsistent regulatory pathways, prolonged approval timelines, and varying scrutiny methodologies (Youssef et al., 2025). For example, autologous exosomes derived from patient samples may face different regulatory requirements than engineered allogeneic exosomes with specific therapeutic payloads. Unification of these classifications is imperative to streamline development and prevent regulatory confusion (Wang et al., 2024a).

24.8.1 Good Manufacturing Practice and Quality Control

Adherence to the GMP standards is critical for producing uniform batches of clinical-grade exosome-based products that are safe and effective (Lin et al., 2022). Some critical manufacturing challenges include identical source cell lines, optimizing cell culture conditions to enhance exosome yield while maintaining consistent cargo profiles, and developing scalable purification methods (Witwer et al., 2021). Techniques such as tangential flow filtration (TFF), size-exclusion chromatography, and affinity purification enhance purity and minimize the co-isolation of contaminants, thereby addressing scalability and reproducibility issues (Wang et al., 2024). Consistent storage and stabilization protocols are also crucial for preserving product integrity over time (Ahmadian et al., 2024).

An in-depth quality control protocol is essential to ensure the identity, purity, potency, and safety of exosome-based preparations (Goo et al., 2024). This encompasses size distribution and cargo profiling, surface marker analysis, and performing functional assays (Théry et al., 2018). Advanced analytical methods, such as single-exosome analysis and multi-omics approaches, are increasingly necessary to achieve thorough characterization and ensure batch-to-batch consistency for clinical-grade products (Louro et al., 2024). According to the Minimal Information for Studies of Extracellular Vesicles (MISEV) guidelines 2023, a standardized framework has been developed emphasizing rigorous nomenclature, reproducibility, and transparent reporting to enhance study quality (Samuels and Giamas 2024; Welsh et al., 2024). These guidelines emphasize the importance of quality control in ensuring reliable, reproducible protocols, which, in turn, support the maintenance of ethical standards and facilitate clinical translation (Samuels and Giamas 2024).

24.8.2 Ethical Considerations

Ethical considerations are central to the responsible development and application of exosome-based therapies. Informed consent is paramount, particularly since exosomes are often derived from patient biofluids or cells (Youssef et al., 2025). Patients must be fully informed about the use of their biological materials, especially in contexts involving genetic modification or biobanking for future research. Clear communication about the experimental nature of exosome therapies is crucial to prevent therapeutic misconceptions, where patients may confuse research participation with established clinical care (Fujita et al., 2024). This is particularly critical in regions where unregistered exosome-based products are marketed (Fujita et al., 2024). High production costs, reliance on specialized infrastructure, and the need for highly trained personnel may restrict access to the products to resource-rich regions only, thereby exacerbating global healthcare disparities (Stawarska et al., 2024; Wang et al., 2024a; Youssef et al., 2025). Strategies to address this include developing cost-effective and scalable production methods tailored to diverse economic contexts (Rai et al., 2024). The growing trend of patients seeking unproven exosome-based products highlights the need for safe, regulated, and globally accessible treatment options to prevent exploitation and ensure equity (Lyons et al., 2022).

The integration of artificial intelligence (AI) in exosome research, particularly for biomarker discovery and cargo analysis, introduces additional ethical complexities (Greenberg et al., 2023). While AI enhances the identification of exosomal biomarkers, it risks perpetuating biases if training datasets lack diversity (Greenberg et al., 2023). For instance, algorithms trained primarily on data from specific ethnic groups may be less effective for other groups, potentially leading to disparities in diagnostic or therapeutic outcomes (Resnik and Hosseini, 2024). Transparent, explainable AI algorithms, accountability for outcomes, and inclusive data collection are crucial to ensuring equitable and accurate AI-driven exosome technologies (Bouhouita-Guermech et al., 2023; Youssef et al., 2025). Additionally, protecting patient data privacy and security in large-scale AI analyses for biomarker discovery is a critical ethical concern that requires robust safeguards (Kumar, 2024).

24.8.3 Fostering Public Trust

Maintaining public trust in exosome-based innovations is crucial, particularly given the proliferation of unproven products in unregulated markets (Youssef et al., 2025). Transparent communication about the scientific rigor, safety, and efficacy of these therapies, coupled with robust public education, is necessary to counter misinformation and differentiate validated research from unproven commercial offerings (Yadav et al., 2024b). A unified framework that prioritizes safety, transparency, and inclusivity, harmonizes global regulatory standards, and fosters interdisciplinary collaboration is essential to realize the transformative potential of exosome-based precision medicine as an equitable healthcare innovation (Samuels and Giamas, 2024; Youssef et al., 2025).

The regulatory and ethical landscape for exosome-based therapies in oncology is complex but pivotal to their safe and effective clinical translation. Harmonized global regulatory standards, robust GMP compliance, and advanced quality control measures are critical to ensure product safety and consistency. Ethical considerations—including informed consent, equitable access, and the responsible use of AI—must be prioritized to protect patients and foster public trust. By addressing these challenges, exosome therapies can fulfill their promise as innovative, accessible, and ethically sound solutions for cancer treatment.

24.9 EMERGING ROLE OF EXOSOMES IN CANCER THERAPY: A CLINICAL PERSPECTIVE

The use of exosome-based treatments in healthcare is gaining popularity for cancer and other pathologies as part of the cell-free therapy approach. This is a significant shift from conventional treatment methods and represents a biomimetic approach. As discussed earlier, exosomes carry a heterogeneous cargo of biologically active molecules and have the potential to deliver chemotherapeutic agents in cancer therapy across biological barriers while maintaining biocompatibility and minimizing immunogenicity (El Andaloussi et al., 2013; Vader et al., 2016). As a result, systems utilizing exosomes are currently being evaluated in various clinical trials targeting different types of cancer and various treatment approaches. This section highlights some clinical trials involving exosome-based cancer therapies. Of the five clinical trials listed here, each one follows a distinct mechanism of action and delivery method, including exosome vaccines from dendritic cells, oral exosome formulations from plants, and customized vesicles loaded with small interfering RNAs or oligonucleotides. These clinical trials provide a comprehensive understanding of exosome-based research in oncology, as summarized in Table 24.2.

A phase II clinical study entitled Exosome Immunotherapy in NSCLC Using IFNγ-Dex (ID: NCT01159288) evaluated the safety and effectiveness of autologous dendritic cell (DC)-derived exosomes treated with interferon-gamma (IFNγ-Dex) in 22 NSCLC patients.

The data showed that exosome-based treatment had a noticeable effect on both the tumor and the nearby normal tissues, compared with the group given curcumin alone. No safety issues were observed, indicating that the intervention was well-tolerated. Moreover, exosome formulation altered metabolic and immune pathways in the colon, suggesting broader therapeutic implications. These data highlight the safety and effectiveness of edible plant-derived exosomes as a natural, targeted drug delivery system for the treatment of gastrointestinal cancer. Continuing in the same vein, a Phase I study (ID: NCT01668849) has successfully used grape-derived exosomes to treat radiotherapy-induced mucositis. The study aimed to demonstrate whether consuming grape-derived exosomes could mitigate the severity of oral mucositis in patients undergoing concurrent chemotherapy and radiotherapy for head and neck cancer. Another Phase 1 clinical trial utilized plant-derived exosomes to deliver curcumin to 35 patients with colon cancer (ID: NCT01294072). A three-arm parallel design was employed, where the first group received curcumin only in capsule form, the second group received curcumin encapsulated in plant-derived exosomes, and the third group served as a control with no intervention. For a week before the surgery, participants in the treatment groups took the designated formulations orally.

The primary objective of the study was to determine the levels of curcumin present in both healthy and cancerous colon tissues, and to assess the safety and tolerability of curcumin. The patients who received platinum-based chemotherapy exhibited no indications of disease progression. Before exosome immunotherapy, patients received metronomic cyclophosphamide (mCTX) to suppress T regulatory cell (Treg) activity temporarily. Patients then received four weekly subcutaneous or intradermal injections of exosomes obtained from IFNγ-matured DCs that contained tumor antigens. The primary finding was that 7 of 22 patients (32%) demonstrated stable disease during the 4-month follow-up period. The median progression-free survival (PFS) was 2.2 months, while the overall survival (OS) was 15 months. Immunomonitoring revealed heightened NK cell activity and increased IFNγ-producing lymphocytes, confirming an activated immune response with a favorable safety profile. This research provides initial clinical evidence that exosomes from DCs can stimulate immune activation, similar to traditional DC-centric immunotherapies, with enhanced feasibility and scalability (Viaud et al., 2010).

Sixty adult participants, aged 20 to 85, participated in a randomized, parallel-group study evaluating the daily use of exosome extract derived from grape powder compared to standard symptomatic therapies during a 35-day radiation treatment period (ID: NCT01668849). Further objectives included assessing immune markers in blood, such as cytokines, T cells, and NK cells, as well as

TABLE 24.2
Summary of the Key Clinical Trials Investigating Exosome-Based Therapies in Cancer Treatment

NCT ID	Intervention (Exosome Source + Payload)	Indication/Patient Population	Phase, Design, and Dosing	Key Outcomes
NCT01159288	Autologous DC-derived exosomes (IFNγ-Dex), preceded by low-dose cyclophosphamide.	Advanced NSCLC after chemotherapy.	Phase II: 22 patients received mCTX, then weekly subcutaneous injections (4 doses).	32% stable disease at 4 months; median PFS 2.2 months, OS 15 months; NK-cell activation; safe.
NCT01294072	Plant-derived exosomes delivering curcumin.	Colon cancer patients' pre-surgical resection.	Phase I, 35 participants; curcumin ± exosomes orally for 7 days.	Higher curcumin in tumor and normal tissues; no adverse events; immune/ metabolic markers improved.
NCT01668849	Grape-derived exosomes (oral powder).	Head and neck cancer patients undergoing chemoradiation.	Phase I, 60 subjects randomized to daily grape exosome vs symptom care during 35 days.	Reduced mucositis pain and severity; modulation of local and systemic immune markers is safe and effective.
NCT03608631	MSC-derived exosomes loaded with KRAS G12D siRNA (iExosomes).	Metastatic pancreatic ductal adenocarcinoma with KRAS G12D mutation.	Phase I dose-escalation, 15 patients; IV infusions on days 1, 4, 10 per cycle.	No dose-limiting toxicity; stable disease in some; KRAS suppression, p-ERK downregulation, and $CD8^+$ T-cell infiltration.
NCT05375604	Engineered exoASO-STAT6 exosomes carrying STAT6 antisense oligonucleotide.	Advanced hepatocellular carcinoma or CRC/gastric cancer liver metastases.	Phase I, up to 9 patients treated; IV escalating doses.	Favorable safety; biomarker modulation of TAM phenotype; vigorous preclinical anti-tumor activity.

Abbreviations: IV: intravenous; NK cells: natural killer cells; OS: overall survival; PFS: progression-free survival.
Source: ClinicalTrials.gov.

in mucosal tissues, including CD3, CD8, CD11b, and F4/80. The data showed a notable reduction in pain and mucositis severity in patients treated with exosomes versus those receiving standard care. This clinical research suggests that exosomes derived from edible plants may serve as a safe, natural, and effective supportive care method to mitigate mucosal toxicity and enhance the quality of life for patients undergoing intensive treatment for head and neck cancer.

A Phase 1 dose-escalation study was conducted using exosomes loaded with KRASg12dg-specific small interfering RNA (siRNA; iExosomes) to treat metastatic pancreatic ductal adenocarcinoma (ID: NCT03608631). The primary objective of the study was to evaluate the safety, tolerability, and efficacy of iExosomes from mesenchymal stromal cells. The study was conducted at MD Anderson Cancer Center in 15 patients with a KRAS G12D oncogenic mutation who had experienced disease progression before conventional therapies. Exosome was administered intravenously on days 1, 4, 10, and 14, with the option to complete up to 3 cycles based on response and tolerance. The primary objective was to determine the optimal dose, with secondary endpoints evaluating pharmacokinetics, disease control rate, and anti-tumor efficacy. The data showed iExosomes significantly reduced KRAS^G12D mRNA levels in both circulating tumor DNA and tumor biopsies, resulting in decreased downstream signaling via the ERK pathway and increased infiltration of $CD8^+$ cytotoxic T cells in the treated lesions. Notably, patients responded favorably to the treatment, and no serious adverse effects were reported. These results support the practical application of exosome-mediated siRNA therapies in specifically targeting oncogenic mutations (Kamerkar et al., 2017; Mendt et al., 2018).

Another Phase 1 study (ID: NCT05375604) used exoASO-STAT6 to reprogram tumor-associated macrophages in the liver and metastases, assessing its safety and efficacy. The study was innovative in using exosome-based antisense oligonucleotides targeting the STAT6 transcription factor in tumor-associated macrophages (TAM). The study focused on patients with advanced HCC or liver metastases due to colorectal or gastric cancer. Given that STAT6 is critical for the M2 (pro-tumor) phenotype of macrophages, facilitating immune evasion and conferring resistance to checkpoint inhibitors, ExoASO-STAT6 induced an M1 (anti-tumor) phenotype in TAMs by inhibiting STAT6 expression, thereby rejuvenating the immune response within the tumor microenvironment. The patients received escalating intravenous doses of ExoASO-STAT6. The data highlight advancements in utilizing engineered exosomes to target tumors and alter the microenvironment in liver and gastrointestinal cancers.

In summary, these selected clinical trials signify the exosome-based therapies in cancer and highlight several vital uses, including boosting immune responses via exosomes from dendritic cells (ID: NCT01159288), increasing bioavailability with plant-derived vesicles (ID: NCT01294072), reducing inflammation in radiation therapy with grape exosomes (ID: NCT01668849), and precise gene silencing using engineered exosomes (ID: NCT03608631). Additionally, the exoASO-STAT6 investigation (NCT05375604) marks a new phase in modifying immune pathways, particularly for liver and gastrointestinal tumors (Cullis et al., 2022; Kamerkar et al., 2017). While preliminary data are encouraging, these studies have their limitations. Numerous studies have faced limitations, including small sample sizes, single-arm designs, and open-label methods, warranting better-designed future studies. Moreover, various techniques for isolating and purifying exosomes—from dendritic cells to edible plants and mesenchymal stromal cells—result in divergent clinical reproducibility, necessitating the development of a uniform protocol (El Andaloussi et al., 2013; Vader et al., 2016). Ren et al. (2024b) indicate that the next phase of clinical translation will require standardized GMP-compliant manufacturing processes, enhanced vesicle characterization, and comprehensive pharmacokinetic evaluation (Ren et al., 2024b). There is a need for extensive, multicenter, randomized studies better to evaluate the clinical effectiveness and safety of this approach (Zhuang et al., 2011; Mendt et al., 2018; Kamerkar et al., 2017).

24.10 EXOSOMAL BIOMARKERS IN CANCER DIAGNOSIS AND PROGNOSIS

Cancer is one of the leading causes of death worldwide, in both men and women, with lung, breast, liver, and stomach malignancies as the most prevalent forms of cancer (Moni et al., 2025). As such, early disease detection is undoubtedly a prominent and growing field of interest. Biomarkers, in particular, are beneficial for early cancer detection and prognosis (Zhu et al., 2024). A key method of cancer detection that has recently gained attention for its high sensitivity, repeatability, and non-invasive nature is liquid biopsy, which uses samples from blood, saliva, cerebrospinal fluid, and other non-solid bodily fluids. Tumor-derived entities are extracted from these body fluids and analyzed for genomic and proteomic data. In this regard, TDEs have transformed the field of cancer diagnosis and prognosis by providing a non-invasive window into tumor biology through the analysis of nanosized EVs secreted by tumor cells, which secrete approximately tenfold more EVs than normal tissues with a heterogeneous array of molecular cargos, including tumor antigens and stemness markers. Collectively, they reflect the tumor's physiological state and genetic alterations (Wang et al., 2022).

The use of exosomal biomarkers in liquid biopsies has garnered significant interest due to their stability in biological fluids and their potential for monitoring. TDEs' biogenesis explains the multifaceted nature of their cargo profile and that of their parent cell, which may cause cancer cells to grow more rapidly, evade the immune system, spread throughout the body, and even increase drug resistance (Wang et al., 2022). They are often isolated from the plasma, saliva, urine, or other bodily fluids of cancer patients using the available methods of isolation, purification, and detection as detailed in earlier sections (Huang et al., 2020; Zhu et al., 2024). TDEs differ discreetly from those of normal cell-derived exosomes due to their molecular cargo, as the former's parent cells are tumorigenic; hence, their contents exhibit oncogenic and pro-tumoral properties (Moni et al., 2025).

Given their regulatory role in cell signaling, which is involved in various cellular processes, exosomal miRNAs (EXO-miR) have attracted attention as promising biomarkers for cancer diagnosis and prognosis and have been extensively studied in this context (Table 24.3). ClinicalTrials.gov lists 44 clinical studies involving exosome-derived entities for cancer diagnosis and prognosis, of which five have been completed (Table 24.4). They possess key properties, including high specificity, sensitivity, stability, and accessibility, and their diseased tissue-specific, fluctuating profiles differ from those of normal tissues (Aramini et al., 2022; Zhu et al., 2024). They are among the best-characterized exosome-derived biomarkers. These small non-coding RNAs regulate gene expression at the post-transcriptional level and play central roles in oncogenic processes, including cell proliferation, evasion of apoptosis, angiogenesis, and metastasis. For example, patients with breast cancer have greater concentrations of exosomes with rich miRNA contents in their peripheral blood compared to healthy subjects (Wu et al., 2020). This suggests that the EXO-miR profile has the potential to serve as a biomarker for breast cancer diagnosis and prognosis. One of the most extensively studied miRs is miR-21, which is consistently overexpressed in exosomes from a wide range of cancers, including breast, lung, colorectal, and gastric cancers. Its oncogenic role has been attributed to the modulation of tumor suppressor genes, the promotion of invasive phenotypes, and resistance to chemotherapy-induced apoptosis, all of which render it a valuable biomarker for early detection and prognosis. EXO-miR-21 also promotes bone metastasis in breast cancer cells. Together with EXO-miR-23b-3p, EXO-miR-21-5p is also associated with the metastasis and progression of lung cancer (Zhu et al., 2024).

Studies have shown that EXO-miR-141 is upregulated in patients with lung cancer, while EXO-miR-144 expression is downregulated (Zhu et al., 2024). Biomarkers for early-stage lung adenocarcinoma include altered expression of EXO-miR-342-5p and EXO-miR-574-5p in cancerous tissues compared to non-cancerous tissues (Moni et al., 2025). Another notable miR is EXO-miR-1246, which is highly upregulated in the exosomes of patients with lung adenocarcinoma, colorectal, and breast cancer. EXO-miR-1246 has been reported to promote tumor growth by targeting tumor suppressor genes and regulating the tumor microenvironment. Its levels in patient plasma correlate strongly with tumor burden, making it an excellent candidate for non-invasive diagnostic assays and monitoring disease progression (Morimoto et al., 2023).

EXO-miR-155 is a standout as a cross-cancer biomarker, as it is upregulated in exosomes from lung cancer, hepatocellular carcinoma (HCC), breast, and gastric cancers. Its role in promoting

TABLE 24.3
Exosomal Cargo Biomarkers in Cancer Diagnosis and Prognosis

Cancer Type	Exosomal Biomarker Type	Key Biomarker	Clinical Role	Citation
Lung Cancer (NSCLC)	lncRNA	MALAT1, HOTAIR, SOX2-OT, H19	Diagnostic, prognostic (all); therapeutic (HOTAIR, H19).	Lin et al. (2023)
Lung cancer (NSCLC & SCLC)	Exosomal lncRNAs, circRNAs, miRNAs	lncRNAs: TBILA, AGAP2-AS1, DLX6-AS1, PCAT6, GAS5, RP11-510M2	Diagnostic, prognostic, therapeutic (via ceRNA axis).	Wang et al. (2024a)
NSCLC (diagnostic)	miRNA/lncRNA/circRNA	>60 miRNAs, 9 lncRNAs, 14 circRNAs	Diagnosis: Sensitivity ≈ 0.83, Specificity ≈ 0.79–0.83; AUC miRNA ≈ 0.90, lncRNA ≈ 0.82, circRNA ≈ 0.86).	Cao et al. (2023)
Prostate cancer (Castration-resistant)	miRNA	miR-222-3p	Promotes CRPC via the mTOR pathway; linked to drug resistance.	Wang et al. (2023b)
Pancreatic Cancer	Plasma-derived exosomal lncRNAs	LINC01268, LINC02802, AC124854.1, AL132657.1	Non-invasive diagnostic markers (AUC 0.8476).	He et al. (2024)
Multiple Myeloma	Exosomal miRNAs, proteins	↓ miR-15a, ↑ oncogenic proteins, cytokines (e.g. IL-6, CCL2)	Promotes tumor growth, angiogenesis, migration, and drug resistance.	Roccaro et al. (2013)
Glioblastoma	miRNA	Specific miRNAs in peripheral blood	Diagnostic biomarker for glioblastoma.	Dong et al. (2014) and Dong et al. (2023)
Breast cancer	Exosomal miRNAs	miR-370-3p	Promotes cancer progression via targeting FBLN5.	Mao et al. (2021)

Abbreviations: NSCLC: non-small cell lung cancer; SCLC: small cell lung cancer; CRPC: castration-resistant prostate cancer; MM: multiple myeloma; GBM: glioblastoma; lncRNA: long noncoding RNA; miRNA: microRNA; circRNA: Circular RNA; AUC: area under the curve; FBLN5: fibulin-5.

tumor growth, invasion, and resistance to apoptosis suggests its diagnostic and prognostic value. In lung cancer, for instance, EXO-miR-155 expression was significantly greater in patients with aggressive tumors and was associated with lymph node metastasis and poor survival rates. These findings hold promise for EXO-miR-155 in informing clinical decision-making for patient risk stratification (Zhang et al., 2024). In HCC, EXO-miRs, such as miR-92a-3p, miR-720, and particularly miR-4661-5p, are significantly overexpressed. The latter shows high diagnostic precision, even in the early stages of the disease, with an AUROC of 0.923 in a large patient cohort. Both EXO-miR-1910-3p and EXO-miR-500a-5p have been found to promote the proliferation and migration of

TABLE 24.4
Use of Exosome-Based Entities as Biomarkers for Diagnosis and Prognosis of Cancer

NCT Number	Study Title	Conditions	Interventions	Sponsor	Study Type
NCT03830619	Serum Exosomal Long Noncoding RNAs as Potential Biomarkers for Lung Cancer Diagnosis.	Lung Cancer (Diagnosis)	DIAGNOSTIC_TEST: collect samples	Wuhan Union Hospital, China	OBSERVATIONAL
NCT02862470	Anaplastic Thyroid Cancer and Follicular Thyroid Cancer-derived Exosomal Analysis Via Treatment of Lovastatin and Vildagliptin and Pilot Prognostic Study Via Urine Exosomal Biological Markers in Thyroid Cancer Patients.	Thyroid Cancer		National Taiwan University Hospital	OBSERVATIONAL
NCT06023121	Use of a Liquid Biopsy Signature to Detect Early-onset Gastric Cancer.	Gastric Cancer	Measurement of levels of an exosome-based liquid biopsy signature.	Chinese PLA General Hospital	OBSERVATIONAL
NCT02921854	Detection of Circulating Biomarkers of Immunogenic Cell Death.	Non-Small Cell Lung Cancer	OTHER: Blood withdrawal	Maastricht Radiation Oncology	INTERVENTIONAL
NCT03488134	Predicting Prognosis and Recurrence of Thyroid Cancer Via New Biomarkers, Urinary Exosomal Thyroglobulin and Galectin-3.	Thyroid Cancer		National Taiwan University Hospital	OBSERVATIONAL

breast cancer cells (Zhu et al., 2024). Additionally, EXO-miR-7641 is overexpressed in patients with highly metastatic breast cancers (Shen et al., 2021).

EXO-miR expression profiles change not only to serve as potential biomarkers for cancer detection but also to predict therapeutic response in cancer patients. For instance, elevated EXO-miR-425-3p and EXO-miR-96 are predictors of cisplatin resistance in lung cancer (Ma et al., 2019), while high levels of EXO-miR-155 and EXO-miR-301 are associated with response to neoadjuvant therapy in breast cancer patients (Wang et al., 2022). In NSCLC and lung adenocarcinoma (LUAD), miR-342-5p, miR-574-5p, miR-1246, and miR-96 were profiled in tumor tissues and patient plasma. LUAD patient plasma contained EXO-miR-4454 and miR-619-5p, which may serve as additional diagnostic markers for distinguishing cancer patients from healthy individuals (Feng et al., 2023). These miRs demonstrated strong sensitivity and specificity, outperforming conventional serum markers, such as carcinoembryonic antigen (CEA).

Another group of exosomal cargo is lncRNA (EXO-IncRNA) (Márton et al., 2025). Despite being newly characterized, EXO-IncRNAs are stable and specific biomarkers in exosomes. In HCC, EXO-lnc-LINC00161 and EXO-lnc-FAM72D-3 have demonstrated high diagnostic accuracy and a correlation with tumor size, vascular invasion, and metastasis. These biomarkers are particularly valuable for early detection, greatly enhancing prognosis in patients in whom it is challenging to detect them with traditional imaging and serum alpha-fetoprotein (AFP) levels. These EXO-lncRNAs regulate gene expression of cell cycle progression and EMT, mechanisms pivotal to cancer invasiveness. The high stability of EXO-lncRNAs, combined with detectable expression levels, makes them ideal candidates for the development of clinical biomarkers.

Exosomal proteins also enhance the diagnostic and prognostic potential of liquid biopsies. Glypican-1 (GPC-1), a membrane-anchored heparan sulfate proteoglycan, has demonstrated extraordinary specificity and sensitivity in pancreatic and breast cancers (Frampton et al., 2018). GPC-1-positive exosomes offer excellent discrimination between patients with pancreatic cancer and healthy individuals, with an area under the receiver operating characteristic curve (AUC) of nearly 1.0 (Melo et al., 2015), making them a gold standard for early cancer detection. Epithelial cell adhesion molecule (EpCAM) and developmental endothelial locus-1 (Del-1) proteins on exosomes have also been shown to successfully differentiate between malignant and benign lesions, facilitating patient stratification. In breast cancer, exosomal biomarkers, specifically polo-like kinases (PLK), are associated with predicting tumor prognosis. Specifically, elevated levels of PLK2, PLK3, and PLK5 are associated with better breast cancer outcomes, whereas elevated levels of PLK1 and PLK4 are associated with poorer survival (Zhu et al., 2024). Gastric cancer patients benefit from serum exosomal biomarkers such as human epidermal growth factor receptor 2 (HER2), which directly correlate with tissue expression and thereby inform patient selection for targeted anti-HER2 therapies. Salivary EXO-miR-151-3p provides a non-invasive avenue for early detection. In contrast, reduced EXO-Gastrokine 1 (GKN1) protein levels have been associated with both early-stage and advanced gastric cancers, providing valuable insights for diagnosis and prognosis (Yoon et al., 2019).

Exosomal DNA (EXO-DNA) from tumors is also crucial for cancer prognosis (Vaidya & Sugaya, 2020). Driver mutations in cancers, which are frequently observed in tumors, including KRAS, EGFR, and TP53, can be identified in circulating EXO-DNA fragments to assist in evaluating tumor progression, identifying minimal residual disease, and recognizing treatment resistance (Yang et al., 2017). Quantitative assessment of circulating exosomes is also linked to tumor burden and disease progression. Elevated serum exosome levels may reflect hypoxic conditions in the tumor microenvironment, increased angiogenesis, and metastatic activity. Profiling exosome-level changes over follow-up can serve as a valuable biomarker for disease status and treatment response.

These findings provide strong evidence of the clinical significance of the fast-emerging exosome-based biomarkers in cancer diagnosis and prognosis. Their specific molecular properties, stability in bodily fluids, and minimally invasive sampling protocols render them suitable for precision oncology. The development of multi-omic profiling strategies, along with their potential integration with artificial intelligence, is expected to enhance the reliability and predictive potential of exosome-based diagnostics. Their routine clinical use as biomarkers is expected to contribute significantly to tumor diagnosis and management.

24.11 FUTURE DIRECTIONS IN EXOSOME RESEARCH AND THERAPY IN ONCOLOGY

The field of oncology is undergoing a transformative shift toward precision diagnostics and therapeutics, driven by the need for enhanced efficacy, specificity, and reduced toxicity (Bhattacharya and Bhirud 2025). EVs, particularly exosomes, have emerged as pivotal mediators of intercellular communication, transitioning from biological curiosities to versatile tools with significant potential in cancer diagnosis and therapy (Muttiah et al., 2024). As foundational discoveries pave the way for advanced applications, future research in exosome-based oncology is poised to integrate cutting-edge engineering, deep biological insights, scalable clinical translation, and solutions to persistent manufacturing and regulatory challenges (Sinha et al., 2021; Wang et al., 2024c). Over the next decade, exosomes are expected to become integral to personalized cancer medicine, propelled by innovations that bridge laboratory research and clinical practice (Sharma et al., 2024).

24.11.1 Molecular Mechanisms of Exosome Biogenesis and Cargo Selection

Elucidating the molecular pathways critical for developing targeted therapeutic strategies. Understanding the roles of ESCRT-independent pathways and lipid-driven mechanisms could enable precise manipulation of exosomal cargo (Han et al., 2022). Factors such as cellular state, environmental cues (e.g., hypoxia, nutrient deprivation), and genetic alterations in cancer cells profoundly influence exosomal content and surface markers. This knowledge is crucial for designing exosomes with targeted therapeutic payloads and specific targeting capabilities. However, exosome heterogeneity, particularly in tumor-derived vesicles, poses a significant challenge to achieving batch-to-batch consistency and therapeutic predictability (Van De Wakker et al., 2023). Future research will utilize single-exosome analysis technologies to dissect this heterogeneity at an unprecedented resolution, moving beyond bulk population analyses to enhance the precision of exosome-based interventions (Xing et al. 2025).

24.11.2 Advances in Isolation Technologies

Scalable and reproducible isolation methods are essential for translating exosome-based therapies into clinical settings. Traditional techniques, such as ultracentrifugation, are limited by low yields, purity issues, and a lack of scalability (Ter-Ovanesyan et al., 2021). Emerging technologies, including microfluidic platforms and tangential flow filtration (TFF), offer promising solutions. Microfluidic systems enable automated, high-throughput separation based on size, density, or surface markers, facilitating rapid and precise isolation (Rahnama et al., 2024). TFF provides a scalable and gentle approach for concentrating and purifying exosomes from large volumes, making it suitable for industrial bioprocessing (Kapoor et al., 2024). Future efforts should focus on refining these methods to ensure high purity and yield while developing integrated, closed-system workflows to minimize contamination and maintain sterility. Hybrid approaches that combine traditional techniques with machine learning (ML) could enhance biomarker discovery and specificity, enabling the high-throughput identification of target exosomes from complex biofluids (Li et al., 2024b; Wu et al., 2022).

24.11.3 Liquid Biopsy Applications

Exosome-based liquid biopsies hold immense potential for non-invasive cancer diagnostics, but challenges in biomarker validation and reproducibility must be addressed (Preethi et al., 2022).

Collaborative consortia could standardize protocols and datasets to accelerate clinical adoption (Dilsiz 2024). Identifying universal biomarkers, such as Clathrin Heavy Chain (CLTC), Ezrin (EZR), Talin-1 (TLN1), Adenylyl Cyclase-Associated Protein 1 (CAP1), and Moesin (MSN), is crucial for early detection across diverse cancer types (Sergazy et al., 2025; Youssef et al., 2025). Next-generation liquid biopsy platforms will likely integrate multi-omics analysis of exosomal cargo (e.g., non-coding RNAs, circulating tumor DNA, proteins, and lipids) with AI-driven algorithms to identify highly sensitive and specific molecular signatures (Lorite, 2025; Xu et al., 2025). These signatures will enable early detection, real-time monitoring of treatment responses, and identification of minimal residual disease, providing a comprehensive molecular snapshot to guide personalized therapeutic decisions and improve patient stratification (Youssef et al., 2025).

24.11.4 Therapeutic Applications

Engineered exosomes offer significant advantages for targeted drug delivery, leveraging their biocompatibility, low immunogenicity, and ability to cross biological barriers, such as the BBB (Dilfy and Abduljaleel 2023). Advances in loading therapeutic agents, such as CRISPR-Cas9 or siRNAs, require efficiency improvements, targeting precision, and scalability. Exosomes outperform synthetic nanocarriers in overcoming drug resistance, a significant challenge in oncology (Youssef et al., 2025). Future strategies will explore delivering sensitizing agents, targeting resistance pathways, or reprogramming the TME to reduce resistance. Hybrid exosomes, which combine natural and synthetic properties, can enhance stability, payload capacity, and release kinetics, offering novel solutions for personalized therapies (Mirgh et al., 2024; Pan et al., 2024).

24.11.5 Immune Modulation and Resistance Mechanisms

Exosome-mediated immune modulation and resistance to immune checkpoint inhibitors (CPIs) remain underexplored. TDEs often carry immunosuppressive cargo, such as PD-L1, FasL, and galectins, which inhibit immune cell function or induce T-cell apoptosis, thereby contributing to immune evasion (Lyu et al., 2024; Wu et al., 2024b). Novel platforms, such as the nanoplasmonic immunoassay integrated cell-on-a-chip (NIIC), enable sensitive, quantitative, and spatial analysis of exosome-mediated immune suppression, revealing mechanisms, such as PD-L1$^+$ exosome activity (Wang et al., 2025b). Future research will focus on strategies to block the release of immunosuppressive exosomes, neutralize their pro-tumorigenic effects, or re-engineer them to be immunostimulatory, thereby enhancing the efficacy of immunotherapies by transforming immunologically "cold" tumors into "hot" ones (Safaei et al., 2024).

24.11.6 Exosomes in Vaccine Development

Engineered exosomes presenting tumor antigens show promise for eliciting robust immune responses, paving the way for personalized cancer vaccines (Zhang et al., 2023). Early-phase clinical trials with dendritic cell-derived exosomes have demonstrated strong anti-tumor T-cell responses in melanoma and NSCLC, mimicking the natural presentation of antigens (Santos & AlMeida, 2021; Youssef et al., 2025). Scalable production methods, such as TFF and immunoaffinity capture, enhance the clinical feasibility of these approaches (Madhan et al., 2024; Orefice et al., 2023). Plant-derived exosomes offer a cost-effective, biocompatible alternative for vaccine delivery. Future efforts should standardize exosome engineering and optimize antigen loading to maximize therapeutic impact, potentially leading to universal, off-the-shelf cancer vaccines (Wang et al., 2025a).

24.11.7 Microbiome–Exosome Interactions

The gut microbiome significantly influences exosomal cargo; however, bidirectional interactions between the microbiota and exosomes in cancer remain underexplored (Ocansey et al., 2020).

Investigating microbiota-derived exosomes as biomarkers or therapeutic agents could yield innovative strategies for managing cancer and related disorders (Zhang et al., 2022). Understanding how microbial dysbiosis affects exosomal content in cancer patients and how host-derived exosomes influence microbial composition could uncover novel diagnostic and therapeutic opportunities, particularly in immunotherapy, where the gut microbiome plays a critical role. Engineered bacterial or host-derived exosomes that carry microbial metabolites or therapeutic agents can modulate the gut-tumor axis, offering new avenues for cancer management (Gabaran et al., 2025; Qiu et al., 2024).

24.11.8 Long-Term Safety and Efficacy

The long-term safety and efficacy of exosome-based therapeutics, particularly in gene editing and immune modulation, require rigorous investigation. Preclinical and clinical studies should address potential immunogenicity, biodistribution, and off-target effects (Li et al., 2019; Liu et al., 2025).As these therapies approach widespread clinical use, robust pharmacovigilance and post-market surveillance will be crucial for monitoring cumulative toxicity and sustained efficacy. Comprehensive clinical trial designs with large cohorts and extended follow-up periods are essential for fully characterizing these aspects (Youssef et al., 2025).

24.11.9 Regulatory and Ethical Frameworks

The integration of EVs into clinical practice necessitates the harmonization of global regulatory standards. Research into ethical implications, particularly regarding patient-derived exosomal material and genetic modifications, will be critical for public trust and widespread implementation (Welsh et al., 2024; Youssef et al., 2025). The unique biological nature of exosomes, which bridge small molecules and cell-based therapies, poses challenges for regulatory bodies such as the FDA and EMA (Shah et al., 2025; Wang et al., 2024b). Future efforts must establish clear guidelines for GMP manufacturing, characterization, and quality control (Wang et al., 2024b). Ethical considerations, including informed consent for patient-derived or genetically modified exosomes, equitable access to high-cost therapies, and prevention of unproven treatments, are critical for maintaining public trust and ensuring responsible implementation (Kalodimou and Muthu 2025).

By tackling these complex challenges, exosome research can unlock its transformative potential for diagnostic and therapeutic uses in oncology. Advances in understanding biogenesis, isolation methods, liquid biopsy applications, therapeutic engineering, immune modulation, vaccine development, microbiome interactions, and regulatory policies will propel the field toward real clinical impact (Youssef et al., 2025). Collaboration among academia, industry, and regulatory agencies will be crucial to turn these developments into effective, accessible, and ethically sound cancer treatments.

24.12 CONCLUSIONS

The evolution of cell-derived exosomes and secretomes represents a paradigm shift in oncology, redefining approaches to cancer diagnosis, prognosis, and treatment. This chapter has explored their transformative potential, highlighting their role in intercellular communication and their emergence as innovative therapeutic platforms. These vesicles offer unique opportunities for targeted drug delivery, immune system modulation, and non-invasive diagnostics. Their inherent biocompatibility and low immunogenicity distinguish them from synthetic nanoparticles and traditional cellular therapies, enhancing safety profiles. Moreover, their ability to cross complex physiological barriers, such as the BBB, positions them as promising candidates for addressing challenging malignancies, including brain tumors.

Advances in engineering/ priming parent cells to produce exosomes with tailored therapeutic payloads or modifying exosomal surfaces have ushered in a new era of precision oncology (Haider,

2024). Envisioning exosomes engineered to deliver potent chemotherapeutic agents directly to tumor cells could minimize systemic toxicity, while those designed to carry immunomodulatory molecules might enhance anti-tumor immune responses. The prospect of personalized medicine, where exosome-based therapies are customized to a patient's tumor molecular profile, is rapidly transitioning from concept to reality.

However, translating these discoveries into routine clinical practice presents significant hurdles. Standardization and scalability of production remain critical. The intrinsic heterogeneity of exosomes, even from identical cell sources, necessitates robust quality control and adherence to GMP standards. Ensuring consistent yield, purity, and cargo composition across batches is crucial for the safety and effectiveness of clinical-grade products (Liang et al., 2025). This involves a significant investment in advanced bioprocessing technologies along with comprehensive characterization methods.

The regulatory landscape for exosome- and secretome-based products is complex and varies globally, creating uncertainty among researchers and developers. International harmonization of regulatory frameworks is urgently needed to streamline development, foster collaboration, and mitigate regulatory arbitrage. Clear, tailored guidelines for safety, efficacy, and manufacturing quality—specifically designed for the unique properties of cell-derived products—are essential for navigating the path from preclinical studies to clinical trials and ultimately to market authorization. Ethical considerations are equally pivotal in the responsible development and use of exosome and secretome therapies. Transparent communication about the experimental nature of these therapies is essential to prevent therapeutic misconceptions, especially given the proliferation of unproven exosome products in specific markets. Strategies such as cost-effective manufacturing, global partnerships, and equitable pricing models are crucial to democratizing access to exosome-based precision medicine, ensuring that benefits reach patients regardless of socioeconomic or geographic constraints.

The integration of artificial intelligence (AI) in exosome and secretome research holds immense promise for accelerating biomarker discovery and cargo analysis. However, it introduces ethical challenges related to data bias and privacy. Proactive measures, including diverse training datasets, transparent AI algorithms, and stringent data protection protocols, are crucial to ensuring the equitable and accurate application of AI-driven technologies. Maintaining public trust requires transparency, rigorous scientific validation, and robust public education to counter misinformation and promote the responsible adoption of these novel therapies.

In summary, cell-derived exosomes and secretomes are poised to revolutionize oncology by leveraging their biological properties and engineered capabilities. Despite challenges in standardization, regulation, and access, ongoing advancements and interdisciplinary collaboration among academia, industry, and regulatory agencies are paving the way for the clinical integration of these technologies. By addressing these obstacles with foresight and ethical accountability, the full transformative potential of exosomes and secretomes can be exploited.

REFERENCES

Ahmadian S, Jafari N, Tamadon A, Ghaffarzadeh A, Rahbarghazi R, Mahdipour M. Different storage and freezing protocols for extracellular vesicles: A systematic review. *Stem Cell Res. Ther.* 2024;15(1):453. https://doi.org/10.1186/s13287-024-04005-7.

Ahmed ZT, Alkahlot MH, Haider KH. MSC-derived exosomes: Advances in cell-free therapy. In: *Handbook of Stem Cell Applications*. Singapore: Springer Nature, 2023, pp. 1–41.

Álvarez-Viejo M, Haider KH. Mesenchymal stem cells. In: Haider, K.H. (eds) *Handbook of Stem Cell Therapy*. Singapore: Springer, 2022. https://doi.org/10.1007/978-981-16-6016-0_6-1.

Aramini B, Haider KH, et al. Future perspectives of exosomal payload of miRNAs in lung cancer. In: Haider, K.H. (eds) *Handbook of Stem Cell Therapy*. Singapore: Springer, 2022. https://doi.org/10.1007/978-981-16-6016-0_28-1.

Aramini B, Masciale V, Haider KH. Defining lung cancer stem cells' exosomal payload of miRNAs from a clinical perspective. *World J Stem Cells.* 2020a Jun 26;12(6):406–21. https://doi.org/10.4252/wjsc.v12.i6.406.

Aramini B, Masciale V, Banchelli F, D'Amico R, Dominici M, Haider KH. Precision medicine. Lung cancer: Challenges and opportunities in diagnostic and therapeutic purposes. In: Park HS-M (eds) *Lung Cancer – Modern Multidisciplinary Management.* London: InTech Open, 2020b.

Basyoni AE, Atta A, Salem MM, Mohamed TM. Harnessing exosomes for targeted drug delivery systems to combat brain cancer. *Cancer Cell Int.* 2025;25(1):150.

Besse B, Charrier M, Lapierre V, Dansin E, Lantz O, Planchard D, et al. Dendritic cell-derived exosomes as maintenance immunotherapy after first-line chemotherapy in NSCLC. *Oncoimmunology.* 2016;5(4):e1071008.

Bhattacharya S, Bhirud D. Exploring the horizon of cancer research: Pioneering breakthroughs in diagnostics and theranostics. *Adv. Cancer Res. Explor. Diagnostics Ther. Break.* 2025;1:1–16.

Bhatti MA, Haider KH. Mesenchymal stem cells as living bio-drugs: Adding a novel dimension to the cardiology and cardiovascular sciences. *J Cardiol Cardiovasc Sciences.* 2025;6(1):30–5.

Bouhouita-Guermech S, Gogognon P, Bélisle-Pipon J-C. Specific challenges posed by artificial intelligence in research ethics. *Front. Artif. Intell.* 2023;6:1149082.

Cao Y, Liu X, Liu J, Su Z, Liu W, Yang L, and Zhang L. Diagnostic value of exosomal noncoding RNA in lung cancer: a meta-analysis. *Front. Oncol.* 2024; 14:1357248. https://doi.org/10.3389/fonc.2024.1357248.

Casari I, Howard JA, Robless EE, Falasca M. Exosomal integrins and their influence on pancreatic cancer progression and metastasis. *Cancer Lett.* 2021;507:124–34.

Colombo M, Raposo G, Théry C. Biogenesis, secretion, and intercellular interactions of exosomes and other extracellular vesicles. *Annu. Rev. Cell Dev. Biol.* 2014;30(1):255–89.

Cullis PR, Hope MJ, Akinc A. Clinical development of lipid nanoparticles for RNA delivery. *Nat. Rev. Drug Discov.* 2022;21(12):915–32. https://doi.org/10.1038/s41573-022-00595-3.

Dilfy S, Abduljaleel R. Extracellular vesicles as promising vectors for CRISPR/Cas9 delivery: Advantages, challenges, and clinical implications. *Trends Pharm. Biotechnol.* 2023;1(1):19–33.

Dilsiz N. A comprehensive review on recent advances in exosome isolation and characterization: Toward clinical applications. *Transl. Oncol.* 2024;50:102121.

Dong L, Li Y, Han C, Wang X, She L, Zhang H. miRNA microarray reveals specific expression in the peripheral blood of glioblastoma patients. *Int J Oncol.* 2014 Aug;45(2):746–56. https://doi.org/10.3892/ijo.2014.2459.

Dong X, Zhang C, Liu X, et al. Peripheral blood exosomal microRNAs as biomarkers for glioblastoma diagnosis: A gene chip analysis. *Mol. Med. Rep.* 2023;27(3):58.

El Andaloussi S, Mäger I, Breakefield XO, Wood MJA. Extracellular vesicles: Biology and emerging therapeutic opportunities. *Nat. Rev. Drug Discov.* 2013;12(5):347–57. https://doi.org/10.1038/nrd3978.

Escudier B, Dorval T, Chaput N, André F, Caby M-P, Novault S, et al. Vaccination of metastatic melanoma patients with autologous dendritic cell (DC) derived-exosomes: Results of the first phase I clinical trial. *J. Transl. Med.* 2005;3:1–13.

European Medicines Agency. EMA Scientific recommendations on classification of ATMP. www.ema.europa.eu/en/human-regulatory-overview/marketing-authorisation/advanced-therapies-marketing-authorisation/scientific-recommendations-classification-advanced-therapy-medicinal-products.

Feng L, Feng Z, Hu J, Gao J, Li A, He X, Liu L, et al. Identification of hsa-miR-619-5p and hsa-miR-4454 in plasma-derived exosomes as a potential biomarker for lung adenocarcinoma. *Front Genet.* 2023;14:1138230. https://doi.org/10.3389/fgene.2023.1138230.

Frampton AE, Prado MM, López-Jiménez E, Fajardo-Puerta AB, Jawad ZAR, Lawton P, Giovannetti E, et al. Glypican-1 is enriched in circulating exosomes in pancreatic cancer and correlates with tumor burden. *Oncotarget.* 2018;9(27):19006–13. https://doi.org/10.18632/oncotarget.24873.

Fujita M, Hatta T, Ikka T, Onishi T. There is an urgent need for clear and concise regulations on exosome-based interventions. *Stem Cell Rep.* 2024;19(11):1517–9.

Gabaran SG, Ghasemzadeh N, Rahnama M, Karatas E, Akbari A, Rezaie J. Functionalized exosomes for targeted therapy in cancer and regenerative medicine: genetic, chemical, and physical modifications. *Cell Commun. Signal.* 2025;23(1):1–19.

Ghosh A, Ramachandran V, Rajan R, et al. Exosomal protein markers as novel diagnostic tools in oncology: A comprehensive review. *Cancer Biomarkers.* 2022;35(2):115–30.

Ghosh S, Rajendran RL, Mahajan AA, Chowdhury A, Bera A, Guha S, Chakraborty K, et al. Harnessing exosomes as cancer biomarkers in clinical oncology. *Front Mol Biosci.* 2023;10:1196728.

Goo J, Lee Y, Lee J, Kim I-S, Jeong C. Extracellular vesicles in therapeutics: A comprehensive review on applications, challenges, and clinical progress. *Pharmaceutics.* 2024;16.

Greenberg ZF, Graim KS, He M. Towards artificial intelligence-enabled precision drug delivery via extracellular vesicles. *Adv. Drug Deliv. Rev.* 2023;199:114974.

Guo W, Li Y, Pang W, Shen H. Exosomes: A potential therapeutic tool targeting communications between tumor cells and macrophages. *Mol. Ther.* 2020;28(9):1953–64.

Guo Y, Ji X, Liu J, Fan D, Zhou Q, Chen C, et al. Effects of exosomes on pre-metastatic niche formation in tumors. *Mol. Cancer.* 2019;18:1–11.

Gurunathan S, Min-Hee K, Kim J-H. A comprehensive review on factors influencing biogenesis, functions, therapeutic and clinical implications of exosomes. *Int. J. Nanomedicine.* 2021;16(null):1281–312. https://doi.org/10.2147/IJN.S291956.

Habib SM, Martini MF, Al Ahmed O, Qassab AR, Haider KH. Mesenchymal stem cell secretomes for cell-free therapy of the broken heart. In: Haider, K.H. (eds) *Handbook of Regenerative Medicine: Cell-Free Therapy.* London: Taylor & Francis Group/CRC Press, 2025a, pp. 274–93.

Habib SM, Martini MF, Al Ahmed O, Qassab AR, Haider KH. The potential of mesenchymal stem cell–derived exosomes in cell-free therapy for myocardial repair. In: Haider, K.H. (eds) *Handbook of Regenerative Medicine: Cell-Free Therapy.* London: Taylor & Francis Group/CRC Press, 2025b, Chapter 24, pp. 372–90.

Habib SY, Martini MF, Hamdan YNA, Shrebaty OMM, Haider KH. Mesenchymal stem cell secretome and exosomes as potential advanced therapy medicinal products for treating a failing heart. *World J Stem Cells.* 2026;18.

Haider KH. Priming mesenchymal stem cells to develop "super stem cells." *World J Stem Cells.* 2024;16(6):623–40. https://doi.org/10.4252/wjsc.v16.i6.623.

Haider KH, Najimi M. Exosome-based Cell-free therapy in regenerative medicine for myocardial repair. In: Haider, K.H. (eds) *Handbook of Stem Cell Therapy.* Singapore: Springer, 2022. https://doi.org/10.1007/978-981-16-6016-0_9-1.

Han Q-F, Li W-J, Hu K-S, Gao J, Zhai W-L, Yang J-H, et al. Exosome biogenesis: Machinery, regulation, and therapeutic implications in cancer. *Mol. Cancer.* 2022;21(1):207.

Han Z, Lv W, Li Y, Chang J, Zhang W, Liu C, et al. Improving tumor targeting of exosomal membrane-coated polymeric nanoparticles by conjugation with aptamers. *ACS Appl. Bio Mater.* 2020;3(5):2666–73.

Hanahan D. Hallmarks of cancer: New dimensions. *Cancer Discov.* 2022;12(1):31–46.

He J, He F, Yang Q, Li Q. Blockade of exosome release sensitizes breast cancer to doxorubicin via inhibiting angiogenesis. *Cancer Med.* 2025;14(8):e70785.

He X, Chen L, Di Y, Li W, Zhang X, Bai Z, Wang Z, et al. Plasma-derived exosomal long noncoding RNAs of pancreatic cancer patients as novel blood-based biomarkers of disease. *BMC Cancer.* 2024 Aug 6;24(1):961. https://doi.org/10.1186/s12885-024-12755-z.

Huang T, Deng CX. Current progress of exosomes as cancer diagnostic and prognostic biomarkers. *Int. J. Biol. Sci.* 2019;15(1):1–11. https://doi.org/10.7150/ijbs.27796.

Huang TY, Wang CY, Chen KY, Huang LT. Urinary exosomal thyroglobulin in thyroid cancer patients with post-ablative therapy: A new biomarker in thyroid cancer. *Front Endocrinol (Lausanne).* 2020;11:382. https://doi.org/10.3389/fendo.2020.00382.

Ji H, Zhang L, Ye L. Exosome, an important transmitter in the drug resistance of non-small cell lung cancer. *Front. Oncol.* 2025;15:1539047.

Jiang M, Zhang K, Meng J, Xu L, Liu Y, Wei R. Engineered exosomes in service of tumor immunotherapy: From optimizing tumor-derived exosomes to delivering CRISPR/Cas9 system. *Int. J. Cancer.* 2025;156(5):898–913.

Jiang Y, Jiang H, Wang K, Liu C, Man X, Fu Q. Hypoxia enhances the production and antitumor effect of exosomes derived from natural killer cells. *Ann. Transl. Med.* 2021;9(6):473.

Kalodimou VE, Muthu S. Regulatory concerns for exosome-and other extracellular vesicle-based diagnostics and medicine products. In: *Extracell. Vesicles Ther. Diagnostic Appl,* 2025, pp. 523–36.

Kalou Y, Al-Khani AM, Haider KH. Bone marrow mesenchymal stem cells for heart failure treatment: A systematic review and meta-analysis. *Heart Lung Circ.* 2023. https://doi.org/10.1016/j.hlc.2023.01.012.

Kamerkar S, LeBleu VS, Sugimoto H, Yang S, Ruivo CF, Melo SA, Lee JJ, et al. Exosomes facilitate targeted therapy for oncogenic KRAS in pancreatic cancer. *Nature.* 2017;546(7659):498–503. https://doi.org/10.1038/nature22341.

Kang S, Mansurov A, Kurtanich T, Chun HR, Slezak AJ, Volpatti LR, et al. Engineered GM-CSF polarizes protumorigenic tumor-associated macrophages to an antitumorigenic phenotype and potently synergizes with IL-12 immunotherapy. *J. Immunother. Cancer.* 2024;12(12):e009541.

Kapoor KS, Harris K, Arian KA, Ma L, Schueng Zancanela B, Church KA, et al. High-throughput and rapid isolation of extracellular vesicles and exosomes with purity using size-exclusion liquid chromatography. *Bioact. Mater.* 2024;40:683–95.

Kaur R, Bhardwaj A, Gupta S. Cancer treatment therapies: Traditional to modern approaches to combat cancers. *Mol. Biol. Rep.* 2023;50(11):9663–76. https://doi.org/10.1007/s11033-023-08809-3.

Kim H, Jang H, Cho H, Choi J, Hwang KY, Choi Y, et al. Recent advances in exosome-based drug delivery for cancer therapy. *Cancers (Basel).* 2021;13(17):4435.

Kumar A. AI-driven precision oncology: Predictive biomarker discovery and personalized treatment optimization using genomic data. *Int. J. Adv. Res. Publ. Rev.* 2024;1(3):21–38.

Lee YJ, Shin KJ, Chae YC. Regulation of cargo selection in exosome biogenesis and its biomedical applications in cancer. *Exp. Mol. Med.* 2024;56(4):877–89. https://doi.org/10.1038/s12276-024-01209-y.

Li B, Kugeratski FG, Kalluri R. A novel machine learning algorithm selects a proteome signature to specifically identify cancer exosomes. *Elife.* 2024b;12:RP90390.

Li J, Kuang X. Global cancer statistics of young adults and their changes in the past decade: Incidence and mortality from GLOBOCAN 2022. *Public Health.* 2024;237:336–43.

Li Q, He G, Yu Y, Li X, Peng X, Yang L. Exosome crosstalk between cancer stem cells and tumor microenvironment: cancer progression and therapeutic strategies. *Stem Cell Res. Ther.* 2024a;15(1):449.

Li X, Corbett AL, Taatizadeh E, Tasnim N, Little JP, Garnis C, et al. Challenges and opportunities in exosome research: Perspectives from biology, engineering, and cancer therapy. *APL Bioeng.* 2019;3(1). https://doi.org/10.1063/1.5087122.

Li X, Guan J, Jiang Z, Cheng S, Hou W, Yao J, Wang Z. Microglial exosome miR-7239-3p promotes glioma progression by regulating circadian genes. *Neurosci Bull.* 2021 Apr;37(4):497–510. https://doi.org/10.1007/s12264-020-00626-z.

Liang S-F, Zuo F-F, Yin B-C, Ye B-C. Delivery of siRNA based on engineered exosomes for glioblastoma therapy by targeting STAT3. *Biomater. Sci.* 2022;10(6):1582–90.

Liang Y, Duan L, Lu J, Xia J. Engineering exosomes for targeted drug delivery. *Theranostics.* 2021;11(7):3183.

Liang Y, Xia J, Xu L, Zhang Y, Xu X, Haider KH. Cell-free and cell-based therapy in regenerative medicine. In: Haider, K.H. (eds) *Handbook of Regenerative Medicine: Cell-Free Therapy Approach* (1st ed.). CRC Press, 2025. Chapter 24. https://doi.org/10.1201/9781003611509.

Lin H, Li J, Wang M, Zhang X, Zhu T. Exosomal long noncoding RNAs in NSCLC: Dysfunctions and clinical potential. *J Cancer.* 2023 Jun 12;14(10):1736–50. https://doi.org/10.7150/jca.84506.

Lin Z, Wu Y, Xu Y, Li G, Li Z, Liu T. Mesenchymal stem cell-derived exosomes in cancer therapy resistance: Recent advances and therapeutic potential. *Mol. Cancer.* 2022;21(1):179. https://doi.org/10.1186/s12943-022-01650-5.

Liu JJJ, Liu D, To SKY, Wong AST. Exosomes in cancer nanomedicine: Biotechnological advancements and innovations. *Mol. Cancer.* 2025;24(1):1–29.

Liu M, Zhang Y, He J, Liu W, Li Z, Zhang Y, et al. Fusion with ARRDC1 or CD63: A strategy to enhance p53 loading into extracellular vesicles for tumor suppression. *Biomolecules.* 2024;14(5):591.

López de Andrés J, Griñán-Lisón C, Jiménez G, Marchal JA. Cancer stem cell secretome in the tumor microenvironment: A key point for an effective personalized cancer treatment. *J. Hematol. Oncol.* 2020;13(1):136. https://doi.org/10.1186/s13045-020-00966-3.

Lorey MB. Secretome analysis of human macrophages activated by microbial stimuli. University of ,Helsinki, Helsinki 2017.

Lorite P, Domínguez JN, Palomeque T, Torres MI. Extracellular vesicles: Advanced tools for disease diagnosis, monitoring, and therapies. *Int. J. Mol. Sci.* 2025;26.

Louro AF, Meliciano A, Alves PM, Costa MHG, Serra M. A roadmap towards manufacturing extracellular vesicles for cardiac repair. *Trends Biotechnol.* 2024;42(10):1035–322.

Lyons S, Salgaonkar S, Flaherty GT. International stem cell tourism: A critical literature review and evidence-based recommendations. *Int. Health.* 2022;14(2):132–41.

Lyu C, Sun H, Sun Z, Liu Y, Wang Q. Roles of exosomes in immunotherapy for solid cancers. *Cell Death Dis.* 2024;15(2):106.

Ma Y, Yuwen D, Chen J, Zheng B, Gao J, Fan M, Xue W, et al. Exosomal transfer of cisplatin-induced miR-425-3p confers cisplatin resistance in NSCLC through activating autophagy. *Int J Nanomedicine.* 2019;14:8121–32. https://doi.org/10.2147/IJN.S221383.

Madhan S, Dhar R, Devi A. Plant-derived exosomes: A green approach for cancer drug delivery. *J. Mater. Chem. B.* 2024;12.

Makler A, Asghar W. Exosomal biomarkers for cancer diagnosis and patient monitoring. *Expert Rev. Mol. Diagn.* 2020;20(4):387–400.

Mao J, Wang L, Wu J, Wang Y, Wen H, Zhu X, Wang B, et al. miR-370-3p as a novel biomarker promotes breast cancer progression by targeting FBLN5. *Stem Cells Int.* 2021 Aug 23;2021:4649890. https://doi.org/10.1155/2021/4649890.

Márton É, Varga A, Domoszlai D, Buglyó G, Balázs A, Penyige A, Balogh I, et al. Non-coding RNAs in cancer: Structure, function, and clinical application. *Cancers (Basel).* 2025;17(4):579. https://doi.org/10.3390/cancers17040579.

Mashouri L, Yousefi H, Aref AR, Ahadi A Mohmad, Molaei F, Alahari SK. Exosomes: Composition, biogenesis, and mechanisms in cancer metastasis and drug resistance. *Mol. Cancer.* 2019;18(1):75. https://doi.org/10.1186/s12943-019-0991-5.

McAndrews KM, Xiao F, Chronopoulos A, LeBleu VS, Kugeratski FG, Kalluri R. Exosome-mediated delivery of CRISPR/Cas9 for targeting of oncogenic KrasG12D in pancreatic cancer. *Life Sci. Alliance.* 2021;4(9).

Melo SA, Luecke LB, Kahlert C, Fernandez AF, Gammon ST, Kaye J, LeBleu VS, et al. Glypican-1 identifies cancer exosomes and detects early pancreatic cancer. *Nature.* 2015;523(7559):177–82. https://doi.org/10.1038/nature14581.

Mendt M, Kamerkar S, Sugimoto H, McAndrews KM, Wu CC, Gagea M, Yang S, et al. Generation and testing of clinical-grade exosomes for pancreatic cancer. *JCI Insight.* 2018;3(8):e99263. https://doi.org/10.1172/jci.insight.

Mirgh D, Sonar S, Ghosh S, Adhikari MD, Subramaniyan V, Gorai S, et al. Landscape of exosomes to modified exosomes: A state of the art in cancer therapy. *RSC Adv.* 2024;14(42):30807–29.

Moeinzadeh L, Razeghian-Jahromi I, Zarei-Behjani Z, Bagheri Z, Razmkhah M. Composition, biogenesis, and role of exosomes in tumor development. *Stem Cells Int.* 2022;2022(1):8392509.

Moni ZA, Hasan Z, Alam MS, Roy N, Islam F. Diagnostic and prognostic significance of exosomes and their components in patients with cancers. *J Cell Commun Signal.* 2022;16(3):267–82.

Moni ZA, Hasan Z, Alam MS, Roy N, Islam F. Diagnostic and prognostic significance of exosomes and their components in patients with cancers. *Cancer Med.* 2025 Jan;14(1):e70569. https://doi.org/10.1002/cam4.70569.

Moon B, Cells SC. Exosome as a delivery vehicle for cancer therapy. mdpi.com. 2022. www.mdpi.com/2073-4409/11/3/316.

Moreno R. Mesenchymal stem cells and oncolytic viruses: Joining forces against cancer. *J. Immunother. Cancer.* 2021;9(2):e001684.

Morimoto M, Maishi N, Tsumita T, Alam MT, Kikuchi H, Hida Y, Yoshioka Y, et al. miRNA-1246 in tumor extracellular vesicles promotes metastasis via increased tumor cell adhesion and endothelial cell destruction. *Front. Oncol.* 2023;13:973871.

Morini M, Vitale C, Ardito M, Dondero A, Cortese K, Bottino C, Castriconi R. Exosomes and immune modulation: Implications for neuroblastoma immunotherapy. *Frontiers in Immunology.* 2025;16:1600062.

Mushtaq M, Zineldeen DH, Mateen MA, Haider KH. Mesenchymal stem cells' "garbage bags" at work: Treating radial nerve injury with mesenchymal stem cell-derived exosomes. *World J Stem Cells* 2024;16(5). https://doi.org/10.4252/wjsc.v16.i5.0000.

Muttiah B, Ng SL, Lokanathan Y, Ng MH, Law JX. Extracellular vesicles in breast cancer: From intercellular communication to therapeutic opportunities. *Pharmaceutics.* 2024;16(5):654.

Najimi M, Haider KH. Extracellular vesicle-based Cell-free therapy for liver regeneration. In: Haider, K.H. (ed.) *Handbook of Stem Cell Therapy.* Singapore: Springer, 2022. https://doi.org/10.1007/978-981-16-6016-0_9-1.

Ocansey DKW, Zhang L, Wang Y, Yan Y, Qian H, Zhang X, et al. Exosome-mediated effects and applications in inflammatory bowel disease. *Biol. Rev.* 2020;95(5):1287–307.

Orefice NS, Di Raimo R, Mizzoni D, Logozzi M, Fais S. Purposing plant-derived exosomes-like nanovesicles for drug delivery: Patents and literature review. *Expert Opin. Ther. Pat.* 2023;33(2):89–100.

Pan W, Miao Q, Yin W, Li X, Ye W, Zhang D, et al. The role and clinical applications of exosomes in cancer drug resistance. *Cancer Drug Resistance.* 2024;7:43.

Pitt JM, Charrier M, Viaud S, André F, Besse B, Chaput N, et al. Dendritic cell–derived exosomes as immunotherapies in the fight against cancer. *J. Immunol.* 2014;193(3):1006–11.

Preethi KA, Selvakumar SC, Ross K, Jayaraman S, Tusubira D. Liquid biopsy: Exosomal microRNAs as novel diagnostic and prognostic biomarkers in cancer. *Cancer Cell Int.* 2022;22(1):56.

Qiu J, Jiang Y, Ye N, Jin G, Shi H, Qian D. Leveraging the intratumoral microbiota to treat human cancer: Are engineered exosomes an effective strategy? *J. Transl. Med.* 2024;22(1):728.

Rahnama M, Ghasemzadeh N, Latifi Z, Kheradmand F, Koukia FA, Sharun K, et al. Menstrual blood and endometrial mesenchymal stem/stromal cells: A frontier in regenerative medicine and cancer therapy. *Eur. J. Pharmacol.* 2025;1000:177726.

Rahnama M, Heidari M, Poursalehi Z, Golchin A. Global trends of exosomes application in clinical trials: A scoping review. *Stem Cell Rev. Reports.* 2024;20(8):2165–93. https://doi.org/10.1007/s12015-024-10791-7.

Rai A, Claridge B, Lozano J, Greening DW. The discovery of extracellular vesicles and their emergence as a next-generation therapy. *Circ. Res.* 2024;135(1):198–221.

Raue R, Frank A-C, Syed SN, Brüne B. Therapeutic targeting of MicroRNAs in the tumor microenvironment. *Int. J. Mol. Sci.* 2021;22(4):2210.

Ren B, Li X, Zhang Z, Tai S, Yu S. Exosomes: A significant medium for regulating drug resistance through cargo delivery. *Front. Mol. Biosci.* 2024a;11:1379822.

Ren M, Wang X, Zhang Y, Wu B, Xu Y, Hu J, Jin H. Emerging roles of extracellular vesicles in targeted cancer therapy: Mechanisms and applications. *Front. Bioeng. Biotechnol.* 2024b;12:1476737. https://doi.org/10.3389/fbioe.2024.1476737.

Resnik DB, Hosseini M. The ethics of using artificial intelligence in scientific research: New guidance needed for a new tool. *AI Ethics.* 2025;5.

Ritchie S, Reed DA, Pereira BA, Timpson P. The cancer cell secretome drives the cooperative manipulation of the tumor microenvironment to accelerate tumorigenesis. *Fac. Rev.* 2021;10:4.

Roccaro AM, Sacco A, Maiso P, Azab AK, Tai YT, Reagan M, Azab F, et al. BM mesenchymal stromal cell-derived exosomes facilitate multiple myeloma progression. *J Clin Invest.* 2013;123(4):1542–55. https://doi.org/10.1172/JCI66517.

Safaei S, Fadaee M, Farzam OR, Yari A, Poursaei E, Aslan C, et al. Exploring the dynamic interplay between exosomes and the immune tumor microenvironment: implications for breast cancer progression and therapeutic strategies. *Breast Cancer Res.* 2024;26(1):57.

Samuels M, Giamas G. MISEV2023: Shaping the future of EV research by enhancing rigour, reproducibility and transparency. *Cancer Gene Ther.* 2024;31(5):649–51.

Sani F, Shojaei S, Tabatabaei SA, Khorraminejad-Shirazi M, Latifi M, Sani M, et al. CAR-T cell-derived exosomes: A new perspective for cancer therapy. *Stem Cell Res. Ther.* 2024;15(1):174.

Santos P, AlMeida F. Exosome-based vaccines: History, current state, and clinical trials. *Front. Immunol.* 2021;711565. https://doi.org/10.3389/fimmu.2021.711565.

Sergazy S, Seydahmetova R, Gulyayev A, Shulgau Z, Aljofan M. The role of exosomes in cancer progression and therapy. *Biology (Basel).* 2025;14: 27.

Shah D, Gandhi S, Shah S, Patel K. Recent breakthroughs in exosome-based drug delivery: A comprehensive review for cancer therapy. *Cancer Biother. Radiopharm.* 2025;40.

Sharma A, Yadav A, Nandy A, Ghatak S. Insight into the functional dynamics and challenges of exosomes in pharmaceutical innovation and precision medicine. *Pharmaceutics.* 2024;16(6):709.

Shen S, Song Y, Zhao B, Xu Y, Ren X, Zhou Y, Sun Q. Cancer-derived exosomal miR-7641 promotes breast cancer progression and metastasis. *Cell Commun. Signaling.* 2021;19(1):20.

Sheykhhasan M, Kalhor N, Sheikholeslami A, Dolati M, Amini E, Fazaeli H. Exosomes of mesenchymal stem cells as a proper vehicle for transfecting miR-145 into the breast cancer cell line and its effect on metastasis. *Biomed. Res. Int.* 2021;2021(1):5516078.

Sinha D, Roy S, Saha P, Chatterjee N, Bishayee A. Trends in research on exosomes in cancer progression and anticancer therapy. *Cancers (Basel).* 2021;13(2):1–31.

Song H, Liu B, Dong B, Xu J, Zhou H, Na S, et al. Exosome-based delivery of natural products in cancer therapy. *Front. Cell Dev. Biol.* 2021;9:650426.

Song Z, Tao Y, Liu Y, Li J. Advances in delivery systems for CRISPR/Cas-mediated cancer treatment: a focus on viral vectors and extracellular vesicles. *Front. Immunol.* 2024;15:1444437.

Stawarska A, Bamburowicz-Klimkowska M, Runden-Pran E, Dusinska M, Cimpan MR, Rios-Mondragon I, et al. Extracellular vesicles as next-generation diagnostics and advanced therapy medicinal products. *Int. J. Mol. Sci.* 2024;25(12):6533.

Tang Y, Liu X, Sun M, Xiong S, Xiao N, Li J, et al. Recent progress in extracellular vesicle-based carriers for targeted drug delivery in cancer therapy. *Pharmaceutics.* 2023;15(7):1902.

Tenchov R, Sasso JM, Wang X, Liaw W-S, Chen C-A, Zhou QA. Exosomes— nature's lipid nanoparticles, a rising star in drug delivery and diagnostics. *ACS Nano.* 2022;16(11):17802–46.

Ter-Ovanesyan D, Norman M, Lazarovits R, Trieu W, Lee J-H, Church GM, et al. Framework for rapid comparison of extracellular vesicle isolation methods. *Elife.* 2021;10.

Théry C, Witwer KW, Aikawa E, Alcaraz MJ, Anderson JD, Andriantsitohaina R, et al. Minimal information for studies of extracellular vesicles 2018 (MISEV2018): A position statement of the International Society for Extracellular Vesicles and update of the MISEV2014 guidelines. *J. Extracell. Vesicles.* 2018;7(1):1535750.

Tian T, Liang R, Erel-Akbaba G, Saad L, Obeid PJ, Gao J, et al. Immune checkpoint inhibition in GBM primed with radiation by engineered extracellular vesicles. *ACS Nano.* 2022;16(2):1940–53.

Tsuchiya A, Terai S, Horiguchi I, Homma Y, Saito A, Nakamura N, Sato Y, Ochiya T, Kino-oka M. Basic points to consider regarding the preparation of extracellular vesicles and their clinical applications in Japan. *Regen.Ther.* 2022;21:19–24.

U.S. Food and Drug Administration. Public safety notification on exosome products. www.fda.gov/vaccines-blood-biologics/safety-availability-biologics/public-safety-notification-exosome-products.

Vader P, Mol EA, Pasterkamp G, Schiffelers RM. Extracellular vesicles for drug delivery. *Adv Drug Delivery Rev.* 2016;106:148–56. https://doi.org/10.1016/j.addr.2016.02.006.

Vagena I-A, Malapani C, Gatou M-A, Lagopati N, Pavlatou EA. Enhancement of EPR effect for passive tumor targeting: current status and future perspectives. *Appl. Sci.* 2025;15(6):3189.

Vaidya M, Sugaya K. DNA associated with circulating exosomes as a biomarker for glioma. *Genes (Basel).* 2020 Oct 29;11(11):1276. https://doi.org/10.3390/genes11111276.

Van De Wakker SI, Meijers FM, Sluijter JPG, Vader P. Extracellular vesicle heterogeneity and its impact for regenerative medicine applications. *Pharmacol. Rev.* 2023;75(5):1043–61.

Viaud S, Théry C, Ploix S, Tursz T, Lapierre V, Lantz O, Zitvogel L, et al. Dendritic cell-derived exosomes for cancer immunotherapy: What's next? *Cancer Res.* 2010 Feb 15;70(4):1281–5. https://doi.org/10.1158/0008-5472.CAN-09-3276.

Wang C, Zhou L, Kang X, Huang C-H, Gao Z, Shen J, et al. A nanoplasmonic cell-on-a-chip for in situ monitoring of PD-L1+ exosome-mediated immune modulation. *Biosens. Bioelectron.* 2025b;277:117293.

Wang C-K, Tsai T-H, Lee C-H. Regulation of exosomes as biologic medicines: Regulatory challenges faced in exosome development and manufacturing processes. *Clin. Transl. Sci.* 2024b;17(8):e13904.

Wang M, Altinoglu S, Takeda YS, Xu Q. Integrating protein engineering and bioorthogonal click conjugation for extracellular vesicle modulation and intracellular delivery. *PLoS One.* 2015;10(11):e0141860.

Wang R, Xu Y, Tong L, Zhang X, Zhang S. Recent progress of exosomal lncRNA/circRNA-miRNA-mRNA axis in lung cancer: implication for clinical application. *Front Mol Biosci.* 2024a Jul 3;11:1417306. https://doi.org/10.3389/fmolb.2024.1417306.

Wang T, Fu Y, Sun S, Huang C, Yi Y, Wang J, et al. Exosome-based drug delivery systems in cancer therapy. *Chinese Chem. Lett.* 2023a;34(2):107508.

Wang W, Kong P, Feng K, Liu C, Gong X, Sun T, Duan X, et al, Tao Z, Liu W. Exosomal miR-222-3p contributes to castration-resistant prostate cancer by activating mTOR signaling. *Cancer Sci.* 2023b Nov;114(11):4252–69. https://doi.org/10.1111/cas.15948.

Wang X, Li A, Wang A, He M, Zeng Y, Li D, et al. Exosome-based vaccines: Pioneering new frontiers in combating infectious diseases and cancer. *Small Methods.* 2025a;2025:2402222.

Wang X, Tian L, Lu J, Ng IO-L. Exosomes and cancer: Diagnostic and prognostic biomarkers and therapeutic vehicles. *Oncogenesis.* 2022;11(1):54. https://doi.org/10.1038/s41389-022-00431-5.

Wang Z, Wang Q, Qin F, Chen J. Exosomes: A promising avenue for cancer diagnosis beyond treatment. *Front. Cell Dev. Biol.* 2024c;12:1344705. www.frontiersin.org/articles/10.3389/fcell.2024.1344705/full.

Waqas MY, Javid MA, Nazir MM, Niaz N, Nisar MF, Manzoor Z, et al. Extracellular vesicles and exosomes: Insight from physiological regulatory perspectives. *J. Physiol. Biochem.* 2022;78(3):573–80.

Welsh JA, Goberdhan DCI, O'Driscoll L, Buzas EI, Blenkiron C, Bussolati B, et al. Minimal information for studies of extracellular vesicles (MISEV2023): From basic to advanced approaches. *J. Extracell. Vesicles.* 2024;13(2):e12404.

Witwer KW, Goberdhan DC, O'Driscoll L, Théry C, Welsh JA, Blenkiron C, et al. Updating MISEV: Evolving the minimal requirements for studies of extracellular vesicles. *J. Extracell. Vesicles.* 2021;10:e12182.

Wu L, Wang J, Chen Y, et al. Comparative analysis of serum and serum exosomal microRNAs for breast cancer diagnosis. *BMC Cancer.* 2020;20:456.

Wu Y, Cao Y, Chen L, Lai X, Zhang S, Wang S. Role of exosomes in cancer and aptamer-modified exosomes as a promising platform for cancer targeted therapy. *Biol. Proced.* 2024a;26(1):15.

Wu Y, Fu H, Hao J, Yang Z, Qiao X, Li Y, et al. Tumor-derived exosomal PD-L1: a new perspective in PD-1/PD-L1 therapy for lung cancer. *Front. Immunol.* 2024b;15:1342728.

Wu Y, Wang Y, Lu Y, Luo X, Huang Y, Xie T, et al. Microfluidic technology for the isolation and analysis of exosomes. *Micromachines.* 2022;13(10):1571.

Xing Y-H, Ren X-S, Li D-H, Liu L. Exosome separation and analysis based on microfluidics technology and its clinical applications. *Chinese J. Chromatogr.* 2025;43(5):455–71.

Xu S, Zhang G, Li G. Therapeutic potential of non-coding RNAs in cancer. *X-Disciplinarity.* 2025;1(2):4.

Yadav A, Xuan Y, Sen CK, Ghatak S. Standardized reporting of research on exosomes to ensure rigor and reproducibility. *Adv. Wound Care.* 2024b;13(11):584–99.

Yadav R, Singh AV, Kushwaha S, Chauhan DS. Emerging role of exosomes as a liquid biopsy tool for diagnosis, prognosis & monitoring treatment response of communicable & non-communicable diseases. *Indian J Med Res.* 2024a;159(2):163–80.

Yang J, Tang S, Saba NF, Shay C, Teng Y. Tumor secretome shapes the immune landscape during cancer progression. *J. Exp Clin Cancer Res.* 2025;44(1):47. https://doi.org/10.1186/s13046-025-03302-0.

Yang Q, Xu J, Gu J, Shi H, Zhang J, Zhang J, et al. Extracellular vesicles in cancer drug resistance: Roles, mechanisms, and implications. *Adv. Sci.* 2022;9(34):e2201609.

Yang S, Che SP, Kurywchak P, Tavormina JL, Gansmo LB, Correa de Sampaio P, Tachezy M, et al. Detection of mutant KRAS and TP53 DNA in circulating exosomes from healthy individuals and patients with pancreatic cancer. *Cancer Biol Ther.* 2017;18(3):158–65. https://doi.org/10.1080/15384047.2017.1281499.

Yang Y, Chen Y, Zhou X, et al. The role of exosomal microRNAs in glioma progression and therapy. *Front Mol Neurosci.* 2024;17:149.

Yoon JH, Park YG, Nam SW, Park WS. The diagnostic value of serum gastrokine 1 (GKN1) protein in gastric cancer. *Cancer Med.* 2019;8(12):5507–14. https://doi.org/10.1002/cam4.2457.

Youssef E, Palmer D, Fletcher B, Vaughn R. Exosomes in precision oncology and beyond: From bench to bedside in diagnostics and therapeutics. *Cancers (Basel).* 2025;17:940.

Yu D, Li Y, Wang M, Gu J, Xu W, Cai H, et al. Exosomes as a new frontier of cancer liquid biopsy. *Mol. Cancer.* 2022;21(1):56.

Zemanek T, Danisovic L, Nicodemou A. Exosomes, their sources, and possible uses in cancer therapy in the era of personalized medicine. *J. Cancer Res. Clin. Oncol.* 2025;151(1):1–14.

Zhang B, Zhao J, Jiang M, Peng D, Dou X, Song Y, et al. The potential role of gut microbial-derived exosomes in metabolic-associated fatty liver disease: Implications for treatment. *Front. Immunol.* 2022;13:893617.

Zhang M, Hu S, Liu L, Dang P, Liu Y, Sun Z, et al. Engineered exosomes from different sources for cancer-targeted therapy. *Signal Transduct. Target. Ther.* 2023;8(1).

Zhou Y, Tao L, Qiu J, Xu J, Yang X, Zhang Y, et al. Tumor biomarkers for diagnosis, prognosis, and targeted therapy. *Signal Transduct. Target. Ther.* 2024;9(1):132. https://doi.org/10.1038/s41392-024-01823-2.

Zhu M, Gao Y, Zhu K, Yuan Y, Bai H, Meng L. Exosomal miRNA as biomarker in cancer diagnosis and prognosis: A review. *Medicine.* 2024;103(42):e40082. https://doi.org/10.1097/MD.0000000000040082.

Zhuang X, Xiang X, Grizzle W, Sun D, Zhang S, Axtell RC, Zhang HG. Treatment of brain inflammatory diseases by delivering exosome-encapsulated anti-inflammatory drugs from the nasal region to the brain. *Mol. Ther.* 2011;19(10):1769–79. https://doi.org/10.1038/mt.2011.164.

25 Exosome-Based Therapy
Therapeutic Potential and Limitations

Mahmood S Choudhery[1], *Taqdees Arif*[1], *Ruhma Mahmood*[2], *and David T. Harris*[3]

[1]Department of Human Genetics and Molecular Biology, University of Health Sciences, Lahore, Pakistan

[2]Allama Iqbal Medical College, Jinnah Hospital, Lahore, Pakistan

[3]Department of Immunobiology and University of Arizona Health Sciences Biorepository, College of Medicine, University of Arizona, Tucson, AZ, USA

25.1 ABBREVIATIONS

ASCs	Adult stem cells
Alix	ALG-2-interacting protein X
BSCB	Blood-spinal cord barrier
CD	Cluster of differentiation
CVDs	Cardiovascular diseases
DCs	Dendritic cells
DLS	Dynamic light scattering
DMOAD	Disease-modifying osteoarthritis drug
EDs	Eye diseases
ESCRT	Endosomal sorting complex transport
FDA	Food and Drug Administration
HSPs	Heat shock proteins
hAMSCs	Human amniotic mesenchymal stem cells
HD-PTP	Histidine domain-containing phosphotyrosine phosphatase
HESCs	Human embryonic stem cells
iPSCs	Induced pluripotent stem cells
IPFPs	Infrapatellar fat pads
IHD	Ischemic heart disease
IRI	Ischemia-reperfusion injury
PEG	Polyethylene glycol
HER-2	Human epidermal growth factor receptor 2
Exo	Exosomes
EVs	Extracellular vesicles
EOPA	Ethyl-N-isopropyl amiloride
ICAM-1	Intercellular adhesion molecule

DOI: 10.1201/9781003653387-25

MVBs	Multivesicular bodies
MNs	Micro needle patches
MHC	Major histocompatibility complex
miRNA	MicroRNA
mRNA	Messenger RNA
MSCs	Mesenchymal stem cells
MI	Myocardial infarction
ALS	Amyotrophic lateral sclerosis
NSCs	Neural stem cells
NTA	Nanoparticle tracking analysis
NGS	Next-generation sequencing
OA	Osteoarthritis
PTX	Paclitaxel
PtdSer	Phosphatidylserine
rRNA	Ribosomal RNA
RPE	Retinal pigment epithelium
SPRM	Surface plasmon resonance microscopy
SMECs	Spinal cord microvascular endothelial cells
TRAIL	Tumor necrosis factor-related apoptosis-inducing ligand
TNF	Tumor necrosis factor
TEM	Transmission electron microscopy
TSG-6	TNF-stimulated gene-6
VEGF	Vascular endothelial growth factor

25.2 INTRODUCTION

25.2.1 Overview of Exosomes

Exosomes, a category of extracellular vesicles, are spherical vesicles encapsulated in lipid membranes. Exosomes are secreted by almost all cell types in multicellular organisms. The most common pathway of their biogenesis is the endosomal pathway. They are nanoscale particles with sizes ranging from 40 nm to 160 nm. The discovery of exosomes in the extracellular environment dates back to the late 1980s. Initially, it was suggested that exosomes are the byproducts of cell homeostasis, serving as cellular trash resulting from cell injury. At that time, their role in affecting the surrounding cells was also not prominent. Recently, these extracellular vesicles have been recognized as functional vehicles that transport a complex array of lipids, proteins, and nucleic acids to target cells. Exosomes facilitate intercellular communication and can potentially influence cellular processes, including cellular development and differentiation, immune system regulation, signal transduction, and gene expression. The stability, biocompatibility, and ability to traverse cellular barriers are unique properties that make exosomes potential candidates for various therapeutic applications (Choudhery et al., 2024). Exosomes can alter their function in response to changes in the body's physiological conditions. Interestingly, exosomes may also serve as biomarkers of specific diseases, thereby opening new avenues for multiparameter diagnostic testing, which involves the simultaneous analysis of multiple biomarkers, proteins, lipids, and nucleic acids, to provide a comprehensive understanding of disease mechanisms, diagnosis, and prognosis (Kalluri and LeBleu, 2020).

Naturally, exosomes are found in several body fluids, including saliva, semen, amniotic fluid, and plasma. Moreover, exosomes are secreted by several cell types, including immune cells (macrophages, dendritic cells), cancer cells, stem cells (adult stem cells, neonatal stem cells, embryonic stem cells), and cells of other adult tissues, including uterine cells, epithelial cells, endometrial cells, and follicular cells (Choudhery et al., 2024). The functional capabilities of exosomes can be changed by manipulating their natural composition to target a specific therapeutic application.

It is possible to engineer exosomes to enhance their inherent qualities for therapeutic purposes. Engineered exosomes may exhibit enhanced stability, selectivity, efficiency, and bioavailability, making them ideal for drug delivery, gene therapy, and immunomodulation (Bahadorani et al., 2024). The varied actions of these engineered exosomes make them advantageous for a wide range of therapeutic applications.

Exosomes are involved in tissue regeneration as they transfer cellular components and replicate the physiological state of donor cells in recipient cells (Bobrie et al., 2011). The contents of exosomes can vary widely depending on the cell type and its state, such as when it is stimulated, differentiated, transformed, or stressed (Liang et al., 2025). The biologically active contents of exosomes have significant roles in the treatment of various diseases, including Osteoarthritis (OA), spinal cord injury, cardiovascular diseases, skin diseases, aesthetics, neurodegenerative diseases, such as Parkinson's disease, Alzheimer's disease, and multiple sclerosis, and eye diseases (Zhang et al., 2019; Ahmed et al., 2023; Ahmed et al., 2025; Safwan et al., 2025). This chapter provides a brief overview of exosome genesis, composition, and mechanism of action. Furthermore, methods for their isolation and characterization are described. This chapter also summarizes the current state of exosome-based therapy, discussing its potential, limitations, and possible future applications.

25.3 EXOSOME SOURCES AND COMPOSITION

Exosomes are naturally released by cells of multicellular organisms (both plant and animal cells), allowing cell-cell communication. Exosomes can originate from various sources, including adult human cells, tumor cells, immune cells, and stem cells. The release of exosomes by dendritic cells (DCs), monocytes, and macrophages helps in the development and activation of adaptive immunity (Choudhery et al., 2024).

The cargo of exosomes consists of proteins, lipids, nucleic acids, and various other signaling molecules, including cytokines, chemokines, and growth factors. Exosome cargo is involved in cell signaling and intercellular communication in both distant and nearby sites. The cell type and physiological condition of the parent cell determine the exact composition of exosomes. The lipid components of exosomes consist of cholesterol, sphingomyelin, ceramides, phosphatidylserine, and other lipid derivatives (such as sphingolipids). These lipids are standard components of exosome membranes (Bahadorani et al., 2024). Their lipid bilayer surrounds an aqueous core containing proteins, nucleic acids, metabolites, and signaling molecules. The drug-delivery capacity of exosomes is enhanced by their amphiphilic nature, which enables them to interact with both lipophilic and hydrophilic molecules. Exosomal protein components are diverse. Cytoskeletal proteins, membrane-associated proteins, various types of enzymes (e.g., proteases, oxidases), tetraspanins (e.g. CD9, CD63, CD81), GTPases, annexins, flotillin, lysosomal proteins, antigen presentation molecules, such as major histocompatibility complex (MHC) class I and II, and intercellular adhesion molecule (ICAM-1) are major protein components of exosomes. Additionally, exosomes contain heat shock proteins (HSPs), such as HSP90 and HSP70. These transmembrane proteins interact with integrins and MHC molecules, playing a crucial role in the development of intricate molecular networks that facilitate cell-to-cell communication. Proteins such as ALIX, TSG101, and VPS4 are involved in exosome production through the endosomal sorting complex required for transport (ESCRT) pathway. In contrast, tetraspanin proteins, specifically CD9, CD37, CD63, CD81, and CD82, are involved in a pathway that is independent of the ESCRT pathway. Some proteins contained in exosomes are specific to their biological origin. Examples include MHC class I and II molecules, as well as proteins specific to various immune cells, such as CD3 (T cells), CD19 (B cells), and CD14 (monocytes and macrophages). Exosomes also transport RNA and proteins across cells and to distant organs, facilitating intercellular communication. In addition, it was demonstrated that exosomes, when delivered to specific cells, have the potential to alter the homeostasis of those cells by altering their lipid profiles, particularly those of cholesterol and sphingomyelin. With over 47,000 entries for proteins, mRNAs, and lipids, ExoCarta is a database that incorporates all published and unpublished data on exosome content (Mathivanan et al., 2012).

25.4 MECHANISM OF EXOSOME BIOGENESIS AND CELLULAR UPTAKE

Cargoes located on the external membrane surface of exosomes can directly interact with recipient cells, either facilitating specific targeting or initiating signaling events. Exosome uptake can lead to the release of internal payloads into recipient cells. Exosomes serve as biological messengers for the transfer of information (Exosomal intercellular communication relies on the combination of internal and external payloads. The function of exosomes depends on their internal contents. Exosome loading with proteins, lipids, and RNAs is selective and depends on the cell type. It is therefore essential to understand the mechanisms of content sorting and loading. Such understanding could lead to advancements in diagnostic and treatment approaches using exosomes (Lee et al., 2024).

25.4.1 Exosomes Biogenesis

Exosome biogenesis refers to the process by which cells produce and release exosomes into the extracellular space. Multivesicular body (MVB) formation is the most common mechanism of exosome biogenesis, as shown in Figure 25.1. First, the cells invaginate extracellular material through a process called endocytosis. A small sac is produced because of this invagination. This sac is called an early endosome. Early endosomes accumulate specific proteins and lipids from the cell

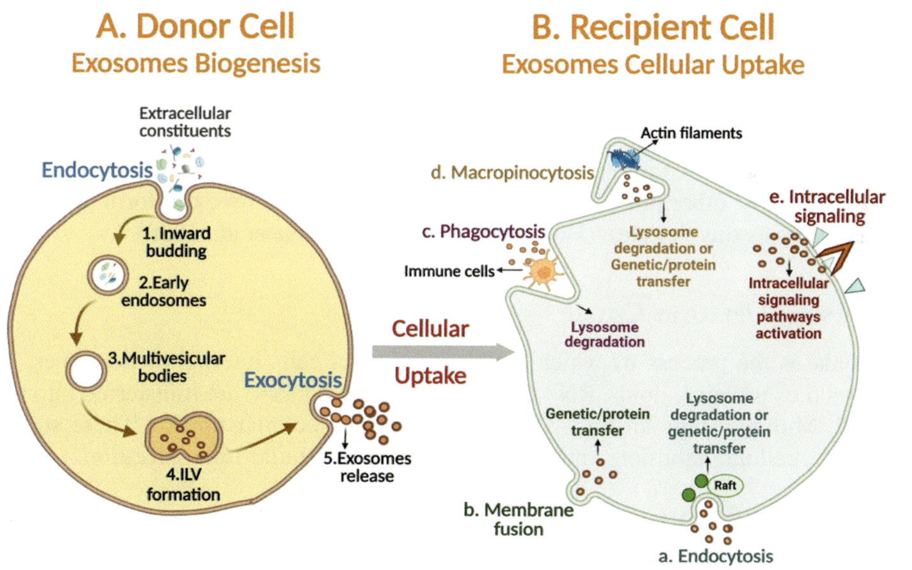

FIGURE 25.1 Exosomes biogenesis and cellular uptake. (A) Donor cell-exosome biogenesis: The plasma membrane undergoes endocytosis to form early endosomes. Early endosomes mature into multivesicular bodies (MVBs). Intraluminal vesicles (ILVs) are formed inside MVBs by inward budding of the membranes. MVBs fuse with the plasma membrane to exocytose ILVs as exosomes. (B) Recipient cell exosome uptake: Exosomes are taken up by cells through multiple mechanisms. a) Endocytosis: Exosomes are internalized into endocytic compartments through raft-receptor-mediated endocytosis. b) Membrane fusion: Exosomes directly fuse with the recipient cell membrane, delivering their contents. c) Phagocytosis: Immune cells engulf exosomes as part of cellular uptake. d) Macropinocytosis: uptake of exosome fluid via actin filament-driven processes. e) Intracellular signaling: exosomal content triggers signaling pathways, affecting the recipient cell's response.

and mature into late endosomes. Late endosomes are also called MVBs. The membrane of MVBs undergoes inward budding, a process known as MVB endocytosis. It results in the formation of intraluminal vesicles (ILVs). These ILVs eventually become exosomes. Cells release exosomes to communicate with other cells (Gruenberg, 2020). The formation of ILV is regulated by ESCRT-dependent and ESCRT-independent mechanisms.

25.4.1.1 ESCRT-Dependent Pathway

In the ESCRT pathway, different ESCRT sub-complexes (0, I, II, III) and ATPases (such as VPS4) control the formation of ILVs. The process proceeds in a stepwise manner (Wang C et al., 2023a). In the first step, ESCRT-0 recognizes ubiquitinated cargo protein on the endosomal membrane. The binding of ESCRT-0 with the ubiquitinated cargo protein recruits ESCRT-I to the endosomal membrane. The recruitment of ESCRT-1 stabilizes the interaction between ESCRT-0 and the endosomal membrane. In the next step, ESCRT-II binds with ESCRT-1 and deforms the endosomal membrane. This process initiates ILV formation inside the endosomes. Finally, ESCRT-III polymerizes to form a spiral filament. It facilitates the inward budding of the endosomal membrane and scission of ILV. The VPS4 disassembles the ESCRT-III filament, completing ILV formation. ILVs are then released into the MVB lumen.

25.4.1.2 ESCRT-Independent Pathway

The ESCRT-independent pathway bypasses the participation of ESCRT sub-complexes. Although the exact mechanism of the ESCRT-independent pathway in ILV formation remains unclear, several proteins are involved. For example, huntingtin-interacting protein D, protein tyrosine phosphatase (HD-PTP), and Alix can recognize and bind to specific cargo proteins of exosomes. These proteins recruit vacuolar protein sorting 4 (VPS4) to the outer membrane of MVBs and facilitate the formation of ILVs within MVBs. This allows for efficient sorting and packaging of cargo proteins into exosomes. Through this pathway, HD-PTP and Alix enable a direct and efficient mechanism for ILV formation. In addition to HD-PTP and Alix, other proteins, particularly those involved in membrane-associated lipid rafts, such as flotillins, may play essential roles in ILV development independent of ESCRT. CD63, for example, may promote ILV formation through ceramide-independent mechanisms. Similarly, other proteins, such as caveolin-1 and flotillins, can form lipid rafts. Lipid rafts can form ILVs as tiny vesicles. (Han et al., 2022, Tschuschke et al., 2020).

25.4.2 Exosome Uptake by Cells

Exosome uptake is the process by which recipient or target cells internalize exosomes. The exosomal contents (i.e., proteins, lipids, RNAs, and mitochondrial DNA) are transferred into the recipient/target cell. Multiple steps are involved in the exosomal cellular uptake. These steps include exosome release, cellular adhesion, antigen recognition, and cellular internalization.

25.4.2.1 Exosomes Release

Donor cells release exosomes into the body fluids. Exosomes have a circulating half-life of approximately 2 to 30 minutes (Parada et al., 2021). The half-life of exosomes is the time required for half of the exosomes in recipient or target cells to be degraded or eliminated by natural processes. Exosomes have been detected in the gut and brain up to 3 hours after intranasal administration (Zhuang et al., 2011; Gotoh et al., 2024). The half-life of exosomes can be detected in the spleen for up to 2 hours (Saunderson et al., 2014). Exosomes are rapidly absorbed by circulating monocytes and macrophages in the liver and spleen. However, it remains unclear whether this uptake primarily serves to clear damaged or foreign materials or to support cellular signaling. The blood vessel randomly distributes exosomes. Exosomes are distributed randomly throughout blood vessels, moving laterally from the center of red blood cells to the arterial wall, where leukocytes are located (Lee et al., 2013). This process facilitates interactions between exosomes and immune cells (such as T cells, monocytes, and macrophages). Exosomes are then systematically transported to distant sites. Moreover, opsonization of circulatory exosomes, similar to liposomes, can enhance their uptake and targeting (van der Meel et al., 2014).

25.4.2.2 Cellular Adhesion and Antigen Recognition

Adhesion refers to the interaction between exosomes and recipient cells. Recipient cells that bind exosomes can be of many types, including neural stem cells, immune cells (e.g., dendritic cells), and mesenchymal stem cells—the exact mechanism by which exosomes target these cells remains unclear. Once in circulation, exosomes may interact with relevant recipient cells via specific surface markers, including integrins, tetraspanins (such as CD82, CD81, CD63, CD53, and CD9), and glycoproteins. The interaction between these markers facilitates the binding and uptake of exosomes by recipient cells through ligand-receptor, protein-protein, and carbohydrate-protein interactions (Maguire et al., 2012). For example, exosomes produced by mature dendritic cells contain the ICAM-1 (CD54) glycoprotein, and T cells express the leukocyte integrin LFA-1 (CD11a/CD18 or αLβ2). T cell initial binding to exosomes is controlled by ICAM-1/LFA-1 interaction. Adhesion can be strengthened by the expression of leukocyte adhesion molecules, including αL (CD11a), α4 (CD49d), CD44, and ICAM-1, as well as by the production of exosomes that contain the tetraspanins CD9 and CD81 (Atay et al., 2011).

The specific binding of exosome surface antigens (e.g., CD44, CD63, and CD81) to corresponding receptors on the surfaces of recipient immune cells, such as dendritic cells, macrophages, and T cells, facilitates exosome uptake into these cells and initiates an immune response. The combination of antigen and MHC class I (Tc cells) and class II molecules (CD4+ T cells) determines the specificity of T-cell recognition for the target cell (Blanchard et al., 2002). The expression of MHC on exosomes depends on the presence of MHC molecules on the exosome-producing cells. Interestingly, microRNAs contained in certain tumor-derived exosomes may be different from those in the parent cell, possibly due to selective packaging mechanisms (de Candia et al., 2013). But the expression of MHC class I and II molecules always closely resembles the originating cells (Wieckowski et al., 2009). MHC class I molecules of exosomes trigger a negative signal through inhibitory receptors on recipient cells. These inhibitory receptors include KIR2DL4, ILT4, and ILT2. This receptor can suppress the responses of natural killer cells and CD8+ T cells. MHC class II molecules generate a stimulatory signal that facilitates the differentiation and proliferation of CD4+ T cells. Exosomes derived from antigen-presenting cells, including B lymphocytes and dendritic cells, exhibit MHC class II expression (Vincent-Schneider et al., 2002).

25.4.2.3 Cellular Internalization of Exosomes

It is the process by which exosomes are transported across the cell membrane, and their contents are released into the cytoplasm, thereby interacting with cellular machinery. The following steps are involved in cellular internalization.

25.4.2.3.1 Endocytosis

Endocytosis is the process by which cells take in substances from outside the cell by engulfing them with a portion of the cell membrane, forming a vesicle that contains the ingested material. Raft-mediated endocytosis and receptor-mediated endocytosis are commonly involved in the cellular uptake of exosomes. The raft-mediated endocytosis relies on the presence of cholesterol and sphingolipid-rich microdomains within the plasma membrane. Receptor-mediated endocytosis requires a ligand on the exosome surface to bind to specific receptors on the plasma membrane of the cell. Exosomes internalized through endocytosis from the external environment or plasma membrane can be directed to lysosomes for degradation or recycled to the plasma membrane (McKelvey et al., 2015).

25.4.2.3.2 Membrane Fusion

Fusion is a potential mechanism for cellular internalization of exosomes. The process by which an exosomal vesicle fuses with a cell's plasma membrane is known as vesicle-cell fusion. The lipid membrane of exosomes fuses with recipient cells. The membranes first come close together, forming a connecting stalk. This stalk then widens, creating a shared membrane layer. Then, a small opening forms, allowing the exosome contents to mix with the cell's interior (Del Conde et al., 2005). The mechanism of cell-cell fusion remains inadequately understood. Tetraspanin complexes on target cells may regulate this phenomenon. Tetraspanins are involved in various processes, including T cell activation, membrane fusion in myoblasts and mononuclear phagocytes, and mammalian viral-cell fusion. Tetraspanin-enriched microdomains facilitate exosome fusion with the target cell (Hemler, 2003).

25.4.2.3.3 Phagocytosis

It is a type of endocytic process in which cells, such as immune cells (e.g., dendritic cells and macrophages), engulf and internalize exosomes by extending membrane protrusions, forming a phagocytic cup that then closes to form a vesicle containing the ingested exosomes. The phagocytic cells of the immune system, in particular, use phagocytosis as the primary mechanism of exosome uptake. Specific receptors and processes, primarily found in specialized cells, are essential for phagocytosis. These include Fc receptors, complement receptors, scavenger receptors, and toll-like receptors. Exosomes are encapsulated in phagosomes by these cells, which subsequently direct the payload to the lysosome. Further research is necessary to determine whether phagocytosis serves as a valid mechanism for exosome internalization in intercellular communication or primarily functions to eliminate them (Gonda et al., 2019).

25.4.2.3.4 Macropinocytosis

Macropinocytosis is a type of endocytosis. In this process, cells engulf large volumes of fluids and solutes from their extracellular environment. In macropinocytosis, actin filaments facilitate the formation of cell membrane protrusions. It results in the formation of an invagination of the plasma membrane. The formation of plasma membrane invagination nonspecifically endocytoses extracellular fluid and small particles—phosphatidylserine (PtdSer) on the surface of oligodendrocyte-derived exosomes has been shown to activate macropinocytosis in a specific subset of microglia/macrophages that lack antigen-presenting capability (McKelvey et al., 2015). Macropinocytosis of exosomes relies on Na+ and PI3K, as evidenced by the reduced exosome uptake following the inhibition of Na+/H+ ion exchange and PI3K activity through the pharmacological inhibitors 5-N-Ethyl-N-isopropyl amiloride (EIPA) and LY294002, respectively (Tian et al., 2014).

25.4.2.3.5 Intracellular Signaling

The internalization of exosomes by the target cells involves highly regulated intracellular signaling pathways. Once the exosomes are taken up by target cells, signaling cascades are activated. Receptor tyrosine kinases and other signaling molecules regulate various signaling pathways. These intracellular signaling pathways facilitate the interaction between exosomes and specific receptors on target cells. For example, juxtacrine signaling necessitates the juxtaposition of ligands and receptors on the surfaces of the exosomes and the target cell. On the other hand, soluble signaling is facilitated by proteolytic cleavage of ligands from the exosomal surface. These intracellular signaling cascades lead to changes in gene expression, protein activity, and cellular behavior (Stenqvist et al., 2013).

25.5 ENGINEERED EXOSOMES

Exosome engineering involves modifying or manipulating exosomes to enhance their properties and functions. The exosome surface is essential for cell-specific targeting, biodistribution, and targeted clinical applications. Internal contents of exosomes, including proteins, lipids, and microRNA, mediate intercellular communication, transfer nucleic acids, and modulate cellular behavior and responses. Pharmacological drugs and other substances may enhance the therapeutic potential of natural exosomes by incorporating them onto their surface and into their internal contents. These engineered/modified exosomes maintain their natural biocompatibility (Choudhery et al., 2024). Including priming of the parent cells (Haider, 2024a and 2024b), some of the standard methods for exosome engineering include, but are not limited to, the following.

25.5.1 Electroporation

Electroporation creates temporary holes in the phospholipid bilayer of the cell membrane, allowing therapeutic cargo to enter exosomes. A lipid extruder is used to transport a combination of an exosome and cargo over a membrane. The rupture of the membrane allows the cargo to enter the exosomes. In addition, the parent cells are incubated with a drug of choice, which promotes drug uptake and eventual encapsulation within exosomes. The drug-loaded exosomes are then spontaneously released by the host body into the specific cell type affected by the disease. The procedure makes it difficult to control the loading efficiency. It is also not possible for exosomes to natively load all RNA and proteins.

25.5.2 Incubation

Incubating the therapeutic drugs (cargo) with exosomes is the most direct method of exosome engineering. Therapeutic drugs diffuse into exosomes via a concentration gradient. Exosomes have both a hydrophobic and lipid-rich plasma membrane and a hydrophilic core. Therefore, both hydrophobic and hydrophilic drugs can interact with and integrate into exosomes. First, exosomes are isolated from body fluids, tissues, or cell culture medium. The required therapeutic drugs are then incubated with the exosomes at a specific temperature for a specified duration. A variety of cargos, including small-molecule drugs, have been successfully loaded into exosomes. For example, the chemotherapeutic drug paclitaxel was incubated with exosomes produced from prostate cancer cells. The resulting drug-loaded exosomes were capable of efficient drug administration to prostate cancer cells, thereby enhancing therapeutic efficacy and potentially treating prostate cancer (Saari et al., 2015).

25.5.3 Heat Shock

Heat shock is a commonly employed method for creating membrane pores and stimulating exosome loading. Heat shock can enhance cellular fluidity and stimulate lipid remodeling of the membrane, resulting in the formation of membrane holes. Heat shock treatment with either $CaCl_2$ or thermal shock ruptures the subcellular membrane, facilitating the loading of therapeutic cargos. However, $CaCl_2$ and thermal shock treatment must be combined for efficient loading of miRNA into exosomes. An exosome-miRNA complex was successfully delivered to the lungs of the animal model. The therapeutic effect was evidenced by elevated miRNA-15a levels in their alveolar macrophages. The loading process involved heating the miRNA-15a and exosome mixture to 42°C for one minute. It was then chilled for 5 minutes on ice in a 0.1 M $CaCl_2$ solution to facilitate membrane permeabilization and effective miRNA loading.

25.5.4 Extrusion

The process of extrusion involves passing the target cells or exosomes through carbon filters with decreasing pore diameters, from 400 nm to 100 nm. When samples are passed through the holes approximately 30 times in both directions, a device is needed that lowers the temperature that rises when force is applied to burst cells or exosomes. This method can produce synthetic exosomes with uniform sizes. Additionally, when loading agents are present, this can result in vesicles loaded with therapeutic cargo. Catalase was successfully incorporated into exosomes by extrusion. They are very effective at preventing, releasing, and encapsulating protease degradation. Similarly, doxorubicin-loaded exosome mimics were created by extrusion. Doxorubicin-encapsulating exosome mimics demonstrated robust anticancer activity both in vitro and in vivo.

25.5.5 Sonication

Sonication involves the use of sound waves (typically in the ultrasonic frequency range). Sonication can break down biological materials such as cell membranes. Sonication can disrupt the exosome membranes. The disruption of exosome membranes by high-frequency waves can lead to their permeabilization. Transient cavitation created by ultrasound waves enhances the fluidity of cell membranes. These changes in the exosome membrane allow exosomes to efficiently uptake proteins, drugs, growth factors, nucleic acids, and other molecules. In some studies, exosomes produced by M1-polarized macrophages were sonicated to introduce an anticancer drug, such as paclitaxel (Wang et al., 2019a). In both in vitro and in vivo settings, the paclitaxel-laden exosomes demonstrated a notable anticancer effect. Additionally, the loaded exosomes increased macrophage-mediated inflammation. This enhances the anticancer effect of paclitaxel. It is a more effective method than electroporation for loading exosomes, as it prevents aggregation (Fu et al., 2020).

25.6 EXOSOME ISOLATION METHODS

Exosomes exhibit diverse structures, sizes, and contents, as well as distinct functions. Maintaining the structure of pure exosomes during isolation is challenging because most exosome separation procedures are labor-intensive and time-consuming, and they require specialized, expensive equipment. Standardizing an exosome isolation methodology for clinical applications in diagnostic laboratories worldwide is a significant challenge. Table 25.1 presents a comparison of different exosome isolation methods. Differential centrifugation, polymer precipitation, sequential filtration, immunomagnetic beads, immunoaffinity, and commercially available exosome isolation kits are among the current methods used to successfully obtain exosomes from various sources.

25.6.1 Differential Ultracentrifugation

Differential ultracentrifugation represents one of the initial techniques employed for exosome isolation. This technique remains the most reliable and frequently utilized method. It requires minimal effort for sample preparation. This technique fundamentally involves separating small particles from larger ones by systematically increasing the centrifugation speed, which is directly proportional to the centrifugal force. The centrifugation time is also increased sequentially as the speed increases.

The size, shape, and density of exosomes are critical determinants in their isolation and purification. The isolation of specific particles in the centrifuge tube depends on the sedimentation rate during centrifugation (Gurunathan et al., 2019). The denser, larger particles initially deposit at the bottom, while the lighter particles remain in the supernatant. The supernatant is separated from the pellet for further processing. For example, the isolation of exosomes released by stem cells follows a specific set of steps. The stem cell culture media is collected and either passed through a 40 μm strainer or centrifuged at a low speed (500 × g) for 10 minutes. It separates cells, large particles, and cellular debris from the culture fluid. The supernatant is subsequently removed and centrifuged at 100,000 × g for 70 minutes. Although increased time and speed increase exosome yield, they also represent a primary drawback of this method. It is pertinent to note that exosome membranes are prone to mechanical degradation after centrifugation for more than 4 hours, potentially leading to

TABLE 25.1
Comparison of Different Methods of Exosome Isolation

Method	Principle	Advantages	Disadvantages
Differential ultracentrifugation	Separation based on size and density	Reliable, widely used, minimal sample preparation	Time-consuming, requires specialized equipment, and has a low yield.
Affinity chromatography	Separation based on specific interactions	High specificity can separate exosomes from other EVs	Requires prior filtration or centrifugation, low yield, dependent on antibody specificity.
Immunomagnetic beads	Separation based on specific antibody interactions	High accuracy, minimal sample contamination	Limited by antibody recognition, not suitable for a clinical setting.
Exosome isolation kits	Size-based separation using membranes	High yield, high purity, short isolation time	Expensive, may not be suitable for downstream uses.
Polymer precipitation	Separation based on solubility differences	Quick, easy, and cost-effective	Low sensitivity, may not be suitable for clinical use, requires preceding centrifugation or filtration.
Ultrafiltration	Separation based on size using filters	Rapid purification, no chemical contamination, no specialized expertise required	Fluid component contamination, exosome deformation, and low exosome yield.
Gel filtration technique	Separation based on hydrodynamic radius	Low contamination levels enable the separation of small vesicles from large ones	Impossible to separate exosomes from similar-sized macromolecules, low yield, and increased dilution of the target product.

the release of exosome contents into the sample. A substantial number of samples is necessary for the time-intensive process of differential ultracentrifugation to ensure reliable, reproducible results. The efficient management of multiple biological samples within a constrained timeframe presents a significant challenge in maintaining sample integrity and experimental consistency (Yu et al., 2018). Additionally, this method requires multiple rounds of ultracentrifugation, resulting in a low yield. Combining ultracentrifugation with microfiltration or sucrose density gradient techniques can significantly enhance the purity of isolated exosomes. Although optimized protocols enhance exosome quality, current methods still struggle to produce a completely homogeneous, contaminant-free final product (Al Abdullah et al., 2024).

25.6.2 Affinity Chromatography

Affinity chromatography is based on the principle that targeted molecules can be separated through the interaction of ligands with specific binding sites on a surface. Antibodies are the commonly used ligands that target specific surface molecules of exosomes. Ligands (such as antibodies) are covalently attached to a solid support matrix to form an affinity column. When the sample suspension passes through the column, specific antibodies (immobilized on the column matrix) bind to their ligands. This technique has the potential to separate exosomes from other extracellular vesicles (EVs), which is an advantage over other separation approaches (Rodriguez et al., 2020). The proteins and contaminants do not bind with the antibodies and are washed away. There are also some drawbacks to this method. This method requires prior filtration or centrifugation of the sample due to the complex nature of biological fluids. Additionally, antibodies cannot detect antigens within membrane-enclosed exosomes. The method's specificity also depends on the antibody's specificity. Moreover, the yield of isolated exosomes is comparatively low compared to other methods (Doyle and Wang, 2019).

25.6.3 Immunomagnetic Beads

The immunomagnetic beads are spherical particles coated with specific antibodies. These antibody-coated immunomagnetic beads facilitate antibody binding to exosomes via particular receptors. The interaction between surface receptors (such as CD9, CD63, and CD81) on exosomes and the antibody-coated magnetic beads enables exosome attachment to the magnetic beads. Exosomes are isolated from the sample using a magnetic field, which induces targeted migration of exosome complexes. This technique is accurate due to a specific antigen-antibody reaction. Additionally, it avoids chemical contamination and requires only a small sample volume. The method, however, may fail to isolate distinct exosome subpopulations from a heterogeneous mixture because they lack specific antigens. This technique is effective for small-scale experiments but cannot be applied to large sample volumes. Therefore, its usage in clinical settings is limited (Yu et al., 2018).

25.6.4 Kit-based Isolation of Exosomes

There are several commercially available exosome isolation kits, including ExoQuick from System Biosciences (Palo Alto, CA, USA), ExoMir kit (Bioo Scientific in Austin, TX, USA), and the total Exosome Isolation Kit from Thermo Fisher Scientific (Waltham, MA, USA), that offer isolation of a pure exosome population. These kits use different strategies to isolate exosomes from the suspension. For example, the ExoMir kit (Bioo Scientific, Austin, TX, USA) isolates exosomes based on size. The kit uses two membranes with 200 nm and 20 nm pore diameters, held in place in a syringe. The material is usually spun at a low speed in a centrifuge to separate intact cells, dead cells, and cell debris. The use of proteinase K breaks down the larger particles, thereby evading the membrane blockage. Samples are filtered using the syringe filter. Syringe filters separate vesicles into two sizes: those larger than 200 nm are retained above the first filter, and those between 20 nm and 200 nm are collected between the two filters. The syringe is designed to eliminate vesicles smaller than 20 nm. This method has several advantages, including high yields and purities of isolated exosomes, a short isolation time, and suitability for small volumes. However, it is not easy to isolate exosomes of enormous size. It is also expensive and may be unsuitable for downstream uses. It is now possible to isolate exosomes using several other commercial kits, including ExoQuick from System Biosciences (Palo Alto, CA, USA) and the total exosome isolation kit from Thermo Fisher Scientific (Waltham, MA, USA) (Liu et al., 2017; Choudhery et al., 2024).

Exosome isolation is a challenging, time-consuming, and non-specific procedure. The total exosome isolation kits rapidly and effectively separate the intact exosomes. In kit-based methods, cells are cultured in serum-free media. The cell culture medium is collected and incubated with the reagents contained in the kit. The incubation time and temperature may range from 12 to 24 hours at 2°F to 8°F. The kit components bind to water molecules, while less-soluble components, including exosomes, are excluded from the solution. Low-speed centrifugation is then applied to separate these vesicles. Exosomes are then precipitated using standard centrifugation at 10,000 × g for 60 minutes. The pellet is resuspended in a compatible buffer (e.g., PBS). Following purification of total RNA and protein, intact exosomes can be used for various downstream or clinical applications. This kit-based method enables the easy isolation of exosomes, requires less time, and maximizes the recovery of intact exosomes.

25.6.5 Polymer Precipitation

The principle of polymer precipitation is that different components can be precipitated from different solutions containing precipitants. It is one of the most effective methods for isolating and separating exosomes. Polyethylene glycol (PEG) is the ideal medium used for polymer precipitation. It attracts water and binds to water molecules, leading to exosome precipitation. After centrifugation, exosomes are collected as a pellet. Exosomes can then be utilized for a range of analyses. This technique can easily handle large amounts of samples (from 100 microliters to several milliliters). Additionally, this method of exosome isolation is simple, effective, and does not require expensive equipment or specialized technical expertise. This approach, however, cannot be used in clinical settings due to its low sensitivity. Minimizing contamination in the final exosome product requires preceding processes such as centrifugation or filtration (Choudhery et al., 2024; Al Abdullah et al., 2024).

25.6.6 Ultrafiltration

Ultrafiltration uses filters with holes of specific sizes. When a fluid, such as cell culture media, is passed through filters, particles are trapped according to their size. Ultrafiltration is performed in multiple steps. Initially, large filters are used to remove large particles. The filter has pore sizes of 0.8 μm, 0.45 μm, 0.22 μm, and 0.1 μm, which progressively remove particles larger than the pore size (Merchant et al., 2010). Exosomes obtained by ultrafiltration are morphologically identical to those obtained by ultracentrifugation. These exosomes have the same number of particles, contamination level, and exosome surface markers. The ultrafiltration method does not require any extra chemicals that could contaminate exosomes. This method allows processing multiple samples due to its rapid purification time. However, these methods also have several limitations, including contamination with fluid components, exosome deformation due to the filter's small pore size, and the isolation of small quantities of exosomes. Filter membranes can capture some exosomes, leading to their loss. On the other hand, proteins and other biopolymers can clog filter pores. It also reduces the purification rate and increases the likelihood of exosome loss (Yakubovich et al., 2022).

25.6.7 Gel Filtration Technique

In this method, exosomes are separated from the solution based on their size. The exosome-containing solution passes through a stationary phase gel. The pores of the gel are so small that they allow large particles to pass through. Exosomes, being small, easily pass through pores. The technique is also helpful in segregating vesicles of different sizes. Additionally, exosomes can be separated from unbound soluble proteins and high-density lipoproteins. Studies indicate that this technique produces exosomes with high purity. The lowest plasma protein contamination levels were observed compared to other approaches (Baranyai et al., 2015). Gel filtration also has some limitations. For example, exosome yields are low, and similar-sized macromolecules cannot be separated from them.

25.7 EXOSOMES CHARACTERIZATION

Exosomes are characterized based on their physiological and chemical properties. These parameters include the shape, size, density, surface charge, and composition of exosomes. Several methods have been employed to characterize exosomes. The primary objective of these methods is to identify exosomes and determine what they transmit to other cells. Morphology, physical qualities, and the nature of contents are the three main criteria used to classify exosomes. Flow cytometry, transmission electron microscopy, and polymerase chain reaction are frequently used for exosome characterization. Additionally, surface plasmon resonance microscopy (SPRM), Raman spectroscopy, next-generation sequencing, and dynamic light scattering are used for exosome characterization. Table 25.2 compares different exosome characterization methods.

25.7.1 Flow Cytometry

Flow cytometry is a powerful technique for analyzing and characterizing exosomes. It enables the detection and quantification of exosomes based on their surface markers. It simultaneously provides information on the size, concentration, and surface proteins of individual exosomes. It enables the

TABLE 25.2
Comparison of Different Characterization Methods

Methods	Principle	Advantages	Disadvantages
Flow cytometry	Analyzes single particles based on size, complexity, surface markers, and cell contents.	Rapid, multi-parameter analysis, high sensitivity.	Size detection limitations require specialized equipment.
Transmission electron microscopy	Uses an electron beam to image the morphology of exosomes.	High-resolution imaging provides information on size, shape, and structure.	Sample preparation is time-consuming and complex, and may alter exosome morphology.
Nanoparticle tracking analysis	Tracks individual particles to determine size, concentration, and distribution.	Provides information on particle size, concentration, and distribution, suitable for polydisperse suspensions.	Requires specialized equipment and expertise, sample preparation is critical.
Raman spectroscopy	Analyzes surface changes and lipid composition based on photon scattering.	High selectivity, label-free, provides information on lipid composition.	Requires specialized equipment and expertise.
Surface plasmon resonance microscopy	Detects exosomal membrane proteins based on changes in reflected light.	Real-time information on protein binding kinetics, efficient identification of membrane proteins.	Requires specialized equipment and expertise.
Dynamic light scattering	Measures particle size and concentration based on scattered light.	Easy to use, rapid measurement, suitable for monodispersed suspensions.	Fails to provide a visual representation, detection of small particles is challenging in polydisperse suspensions.
Next-generation sequencing	Evaluates the sequence of the genetic composition of exosomes.	Provides high-throughput information on DNA and RNA content, suitable for complex genetic analysis.	Requires specialized equipment and expertise, and data analysis is complex.

rapid and accurate quantification of a large number of exosomes as well. It is common practice to use flow cytometry while investigating the physical and chemical properties of exosomes.

The flow cytometry procedure involves the following steps.

- The target exosomes are prepared by fixing, washing, and fluorescently staining with specific antibodies against exosome surface markers (CD9, CD81, or CD63).
- The sample is introduced to the flow chamber, where particles pass one by one. At a rate of thousands of pulses per second, the exosome sample in the flow chamber is rapidly scanned by the laser beam.
- The photomultiplier tube then collects and converts these fluorescence signals into electrical impulses.
- Flow cytometry enables multi-parameter quantitative analysis by simultaneously recording fluorescence intensity, size, and shape of a single exosome.
- The data is analyzed using specialized software, and the results can be expressed in various formats, such as graphs or percentages.

Flow cytometry analyzes and classifies exosomes based on the expression of specific antigens. It is possible to analyze exosomes from samples at low concentrations. Specialized flow cytometers have been designed to differentiate exosomes from background contaminants. The latest generation of flow cytometers uses multiple angles for FSC detection, improving particle resolution. This enhanced capability offers several benefits. It enables rapid measurement of suspended exosomes. It also allows for the detection of extremely small EVs.

25.7.2 Transmission Electron Microscopy

Transmission electron microscopy (TEM) is used to examine the morphology of small biological specimens, such as exosomes. TEM utilizes an accelerated electron beam for high resolution. This technique provides detailed nanoscale structural information, enabling visualization of the objects' morphology. TEM can analyze the size, morphology, and surface features of exosomes. With this technique, sub-molecular structures of exosomes can also be studied. Before TEM examination, glutaraldehyde is used to fix the samples. When an electron beam passes through a material, a primary signal is produced. Secondary signals are produced when the objects interact with the primary electron beam. These signals are emitted as electrons and are collected by the detectors. The detected electrons are collected and converted into an electronic signal. These signals are amplified and processed to create an image. TEM images of isolated exosomes reveal a cup-shaped morphology. Sample preparation is a significant factor to consider when utilizing TEM. The procedure is time-consuming, involves multiple steps, and can alter exosome shape (Wu et al., 2015; Choudhery et al., 2024).

25.7.3 Nanoparticle Tracking Analysis

Nanoparticle tracking analysis (NTA) is a method used to characterize nanoparticles, including exosomes. NTA can measure the concentration and size of exosomes. It is a light-scattering technique that uses a laser beam to illuminate suspended particles (exosomes). When the laser light hits the particles, it scatters in all directions. A camera system captures the scattered light to track the movement of individual particles. NTA can track exosome movement and provide information on their size, concentration, and distribution. NTA can detect particles as small as 40 nm and identify exosomes from various extracellular vesicle types. The significant advantage of this method is that the sample can be recovered after analysis. However, proper sample preparation and dilution are essential for successful NTA analysis (Doyle and Wang, 2019; Gurunathan et al., 2019).

25.7.4 Raman Spectroscopy

Raman spectroscopy uses photon scattering to probe the rotational and vibrational properties of molecules. It can differentiate exosomes based on their molecular signatures and lipid composition. The technique involves pointing a powerful laser at a sample. The incident light scatters, causing the molecules to vibrate. Light deflects based on molecular vibrations, creating a unique pattern. Some light scatters at wavelengths other than the laser's. This phenomenon is called Raman scattering. Generally, the intensity of the light scattering depends on the chemical composition of the substance. Intensity peaks reveal the unique vibrations of specific chemical bonds in a molecule. This method does not require labeling with dyes or antibodies. It shows a high degree of selectivity. It can distinguish different types of lipids in exosomes (Li et al., 2021a).

25.7.5 Surface Plasmon Resonance Microscopy

Surface plasmon resonance microscopy (SPRM) provides information about protein binding. SPRM recognizes exosomal membrane proteins in real time. Specialized plasmon sensors are used to detect these proteins. The angle of incidence of light determines the amount of light reflected or absorbed. A camera records the reflected light and detects whether proteins are present or absent on the membrane by analyzing the light pattern. First, the mass bound to the sensor surface is determined from the SPR response, and then the suspended exosome concentration is calculated. A mathematical method is used to determine a correlation between time-dependent changes in mass absorption and exosome concentration (Gool et al., 2017).

25.7.6 Dynamic Light Scattering

Dynamic light scattering (DLS) is also used to characterize exosomes and other nanoparticles in suspension. It is used to measure the size and concentration of exosomes in suspension by analyzing the scattered light from their Brownian motion. It measures particles ranging from 1 nm to 6 μm (Szatanek et al., 2017). A monochromatic coherent laser beam is used to scan the sample and estimate the size of suspended exosomes. First, an exosome sample is prepared for measurement. Then, a monochromatic, coherent laser beam is bombarded onto the sample containing the exosomes. The relative Brownian movements scatter the light. This scattered light is used to estimate particle size and concentration. This method is simple to use, especially for solutions that contain only one type of particle. However, this method fails to provide a visual representation of the particles. Moreover, it is challenging to detect tiny particles when large vesicles are present in the suspension (Gurunathan et al., 2019). In addition, it cannot accurately measure complex mixtures that contain large molecules, even at low levels (Zhang et al., 2020).

25.7.7 Next Generation Sequencing

Next-generation sequencing (NGS) is a novel method for sequencing nucleic acids (RNA and DNA). In exosomes, this technique can be used to evaluate their genetic composition. The concept of using NGS was first implemented in 2013, when exosomal RNA was sequenced using this technology. NGS was used to identify the fragments of 28S and 18S ribosomal RNA (rRNA) in human breast cancer cell exosomes. Additionally, short RNA transcriptomes were identified simultaneously in exosomes derived from human saliva. Exosomal RNA comprises sRNA, piwi-interacting RNA, messenger RNA (mRNA), and microRNA (miRNA). It is also crucial to evaluate the quantity of different types of RNAs in exosomes using quantitative and qualitative approaches, since exosomes contain several types of RNAs.

25.8 EXOSOMES AS STEM CELL-BASED ACELLULAR THERAPY

Acellular therapy utilizes bioactive molecules secreted by stem cells rather than stem cells themselves. Exosomes are one of the major therapeutic components secreted by stem cells. The immunomodulatory properties, small size, biocompatibility, regenerative potential, targeted delivery, off-the-shelf storage, and lower toxicity of exosomes make them ideal for the development of acellular-based treatments (Choudhery et al., 2024). Exosomes offer a potential alternative to conventional stem cell-based therapies that require invasive operations or surgical interventions. Additionally, exosome administration is a relatively simple and less invasive procedure. Exosomes are administered by aerosol spray, intramuscular injection, subcutaneous injection, and intravenous injection for treatment. Therapeutic molecules (including mitochondrial DNA, RNA, and proteins) are encapsulated in exosomes, which also modulate the immune system, target specific cell types, and prevent destruction. Exosomes possess desirable qualities, making them a promising approach for the treatment of a wide range of disorders (Ghafouri-Fard et al., 2021).

Stem cells can be obtained from a wide range of easily accessible tissues, including bone marrow, fat, dental pulp, cord blood, and the umbilical cord (Kamal et al., 2022). Large-scale stem cell culture facilitates exosome isolation. Exosomes are released not only by adult stem cells, but also

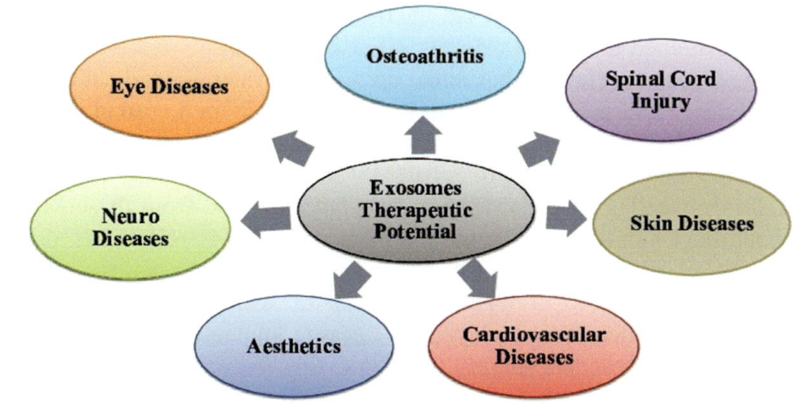

FIGURE 25.2 Therapeutic potential of exosomes. Exosomes have a significant impact on the treatment of osteoarthritis, spinal cord injury, skin diseases, cardiovascular diseases, aesthetics, neurodegenerative diseases, and eye diseases.

by induced pluripotent stem cells (iPSCs) and human embryonic stem cells (hESCs). Compared with adult stem cells, these cells exhibit rapid self-renewal, differentiation, and proliferation. The use of these cells as a cell-based therapy is currently limited in regenerative medicine; however, the exosomes they generate can be used as substitutes for various clinical applications. Exosomes also alleviate concerns about the potential for teratoma formation following stem cell transplantation (Wang, 2021). Exosomes can be lyophilized at room temperature or stored at low temperatures for extended periods. This makes their storage and transportation more convenient. Their compact size also enables filtration-based sterilization. This allows for safe and efficient removal of contaminants. Inhalation of lyophilized or nebulized exosomes produced from lung stem cells for the treatment of lung disease is one example of several methods for delivering exosomes to injury sites. This method shows promise for targeted and non-invasive therapy (Zhang and Cheng, 2023).

Exosomes are an excellent choice for acellular stem cell therapy. However, multiple challenges prevent the practical deployment of exosome-based therapeutics in clinical settings. Concerns include the safety of delivering exosomes to humans, the necessity to standardize processes for isolating and purifying exosomes, and the advancement of a better understanding of exosome absorption strategies and target cell interactions. Despite these obstacles, the therapeutic promise of exosomes remains an effective inducer for research and development (Choudhery et al., 2024).

25.9 THERAPEUTIC POTENTIAL OF EXOSOMES

The precise tailoring of exosomes according to unique molecular signatures promises to improve treatment efficacy. Exosomes possess the therapeutic properties of their progenitor cells, specifically stem cells, which can repair, regenerate, and replace damaged tissue or organs. Given that stem cells exhibit anti-inflammatory, immunomodulatory, and tissue-regenerative properties, exosomes derived from these cells may similarly exhibit these effects. Investigating these possibilities facilitates the development of innovative treatment options for a wide range of disorders. This section explores the applications of exosomes in various diseases, including osteoarthritis, spinal cord injury, cardiovascular diseases, skin diseases, aesthetic applications, neurodegenerative diseases, and eye diseases (Figure 25.2).

25.9.1 Osteoarthritis

Osteoarthritis (OA) is defined by the degeneration of joint cartilage and the underlying bone. It induces discomfort and rigidity, limiting joint mobility. It affects approximately 240 million people worldwide. It is the most prevalent joint disease and the most common cause of adult activity limitation. It was initially classified as a cartilage-only disease. However, it affects the bone, synovium, muscle, ligament, and periarticular fat that make up a joint. Systemic medications are used only to reduce structural degeneration (e.g., cathepsin K and Wnt inhibitors) or to alleviate OA discomfort (e.g., nerve growth factor inhibitors). Stem cell-derived exosomes have also been utilized to treat OA. For example, mesenchymal stem cells (MSCs) were isolated from synovium, gingiva, bone marrow, and infrapatellar fat pads (IPFPs). Exosomes derived from these MSCs were then used to target chondrocytes in in vitro OA models (Tan et al., 2024). miRNA is also a key player in exosome-based acellular OA treatment. MSC-derived exosomes suppress the mTOR-autophagy pathway in OA mice through miR-100-5p, which protects the articular cartilage from degradation and improves gait abnormalities (Wu et al., 2019). Because human IPFP can be easily extracted from OA patients during arthroscopic surgery in a clinic, this exosome therapy has the potential to streamline and simplify the transition from the laboratory to patient care, making it a promising, clinically translatable treatment approach (Wu et al., 2019).

Osteochondritis is a similar disorder characterized by a single lesion in one joint, typically the knee. Exosomes have also been used to treat osteochondral defects. This treatment changed the immune response to promote healing. It increased the number of beneficial immune cells, known as CD163+ M2 macrophages, while reducing the number of harmful immune cells, known as CD86+ M1 macrophages. Additionally, the treatment decreased inflammation by lowering levels of proinflammatory synovial cytokines, such as IL-1β and TNF-α (Zhang et al., 2018). Exosomes derived from gingival MSCs were also used to prevent collagen-induced arthritis. These exosomes functioned as immunosuppressants, inhibiting the IL-17RA-Act1-TRAF6-NF-κB signaling pathway. This effect, in turn, reduced the occurrence of arthritis and bone degradation by elevating IL-10 levels (Tian et al., 2022). Different exosome sources are chosen based on the type and condition of OA. Currently, there is no singular therapy that is universally applicable to all patients with OA. A disease-modifying OA drug (DMOAD) is a pharmacological agent designed to inhibit or potentially reverse the development of OA. The primary characteristic of OA is the degradation of cartilage. Thus, an ideal DMOAD would inhibit cartilage loss and facilitate its regeneration. Alternative DMOADs can attempt to promote the restoration of surrounding tissues by reducing inflammation. It is anticipated that a successful DMOAD would demonstrate improvements in joint tissue health, reductions in patient pain, and improved function. Both MSC-derived exosome therapy and DMOADs can target inflammatory cytokines, matrix-degrading enzymes, and the Wnt pathway. Therefore, MSC-based exosome therapy for OA can be enhanced by combining it with DMOAD (Cho et al., 2021).

25.9.2 Spinal Cord Injury

A spinal cord injury is a condition in which the spinal cord is damaged, resulting in temporary or permanent alterations in its function. It is a devastating neurological disorder that disrupts autonomic nervous system function, sensation, and movement. It is a global health concern that imposes a substantial economic and functional burden on both patients and society (Ding et al., 2022). Mechanical trauma causes initial injuries in the spinal cord. It is followed by a cascade of other symptoms, including edema, ischemia, inflammatory cell infiltration, excitotoxicity, and death (Ahuja et al., 2017). Traditional treatment options include neuroprotective or neuroregenerative pharmaceutical drugs and surgery. Interestingly, exosomes derived from either MSCs or neural stem cells (NSCs) can enhance spinal cord regeneration and protection (Hosseini et al., 2018). There is a risk of neural cell death following exosome-mediated treatment with neural stem cells (NSCs). A specific pathway, miR-219a-2-3p/YY1, was found to play a key role in promoting recovery. This pathway is activated by NSC-derived exosomes and triggered by the growth factor IGF-1. By activating this pathway, these exosomes can promote functional recovery after spinal cord injury and reduce neuronal death (Ma et al., 2019).

Angiogenesis is the process by which new blood vessels form from pre-existing ones. After spinal cord injury, the affected area experiences ischemia (a lack of oxygen and nutrients). The nutrient and oxygen deficiencies in the affected regions result in tissue damage. Angiogenesis can restore blood flow to the injured area. The delivery of oxygen and nutrients to the damaged regions ultimately restores tissue repair and regeneration. Angiogenesis also maintains the integrity of the blood-spinal cord barrier (BSCB) in spinal cord injury. Exosomes have been found to enhance the angiogenic activity of transplanted cells. For example, the angiogenic potential of spinal cord microvascular endothelial cells (SCMECs) was enhanced following exosome pretreatment. Similarly, NSC-derived exosomes enriched in vascular endothelial growth factor (VEGF)-A were used to improve spinal cord angiogenic activity (Zhong et al., 2020). Moreover, hind limb function was enhanced by modifying NSC-derived exosomes with FTY720, a modulator of the sphingosine 1-phosphate receptor, thereby improving their therapeutic potential for spinal cord injury repair. It is an immunomodulatory and microvascular regulator. It protects the SCMEC barrier function by activating the PTEN/Akt pathway and improving outcomes in spinal cord injury (Chen et al., 2021).

The balance between the functions of neurotoxic astrocytes and the sustainability of endogenous neural stem cells (NSCs) determines the outcome of spinal cord injuries (Parekh et al., 2022). Harmful substances released by neurotoxic astrocytes worsen spinal cord damage. Sustainable endogenous neural stem cells (NSCs) repair spinal cord tissues by promoting neuroprotection and neuroregeneration. Thus, NSCs support the repair and recovery of damaged spinal cord tissues. Exosomes derived from modified human umbilical cord mesenchymal stem cells were found to target and modulate neurotoxic astrocytes. The goal of exosome modification is to promote spinal cord function restoration after injury. Axonal regeneration and neuronal differentiation of NSCs were both improved by these modified exosomes (Lai et al., 2022). Research has also examined the use of exosomes to enhance locomotor function in patients with spinal cord injury. For instance, exosomes derived from MSCs could activate endogenous neural stem cells (NSCs) in a proliferative state, thereby enhancing locomotor activity (Perez et al., 2022).

25.9.3 Cardiovascular Diseases

Cardiovascular diseases (CVDs) continue to be the leading cause of mortality and comorbidity in developed nations, even though there has been significant improvement in prevention and treatment (Netala et al., 2024). Ischemic heart disease (IHD) is the most common cardiovascular disease and the leading cause of death globally. In IHD, blood flow to the heart muscle is reduced, leading to hypoxia, or an inadequate oxygen supply. IHD is mainly caused by obstructive coronary atherosclerosis due to the gradual closing or narrowing of arteries that supply blood to the heart. A myocardial infarction (MI) occurs when blood flow to a coronary artery is interrupted or significantly reduced, leading to cardiac muscle necrosis (infarction). Myocardial ischemia-reperfusion injury (IRI) is the damage that occurs to the heart muscle (myocardium) when blood flow is restored after a period of ischemia (lack of blood flow). Myocardial cardiomyocyte death can occur after an extended period of oxygen deprivation. Reperfusion therapy is a medical intervention designed to restore blood flow through or around occluded arteries, typically following a MI. However, IRI limits the beneficial effects of reperfusion. Heart failure after an acute MI continues to cause significant death and disability, even with current coronary reperfusion. This observation highlights the need for next-generation cardioprotective treatments, such as stem cell-derived exosomes (Heusch, 2020; Haider and Najimi, 2022). Exosomes derived from MSCs were modified with miR-25-3p. These modified exosomes were used to target pro-apoptotic proteins in animal models (Peng et al., 2020). Moreover, exosomes modified with miR-144 were used to target the PTEN/Akt pathway. This approach demonstrated promising results in reducing cardiomyocyte apoptosis under low-oxygen conditions (Wen et al., 2020; Santoso et al., 2020).

Exosomes derived from endothelial progenitor cells have the potential to treat cardiac diseases (Al-Omar et al., 2022). Exosomal miR-294 could enhance cardiac function after MI. Exosomes protect the cardiomyocytes by promoting neovascularization and inhibiting fibrosis (Khan et al., 2015). Moreover, exosomes produced from endothelial stem cells can reduce doxorubicin-induced cardiotoxicity. These exosomes block TLR4-NLRP3-mediated pyroptotic cell death in cardiomyocytes, besides promoting the development of anti-inflammatory M2 macrophages, and inhibit inflammation-induced pyroptosis, leading to cardiac remodeling. Stem cell-derived exosomes can also restore cardiac function, such as decreasing the end-systolic and end-diastolic volumes of the left ventricle (Kervadec et al., 2016).

Exosomes derived from induced pluripotent stem cells (iPSCs) and hematopoietic stem cells (HSCs) have been found to protect cardiomyocytes in vitro and in vivo. In one study, iPSC-derived exosomes increased cardiomyocyte survival and preserved intracellular Ca^{2+} homeostasis. This can enhance the MI recovery (Li et al., 2023b). HSC-derived exosomes can also improve heart function by reducing scarring (Li et al., 2023a). Furthermore, systemic infusion of HSC-derived exosomes can also treat ischemic cardiomyopathy in acute MI by promoting angiogenesis, reducing apoptosis, and enhancing cardiac function. These exosomes can also mitigate adverse effects, such as renal injury, oxidative stress, and inflammation, thereby improving overall survival and quality of life (Angulski et al., 2019). HSCs can also be modified with the angiogenic factor sonic hedgehog (Shh). These modified exosomes can maintain heart function after an acute MI. This modification would include delivering exosomal Shh to the ischemic myocardium to promote angiogenesis and protect the heart muscle from damage (Mackie et al., 2012).

25.9.4 Skin Diseases

The skin is the body's outermost structure, comprising approximately 70% of its surface area. It simultaneously perceives temperature changes, sweats, and protects the body from external pathogens. Skin injuries can be either acute or chronic, including burns, dermatitis, and more. The person's physical and mental well-being is impacted by the scars left behind by severe skin injury. Several skin issues, including pigmentation and fine wrinkles, also worsen with age. Exosomes derived from stem cells are effective in treating and repairing skin disorders. A wide selection of skin care products, including therapeutic exosomes, is now available. Wound healing with therapeutic exosomes involves multiple steps in the natural repair and regeneration processes of skin. These therapeutic exosomes can modulate immunological responses, enhance cell migration and proliferation, and exert other beneficial effects. Wound healing following cosmetic surgery is another common application of exosomes, which leverages their capabilities for repair and regeneration. Therapeutic exosomes can actively regulate collagen production and restore the skin to a normal state from its pathological condition (Yang et al., 2021).

Exosomes were isolated from adult stem cells (ASCs) and modified with an alginate-based hydrogel (alg-EXO). These modified exosomes have been employed to treat skin conditions. Alg-EXO was applied to one group of animals, while the other group received no alginate treatment. After 15 days, the in vivo data showed that wound closure was significantly better in the alg-EXO group than in the other groups. Exosomes induced collagen deposition and vascularization, promoting wound healing (Shafei et al., 2020). Self-healing hydrogel (FHE) made of Pluronic F127, oxidative hyaluronic acid (OHA), and poly-ε-L-lysine (EPL) was also injected into diabetic mice. Exosomes released from these components reduced the period of wound healing, facilitated by re-epithelialization, collagen remodeling, accelerated granulation tissue formation, and increased cell proliferation. Consequently, wounds healed more quickly and with less scar tissue (Wang et al., 2019a).

25.9.5 Aesthetics

Aesthetics deals with the nature of beauty, taste, and art. It involves the study of how we perceive and appreciate beauty, as well as the principles that govern artistic expression and creativity. Medical aesthetics is the branch of medicine that focuses on improving patients' physical appearance and aesthetic appeal through non-invasive or minimally invasive medical treatments and procedures. Medical aesthetics primarily aims to cure wounds, prevent aging, reduce hyperpigmentation, and stop hair loss. Aesthetic medicine and regenerative medicine work hand in hand in medical aesthetics. It utilizes regenerative medicine techniques, including stem cells, natural or synthetic scaffolds, and growth hormones, to repair, replace, or regenerate human cells and tissues. It is used in medical aesthetics to restore, alter, and enhance human appearance, form, and function, thereby achieving balance and improvement in these areas (Zhang et al., 2022).

Currently, the primary applications of exosomes in medical aesthetics research include topical application and local injection. It is possible to enhance the transdermal action of exosomes by applying them topically, either using micro-needles or transdermal drug administration, or by simply applying them locally to the skin. Most local injections are administered subcutaneously. Identifying the most effective method for administering exosomes remains challenging because no universally accepted criteria exist for this purpose. In addition, various methods of administration have varied effects (Zhang et al., 2022).

The effects of topical application can be tailored by applying it to different sites, and the dosage can be easily managed. Using a skin injury model in mice, it was discovered that both normal and diabetic animals may repair wounds more quickly when endothelial progenitor cells secrete exosomes or miRNA-221-3p. Transdermal drug delivery, on the other hand, can bypass the stratum corneum barrier, allowing active components to be delivered directly under the epidermis. This increases the rate of absorption and decreases the loss of active compounds (Xu et al., 2020). A new method has been developed to deliver hair follicle stem cell activators more effectively. This method uses a detachable micro-needle device to provide the activators. The device is part of a drug delivery system. This method was used to treat hair loss, combining micro-needle technology with a small-molecule drug, UK5099, and exosomes generated from MSCs. After two rounds of treatment, the results showed improved efficacy at a lower dose. Pigmentation and hair regrowth were also stimulated within six days. Better results were also obtained with this transdermal delivery method using micro-needles than with topical UK5099 administration or subcutaneous exosome injection (Yang et al., 2017).

Soluble microneedle patches (MNs) were combined with hair nanoparticles and human amniotic MSC-derived exosomes (hAMSCs). HNP-modified microneedle patches (HMNs) effectively penetrated the stratum corneum and delivered hAMSC exosomes into the dermis (Hong et al., 2021). This process induces the hair follicle to transition from telogen to anagen and stimulates stem cell proliferation within the hair follicle, thereby promoting hair regeneration. The efficacy and safety of these transdermal combination treatments surpass those of direct administration (Shang et al., 2022). The studies exemplified here suggest that exosomes, when administered in various ways, can contribute to medical aesthetics. However, transdermal delivery is a better option for medical aesthetics. Transdermal medicine delivery has several advantages over direct exosome application, including the ability to cross the cuticle, increasing the rate of effective substance absorption by the skin, and reducing the loss of practical components. The transdermal distribution of drugs is safer, more convenient, and less prone to adverse effects than injecting exosomes. It has enormous potential for use in the future medical beauty sector (Yang et al., 2021).

25.9.6 Neurodegenerative Diseases

A neurodegenerative disease results from the gradual deterioration of neurons and nervous tissues. This process is also called neurodegeneration. Ultimately, neuronal damage may lead to neural death. Neurodegenerative diseases include stroke, amyotrophic lateral sclerosis, multiple sclerosis, Alzheimer's disease, and Parkinson's disease. Brain-derived exosomes have been shown to carry specific payloads, facilitating easier communication between brain regions and enabling them to coordinate complex brain processes. The cargo of the exosome can mediate neuronal development and protection, nerve regeneration, and synaptic plasticity. Additionally, exosomes have been observed to facilitate protein production and tissue regeneration by delivering regulatory components to areas of neurological injury (Wei et al., 2016).

A stroke occurs when the blood supply to the brain is interrupted, either due to a blockage (ischemic stroke) or a rupture (hemorrhagic stroke) of the blood vessels, resulting in inadequate blood flow to the brain. Exosomes produced from mesenchymal stem cells (MSCs) can preserve tissues after neurological injury caused by middle cerebral artery occlusion in stroke models (Xin et al., 2012). The number and length of neurite branches were shown to grow considerably after exosomes generated from MSCs were delivered. It has been demonstrated that the exosomal transfer of microRNA-133b facilitates neuronal regeneration following spinal cord damage (Yu et al., 2011). Amyotrophic lateral sclerosis (ALS) is a neurodegenerative disease that primarily affects the central nervous system. Muscle control is lost in amyotrophic lateral sclerosis. ALS primarily affects the brain's higher and lower motor neurons in the spinal cord and brain, respectively. It is a fatal, progressive disease that typically begins in adulthood. Neuroinflammatory reactions, such as microglial activation in the central nervous system, and further involvement of the peripheral immune system, underlie this neurodegeneration (Zondler et al., 2016). It was found that peripheral monocytes from ALS donors secrete fewer pro-inflammatory cytokines after exosomal activation. Moreover, exosomal transactive response DNA-binding protein 43 kDa (TDP-43) is prone to aggregation in ALS histopathology and increases monocyte activation. Peripheral monocytes can invade the central nervous system in ALS. Exosomes may mediate TDP-43 invasion of the central nervous system, as monocytes have been shown to preferentially uptake exosomal TDP-43 over free TDP-43 (Zondler et al., 2017).

Alzheimer's disease is characterized by neurodegeneration, neurofibrillary tangles, and an excess of beta-amyloid peptide (Aβ) in the brain that fails to break down, resulting in amyloid plaques. Exosomes are a promising therapeutic approach for treating Alzheimer's disease. Exosomes were isolated from MSCs derived from adipose tissue. Exosomes were engineered to express neprilysin, thereby addressing the imbalance in Aβ accumulation. This approach could mitigate Aβ buildup and potentially alleviate symptoms of Alzheimer's (Hickman et al., 2016). Exosome-based treatments for Alzheimer's disease have been employed in mouse models. Dendritic cell-derived exosomes were loaded with short interfering RNA (siRNA). These exosomes crossed the blood-brain barrier. BACE1 is a protease that causes Aβ aggregations through the N-terminal cleavage of amyloid precursor proteins. The mRNA and protein levels of BACE1 were significantly reduced following exosome-delivered siRNA. Targeted exosomal delivery of siRNA can knock down specific genes in the brain and improve the symptoms of Alzheimer's disease (Alvarez-Erviti et al., 2011).

Parkinson's disease is a neurological disease characterized by involuntary or uncontrollable movements, including tremors, rigidity, and challenges with coordination and balance. It has also been treated with exosome-based methods. The pathological hallmarks of Parkinson's disease include neuroinflammation mediated by microglia, the degeneration of dopaminergic neurons in the substantia nigra of the midbrain, and the buildup of Lewy bodies, which are protein aggregates containing α-synuclein (Fraser et al., 2016). Exosomes from bone marrow MSCs contain TNF-stimulated gene-6 (TSG-6), which can help reduce damage to brain cells caused by a toxic substance called MPP+, which is related to the neurotoxin MPTP. It was also discovered that the impact of exosomal anti-PD progression on an in vitro PD model, including SH-SY5Y and SK-N-SH cells, is mediated by the STAT3/miR-7/NEDD4 axis (Huang et al., 2022). NSC-derived exosomes treatment prevented the progression of Parkinson's disease (PD). These exosomes inhibited apoptotic pathways and intracellular reactive oxygen species (ROS) in vitro using SH-SY5Y and BV2 cells. Exosomes also diminish dopaminergic neuronal loss and downregulate pro-inflammatory markers in 6-hydroxydopamine-induced PD rats. Furthermore, the presence of NSC-specific microRNAs, including let-7, miR-182-5p, miR-183-5p, and miR-9, impacts cell differentiation, neurotropic function, and immunological regulation (Lee et al., 2022).

25.9.7 Eye Diseases

Eyes are vital human sensory organs. Humans can only perceive the world through their own eyes. Some of the most common causes of eye diseases (EDs) include inadequate eye care, prolonged periods of screen time, hereditary factors, aging, and excessive eye strain. It is easy to become blind from an ED if it is not treated promptly or properly. Present medications do not adequately address ED. Exosomes have been shown in multiple studies to transport drugs to specific cells while preventing them from being broken down outside of cells. The therapeutic use of exosomes in EDs has been the subject of multiple clinical trials. Dry eyes are the primary symptom of dry eye syndrome, an ocular condition resulting from a problem with tear production. Damage to the cornea and potential blindness can result from severe dry eye. Current research suggests that therapeutic exosomes applied topically to the eye can alleviate dryness by altering the ocular surface microenvironment and enhancing the cornea's immune microenvironment. One potential future application of therapeutic exosomes is daily treatment for dry eye. Inside the eye, on the retina, is a thin, transparent membrane that acts as a sensor for imaging and light stimulation. Blindness, impaired vision, or dark circles around the eyes may result from retinal illness or injury. Regeneration of the retina is not routinely possible, unfortunately, although significant advances have been made using laboratory-grown iPSC-derived retinal pigment epithelium (RPE) tissues. Nevertheless, therapeutic exosomes could aid in retinal repair by preventing cell death, reducing oxidative stress in retinal pigment epithelium cells (RPECs), and enhancing their proliferation when injected intravitreally or retrobulbarly. This provides new opportunities for cell-free treatment of eye disorders (Rad et al., 2022).

Diabetic retinopathy (DR), a severe complication of diabetes, is the most significant manifestation of diabetic microangiopathy. It is characterized by a fundus lesion that undergoes specific modifications. The three primary pathogenic hallmarks of DR are non-proliferative, pre-proliferative, and proliferative DR. Diabetic macular edema is a type of non-proliferative diabetic retinopathy, characterized by macular and vascular leakage (Li et al., 2021b).

Intravitreal injection of MAC-secreted exosomes into the eye can enhance the treatment of diabetic retinopathy (DR). Specifically, a study using oxygen-induced retinopathy (OIR) mouse models has demonstrated that exosomes can reduce retinal ischemia. The researchers injected hypoxia-cultured exosomes containing angiogenic components from human MSCs (hMSCs) or control saline into the animals' eyes. Treatment with exosomes reduced retinal ischemia and prevented retinal thinning as compared to the control group (Li et al., 2022).

Age-related macular degeneration (AMD) is the primary cause of blindness among the elderly in developed countries. This condition predominantly affects individuals over 50. Dry AMD and wet AMD are classified based on clinical symptoms and changes in the disease's pathology. Choroidal neovascularization (CNV) and retinal and choroidal hemorrhage are hallmarks of wet AMD. The visual impairment caused by wet AMD is more severe than that caused by dry AMD. Dry AMD is characterized by the formation of yellow drusen beneath the retinal pigmented epithelium (RPE) rather than neovascularization (Xu et al., 2023). Some retinal ganglion cells (RGCs) and astroglial cells (RACs) can produce and secrete exosomes. These cells can play a role in the degenerative process of AMD. Exosomes produced by RACs contain anti-angiogenic components that can limit CNV and retinal vascular leakage (Li et al., 2020).

25.10 LIMITATIONS AND FUTURE DIRECTION OF EXOSOME-BASED THERAPY

The diagnostic and therapeutic potential of exosomes has begun to attract considerable interest. Exosomes are natural, biocompatible, and low-immunogenic. The diversity of exosomes in terms of size, cargo, function, and source is one of the many choices that researchers must consider. There is currently no unified standard protocol for the isolation, characterization, and manipulation of exosomes for clinical use. Differences in methodology can lead to variations in exosome quantity, composition, and even therapeutic activity, making it challenging to replicate results across studies and practical applications (Zhang et al., 2020). The limitations of existing technology for isolating and characterizing exosomes are impeding the development of novel exosome-based diagnostic tests and treatments (Table 25.3). The density ranges of exosomes are very similar. The techniques used for sample separation and the strategies for sample preparation and characterization influence exosome structure. Exosomes produced by the same cell type exhibit distinct characteristics depending on the physical conditions or other cells involved in the process (Wang X et al., 2023b).

Another potential challenge in employing exosomes as acellular therapeutics is the need to isolate, process, and administer them under sterile conditions. During their biogenesis phase, exosomes can transport retroviruses, such as human immunodeficiency virus (HIV)-1 and human T-lymphotropic virus type 1, allowing them to replicate in different parts of the body undetected by the immune system. The slight size disparity between exosomes and viruses makes it easy for viral material (such as retroviruses) to contaminate exosomes. Therefore, to minimize hazards, it is essential to monitor any pathogens closely before their clinical use. The potential for oncogenic payloads is another major drawback of exosome use. Exosomes include miRNAs, which are among their primary components. Tumors may arise when some miRNAs encourage cell growth, a process known as oncogenic potential. Exosomes may also include enzymes and proteins that can activate pathways associated with cancer. Additionally, the long-term effects of exosomes in humans remain unclear. It is crucial to investigate safer and more standardized methods for purifying exosomes, as well as strategies to enhance the isolation process (Rezaie et al., 2022).

The intrinsic limitations of exosomes include inadequate targeting specificity, low yields, and restricted encapsulation efficiency of therapeutic cargo, which have limited their therapeutic applications. Therefore, it is necessary to modify and design exosomes targeted to specific targets to reduce these challenges. Gene transfection is a challenging procedure for producing exosomes. Concerns involve safety, variability in transfection efficiency, mutagenesis, and inadequate targeting. The therapeutic potential of artificial exosomes is highly dependent on the yield and purity of the exosomes isolated from their host. Artificial exosomes must have high purity and yield to meet the standards for therapeutic applications. The biocompatibility of artificial exosomes is equally essential. Any unfavorable immune reactions or toxicities that occur in vivo can hinder the efficacy and safety of artificial exosomes. It is necessary to address these critical issues to fully leverage

TABLE 25.3
Limitations of Exosome-Based Acellular Therapy

Limitations	Description
Heterogeneity	Diversity in size, cargo, function, and donor cells.
Isolation methods	Lack of standardized methods, each with its pros and cons.
Purity and yield	Variability in isolation procedures affects therapeutic potential.
Characterization	Limited understanding of exosome composition, structure, and function.
Targeting	Inadequate targeting, low yields, and restricted encapsulation efficiency.
Cargo loading	Limited understanding of cargo loading mechanisms and efficiencies.
Immune response	Potential for adverse immune responses or toxicities.
Oncogenic potential	Risk of oncogenic payloads, including miRNAs and proteins.
Scalability	Difficulty in scaling up exosome production for clinical applications.
Regulatory approval	Challenges in obtaining FDA approval due to a lack of standardized methods and characterization.
Storage and stability	Limited understanding of exosome stability and storage conditions.
Ethical considerations	Concerns regarding production methods, origin, and potential misuse.

the potential of artificial exosomes in clinical settings. Moreover, custom-designed approaches are required to account for the distinctive characteristics of various biological targets, including specific cellular receptors and microenvironments. There is an urgent need to develop standardized guidelines for evaluating cargo loading and encapsulation efficiencies across various preclinical and clinical applications. Additionally, understanding the intracellular fate of exosomes is essential to improve their function and therapeutic efficiency (Bahadorani et al., 2024).

Exosomes are challenging to maintain because these vesicles are only stable when stored for an extended period at −80°C or below. Since such low temperatures can reduce translational activity, this requirement poses a secondary difficulty. A solution to this problem is vacuum-assisted freeze-drying of exosomes, which involves the selective removal of water from a frozen exosome sample. However, different studies have reported different results. The accepted standards for exosome preservation have not been validated yet (Sokolova et al., 2011).

From basic science to the clinic, numerous concerns exist regarding the utilization of exosomes. One significant issue is exosome scalability. Good manufacturing practices are crucial for the clinical applications of exosomes. Exosomes must be better characterized to analyze their heterogeneity, cargo, and functions. Factors such as dosage, duration, and administration method should be evaluated when using supraphysiological amounts of exosomes. Developing more accurate, early-phase biomarkers to inform clinical decision-making from the various types of information exosomes convey remains an open question. Obtaining the necessary approvals from the Food and Drug Administration (FDA) for clinical applications is a crucial step in utilizing exosomes as therapeutic agents. This involves determining optimal producer cells, culture conditions, and stimulation conditions, followed by the large-scale generation of homogeneous exosomes. Numerous studies have demonstrated that exosomes, and more specifically MSC-exosomes, play crucial roles in tumors, immunomodulation, and tissue regeneration. Exosomes from "old" MSC, for instance, have little influence on skin regeneration, whereas exosomes from other sources may have the reverse impact. Further research is needed to fully understand the intricate role of exosomes (Conroy, 2021).

The ethical implications of the origin and production of artificial or engineered exosomes are also significant. The lack of uniform standards for evaluating exosome products, which leads to variations in quality assessment methods across different companies, exacerbates these issues.

Obtaining regulatory approval for exosome use in clinical applications is challenging due to their complexity.

Collaborative efforts and interdisciplinary cooperation are necessary to streamline the process from exosome production to their clinical implementation. This includes pharmaceutical designs, cellular response dynamics, careful toxicological evaluations, thorough preclinical investigations, and robust clinical evaluations (Bahadorani et al., 2024).

25.11 CONCLUSION

The use of exosomes is an ideal option for cell-based acellular therapy. The therapeutic effects of exosomes are attributed to their biocompatibility, precise targeting, low toxicity, and immunogenicity, among other factors. Exosomes inherit the molecular characteristics of their parent cells. The exosomes derived from stem cells have a dual function. They not only replicate the functions of stem cells but also mitigate the adverse effects associated with stem cell-based therapies. Interestingly, exosomes can be engineered to load and deliver personalized therapeutic molecules or drugs to the targeted cells and organs. Despite their therapeutic significance, the use of exosomes presents several challenges, including exosome heterogeneity, uncertainty regarding the underlying mechanism of action, a lack of standardized isolation protocols, poorly understood and costly characterization methods, and insufficient published data from clinical trials. Cooperative initiatives among researchers, physicians, and regulatory agencies are crucial for addressing complications and facilitating the smooth translation of laboratory research into patient care.

REFERENCES

Ahmed LA, Al-Massri KF, Haider KH. Therapeutic efficacy of mesenchymal stem cells and their derived exosomes in different neurodegenerative disorders: Current status and future perspectives. In: *Handbook of Regenerative Medicine. Cell-free Therapy Approach*. 2025, pp. 55–67.

Ahmed ZT, Alkahlot MH, Haider KH. MSC-derived exosomes: advances in cell-free therapy. In: *Handbook of Stem Cell Applications*. Singapore: Springer Nature, 2023 Aug 18, pp. 1–41.

Ahuja CS, Wilson JR, Nori S, Kotter M, Druschel C, Curt A, et al. Traumatic spinal cord injury. *Nature Reviews Disease Primers*. 2017;3(1):1–21.

Al Abdullah S, Cocklereece I, Dellinger K. Unlocking the potential of circulating small extracellular vesicles in neurodegenerative disease through targeted biomarkers and advancements in biosensing. *Exploration of BioMat-X*. 2024;1(2):100–23.

Al-Omar MT, Alnajjar MT, Ahmed ZT, Salaas FMI, Alrefaei TSM, Haider KH. Endothelial progenitor cell-derived small extracellular vesicles for myocardial angiogenesis and revascularization. *Journal of Clinical and Translational Research*. 2022; 8(6): 1–12

Alvarez-Erviti L, Seow Y, Yin H, Betts C, Lakhal S, Wood MJ. Delivery of siRNA to the mouse brain by systemic injection of targeted exosomes. *Nature Biotechnology*. 2011;29(4):341–5.

Angulski AB, Capriglione LGA, Barchiki F, Brofman P, Stimamiglio MA, Senegaglia AC, et al. Systemic Infusion of expanded CD133+ cells and expanded CD133+ cell-derived EVs for the treatment of ischemic cardiomyopathy in a rat model of AMI. *Stem Cells International*. 2019;2019(1):4802578.

Atay S, Gercel-Taylor C, Kesimer M, Taylor DD. Morphologic and proteomic characterization of exosomes released by cultured extravillous trophoblast cells. *Experimental Cell Research*. 2011;317(8):1192–202.

Bahadorani M, Nasiri M, Dellinger K, Aravamudhan S, Zadegan R. Engineering exosomes for therapeutic applications: Decoding biogenesis, content modification, and cargo loading strategies. *International Journal of Nanomedicine*. 2024:7137–64.

Baranyai T, Herczeg K, Onódi Z, Voszka I, Módos K, Marton N, et al. Isolation of exosomes from blood plasma: Qualitative and quantitative comparison of ultracentrifugation and size exclusion chromatography methods. *PLOS ONE*. 2015;10(12):e0145686.

Blanchard N, Lankar D, Faure F, Regnault A, Dumont C, Raposo G, et al. TCR activation of human T cells induces the production of exosomes bearing the TCR/CD3/ζ complex. *The Journal of Immunology*. 2002;168(7):3235–41.

Bobrie A, Colombo M, Raposo G, Théry C. Exosome secretion: Molecular mechanisms and roles in immune responses. *Traffic*. 2011;12(12):1659–68.

Chen J, Zhang C, Li S, Li Z, Lai X, Xia Q. Exosomes derived from nerve stem cells loaded with FTY720 promote the recovery after spinal cord injury in rats by the PTEN/AKT signal pathway. *Journal of Immunology Research*. 2021;2021(1):8100298.

Cho Y, Jeong S, Kim H, Kang D, Lee J, Kang S-B, et al. Disease-modifying therapeutic strategies in osteoarthritis: current status and future directions. *Experimental & Molecular Medicine*. 2021;53(11):1689–96.

Choudhery MS, Arif T, Mahmood R, Harris DT. Stem cell-based acellular therapy: Insight into biogenesis, bioengineering, and therapeutic applications of exosomes. *Biomolecules*. 2024;14(7):792.

Conroy CM. *Isolation and Characterisation of Immune Modulatory Extracellular Vesicles from CD362-Selected Human Umbilical Cord Mesenchymal Stromal Cells*. National University of Ireland, Galway, 2021.

De Candia P, Torri A, Gorletta T, Fedeli M, Bulgheroni E, Cheroni C, et al. Intracellular modulation, extracellular disposal, and serum increase of MiR-150 mark lymphocyte activation. *PLOS ONE*. 2013;8(9):e75348.

Del Conde I, Shrimpton CN, Thiagarajan P, López JA. Tissue-factor–bearing microvesicles arise from lipid rafts and fuse with activated platelets to initiate coagulation. *Blood*. 2005;106(5):1604–11.

Ding W, Hu S, Wang P, Kang H, Peng R, Dong Y, et al. Spinal cord injury: The global incidence, prevalence, and disability from the Global Burden of Disease Study 2019. *Spine*. 2022;47(21):1532–40.

Doyle LM, Wang MZ. Overview of extracellular vesicles, their origin, composition, purpose, and methods for exosome isolation and analysis. *Cells*. 2019;8(7):727.

Fraser KB, Rawlins AB, Clark RG, Alcalay RN, Standaert DG, Liu N, et al. Ser (P)-1292 LRRK2 in urinary exosomes is elevated in idiopathic Parkinson's disease. *Movement Disorders*. 2016;31(10):1543–50.

Fu S, Wang Y, Xia X, Zheng JC. Exosome engineering: Current progress in cargo loading and targeted delivery. *NanoImpact*. 2020;20:100261.

Ghafouri-Fard S, Niazi V, Hussen BM, Omrani MD, Taheri M, Basiri A. The emerging role of exosomes in the treatment of human disorders with a special focus on mesenchymal stem cell-derived exosomes. *Frontiers in Cell and Developmental Biology*. 2021;9:653296.

Gonda A, Kabagwira J, Senthil GN, Wall NR. Internalization of exosomes through receptor-mediated endocytosis. *Molecular Cancer Research.* 2019;17(2):337–47.

Gool EL, Stojanovic I, Schasfoort RB, Sturk A, van Leeuwen TG, Nieuwland R, et al. Surface plasmon resonance is a highly sensitive analytical method for profiling antigens on extracellular vesicles. *Clinical Chemistry.* 2017;63(10):1633–41.

Gotoh S, Kawabori M, Fujimura M. Intranasal administration of stem cell-derived exosomes for central nervous system diseases. *Neural Regeneration Research.* 2024;19(6):1249–55.

Gruenberg J. Life in the lumen: The multivesicular endosome. *Traffic.* 2020;21(1):76–93.

Gurunathan S, Kang M-H, Jeyaraj M, Qasim M, Kim J-H. Review of the isolation, characterization, biological function, and multifarious therapeutic approaches of exosomes. *Cells.* 2019;8(4):307.

Haider KH. Melatonin-based priming of stem cells to alleviate oxidative stress. *World Journal of Stem Cells.* 2024b;16(11):985–89. https://doi.org/10.4252/wjsc.v16.i11.985.

Haider KH. Priming mesenchymal stem cells to develop "super stem cells." *World Journal of Stem Cells.* 2024a;16(6):623–40. https://doi.org/10.4252/wjsc.v16.i6.623.

Haider KH, Najimi M. Exosome-based Cell-free therapy in regenerative medicine for myocardial repair. In: Haider, K.H. (eds) *Handbook of Stem Cell Therapy.* Singapore: Springer, 2022, pp. 1007–34. https://doi.org/10.1007/978-981-16-6016-0_9-1.

Han Q-F, Li W-J, Hu K-S, Gao J, Zhai W-L, Yang J-H, et al. Exosome biogenesis: Machinery, regulation, and therapeutic implications in cancer. *Molecular Cancer.* 2022;21(1):207.

Hemler ME. Tetraspanin proteins mediate cellular penetration, invasion, and fusion events and define a novel type of membrane microdomain. *Annual Review of Cell and Developmental Biology.* 2003;19(1):397–422.

Heusch G. Myocardial ischaemia–reperfusion injury and cardioprotection in perspective. *Nature Reviews Cardiology.* 2020;17(12):773–89.

Hickman RA, Faustin A, Wisniewski T. Alzheimer's disease and its growing epidemic: Risk factors, biomarkers, and the urgent need for therapeutics. *Neurologic Clinics.* 2016;34(4):941–53.

Hong C, Zhang G, Zhang W, Liu J, Zhang J, Chen Y, et al. Hair grows hair: Dual-effective hair regrowth through a hair-enhanced dissolvable microneedle patch, cooperated with the pure yellow light irradiation. *Applied Materials Today.* 2021;25:101188.

Hosseini SM, Sani M, Haider KH, Dorvash MR, Ziaee SM, Karimi A. Concomitant use of mesenchymal stem cells and neural stem cells for treatment of spinal cord injury: A combo cell therapy approach. *Neurosc Lett.* 2018;668:138–46.

Huang D, Zhang M, Tan Z. Bone marrow stem cell-exo-derived TSG-6 attenuates 1-methyl-4-phenylpyridinium+-induced neurotoxicity via the STAT3/miR-7/NEDD4/LRRK2 axis. *Journal of Neuropathology & Experimental Neurology.* 2022;81(8):621–34.

Kalluri R, LeBleu VS. The biology, function, and biomedical applications of exosomes. *Science.* 2020;367(6478):eaau6977.

Kamal M, Kassem D, Haider KH. Sources and therapeutic strategies of mesenchymal stem cells in regenerative medicine. In: Haider, K.H. (eds) *Handbook of Stem Cell Therapy.* Singapore: Springer, 2022, pp. 1–28. https://doi.org/10.1007/978-981-16-6016-0_2-1.

Kervadec A, Bellamy V, El Harane N, Arakélian L, Vanneaux V, Cacciapuoti I, et al. Cardiovascular progenitor–derived extracellular vesicles recapitulate the beneficial effects of their parent cells in the treatment of chronic heart failure. *The Journal of Heart and Lung Transplantation.* 2016;35(6):795–807.

Khan M, Nickoloff E, Abramova T, Johnson J, Verma SK, Krishnamurthy P, et al. Embryonic stem cell–derived exosomes promote endogenous repair mechanisms and enhance cardiac function following myocardial infarction. *Circulation Research.* 2015;117(1):52–64

Lai X, Wang Y, Wang X, Liu B, Rong L. miR-146a-5p-modified hUCMSC-derived exosomes facilitate spinal cord function recovery by targeting neurotoxic astrocytes. *Stem Cell Research & Therapy.* 2022;13(1):487.

Lee EJ, Choi Y, Lee HJ, Hwang DW, Lee DS. Human neural stem cell-derived extracellular vesicles protect against Parkinson's disease pathologies. *Journal of Nanobiotechnology.* 2022;20(1):198.

Lee T-R, Choi M, Kopacz AM, Yun S-H, Liu WK, Decuzzi P. On the near-wall accumulation of injectable particles in the microcirculation: Smaller is not better. *Scientific Reports.* 2013;3(1):2079.

Lee YJ, Shin KJ, Chae YC. Regulation of cargo selection in exosome biogenesis and its biomedical applications in cancer. *Experimental & Molecular Medicine.* 2024;56(4):877–89.

Li C, Li X, Shi Z, Wu P, Fu J, Tang J, et al. Exosomes from LPS-preconditioned bone marrow MSCs accelerated peripheral nerve regeneration via M2 macrophage polarization: Involvement of TSG-6/NF-κB/NLRP3 signaling pathway. *Experimental Neurology.* 2022;356:114139.

Li G, Zhu N, Zhou J, Kang K, Zhou X, Ying B, et al. A magnetic surface-enhanced Raman scattering platform for performing successive breast cancer exosome isolation and analysis. *Journal of Materials Chemistry B*. 2021;9(11):2709–16.

Li H, Gu J, Sun X, Zuo Q, Li B, Gu X. Isolation of swine bone marrow lin-/CD45-/CD133+ cells and cardioprotective effects of its exosomes. *Stem Cell Reviews and Reports*. 2023a;19(1):213–29.

Li H, Wang L, Ma T, Liu Z, Gao L. Exosomes secreted by endothelial cells derived from human induced pluripotent stem cells improve recovery from myocardial infarction in mice. *Stem Cell Research & Therapy*. 2023b;14(1):278.

Li Y, Wang J, Chen S, Wu P, Xu S, Wang C, et al. miR-137 boosts the neuroprotective effect of endothelial progenitor cell-derived exosomes in oxyhemoglobin-treated SH-SY5Y cells partially via COX2/PGE2 pathway. *Stem Cell Research & Therapy*. 2020;11:1–18.

Liang Y, Xia J, Xu L, Zhang Y, Xu X, Haider KH. Cell-free and cell-based therapy in regenerative medicine. In: Haider, K.H. (eds) *Handbook of Regenerative Medicine: Cell-Free Therapy*. New York: Taylor & Francis Group CRC Press, 2025, pp.134–53.

Liu F, Vermesh O, Mani V, Ge TJ, Madsen SJ, Sabour A, et al. The exosome total isolation chip. *ACS Nano*. 2017;11(11):10712–23.

Ma K, Xu H, Zhang J, Zhao F, Liang H, Sun H, et al. Insulin-like growth factor-1 enhances neuroprotective effects of neural stem cell exosomes after spinal cord injury via an miR-219a-2-3p/YY1 mechanism. *Aging (Albany, NY)*. 2019;11(24):12278.

Mackie AR, Klyachko E, Thorne T, Schultz KM, Millay M, Ito A, et al. Sonic hedgehog–modified human CD34+ cells preserve cardiac function after acute myocardial infarction. *Circulation Research*. 2012;111(3):312–21.

Maguire CA, Balaj L, Sivaraman S, Crommentuijn MH, Ericsson M, Mincheva-Nilsson L, et al. Microvesicle-associated AAV vector as a novel gene delivery system. *Molecular Therapy*. 2012;20(5):960–71.

Mathivanan S, Fahner CJ, Reid GE, Simpson RJ. ExoCarta 2012: Database of exosomal proteins, RNA, and lipids. *Nucleic Acids Research*. 2012;40(D1):D1241–4.

McKelvey KJ, Powell KL, Ashton AW, Morris JM, McCracken SA. Exosomes: Mechanisms of uptake. *Journal of Circulating Biomarkers*. 2015;4:7.

Merchant ML, Powell DW, Wilkey DW, Cummins TD, Deegens JK, Rood IM, et al. Microfiltration isolation of human urinary exosomes for characterization by MS. *PROTEOMICS–Clinical Applications*. 2010;4(1):84–96.

Netala VR, Teertam SK, Li H, Zhang Z. A comprehensive review of cardiovascular disease management: Cardiac biomarkers, imaging modalities, pharmacotherapy, surgical interventions, and herbal remedies. *Cells*. 2024;13(17):1471.

Parada N, Romero-Trujillo A, Georges N, Alcayaga-Miranda F. Camouflage strategies for therapeutic exosomes evasion from phagocytosis. *Journal of Advanced Research*. 2021;31:61–74.

Parekh Y, Bokara KK, Haider KH. Neural stem cells: From bench to bedside. In: Haider, K.H. (eds) *Handbook of Stem Cell Therapy*. Singapore: Springer, 2022. https://doi.org/10.1007/978-981-16-6016-0_28-1.

Peng Y, Zhao J-L, Peng Z-Y, Xu W-F, Yu G-L. Exosomal miR-25-3p from mesenchymal stem cells alleviates myocardial infarction by targeting pro-apoptotic proteins and the epigenetic regulator EZH2. *Cell Death & Disease*. 2020;11(5):317.

Perez NE, Godbole NP, Amin K, Syan R, Gater Jr DR. Neurogenic bladder physiology, pathogenesis, and management after spinal cord injury. *Journal of Personalized Medicine*. 2022;12(6):968.

Rad LM, Yumashev AV, Hussen BM, Jamad HH, Ghafouri-Fard S, Taheri M, et al. Therapeutic potential of microvesicles in cell therapy and regenerative medicine of ocular diseases with a special focus on mesenchymal stem cell-derived microvesicles. *Frontiers in Genetics*. 2022;13:847679.

Rezaie J, Feghhi M, Etemadi T. A review on exosomes application in clinical trials: Perspective, questions, and challenges. *Cell Communication and Signaling*. 2022;20(1):145.

Rodriguez EL, Poddar S, Iftekhar S, Suh K, Woolfork AG, Ovbude S, et al. Affinity chromatography: A review of trends and developments over the past 50 years. *Journal of Chromatography B*. 2020;1157:122332.

Saari H, Lázaro-Ibáñez E, Viitala T, Vuorimaa-Laukkanen E, Siljander P, Yliperttula M. Microvesicle- and exosome-mediated drug delivery enhances the cytotoxicity of Paclitaxel in autologous prostate cancer cells. *Journal of Controlled Release*. 2015;220:727–37.

Safwan M, Bourgleh MS, Baroudi L, Almesned A, Albalawi R, Aybour B, Shahad Alfawaz, Bourgleh SM, Haider KH. Entering the era of living biopharmaceuticals for treating knee osteoarthritis: Systematic review and network meta-analysis to compare allogenic and autologous MSCs. *World Journal of Stem Cells*. 2025 August 26;17(8):107076. https://doi.org/10.4252/wjsc.v17.i8.107076. ISSN 1948.

Santoso MR, Ikeda G, Tada Y, Jung JH, Vaskova E, Sierra RG, et al. Exosomes from induced pluripotent stem cell–derived cardiomyocytes promote autophagy for myocardial repair. *Journal of the American Heart Association*. 2020;9(6):e014345.

Saunderson SC, Dunn AC, Crocker PR, McLellan AD. CD169 mediates the capture of exosomes in the spleen and lymph nodes. *Blood, The Journal of the American Society of Hematology*. 2014;123(2):208–16.

Shafei S, Khanmohammadi M, Heidari R, Ghanbari H, Taghdiri Nooshabadi V, Farzamfar S, et al. Exosome-loaded alginate hydrogel promotes tissue regeneration in full-thickness skin wounds: An in vivo study. *Journal of Biomedical Materials Research Part A*. 2020;108(3):545–56.

Shang H, Younas A, Zhang N. Recent advances on transdermal delivery systems for the treatment of arthritic injuries: From classical treatment to nanomedicines. *Wiley Interdisciplinary Reviews: Nanomedicine and Nanobiotechnology*. 2022;14(3):e1778.

Sokolova V, Ludwig A-K, Hornung S, Rotan O, Horn PA, Epple M, et al. Characterisation of exosomes derived from human cells by nanoparticle tracking analysis and scanning electron microscopy. *Colloids and Surfaces B: Biointerfaces*. 2011;87(1):146–50.

Stenqvist A-C, Nagaeva O, Baranov V, Mincheva-Nilsson L. Exosomes secreted by human placenta carry functional Fas ligand and TRAIL molecules and convey apoptosis in activated immune cells, suggesting exosome-mediated immune privilege of the fetus. *The Journal of Immunology*. 2013;191(11):5515–23.

Szatanek R, Baj-Krzyworzeka M, Zimoch J, Lekka M, Siedlar M, Baran J. The methods of choice for characterizing extracellular vesicles (EVs). *International Journal of Molecular Sciences*. 2017;18(6):1153.

Tan F, Li X, Wang Z, Li J, Shahzad K, Zheng J. Clinical applications of stem cell-derived exosomes. *Signal Transduction and Targeted Therapy*. 2024;9(1):17.

Tian T, Zhu Y-L, Zhou Y-Y, Liang G-F, Wang Y-Y, Hu F-H, et al. Exosome uptake through clathrin-mediated endocytosis and macropinocytosis mediates miR-21 delivery. *Journal of Biological Chemistry*. 2014;289(32):22258–67.

Tian X, Wei W, Cao Y, Ao T, Huang F, Javed R, et al. Gingival mesenchymal stem cell-derived exosomes are immunosuppressive in preventing collagen-induced arthritis. *Journal of Cellular and Molecular Medicine*. 2022;26(3):693–708.

Tschuschke M, Kocherova I, Bryja A, Mozdziak P, Angelova Volponi A, Janowicz K, et al. Inclusion biogenesis, methods of isolation, and clinical application of human cellular exosomes. *Journal of Clinical Medicine*. 2020;9(2):436.

van der Meel R, Fens MH, Vader P, Van Solinge WW, Eniola-Adefeso O, Schiffelers RM. Extracellular vesicles as drug delivery systems: Lessons from the liposome field. *Journal of Controlled Release*. 2014;195:72–85.

Vincent-Schneider H, Stumptner-Cuvelette P, Lankar D, Pain S, Raposo G, Benaroch P, et al. Exosomes bearing HLA-DR1 molecules require dendritic cells to efficiently stimulate specific T cells. *International Immunology*. 2002;14(7):713–22.

Wang AYL. Human-induced pluripotent stem cell-derived exosomes as a novel therapeutic strategy for various diseases. *International Journal of Molecular Sciences*. 2021;22(4):1769.

Wang C, Chen Y, Hu S, Liu X. Insights into the function of ESCRT and its role in enveloped virus infection. *Frontiers in Microbiology*. 2023;14:1261651.

Wang C, Wang M, Xu T, Zhang X, Lin C, Gao W, et al. Engineering bioactive self-healing antibacterial exosomes hydrogel for promoting chronic diabetic wound healing and complete skin regeneration. *Theranostics*. 2019;9(1):65.

Wang X, Xia J, Yang L, Dai J, He L. Recent progress in exosome research: Isolation, characterization, and clinical applications. *Cancer Gene Therapy*. 2023;30(8):1051–65.

Wei J, Chen Y, Xue C, Ma B, Shen Y, Guan J, et al. Protection of nerve injury with exosomes extracted from mesenchymal stem cells. Zhongguo yi xue ke xue yuan xue bao *Acta Academiae Medicinae Sinicae*. 2016;38(1):33–6.

Wen Z, Mai Z, Zhu X, Wu T, Chen Y, Geng D, et al. Mesenchymal stem cell-derived exosomes ameliorate cardiomyocyte apoptosis in hypoxic conditions through microRNA144 by targeting the PTEN/AKT pathway. *Stem Cell Research & Therapy*. 2020;11:1–17.

Wieckowski EU, Visus C, Szajnik M, Szczepanski MJ, Storkus WJ, Whiteside TL. Tumor-derived microvesicles promote regulatory T cell expansion and induce apoptosis in tumor-reactive activated CD8+ T lymphocytes. *The Journal of Immunology*. 2009;183(6):3720–30.

Wu J, Kuang L, Chen C, Yang J, Zeng W-N, Li T, et al. miR-100-5p-abundant exosomes derived from infrapatellar fat pad MSCs protect articular cartilage and ameliorate gait abnormalities via inhibition of mTOR in osteoarthritis. *Biomaterials*. 2019;206:87–100.

Wu Y, Deng W, Klinke II DJ. Exosomes: improved methods to characterize their morphology, RNA content, and surface protein biomarkers. *Analyst.* 2015;140(19):6631–42.

Xin H, Li Y, Buller B, Katakowski M, Zhang Y, Wang X, et al. Exosome-mediated transfer of miR-133b from multipotent mesenchymal stromal cells to neural cells contributes to neurite outgrowth. *Stem cells.* 2012;30(7):1556–64.

Xu J, Bai S, Cao Y, Liu L, Fang Y, Du J, et al. miRNA-221-3p in endothelial progenitor cell-derived exosomes accelerates skin wound healing in diabetic mice. *Diabetes, Metabolic Syndrome, and Obesity.* 2020;13:1259–70.

Xu X, Zhang H, Li J, Chen Y, Zhong W, Chen Y, et al. A combination of EPC-EXs and NPC-EXs, overexpressing miR-126 and miR-210, produces better therapeutic effects on ischemic stroke by protecting neurons through the Nox2/ROS and BDNF/TrkB pathways. *Experimental Neurology.* 2023;359:114235.

Yakubovich E, Polischouk A, Evtushenko V. Principles and problems of exosome isolation from biological fluids. *Biochemistry (Moscow), Supplement Series A: Membrane and Cell Biology.* 2022;16(2):115–26.

Yang GH, Lee YB, Kang D, Choi E, Nam Y, Lee KH, et al. Overcome the barriers of the skin: Exosome therapy. *Biomaterials Research.* 2021;25(1):22.

Yang JS, Lee JC, Byeon SK, Rha KH, Moon MH. Size-dependent lipidomic analysis of urinary exosomes from patients with prostate cancer by flow field-flow fractionation and nanoflow liquid chromatography-tandem mass spectrometry. *Analytical Chemistry.* 2017;89(4):2488–96.

Yu L-L, Zhu J, Liu J-X, Jiang F, Ni W-K, Qu L-S, et al. A comparison of traditional and novel methods for the separation of exosomes from human samples. *BioMed Research International.* 2018;2018.

Yu YM, Gibbs KM, Davila J, Campbell N, Sung S, Todorova TI, et al. MicroRNA miR-133b is essential for functional recovery after spinal cord injury in adult zebrafish. *European Journal of Neuroscience.* 2011;33(9):1587–97.

Zhang B, Gong J, He L, Khan A, Xiong T, Shen H, et al. Exosome-based advancements for application in medical aesthetics. *Frontiers in Bioengineering and Biotechnology.* 2022;10:1083640.

Zhang K, Cheng K. Stem cell-derived exosome versus stem cell therapy. *Nature Reviews Bioengineering.* 2023;1(9):608–9.

Zhang S, Chuah SJ, Lai RC, Hui JHP, Lim SK, Toh WS. MSC exosomes facilitate cartilage repair by promoting proliferation, inhibiting apoptosis, and modulating immune reactivity. *Biomaterials.* 2018;156:16–27.

Zhang Y, Bi J, Huang J, Tang Y, Du S, Li P. Exosome: A review of its classification, isolation techniques, storage, diagnostic and targeted therapy applications. *International Journal of Nanomedicine.* 2020;15:6917–34.

Zhang Y, Liu Y, Liu H, Tang WH. Exosomes: biogenesis, biological function and clinical potential. *Cell & Bioscience.* 2019;9:1–18.

Zhong D, Cao Y, Li C-J, Li M, Rong Z-J, Jiang L, et al. Neural stem cell-derived exosomes facilitate spinal cord functional recovery after injury by promoting angiogenesis. *Experimental Biology and Medicine.* 2020;245(1):54–65.

Zhuang X, Xiang X, Grizzle W, Sun D, Zhang S, Axtell RC, et al. Treatment of brain inflammatory diseases by delivering exosome encapsulated anti-inflammatory drugs from the nasal region to the brain. *Molecular Therapy.* 2011;19(10):1769–79.

Zondler L, Feiler MS, Freischmidt A, Ruf WP, Ludolph AC, Danzer KM, et al. Impaired activation of ALS monocytes by exosomes. *Immunology and Cell Biology.* 2017;95(2):207–14.

Zondler L, Müller K, Khalaji S, Bliederhäuser C, Ruf WP, Grozdanov V, et al. Peripheral monocytes are functionally altered and invade the CNS in patients with ALS. *Acta Neuropathologica.* 2016;132:391–411.

26 The Use of Secretome and Extracellular Vesicles in Regenerative Medicine

Krish V. Prajapati[1], Allison M. Lockhart[1], Prakhya S. Ganti[1], Amritha T.M. Seetharaman[1], and Rajashekhar Gangaraju[1,2]

[1]Department of Ophthalmology, University of Tennessee Health Science Center, Memphis, TN, USA

[2]Department of Anatomy and Neurobiology, University of Tennessee Health Science Center, Memphis, TN, USA

26.1 LIST OF ABBREVIATIONS

6-OHDA	6-hydroxydopamine
AD	Alzheimer's disease
AKT	Protein kinase B
ALS	Amyotrophic lateral sclerosis
ANG1	Angiopoietin-1
ANG2	Angiopoietin-2
ApoBDs	Apoptotic bodies
APP/PS1	Amyloid precursor protein/presenilin 1
ASC-CCM	Adipose tissue-derived mesenchymal stem cells
ASC-CM	Adipose stem cell-conditioned media
BM-MSC	Bone marrow MSCs
CAR-T	Chimeric antigen receptor T-cell
CCL2	Chemokine ligand 2
CCR2	C-C chemokine receptor type 2
CD44	Cluster of differentiation 44
CNS	Central nervous system
COL1	Collagen Type I ALP alkaline phosphatase
DNA	Deoxyribonucleic acid
EGF	Epidermal growth factor
EMA	European Medical Association
EPC	Endothelial progenitor cells
ERBB2	Erb-B2 receptor tyrosine kinase 2
ERK	Extracellular regulated kinase
ESC	Embryonic stem cells
EVs	Extracellular vesicles
FDA	Food and Drug Administration
GFAP	Glial fibrillary acidic protein

DOI: 10.1201/9781003653387-26

DSS	Dextran sodium sulfate
GMP	Good Manufacturing Practice
GPCR	G-protein-coupled receptor
GvHD	Graft-versus-host disease
HaCaT	Human keratinocyte cell line
HEK293	Human embryonic kidney epithelial cells 293
hPDLSCs	Human periodontal ligament stem cells
HSCs	Hematopoietic stem cells
hUC-MSCs	Human umbilical cord mesenchymal stem cells
ISEV	International Society for Extracellular Vesicles
microRNA	Micro ribonucleic acid
MISEV	Minimal information for studies of extracellular vesicles
MRSA	Methicillin-resistant staphylococcus aureus
MSCs	Mesenchymal stem cells
MVBs	Multivesicular bodies
NTA	Nanoparticle tracking analysis
OCN	Osteocalcin (also known as Bone Gla Protein)
OPN	Osteopontin
PAD	Peripheral artery disease
PDGFs	Platelet-derived growth factors
PRRs	Pattern recognition receptors
RGD	Arginine-glycine-aspartic acid
SEC	Size exclusion chromatography
SHED	Human exfoliated deciduous teeth
TBI	Traumatic brain injuries
TGF-β	Transforming growth factor beta
TSG-6	Tumor necrosis factor-stimulated gene-6
VEGF	Vascular endothelial growth factor
VEGF	Vascular endothelial growth factor

26.2 INTRODUCTION

26.2.1 Cell-Based Therapies

Cell-based therapies, particularly those involving stem cells, present considerable advantages in managing various medical conditions. Stem cells intrinsically secrete growth factors that enhance the body's regenerative mechanisms, facilitating tissue repair while alleviating pain without the dependency on analgesic medications (Anwar et al., 2025). A significant benefit of these therapies is their ability to relieve pain and inflammation in targeted areas, providing practical and sustained relief. Additionally, they are associated with reduced post-procedural recovery time, enabling patients to resume their daily activities more promptly. These therapies symbolize responsible medical innovation by fostering the repair and regeneration of damaged tissues—nerves, cartilage, and muscles—ultimately augmenting functionality, range of motion, and flexibility for conditions characterized by extensive tissue damage, such as heart disease, spinal cord injuries, and degenerative disorders. In numerous instances, they restore impaired tissues to their pre-damaged condition (Safwan et al., 2025). Importantly, cell-based treatments offer a non-surgical, minimally invasive alternative to conventional interventions, thereby reducing the risks associated with surgical procedures—such as complications and nerve damage—and eliminating the need for general anesthesia. Moreover, using autologous biological materials, cells harvested from the patient, substantially reduces the risk of immune rejection and adverse reactions. By reducing reliance on pharmaceutical interventions, these therapies enhance overall physiological function while minimizing the risk of disease transmission. Ultimately, these therapies can be tailored to meet the specific needs of individual patients, resulting in more effective and personalized treatment plans.

26.2.2 Historical Context and Evolution of Cell-Free Therapies

The concept of cell-free therapies emerged from the extensive history of gene therapy and stem cell research. The initial conceptual history of gene therapy began with the identification of DNA as the transforming substance capable of altering the phenotypic characteristics of bacteria (Malech et al., 2022). This understanding laid the groundwork for investigating how genetic material could be utilized to modify and treat various conditions. As research advanced, scientists found that the therapeutic effects of hematopoietic stem cells (HSCs), mesenchymal stem cells (MSCs), and non-stem cell-based therapies like CAR-T cells and immune cells were primarily elicited by their secreted products, known as the secretome, rather than by cellular differentiation (Sreenivas and Jha, 2025). This realization led to a paradigm shift from cell-based therapies to cell-free approaches (Liang et al., 2025). The evolution of cell-free therapies has gained momentum in recent years, driven by the need to address ethical issues associated with cellular transplantation and prevent complications from arising due to the incorrect differentiation of cells in host tissue (Drobiova et al., 2023). The overall use of cells' secretomes, or just extracellular vesicles (EVs), for treating diseases is referred to as cell-free therapy; this is particularly important with MSCs (Habib et al., 2025). This approach preserves the therapeutic potential of stem cells while circumventing some of the challenges associated with direct cell transplantation. This chapter will primarily focus on secretome and EV-based methods, especially those involving MSCs.

26.2.3 Overview of the Secretome and Extracellular Vesicles

The secretome refers to the collection of bioactive substances released by cells, particularly MSCs. It comprises free soluble factors and EVs (Ahmed et al., 2023). The secretome plays a pivotal role in MSCs' therapeutic effects, encompassing their neuroregenerative, neuroprotective, and immunomodulatory attributes. EVs, integral components of the secretome, are membrane-bound structures capable of transporting and delivering a diverse array of bioactive molecules, including microRNAs, non-coding RNAs, messenger RNAs, and proteins, to target cells. EVs derived from MSCs have demonstrated therapeutic potential in various conditions, including inflammatory diseases, osteoarthritis, and rheumatoid arthritis (Mushtaq et al., 2024). The secretome and EVs are receiving significant attention in regenerative medicine due to their ability to influence various biological processes. Empirical evidence indicates that they promote angiogenesis, facilitate tissue repair, and induce immunomodulation. The application of secretomes and EVs in cell-free therapies offers numerous advantages over traditional cell-based methodologies, including simplified storage and handling, reduced risk of tumorigenicity, and the potential for off-the-shelf treatments. The investigation of the secretome and EVs has proliferated rapidly, with research delving into their potential for addressing neurodegenerative diseases, cardiovascular disorders, and other chronic conditions. As our comprehension of these complex biological entities expands, they will likely assume an increasingly significant role in the future of regenerative medicine and personalized healthcare therapies.

26.3 COMPOSITION AND BIOLOGY OF THE SECRETOME

26.3.1 Soluble Bioactive Molecules and Their Role in Cell-to-Cell Communication and Tissue Homeostasis

The secretome refers to the set of bioactive molecules secreted by a cell that induce changes in cellular mechanisms via various pathways (Xia et al., 2023). These molecules, including cytokines, chemokines, growth factors, and bioactive lipids, vary in function and structure, yet they collaborate to maintain cellular homeostasis. Cytokines are signaling molecules secreted by both immune and non-immune cells, playing a vital role in cellular communication. Typically, cytokines are secreted when a specific disease, infection, or cancer develops, helping modulate inflammatory pathways in the immune response. Cytokines have a helical shape; four amphipathic helices form the core of the helical bundle (Hughes and Nibbs, 2018). This molecule is too small to enter the cell membrane. Thus, cytokines bind to an extracellular receptor, such as a G-protein-coupled receptor (GPCR), to activate a cellular response. Cytokines are widely used in paracrine signaling to co-stimulate immune cells during an immune response, inducing clonal selection and promoting pathogen clearance. Cytokines can be further classified into more specific types, including interleukins and interferons. Interleukins are cytokines that regulate cellular growth and differentiation, specifically in leukocytes, and are produced when the immune system detects antigens or pathogens (Vaillant, 2025). Another example is interferons, small proteins that, during a viral infection, induce cells to inhibit viral replication (Urry et al., 2021).

Chemokines are a subcategory of cytokines, distinguished by their unique overall structure and function. Chemokines contain a central three-stranded β-sheet, an overlying C-terminal α-helix, and a short unstructured N terminus (Hughes and Nibbs, 2018). They signal to other cells through GPCRs, which, along with chemokines' polar structures, implies that chemokines are polar and cannot freely enter the cell membrane. Some of their functions include angiogenesis and cell-to-cell adhesion. They are also responsible for stimulating cellular migration, such as haptotaxis, chemokinesis, and haptokinesis, as well as recruiting leukocytes. For example, chemokine ligand 2 (CCL2) via its functional receptor, CCR2, is responsible for prostate cancer cell survival (Zhang et al., 2010).

Growth factors are substances produced by the body that regulate cell division and the cell cycle. They affect a wide variety of cell types, particularly because each growth factor is specific for its receptors. For example, some growth factors can act only on hematopoietic cells, which are bone marrow cells responsible for hematopoiesis (Stone et al., 2025). When growth factors bind to their specific transmembrane receptors, tyrosine kinase activity increases, triggering a signal transduction cascade that modulates protein synthesis and gene expression. Growth factors have varying structures based on their unique functions, including differences in polarity; lipid-soluble growth factors can pass through the cell membrane, whereas those that cannot pass bind to cell-surface receptors. Growth factors are also involved in wound healing, where platelet-derived growth factors (PDGFs) and Transforming Growth Factor Beta (TGF-β) are released from platelets to initiate the inflammatory phase and attract neutrophils and macrophages. However, growth factors can also have detrimental effects. Vascular endothelial growth factor (VEGF) promotes neovascularization, or the abnormal growth of new retinal blood vessels, leading to macular degeneration in ischemic retinal disorders (Penn et al., 2008).

Bioactive lipids are molecules that mediate biological phenomena such as inflammation, immune regulation, and homeostasis (Casati et al., 2021). Due to their non-polar nature, bioactive lipids can cross the cell membrane and interact with intracellular nuclear receptors. They can also work extracellularly on GPCRs. There are three broad categories of bioactive lipids: eicosanoids, endocannabinoids, and lysophospholipids, each of which generates distinct effects on inflammatory pathways (Casati et al., 2021). Bioactive lipids can impair cellular function when secreted in large quantities. For example, one class of eicosanoids has been implicated in the development of cardiovascular disease in preclinical studies (Wasserman et al., 2020). Bioactive lipids can also induce inflammation, leading to the development of several chronic diseases, including atherosclerosis, cancer, and rheumatoid arthritis (Chiurchiu et al., 2018).

26.3.2 Similarities and Differences in Various Cell Types' Secretome

Cell types exhibit distinct secretomes due to their unique origins and functions. This section will emphasize the secretomes of three well-studied cell types: Mesenchymal Stem Cells (MSCs),

Immune Cells, and Endothelial Progenitor Cells. The secretome of MSCs plays a crucial role in facilitating cell-to-cell communication and mediating immune modulation and regeneration. It contains a soluble portion and a vesicular portion in its secretome (Ghasemi et al., 2023). The soluble portion includes proteins such as cytokines, chemokines, and growth factors, whereas the vesicular portion contains microvesicles and exosomes (Ghasemi et al., 2023). Immune cells are a broad category encompassing specific cell types with distinct secretomes. Cytokines are a significant component of the immune cell secretome and are widely used to co-stimulate specific immune cells during infection (Urry et al., 2021). Interleukins, a family of cytokines, are produced by immune cells and play a crucial role in stimulating the proliferation and migration of these cells. Other molecules include interferons and immunoglobulins, both of which play significant roles in the immune response (Urry et al., 2021). Endothelial progenitor cells secrete numerous proteins, ranging from growth factors that promote proliferation to cytokines that facilitate signaling (Al-Omar et al., 2022). Such proteins include angiogenin, platelet-derived growth factor B, thymidine phosphorylase, and granulocyte-macrophage colony-stimulating factor, among others (Alwjwaj et al., 2021). Many of these proteins are responsible for angiogenesis, the process by which blood vessels form.

26.4 BIOGENESIS AND CLASSIFICATION OF EXTRACELLULAR VESICLES

Cells secrete EVs and pack them with material for shipment, either within the cell or between cells. Three of the EVs that are under examination are exosomes, microvesicles, and apoptotic bodies. Exosomes, which are part of the small EVs (< 200 nm), carry nucleic acids, proteins, lipids, and metabolites (Kalluri and LeBleu, 2020; Welsh et al., 2024). They are widely used in medicine for their regenerative capabilities and their support for near- and long-distance intercellular communication and cell maintenance (Doyle and Wang, 2019). Exosomes interact with cells via surface receptors and are internalized via micropinocytosis and phagocytosis, providing therapeutic benefits in the treatment of various diseases, including myocardial infarction, diabetes mellitus, infertility, COVID-19, and spinal cord injury (Muthu et al., 2021; Habib et al., 2026). Exosomes form from the inward budding of an early endosome's limiting membrane (Doyle and Wang, 2019). This then matures into multivesicular bodies (MVBs), which are organelles responsible for protein sorting, recycling, storage, transport, and release of materials. These MVBs have two fates: either they are sent to a lysosome to be degraded or are sent to the cell's plasma membrane, where they fuse to release the content, including exosomes, into the cell's extracellular space (Hessvik and Llorente, 2018).

Microvesicles are a type of vesicle that are approximately 100–1000 nanometers in size, making them larger than exosomes, and transport proteins, nucleic acids, and lipids (Stahl et al., 2019). Microvesicles transport membrane-derived receptors, lipids, carbohydrates, mRNA, microRNA, cytokines, and chemokines, all of which may aid in cellular signaling and migration. In addition to playing a key role in cellular communication by transporting biomolecules, microvesicles can serve as markers of disease progression, as they are found at elevated concentrations in diseased tissue (Schwager and Reinhart-King, 2020). Microvesicles are formed by budding directly from the plasma membrane; subsequently, a process of severe pinching and scission releases them into the extracellular space. This process is facilitated by changes in calcium ion concentration and by a stimulus that triggers the release of microvesicles (Stahl et al., 2019).

Apoptotic bodies (ApoBDs), as the name suggests, are vesicles that are 1–5 μm in diameter and are formed when cells undergo apoptosis or programmed cell death (Yu et al., 2023). Apoptotic bodies form from the disassembly of cells, but each cell type disassembles distinctly. This means that the quality and quantity of ApoBDs formed will vary according to the cell type undergoing apoptosis (Elmore, 2007). These vesicles can contain cellular components such as micronuclei, degraded proteins, DNA fragments, or cytosolic components (Battistelli and Falcieri, 2020). ApoBDs promote the removal of cell debris by attracting nearby phagocytes, delivering essential nutrients to healthy cells, and harboring biomolecules such as DNA and microRNAs to regulate intercellular communication (Battistelli and Falcieri, 2020). These molecules in ApoBDs stimulate tissue regeneration and the immune response; after phagocytes engulf ApoBDs, these molecules are released and begin regulating receptors involved in intercellular communication (Li et al., 2022b). Apoptotic bodies are widely used in regenerative medicine and help regulate diseases such as atherosclerosis and hepatic fibrosis (Li et al., 2022b).

26.5 THERAPEUTIC APPLICATIONS OF THE SECRETOME

The therapeutic applications of the secretome are vast and varied, with ongoing research continually uncovering new potential uses. Its ability to influence multiple biological processes and its versatility across diverse medical contexts highlight its significance in advancing modern medicine. The following sections provide a brief overview.

26.5.1 Wound Healing and Tissue Repair

Chronic wounds are characterized by persistent inflammation, infection, and cellular necrosis, resulting from disruptions in normal tissue healing (Raziyeva et al., 2021). The cell secretome can be utilized for cell regeneration and expedite wound healing by reducing inflammation, supporting angiogenesis, and stimulating the proliferation of keratinocytes and fibroblasts, all essential components of wound healing. For instance, in a human keratinocyte cell line (HaCaT), a human umbilical cord mesenchymal stem cell (hUCMSC)-conditioned medium was applied, positively affecting keratinocyte proliferation and migration (Sriramulu et al., 2020). In another *in vitro* study examining the application of umbilical cord-derived mesenchymal stem cell-conditioned medium to dermal fibroblasts, the results showed enhanced fibroblast proliferation and migration, improved wound healing, and reduced scar formation (Li et al., 2017). These findings were confirmed in an *in vivo* C57BL mouse model, which demonstrated a heightened, faster wound-healing response and decreased collagen deposition, resulting in significantly reduced scar formation in C57BL mice treated with UC-MSC conditioned medium (Li et al., 2017).

Secretome uses also include improvements in vascularity and the dampening of immune responses, thereby enhancing wound-healing processes. For example, a human umbilical cord Wharton's jelly-conditioned medium applied to murine excisional and diabetic wounds found that the treated wounds exhibited significantly greater vascularity, cellular density, and epithelialization, resulting in faster healing rates than control groups (Fong et al., 2014). In mouse models infected with Methicillin-resistant Staphylococcus aureus (MRSA), BM-MSC-conditioned medium increased granulation tissue, vasculature, and hair follicles, accelerating wound closure in treated mice (Rajesh et al., 2024). Additionally, the bacterial load in MRSA-infected wounds was significantly reduced, accompanied by an increase in M2 macrophages, which support healing.

26.5.2 Neuroprotection in Neurodegenerative Diseases and Traumatic Brain Injuries

Mesenchymal stem cells have the potential to secrete neuroprotective agents with immunomodulatory properties, which can be used for neuroregeneration and neuroprotection in a wide range of neurodegenerative disorders and traumatic brain injuries (TBI). The conditioned medium of dental pulp stem cells (DPSC-CM) in an mSOD1G93A mouse model of Amyotrophic Lateral Sclerosis (ALS) showed increased survival in the mice during DPSC-CM administration across all stages of ALS (Wang et al., 2019). In a 6-hydroxydopamine (6-OHDA) rat model of Parkinson's disease (PD), the application of bone marrow stem cell secretome resulted in reduced inflammation, oxidative stress, neurobehavioral impairments, and apoptotic markers in secretome-treated rats compared to controls (Mahendru et al., 2021). The secretome has also demonstrated promising results in treating Alzheimer's Disease (AD). In APP/PS1 AD mice, MSC secretome treatment reduced plaque accumulation, lowered AβO levels, and diminished brain atrophy (Santamaria et al., 2021). Additionally, weekly intranasal administration of MSC secretome induced persistent memory recovery in treated APP/PS1 AD mice, whereas single administrations only produced transient memory recovery.

When brain structures are damaged during traumatic brain injury (TBI), damage-associated molecular patterns (DAMPs) and alarmins are released from the meninges, glial limitans, and brain parenchyma, activating recognition by pattern recognition receptors (PRRs) on central nervous system (CNS) glial cells (Pischiutta et al., 2022). This recognition of stress signals initiates a cascade of inflammatory events, including the activation of inflammasomes and the release of cytokines and chemokines by CNS glial cells. In a study demonstrating the impact of secretome on TBI, a rat model of severe TBI was administered UC-MSC secretome, which improved cognition (Liu et al., 2020b). The utility of the secretome was also established in visual deficits following TBI. In a study investigating mice with mild traumatic brain injury (mTBI), vision loss was prevented by administering concentrated conditioned media from adipose tissue-derived mesenchymal stem cells (ASC-CCM) (Jha et al., 2019). The ASC-CCM showed that TSG-6 secretion suppressed microglial activation, a key event in vision loss, and inhibited increases in endothelial permeability. In a similar study, mice subjected to repetitive ocular blast injury (rOBI) showed positive results from intravitreal administration of adipose-derived stem cell concentrated conditioned medium (ASC-CCM), leading to decreased neuroinflammation, reduced oxidative stress, and lowered levels of retinal Glial Fibrillary Acidic Protein (GFAP), a prominent retinal injury marker (Jha et al., 2022).

26.5.3 Modulation of Immune Responses in Autoimmune Diseases

It is now well established that MSCs mediate immune homeostasis by secreting cytokines, chemokines, growth factors, and signaling molecules that modulate both adaptive and innate immune responses (Ghasemi et al., 2023). The secretome itself is less immunogenic than MSC implantation due to its unique cell-free nature, showing great promise for therapeutic use in autoimmune disorders, including inflammatory bowel diseases such as Crohn's Disease and Ulcerative Colitis (Muzes and Sipos, 2022). For instance, a study intraperitoneally injected conditioned media from spindle-shaped MSCs (SS-AF-MSCs) derived from second-trimester amniotic fluid into Dextran Sodium Sulfate (DSS)-treated mice, which induced colitis (Legaki et al., 2016; Lee et al., 2025). In a separate study investigating the effect of intraperitoneal injection of conditioned mouse adipose-derived stem cell (mADSC) secretome in DSS-induced mice, researchers observed improvements in body weight, colon length, and histopathological scores. In another example, conditioned medium from hypoxic human periodontal ligament stem cells (H-hPDLSCs-CM) was administered to mice with autoimmune encephalomyelitis (EAE), a murine model of multiple sclerosis (Giacoppo et al., 2017). In a similar study, conditioned medium from human exfoliated deciduous tooth stem cells (SHED-CM) was injected into mice with EAE; the results showed significantly improved disease scores, reduced demyelination, reduced axonal injury, and reduced levels of pro-inflammatory cytokines (Shimojima et al., 2016). These studies, along with others, suggest the potential use of stem cell-conditioned medium as a treatment for several autoimmune disorders.

26.5.4 Angiogenesis in Ischemic Conditions

Ischemic heart disease is caused by the narrowing of the coronary arteries due to clots, atherosclerosis, and vessel constriction, reducing blood flow to the heart muscle and increasing the risk of myocardial infarction and ischemic stroke. In porcine models of myocardial infarction, intravenous administration of MSC-CM stimulated angiogenesis, preserving cardiac function and increasing capillary density (Timmers et al., 2011). In a mouse model of ischemia-reperfusion (I/R) injury, the conditioned medium from intravenously administered stem cells derived from human exfoliated deciduous teeth (SHED) significantly suppressed apoptosis and inflammation, protecting against acute ischemic injury (Yamaguchi et al., 2015). Conditioned medium from human embryonic MSCs (hMSC-CM), administered to adult male Wistar rats with middle cerebral artery occlusion, was shown to alleviate some effects of ischemic stroke (Asgari Taei et al., 2021). MSC-CM has also been helpful in other conditions leading to ischemia, including retinal ischemia and Peripheral Artery Disease (PAD) (Elshaer et al., 2021). For example, intravitreal delivery of Adipose Stem Cell-Conditioned Media (ASC-CM) to Ins 2 Akita mouse models with retinal ischemia increased retention of neurovascular architecture, reduced vascular leakage, and decreased inflammation, indicating improved vision (Elshaer et al., 2018). Administration of the secretome from hypoxia-induced mesenchymal stem cells (S-MSCs) to diabetic rats with peripheral artery disease (PAD) increased VEGF and CD31 levels, indicating upregulated angiogenesis in the affected area (Sazli et al., 2023).

26.6 THERAPEUTIC APPLICATIONS OF EVS

EVs have garnered significant attention as potential diagnostic and prognostic biomarkers due to their ability to carry diverse molecular cargos, including proteins, lipids, and nucleic acids. Additionally, both natural and engineered EVs serve as drug-delivery vehicles. In the following sections, representative examples of various diseases are provided to exemplify this.

26.6.1 Anti-inflammatory and Regenerative Roles in Cardiovascular and Metabolic Diseases

EVs have the potential to alleviate various cardiac pathologies associated with ischemia and inflammation (Haider and Najimi, 2022). For example, MSC-derived exosomes inhibit apoptosis and improve cardiomyocyte conditions by delivering miR-451 and miR-21, which inhibit caspase 3/7 and miR-210, inhibiting pro-apoptotic proteins (Xiao et al., 2016). In addition, MSC-derived exosomes can reduce cardiac hypertrophy and inflammation by inhibiting the proliferation of cardiac fibroblasts and pro-inflammatory cytokines (Mani et al., 2023; Haider and Aramini, 2020). Additionally, the therapeutic potential of MSC-derived EVs in reducing lipid accumulation and liver injury, and immune cell-derived EVs in mitigating inflammation and fibrosis, has been described for metabolic diseases (Song et al., 2020).

26.6.2 Applications in Cancer Therapy: Immune Modulation and Chemotherapy Sensitization

EVs effectively target cells in RNA-based cancer gene therapy by altering surface molecules and modulating immune responses. They prime cytolytic T-cell responses using dendritic cell-derived EVs pulsed with tumor peptides, enabling T-cell-dependent tumor attacks. Tumor-derived EVs expose dendritic cells to neoantigens, thereby priming CD8+ T-cell responses (Ruan et al., 2022). Additionally, EVs present tumor-associated antigens, immune adjuvants, immune checkpoint inhibitors, and immunomodulators, enhancing immune responses against cancer cells (Poggio et al., 2019). Combining EVs with other therapies, such as oncolytic viruses, CAR-T cells, and CRISPR/Cas9, is being explored to enhance their effectiveness, thereby opening new avenues for research (Chen et al., 2024). Research has demonstrated that EVs can be engineered to deliver anti-cancer drugs directly to tumor cells, thereby enhancing treatment efficacy while minimizing side effects. For example, studies have explored the use of EVs to deliver chemotherapeutic agents, such as doxorubicin, to cancer cells, thereby improving drug delivery and reducing off-target toxicity (Chen et al., 2022).

26.6.3 Applications for Neurodegenerative Diseases

EVs are emerging as promising therapeutic tools for neurodegenerative diseases like Alzheimer's, Parkinson's, Huntington's, and brain injuries (Wang et al., 2024c). EVs can potentially deliver therapeutic agents across the blood-brain barrier, offering new treatment avenues. These vesicles can deliver bioactive molecules, such as proteins, lipids, and nucleic acids, directly to affected neurons, promoting neuroprotection and regeneration. They can be engineered to deliver therapeutic molecules, such as RNA interference or small molecules, directly to affected neurons, potentially halting or reversing disease progression (Wang et al., 2024c). EVs also modulate neuroinflammation and facilitate the removal of toxic protein aggregates, thereby slowing disease progression and improving neuronal function (Pinho et al., 2020). Additionally, EVs have shown potential in treating TBI by enhancing neurovascular regeneration, reducing inflammation, and supporting neuronal survival. These properties help restore neural networks and improve cognitive functions, offering a potential therapeutic strategy for these challenging conditions.

26.7 MECHANISTIC INSIGHTS INTO SECRETOME AND EV THERAPIES

By leveraging several key molecular pathways, EV- and secretome-based therapies hold promise for regenerative medicine, including bone and cartilage repair, immunomodulation, cancer immunosurveillance, and neuroprotection in ocular and neurodegenerative conditions (Table 26.1).

26.7.1 Molecular Pathways Influenced by the Secretome

The secretome encompasses all soluble factors released by cells, including cytokines, growth factors, and other proteins that regulate key biological processes. By acting through autocrine, paracrine, and endocrine signaling, these molecules orchestrate tissue homeostasis, inflammation, and regenerative responses (Wu and Krijgsveld, 2024). For instance, in cancer, the secretome can be biphasic, with secreted molecules like TGFβ initially inhibiting tumor growth yet later driving invasiveness (Massague, 2008). In regenerative medicine, secreted factors from mesenchymal stem cells (MSCs) have been shown to stimulate angiogenesis, osteogenesis, and tissue repair. In addition, several cytokines and chemokines, like VEGF and IGF-1, in varying concentrations have been reported in MSC-CM, and these also induce bone regeneration by endogenous stem cell stimulation (Osugi et al., 2012).

Studies on the cancer cell secretome have also identified specific factors that could be potential biomarkers of cancer. In the pancreatic cancer cell secretome, some proteins like extracellular regulated kinase (ERK), protein kinase B (AKT), epidermal growth factor (EGF), and erb-B2 Receptor Tyrosine Kinase 2 (ERBB2) were considered as biomarkers because of their role in tumor proliferation, migration, and inhibition of apoptosis (Li et al., 2022a). Biotinidase, clusterin, enolase 1, and PTMA were identified in thyroid cancer cell lines, including TPC-1 and CAL 62, and have been proposed as potential biomarkers for thyroid cancer detection (Kashat et al., 2010). Thus, the secretome's versatility enables it to modulate complex molecular pathways, such as the MAPK/ERK, PI3K/AKT, and Wnt/β-catenin pathways, under both physiological and pathological conditions.

26.7.2 Molecular Pathways Influenced by EVs

EVs influence a variety of cell signaling pathways, enabling the precise and targeted delivery of critical factors that regulate immune function, stem cell differentiation, and disease progression. For example, EVs from MSCs are enriched with osteogenic and angiogenic molecules, aiding bone repair and vascularization by triggering the expression of various angiogenesis-inducing factors, such as VEGF, ANG1, ANG2, and COL1, as well as the gene expression of ALP, OCN, and OPN associated with osteogenesis (Takeuchi et al., 2019). In another example, the enriched EV fraction from MSCs, containing TSG-6, regulates microglial dynamics through a feedback loop with CD44, thus mitigating excessive inflammation or immune dysfunction (Hossain et al., 2025). In cancer models, EVs can transport oncogenic β-catenin, stimulating Wnt signaling in recipient cancer cells and promoting cancer progression (Kalra et al., 2019). In neuroregeneration, EVs transport

TABLE 26.1
Multifaceted Mechanisms Offered by Secretome and EV Therapies

Mechanism	Secretome	EVs
Paracrine and autocrine signaling	• Soluble proteins TGFβ, IFNγ, and VEGF regulate proliferation (Gao et al., 2016), differentiation (Ge et al., 2018), and immune modulation (English et al., 2007). • Autocrine feedback loops control cell survival or invasiveness (Wu & Krijgsveld, 2024). • Maintains tissue homeostasis via stress response and repair (Wu and Krijgsveld, 2024).	• EVs deliver proteins, lipids, and nucleic acids (mRNA, miRNA, lncRNA) for targeted intercellular communication (Kumar et al., 2024). • Cargo varies with physiological/pathological state (Yates et al., 2022). • Modulates recipient cell signaling via direct transfer of functional molecules (β-catenin, YAP1, transcription factors) (Forero et al., 2024).
Immune modulation	• Inflammatory cytokines (IFNγ, IL-10) can trigger or suppress immune responses (Wu et al., 2024). • Immunosuppressive factors in MSC secretome aid tissue repair and inflammation control (Cahill et al., 2016).	• EVs carry cytokines and antigens that regulate T-cell and B-cell function (Zhou et al., 2020). • Tumor-derived EVs facilitate immune evasion by transferring immunosuppressive molecules (Clarkson et al., 2022).
Tissue regeneration and repair	• MSC-derived secretome contains VEGF, IGF-1, and cytokines to promote angiogenesis, osteogenesis, and tissue healing (Osugi et al., 2012). • Chemotactic recruitment of stem cells or immune cells for regeneration (González-González et al., 2020).	• EVs transfer pluripotency factors (Oct4, Sox2, Nanog) to stimulate cell differentiation (Katsman et al., 2012). • Delivers survival signals (e.g., AKT pathway activators) to enhance cell viability under stress (Arslan et al., 2013).
Oncogenic and disease pathways	• Dual role in cancer: TGFβ can suppress or promote metastasis (Massagué, 2008). • Oncoproteins (ERK, AKT, EGF) in the secretome drive tumor progression (Li et al., 2022a). • Disease-specific secreted proteins serve as diagnostic biomarkers (Kashat et al., 2010).	• EVs transfer oncogenes (e.g., β-catenin, mutated receptors) to recipient cells, driving tumor progression (Kalra et al., 2019). • Theranostic potential: EV cargo (proteins, miRNAs) used as biomarkers and drug carriers (Carpi et al., 2020).
Key signaling pathways	• TGFβ pathway regulates apoptosis (Herr et al., 2009), fibrosis (An et al., 2017), and cancer progression (Giusti et al., 2018). • Wnt/β-catenin, Hippo/YAP (Ji et al., 2020), and MAPK/ERK pathways (Li et al., 2022a) influence cell proliferation and regeneration. • PI3K/AKT and FAK/JNK pathways modulate survival, migration, and adhesion (Desrochers et al., 2016).	• EVs deliver β-catenin, YAP (Ji et al., 2020), and PI3K/AKT (Liem et al., 2017) Activators to regulate recipient cell fate. • EV-mediated signaling affects tissue regeneration, immune modulation, and tumor progression (Cerrotti et al., 2024).
Clinical applications	• Regenerative Medicine: Conditioned media used for tissue repair (bone (Osugi et al., 2012), cartilage, cardiac). • Immunomodulation: Modifies immune response in autoimmune diseases. • Diagnostics: Disease-specific proteins used as biomarkers.	• EV Therapies: Engineered EVs for drug/gene delivery in regenerative medicine (McBride et al., 2018). • Cancer Immunotherapy: EV-based vaccines activate anti-tumor responses (Luo et al., 2025). • Biomarkers: EV-associated miRNAs and proteins for non-invasive diagnostics (Xu et al., 2022).

Abbreviations: EV: extracellular vesicle; TGFβ: transforming growth factor beta; IFNγ: interferon gamma; VEGF: vascular endothelial growth factor; IL-10: interleukin-10; IGF-1: insulin-like growth factor 1; ERK: extracellular signal-regulated kinase; AKT: protein kinase B (also known as AKT); EGF: epidermal growth factor; mRNA: messenger ribonucleic acid; miRNA: microRNA; lncRNA: long non-coding RNA; YAP1: yes-associated protein 1; Oct4: octamer-binding transcription factor 4; Sox2: SRY-Box transcription factor 2; Nanog: homeobox protein NANOG; MAPK: mitogen-activated protein kinase; PI3K: phosphoinositide 3-kinase; FAK: focal adhesion kinase; JNK: c-Jun N-terminal kinase.

critical molecules, including transcription factors, that influence cell fate and differentiation (Forero et al., 2024). For example, YAP1, a crucial transcription factor in human brain development, is transported and trafficked between cells to exert its physiological function (Forero et al., 2024). ESC-MVs can trigger Müller cell dedifferentiation by activating the retinal regeneration program, delivering pluripotency factors such as Oct4, Nanog, Sox2, and miR-290 (Katsman et al., 2012). Moreover, exosomes and autophagy share standard molecular mechanisms, influencing pathways such as the regulation of inflammatory cytokines and the control of autophagic flux. They modulate conditions such as diabetic retinopathy and keratoconus by affecting cell survival, proliferation, and immune responses through their secreted cargo (Ma et al., 2022). Thus, the EV's versatility allows it to modulate complex molecular pathways under both physiological and pathological conditions, making it a valuable target for therapeutic intervention and biomarker discovery.

26.8 ISOLATION AND CHARACTERIZATION TECHNIQUES

26.8.1 Methods for Secretome Collection

Secretome collection is crucial for studying cell-secreted factors, which play vital roles in cell communication and the development of therapeutic agents. Conditioned media (CM) is a widely used method for collecting and storing secreted factors from cells cultured in a growing medium, then washed and incubated for a period of 24 to 48 hours (Pinho et al., 2020). Today, there are various kinds of media: serum-free, animal component-free, chemically defined, and media for vaccine processes are some notable examples (Whitford et al., 2018). Another commonly used method for secretome collection is the application of bioreactors. Bioreactors are systems or vessels that provide a carefully controlled environment for collecting products from bioprocesses. They offer advantages such as scalability and higher yield compared to traditional static cultures. Perfusion-based bioreactors enable continuous collection, reducing cell stress and enhancing secretome integrity (Schirmaier et al., 2014). These collection methods are fundamental to the isolation and characterization of EVs.

26.8.2 EV Isolation Techniques: Ultracentrifugation, Precipitation, and Chromatography

EVs' isolation is a critical step in studying their composition and function. While complete isolation is not possible, various techniques are used to achieve high purity and yield. One such technique is ultracentrifugation, which separates EVs from cell culture supernatants or biological fluids through high-speed centrifugation. Different types of centrifugation can be employed, depending on the desired yield and specificity (Théry et al., 2018). However, this method requires expensive equipment and is time-consuming. Precipitation-based isolation methods, such as polyethylene glycol (PEG)-mediated precipitation, offer a cost-effective and scalable alternative. Chromatography techniques, such as size-exclusion chromatography (SEC), provide high yields and purity while maintaining the integrity of the EVs and are also viable for a more extended period (Gamez-Valero et al., 2016). A combination of these techniques is often used to optimize purity and recovery, depending on the research application.

26.9 CHARACTERIZATION TOOLS: NANOPARTICLE TRACKING ANALYSIS, FLOW CYTOMETRY, AND OMICS

Characterizing EVs is crucial for verifying their size, concentration, and molecular composition. Nanoparticle tracking analysis (NTA) is widely used to measure EV size distribution and concentration in a suspension by tracking Brownian motion (Dragovic et al., 2011). Flow cytometry can characterize the size, amount, and phenotype of an EV and is used for surface-level markers, as it is less accurate at the range required to analyze smaller particles (Gul et al., 2022). Additionally, omics-based techniques, including proteomics, transcriptomics, metabolomics, and lipidomics, offer valuable insights into the molecular composition and functional properties of EVs by identifying their associated proteins, RNA sequences, lipids, and metabolic profiles (Shaba et al., 2022). Integrating multiple characterization techniques enhances the reliability of EV studies and supports their therapeutic and diagnostic applications.

26.10 CHALLENGES AND LIMITATIONS

26.10.1 Standardization of Secretome and EV Preparation

Reliability and reproducibility in therapeutic applications of exosomes hinge on the standardization of secretome and EV preparation. Recent studies highlight various methodologies and challenges in achieving consistent quality across different preparations. A better understanding of key aspects of secretome and EV standardization—including isolation techniques, characterization methods, and the importance of quality control—is essential.

Regarding isolation techniques, Differential Ultracentrifugation is considered the gold standard for EV isolation. Still, other methods, such as PEG precipitation and size-exclusion chromatography (SEC), are also used (Tiwari et al., 2021). While ultracentrifugation methods may yield pure EVs, they are not scalable. Therefore, SEC columns that are commercially available have become popular, as they can produce yields at industrial scales. The International Society for Extracellular Vesicles (ISEV) emphasizes the need for standardized methods to enhance the reliability of EV research (Nieuwland et al., 2020), thereby contributing to the rigor and reproducibility of EV-based research. As for characterization methods, they can be categorized as biochemical analysis, which involves techniques such as flow cytometry and mass spectrometry, utilized to analyze the protein and lipid composition of EVs (Tiwari et al., 2021), and through the Raman Spectroscopy method, which provides insights into the composition of secretome, confirming the presence of key therapeutic factors (Giannasi et al., 2021). Indeed, harmonization efforts that follow ISEV recommendations for exosome isolation and processing will help minimize contamination and variability in downstream characterization processes, such as proteomic analyses (Abramowicz et al., 2018), as well as in preclinical and clinical studies. While significant progress has been made to date in standardizing secretome and EV preparations, challenges remain, particularly regarding the influence of pre-analytical variables and the need for dedicated reference materials. Focusing on these issues is essential for advancing the clinical application of EVs and their therapeutic potential.

26.10.2 Scalability and Cost of Production

The scalability and production costs of EVs are critical factors influencing their therapeutic applications. Despite advancements, the field continues to struggle with the need for robust, cost-effective manufacturing solutions to meet clinical demands. The ongoing research aims to streamline these processes, potentially leading to more accessible EV-based therapies. Current manufacturing processes face challenges such as low yields, high heterogeneity, and significant labor and consumable costs (Paganini et al., 2019). Scalability in EV production is influenced by lot size and technology choice. The costs of labor and consumables during EV harvesting represent significant financial constraints, necessitating the development of automated, high-throughput technologies. The decision support tool identifies optimal technologies, emphasizing the need for higher efficiency automated methods to reduce costs (Ng et al., 2019). The scalable production of EVs can be standardized by bioreactors, leading to consistent yields and increased therapeutic efficacy. The use of micro-carrier-based bioreactor systems, such as the Vertical-Wheel™ bioreactor, has been shown to significantly increase EV yield and productivity compared to traditional static cultures (de Almeida Fuzeta et al., 2020). Upstream optimization can enhance EV secretion from cells through strategies like fed-batch suspension cultures, cytokine stimulation, and genetic modifications (Ng et al., 2022).

Downstream processing is essential for obtaining ultra-pure EVs and critical for clinical applications, such as employing fed-batch suspension HEK293 cell culture and chromatographic purification, addressing scalability challenges, and enabling potential cost-effective GMP production of ultra-pure EVs for clinical purposes (Yang et al., 2024).

26.10.3 Safety Concerns, Including Immune Reactions and Off-Target Effects

EVs have emerged as high-potential therapeutic agents, but they also pose several safety concerns, including immune responses, off-target effects, and compositional variability. EVs from specific sources have been shown to increase production of IL-1β, IL-6, and TNF, potentially leading to inflammation. Some EVs can activate microglia and other immune cells, potentially leading to neuroinflammation or other immune-mediated effects. Additionally, EVs can influence B-cell activation and function, potentially altering antibody production and immune responses (Wang et al., 2024d). EVs can also have unintended effects on non-target cells and tissues. Since EVs can circulate throughout the body, they can affect multiple organ systems beyond their intended target (Kumar et al., 2024), making it challenging to control their biodistribution. EVs can exert unintended biological effects by altering the gene expression and signaling pathways of recipient cells (Kim et al., 2024). While EVs do not proliferate, their contents may induce the proliferation of native cells, raising concerns about their safety. EVs contain a diverse array of biomolecules, making it challenging to predict their exact effects and ensure consistent therapeutic outcomes. Although EVs generally have a better safety profile than cell therapies, they may still carry risks of toxicity that require thorough evaluation. For example, intraocularly delivered EVs may aggregate, leading to unintended neovascularization (Anvarinia et al., 2024). Furthermore, the long-term consequences of EV-based therapies are not yet fully understood and require further investigation to verify the suitability of vesicles before they are tested in clinical trials (Shahi et al., 2024). Despite these concerns, recent meta-analyses of clinical trials have shown that EV-based therapies have a relatively low incidence of serious adverse events (0.7%) and adverse events (4.4%), suggesting a generally favorable safety profile (Van Delen et al., 2024). However, the potential risks associated with therapeutics based on EVs must be thoroughly understood and mitigated through ongoing research and rigorous preclinical and clinical studies. These studies must transparently report the methods of EV isolation and characterization, along with any adverse outcomes associated with their application in preclinical and clinical trials.

26.11 REGULATORY AND ETHICAL CONSIDERATIONS

26.11.1 Current Regulatory Frameworks for Secretome and EV-Based Therapies

The regulatory frameworks for secretome and EV-based therapies are evolving, highlighting the complexities of their biological mechanisms and applications. These frameworks differ significantly across countries, requiring a nuanced understanding of classification, manufacturing, and clinical investigation requirements. A role-based classification of biologics, such as active drug constituents or carriers, shapes regulatory requirements (Lener et al., 2015). This classification also dictates necessary quality control and manufacturing processes, underscoring the need for clear definitions in regulatory guidelines (Lener et al., 2015). Standardizing manufacturing techniques poses regulatory hurdles due to the diversity of EVs and secretomes, complicating oversight. Regulatory agencies mandate detailed assessments of EVs' molecular composition, particle size, and pharmacokinetics for safety and efficacy (Wang et al., 2024a). Comprehensive regulations are essential to prevent adverse outcomes, as unclear laws have harmed patients. Safe translation into clinical practice relies on collaboration among researchers, clinicians, and regulatory bodies (Lener et al., 2015). Although current frameworks aim to ensure safety and efficacy, rapid advancements in EV and secretome research may outpace regulatory adaptations, potentially creating oversight gaps. Exosome, secretome, or EV-based therapies remain unapproved globally. However, regulatory frameworks are adapting to these emerging therapies.

26.11.2 Quality Control and Clinical-Grade Production Standards

The therapeutic efficacy of secretome or EVs depends on quality control and adherence to clinical-grade production standards. Recent studies emphasize the need for standardized production processes and robust quality control measures to protect the safety and efficacy of EV therapeutics. Quality control and clinical-grade production standards for secretome-based therapeutics worldwide underscore the importance of adhering to stringent Good Manufacturing Practice (GMP) compliance to ensure consistency, safety, and efficacy. This involves meticulously characterizing bioactive components, including cytokines, lipids, and EVs, and rigorously testing for batch-to-batch reproducibility. Standardization or consistency across studies can be achieved by adhering to standard guidelines, such as the Minimal Information for Studies of Extracellular Vesicles (MISEV) (Patel et al., 2024). RNA-seq and proteomic analyses are used to ensure the reproducibility of EV products (Wang et al., 2024b). Stability assessments are crucial, ensuring that secretomes maintain their therapeutic properties under various storage conditions. Studies indicate that EVs can remain stable under specific storage conditions, which are vital for clinical applications (Wang et al., 2024b). Standardization of production protocols and quality control parameters is essential for meeting regulatory requirements and facilitating the clinical translation of findings. These standards guarantee that secretome-based therapies are reliable and effective for global patient use. Manufacturing processes must meet stringent FDA requirements for products in the USA, necessitating close collaboration between researchers and regulatory bodies (Cheng and Kalluri, 2023). While EV production and quality control advancements are promising, challenges remain, particularly in establishing universal guidelines and overcoming manufacturing hurdles. Continued dialogue between researchers and regulatory agencies will be crucial for effectively commercializing EV-based therapeutics.

26.11.3 Ethical Implications of Cell-Free Therapies

The ethical implications of cell-free therapies, such as those using EVs or secretomes, revolve around several key considerations (van Hooff et al., 2025). These therapies promise significant medical advancements but raise concerns about equitable access, informed consent, and long-term safety. One of the primary ethical considerations is the potential for unintended consequences. As with any new medical intervention, there are risks of adverse effects that may not be immediately apparent. The long-term health effects remain largely unknown, raising ethical concerns about exposing patients to potential hazards. Beyond the immediate medical concerns, cell-free therapies may have broader societal implications. The potentially higher cost of these therapies could create or exacerbate healthcare disparities, raising questions of social justice and equitable access to treatment (Assen et al., 2021). Ensuring that patients fully understand the potential risks and benefits is crucial, as is addressing the affordability and availability of these treatments. Additionally, how these cell-free therapies are perceived and understood by the public could influence their acceptance and integration into healthcare systems. The focus on developing cell-free therapies might divert funding and attention from other areas of medical research, potentially affecting overall healthcare advancements (Assen et al., 2021). The rapid pace of innovation in this field necessitates robust regulatory oversight to prevent misuse and ensure ethical standards are maintained throughout the development and clinical application processes.

26.12 CLINICAL ADVANCES

As of this chapter writing, there are over 200 clinical studies (clinicaltrials.gov) on EVs, over 400 that include the term "exosomes," and fewer than a hundred focusing on the secretome, 33 of which incorporate the term "mesenchymal stem cells" (Table 26.2).

TABLE 26.2
Clinical trials utilizing Secretome and EV therapies.

Disease	Subjects (n)	Trial Phase	Application Site or Evaluation Procedure	EV/Secretome Used; Source of EVs	Results or Purpose (if study ongoing or no results given)	Reference
Diabetic foot ulcers (from diabetes mellitus)	10	1	Perilesional injection into diabetic foot ulcers	Microvesicles; Auto-human platelet granules	Determine the effectiveness of EVs on healing diabetic foot ulcers	NCT06825884
Deep Second-degree burns of the skin	1	1	Administered directly onto the burn	N/A; Allo-BM-MSCs	Determine the safety of administering EVs to second-degree burns.	NCT05078385
Androgenetic Alopecia	30	3	Injection into the scalp	Foreskin-derived mesenchymal stromal cells-derived EXO	Determine the effects of EXO on the hair density of patients with androgenetic alopecia.	NCT06539273
Locally advanced breast cancer	200	N/A	Measuring EXO and ctDNA between neoadjuvant chemotherapy and surveillance after curative surgery	EXO; Auto	Determine if EXO can be used as a prognostic biomarker for early-stage breast cancer.	NCT05955521
Oral leukoplakia	225	N/A	Saliva and plasma samples from each subject were analyzed for exosomal miR-185, a microRNA that suppresses tumor proliferation	EXO; Auto	To determine the cancer risk of oral leukoplakia in its early stages through levels of salivary and plasma exosomal miR-185.	NCT06469892
COVID-19	102	2	Intravenous administration	BM-MSCs-derived EXO	Over 60 days, the group with the highest dosage of EXO had a higher survival rate (69.6%) compared to the lower dosage/saline (53.4% and 51.6%, respectively).	NCT04493242

(Continued)

TABLE 26.2 CONTINUED
Clinical trials utilizing Secretome and EV therapies.

Disease	Subjects (n)	Trial Phase	Application Site or Evaluation Procedure	EV/Secretome Used; Source of EVs	Results or Purpose (if study ongoing or no results given)	Reference
COVID-19	24	1	Inhalation of adipose MSCs-derived EXO	5 times aerosol inhalation of MSCs-derived EXO (2.0*10E8 nano vesicles/3 ml at Days 1, 2, 3, 4, and 5).	Up to a total amount of 2.0×10^9 nano vesicles was feasible and well tolerated.	NCT04276987
Topical and regenerative application	10	1	Topical	MSC-derived EXO ointment combined with liposome cream for skin applications.	Safety and tolerability studies over 21 days.	NCT05523011
Multiple sclerosis	48	2	Either intrathecally or intravenously	Secretome of Auto-BM-MSCs; N/A.	Treatment was well-tolerated in progressive multiple sclerosis and conferred short-term benefits.	NCT02166021
COVID-19	36	1 and 2	Intravenous infusion	Allo-MB stromal cells; N/A	Reversal of hypoxia, immune reconstitution, and downregulation of cytokine storm. No treatment-related adverse effects were observed.	NCT05019287
Liver cirrhosis	54	1	Injection of secretome (unspecified route)	hUC-MSC secretome; N/A	Determine if secretome injection improves liver function, if it is a viable treatment for liver cirrhosis, and if it improves the subject's quality of life.	NCT06629909

Abbreviations: Auto: autologous; Allo: allogenic; BM-MSCs: bone marrow-derived MSCs; EXO: Exosomes; hUC-MSCs: human umbilical cord-derived MSCs; MB: menstrual blood; MSCs: mesenchymal stem cells;

This suggests that fewer studies have investigated the potential of cell secretomes, while highlighting the immense promise of EVs in clinical contexts. The search results indicate that multiple clinical trials have investigated secretome-derived products from MSCs and other cell types for various conditions, particularly for immunomodulation in COVID-19 and for potential applications in cardiac conditions. These trials have demonstrated the safety and preliminary efficacy of the secretome across multiple applications, suggesting its potential as an "off-the-shelf" therapeutic solution, free of many of the complexities associated with direct cell transplantation.

EVs, exosomes derived from MSCs, and other cell types are actively investigated in clinical trials for diverse therapeutic applications. These trials leverage the immunomodulatory, regenerative, and target-delivery properties of EVs while addressing challenges of standardization and scalability. Some of the key trials focus on COVID-19 and respiratory diseases, topical and regenerative applications, oncology, immunomodulation, and more. Completed and ongoing trials use MSC-derived exosomes for conditions such as graft-versus-host disease (GvHD), leveraging their ability to modulate immune responses without the risks of cell-based therapies. The recent approval of Ryoncil (remestemcel-L) by the US Food and Drug Administration (USFDA) for the treatment of steroid-refractory acute graft-versus-host disease (aGvHD) in children marks a substantial advancement for MSC-based treatments. It offers essential guidance for advancing MSC-EV therapies (Bernd, 2025). Current trials highlight the potential of EV-based therapies as safer, scalable alternatives to cell transplantation, with ongoing research addressing remaining challenges, such as enhancing targeting efficiency and cargo loading (e.g., miRNAs, growth factors) for precision medicine, as well as developing consensus on characterization, potency testing, and biodistribution to meet FDA/EMA requirements for clinical adoption (Lotfy et al., 2023). It is also worth noting that most clinical trials testing the secretome of EV-based therapies are conducted in rare or end-stage diseases, highlighting the risk aversion associated with cell therapies.

26.13 FUTURE DIRECTIONS IN SECRETOME AND EV RESEARCH

The future of research on secretomes and EVs is auspicious, especially with the incorporation of bioengineering techniques that can enhance the accuracy and effectiveness of these biological tools. By modifying genetic and chemical factors to influence secretome composition and EV yield, researchers can improve the therapeutic capabilities of these vesicles (Haider, 2024). Additionally, integrating secretome- and EV-based therapies with current treatments could transform regenerative medicine and disease management, creating synergistic effects that improve patient outcomes. These developments underscore the groundbreaking potential of secretome and EV research in the years ahead.

26.13.1 Integration with Bioengineering

The integration of MSC secretome and EVs with bioengineering approaches holds great promise for enhancing therapeutic efficacy in tissue repair and regeneration. A promising approach involves culturing cells within synthetic biomimetic matrices, such as polyisocyanide (PIC)-based hydrogels, which serve as a powerful tool for directing the composition of the secretome. For example, cells in non-adherent PIC gels secrete increased levels of anti-inflammatory factors, such as IL-10, and the conditioned media from these 3D cultures can accelerate wound closure (Liu et al., 2020a). Moreover, these synthetic matrices can be designed with specific properties, such as the incorporation of RGD motifs that enable cell adhesion via integrin-mediated binding. Minimal, straightforward PIC-based systems have the potential to modulate cell adhesion, motility, proliferation, vascularization, and cytocompatibility, thereby enabling the production of distinct microenvironments that influence secretome composition (Wechsler et al., 2021). Much like secretome engineering, biopolymeric scaffolds, such as hydrogels, are being explored as vehicles for EV delivery, offering several advantages (Leung et al., 2022). These scaffolds mimic the natural extracellular matrix, providing a supportive environment for EVs and promoting sustained release over time (Li et al., 2024). Natural biopolymers, such as alginate, silk fibroin, collagen, gelatin, and chitosan, are particularly attractive for this purpose due to their biocompatibility, biodegradability, and ability to promote cell adhesion and migration (Leung et al., 2022). When combined with EVs, these scaffolds have demonstrated positive therapeutic effects in various tissues and disease models (Leung et al., 2022). Recent research has focused on developing multifunctional hydrogel-based engineered EVs for applications such as wound healing (Li et al., 2024). These functionalized hydrogels not only protect EVs but also ensure their sustained release and maintain their bioactivity during the healing process (Li et al., 2024). The release profiles of EVs from biodegradable hydrogels can be fine-tuned, ranging from 6 to 27 days, allowing for prolonged therapeutic effects (Li et al., 2024). Future research in this area will likely focus on optimizing the integration of EVs with hydrogels, exploring different crosslinking methods, and investigating the timing of EV incorporation to maximize their therapeutic potential (Li et al., 2024). Additionally, the development of intelligent, responsive hydrogels that can release EVs in response to specific physiological cues may further enhance the precision and efficacy of these treatments.

26.13.2 Enhancing Secretome Composition and EV Yield via Targeting through Genetic and Chemical Modifications

Secretome therapies leverage diverse engineering strategies to optimize therapeutic potential. Genetic approaches involve modifying stem cells to overexpress key factors (e.g., VEGF, IGF-1, HGF) or using microparticles to transiently regulate pathways, such as NF-κB, thereby reducing proinflammatory secretions without requiring permanent genetic alterations. Biochemical methods include priming cells with cytokines (e.g., IL-1α) or hypoxic conditioning to enhance both angiogenic and anti-inflammatory profiles, as well as hydrogel-based systems that incorporate peptides or small molecules to direct secretory outputs. Mechanical strategies exploit mechanotransduction through 3D culture systems (e.g., hydrogels or spheroids) and applied forces to modulate secretion, such as increased nitric oxide production in bone cells under mechanical loading or the release of cartilage-specific factors during chondrogenic induction. These approaches synergistically enhance secretome efficacy for applications like angiogenesis, tissue repair, and immunomodulation (Daneshmandi et al., 2020; Gaddam et al., 2019).

Enhancing the composition, yield, and targeting capabilities of EVs involves strategies similar to those used in secretome therapeutics. EVs can be loaded with therapeutics either during cell replication or by adding them to isolated EVs through diffusion or membrane disruption. Despite their advantages, EVs encounter issues such as rapid systemic clearance and poor tissue targeting, often leading to accumulation in the liver and spleen, where the mononuclear phagocyte system degrades them (Zhao et al., 2024). To address these challenges, engineered exosomes with modified surfaces are being developed using various methods to enhance targeting, prolong circulation time, and improve therapeutic potential. Genetic engineering of EV-secreting cells is one method under exploration, though it carries the risk of horizontal gene transfer (N'Diaye et al., 2022). Researchers are also focusing on modifying EVs post-secretion, utilizing techniques such as click chemistry to create tailored vesicles with enhanced targeting. Chemical modifications, including the use of RGD peptides, Lamp2b, tetraspanins, and nucleic acid aptamers, have been explored to improve targeting and efficiency (Zhao et al., 2024). These innovations aim to enable selective delivery of therapeutics, such as small molecules and proteins, to target tissues. Future research will likely focus on more sophisticated modification techniques to improve EV yield and stability while preserving their natural properties, including synthetic biology and high-throughput screening to identify optimal targeting moieties.

26.13.3 Combining Secretome or EV-Based Therapies with Existing Treatments

Combining MSC secretome or EV-based therapies with existing treatments represents a promising approach to enhance therapeutic outcomes. As cell-free alternatives, secretome and EV-based approaches offer advantages over whole-cell therapies, including improved safety profiles and easier storage and handling. One potential direction is the integration of secretome or EV-based therapies with current pharmacological treatments. Moreover, the immunomodulatory and regenerative properties of MSC-derived EVs could be leveraged to enhance the effectiveness of existing therapies in various diseases, particularly in neurodegenerative disorders and cancer (Zhao et al., 2024). Investigating the potential of secretome and EV-based therapies to overcome drug resistance or enhance the efficacy of immunotherapies could open new avenues for treating challenging diseases. Future research will likely focus on optimizing these combination therapies, identifying synergistic effects between secretome/EV-based treatments and existing therapies, and developing standardized protocols for their clinical application.

26.14 CONCLUSIONS AND PERSPECTIVES

The secretome and EVs have emerged as powerful tools in regenerative medicine, offering multifaceted mechanisms to influence cell behavior. Primarily, they deliver bioactive molecules that modulate key signaling pathways, promote tissue repair, regulate immune responses, and potentially target disease-specific pathways. Their ability to enhance cellular communication and healing processes has been demonstrated in various applications, including wound healing, cardiac repair, and neuroregeneration. The therapeutic promise of the secretome and EVs lies in their versatility, biocompatibility, and potential for personalized medicine, making them invaluable assets in translational medicine. By refining our understanding of their cargo composition and delivery mechanisms, we can harness secretome and EVs for precise and effective treatments in regenerative medicine, oncology, immunotherapy, and beyond (Figure 26.1).

Looking ahead, integrating secretome and EV-based therapies with cutting-edge technologies will pave the way for next-generation regenerative medicine. Innovations in bioengineering, such as genetic and chemical modifications, will enable precise control over secretome composition and EV yield, thereby optimizing their therapeutic potential. The molecular pathways involved in EV biogenesis, cargo sorting, and release differ significantly from those involved in the secretion and processing of non-vesicular components from stem cells. Understanding these differences is crucial for developing targeted therapies and regenerative medicine approaches based on components of the

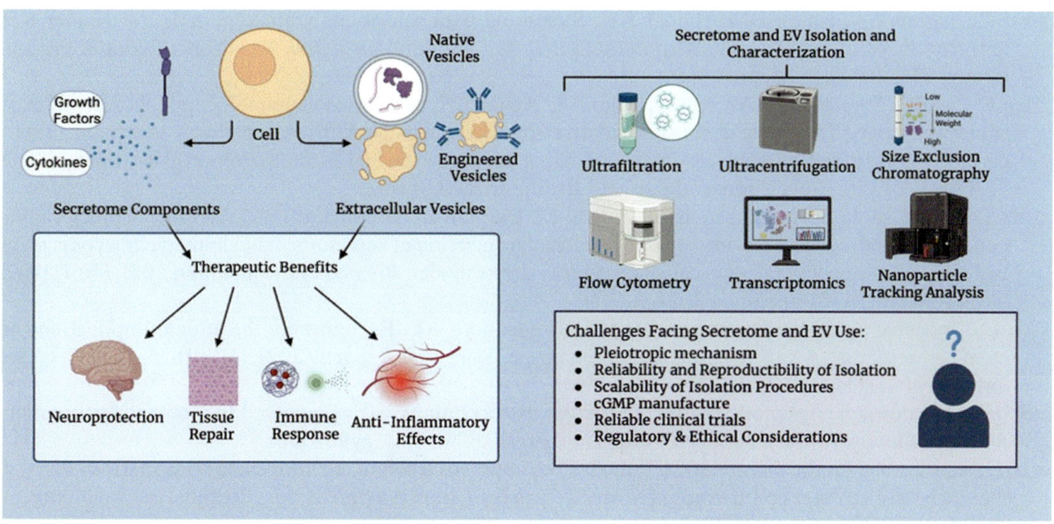

FIGURE 26.1 Overview of cell-derived secretome and extracellular vesicles (EVs) for therapeutic applications and associated challenges.

stem cell secretome. The future of regenerative medicine is likely to see a convergence of multidisciplinary approaches, leveraging the strengths of secretome and EV research to address complex medical challenges and improve patients' quality of life worldwide.

Declaration: During the preparation of this work, the author(s) used regenerative AI tools to organize and improve the readability and language of the manuscript. After using this tool/service, the author(s) reviewed and edited the content as needed and take(s) full responsibility for the content of the published article.

Conflicts of interest: None of the authors declares any conflicts of interest.

Acknowledgments: The authors wish to acknowledge editorial assistance from Vaishnavi Sowmya Narayn.

Funding Sources: Authors gratefully acknowledge funding support from CDMRP—W81XWH-22-1-0697 (RG); R01EY034716, National Eye Institute (RG), National Institute of Neurological Disorders and Stroke, R56NS127924 (RG), and Research to Prevent Blindness (Hamilton Eye Institute).

REFERENCES

Abramowicz A, Marczak L, Wojakowska A, Zapotoczny S, Whiteside TL, Widlak P, Pietrowska M. Harmonization of exosome isolation from culture supernatants for optimized proteomics analysis. *PLoS One*. 2018;13(10):e0205496. https://doi.org/10.1371/journal.pone.0205496.

Ahmed ZT, Alkahlot MH, Haider KH. MSC-derived exosomes: advances in cell-free therapy. In: Haider KH (ed) *Handbook of Stem Cell Applications*. Singapore: Springer; 2023, pp. 1–41.

Al-Omar MT, Alnajjar MT, Ahmed ZT, Salaas FMI, Alrefaei TSM, Haider KH. Endothelial progenitor cell-derived small extracellular vesicles for myocardial angiogenesis and revascularization. *J Clin Transl Res*. 2022;8(6):1–12.

Alwjwaj M, Kadir RRA, Bayraktutan U. The secretome of endothelial progenitor cells: A potential therapeutic strategy for ischemic stroke. *Neural Regen Res*. 2021;16(8):1483–9. https://doi.org/10.4103/1673-5374.303012.

An SY, Jang YJ, Lim HJ, Han J, Lee J, Lee G, et al. Milk fat globule-EGF factor 8, secreted by mesenchymal stem cells, protects against liver fibrosis in mice. *Gastroenterology*. 2017;152(5):1174–86. https://doi.org/1053/j.gastro.2016.12.003.

Anvarinia Y, Del Mar NA, Awad AM, Hossain S, Seetharaman AT, Ravindran S, Roth S, et al. MicroRNA-based engineered mesenchymal stem cell extracellular vesicles to treat visual deficits after blast-induced trauma. *Exp Eye Res*. 2024;247:110031. https://doi.org/10.1016/j.exer.2024.110031.

Anwar EI, Fatima SA, Ismail DM, Haider KH. Stemming pain sensations with stem cells. In: Haider KH (ed) *Handbook of Regenerative Medicine: Cell-Based Therapy Approach*. New York: Taylor & Francis Group, CRC Press, 2025, pp. 83–99.

Arslan F, Lai RC, Smeets MB, Akeroyd L, Choo A, Aguor EN, et al. Mesenchymal stem cell-derived exosomes increase ATP levels, decrease oxidative stress and activate PI3K/Akt pathway to enhance myocardial viability and prevent adverse remodeling after myocardial ischemia/reperfusion injury. *Stem Cell Res*. 2013;10:301–12. https://doi.org/10.1016/j.scr.2013.01.002.

Asgari Taei A, Nasoohi S, Hassanzadeh G, Kadivar M, Dargahi L, Farahmandfar M. Enhancement of angiogenesis and neurogenesis by intracerebroventricular injection of secretome from human embryonic stem cell-derived mesenchymal stem cells in ischemic stroke model. *Biomed Pharmacother*. 2021;140:111709. https://doi.org/10.1016/j.biopha.2021.111709.

Assen LS, Jongsma KR, Isasi R, Tryfonidou MA, Bredenoord AL. Recognizing the ethical implications of stem cell research: A call for broadening the scope. *Stem Cell Reports*. 2021;16(7):1656–61. https://doi.org/10.1016/j.stemcr.2021.05.021.

Battistelli M, Falcieri E. Apoptotic bodies: Particular extracellular vesicles involved in intercellular communication. *Biology (Basel)*. 2020;9(1). https://doi.org/10.3390/biology9010021.

Bernd G. A milestone for the therapeutic EV field: FDA approves Ryoncil, an allogeneic bone marrow-derived mesenchymal stromal cell therapy. *Extracell Vesicles Circul Nucleic Acids*. 2025;6(1):183–90. https://doi.org/10.20517/evcna.2025.02.

Cahill EF, Kennelly H, Carty F, Mahon BP, English K. Hepatocyte growth factor is required for mesenchymal stromal cell protection against bleomycin-induced pulmonary fibrosis. *Stem Cells Transl Med*. 2016;5(10):1307–18. https://doi.org/10.5966/sctm.2015-0337.

Carpi S, Polini B, Nieri D, Dubbini N, Celi A, Nieri P, et al. Expression analysis of muscle-specific miRNAs in plasma-derived extracellular vesicles from patients with chronic obstructive pulmonary disease. *Diagnostics*. 2020;10:502. https://doi.org/10.3390/diagnostics10070502.

Casati S, Giannasi C, Niada S, Bergamaschi RF, Orioli M, Brini AT. Bioactive lipids in MSCs biology: State of the art and role in inflammation. *Int J Mol Sci*. 2021;22(3):1481. https://doi.org/10.3390/ijms22031481.

Cerrotti G, Buratta S, Latella R, Calzoni E, Cusumano G, Bertoldi A, et al. Hitting the target: cell signaling pathways modulation by extracellular vesicles. *Extracell Vesicles Circ Nucleic Acids*. 2024;5:527. https://doi.org/10.20517/evcna.2024.XX.

Chen J, Tan Q, Yang Z, Jin Y. Engineered extracellular vesicles: Potentials in cancer combination therapy. *J Nanobiotechnology*. 2022;20(1):132. https://doi.org/10.1186/s12951-022-01330-y.

Chen Y, Tang S, Cai F, Wan Y. Strategies for small extracellular vesicle-based cancer immunotherapy. *Research (Wash D C)*. 2024;7:0421. https://doi.org/10.34133/research.0421.

Cheng K, Kalluri R. Guidelines for clinical translation and commercialization of extracellular vesicles and exosomes based therapeutics. *Extracellular Vesicle*. 2023;2:100029.

Chiurchiu V, Leuti A, Maccarrone M. Bioactive lipids and chronic inflammation: Managing the fire within. *Front Immunol*. 2018;9:38. https://doi.org/10.3389/fimmu.2018.00038.

Clarkson BD, Grund E, David K, Johnson RK, Howe CL. ISGylation is induced in neurons by demyelination driving ISG15-dependent microglial activation. *J Neuroinflammation*. 2022;19:258. https://doi.org/10.1186/s12974-022-02593-6.

Daneshmandi L, Shah S, Jafari T, Bhattacharjee M, Momah D, Saveh-Shemshaki N, Lo KW, et al. Emergence of the stem cell secretome in regenerative engineering. *Trends Biotechnol*. 2020;38(12):1373–84. https://doi.org/10.1016/j.tibtech.2020.04.013.

de Almeida Fuzeta M, Bernardes N, Oliveira FD, Costa AC, Fernandes-Platzgummer A, Farinha JP, Rodrigues CAV, et al. Scalable production of human mesenchymal stromal cell-derived extracellular vesicles under serum-/xeno-free conditions in a microcarrier-based bioreactor culture system. *Front Cell Dev Biol*. 2020;8:553444. https://doi.org/10.3389/fcell.2020.553444.

Desrochers LM, Bordeleau F, Reinhart-King CA, Cerione RA, Antonyak MA. Microvesicles provide a mechanism for intercellular communication by embryonic stem cells during embryo implantation. *Nat Commun*. 2016;7:11958. https://doi.org/10.1038/ncomms11958.

Doyle LM, Wang MZ. Overview of extracellular vesicles, their origin, composition, purpose, and methods for exosome isolation and analysis. *Cells*. 2019;8(7). https://doi.org/10.3390/cells8070727.

Dragovic RA, Gardiner C, Brooks AS, Tannetta DS, Ferguson DJ, Hole P, Carr B, et al. Sizing and phenotyping of cellular vesicles using nanoparticle tracking analysis. *Nanomed Nanotechnol Biol Med*. 2011;7(6):780–8. https://doi.org/10.1016/j.nano.2011.04.003.

Drobiova H, Sindhu S, Ahmad R, Haddad D, Al-Mulla F, Al Madhoun A. Wharton's jelly mesenchymal stem cells: A concise review of their secretome and prospective clinical applications. *Front Cell Dev Biol*. 2023;11:1211217. https://doi.org/10.3389/fcell.2023.1211217.

Elmore S. Apoptosis: A review of programmed cell death. *Toxicol Pathol*. 2007;35(4):495–516. https://doi.org/10.1080/01926230701320337.

Elshaer SL, Bahram SH, Rajashekar P, Gangaraju R, El-Remessy AB. Modulation of mesenchymal stem cells for enhanced therapeutic utility in ischemic vascular diseases. *Int J Mol Sci*. 2021;23(1). https://doi.org/10.3390/ijms23010249.

Elshaer SL, Evans W, Pentecost M, Lenin R, Periasamy R, Jha KA, Alli S, et al. Adipose stem cells and their paracrine factors are therapeutic for early retinal complications of diabetes in the Ins2(Akita) mouse. *Stem Cell Res Therapy*. 2018;9(1):322. https://doi.org/10.1186/s13287-018-1059-y.

English K, Barry FP, Field-Corbett CP, Mahon BP. IFN-γ and TNF-α differentially regulate immunomodulation by murine mesenchymal stem cells. *Immunol Lett*. 2007;110:91–100.

Fong CY, Tam K, Cheyyatraivendran S, Gan SU, Gauthaman K, Armugam A, Jeyaseelan K, et al. Human Wharton's jelly stem cells, and its conditioned medium, enhance the healing of excisional and diabetic wounds. *J Cell Biochem*. 2014;115(2):290–302. https://doi.org/10.1002/jcb.24661.

Forero A, Pipicelli F, Moser S, Baumann N, Gratz C, Gonzalez Pisfil M, Pfaffl MW, et al. Extracellular vesicle-mediated trafficking of molecular cues during human brain development. *Cell Rep*. 2024;43(10):114755. https://doi.org/10.1016/j.celrep.2024.114755.

Gaddam S, Periasamy R, Gangaraju R. Adult stem cell therapeutics in diabetic retinopathy. *Int J Mol Sci*. 2019;20(19). https://doi.org/10.3390/ijms20194876.

Gamez-Valero A, Monguio-Tortajada M, Carreras-Planella L, Franquesa M, Beyer K, Borras FE. Size-exclusion chromatography-based isolation minimally alters Extracellular Vesicle characteristics compared to precipitating agents. *Sci Rep*. 2016;6:33641. https://doi.org/10.1038/srep33641.

Gao F, Chiu SM, Motan DAL, Zhang Z, Chen L, Ji HL, et al. Mesenchymal stem cells and immunomodulation: current status and future prospects. *Cell Death Dis.* 2016;7:e2062. https://doi.org/10.1038/cddis.2015.327.

Ge Q, Zhang H, Hou J, Wan L, Cheng W, Wang X, et al. VEGF secreted by mesenchymal stem cells mediates the differentiation of endothelial progenitor cells into endothelial cells via paracrine mechanisms. *Mol Med Rep.* 2018;17:1667–75.

Ghasemi M, Roshandel E, Mohammadian M, Farhadihosseinabadi B, Akbarzadehlaleh P, Shamsasenjan K. Mesenchymal stromal cell-derived secretome-based therapy for neurodegenerative diseases: Overview of clinical trials. *Stem Cell Res Therapy.* 2023;14(1):122. https://doi.org/10.1186/s13287-023-03264-0.

Giacoppo S, Thangavelu SR, Diomede F, Bramanti P, Conti P, Trubiani O, Mazzon E. Anti-inflammatory effects of hypoxia-preconditioned human periodontal ligament cell secretome in an experimental model of multiple sclerosis: A key role of IL-37. *FASEB J.* 2017;31(12):5592–608. https://doi.org/10.1096/fj.201700524R.

Giannasi C, Niada S, Della Morte E, Casati S, Orioli M, Gualerzi A, Brini AT. Towards secretome standardization: Identifying key ingredients of MSC-derived therapeutic cocktail. *Stem Cells Int.* 2021;2021(1):3086122. https://doi.org/10.1155/2021/3086122.

Giusti I, Di Francesco M, D'Ascenzo S, Palmerini MG, Macchiarelli G, Carta G, Dolo V. Ovarian cancer-derived extracellular vesicles affect normal human fibroblast behavior. *Cancer Biol Ther.* 2018;19(8):722–34. https://doi.org/10.1080/15384047.2018.1451286.

González-González A, García-Sánchez D, Dotta M, Rodríguez-Rey JC, Pérez-Campo FM. Mesenchymal stem cells secretome: the cornerstone of cell-free regenerative medicine. *World J Stem Cells.* 2020;12:1529–52. https://doi.org/10.4252/wjsc.v12.i12.1529.

Gul B, Syed F, Khan S, Iqbal A, Ahmad I. Characterization of extracellular vesicles by flow cytometry: Challenges and promises. *Micron.* 2022;161:103341. https://doi.org/10.1016/j.micron.2022.103341.

Habib SM, Martini MF, Al Ahmed O, Qassab AR, Haider KH. Mesenchymal stem cell secretomes for cell-free therapy of the broken heart. In: Haider, KH (ed) *Handbook of Regenerative Medicine: Cell-Free Therapy.* New York: Taylor & Francis Group/CRC Press, 2025, Chapter 18, pp. 274–93.

Habib SY, Martini MF, Hamdan YNA, Shrebaty OMM, Haider KH. Mesenchymal stem cell secretome and exosomes as potential advanced therapy medicinal products for treating a failing heart. *World J Stem Cells.* 2026;18.

Haider KH. Priming mesenchymal stem cells to develop "super stem cells." *World J Stem Cells* 2024;16(6):623–40. https://doi.org/10.4252/wjsc.v16.i6.623.

Haider KH, Aramini B. Micrining the injured heart with stem cell-derived exosomes: An emerging strategy of cell-free therapy. *Stem Cell Res Ther.* 2020; 11(1):23. https://doi.org/10.1186/s13287-019-1548-7.

Haider KH, Najimi H. Exosome-based cell-free therapy in regenerative medicine for myocardial repair. In: Haider, K.H. (ed) *Handbook of Stem Cell Therapy.* Singapore: Springer, 2022. https://doi.org/10.1007/978-981-16-6016-0_9-1.

Herr B, Zhou J, Werno C, Menrad H, Namgaladze D, Weigert A, Dehne N, Brüne B. The supernatant of apoptotic cells causes transcriptional activation of hypoxia-inducible factor-1alpha in macrophages via sphingosine-1-phosphate and transforming growth factor-beta. *Blood.* 2009;114(10):2140–48. https://doi.org/10.1182/blood-2009-01-201889.

Hessvik NP, Llorente A. Current knowledge on exosome biogenesis and release. *Cell Mol Life Sci.* 2018;75(2):193–208. https://doi.org/10.1007/s00018-017-2595-9.

Hossain MS, Rasiah PK, Seetharaman ATM, Alvarado D, Luo M, Wohlschlegel JA, Pentecost M, et al. TNF-stimulated gene-6, part of extracellular vesicles in adipose tissue-derived mesenchymal stem cell concentrated conditioned medium, affects microglial activity. *J Neuroimmune Pharmacol.* 2025;20(1):60. https://doi.org/10.1007/s11481-025-10216-3.

Hughes CE, Nibbs RJB. A guide to chemokines and their receptors. *FEBS J.* 2018;285(16):2944–71. https://doi.org/10.1111/febs.14466.

Jha KA, Pentecost M, Lenin R, Gentry J, Klaic L, Del Mar N, Reiner A, et al. TSG-6 in conditioned media from adipose mesenchymal stem cells protects against visual deficits in mild traumatic brain injury model through neurovascular modulation. *Stem Cell Res Therapy.* 2019;10(1):318. https://doi.org/10.1186/s13287-019-1436-1.

Jha KA, Rasiah PK, Gentry J, Del Mar NA, Kumar R, Adebiyi A, Reiner A, et al. Mesenchymal stem cell secretome protects against oxidative stress-induced ocular blast visual pathologies. *Exp Eye Res.* 2022;215:108930. https://doi.org/10.1016/j.exer.2022.108930.

Ji C, Zhang J, Zhu Y, Shi H, Yin S, Sun F, et al. Exosomes derived from hucMSC attenuate renal fibrosis through CK1δ/β-TRCP-mediated YAP degradation. *Cell Death Dis*. 2020;11:327. https://doi.org/10.1038/s41419-020-2470-0.

Kalluri R, LeBleu VS. The biology, function, and biomedical applications of exosomes. *Science*. 2020;367(6478). https://doi.org/10.1126/science.aau6977.

Kalra H, Gangoda L, Fonseka P, Chitti SV, Liem M, Keerthikumar S, Samuel M, et al. Extracellular vesicles containing an oncogenic mutant of beta-catenin activate the Wnt signalling pathway in recipient cells. *J Extracell Vesicles*. 2019;8(1):1690217. https://doi.org/10.1080/20013078.2019.1690217.

Kashat L, So AK, Masui O, Wang XS, Cao J, Meng X, Macmillan C, et al. Secretome-based identification and characterization of potential biomarkers in thyroid cancer. *J Proteome Res*. 2010;9(11):5757–69. https://doi.org/10.1021/pr100529t.

Katsman D, Stackpole EJ, Domin DR, Farber DB. Embryonic stem cell-derived microvesicles induce gene expression changes in Müller cells of the retina. *PLoS One*. 2012;7(11):e50417. Epub 2012/12/12. https://doi.org/10.1371/journal.pone.0050417.

Kim HI, Park J, Zhu Y, Wang X, Han Y, Zhang D. Recent advances in extracellular vesicles for therapeutic cargo delivery. *Exp Mol Med*. 2024;56(4):836–49. https://doi.org/10.1038/s12276-024-01201-6.

Kumar MA, Baba SK, Sadida HQ, Marzooqi SA, Jerobin J, Altemani FH, Algehainy N, et al. Extracellular vesicles as tools and targets in disease therapy. *Signal Transduct Target Therapy*. 2024;9(1):27.

Lee S, Heo J, Ahn EK, Kim JH, Kim YH, Chang HK, Lee SJ, et al. Conditioned secretome of adipose-derived stem cells improves dextran sulfate sodium-induced colitis in mice. *World J Gastroenterol*. 2021;27(23):3342–56. https://doi.org/10.3748/wjg.v27.i23.3342.

Legaki E, Roubelakis MG, Theodoropoulos GE, Lazaris A, Kollia A, Karamanolis G, Marinos E, et al. Therapeutic potential of secreted molecules derived from human amniotic fluid mesenchymal stem/stroma cells in a mice model of colitis. *Stem Cell Rev Rep*. 2016;12(5):604–12. https://doi.org/10.1007/s12015-016-9677-1.

Lener T, Gimona M, Aigner L, Borger V, Buzas E, Camussi G, Chaput N, et al. Applying extracellular vesicles based therapeutics in clinical trials – An ISEV position paper. *J Extracell Vesicles*. 2015;4(1):30087. https://doi.org/10.3402/jev.v4.30087.

Leung KS, Shirazi S, Cooper LF, Ravindran S. Biomaterials and extracellular vesicle delivery: Current status, applications and challenges. *Cells*. 2022;11(18). https://doi.org/10.3390/cells11182851.

Li M, Luan F, Zhao Y, Hao H, Liu J, Dong L, Fu X, et al. Mesenchymal stem cell-conditioned medium accelerates wound healing with fewer scars. *Int Wound J*. 2017;14(1):64–73. https://doi.org/10.1111/iwj.12551.

Li X, Liu H, Dun MD, Faulkner S, Liu X, Jiang CC, Hondermarck H. Proteome and secretome analysis of pancreatic cancer cells. *Proteomics*. 2022a;22(13–14):e2100320. https://doi.org/10.1002/pmic.202100320.

Li X, Liu Y, Liu X, Du J, Bhawal UK, Xu J, Guo L, et al. Advances in the therapeutic effects of apoptotic bodies on systemic diseases. *Int J Mol Sci*. 2022b;23(15). Epub 2022/07/29. https://doi.org/10.3390/ijms23158202.

Li Z, Liu J, Song J, Yin Z, Zhou F, Shen H, Wang G, et al. Multifunctional hydrogel-based engineered extracellular vesicles delivery for complicated wound healing. *Theranostics*. 2024;14(11):4198–217. https://doi.org/10.7150/thno.97317.

Liang Y, Xia J, Xu L, Zhang Y, Xu X, Haider KH. Cell-free and cell-based therapy in regenerative medicine. In: Haider, K.H. (ed) *Handbook of Regenerative Medicine* Boca Raton: CRC Press, 2025, pp. 134–52.

Liem M, Ang CS, Mathivanan S. Insulin-mediated activation of PI3K/Akt signalling pathway modifies the proteomic cargo of extracellular vesicles. *Proteomics*. 2017;17:1600165. https://doi.org/10.1002/pmic.201600165.

Liu K, Veenendaal T, Wiendels M, Ruiz-Zapata AM, van Laar J, Kyranas R, Enting H, et al. Synthetic extracellular matrices as a toolbox to tune stem cell secretome. *ACS Appl Mater Interfaces*. 2020a;12(51):56723–30. https://doi.org/10.1021/acsami.0c16208.

Liu XY, Wei MG, Liang J, Xu HH, Wang JJ, Wang J, Yang XP, et al. The injury-preconditioning secretome of umbilical cord mesenchymal stem cells enhanced neurogenesis and cognitive recovery after severe traumatic brain injury in rats. *J Neurochem*. 2020b;153(2):230-51. https://doi.org/10.1111/jnc.14859.

Lotfy A, AboQuella NM, Wang H. Mesenchymal stromal/stem cell (MSC)-derived exosomes in clinical trials. *Stem Cell Res Therapy*. 2023;14(1):66. https://doi.org/10.1186/s13287-023-03287-7.

Luo X, Kugeratski FG, Dowlatshahi DP, Sugimoto H, Arian KA, Fan Y, et al. Engineered immunomodulatory extracellular vesicles from epithelial cells with the capacity for stimulation of innate and adaptive immunity in cancer and autoimmunity. *ACS Nano*. 2025;19(5):5193–216. https://doi.org/10.1021/acsnano.4c09688.

Ma S, Liu X, Yin J, Hao L, Diao Y, Zhong J. Exosomes and autophagy in ocular surface and retinal diseases: new insights into pathophysiology and treatment. *Stem Cell Res Therapy.* 2022;13(1):174.

Mahendru D, Jain A, Bansal S, Malik D, Dhir N, Sharma AR, Sarma P, et al. Neuroprotective effect of bone marrow-derived mesenchymal stem cell secretome in 6-OHDA-induced Parkinson's disease. *Regen Med.* 2021;16(10):915–30. https://doi.org/10.2217/rme-2021-0018.

Malech HL, Garabedian EK, Hsieh MM. Evolution of gene therapy: Historical perspective. *Hematol Oncol Clin North Am.* 2022;36(4):627–45. https://doi.org/10.1016/j.hoc.2022.05.001.

Mani S, Gurusamy N, Ulaganathan T, Paluck AJ, Ramalingam S, Rajasingh J. Therapeutic potentials of stem cell-derived exosomes in cardiovascular diseases. *Exp Biol Med (Maywood).* 2023;248(5):434–44. https://doi.org/10.1177/15353702231151960.

Massague J. TGFbeta in cancer. Cell. 2008;134(2):215–30. https://doi.org/10.1016/j.cell.2008.07.001.

McBride JD, Rodriguez-Menocal L, Candanedo A, Guzman W, Garcia-Contreras M, Badiavas EV. Dual mechanism of type VII collagen transfer by bone marrow mesenchymal stem cell extracellular vesicles to recessive dystrophic epidermolysis bullosa fibroblasts. *Biochimie.* 2018;155:50–58. https://doi.org/10.1016/j.biochi.2018.08.009.

Mushtaq M, Zineldeen DH, Mateen MA, Haider KH. Mesenchymal stem cells' "garbage bags" at work: Treating radial nerve injury with mesenchymal stem cell-derived exosomes. *World J Stem Cells* 2024;16(5). https://dx.doi.org/10.4252/wjsc.v16.i5.0000.

Muthu S, Bapat A, Jain R, Jeyaraman N, Jeyaraman M. Exosomal therapy – A new frontier in regenerative medicine. *Stem Cell Investig.* 2021;8:7. https://doi.org/10.21037/sci-2020-037.

Muzes G, Sipos F. Mesenchymal stem cell-derived secretome: A potential therapeutic option for autoimmune and immune-mediated inflammatory diseases. *Cells.* 2022;11(15). https://doi.org/10.3390/cells11152300.

N'Diaye ER, Orefice NS, Ghezzi C, Boumendjel A. Chemically modified extracellular vesicles and applications in radiolabeling and drug delivery. *Pharmaceutics.* 2022;14(3). https://doi.org/10.3390/pharmaceutics14030653.

Ng CY, Kee LT, Al-Masawa ME, Lee QH, Subramaniam T, Kok D, Ng MH, et al. Scalable production of extracellular vesicles and its therapeutic values: A review. *Int J Mol Sci.* 2022;23(14):7986. https://doi.org/10.3390/ijms23147986.

Ng KS, Smith JA, McAteer MP, Mead BE, Ware J, Jackson FO, Carter A, et al. Bioprocess decision support tool for scalable manufacture of extracellular vesicles. *Biotechnol Bioeng.* 2019;116(2):307–19. https://doi.org/10.1002/bit.26809.

Nieuwland R, Falcon-Perez JM, Thery C, Witwer KW. Rigor and standardization of extracellular vesicle research: Paving the road towards robustness. *J Extracell Vesicles.* 2020;10(2):e12037. https://doi.org/10.1002/jev2.12037.

Osugi M, Katagiri W, Yoshimi R, Inukai T, Hibi H, Ueda M. Conditioned media from mesenchymal stem cells enhanced bone regeneration in rat calvarial bone defects. *Tissue Eng Part A.* 2012;18(13–14):1479–89. https://doi.org/10.1089/ten.TEA.2011.0325.

Paganini C, Capasso Palmiero U, Pocsfalvi G, Touzet N, Bongiovanni A, Arosio P. Scalable production and isolation of extracellular vesicles: Available sources and lessons from current industrial bioprocesses. *Biotechnol J.* 2019;14(10):e1800528. https://doi.org/10.1002/biot.201800528.

Patel U, Susman D, Allan AL. Quality control and validation of extracellular vesicles isolated from cultured human breast cancer cells. *BMC Res Notes.* 2024;17(1):202. https://doi.org/10.1186/s13104-024-06865-x.

Penn JS, Madan A, Caldwell RB, Bartoli M, Caldwell RW, Hartnett ME. Vascular endothelial growth factor in eye disease. *Prog Retin Eye Res.* 2008;27(4):331–71. https://doi.org/10.1016/j.preteyeres.2008.05.001.

Pinho AG, Cibrao JR, Silva NA, Monteiro S, Salgado AJ. Cell secretome: Basic insights and therapeutic opportunities for CNS disorders. *Pharmaceuticals (Basel).* 2020;13(2):31. https://doi.org/10.3390/ph13020031.

Pischiutta F, Caruso E, Cavaleiro H, Salgado AJ, Loane DJ, Zanier ER. Mesenchymal stromal cell secretome for traumatic brain injury: Focus on immunomodulatory action. *Exp Neurol.* 2022;357:114199. https://doi.org/10.1016/j.expneurol.2022.114199.

Poggio M, Hu T, Pai CC, Chu B, Belair CD, Chang A, Montabana E, et al. Suppression of exosomal PD-L1 induces systemic anti-tumor immunity and memory. *Cell.* 2019;177(2):414–27 e13. https://doi.org/10.1016/j.cell.2019.02.016.

Rajesh A, Ju EDE, Oxford KA, Harman RM, Van de Walle GR. The mesenchymal stromal cell secretome promotes tissue regeneration and increases macrophage infiltration in acute and methicillin-resistant Staphylococcus aureus-infected skin wounds in vivo. *Cytotherapy.* 2024;26(11):1400–10. https://doi.org/10.1016/j.jcyt.2024.06.007.

Raziyeva K, Kim Y, Zharkinbekov Z, Kassymbek K, Jimi S, Saparov A. Immunology of acute and chronic wound healing. *Biomolecules*. 2021;11(5). https://doi.org/10.3390/biom11050700.

Ruan S, Greenberg Z, Pan X, Zhuang P, Erwin N, He M. Extracellular vesicles as an advanced delivery biomaterial for precision cancer immunotherapy. *Adv Healthc Mater.* 2022;11(5):e2100650. https://doi.org/10.1002/adhm.202100650.

Safwan M, Bourgleh MS, Baroudi L, Almesned A, Albalawi R, Aybour B, Shahad Alfawaz, Bourgleh SM, Haider KH. Entering the era of living biopharmaceuticals for treating knee osteoarthritis: Systematic review and network meta-analysis to compare allogenic and autologous MSCs. *World J Stem Cells* 2025;17(8):107076. https://doi.org/10.4252/wjsc.v17.i8.107076.

Santamaria G, Brandi E, Vitola P, Grandi F, Ferrara G, Pischiutta F, Vegliante G, et al. Intranasal delivery of mesenchymal stem cell secretome restores brain function in Alzheimer's mice. *Cell Death Differentiation*. 2021;28(1):203–18. https://doi.org/10.1038/s41418-020-0592-2.

Sazli BI, Lindarto D, Hasan R, Putra A, Pranoto A, Sembiring RJ, Ilyas S, et al. Secretome of hypoxia-preconditioned mesenchymal stem cells enhances angiogenesis in diabetic rats with peripheral artery disease. *Med Arch*. 2023;77(2):90–6. https://doi.org/10.5455/medarh. 2023.77.90-96.

Schirmaier C, Jossen V, Kaiser SC, Jüngerkes F, Brill S, Safavi-Nab A, Siehoff A, et al. Scale-up of adipose tissue-derived mesenchymal stem cell production in stirred single-use bioreactors under low-serum conditions. *Eng Life Sci.* 2014;14(3):292–303. https://doi.org/10.1002/elsc.201300134.

Schwager SC, Reinhart-King CA. Chapter 7 – Mechanobiology of microvesicle release, uptake, and microvesicle-mediated activation. In: ILevitan , Trache A (eds) *Current Topics in Membranes*. Vol. 86. New York: Academic Press, 2020, pp. 255–78.

Shaba E, Vantaggiato L, Governini L, Haxhiu A, Sebastiani G, Fignani D, Grieco GE, et al. Multi-omics integrative approach of extracellular vesicles: A future challenging milestone. *Proteomes*. 2022;10(2). https://doi.org/10.3390/proteomes10020012.

Shahi S, Kang T, Fonseka P. Extracellular vesicles in pathophysiology: A prudent target that requires careful consideration. *Cells*. 2024;13(9):754.

Shimojima C, Takeuchi H, Jin S, Parajuli B, Hattori H, Suzumura A, Hibi H, et al. Conditioned medium from the stem cells of human exfoliated deciduous teeth ameliorates experimental autoimmune encephalomyelitis. *J Immunol*. 2016;196(10):4164–71. https://doi.org/10.4049/jimmunol.1501457.

Song H, Canup BSB, Ngo VL, Denning TL, Garg P, Laroui H. Internalization of garlic-derived nanovesicles on liver cells is triggered by interaction with CD98. *ACS Omega*. 2020;5(36):23118–28. https://doi.org/10.1021/acsomega.0c02893.

Sreenivas A, K. Jha D. Cell-free therapies: Revolutionizing the approach to cellular treatments. In: Paul M, Bisht B, Singh BN (eds) *Advances in Regenerative Medicine and Tissue Engineering*. Rijeka: IntechOpen, 2025.

Sriramulu S, Banerjee A, Jothimani G, Pathak S. Conditioned medium from the human umbilical cord-mesenchymal stem cells stimulates the proliferation of human keratinocytes. *J Basic Clin Physiol Pharmacol.* 2020;32(2):51–6. https://doi.org/10.1515/jbcpp-2019-0283.

Stahl AL, Johansson K, Mossberg M, Kahn R, Karpman D. Exosomes and microvesicles in normal physiology, pathophysiology, and renal diseases. *Pediatr Nephrol.* 2019;34(1):11–30. https://doi.org/10.1007/s00467-017-3816-z.

Stone WL, Leavitt L, Varacallo MA. *Physiology, Growth Factor.* Treasure Island, FL: StatPearls Publishing, 2025.

Takeuchi R, Katagiri W, Endo S, Kobayashi T. Exosomes from conditioned media of bone marrow-derived mesenchymal stem cells promote bone regeneration by enhancing angiogenesis. *PLoS One.* 2019;14(11):e0225472. https://doi.org/10.1371/journal.pone.0225472.

Théry C, Witwer KW, Aikawa E, Alcaraz MJ, Anderson JD, Andriantsitohaina R, Antoniou A, et al. Minimal information for studies of extracellular vesicles 2018 (MISEV2018): A position statement of the International Society for Extracellular Vesicles and update of the MISEV2014 guidelines. *J Extracell Vesicles*. 2018;7(1):1535750. https://doi.org/10.1080/20013078.2018.1535750.

Timmers L, Lim SK, Hoefer IE, Arslan F, Lai RC, van Oorschot AA, Goumans MJ, et al. Human mesenchymal stem cell-conditioned medium improves cardiac function following myocardial infarction. *Stem Cell Res*. 2011;6(3):206–14. https://doi.org/10.1016/j.scr.2011.01.001.

Tiwari S, Kumar V, Randhawa S, Verma SK. Preparation and characterization of extracellular vesicles. *Am J Reprod Immunol.* 2021;85(2):e13367. https://doi.org/10.1111/aji.13367.

Urry LA, Cain ML, Wasserman SA, Minorsky PV, Orr RB. Chapter 43 – The immune system. In: Campbell NA (ed) *Campbell Biology.* New York: Pearson, 2021, pp. 952–77.

Vaillant AA JQA. *Interleukin*. [Updated 2022 Aug 22]. Treasure Island, FL: StatPearls Publishing, 2025.

Van Delen M, Derdelinckx J, Wouters K, Nelissen I, Cools N. A systematic review and meta-analysis of clinical trials assessing the safety and efficacy of human extracellular vesicle-based therapy. *J Extracell Vesicles*. 2024;13(7):e12458. https://doi.org/10.1002/jev2.12458.

van Hooff LC, Merz EM, Kidane Gebremeskel AS, de Jong JA, Burchell GL, Lunshof JE. Balancing benefits and burdens: A systematic review on ethical and social dimensions of gene and cell therapies for hereditary blood diseases. *BMC Med Ethics*. 2025;26(1):36. https://doi.org/10.1186/s12910-025-01188-3.

Wang CK, Tsai TH, Lee CH. Regulation of exosomes as biologic medicines: Regulatory challenges faced in exosome development and manufacturing processes. *Clin Transl Sci*. 2024a;17(8):e13904.

Wang J, Chen ZJ, Zhang ZY, Shen MP, Zhao B, Zhang W, Zhang Y, et al. Manufacturing, quality control, and GLP-grade preclinical study of nebulized allogenic adipose mesenchymal stromal cells-derived extracellular vesicles. *Stem Cell Res Therapy*. 2024b;15(1):95. https://doi.org/10.1186/s13287-024-03708-1.

Wang J, Zuzzio K, Walker CL. Systemic dental pulp stem cell secretome therapy in a mouse model of amyotrophic lateral sclerosis. *Brain Sci*. 2019;9(7). https://doi.org/10.3390/brainsci9070165.

Wang L, Zhang X, Yang Z, Wang B, Gong H, Zhang K, Lin Y, et al. Extracellular vesicles: biological mechanisms and emerging therapeutic opportunities in neurodegenerative diseases. *Transl Neurodegener*. 2024c;13(1):60. https://doi.org/10.1186/s40035-024-00453-6.

Wang Y, Li C, Wu F, Mao J, Zhu J, Xie H, Zhou X, et al. The negative effects of extracellular vesicles on the immune system. *Front Immunol*. 2024d;15:1410273. https://doi.org/10.3389/fimmu.2024.1410273.

Wasserman AH, Venkatesan M, Aguirre A. Bioactive lipid signaling in cardiovascular disease, development, and regeneration. *Cells*. 2020;9(6). https://doi.org/10.3390/cells9061391.

Wechsler ME, Rao VV, Borelli AN, Anseth KS. Engineering the MSC secretome: A hydrogel focused approach. *Adv Healthc Mater*. 2021;10(7):e2001948. https://doi.org/10.1002/adhm.202001948.

Welsh JA, Goberdhan DCI, O'Driscoll L, Buzas EI, Blenkiron C, Bussolati B, Cai H, et al. Minimal information for studies of extracellular vesicles (MISEV2023): From basic to advanced approaches. *J Extracellular Vesicles*. 2024;13(2):e12404. https://doi.org/10.1002/jev2.12404.

Whitford WG, Lundgren M, Fairbank A. Cell culture media in bioprocessing. In: Jagschies G, Lindskog E, Łącki K, Galliher P (eds) *Biopharmaceutical Processing*. Amsterdam: Elsevier, 2018, pp. 147–62.

Wu W, Krijgsveld J. Secretome analysis: Reading cellular sign language to understand intercellular communication. *Mol Cell Proteomics*. 2024;23(1):100692. https://doi.org/10.1016/j.mcpro.2023.100692.

Xia J, Arai S, Haider KH. Research and applications of stem cell secretome. In: Haider, K.H. (ed) *Stem Cells: From Potential to Promise*. Singapore: Springer, 2021. https://doi.org/10.1007/978-981-16-0301-3_7.

Xiao J, Pan Y, Li XH, Yang XY, Feng YL, Tan HH, Jiang L, et al. Cardiac progenitor cell-derived exosomes prevent cardiomyocyte apoptosis through exosomal miR-21 by targeting PDCD4. *Cell Death Disease*. 2016;7(6):e2277. https://doi.org/10.1038/cddis.2016.181.

Xu D, Di K, Fan B, Wu J, Gu X, Sun Y, et al. MicroRNAs in extracellular vesicles: sorting mechanisms, diagnostic value, isolation, and detection technology. *Front Bioeng Biotechnol*. 2022;10:948959. https://doi.org/10.3389/fbioe.2022.948959.

Yamaguchi S, Shibata R, Yamamoto N, Nishikawa M, Hibi H, Tanigawa T, Ueda M, et al. Dental pulp-derived stem cell conditioned medium reduces cardiac injury following ischemia-reperfusion. *Sci Rep*. 2015;5(1):16295. https://doi.org/10.1038/srep16295.

Yang S, Zhang L, Zhou X, Peng X, Zhang X, Wang W, Zhao J, et al. A new large-scale extracellular vesicle production strategy for biomedical drug development. *bioRxiv*. 2024:2024.04. 15.589541.

Yates AG, Pink RC, Erdbrügger U, Siljander PRM, Dellar ER, Pantazi P, et al. In sickness and in health: The functional role of extracellular vesicles in physiology and pathology in vivo: part II: pathology. *J Extracell Vesicles*. 2022;11:e12190. https://doi.org/10.1002/jev2.12190.

Yu L, Zhu G, Zhang Z, Yu Y, Zeng L, Xu Z, Weng J, et al. Apoptotic bodies: bioactive treasure left behind by the dying cells with robust diagnostic and therapeutic application potentials. *J Nanobiotechnology*. 2023;21(1):218. https://doi.org/10.1186/s12951-023-01969-1.

Zhang J, Patel L, Pienta KJ. Targeting chemokine (C-C motif) ligand 2 (CCL2) as an example of translation of cancer molecular biology to the clinic. *Prog Mol Biol Transl Sci*. 2010;95:31–53. https://doi.org/10.1016/B978-0-12-385071-3.00003-4.

Zhao S, Di Y, Fan H, Xu C, Li H, Wang Y, Wang W, et al. Targeted delivery of extracellular vesicles: The mechanisms, techniques and therapeutic applications. *Mol Biomed*. 2024;5(1):60. https://doi.org/10.1186/s43556-024-00230-x.

Zhou X, Xie F, Wang L, Zhang L, Zhang S, Fang M. et al. The function and clinical application of extracellular vesicles in innate immune regulation. *Cell Mol Immunol*. 2020;17:323–34. https://doi.org/10.1038/s41423-019-0301-7.

27 Diabetes Mellitus
Treatment with MSCs and their Secretome

Mohamed Abdullah alJarroud[1], Syed Abubakar Habib[1], Hassan Alnatour[1], Farras Trityan Aqma[1], Ahmad Mohammed Adeb Adi[1], and Khawaja H. Haider[1]

[1]Department of Basic Sciences, Sulaiman AlRajhi University, AlQaseem, Kingdom of Saudi Arabia

27.1 LIST OF ABBREVIATIONS

AD-MSCs	Adipose-derived mesenchymal stem cells
AFP	Alpha-fetoprotein
AMMSCs	Amniotic membrane-derived mesenchymal stem cells
ASCs	Adipose tissue-derived mesenchymal stem cells
BM-MSCs	Bone marrow-derived mesenchymal stem cells
CA-199	Carbohydrate antigen 19-9
CD	Cluster of differentiation
CD4+	Cluster of differentiation 4 positive T cells
CD8+	Cluster of differentiation 8 positive T cells
CD8 T-cells	cluster of differentiation 8 T cells
CEA	Carcinoembryonic antigen
DP-MSCs	Dental pulp mesenchymal stem cell
DMT1	Diabetes mellitus type 1
DMT2	Diabetes mellitus type 2
DKA	Diabetic ketoacidosis
EVs	Extracellular vesicles
ESCs	Embryonic stem cells
FBS	Fetal bovine serum
FBG	Fasting blood glucose
FGF21	Fibroblast growth factor 21
GADA	Glutamic acid decarboxylase antibodies
GLP1	Glucagon-like peptide 1
GLUT4	Glucose transporter type 4
HbA1c	Glycated blood glucose
HFD	High-fat diet

DOI: 10.1201/9781003653387-27

HGF	Hepatocyte growth factor
HLA-DR	Human leukocyte antigen-DR isotype
HOMA-B	Homeostatic model assessment of beta-cell function
HOMA-IR	Model Assessment of Insulin Resistance
hUC-MSC	Human umbilical cord mesenchymal stem cell
IA-2	Insulinoma-associated protein-2 antibodies
IDO	Indoleamine 2,3-dioxygenase
IGF-1	Insulin-like growth factor 1
IHC	Immunohistochemistry
IL-10	Interleukin-10
IL-1β	Interleukin-1 beta
IL-6	Interleukin-6
iPSCs	Induced pluripotent stem cells
IPCs	Insulin-producing cells
ISCT	International Society for Cellular Therapy
M1	Pro-inflammatory macrophage phenotype
M2	Anti-inflammatory macrophage phenotype
MHC	Major histocompatibility complex
miRNAs	MicroRNAs
MSC	Mesenchymal stem cell
MSC-	EXOs: MSCs derivative exosomes
MSC-GM	Mesenchymal stem cell growth medium
NGF	Nerve growth factor
NGN3	Neurogenin 3
NKX6.1	NK6 Homeobox 1
PBS	Phosphate buffered saline
PGE2	Prostaglandin E2
PDX1	Pancreatic and duodenal homeobox 1
STZ	Streptozotocin
T1DM	Type 1 diabetes mellitus
T2DM	Type 2 diabetes mellitus
TGF-β	Transforming growth factor beta
TNF-α	Tumor necrosis factor alpha
Tregs	Regulatory T cells
UC-MSC	Umbilical cord mesenchymal stem cell
VEGF	Vascular endothelial growth factor
WJ-MSCs	Wharton's jelly mesenchymal stem cells
ZNT8	Zinc transporter 8
β-cells	Beta cells
Ig	Immunoglobulin

27.2 INTRODUCTION

Diabetes Mellitus (DM) is a chronic metabolic disorder characterized by persistent hyperglycemia due to defective insulin secretion, insulin resistance, or both (Päth et al. 2019). This disease is mainly divided into DMT1 and DMT2. The first one is due to the autoimmune destruction of β-cells in the pancreas that produce insulin, resulting in deficient production and thus requiring the person to take daily dosages of insulin (El-Badri & Ghoneim 2013). Meanwhile, the second type involves insulin resistance and relative impairment in its secretion, which can lead to elevated levels of blood sugar if left untreated (Mikłosz and Chabowski 2023; Dludla et al., 2023). As time passes, chronic hyperglycemia can result in microvascular complications and may eventually cause neuropathy, nephropathy, and retinopathy (El-Badri and Ghoneim 2013; Päth et al. 2019). According to the free fatty acid hypothesis, faulty fatty acid metabolism is a primary contributor to DMT2 (Musleh et al., 2018). According to data from the eleventh edition of the Diabetes Atlas, published by the International Diabetes Federation in 2025, the estimated number of patients with diabetes worldwide in the 20–79 age group in 2024 was approximately 588.7 million, 98% of whom had DMT2 (https://idf.org/resources/idf-diabetes-atlas-2025/; IDF reports, 2025). This number is expected to increase by 45%, thus reaching a staggering 852.5 million patients by 2050 (https://idf.org/resources/idf-diabetes-atlas-2025/ IDF report, 2025). This shows that diabetes has become a critical global health concern, escalating to an alarming degree. The published data highlight diabetes as one of the fastest-growing health crises worldwide in the 21st century (Hossain et al., 2024).

Current treatment options primarily focus on symptomatic management, primarily by controlling hyperglycemia to prevent its acute and chronic complications, rather than addressing the underlying pathophysiological root cause of the disease (Weinberg et al., 2023). Insulin injections, with an expanding variety of biosynthetic human insulin analogs, remain the primary therapy for controlling blood sugar levels in DMT1; however, they do not alleviate the autoimmune attack on β-cells (Nally et al., 2019). In DMT2, available medications either improve insulin sensitivity, such as biguanides and thiazolidinediones, or stimulate increased insulin secretion, including sulfonylureas and GLP-1 agonists (Mikłosz & Chabowski, 2023). Although improving insulin sensitivity can be achieved through weight loss and lifestyle modification, these measures do not fully restore β-cell function, especially in cases of severe dysfunction (Andreadi et al., 2023).

With that being said, the search for new therapeutic strategies has remained an area of special research interest over the past decade, encompassing stem cell-based regenerative therapy and the targeted preservation and restoration of beta-cell function (Boughton et al., 2017; Saima & Haider, 2021). One that has attracted considerable attention is the use of mesenchymal stem cell (MSC)-based therapy due to their excellent reparative and regenerative potential (Alvarez and Haider, 2022; Zeinhom et al., 2024). These are multipotent stem cells. MSCs are found in mesenchymal tissues, including the bone marrow, adipose tissue, placenta, and umbilical cord (Kamal et al., 2022). Given their near-ideal characteristics and cell biology, they have already established a presence in the biopharmaceutical market as living biodrugs for routine use in clinical settings (Bhatti and Haider, 2025).

Besides differentiation potential, MSCs reparability and regenerative potential have also been attributed to their paracrine activity and anti-inflammatory and immunomodulatory effects due to the release of numerous bioactive substances such as growth factors, cytokines, and chemokines, facilitating tissue repair, promoting angiogenesis, and supporting cell survival (Haider & Aziz, 2017; Xu et al., 2020). MSCs can also modulate the immune response by interacting with innate and adaptive immune cells, leading to downregulation of pro-inflammatory cytokines and upregulation of anti-inflammatory cytokines (Ma et al., 2021; Rajab et al., 2019).

In the context of MSCs' use for diabetes treatment, several preclinical studies in experimental animal models and early phase clinical trials have shown their potential in facilitating endogenous regeneration of β-cells, helping to sustain the remaining β-cell mass, and regulating the immune system, addressing the autoimmune attack on β-cells (Ma et al. 2021; Gao et al., 2022). In DMT2, MSCs may enhance insulin sensitivity in peripheral tissues and reduce chronic low-grade inflammation, both of which are key factors in insulin resistance and β-cell dysfunction. They are also being used to support islet grafts post-engraftment (Mou et al., 2024). This multimodal mechanism of action of MSCs renders them a suitable cell type for ameliorating DM-associated complications and as an emerging treatment option in clinical settings. This review aims to systematically evaluate the therapeutic role of mesenchymal stem cells (MSCs) in the treatment of both DMT1 and DMT2.

27.3 PATHOPHYSIOLOGY OF DM1 AND DM 2

27.3.1 Pathophysiology of DMT1

DMT1 is a chronic autoimmune condition characterized by the selective autodestruction of insulin-producing β-cells in the islets of Langerhans of the pancreas, resulting in absolute insulin deficiency (Speight & Pouwer, 2019; Cotran et al., 2021). This condition necessitates lifelong dependency of the patient on exogenous insulin to avoid hyperglycemia and fatal ketoacidosis. Unlike DMT2, which is more prevalent in the elderly, DMT1 is more common in younger patients. T-lymphocytes primarily mediate chronic β-cell destruction in the pancreas (Roep, 2003; Lin et al., 2025). The problem is exacerbated by the humoral response, which involves B-lymphocytes that produce specific autoantibodies against several antigens, including islet cell cytoplasm (ICC), insulin, insulinoma antigen 2 (IA-2), and the zinc transporter isoform 8 (ZNT-8) (Bonilla & Oettgen, 2010). The detection of these autoantibodies constitutes an integral part of the differential diagnosis of DMT1 (Lucier & Mathias, 2024). In DMT1 cases, they directly attack their own body cells after infiltrating beta cells, causing insulitis by targeting beta-cell antigens expressed on their cell surface (Sun et al. 2020). This process is considered the primary pathologic event in the development of DMT1, which typically occurs in three stages (Cotran et al., 2021). These stages could be specified based on the presence of pancreatic autoantibodies (stage 1 of autoimmunity), laboratory result findings of fasting blood glucose (FBG) and Glycated Blood Glucose (HbA1c) as the second stage, and the hyperglycemic symptoms due to prolonged destruction of β-cells and seriously declined insulin production (Lin et al., 2025). The first two stages are preclinical, meaning that patients remain asymptomatic. The third stage represents the clinical onset of the disease process. Some recent epidemiological studies have observed that in patients with higher blood autoantibody titers, the preclinical stage may directly progress to the third stage, resulting in a rapid shift from asymptomatic to symptomatic, characterized by symptoms such as fatigue, weakness, polyphagia, polydipsia, polyuria, and unintentional weight loss (Insel et al. 2015; Speight & Pouwer 2019). If kept untreated, DMT1 may lead to a poor prognosis for Diabetic Ketoacidosis (DKA) with life-threatening complications.

27.3.2 Pathophysiology of Diabetes Mellitus Type 2

DMT2, also known as non-insulin-dependent DM, is a multifactorial metabolic disorder that primarily results from defective/low insulin secretion, increased insulin resistance at the tissue and organ levels, predominantly affecting skeletal muscles, adipose tissue, and the liver in terms of glucose uptake from the blood circulation or a combined result of these conditions (Galicia-García et al., 2020). Insulin is a protein hormone that interacts with insulin receptors on the surface of target tissue cells to initiate intracellular signaling, leading to molecular-level changes that facilitate glucose uptake from the circulation and maintain glucose homeostasis. The most specific cell surface receptor for insulin is the tyrosine kinase (Lee et al., 2022). As part of the insulin resistance mechanism, the insulin receptors in peripheral tissues and organs become impaired, failing to respond to the bound insulin, thus resulting in dysregulated glucose homeostasis (Młynarska et al., 2025; Cotran et al., 2021). The body's response to hyperglycemia involves increased insulin production, resulting in hyperinsulinemia. However, a prolonged compensatory increase in insulin production leads to β-cell exhaustion and ultimately complete dysfunction, resulting in a hyperglycemic state that requires pharmacological intervention (Haider et al., 2025). Although DMT2 causes hyperglycemia similar to DMT1, and they share the same symptoms of hyperglycemia—fatigue, weakness, unintentional weight loss, polyphagia, and polydipsia—DMT2 may present with additional peripheral vascular complications, including microvascular and macrovascular complications that affect small and large vessels. In summary, DMT1 and DMT2 are diverse diseases in their pathophysiological mechanism, but they share the same consequence of dysregulated glucose homeostasis as summarized in Figure 27.1. In summary, DMT1 and DMT2 are diverse diseases in

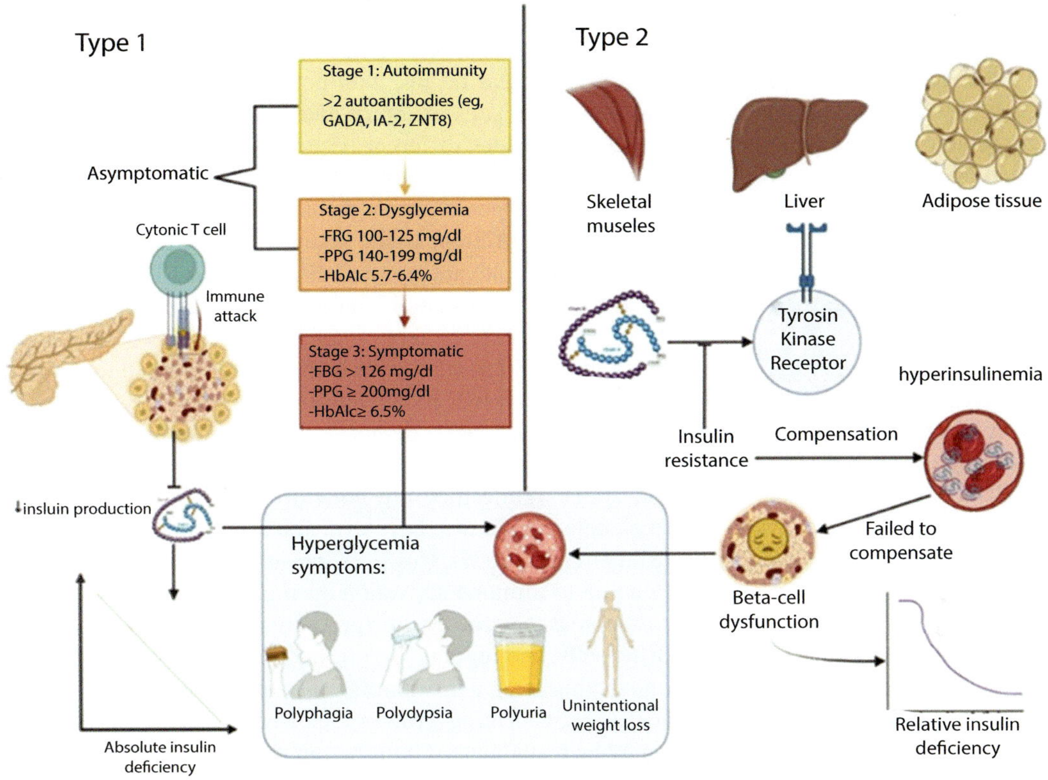

FIGURE 27.1 Pathophysiology of diabetes mellitus type 1 and type 2.

their pathophysiological mechanism, but they share the same consequence of dysregulated glucose homeostasis as summarized in Figure 27.1.

27.4 MSCS ARE A VIABLE OPTION FOR CELL-BASED THERAPY FOR DMT1 AND DMT2

Among the possible cellular options for a cell-based therapy approach, MSCs are a subtype of adult stem cells that are the leading cell choice for diabetes treatment (Haider et al., 2025). They are characterized by a fibroblast-like appearance, plastic adherence, and the potential to differentiate into osteogenic, chondrogenic, and adipogenic lineages under well-defined culture conditions. Based on the International Society for Cellular Therapy (ISCT)'s criterion, MSCs are defined by plastic adherence in culture, express surface antigens CD73, CD90, and CD105, and lack the expression of CD34, CD45, CD14, and HLA-DR, as well as demonstrate differentiation into mesodermal lineages, including adipocytes, chondrocytes, and osteocytes (Dominici et al., 2006). This criterion has helped in defining the purity of MSCs in culture for experimental use.

MSCs were initially isolated from bone marrow for experimental and clinical characterization; however, as the research progressed, MSCs have been procured from different mesenchymal tissues, notable amongst which include adipose tissue (AD-MSCs), Wharton's jelly (WJ-MSCs), umbilical cord (UC-MSCs), umbilical cord blood (UCB-MSCs), dental pulp (DP-MSCs), etc (Kamal et al., 2022), MSCs have been procured from different mesenchymal tissues, notable amongst which include adipose tissue (AD-MSCs), Wharton's jelly (WJ-MSCs), umbilical cord (UC-MSCs), umbilical cord blood (UCB-MSCs), and dental pulp (DP-MSCs) (Kamal et al., 2022). First identified and used for skeletal repair and restoration of the hematopoietic microenvironment (Saeed et al., 2016; Hass et al., 2012), restoration of the hematopoietic microenvironment (Saeed et al., 2016; Hass et al., 2012). MSCs also act as pericytes of perivascular origin, with potent paracrine and immunomodulatory effects, and the capacity to migrate to sites of injury and secrete bioactive molecules that promote tissue repair (Feng et al., 2011). Besides different tissue sources, they are available from allogenic and autologous sources, each with its own advantages and disadvantages. Therefore, there is ongoing debate over the superiority of allogenic versus autologous MSCs for clinical use (Ahmed et al., 2025; Feng et al., 2011). Besides different tissue sources, they are available from allogenic and autologous sources, each with its own advantages and disadvantages. Therefore, there is ongoing debate over the superiority of allogenic versus autologous MSCs for clinical use (Ahmed et al., 2025).

What distinguishes MSCs from the remaining adult stem cells is their demonstrated low immunogenicity due to the limited expression of MHC class II molecules and absence of co-stimulatory markers CD80 and CD86 on their surface (Ma et al., 2021) demonstrated low immunogenicity due to the limited expression of MHC class II molecules and absence of co-stimulatory markers CD80 and CD86 on their surface (Ma et al., 2021). While they are not fully immune-privileged, this property enables them to evade allogeneic rejection in the short term and exert immunosuppressive effects through paracrine signaling (Müller et al., 2021). However, pro-inflammatory stimuli may upregulate MHC class II expression, leading to immune activation during prolonged exposure (Oh et al., 2022). MSCs have been shown to suppress the proliferation and activation of CD4+ and CD8+ T-cells, inhibit the maturation of dendritic cells, and regulate β-cell function (Ma et al., 2021). They promote the development of Tregs, contributing to an anti-inflammatory environment.

The secretion of interleukin-10 (IL-10), transforming growth factor-beta (TGF-β), and prostaglandin E2 has been observed, which may help attenuate inflammation and autoimmune responses (Ma et al., 2021). They also inhibit proinflammatory cytokines, including IL-1β, tumor necrosis factor-alpha (TNF-α), and IL-6, thereby reducing inflammatory activity in pancreatic islets. Zhou et al. demonstrated that wild-type p53 regulates MSC immunomodulatory properties via interferon-alpha and bone marrow stromal cell antigen-2, thereby contributing to the effectiveness of DMT1 (Zhou et al., 2021). They also promote a microenvironment favorable to increased insulin sensitivity and glucose uptake by repolarizing macrophages from an M1 (pro-inflammatory) to an M2 (anti-inflammatory) phenotype (Jin et al., 2019). A recent systematic review and meta-analysis report that MSC-based therapy is an efficacious glycemia-lowering approach for both DMT1 and DMT2, compared to conventional treatment options (Kashbour et al., 2025). The study involved 507 patients from 13 clinical trials (DMT1= 199 patients and DMT2 = 308 patients). The authors observed a significant reduction in daily insulin requirements and a corresponding decrease in HbA1c levels in the cell therapy group. Despite encouraging data in experimental and clinical settings, cell-based therapy has some issues that can be addressed by using stem cell-derived secretomes as an alternative cell-free therapy approach, as discussed in the following sections of the chapter.

27.5 SUPPORT AND PRESERVATION OF THE B-CELL

The progressive decline over time in the number of morphofunctionally competent β-cells and their ability to respond to glycemic changes by insulin secretion to maintain glycemic homeostasis represents a hallmark of both DMT1 and advanced-stage DMT2. Hence, supporting β-cell survival and regeneration while increasing their capacity to respond to hyperglycemia through insulin secretion is needed to address these issues. MSC-based therapies have been shown to exert multiple factorial supportive effects on the pancreatic microenvironment, facilitating the preservation and regeneration of endogenous β-cell populations (Azizi et al., 2022). MSCs enhance β-cell survival by secreting factors that exhibit anti-apoptotic, immunomodulatory, and angiogenic properties. These factors may include, but are not limited to, IGF-1, VEGF, PDGF, HGF-1, SDF-1, and TGF-β, which provide cytoprotection against hypoxic conditions in pancreatic islets (Gao et al., 2014) (Figure 27.2). Additionally, IL-10 and TGF-β, as humoral factors, synergistically protect against immune-mediated β-cell destruction by suppressing T and B cells (Komai et al., 2018).

The route of MSC administration significantly affects therapeutic outcomes (Kalou et al., 2021; Jihwaprani et al., 2024). Direct intrapancreatic administration is more effective than systemic administration at protecting and regenerating β-cells in experimental models. This is due to cell-cell interactions and paracrine signals that maintain islet structure and function. MSCs provide direct protective effects for pancreatic islets. Adopting a therapeutic combinatorial approach, combining islet transplantation with MSC-based cell therapy has been shown to significantly enhance therapeutic outcomes, leading to improved survival of the transplanted islets (Arzouni et al., 2018). This combined approach enhances graft revascularization, reduces immune cell infiltration, and improves glucose-stimulated insulin secretion. In most cases, MSC delivery is carried out as an adjuvant therapy; hence, MSCs do not co-engraft with islets, which are delivered exclusively via the hepatic portal vein. The beneficial effects of MSCs are therefore attributed to paracrine mechanisms rather than to direct interactions between the islets and the adjuvant MSCs. Furthermore, MSC co-transplantation may reduce the number of donor islets required per transplant, thereby overcoming the limited availability of donor pancreatic tissue.

Cellular and subcellular preconditioning strategies have been developed to enhance the survival and functionality of stem cells after engraftment (Haider and Ashraf, 2010; Lu et al., 2012). Various strategies have been employed to precondition cells before transplantation (Haider, 2024). Given the effectiveness of this approach, an alternative to using adjuvant MSCs is preconditioning islets by pre-incubating them with MSCs, a method that has been used to enhance the survival and function of donor islets before transplantation (Gamble et al., 2018) (Figure 27.3).

In experimental settings, mouse islets co-cultured with mouse Ad-MSCs for 48 hours performed significantly better in secreting glucose-stimulated insulin in vitro and successfully reverted blood glucose to euglycemia as compared to the non-cocultured islets in a mouse model. In this regard,

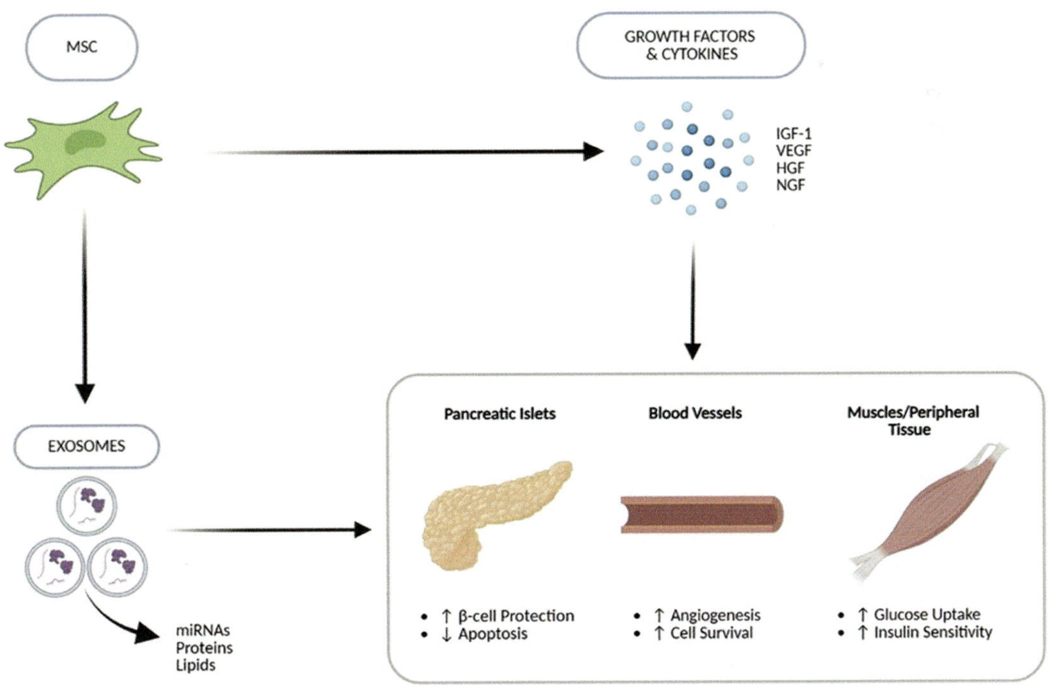

FIGURE 27.2 Paracrine signaling of bioactive molecules leading to the release of secretomes and exsomes from mesenchymal stem cells.

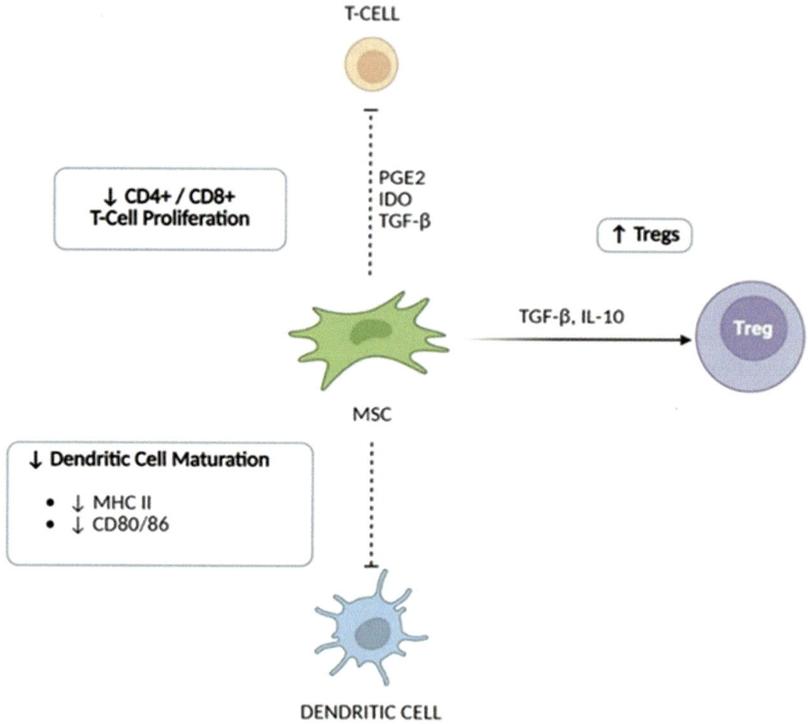

FIGURE 27.3 Immunomodulatory and anti-inflammatory properties of mesenchymal stem cells. 27.3 Paracrine signaling and exosomes derived from MSCs.

direct contact co-culture between monolayered MSCs and islets was more effective than indirect co-culture conditions (Rackham et al., 2013). This preconditioning strategy can improve islet viability by mitigating oxidative stress and inflammatory damage during the immediate post-transplantation period. These results highlight the synergistic protective and regenerative effects of MSCs on pancreatic islet grafts.

In addition to their β-cell-supportive effects, MSCs also affect peripheral glucose metabolism. Published data suggest that MSC-based therapy reduces gluconeogenesis while promoting glycogen storage in hepatic tissues in experimental settings. Si et al. have reported that MSC infusion ameliorated insulin resistance in peripheral target tissues in an experimental rat model of high-fat diet/streptozotocin-induced DMT2 (Si et al., 2012). Cell administration was performed on days 7 and 21 after streptozotocin injection. While exploring the mechanism, the authors observed enhanced B-cell function and amelioration of insulin resistance, accompanied by increased GLUT4 expression and elevated phosphorylated IRS-1 and Akt in insulin target tissues (Si et al., 2012). Subsequent studies have substantiated these data. For example, Kotikalpudi et al. showed that intramuscular treatment with placenta-derived MSCs restored the dysregulated cytokine profile in a rodent obesity model. This was achieved at the molecular level by activating Akt signaling and GLUT4-dependent glucose uptake in comparison with control animals (Kotikalpudi et al., 2021) (Figure 27.4).

In another study, MSCs were genetically modified to overexpress fibroblast growth factor 21 (FGF21) or glucagon-like peptide-1 (GLP-1) to enhance their therapeutic effects on glucose regulation and insulin sensitivity (Xue et al., 2021). Treatment with genetically modified MSCs enhanced insulin secretion, supported standard liver structure, and improved the lipid profile in a DMT1 mouse model. These data have been extrapolated to clinical use in multiple studies. In a recent meta-analysis, Ranjbaran et al. reported an association between MSC-based therapy and glucose

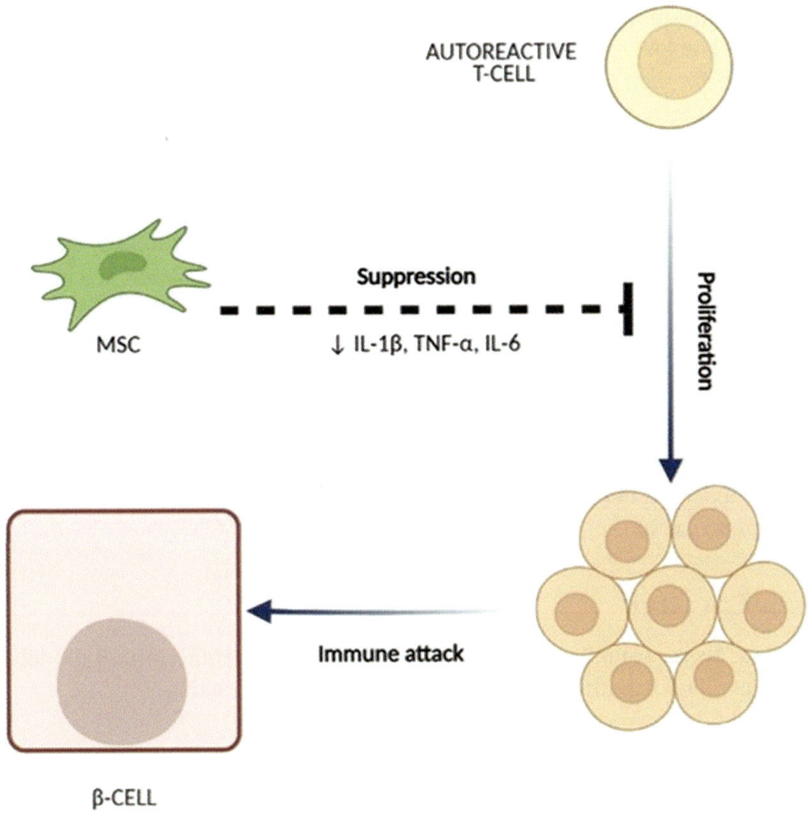

FIGURE 27.4 Support and preservation of the β-cell.

regulation in patients with DMT2 (Ranjbaran et al., 2021). These approaches expand the potential utility of MSCs for metabolic modulation in diabetes management.

27.6 MSC DIFFERENTIATION INTO INSULIN-PRODUCING CELLS

Trilineage differentiation potential—i.e., adipogenic, osteogenic, and chondrogenic—under specific culture conditions was identified as the primary mechanism of action for MSCs as living biodrugs. MSCs have also demonstrated significant potential to differentiate into insulin-producing cells (IPCs) for the treatment of diabetes (Chen et al., 2004). Although complex, MSCs can be guided to acquire the morphofunctional features of pancreatic β-cells by triggering key transcription factors, such as PDX1, NGN3, and NKX6.1, in combination with specific chemical cues in the culture medium. The optimized differentiation protocols enable MSC-derived β-cells to release insulin and C-peptide in response to changes in glucose levels, as natural β-cells do. The source of the MSCs used remains a critical contributing factor to the success of the differentiation protocol (Xu et al., 2017; Safwan et al., 2024).

Among the different MSC tissue sources, AD-MSCs, along with BM-MSCs and UC-MSCs, are especially useful because they are easily collected and grow well in culture. Studies have found that IPCs made from AD-MSCs can lower blood sugar in diabetic animals, although they don't yet produce as much insulin as fully developed human islets. Despite the superiority of pluripotent stem cells, specifically embryonic stem cells (ESCs) and induced pluripotent stem cells (iPSCs), in terms of β-cell differentiation rates, MSCs are preferable due to their safety in use and the absence of ethical concerns and moral issues (Ibrahim et al., 2016). Although iPSCs mitigate ethical concerns, they pose risks of genetic instability and oncogenesis due to reprogramming errors (Ahmed et al., 2011; Omole et al., 2022). The low immunogenicity of MSCs, particularly those derived from autologous tissue sources, allows for their safe use across donors, even in allogeneic applications. This also adds to the logistical advantages, with ready-to-use off-the-shelf availability. MSCs are also compatible with bioengineering platforms, including encapsulation devices and hydrogel matrices, to improve therapeutic persistence and heterogeneity. Furthermore, MSCs' proven safety record spanning more than 2 decades of clinical experience in other disorders, and their market launch as living biodrugs for treating various pathologies (Bhatti and Haider 2025), provide a solid platform for their use in diabetes. In sum, these findings underscore that MSCs are not only a feasible but also a strategically preferred option for regenerative approaches to diabetes treatment.

27.7 EXPERIMENTAL STUDIES ON ANIMAL MODELS IN DMT2

MSCs have been extensively studied as a potential treatment modality in various experimental animal models (Table 27.1). These studies have utilized a range of MSC sources, refined their preparation protocols, and assessed the therapeutic impact of these cells in well-established rodent models of DMT2, as summarized in the following sections of the chapter.

27.7.1 Type of Cells Used

MSCs from various tissue sources have been utilized in experimental animal models across multiple studies to treat diabetic pathology. The primary cell types include:

- BM-MSCs: MSCs isolated from rat femurs and expanded in low-glucose DMEM were characterized for their therapeutic efficacy. These cells were selected for their multipotency and paracrine effects for mitigating diabetic complications (Wang et al., 2017; Zhao et al., 2015).
- UC-MSCs and AM-MSCs from human tissues were studied for their potent differentiation and immunomodulatory properties and ease of isolation from postpartum tissues (Wang et al. 2022; Sun et al. 2017; Xie et al. 2016).
- AD-MSCs: MSCs can be genetically modified using lentiviral particles to overexpress FGF21 and GLP1 to enhance insulin sensitivity and metabolic regulation (Xie et al. 2017; Xue et al. 2021). Rat AD-MSCs from different mouse models, including those with DMT2 and leptin receptor-deficient (db/db) mice, were used with adult stem cells from healthy mice as controls (Wang et al. 2018).
- hUC-MSCs and their derived Exosomes (MSC-EXO): hUC-MSCs were used to produce therapeutic exosomes. In one case, these exosomes were further modified with a β-cell-targeting aptamer and PEG to improve their circulation time and specificity, to enhance pancreatic β-cell function (Xia et al. 2024). Another report employed hUC-MSCs directly as treatment to corroborate their use in modulating diabetic endpoints (Song et al. 2024).
- Primary rat BM-MSCs-derived conditioned medium and mouse pancreatic α-cell lines (αTC1-6 and InR1G9) to study glucagon secretion in vitro (Song et al. 2021).
- hUC-MSCs from consenting donors were characterized for MSC-specific markers (positive for CD90, CD105, CD73; negative for CD45 and CD34, and trilineage differentiation potential (Li et al. 2021). Another study used hUC-MSC-derived exosomes instead of cells for potential tissue regeneration and metabolic regulation (Sun et al., 2018).

27.7.2 Preparation of Cells

Across various studies, standardized cell preparation protocols have been employed to ensure high-quality cell preparation, thereby enhancing cell viability and functionality.

TABLE 27.1
Therapeutic Efficacy of Mesenchymal Stem Cells in Experimental Studies

#	Title	MSC Type	Animal Model	Disease Model	Outcome	Mechanism	Reference
1	Protective role of MSCs in islet microvascular endothelium under oxidative stress.	Rat BM-MSCs	Male Sprague-Dawley rats	DMT2	The infusion of BM-MSCs significantly reduced fasting glucose levels.	MSCs activated a signaling pathway that protected MS-1 cells, with Wnt4 being protective and Wnt5a being destructive.	(Wang et al. 2017)
2	Restored High-Fat-Fed Induced Hyperinsulinemia in Rats at the Early Stage of DMT2.	BM-MSCs	Male Sprague-Dawley rats	DMT2, an early-stage induced by a high-fat diet	Significant decrease in both insulin and glucose levels.	BM-MSCs resulted in increased GLUT4 expression and enhanced p-IRS-1, while lowering inflammatory cytokines TNFα and IL-6.	(Li et al. 2021)
3	Therapeutic Effects of Amniotic Membrane and UC-MSCs for the Treatment of DMT2.	Amniotic Membrane-MSCs and UC-MSCs	db/db mice	DMT2	Lowering DMT2 symptoms—specifically hyperglycaemia and insulin resistance—improves glycolipid metabolism..	Increased Akt phosphorylation and reduction of inflammation in the liver increased expression of (G6PC), decreased expression of (FASN), and (SREBP-1).	(Wang et al. 2022)
4	MSCs modified by FGF21 and GLP-1 ameliorate lipid metabolism and lower blood glucose levels.	AD-MSCs	KS.Cg-Dock7 m+/+Leprdb/Nju mice (DMT2 mouse model)	DMT2	Significantly improved blood glucose levels and body weight in the diabetic mice.	MSCs modified to express FGF21 and GLP1 and modulation of lipid metabolism genes, SREBP1 and SREBP2.	(Xue et al. 2021)

(Continued)

TABLE 27.1 CONTINUED
Therapeutic Efficacy of Mesenchymal Stem Cells in Experimental Studies

#	Title	MSC Type	Animal Model	Disease Model	Outcome	Mechanism	Reference
5	β-Cell-Targeting Exosomes from MSCs Improve β-Cell Function by Suppressing NRF2-Mediated Ferroptosis.	MSC-EXOs	C57BL/6J mice	DMT2	Reduction of random blood glucose levels, enhancement of glucose and insulin tolerance, and increased insulin secretion.	Effective by increasing β-cell mass through suppression of NRF2-mediated ferroptosis and regulating Akt/ERK signaling to improve β-cell function.	(Xia et al. 2024)
6	MSCs ameliorate mitochondrial dysfunction in α cells and hyperglucagonemia in DMT2 by SIRT1/FoxO3a signaling.	HucMSCs	C57BL/6J mice	DMT2	HucMSC infusion improved glucose and insulin tolerance, as well as hyperglycemia and hyperglucagonemia.	Activating SIRT1/FoxO3a signaling pathways enhanced glucose uptake and utilization, improved mitochondrial function, and increased ATP production.	(Song et al. 2024)
7	BM-MSCs ameliorate chronic high glucose-induced β-cell injury through autophagy modulation.	BM-MSCs	high-fat diet/STZ rats	DMT2	lowered hyperglycemia and promoted the restoration of pancreatic β-cells in DMT2 rats.	Reduced β-cell apoptosis and increased the number of insulin granules.	(Zhao et al. 2015)
8	HucMSCs Elicit Macrophages into an Anti-Inflammatory Phenotype to Alleviate Insulin Resistance in DMT2 Rats.	HucMSCs	A high-fat diet induced Sprague-Dawley rats	DMT2	Improved glucose homeostasis and insulin sensitivity.	UC-MSCs Induced M2 Macrophages Polarization in Adipose Tissue and Regulated the Inflammation State via Secretion of IL-6.	(Xie et al. 2016)

#	Title	MSC Type	Animal Model	Disease Model	Outcome	Mechanism	Reference
9	AD-MSCs ameliorate hyperglycemia through regulating hepatic glucose metabolism in DMT2 rats.	AD-MSCs	STZ-induced DMT2 rat models	DMT2	significant decrease in blood glucose levels Glucose tolerance was significantly alleviated in the treated group. improvement of insulin sensitivity.	Regulates hepatic glucose metabolism by promoting phosphorylation of AMPK.	(Xie et al. 2017)
	HUC-MSCs ameliorate insulin resistance by suppressing NLRP3 inflammasome-mediated inflammation.	HucMSCs	HFD/STZ-induced DMT2 model	DMT2	UC-MSC infusion significantly ameliorated hyperglycemia in DMT2 rats and reduced inflammatory activity, thereby improving insulin sensitivity.	Removal of NLRP3/ IL-1β partially improved PA and LPS-induced insulin signaling impairments by infusion of UC-MSCs.	(Sun et al. 2017)
	Therapeutic Effects of AD-MSCs from Diabetic Mice for the Treatment of DMT2	AD-MSCs	db/db mice	DMT2	Improved insulin sensitivity and reduced β-cell death were observed in DMT2 mice receiving MSC therapy.	Reduced mRNA expression of TNF-α compared with T2D control pancreases. NOD2 and F4/80 were also reduced, but not significantly.	(Wang et al. 2018)
	Mesenchymal stem cells ameliorate β-cell dysfunction of human DMT2 islets by reversing β-cell dedifferentiation.	HucMSCs	db/dbmice	DMT2	Reversal of β-cell dedifferentiation and improved glycemic control.	MSC secretion of IL-1R antagonist (IL-1Ra), which acted on the inflamed islets and reversed cell differentiation.	(Wang et al. 2020a)

(Continued)

TABLE 27.1 CONTINUED
Therapeutic Efficacy of Mesenchymal Stem Cells in Experimental Studies

#	Title	MSC Type	Animal Model	Disease Model	Outcome	Mechanism	Reference
	Exendin-4 gene modification and micro-scaffold encapsulation promote self-persistence and antidiabetic activity of MSCs.	AD-MSCs	HFD mice	DMT2	Ameliorated insulin resistance Ameliorated lipid Accumulation in vivo promoted insulin sensitivity and inhibited lipid accumulation.	Improved survival under high glucose stress via autocrine activation of the GLP–1R–mediated AMPK signaling pathway.	(Zhang et al. 2021)
	MSC-conditioned medium alleviates high-fat-induced hyperglucagonemia via miR-181a-5p and its target, the PTEN/Akt signaling pathway.	BM-MSCs	C57BL/6J mice	DMT2	improved glucose and insulin tolerance, protected against induced hyperglycemia/ hyperglucagonemia.	miR-181a-5p mitigates glucagon secretion of α-cells by regulating PTEN/AKt signaling.	(Song et al. 2021)
	Reversion of early- and late-stage B-cell dedifferentiation by HUC-MSCs in DMT2 mice.	UC-MSCs	(HFD)/ streptozotocin (STZ)-induced DMT2 mice	DMT2	Improved glucose homeostasis and reversed β-cell dedifferentiation.	Expression of ChrA Preserved transcription factors like Pdx1, MafA, and ALDH1A3.	(Li et al. 2021)
	Human MSCs' Exosomes Alleviate DMT2 by Reversing Peripheral Insulin Resistance and Relieving β-Cell Destruction.	HucMSC-ex	(HFD)/ streptozotocin (STZ)-induced DMT2 mice	DMT2	Partially reversed insulin resistance in DMT2 indirectly accelerates glucose metabolism.	Reversing peripheral insulin resistance and relieving β-cell destruction. Restored the phosphorylation (tyrosine site) of the insulin receptor substrate one and protein kinase B in DMT2.	(Sun et al. 2018)

Abbreviations: AD-MSCs: adipose-derived MSCs; BM-MSCs: bone marrow-derived mesenchymal cells; MSCs: HUC-MSCs: human umbilical cord MSC mesenchymal stem cells; DMT2: type 2 diabetes mellitus; UC-MSCs: Umbilical cord-derived mesenchymal stem cells.

27.7.2.1 Isolation and Culture Conditions

- BM-MSCs from rat femur and cultured in low-glucose DMEM supplemented with FBS (Wang et al. 2017; Zhao et al. 2015).
- AM-MSCs and UC-MSCs from their respective tissues, cultured in MSC NutriStem XF Medium supplemented with knockout serum replacement, to support cell expansion and maintain stemness (Wang et al. 2022).
- AD-MSCs were isolated by enzymatic digestion and cultured in α-MEM with 10% FBS. The adherent cells were expanded in vitro (Xue et al. 2021).
- hUC-MSCs cultured in α-MEM with FBS and antibiotics until 80% confluence and functionally ready for use (Song et al. 2024; Xia et al. 2024; Li et al., 2021).

27.7.2.2 Co-culture and Conditioned Medium Approaches

- BM-MSCs isolated by flushing rat femurs were cultured in DMEM/F12 supplemented with 20% FBS and expanded until 80–90% confluency in 10% medium. The conditioned medium was collected from refined MSCs in serum-free medium for 24 hours, concentrated 20-fold by ultrafiltration, and sterilized by filtration. α-cell lines were refined in suitable media to 70–80% confluency and treated with palmitate (Pa) to induce dysfunction, with or without BM-MSC-derived conditioned medium (Song et al. 2021).
- Some studies have used the transwell co-culture system approach to assess paracrine effects without direct cell-to-cell contact on endothelial and β-cells (Wang et al., 2017).
- MSCs were cultured in CMRL-1066 medium, either alone or in islet-conditioned medium, to assess their impact on β-cell function. The cells were maintained at approximately 80% confluency (Wang et al. 2020a).
- hUC-MSC-Exo isolated from the culture supernatant of hUC-MSCs using the ultracentrifugation protocol set. The exosomes were characterized for their morphology, size analysis, and exosome-specific markers, including CD9 and CD81. Protein content was further analyzed by LC-MS/MS, confirming the enrichment of glucose metabolism-associated proteins, such as GLUT4 (Sun et al., 2018).

27.7.2.3 Genetic Modifications

While the genetic modification step is not a routine part of cell preparation protocols, as most studies prefer to use naive MSCs, a select group of studies has innovated this approach to enhance the efficacy of MSC-based interventions. For example,

- AD-MSCs were engineered to overexpress the genes FGF21 and GLP1, aiming to enhance their intrinsic metabolic regulatory functions.
- UC-MSCs were genetically modified using small interfering RNA (siRNA) to knock down interleukin-6 (IL-6) to study its role in the anti-inflammatory effects of UC-MSCs and M2 macrophage polarization (Xie et al. 2016).
- Knockdown of IL-1 receptor antagonist in MSCs using a recombinant lentivirus containing shRNAs to study the role of IL-1Ra in mediating the effects of MSCs on β-cell dedifferentiation (Wang et al. 2020a).
- MSCs were engineered to overexpress Exendin-4 to enhance their antidiabetic efficacy and survive under hyperglycemic stress by activation of the GLP-1 receptor (GLP-1R) and the AMPK signaling pathway (Zhang et al. 2021),
- MSC-derived exosomes were engineered by incorporating a β-cell-targeting aptamer with PEG, thereby improving targeting specificity and circulation time, which potentiated the therapeutic outcomes on β-cell viability and function (Xia et al. 2024; Xue et al. 2021).

27.7.2.4 Use of Cells

In the studies analyzed, the primary role of the cells was to serve as a treatment modality:

- Treatment Intervention Only: MSCs or their derivatives (e.g., exosomes) were administered as the sole therapeutic intervention via intravenous injection or direct tissue infusion to invoke regeneration, reduce inflammation, and restore insulin sensitivity (Li et al., 2021; Song et al., 2024; Wang et al., 2017, 2022; Xia et al., 2024; Xue et al., 2021).
- In addition to using UC-MSCs as treatment, cells were cocultured with insulin-resistant HepG2 cells to investigate their anti-inflammatory effects. This allowed for the examination of the paracrine mechanisms through which UC-MSCs could alleviate insulin resistance (Sun et al. 2017).
- MSC-derived exosomes were used as the primary treatment intervention in DMT2 rats by intravenous infusions via the tail vein. Control groups included DMT2 mice treated with PBS and a group treated with hUC-MSCs for comparison (Sun et al. 2017).

27.7.2.5 Experimental Animal Models Used

Rodent models are the most commonly used and indispensable for studying DMT2:

27.7.2.5.1 Experimental Rat Model

The BM-MSC study employed a high-fat-fed Sprague-Dawley rat model to simulate early-stage DMT2, with the animals exhibiting diet-induced insulin resistance, closely mimicking the metabolic syndrome conditions observed in humans (Wang et al. 2017).

27.7.2.5.2 Experimental Mouse Model

- The db/db mouse model, which is genetically predisposed to obesity and severe DMT2-like symptoms, was used to study the therapeutic potential of AM-MSCs, UC-MSCs, and AD-MSCs (Wang et al. 2018, 2022).
- A model of a high-fat diet combined with STZ in C57BL/6J mice to replicate both insulin resistance and β-cell dysfunction (Song et al., 2024; Wang et al., 2018, 2020; Xia et al., 2024; Xie et al., 2016; Zhao et al., 2015).
- BKS.Cg-Dock7^m+/+Lepr^db/Nju mice model to substantiate the impact of genetically modified AD-MSCs on diabetic conditions (Xue et al. 2021).
- Male C57BL/6J mice models fed a fat-rich diet (45% fat) for seven months to cause metabolic dysfunction of DMT2, including hyperglycemia and hypertension (Xue et al. 2021).
- Male C57BL/6J (8 weeks old) was used for the HFD/STZ model, and C57BL/6 DB/DB male mice (8 weeks old) and wild littermates (DB/+) have been used for genetic models (Li et al. 2021).

27.7.2.6 Summary Results and Significance of Treatment

Collectively, the studies demonstrated notable therapeutic benefits from MSC-based interventions as summarized below:

27.7.2.6.1 Metabolic and Functional Improvement

- **Glucose and Insulin Regulation:** Treatment with MSCs, whether delivered in their cellular form or via paracrine secretions, consistently resulted in significant reductions in blood glucose levels and improvements in insulin sensitivity, associated with enhanced insulin signaling (Song et al., 2024; Wang et al., 2017, 2018).
- **Restoration of Lipid Metabolism and Organ Function**: In addition to improved glycemic control, there were broad metabolic benefits in terms of restoring disrupted lipid profiles and liver function (Wang et al. 2017; Xue et al. 2021).

- **Enhanced autophagic activity in pancreatic beta cells:** This led to reduced apoptosis and improved mitochondrial function in INS-1 cells exposed to high glucose levels (Zhao et al., 2015).

27.7.2.6.2 Anti-inflammatory Effects

MSCs promoted the polarization of macrophages into the M2 phenotype, which is associated with anti-inflammatory effects. This was associated with increased IL-6 expression.

27.7.2.6.3 Suppressing Inflammasome-Mediated Inflammation

These molecular-level changes significantly alleviated insulin resistance in DMT2 rats by suppressing NLRP3 inflammasome-mediated inflammation. Treatment also reduced inflammatory cytokines and improved insulin signaling (Sun et al. 2017).

27.7.2.6.4 Reduced Fat Deposition

Decreased fat deposition in adipose tissue and liver, and increased β-cells over long-term follow-up, besides improved insulin sensitivity and reduced inflammation (Wang et al. 2018). Repeated treatment with conditioned medium (200 μL, 10 mg/mL, every 3 days for 10 cycles) significantly improved glucose tolerance and insulin sensitivity in diabetes induced by a high-fat diet. Treatment inhibited glucagon secretion from α cells (Song et al. 2021).

27.7.2.6.5 Reversal of Cellular Dysfunction

Marked by the restoration of insulin expression and reduction of the precursor (NGN3+) signs in pancreatic islets. The reverse effect is more pronounced at the early stage of β-cell dedifferentiation than at the later stage (Song et al., 2021).

27.7.3 Mechanistic Insights

29.7.3.1 Signal Transduction Pathways

MSC-based therapy activates the Wnt/β-catenin signaling cascade, especially Wnt4, with a differential influence compared to the detrimental effects of Wnt5a (Wang et al., 2017).

27.7.3.2 Protection against Oxidative Stress and Inflammation

Reduced oxidative stress-induced apoptosis and low expression of adhesion molecules, such as VCAM-1, MSC therapy promoted endothelial cell survival, thereby restoring vascular integrity, besides a reduction in inflammation (Song et al., 2024; Wang et al., 2017; Xia et al., 2024).

27.7.3 Enhanced β-Cell Function

Improved pancreatic islet morphology and function were observed after MSC-EXOs treatment, which led to enhanced β-cell viability and reduced ferroptosis (Xia et al. 2024).

- hUC-MSC-EXO treatment significantly improved blood sugar in DMT2 mice by reversing peripheral insulin resistance, restoring insulin receptor substrate-1 (IRS-1) and Akt phosphorylation, which promoted GLUT4 expression in muscles and increased hepatic glycogen storage (Sun et al. 2018).
- Overexpression of MIR-181A-5P in α cells has reduced the secretion of Glucagon, indicating that MSC-derived miRNAs mediate therapeutic effects through the PTEN/Akt signaling pathway (Song et al. 2021).

The integration of experimental animal data underscores the robust potential of MSC-based therapies. Regardless of tissue source, MSCs and their secretomes offer significant therapeutic benefits by modulating metabolic dysfunction, reducing systemic inflammation, and enhancing β-cell survival and function. Although some studies have explored a combinatorial approach of preconditioning and genetic modulation of MSCs before use, optimization of the protocols for clinical use is warranted. Continued research using refined and optimized protocols is necessary to translate these preclinical data into clinical applications (Song et al., 2024; Wang et al., 2017, 2022; Xia et al., 2024; Xie et al., 2017; Xue et al., 2021; Zhao et al., 2015).

27.8 MSCS FOR DM TREATMENT FROM A CLINICAL PERSPECTIVE

Our literature search was conducted using available clinical trials from various databases, including clinicaltrials.gov, PubMed, and the National Library of Medicine, using the following keywords: type 2 diabetes mellitus, mesenchymal stem cells, randomized controlled trials, and autologous. We determined the efficacy and safety of MSC therapy for DMT2, including the methods used to prepare stem cell samples, with or without genetic modifications, and the preconditioning protocols.

UC-MSCs, AD-MSCs, and autologous BM-MSCs remain the most commonly used cells due to their availability and fewer ethical considerations and restrictions. Other MSC types, such as placenta-derived MSCs (Jiang et al., 2011), have also been used, but not very commonly. A key conclusion from the clinical trial data is that the MSC-based treatment may be a promising solution for patients with DMT2. For example, astudy comparing a UC-MSC-based intervention to a placebo group showed that 20% of patients in the intervention group achieved <7.0% HbA1c, and 13.5% of the treated patients became insulin-free during the follow-up period (Zang et al., 2022). Another study, using Wharton's jelly-derived MSCs (WJ-MSCs) at Peking University Shenzhen Hospital, China, reported that one patient in the treatment group permanently discontinued hypoglycemic agents (Lian et al., 2022). This study offered hope to DMT2 patients, enabling them to improve their quality of life and potentially reduce or eliminate the need for hypoglycemic agents or insulin treatment.

In a single-blinded, randomized study involving 21 patients on oral anti-hyperglycemic medication, Bhansali et al. (2014) reported a significant increase in C-peptide release, a serum marker of insulin secretion and pancreatic function, in the autologous BM-MSC-treated group (Bhansali et al. 2014). The intervention group (n = 11 patients) received two doses of BM-MSCs and were followed up for 12 months. Serum C-peptide doubled from baseline during the follow-up period. Although the data were promising, they need to be substantiated in a larger patient population. Another treatment perspective involves the immune-mediated destructive effects of DM on β-cells. Lian et al. (2022) demonstrated that Homeostatic Model Assessment for β-cell Function (HOMA-β) serum levels improved by day 28 ± 3 and remained stable throughout the entire study. The authors also measured HOMA-IR, but it was not statistically significant during the exact observation period. Hu et al. (2016) at the Hospital of Qingdao University, China, demonstrated marked increases in HOMA-β serum levels by the 3rd month of the trial, which persisted throughout the remainder of the follow-up. Contrary to the previous study, HOMA-IR did not show any significant change.

Regarding the complications and adverse effects of cell-based treatment, Lian et al. (2022) in China measured serum tumor markers (AFP, CEA, and CA-199), liver and renal function, blood lipid indices, and immunoglobulin levels. The data showed an increase in D-dimer levels throughout the first three weeks of therapy, which returned to baseline thereafter and remained stable until the end of follow-up. The same group later reported an association with an increased incidence of microvascular umbilication, but with negative serum tumor markers (Lian et al., 2023). Zang et al. employed a Continuous Glucose Monitoring approach to measure the current glucose level, as well as the direction and rate of change in glucose levels (Zang et al., 2023). By the 48th week, there was a substantial increase in all parameters compared with baseline, except TBR (time below range). There is also a risk of an immune reaction to MSC intervention or the development of autoantibodies, such as insulin autoantibodies (IAA), during the trial. Still, very few studies have reported this aspect (Zang et al., 2022). Bhansali et al. reported 19 hypoglycemic episodes, most of which occurred within the first 3 months of treatment. In contrast, most other studies reported only one case of hypoglycemia after the initiation of cell therapy (Bhansali et al., 2014). These findings, in addition to indicating the treatment's efficacy, also raise a note of caution about the need for continuous, immediate patient follow-up.

Most clinical trials have used naïve MSCs, without preconditioning or genetic modification. However, given the robust nature of MSCs to withstand the rigors of genetic modification protocols and the effectiveness of genetically modified MSCs, clinical trials are being designed using this combinatorial approach. A recent study at Shanghai Hospital, China (NCT06727721), has utilized OCN19-overexpressing hUC-MSCs to assess their safety and efficacy. The encouraging safety data from this study offer a promising opportunity to design future clinical trials that enhance treatment outcomes, as the combined approach leverages the best aspects of both gene and cell-based therapies.

We sought the optimal method of sample preparation and the cultured media to enhance treatment outcomes. The most commonly used medium is MSC-GM medium, provided by MSC-GM BulletKit (Germany) and MesenCult from Stem Cells Technologies (Germany) as a growth promoter. Another medium used was serum-free Dulbecco's modified Eagle's medium (DMEM). Table 27.2 summarizes the clinical studies that used MCS therapy in DMT2 patients.

27.8.1 MSC-Derived Paracrine Secretions as Cell-Free Therapy

By the paracrine hypothesis (Haider and Aziz, 2017), MSCs exert their therapeutic effects through their soluble (secretome) and insoluble (exosomes) paracrine secretions, which comprise bioactive molecules (Xia et al., 2021). The secretome typically contains a wide range of bioactive molecules (Habib et al., 2025). This paracrine mechanism allows MSCs to mediate systemic therapeutic effects without requiring engraftment into target tissues. It may involve an intrinsic pool of stem/progenitor cells participating in the repair process. MSC-derived exosomes are endosome-derived nanovesicles that carry miRNAs, mRNAs, proteins, and lipids, delivering them to recipient cells to support intercellular communication (Mushtaq et al., 2024). They provide a natural platform for delivering their payload, which can also be manipulated/engineered to achieve the desired bioactive molecular profile required for a specific therapeutic outcome (Liang et al., 2025). For instance, BM-MSC-derived exosomes enriched with miR-146a reversed diabetic β-cell dedifferentiation and improved β-cell insulin secretion in both in vivo and in vitro experimental models of the apoptotic signaling pathway (Chen et al., 2020). Mechanistically, they reported that exosome treatment successfully suppressed p38/mitogen-activated protein kinase DMT2 in rats and in high-glucose-treated primary islets via the NUMB/b-catenin signaling pathway (He et al., 2021). Similarly, Chen et al. have reported that hUC-MSC-derived exosomes protected β-cells cultured under hypoxic conditions by suppressing the pro-(p38/MAPK) pathway in β-cells, which was facilitated by miRNA-21 and let-7g.

These data showed that therapeutic signals expressed by these EVs exhibited a substantial overlap with those produced by MSCs. It is now being considered that the therapeutic benefits of MSC-based therapy are primarily due to paracrine mechanisms (Park et al., 2012). Secreted growth factors in the soluble secretome, including IGF-1, VEGF, NGF, and HGF, promoted angiogenesis, cytoprotection, cell proliferation, and survival. These factors also contributed to myogenesis in ischemic pancreatic tissues and helped to maintain residual β-cell function by modulating apoptotic signaling. Additionally, they served as potential chemical cues for the intrinsic pool of stem/progenitor cells to home in on and sustain their presence at the injury site, thereby participating in the repair process. Cell-free therapy approach using MSC-derived conditioned medium or exosome isolates may offer an alternative treatment strategy with several advantages over cell-based therapy, including reduced risks of immunogenicity, tumorigenesis, and vascular occlusion. Preclinical studies and

TABLE 27.2
Therapeutic Efficacy of Mesenchymal Stem Cells in Clinical Studies

MSC source	Route of Administration /Sample Preparation	Genetic Modulation	Results	Reference
BM-MSC	Intravenously through the pancreatic artery	None	Not completed	NCT01142050
PD-MSCs	Placenta tissues were treated with collagenase and trypsin, and the cells were cultured in D-MEM/F-12 medium supplemented with Epidermal growth factor.	None	By the 6th month, a significant reduction in the required insulin dose and a reduction in the glycosylated hemoglobin values.	Jiang et al., 2011
Human Menstrual Blood-MSCs	Unknown	Unknown	Unknown	NCT01496339
Allogeneic hMSCs	Intravenous infusion	None	EPC Colony Forming Units (EPC-CFU)/a significant increase from the baseline, which was drastically reduced by the end. Negative CRP, Vascular Endothelial Growth Factor (VEGF), Stem Cell Factor (SCF), and Stromal Cell-Derived Factor 1 (SDF-1). Slight increase in Flow-mediated diameter% No change in TNF from the baseline One case developed an infection, and one case experienced side effects, including benign, malignant, and unspecified neoplasms.	NCT02886884
BM-MSCs	Token from the anterior iliac crest, BM-MSCC were separated using density gradient, cultured in Xeno-free and serum-free medium, then expanded.	None	After 12 months, 2 of 4 patients reported reducing their insulin dose by an average of 6 IU/day. Five out of 10 patients reduced their oral medication by an average of 868 mg. One out of three reduced his drug intake but increased his insulin intake.	(Nguyen et al., 2021)
UC-MSCs	A combination of stem cell therapy and the liraglutide (LIRA) drug was used; MSCs were administered via the pancreatic artery.	None	Group B patients (MSC+LIRA) exhibited an increase in ΔCP30/ΔG30 and AUCCP180, accompanied by a significant decrease in HOMA-IR, FPG, 2hPG, and HbA1c levels.	(Chen et al., 2016)
WJ-MSCs	Serum-free NutriStem® MSC XF medium was used for cell culture, and the sample was then collected and expanded in a laminar-flow laboratory.	None	Significant decrease in the levels of both postprandial plasma glucose and HbA1c at the 6th month, with a marked increase in fasting C-peptide and CPGR during the follow-up. HOMA-B showed a rise, whereas HOMA-IR showed a reduction in the insulin dose.	Hu et al., 2016

(Continued)

TABLE 27.2 CONTINUED
Therapeutic Efficacy of Mesenchymal Stem Cells in Clinical Studies

MSC source	Route of Administration /Sample Preparation	Genetic Modulation	Results	Reference
UC-MSCs	Unknown	OCN19	Recruiting	NCT06727721
Autologous AD-MSCs	The sample will be gained through a tumescent liposuction procedure. Delivered intravenously in a dosage of 1 × 10^6 MSCs/kg body weight.	None	Enrollment	NCT06789302
UC-MSCs	Collected from an unborn full-term placenta, cultured using MSCGM medium and MesenCult as growth-promotion medium, plates were coated with FCS, and samples were stored at 4°C until further use.	None	20% of the patients in the UC-MSCs group achieved the goal of HbA1c <7.0% and daily insulin reduction ≥50%	(Zang et al., 2022)
UC-MSCs	Collected from an unborn full-term delivery placenta, isolated using the Density gradient centrifugation method, cultured using MSCGM medium and MesenCult.	None	Improvement in all CGM parameters* at week 48, except TBR, insulin requirement, and HbA1c, was lower in the intervention group. The valid group** had lower HbA1c at 48 weeks.	(Zang et al., 2023)
WJ-MSCs	Cultured on DMEM media, then stored in −196°C DMSO cryopreservation solution for later use.	None	Significant reduction in lymphocytes and an increase in neutrophil ratio in 2–3 weeks, which returns to normal during follow-up. Increase both D-dimer and immunoglobulin levels in the intervention group, with negative AFP, CEA, and CA199 tumor markers.	(Lian et al., 2023)
WJ-MSCs	Serum-free NutriStem® MSC XF medium was used for cell culture, and the sample was then collected and expanded in a laminar-flow laboratory.	None	Significant decreases in postprandial plasma glucose and HbA1c at the 6th month, accompanied by a marked increase in fasting C-peptide and CPGR throughout the entire follow-up period. HOMA-B showed a marked increase, whereas HOMA-IR showed a significant reduction in the required insulin dose.	(Hu et al., 2016)
BM-MSCs	The cells were aspirated from PSIS stem cells, separated by ultracentrifugation, and then layered onto a density gradient medium (Ficoll–Hypaque).	None	7 of 10 patients achieved the study goal of reducing insulin requirements by more than 50%. Two patients were able to stop insulin entirely on days 7 and 41. C-peptide levels increase significantly in the responder group.	(Bhansali et al., 2009)

(*Continued*)

TABLE 27.2 CONTINUED
Therapeutic Efficacy of Mesenchymal Stem Cells in Clinical Studies

MSC source	Route of Administration /Sample Preparation	Genetic Modulation	Results	Reference
BM-MSCs	Aspirated from PSIS and MNC isolated using centrifugation and layered over density-gradient medium (Ficoll-Hypaque)	None	Six out of seven patients achieved the target of reducing their daily insulin requirement by 51% at the 6-month mark. Significant increase in FPG and the AUC of 2nd-phase C-peptide by the end of the study. There was an eight-episode reduction per patient per month at the 3rd month, which continued to decrease significantly by the 6th month.	(Bhansali et al., 2017)
ABM-MSCs	The sample was collected from PSIS and cultured on density-gradient medium.	None	Reduction in insulin dose by ≥50% after 3 months, significant increase in glucagon-stimulated C-peptide by the 6th month.	Bhansali et al., 2009
ABM-MSCs	Collected from the PSIS bone, the MNCs were suspended in α-MEM and supplemented with 10% pooled human platelet lysate until they reached 80% confluency, after which they were expanded in T-225 flasks.	None	A > 50% reduction in insulin requirement was observed in the 9th week; however, the decrease in HbA1c was not significant.	Bhansali et al., 2017b
ABM-MSCs	Collected from the PSIS bone and cultured on a density gradient medium.	None	A Significant reduction in insulin dose correlates with an increase in C-peptide level, but there is an insignificant increase in HbA1c at 12 months.	Bhansali et al., 2014
WJ-MSCs	Cultured on DMEM media, then stored in −196°C DMSO cryopreservation solution for later use.	None	Significant decrease in FPG level at day 14 and HbA1c at day 84. improvement in the HOMA-β index on day 28.	Lian et al., 2022

Abbreviations: UC-MSCs: umbilical cord mesenchymal stem cells; FCS: fetal calf serum; MSCGM: mesenchymal stem cell growth medium; ABM-MSCs: autologous bone marrow mesenchymal stem cells; PSIS: posterior superior iliac spine; WJ-MSC: Wharton jelly MSCs; DMEM: serum-free Dulbecco's modified Eagle's medium; DMSO: dimethyl sulfoxide; FPG: fasting plasma glucose; HOMA-β: homeostasis model assessment of beta cell function; AFP: alpha fetoprotein; CEA: cancer embryonic antigen; CA199: cancer antigen 199; MSCs XF: mesenchymal cell xeno-free; CPGR: C peptide glucose ratio; HOMA-ir: homeostasis model assessment of insulin resistance; MNCs: mononuclear cells; PD-MSCs: placenta derived mesenchymal stem cells; PD-MSCs: placenta-derived mesenchymal stem cells; 2hPG: postload glucose; FPG: fasting plasma glucose; AUCCP180: amount of C-peptide secretion function; HbA1c: hemoglobin A1c.

*CGM parameters (glucose coefficient of variation (CV, mean amplitude of glycemic excursions (MAGE, mean glucose (MG, standard deviation (SD) of blood glucose values, time in range 70–179 mg/dL (TIR), time above range (TAR), time below range (TBR)).

**The valid group (those patients from the intervention group who achieved the target of time in range ≥ 70% and HbA1c < 7%).

early-phase clinical trials have demonstrated potential efficacy in improving glycemic control and insulin sensitivity (Table 27.3).

27.8.2 Extracellular Vesicles Therapy in Diabetes Mellitus

As a part of the cell-free therapy approach, EVs are fast-emerging as transformative therapeutic agents for DM, addressing the long-standing unresolved issues of β-cell dysfunction, insulin resistance, and related diabetic complications. EVs are key components of MSC-derived paracrine secretions and exhibit regenerative, anti-inflammatory, and immunomodulatory properties, addressing the core pathophysiological issues in both DMT1 and DMT2 (Soltani et al. 2022a & 2022b; Habib et al., 2026). This cell-free approach circumvents the limitations associated with contemporary pharmacological and insulin-based treatments, positioning EVs as a next-generation biopharmaceutical for diabetes management. Also, given their cell-specific payload profile, they are being explored for the diagnosis of DM and its complications (Sun et al., 2021).

27.8.3 Current Therapeutic Limitations and Advantages of EV-Based Therapy

Contemporary DM management and treatment emphasize glycemic control in patients, but are unable to halt β-cell loss or reverse insulin resistance (Soltani et al., 2022a and 2022b). In DMT1, lifelong insulin dependence persists, continually contributing to autoimmune β-cell destruction. In DMT2, treatments are insufficient to combat chronic inflammation, which drives disease progression (Hu et al., 2020; Liu et al., 2021). MSC transplantation, despite its promise, faces challenges such as poor cell survival, low rate of differentiation, and risks of tumorigenicity (Soltani et al., 2022a). EVs overcome these limitations through their inherent stability, low immunogenicity, and targeted delivery of bioactive molecular payloads, such as miRNAs and proteins, to pancreatic β-cells and peripheral tissues (Hu et al., 2020). Moreover, they can also be primed and engineered to modify their payload profile for targeted delivery to the tissue of interest (Shi et al., 2017; Li et al., 2017; Liang et al., 2025).

27.8.4 Mechanistic Insights into EV Therapeutics

In diabetes, MSC-derived EVs have been shown to modulate key pathways involved in metabolic balance, pancreatic function, and immune regulation through multifactorial mechanisms. One reported mechanism is that MSC-derived EVs upregulate pancreatic duodenal homeobox 1 (PDX1) and TGF-β (both are key to β-cell development and survival), enhance β-cell proliferation, and promote insulin secretion (Soltani et al. 2022a). In addition, adipose MSC-derived EVs promote macrophage polarization toward an anti-inflammatory (M2-like) phenotype, reducing systemic inflammation in models of DMT2 (Liu et al., 2021). In DMT1, EVs have demonstrated immunomodulatory effects by inhibiting T-cell activation and promoting T-cell (Treg) populations (Hu et al., 2020; Soltani et al., 2022a). In DMT1 models, EV-based therapy has been associated with reducing autoimmune destruction and stimulating β-cell regeneration. For example, MSC-derived exosomes increased insulin-positive β-cell numbers by 40% in streptozotocin (STZ)-induced diabetic mice compared with parental MSCs, thereby promoting β-cell regeneration (Soltani et al., 2022a). Furthermore, BM-MSC-derived EVs deliver miR-106b-5p/miR-222-3p, which silence cell cycle inhibitors (p21Cip1, p27Kip1) to boost β-cell replication. In DMT2, EV-based treatment improved insulin sensitivity in the liver, muscle, and adipose tissue. Adipocyte-derived EVs were also reported to reduce hepatic gluconeogenesis by approximately 30% using the AMP-activated protein kinase (AMPK) activation. On the same note, endothelial cell-derived EVs enhance glucose uptake in skeletal muscle. Furthermore, EVs have also shown protective effects in diabetic nephropathy (Man et al., 2025).

TABLE 27.3
Role of Exosomes and Secretomes in the Treatment of DMT2 in Experimental Animal Studies.

Title	Therapeutic Agent	Animal Model	Disease Model	Outcome	Mechanism	Reference
Exosomes From Adipose-Derived Stem Cells Attenuate Adipose Inflammation and Obesity Through Polarizing M2 Macrophages and Beiging in White Adipose Tissue.	ADSC-derived exosomes	C57BL/6 male mice	Diet-induced obesity (DMT2)	Improved insulin sensitivity (27.8%), reduced obesity, alleviated hepatic steatosis, reduced WAT inflammation, promoted WAT beiging.	Exosomes transferred active STAT3 to macrophages → induced M2 polarization → increased catecholamine release → promoted beiging.	Zhao et al. Diabetes 2018;67:235–247
Human Mesenchymal Stem Cell-Derived Exosomes Alleviate Type 2 Diabetes Mellitus by Reversing Peripheral Insulin Resistance and Relieving β-Cell Destruction.	hUCMSC-derived exosomes	HFD+STZ-induced rats	Type 2 diabetes mellitus	Reversed insulin resistance, reduced β-cell apoptosis, restored insulin secretion, and improved glucose homeostasis.	Restored phosphorylation of Akt/IRS-1 → increased GLUT4 translocation → enhanced glucose uptake.	Sun et al. ACS Nano 2018;12:7613–28
Exosomes secreted by human urine-derived stem cells could prevent kidney complications in rats with type 1 diabetes.	USCs-derived exosomes (USCs-Exo)	Sprague-Dawley rats	Diabetic nephropathy	Reduced urine volume, urinary microalbumin excretion; prevented podocyte/tubular apoptosis; suppressed caspase-3; increased endothelial proliferation.	Inhibited podocyte apoptosis via BMP-7; promoted angiogenesis via VEGF/TGF-β1/ angiogenin; contained regenerative factors.	Jiang et al. Stem Cell Res Ther (2016) 7:24

Title	Therapeutic Agent	Animal Model	Disease Model	Outcome	Mechanism	Reference
Secretome from Human Mesenchymal Stem Cells-Derived Endothelial Cells Promotes Wound Healing in a Type-2 Diabetes Mouse Model.	hMSC-EC Secretome + protein cocktail (Ang-1/Ang-2/FGF-7/MMP-9/VEGF-C)	C57Bl/6J mice	Impaired wound healing	Accelerated wound closure; enhanced re-epithelialization and angiogenesis in hyperglycemic conditions.	Promoted endothelial proliferation; enriched angiogenic/regenerative factors; matrix remodelling via MMP-9.	Ormazabal et al. Int J Mol Sci (2022) 23:941
MicroRNA-146a Mimics Reduce the Peripheral Neuropathy in Type 2 Diabetic Mice.	microRNA-146a (miR-146a)	Age-matched heterozygote mice (db/m)	diabetic peripheral neuropathy	miR-146a mimics greatly suppressed the expression of many proinflammatory genes and downstream related cytokines.	Ability to suppress hyperglycaemia-induced proinflammatory gene expression and improve neurovascular function.	Liu et al. Diabetes 2017;66:3111–3121
Exosomes Derived From Schwann Cells Ameliorate Peripheral Neuropathy in Type 2 Diabetic Mice.	Schwann cell–derived exosomes	type 2 diabetic db/db	diabetic peripheral neuropathy (DPN)	SC-Exos to diabetic db/db mice significantly ameliorated DPN.	C-Exos promoted neurite outgrowth of diabetic DRG neurons and migration of Schwann cells challenged by high glucose.	Wang et al. Diabetes 2020b;69:749–759
Human Mesenchymal Stem Cell-Derived Exosomes Alleviate Type 2 Diabetes Mellitus by Reversing Peripheral Insulin Resistance and Relieving β-Cell Destruction.	HucMSC-ex	(HFD)/streptozotocin (STZ)-induced DMT2 mice	DMT2	Partially reversed insulin resistance in DMT2 indirectly accelerates glucose metabolism.	Reversing peripheral insulin resistance and relieving β-cell destruction. Restored the phosphorylation (tyrosine site) of the insulin receptor substrate one and protein kinase B in DMT2.	Sun et al. ACS Nano 2018;12:7613–7624

(*Continued*)

TABLE 27.3 CONTINUED

Role of Exosomes and Secretomes in the Treatment of DMT2 in Experimental Animal Studies.

Title	Therapeutic Agent	Animal Model	Disease Model	Outcome	Mechanism	Reference
Mesenchymal stem cell-derived exosomal miR-146a reverses diabetic β-cell dedifferentiation.	Bone marrow MSCs (bmMSCs)	HFD + STZ-induced DMT2	DMT2	Reversed β-cell dedifferentiation, improved insulin secretion, reduced hyperglycaemia, restored β-cell mass, enhanced glucose tolerance, and insulin sensitivity.	Exosomal miR-146a targets NUMB* → activates β-catenin** signalling → inhibits β-cell dedifferentiation.	He et al. Stem Cell Res Ther (2021) 12:449
Human Umbilical Cord Mesenchymal Stem Cell-Derived Small Extracellular Vesicles Ameliorated Insulin Resistance in DMT2 Rats.	Umbilical cord MSCs (hUC-MSCs)	HFD + nicotinamide +STZ-induced DMT2	DMT2	Improved glucose tolerance, reduced HbA1c, lowered HOMA-IR, enhanced QUICKI, restored pancreatic islet/kidney/ liver architecture, improved blood parameters.	Promotes glucose uptake via GLUT1-4 transporters; organ-protective effects.	Yap et al. Pharmaceutics (2022) 14:649

Abbreviation: DMT2: type 2 diabetes mellitus; DMT1: type 1 diabetes mellitus; STZ: streptozotocin; HFD: high-fat diet; ADSC: adipose-derived stem cell; hUCMSC: human umbilical cord mesenchymal stem cell; USC: urine-derived stem cell; hMSC: human mesenchymal stem cell; EC: endothelial cell; DPN: diabetic peripheral neuropathy; DRG: dorsal root ganglion; STAT3: signal transducer and activator of transcription 3; IRS-1: insulin receptor substrate-1; Akt: protein kinase B (also known as PKB); GLUT4: glucose transporter type 4; BMP-7: bone morphogenetic protein 7; VEGF: vascular endothelial growth factor; TGF-β1: transforming growth factor beta 1; Ang-1: angiopoietin-1; Ang-2: angiopoietin-2; FGF-7: fibroblast growth factor 7; MMP-9: matrix metalloproteinase 9; miR-146a: microRNA-146a; HOMA-IR: homeostatic model assessment of insulin resistance; QUICKI: quantitative insulin sensitivity check index; TC: total cholesterol; TG: triglycerides.

* NUMB is a signalling adapter protein involved in β-cell differentiation (targeted by miR-146a).

** β-catenin is a key signalling molecule in the Wnt signalling pathway.

Despite encouraging data, the field is still having challenges that need to be addressed to translate into clinical use. For example, selecting the optimal EV source remains an active area of research. Similarly, methodologies need to be optimized to enhance their functionality via molecular engineering—for example, by loading EVs with antidiabetic miRNAs (e.g., miR-146a to suppress inflammation) or glucose-responsive insulin genes—to mitigate diabetes-related complications, accelerate wound healing, and reduce myocardial fibrosis in diabetic cardiomyopathy. Clinical trials are investigating the safety and efficacy of MSC-derived EVs for preserving β-cell mass in patients with recent-onset DMT1. Current findings indicate improved C-peptide levels; however, long-term outcomes remain under investigation. Furthermore, long-term security profiles and large-scale production protocols require further approval and validation. Additionally, combination strategies combining EVs with immunomodulatory drugs or GLP-1 agonists may yield additive or synergistic effects.

27.9 CONCLUSION

EV therapy represents a significant shift in diabetes treatment, focusing on the disease etiology rather than its side effects. Although preclinical data support the role of EVs in β-cell regeneration, immune modulation, and metabolic improvement, significant challenges in manufacturing and delivery need to be addressed in future research. As research advances, EV-based interventions may redefine care guidelines, offering durable, regulatable glycemic control with low complication burden and safety risks.

REFERENCES

Ahmed OTF, Ahmed ZT, Dairi AW, Zain Al-Abeden MS, Alkahlot MH, Alkahlot RH, Al Jowf GI, et al. The inconclusive superiority debate of allogeneic versus autologous MSCs in treating patients with HFrEF: A systematic review and meta-analysis of RCTs. *Stem Cell Res Ther.* 2025;16(1):175. https://doi.org/10.1186/s13287-025-04209-5.

Ahmed RPH, Ashraf M, Buccini S, Shujia J, Haider KH. Cardiac tumorigenic potential of induced pluripotent stem cells in immunocompetent host: A note of caution. *Regenerative Med.* 2011;6:171–8.

Álvarez-Viejo M, Haider KH. Mesenchymal stem cells. In: Haider, K.H. (eds) *Handbook of Stem Cell Therapy.* Singapore: Springer, 2022. https://doi.org/10.1007/978-981-16-6016-0_6-1

Andreadi A, Muscoli S, Tajmir R, Meloni M, Muscoli C, Ilari S, Mollace V, et al. Recent pharmacological options in type 2 diabetes and synergic mechanism in cardiovascular disease. *Int J Mol Sci.* 2023 Jan 13;24(2):1646. https://doi.org/10.3390/ijms24021646.

Arzouni AA, Vargas-Seymour A, Nardi N, JF King A, Jones PM. Using mesenchymal stromal cells in islet transplantation. *Stem Cells Transl Med.* 2018;7(8):559–63. https://doi.org/10.1002/sctm.18-0033.

Azizi Z, Abbaszadeh R, Sahebnasagh R, Norouzy A, Motevaseli E, Maedler K. Bone marrow mesenchymal stromal cells for diabetes therapy: Touch, fuse, and fix? *Stem Cell Res Ther.* 2022;13:348. https://doi.org/10.1186/s13287-022-03028-2

Bhansali A, Asokumar P, Walia R, Bhansali S, Gupta V, Jain A, Sachdeva N, et al. Efficacy and safety of autologous bone marrow-derived stem cell transplantation in patients with type 2 diabetes mellitus: A randomized placebo-controlled study. *Cell Transplant.* 2014;23:1075–85. https://doi.org/10.3727/096368913X665576.

Bhansali A, Upreti V, Khandelwal N, Marwaha N, Gupta V, Sachdeva N, Sharma RR, et al. Efficacy of autologous bone marrow-derived stem cell transplantation in patients with type 2 diabetes mellitus. *Stem Cells Develop.* 2009;18:1407–15. https://doi.org/10.1089/scd.2009.0164.

Bhansali S, Dutta P, Yadav MK, Jain A, Mudaliar S, Hawkins M, Kurpad AV, et al. Efficacy of autologous bone marrow-derived mesenchymal stem cell and mononuclear cell transplantation in type 2 diabetes mellitus: A randomized, placebo-controlled comparative study. *Stem Cells Develop.* 2017;26:471–81. https://doi.org/10.1089/scd.2016.0275.

Bhansali S, Dutta P, Yadav MK, Jain A, Mudaliar S, Hawkins M, Kurpad AV, et al. Autologous bone marrow-derived mononuclear cells transplantation in type 2 diabetes mellitus: Effect on β-cell function and insulin sensitivity NCT01759823 NCT. *Diabetol Metab Syndrome.* 2017b;9(1). https://doi.org/10.1186/s13098-017-0248-7.

Bhatti MA, Haider KH. Mesenchymal stem cells as living bio-drugs: Adding a novel dimension to the cardiology and cardiovascular sciences. *J Cardiol Cardiovasc Sci.* 2025;6(1):30–5.

Bonilla FA, Oettgen HC. Adaptive immunity. *J Allergy Clin Immunol.* 2010;125. https://doi.org/10.1016/j.jaci.2009.09.017.

Boughton CK, Munro N, Whyte M. Targeting beta-cell preservation in the management of type 2 diabetes. *Br J Diabetes* 2017;17(4):134–44. https://doi.org/10.15277 /bjd.2017.148

Chen J, Chen J, Cheng Y, Fu Y, Zhao H, Tang M, et al. Mesenchymal stem cell-derived exosomes protect beta cells against hypoxia-induced apoptosis via miR-21 by alleviating ER stress and inhibiting p38 MAPK phosphorylation. *Stem Cell Res Ther.* 2020;11:97.

Chen LB, Jiang XB, Yang L. Differentiation of rat marrow mesenchymal stem cells into pancreatic islet beta-cells. *World J Gastroenterol.* 2004 Oct 15;10(20):3016–20. https://doi.org/10.3748/wjg.v10.i20.3016.

Chen P, Huang Q, Xu XJ, Shao ZL, Huang LH, Yang XZ, Guo W, et al. The effect of liraglutide in combination with human umbilical cord mesenchymal stem cells treatment on glucose metabolism and β-cell function in type 2 diabetes mellitus. *Zhonghua Nei Ke Za Zhi.* 2016;55:349–54. https://doi.org/10.3760/cma.j.issn.0578-1426.2016.05.004.

Cotran RS, Kumar V, Fausto N. *Robbins & Cotran pathologic basis of disease.* Philadelphia: Elsevier, 2021.

Dludla PV, Mabhida SE, Ziqubu K, Nkambule BB, Mazibuko-Mbeje SE, Hanser S, Basson AK, et al. Pancreatic β-cell dysfunction in type 2 diabetes: Implications of inflammation and oxidative stress. *World J Diabetes.* 2023 Mar 15;14(3):130–46. https://doi.org/10.4239/wjd.v14.i3.130.

Dominici M, Le Blanc K, Mueller I, Slaper-Cortenbach I, Marini F, Krause D, Deans R, et al. Horwitz E. Minimal criteria for defining multipotent mesenchymal stromal cells. The International Society for Cellular Therapy position statement. *Cytotherapy.* 2006;8(4):315–7. https://doi.org/10.1080/14653240600855905.

El-Badri N, Ghoneim MA. Mesenchymal stem cell therapy in diabetes mellitus: Progress and challenges. *J Nucleic Acids.* 2013;2013:194858.

Feng J, Mantesso A, De Bari C, Nishiyama A, Sharpe PT. Dual origin of mesenchymal stem cells contributing to organ growth and repair. *Proc Natl Acad Sci U S A.* 2011 Apr 19;108(16):6503–8. https://doi.org/10.1073/pnas.1015449108.

Galicia-García U, Benito-Vicente A, Jebari S, Larrea-Sebal A, Siddiqi H, Uribe KB, Ostolaza H, et al. Pathophysiology of type 2 diabetes mellitus. *Int J Mol Sci.* 2020 Aug 30;21(17):6275. https://doi.org/10.3390/ijms21176275.

Gamble A, Pawlick R, Pepper AR, Bruni A, Adesida A, Senior PA, Korbutt GS, et al. Improved islet recovery and efficacy through co-culture and co-transplantation of islets with human adipose-derived mesenchymal stem cells. *PLoS One.* 2018 Nov 12;13(11):e0206449. https://doi.org/10.1371/journal.pone.0206449.

Gao S, Zhang Y, Liang K, Bi R, Du Y. Mesenchymal stem cells (MSCs): A novel therapy for type 2 diabetes. *Stem Cells Int.* 2022 Aug 22;2022:8637493. https://doi.org/10.1155/2022/8637493.

Gao X, Song L, Shen K, Wang H, Qian M, Niu Q, Qin X. Bone marrow mesenchymal stem cells promote the repair of islets from diabetic mice through paracrine actions. *Mol Cell Endocrinol.* 2014;388(102):41–50. https://doi.org/10.1016/j.mce.2014.03.004.

Habib SM, Haider KH, Martini MF. The potential of mesenchymal stem cell-derived exosomes in cell-free therapy for myocardial repair. In: Haider, K.H. (ed.) *Handbook of Regenerative Medicine: Cell-Free Therapy Approach* (1st ed.). CRC Press, 2025. https://doi.org/10.1201/9781003611509.

Habib SY, Martini MF, Hamdan YNA, Shrebaty OMM, Haider KH. Mesenchymal stem cell secretome and exosomes as potential advanced therapy medicinal products for treating a failing heart. *World J Stem Cells.* 2026;18.

Haider KH. Priming mesenchymal stem cells to develop "super stem cells." *World J Stem Cells* 2024;16(6):623–40. https://doi.org/10.4252/wjsc.v16.i6.623

Haider KH, Alreshidi MA, Al-Rasheedi M, Aldmasi KS. Reversing insulin resistance by novel cell-based and cell-free therapy approaches with skeletal myoblasts. In: Haider, K.H. (eds) *Handbook of Multisystem Applications of Stem Cell-based Therapy*, Chapter 19. New York: Taylor & Francis, 2025.

Haider KH, Ashraf M. Preconditioning and stem cell survival. *J Cardiovasc Trans Res.* 2010;3:89–102.

Haider KH, Aziz S. Paracrine hypothesis and cardiac repair. *Int J Stem Cell Res Transplant.* 2017;5(1):265–7.

Hass R, Otte A. Mesenchymal stem cells as all-round supporters in a normal and neoplastic microenvironment. *Cell Commun Signal.* 2012 Sep 3;10(1):26. https://doi.org/10.1186/1478-811X-10-26.

He Q, Song J, Cui C, Wang J, Hu H, Guo X, Yang M, et al. Mesenchymal stem cell-derived exosomal miR-146a reverses diabetic β-cell dedifferentiation. *Stem Cell Res Ther.* 2021 Aug 11;12(1):449. https://doi.org/10.1186/s13287-021-02371-0.

Hossain MJ, Al-Mamun M, Islam MR. Diabetes mellitus is the fastest-growing global public health concern: Early detection should be a priority. *Health Sci Rep.* 2024;7(3):e2004. https://doi.org/10.1002/hsr2.2004.

Hu J, Wang Y, Gong H, Yu C, Guo C, Wang F, Yan S, et al. Long-term effect and safety of wharton's jelly-derived mesenchymal stem cells on type 2 diabetes. *Exp Therap Med.* 2016;12:1857–66. https://doi.org/10.3892/etm.2016.3544.

Hu W, Song X, Yu H, Sun J, Zhao Y. Therapeutic potentials of extracellular vesicles for the treatment of diabetes and diabetic complications. *Int J Mol Sci.* 2020;21:1–24.

Ibrahim AY, Mehdi Q, Abbas AO, Alashkar A, Haider KH. Induced pluripotent stem cells: Next-generation cells for tissue regeneration. *J Biomed Sci Eng.* 2016;9(4):226–44.

Insel RA, Dunne JL, Atkinson MA, Chiang JL, Dabelea D, Gottlieb PA, et al. Staging presymptomatic type 1 diabetes: A scientific statement of JDRF, the Endocrine Society, and the American Diabetes Association. *Diabetes Care.* 2015;38:1964–74. https://doi.org/10.2337/dc15-1419.

International Diabetes Federation (IDF). *Diabetes Atlas.* Brussels: International Diabetes Federation, 2025.

Jiang R, Han Z, Zhuo G, Qu X, Li X, Wang X, Shao Y, et al. Transplantation of placenta-derived mesenchymal stem cells in type 2 diabetes: A pilot study. *Front Med China.* 2011;5:94–100. https://doi.org/10.1007/s11684-011-0116-z.

Jiang ZZ, Liu YM, Niu X, Yin JY, Hu B, Guo SC, et al. Exosomes secreted by human urine-derived stem cells could prevent kidney complications in rats with type I diabetes. *Stem Cell Res Therapy* 2016;7(1):24.

Jihwaprani MC, Sula I, Charbat MA, Haider KH. Establishing delivery route-dependent safety and efficacy of living biodrug mesenchymal stem cells in heart failure patients. *World J Cardiol.* 2024 Jun 26;16(6):339–54. https://doi.org/10.4330/wjc.v16.i6.339.

Jin L, Deng Z, Zhang J, Yang C, Liu J, Han W, Ye P, Si Y, Chen G. Mesenchymal stem cells promote type 2 macrophage polarization to ameliorate the myocardial injury caused by diabetic cardiomyopathy. *J Transl Med.* 2019;17(1):251. https://doi.org/10.1186/s12967-019-1999-8.

Kalou YMN., Hashemi ASA, Joudeh RM, Aramini B, Haider KH. Mesenchymal stem cell-based heart cell therapy: The effect of route of cell delivery in the clinical perspective. In: Haider K.H. (eds) *Stem Cells: From Potential to Promise.* Singapore: Springer, 2021. pp. 151–91. https://doi.org/10.1007/978-981-16-0301-36.

Kamal M, Kassem D, Haider KH. (2022). Sources and therapeutic uses of mesenchymal stem cells in regenerative medicine. In: Haider K.H. (eds) *Handbook of Stem Cell Therapy*, Springer, Singapore. http://doi.org/10.1007/978-981-19-2655.

Kashbour M, Abdelmalik A, Yassin MNA, Yassin MNA, Abed M, Aldieb E, Milad D, et al. Mesenchymal stem cell-based therapy for type 1 & 2 diabetes mellitus patients: A systematic review and meta-analysis of randomized controlled trials. *Diabetol Metab Syndr.* 2025;17:189. https://doi.org/10.1186/s13098-025-01619-6.

Komai T, Inoue M, Okamura T, Morita K, Iwasaki Y, Sumitomo S, Shoda H, et al. Transforming growth factor-β and interleukin-10 synergistically regulate humoral immunity *via* modulating metabolic signals. *Front Immunol.* 2018 Jun 14;9:1364. https://doi.org/10.3389/fimmu.2018.01364.

Kotikalapudi N, Sampath SJP, Sukesh Narayan SRB, Nemani H, Mungamuri SK, Venkatesan V. The promise(s) of mesenchymal stem cell therapy in averting preclinical diabetes: Lessons from in vivo and in vitro model systems. *Sci Rep.* 2021;11(1):16983. https://doi.org/10.1038/s41598-021-96121-0.

Lee SH, Park SY, Choi CS. Insulin resistance: From mechanisms to therapeutic strategies. *Diabetes Metab J.* 2022;46:15–37.

Liang Y, Xia J, Xu L, Zhang Y, Xu X, Haider KH. Cell-free and cell-based therapy in regenerative medicine. In: Haider K.H. (eds) *Handbook of Regenerative Medicine: Cell-Free Therapy Approach* (1st ed.). CRC Press, 2025. https://doi.org/10.1201/9781003611509.

Li B, Yu Cheng Y, Yin Y, Xue J, Yu S, Gao J, Liu J, Li Zang, et al. Reversion of early- and late-stage β-cell dedifferentiation by human umbilical cord-derived mesenchymal stem cells in type 2 diabetic mice. *Cytotherapy.* 2021;23:510 20. https://doi.org/10.1016/j.jcyt.2021.01.005.

Li G, Peng H, Qian S, Zou X, Du Y, Wang Z, Zou L, et al. Bone marrow-derived mesenchymal stem cells restored high-fat-fed induced hyperinsulinemia in rats at the early stage of type 2 diabetes mellitus. *Cell Transplant.* 2020;29. https://doi.org/10.1177/0963689720904628.

Li H, Liu D, Li C, Zhou S, Tian D, Xiao D, et al. Exosomes secreted from mutant-HIF-1α-modified bone-marrow-derived mesenchymal stem cells attenuate early steroid-induced avascular necrosis of the femoral head in rabbits. *Cell Biol Int.* 2017;41(12):1379–90. https://doi.org/10.1002/cbin.10869.

Lian X-F, Lu D-H, Liu H-L, Liu Y-J, Yang Y, Lin Y, Xie F, et al. Effectiveness and safety of human umbilical cord-mesenchymal stem cells for treating type 2 diabetes mellitus. *World J Diabetes.* 2022;13:877–87. https://doi.org/10.4239/wjd.v13.i10.877.

Lian X-F, Lu D-H, Liu H-L, Liu Y-J, Yang Y, Lin Y, Xie F, et al. Safety evaluation of human umbilical cord-mesenchymal stem cells in type 2 diabetes mellitus treatment: A phase 2 clinical trial. *World J Clin Cases.* 2023;11:5083–96. https://doi.org/10.12998/wjcc.v11.i21.5083.

Lin, T-C, Lacorcia M, Mannering SI. Current and emerging assays for measuring human T-cell responses against beta-cell antigens in type 1 diabetes. *Biomolecules.* 2025;15. https://doi.org/10.3390/biom15030384.

Liu J, Sun X, Zhang FL, Jin H, Yan XL, Huang S, Guo ZN, Yang Y. Clinical potential of extracellular vesicles in type 2 diabetes. *Front Endocrinol.* 2021;11.

Liu XS, Fan B, Szalad A, Jia L, Wang L, Wang X, Pan W, et al. MicroRNA-146a mimics reduce the peripheral neuropathy in type 2 diabetic mice. *Diabetes.* 2017 Dec;66(12):3111–21. https://doi.org/10.2337/db16-1182.

Lu G, Jiang S, Ashraf M, Haider KH. Mitochondria-specific connexin-43 transgene expression protects stem cells in the ischemic heart via shifting the balance between Bak and Bcl-xL. *Regen Med.* 2012;7(3):323–34.

Lucier J, Mathias PM. *Type 1 Diabetes.* Treasure Island, FL: StatPearls Publishing; 2024.

Ma X, Xiao L, Liu L, Ye L, Su P, Bi E, Wang Q, et al. CD36-mediated ferroptosis dampens intratumoral CD8+ T cell effector function and impairs their antitumor ability. *Cell Metab.* 2021;33:1001–12.e5. https://doi.org/10.1016/j.cmet.2021.02.015.

Man X, Lin T, Xie Z, Jin J, He Q. Beneficial effects of cell-derived exosomes on diabetic nephropathy: A systematic review and meta-analysis of preclinical evidence. *Acta Diabetol.* 2025 May;62(5):607–20. https://doi.org/10.1007/s00592-025-02473-8.

Mikłosz A, Chabowski A. Adipose-derived mesenchymal stem cells therapy as a new treatment option for diabetes mellitus. *J Clin Endocrinol Metab.* 2023;108:1889–97. https://doi.org/10.1210/clinem/dgad142.

Młynarska E, Czarnik W, Dzieża N, Jędraszak W, Majchrowicz G, Prusinowski F, Stabrawa M, et al. Type 2 diabetes mellitus: New pathogenetic mechanisms, treatment and the most important complications. *Int J Mol Sci. 2025;*26. https://doi.org/10.3390/ijms26031094.

Mou L, Wang TB, Wang X, Pu Z. Advancing diabetes treatment: The role of mesenchymal stem cells in islet transplantation. *Front Immunol.* 2024;15:1389134. https://doi.org/10.3389/fimmu.2024.1389134.

Müller L, Tunger A, Wobus M, von Bonin M, Towers R, Bornhäuser M, Dazzi F, et al. Immunomodulatory properties of mesenchymal stromal cells: An update. *Front Cell Dev Biol.* 2021 Feb 9;9:637725. https://doi.org/10.3389/fcell.2021.637725.

Mushtaq M, Zineldeen DH, Mateen MA, Haider KH. Mesenchymal stem cells' "garbage bags" at work: Treating radial nerve injury with mesenchymal stem cell-derived exosomes. *World J Stem Cells.* 2024 May 26;16(5):467–78. https://doi.org/10.4252/wjsc.v16.i5.467.

Musleh AR, Al-Kasssas W, Haider KH. Fatty acid escape hypothesis: The pathway to type-II diabetes. *Acta Endocrinologica* 2018;5:201.

Nally LM, Sherr JL, Van Name MA, Patel AD, Tamborlane WV. Pharmacologic treatment options for type 1 diabetes: What's new? *Expert Rev Clin Pharmacol.* 2019 May;12(5):471–9. https://doi.org/10.1080/17512433.2019.1597705.

Nguyen LT, Hoang DM, Nguyen KT, Bui DM, Nguyen HT, Le HTA, Hoang VT, et al. Type 2 diabetes mellitus duration and obesity alter the efficacy of autologously transplanted bone marrow-derived mesenchymal stem/stromal cells. *Stem Cells Transl Med.* 2021;10:1266–78. https://doi.org/10.1002/sctm.20-0506.

Oh JY, Kim H, Lee HJ, Lee K, Barreda H, Kim HJ, Shin E, et al. MHC class I enables MSCs to evade NK-cell-mediated cytotoxicity and exert immunosuppressive activity. *Stem Cells.* 2022 Sep 26;40(9):870–82. https://doi.org/10.1093/stmcls/sxac043.

Omole AE, Fakoya AOJ, Nnawuba KC, Haider KH. Common ethical considerations of human-induced pluripotent stem cell research. In: Haider K.H. (eds) *Handbook of Stem Cell Therapy.* Singapore: Springer, 2022. https://doi.org/10.1007/978-981-16-6016-0_40-1.

Ormazabal V, Nova-Lampeti E, Rojas D, Zúñiga FA, Escudero C, Lagos P, Moreno A, et al. Secretome from human mesenchymal stem cells-derived endothelial cells promotes wound healing in a type-2 diabetes mouse model. *Int J Mol Sci.* 2022 Jan 15;23(2):941. https://doi.org/10.3390/ijms23020941.

Park JH, Hwang I, Hwang SH, Han H, Ha H. Human umbilical cord blood-derived mesenchymal stem cells prevent diabetic renal injury through paracrine action. *Diabetes Res Clin Pract.* 2012;98(3):465–73. https://doi.org/10.1016/j.diabres.2012.09.034.

Päth G, Perakakis N, Mantzoros CS, Seufert J. Stem cells in the treatment of diabetes mellitus—Focus on mesenchymal stem cells. *Metab Clin Exp.* 2019;90:1–15.

Rackham CL, Dhadda PK, Chagastelles PC, Simpson SJ, Dattani AA, Bowe JE, Jones PM, et al. Pre-culturing islets with mesenchymal stromal cells in a direct-contact configuration is beneficial for transplantation outcomes in diabetic mice. *Cytotherapy.* 2013 Apr;15(4):449–59. https://doi.org/10.1016/j.jcyt.2012.11.008.

Rajab AM, Rajab TM, AlJundi S, Haider KH. Bone stem cell therapy in the clinical perspective: A focus on nonrandomized and randomized trials. In: Haider K.H. (eds) *Stem Cells: From Myth to Reality and Evolving.* Berlin: Medicine & Life Sciences, De Gruyter, 2019.

Ranjbaran H, Jobani BM, Amirfakhrian E, Alizadeh-Navaei R. Efficacy of mesenchymal stem cell therapy on glucose levels in type 2 diabetes mellitus: A systematic review and meta-analysis. *J Diabetes Investig.* 2021;12:803–10.

Roep BO. The role of T-cells in the pathogenesis of type 1 diabetes: From cause to cure. *Diabetologia* 2003;46:305–21.

Saeed H, Ahsan M, Saleem Z, Iqtedar M, Islam M, Danish Z, Khan AM. Mesenchymal stem cells (MSCs) as skeletal therapeutics – An update. *J Biomed Sci.* 2016 Apr 16;23:41. https://doi.org/10.1186/s12929-016-0254-3.

Safwan M, Bourgleh MS, Aldoush M, Haider KH. Tissue-source effect on mesenchymal stem cells as living biodrugs for heart failure: Systematic review and meta-analysis. *World J Cardiol.* 2024;16(8):469–83. https://doi.org/10.4330/wjc.v16.i8.469.

Saima Kh, Haider KH. Stem cells: A renewable source of pancreatic β-cells and future for diabetes treatment. In: Haider K.H. (eds) *Stem Cells.* Cham: Springer, 2021, pp. 185–202. https://doi.org/10.1007/978-3-030-77052-5_12.

Shi H, Xu X, Zhang B, Xu J, Pan Z, Gong A, et al. 3, 3′-diindolylmethane stimulates exosomal Wnt11 autocrine signaling in human umbilical cord mesenchymal stem cells to enhance wound healing. *Theranostics.* 2017;7(6):1674. https://doi.org/10.7150/thno.18082.

Si Y, Zhao Y, Hao H, Liu J, Guo Y, Mu Y, Shen J, et al. Infusion of mesenchymal stem cells ameliorates hyperglycemia in type 2 diabetic rats: Identification of a novel role in improving insulin sensitivity. *Diabetes.* 2012;61(6):1616–25. https://doi.org/10.2337/db11-1141.

Soltani S, Mansouri K, Aleagha MSE, Moasefi N, Yavari N, Shakouri SK, Notararigo S, et al. Extracellular vesicle therapy for type 1 diabetes. *Front Immunol.* 2022a;13.

Soltani S, Mansouri K, Parvaneh S, Thakor AS, Pociot F, Yarani R. Diabetes complications and extracellular vesicle therapy. *Rev Endocrine Metab Dis.* 2022b;23:357–85.

Song J, He Q, Guo X, Wang L, Wang J, Cui C, Hu, et al. Mesenchymal stem cell-conditioned medium alleviates high fat-induced hyperglucagonemia via MiR-181a-5p and its target PTEN/Akt signaling. *Mol Cell Endocrinol.* 2021;537. https://doi.org/10.1016/j.mce.2021.111445.

Song J, Wang L, Wang L, Guo X, He Q, Cui C, Hu H, et al. Mesenchymal stromal cells ameliorate mitochondrial dysfunction in α cells and hyperglucagonemia in type 2 diabetes via SIRT1/FoxO3a signaling. *Stem Cells Translational Med.* 2024;413:776–90. https://doi.org/10.1093/stcltm/szae038.

Speight J, Pouwer F. Diabetes mellitus, type 1. In: *Cambridge Handbook of Psychology, Health and Medicine.* Cambridge: Cambridge University Press, 2019, pp. 477–480.

Sun L, Xi S, He G, Li Z, Gang X, Sun C, Guo W, Wang G. Two to tango: Dialogue between adaptive and innate immunity in type 1 diabetes. *J Diabetes Res.* 2020 Jul 30;2020:4106518. https://doi.org/10.1155/2020/4106518.

Sun X, Hao H, Han Q, Song X, Liu J, Dong L, Han W, Mu Y. Human umbilical cord-derived mesenchymal stem cells ameliorate insulin resistance by suppressing NLRP3 inflammasome-mediated inflammation in type 2 diabetes rats. *Stem Cell Res Ther.* 2017 Nov 2;8(1):241. https://doi.org/10.1186/s13287-017-0668-1

Sun Y, Shi H, Yin S, Ji C, Zhang X, Zhang B, Wu P, et al. Human mesenchymal stem cell-derived exosomes alleviate type 2 diabetes mellitus by reversing peripheral insulin resistance and relieving β-cell destruction. *ACS Nano.* 2018;12:7613–28. https://doi.org/10.1021/acsnano.7b07643.

Sun Y, Tao Q, Wu X, Zhang L, Liu Q, Wang L. The utility of exosomes in the diagnosis and therapy of diabetes mellitus and associated complications. *Front Endocrinol (Lausanne).* 2021 Oct 26;12:756581. https://doi.org/10.3389/fendo.2021.756581.

Wang L, Chopp M, Szalad A, Lu X, Zhang Y, Wang X, Cepparulo P, Lu M, Li C, Zhang ZG. Exosomes derived from Schwann cells ameliorate peripheral neuropathy in type 2 diabetic mice. *Diabetes.* 2020b Apr;69(4):749–59. https://doi.org/10.2337/db19-0432.

Wang L, Liu T, Liang R, Wang G, Liu Y, Zou J, Liu N, et al. Mesenchymal stem cells ameliorate β cell dysfunction of human type 2 diabetic islets by reversing β cell dedifferentiation." *EBioMedicine*. 2020a;51. https://doi.org/10.1016/j.ebiom.2019.102615.

Wang L, Qing L, Liu H, Liu N, Qiao J, Cui C, He T, Zhao R, et al. Mesenchymal stromal cells ameliorate oxidative stress-induced islet endothelium apoptosis and functional impairment via Wnt4-β-catenin signaling. *Stem Cell Res Therapy.* 2017;8. https://doi.org/10.1186/s13287-017-0640-0.

Wang M, Song L, Strange C, Dong X, Wang H. Therapeutic effects of adipose stem cells from diabetic mice for the treatment of type 2 diabetes. *Mol Therapy.* 2018;26:1921–30. https://doi.org/10.1016/j.ymthe.2018.06.013.

Wang Z, Li H, Fang J, Wang X, Dai S, Cao W, et al. Comparative analysis of the therapeutic effects of amniotic membrane and umbilical cord-derived mesenchymal stem cells for the treatment of type 2 diabetes. *Stem Cell Rev Rep. 2022;*18:1193–206. https://doi.org/10.1007/s12015-021-10320-w.

Weinberg Sibony R, Segev O, Dor S, Raz I. Drug therapies for diabetes. *Int J Mol Sci.* 2023 Dec 5;24(24):17147. https://doi.org/10.3390/ijms242417147.

Xia J, Arai S, Haider KH. Research and applications of stem cell secretome. In: Haider K.H. (eds) *Stem Cells: From Potential to Promise*. Singapore: Springer, 2021. https://doi.org/10.1007/978-981-16-0301-3_7.

Xia MY, Zang N, Song S, Chen J, Hu H, Wang K, et al. PEGylated β-cell-targeting exosomes from mesenchymal stem cells improve β-cell function and quantity by suppressing NRF2-mediated ferroptosis. *Int J Nanomedi.* 2024;19:9575–96. https://doi.org/10.2147/IJN.S459077.

Xie M, Hao HJ, Cheng Y, Xie ZY, Yin YQ, Zhang Q, Gao JQ, Liu HY, Mu YM, Han WD. Adipose-derived mesenchymal stem cells ameliorate hyperglycemia through regulating hepatic glucose metabolism in type 2 diabetic rats. *Biochem Biophys Res Commun.* 2017;483:435–41. https://doi.org/10.1016/j.bbrc.2016.12.125.

Xie Z, Haojie H, Tong C, Cheng Y, Liu J, Pang Y, Si Y, et al. Human umbilical cord-derived mesenchymal stem cells elicit macrophages into an anti-inflammatory phenotype to alleviate insulin resistance in type 2 diabetic rats. *Stem Cells* 2016;34:627–39. https://doi.org/10.1002/stem.2238.

Xu L, Liu Y, Sun Y, Wang B, Xiong Y, Lin W, Wei Q, et al. Tissue source determines the differentiation potentials of mesenchymal stem cells: A comparative study of human mesenchymal stem cells from bone marrow and adipose tissue. *Stem Cell Res Ther.* 2017;8(1):275. https://doi.org/10.1186/s13287-017-0716-x.

Xu J, Liu X, Zhao F, Zhang Y, Wang Z. HIF1α overexpression enhances diabetic wound closure in high glucose and low oxygen conditions by promoting adipose-derived stem cell paracrine function and survival. *Stem Cell Res Therapy.* 2020;11. https://doi.org/10.1186/s13287-020-01654-2.

Xue B, Xiao X., Yu T, Xiao X, Xie J, Ji Q, Wang L, et al. Mesenchymal stem cells modified by FGF21 and GLP1 ameliorate lipid metabolism while reducing blood glucose in type 2 diabetic mice. *Stem Cell Res Ther.* 2021;12:133. https://doi.org/10.1186/s13287-021-02205-z.

Yap SK, Tan KL, Abd Rahaman NY, Saulol Hamid NF, Ooi J, Tor YS, Daniel Looi QH, et al. Human umbilical cord mesenchymal stem cell-derived small extracellular vesicles ameliorated insulin resistance in type 2 diabetes mellitus rats. *Pharmaceutics.* 2022 Mar 16;14(3):649. https://doi.org/10.3390/pharmaceutics14030649.

Zang L, Li Y, Hao H, Liu J, Zhang Q, Gao F, Wang H, et al. Efficacy of umbilical cord-derived mesenchymal stem cells in the treatment of type 2 diabetes assessed by retrospective continuous glucose monitoring. *Stem Cells Translational Med. 2023;*12:775–82. https://doi.org/10.1093/stcltm/szad060.

Zang L, Yijun L, Haojie H, Jiejie L, Yu C, Bing L, et al. Efficacy and safety of umbilical cord-derived mesenchymal stem cells in Chinese adults with type 2 diabetes: A single-center, double-blinded, randomized, placebo-controlled phase II trial. *Stem Cell Res Ther.* 2022;13(1). https://doi.org/10.1186/s13287-022-02848-6.

Zeinhom A, Fadallah SA, Mahmoud M. Human mesenchymal stem/stromal cell-based therapy in diabetes mellitus: Experimental and clinical perspectives. *Stem Cell Res Ther.* 2024;15:384. https://doi.org/10.1186/s13287-024-03974-z.

Zhang Y, Gao S, Liang K, Wu Z, Yan X, Liu W, Li J, Wu B, Du Y. Exendin-4 gene modification and microscaffold encapsulation promote self-persistence and antidiabetic activity of MSCs. *Science Adv.* 2021;7. https://doi.org/10.1126/sciadv.abi4379.

Zhao H, Shang Q, Pan Z, Bai Y, Li Z, Zhang H, Zhang Q, et al. Exosomes from adipose-derived stem cells attenuate adipose inflammation and obesity through polarizing M2 macrophages and beiging in white adipose tissue. *Diabetes.* 2018 Feb;67(2):235–47. https://doi.org/10.2337/db17-0356.

Zhao KH. Liu HJ, C. Cheng TY, Z. Zang XL, Mu Y, Han W. Bone marrow-derived mesenchymal stem cells ameliorate chronic high glucose-induced ss-cell injury through modulation of autophagy. *Cell Death Dis.* 2015;6. https://doi.org/10.1038/cddis. 2015.230.

Zhou N, Liu W, Zhang W, Liu Y, Li X, Wang Y, Zheng R, Zhang Y. Wip1 regulates the immunomodulatory effects of murine mesenchymal stem cells in type 1 diabetes mellitus via targeting IFN-α/BST2. *Cell Death Discov.* 2021 Oct 29;7(1):326. https://doi.org/10.1038/s41420-021-00728-1.

Index

F

G

H

P

R

S

T

U

V

W

For Product Safety Concerns and Information please contact our
EU representative GPSR@taylorandfrancis.com Taylor & Francis
Verlag GmbH, Kaufingerstraße 24, 80331 München, Germany